KU-095-507

NUCLEAR REACTOR SAFETY HEAT TRANSFER

PROCEEDINGS OF THE INTERNATIONAL CENTRE FOR HEAT AND MASS TRANSFER
ZORAN P. ZARIĆ, EDITOR
C. L. TIEN, SENIOR CONSULTING EDITOR

---

1. Afgan and Schlünder **Heat Exchangers: Design and Theory Sourcebook** 1974
2. Afgan and Beer **Heat Transfer in Flames** 1974
3. Blackshear **Heat Transfer in Fires: Thermophysics, Social Aspects, Economic Impact** 1974
4. deVries and Afgan **Heat and Mass Transfer in the Biosphere: Transfer Processes in Plant Environment** 1975
5. Denton and Afgan **Future Energy Production Systems: Heat and Mass Transfer Processes** 1976
6. Hartnett **Alternative Energy Sources** 1976
7. Spalding and Afgan **Heat Transfer and Turbulent Buoyant Convection: Studies and Applications for Natural Environment, Buildings, Engineering Systems** 1977
8. Zarić **Thermal Effluent Disposal from Power Generation** 1978
9. Hoogendoorn and Afgan **Energy Conservation in Heating, Cooling, and Ventilating Buildings: Heat and Mass Transfer Techniques and Alternatives** 1978
10. Durst, Tsiklauri, and Afgan **Two-Phase Momentum, Heat and Mass Transfer in Chemical, Process, and Energy Engineering Systems** 1979
11. Spalding and Afgan **Heat and Mass Transfer in Metallurgical Systems** 1981
12. Jones **Nuclear Reactor Safety Heat Transfer** 1981
13. Bankoff and Afgan **Heat Transfer in Nuclear Reactor Safety** 1981
14. Zarić **Structure of Turbulence in Heat and Mass Transfer** 1981
15. Hewitt, Taborek, and Afgan **Advancement in Heat Exchangers** 198
16. Metzger and Afgan **Heat and Mass Transfer in Rotating Machinery** 1983

NUCLEAR REACTOR SAFETY HEAT TRANSFER

Edited by
OWEN C. JONES, JR.
Department of Nuclear Energy
Brookhaven National Laboratory

HEMISPHERE PUBLISHING CORPORATION
Washington New York London

DISTRIBUTION OUTSIDE THE UNITED STATES
McGRAW-HILL INTERNATIONAL BOOK COMPANY
Auckland Bogotá Guatemala Hamburg Johannesburg Lisbon
London Madrid Mexico Montreal New Delhi Panama Paris
San Juan São Paulo Singapore Sydney Tokyo Toronto

05Y54519

NUCLEAR REACTOR SAFETY HEAT TRANSFER

Copyright © 1981 by Hemisphere Publishing Corporation. All rights reserved. Printed in the United States of America. No part of this publication may be reproduced, stored in a retrieval system, or transmitted, in any form or by any means, electronic, mechanical, photocopying, recording, or otherwise, without the prior written permission of the publisher.

1 2 3 4 5 6 7 8 9 0 B C B C 8 9 8 7 6 5 4 3 2 1

Library of Congress Cataloging in Publication Data
Main entry under title:

Nuclear reactor safety heat transfer.

(Proceedings of the International Centre for Heat and Mass Transfer ; 12)
Includes bibliographical references and index.
1. Nuclear reactors—Safety measures—Congresses. 2. Nuclear reactors—Cooling—Congresses. 3. Heat—Transmission—Congresses. I. Jones, Owen C., date. II. Series.
TK9152.N7956 621.48'35 81-2932
ISBN 0-89116-224-0 AACR2
ISSN 0272-880X

D
621.4835
NUC

CONTENTS

PART 2 FUNDAMENTAL CONCEPTS

PREFACE

Ideally one would like to write a textbook on a field that has achieved its majority, However, in today's rapidly changing technological world, it is not always possible to do so and at the same time to provide timely instructive material for the serious student. This is certainly true in the field of nuclear safety heat transfer, which is the subject of this book. On the other hand, instructive material must be provided for students just entering the field, as well as seasoned engineers who need an intensive introduction or detailed refresher. This is the intended purpose of this textbook.

The material in this textbook was compiled through the generous assistance of several internationally active recognized authorities in the field of nuclear heat transfer. The material was originally assembled for use in an intensive, 40-hour summer school at the International Centre for Heat and Mass Transfer in Dubrovnik, Yugoslavia, August 25–29, 1980. The purpose of the school was to bring together in one course the major areas of concern in the field of nuclear safety heat transfer and to describe the state-of-the-art at the turn of the decade.

It had been the intention that a thorough editing, integration, unification of viewpoint, and standardization of each part of the text would be undertaken prior to its first release at the 1980 Summer School. This turned out to be impractical on the time scales over which this project was undertaken. It had also been hoped that numerous examples and problems would be developed for each chapter. This also turned out to be impractical. Had this book been written by one person in the process of developing a course over a number of years it might have been otherwise. However, it was instead prepared mostly by researchers, specifically for the summer school, who gave freely of their time generally outside of their daily schedules during a period in the history of nuclear power when intensive efforts were required for input into the regulatory and political processes. The result, then, rather than being in the nature of a textbook, is instead more of an organized composite summary of nuclear safety heat transfer technology.

The material in the text is presented in five major sections comprising twenty-four chapters. The first section, entitled "Overview," presents the background material placing nuclear power in perspective. Starting with a historical overview, followed by fundamental concepts of nuclear energy and the philosophy of risk, the first three chapters: give the reader a brief but thorough introduction to nuclear power generation; describe the different types of nuclear reactors built (PWR, CANDU, HTGR, LMFBR, etc.); discuss the factors that couple the physics to the thermal hydraulics; and place the safety issues of nuclear power in the context of the thermohydraulics.

The second section of the text, "Fundamental Concepts," describes the startup and shutdown procedures, typical maneuvering transients, faults of moderate and remote frequency, and the concept of design basis accidents. This is done for both water reactor systems (LWR) and liquid metal fast breeder reactor systems (LMFBR). The balance of the second section is devoted to presenting fundamental concepts of single- and two-phase flow and heat transfer and of system safety modelling.

The balance of the textbook is concerned with specific features of LWRs and LMFBRs, specifically concerning the design basis accident—large break guillotine loss of coolant accident (LOCA) in LWRs and the core disruptive accident (CDA) in LMFBRs—and topics of a more general nature. The third section, "Design Basis Accident: Light Water Reactors," covers the four major phases of a large break LOCA in detail, including a full description of a possible sequence of conditions during a LOCA and then details of the blowdown phase, emergency cooling injection, and reflood and rewet heat transfer. A detailed discussion of system analysis methods by computer calculation is presented, which includes typical comparisons of computations and experimental results.

Then the LMFBR is described: the complete sequences that may progress through a core disruptive accident, followed by discussions of the initiation phase up through clad melting and motion, transition phase through the possible sequences of core motion to permanent nuclear shutdown, and the heat removal phases both in the reactor vessel itself and ex-vessel following an accident. Computational methods and comparisons are also given in this fourth section, entitled "Design Basis Accident: Liquid Metal Fast Breeder Reactors."

Finally, certain areas of nuclear system thermal hydraulics, while not major concerns insofar as the design basis accident for a particular reactor concept, are nevertheless receiving intensive study since they are important for more moderate fault conditions or possibly for advanced reactor design steady state or maneuvering conditions. Others are applicable to both LWRs and LMFBRs. These are discussed in the fifth section, "Special Topics," which includes vapor explosions, natural convective cooling, blockages, sodium boiling, and experimental techniques. The section is completed by a brief but thorough description of the sequence of events that occurred during the accident at Three Mile Island.

ACKNOWLEDGMENTS

The editor is indebted to many people, too numerous to mention, who helped in the preparation of this book. Special thanks, however, must go to the persons who played key roles.

First, thanks must go to Professor S. George Bankoff of Northwestern University in Evanston, Illinois, U.S.A., who was chairman of the International Seminar on Nuclear Reactor Safety Heat Transfer in Dubrovnik, Yugoslavia. The seminar was held the week immediately following the summer school wherein this material was first presented. This editor acted as the school's director. The planning and development of the direction the school and seminar would take was a cooperative effort on both parts. Professor Bankoff thus played an instrumental role in the formulation of this book as well as contributing one of its chapters. His help and friendship are gratefully appreciated.

In addition to Professor Bankoff, thanks must also be extended to the authors of the chapters, without whose support there would be no book. These include J. G. Collier of the Atomic Energy Research Establishment in Harwell, U.K.; J. Costa and D. Grand of Centre d'Études Nucléaires de Grenoble, France; M. Epstein and H. K. Fauske of Fauske and Associates, Willowbrook, Illinois, U.S.A.; J. F. Jackson of Los Alamos Scientific Laboratory, New Mexico, U.S.A.; H. M. Kottowski of the Joint Research Center in Ispra, Italy; R. T. Lahey of the Rensselaer Polytechnic Institute in Troy, New York, U.S.A.; F. Mayinger of the University of Hannover in the Federal Republic of Germany; R. Peckover, Culham Laboratory in Abbingdon, U.K.; and W. Riebold of the Joint Research Center in Ispra, Italy. Thanks must also go to their coauthors for the additional support provided.

The generous assistance of several other people must also be acknowledged with thanks. The continued support of L. Reilly of Hemisphere Publishing Corporation during the preparation and duplication of the original manuscripts and presentation of the Dubrovnik summer school was deeply appreciated. T. Rowland and M. McGrath were both most helpful in preparation of the editor's own manuscripts. Assistance in compilation of the index was obtained from E. Gialdi of AGIP Nucleare in Milan, Italy; J. Kilpi of Teollisuuden in Olkiluoto, Finland; H. Tuomisto of Imatran Voima OY in Helsinki, Finland; S. Dedović and A. Jovanović of the Institute Goša in Beograd, Yugoslavia; and L. Peddicord of Oregon State University in Corvallis, Oregon, U.S.A.; and their assistance is also gratefully acknowledged. Frances M. Scheffel and her staff, at the National Nuclear Data Center at Brookhaven National Laboratory in Upton, New York, U.S.A., were most helpful in organizing this index.

As a final note, this editor gives grateful appreciation for the continued support and assistance of Professors Zarin Zarić, secretary general of the International Centre for Heat and Mass Transfer, and Naim Afgan, scientific secretary for the Centre. Their efforts and the support of the Centre were instrumental in the conduct of the school and the preparation of this book.

The editor's work in the preparation of the book was undertaken under the auspices of the U.S. Nuclear Regulatory Commission.

Owen C. Jones, Jr.

PART 1

OVERVIEW

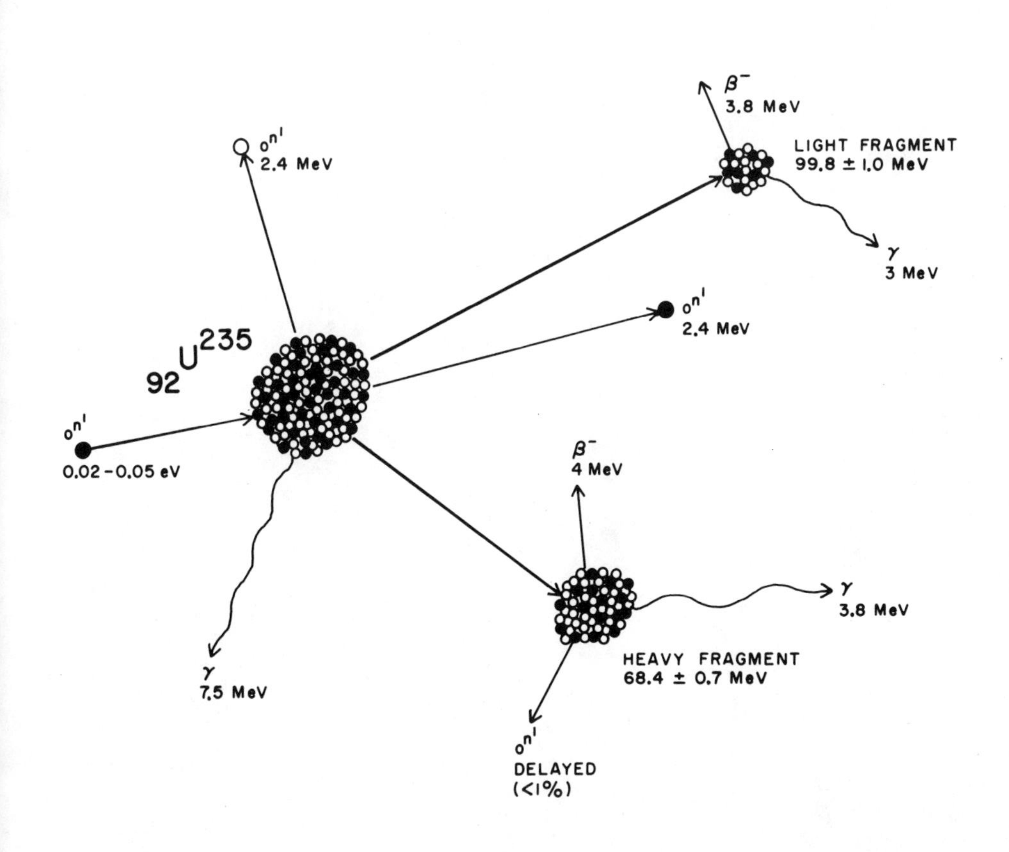

1 INTRODUCTION

by

O.C. Jones, Jr.

BNL – AUI, Upton, NY, USA

In this chapter an overview of nuclear technology is presented starting first with an historical synopsis of the important scientific events leading up to the first criticality. Reactor physics fundamentals are presented in the context of the thermal hydraulic concerns and a short section on the philosophy and practice of nuclear safety including consideration of risk is included at the end.

1. TOPICAL OVERVIEW

It has been barely a century since Hittdorf obtained and recognized man's first physical evidence of the existence of an atomic structure. Not even half that time has elapsed since the neutron was discovered. Yet in this relatively short period of time nuclear energy has been harnessed through use of the fission process, and intensive efforts promise to soon tame the mechanics of the sun's energy themselves to make power by fusion. There has not been, in the entire modern history of mankind, the extraordinary amount of work placed on any other technological achievement to make it safe for human utilization. Yet, while we stand on the doorstep of achieving the realization of a safe and virtually inexhaustible supply of energy, our own fears, born of the trauma of Hiroshima and Nagasaki and nurtured by modern soothsayers of nuclear doom, threaten to rob us of this realization.

To put current concerns into perspective, however, the following quotation taken originally from the Congressional Record in 1875 as given by El-Wakil[5] may prove enlightening:

> "A new source of power, which burns a distillate of kerosene called gasoline, has been produced by a Boston engineer. Instead of burning the fuel under a boiler, it is exploded inside the cylinder of an engine. This so-called internal combustion engine ... begins a new era in the history of civilization ... Never in history has society been confronted with a power so full of potential danger and at the same time so full of promise for the future of man and for the peace of the world.
>
> "The dangers are obvious. Stores of gasoline in the hands of people interested primarily in profit would constitute a fire and explosive hazard of the first rank. Horseless carriages propelled by gasoline engines might attain speeds of 14 or even 20 miles per

hour. The menace to our people of vehicles of this type hurtling through our streets and along our roads and poisoning the atmosphere would call for prompt legislative action, even if the military and economic implications were not so overwhelming ..."

Sound familiar? We are in a similar situation today with nuclear energy. Nevertheless, a small but vocal group of concerned, probably well intentioned but nevertheless ill-informed people, have made and continue to make a large impression on the overall political process through an extremely efficient amplification process known as the free press. The net result is that the governmental legislative bodies are in such a state of confusion, and the regulatory process in such a shambles, that their combined effects seem on the verge of destroying the very thing that has the potential of providing energy for the world for centuries to come--well after all our known reserves of oil and even coal have been consumed.

But nuclear steam supply systems are safe. This has been demonstrated many times over, including rather severe accidents at Brown's Ferry and Three Mile Island. Although the full consequences in terms of system damage to the latter power plant will not be known for some time, they may result in a four-year outage and complete core replacement at a cost of well over one billion of today's dollars. Yet there has not been one nuclear-related fatality in any of these mishaps, in spite of well over 2,000 reactor years of operation.[1] No other of man's advanced technology systems can boast near equal performance.

Nuclear safety continues to receive intensive attention the world over. This includes research dealing with interrelated systems behavior, human factors and man-machine interfacing, and research related to specific technological questions, among which thermal hydraulics remains prominent.

Several years ago the Rasmussen study[2] placed probabilistic estimates on the frequency and consequences of certain types of accidents. These probabilities were based on early 1970 technology, but the technology continues to improve through research as well as evaluation of operational experience including such severe incidents as occurred at Three Mile Island. Indeed, probably the heaviest fallout from this accident will be the technological improvements leading to much improved safety in future designs, and yet still lower probabilistic estimates of accidents with severe consequences.

A vast majority of the safety considerations of modern nuclear power generation systems hinge directly on fluid, mechanical or thermal considerations. This is because most of the power systems in operation today, or on the drawing board, utilize the reactor to generate heat and a fluid to transfer thermal energy from the reactor to a prime mover. In some cases this energy is transferred through an intermediate device, while in others the fluid is used directly.

While great strides have been made in studies of phenomena involved in these various transfer processes, still more is yet to be accomplished. It is the purpose of this text to provide an orderly survey of the current safety aspects of nuclear system heat transfer relative to commercial generation of electrical power. While descriptions of most of the different types of nuclear systems will be provided, the emphasis shall be on light water reactor (LWR) systems and on liquid metal fast breeder reactor (LMFBR) systems. The former are emphasized because of the 530 nuclear power units operating (229), or on order or under construction (301), representing a total of 405,768 megawatts of electrical generating capability (124,750 MWe installed), by far the majority use light water as a coolant (Table 1.1).[3] On the other hand, intensive thermal hydraulic research activities in recent years have centered on problems dealing with the liquid metal fast breeder reactor. This has been due to the fact that in LMFBRs, contrary to most water-based systems, the nuclear heat source is not in its most reactive configuration and some accident situations could conceivably lead to destructive high power behavior and possibly energetic fuel dispersal. The LMFBR research has thus been directed toward the processes which might lead to the reactor as a source of radionuclides to the surrounding population and to the mitigation of such behavior and/or containment of resultant source material.

This text is divided into five parts: Overview; Fundamental Concepts; Design Basis Accident--Light Water Reactors; Design Basis Accident--Liquid Metal Fast Breeder Reactors; Special Topics. In the first part, the Overview, the history and fundamental physical principles coupling reactor physics to thermal hydraulics are discussed and the overall philosophy of risk put into context. Different power reactor concepts are described and the thermal hydraulic considerations in various safety issues are identified. In the second part, the details relating to maneuvering transients and operational faults are covered and the reader is given the fundamental background in single- and two-phase fluid mechanics, heat transfer, and analysis techniques required to understand the phenomena described in the following chapters. The design basis accident for LWRs discussed in the third part of this text emphasizes the loss of coolant accident, LOCA, including a description of the whole accident itself, followed by detailed coverage of blowdown, emergency cooling water injection, reflood and rewet heat transfer, and computational methods and comparisons. A similar detailed discussion for LMFBRs will center on the core disruptive accident (CDA), including the various phases of initiation and start of clad motion, gross clad and fuel motion up to the termination of neutronic activity, post-accident heat removal both inside the vessel and in the containment following vessel melt-through, and computational methods and comparisons. Finally, the last section of this text discusses topics that are either of interest across the board for both reactor types including vapor explosions, natural convection, blockage effects and experimental techniques, as well as the area of sodium boiling which is important in long-term integral core coolability for the LMFBR, but generally not treated in detail relative to design basis accidents. The final chapter is a comprehensive review of the events which occurred during the course of the accident at Three Mile Island, with attention to particular items which may have been overlooked in the design and operation of the power plant.

Table 1.1

World List of Nuclear Power Plants Operable, Under Construction, or On Order (30 MWe and Over) as of December 31, 1979
(Source: Nuclear News, 23, 2, Feb., 1980)

Pressurized water reactor	PWR	313
Boiling water reactor	BWR	120
Gas-cooled reactor	GCR	36
Pressurized heavy water reactor	PHWR	36
Advanced gas-cooled reactor	AGR	11
Liquid metal fast breeder reactor	LMFBR	7
Gas-cooled heavy water moderated reactor	GCHWR	2
Light water-cooled heavy water moderated reactor	LWCHW	2
High-temperature gas-cooled reactor	HTGR	1
Heavy water moderated boiling light water reactor	HWLWR	1
Thorium high-temperature reactor	THTR	1

1.1 Historical Perspective

It won't be long before well over 500 nuclear steam supply systems capable of generating more than 400 million kilowatts of electricity will be in operation. To understand the attractiveness of these types of systems, one must realize the economics involved: economy of bulk, economy of transportation, economy of energy liberation.

Complete combustion of a carbon atom by oxidation will yield approximately 4 electron volts of energy. By contrast, complete "burning" of one uranium atom by fission liberates 200 MeV (50 million times the energy obtained in combusting the carbon atom.[4] In terms of our everyday experience, nearly 2500 metric tons of coal must be burned to produce the 81 trillion Joules of energy which could be given off by complete fissioning of one kilogram ($\sim$ 100 cc) of uranium. Even with only the few percent burnup achieved (% of total possible utilization) in modern nuclear power plants, the advantages appear enormous from many aspects including the mining, the transportation, the waste, even the pollution. These considerations do not even consider the differences in human suffering and tragedy currently experienced in mining or the continual radioactivity release of coal combustion--many, many times that expected under worst daily happenings in a nuclear plant of equivalent size--or the costs in terms of property damage, human lives, or environmental havoc and destruction caused by release of sulfur into the atmosphere. By nearly any reasonable standard it would seem that the deep concerns expressed by conservationists regarding nuclear energy production are misplaced at best.

While understanding the mechanics of nuclear power does not require one to know the details of its development, and indeed few texts on nuclear energy provide this insight, an interesting historical perspective is achieved through a brief perusal of the ideas, philosophy and discoveries which have led to its current status as a source of electrical power.

Several sources have provided viewpoints of varying degrees of depth, the most comprehensive being a brief monograph on Nuclear Physics authored by Heisenberg and translated from the original German by Gagnor, von Zeppelin and Wilson.[6] Various other texts have been utilized also to provide the historical overview that follows, especially those by Profio,[4] Oldenberg,[7] Burcham,[8] Soodak and Campbell,[9] El-Wakil,[5] and Massimo.[10]

The important ideas, philosophies or events having an impact on the development of nuclear energy are itemized briefly in Table 1.2. The early Greek philosophers began to ponder the fundamental makeup of matter as early as the sixth century before the time of Christ. Perhaps it is a blessing that the early ideas of Democritus and Plato which were quite accurate were not followed up for nearly 2,000 years. If the Christians had had the atomic bomb during the Crusades, this book might never have been written.

In direct opposition to the ideas of the alchemists during the sixteenth century, the scientists in the following three centuries showed incontrovertibly that matter could not be transmuted by chemical means. By the mid to latter part of the seventeenth century chemists had formulated theories which revolved around the idea proposed by Cassendi that all matter was composed of indivisible building blocks, or atoms, and that according to Boyle these building blocks were of limited number.

It would not be until a century later, however, that quantitative means would be used to build upon these ideas. But in a startling flash of insight which was to be equalled only later by Avogadro and Einstein, Lavoisier postulated what was to later become known as the law of conservation of mass.

Table 1.2

Chronological Summary of Important Events in the History of Nuclear Energy

Dates	People	Description of Major Event, Idea, or Philosophy Having an Impact
~600 BC	Thales	Water is the source of all things.
~500 BC	Anaxagoras	An infinite of basic but indestructable substances could combine to produce the "variety of world processes."
~500 BC	Empedocles	There are only four fundamental substances from which all other things are derived: earth, air, fire, water.
~500 BC ~500 BC	Leucippus and Democritus	Concept of "full" and "empty" was stated and extended to idea of smallest ultimate individual particles (atoms) as fundamental building blocks of matter.
~500 BC	Plato	Coupled ideas of Empedocles and Pythagoras (number theory) to identify "atoms" with symmetrical shapes: cube, octahedron, tetrahedron...
		Antiqual Hiatus Between Early Greek Philosophical Ideas and Modern Theory
1500's	Alchemists	Every substance reducible to one common substance - should be possible to transmute one into another.
~1625-1655	Cassendi	Abandoned Aristotlian philosophy to re-embrace that of Democritus. Physical variety resulted from varied arrangement of building blocks or atoms.
~1650-1691	Boyle	First to develop idea that there were basic "chemical" substances or "elements" from which remaining infinite variety of substances are composed.
~1764-1794	Lavoisier	Founded quantitative chemistry. Explained "fire" as oxidation and proved mass gain. Formulated law of conservation of mass.
1792	Richter	Discovered proportionality between elements in chemical compounds.
1803	Dalton	"Law of constant proportions" was proposed and utilized to provide geometric interpretation to atomic combinations.
1811	Avogadro	Hypothesized that any gas in a given volume at a specified temperature would be composed of the same number of molecules. Provided a permanent base for Dalton's theory and lit the way for later determination of atomic and molecular weights.
1812	Berzelius	Determined atomic weights of many molecules and introduced notion of valency in connection with binding forces.
1815	Prout	Hypothesized that all elements were simply integral multiples of hydrogen atoms. i.e.: further divisible.
1834	Faraday	Correlated facts observed in experiments on electrolytic conduction to determine the "quantity of electricity" in a mole of matter.

Table 1.2 (Continued)

Dates	People	Description of Major Event, Idea, or Philosophy Having an Impact
1848	Weber	Based on general ideas of Faraday relating atomic and electrical theory, Weber stated that a fixed quantity of matter was associated with a fixed quantity of electricity in a way as yet undetermined.
~1860	Maxwell Bolzmann Clausius	Quantized behavior of gases as molecules in motion. Developed mathematical principles which provided a firm foundation for this behavior.
1865	Loschmidt	Provided first order of magnitude estimate of the physical size of an atom based on studies of internal friction of gases.
1869	Hittorf	Discovered free atoms of electricity (later termed "electrons" by Stoney in 1874) and determined their approximate mass in cathode ray experiments.
1895	Roentgen	Discovered x-rays when platinum barium cyanide outside a blackened discharge tube was found to fluoresce.
1896	Becquerel	Discovered radiation occurred naturally from pitchblend (U_2O_8) in his ultra-violet fluorescence experiments.
1898-1900	Curies, M.&P.	Chemically separated radium from pitchblend and also identified thorium and polonium as radioactive.
1902 etc.	Rutherford	Classified natural radiation from uranium into α, β, and γ categories according to their penetrating power. Identified decay constants we now term half-life and determined energy levels too low to use as an energy source (~30 mW/Ci).
1913	Bohr	Determined that atomic quantum states were characterized by the energy level of electrons in their orbits.
1917-1919	Rutherford/Soddy	Concluded that emission of α-particles was associated with transmutation of one chemical element into another using the nitrogen-to-oxygen reaction with α-bombardment. Subsequently showed protons are fundamental nuclear particles (see Prout 1815).
1924	de Broglie	Found that matter showed particle-wave duality, a cornerstone discovery which lead to the development of quantum mechanics.
1930	Bothe and Becker	Bombarded beryllium and boron with 5 MeV alphas and found penetrating radiation resulted which they thought to be gammas.
1931	Joliot-Curie	Repeated experiments of Bothe and Becker and thought the 50 MeV gammas behaved in wrong manner when considering proton scattering in paraffin.
1932	Chadwick	Reasoned that the 50 MeV gammas were instead heavy uncharged particles to explain their penetrating power. Termed them neutrons.

Table 1.2 (Continued)

Dates	People	Description of Major Event, Idea, or Philosophy Having an Impact
1933-1935	Fermi	U. Rome experiments of orderly bombardment of chemical elements found secondary emissions from over 40 of these including uranium which he suggested transmuted to higher atomic member.
1934	Pontecarro	Discovered slowing down of neutrons through collisions with hydrogen in paraffin and water.
1935-1937	Fermi	Developed ideas of quantum nuclear states to explain strong energy selectivity in neutron absorption. Later clarified by Bohr and by Bret and Wigner to be selective metastable states of nuclear excitation similar to electron levels.
1939	Hahn/Strassmann	Explained many confusing chemical separation results following neutron activation of uranium in terms of fission. Followed immediately by Meitner and Frisch who calculated an $\sim$200 MeV energy release by mass defect. Determined U^{235} fissions only due to low energy neutrons while both thorium and U^{238} could be fissioned by high energy neutrons.
1939	von Halban	Discovered secondary neutrons released due to fission in uranyl-nitrate solution. Also showed existance of $\sim$1% delayed neutrons.
1939	Fermi	Tried and failed to construct a device to develop a chain reaction using natural uranium and light water. Further tests showing carbon to be 100 times less absorbing than hydrogen, coupled with Fermi's idea of minimizing resonant capture in the heavy uranium isotope by localizing the uranium, led directly to the first successful critical assembly, CP-1.
Dec. 2, 1942	Fermi	CP-1 went critical.

This postulation led directly to the initiation of quantitative chemistry without which further progress would have been impossible.

Following the leads given by Lavoisier, Richter and Dalton reasoned that there was a proportional basis for the way in which atoms combined in chemical compounds. However, with little clues, Avogadro's quantum leap to suppose that all gases of equal temperature and volume would comprise identical numbers of molecules formed the basis upon which the following determinations of molecular and atomic weights by Berzelius and others were made.

By the early nineteenth century, the physical chemistry of substances had come full circle from the early days of the Greeks when Thales, six centuries before Christ, hypothesized that water was the source of all things. In 1815, Prout noted that the atomic weights in those days all seemed to be integral multiples of that of hydrogen. He then postulated that all atoms of various elements recognized up to that time could be further subdivided but that hydrogen was the basic, indivisible element from which all others were made.

Until the middle of the nineteenth century the study of matter and chemistry had proceeded separately from the infant field of electricity. This was true in spite of the introduction by Berzelius of the notions of valency in 1812. However, in 1848 Weber conceived that the general ideas of Faraday could be stated more firmly. He viewed the relationship between matter and electricity to be fixed so that a specific quantity of one was always associated with a fixed quantity of another in a way as yet undetermined.

During the next decade, Boltzman, Clausius and Maxwell developed the ideas on which the molecular theories of gases are founded, and provided a firm mathematical theory to explain this behavior. This was followed quickly by the work of Loschmidt who, in his studies of the internal friction of gases, obtained the first approximate estimate for the physical size of an atom. In fact, it has been only recently that Loschmidt's number, the number of molecules in a mole of a substance, has become to be known instead as Avogadro's number. This concept was extremely important since it allowed one to count molecules of a substance by weighing them. Thus, 28 grams of nitrogen would contain (as later accurately determined) 6.025×10^{23} molecules and 2 grams of hydrogen the same. The atomic weight of a hydrogen atom was therefore determinable as 1.66×10^{-24} gm, now called the "atomic mass unit."

It was not long after Loschmidt's first rough determination of atomic size that Hittdorf, in a series of cathode ray experiments, discovered "free atoms of electricity" and determined their approximate mass, thus providing the first physical evidence of an atomic structure. Earlier, however, Faraday had observed the relationship between quantities of electricity passed through an electrolyte solution and the amount of matter liberated in terms of atomic weight and valency. In 1834 his observations coalesced to become later known as Faraday's Laws, the one most important being that the quantity of electrical charge per mole was an even multiple of a constant value depending on the valence. This constant which today bears his name is 96,520 coulombs per gram atom. Stony then combined these results to conclude that the minimum subdivision of electrical charge of the "free atoms" observed by Hittdorf, which Stony now termed the "electron," would be the charge carried by a single electron. This quantity as we now know it is 96,520 coulombs per gram atom divided by 6.0247×10^{23} atoms per gram atom, or 1.602×10^{-19} coulombs per electron.

It had been over 2500 years since the basic nature of matter had been seriously considered. The existence of atoms, their numbers, sizes, and weights had been determined, and the electrical nature of atomic structure had for the first time been unlocked and quantified. Now, over little more than hal a century since these first rudimentary cornerstone discoveries, the rush to unlock the secrets of the nucleus would result in unleashing the power of the atom for production of energy.

In 1895 and 1896, Roentgen and Becquerel discovered x-rays and natural radiation from pitchblend, respectively. Within four years Marie and Pierre Curie succeeded in chemically separating radium from pitchblend and identified thorium and polonium as radionuclides. Rutherford followed two years later in his classification of radiation from uranium by their penetrating power, alphas, betas and gammas, and identified the decay constants proportional to what we now call half-lives. Ten years after this, Neils Bohr determined the relationship between energy quantum states and electron levels and the "Bohr Atom" was born. But it was Rutherford and Soddy who intuited that transmutation of the elements was not only possible but was the natural event which led to the discharge of an alpha particle. In just over one hundred years since Prout had hypothesized that hydrogen was the root of all matter, they showed by 1919 that protons in fact were fundamental nuclear particles.

In a discovery that was later to play a major role in nuclear theory, de Broglie in 1924 found that there was a duality between matter and energy which was to lay the foundation for development of quantum mechanics. Meanwhile, numerous experiments were being undertaken with pitchblend and its chemical constituents. A major quandry was reached in 1930 when Boethe and Becker bombarded beryllium and boron with 5 MeV alpha particles and noticed extremely penetrating radiation which they thought to be gammas. Calculations, however, showed the necessary energy to be near 50 MeV. A year later the Joliot-Curies repeated the experiments and found that a sheet of paraffin wax between their higher intensity polonium source and their detector substantially increased the activity where heavier sheets had much less or little effect--an apparent paradox.

During the next year Chadwick repeated similar experiments and through his measurements showed that the results could not be explained in terms of Compton scattering of gamma rays. He reasoned in 1932 that only a relatively heavy particle of the same order as the mass of the proton, but without charge, could have the penetrating power observed. He thus termed this particle the "neutron."

In quick succession Fermi, in his work at the University of Rome, subjected sequentially heavier elements to neutron bombardment and identified more than 40 which gave secondary emissions. These included uranium, which he suggested transmuted to an element of higher atomic number. At about the same time Pontecarro found that neutrons were slowed down selectively by the hydrogen atom in experiments with paraffin and water. Fermi continued his work developing ideas of quantum nuclear states to explain the strong selectivity in neutron absorption experiments. These were later amplified by Bohr and by Bret and Wigner in terms of selective metastabilities similar to those exhibited by electrons in their shells.

In the meantime, many workers were applying chemical analysis techniques on pitchblend following neutron bombardment. The emphasis was directed towards identification of new transuranium elements through chemical separation. Many different activities were identified which could not be reconciled with ideas or evidence of the existence of these heavy elements. The confusion was, however, ended in 1939 by Hahn and Strassman who identified a lighter element, barium, which was decaying into lanthanum, and immediately Meitner and Frisch concluded that the uranium was splitting into lighter elements. Their calculations showed a mass defect of 200 MeV which would be released by this "fissioning." Once the von Halbon experiments in 1939 showed more neutrons were released in the fission process, the beginning of the nuclear age was in sight.

It would only be a short time before Fermi's first experiment attempting to devise a self-sustaining chain reaction using water failed due to selective absorption by the heavier isotope of uranium and hydrogen absorption. But finding that carbon was 100 times less absorbing of neutrons than water, he immediately laid plans for another experiment.

The work leading up to the first successful critical assembly is well described by Profio.[4]

> "In October 1942, enough material became available to begin assembly of the full-scale, chain-reacting pile, so called because it was to consist of layers of graphite blocks, and graphite blocks containing uranium lumps, piled on top of one another. There were only six tons of pure uranium metal available. This was cast into 6-lb. cylinders, 2.25 inches in diameter, placed near the center of the pile. The rest of the lattice was filled out with UO_2 powder compressed into 'pseudospheres' about 3.25 inches in diameter. The basic lattice cell was 8.25 inches on a side. The purest graphite was used at the center, with less pure graphite farther out, ending with a foot of graphite alone as a neutron reflector. This, the world's first reactor, CP-1 (Chicago Pile No. 1), was built in a squash court under the West Stands of the unused football stadium at the University of Chicago. The original plan was to build a sphere 26 feet in diameter. However, as the piling progressed, it became evident that criticality would be achieved at a lower level, and the final shape approximated an ellipsoid ....
>
> "Interspersed in the pile were a number of control rods, consisting of cadmium sheets nailed to wood. Shim or coarse control rods were either pulled out completely, or inserted and locked in place. In addition, there were cocked safety rods, which would be released to fall into the core automatically, should the neutron intensity as measured by a BF_3 filled ionization chamber rise too high. Another rod, for fine regulation, could be controlled manually or through a servo system. The neutron multiplication was monitored by a BF_3 proportional counter located at the center of layer 11. The approach to critical was also monitored by the activation of a standard indium foil irradiated at the center of the pile.
>
> "On December 2, 1942, under Fermi's direction, the rods were pulled out one by one, the last in steps, each time predicting the critical rod position from the counting rate of the BF_3 detector. Finally, the first self-sustaining chain reaction in history was achieved. The pile was operated for 28 minutes at less than one-half watt, and then shut down, or "scrammed," by dropping the safety rods. The atomic age had arrived."

In rapid succession several other reactors were constructed, driven primarily by the military implications and the need to obtain plutonium. These are summarized in Table 1.3, including the two first nuclear reactors, one of water coolant design and the other using sodium, utilized for naval propulsion purposes.[11]

Table 1.3

Tabulation of Characteristics of Early Nuclear Reactors[A,B]

Year	Reactor	Country	Description*	Type	Moderator	Coolant	Power
1942	CP-1	USA	Natural uranium, graphite, ellipsoidal 3.88-m dia x 3.08-m high	Pile	C	-	200Wt
1943	CP-2	USA	Natural uranium, cubic lattice forming a roughly cubic core 6 meters high	Pile	C	Air	10kWt
1943	X-10	USA	Natural uranium fuel slugs 10-cm long x 2.5-cm dia in graphite matrix forming a cubic core ∿ 6 m on a side	Tank	C	Air	3.5MWt
1944	CP-3	USA	Natural uranium metal rods 2-m long in a square lattis forming a core of 2-m diameter with C and D_2O reflector	Tank	D_2O	D_2O	300kWt
1944	LOPO	USA	$UO_2SO_4 \cdot H_2O$ solution in a 30-cm sphere with beryllium oxide and graphite reflector	Aqu. Homog.	H_2O	-	-
1944	HYPO	USA	Similar to LOPO but with cooling coils inside sphere and higher concentration fuel in water	Aqu. Homog.	H_2O	H_2O	5.5kWt
1944	HEW-305	USA	Natural uranium arranged in a sphere lattice in a graphite matrix forming an elongated cube ∿4x4x5-m high	Pile	C	-	30Wt
1945	ZEEP	Canada	Natural uranium, plutonium, and U-235 fueled reactor having variable dimensioned elements in a tank of 200-cm diameter x 250-cm high	Tank	D_2O	-	100Wt
1946	Clementine	USA	Plutonium slugs with uranium reflectors of ∿2-cm diameter in close-packed peripheral loading forming a core 15-cm diameter x 15-cm high	Fast	-	Hg	25kWt
1947	NRX	Canada	Mixed natural and enriched fuel with plutonium in 3.5-cm diameter rods 3-m long in a hexagonal lattice forming a core 2.7-m diameter x 3 meters high	Tank	D_2O	H_2O	40MWt
1947	GLEEP	UK	A graphite, barn-shaped stack having natural uranium cylinders inserted horizontally in a square matrix forming a 5x5-m cylindrical core	Pile	C	Air	100kWt
1948	ZOE	France	Natural uranium slugs inside tubes in a hexagonal lattice forming a core 1.8-m diameter x 1.5-m high surrounded by a graphite reflector	Tank	D_2O	D_2O	150kWt
1948	BEPO	UK	30-cm long, 2-cm diameter rods with spiral aluminum fins and a graphite matrix with square lattice cylindrical core of ∿6-meter dimensions	Pile	C	Air	65MWt

Table 1.3 (Continued)

Year	Reactor	Country	Description*	Type	Moderator	Coolant	Power
1949	TR	SSSR	Slightly enriched uranium slugs in aluminum cladding in 64 fuel channels forming a square lattice in a core 1.1-m diameter x 1.22-m high	Tank	D_2O	D_2O	2.5MWt
1950	BGRR	USA	Bent fuel plates of highly enriched U-Al alloy on a square pitch in a graphite matrix forming a core of 4x5x6-m dimensions	Pile	C	Air	20MWt
1950	BSR-1	USA	MTR-type fuel plates 1.5-mm thick x 7-cm wide x 60-cm long, highly enriched, separated by rectangular coolant channels ∿4-mm thick forming a core 0.4x 0.45x0.6-m in size with a beryllium reflector	Pool	H_2O	H_2O	2MWt
1950	LITR	USA	MTR type fuel plates (see BSR-1 description) in a rectangular grid in a core of variable shape typically 38x70x61-cm with a beryllium reflector	Tank	H_2O	H_2O	3MWt
1951	TTR	USA	Fully enriched uranium disks on aluminum rods arranged in a circular lattice forming a hollow cylindrical core of 45-cm diameter x 45-cm high	Tank	H_2O C	-	100Wt
1951	EBR-1	USA	7.5-mm diameter fuel rods of plutonium arranged in a centered hexagonal lattice providing a cylindrical core 18-cm in diameter, 21-cm high	Fast	-	NaK	1.2MWt
1951	JEEP-1	Norway	Natural uranium slugs in cylinders on a square matrix forming a core 1.7-m diameter x 1.9-m high in a graphite reflector	Tank	D_2O	D_2O	450kWt
1951	WWR-2	SSSR	Slightly enriched uranium rods 0.5-cm dia x 0.5-m long in a square lattice forming a 40-cm-diameter cylinder 50-cm high	Tank	H_2O	H_2O	3MWt
1951	S1W	USA	Light water cooled and moderated PWR for naval propulsion in USS Nautilus, SSN 571	PWR	H_2O	H_2O	-
1952	MTR	USA	Rectangular fuel elements (see BSR-1) in a rectangular lattice forming a core 22x67x60-cm size with a beryllium reflector	Tank	H_2O	H_2O	40MWt
1952	HRE-1	USA	$UO_2SO_4 \cdot H_2O$ solution in a 45-cm diameter sphere cooled by circulating fuel solution set within a D_2O reflector	Aqu. Homog.	H_2O	H_2O	1MWt
1952	EL-2	France	Natural uranium slugs in 2.6-cm diameter rods 53-cm long forming a core diameter of 2-meters, 2.2-m high in a graphite reflector	Tank	D_2O	D_2O	2MWt

Table 1.3 (Continued)

Year	Reactor	Country	Description*	Type	Moderator	Coolant	Power
1952	ZETR	UK	Plutonium or highly enriched uranium as nitrate or fluoride in water. Cylindrical vessel core 12 or 30-cm diameter, 1-m high	Aqu. Homog.	H_2O	H_2O	-
1952	RPT	SSSR	Fully enriched fuel elements 102-cm long in a square lattice forming a core 100-cm diameter	Tank	C	H_2O	40MWt
1952	SR305	USA	Natural uranium slugs in a square graphite lattice forming an elongated 3-meter cube with graphite reflector	Pile	C	-	1kWt
1952	AE-6	USA	Highly enriched uranyl sulfate solution in a 30-cm sphere held within a graphite reflector. Cooling coil within core	Aqu. Homog.	H_2O	H_2O	3kWt
1953	SP	USA	Fully enriched uranium disks 38-cm long rods arranged in a circular lattice forming a hollow 45-cm O.D. cylindrical core 45-cm long	Tank	C	H_2O	10kWt
1953	NCSR-L	USA	First university owned and operated reactor. Highly enriched uranyl sulfate solution in a 26-cm diameter, 22-cm high core	Aqu. Homog.	H_2O	H_2O	10kWt
1953	BORAX-1	USA	MTR type fuel plates (see BSR-1) in a rectangular lattice forming a square core 50x50x60-cm high	Tank	H_2O	H_2O	1.4MWt
1953	PDP	USA	A process development reactor of variable components and geometry in a reactor tank of 5-m diameter, 5-m high	Tank	D_2O	D_2O	1kWt
1960	S1G	USA	Sodium cooled reactor for naval propulsion in USS Seawolf, SSN 575	Tank	-	Na	-

*All dimensions approximate.

[A] I.A.E.A. Directory of Nuclear Reactors, Vols. II-VI, Vienna, 1959-1964.

[B] U.S. Government Printing Office, Tech. Info. Center, "Nuclear Reactors - Built, Planned, or Being Built," TID-8200-R39, March 1979.

2. FUNDAMENTAL CONCEPTS OF NUCLEAR ENERGY

While this course does not emphasize the nuclear aspects of nuclear power production, there are certain factors relating to the physical behavior of neutrons in the environment of a nuclear reactor that should be grasped in order for the reader to understand the thermal hydraulic behavior of such a system. This is because the nuclear and thermal hydraulic processes are coupled. For instance, the presence or absence of a coolant vapor region in a liquid-cooled reactor would affect the neutron energies in this zone, which in turn affects the local fission probability. This then has an effect on the power level and, as a result, the extent or behavior of the vapor region may change. To understand the importance of these interactions and the associated thermal hydraulic interactions, one must gain a feel for such basic terms as "reactivity," "cross section," "thermal utilization," etc. The purpose of this section is to provide the reader with a very brief idea of the physics of nuclear interactions in a mostly qualitative manner.

2.1 The Fission Process

The basic idea exploited in production of energy through the nuclear reaction is the release of energy through the process of splitting or fissioning a heavy atom to yield two lighter ones having less total mass than the original nucleus. The difference, or mass defect, of the combined resulting system of fragments and particles appears as energy.

Aston originally made a systematic study of the mass defect of nuclides in terms of the ratio of the mass defect per unit mass number (sum of neutrons and protons) which he called packing density.[7] His results, shown in Figure 1.1, were based on the standard of 16.0000 for the atomic weight of oxygen. The packing density data thus have a zero (relative) defect at this value of mass number.

A more recent curve of binding energy per unit mass number, shown in Figure 1.2, may be compared with Figure 1.1. The message contained in these two figures is that there is energy to be obtained through nuclear reactions on either side of the minimum. On the light side, combining two nuclides to form a larger one will liberate energy. This process is called fusion, itself the current subject of intensive physical investigations. On the heavy side of the minimum, large nuclides can break apart (fission), yielding two lighter atoms with a similar release of energy.

In the case of the rare ($\sim$ 0.7%) isotope of uranium $_{92}U^{235}$ (which has 92 protons and 143 neutrons in the nucleus for a total of 235 nucleons), the fission process occurs schematically as shown in Figure 1.3.[4] A neutron of low energy generally at equilibrium with its surroundings (thermal) is absorbed by an atom of uranium. In an extremely short time (order of microseconds) this unstable nuclear configuration fissions, breaking into two lighter nuclides* called "fission fragments" or "fission products," some of which may themselves be unstable. In addition, beta and gamma rays are given off during fission and more neutrons, between two and three on the average, are produced directly by the fission process. In addition, the unstable fission products themselves decay over a period of time following their formation, yielding more betas,

---

* The case where three separate nuclides form rather than two is approximately 1,000 times less probable than binary fission, but does occur.

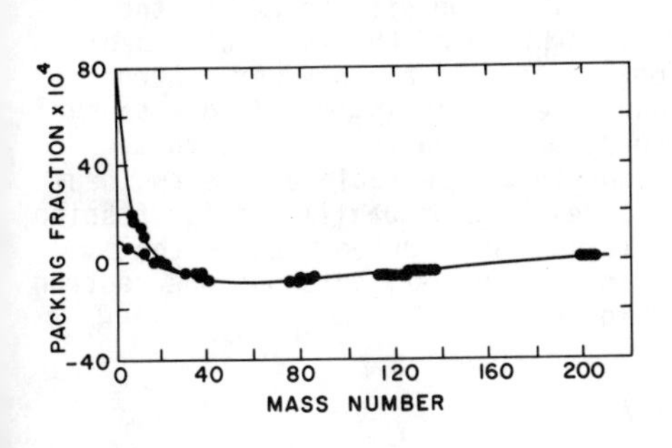

Figure 1.1. Packing fractions of various nuclides determined by Aston.[7] (BNL Neg. 7-875-80)

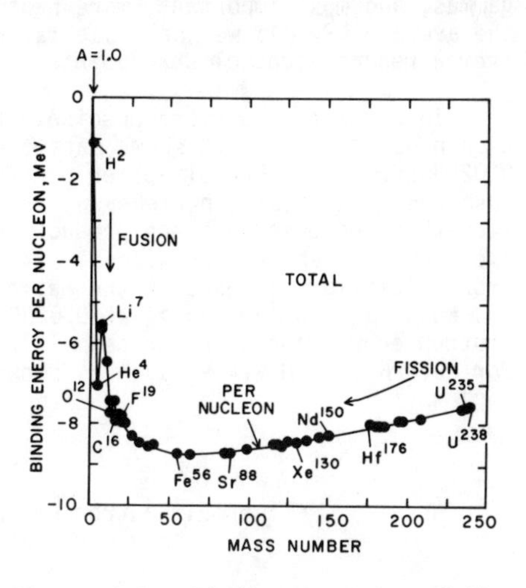

Figure 1.2. Binding energies of the stable nuclides. (BNL Neg. 7-871-80)

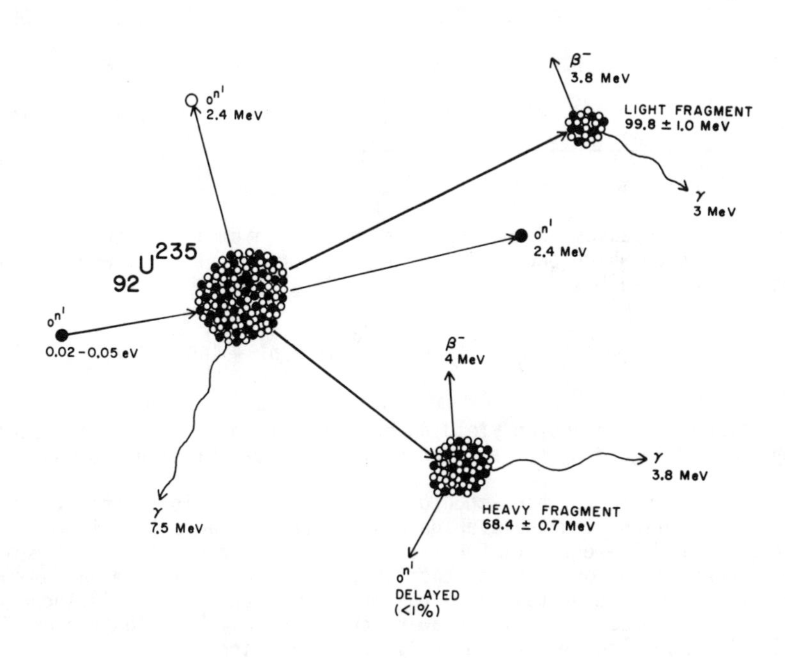

Figure 1.3. Schematic of nuclear fission of $_{92}U^{235}$ including both prompt and secondary emissions and energies. (BNL Neg. 7-862-80)

gammas, and most important, more neutrons, approximately another 1% neutrons on the average.[5] As we shall see later, these neutrons are most important from a reactor control standpoint.

In order to show more precisely the multiplication effects due to the fission process, Table 1.4 shows data for the absorption of thermal neutrons of 0.0253 eV. Both the multiplier per fission, ν, as well as that per neutron absorbed by a fissile nucleus, η, are shown. The latter is smaller due to the fact that there is some non productive absorption of neutrons by the fuel, i.e., neutron absorption without fission occurring. The ratio of the two neutron multipliers is exactly the ratio of the relative probabilities for fission and total absorption, σ_f/σ_a at 0.0253 eV. It should be noted that as the neutron energy increases these multipliers increase slowly by about one neutron for each 6 to 7 MeV,[16] such as shown in Figure 1.4.

Table 1.4

Thermal Neutron Production Ratio per Fission and per Neutron Absorbed[15]

Nucleus	Number of Fission Neutrons Produced per Fission - ν	Number of Fission Neutrons Produced per Thermal Neutron Absorbed - η
$_{92}U^{233}$	2.492 ± 0.008	2.287 ± 0.007
$_{91}U^{235}$	2.418 ± 0.008	2.068 ± 0.009
$_{94}Pu^{239}$	2.871 ± 0.006	2.108 ± 0.008
$_{94}Pu^{241}$	2.927 ± 0.014	2.145 ± 0.014

The energy releases of both uranium and plutonium are given in Table 1.5. A typical reaction which occurs in the fission process is that where barium and krypton are released,[5,8]

$$ {}_0n^1 + {}_{92}U^{235} \rightarrow {}_{56}Ba^{137} + {}_{36}Kr^{97} + 2\,{}_0n^1 + 189.6 \text{ MeV}. $$

In fact, it was this reaction yielding barium that led to Hahn's original conclusion in 1939 regarding the existence of the actual fission process.

There are two facts which should be noticed from this reaction. The first, by comparison with the average quantitites in Table 1.5, is that the energy release is different from the average. This is because there are many possible combinations of fission reactions, one of which is the barium-krypton product reaction. The actual distribution of fission products is shown in Figure 1.5. (The reader should be aware that the normalization factor for the yield is 200% since there are two nuclides per fission.)

The second fact that should be kept in mind is that by far the majority of the fission energy is associated with the large fission fragments as shown in Table 1.5, and very little with the neutrons or other radiation. The fission

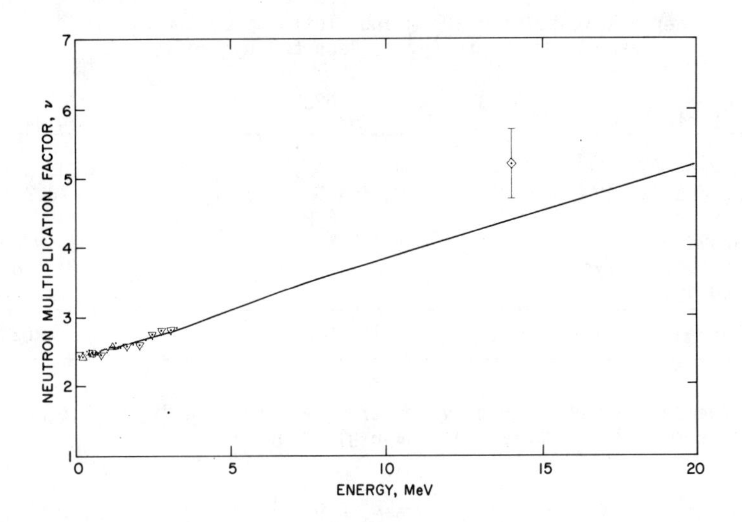

Figure 1.4. Neutron production ratio per fission, ν, for $_{92}U^{235}$, (Reference 23). (BNL Neg. 10-769-80)

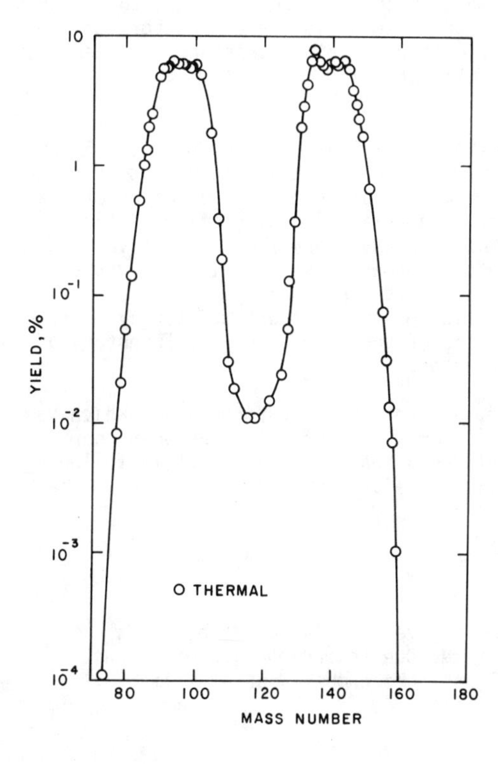

Figure 1.5. Fission product yield as a function of mass number for $_{92}U^{235}$ for thermal neutrons (0.0253 MeV).[13] (BNL Neg. 7-873-80)

Table 1.5

Energy Release of Uranium and Plutonium During Fission Taken From Keepin[12] as Reported by Profio[4]

Particle	${}_{92}U^{235}$	${}_{94}Pu^{239}$
Light Fragment (av.)	99.8 ± 1 MeV	101.8 ± 1 MeV
Heavy Fragment (av.)	68.4 ± 0.7 MeV	73.2 ± 0.7
Prompt Neutron	4.8 MeV	5.8
Prompt γ-rays	7.5	∿ 7
Fission Product β-rays	7.8	∿ 8
Fission Product γ-rays	6.8	∿ 6.2
Total Energy	195 MeV	202 MeV

neutrons themselves have an energy spectrum, as shown in Figure 1.6, which is well represented by the Maxwellian shape given by

$$n(E) = \frac{2}{\sqrt{\pi}\, T^{3/2}} E^{\frac{1}{2}} \exp\left(-\frac{E}{T}\right)$$

with an average energy given by the nuclear temperature $\overline{E}=3T/2=1.94\pm.05$MeV. Thus, while the neutrons, being extremely penetrating due to their lack of charge, are free to move to areas outside the reaction and there to give up their energy, the fission fragments instead deposit their energy, 168.2 MeV on the average, directly in the fuel regions. In solid fueled reactors this generally presents the biggest source of difficulty since this energy appears as a heat source, requiring that the fuel elements must be continuously cooled to maintain their integrity.

A third fact not immediately obvious from Table 1.5 and Figure 1.5 but nevertheless as important from the viewpoint of nuclear safety heat transfer, is the fact that a large fraction of the fission products are radionuclides. These nuclides decay according to their half-lives* giving off radiation and thermal energy. Some of these nuclides have very short half-lives, in the order of milliseconds or less, while some others have half-lives measured in thousands of years. While a steady state fission process is being experienced in a neutron-fuel environment, these fission products are constantly being generated, their radioactivity forming part of the overall energy inventory of the nuclear reaction system or nuclear reactor.

Once the reaction is terminated, however, these radionuclides which have been formed continue to release heat. For a reactor which has been operating for some length of time, this initial decay heat level immediately following shutdown of the reaction may be 5-10% of the preshutdown power level. The

---

* The half-life is the length of time a given population of nucleons will have 50% of their population altered due to radioactive emissions. If N_0 is the original number of nucleons, the number N at any time is $N = N_0 \exp(-0.693t/t_{\frac{1}{2}})$ when $t_{\frac{1}{2}}$ is the half-life.

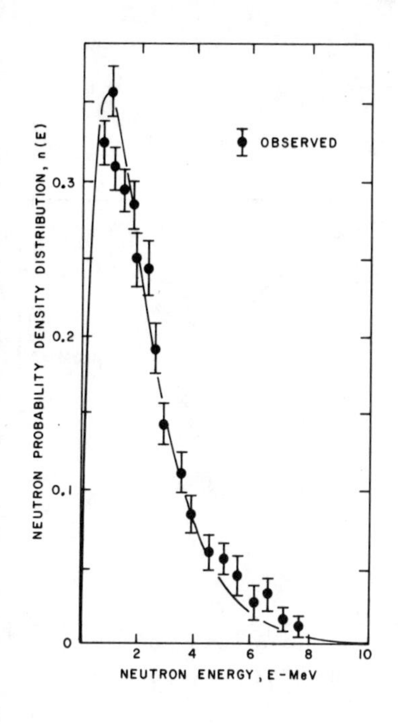

Figure 1.6. Prompt neutron energy spectrum, (data from Reference 13). (BNL Neg. 7-874-80)

actual value varies with length of operation prior to shutdown (Figure 1.7 a-d), and also with the type of fuel used. Even two months after shutdown, a 3000 MWt reactor (i.e., one generating 3000 megawatts of thermal energy) may still be generating 3 MWt (Figure 1.8). Unlike conventional power sources which stop when they are turned off, a nuclear power source does not. Hence, provisions must be made to continuously remove this "decay heat" energy following termination of operation to keep the fuel elements from overheating and becoming damaged.

Finally, the radioactivity itself in the fission products inventory built up in a reactor during operation presents a substantial potential safety hazard. This is because all penetrating radiation can interact with the atomic structure in the tissues of the body, causing mutation and damage which may, in sufficiently large doses, prove fatal. Therefore, the operational and systems safety design of any nuclear power system must take into account the previously mentioned factors in order to provide a workable, technically and economically feasible device which should be virtually unbreachable under all normal and abnormal conditions which may be encountered. It is toward this end that such heavy emphasis is being placed on nuclear safety research in general and thermal hydraulic research in particular. It is also mainly for this reason that the uranium fuel of a nuclear reactor is usually confined inside a metallic container, usually in the form of small diameter cylindrical rods or plates perhaps a centimeter thick or in diameter and one or more meters long. These fuel elements are then typically arranged in a lattice occupying about 50-60% of the total reactive volume.

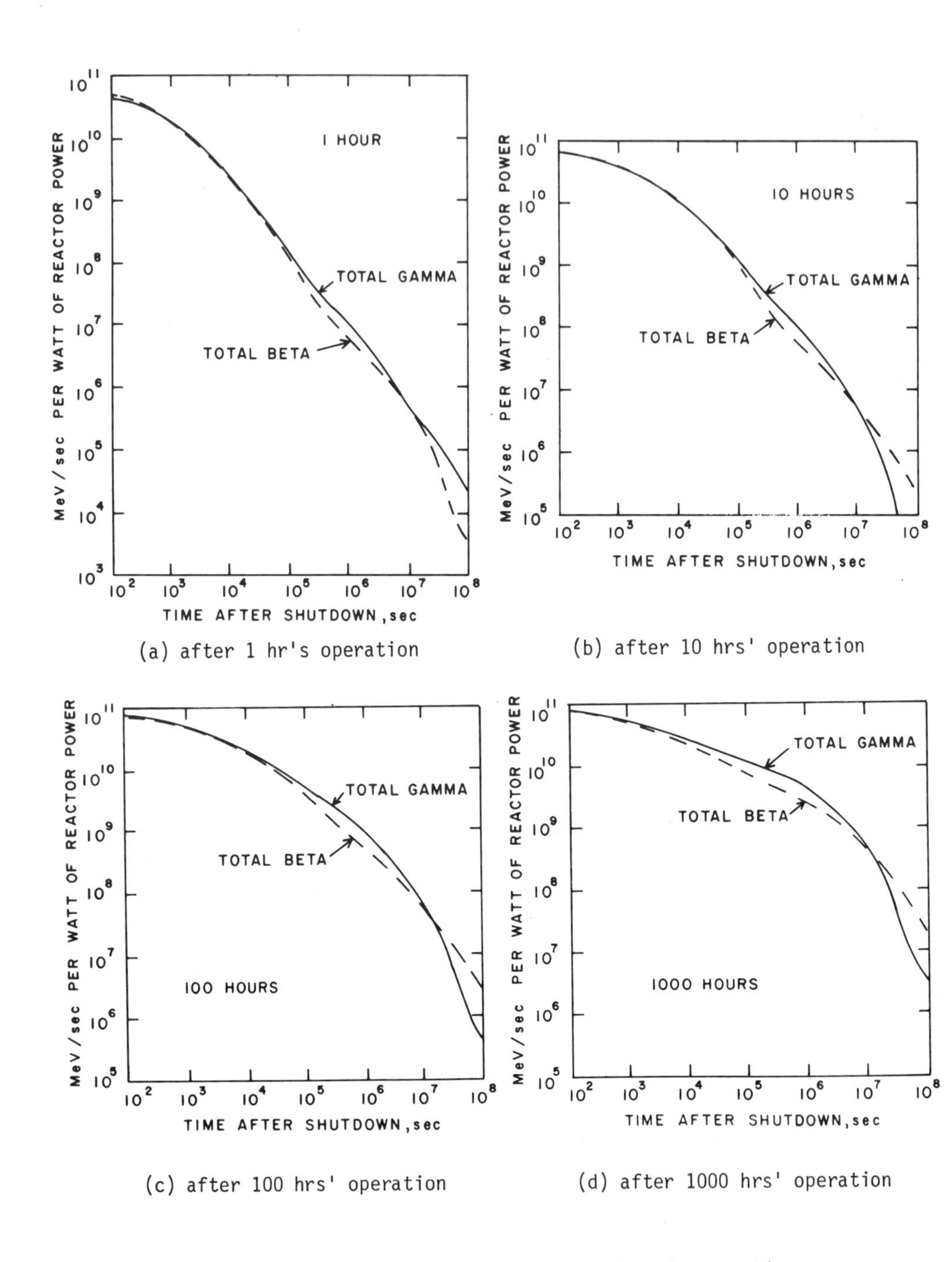

(a) after 1 hr's operation

(b) after 10 hrs' operation

(c) after 100 hrs' operation

(d) after 1000 hrs' operation

Figure 1.7. Decay rates after reactor shutdown for reaction operation of: (a) 1 hr; (b) 10 hrs; (c) 100 hrs; (d) 1000 hrs.[13] (BNL Neg. 7-866-80)

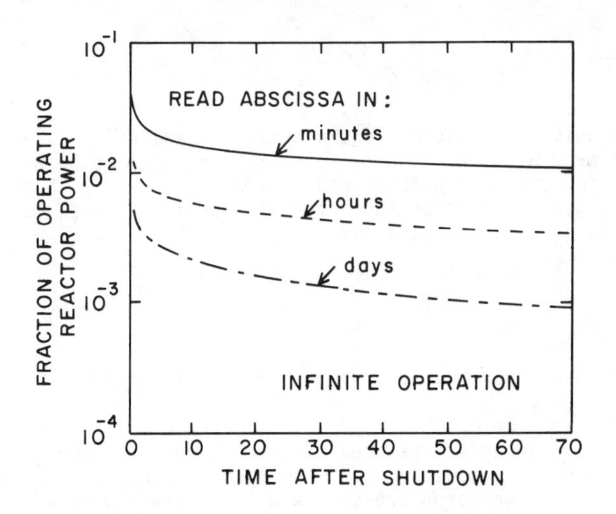

Figure 1.8. Decay heat power level due to fission products after shutdown of a reactor following "infinite" operation.[13] (BNL Neg. 7-872-80)

2.2 Neutron Interactions

The previous section described the fission process, the effect of neutron multiplication, and the energies involved if such an event were to take place. However, nothing has yet been said regarding the likelihood of a fission event actually happening. There are many competing processes which tend to remove neutrons from the neutron inventory of a reactor in a nonproductive manner, i.e., without leading to a fission reaction. From Table 1.4 it is obvious that a reactor using ${}_{92}U^{235}$ as a fuel must lose no more than 1.47 neutrons per fission interaction to operate at steady state.

Once a neutron is born only three things can occur which will directly affect the neutron population as a whole:

a) The neutron can be absorbed productively, yielding more neutrons.

b) The neutron can be absorbed unproductively.

c) The neutron can leak out of the reactor.

In order for a neutron to be lost by leakage out of a reactor environment, the reactor must be finite in size. In considering the behavior of a given system, it becomes convenient to consider first an infinitely large geometry, from which no leakage can occur. Then, the only considerations which become important are the relative neutron interaction probabilities. These can be specified in terms of what have become known as cross sections, as discussed in the next sections.

Microscopic Cross Sections. Consider a beam of neutrons expressed in terms of a neutron flux vector $\hat{\phi}$ neutrons per second per cm^2. If an arbitrary area is placed to intersect the neutron beam, the neutron crossing rate is

$$C = \int_A \hat{\phi} \cdot \hat{n} dA = \phi_n A \quad (1)$$

where $\hat{n}$ is the unit normal to A and the nonvector value ϕ_n is taken as the average of $\hat{\phi} \cdot n$ over A. Since the crossing rate is seen to be area-dependent, it becomes obvious that for spatial scales small compared with those where significant changes in ϕ_n occur the ratio of crossing rate C_j for a locally nonspecific area A_j which is part of A,

$$p_j = \frac{C_j}{C} = \frac{A_j}{A} \quad (2)$$

is simply the probability that a given neutron will cross A_j.

By analogy, an atomic nucleus presents an effective area σ for a nuclear interaction. It is an effective area because it seems to change in value under different circumstances, neutron energy being one major factor. Nevertheless, the probability that a neutron crossing A will interact with a given nucleus in a particular reaction χ is

$$p_\chi = \frac{\sigma_\chi}{A} \quad (3)$$

While physical nuclear areas are all on the order of 10^{-24} cm^2, the effective areas, σ_χ's, may in many cases be orders of magnitude larger. Rather than carry the exponent, it has become standard practice to use another, more convenient term, the barn, where 1.0 b = 10^{-24} cm^2.

Considering mutually exclusive interactions, the total probability that an interaction of any kind will occur is

$$p = \sum_j p_j = \frac{\sigma_1}{A} + \frac{\sigma_2}{A} + \ldots + \frac{\sigma_\chi}{A} + \ldots \quad (4)$$

from which the total effective area or "cross section," σ_t, is determined to be

$$\sigma_t = \sum_j \sigma_j \quad . \quad (5)$$

The different types of interactions which may occur lead to the following cross section definitions:

1. σ_s = elastic scattering cross section
2. σ_i = inelastic scattering cross section
3. σ_γ = capture cross section (n,γ reaction)
4. σ_p = capture cross section (n,p reaction)
5. σ_α = capture cross section (n,α reaction)
6. σ_f = fission cross section

The total absorptions include the latter four interactions so that an absorption cross section is usually defined as

$$\sigma_a = \sigma_f + \sigma_\gamma + \sigma_p + \sigma_\alpha + \ldots \tag{6}$$

Other absorptive reactions which can occur are simply added to (6).

It is instructive to examine the relative magnitudes of cross sections of selected elements or isotopes as shown in Table 1.6. It is seen that boron is 16 times more likely to absorb a thermal neutron per nuclide than is hydrogen, while carbon is 10 times less likely. Similarly, $_{92}U^{235}$ is 13 times more likely and a nucleus of $_{94}Pu^{241}$ is 26 times more absorptive of thermal neutrons than is hydrogen. On the other hand, it is interesting to note that at neutron energies of 1.0 MeV all the total cross sections noted are of approximately the same magnitude.

Table 1.6

Approximate Total Cross Sections for Selected Naturally Occurring Elements (No Superscripts) or Isotopes as Read From Charts in Reference 16

Nuclide	Total Thermal Cross Section σ_t(barns)	Total 1.0 MeV Cross Section σ_t(barns)
$_1H$	53	4.2
$_5B$	800	1.7
$_6C$	5	2.6
$_8O$	4.3	8.0
$_{13}Al^{27}$	1.5	3.1
$_{26}Fe$	14	3
$_{28}Ni$	21.5	3
$_{40}Zr$	6.4	6.3
$_{54}Xe$	28	6.8
$_{92}U^{233}$	600	6.8
$_{92}U^{235}$	690	6.8
$_{92}U^{238}$	15	6.8
$_{94}Pu^{238}$	600	2.2
$_{94}Pu^{239}$	1000	7.0
$_{94}Pu^{241}$	1400	≥ 1.5 (σ_f only)

Point behavior, however, is not necessarily a good indication of the cross section for a given material, as Figures 1.9-1.11 indicate for the elements hydrogen and boron, and the isotope plutonium-239. Any given isotope can exhibit regions where the total cross section falls rapidly with energy increase, remain constant, or even show resonant peaks and valleys due to its quantum state values. These peaks in the cross sections are termed "resonances."

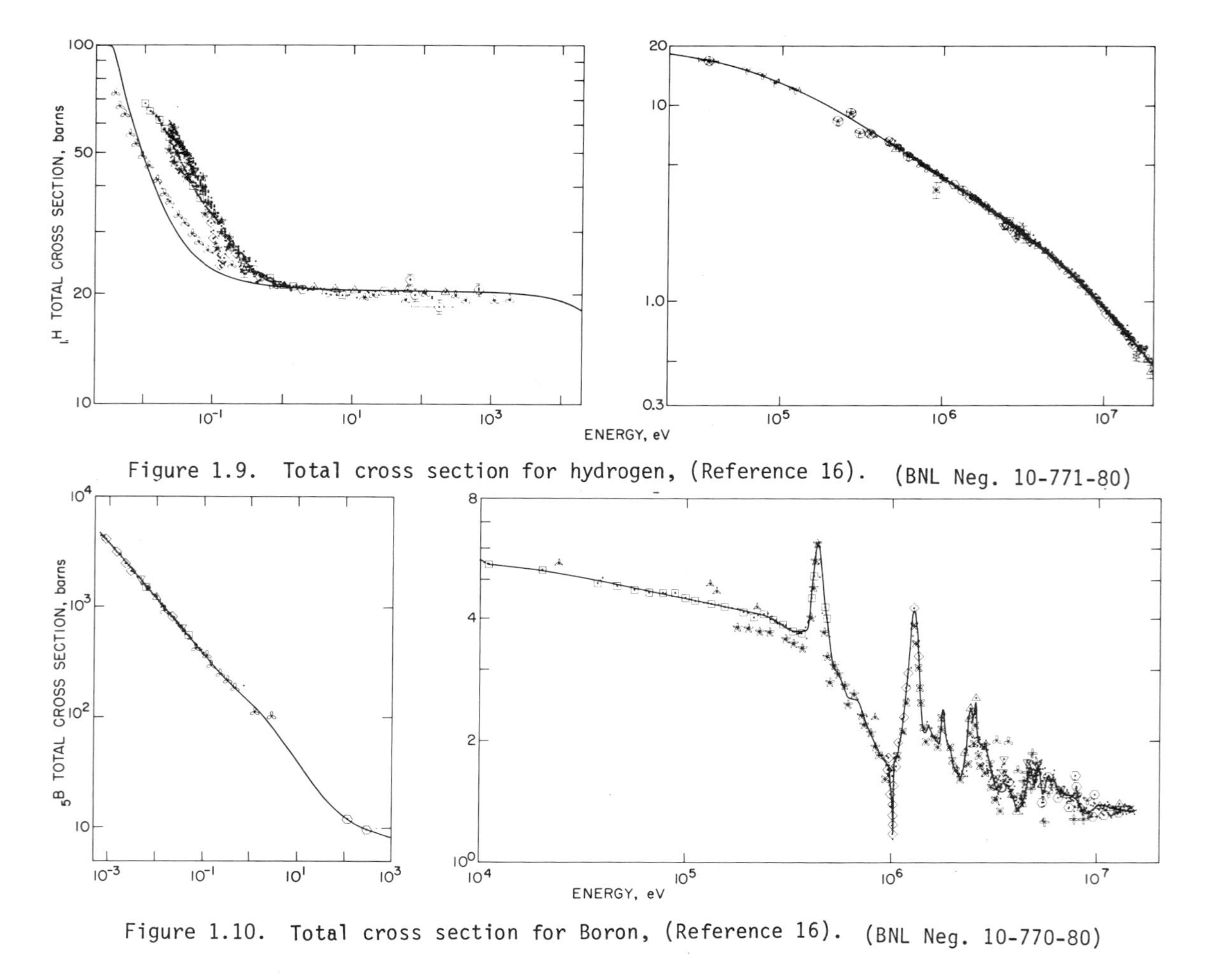

Figure 1.9. Total cross section for hydrogen, (Reference 16). (BNL Neg. 10-771-80)

Figure 1.10. Total cross section for Boron, (Reference 16). (BNL Neg. 10-770-80)

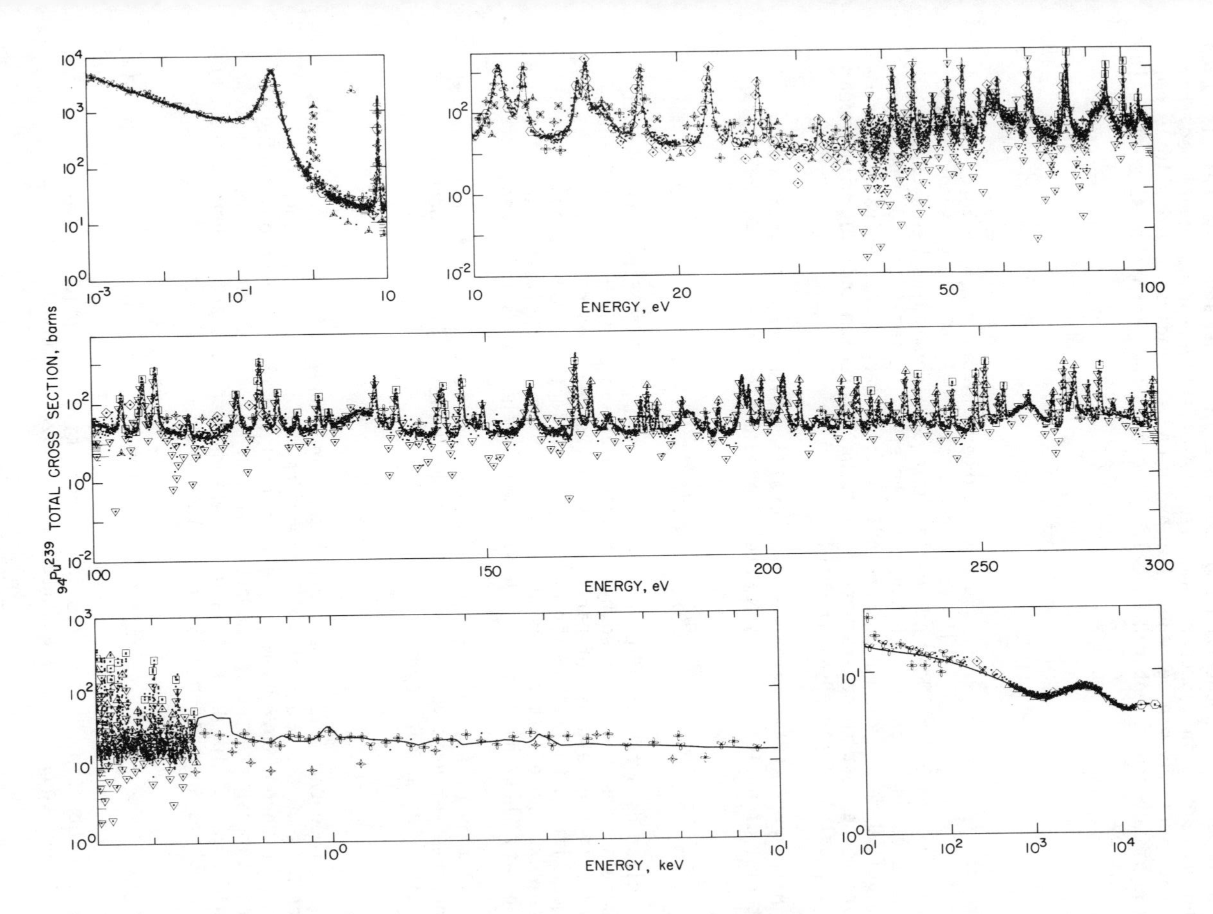

Figure 1.11. Total cross section for plutonium-239, (Reference 16). (BNL Neg. 10-772-80)

A useful rule of thumb developed by Neils Bohr for calculating the radius at which the Coulomb forces just balance the short range nuclear attractive forces is given by[18]

$$r = 1.4 \times 10^{-13} A^{1/3} \tag{7}$$

in centimeters. A is the number of nucleons in the nucleus. If this is taken as the nuclear radius, comparisons with the effective radius determined from the cross sections can be made. For hydrogen, the radius is 1.4×10^{-13} cm, whereas if the equivalent radius calculated from the thermal cross section in Table 1.6 by

$$r_e = \sqrt{\frac{\sigma_t}{\pi}} \tag{8}$$

the result is 41×10^{-13} or ~ 30 times larger. Similarly, for boron $r = 3.1 \times 10^{-13}$ cm and $r_e = 1.6 \times 10^{-11}$ cm, or approximately 50 times larger. Also, for $_{94}Pu^{241}$ $r = 8.7 \times 10^{-13}$ cm, while the equivalent radius based on the total thermal cross section is ~ 25 times larger at 2.1×10^{-11} cm. On the other hand, at 1.0 MeV, hydrogen appears only 8 times larger than actual; boron appears to be just twice its actual size, while uranium and plutonium appear nearly equal to their physical size.

Macroscopic Cross Section. The total reaction rate for interaction χ in a material of nuclear density N, area A, and differential thickness dx, is

$$dR_\chi = (\phi_n \sigma_\chi)(NAdx) \quad . \tag{9}$$

The nuclear density is, of course, given by the product of physical density, ρ, Avagadro's number N_A, divided by the molecular weight M; $N = \rho N_A/M$. The quantity in the left set of parentheses is the reaction rate per nucleus, and the quantity in the right pair of parentheses is the number of nuclei in the volume Adx. Note that the product $\sigma_\chi NAdx$ divided by the target area A is the probability that the reaction χ will occur in the target thickness, dx.

For many cases, workers find it convenient to discuss effective interaction area per unit volume rather than per nuclei. Thus, the product $N\sigma_\chi$ has been given the special symbol

$$\Sigma_\chi \equiv N\sigma_\chi \tag{10}$$

and is termed the "macroscopic cross section," conventionally having units (since N is of the order of 10^{24}) of cm^2/cm^3 or cm^{-1}. The reasons for this usage become obvious when one considers the flux of neutrons moving in a direction represented by the coordinate x. If a neutron balance is made on an element of length in this direction when changes are occurring due to nonproductive absorption only, it is seen that

$$\phi_n(x) = \phi_n(x + dx) + \frac{R_{na}(x)}{A} \tag{11}$$

where the subscript "na" refers to nonproductive absorption. When ϕ_n is expanded and Equations (9) and (10) are utilized, (11) becomes

$$\frac{d\phi_n}{dx} = - \Sigma_{na}\phi_n \quad . \tag{12}$$

Integrating with the initial condition $\phi_n = \phi_{no}$ at $x = x_o = 0$ for convenience yields

$$\phi_n = \phi_{no} e^{-\Sigma_{na} x} . \tag{13}$$

Note that this equation holds only when the neutron-nuclide interactions result in nonproductive absorption, and does not consider the case where neutrons are simply scattered as in the case of shielding or a reflector, nor the case where fission occurs as a result of the absorption, thereby yielding more neutrons. From (13) one sees that a characteristic absorption length where neutron flux is decreased one e-fold is identically the inverse of the macroscopic absorption cross section,

$$\lambda_{na} = \frac{1}{\Sigma_{na}} . \tag{14}$$

Noting that Equation (12) can be rearranged to yield

$$\frac{-\,d\phi_n}{\phi_n} = \Sigma_{na} dx , \tag{15}$$

it is immediately recognized that the quantity Σ_{na} can be considered the probability per unit neutron path length that an interaction will occur in dx. Stated mathematically, since $\phi_n(x)$ is the intensity of noninteracted neutrons at x, and $\phi_n(x)/\phi_o = \exp(-\Sigma_{na}x)$ is therefore the probability that a neutron will survive up to the point x, the product of the two probabilities yields the probability that an absorption will occur within dx only. Thus

$$p(x)dx = e^{-\Sigma_{na} x} \, \Sigma_{na} dx \tag{16}$$

represents this differential probability. The expectation that a neutron will survive a distance x without interaction can be expressed in terms of the mean of the distances traveled or "mean free path" given by

$$\lambda_{na} = \int_0^\infty xp(x)dx = \Sigma_{na} \int_0^\infty xe^{-\Sigma_{na} x} \, dx = \frac{1}{\Sigma_{na}} . \tag{17}$$

Comparison with Equation (14) shows that the mean free path, or the expected distance a neutron will travel without absorption occurring, is exactly given by the inverse of the macroscopic absorption cross section, and is exactly the distance over which an e-fold decrease in neutron flux would occur. It can be shown that this result also applies to all interactions, including scattering and fission, so that in general,

$$\lambda_t = \sum_j \frac{1}{\Sigma_j} \tag{18}$$

where λ_t may be considered the mean free path for all neutron interactions. Note that similarly mean free paths can be defined for scattering--λ_s, fission--λ_f, etc., and a comparison of these yields a good indication of the likelihood of one interaction occurring versus another.

Elastic and Inelastic Scattering. In addition to absorptive interactions between neutrons and other nuclei, neutrons can effectively impact other nuclei in a way such that both recoil and move off from each other, conserving momentum and energy through both their velocities and possible exitation of the target nucleus. Such a situation is diagramed in Figure 1.12. Ignoring initial velocity for the Ψ-nuclide, the momentum and energy conservation equations are

$$\hat{P}_o = \hat{P}_n + \hat{P}_\Psi \tag{19}$$

and

$$E_o = E_n + E_\Psi + E^* \tag{20}$$

where E* is the additional energy carried by the reactant nucleon as exitation energy. If the exitation energy E* is nonzero, the scattering is inelastic. Since ground states for nuclear absorptions usually require neutron energies to be several MeV or more, the elastic scattering case is much more probable. Recalling that $P^2 = 2mE$ and letting $m_\Psi/m_n = A$, Equations (19) and (20) may be combined to yield in the elastic case ($E^* = 0$)

$$(A+1)E_n - 2\sqrt{E_n}\sqrt{E_\Psi}\cos\Theta_n - (A-1)E_\Psi = 0 \tag{21}$$

which may be solved for E_n quadratically to yield[14]

$$E_n = \frac{E_o}{(A+1)^2}\left[\cos\Theta_n + \sqrt{A^2 - \sin^2\Theta_n}\right]^2. \tag{22}$$

The maximum fractional energy loss per collision occurs when $\Theta_n = \pi$ and is

$$\left(\frac{E_n}{E_o}\right)_{min} = \left(\frac{A-1}{A+1}\right)^2 \equiv \alpha_\ell \tag{23}$$

where α_ℓ is defined as the "collision parameter." It is obvious that to be most effective at reducing neutron energies (or slowing down neutrons), a nucleon α_ℓ must be light. Typical values of α are given in Table 1.7.

While Equation (23) indicates the maximum energy loss in an elastic collision, it says nothing about the average value, usually given in terms of the logarithmic energy decrement ξ, or "lethargy." This decrement, defined as

$$\xi \equiv \ell n\, E_o - \overline{\ell n\, E} \tag{24}$$

may be shown to be given by

$$\xi = 1 - \frac{(A-1)^2}{2A}\,\ell n\left[\frac{A+1}{A-1}\right] \tag{25}$$

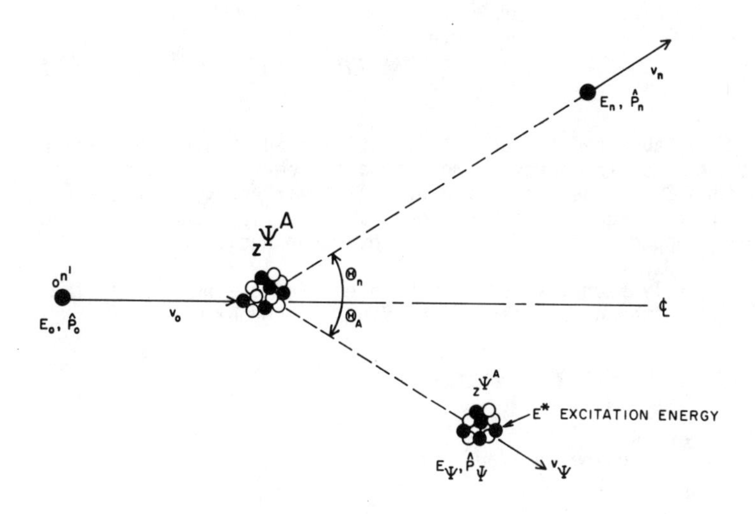

Figure 1.12. Diagram for neutron scattering by a nucleus $_Z\Psi^A$. (BNL Neg. 7-878-80)

Table 1.7

Collision Parameters for Neutrons Slowing Down by Elastic Scattering Collisions With Various Nucleons

Nucleus	A	α_ℓ	$\overline{\Delta E}/E_0$	n_c
Hydrogen	1	0	0.6321	18
H_2O		*	0.601**	20
Duterium	2	0.111	0.516	25
D_2O		*	0.399**	36
Beryllium	9	0.640	0.187	88
Carbon	12	0.716	0.146	115
Oxygen	16	0.779	0.113	152
Sodium	23	0.840	0.0810	215
Zirconium	40	0.905	0.0480	362
Iron	56	0.931	0.0347	515
Uranium	238	0.983	0.00834	2171

*Not defined.
**An appropriately averaged value.[14]

This relationship may be used to determine the average fractional energy loss in an elastic collision to be

$$\frac{\overline{\Delta E}}{E_o} = 1 - e^{-\xi} \tag{26}$$

which is also tabulated in Table 1.7. It is readily seen that while neutrons lose only 2% of their energy at most in a scattering interaction with uranium and less than 1% on the average, over half their energy on the average is lost through interactions with hydrogen and, in fact, the neutron could be completely stopped by colliding with a hydrogen nucleus.

It is easily seen that since $\overline{E/E_o} = e^{-\xi}$ for a single collision on the average, that after n collisions, the final energy E_f will be

$$E_f = E_o e^{-n\xi} \tag{27}$$

so that it will require an average of

$$n_c = \frac{\ell n\ E_o/E_f}{\xi} \tag{28}$$

collisions to slow a neutron from energy E_o to some final energy E_f. For hydrogen, approximately 18 collisions would be required to slow a neutron down from 2 MeV to 0.0253 thermal energies. On the other hand, over 2000 would be required in uranium. Other values are shown in Table 1.7.

The process of slowing down of neutrons is termed "moderation," and the nuclides most effective in the slowing down process are called moderators. It is seen from Table 1.7 why the best moderators are those which contain hydrogen (water and duterium) or carbon or beryllium. Note that the relative probability for the scattering of neutrons is given by the scattering microscopic or macroscopic cross sections, σ_s and Σ_s, and the scattering mean free path is given by

$$\lambda_s = \frac{1}{\Sigma_s} \ . \tag{29}$$

Comparative values of thermal scattering and absorption figures are given in Table 1.8. It is obvious that at thermal energies the common moderators are all much better scatterers than absorbers, while iron tends to parasitically absorb neutrons quite readily in comparison with its scattering ability. Interestingly enough, uranium is just as likely to scatter thermal neutrons as to absorb them. It is also of interest that the mean free path for neutron scattering in water, shown in Table 1.8, is also of the same order as the subchannel sizes in the rod bundle geometries of many nuclear reactors. Finally, comparison of the mean free paths for scattering vs. absorption for iron and zirconium indicates that the latter has a much less relative absorption probability which is why, in spite of the mechanical difficulties of working with the material, its use as a fuel element containment (clad) material has been pursued. Nevertheless, stainless steel remains a predominant cladding material in use today.

Table 1.8

Thermal Scattering and Absorption Comparisons for Selected Materials (0.0254 eV).[14]

Material	σ_a b	σ_s b	Σ_a cm^{-1}	Σ_s cm^{-1}	λ_a cm	λ_s cm
Water	0.664	103	0.0222	3.443	45	0.29
Heavy Water	0.00133	13.6	4.4×10^{-5}	0.452	2.3×10^4	2.21
Carbon	0.0034	4.75	2.7×10^{-4}	0.3811	4.2×10^3	2.62
Beryllium	0.0092	6.14	0.00114	0.759	877	1.32
Zirconium	0.185	6.40	0.00794	0.2746	126	3.64
Iron	2.55	10.9	0.2164	0.9251	4.62	1.08
Uranium	7.59*	8.90	0.3668	0.4301	2.73	2.33

*σ_f = 4.19 b.

Multigroup Concepts. The previous discussions have shown that cross sections, and hence interaction probabilities, are energy-dependent. On the other hand, all the equations so far considered have dealt with single energy concepts only. While the consideration of energy as a variable is much beyond the scope of this chapter, and certainly outside the range of this book, it is worthwhile to discuss energy concepts briefly in order to place the ideas developed into proper context. This discussion will follow closely that given by Lamarsh.[14]

If n(E)dE represents the neutron density in the energy band dE, then the differential energy flux is simply

$$d\phi(E) = n(E)v(E)dE \quad . \tag{30}$$

From Equations (9) and (10), the reaction rate density in this energy band is given by

$$dR_\chi(E) = n(E)\Sigma_\chi(E)v(E)dE \tag{31}$$

so that the overall reaction rate is

$$dR_\chi = \int_V \left[\int_E n(E)\Sigma_\chi(E)v(E)dE \right] dV \tag{32}$$

where integrations are undertaken spatially over the entire energy spectrum E in the range $(0, E_{max})$ within the volume V.

Now the neutron flux having energies contained within a region E_k is simply

$$\phi_k = \int_{E_k} n(E)v(E)dE \tag{33}$$

where ϕ_k can be considered to be the kth "group" of neutrons, and the entire energy spectrum can be considered to be broken down into k energy groups. The quantity within brackets in Equation (32) may be written as

$$dR_X = \int_E n(E)\Sigma_X(E)v(E)dE = \sum_{k=1} \int_{E_k} n(E)\Sigma_X(E)v(E)dE \qquad (34)$$

or

$$dR_{Xk} = \int_{E_k} n(E)\Sigma_X(E)v(E)dE \qquad (35)$$

but Equation (35) may be written as

$$dR_{Xk} = \phi_k \frac{\int_{E_k} n(E)\Sigma_X(E)v(E)dE}{\int_{E_k} n(E)v(E)dE} \qquad (36)$$

or

$$dR_{Xk} = \phi_k\Sigma_{Xk} \qquad (37)$$

where the macroscopic cross section Σ_{Xk} represents the reaction-X probability per unit length that neutrons having energies within the E_k range will undergo interaction-X. Σ_{Xk} is thus termed the group cross section and Equation (36) shows it to be a weighted average within the energy group E_k, the weight kernal being the energy-dependent neutron flux. The total interaction rate is simply

$$R_X = \sum_{k=1} \int_V \phi_k\Sigma_{Xk}dV \qquad (38)$$

which is seen to be the sum of the individual energy group interactions.

Because of the complicacy of using many groups of neutrons in calculations, it has become standard practice in the past to collapse all the neutron interaction variables into one or two groups, thermal and fast, or sometimes three groups where an intermediate energy range of "epithermal" neutrons is taken into account. Generally, in cases involving more than one or two groups, even for simple geometries, digital computers must be utilized to obtain reasonable results. Needless to say, the process of properly collapsing cross section data to obtain appropriate group-averaged values is not a trivial task in itself. In addition, intergroup transfer cross sections must also be considered in order to accurately specify group population.

A special case is worthy of note. At low energies, many nuclei exhibit "1/v" absorption characteristics where

$$\Sigma_a(E) = \Sigma_a(E_0)\frac{v_0}{v(E)} \qquad (39)$$

where E_o is arbitrarily chosen. Introducing (39) into Equation (35) yields

$$dR_{ak} = \Sigma_a(E_o)v_o \int_{E_k} n(E)dE \tag{40}$$

or

$$dR_{ak} = n_k \Sigma_a(E_o)v_o \tag{41}$$

Thus, for a 1/v absorber, the absorption rate is independent of the energy distribution of the neutrons and determined only by the absorption cross section at a given but arbitrary energy identical to that of a monoenergetic beam in E_k. For this reason, then, it has become standard practice to specify <u>all</u> absorption cross sections regardless of their actual characteristics at the single energy E_o = 0.0253 eV, where the Maxwellian velocity is 2200 m/sec. Such cross sections are generally termed "thermal cross sections."

For the case of a non-1/v absorber

$$dR_k = g_k(E)\Sigma_a(E_o)\phi_o \tag{42}$$

where the 2200 m/sec flux is ϕ_o and $g_k(E)$ is a tabulated non-1/v factor in terms of energy or its Maxwellian counterpart, temperature.

2.3 Neutron Diffusion

We shall consider the arbitrary control volume shown in Figure 1.13. Relative to this control volume V_{cv}, having an external bounding surface S_{cv}, we can identify the following neutron events within an element dV or through a surface element dA:

1. Rate of increase of neutrons in V -- $\int_V \frac{\partial N}{\partial t} dV$ (43a)

2. Production of neutrons in V -- $\int_V s_p dV$ (43b)

3. Absorption of neutrons in V -- $\int_V s_a dV$ (43c)

4. Efflux of neutrons through S_o -- $\int_{S_o} \hat{n} \cdot \hat{J} dA$ (43d)

The quantities s_p and s_a are production and absorption densities having dimensions of neutrons per unit time per unit volume; $\hat{J}$ is the net neutron current density expressed as neutrons per unit time per unit area, and $\hat{n}$ is the unit outward normal to the surface S_o at dA.

Since the neutron flux is generally considered to be an undirected quantity representing the product of the neutron density n, and the neutron <u>speed</u> v, we use the separate term $\hat{J}$ to denote the <u>net</u> directed quantity. Thus, while ϕ would represent the average random motion of the neutrons as they move in

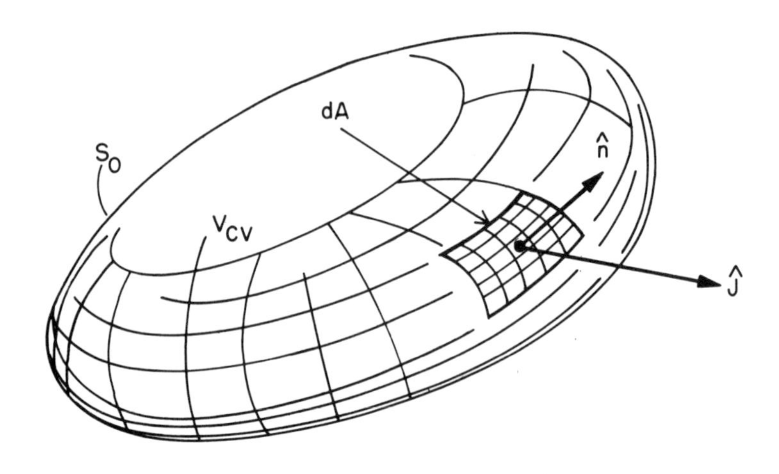

Figure 1.13. Control volume for neutron balance.
(BNL Neg. 7-880-78)

space, i.e., how many would cross an area in a given time regardless of direction--$\hat{J}$ would specify the preferential motion which would constitute a net migration of matter from one location to another. Mathematically, then for our one group space

$$\phi = \int_E n(E)v(E)dE \tag{44}$$

whereas

$$\hat{J} = - D \hat{\nabla} \phi \tag{45}$$

expresses the net neutron current density through Fick's law of diffusion due to the gradient of the neutron flux. Thus, ϕ can be a nonzero constant representing uniformly random motion of a neutron field with a net current density of zero. It is emphasized here that Fick's law is only an approximation valid if $\lambda^2 d^3N/d\zeta^3 << dN/d\zeta$[14] where ζ is the gradient coordinate. In particular, it is invalid (a) if tne medium is strongly absorbing, (b) near (within approximately three mean free paths) a neutron source, sink or surface of a medium, (c) if the medium is strongly anisotropic. In such cases, more exact theory based specifically on the neutron-nuclide interactions must be applied to yield accurate results.

Writing a neutron balance on V_{cv} using the terms in (43) yields

$$\int_{V_{cv}} \frac{\partial N}{\partial t} dV = \int_{V_{cv}} s_p dV - \int_{V_{cv}} s_a dV - \int_{S_o} \hat{n} \cdot \hat{J} dA \quad . \tag{46}$$

Recalling that the absorption rate per unit volume is given in terms of the macroscopic absorption cross section as

$$s_a = \Sigma_a \phi \qquad (47)$$

and applying the divergence theorem along with (45) to Equation (46) results in the null integral

$$\int_{V_{cv}} \left[\frac{\partial N}{\partial t} - \hat{\nabla} \cdot (D\hat{\nabla}\phi) - s_p + \Sigma_a \phi \right] dV = 0 \ . \qquad (48)$$

Since the control volume is completely arbitrary, the integral must vanish everywhere in and on V_{CV} yielding

$$\frac{\partial N}{\partial t} = \hat{\nabla} \cdot (D\hat{\nabla}\phi) + s_p - \Sigma_a \phi \ . \qquad (49)$$

Since we are concerned with only a single group of neutrons, for isotropic media (49) with (44) therefore becomes the neutron diffusion equation,

$$\frac{1}{v}\frac{\partial \phi}{\partial t} - D\nabla^2 \phi + \Sigma_a \phi = s_p \ . \qquad (50)$$

Steady State Diffusion. Equation (50) is the basic equation which describes, together with appropriate boundary conditions, the neutron economy in a nuclear power generation system. In the case of a steady or quasistatic configuration, the steady state diffusion equation may be written as

$$\nabla^2 \phi - \frac{1}{L^2} \phi = - \frac{S}{D} \ . \qquad (51)$$

The quantity L given by

$$L^2 = \frac{D}{\Sigma a} \qquad (52)$$

is generally known as the diffusion area and its positive root, L, is called the diffusion length for reasons which will become clear presently.

Boundary Conditions. Boundary conditions which are generally applicable include the interface matching conditions where flux and current density are taken to be identical on each side of the interface in the absence of a source. If, on the other hand, a source is present, the difference between current densities on either side of the source must be the source strength. Therefore, the interface matching conditions become

1. $\lim_{\Delta x \to 0} \{\phi(x_i + \Delta x) - \phi(x_i - \Delta x)\} \to 0$ (53a)

2. $\lim_{\Delta x \to 0} \{J(x_i + \Delta x) - J(x_i - \Delta x)\} \to S$ (53b)

where S is the source strength in units of neutron current density. In this case, x_i can be considered as the generalized coordinate specifying the interface location.

Similary, point sources require conservation of neutrons in a vanishingly small sphere around the source. Thus,

$$\lim_{\Delta\hat{r}\to 0} 4\pi r^2 J(\hat{r}+\Delta\hat{r}) \to S(\hat{r}) \tag{54}$$

where S is the total source strength in neutrons per unit time at coordinate $\hat{r}$.

At the external surface of a diffusing solid, the neutrons continue to be scattered and, hence, exit the surface. In this case, Fick's law is not a good approximation and so diffusion theory will generally underestimate the flux gradient approaching this surface. A useful approximation which provides reasonable results inside the solid, at least three mean free paths away from the surface, is to use an "extrapolation distance," d, outside the medium at which the neutron flux would vanish. Confirmed theory has shown that for most cases of interest the extrapolation distance is 71% of the transport mean free path λ_{tr}, or

$$d = 0.71\ \lambda_{tr} \tag{55}$$

where λ_{tr} is generally taken to be approximately three times the diffusion coefficient,

$$\lambda_{tr} \sim 3D\quad . \tag{56}$$

Various thermal diffusion parameters are tabulated for common moderators in Table 1.9. In this case we have approximate bounding surface conditions

$$\lim_{x\to x_e} \phi(x + d) \to 0 \tag{57}$$

where x_e is the generalized coordinate of the external bounding surface.

Table 1.9

Thermal neutron diffusion parameters[14]

Moderator	Σ_a cm^{-1}	D cm	L^2 cm^2	L cm	λ_{tr} cm
Light Water	0.0222	0.16	7.21	2.7	0.48
Heavy Water	4.4×10^{-5}	0.87	19772	141	2.61
Beryllium	0.00114	0.50	439	21	1.50
Carbon	2.7×10^{-4}	0.84	3111	56	2.52

Simple Solutions. For purposes of describing the expected behavior in simple homogeneous media, the following examples are included:

(a) Point source of strength S in a finite spherical medium of radius r_o.

Expressing Equation (51) in spherical, one-dimensional coordinates yields

$$\frac{1}{r^2}\frac{d}{dr}\left(r^2\frac{d\phi}{dr}\right) - \frac{1}{L^2}\phi = 0 \tag{58}$$

where the source strength is zero everywhere except at r=0. The appropriate boundary conditions are

$$\text{BC 1:} \quad \lim_{r \to 0} -4\pi D r^2 \frac{d\phi}{dr} \to S \tag{59a}$$

$$\text{BC 2:} \quad \phi\,(r_o^+) = 0 \tag{59b}$$

where $r_o^+ = r_o + d$. The solution to (58) with (59) may be written conveniently as

$$\phi(r) = \frac{S}{4\pi D r}\,\frac{\sinh\left(\frac{r_o^+ - r}{L}\right)}{\sinh\frac{r_o^+}{L}} \tag{60}$$

which may also be written as

$$\phi(r) = \frac{S}{4\pi D r}\left[e^{-\frac{r}{L}} - e^{\frac{r}{L}}\,e^{-2\frac{r_o^+}{L}}\right]\Bigg/\left[1 - e^{-2\frac{r_o^+}{L}}\right]. \tag{61}$$

In the case where $r \ll r_o^+$, the behavior is nearly identical to the behavior of a point source in an infinite medium given by

$$\lim_{r_o^+ \to \infty} \phi(r) \to \frac{S}{4\pi D r}\,e^{-\frac{r}{L}}. \tag{62}$$

It is quite straightforward to see this is the solution to (58) with BC2 changed to $\phi \to 0$ as $r \to \infty$.

(b) Plane source of strength S in a symmetrical, semi-infinite homogeneous medium of thickness 2t.

In this case, the diffusion equation is

$$\frac{d^2\phi}{dx^2} - \frac{1}{L^2}\phi = 0 \tag{63}$$

with the boundary conditions

$$\text{BC 1:} \quad \lim_{x \to 0} J(x) \to \frac{S}{2} \tag{64a}$$

$$\text{BC 2:} \quad \phi\,(t+d) = 0 \quad . \tag{64b}$$

The solution in this case becomes ($t^{+} = t+d$)

$$\phi(x) = \frac{SL}{2D} \; \frac{\sinh(\frac{|t^{+}-x|}{L})}{\sinh(\frac{t^{+}}{L})} \tag{65}$$

For the limiting case of a plane source in an infinite homogeneous medium, (65) becomes

$$\lim_{t^{+} \to \infty} \phi(x) = \frac{SL}{2D}\, e^{-\frac{x}{L}} \tag{66}$$

(c) Plane slab of thickness 2t with a uniform source of strength S neutrons per unit volume per unit time.

In this case Equation (51) becomes

$$\frac{d^2\phi}{dx^2} - \frac{1}{L^2}\,\phi = -\frac{S}{D} \tag{67}$$

and the boundary conditions become

$$\text{BC 1:} \quad \text{at } x=0 \quad \frac{d\phi}{dx} = 0 \quad \text{(symmetric)} \tag{68a}$$

$$\text{BC 2:} \quad \text{at } x=t+d \quad \phi = 0 \quad . \tag{68b}$$

The solution becomes

$$\phi(x) = \frac{SL^2}{D} \left[1 - \frac{\cosh \frac{x}{L}}{\cosh(\frac{t+d}{L})} \right] . \tag{69}$$

It is easily seen that the neutron current at the external surface of the slab is

$$J(t) = SL \, \frac{\sinh \frac{t}{L}}{\cosh \frac{t+d}{L}} \quad . \tag{70}$$

On the other hand, the neutron generation per unit surface area is 2tS. Since the slab is symmetric, the total neutron current at $x = \pm t$ is twice that given by (70) so that the leakage probability or fraction of neutrons generated which leak out through the slab surfaces is just

$$P(\text{Leakage}) = \frac{L}{t} \frac{\sinh \frac{t}{L}}{\cosh \frac{t+d}{L}} . \tag{71}$$

It is easily seen that as the slab half thickness t becomes very large compared with the diffusion length and transport mean free path, that the leakage probability decreases linearly with increasing slab thickness. Quite similar results are obtained for the sphere with uniform neutron generation.

(d) Finite cylinder with a uniform source of strength S per unit volume per unit time.

The geometry for this problem is shown in Figure 1.14. Note that the extrapolated boundary conditions are treated equivalently to a cylinder of radius $a^+=a+d$ and height $2h^+=2(h+d)$. The neutron flux is taken to be symmetric about r=0 and z=0. In this case, the diffusion Equation (51) is written as

$$\frac{1}{r} \frac{\partial}{\partial r} \left[r \frac{\partial \phi}{\partial r} \right] + \frac{\partial^2 \phi}{\partial z^2} - \frac{1}{L^2} \phi = - \frac{S}{D} \tag{72}$$

with the following boundary conditions:

BC 1: $\lim_{r \to 0} \phi(r,z)$ is finite

BC 2: $\phi(a^+,z) = 0$

BC 3: $\frac{\partial \phi}{\partial z}(r,0) = 0$ (symmetric in z)

BC 4: $\phi(r,h^+) = 0$.

The neutron flux for this problem may be found once the particular solution satisfying the inhomogeneity is determined to be $-SL^2/D$. As shown in Appendix A the flux distribution is

$$\phi(r,z) = \frac{SL^2}{D} \left\{ 1 - \frac{2}{h^+} \sum_{n=0}^{\infty} \frac{(-1)^n}{k_n} \frac{I_o(\beta_n r)}{I_o(\beta_n a^+)} \cos k_n z - \ldots \right.$$

$$\left. \ldots - \frac{2}{a^+} \sum_{n=0}^{\infty} \frac{J_o(j_n r)}{j_n J_1(j_n a^+)} \frac{\cosh \gamma_n z}{\cosh \gamma_n h^+} \right\} \tag{73}$$

where the following definitions apply:

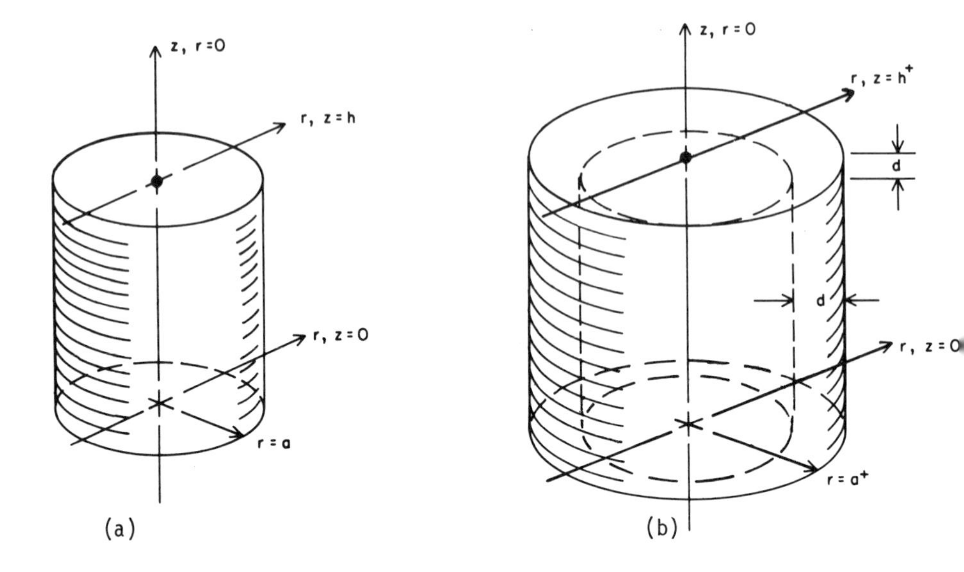

Figure 1.14. Coordinate diagram for finite cylinder, symmetric about z=0 (upper half shown). (a) Actual geometry; (b) Extrapolated geometry. (BNL Neg. 7-867-80)

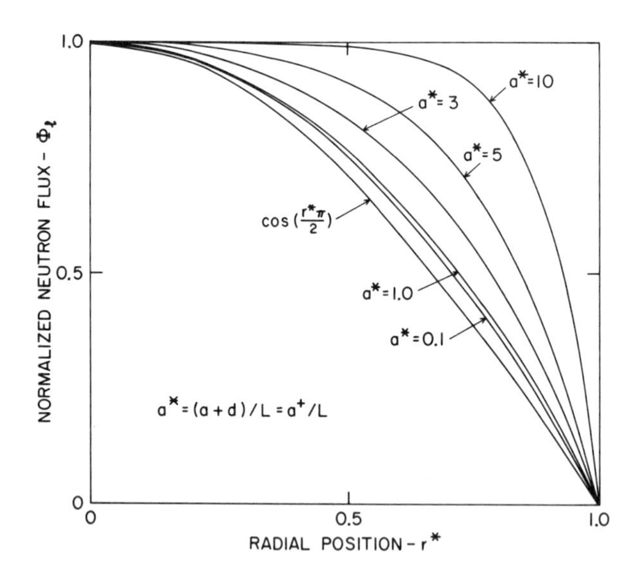

Figure 1.15. Dimensionless radial neutron flux distribution for an infinite cylinder of radius a, extrapolated radius a^+, and dimensionless extrapolated radius a* diffusion lengths. (BNL Neg. 7-869-80)

$$k_n = \frac{(2n+1)\pi}{2h^+} \tag{74}$$

$$\beta_n^2 = k_n^2 + \frac{1}{L^2} \tag{75}$$

$$j_n = \text{eigenvalues of } J_o(j_n a^+) = 0 \tag{76}$$

$$\gamma_n^2 = j_n^2 + \frac{1}{L^2} \quad . \tag{77}$$

It may easily be seen that for the infinitely long cylinder, $h^+ \to \infty$ with $z \sim 0$, the result (73) becomes

$$\lim_{\substack{h^+ \to \infty \\ z \sim 0}} \phi(r,z) \to \frac{SL^2}{D} \left\{ 1 - \frac{I_o(r/L)}{I_o(a^+/L)} \frac{4}{\pi} \sum_{n=0}^{\infty} \frac{(-1)^n}{2n+1} \right\} \quad . \tag{78}$$

It may be shown that the summation in (78) approaches $\pi/4$ so that the solution becomes identical with that which would be obtained from the one-dimensional diffusion equation in cylindrical coordinates, Equation (72) with $\phi_{zz} = 0$,

$$\lim_{\substack{h^+ \to \infty \\ z \sim 0}} \phi(r,z) \to \frac{SL^2}{D} \left\{ 1 - \frac{I_o(r/L)}{I_o(a^+/L)} \right\} \quad . \tag{79}$$

If the limiting neutron flux distribution is denoted $\phi_\ell(r)$ we may normalize the result to the center flux to obtain

$$\Phi_\ell \equiv \frac{\phi_\ell}{\phi_{\ell o}} = \frac{I_o(a^+/L) - I_o(r/L)}{I_o(a^+/L) - 1} \tag{80}$$

which may be written as

$$\Phi_\ell = \frac{I_o(a^*) - I_o(a^* r^*)}{I_o(a^*) - 1} \tag{81}$$

where $a^* = a^+/L$ and $r^* = r/a^+$. This function is plotted in Figure 1.15. It is seen that a leaky cylinder where the radius a^+ is small compared with the diffusion length L, the profile is very nearly a cosine shape. As the size increases, the neutron flux becomes flater as the leakage effects are confined to a relatively smaller and smaller zone near the boundary.

2.4 The Steady State Nuclear Reactor

In this section we shall develop approximate relations for nuclear reactor systems. We shall base the development almost entirely without attention to the details of neutron energy variations or transport, the purpose being to provide, perhaps in some cases somewhat heuristically, a simple picture of a reactor system upon which the thermal hydraulic development shall be based. We shall return now to the single group neutron diffusion Equation (50) for isotropic media

$$\frac{1}{v}\frac{\partial\phi}{\partial t} - D\nabla^2\phi + \Sigma_a\phi = s_p. \tag{50}$$

Recalling Equations (9) and (10) indicates that the reaction rate in a volume dV = dxdydz of a process χ is given by

$$dR_\chi = \phi_n \, \Sigma_\chi \, dx \, dy \, dz \tag{82}$$

For all neutrons in dV, if Σ_f represents the absorptions leading to the fission process only, then the fission reaction rate r_F fissions per unit time per unit volume is

$$r_F = \Sigma_f \, \phi. \tag{83}$$

If each fission process yields ν neutrons per fission then the source of neutrons per unit time per unit volume s_p is

$$s_p = \nu \, \Sigma_f \, \phi \tag{84}$$

where ν is the neutron production ratio per fission specified in Table 1.4. On the other hand, the total number of neutrons absorbed in the fuel is given by $\Sigma_{aF}\,\phi$ neutron per unit time and volume and the ratio of neutrons absorbed in the fuel to those absorbed in the reactor system as a whole is Σ_{aF}/Σ_a. We may therefore rewrite (84) in terms of the reactor flux ϕ and the total reactor absorption cross section Σ_a as

$$s_p = \nu \, \Sigma_a \, \frac{\Sigma_{aF}}{\Sigma_a} \, \frac{\Sigma_f}{\Sigma_{aF}} \, \phi. \tag{85}$$

Now Σ_f/Σ_{aF} is simply the fission-to-capture ratio for the fuel. Recalling the discussion regarding the differences between ν and η in Table 1.4 we recognize that

$$\eta = \nu \, \frac{\Sigma_f}{\Sigma_{aF}} \tag{86}$$

so that (85) may be written in terms of the total reactor flux and absorption cross section Σ_a as

$$s_p = \eta f \Sigma_a \phi. \tag{87}$$

In Equation (87) the factor f is defined as

$$f \equiv \frac{\Sigma_{aF}}{\Sigma_a} \tag{88}$$

and represents the ratio of neutrons absorbed in the fuel to those absorbed in the entire reactor termed the "fuel utilization" factor. Thus, Equation (50) may be written as

$$\frac{1}{v}\frac{\partial \phi}{\partial t} - D\nabla^2\phi + (1-\eta f)\Sigma_a\phi = 0 \tag{89}$$

which describes approximately the neutron economy in a system where neutrons are produced by fission only, a nuclear reactor system.

The Infinite Reactor

In the case of a reactor of infinite size, the curvature of the neutron flux vanishes so that ϕ is constant instantaneously everywhere in the reactor. In this case we have

$$\frac{d\ell n\phi}{dt} = -(1-\eta f)\Sigma_a v. \tag{90}$$

For this reactor to operate at steady state conditions, (i.e., to be "critical"), Equation (90) tells us that

$$\eta f = 1, \qquad \text{(for steady state).} \tag{91}$$

But since all neutrons will be eventually absorbed in the infinite reactor, $\Sigma_a\phi$ representing this rate for a unit volume, and since $\Sigma_a\phi$ neutrons absorbed in one neutron generation will lead to $\eta f\Sigma_a\phi$ neutrons in the next, we define the neutron "multiplication factor" for the infinite reactor as

$$k_\infty \equiv \eta f = \frac{\eta f\Sigma_a\phi}{\Sigma_a\phi}. \tag{92}$$

Equation (91) shows us that if an infinite reactor is to operate in a steady state mode, the multiplication factor must be unity: $k_\infty = 1.0$ for steady state. That is, on the average there must be only as many thermal neutrons generated in one generation as were absorbed in the previous.

The Finite Reactor

In a reactor of finite size, there will be some curvature in the neutron flux profile. If we denote the instantaneous curvature as B^2, it may be written as

$$B^2 = -\frac{1}{\phi}\nabla^2\phi. \tag{93}$$

In terms of the curvature, the neutron diffusion Equation (89) using both (92) and (93) is

$$\frac{1}{Dv}\frac{d\ell n\phi}{dt} = \frac{k_\infty - 1}{L^2} - \beta^2 \tag{94}$$

which shows us that for steady state operation the curvature of the neutron flux must be given wholly in terms of the infinite reactor multiplication factor k_∞, and the diffusion length $L^2 = D/\Sigma_a$. But we know that Equation (93) written as

$$\nabla^2\phi + \beta^2\phi = 0 \tag{95}$$

has no unique solution for positive β^2 but instead has an infinite set of solutions (eigenfunctions)

$$\phi_n\ (\vec{x},t) = \phi_n\ (\vec{x},t;\beta_n) \tag{96}$$

where the β_n are the characteristic values or eigenvalues of a characteristic equation

$$G(\beta_n) = 0. \tag{97}$$

The general solution becomes a linear superposition of all the eigenfunctions

$$\phi(\vec{x},t) = \sum_{n=1}^{\infty} C_n\phi_n(\vec{x},t;\beta_n) \tag{98}$$

If the diffusion Equation (89) is written in terms of (98) where the first eigenfunction is separated from all the rest, we have

$$C_1\left[\frac{1}{Dv}\frac{\partial\phi_1}{\partial t} - \frac{\nabla^2\phi_1}{\phi_1}\phi_1 - \frac{k_\infty-1}{L^2}\phi_1\right] = -\sum_{n=2}^{\infty} C_n\left\{\frac{1}{Dv}\frac{\partial\phi_n}{\partial t} - \frac{\nabla^2\phi_n}{\phi_n}\phi_n - \frac{k_\infty-1}{L^2}\phi_n\right\} \tag{99}$$

or instantaneously as before

$$C_1\left[\frac{1}{Dv}\frac{\partial\phi_1}{\partial t} + \left(\beta_1^2 - \frac{k_\infty-1}{L^2}\right)\phi_1\right] = -\sum_{n=2}^{\infty} C_n\left\{\frac{1}{Dv}\frac{\partial\phi_n}{\partial t} + \left(\beta_n^2 - \frac{k_\infty-1}{L^2}\right)\phi_n\right\} \tag{100}$$

where we have assumed that the instantaneous curvature of each characteristic function specifying the flux shape is represented by the eigenvalues β_n, n = 2,3,.... We know that eigenfunctions can always be specified in terms of ascending values of their eigenvalues, β_n such that $\beta_2 > \beta_1$, $\beta_3 > \beta_2$, ... $\beta_{n+1} > \beta_n$. Thus, if the reactor configuration represented by Equation (100) is adjusted so that $\partial\phi_1/\partial t = 0$ and $(C_1 \neq 0)$

$$\frac{k_\infty-1}{L^2} = \beta_1^{\,2} \tag{101}$$

then for all the higher eigenfunctions

$$\beta_n^2 > \frac{k_\infty - 1}{L^2} \qquad n = 2,3, \ldots \tag{102}$$

That is, the first mode flux shape remains constant. The higher mode shapes are represented by

$$\sum_{n=2}^{\infty} C_n\left(\frac{\partial \phi_n}{\partial t} + \mu_n \phi_n\right) = 0 \tag{103}$$

where

$$\mu_n = \beta_n^2 - \frac{k_\infty - 1}{L^2} > 0. \tag{104}$$

Assuming a solution for the ϕ_n in terms of an exponential in time of the form

$$\phi_n(\vec{x},t) = \psi_n(\vec{x})\, e^{t/\tau_n} \tag{105}$$

and substituting into (103) yields

$$\sum_{n=2}^{\infty} C_n\left(\frac{1}{Dv\tau_n} + \mu_n\right) \psi_n(\vec{x})\, e^{t/\tau_n} = 0. \tag{106}$$

Since $\phi_n(\vec{x},t)$ and hence $\psi_n(\vec{x})$ must be orthogonal within and on the reactor geometry we may multiply (106) by $K_m(\vec{x})\psi_m(\vec{x})$ where $K_m(x)$ is the appropriate kernel and integrate over the region to the boundaries T of the reactor. Due to the orthogonality, all integrations will identically vanish except where m = n so that

$$e^{t/t_m}\left(\frac{1}{Dv\tau_m} + \mu_m\right)\int_T \psi_m^2(\vec{\xi})\, K_m(\xi)\, d\xi = 0 \tag{107}$$

from which the only nontrivial solution is for

$$\tau_m = -\frac{1}{Dv\mu_m} < 0 \quad . \tag{108}$$

Thus, a reactor which is steady in its first mode has the characteristic that all subsequent higher modes <u>decay</u> in time rather than grow, and are thus unimportant insofar as steady state operation is concerned.

The eigenvalue of the first mode flux shape is known as the reactor "buckling" since it represents the way the flux shape curves or "buckles," and is given the special symbol

$$B^2 = \frac{k_\infty - 1}{L^2} \; . \tag{109}$$

Simple Solutions. The steady state finite reactor thus has a neutron flux shape satisfied by the equation

$$\nabla^2\phi + B^2\phi = 0 \tag{110}$$

where B^2 is the first eigenvalue of (110)

$$B^2 = \beta_1 \tag{111}$$

which satisfies the appropriate boundary conditions of the Sturm-Liouville type which result in the characteristic equation given by (97).

In order to operate at steady state the reactor configuration must be adjusted so that

$$\frac{k_\infty-1}{L^2} = B^2 \tag{112}$$

and any uncompensated for change in the reactor configuration with constant properties, or any noncompensated change in reactor properties with constant configuration will result in either the reactor neutron population growing or decreasing generally in an exponential manner. Examples of specific reactor geometries and their eigenvalues (buckling) are given below

(a) Spherical reactor of radius a.

$$\nabla^2\phi + B^2\phi = \frac{1}{r^2}\frac{d}{dr}\left(r^2\frac{d\phi}{dr}\right) + B^2\phi = 0 \tag{113}$$

with

BC 1: ϕ everywhere finite
BC 2: $\phi(a^+) = 0$

where $a^+ = a+d$, the extrapolated reactor boundary.

Since $B^2 > 0$ the solutions to (113) are sines and cosines divided by r or

$$\phi(r) = A\frac{\text{Sin } Br}{r} + C\frac{\text{Cos } Br}{r} \tag{114}$$

BC 1 requires that C = 0 whereas BC 2 requires that

$$B = \frac{n\pi}{a^+}, \quad n = 1 \tag{115}$$

for the primary configuration. Thus, the spherical reactor must be built so that its nuclear geometry satisfies $B^2=(\pi/a^+)^2=(k_\infty-1)/L^2$, or

$$a = \frac{\pi L}{\sqrt{k_\infty-1}} - d \quad . \tag{116}$$

While any spherical reactor having the given flux profile will satisfy the field Equation (113) and its boundary conditions so that

$$\phi(r) = \frac{A}{r} \sin \frac{\pi r}{a^+} \tag{117}$$

there is only one flux shape which will produce power at a specified value P. Since $\Sigma_f \phi$ is the fission rate per unit volume, if Λ is the energy released per fission, the total power yield from the reactor is

$$P = \Lambda \Sigma_f \int_V \phi(r) dV. \tag{118}$$

Substitution of (117) into (118) and performing the appropriate integration yields

$$\phi(r) = \frac{P}{4\Lambda\ \Sigma_{f.}\ a^{+2}\ \frac{r}{\pi} \left[\sin \frac{\pi a}{a^+} - \frac{\pi a}{a^+} \cos \frac{\pi a}{a^+} \right]} \sin \frac{\pi r}{a^+}\ . \tag{119}$$

In the limiting case where d/a << 1 we have

$$\lim_{\frac{d}{a} \to 0} \phi(r) \to \frac{P}{4\Lambda\Sigma_f a^2} \frac{\sin \frac{\pi r}{a}}{r} \tag{120}$$

To see the effect of increasing the reactor size, the neutron flux profiles are shown in Figure 1.16. In this case they are normalized to the volume-averaged value. This average is easily determined since the power P is simply $\Lambda\Sigma_f\phi_{avg}V$. It is seen that the effect of neutron leakage is to flatten the profile, just opposite to the case of a constant, uniform source of neutrons. This is, of course, because leakages have the effect of reducing the source nearer the edge region. Note that since the local power density is directly proportional to the local neutron flux, these profiles also represent power density profiles with constant properties. These profiles would be needed if localized coolant rates and/or temperatures would be required since the power density is directly related to q''' required to calculate temperatures in the Fourier conduction equation. i.e.: $\overline{q'''_{tot}} = P/V$ whereas $q'''(r) = \Lambda\Sigma_f\phi(r)$.

(b) Cylindrical reactor of radius a, height 2h.

$$\frac{1}{r} \frac{\partial}{\partial r} (r \frac{\partial \phi}{\partial r}) + \frac{\partial^2 \phi}{\partial z^2} + B^2 \phi = 0 \tag{121}$$

with

BC 1 ϕ everywhere finite

BC 2 $\phi(a^+, z) = 0$

BC 3 $\phi_z(r, o) = 0$ (symmetric)

BC 4 $\phi(r, h^+) = 0$

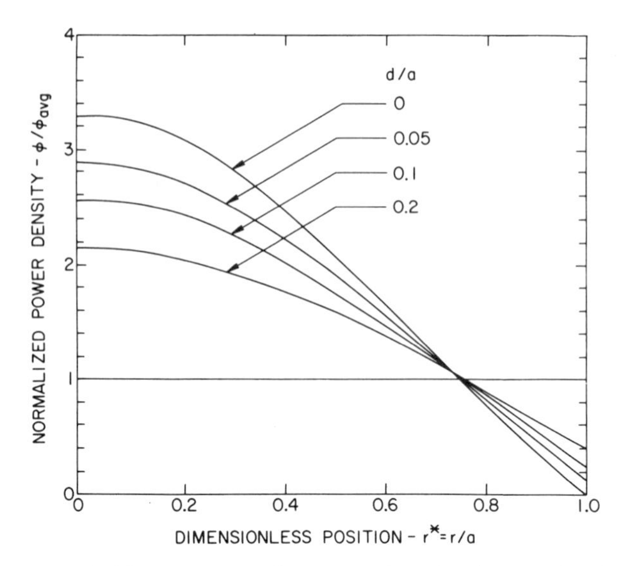

Figure 1.16. Neutron flux profiles for bare spherical reactor normalized to the volume averaged value. (BNL Neg. 7-868-80)

Following methods similar to those used in Appendix A we obtain

$$\phi(r,z) = A\, J_0(\beta r) \cos \kappa z \tag{122}$$

where the only eigenvalues of importance for the steady reactor are

$$\kappa = \frac{\pi}{2h^+} \tag{123}$$

and

$$\beta = \frac{2.405}{a^+} \tag{124}$$

and where the buckling is

$$B^2 = \left(\frac{2.405}{a^+}\right)^2 + \left(\frac{\pi}{2h^+}\right)^2 . \tag{125}$$

As for the spherical reactor, the power is the integral of $\Lambda\Sigma_f\phi$ over the entire volume or

$$P = 4\pi\, \Lambda\, \Sigma_f\, A \int_0^h \int_0^a r\, J_0(\beta r) \cos \kappa z\, dr dz \tag{126}$$

where it is recalled that the reactor is symmetric in an overall height of 2h. For the case of the large reactor $h \to h^+$ and $a \to a^+$

$$\int_0^{h^+} \cos \kappa z\, dz \to \frac{2h}{\pi} \tag{127}$$

while

$$\int_0^{a^+} r\, J_o(\beta r)dr \rightarrow \frac{a^2}{2.405} J_1(2.405). \tag{128}$$

For a large reactor ($J_1[2.405] = 0.5184$), the flux shape becomes

$$\phi(r,z) \rightarrow 3.643 \frac{P}{V\Lambda\Sigma_f} J_o(\beta r) \cos \kappa z \quad \begin{cases} a \rightarrow a^+ \\ h \rightarrow h^+ \end{cases} \tag{129}$$

where V is the reactor volume $2\pi ha^2$.

This flux profile is plotted in Figure 1.17 normalized to the average value given by $\phi_{avg}=P/V\Lambda\Sigma_a$, which also represents the localized power density profiles if nuclear properties are constant. It is seen that if a coolant were to be forced parallel to the z-axis, the most rapid rate of heat input to the coolant would occur where z=0, at the center of the reactor. The axially-averaged rate of temperature rise would decrease with increasing distance from the core centerline toward the outer edge. The net result would be large temperature gradients throughout the core which is an undesirable situation. In order to eliminate or substantially reduce these temperature gradients, the neutron flux profiles are made flatter in the radial direction by zoning the fuel to make it more reactive at the periphery and adding a neutron reflector. Both have the effect of increasing the neutron flux near the outer periphery. In addition, where insufficient radial flux profile flattening is not possible, the axial coolant flow rates through the core are varied radially, usually by orificing the inlet regions. Looking again at the neutron flux shape, in the case where the reactor height becomes very large, (125) approaches the shape of that for an infinite circular cylinder given by

$$\lim_{\substack{h^+ \rightarrow \infty \\ z \text{ finite}}} \phi(r) \rightarrow 0.736 \frac{(dP/dz)}{\Lambda\Sigma_f a^2} J_o(\beta r). \tag{130}$$

The buckling becomes

$$B^2 \rightarrow \beta^2 = \left(\frac{2.405}{a}\right)^2 \tag{131}$$

which would specify the radius in terms of the nuclear properties.

(c) Rectangular parallelepiped reactor of sides 2a, 2b, 2c.

$$\frac{\partial^2\phi}{\partial x^2} + \frac{\partial^2\phi}{\partial y^2} + \frac{\partial^2\phi}{\partial z^2} + B^2\phi = 0 \tag{132}$$

where

BC 1 $\quad \phi(a^+,y,z) = 0$

BC 2 $\quad \phi_x(0,y,z) = 0$

BC 3 $\quad \phi(x,b^+,z) = 0$

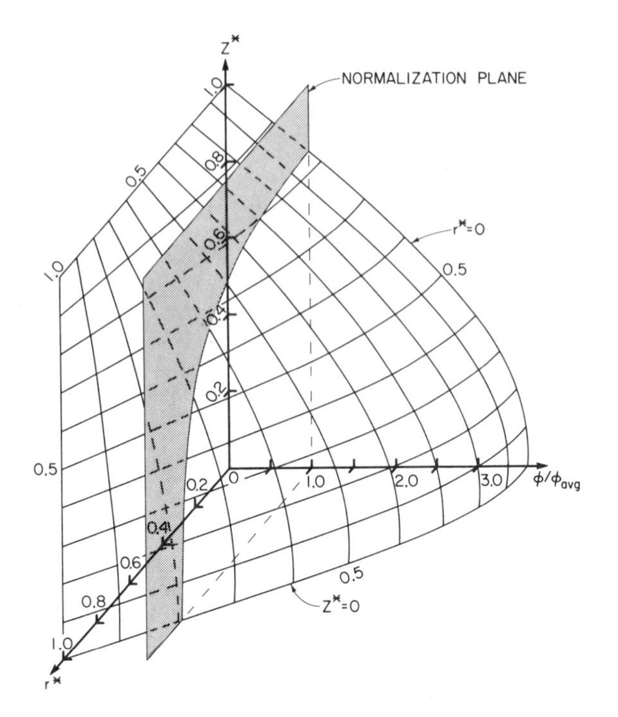

Figure 1.17. Normalized neutron flux profiles for the large bare cylindrical reactor. (BNL Neg. 7-879-80)

BC 4 $\phi_y(x,0,z) = 0$

BC 5 $\phi(x,y,c^+) = 0$

BC 6 $\phi_z(x,y,0) = 0$.

Choosing a product solution for ϕ in the form $\phi = \psi(y,z)\,X(x)$ yields

$$\frac{1}{\psi}\left(\frac{\partial^2\psi}{\partial y^2} + \frac{\partial^2\psi}{\partial z^2}\right) + B^2 = -\frac{1}{X}\frac{d^2X}{d_x^{\;2}} = \mu \tag{133}$$

It is quickly seen that $\mu < 0$ and $\mu = 0$ leads to trivial results whereas $\mu > 0$ yields

$$X(x) = A_x \cos\frac{\pi x}{2a^+} \;. \tag{134}$$

In this case the lowest eigenvalue was $\sqrt{\mu} = \pi/2a^+$. For the ψ-product we choose $\psi(y,z) = Y(y)Z(z)$ so that

$$Y(y) = A_y \cos\frac{\pi y}{2b^+} \tag{135}$$

and

$$Z(z) = A_z \cos \frac{\pi z}{2c^+} \quad . \tag{136}$$

The resulting neutron flux profiles are given by

$$\phi(x,y,z) = A \cos \frac{\pi x}{2a^+} \cos \frac{\pi y}{2b^+} \cos \frac{\pi z}{2c^+} \quad . \tag{137}$$

The buckling for a critical assembly is

$$B^2 = \left(\frac{\pi}{2a^+}\right)^2 + \left(\frac{\pi}{2b^+}\right)^2 + \left(\frac{\pi}{2c^+}\right)^2 \tag{138}$$

and the coefficients obtained from the power may be specified as

$$A = \frac{(\pi/4)^3}{a^+b^+c^+\Lambda\Sigma_f} \left(\sin \frac{\pi a}{2a^+}\right)^{-1} \left(\sin \frac{\pi b}{2b^+}\right)^{-1} \left(\sin \frac{\pi c}{2c^+}\right)^{-1} \quad . \tag{139}$$

For the large reactor, (137) and (139) yield

$$\lim_{\substack{a \to a^+ \\ b \to b^+ \\ c \to c^+}} \phi(x,y,z) \to \frac{(\pi/2)^3 P}{V\Lambda\Sigma_f} \cos \frac{\pi x}{2a} \cos \frac{\pi y}{2b} \cos \frac{\pi z}{2c} \quad . \tag{140}$$

For the case where the reactor is stretched infinitely in the z-direction we obtain the solution for an infinite rectangular rod

$$A = \frac{(\pi/4)^2 \ (dP/dz)}{a^+b^+\Lambda\Sigma_f} \left(\sin \frac{\pi a}{2a^+}\right)^{-1} \left(\sin \frac{\pi b}{2b^+}\right)^{-1} \tag{141}$$

with the buckling given by

$$B^2 = \left(\frac{\pi}{2a^+}\right)^2 + \left(\frac{\pi}{2b^+}\right)^2 \tag{142}$$

and the limiting flux shape given by

$$\lim_{\substack{a \to a^+ \\ b \to b^+}} \phi(x,y) \to \frac{(\pi/2)^2 (dP/dz)}{A_x\Lambda\Sigma_f} \cos \frac{\pi x}{2a} \cos \frac{\pi y}{2b} \equiv \phi_{xy}(x,y) \tag{143}$$

where A_x is the cross sectional area of the reactor. dP/dx is the power per unit length generated by the reactor.

For the case of the infinite rectangular slab the coefficient is simply

$$A = \frac{(\pi/4)\ (dP/dA_x)}{a^+\ \Lambda\ \Sigma_f} \left(\sin \frac{\pi a}{2a^+}\right)^{-1} \tag{144}$$

and the limiting case of a very thick slab yields

$$\lim_{a \to a^+} \phi(x) \to \frac{(\pi/2)\ (dP/dA_x)}{2a\Lambda\Sigma_f} \cos \frac{\pi x}{2a} \equiv \phi_x(x) \tag{145}$$

dP/dA_x is the power which is generated in an element of area dA_x and having a thickness 2a.

Now the average neutron flux in the x-direction for the infinite slab is given by

$$\overline{\Phi}_x = \frac{1}{2a}\int_{-a}^{a} \phi(x)dx = \frac{2}{\pi}\ \phi_x(0) \tag{146}$$

so that the ratio of maximum-to-average neutron flux for the cosine shape is $\pi/2 = 1.571$ in the infinite slab. Similarly, for the infinite rectangular rod reactor the ratio of maximum to average neutron flux is $(\pi/2)^2 = 2.467$ whereas for the rectangular finite reactor the ratio is $(\pi/2)^3 = 3.876$.

The results for these various reactor shapes are summarized in Table 1.10. It is easily seen why the axial nonuniform flux profiles most encountered in thermal hydraulic tests are chopped cosine power distributions having peak-to-average heat flux profiles of $\sim$ 1.5:1. (Note that the local power density is directly proportional to the local neutron flux density presuming Σ_f is not space dependent. As we shall see later, however, Σ_f may change spatially due to heterogeneities, fuel and poison (absorber) distributions, and specifically due to control rod locations (movable poison).)

The size of any critical assembly is given in terms of the buckling and the utilization of Equation (112). For instance, for a finite cylindrical geometry, the dimensions must be arranged such that

$$\left(\frac{2.405}{a+d}\right)^2 + \left(\frac{\pi/2}{h+d}\right)^2 = \frac{k_\infty - 1}{L^2} = \frac{\eta\Sigma_{af} - \Sigma_a}{D}\ . \tag{147}$$

Thus, if the particular constituent makeup of the reactor is specified, (147) will indicate what combination of dimensions will result in a critical assembly, if any. Similarly, if a given size is desired, Equation (147) will allow the constituent to be selected, if any are possible, to make the desired reactor.

Finally, we will inquire regarding the general effect of neutron leakage from a reactor assembly. Recalling that the net neutron current is $\hat{J} = -D\hat{\nabla}\phi$, we may sum up the neutrons exiting a given finite reactor at the physical surfaces as

$$R_L = \text{Leakage Rate} = \int_A \hat{J} \cdot \hat{n}\ dA = -D\int_A (\hat{\nabla}\phi) \cdot \hat{n}\ dA\ . \tag{148}$$

Table 1.10

Neutron Flux Profile Factors for the Bare, Homogeneous Reactors of Shapes Given
Values for the Energy per Fission are Given in Table 1.5 for Both $_{92}U^{235}$ and $_{94}Pu^{239}$

Geometry	Dimensions	Buckling	Flux Profile	Coefficient A*	ϕ_{max}/ϕ_{avg} *
Infinite	None	0	Constant = A	$A = \frac{dP}{dV}$ given by power density	1.0
Sphere	Radius a	$\left(\frac{\pi}{a^+}\right)^2$	$A \frac{\sin \frac{\pi r}{a^+}}{r}$	$\frac{P}{4\Lambda\Sigma_f a^2}$	$\frac{\pi^2}{3} = 3.29$
Infinite Rectangular Rod	Thickness 2a	$\left(\frac{\pi}{2a^+}\right)^2$	$A \cos\left(\frac{\pi x}{2a^+}\right)$	$\frac{(\pi/2)(dP/dA_x)}{2a\Lambda\Sigma_f}$	$\frac{\pi}{2} = 1.57$
Infinite Slab	Thickness 2a Width 2b	$\left(\frac{\pi}{2a^+}\right)^2+\left(\frac{\pi}{2b^+}\right)^2$	$A \cos\left(\frac{\pi x}{2a^+}\right) \cos\left(\frac{\pi y}{2b^+}\right)$	$\frac{4(\pi/4)^2(dP/dz)}{A_x\Lambda\Sigma_f}$	$\left(\frac{\pi}{2}\right)^2 = 2.47$
Rectangular Parallel Piped	2ax2bx2c	$\left(\frac{\pi}{2a^+}\right)^2+\left(\frac{\pi}{2b^+}\right)^2+\left(\frac{\pi}{2c^+}\right)^2$	$A \cos\left(\frac{\pi x}{2a^+}\right)\cos\left(\frac{\pi y}{2b^+}\right)\cos\left(\frac{\pi z}{2c^+}\right)$	$\frac{8(\pi/4)^3\ P}{V\Lambda\Sigma_f}$	$\left(\frac{\pi}{2}\right)^3 = 3.88$
Infinite Cylinder	Radius a	$\left(\frac{2.405}{a^+}\right)^2$	$A\ J_o\left(\frac{2.405r}{a^+}\right)$	$\frac{0.736(dP/dz)}{\Lambda\Sigma_f a^2}$	$\frac{2.405}{1.037} = 2.32$
Finite Cylinder	Radius a Height 2h	$\left(\frac{2.405}{a^+}\right)^2 + \left(\frac{\pi}{2h^+}\right)^2$	$A\ J_o\left(2.405\ \frac{r}{a^+}\right)\cos\left(\frac{\pi z}{2h^+}\right)$	$\frac{3.64\ P}{V\Lambda\Sigma_f}$	$\frac{\pi}{2}\left(\frac{2.405}{1.037}\right) = 3.64$

*For very large reactor where the geometric size a or equivalent is much larger than d.

Using Gauss' theorem (148) may be written as

$$R_L = - D\int_V \nabla^2\phi dV = + DB^2\int_V \phi\ dV \tag{149}$$

where (110) was used to express the Laplacian in terms of the buckling. The neutron absorption rate in the reactor is simply

$$R_A = \Sigma_a\int_V \phi\ dV. \tag{150}$$

The fraction of all neutrons in the reactor which leak from the reactor is thus the probability of leakage, P_L, given by

$$p_L = \frac{R_L}{R_L+R_A} = \frac{L^2B^2}{1+L^2B^2}. \tag{151}$$

The nonleakage probability is simply $1-P_L$

$$P_{NL} = \frac{1}{1+L^2B^2}. \tag{152}$$

Examining the average neutron production inventory, the neutron absorption rate is $\Sigma_a vN_{avg}$ resulting in $k_\infty\Sigma_a vN_{avg}$ of which a fraction P_{NL} will be absorbed in the following cycle. The effective finite reactor multiplication factor k_{eff} is thus

$$k_{eff} = \frac{p_{NL}k_\infty\Sigma_a vN_{avg}}{\Sigma_a vN_{avg}} = p_{NL}k_\infty. \tag{153}$$

Now, recalling (112) together with (153), Equation (152) yields

$$k_{eff} = \frac{k_\infty}{1+L^2B^2} = 1 \tag{154}$$

showing that the effective neutron multiplication factor in a finite reactor must be unity for steady state critical operation, and that the reactor materials must be chosen such that $k_\infty = 1/p_{NL}$.

Extensions of the Simple Case. The foregoing discussion has all been developed from the viewpoint of constant nuclear properties and without consideration of variable neutron energy. Neutrons are born from the fission process, however, with an average energy of approximately 2 MeV. On the other hand, the energies most likely to cause fission in $_{92}U^{235}$ are in the thermal zone.

In an actual reactor, then, consideration must be taken of the fact that not all fissions occur at thermal energies but some are produced at higher energies resulting in a fraction $(\varepsilon-1)$ more neutrons than expected, ε being termed the "fast fission factor." On the other hand, some of the neutrons born at high energies will be absorbed while slowing down. The probability that a neutron will not be absorbed is termed the "resonance escape probability," p. As a result, the actual multiplication factor for the infinite reactor would be

$$k_\infty = \eta_T p\varepsilon f \tag{155}$$

where the subscript on η is to indicate that this is a value for thermal energies only. The fuel utilization factor is effectively averaged over the entire neutron spectrum rather than using values at a given energy. Equation (155) is usually termed the "four factor formula."

If neutrons are considered to slow down from higher energies, this is accomplished through successive scattering interactions with the constituents in the reactor. During this period, there is a finite probability that some neutrons will escape from the reactor. If the fast nonescape probability is denoted p_{NLF}, then the effective multiplication factor for the finite reactor becomes

$$k_{eff} = k_\infty \, p_{NLT} \, p_{NLF} = \eta_T p\varepsilon f p_{NLT} p_{NLF} \tag{156}$$

where k_∞ was obtained from (155). It is reasonably straightforward to show that[14]

$$p_{NLF} = \frac{1}{1+B^2\tau_T} \tag{157}$$

where

$$\tau_T = \frac{D_1}{\Sigma_1} . \tag{158}$$

This quantity termed "neutron age" is the ratio of the fast neutron diffusion coefficient to the slowing down cross section Σ_1 where $\Sigma_1\phi_1$ represents the rate of neutrons per unit volume scattered from the fast flux ϕ_1 into the thermal range. In actual practice for most reactors both p_{NLT} and p_{NLF} are close to unity. Examples of reactor sizing calculations using these variables are given by Lamarsh.[14]

As was mentioned earlier, it is undesirable to have radial variations in neutron flux due to the potential they provide for thermal stress in nuclear systems. To elminate or significantly reduce these gradients, nonproducing materials are usually placed around the reactor to scatter some neutrons back into the core which otherwise would be lost to the reactor. This "reflector" has the effect of peaking the thermal flux profile just outside the core boundary due to the slowing down of fast neutrons which enter the reflector. Further, the extrapolated zero flux location is forced farther from the reactor boundary. Simple one group calculations may be made by analyzing a two-region problem, the inner region, or core, being a producing region when Equation (110) applies, and the outer reflector region being analyzed by the diffusion equation with no source. Standard flux and current matching conditions are utilized at the interface between the two.

Finally, since the scattering cross section for uranium is much smaller than for a moderator while the resonance capture probability is higher, it has been found that by heterogeneously localizing zones of fuel and surrounding these zones by moderator that less nonproductive absorption occurs in the reactor assembly on a whole. Thus, most slowing down of the neutrons is accomplished in fuel-free zones where the neutrons, once thermalized, then are scattered back into the fuel regions to be absorbed causing fission.

Typical heterogeneous designs include plate-type geometries shown schematically in Figure 1.18, and rod-type geometries such as shown in Figure 1.19. Plate-type geometries have been utilized in many test reactors (TR's) of the

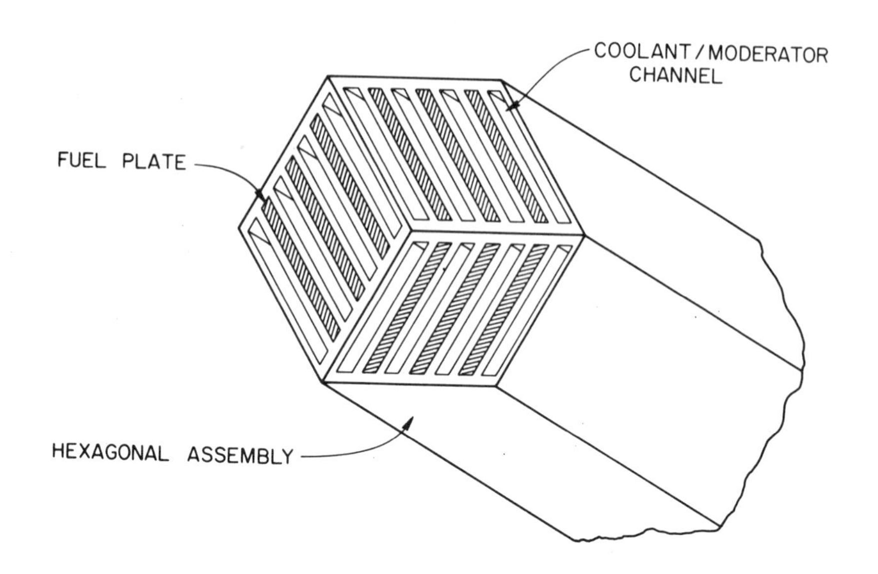

Figure 1.18. Typical schematic arrangement of fuel and coolant/moderator ducts in plate geometry arranged in a hexagonal cell assembly. (BNL Neg. 7-876-80)

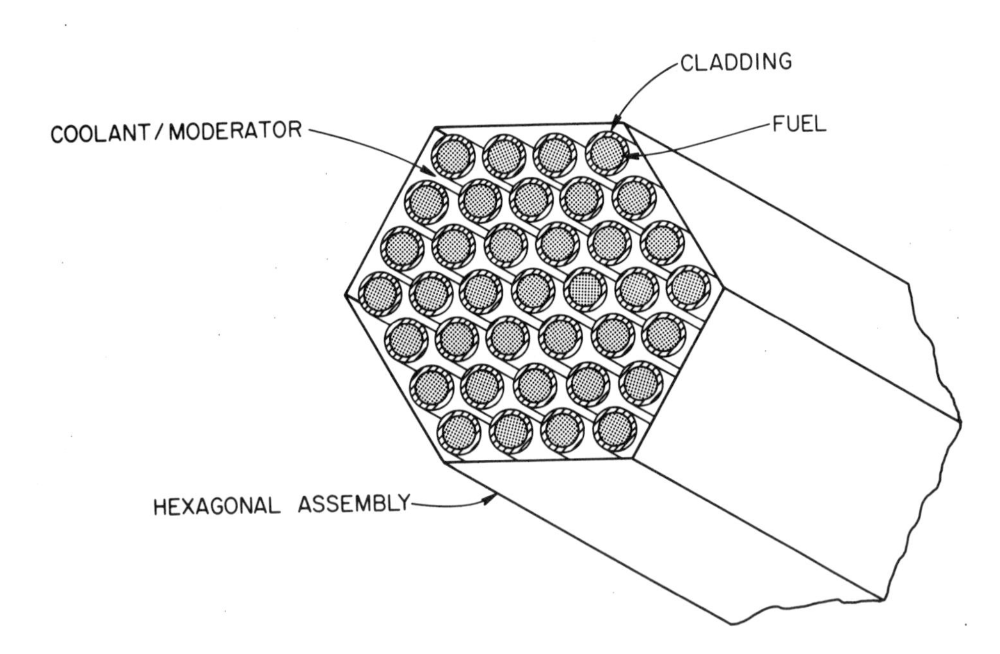

Figure 1.19. Typical schematic arrangement for a rod-bundle reactor geometry arranged in a hexagonal lattice. (BNL Neg. 7-877-80)

ETR (Experimental) or MTR (Material) type. Fuel thickness and cladding thickness may typically be 0.5 mm to 1 mm whereas the coolant channels may be up to 5-6 mm. The hexagonal subassembly may be 10 cm or more across the flats. Fuel itself is usually uranium oxide UO_2. Rod-type geometries are those commonly found in commercial power reactor systems, and, as in current usage, are generally built on a square lattice rather than a hexagonal lattice. Fuel rods may be up to $\sim$ 10 mm in diameter with clearances between them of 0.5-1 mm. They are usually formed of zirconium tubes having a thickness of 0.5-1 mm with UO_2 pellets stacked inside. Axial reflectors may simply be solid metal exteriors of the fuel rods on either end. The rods may be positioned in a hexagonal array by means of wires wrapped in a spiral around the rods, as in certain liquid metal breeder designs, or held in place by means of grid spacers. In some cases the fuel-pin array forms a core-wide open lattice without individually closed subassemblies. In others they may be encased in a square or hexagonal "can" forming an assembly unit for ease of handling. Typical linear power densities in commercial reactors are $\sim$ 15 kW per meter per rod. Thus, a 3000 MWt (megawatt-thermal) power plant generating $\sim$ 1000 MWe of electrical energy with fuel rods 4-meters long would be constructed of 50,000 fuel rods. Other reactor geometries are discussed in later chapters.

2.5 The Transient Nuclear Reactor

The previous section discussed neutron inventory in a nuclear power generating system under steady state operating situations. Under many circumstances, however, time variations occur due to changing operating requirements such as start-up, shutdown, or changing power levels. Time variations may also occur normally without operator intervention such as when the nuclear steam supply system responds to a change in turbine steam demand due to varying electrical load. Abnormal, or infrequent but expected sequences can also cause transient behavior of the nuclear reactor. Things such as a storm causing the power plant breakers to trip, or a bearing seizure in a pump, or perhaps a control rod malfunction or inadvertent actuation may all occur at one time or another in the life of a nuclear power plant.

Nuclear systems are designed to provide stable operation or shutdown under all conceivable circumstances. In many cases significant thermal hydraulic-nuclear interaction may occur as a result of coolant/moderator behavior. This is especially true where, as is the case with the vast majority of operating reactors today, the coolant also serves as the moderator. While it is impossible in the brief scope provided herein to provide a detailed treatment of reactor kinetics (as transient behavior has come to be called), we will attempt to provide sufficient background material to enable the reader to identify where and why fluid-nuclear interactions occur, and why they are important.

As was mentioned in the previous section, the size of a reactor may be specified once the material constituent makeup has been selected. Part of the steady state operation of a reactor, however, involves the use or "burning" of some of the fuel which results in a constantly changing fuel inventory. Provisions must therefore be made to account for the gradual reduction of the quantity of nuclear fuel along with the generation of fission products which would otherwise result in the reactor becoming subcritical and shutting down.

Similarly, some of the fission products of the nuclear reactore are partucularly good in absorbing neutrons unproductively and are termed "poisons." By far the largest effect is due to xenon having resonant cross sections exceeding 10^4 barns. A particular short lived isotope $_{54}Xe^{135}$ is formed by decay of iodine and has a half life of approximately 9 hours. This xenon isotope has a cross section over a broad range of energies of close to 500,000 barns, and at thermal neutron energies has a cross section of 2.6×10^6 barns! Under steady state reactor operation its production rate is generally balanced by its fission rate. But following shutdown, its continued production due to decay of iodine followed by its own radiodecay results in a peaking of the xenon inventory approximately 10 hours after shutdown of a reactor.

In addition to the generally changing constituent makeup of a reactor as specified above, provisions must also be made to allow for gradually increasing or decreasing the reactor power levels by increasing or decreasing the neutron inventory. A nuclear reactor must therefore be designed to have continued variability in its geometry or material inventory to allow for all these eventualities to occur and still operate in a stable and power producing fashion. Such control is usually provided through application of movable (and/or removable) materials. At their fullest concentration they would positively shut down the reactor under all circumstances during the life of the reactor but would be removable to the extent required to allow operation at the end of reactor life where fuel inventory is at a minimum. They may also provide some fraction of peak xenon override (ability to startup in spite of some xenon inventory). Where reactors are required on demand full peak xenon override is necessary. Generally such control is provided by control elements or rods which may be moved into or out of the reactor assembly at will. These usually are made of boron clad with stainless steel. In some instances, small changes in poison level in a reactor is provided by boron "shim"--small amounts of boron added to or extracted from the reactor coolant (usually water). In some cases a supply of heavily borated water is kept on hand to provide positive reactor shutdown by injection if emergency situations require.

While many of these situations occur rapidly enough or locally enough in a reactor to substantially alter the neutron flux profiles, significant time behavior does occur in ways which maintain a relatively fixed flux profile. In such cases, the changes occur such that the neutron diffusion equation remains linear and a product solution for a function of time and a function of space will allow separation of the spatial terms satisfying Equation (110) identically. For purposes of this discussion we shall consider only the latter case while realizing that under some circumstances, what may suffice for our purpose here may be substantially in error.

Preliminary Concepts. We shall first incorporate both the fast nonleakage factor p_{NLF} with the thermal nonleakage factor p_{NLT} as

$$p_{NL} \equiv p_{NLT}\, p_{NLF} = \frac{1}{1+L^2B^2} \cdot \frac{1}{1+\tau B^2} \sim \frac{1}{1+M^2B^2} \tag{159}$$

where since both probabilities are near unity $B^4L^2\tau \ll M^2B^2$ and is neglected. The factor

$$M^2 = L^2 + \tau_T \tag{160}$$

is termed the thermal migration area. Recalling (156) we see that

$$(1 + M^2B^2) = \frac{k_\infty}{k_{eff}} \tag{161}$$

so that the diffusion equation for constant spatial shape may be written as

$$\frac{dN}{d\tau} + k_\infty \frac{1-k_{eff}}{k_{eff}} N = 0 \; . \tag{162}$$

Note that (162) has been nondimensionalized through the use of the average neutron lifetime given by

$$\ell_o = \frac{\lambda_s}{v_F} + \frac{\lambda_a}{v_T} \tag{163}$$

where quantities have been suitably averaged. ℓ_o thus represents the neutron slowing down time as well as the thermal migration time for an equivalent infinite reactor. Since v_F is much larger than the mean thermal velocity v_T as the neutron slows down while the slowing down and absorption mean free paths are similar, the thermal diffusion time is much larger than the slowing down time. Note that the mean lifetime of a neutron for thermal reactors is of the order of 10^{-4}-10^{-3} sec while for fast reactors it is nearer to 10^{-8} sec.[5] As will be seen this represents a substantial difference in reactor control. It should also be noted that in a finite reactor the lifetime is actually shorter since some neutrons leak out of the system. The finite neutron lifetime is reduced from ℓ_o by the product of fast and thermal nonleakage probabilities given by (159).

An instructive exercise proceeds by assuming the reactor starts from a steady state where $N = N_o$ and undergoes a transient perturbation due to a small change in $k_{eff} = 1 + \delta k$. The perturbation in neutron density δN is thus represented by

$$\frac{d\delta N}{d\tau} - k_\infty \rho \delta N = k_\infty \rho N_o \ . \tag{164}$$

Note that a new quantity has been introduced which is called the "reactivity" defined as

$$\rho = \frac{\delta k}{k_{eff}} \ . \tag{165}$$

The perturbation behavior is easily found to be

$$\frac{\delta N}{N_o} = e^{k_\infty \rho \tau} - 1 = e^{\tau/T} - 1 \ . \tag{166}$$

For this simple example the "reactor period" is given as

$$T = \frac{\ell_o}{k_\infty \rho} = \frac{1}{k_\infty \rho}\left[\frac{\lambda_s}{v_F} + \frac{\lambda_a}{v_T}\right]. \tag{167}$$

From the above results we see that the smaller a reactor is (i.e., the larger the buckling and hence the larger k_∞) the more rapidly it would respond to an upset in k_{eff}--a reactivity insertion represented by the relative perturbation $\rho = \delta k/k_{eff}$. The reactor period is also a function of the reactivity ρ itself so that the larger the upset, the more rapidly the reactor will respond. In addition, a significant feature is the stabilizing effect found in an increase in ℓ_o. While we have heretofore tacitly assumed that neutrons are born virtually instantaneously, they are not. As we shall see in the next section there is a small group of delayed neutrons which occur due to decay of fission products and which play an exceedingly important role in nuclear reactor control.

<u>Effects of Delayed Neutrons</u>. Returning to the neutron diffusion equation

$$\frac{1}{v}\frac{\partial N}{\partial t} - D\nabla^2\phi + \Sigma_a\phi = s_p \tag{50}$$

we previously showed that the source for neutrons could be stated as

$$s_p = k_\infty \Sigma_a \phi \ . \tag{168}$$

For the finite reactor including slowing down leakage, thermal leakage, and fast fission, we had the four factor formula (155) for k_∞.

In the actual case, only the fraction $(1-\beta)$ are prompt neutrons which appear almost immediately after a neutron is absorbed causing fission. A fraction β arrives instead as a result of radioactive nucleonic decay of fission products. If β is taken to be the total fraction delayed in a $_{92}U^{235}$ reactor than

$$\beta = \sum_{i=1}^{6} \beta_i = 0.0065 \tag{169}$$

where the β_i are shown in Table 1.11. As is seen in this table, the half life for decay in some cases is very large compared with the total prompt neutron life times.

Table 1.11

Groups of Delayed Neutrons for $_{92}U^{235}$[14]

Group	Half-Life sec	Decay Constant sec^{-1}	Fraction
1	55.7	0.0124	0.000215
2	22.7	0.0305	0.001424
3	6.22	0.111	0.001274
4	2.30	0.301	0.002568
5	0.610	1.14	0.000748
6	0.230	3.01	0.000273
			0.0065

As before, we shall consider a simple one group field consisting entirely of thermal neutrons. The source of <u>thermal</u> neutrons may thus be written in terms of the prompt neutron source s_{pn} and the sum of the delay groups s_{dn_i}. Thus

$$s_p = s_{pn} + \sum_{i=1}^{6} s_{dn_i} \ . \tag{170}$$

In this case since only $(1-\beta)$ of those given by (168) are really prompt we have

$$s_{pn} = (1-\beta)p_{NLF}k_\infty\Sigma_a vN \tag{171}$$

where p_{NLF} [Equation (157)], accounts for the probability that a fast neutron remains in the system to become thermal.

On the other hand, if C_i is the nuclear concentration of the ith group of neutron precursors (fission products which decay to produce the β_i fraction), the delayed neutron source from the ith group is simply

$$s_{dn_i} = \lambda_i C_i p p_{NLF} , \tag{172}$$

where λ_i is the decay constant of the precursor. Note that the factors p and p_{NLF} account for the fact that not all neutrons generated at elevated energies survive. Only a fraction p escape resonance absorption and another fraction p_{NLF} do not leak out of the reactor while being thermalized.

Similar to the neutrons, while the delayed neutron precursors do not significantly diffuse during time scales of interest, they do follow a material balance given by

$$\frac{dC_i}{dt} = \text{production - decay} . \tag{173}$$

The production rate is identical to the thermal neutron production rate except only a fraction β_i are generated and there is much less concern regarding leakage or resonance absorption. Thus,

$$\text{i-species production} = \beta_i \frac{k_\infty}{p} \Sigma_a vN, \tag{174}$$

whereas the decay is specified in terms of the decay rate λ_i so that

$$\text{i-species decay} = \lambda_i C_i . \tag{175}$$

By using (171) and (172) with (49), allowing the spatial variations of the flux to be specified in terms of the buckling, using (174) and (175) in (173) and rearranging, the following system of seven equations in seven unknowns is obtained:

$$\frac{dN}{d\tau} + \Omega N = \sum_{i=1}^{6} \Lambda_i C_i \tag{176}$$

$$\frac{dC_i}{d\tau} + \Lambda_i^* C_i = B_i^* N, \qquad i = 1,6 . \tag{177}$$

The parameters are defined as follows:

$$\Omega = \beta - \rho \tag{178}$$

$$\Lambda_i = \mu \lambda_i \tag{179}$$

$$\mu = \frac{\rho}{v\Sigma_a k_\infty} \tag{180}$$

$$\Lambda_i^* = \Lambda_i/pp_{NLF} \tag{181}$$

$$\beta_i^* = \beta_i/pp_{NLF} \tag{182}$$

$$\tau = \frac{k_{eff}v\Sigma}{p_{NLT}} t = k_{eff}\frac{t}{\ell} \quad . \tag{183}$$

Note that we are concerned here with situations where changes in any of the parameters included in (180)-(183), and especially k_{eff}, are small and can be ignored. Several conclusions can be drawn from these equations as follows.

(a) Small perturbation.

If the neutron density and precursor concentrations can be considered to be only slightly perturbed due to a reactivity insertion ρ from the critical state, Equations (176) and (177) become

$$\frac{d\delta N}{d\tau} + \Omega\delta N = \sum_{i=1}^{6} \Lambda_i C_{i_o} - \Omega N_o \equiv K_N \tag{184}$$

and

$$\frac{d\delta C_i}{d\tau} + \Lambda_i^*\delta C_i = \beta_i^* N_o - \Lambda_i^* C_o \equiv K_{C_i} \qquad i=1,6 \tag{185}$$

where $\delta N << N_o$ and $\delta C_i << C_{i_o}$ and were thus neglected by comparison in the Λ_i and β_i^* terms.

Solutions to (184) and (185) are immediately seen to be

$$\delta N = \frac{K_n}{\Omega} + (\delta N_o - \frac{K_n}{\Omega})e^{-\Omega\tau} \tag{186}$$

and

$$\delta C_i = \frac{K_{C_i}}{\Lambda_i^*} + (\delta C_{i_o} - \frac{K_{C_i}}{\Lambda_i^*})e^{-\Lambda_i^*\tau} \tag{187}$$

where δN_o and δC_{i_o} are initial perturbations which will be discussed shortly.

It is obvious that during the period of time where changes are small, the effect of a reactivity insertion dies out if

$$\Omega = (\beta - \rho) > 0 \quad . \tag{188}$$

Thus, unless a positive reactivity insertion is greater than the fraction of delayed neutrons, the reactor remains "prompt subcritical" ($\rho<\beta$), for some period of time.

If sufficient reactivity is added ($\rho > \beta$) to the nuclear system to cause the reactor to immediately begin to increase its power exponentially the reactor is then supercritical on the prompt neutrons alone and the perturbation is immediately excursive.

The condition where the reactivity insertion is just sufficient to be critical on the prompt neutrons alone ($\rho = \beta$), is termed "prompt critical." This amount of reactivity is defined as "one dollar" of reactivity. Thus,

$$\$ = \frac{\rho}{\beta} \; ; \; ¢ = 100 \frac{\rho}{\beta} \tag{189}$$

where hundredths of a dollar are termed "cents."

If the reaction frequencies for neutron and precursor generation are compared we find from (178) and (180) that

$$\frac{\Lambda_i^*}{\Omega} = \frac{p\lambda_i}{v\Sigma_a k_\infty(\beta-\rho)} \; . \tag{190}$$

Thus in a light-water moderated reactor even for the precursor with the shortest half life, the precursor concentrations change at a much slower rate than that of the neutron population density. This is not true in a graphite-moderated reactor, however, or one moderated by heavy water, because the prompt neutron life times (represented approximately by $\ell_o \sim 1/v\Sigma_a$) are themselves so long.

The much slower rate of change of precursor density causes another phenomenon to occur which is important from a control standpoint. On time scales short enough with small perturbations where C_i can be assumed initially constant, $|dN/d\tau| << \Omega N$ and we have

$$N_o^+ = \frac{1}{\Omega^+} \sum_{i=1}^{6} \Lambda_i^+ C_{oi}^- \tag{191}$$

where the + and - signs refer to just after and just before the perturbation in reactivity. But

$$C_{oi}^- = \frac{\beta_i^{*-}}{\Lambda_i^{*-}} N_o^- \tag{192}$$

so that

$$N_o^+ = \frac{\beta(1-\rho)}{\beta-\rho} N_o^- \; , \tag{193}$$

showing that the initial perturbation is given by

$$\frac{\delta N_o}{N_o^-} = \frac{\rho(1-\beta)}{(\beta-\rho)} \; . \tag{194}$$

This shows that almost immediately following a reactivity insertion (order of prompt neutron life time) there is a "prompt jump" in the neutron density, (193) being known as the prompt jump approximation.

We shall see presently that small reactivity insertions can occur due to the sudden appearance or disappearance of vapor zones in a reactor, passage of cold or hot water slugs relative to the existing fluid temperature, pressure changes in the moderator, or even thermal expansion effects. Changes in reactivity may also occur if some sudden material or configuration changes occur in the assembly itself.

(b) Stable period.

The neutron kinetics equation set (176) and (177) may be written in matrix form as

$$\underline{I}\dot{Y} - \underline{D}Y = 0 \tag{195}$$

where $\underline{I}$ is the identity matrix and $\underline{D}$ is the coefficient matrix for all the zeroth-order times in the set. The vector Y is the transpose of $[N \;\; C_1 \;\; C_2 \;\; \ldots \;\; C_6]$. If a solution of the form

$$Y = Ae^{\omega^* \tau} \tag{196}$$

is assumed (195) becomes

$$(\omega\underline{I} - \underline{D})\; Y = 0 \tag{197}$$

which has nontrivial solutions only for the case when

$$|\omega^*\underline{I} - \underline{D}| = 0 \; . \tag{198}$$

Being a 7x7 matrix for which we wish the determinant we expect there to be seven eigenvalues which satisfy (198): $\omega_0^*, \omega_1^*, \ldots \omega_6^*$. Thus each element of the solution vector has the solution

$$Y_i = \sum_{j=1}^{6} A_{ij} \; e^{\omega_j^* \tau} \; . \tag{199}$$

By observing that each C_i is independent of all other C_i (196) may be substituted directly into (177) allowing specification of A_{ij} for a given ω_i^* in terms of A_{oj}, the coefficient of the N-behavior. Substitution of the result into (176) yields after some rearrangement

$$\rho = \omega^* + pp_{NLF} \sum_{i=1}^{6} \frac{\omega^* \beta_i^*}{\omega^* + \Lambda_i^*} \; . \tag{200}$$

By substituting the definitions (178)-(183) into (200) and identifying

$$\omega^*\tau = \omega t \tag{201}$$

leads immediately to

$$\rho = \frac{\ell\omega}{1+\ell\omega} + \frac{1}{1+\ell\omega} \sum_{i=1}^{6} \frac{\omega\beta_i}{\lambda_i + \omega} \equiv f(\omega) \tag{202}$$

where ℓ is the prompt neutron lifetime given by

$$\ell = \frac{p_{NLT}}{v\Sigma_a} \sim p_{NLT}\ell_o \quad . \tag{203}$$

Typical values of ℓ_o are shown in Table 1.12.

Table 1.12

Approximate Prompt Neutron Lifetimes for Thermal Reactors[14]

Moderator	ℓ_o-sec
H_2O	2.1×10^{-4}
D_2O	0.14
Be	3.9×10^{-3}
C	0.017

Examination of Equation (202) shows the following. For a specified step reactivity insertion ρ is constant. Thus (202) shows $f(\omega)$ is a finite constant having seven poles, one at $\omega_6 = -1/\ell$, and one each at $\omega_i = -\lambda_i$. The traverse in ω from $\omega = -\lambda_i - \varepsilon$ to $\omega = -\lambda_i + \varepsilon$ shows that $f(\omega)$ switches from positive to negative infinity at each λ_i pole as ω increases through the pole. Thus, between each λ_i there must be a crossing where $f(\omega) = 0$. At $\omega = 0$, $f(\omega) = 0$ also and for large $|\omega|$ the limit of $f(\omega)$ is $1+\beta/1+\ell\omega$, or unity, ($\sim$ \$140 of reactivity).

The function $f(\omega)$ is shown schematically in Figure 1.20[20] as the bold lines. The roots of the Equation (202) are the intersections of lines of constant $\rho = f(\omega)$ with the bold curves--the large dots shown in the figure. For negative reactivity there are no positive roots and the reactor is unconditionally stable. For positive insertions there is only one positive root, the rest being negative.

Note that the largest negative root is near $\omega_6 \sim -1/\ell$. Thus, for either positive or negative reactivity insertions a very rapid decay occurs for one term in the seven-term neutron density equation. For a water reactor Table 1.12 shows us that this exponential is $\exp[-t/2.1 \times 10^{-4}]$ or approximately 95% in three prompt neutron lifetimes of slightly over 600 microseconds, ($p_{NLT} \sim 1$)! This is the source of the prompt jump described previously. If a relatively large negative reactivity insertion of, say 0.2 or $\sim$ \$30 occurs, (193) shows that the neutron density (power) decreases in this initial short period to approximately 3% of its original value.

For any reactivity, the terms with most negative ω_j's die out quickly. With negative reactivity the reactor period rapidly settles to something between 0 and $-1/\lambda_1$ or, from Table 1.11, approximately 80 seconds--a remarkedly long period considering the very short time between neutron generations represented by ℓ--and demonstrates the strong stabilizing effect of the delayed neutrons.

The reactor simply decreases in its power level by a factor of 1/e from the initial prompt jump value once every 80 seconds.

Since the largest root of Equation (202) for zero reactivity is zero, it follows that for small positive reactivity insertions the largest root ω_0 is only slightly positive, near zero. This being the case, we see from (202) that the lead term can be ignored and (202) yields a reactor period given by

$$T_{aprox} = \frac{1}{\omega_0} \sim \frac{1}{\rho} \sum_{i=1}^{6} \frac{\beta_i}{\lambda_i} = \frac{\overline{t}_d}{\$} \tag{204}$$

where

$$\overline{t}_d = \frac{\sum_{i=1}^{6} \frac{\beta_i}{\lambda_i}}{\sum_{i=1}^{6} \beta_i} \tag{205}$$

represents the precursor-averaged decay period for the particular fuel, tabulated in Table 1.13.

Table 1.13

Delayed Neutron Reactor Period Data for Three Fissile Isotopes[14]

Fuel	$\Sigma\beta_i/\lambda_i$ sec	$\overline{t}_d$ sec
$_{92}U^{233}$	0.0479	7.4
$_{92}U^{235}$	0.0848	13.0
$_{94}Pu^{239}$	0.0324	5.0

For positive reactivities of a few cents, say 1-10¢, it is seen that the reactor periods are on the order of 50-1200 sec, exceedingly long stable periods due to the effects of the delayed neutrons.

It is easily seen how a reactor power level may be changed by simply inserting a small amount of reactivity. This can be done mechanically by moving (shimming) control elements (usually rods containing boron) or chemically by altering the concentration of a poison (boron) in the coolant (called chemical shim). The reactor power level can then be monitored while it changes slowly according to the period T, and the power ramp terminated by the insertion of a similarly small reactivity of opposite sign as that which began the ramp. Reactor periods of $\sim$ 2 min or longer are generally used for power range maneuvering corresponding to $\sim$ 10¢ of reactivity to initiate the maneuver.

(c) Generalized behavior.

The actual solution to Equation (202) depends both on the type of fuel used through β_i and λ_i, and on the moderator through ℓ. The solution for ω_0 for the case of $_{92}U^{235}$ fuel is shown in Figure 1.21.[14] It should be noticed that for a fast reactor, i.e., a

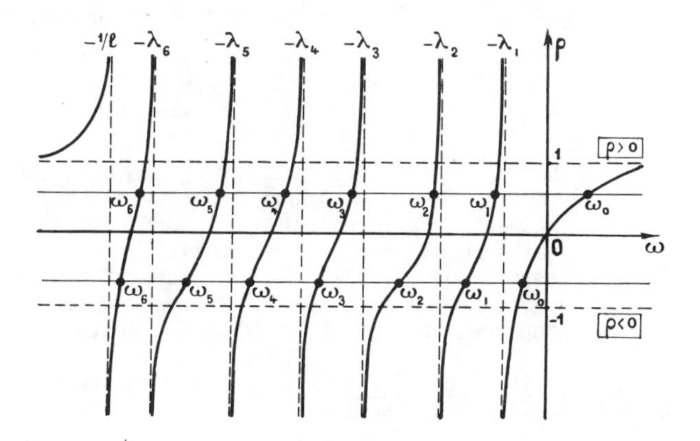

Figure 1.20. Reaction frequency solutions for six groups of delayed neutrons, (Reference 21, reproduced by permission). (BNL Neg. 7-834-80)

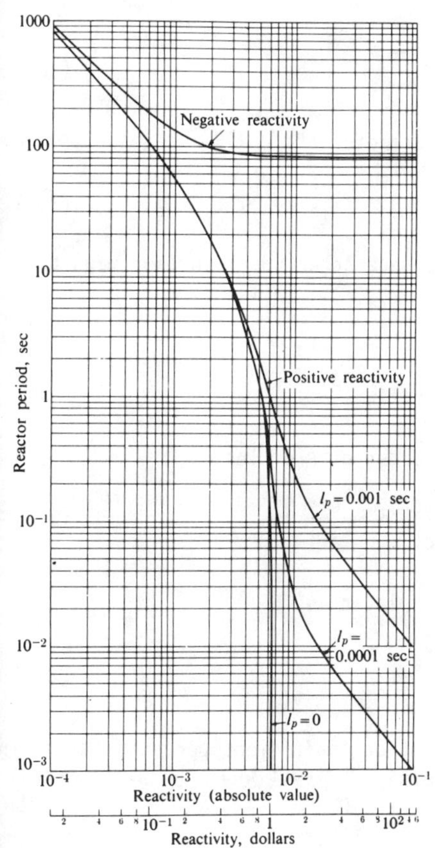

Figure 1.21. Reactor period for a $_{92}U^{235}$-fueled reactor as a function of the reactivity insertion, (Reference 14, reproduced by permission). (BNL Neg. 7-833-80)

reactor where the majority of the fissions occur in the high neutron energy ranges, the neutron lifetime is very short, on the order of 10^{-7}sec.[14] In such a case the characteristic equation becomes

$$\rho \sim \omega \sum_{i=1}^{6} \frac{\beta_i}{\omega+\lambda_i} \ . \tag{206}$$

Solutions for this case for ω_0 are thus obtained from Figure 1.21 for the curve $\ell = 0$, and indicate the major control differences between a fast reactor and a thermal reactor, the former having quite short periods for reactivity insertions approaching or exceeding one dollar. "Prompt bursts" are thus obtained on a much shorter time scale in a fast reactor due to the loss of the slowing down and thermal diffusion times.

Reactivity Coefficients. Let us now return to Equation (176) using the definition (178) for Ω so that

$$\frac{dN}{d\tau} + (\beta-\rho)N = \sum_{i=1}^{6} \Lambda_i C_i \ . \tag{207}$$

If the reactivity insertion ρ is due to some change in paramter ξ we may write for small changes in reactivity

$$\frac{dN}{d\tau} + (\beta-\alpha_{\rho\xi}\Delta\xi)N = \sum_{i=1}^{6} \Lambda_i C_i \tag{208}$$

where the "reactivity coefficient for the parameter ξ"

$$\alpha_{\rho\xi} = \frac{d\rho}{d\xi} \tag{209}$$

reflects the sensitivity of the reactor response to an upset in ξ of magnitude $\Delta\xi$.

Recalling that

$$\rho = \frac{k_{eff}-1}{k_{eff}} \quad \text{and} \quad k_{eff} = \frac{1}{1-\rho} \tag{210}$$

we may write for small reactivities of the order of β or less

$$d\rho = \frac{dk_{eff}}{k_{eff}^2} = (1-\rho)\frac{dk_{eff}}{k_{eff}} \sim \frac{dk_{eff}}{k_{eff}} \ . \tag{211}$$

The reactivity coefficient is thus usually written as

$$\alpha_\rho = \frac{dk_{eff}/k_{eff}}{d\xi} \tag{212}$$

which is the fractional change in multiplication coefficient per unit change in the parameter ξ. From the definition of k_{eff}, the fractional change in k_{eff} may be approximated by[20]

$$\frac{dk_{eff}}{k_{eff}} \sim \frac{d\eta_T}{\eta_T} + \frac{dp}{p} + \frac{d\varepsilon}{\varepsilon} + \frac{df}{f} - \frac{B^2 dM^2}{k_{eff}} - \frac{M^2 dB^2}{k_{eff}} \tag{213}$$

The areas of most concern from a thermal-hydraulic-neutronic viewpoint include

(a) temperature coefficient of reactivity, $\alpha_{\rho T}$
(b) void coefficient of reactivity, $\alpha_{\rho V}$
(c) pressure coefficient of reactivity, $\alpha_{\rho p}$.

Each shall be discussed briefly in what follows. Since an increase in reactor power would usually lead to an increase in each of these variables, it is generally desirable to have each coefficient be negative so that an up-power perturbation be self-limiting.

(a) Temperature coefficient, $\alpha_{\rho T} = \frac{\partial \rho}{\partial T}$.

The temperature coefficient includes several effects all of which contribute to the overall temperature effect such that $\alpha_{\rho T} = \alpha_{\rho T1} + \alpha_{\rho T2} + \ldots$. The following terms are included and are generally considered of most significance.[5,22]

1. Nuclear-temperature coefficient, $\alpha_{\rho Tn} = \partial \rho_n / \partial T$.

This coefficient concentrates on the temperature effects in the thermal migration area M^2. The largest effect is on the thermal diffusion length L such that

$$\alpha_{\rho Tn} = - \frac{B^2}{k_\infty} \frac{\partial L^2}{\partial \Sigma_a} \frac{\partial \Sigma_a}{\partial T} \tag{214}$$

Since the largest effect is on the absorption cross section σ_a, application of the 1/v hypothesis for the σ_a behavior yields

$$\alpha_{\rho Tn} \sim - \frac{B^2 L^2}{2 k_\infty T} \tag{215}$$

which arises mainly due to square root dependency of v on T, and which is applicable to small temperature changes. This shows that the nuclear coefficient is negative causing a negative reactivity insertion for any up-power upset.

2. Density-temperature coefficient, $\alpha_{\rho Td} = \partial \rho_d / \partial T$.

This effect is strictly due to change in nuclear density and is mainly felt in the migration area so that

$$\alpha_{\rho Td} = \frac{B^2}{k_\infty} \frac{\partial M^2}{\partial N} \frac{\partial N}{\partial T} \quad . \tag{216}$$

The nuclear density is inversely proportional to the cube of the linear dimension. By using the linear thermal expansion coefficient $\overline{\alpha}_{C1}$ for the average of the core materials it is found that (recalling $B^2 M^2 = k_\infty - 1$)

$$\alpha_{\rho Td} = - 6\bar{\alpha}_{C1} \frac{k_\infty - 1}{k_\infty} \tag{217}$$

which is also self-compensating for up-power transients.

3. Volume-temperature coefficient, $\alpha_{\rho Tv} = \partial\rho_v/\partial T$.

This effect is due to a change in the reactor buckling due to temperature changes through

$$\alpha_{\rho Tv} = - \frac{M^2}{k_\infty} \frac{\partial B^2}{\partial T} \tag{218}$$

The buckling is usually in the form (Table 1.10)

$$B^2 = \sum_k \left(\frac{a_k}{x_k}\right)^2 \tag{219}$$

so that for small temperature increases

$$\alpha_{\rho Tv} = 2\bar{\alpha}_{C2} \frac{k_\infty - 1}{k_\infty} \tag{220}$$

where $\bar{\alpha}_{C2}$ is the core-averaged thermal expansion coefficient including the structural materials which govern the geometric sizes. While this term provides autocatalysis for a positive power surge, its effect is generally much smaller than the other two effects.

A first approximation to the temperature coefficient for a reactor is

$$\alpha_{\rho T} = - \frac{B^2 L^2}{2k_\infty T} - \frac{k_\infty - 1}{k_\infty} (6\bar{\alpha}_{C1} - 2\bar{\alpha}_{C2}). \tag{221}$$

Only very slight effects are encountered from the other terms in Equation (213) and are neglected. Typical temperature coefficients for experimental and commercial nuclear reactors have been found to have magnitudes of between $\sim$ -0.1¢/°K and $\sim$ -30¢/°K.

4. Nuclear Doppler Effect, $\alpha_{\rho Tp} = \partial\rho_p/\partial T$

While the previous factors apply to the neutron population as a whole, there is an additional factor which is extremely important since it acts on the prompt neutrons as they slow down through the resonance absorption regions. This factor, termed the Doppler effect, yields a negative prompt neutron temperature coefficient in most reactors, a significant factor for reactor control in the event of a positive reactivity insertion of over \$1. Since it acts strictly on the resonance escape, Equation (213) shows that the prompt coefficient is given by

$$\alpha_{\rho Tp} = \frac{1}{p}\frac{\partial p}{\partial T}. \tag{222}$$

In the neighborhood of an isolated resonance of energy E_r, the energy dependence of the cross section is given by the Breit-Wigner formula

$$\frac{\sigma}{\sigma_{max}} = \frac{\Gamma^2/4}{(E - E_r)^2 + \Gamma^2/4}. \tag{223}$$

It is easily seen that Γ is the total width of the resonance, $2|E_{\frac{1}{2}}-E_r|$, where the cross section is half its maximum value. This quantity is simply termed the resonance width. Equation (223), however, applies under the tacit assumption that the nuclei with which the neutrons interact are at rest in the observational reference frame. As a result, it applies strictly to the case of absolute zero temperature of the absorber. However, as the temperature of the material increases, the atoms and their nuclei become agitated so that they exist in a continual state of motion. This causes the relative energy difference $(E-E_r)$ to appear as a continuous spectrum even for monoenergetic neutron population. The effective absorption cross section at a particular energy E near E_r thus seems to decrease, and the width of the resonance, Γ, appears to increase. Similar to a family of Gaussian curves of increasing standard deviation, the area under the absorption resonance curves remain the same with increasing widths, Γ.

At any temperature T, the resonance width Γ, which is greater than the value at absolute zero temperature Γ_0, will also represent an energy range within which the resonance absorption integral remains constant. Thus,

$$I_r \equiv \int_{E_r-\Gamma}^{E_r+\Gamma} \Sigma_a(E;T)dE = \int_{E_r-\Gamma_0}^{E_r+\Gamma_0} \Sigma_{ao}(E;0)dE \tag{224}$$

Considering absorption in the resonance band to be expressed as previously done for multigroup situations, (Equation (30) etc.), the resonance absorption rate is given by

$$dR_{ar} = \int_{E_r-\Gamma}^{E_r+\Gamma} \phi(E)\Sigma(E;T)dE \tag{225}$$

This equation may be written as in Equation (36) while at the same time multiplying and dividing by the resonance absorption integral (224) to yield

$$dR_{ar} = \phi_r \int_{E_r-\Gamma}^{E_r+\Gamma} \Sigma_a(E;T)dE \; \frac{\displaystyle\int_{E_r-\Gamma}^{E_r+\Gamma} \phi(E)\Sigma_a(E;T)dE}{\displaystyle\int_{E_r-\Gamma}^{E_r+\Gamma} \phi(E)dE \int_{E_r-\Gamma}^{E_r+\Gamma} \Sigma_a(E;T)dE} \tag{226}$$

The integral ratio on the right hand side of (226) is simply the covariance function of the flux and the cross section which shall be termed $G(\phi,\Sigma_a)=cov(\phi,\Sigma_a)$. The resonance group flux ϕ_r is

$$\phi_r = \int_{E_r-\Gamma}^{E_r+\Gamma} \phi(E)dE \qquad (227)$$

so that the resonance zone absorption rate may be written as

$$dR_{ar} = G(\phi,\Sigma_a)\ \phi_r\ I_r \qquad (228)$$

Consider what occurs as the temperature increases. First, except near absolute zero temperature, the G-function is near unity, and the resonance absorption integral, I_r, is a constant by definition. Since the resonance width $\Gamma(T)$ increases with temperature the resonance flux ϕ_r given by (227) must also increase causing the absorption rate to increase. Thus, Doppler broadening is seen to cause increasing neutron absorption with generally negative feedback effects on reactor power. An expression which approximates the reactivity feedback coefficient in terms of the absolute temperature may be derived[14] to yield

$$\alpha_{\rho Tp} = -\frac{\beta_p}{\sqrt{T}} \qquad (229)$$

In this case, β_p is a material and geometry dependent constant.

The physical explanation for this effect is quite simple. Since a finite energy loss occurs during each scattering collision, a broader resonance region yields a higher possibility for multiple collisions to occur while slowing down within the energy range of the resonance. A higher probability for capture therefore results. Complicating factors from a mathematical viewpoint arise due to overlap of adjacent resonance peaks when broadening occurs. This may happen in a single material having closely spaced resonance regions, or in differing materials having resonances at similar but unequal energies. The underlying physics underlying the Doppler effect, and the resultant effect on the prompt temperature coefficient remains the same.

(b) Void reactivity coefficient, $\alpha_{\rho\alpha} = \frac{\partial \rho}{\partial \alpha}$.

The void reactivity generally refers to the effects of vapor generation and resultant vapor volume fraction α, in a liquid-cooled/moderated reactor. In most liquid moderated thermal reactors the void coefficient is strongly negative, generally because of the increase in the migration area and hence the direct increase in neutron lifetime caused by removal of slowing-down capability. In single terms, a decrease in the neutron slowing down rate is felt basically as a decrease in the nuclear source, offset slightly in the fuel utilization factor, f. The strong negative void coefficient in boiling water reactors, for instance, gives these reactors very attractive self-load-following characteristics without operator intervention. In some cases, however, a D_2O moderated H_2O cooled reactor may have a slight positive void coefficient since, being overmoderated, the removal of H_2O removes a parasitic absorber without strongly reducing the moderation effects. This is the case of some CANDU (CANadian DeUterium) designs.

The situation is slightly more complex in fast reactors since little moderation occurs and the effects of moderation are felt mainly in the decrease in the value of η with neutron energy. For an infinite reactor or very large reactor, the increase in vapor volume would decrease moderation and cause a hardening of the neutron spectrum towards higher average energies. The resultant increase in η is dominant and the void coefficient may be strongly positive. For a finite reactor, leakage effects become important. The smaller the reactor, the larger the leakage effects would be in general. Since voiding of the moderator would in effect increase the leakage, this effect is felt more strongly the smaller the reactor geometry. This is why some sodium-cooled fast reactors such as EBR-II and FFTF have negative void coefficients while larger but otherwise similar core designs such as CRBR and SUPER PHOENIX have positive void coefficients.

(c) Pressure coefficient of reactivity, $\alpha_{\rho p} = \frac{\partial \rho}{\partial p}$.

Since liquids and solids are generally of very low compressibilities, pressure effects on systems where no voids exist are negligible. Where voids do exist, the effects of increasing pressures are to decrease the void fraction (vapor volume) so that

$$\alpha_{\rho p} = \frac{\partial \rho}{\partial \alpha}\frac{\partial \alpha}{\partial p} = \alpha_{\rho\alpha}\frac{\partial \alpha}{\partial p} \tag{230}$$

leading to effects as previously discussed amplified by the factor $\partial \alpha/\partial p$. Since α, the void fraction, decreases with increasing pressure, the pressure reactivities are usually positive for reasons previously mentioned. Since $\partial \alpha/\partial p$ is strongly negative at low pressures, boiling water reactors (BWR) may sometimes experience instabilities at low pressure if proper precautions in design and operation are not observed. Similarly, positive pressure pulses due, say, to a turbine trip in a BWR, may lead to sufficient void collapse to cause a considerable power surge until the increasing void fraction due to increased power turns the power around. Such was recently verified to be the case in special tests conducted at the Peach Bottom plant. The strong self-limiting characteristics of the positive power-to-void transfer function coupled with negative void coefficients eliminated any problems which might otherwise have occurred.

3. PHILOSOPHY AND PRACTICE OF NUCLEAR SAFETY

Any activity of human endeavor, whether it be active or passive from an individual standpoint, carried with it some risk. The simple act of inactivity to escape risk carries with it some risk in itself if for no other reason than atrophy. This section is concerned with the practical aspects of risk, and its minimization, insofar as the utilization of nuclear energy for power production is concerned. Arguments in this area can be both philosophical and practical. For example, how much risk are we, as individuals, willing to accept? How much are we willing to pay to minimize this risk? To what extent is our individual and collective safety affected by the presence of nuclear power systems? Are these risks reasonable? Tolerable? Increasing or decreasing in magnitude? In order to answer some of these questions we must understand some of the differences between nuclear systems and conventional power production systems.

3.1 Nuclear vs. Conventional Power Systems

Radioactivity. The first and most obvious difference is the fact that the nuclear fuel and its waste are radioactive, which is to say they are poisonous on an atomic scale. Either ingestion or absorption of radioactive material can be harmful or fatal if the dose level is sufficiently high, and proximate exposure, unlike other poisons which act on a chemical basis, can also have an effect. Therefore, the radioactive materials stored or generated in the nuclear power cycle must be controlled in such a way to minimize exposure of individuals to acceptable levels. This is the single most important difference between conventional power systems and nuclear systems and is the factor which has driven and continues to drive the intense safety research work in the world. Radioactivity must be accepted as a cost of doing business in the same way that oil retention and cleanup, coal slag removal, and sulphur and carbon dioxide release must be accepted if coal or oil fuels are used to generate power.

Decay Heat. Unlike conventional power sources which stop when they are turned off, nuclear systems do not. As mentioned previously and shown in Figure 1.8 , nuclear systems continued to generate power due to radiodecay, approximately 15 billion curies per 1000 MWe,[1] of fission products over long periods of time. Methods must thus be devised for maintaining long term cooling of the nuclear fuel in such a manner that inadvertent disruption of the normal cooling modes would not lead to release of radioactivity. Thus, nuclear steam supply systems are generally equipped with residual heat removal, (RHR), systems.

Reaction Control. It is quite difficult to imagine a conventional power system suffering a substantially harmful up-power transient resulting from inadvertent miss control or loss of control. On the other hand, nuclear systems operate in a metastable state where a slight perturbation due to a positive or negative reactivity insertion gives an effective multiplication factor, k_{eff}, slightly different from unity. This difference yields an exponential growth or decay in the neutron inventory accompanied by a like change in the reactor power level. Exponential periods in the fractional millisecond range are indeed possible and precautions in design and operation must be taken to avoid or mitigate these situations. As we have seen earlier the effect of delayed neutrons makes the task of controlling the multiplication factor to exact unity a relatively simple task, in spite of the very short prompt neutron lifetimes.

In addition, since the nuclear configuration is constantly changing due to fuel burnup, the reactor geometry must be constantly changeable through the buckling. This is generally accomplished by a combination of burnable and removable poisons, some mechanically operated. Provisions must be made for rapidly shutting down a reactor in the event of an inadvertent excess of poison removal which would otherwise result in an up-power excursion. Such reactor "SCRAM" capability is usually obtained by providing control rods with a rapid insertion capability. The resultant prompt jump due to substantial negative reactivity insertion acts as an extremely effective immediate shutdown mechanism. As such it is utilized under any off-normal situations which are indicative of a potential instability.

Fuel Cooling. Since the fission fragments carry the majority of the energy released in the fission process, and since they are generally contained inside a barrier for both safety and physical reasons, the fuel itself must be cooled. In addition, due to the possible disruption of the cooling process, emergency cooling measures must be provided to maintain continued fuel element integrity. This is usually accomplished by redundant emergency core cooling systems (ECCS) which are separate from the main cooling system both in source of coolant and nature of the supply. Provisions for positive nuclear shutdown may also be provided in the ECCS by the inclusion of the poison boron in the coolant.

Reaction Rates. The nuclear reaction rates are dependent on nuclear interaction parameters which may be altered by changes in the physical environment. Therefore, insofar as possible, all environmental factors must result in a negative power feedback effect where possible. This is especially true of the fuel temperature coefficient exclusive of moderator effects since fuel reactivities are effective on the time scale of prompt neutron lifetimes.

3.2 Environmental Interaction

A nuclear power system may be considered to be a control volume in the same sense that a thermodynamic system may be considered. To examine the safety and risks associated with such a plant, one must examine the interactions between the control volume and its surroundings. These include thermal interactions, electrical interactions, and material interactions.

Thermal Interactions. These interactions are due to the Carnot limitations inherent in any energy conversion system. They are identical in nature to those which occur in other types of power plants. They may represent a somewhat greater percentage of the total energy inventory of the system because the thermal efficiencies of nuclear systems are somewhat lower than conventional systems due to lower operating temperatures--30-35% rather than 35-45%.

Electrical Interactions. These also are no different in a nuclear system than in a fossile-fueled power plant. They thus deserve no special considerations except for the possible consequences of electrical supply failure.

Material Interactions. These are the interactions between a nuclear power system and the environment which deserve and indeed require the majority attention. They may be of several types including personnel, operating and maintenance materials and supplies exclusive of fuel, air, services, coolant, nuclear fuel, and nuclear waste material. The services such as water and non-nuclear waste require attention because waste may carry radioactivity which has been purged or leaked from the system. Similarly, air venting or release during physical access to the plant may carry radioactive gases. The coolant paths must be considered because of potential leakage and subsequent release of radioactivity carried in the coolant. Nuclear material transport to and from the plant is of concern from both a transport safety as well as a storage aspect. Indeed, the latter is receiving intense attention and study at this time. By far, though, the major concern has been and continues to be the inadvertent accidental release of radioactive material and the entire reactor system is designed to prevent such an eventuality. Virtually all nuclear safety research relating to the nuclear power system itself relates to the questions of safeguarding public health and welfare by preventing release of radioactive materials.

3.3 Nuclear Safety and System Configuration

There are many ways in which nuclear system safety is assured including design control, quality specification and control, administrative control, and operational control.

Design Control. Design control is the control exerted by the system designer to insure sufficiently benign behavioral response to any foreseen and unforeseen eventuality. The nuclear system must be designed to insure absolute control against the release of excess radioactive material under all circumstances. A system has evolved which relies on three barriers or levels of containment, all three of which must be breached before substantial levels of radioactivity are released. These levels include the fuel cladding, the nuclear heat supply system itself, and the containment building within which the system is placed. The ceramic fuel matrix itself acts implicitly as a barrier also.

The fuel cladding itself represents the first barrier to radioactivity release. Current design practice is to design the nuclear system to try to insure the temperature of the cladding never exceeds a limit which would otherwise lead to failure due to melting, cracking, rupturing, or oxidizing. This limit is usually taken as $\sim$ 1200°C (2200°F), and extensive calculation and research is expended to attempt to insure these temperatures are not exceeded under normal or off-normal sequences.

The nuclear power system itself constitutes the second barrier to the release of radioactivity, usually consisting of the reactor vessel, coolant piping, and heat exchange system or in some cases, the prime mover itself. Even if one or more fuel elements were to be breached, the system barrier would also have to be breached to release the resultant radioactivity. This may happen in several ways--rupture of a pipe being the least expected and most difficult to control. Valve steam leaks, bearing leaks in pumps, instrument line leaks, also play a role as do intentional or unintentional purging through vent valves or relief valves. It was the latter, in fact, which led to the difficulties recently experienced in the Three Mile Island Accident.

The reactor system containment bulding represents the third and sometimes the fourth barrier to the release of radioactive materials. The nuclear system is placed itself inside a large container which is generally held at subatmospheric pressures to insure any atmospheric leakage is inward. In the case of some BWR's there is an internal containment for the reactor vessel itself termed the "dry well" which also houses a liquid pool system designed for pressure suppression in the event of a pipe rupture. The overall containment structures are designed to hold the contents of a nuclear reactor system in the event of a rupture leading to escape of the coolant into the containment volume and will take moderate overpressures with respect to atmosphere. Detailed descriptions of reactor systems are given in the following chapters.

In addition to the physical design of the nuclear system and containment, the system instrument and control design is carefully undertaken with sufficient redundancy to insure continuous operation in the event of a failure. In addition, separate auxilliary electrical power generation systems are kept on standby to provide instrument and control power in the event a loss of off-site power occurs.

Quality Specification and Control. The highest and most up-to-date standards are utilized in the design and construction of nuclear power systems. Applicable formal Codes and Standards such as the ASME standards for nuclear piping are usually met or exceeded in a system design. These standards not only give material and geometric specifications, but also give specifications for quality control and inspection. For instance, material certifications specifying the heat and spectrographic analysis of the melt may be required for nuclear system components as well as full radiographic and/or ultrasonic inspection of welds or castings for flaws.

Administrative Controls. Administrative controls are exercised throughout all stages of design, construction and operation of a nuclear reactor power system. While these are exercised by architectural engineers, system vendors, and utilities, the final responsibility for nuclear system quality and safety in all phases rests with governmental bodies. In the United States, controls are administered on various levels--local, state, and federal--with reviews being undertaken through public hearings as well as by regulatory bodies. The ultimate authority for regulation in the U.S. is the Nuclear Regulatory Commission (NRC) and reviews are undertaken by various departments within the NRC regarding different aspects of plant siting, design, operation, and emergency evacuation. In addition, reviews are undertaken by the Advisory Committee on Reactor Safety (ACRS) with their reports going directly to the NRC commissioners. Similar and generally no less stringent review and control procedures exist in most countries utilizing nuclear power today.

In addition to regulations promulgated by various agencies to insure that the highest standards of safety are maintained, laws themselves are usually enacted which specify certain requirements which must be met as well as to provide for regulatory authority. In the United States these are included in the Code of Federal Regulations Title 10. The different parts of this Title cover different aspects. For instance, part 10 pertains to radiation protection while part 50 refers to licensing of nuclear power plants. General design criteria for nuclear reactors are contained in 10CFR50 Appendix A.

Operational Controls. The operation of a nuclear power plant is governed by the designed control system and operational manuals providing specifications for control procedures. The operational criteria are subject to review as are other aspects of nuclear system design. Certain functions of the reactor control system actuate automatically and cannot be overridden such as the period meter tied into the SCRAM system to rapidly shut down the reactor by control rod insertion in the event of an unacceptably rapid up-power excursion. The balance of the operational controls are either manual or automatic with manual override. An example of the latter is the ECCS. Manual termination at Three Mile Island Unit Number 2 contributed to the accident. This latter fact brings up the controls applied to reactor operation itself. Operators must receive acceptable training and qualify by both written and demonstrative examination.

There are three levels of safety for nuclear power systems parallel to the three barriers. These three levels are designed to provide "defense in depth" against the possibilty that the containment barriers might be breached.

Safety Level One. The nuclear system designer must incorporate features for maximum safety and also give the system the maximum tolerance for malfunction. The features incorporated include:

(a) prompt negative temperature coefficient
(b) use of materials only of known and acceptable behavior
(c) instrumentation redundancy to provide operators with continuous plant performance data
(d) use of highest current engineering practice and standards
(e) monitoring of system and components as required to detect wear or incipient failures.

Safety Level Two. Design to prevent or minimize damage and protect operational and civilian personnel in the event of an anticipated potential incident. Design should include:

(a) emergency core cooling system (ECCS)
(b) redundant SCRAM systems
(c) separate independent emergency power systems for ECCS, instruments, and control devices.

Safety Level Three. Develop all possible hypothetical accident sequences including the possibility of simultaneous failure. Add a safety margin for unforeseen events and include additional system protections and controls to insure stable and safe operation in the event such a sequence should occur. Such events utilized to analyze and evaluate the nuclear system performance are termed "design basis accidents." For the light water reactor the DBA is a guillotine rupture of one of the main cooling lines (PWR) or recirculation line (BWR) leading to a loss of coolant accident (LOCA). For an LMFBR the DBA is the core disruptive accident (CDA) which may result from a loss of flow without SCRAM. The overall LOCA and its detailed sequences are examined in detail in Part 3 of this book. The CDA is similarly examined in Part 4. For the High Temperature Gas-Cooled Reactor (HTGR) the DBA is the break of a reactor vessel penetration leading to a depressurization.

The potential for radioactivity release from a nuclear reactor system must also be considered both for expected releases such as sump clearing, fission gas venting, or condensor purging, and for the unexpected release as may ocur as a result of an accident. Factors such as population density, meteorological data, earthquake expectancy, etc., must all be considered and taken into account in plant siting and design.

Reactor accidents have been classified according to the increasing likelihood for the release of radioactivity, and are numbered from 1 to 9 in order of increasing severity. Table 1.14 itemizes each class and these are described in more detail in Chapters 3 and 4. Accident classes 1-8 must be considered in the design of a reactor plant and the class 8 accidents must be analyzed in detail and included for review in a detailed safety analysis report (SAR) subject to legal review.

The class 9 accidents, while being potentially of the greatest severity, are so remote that they currently need not be analyzed or protected against through design in LWR's, although this may change as a result of the TMI accident. This points out that the risks associated with certain events are so small that the economics of protection, mitigation, or prevention are not considered justifiably important. The question of risk leads us directly into the final section of this chapter.

Table 1.14

Nuclear System Safety Accident Classification

Class	Description
1	Trivial accidents
2	Small releases outside containment
3	Failure of radioactive waste storage or handling system
4	Escape of fission products from fuel elements into primary (BWR) systems
5	Escape of fission products from fuel elements to the primary and secondary systems (PWR)
6	Refueling accidents
7	Spent fuel handling accidents
8	Accident initiators considered in DBA's.
9	Extremely serious events so unlikely that their likelihood can be "ignored in analyzing the safety and environmental aspects of the plant."[14]

3.4 Risk

The concept of risk appears to be a combination of frequency and severity of events which if "low" is tolerable, but if "high" is intolerable. Obviously the terms "low" and "high" are quite subjective and their perception is very much an individual psychological factor. Risks of a given type can be quantified and, when compared in consistent units, may provide quantifiable indices of acceptability. Thus, if a person were to consider an undertaking which if executed would have a 10^{-5} possibility of death, would probably proceed if minimal incentive were to exist. Such an undertaking might include the use of a gun or taking a boat ride. Both could be considered pleasurable experiences if undertaken for recreational purposes and people would usually not give such activities a second thought.

On the other hand, if the probability of death becomes higher, say one in 3000 per person per year, perhaps a higher psychological motivation might be required to induce the individual to proceed. Such is the case of riding in an automobile for average distances on a continuing basis. High risk would, outside of the psychological catharsis which may result, generally require sufficient inducement for the undertaking to proceed, such inducement perhaps being an equally high or higher risk due to not proceeding.

Risk can be considered the product of frequency of occurrence and severity per event, either on an individual or a group basis. Thus,

$$\left(\frac{\text{consequences}}{\text{p.u. time}}\right) = \left(\frac{\text{events}}{\text{p.u. time}}\right) \times \left(\frac{\text{consequences}}{\text{per event}}\right) .$$

For instance the risk could be stated simply in terms of impersonal events such as automobile accidents per year in the U.S.,

$$10^4 \,\frac{\text{miles}}{\text{year-auto}} \times 0.75\text{x}10^{-4} \,\frac{\text{accidents}}{\text{mile}} \times 2\text{x}10^7 \text{ autos} = 15\text{x}10^6 \,\frac{\text{accidents}}{\text{year}} .$$

Or the risk could be monitory loss as for automobiles for the United States population as a whole,

$$15\times10^{6}\ \frac{\text{accidents}}{\text{year}} \times \$300/\text{accident} = \$4.5\times10^{9} \text{ per year cost}\ .$$

On the other hand, the risk could be in terms of quite personal events but still for whole populations. Then

$$15\times10^{6}\ \frac{\text{accidents}}{\text{year}} \times \frac{1 \text{ death}}{300 \text{ accidents}} = 50{,}000\ \frac{\text{deaths}}{\text{year}}\ .$$

Risk may instead be stated in terms of individual risk. For the above, figuring a population of 200 million the individual risks would be

0.075 accidents per person per year
$22.50 per person per year
1 in 4000 deaths per person per year

Obviously these risks of automobile driving are deemed acceptable in the U.S. People don't really consider that in a 40-year driving career at 30,000 miles per year their risk might be one chance in 30 of being killed. This writer certainly accepts these risks in order to continue his way of life.

Individual risks for various eventualities are given in Table 1.15. It is obvious that for risks above one in a million, people are willing collectively to spend considerable money and time on preventative measures. A study based generally on late '60's technology and reported first in 1974 by N. Rasmussen and his coworkers[24] for the U.S. Atomic Energy Commission, evaluated the risks involved in reactor operation. They estimated based on fault tree and event tree probabilistic analysis techniques the various risks associated with different circumstances to arrive at societal and individual risk estimates. Their conclusions showed that while it could be expected that 115,000 deaths would occur each year in the U.S. due to accidents of all types, deaths would

Table 1.15

1969 Approximate Risk Table for Individual Fatalities[24]

Accident	Individual Risk Death/Person-Year
Auto	3×10^{-4}
Falls	9×10^{-5}
Fire and Heat	4×10^{-5}
Drowning	3×10^{-5}
Poison	2×10^{-5}
Guns	10^{-5}
Machinery	10^{-5}
Boats	9×10^{-6}
Airplanes	9×10^{-6}
Falling Objects	6×10^{-6}
Electrocution	6×10^{-6}
Trains	4×10^{-6}
Lightening	5×10^{-7}
Hurricanes	4×10^{-7}
Tornados	4×10^{-7}
All Others	4×10^{-5}
All Accidents	6×10^{-4}

occur at a rate of 4 per hundred years due to nuclear accidents if 100 reactors were in operation. The individual risk was shown to be approximately 2×10^{-10} fatality per year per person--a very minor possibility considering the risks shown in Table 1.15. Stated another way, the possibility of 1000 deaths occurring as a result of 100 operating nuclear power plants in the U.S. was seen to be identical to the potential for loss of life due to the impact of a meteor. The author finds it difficult to recall the last time he learned of such an event. Loss of similar numbers due to earthquakes is 30,000 times more likely.

It is thus seen that the risk associated with nuclear power generation by any standards should be considered negligible when viewed in the light of 1970 technology. The technology is continually improving. By the same token our safety methods are also continually improving due to learning experiences associated with low class accidents. The resultant risk thus continues to decrease. Certainly the largest benefit by far of the Three Mile Island incident was the resultant introspection by workers in the fields of regulation and safety which led to substantial improvements in existing and forthcoming nuclear system designs. It is difficult to conceive of an accident so severe that it caused hundreds of millions of dollars of damage--with no loss of life and no personal injury! It is testimony to the intense efforts expended on nuclear safety technology that this was true in the nuclear industry. Reactor power systems are safe and through continued efforts will become safer still.

4. SUMMARY

In this chapter the reader was provided with an overview of nuclear power technology as it pertains to the thermal hydraulic aspects. A brief introductory review was given of the significant events leading up to the beginning of nuclear power--from Thales to Fermi. In the second section nuclear energy fundamentals were presented while concepts of nuclear safety were briefly presented in the third section.

In the second section the concepts of nuclear interactions were discussed and certain definitions such as cross section and mean free path were introduced. Neutron scattering was presented and it was shown why one material is a good moderator and another is not. The fission process was described and it was shown that the majority of the reactants of the process carried the energy and radioactivity--hence the need for fuel isolation, containment, and cooling. The neutron diffusion approximation was presented and solutions to the neutron diffusion problem with point, plane, and volumetric sources were presented for simple geometries. The special case where the neutron source is the absorber itself accompanied by fission, release of energy, and birth of more than one new neutron was reviewed and solutions for simple geometries given. It was shown why certain power profiles exist in today's reactors and why the need for cooling, flux shaping, and perhaps coolant flow proportioning to different regions of the reactor inlet may exist. Extensions to the transient case where the buckling--power shape--remains constant were then based on the transient diffusion equation. It was shown why the fortuitous existence of delayed neutron groups arising from the radiodecay of fission products is an extremely beneficial factor for reactor control, and why the control of fast reactors presents a more difficult challenge than the control of thermal reactors. The effects of perturbations in the thermal hydraulic parameters such as temperature, pressure, and void fraction was described through the introduction of reactivity coefficients.

In the third and final section of this chapter, the philosophy and practice of nuclear safety was briefly discussed. It was mentioned that the continued striving for safe nuclear power systems has lead to a dual trilogy--the trilogy of containment barriers and the trilogy of safety levels--which provide a defense in depth against the primary concern of the nuclear industry, the release of radioactivity harmful to public health and welfare. These were briefly reviewed and the risk associated with nuclear plant operation was presented. It was mentioned that the expectancy of the loss of 1000 lives due to a reactor accident was approximately 30,000 times more remote than similar loss from earthquake.

Finally, a bit of personal philosophy was introduced by the author and will be expanded on here. The risk figures mentioned in Section 3.4 show nuclear power to be by far the safest industrial activity of its size the world has known. Probabilities for a catastrophic accident based on 1970 technology are so remote as to be almost an impossibility. In order to maintain this enviable record--no fatalities in the commercial field in over 2000 reactor years of power plant operation--several things are needed.

1. The utility of accidents of slight or moderate frequency or severity of occurrence which would be inordinately expensive to eliminate must be tolerated, accepted, and recognized for the potential learning experience they provide.

2. The learning experiences provided by such accidents must be capitalized upon for what they can teach us about improved power plant and component design.

3. The rate of decrease of risk associated with incorporation of new technology must be larger than the rate of increase of risk of accidents associated with increased numbers of power plants and lack of a severe accident itself.

4. As a public we must be extraordinarily careful that extreme positions do not have the effect of paralyzing the effectiveness of our regulatory and safety research.

5. As regulators we must have the courage to place our vision forward and be ever mindful of our true mission--the continued and continually improving safety of nuclear power systems.

Regarding the latter we must keep in mind the old adage, "It is hard when surrounded by alligators to recall that the primary mission is to drain the swamp." It is easy for a regulatory body to succumb to backbiting, bureaucratic infighting, and bungling in the face of a difficult situation such as occurred at Three Mile Island. But boldness and leadership rather than rear end protectiveness must be exercised if the nuclear industry is to survive to recognize its potential--a truly cheap and abundant power source for thousands of years to come.

ACKNOWLEDGEMENTS

The author is indebted to Drs. S. Pearlstein, M. Levine, and D. Diamond for their careful review of the manuscript and the helpful suggestions they gave, and especially to Dr. Pearlstein for the information he and members of his staff of the National Nuclear Data Center provided. A special thanks goes to M. McGrath and D. Thompson for their preparation of the manuscript. Permission to reproduce material for Figures 1.20 and 1.21 granted by Addison-Wesley and Dunod Publishing respectively is gratefully acknowledged.

APPENDIX A. SOLUTION TO THE PROBLEM OF DIFFUSION IN A FINITE CYLINDER WITH A CONSTANT SOURCE

The equation to be solved is given by Equation (72)

$$\frac{1}{r}\frac{\partial}{\partial r}\left(r\frac{\partial\phi}{\partial r}\right) + \frac{\partial^2\phi}{\partial z^2} - \frac{1}{L^2}\phi = -\frac{S}{D} \tag{72}$$

subject to the conditions

BC 1 $\lim_{r \to 0} \phi(r,z)$ is finite

BC 2 $\phi(a^+,z) = 0$

BC 3 $\frac{\partial\phi}{\partial z}(r,0) = 0$ (symmetric in z)

BC 4 $\phi(r,h^+) = 0$.

A particular solution to this problem is easily obtained by inspection. We thus look for a solution of the form

$$\phi(r,z) = \frac{SL^2}{D} + u(r,z) + v(r,z) \tag{A1}$$

where both $u(r,z)$ and $v(r,z)$ satisfy the homogeneous equation

$$\frac{\partial^2\psi}{\partial z^2} + \frac{1}{r}\frac{\partial}{\partial r}\left(r\frac{\partial\psi}{\partial r}\right) - \frac{1}{L^2}\psi = 0 \tag{A2}$$

with boundary conditions separated between u and v such that

	u-conditions		v-conditions
BC 1:	$u(0,z)$ finite	;	$v(0,z)$ finite
BC 2:	$u(a^+,z) = -SL^2/D = u_1$	;	$v(a^+,z) = 0$
BC 3:	$u_z(r,0) = 0$	;	$v_z(r,0) = 0$
BC 4:	$u(r,h^+) = 0$	;	$v(r,h^+) = -SL^2/D = v_1$

It is readily verified that both functions and boundary conditions sum to the appropriate values.

A.1 Solution for u

The equation to be solved is

$$\frac{1}{r}\frac{\partial}{\partial r}\left(r\frac{\partial u}{\partial r}\right) + \frac{\partial^2 u}{\partial z^2} - \frac{1}{L^2}u = 0 \tag{A3}$$

subject to the above boundary conditions.

A product solution assumed is of the form

$$u = R_u(r)\, Z_u(z) \tag{A4}$$

which upon being substituted into (A3) yields

$$\frac{1}{R_u r}\frac{d}{dr}\left(r\frac{dR_u}{dr}\right) - \frac{1}{L^2} = -\frac{1}{Z_u}\frac{d^2Z_u}{dz^2} = \mu_r. \tag{A5}$$

The choice of leaving the diffusion length term with the r-solution is arbitrary and somewhat simplifies the solution. The three possible cases for μ_r--positive, negative, and zero--will be examined separately.

Case 1 - $\mu_r < 0$. In this case ($\mu_r = -k^2$) the z-solution becomes

$$Z_u(z) = ae^{kz} + be^{-kz} \tag{A6}$$

whereas

$$Z_u'(z) = k(ae^{kz} - be^{-kz}). \tag{A7}$$

From condition BC 3 we find that at $z = 0$

$$\left.\frac{\partial u}{\partial z}\right|_{z=0} = R_u Z_u'\Big|_{z=0} = kR_u(a-b) = 0 \tag{A8}$$

yielding

$$a = b\ . \tag{A9}$$

Similarly, BC 4 yields

$$u(r,a^+) = 0 = R_u(e^{kh^+} + e^{-kh^+})\, a \tag{A10}$$

from which we obtain the trivial solution $a = 0$.

Case 2 - $\mu_r = 0$. In this case the z-solution is

$$Z(z) = a + bz. \tag{A11}$$

Application of conditions BC 3 and BC 4 again yield the trivial result $a = b = 0$.

Case 3 - $\mu_r > 0$. For positive $\mu_r = +k^2$ we have

$$\frac{d^2Z_u}{dz^2} + k^2 Z_u = 0 \tag{A12}$$

from which

$$Z_u(z) = a \cos kz + b \sin kz. \tag{A13}$$

and

$$Z_u'(z) = - k \, [a \sin kz - b \cos kz] \tag{A14}$$

The symmetricity condition (BC 3) shows that b = 0 whereas BC 4 results in

$$a \cos kh^+ = 0. \tag{A15}$$

The only nontrivial solutions to (A15) are the eigenfunctions produced by letting kh^+ take an odd multiples of $\pi/2$. Thus

$$k = k_n = \frac{(2n+1)\pi}{2h^+} \qquad n = 0,1,2, \ldots \tag{A16}$$

yielding an infinity of solutions

$$Z_{un}(z) = a_n \cos k_n z \qquad n = 0,1,2, \ldots \tag{A17}$$

The r-solution is readily obtained from the infinite set of equations

$$\frac{d^2R_{un}}{dr^2} + \frac{1}{r}\frac{dR_{un}}{dr} - \beta_n^2 R_{un} = 0 \qquad n = 0,1,2, \ldots \tag{A18}$$

where

$$\beta_n^2 = k_n^2 + \frac{1}{L^2} \qquad n = 0,1,2, \ldots \tag{A19}$$

is always real and positive. The solutions to (A18) are modified Bessel functions

$$R_{un}(r) = C \, I_0(\beta_n r) + d \, K_0(\beta_n r), \qquad n = 0,1,2, \ldots \tag{A20}$$

Utilization of the condition at r = 0, since K_0 has a logarithmic discontinuity at the origin, gives d = 0. We thus find that

$$u(r,z) = \sum_{n=0}^{\infty} C_n \, I_0(\beta_n r) \cos k_n z \quad . \tag{A21}$$

In order to determine the coefficients c_n we utilize the condition at $r = a^+$ to give

$$u(a^+,z) = u_1 = \sum_{n=0}^{\infty} C_n \, I_0(\beta_n a^+) \cos k_n z \tag{A22}$$

It is obvious that for (A21) to be valid the series must be a convergent representation of the constant u_1.

Since Z(z) and its boundary conditions satisfy the Sturm-Liouville conditions in $[0,h^+]$, cos $k_n z$ is orthogonal and we can thus separate out C_n by multiplying both sides of (A21) by cos $k_m z$ dz and integrating over the range of z. Thus

$$\int_0^{h^+} u_1 \cos k_m z \, dz = \int_0^{h^+} \sum_{n=0}^{\infty} C_n I_o(\beta_n a) \cos k_n z \cos k_m z \, dz. \qquad (A23)$$

We may interchange summation and integration since the sum is uniformly convergent so that the only nonzero term on the right hand side is for m = n yielding

$$(-1)^n \frac{u_1}{k_n} = C_n I_o(\beta_n a^+) \frac{h^+}{2} . \qquad (A24)$$

Thus, the z-homogeneous solution is, recapitulating,

$$u(r,z) = \sum_{n=0}^{\infty} C_n I_o(\beta_n r) \cos k_n z \qquad (A21)$$

$$C_n = (-1)^n u_1 \frac{2}{h^+} \frac{1}{k_n I_o(\beta_n a^+)} \qquad (A25)$$

$$k_n = \frac{(2n+1) \pi}{2h^+} \qquad (A16)$$

$$\beta_n^2 = k_n^2 + \frac{1}{L^2} \qquad (A19)$$

A.2 Solution for v

The equation we wish to solve is

$$\frac{1}{r} \frac{\partial}{\partial r} (r \frac{\partial v}{\partial r}) + \frac{\partial^2 v}{\partial z^2} - \frac{1}{L^2} v = 0 \qquad (A26)$$

subject to the previously specified conditions for v(r,z). Choosing a product solution

$$v = R_v(r) Z_v(z) \qquad (A27)$$

which yields

$$\frac{1}{R_v r}\frac{d}{dr}\left(r\frac{dR_v}{dr}\right) = -\frac{1}{Z_v}\frac{d^2Z_v}{dz^2} + \frac{1}{L^2} = \mu_v \tag{A28}$$

where we have chosen the diffusion length to be incorporated into the z-solution for convenience. The three cases for μ_v are considered as follows

Case 1 - $\mu_v > 0$. Letting $\mu_v = +j^2$ we obtain for the r-case

$$\frac{d^2R_v}{dr^2} + \frac{1}{r}\frac{dR_v}{dr} - j^2R_v = 0 \tag{A29}$$

solutions of which are modified Bessel functions of first and second kind. Thus

$$R_v(r) = a\ I_o(jr) + b\ K_o(jr) \tag{A30}$$

Since K_o has a logarithmic discontinuity at the origin b = 0 due to condition BC 1. On the other hand, condition BC 2 yields

$$R_v(o) = a\ I_o(ja^+). \tag{A31}$$

Since $I_o(u)$ increases monotonically from unity at the origin (being the cylindrical equivalent of the hyperbolic cosine), it has no zeros so (A31) yields the trivial solution a = 0.

Case 2 - $\mu_v = 0$. This choice leads to the result

$$R_v(r) = a\ \ell n\ r + b \tag{A32}$$

which is seen to be the solution to (A29) if $j^2 = 0$. Condition BC 1 requires a = 0 whereas BC 2 then leads to the trivial solution b = 0.

Case 3 - $\mu_v < 0$. Letting $\mu_v = -j^2$ yields the r-equation

$$\frac{d^2R_v}{dr^2} + \frac{1}{r}\frac{dR_v}{dr} + j^2 R_v = 0 \tag{A33}$$

which has first and second kind Bessel function solutions resulting in

$$R_v = a\ J_o(jr) + b\ Y_o(jr) \tag{A34}$$

BC 1 immediately requires that b = 0 since $Y_0(w)$ is assymptotic to $(2/\pi)\,\ell n\, w$ as $u \to 0$. The second boundary condition requires

$$R_v(a^+) = a\, J_o(ja^+) = 0 \quad . \tag{A35}$$

Since $J_0(w)$ is the cylindrical equivalent of the cosine function it has an infinity of zeros yielding the set of nontrivial solutions

$$R_v(r) = a_n\, J_o(j_n r) \qquad n = 0,1,2, \ldots \tag{A36}$$

where the k_n are eigenvalues corresponding to the zeros of

$$J(j_n a^+) = 0 \qquad n = 0,1,2, \ldots \tag{A37}$$

The equation for Z_v becomes

$$\frac{d^2 Z_{vn}}{dz^2} - \gamma_n^{\,2}\, Z_{vn} = 0 \qquad n = 0,1,2, \ldots \tag{A38}$$

where

$$\gamma_n^{\,2} = j_n^{\,2} + \frac{1}{L^2} \qquad n = 0,1,2, \ldots \tag{A39}$$

Solutions are thus exponential in nature. The hyperbolic functions are chosen for convenience yielding

$$Z_{vn}(z) = c \cosh \gamma_n z + d \sinh \gamma_n z \tag{A40}$$

and

$$\frac{dZ_{vn}}{dz} = \gamma_n\, (c \sinh \gamma_n z + d \cosh \gamma_n z) \; . \tag{A41}$$

The boundary condition at the plane of symmetry, BC 3, requires that d = 0. We thus find the r-solution to be the linear combination of the infinite set resulting from the product of (A40) (with d = 0) and (A36) or

$$v(r,z) = \sum_{n=0}^{\infty} d_n\, J_o(j_n r) \cosh \gamma_n z. \tag{A42}$$

Condition BC 4 when imposed gives immediately

$$v(r,h^+) = v_1 = \sum_{n=0}^{\infty} d_n \cosh \gamma_n h^+\, J_o(j_n r). \tag{A43}$$

We may now take advantage of the orthogonality of the Bessel function in $[0,a^+]$, recognizing that (A43) represents the Fourier-Bessel series representation of the constant v_1. Multiplying (A43) by $rJ_o(j_m r)dr$ and integrating over the range of r yields

$$v_1 \int_0^{a^+} r\, J_o(j_m r)dr = \sum_{n=0}^{\infty} d_n \cosh \gamma_n h^+ \int_0^{a^+} r\, J_o(j_n r)\, J_o(j_m r)dr \ . \tag{A44}$$

Note that we take advantage of the uniformity of convergence of (A43) to interchange the two operators. Now

$$\int_0^{a^+} r\, J_o(j_m r)\, J_o(j_n r)dr = \begin{cases} 0 & n \neq m \\ \frac{a^{+2}}{2} J_1^{\,2}(j_m a^+), & m = n \end{cases} \tag{A45}$$

whereas

$$\int_0^{a^+} r\, J_o(j_m r)dr = \frac{a^+}{j_m} J_1(j_m a^+) \tag{A46}$$

resulting in the identity

$$\frac{v_1 a^+}{j_m} J_1(j_m a^+) = (d_m \cosh \gamma_m h^+)\, \frac{a^{+2}}{2} J_1^{\,2}(j_m a^+) \tag{A47}$$

from which the d_m are determined. Recapitulating the result for v we have for the r-homogeneous solution

$$v(r,z) = \sum_{n=0}^{\infty} d_n\, J_o(j_n r) \cosh \gamma_n z \tag{A42}$$

$$d_n = v_1 \frac{2}{a^+} \frac{1}{j_n\, J_1(j_n a^+) \cosh \gamma_n h^+} \tag{A48}$$

$$j_n = \text{eigenvalues of } J_o(j_n a^+) = 0 \tag{A37}$$

$$\gamma_n^{\,2} = j_n^{\,2} + \frac{1}{L^2} \tag{A39}$$

A.3 Solution for the Neutron Flux

Recalling Equation (A1) with (A21), (A42), and the definitions for u_1 and v_1 identified in the tabulation of boundary conditions we find the solution as

$$\phi(r,z) = \frac{SL^2}{D}\left\{1 - \frac{2}{h^+}\sum_{n=0}^{\infty}\frac{(-1)^n}{k_n}\frac{I_o(\beta_n r)}{I_o(\beta_n a^+)}\cos k_n z - \ldots\right. \tag{A49}$$

$$\left. \ldots - \frac{2}{a^+}\sum_{n=0}^{\infty}\frac{J_o(j_n r)}{j_n\, J_1(j_n a^+)}\frac{\cosh \gamma_n z}{\cosh \gamma_n h^+}\right\}$$

where Equations (A16), (A19), (A37), and (A39) specify k_n, β_n, j_n, and γ_n, respectively. It is straightforward to verify that (A49) is the solution to Equation (72) and its boundary conditions since $I_o(u)$, $J_o(u)$, cos(u) and cosh(u) all satisfy the homogeneous equation, SL^2/D is the inhomogeniety, and the following conditions hold for each sum in (A49)

a) First Sum:

1. vanishes identically at $z = h^+$
2. sums to 1.0 at $z = 0$, $r = a^+$
3. sums to 1.0 at $r = a^+$, all $z \neq h^+$.

b) Second Sum:

1. vanishes identically at $r = a^+$
2. sums to 1.0 at $r = 0$, $z = h^+$
3. sums to 1.0 at $z = h^+$, all $r \neq a^+$.

NOMENCLATURE

English

A	area; number of nuclides in nucleus
a	radius
B^2	buckling
C	nuclide volumetric concentration; neutron beam crossing rate
D	diffusion coefficient
d	extrapolation distance
E	energy
f	fuel utilization factor
g	non-1/v factor (Equation 42)
h	height
J	neutron current density (Equation 45)
J_n, Y_n, I_n, K_n	Bessel functions
K	integration kernal
k	neutron multiplication factor
L	diffusion length
M	migration length $\sqrt{L^2+\tau_T}$
m	nuclear mass
N	neutron density = $\int_E n(E)dE$
N	nuclear density
N	total number of nucleons
n	neutron spectrum or neutron energy density
$\hat{n}$	unit normal
P	momentum; probability; power
p	probability density; resonance escape probability; pressure
R	total reaction rate
r	radial coordinate
r	nuclear radius; reactor rate density
S	source strength
S_o	external control volume surface
s	source or sink term
T	temperature
t	time; thickness
$t_{1/2}$	half-life
V	volume
v	velocity
x	coordinate-general material direction
y	coordinate
z	coordinate
ℓ_o	thermal neutron lifetime-infinite reactor
ℓ	thermal neutron lifetime-finite reactor
$	ρ/β-dollars of reactivity
¢	100 ρ/β-cents of reactivity

Greek

α_ℓ	maximum fractional energy loss per collision
α_ρ	reactivity coefficient
α	void fraction
β	delayed neutron fraction-total if unsubscripted
β^2	generalized curvature
Δ	difference in quantity defined by following symbol
ε	fast fission factor
ζ	flow coordinate
η	neutrons produced per neutron absorbed in fuel

Θ	scattering angle
Λ	energy release per fission constant
λ	decay constant, mean free path
ν	neutrons produced per fission
ξ	refers to type of interaction ; lethargy
ρ	reactivity
Σ	macroscopic cross section
σ	microscopic cross section
τ	time constant; dimensionless time
ϕ	neutron flux
Ω	dimensionless frequency
ω	reaction frequency (inverse period)
∇	gradient operation $\hat{i}\frac{\partial}{\partial x} + \hat{j}\frac{\partial}{\partial y} + \hat{k}\frac{\partial}{\partial z}$

Subscripts

a	absorption
c	collisions
d	density
e	equivalent; extrapolated
eff	effective for finite reactor
F	fuel
f	fission; final
i	inelastic; interface
i	refers to species i
k	subregion-k; index
lo	limiting function evaluated at origin
min	minimum
n	normal, neutron, referring to characteristics
n	nuclear
na	nonproductive absorption
NL	nonleakage (T-thermal, F-fast)
o	initial
p	neutron-proton reaction; production
p	pressure
s	scattering
T	temperature
T	thermal
t	total
tr	transport
v	volume
α	neutron-alpha reaction
α	void
γ	neutron-gamma reaction
ξ	refers to interaction
Ψ	nuclide Ψ
∞	referring to an infinite reactor

Superscripts and Other Operators

*	excitation; dimensionless
$\bar{}$	averaged
$\hat{}$	vector
$\underline{}$	matrix
$\mp$	extrapolated

REFERENCES

1. Nero, N. V., Jr., A Guidebook to Nuclear Reactors, U. Cal. Press, Berkeley, 1979.

2. Rasmussen, N., "Reactor Safety Study: An Assessment of Accident Risks in U.S. Commercial Nuclear Power Plants," U.S.N.R.C. Report WASH-1400, NUREG-75/014, October, 1975.

3. Nuclear News, February 1980, 23, 2, p. 30.

4. Profio, E. A., Experimental Reactor Physics, John Wiley and Sons, New York, 1976.

5. El-Wakil, M. M., Nuclear Power Engineering, McGraw-Hill Book Company, New York, 1962.

6. Heisenberg, W., Nuclear Physics, Methuen & Co. Ltd., London, Translated from the German (Die Physik der Atomkerne) by Gaynor, F., von Zeppelin, A., and Wilson, W., 1953.

7. Oldenberg, O., Introduction to Atomic Physics, McGraw-Hill Book Company, Inc., New York, 1954.

8. Burcham, W. E., Nuclear Physics; An Introduction, Longmans Greene and Co. Ltd., London, 1963.

9. Soodak, H., and Campbell, E. C., Elementary Pile Theory, John Wiley & Sons, Inc., New York, Chapman & Hall, Ltd., London, 1950.

10. Massimo, L., Physics of High-Temperature Reactors, Pergamon Press, Ltd., Oxford, 1976.

11. "Nuclear Reactors, Build, Being Built, or Planned in the U.S. as of June 30, 1979," DOE/TIC-8200-R40.

12. Keepin, G. R., Physics of Nuclear Kinetics, Addison-Wesley, Reading, Massachusetts, 1965.

13. Argonne National Laboratory, "Reactor Physics Constants," ANL-5800, 1958.

14. Lamarsh, J. R., Introduction to Nuclear Engineering, Addison-Wesley Publishing Company, Reading, Massachusetts, 1977.

15. Mughabghab, S. F., and Garber, D. I., "Neutron Cross Sections - Volume 1, Resonance Parameters," BNL-325, Vol. I, 3rd Ed., June, 1973.

16. Garber, D. I., and Kinsey, R. R., "Neutron Cross Sections - Volume 2, Curves," BNL-325, Vol. II, 3rd Ed., January, 1976.

17. Garber, D. I., and Brewster, C., "ENDF/B Cross Sections," BNL-17100, 2nd Ed., October, 1975.

18. Murray, R. L., Nuclear Energy, Pergamon Press Inc., New York, 1975.

19. Goodman, C. D., Ed., Introduction to Pile Theory, Addison-Wesley Press Inc., Cambridge, Mass., 1952.

20. Kahan, T., and Gauzit, M., Introduction au Genie Nucleaire, Vol. 1- Physique et Calcul des Reacteurs Nucleaires, Dunod, Paris, (1957).

21. Soodak, H., Ed., Reactor Handbok, Vol. 3, Part A - Physics, Interscience, New York, 1962.

22. Galanin, A. D., Thermal Reactor Theory, translated from the Russian edition by J. B. Sykes, 2nd Ed., Pergamon Press Ltd., New York, 1960.

23. Kinsey, R.R., Brookhaven National Laboratory, National Nuclear Data Center, Private Communication, 1980.

24. Rasmussen, N., et al., "Nuclear Safety Study," WASH-1400, 1974.

2 POWER REACTOR CONCEPTS AND SYSTEMS OVERVIEW

by
R.T. Lahey, Jr.
RPI, Troy, NY, USA

The purpose of this chapter is to provide an overview of some selected nuclear reactor systems which have been constructed or proposed. While this chapter will not be comprehensive, it should provide the reader with a good understanding of the evolution of the various converter and breeder reactor types in use today. Subsequent chapters will go into more detail on these reactor types, and in particular, will stress the unique safety problems associated with each design.

As discussed in the preceding chapter, in the United States, commercial nuclear power sprang from the naval nuclear reactor programs (PWRs), and the early Argonne National Laboratory work (BWRs and Breeders). A similar history can be seen in the power reactor programs in most other developed nations, although in some cases, the technology was transferred through licensee agreements.

1. LIGHT WATER NUCLEAR REACTORS (LWR)

There are two basic LWR types used to produce power today. These are the pressurized water nuclear reactor (PWR) and the boiling water nuclear reactor (BWR). Both reactor types are based on a Rankine cycle; however, it will be convenient to consider them one at a time.

1.1 Pressurized Water Nuclear Reactors

A typical first generation pressurized light water power reactor (e.g., Shippingport) is shown schematically in Fig. 1. It can be seen that there is a primary loop and a secondary loop; thus, the reactor coolant is isolated from the steam turbine. This type of arrangement is referred to as an indirect cycle. Current generation PWRs are typical of the system shown in Fig. 2. It can be seen that the system is basically the same as earlier systems; however, certain safety advances have been made. Specifically, passive emergency core coolant system (ECCS) accumulators are installed as standard equipment, and the cold leg penetration has been moved to the top of the downcomer to enhance core reflood in the event of a cold leg break.

The power control for PWRs is through control rod motion, change in boron content in the coolant and through core inlet subcooling control. The latter means is frequently called "power on demand" control, and occurs when the load on the steam generator is varied causing a different coolant temperature to

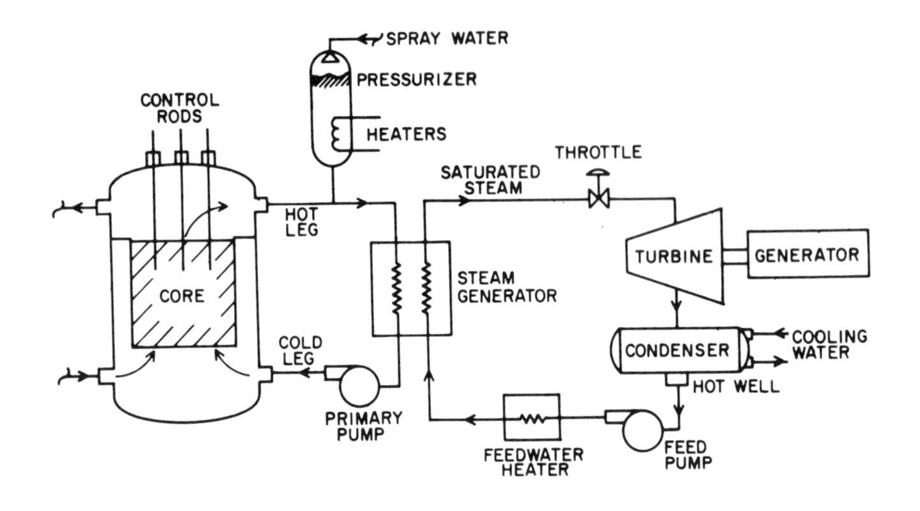

Fig. 1 First Generation PWR

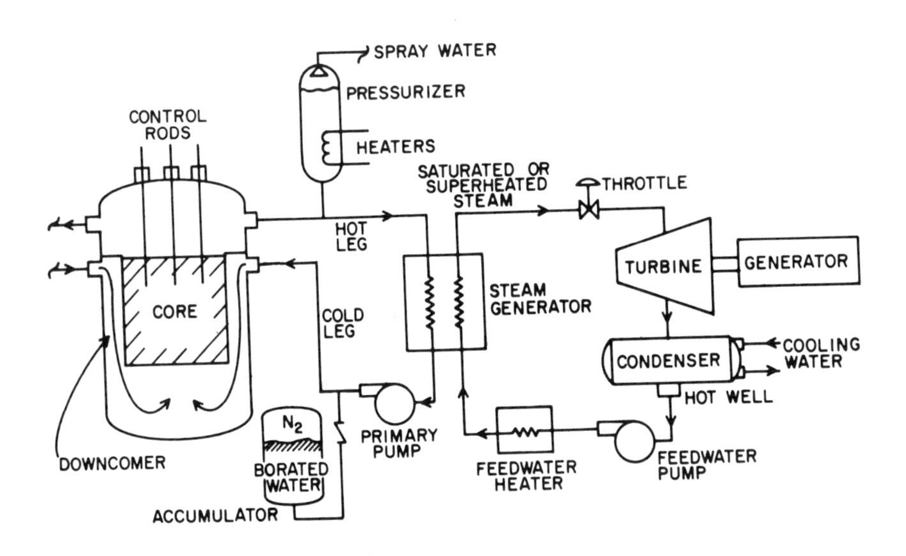

Fig. 2 Modern PWR

enter the core. This means of power control is particularly effective for small (e.g., naval) nuclear reactors.

Not all PWRs use light water as the coolant and moderator, nor are they all of the design shown in Figs. 1 and 2. The Canadian PWR (CANDU) uses heavy water (D_2O) for the coolant and moderator. Figure 3 is the schematic of a typical CANDU reactor (e.g., Bruce-A). It should be noted that the natural uranium fuel rod bundles are located in horizontal pressure tubes, which, in turn, are located inside nitrogen pressurized Calandria tubes. This design feature eliminates the need for a pressure vessel and allows for on-line refueling. On the other hand, due to the positive power reactivity coefficient of this type design, sophisticated control devices and safety systems are normally required.

The British have proposed a reactor similar to the CANDU design; however, in order to minimize the use of D_2O (which is quite expensive), light water is the coolant. In addition, unlike CANDU, in the British design, the coolant is allowed to boil in the pressure tubes. Thus, the British Steam Generating Heavy Water Reactor (SGHWR) is a type of boiling water nuclear reactor (BWR). It should be mentioned in passing that the Japanese (PNC) have also developed a similar system; however, unlike the schematic in Fig. 3, in their design the pressure tubes are mounted vertically and the fuel is enriched.

1.2 Boiling Water Nuclear Reactors

A number of different boiling water nuclear reactor (BWR) type designs have been built and operated. Let us focus here on the design types which involve the use of a pressure vessel, since this type is what is normally meant when one refers to a BWR.

Figure 4 is a schematic of a typical dual-cycle BWR/1 (e.g., Dresden-1). It can be seen that this system is a mixture of indirect cycle PWR and direct cycle BWR design features. It was built this way to obtain operational experience, since the stability of direct cycle BWRs was not yet proven.

The control of a dual cycle BWR is by control rod motion and subcooling control. That is, if during normal operation more power is demanded by the generator, the turbine slows causing the governor to open the throttle between the steam generator and the low pressure (LP) stages of the turbine. As more energy is extracted from the steam generator, the inlet water to the core is cooled. When this subcooled water enters the core, it collapses core voids, which increases the core power level due to void-reactivity feedback. This method of load following is a form of "power on demand" control and works because the reactor is basically slaved to the turbine. It should be noted that the steam to the high pressure (HP) stages of the turbine is virually unaffected by this change in load; thus, over a certain range, all load changes are accommodated by the LP stages.

Figures 5 and 6 are schematics of indirect cycle BWRs (e.g., KAHL). Figure 5 shows a typical forced circulation plant, while Fig. 6 is a natural circulation plant. This type of plant was built to gain operational experience and to "hedge our bets" in the event that internal steam separation did not work. The indirect cycle BWR design is not economical and actually includes the worst design features of both PWRs and BWRs. Plants of this type are no longer considered viable.

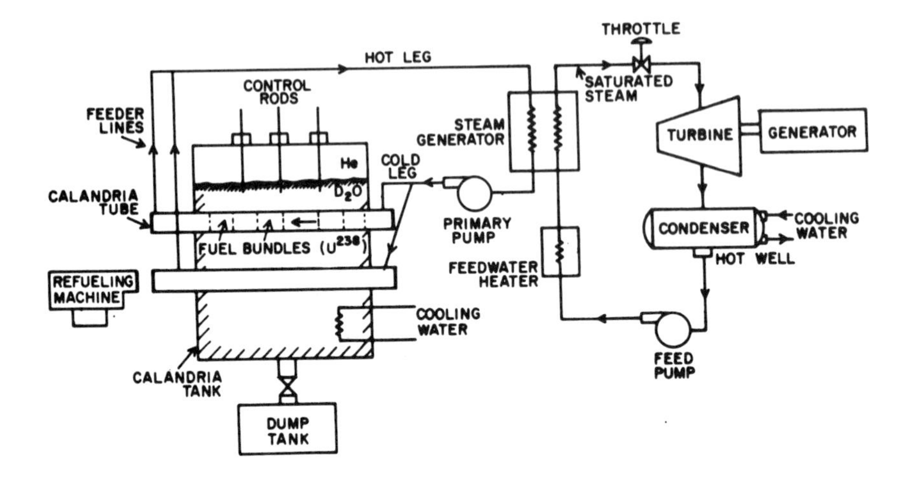

Fig. 3 CANDU (HWR)

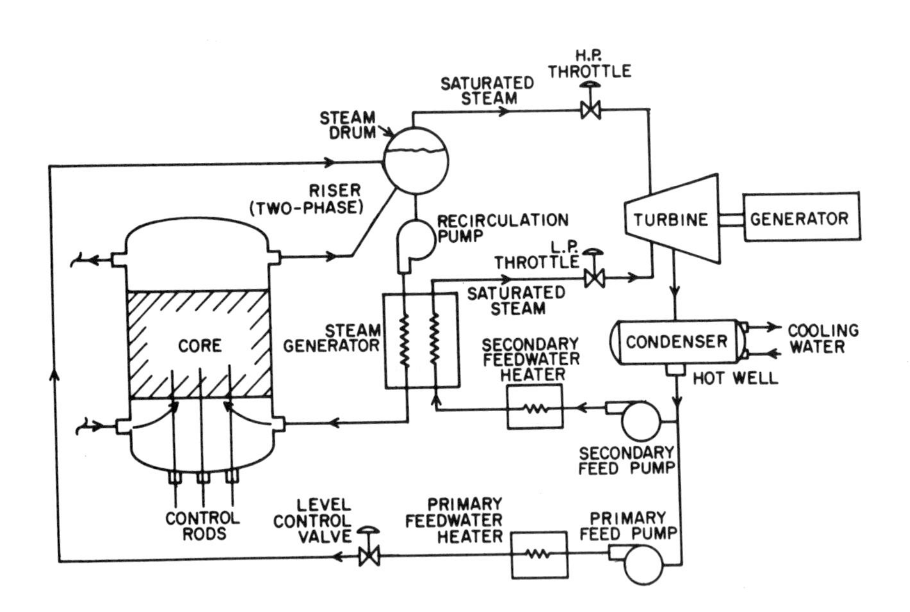

Fig. 4 Dual Cycle BWR

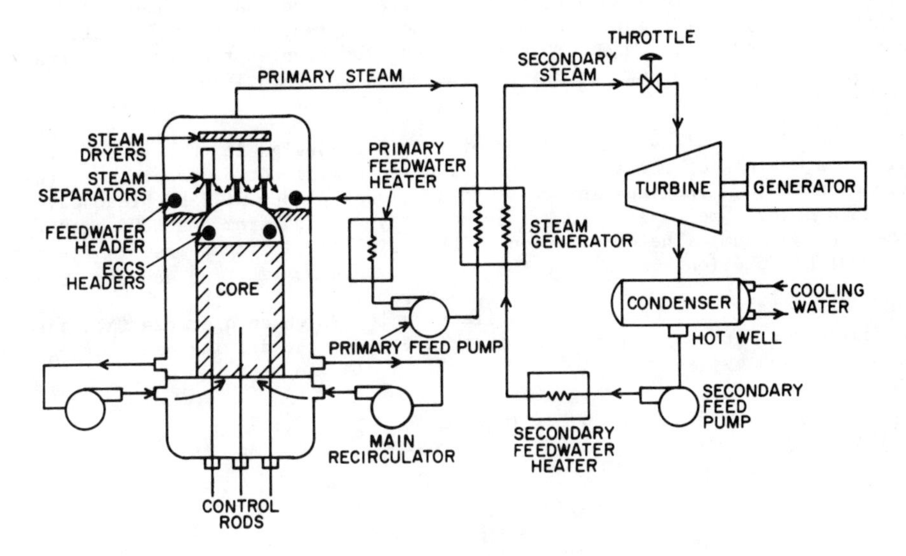

Fig. 5 Indirect Cycle BWR (Forced Circulation)

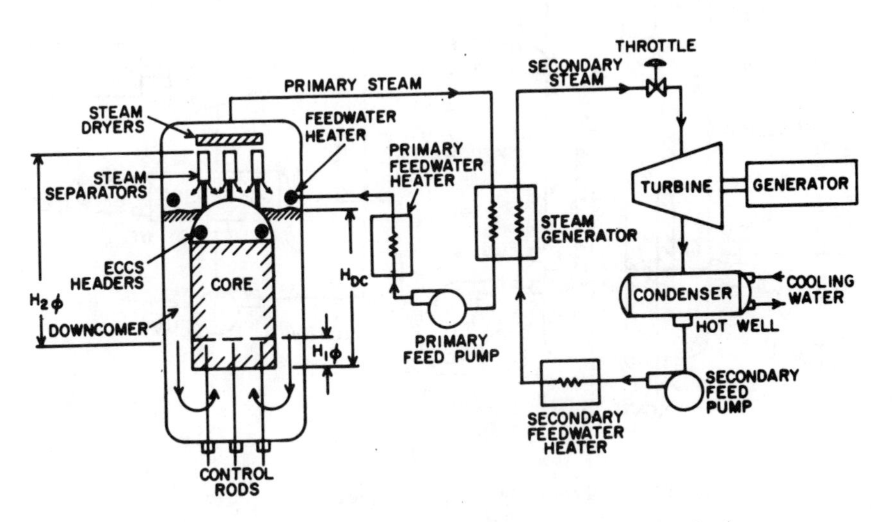

Fig. 6 Indirect Cycle BWR (Natural Circulation)

Early direct cycle BWRs were of the natural circulation type (e.g., Bodega Bay) or the forced circulation type (Oyster Creek). Figure 7 is the schematic of a typical BWR/2. It should be noted that in this design, the full core flow passed through the recirculation pumps. Power control is by control rod position and flow control. The latter method is used for maneuvering transients. For example, as the generator load changes, the motor-generator (MG) sets change the speed of the main recirculators which leads to a change in core flow rate. This change in flow changes the core void fraction, resulting in a new operating state which is compatible with the new generator load.

One of the safety consequences of this type design is that in the event of a large break of the recirculation lines, it will not be possible to reflood the core, since the integrity of the lower plenum will be breached. In this case, core cooling must rely on spray cooling from the ECCS headers in the upper plenum. The elimination of the main circulation pumps, in favor of natural circulation, was an attempt to improve the safety and economy of BWRs.

Figure 8 is a schematic of a direct cycle BWR which operates in natural circulation. It can be noted that, as in Fig. 6, the driving head through the core is given by:

$$\Delta p_{core} = \rho_f \frac{g}{g_c} H_{DC} - \rho_f \frac{g}{g_c} H_{1\phi} - [\rho_f(1 - \langle\bar{\alpha}\rangle) + \rho_g\langle\bar{\alpha}\rangle] \frac{g}{g_c} H_{2\phi} \quad (1)$$

where,

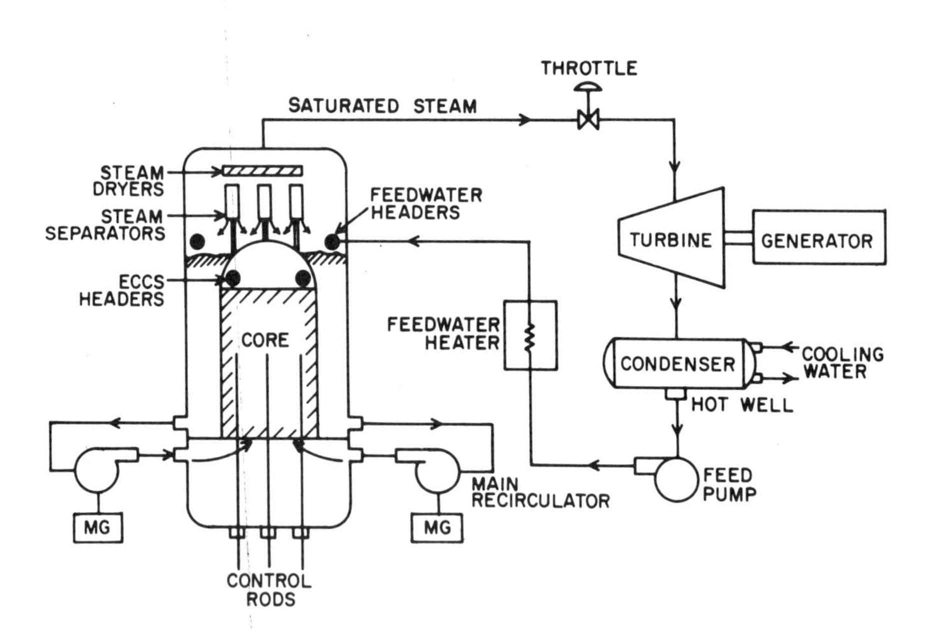

Fig. 7 Direct Cycle BWR/2 (Forced Circulation)

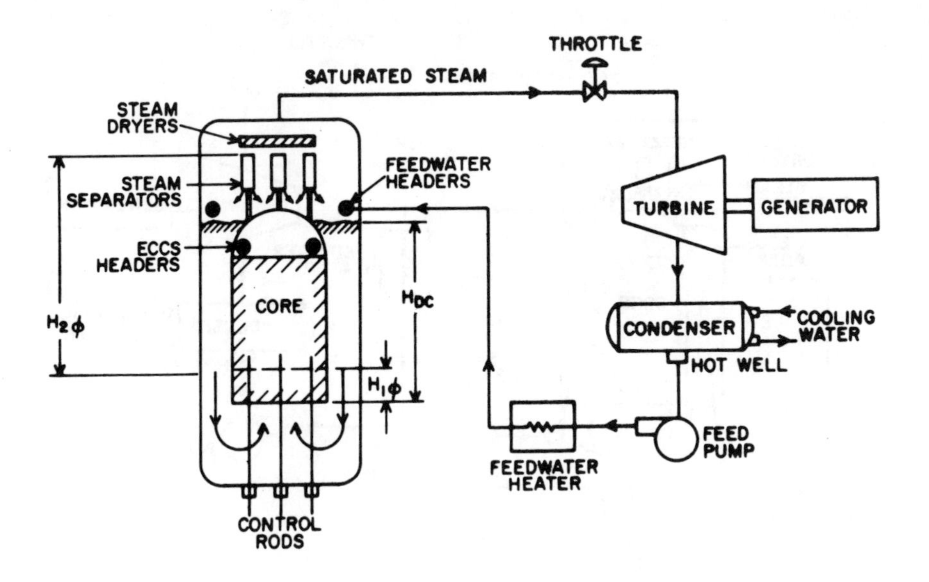

Fig. 8 Early Direct Cycle BWR (Natural Circulation)

$$\langle\overline{\alpha}\rangle = \frac{1}{H_{2\phi}}\int_0^{H_{2\phi}} \langle\alpha\rangle\, dz, \text{ axial average void fraction} \tag{2}$$

$H_{1\phi}$ = Non-boiling height in the core (3)

$H_{2\phi}$ = Two-phase height in the core, plenum and separators (4)

H_{DC} = Height of liquid in downcomer (5)

Thus, the maximum flow rate is determined by the relative elevations in the core and downcomer. Operational experience has indicated that achievable flows are rather low and thus reactors of this type are not being built today because of inherent limitations on control, stability and core power density.

Modern BWRs use either the G.E. jet pump design (BWR/3-BWR/6) or the KWU/ASEA-ATOM internal pump design.

Figure 9 is a schematic of a typical BWR/5 or BWR/6. For this type design, a flow control valve is used to provide flow control, while for the BWR/3 and BWR/4 designs, MG sets are used to control the speed of the recirculation pumps. It should be noted that unlike the forced circulation BWR/2 design, in Fig. 7, in which all the core flow had to be passed through the recirculation pumps, in the jet pump design shown in Fig. 9, only about one-third of the core flow goes through the recirculation pumps (as the jet pump

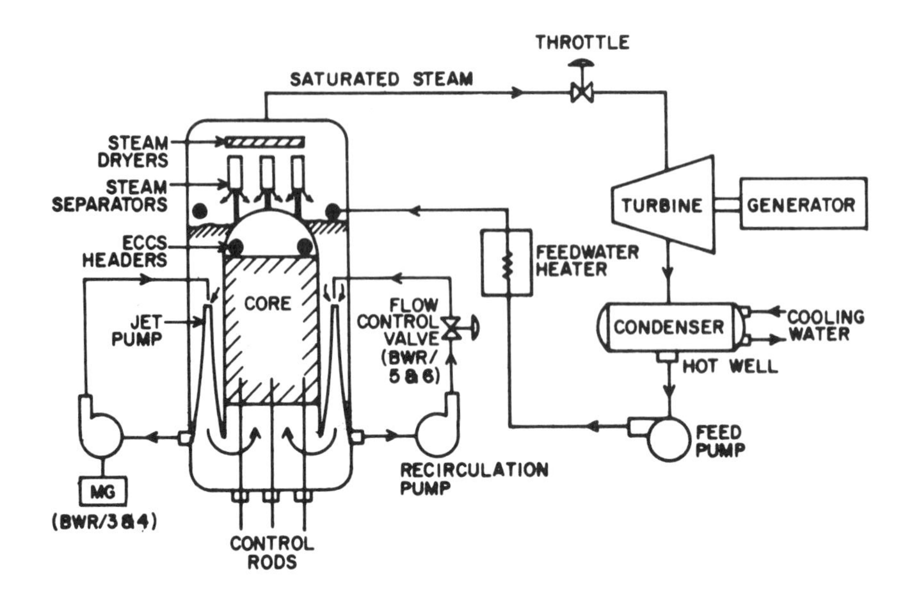

Fig. 9 Modern BWR (G.E. Design)

drive flow). This design innovation dramatically reduces the recirculation pipe size (and thus reduces LOCA* discharge rates). More importantly, however, it can be noted that if the recirculation line breaks in a jet pump plant, reflood of the core is still possible since the lower plenum is still intact.

Figure 10 is a schematic of a modern KWU (German) or ASEA-ATOM (Swedish) type direct cycle BWR in which internal pumps have been used to eliminate the possibility of a recirculation pipe break. This type of design has numerous inherent safety advantages; unfortunately however, European licensing rules have evolved to require an analysis of lower plenum breaks, in which the consequence of such breaks are similar to (if not worse than) classical recirculation line breaks. It is truly unfortunate that design innovations which clearly improve reactor safety are so frequently countered by escalating licensing requirements. It is exactly this type of "ratching" which discourages reactor vendors from making major investments in design innovations.

2. GAS-COOLED REACTORS

Let us now turn our attention to reactor types in which the coolant is a gas, such as air, helium or carbon dioxide. Such reactors are normally moder-

*Loss-of-Coolant-Accident (see Chapter 4).

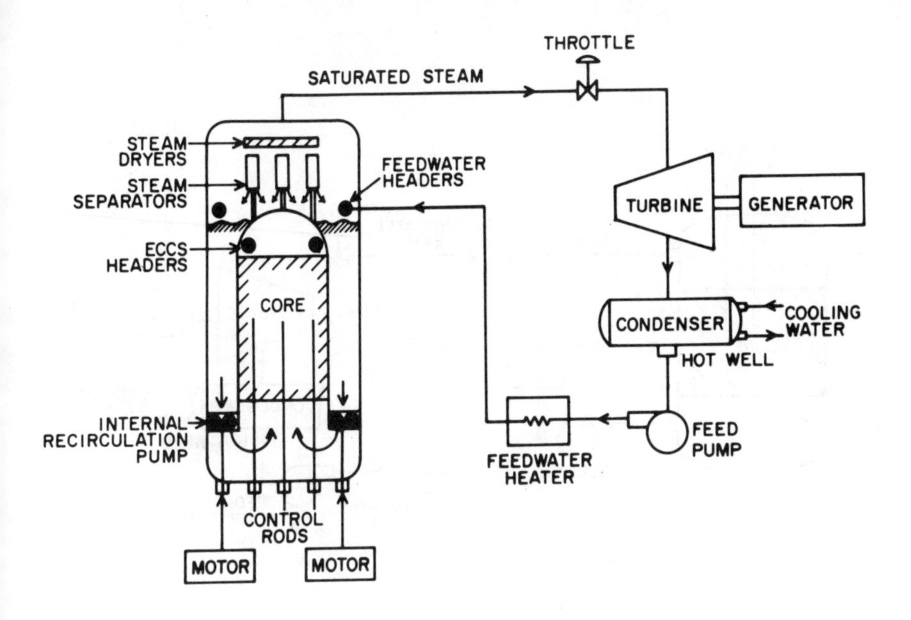

Fig. 10 Modern BWR (KWU/ASEA-ATOM Design)

ated by a high temperature material such as graphite. Reactors of this type have been operated for many years in Great Britain, the United States and elsewhere.

Figure 11 is a schematic of an indirect closed cycle high temperature gas-cooled reactor (HTGR). Large power reactors of this type have been built by Gulf General Atomics (GGA) in the United States; however, they have been plagued with mechanical difficulties (e.g., helium blower problems). Due to its high operating temperature, and thus high thermodynamic efficiency, this type of reactor has potential economic advantages compared to LWRs. In addition, it appears to have safety advantages; however, safety studies have been comparatively few on this reactor type.

Although the market for HTGRs is currently depressed, advanced generation type systems have been proposed. Figure 12 is a schematic of a direct, closed cycle, HTGR. It can be seen in this case we have a Brayton rather than a Rankine (steam) cycle. This design eliminates the need for a steam generator and the associated equipment, but does require the development of a large high temperature gas turbine.

It should be noted that this type design is not all together new. Indeed, the direct open cycle system shown schematically in Fig. 13 was the basis for the nuclear aircraft, which was cancelled for obvious safety reasons. The ex-

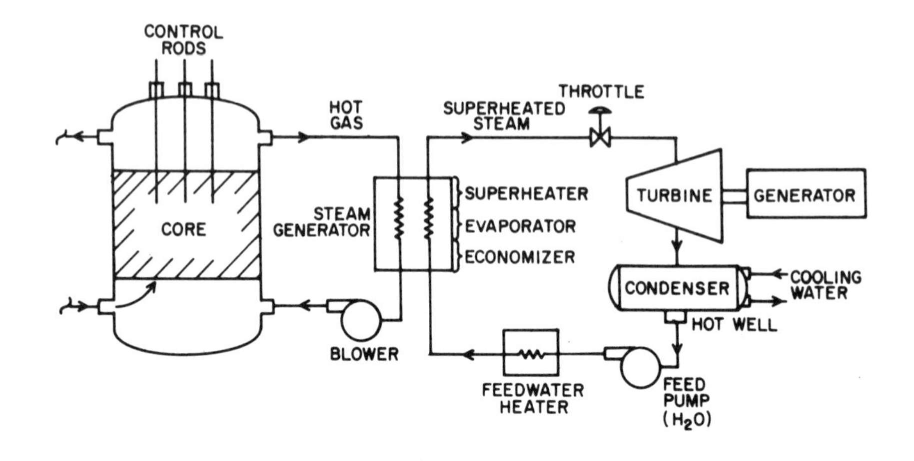

Fig. 11 Indirect, Closed Cycle High Temperature Gas-Cooled Reactor (HTGR)

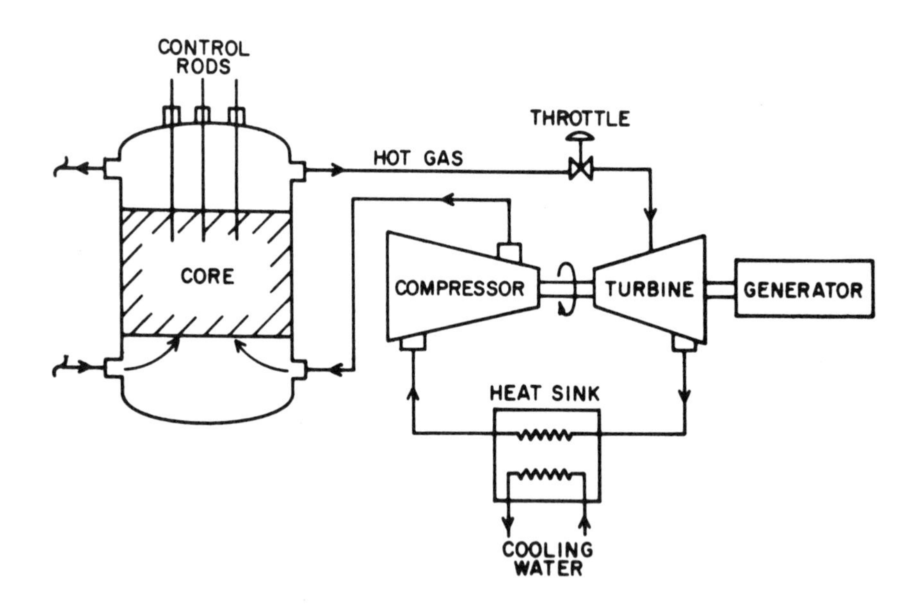

Fig. 12 Direct, Closed Cycle High Temperature Gas-Cooled Reactor

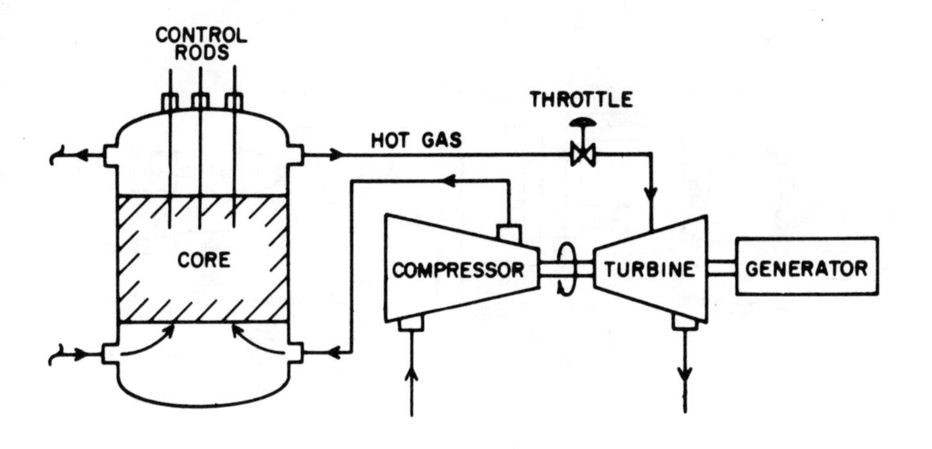

Fig. 13 Direct, Open Cycle High Temperature Gas-Cooled Reactor

perience gained in this project was valuable, however, since it indicated that a direct closed cycle HTGR was both possible and practical.

3. BREEDER REACTORS

The reactor types that have been discussed so far in this chapter have been of the converter type. That is, they burn more fissile isotope than they produce. It is possible to design reactors in which one makes more nuclear fuel than is consumed through a process called breeding. Examples include the Uranium-Plutonium and Thorium-Uranium fuel cycles in which fissile isotopes of Plutonium ($Pu^{239 \text{ and } 241}$) and Uranium ($U^{233}$), respectively, are produced in a blanket surrrounding the core. Obviously, breeder reactors multiply our energy resources and have many important economic advantages. Unfortunately, the fuel which they produce may be made into a nuclear weapon, and thus there is considerable controversy as to the role of breeder reactors for domentic power production.

Breeder reactors can be of many types, including the LWR, HTGR and liquid metal fast breeder reactor (LMFBR) type. Since we have previously considered LWR and HTGR type cycles, we will concentrate here on LMFBR cycles. This emphasis is appropriate since most breeder reactors being planned or operated are of two basic LMFBR types; loop type LMFBRs and pool type LMFBRs.

Figure 14 is a schematic of a typical loop type LMFBR (e.g., Clinch River). It can be seen that there is both a primary and secondary sodium (Na) loop. This design insures that any sodium fires, due for instance to leakage

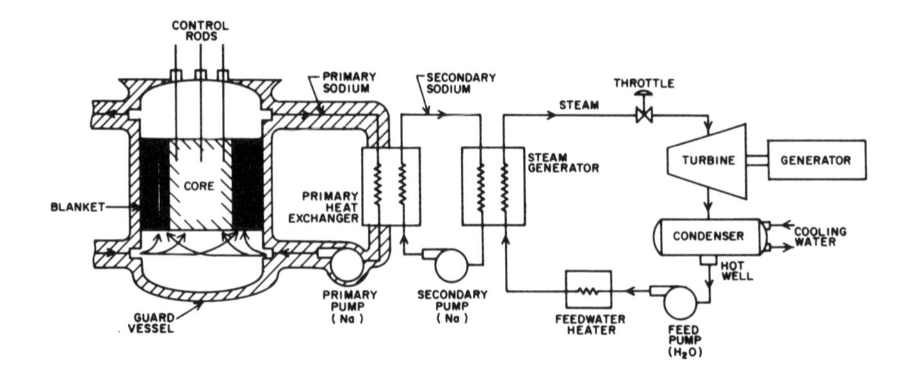

Fig. 14 Loop Type LMFBR

in the sodium/water steam generator, will not result in the release of large amounts of radioactivity.

LMFBRs have inherent safety features due to the fact that the primary loop operates at a low pressure (∿2 atmospheres) and thus a pipe break will produce only a modest blowdown transient. In addition, loop type designs are normally equipped with guard vessel and piping to retain the radioactive sodium in the event of a LOCA. It is significant to note, however, that loop type LMFBRs normally have a positive void-reactivity coefficient and thus some postulated accidents may lead to core melting and disassembly. Once the core has disassembled, one must be concerned about the possibility of recriticality, since the enrichment in LMFBR is rather high (∿20%). Thus the consequences of hypothetical accidents may be more serious in an LMFBR than an LWR.

Other LMFBR designs have developed which appear to have more inherent safety. A typical pool type LMFBR (e.g., Phenix) is shown schematically in Fig. 15. In this type design, the entire core and primary heat exchanger is immersed in a large pool of molten sodium. This design feature leads to some obvious safety advantages, but has some disadvantages concerning equipment inspection and maintenance.

4. FLUID-FUELED REACTORS

The reactor types discussed so far (i.e., LWR, HTGR, LMFBR) are in relatively wide use throughout the world. Other reactor types are possible, and indeed many have been built in various laboratories. A typical fluid-fueled reactor is shown schematically in Fig. 16. Reactors of this type may use the

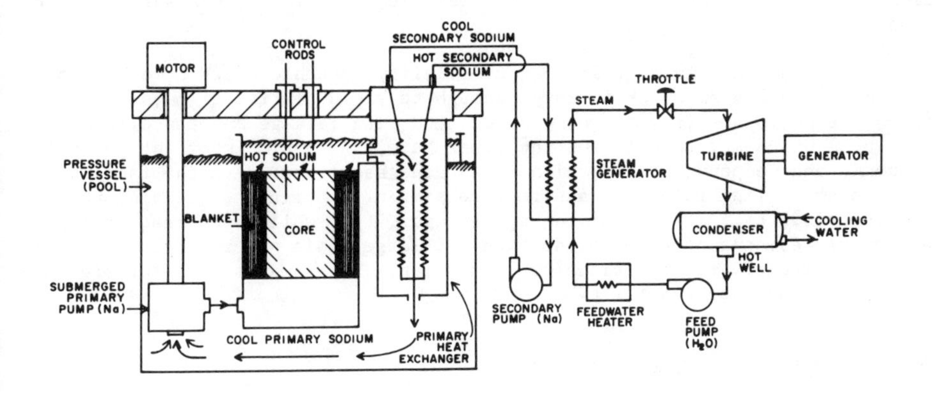

Fig. 15 Pool Type LMFBR

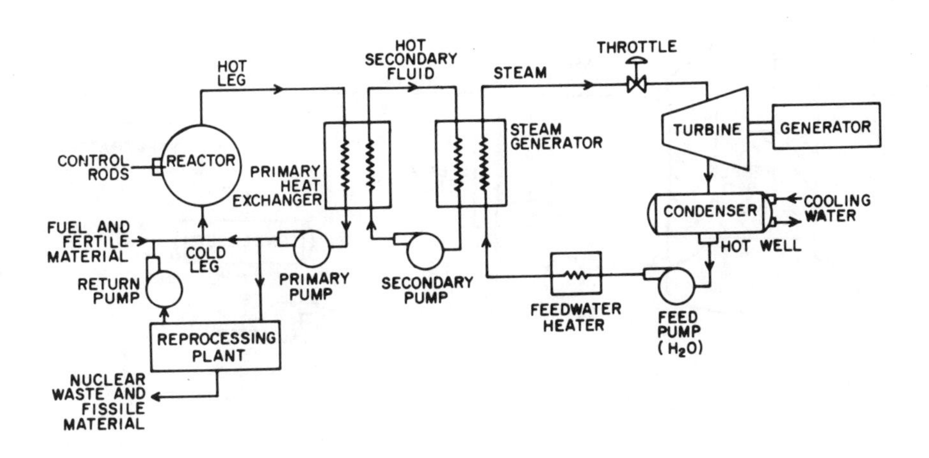

Fig. 16 Fluid-Fueled Reactor

following coolant-fuel configurations:

- Aqueous fuel (solution or slurry)
- Liquid metal (e.g., uranium in molten bismuth)
- Gaseous suspension
- Molten salt

In the past, Oak Ridge National Laboratory (ORNL) has been particularly aggressive in promoting the use of a thorium tetrafluoride molten salt reactor type.

It can be seen in Fig. 16 that criticality is achieved when we have a volume in the system which is large enough to yield a critical mass. Thus there are obvious safety concerns associated with leakage and the criticality of a pool which may result. On the other hand, there are important potential economic and weapons proliferation advantages associated with the fact that one can do on-line refueling and reprocessing of the fluid containing the fission products. It is fair to say, however, that while such reactor types are possible means of power production, no large scale power reactor of this type is currently under serious consideration. The same is true for organic-cooled and moderate reactor types.

5. PEBBLE BED TYPE REACTORS

There are other types of reactors which have been, or are being considered as possible candidates for power production. Examples include the fluidized bed and single and binary (size) pebble bed reactor concepts being developed in Europe. Figure 17 is a schematic of a typical gas-cooled

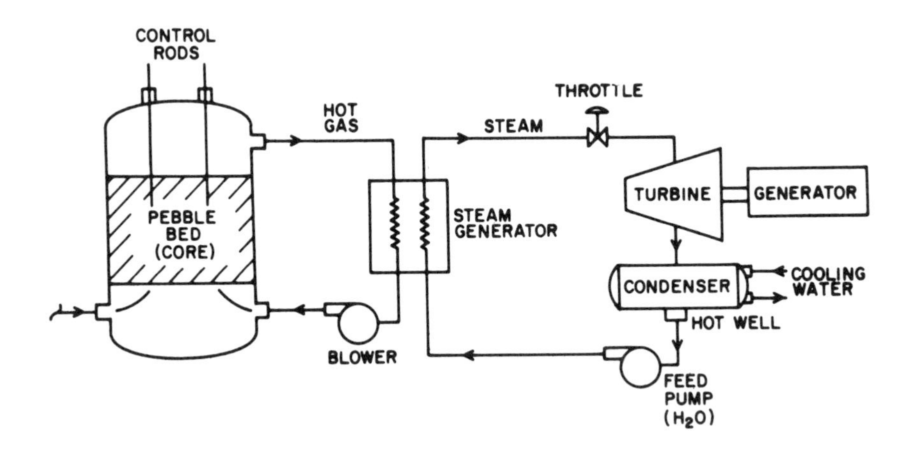

Fig. 17 Pebble Bed Reactor

pebble bed reactor. It can be seen that reactors of this type have a cycle diagram which is essentially the same as the HTGR in Fig. 11. The status of safety technology of such reactors is also similar. That is, both have inherent safety features; however, neither have been studied in great depth.

6. SUMMARY AND CONCLUSIONS

An overview of various reactor types has been given. The point of view which has been adopted has been to focus on the essential features of the various power cycles employed and to describe some of the inherent economic and safety features of each. Subsequent chapters will be concerned with a detailed discussion of the safety technology and considerations of the most significant of these reactor types. In particular, considerable material will be presented on LWR and LMFBR safety technology.

3 SAFETY ISSUES–THERMAL HYDRAULIC CONSIDERATIONS

by
J.G. Collier
UKAEA, Harwell, UK

Nuclear reactors may experience a wide range of plant conditions ranging from normal operational states to highly improbable accident conditions. This Chapter discusses the classification of these plant conditions and the associated limitations to be placed on the condition of the fuel. The different thermo-hydraulics characteristics of gas-cooled, water-cooled and liquid metal-cooled reactors are delineated. A brief summary of the various types of faults including those of moderate frequency and infrequent faults or emergency conditions is included, as well as a description of limiting faults or "design basis accidents." This overview is intended to serve as an introduction to later chapters.

1. INTRODUCTION

This chapter discusses the central role that thermohydraulic considerations play in determining the safety of all types of nuclear reactor. The discussion starts with a brief description of the basic thermal characteristics of gas-cooled, water-cooled and liquid metal-cooled reactors. This is followed by a classification and discussion of the various states of the plant, including fault and accident conditions, which must be taken into account in the design of a reactor system. This chapter thus serves as an introduction to the more detailed presentations to follow both on specific reactor types and on specific safety issues.

2. BASIC THERMAL CHARACTERISTICS OF NUCLEAR REACTORS

Heat in a nuclear reactor is generated in the uranium fuel as a result of the fission process, the rate of generation being basically proportional to the neutron density. To a first approximation, the rate of heat generation is independent of fuel temperature. Thus, for specified values of neutron density and coolant flow, the temperature of the fuel will adjust itself until the heat transfer from the fuel balances the heat generation within the fuel and steady conditions are achieved.

The limiting rate of heat release is determined by the heat transfer process together with the maximum permissible temperature at which the fuel may operate. This maximum fuel temperature will vary with reactor type and with the form of the fuel, e.g. U metal, UO_2, UC. It will usually be determined by metallurgical considerations. In most designs of reactor the fuel must be protected from reacting chemically with the coolant. In addition, the radioactive fission products formed in the fuel must be prevented from escaping into the coolant. This is done by providing some form of barrier around the fuel; usually in the form of a metallic sheath but, in the case of the high temperature gas-cooled reactor (HTR), an impermeable ceramic coating around individual fuel particles. As we shall see later it is the necessity to maintain the integrity of this barrier under all conceivable states of the plant which imposes a limit on the steady state heat generation.

Another basic aspect of the nuclear fission reaction stems from the fact that not all the energy from fission is released promptly. Some heat continues to be generated over long periods of time by the radioactive fission products as they transmute or decay to a stable species. At the time when the fission reaction is stopped, i.e. at reactor shutdown, this decay heat represents 6-7% of the heat generation occurring when the reactor is in operation. However, it falls progressively with time after shutdown as shown in Table 3.1. The presence of this "decay heat" means that it is essential to continue to provide adequate cooling of the fuel for significant periods of time after the reactor has been shut down.

Table 3.1

Decay heat rates following shutdown for an end-of-cycle equilibrium PWR core

Cooling Time sec(h)	% of Steady Power at Shutdown
1 sec	6.5
10 sec	5.1
100 sec	3.2
1000 sec	1.9
1 hr	1.4
10 hr	0.74
100 hr = 4.17 d	0.33
1000 hr = 1.39 m	0.11
8760 hr = 1 year	0.023

3. THE CHARACTERISTICS OF SPECIFIC REACTOR SYSTEMS

Table 3.2 gives a comparison for different reactor types in terms of the core dimensions, the average power output per unit volume of core (volumetric power density), the average power output per unit mass of fuel (fuel rating) and the average power output per unit length of fuel (linear rating).

3.1 Gas-Cooled Reactors

Gas-cooled reactors tend to have large physical dimensions and correspondingly low volumetric power densities. They also have somewhat lower fuel ratings than liquid-cooled reactors because gases have relatively poor heat transfer characteristics. In an attempt to overcome this deficiency use is often made of either extended - "finned" - or enhanced - "roughened" - surfaces for the fuel sheath and steam generator elements. Even so relatively high temperature differences are needed to transfer the heat from the fuel to the gas

Table 3.2

Volumetric power densities and linear fuel ratings for various reactor systems

	Reactor	Thermal Power MW(t)	Core Diameter m	Core Height m	Core Volume m^3	Average Vol. Power Density MW/m^3	Average Fuel Rating MW/tonne	Average Linear Fuel Rating kW/m
MAGNOX	Calder Hall	225	9.45	6.40	449	0.50	-	-
	Bradwell	538	12.19	7.82	913	0.59	2.20	26.2
	Wylfa	1875	17.37	9.14	2166	0.865	3.15	33.0
AGR	Hinkley - 'B'	1500	9.1	8.3	540	2.78	11.0	16.9
	Hartlepool	1507	9.3	8.2	557	2.70	11.5	16.1
HTR	High Temperature Reactor	3360	9.71	6.34	470	7.15	115	-
HWR	CANDU	3425	7.74	5.94	280	12.2	26.4	27.9
	Winfrith SGHWR	308.2	3.12	3.66	28	11.0	14.3	15.8
LWR	PWR	3800	3.6	3.81	40	95	38.8	17.5
	BWR	3800	5.01	3.81	75	51	24.6	19.0
FAST REACTOR	Phenix	563	1.39	0.85	1.38	406	149	27.0
	PFR	612	1.47	0.91	1.61	380	153	27.0

and from the gas to the steam generator. To achieve the required heat transport capability with a moderate expenditure of pumping power - 2-3% of the overall power output - it is necessary to increase the gas density by operating at high pressures, 4-7 MPa. The heat removal capacity falls off essentially in proportion to the gas density - thus this may reduce by a factor of approximately 40 during a depressurisation from 4 MPa to atmospheric pressure. However, over the timescale of a depressurisation this reduction in heat removal capacity is matched by the corresponding reduction in heat generation in the fuel in going from normal operation to a decay heat (shutdown) situation. Thus provided circulation of the gas is maintained and the reactor is safely shut down it should not be necessary for fuel temperatures to rise significantly above their normal operating values during a depressurisation accident.

Because no phase change occurs as a result of changes of pressure or temperature, there are no discontinuities in the cooling process under accident conditions. Consequently flows and temperatures are reasonably predictable. In addition with a gaseous coolant there is no risk of explosions which are postulated to occur in certain circumstances should molten fuel become dispersed in a liquid coolant. Gases also carry a relatively low burden of radioactive particulate matter. The presence or absence of the gas does not significantly influence the neutron density in the reactor. Therefore, unlike a water-cooled reactor, a loss of coolant in a gas-cooled reactor does not inherently lead to the nuclear fission process being shut down. There is therefore a need for a diversity of highly reliable methods of shutting the reactor down in the event of an accident.

3.2 Water-Cooled Reactors

Water-cooled reactor cores when considered as heat sources are relatively highly rated. Volumetric heat generation rates are in the range 50 to 100 MW/m^3 with the lower figure appropriate for a Boiling Water Reactor (BWR) and the higher figure for a Pressurised Water Reactor (PWR). To achieve the necessary temperatures to drive a steam turbine the water has to be pressurised to 7 MPa for a BWR and to 16 MPa for a PWR. The surface heat flux - that is the amount of heat passing through unit surface area of the fuel element in unit time - in an operating water reactor is up to 1 MW/m^2. Such heat fluxes can be removed quite satisfactorily using flowing water at velocities of a few metres per second. In these circumstances the difference in temperature between the fuel element surface and the cooling water will be quite low - a few tens of degrees. Should the temperature of the surface of the fuel element exceed the boiling point then boiling will occur. Again, heat fluxes of the order of 1 MW/m^2 can be safety dissipated by flowing boiling water provided the system remains pressurised.

If, however, the surface heat flux becomes too high or alternatively the water flow too low (or the steam content too high in a boiling system) then overheating of the fuel can occur. This overheating occurs quite suddenly at a particular set of thermal and hydraulic conditions. The threshold at which it occurs is usually described in terms of the maximum heat flux which can be sustained without overheating for a particular set of conditions - the "critical heat flux" (CHF) or point of "departure from nucleate boiling" (DNB) or colloquially the "burnout heat flux" because damage to the fuel can occur if the heat flux is not removed. If no action is taken, for example, to shut down the reactor and reduce the heat generation in the fuel, then the fuel element will heat up until the heat flux can be dissipated through the blanket of steam which forms adjacent to the surface and which tends to act as an insulator. The surface temperature reached in achieving this new equilibrium state would be such that the zirconium alloy cladding used in water-cooled reactors would become hydrided and brittle and indeed may rise to the point where an exothermic reaction with the steam to form hydrogen is initiated. This chemical reaction starts at about 1100-1200°C. As a result the fuel element will be badly damaged. It is therefore important to know the "critical" or "burnout" heat flux for every likely condition within a water-cooled reactor and this topic will be returned to later in this Chapter and in the following Chapters, particularly Chapter 7.

A further characteristic of light water cooled reactors which is important and which was first demonstrated in the so-called BORAX experiments undertaken at Idaho in 1953 is that loss of water from the core shuts the reactor down. This is because the water acts as both coolant and moderator and the thermal neutron population will decrease in the absence of a moderator. This is a very important "fail safe" characteristic.

Even when the reactor is shut down the power levels from the "decay heating" are such that the core must be kept covered with water to prevent it overheating.

3.3 Liquid Metal-Cooled Reactors

Liquid metal-cooled fast reactors are characterised by still higher volumetric power densities and fuel ratings. This is made possible by the use of an excellent heat transfer agent like sodium. Although about five times as much sodium is needed to transport the same amount of heat as compared with water, its heat transfer characteristics are considerably superior to those of water. In contrast to other coolants liquid sodium has the considerable advantage that it can be used at pressures close to atmospheric because of its relatively high boiling point (880°C). Moreover, in designs in which the core is immersed in a liquid pool, the very large heat capacity of the sodium in the pool acts as an effective heat sink for the removal of the decay heat from the core over many hours.

Unlike a thermal reactor, the neutron density in a fast reactor is greatly affected by geometrical rearrangement of the core. There exists at least a theoretical possibility that the significant increase in reactivity resulting from a coherent compaction of the core could cause an increase in fuel temperature to the point where the fuel begins to melt or even to boil. At this stage there would be powerful disruptive forces within the fuel which would cause a disassembly of the core and a permanent reduction of the neutron population leading to a termination of the fission reaction. Increases in reactivity can also come about as a result of removing the neutron absorbing sodium coolant from the core by boiling. Fast reactor designs where this possibility exists are said to have "positive void coefficients". Alternative fast reactor designs are available which do not have "positive void coefficients".

Following an accident sequence in which quantities of molten fuel are produced, the further possibility exists with both water cooled- and liquid metal-cooled reactors that the enthalpy present in the molten fuel could be rapidly converted to a pressure shock wave and cause a vapour explosion. This topic is considered further in Chapter 14 and discussed in detail in Chapter 20.

The final phase of an accident to any reactor in which there is serious fuel damage involves the long term removal of the decay heat from the resulting fuel debris. The location and form of this debris and the extent of the measures which need to be taken to provide adequate cooling vary with reactor type, size and design. This topic is discussed in Chapters 17 and 18.

4. REACTOR OPERATING STATES

The design of a reactor system must be such that it can withstand a wide spectrum of operational transients without any damage to the fuel. In addition the reactor designer must demonstrate that the reactor can cope with various specified but less frequent plant conditions and accident sequences without the release of significant quantities of radioactive materials which could be a danger to the general public as well as to the plant operators.

It is helpful to distinguish the different plant states in terms of the frequency of occurrence of the various possible transients. Thus lists of various foreseeable transients and plant states can be usefully categorised under the following headings:

1. Normal operation and operational transients
2. Faults of moderate frequency or upset conditions
3. Infrequent faults or emergency conditions
4. Limiting fault conditions.

An inherent assumption is made in drawing up this list that transients coming into categories 1 and 2 can be accommodated by the design of the plant without any damage to the fuel and that the fuel damage which might occur during accident or fault sequences falling into categories 3 and 4 would be known and limited within certain specified bounds either in terms of the temperatures reached or the physical distortion which results.

5. NORMAL OPERATION AND OPERATIONAL TRANSIENTS

This category includes conditions that occur frequently or regularly in the course of normal operation, manoeuvring at power, refuelling or maintenance. Such operations must be undertaken in such a manner that there exists a comfortable margin between any operating or plant design variables and the value of that parameter which necessitates protective action in order to prevent damage to the plant. This category can be sub-divided further into:

5.1 Normal operation

This sub category covers steady state operation over a range of power levels (typically from about 10% to 100% design power) start-up, hot stand-by, plant cold shutdown, and normal refuelling (either off-load as in the case of light water reactors and liquid metal-cooled fast breeder reactors or on-load as in the case of heavy water reactors and gas-cooled reactors).

5.2 Variations in normal operation

Plant technical specifications will usually provide for specified departures from normal operating conditions. These departures would include such plant conditions as operation with components out of service, operation with a limited number of failed fuel pins, operation with a limited leakage of primary coolant, etc.

5.3 Operational transients

The process of passing from one steady plant condition to another is referred to as a "transient". Normal operational transients would include plant heat up and cool down, step changes in load and ramp changes in load. Limitations are placed on the rates of change of temperature, pressure, flow and reactor power in order not to exceed a particular plant design limitation, for example to avoid overpressurisation of the reactor pressure vessel at low temperatures in the case of a light water reactor. Limitations on the rate of change of reactor power with time are usually imposed to ensure the integrity of the fuel and to achieve low fuel failure rates.

Because normal operation and the transient states associated with normal operation form the initiating plant conditions for more serious fault conditions, each of these frequent plant operation states must be analysed very fully and the operational limitations set with care.

More detailed discussion of these conditions for light water reactors and liquid metal-cooled reactors is given in Chapters 4 and 5 respectively.

6. FAULTS OF MODERATE FREQUENCY OR UPSET CONDITIONS

Included in this category are all faults not expected during normal operation but which can be reasonably expected during the lifetime of the plant. Plant operation should be such that the worst set of conditions resulting in reactor shutdown should still permit the plant to subsequently return to power. Such transients should be accommodated without any propagation to more serious fault conditions. The design intent in meeting "upset" plant conditions is that no fission product containment barrier should be breached as a consequence, i.e. fuel pin cladding, primary coolant system or containment building.

Such transients can be listed under a series of sub-categories depending on the consequence thus

(a) increases in heat removal from secondary (steam) system
(b) decreases in heat removal from secondary (steam) system
(c) decrease in reactor coolant system flow rate
(d) reactivity and power distribution anomalies
(e) increases in reactor coolant inventory
(f) decrease in reactor coolant inventory

Typical initiating events for light water-cooled reactors which fall under these headings are given in Table 3.3. A similar list can be drawn up for gas-cooled reactors and would include such items as failure of boiler feed water, failure of one gas circulator, single control rod withdrawal at power etc. To ensure that none of the fission product containment barriers is breached it is necessary to analyse such faults to establish that no fuel damage criterion is exceeded and that the reactor pressure, temperature, heat generation etc. do not fall outside acceptable limits.

Table 3.3

Initiating Events for Faults of Moderate Frequency for LWRs

(a) increases in heat removal from secondary (steam) system

- (i) feedwater system malfunctions that result in a decrease in feedwater temperature
- (ii) feedwater system malfunctions that result in an increase in feedwater flow
- (iii) steam pressure regulator malfunction or failure that results in increasing steam flow
- (iv) inadvertent opening of a steam generator relief or safety valve

(b) decreases in heat removal from secondary (steam) system

- (i) steam pressure regulator malfunction or failure that results in decreasing steam flow
- (ii) loss of external electrical load
- (iii) turbine trip (stop valve closure)
- (iv) inadvertent closure of main steam isolation valves
- (v) loss of condenser vacuum
- (vi) coincident loss of on-site and external (off-site) a.c. power to station
- (vii) loss of normal feedwater

(c) decrease in reactor coolant system flow rate

- (i) single reactor coolant pump trip resulting in partial loss of flow
- (ii) BWR recirculation loop controller malfunction that results in decreasing flow rate

(d) reactivity and power distribution anomalies

- (i) uncontrolled control rod assembly withdrawal from a subcritical or low power start-up condition or from a particular power condition
- (ii) control rod maloperation
- (iii) start-up of an inactive reactor coolant loop or recirculating loop at an incorrect temperature
- (iv) a malfunction or failure of the flow controller in a BWR loop that results in an increased reactor coolant flow rate
- (v) chemical and volume control system malfunction resulting in a decrease in boron concentration in the reactor coolant of a BWR

(e) increases in reactor coolant inventory

- (i) inadvertent operation of the ECCS during power operation
- (ii) chemical and volume control system malfunction (or operator error that increases reactor coolant inventory)

(f) decreases in reactor coolant inventory

- (i) inadvertent opening of a pressuriser safety or relief valve in a PWR or safety or relief valve in a BWR.

7. INFREQUENT FAULTS OR EMERGENCY CONDITIONS

In this category fall departures from normal plant conditions that are expected to occur very infrequently in the lifetime of any particular plant but which may be expected to occur a few times within a large number of such plants over a 30-40 year period. Because such conditions are very infrequent some limited amount of fuel failure is permitted to the extent that resumption of power operation might be precluded for a significant period of time. However, any radioactive release from the plant must be such as to not interfere with the activities of the general public beyond the confines of the site on which the plant is situated. For light water reactors typical faults which come into this category are

(a) small break loss of coolant accidents resulting from the rupture of small bore pipes or from leakages of large pipes and involving actuation of the Emergency Core Cooling System (ECCS).

(b) small breaks in the secondary (steam) system either in the steam lines or in the feedwater lines.

(c) inadvertent loading of a fuel subassembly into an incorrect location in the core.

(d) a complete loss of reactor coolant flow as a result of multiple pump trips or other causes.

(e) radioactive gas waste or liquid waste tank leak or failure.

For gas-cooled reactors typical faults in this category might be

(a) depressurisation following a breach of the primary circuit outside the prestressed concrete pressure vessel, e.g. through a safety valve or a break in the pipework in the gas purification plant.

(b) loss of reactor coolant flow due to failure of all gas circulators.

(c) withdrawal of a group of control rods either at power or with the reactor shut down.

(d) single channel faults resulting from inlet blockages or fracture of the graphite sleeves surrounding the fuel element.

For the LMFBR, in particular, considerable attention is given to faults which may develop within individual channels or sub-assemblies. Fuel failures incur an economic penalty but as such are not of importance in relation to safety unless there is a risk of the fault escalating and involving other sub-assemblies or ultimately the whole core. In the LMFBR sub-assembly faults have the potential to initiate more serious accident sequences, such as a loss of flow (LOF) or transient over-power (TOP) (see sections 8 and 10) only if the fuel overheats as a result, for example, of a blockage.

It is accepted that in this class of faults the plant design criteria set specifically for fuel damage or for reactor pressure, temperature, and heat generation limits may be reached. However the limiting condition for this set of infrequent plant conditions is that the off-site dose limits set by the licensing body should not be exceeded.

8. LIMITING FAULT CONDITIONS

Under this category come those faults or plant conditions considered to be very improbable; that is occurring less than once in, say, 1000 reactor operating years but whose consequences include the potential for serious injury or damage to the general public. The possibility of serious plant damage is accepted but it is required that in the event of such a fault the off-site radioactivity release must be contained within limits specified by national licensing authorities. These faults are very severe but must be catered for in the design process. Because such plant conditions represent the limiting design case they are often referred to as "Design Basis Accidents". Such accidents vary with reactor type and between one reactor vendor and another. However, as an illustration the following limiting faults need to be considered during the design of a PWR reactor system.

(a) Rupture of a main coolant pipe of the primary system (Loss of Coolant Accident - LOCA).

(b) Rupture of main secondary system pipe (steam line break - SLB).

(c) Steam generator tube rupture.

(d) Single reactor coolant pump locked rotor.

(e) Fuel handling accidents.

(f) Rupture of a control rod mechanism housing (control rod ejection accident).

A more detailed description of these limiting faults for light water reactors is given in Chapters 4 and 9.

In the case of the LMFBR two major classes of accident need to be considered.

(a) Loss of Flow (LOF) - this will occur if electric power is lost to all the primary sodium pumps. This failure will normally result in the tripping of the reactor but consideration is also given to the improbable situation where the reactor fails to trip automatically (see section 10).

(b) Transient overpower (TOP) - this might be initiated by a control drive failure which removed one rod from the core. A rate of change of reactivity above the design rate would result in the reactor being tripped but again consideration is given to situations where the protection system fails to shut the reactor down (see section 10).

Further discussion of these accident situations is to be found in Chapters 5 and 14.

In practice, detailed consideration of the consequences of these limiting "design basis accidents" determines ultimately the limits which are set for maximum heat generation rates in the fuel under normal operation. In particular, the peak fuel temperature calculated to occur during a loss of coolant accident is a strong function of the linear fuel rating (kW/m). This can be reduced if necessary (thus reducing peak fuel temperatures during a LOCA) whilst maintaining the same core volumetric power density by sub-dividing the fuel further, i.e. by increasing the number of fuel pins and reducing their diameter.

9. CONTROL AND PROTECTION SYSTEMS

In a nuclear power plant the electrical power output from the turbine-generator is directly related to the neutron flux density in the reactor via the steam flow to the turbine and the heat generation in the fuel. Controlling the neutron flux to meet the electrical power demand is the function of the reactor control system. In addition to dealing with normal operation the control system must be capable of handling unexpected non-routine plant conditions such as those described above, in such a way that the design safety limits are not exceeded. To ensure this is done the reactor is equipped with a protection system, the task of which is to monitor essential plant variables and to take automatic protective action as soon as these plant variables reach certain specified limiting values. The protection system would thus be called into play to protect the plant under conditions other than normal operation or during an operational transient.

To guarantee a reliable detection of deviations from normal operation each possible transient is monitored by measuring more than one plant variable. Thus at least two channels of protection exist for each protection function. In addition, by using a combination of redundancy - that is the coincidence of two out of three measurements or two out of four measurements for each variable - diversity, in service testing and self checking modes of operation the reactor protection system achieves very high reliability.

10. ENGINEERED SAFETY SYSTEMS

The reactor protection system can bring a number of specially provided safety systems or "engineered" safety systems into play. These systems again differ from one reactor type to another. For the purposes of illustration those systems provided on a pressurised water reactor will be discussed although similar principles pertain to gas-cooled and liquid metal-cooled reactors.

10.1 Reactor Trip

In order to reduce the neutron flux and stop the fission reaction a system of control rods entering the core from above act as a fast shutdown system. If for any reason this primary shutdown system fails to work a shutdown can be achieved by increasing the boron (neutron poison) concentration of the primary coolant.

10.2 Emergency Core Cooling (ECC) System

In the event of a loss of coolant from the primary circuit a series of components are provided to inject sufficient water into the core to prevent excessive overheating of the fuel. To ensure an adequately high reliability of such a system it usually consists of 3 or 4 identical but completely independent sub-systems. Sufficient water can be provided with just two of the four sub-systems in operation. Each sub-system itself consists of both "active" and "passive" components. The "active" components comprise a high pressure low flow injection pump to cope with small leaks occurring at high primary circuit pressures and a low pressure high flow pump to handle large leaks at low pressure. The "passive" component consists of a vessel or vessels filled with borated water held at a pressure of 3 MPa by a nitrogen overpressure and released into the primary circuit via a check or non-return valve. It is called a "passive" system because its operation does not require any independant action on the part of the operator or control system.

10.3 Auxillary Feedwater System

This system provides feedwater to the steam generators in the case of a loss of the main feedwater supply. Alternative sources of feedwater are used and the auxillary feed water pumps may be driven either by electric motors or by steam turbines.

10.4 Containment

A containment building is normally provided around the primary circuit of the reactor. It serves as the last of three separate barriers to the release of radioactivity to the environment, the other two being the fuel pin sheath and the primary circuit vessels and pipework. The containment building may be "strong", i.e. capable of withstanding significant internal pressure or "weak" or alternatively "vented", in which case the aim is to control rather than prevent the release of radioactivity to the environment. Even with a "sealed" containment it is necessary to penetrate the containment building walls to provide essential services to the reactor even under extreme limiting fault conditions. The containment building will normally only be called to undertake its function as a barrier under Infrequent and Limiting fault conditions. The containment may be fitted with additional engineered safety features such as

post accident heat removal (PAHR) capacity - to remove decay heat from within the containment thereby preventing overpressurisation of the containment.

post accident radioactivity removal (PARR) capacity - to remove radioactivity released from the core to the containment atmosphere.

hydrogen recombiners - to lower the hydrogen levels formed by zirconium water reactions, by radiolysis and chemical corrosion.

Although not strictly part of the engineered safety features, mention should be made of a number of other plant features. To ensure that the reactor has a guaranteed heat sink the turbine can be bypassed and steam from the steam generators passed directly to the condenser. In addition the secondary (steam) circuit can be vented by means of the steam line safety valves. Furthermore most reactors are provided with a separate decay heat removal circuit which provides an alternative to the use of the steam generators as a means of removing decay heating produced when the reactor is shut down. Finally, the primary circuit of gas-cooled and water-cooled reactors are protected from undue overpressure by relief and safety valves.

11. PROTECTED AND UNPROTECTED ACCIDENTS

Plant safety analyses are undertaken on the basis that the reactor protection system functions correctly and will activate the necessary engineered safety features. In particular the reactor protection system must ensure that the reactor is tripped and the heat generation in the fuel greatly reduced. Accident sequences in which the protection system functions correctly and the engineered safety features perform as intended are referred to as "protected accidents". It is possible to estimate the frequency of the initiating event in say a limiting fault condition, e.g. a rupture of a large diameter primary circuit pipe might occur once in 10^4 reactor operating years. It is also possible to put a figure on the reliability of operation of the engineered safety feature, e.g. the 4 train ECCS may fail once in 100 demands to operate. On this basis a major LOCA with a failure of the ECCS might occur once in 10^6 reactor operating years. Such an accident would be referred to as an "unprotected accident".

The question must be asked "what happens if the reactor protection system fails?" This could come about because a component failure might remain undetected and uncorrected or might result from the initiating transient itself. Because the reactor protection systems are very reliable indeed estimating failure rates from actual operating experience is very difficult. Although component and sub-system failures have occurred and design manufacturing and operating errors have been observed, the redundancy feature in the design of the reactor protection system means that it will perform its function even with the infrequent occurrence of single failures. However multiple failures are conceivable because of a common defective manufacturing process or caused by a common external event such as a fire. These "common mode" failures as they are called do, however, have the potential to cause failure of the reactor protection system.

Typically, a nuclear reactor might experience about 10 "upset" transients per year. About half of these might require the reactor to be tripped. To ensure that unprotected accidents involving a failure to trip - so called "Anticipated Transient without Scram" (ATWS) - occur with a frequency less than say once in 10^6 reactor operating years it would be necessry to demonstrate a failure rate for the reactor protection system of less than once in 10^6 demands. Such low values can be achieved by using highly reliable equipment and by providing redundancy but there remains the uncertainty of the common mode failure. As a result it is difficult to be confident about failure rates lower than once in 10^4 demands.

It is difficult to reduce the number of "upset" transients experienced by a plant significantly and the provision of further separate diverse and fully independant means of reactivity shutdown may not always be possible. Therefore reactor designers have given some consideration to what happens during an ATWS event and to the mitigation of the consequences. With gas-cooled reactors the time constants of the system are long and the operator can usually activate the reactor trip manually. Even so secondary and sometimes tertiary shutdown systems are provided as a back-up. With water-cooled reactors the mismatch between the continuing heat generation in the reactor core and the reduced heat removal capacity of the secondary system means that the primary coolant pressure will rise. In a PWR the magnitude of this rise is set by the value of the moderator temperature coefficient, the primary circuit relief valve capacity, the rate of heat removal in the steam generators and other factors which vary from one design to another. In a BWR power can be decreased by manually tripping the main coolant recirculation pumps.

In the case of liquid metal-cooled reactors the large thermal capacity of the primary liquid sodium makes the reactor much less responsive to secondary circuit faults and places the emphasis on understanding primary circuit and core faults which could lead to whole core disrupture accidents (CDAs).

Such whole core accidents are usually postulated to come about as a result of i) loss of coolant or loss of flow (LOF) without reactor trip and ii) reactivity-insertion or transient overpower (TOP) without reactor trip. A loss of cooling may stem from an initiating event in a single fuel channel leading initially to a blockage and propagation to a whole core accident. Alternatively a complete loss of forced flow may occur as a result of the tripping of the primary pumps. In this case there is some evidence that for small fast reactors at least the consequences of a loss of forced flow without a reactor trip may be tolerable(4) and that for large fast reactors the large negative temperature coefficients may delay sodium boiling for several minutes sufficient to allow manual operator action.

12. LIMITING PLANT PARAMETERS

In the discussion of the various plant conditions or various types of transient reference was made to limiting values of plant parameters beyond which action by the reactor protection system is necessary. These limiting values of plant parameters are usually chosen by consideration as to whether a particular set of conditions could lead to fuel damage. In practice, this means a limitation on the peak fuel rating and the setting of a peak clad temperature in the case of gas-cooled or liquid metal-cooled reactors or a minimum margin between the operating heat flux and the "critical heat flux" in the case of water-cooled reactors.

Establishing allowable core power densities for the various reactor types is a complicated process requiring consideration of many design restraints. A simplified treatment would consider just two sets of limits; limits set so that fuel does not fail during normal operation, or during operational and upset transients and limits set so as to establish the extent of the permissible damage during infrequent and limiting fault conditions.

12.1 Limiting Parameters to Prevent Fuel Damage During Normal Operation, Operational Transients and Upset Conditions

Such limits are often set by economic as much as safety considerations. In addition they may vary with operating practice from one country to another depending upon licensing considerations and radioactive discharge limits. As an example in the United States for light water reactors the following limitations are set.

(a) minimum ratio between the critical heat flux and the peak local operating heat flux to be equal to or greater than 1.3

(b) fuel centre temperature to be below the melting point of UO_2 (2800°C)

(c) the internal fission gas pressure to be less than the nominal external pressure at the end of core life.

(d) fuel sheath stresses to be less than yield stress

(e) fuel sheath strain to be less than 1%

(f) cumulative strain fatigue cycles to be less than 80% of design strain fatigue life.

In the case of gas-cooled reactors a peak clad temperature limit of 1100°C is set to prevent fuel failures during normal and upset conditions.

The design of the fuel is usually optimised so that no single limit dictates the fuel linear power rating but that all limitations occur at around the same value. The reactor protection system ensures these limitations are not exceeded.

12.2 Limiting Parameters to Ensure that Fuel Damage During Infrequent and Limiting Fault Conditions is Predictable

Although during an accident sequence it is expected that plant parameters will exceed those at which fuel failure is initiated, it is important that the core should remain in a coolable geometry and that requires the extent of the fuel damage to be known. Again as an example the following limitations are in use in the United States for light water reactors.

(a) peak fuel sheath temperatures to remain below 1200°C

(b) oxidation of the fuel sheath to be less than 17% of the fuel sheath thickness

(c) the amount of zirconium to react with steam to form hydrogen is limited to 1%

(d) the core geometry must remain coolable both during the course of the accident and during the subsequent recovery period.

In the case of gas-cooled reactors fuel sheaths are unlikely to fail in the short term until peak temperatures of 1400°C are reached.

13. IMPLICATIONS OF THE ACCIDENT AT THREE MILE ISLAND NO.2, MARCH 1979

On March 28, 1979 what has been called the worst accident in the history of commercial nuclear power generation occurred at the Three Mile Island No.2 PWR unit in Pennsylvannia. It is interesting to compare what actually happened during the course of this incident with the foregoing description since a number of differences do emerge.

Firstly the initiating event was common enough - a loss of feedwater transient which would be classified above as a "fault of moderate frequency". The reactor protection system functioned correctly in bringing the various engineered safety features into operation. The turbine was tripped and the steam bypass to the condenser was initiated. The auxiliary feed water pumps started but due to isolation valves being closed, initially at least, no feedwater was delivered to the steam generators. The relief valves on the pressuriser opened, the reactor was tripped, but unknown to the operators the relief valve failed to reseat. Thus the design intention that "faults of moderate frequency" should be terminated without propagation to a more serious fault condition was not achieved in this instance. According to the above classification the TMI-2 incident had escalated to an "infrequent fault or emergency condition" in that the plant was then experiencing a "small break loss of coolant accident". Again the reactor protection system worked and the emergency core cooling system was automatically activated only to be throttled back by the operators because the indicated pressuriser level suggested the primary circuit was full of water.

About 100 minutes after the accident started the primary pumps were tripped as a result of excessive vibration. The operators expected the plant to enter a condition in which the decay heat from the core would be transported to the heat sink - the steam generators - by natural convection. In practice the plant experienced a condition which was not foreseen by the designers, namely a situation where the upper parts of the primary circuit were filled with vapour or non-condensible gas and natural circulation was not possible. Furthermore the condition of the plant was such as to preclude the bringing into use of the separate decay heat removal circuit. The damage to the fuel in the core went significantly beyond the limitations discussed above, although the debris remained coolable. Other engineered safety features such as the containment building isolation and post accident heat removal systems worked either partially or completely. The release of radioactivity off-site was minimal and the overall safety intention was met despite many of the engineered safety features being prevented from carrying out their allotted functions either partially or completely by inappropriate operator action or by deficiencies in systems design.

Although the categorisation of transient and fault conditions and the setting of limiting fuel parameters as discussed in this Chapter are indeed helpful to designers in ensuring the safety of nuclear power plants, the accident at TMI suggests that consideration should also be given to a wider spectrum of faults initiated from operational or upset transients and also to accidents in which fuel damage significantly exceeds that of the previously set limits.

ACRONYMS

AGR - Advanced Gas-cooled Reactor
ATWS - Anticipated Transient without Scram
BWR - Boiling Water Reactor
CANDU - Canadian Deuterium Uranium Reactor
CDA - Core Disruptive Accident
CHF - Critical Heat Flux
HTR - High Temperature Reactor
HWR - Heavy Water Reactor
LMFBR - Liquid Metal Fast Breeder Reactor
LOF - Loss of Flow
LWR - Light Water Reactor
MAGNOX - Magnesium alloy used as cladding material in British natural uranium gas-cooled reactors
PFR - Prototype Fast Reactor
PWR - Pressurised Water Reactor
SGHWR - Steam Generating Heavy Water Reactor
TOP - Transient Overpower

REFERENCES

1. L.S. Tong and J. Weisman. "Thermal Analysis of Pressurised Water Reactors", American Nuclear Society (1970).

2. R.T. Lahey and F.J. Moody. "The Thermal-Hydraulics of a Boiling Water Nuclear Reactor", American Nuclear Society (1977).

3. Y.Y. Hsu and H. Sullivan. "Thermal Hydraulic Aspects of PWR Safety Research", Symp. on the Thermal and Hydraulic Aspects of Nuclear Reactor Safety, Vol.I, Winter Annual Meeting of ASME, Atlanta Georgia Nov-Dec 1977.

4. C.V. Gregory et al. "Natural circulation boosts safety of pool type fast reactor", Nuclear Engineering International, Dec. 1979 p.29-33.

PART 2

FUNDAMENTAL CONCEPTS

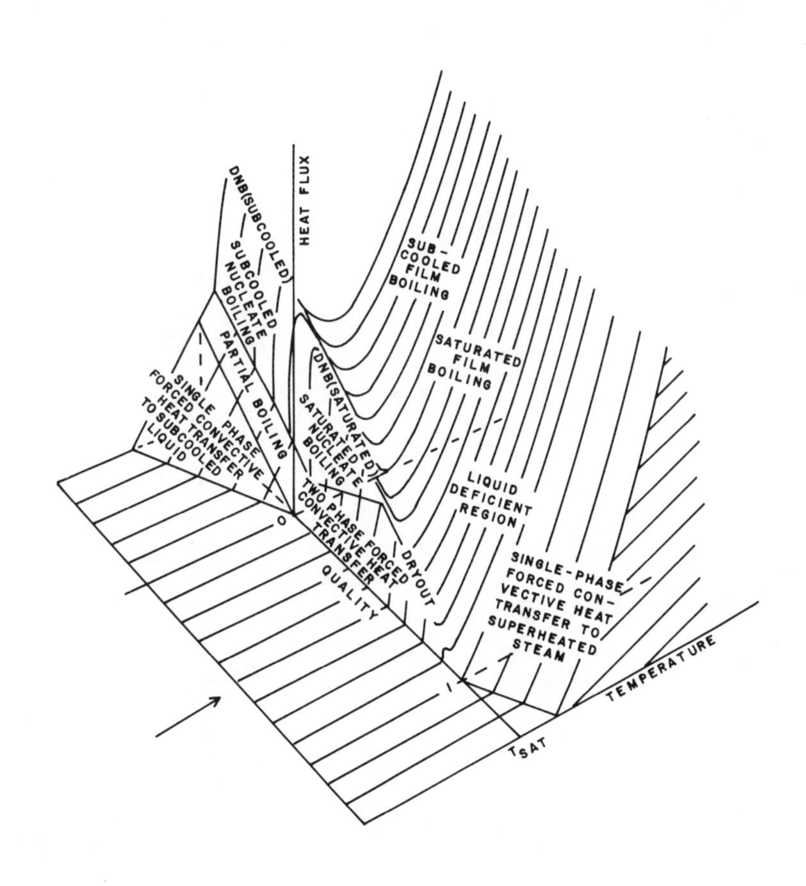

4 TRANSIENT RESPONSE OF LIGHT WATER REACTORS

R.T. Lahey, Jr.
RPI, Troy, NY, USA

The purpose of this chapter is to provide an overview of the thermal-hydraulic considerations associated with light water nuclear reactor (LWR) safety issues. Subsequent chapters will provide more details on the various accident scenarios and the technology involved in LWR safety analysis.

For an LWR to be granted an operating license, the design of the reactor must be such that the reactor is not only able to perform satisfactorily at rated power, but is able to withstand various postulated accidents without release of significant amounts of radioactive material which may affect public health and safety. The full spectrum of plant operational and fault conditions can be divided into four categories, in accordance with their anticipated frequence of occurrence. In considering the plant safety and design limits on power densities, the following operational and fault categories can be specified:

- Category 1 Normal Operation and Operational Transients
- Category 2 Faults of Moderate Frequency
- Category 3 Infrequent Faults
- Category 4 Limiting Faults

The basic principle applied in relating design requirements to each of the above conditions is that the most frequent occurrences yield little or no fuel damage problems, and those extreme situations, having the potential for the greatest risk of causing fuel damage, shall be those least likely to occur.

1. PWR SAFETY ISSUES

Both pressurized and boiling water nuclear reactor safety concerns will be treated in this chapter; however, let us begin with a consideration of PWR concerns.

1.1 PWR Normal Operation and Operational Transients

The first category, Normal Operation and Operational Transients, includes those which are expected frequently in the course of power operation, refueling, maintenance, or maneuvering of the reactors. This category should be accommodated with some margin between any operating parameter, or plant design parameter, and the value of that parameter which would require protective action. A list of normal operation and operational transients usually considered for a PWR can be divided into the following classes:

- Normal (Steady-State) Operation
- Operation at Permissible Limits
- Operational Transients

The first class of conditions consist of steady-state operating conditions such as power operation (reactor power level from about 10 to 100 percent), start-up, hot shutdown, cold shutdown, and normal refueling.

Several deviations from steady-state operation are permitted (specification of these deviations are provided by plant technical specifications). Operations at permissible limits include such conditions as operation with some components of the system out of service, and leakage from fuel with cladding defects causing high activity levels in the reactor coolant.

Operational transients are concerned with system load changes and include, step load changes and ramp load change. Because of the possibility of operation outside the plant design specification, limits are defined for such transients. These transients normally involve primary system temperature changes, but the rate of such changes should be no larger than a specified value. Ramp load changes are generally also limited to specified value, usually to $\leq 5\%$ per minute.

The reason that each of these operational transients must be carefully analyzed is that they constitute initial conditions for the more adverse fault conditions. Initial conditions of categories two through four, are devised from the most adverse conditions determined in the normal operation and operational transients category. Thus the analysis of this category should be performed with precision to insure proper initial conditions for all other fault conditions.

1.2 PWR Faults of Moderate Frequency

The second category of plant operation is chosen so that the worst possible condition that could result in reactor shutdown would still allow a return to power. These transients are to be accommodated with no failure mode propagation which could cause more serious fault conditions. Transients of this type are not expected during normal operation but can reasonably be expected during the lifetime of the plant. The design criteria for this category is that such faults will not lead to a breach of any fission product containment barrier, i.e., fuel rod cladding, reactor coolant barrier (piping or vessel) and containment building. Transients normally considered as Faults of Moderate Frequency are:

- Uncontrolled Reactor Control Rod Assembly Withdrawal
- Reactor Control Rod Assembly Misalignment
- Chemical and Volume Control System Malfunction
- Partial Loss of Forced Reactor Coolant Flow
- Startup of an Inactive Reactor Coolant Loop
- Loss of External Electrical Load and/or Turbine
- Loss of Normal Feedwater
- Loss of All AC Power to the Station Auxiliaries (Station Blackout)
- Excessive Heat Removal Due to Feedwater System Malfunctions
- Excessive Load Increase

Design parameters for not breaching any fission product containment barriers require that no fuel safety limit be exceeded or unacceptable pressure, temperature, and reactor power transients result from these transients.

1.3 Infrequent PWR Faults

The third category which must be considered is Infrequent Faults. This category of transients, or operating conditions, should be accomodated with limited fuel failure even though the fuel damage might be of the extent to preclude resumption of operation. Radioactive release should not be to the extent that public activities are restricted beyond the exclusion radius defined for the plant. The reactor transient and operating conditions considered as Infrequent Faults include:

- Loss of Reactor Coolant from Small Ruptured Pipes, Cracks in Large Pipes, or Stuck Open Relief Valves, which Actuates the Emergency Core Cooling System (ECCS)
- Minor Secondary System Pipe Breaks
- Inadvertent Loading of a Fuel Assembly into an Improper Position
- Complete Loss of Forced Reactor Coolant Flow
- Inadvertent Boron Dilution
- Waste Gas Decay Tank Rupture
- Various Anticipated Transients Without Scram (ATWS)

Design parameters for fuel safety limits and pressure/temperature/power limits may be exceeded as long as the result is not more than a limited amount of fuel rod failure,with offsite dose limits not being exceeded.

Since the occurence of the incident at Three Mile Island - Unit #2 (TMI-2), this category has taken on even greater importance. Prior to this incident the effect of improper operator action was not fully appreciated. Current analyses are attempting to model this aspect of the problem.

1.3.1 PWR Anticipated Transients Without SCRAM (ATWS)

This infrequent fault has been the subject of intense controversy for many years, thus it merits further discussion. The USNRC has treated this subject in great detail in recent Staff Topical Reports [1]. The basic controversy is concerned with the role of common mode failures in reducing the reliability of the Reactor Protection System. The frequency of ATWS events is the product of the frequency of various anticipated transients and the conditional probability of a SCRAM failure (given the occurence of a particular transient). This failure may be prior to the transient (ie: an undetected failure) or, less likely, it may occur during the transient (ie: a failure caused by the transient).

Typical PWR ATWS events which have been analyzed include:

- Station Blackout
- Reactivity Insertion due to:
  - Rod withdrawal (from subcritical)
  - Rod cluster control bank withdrawal
- Boron Dilution Transients
- Loss of Core Flow Transients
- Cold Water Incidents (eg: Startup of an Idle Loop)
- Turbine trip
- Loss of Feed Water Flow

The PWR ATWS which cause the largest pressure increase are those associated with a loss of "heat sink". Both the turbine trip and loss of feed water flow ATWS events deprive the primary side of an effective heat sink. This causes the primary coolant to heat-up and expand. On some plants this expansion may be great enough to completely fill the pressurizer, causing the discharge of liquid through the safety/relief valve(s). Such transients have been calculated [1] to give peak system pressures in excess of 34.5Mpa (5,000 PSIA). Clearly such high pressures are unacceptable.

There are various ways which have been proposed by the PWR vendors to mitigate ATWS transients, including the rapid injection of a high concentration of boron solution through the ECCS. While this controversy is not completely resolved, work is well underway to achieve an acceptable solution.

1.4 Limiting PWR Faults

Limiting fault condition occurrences are faults which are not expected to take place, but are postulated because their consequences include the potential for the release of significant amounts of radioactive material. They are the most drastic and must be designed against since they represent the limiting design case. For the purpose of discussion, the limiting faults for a PWR reactor system have been chosen and classified in the following categories.

- Major Rupture of Pipes Containing Reactor Coolant (Loss-of-Coolant-Accident, LOCA)
- Major Secondary System Pipe Rupture (Steam Pipe Break)
- Steam Generator Tube Rupture
- Single Reactor Coolant Pump Locked Rotor
- Fuel Handling Accident
- Rupture of a Control Rod Mechanism Housing (RCCA Ejection)

These transients represent the postulate accident conditions which ultimately set the power limits of the reactor. The first of these, the Loss-of-Coolant Accident, is the design basis accident (DBA) and thus establishes core power limits. For completeness a brief discussion of this event will be given here. A subsequent chapter will treat this important subject in more detail.

1.5 PWR LOCA

The Loss-of-Coolant-Accident is postulated to be a double-ended break of the main coolant pipes of the primary loop, either in the cold leg or hot leg of the reactor. These are the cases under which very rapid depletion of coolant can result. Figure 1 shows the time history of a typical PWR LOCA.

1.5.1 The PWR Blowdown Stage

SUBCOOLED: The water is discharged through the break, resulting in rapid depressurization of the reactor vessel from about 150 bar, to the saturation pressure corresponding to coolant temperature, which is about 100 bar at 315°C. This is called the subcooled depressurization period.

SATURATED: Pressure continues to drop, but not as rapidly as in the first stage (on the order of 3-30 bar/sec). Flashing of liquid occurs, creating a large void in the core. The core flow can go up or down, or both, depending upon location of the break and the relative hydraulic resistance of each path. That is, the flow rates are determined by the resistance due to the various components in the broken loop. During the blowdown stage, although the fission

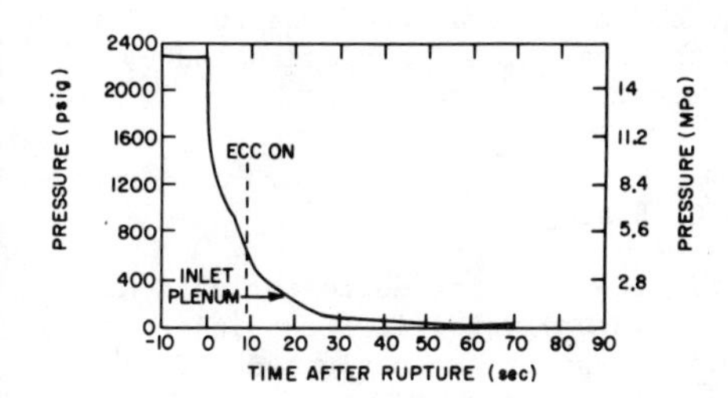

Fig. 1a PWR Pressure History During Blowdown

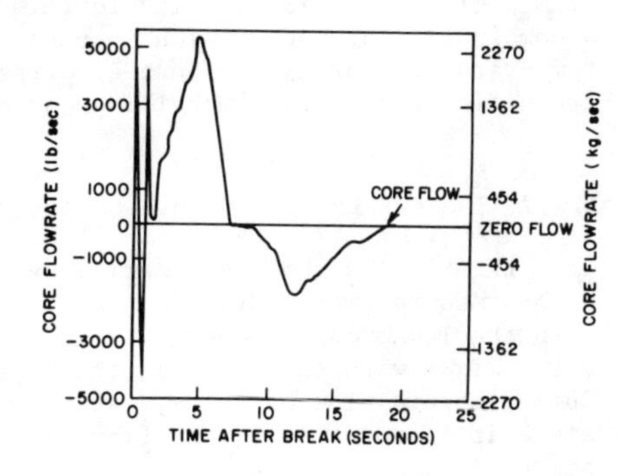

Fig. 1b BWR Core Flow During LOCA (Cold Leg Break)

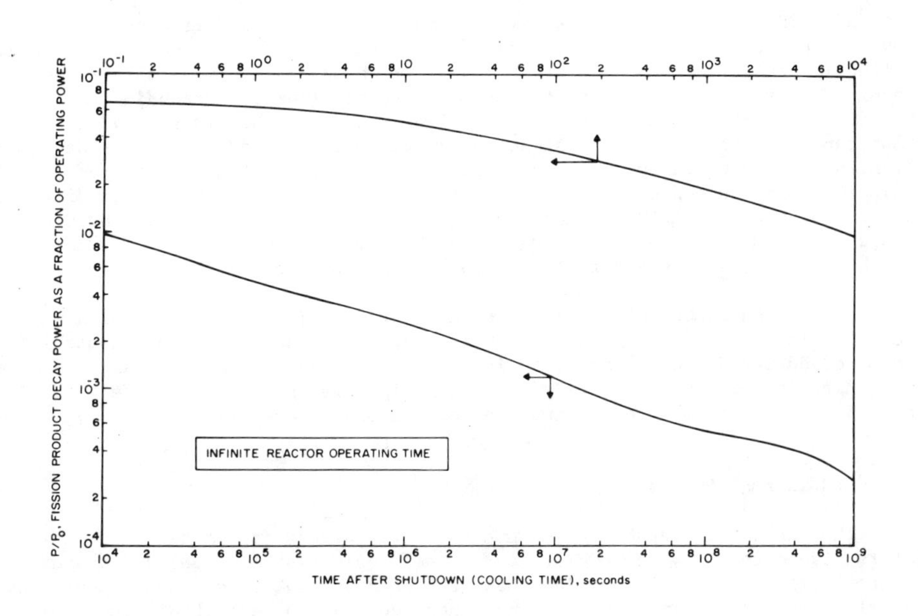

Fig. 2 Decay Heat Curve

process has ceased, due to both insertion of control rods (ie: SCRAM) and the generation of void, there is a significant amount of decay heat, which decays exponentially but, as shown in Figure 2, levels off at a certain value. The decrease of coolant flow together with generation of void, cause the onset of boiling transition (BT). The BT condition can be either of DNB-type (departure from nucleate boiling) or liquid film dry-out type (at high void fraction), or a combination of both, which in turn, causes the excursion of the fuel clad temperature. During the post-BT period, the fuel clad is being cooled by transition boiling, film boiling, or dry steam.

1.5.2 The PWR Emergency Core Cooling (ECC) Stage

BYPASS: As shown schematically in Figure-3, during depressurization, the high-pressure emergency core cooling (accumulator) water is first injected and then the low pressure emergency core cooling systems are activated. Cooling water mixes with the steam in the injection line and then goes to the downcomer. Due to upward flow of steam, water cannot readily go into the lower plenum since it is counter-current flow limited (CCFL). As a result, some of the emergency core coolant may bypass the downcomer and, instead of draining to the lower plenum, will be discharged out the breach. This situation is shown schematically in Figure 4.

It is interesting to note in Figure 5 that considerable vapor may be produced by boiling of the ECC fluid on the elevated temperature pressure vessel walls and internal structures. This vapor may tend to extend the period of steam by-pass, a phenomena known as "hot-wall delay."

REFLOOD: Subsequent to ECC bypass, most ECC water is delivered to the lower plenum to refill it. The level then starts to rise, reflooding the core. Reflooding of the core and quenching of the rods may be met with some resistance. One is the thermal resistance of the vapor-blanket surrounding the very hot fuel rods which delays quenching of fuel rods. Another is the hydraulic resistance due to the flow of steam and droplets downstream of the reflood front which can build up a pressure drop along its path towards the break. This phenomena, shown schematically in Figure 6, is known as "steam binding". It should be noted in this figure that secondary-to-primary leakage through ruptured steam generator tubes aggravates "steam binding", while B&W-type check valves in the core barrel tend to minimize this concern.

Another phenomena of interest is the hydrodynamic instability which may occur during reflood. Figure 7 is typical of the downcomer-to-core instability observed during German experiments in PKL. The effect of this instability mode on PWR reflood is not completely clear at this time, however it is felt that oscillations of this type may improve reflood heat transfer.

1.6 PWR Thermal Limits

The generating capability of nuclear power plants is limited by a series of requirements whicy include nuclear steam supply system (NSSS) safety and factors affecting fuel integrity. For PWR core design, these requirements can be simplified to a limit on the linear heat generation rate (LHGR) and a minimum value of the departure from nucleate boiling ratio (DNBR) for steady-state operation and certain transients.

Determining permissible power limits of the core is a complicated process involving many design constraints [2]. The attempt here is to explain these

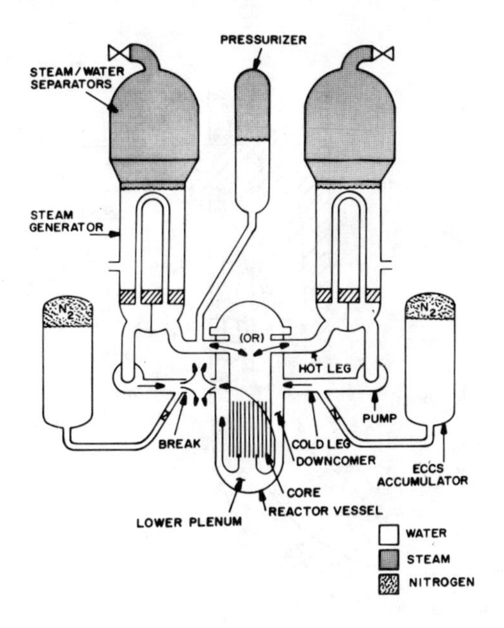

Fig. 3 PWR LOCA (Time of Initiation)

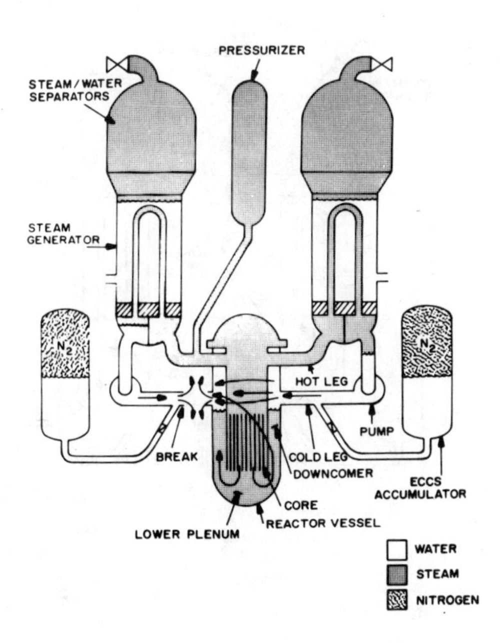

Fig. 4 PWR Bypass Phenomena

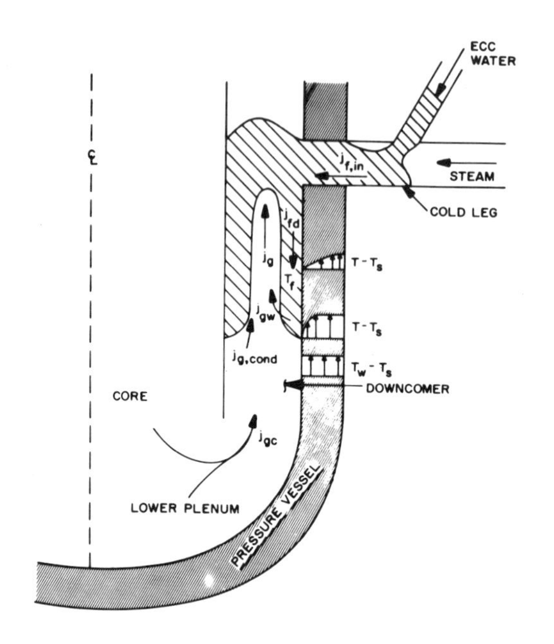

Fig. 5 "Hot Wall Delay"

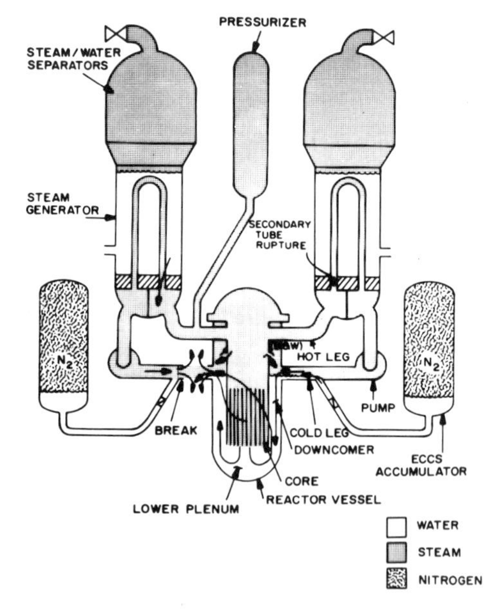

Fig. 6 PWR Steam Binding

limits in relatively simple terms. Two terms used in this discussion will be the safety limits and accident limits. The safety limits are the maximum allowable LHGR in normal operations such that if an abnormal transient or fault condition occurs, the fuel will not exceed the appropriate limits during the course of the transient. In determining the limiting design parameters, each of these will be discussed.

1.7.1 PWR Safety Limits

Safety limits which are determined in relation to fuel safety represent an actual limit at which fuel damage is assumed to occur. Fuel safety limits have been set at levels relating to:

- Departure from nucleate boiling ratio (DNBR)
- Fuel centerline temperature at melting conditions (2881K)
- Internal fission gas pressure less than the nominal reactor coolant pressure
- Clad strain less than 1 percent
- Cumulative strain fatigue cycles less than 80 percent of design strain fatigue life of the irradiated fuel

Each of these may be considered as a safety limit in terms of the LHGR. Since the DNBR is a key parameter in establishing PWR thermal limits let us briefly consider the implications of this operational parameter. As shown in Figure 8, DNBR is defined for a given local flow rate, pressure and enthalpy as, the critical heat flux given by a design correlation (eg: the W-3 correlation corrected for axial and spacer effects [2]) divided by the local heat flux from the reactor fuel. That is,

$$\mathrm{DNBR}(z/L_H) = \frac{q''_c(z/L_H)}{q''(z/L_H)} \tag{1}$$

The axial (and radial) position in the core where the DNB ratio is a minimum is denoted by the minimum DNR ratio (MDNBR). That is,

$$\mathrm{MDNBR} = \frac{q''_c(z_*/L_H)}{q''(z_*/L_H)} \tag{2}$$

It is the MDNBR which establishes the core safety limit. The calculated MDNBR is compared with a lower limit which has been established to account for the effects of fuel geometric distortion, and thermal-hydraulic uncertainties. The calculated MDNBR must be no less than this lower limit, which in general, is plant dependent.

The calculation of MDNBR in a PWR core is normally performed using subchannel codes such as THINC or COBRA. The basic procedure is to perform a core-wide subchannel analysis in which each subchannel may contain a number of fuel assemblies. Once the subchannel having the most limiting DNBR is located, then another subchannel analysis is performed on the limiting subchannel. That is, the large subchannel is subdivided into many smaller ones to get finer resolution of the local thermal-hydraulic conditions. The flow and enthalpy cross-flow calculated for the limiting large subchannel are normally imposed as boundary conditions on the peripheral subchannels in the next run. In this way we obtain overall conservation of mass and energy. Current versions of the Westinghouse THINC-CHAIN approach have automated this procedure.

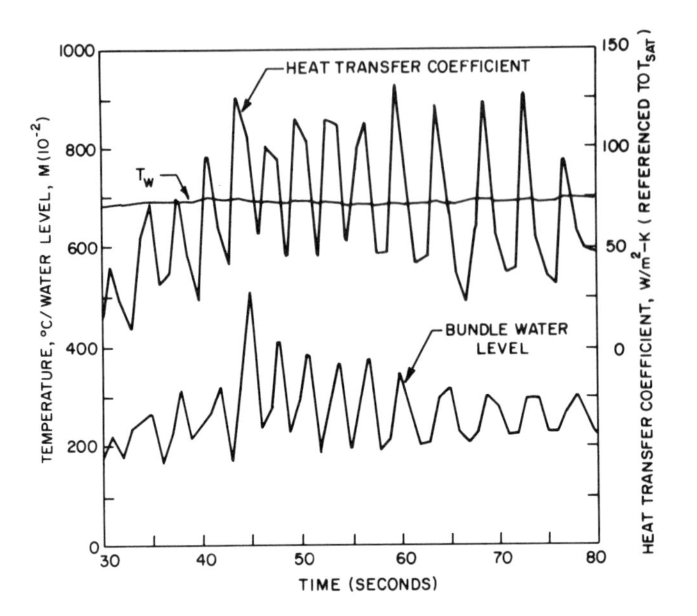

Fig. 7 PKL Reflood Oscillations

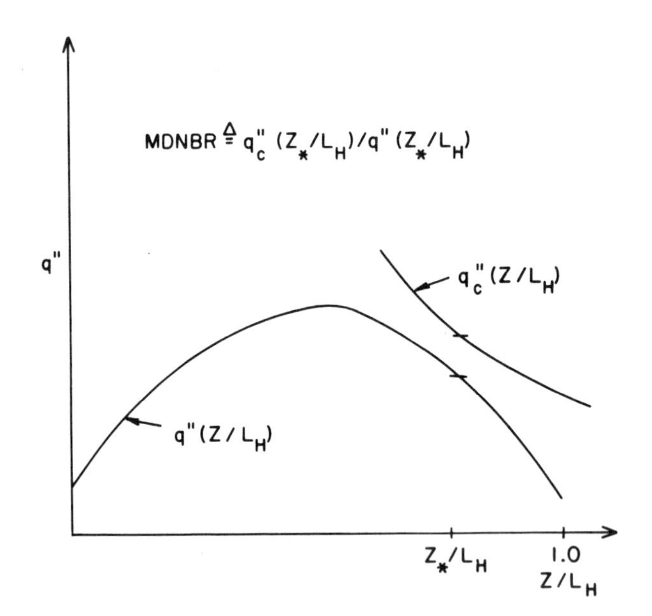

Fig. 8 PWR Thermal Margin

Automation is valuable since, in general, a large number of calculations are required to determine the appropriate MDNBR for each time in core life, control rod position, etc. It is generally true, however, that the DNBR criteria is the most limiting of any of the safety limits.

For example, if we consider a typical Westinghouse 15 x 15 fuel design, the USNRC approved MDNBR limit of 1.3 establishes a maximum LHGR of approximately 69 KW/m. Fuel centerline melting is usually larger than 69 KW/m but can be slightly lower for densified fuel, and is therefore plant dependent. A cladding strain limit is not directly considered since the 1 percent limit normally imposed occurs at a greater LHGR than fuel centerline melt. The reactor protection system is designed to insure that the core limits are not exceeded.

1.7.2 PWR Accident Limits

Accident limits are set so that if an abnormal transient or fault condition occurs, the fuel will not exceed the specified limits during the course of the transient. Accident limits are set by specific criteria for each accident. One of the best known of these criteria is specified by 10 CFR 50, Appendix-K. Appendix-K limits the fuel cladding temperature to below 1478K (2200°F for a loss-of-coolant-accident (LOCA). Lowering the peak LHGR at which the LOCA analysis is initiated lowers the maximum cladding temperature. Therefore, there exists a linear power density which corresponds to a peak clad temperature (PCT) of 1478K (2200°F). The accident limit for a LOCA is plant dependent, but for a typical W 4-loop 15 x 15 fuel design, the LHGR limit is approximately 51 KW/m. By comparing the DNBR and PCT limits, it can be seen that the LOCA limits are the most restrictive.

Only for relatively short periods of time during load-following transients does the potential exist for approaching limits, and most load-following transients do not follow the extreme paths assumed above, the reactor is normally operated such that the peak LHGR is within a range of 60 percent to 85 percent of the allowable LHGR.

2. BWR SAFETY ISSUES

As in PWR technology, a boiling water nuclear reactor (BWR) is designed not only to operate in a prescribed range of steady-state conditions, but also to successfully undergo changes between different operating conditions without exceeding established limits. Moreover, the design must also accomodate various anticipated abnormal operating conditions. These abnormal conditions are associated with some deviation from normal operating conditions and are anticipated to occur often enough that the design should include the capability to withstand these conditions without operational impairment. Included are transients that result from any single operator error or equipment malfunction, transients caused by a fault in a system component requiring its isolation from the system, transients caused by loss of load or power, and any system upset not resulting in a forced outage.

A single operator error is the set of actions that is a direct consequence of a single reasonably expected erroneous decision by a reactor plant operator.

In addition to anticipated abnormal conditions, the BWR design must withstand various postulated accidents without releasing an unacceptable amount of radioactivity to the environment. An accident is defined as a single event, not reasonably expected during the course of plant life, that has been postulated from unlikely but possible situations, that has the potential to cause the release of an amount of radioactive material in excess of prescribed limits.

The general safety features of the BWR and its containment are designed to handle anticipated and abnormal transients as well as postulated accidents. A safety function is identified for each event and is incorporated into the design of one or more engineered safety systems. These safety systems provide a safety function to mitigate the consequences of accidents which may cause major fuel damage. The purpose of these engineered safety features is to prevent the release of radioactive material in excess of prescribed limits, and to insure that the radiation exposure to plant personnel does not exceed allowable limits.

It is significant to note that the "multiple barrier" concept is employed in the design of all light water nuclear reactors. That is, the radioactive fission products are normally contained within the fuel rod cladding (first barrier). If this barrier is breached for some reason, then the reactor pressure vessel serves as an effective barrier to the release of radioactivity to the environment (second barrier). Finally, the containment system itself (third barrier) is engineered to prevent the release of an unacceptable amount of reactivity to the environment in the event that the first two barriers are breached.

As in all light water reactors (LWR), it is the various safety issues which ultimately determine the power rating of the core. The safety issues of greatest interest today in BWR technology are concerned with the thermal-hydraulic response of the reactor, given that a particular accident has occured. Since all BWRs manufactured in "free world" countries are made in accordance with General Electric (G.E.) licensee agreements, G.E. design features will be stressed in this paper.

Current BWR safety issues include:

- Nuclear boiler system safety issues
- BWR containment system safety issues

In the following discussion, each of these categories will be explored from the point of view of providing the reader with an appreciation of the particular safety-related issues involved. BWR containment system safety issues have been included, not because they directly effect the core rating, but rather that they demonstrate the scope of thermal-hydraulic phenomena which must be considered in the safety analysis of BWRs. A corresponding discussion could have been given for PWR containment systems, however, with the exception of PWR ice containments, these systems tend to have less interesting problems.

2.1 BWR Nuclear Boiler System Safety Issues

The safety related issues relevant to BWR nuclear boiler systems are normally classified as:

- Single operator error or equipment malfunction events
- Anticipated transients without scram (ATWS) events
- Hypothetical Loss-of-Coolant-Accidents (LOCA) events

These postulated events are discussed below for current generation BWRs.

2.2 Single Operator Error or Equipment Malfunction

In order to assure compliance with the established limits, specific abnormal conditions which could result from a possible operator error or equipment malfunction must be analyzed. The following describes abnormal conditions which are considered in the design of a BWR:

1) Nuclear reactor system pressure increases caused by:
   - Generator trip
   - Turbine trip (with or without bypass)
   - Loss of condenser vacuum
   - Isolation of one main steam line
   - Isolation of all main steam lines

2) Positive reactivity insertion caused by moderator temperature decrease, due to:
   - Loss of feedwater heating
   - Inadvertent recirculation pump start

3) Positive reactivity insertion, caused by:
   - Control rod withdrawal error
   - Improper fuel assembly insertion or drop*
   - Control rod removal or dropout

4) Loss-of-coolant inventory, caused by:
   - Pressure regulator failure
   - Inadvertent opening of relief or safety valve(s)
   - Loss of feedwater flow
   - Total loss of offsite power (i.e., loss of pumping power)

5) Core coolant flow decrease, caused by:
   - Recirculation flow control failure
   - Trip of one recirculaiton pump
   - Trip of two recirculation pumps
   - Recirculation pump seizure*

6) Positive reactivity insertion caused by core coolant flow increase, due to:
   - Recirculation flow control failure
   - Inadvertent startup of idle recirculation pump

7) Core coolant temperature increase, caused by:
   - Feedwater controller failure

8) Excess of coolant inventory, caused by:
   - Feedwater controller failure

In addition to human (operator) error and equipment malfunctions, other postulated events which are analyzed include natural phenomena such as earthquakes, tornadoes, tropical storms and/or hurricanes, floods, drought, excessive rain, ice and snow. Other potential hazards which must be considered

---

* Currently classified as an accident.

include the plant proximity to airports, ordinance plants, chemical plants, and transportation routes for shipping explosive or corrosive materials.

These areas of potential concern have been extensively analyzed in the past and continue to be analyzed to insure that adequate thermal margins exist to accomodate these postulated events.

2.3 BWR Thermal Limits

The thermal margin is a measure of how close a fuel rod is to a boiling transition situation. Currently, critical power ratio (CPR) is used in the United States as the figure of merit for BWR thermal margin. This ratio is defined for a given bundle thermal-hydraulic conditions as the critical power of a bundle divided by the power of that bundle, that is,

$$CPR = \frac{P_{crit}}{P} \qquad (3)$$

The plant must operate so that the minimum critical power ratio (MCPR) is at, or above, a specified value (normally around 1.2). This value is established to insure that nonaccident type transients can be accomodated without experiencing a boiling transition. That is, as shown schematically in Figure-9, margin exists for geometric, thermal-hydraulic and BT data uncertainty. In practice MCPR > 1.2 to allow manuvering room.

In the 1970's a major effort was made by the General Electric Company to improve their ability to evaluate both steady-state and transient thermal margins. This effort resulted in the new General Electric Thermal Analysis Basis (GETAB) [3].

The General Electric critical quality - boiling length correlation (GEXL), contained in GETAB, is empirical and is based on a large amount of boiling transition data. Many of these data were taken in full scale, electrically heated rod bundles having various axial and local (i.e., rod-to-rod) peaking patterns typical of BWR operating conditions. The GEXL correlation is of the form,

$$\langle X_e \rangle_c = f(L_{B_c}, G, p, L_H, D_q, R) \qquad (4)$$

where,

$\langle X_e \rangle_c$ = the critical equilibrium quality

L_{B_c} = The critical boiling length

G = the mass flux

L_H = heated length

D_q = thermal diameter $\left(\frac{4A_{x-s}}{P_{rods}} \right)$

P_{rods} = perimeter of the heated rods

R = generalized local peaking factor

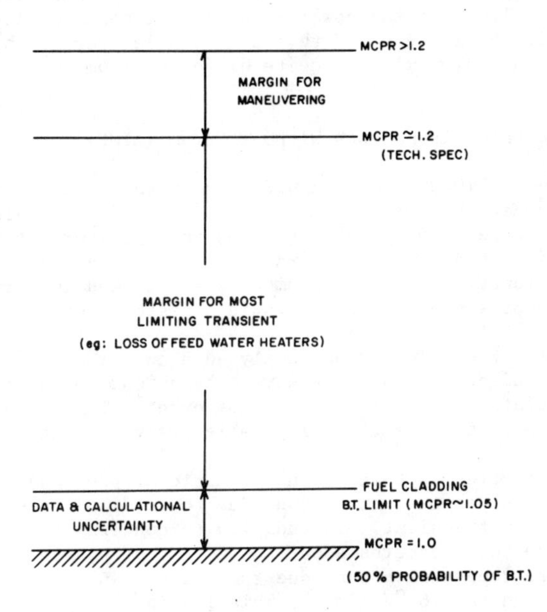

Fig. 9 Thermal Margin Evaluation

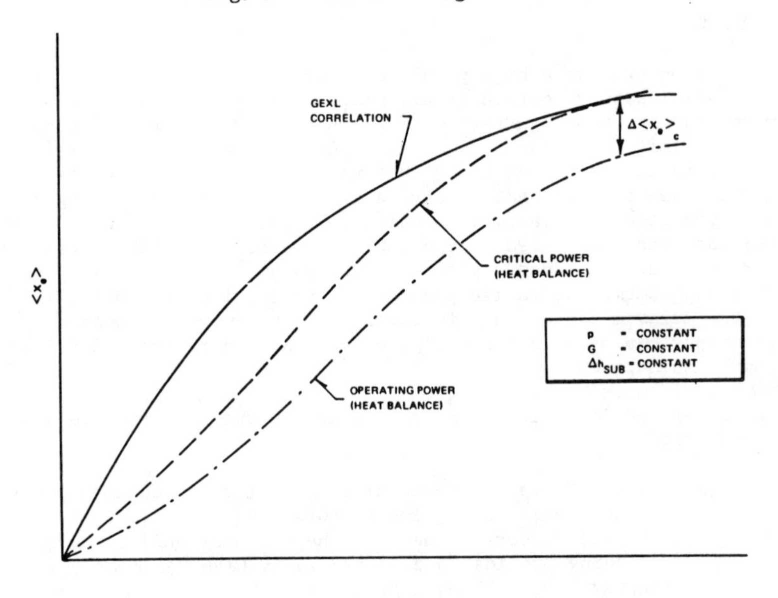

Fig. 10 Graphical Display of GEXL Correlation and BWR Heat Balance Curves

The generalized local peaking factor, R, is an empirical factor used to determine the effect of the local peaking pattern on the critical power of a BWR rod bundle. Figure 10 is a schematic of how GEXL is used in practice. At given bundle power level there is a positive critical quality defect (ΔX_c) and thus MCPR > 1.0. As the power level is increased we achieve a boiling transition. It should be noted that this bundle-average approach to the evaluation of thermal limits is quite different from PWR (subchannel) methods.

2.4 BWR Anticipated Transients Without Scram (ATWS)

In September, 1973, the then U.S. Atomic Energy Commission published WASH-1270 which established the acceptance criterion for ATWS transients for LWR's. These criteria were developed due to regulatory staff concern over the reliability of methodology to analyze common mode failure (which could take out the SCRAM function). This document was followed in April, 1978 by NUREG-0460 [1] which presented a probabilistic assessment of ATWS transients.

The General Electric Company analyzed a number of postulated ATWS transients and concluded that the most severe transient was an inadvertent main steam line isolation valve (MSIV) closure event. This transient caused the vessel pressure to increase, causing relief valve actuation.

Current practice is to have an automatic trip of the recirculation pumps to prevent excessive power excursions due to void collapse during the postulated pressurization transient. In addition to concerns about power excursions, there have also been concerns about the effect of possible condensation loads in the pressure suppression tank due to pool heat-up (because of relief valve discharge) beyond the 160°F limit. This loading mechanism will be discussed more fully later in the chapter when BWR containments issues are considered.

2.5 BWR LOCA

Protection against a highly unlikely loss-of-coolant accident (LOCA) is an essential safety feature of all nuclear reactors. The main purpose of the Emergency Core Cooling System (ECCS) is to provide sufficient cooling of the core to limit peak clad temperature (PCT), thereby limiting release of radioactive materials and ensuring that the core maintains a coolable geometry. In order to assure sufficient safety margin and to avoid exceeding cladding fragmentation limits, a maximum cladding temperature criterion of 1478K (2200°F) has been legislated by the USNRC for acceptable ECC system performance. In order to insure the adequacy of an ECCS design, it is necessary to determine the important thermal-hydraulic phenomena, and consider the overall effect of each on the maximum temperature in the core. The important considerations and philosophy employed in the design of current generation BWR ECC systems will now be summarized.

The major considerations that contribute to the conservative design of a BWR ECCS are:

(1) Stored Energy - The assumption of 102% of maximum power, the highest allowed peaking factors, and conservatively calculated thermal resistance between the UO_2 fuel pellets and zircaloy cladding provides a conservatively high value of the stored energy.

(2) Blowdown Heat Transfer - Heat transfer during the blowdown phase of a LOCA is also calculated in a very conservative manner. There is strong evidence [4] that more stored energy would be removed during blowdown that current design assumptions permit, resulting in conservatively high calculated cladding temperatures.

(3) Decay Heat - It is conservatively assumed that the heat generation rate from the radioactive decay of fission products is 20% higher than the proposed ANS standard [5], and that the fuel has been irradiated at full power for infinite time.

(4) The Peak Clad Temperature Criterion - The limitation of the calculated peak clad temperature to 1478K (2200°F) on the hottest fuel rod provides substantial conservatism to insure that the core would suffer little damage in the event of a LOCA. Indeed, this temperature limit is well below the melting and clad fragmentation limit, and the core-average temperature rise would be much less than the "hot spot" temperature on the peak power rod.

(5) Blowdown Flow Rate - There is considerable evidence [6] that the critical flow rates assumed are conservatively high (ie: the homogeneous equilibrium model is appropriate rather than Moody's slip model), and thus the calculated blowdown transient is more rapid than would actually occur.

(6) LOCA Assumptions - There are many conservative assumptions required in licensing calculations such as, double-ended guillotine breaks, simultaneous loss of off-site power, etc.

These and other conservations are incorporated into the design basis for BWR ECC systems.

2.6 BWR ECC System Design Criteria

The overall objective of the BWR ECCS, in conjunction with the containment, is to limit the release of radioactive material following a hypothetical LOCA so that resulting radiation exposures are within the established guidelines. Moreover, the ECCS must be designed to meet its objective even if certain associated equipment is damaged. Ground rules for the postulated equipment damage are summarized by the so-called "single failure" criterion [7] which states:

> "An analysis of possible failure modes of ECCS equipment and of their effects on ECCS performance must be made. In carrying out the accident evaluation, the combination of ECCS subsystems assumed to be operating shall be those available after the worst damaging single failure of ECCS equipment has taken place".

In order to satisfy the single failure criterion, evaluation of the BWR loss-of-coolant accident is performed assuming the active component failure that results in the most severe consequences, and also assuming loss of normal auxiliary power. The combination of ECC subsystems assumed to be operating are those remaining after the component failure has occurred.

The term "active component" means a mechanical component in which moving parts must operate in order to accomplish its safety function. In addition, active and passive electrical failures must be considered. A single failure may involve only one active component, or it may include the failures of several components resulting from the cascading effect of an initial failure. For either a large main steam line or recirculation line break, the worst single failure which can be hypothesized for a modern BWR (BWR/6) is failure of the low pressure coolant injection (LPIC) diesel-generator (diesel-B; Figure 11).

Therefore, following the assumed LOCA, and assuming the most damaging single failure, the BWR ECCS is designed to:

1) Provide adequate core cooling in the event of any size break or leak in any pipe up to, and including, a double-ended recirculation line break.

2) Remove both residual stored energy and radioactive decay heat from the reactor core at a rate that will limit the calculated maximum fuel cladding temperature to a value less than the established limit of 1478K (2200°F).

3) Have sufficient capacity, diversity, reliability, and redundancy to cool the reactor core under all accident conditions currently postulated.

4) Be automatically initiated by conditions that sense that an accident has occurred.

5) Be capable of startup and operation regardless of the availability of offsite power, and the normal generation systems of the plant.

6) Operate independently of containment back pressure.

7) Have components which are designed to withstand transient mechanical loads during a LOCA.

8) Have essential components which can withstand such effects as missiles (i.e., hurled objects), fluid jets, pipe whip, high temperature, pressure, humidity, and seismic acceleration.

9) Be able to obtain cooling water from a stored source located within the containment barrier.

10) Have flow rate and sensing networks which are testable during normal reactor operation to insure that all active components are operational.

The General Electric BWR/6 ECCS is a good example of a modern engineered safeguard. It is composed of four separate subsystems: the High Pressure Core Spray System (HPCS); the Automatic Depressurization System (ADS); the Low Pressure Core Spray System (LPCS); and the Low Pressure Coolant Injection System (LPCI). These systems are shown schematically in Figure 11 for a BWR/6 plant with a Mark III pressure suppression containment system.

The HPCS pump obtains suction from the condensate storage tank and/or the pressure suppression pool. Injection piping enters the vessel near the top of the shroud, and feeds two semi-circular spargers which are designed to spray water radially over the core into the fuel assemblies. The system functions over the full range of reactor pressure. For smaller breaks, the HPCS system is activated by either a low reactor water level signal or high drywell pressure.

If the HPCS system cannot maintain water level, or if HPCS failure occurs, reactor pressure can be reduced by the independent actuation of the ADS so that flow from the LPCI and LPCS systems can provide sufficient cooling. The ADS employs pressure relief-valves for steam discharge to the pressure suppression pool.

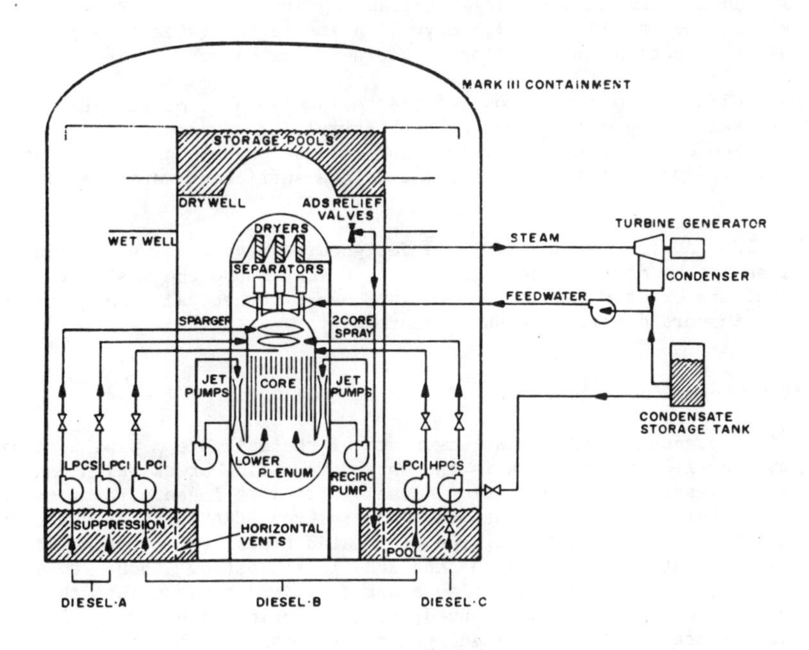

Fig. 11 BWR/6 ECCS Schematic

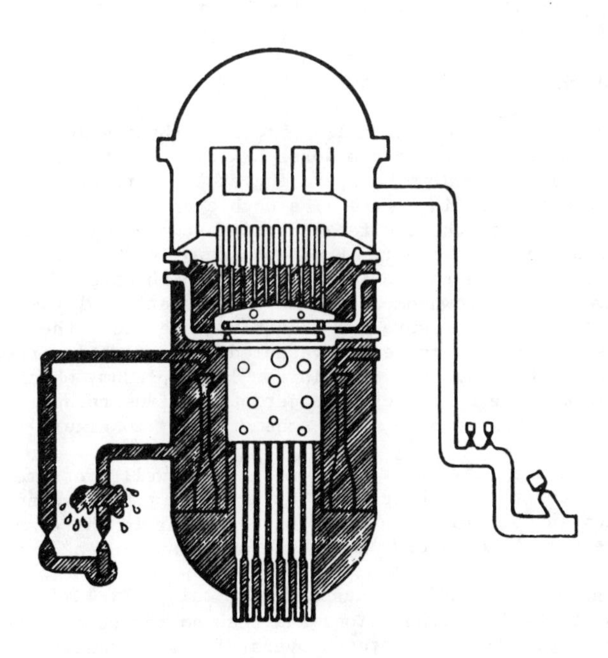

Fig. 12 Hypothetical BWR LOCA Event - Time of Initiation

The LPCS pump draws suction flow from the suppression pool and discharges from a second circular spray sparger in the top of the reactor vessel above the core. Low water level or high drywell pressure activates this system, which begins injection when reactor pressure is low enough.

The LPCI is an operating mode of the residual heat removal (RHR) system. It is actuated by low water level or high drywell pressure and, in conjunction with other ECC sub-systems, can reflood the core before cladding temperatures reach 1478K (2200°F), and thereafter maintain a sufficient water level in the core.

The HPCS, ADS, LPCS, and LPCI are designed to accomodate steam line and liquid line breaks of any size. The breaks which impose the most severe demands on the ECCS are briefly considered next. These hypothetical LOCA events will be treated in more detail in a subsequent chapter.

2.6.1 BWR Steam Line Breaks

An instantaneous, guillotine severance of the main steam pipe upstream of the steam line flow restrictors is the worst steam line break which can be postulated. Vessel depressurization causes sufficient in-core voids to shut down the reactor, although an automatic control rod SCRAM also occurs to insure shutdown. About 20 seconds after the postulated break, the emergency diesels are running at rated conditions and the HPCS is activated. About 20 seconds later, the LPCS and LPCI systems also start to inject coolant into the vessel. The worst single failure to be assumed, in conjunction with a large steam line break, is failure of the LPCI diesel-generator. Even in this severe case, the core is always submerged and the minimum critical power ratio (MCPR) is always larger than unity, thus, boiling transition would not be expected and fuel cladding temperatures would not rise.

2.6.2 BWR Liquid Line Breaks

The double-ended recirculation line break is the design basis accident (DBA)for the ECCS; that is, it is the worst BWR pipe break which can be postulated. The reactor is assumed to be operating at 102 percent of rated power when, as shown schematically in Figure 12 a double-ended circumferential rupture instantly occurs, in one of the two recirculation pump suction lines, simultaneously with a loss of off-site power. Pump coastdown in the intact loop, and natural circulation continue to provide relatively high core flow until the falling water level in the downcomer reaches the elevation of the jet pump suction, shown schematically in Figure 13. Shortly thereafter, the break discharge flow changes to steam, and an increased vessel depressurization rate causes vigorous flashing of the residual water in the lower plenum, which in turn, forces a two-phase mixture up through the jet pump diffusers and core. This lower plenum flashing phenomena is shown schematically in Figure 14.

ECCS is initiated by low water level or high drywell pressure within about 30 seconds after the break occurs. Using current NRC ground rules [7], boiling transition is calculated to occur much earlier in the transient, typically in less than 5-10 seconds after the postulated break.

It is currently conservatively assumed that only very limited heat transfer occurs subsequent to boiling transition, and that no rewetting occurs during lower plenum flashing. When the core uncovers, it is currently assumed that there is no convective heat transfer until the core is reflooded, at which time appropriate reflood heat transfer coefficients are applied [7].

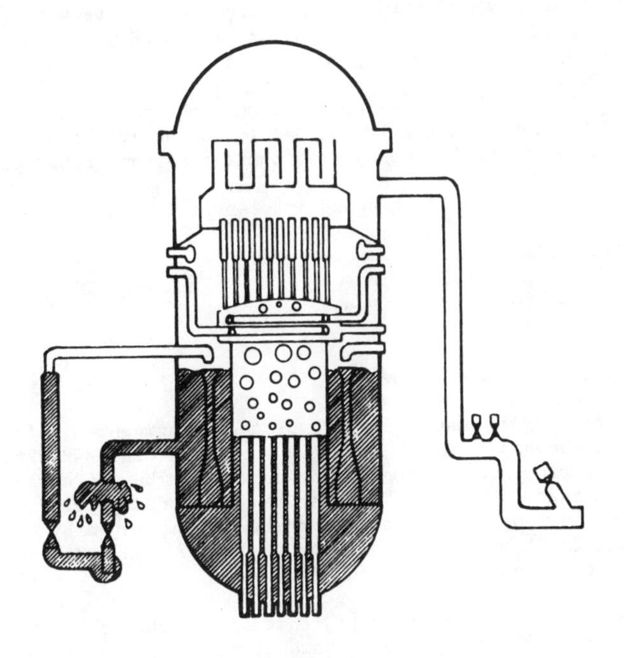

Fig. 13 Hypothetical BWR LOCA Event - Time of Jet Pump Suction Uncovery

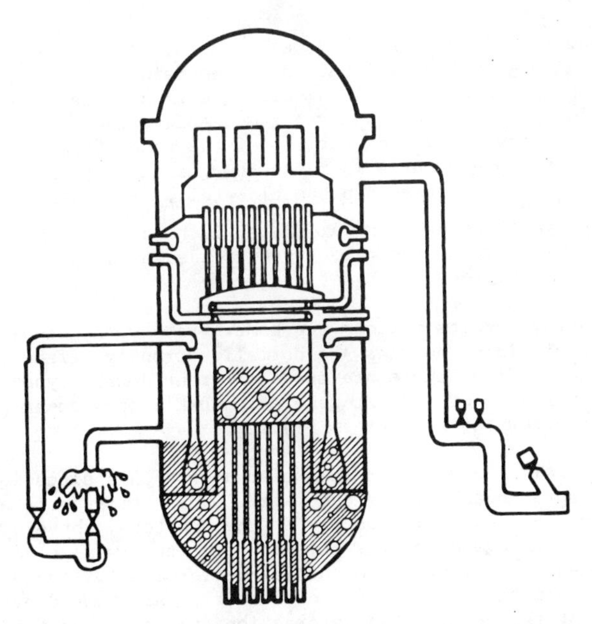

Fig. 14 Hypothetical BWR LOCA Event - Lower Plenum Flashing

The heat transfer assumptions currently applied in licensing are in conflict with the heat transfer seen in the BWR Blowdown Heat transfer (BDHT) test program [4]. As an example, a calculation using current G.E. safety analysis techniques, with Appendix-K heat transfer assumptions, is compared with some BDHT data in Figure 15. The degree of conservatism is obvious.

The items discussed above have had to do with the blowdown phase of the hypothetical LOCA event. Later in the DBA (t>30 sec.), there are a number of issues concerned with the Emergency Core Cooling System (ECCS) performance.

These can be conveniently grouped into several categories:

- Spray distribution issues
- CCFL reflood delays
- BWR "steam binding" issues

2.7 BWR Spray Distribution Issues

All operating BWR's employ as part of their engineered safeguards a core spray system in the upper plenum region. In modern G.E. BWR's this consists of two redundant ring headers containing a large number of spray nozzles. In order to insure that each bundle gets at least the design minimum flow for decay heat cooling, nozzle aiming is normally accomplished by trial-and-error in a full scale facility. This facility is typically an atmospheric pressure facility which contains no provisions for steam.

It has been recently observed [8] that spray nozzles behave differently in air than in a pressurized steam environment; moreover, their spray distribution is also effected by steam up-draft [9]. The presence of steam and elevated pressures may reduce the spray cone angle, and thus distort the reference distribution measured in air.

This phenomena is qualitatively well understood [8]. As shown schematically in Figure 16 the high velocity liquid drops entrain the surrounding vapor by a momentum exchange mechanism. This pumping action causes the vapor to move into the liquid spray jet. As the vapor moves into the spray, it drags the outer droplets in towards the centerline of the cone, thus reducing the cone angle. Higher system pressures increase the gas density and thus the drag force ($1/2\ c_d \rho_g v_g{}^2$) increases proportionately. The effect of a steam environment is even more pronounced, since, in this case, the steam rushes toward the spray jet at a high velocity to condense on the cold droplets, thus shrinking the cone angle.

Not all spray nozzles perform the same. Very small droplets from hollow cone (VNC) and atomizing nozzles are normally strongly effected, while coarse droplets are not. Since there are many different nozzle types which have been used in the ECC system of current generation BWR's the actual performance of a given spray header requires steam testing.

The potential safety concern associated with this situation is due to the fact that the spray headers have been sized in an atmospheric air facility to provide coolant to each fuel bundle. In the reactors, these systems will only be needed if a LOCA event occurs, at which time they will be in a pressurized steam environment. Evaluating the actual performance of these systems is exceedingly complicated. The individual nozzles are spaced around the ring header such that there is strong interaction of the sprays from the various

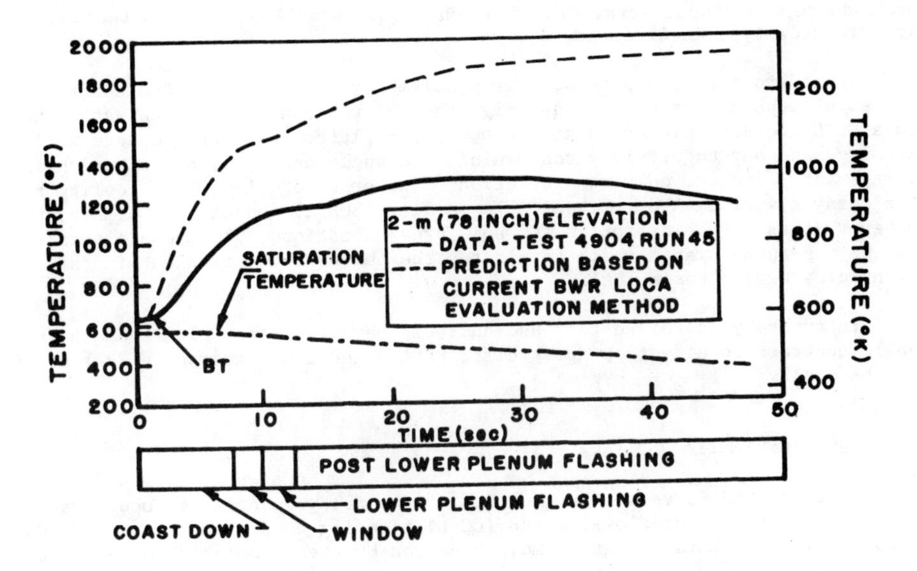

Fig. 15 Predicted and Measured Rod Temperatures

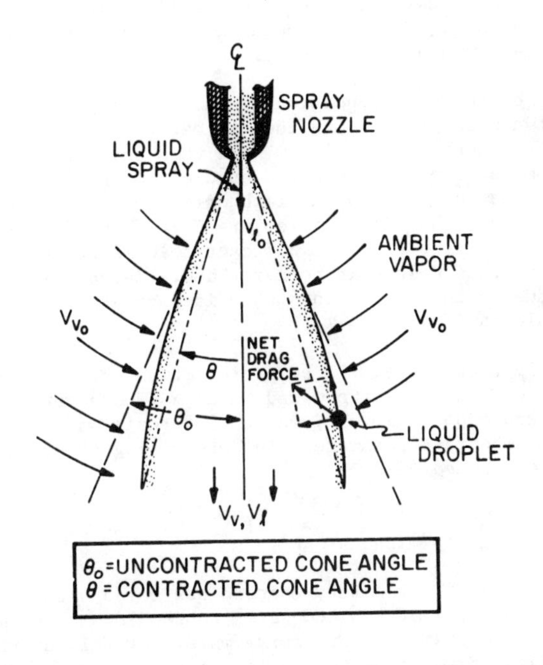

Fig. 16 Schematic of Spray Cone Angle Contraction Mechanisms

nozzles. Moreover, as shown schematically in Figure 17, for the in-reactor application, a vortex may be formed in the upper plenum which could also affect the spray distribution. Thus, evaluation of what a local cone angle reduction will do to the global spray distribution requires multiple nozzle testing in in a reactor type environment.

In modern jet pump plants, this problem may only be of academic interest. If, as will be discussed subsequently, the upper plenum fills with water due to a CCFL flooding situation at the upper tie plate, the spray headers will submerge and any uncertainty concerning cone angle and spray distribution is irrelevant. Indeed, only consideration of upper bundle long-term-cooling by the spray system may be relevant. In earlier (BWR/2) nonjet pump plants, however, this may be a problem for some break locations. In these plants, it is the top spray system which terminates the temperature transient in the event of a design basis LOCA.

Fortunately, detailed test and analysis have indicated [10] that the cone angle contraction effect on spray distribution and the resultant effect of PCT, may be small.

2.8 BWR CCFL Reflood Delays

Another ECCS issue is the question of the delay in core reflood caused by the counter-current-flow-limited (CCFL) flooding condition that occurs during ECCS operation. This situation may be caused by steam coming through the upper tie plate (top of the fuel rod bundle) at such a high velocity that the liquid in the upper plenum cannot easily run down into the core. This steam is produced in the lower plenum due to the depressurization process (i.e., flashing) and the sensible heat stored in the vessel walls. In addition, there is evaporation in the rod bundles of the residual water and the spray water that is able to penetrate down into the bundle.

The effect of the CCFL condition is that the spray water, which enters the upper plenum at a much higher rate that it can run down into the bundles, must either accumulate in the upper plenum (and thus submerge the top spray headers), or shunt down some of the lower powered bundles to the lower plenum. Since all testing to date has been with a single bundle, current licensing calculations conservatively assume that all bundles are CCFL flooded and shunting does not occur. Naturally this leads to significant calculated delays in refilling the lower plenum, and thus delays in core reflood. In addition, the convective heat transfer due to the updrafting steam is currently conservatively ignored in licensing calculations.

It is likely that there will be preferential shunting of the spray water down through some of the lower powered bundles. Moreover, it can be anticipated that relative bundle power levels, inlet flow resistance and spray water subcooling will be important parameters in determining parallel channel phenomena during the reflood phase of a DBA.

2.9 BWR "Steam Binding" Issues

One recent area of BWR safety concern has to do with the possibility that there may be no shunting of the ECC water to the lower plenum. In this situation it can be speculated that the core spray water may fill up the upper plenum and the steam separator standpipes to the top of the steam separators. As the lower plenum gradually fills up with water to the bottom of the jet pump diffuser

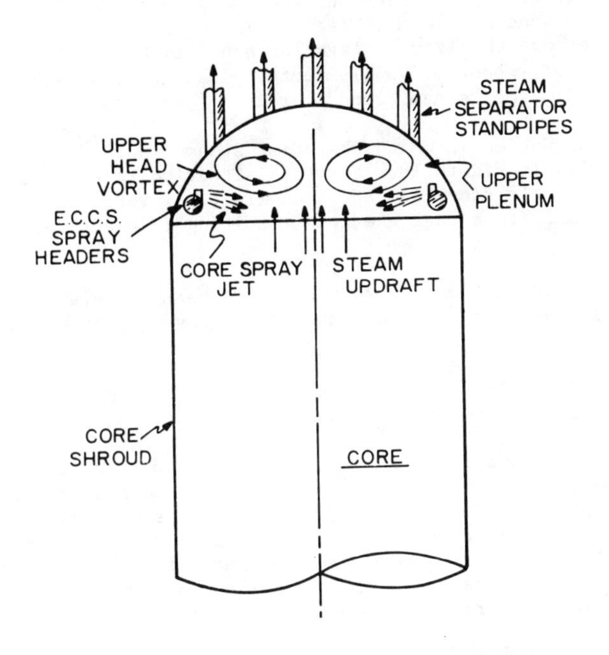

Fig. 17 Upper Plenum Vortex Structure

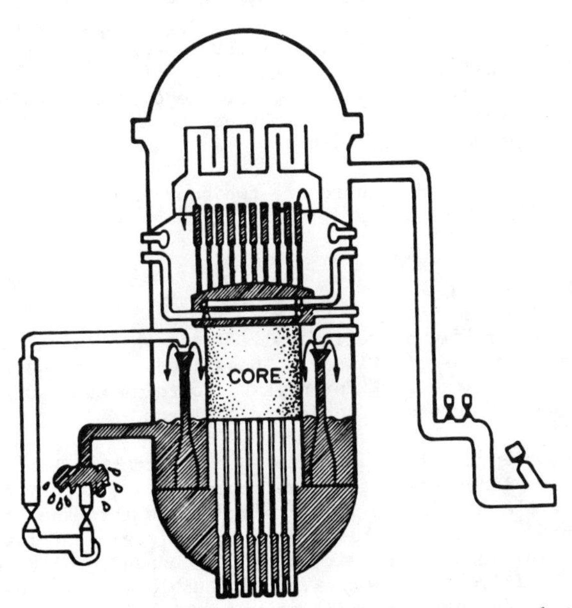

Fig. 18 "Steam Binding" in a BWR During a Hypothetical LOCA Event

tail pipes, the water will seek the path of least hydraulic resistance to the break. One path is through the jet pumps; the pressure head here is due to a liquid elevation head of 2.44m (8 feet). In contrast, to flow up through the core, it must overcome the liquid elevation head of the upper plenum and steam separator standpipes, about 5.79m (19 feet), plus the 3.66m (12 foot) elevation head and pressure in the "steam bound" core.

A possible situation in which there is "steam binding" in the core is shown schematically in Figure 18. For this situation to persist, there would have to be considerable steam and spray cooling; however, there is a potential for the reflood water to bypass the core resulting in a net reduction in core cooling.

Having now discussed some current BWR LOCA heat transfer concerns and conservatisms, let us consider some important pressure suppression containment issues.

2.10 BWR Containment System Safety Issues

Modern BWR containment systems are of the pressure suppression type, in which a large pool of water is used to condense the LOCA steam, thus keeping the containment pressure levels low. Three basic systems have been used by General Electric. As indicated schematically in Figure 19, Mark I, II, and III type designs all work on the same principle. Mark III, which is the system currently offered by General Electric, is unique in that personnel access to the wet well region is permitted during normal plant operation.

In pressure suppression containment systems, there are two main questions of concern. First, do these systems quench all the steam, thus preventing excessive wet well pressures? Secondly, are there excessive loads during system operation? The answer to the first question is based on a rather extensive large scale test program conducted by G.E. which showed that very little steam bypass occurs during system operation. Apparently the only possibility for substantial steam bypassing of the pressure suppression pool is when there is a seismic event causing wave action which is sufficient to periodically uncover some of the steam vents in the pool. Model tests conducted by SWRI have indicated that seismic vent uncovering may not be a problem, and, if necessary, can be easily controlled with baffling.

On the question of loads, there are two sources which must be considered: relief valve loads and LOCA loads.

2.10.1 BWR Relief Valve Loads

There are two significant relief valve loads which have been observed: clearing loads and condensation loads. These loads may be quite strong. In fact, one of the early Kraftwerk Union (KWU) containments (Wurgassen, Germany) was damaged because of excessive condensation loads.

Relief valve clearing loads are due to the fact that the air, which was in the relief valve lines prior to valve actuation, is highly compressed by the high pressure steam. As the liquid plug clears into the pressure suppression pool, the high pressure air "bubble" moves into a region of low, hydrostatic only, pressure. This causes the air to rapidly expand, which loads the containment walls in a positive direction. The air bubble overexpands in the pool and, as it contracts, sends out a rarefaction wave which loads the containment walls

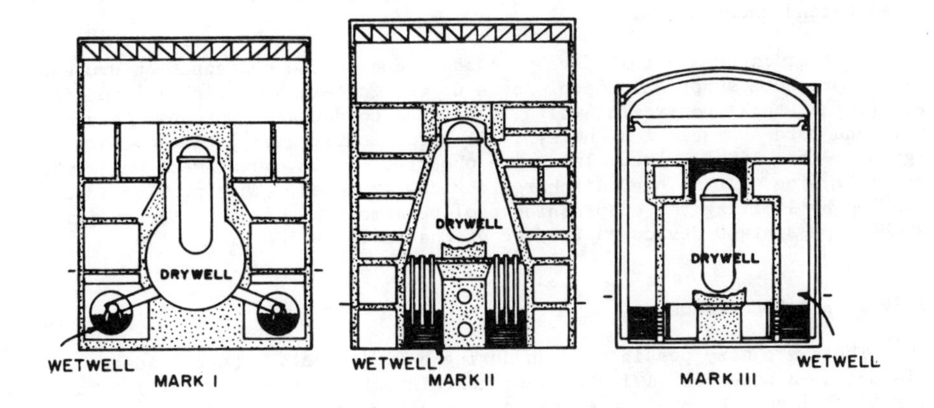

Fig. 19 Pressure Suppression Containment Systems

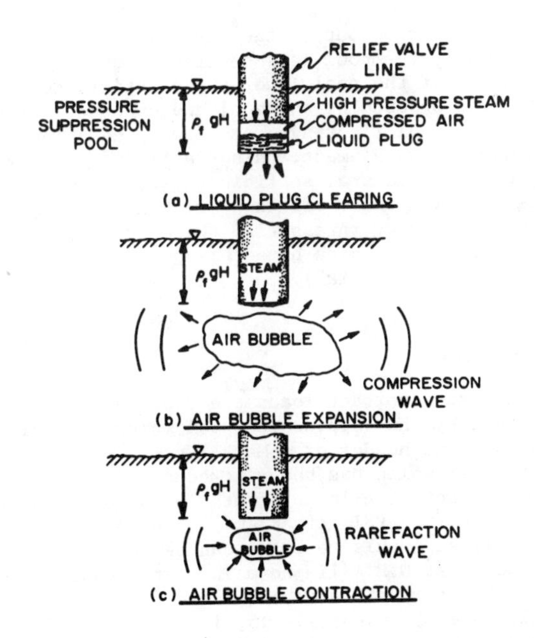

Fig. 20 Relief Valve Clearing Phenomena

in a negative direction. This sequence of events is shown shcematically in Figure 20. The mixed polarity loading, discussed above, can do, and has done, damage to containment walls and internal baffling. Current BWR design practice is to install a KWU-type sparger devices at the end of the relief valve lines to attenuate the clearing loads in containment systems which are not designed to withstand these loads.

Another important relief valve load is due to the condensation process. As the pressure suppression pool heats up as, for example, might happen in the event of a stuck-open relief valve, the steam condensation process becomes more and more sporadic and "noisy". At sufficiently high pool temperatures (greater than 160°F), large loads are observed on the containment walls when straight pipe or rams head discharge devices are used. This problem is currently handled by limiting the suppression pool temperature and/or by the addition of a KWU-type sparger device on the end of the relief valve pipe.

2.10.2 BWR LOCA Loads

There are many possible loads during a LOCA event. These loads are discussed in some detail in relevant containment documentation (11). The most important loads, however, are the pool acceleration loads (i.e., vent clearing loads) and the impact loads.

2.10.2.1 BWR Vent Clearing Loads

As the air which is purged out of the dry well region is rapidly introduced into the suppression pool, the pool is accelerated upwards. A reaction force is imposed on the bottom of the pool in order to satisfy conservation of linear momentum. This force is opposed by an equal and opposite force from the structure supporting the pool flooring. As shown shcematically in Figure 21, in the case of the Mark I containment, this reaction force may cause an unsecured torus to "jump" up off the pedestal sufficiently to damage the vent pipes connecting the dry well and wet well. In this unlikely event, steam can bypass the suppression pool and may overpressurize the containment building. Current practice is to retrofit plants which have this potential problem with a hardware fix (e.g., "Vermont Yankee Fix").

2.10.2.2 BWR Impact Loads

The largest and most important loads are those due to the impacting of a containment structure by the suppression pool which is moving rapidly upwards. The impact load on the ring header of a Mark-I containment is shown schematically in Figure 21. If the air which has been introduced into the pool breaks through the surface prior to impact, the loads (froth loads) are small, fairly long duration impulsive loads ($\sim$15 PSI, $\Delta t > 1$ ms). In contrast, if liquid impacts a structure, the localized impulsive loads can be quite high (200-600 PSI, $\Delta t < 1$ ms). Experiments at G.E. [11] have indicated that a good-rule-of-thumb is that "breakthrough" occurs at between 1.5 and 2 times the initial vent submergence in Mark-III (somewhat less, 1-1.5, in Mark I and II). Thus if a structure is greater than twice the initial vent submergence above the pool free surface, LOCA impact loads will be small. If it is closer to the pool surface, these loads may be much higher, and the structure must be designed to withstand them.

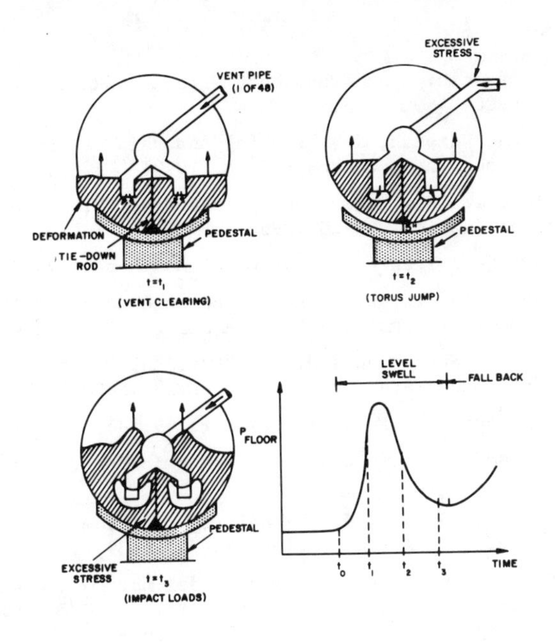

Fig. 21 Mark I Inertial Loads

3. SUMMARY AND CONCLUSION

An overview of the thermal-hydraulic considerations of some of the most important current LWR safety issues has been given in this chapter. It has been shown that LWRs are designed to accommodate a whole host of events and accidents. Unfortunately, as was clearly demonstrated during the incident at TMI-2, nothing is fool-proof and unanticipated situations can occur which may lead to core damage.

Fortunately the "defense-in-depth" design philosophy of LWRs is such that even severe incidents can be accommodated without affecting the health and welfare of the general public. Nevertheless, it is clear that if LWRs are to be a viable energy alternative, we must strive to increase our understanding of reactor thermal-hydraulic phenomena so that LWR safety can be improved and/or verified. It is hoped that the material presented in this chapter will contribute to this goal.

ACKNOWLEDGMENTS

The permission of Drs. Y. Y. Hsu and H. Sullivan (USNRC) to use some of the material that they had previously published [12] on PWR safety is gratefully acknowledged.

REFERENCES

[1] USNRC Staff, "Anticipated Transients Without SCRAM for Light Water Reactors", NUREG-0460, 1978.

[2] Tong, L. S. and Weisman, J., "Thermal Analysis of Pressurized Water Reactors", ANS Monograph, Second Editon, 1979.

[3] "General Electric LWR Thermal Analysis Basis (GETAB); Data, Correlation and Design Application", NEDO-10958, 1973.

[4] "BWR Blwodown Heat Transfer Program - Final Report", GEAP-21214, 1976.

[5] "Proposed ANS Standard, Decay Energy Release Rates Following Shutdown of Uranium Fueled Thermal Reactor", ANS-5 Sub-Committee, 1971.

[6] Sozzi, G. L. and Sutherland, W. A., "Critical Flow Measurements of Saturated and Subcooled Water at High Pressure", NEDO-13418, 1975.

[7] "10CFR50, Appendix-K", Federal Register, 39, No. 3, 1974.

[8] Block, J. A. and Rothe, P. H., "Aerodynamic Behavior of Liquid Sprays", Creare TN-206, 1975.

[9] Sandoz, S. A. and Sun, K. H., "Modelling Environmental Effects on Nozzle Spray Distribution", ASME Preprint, 76-WA/FE-39, 1976.

[10] "General Electric Company Analytical Model for LOCA Analysis in Accordance with 10CFR50 Appendix K, Amendment No. 3 - Effect of Steam Environment on BWR Spray Distribution", NEDO-20566-3, 1977.

[11] James, A. J. et al, "Information Report - Mark III Containment Dynamic Loading Conditions", NEDO-11314-08, 1975. (Also, see Gessar, Appendix 3B, Part-f).

[12] "Hsu, Y. Y., and Sullivan, H., "Thermal-Hydraulic Aspects of PWR Safety Research", ASME Symposium: The Thermal and Hydraulic Aspects of Nuclear Reactor Safety, Volume-1: LWRs, 1976.

5 TRANSIENT RESPONSE OF LIQUID METAL FAST BREEDER REACTORS

J.I. Sackett,* G.H. Golden, R.R. Smith
H.K. Fauske
ANL, Argonne, IL, USA

The purpose of this chapter is to provide an overview of the thermal-hydraulic considerations associated with liquid metal fast breeder reactor (LMFBR) safety issues. Similar to that presented in the previous chapter for LWR's, this chapter serves as an introduction to chapters which follow giving much more detail.

1. INTRODUCTION

One of the most important lessons from the TMI-2 accident is that relatively innocuous plant upsets may, through subsequent faults and operator error, cascade into serious reactor accidents. As a result, increased attention is being given to lower-level anticipated upsets that, if ignored, could lead to more serious consequences. This paper considers the response of LMFBRs to anticipated upsets.

Clearly desirable are systems that self-protect under emergency conditions Such systems should be able to tolerate anticipated malfunctions and operator indifference in such a way that any ensuing damage would be limited to conventional replaceable components outside the nuclear zone. Such systems should be so designed that component malfunction compounded by operator error cannot, under any circumstances, lead to fuel damage. LMFBRs offer the best opportunity for achieving this goal.

Anticipated upsets for the purposes of this paper, have been classified as follows:

A. Loss of electric power (and flow coastdown)

B. Inadvertent reactivity insertion

C. Local damage within a subassembly

D. Loss of heat removal from the secondary sodium or steam systems.

Each of these is considered, herein, with an eye toward inherent response of an LMFBR to determine whether reliance on active protective systems, and operator action, can be minimized.

The most likely event is loss of electric power leading to coastdown of the primary-coolant pumps. It, therefore, requires the most attention in plant design and has historically represented the path by which the occurrence of core disruption has been postulated (the LOF driven TOP). However, more recent analysis suggest that even if one neglects the many active mechanical features intended to protect against loss of flow (such as emergency power, pony motors

*Present Address: Fauske and Associates, Incorporated, 627 Executive Drive, Willowbrook, Illinois 60521.

and fast-acting shutdown systems) for some of the current designs the reactor apparently can still be protected by inherent reactivity feedback and natural convection cooling. This point is discussed in more detail in Section 2.

Inadvertent reactivity insertion is much less likely to occur than loss of flow, requiring specific operation error or multiple mechanical failures. Also, reactivity insertion rates possible for control-rod motion are extremely slow and controllable by design. Rapid reactivity insertion is possible from core motion during seismic events, but proper core design can limit its magnitude to values well within the capability of protection system response.

Protection against inadvertent reactivity insertion requires emphasis on mechanical reliability of shutdown systems. In order to provide "inherent" capability, attention is being given to self-actuated shutdown systems, directly triggered by high temperatures in the core and requiring no out-of-reactor mechanisms. Attention is also being given to possible annular fuel pin designs that will provide for fuel melting and movement away from the core centerline without breach of cladding. These are discussed in Section 3.

Local damage within a subassembly is an event which gained wide attention with the flow blockage at the FERMI reactor. After much consideration and modification in subassembly designs, this issue has been resolved. Fuel elements, if they do deteriorate, will do so only slowly and may be reliably detected. Oxide fuel has been found to react with the sodium once breached, limiting its useful lifetime, but represents only an operational difficulty. The existing question is the extent to which present fuel-element designs can remain in the core once the cladding is breached. These issues are discussed in Section 4.

The response of an LMFBR to loss of heat removal from the secondary sodium or steam systems is both predictable and benign. Here, the advantages of low pressure of the sodium coolant, the large margin to sodium boiling, and reliable cooling of the core by natural convection of sodium provide inherent protection. In the same way as for the loss-of-flow event discussed earlier, it may not even be necessary for active fast-shutdown systems to operate in order to provide protection. As the primary sodium heats up, the reactor will shut itself down from reactivity feedback. Subsequent rejection of decay heat can be accomplished with systems utilizing natural convective cooling. These issues are discussed more fully in Section 5.

An interesting conclusion, which relates to these points, was reached by the International Nuclear Fuel Cycle Evaluation (INFCE) working group 5 (as reported in Ref. [1]). They concluded that *Risks associated with fast breeders amount to only a few percent of the total hazard of the LWR system.* According to the INFCE group, use of sodium *has, in fact, turned out to be a new progressive solution, enhancing heat transfer at low perssures; the chemical interaction of water and air with sodium and sodium activation in a core are problems which have been solved during the development and operation of fast reactors.* Further, while the reactivity margin is indeed smaller than for LWRs, and that causes concern over 'prompt supercriticality'; in a larger power reactor, *the Doppler effect prevents such accidents from happening, to a considerable extent.*

2. RESPONSE TO TOTAL LOSS OF ELECTRIC POWER AND FLOW COASTDOWN

2.1 Background

Loss of power to the primary pumps is the most likely event that will result in abnormally high temperatures in an LMFBR. Therefore, in the design of an

LMFBR power plant, great attention is paid to providing: (1) diversity and redundancy of reactor trip for low flow conditions; (2) an independent source of forced circulation at a minimally acceptable flow rate; and (3) inherent natural circulation cooling of the shutdown plant. Reactor trip signals can come from indications of low voltage across the primary pumps, low flow rate through primary system flowmeters, high reactor outlet temperature, etc. Independent forced flow capability is needed primarily to provide an initial backup to the primary pumps which must fail prior to reliance upon natural circulation cooling. Ultimate rejection of heat is through all or part of the regular cooling system, or through a backup shutdown heat removal system.

Licensing of a commercial LMFBR power plant would be greatly facilitated if it could be shown to survive a loss of primary pumping power without scram, a highly unlikely fault condition. Recently reported analyses indicate that SUPER PHENIX could survive such an extreme condition without any sodium boiling.[2,3] Preliminary analysis for both the Prototype Fast Reactor (PFR) and EBR-II suggests the same result. The nonboiling result at SUPER PHENIX is due to a strong negative net reactivity feedback, two major components of which result from restrained radial expansion of subassemblies and differential expansion of control rod drives causing some movement of the rods into the core. Sufficiently high temperatures would be reached, however, to cause structural damage in the primary system if they were not reduced within a certain period of time -- about 20 min for the postulated conditions in SUPER PHENIX.[4] To the extent that this nonboiling effect can be demonstrated, it converts the unprotected loss-of-flow event from an issue of public safety to one of plant availability.

The thermal hydrualic effects of loss of primary pumping power are being carefully studied by every country that has an LMFBR program. Of particular importance in these studies is the characterization of both the transition from forced to natural circulation and the equilibrium natural circulation. The latter may depend to a great extent on the transient behavior of the secondary flow rate. Although these effects are strongly dependent upon the specific features of a given plant, the fundamental phenomena involved are common to all LMFBR plants. Therefore, considerable effort has been spent at EBR-II, a small but complete LMFBR power plant, on running tests and modeling loss-of-flow events. Discussed below are results from convective flow testing and analysis at EBR-II, FFTF, and PFR.

2.2 EBR-II Work

EBR-II is a pool-type LMFBR plant that operates at a nominal power of 62.5 MWt. Figure 1 is a simplified flow schematic of the EBR-II plant. In the primary system two centrifugal pumps in parallel provide cooling for the core and an auxiliary electromagnetic pump having its own power supply, located on the outlet pipe and provides about 5% of rated flow when the main pumps are off. The rated reactor inlet (primary tank) temperature is 371°C and the reactor ΔT = 102°C. The intermediate heat exchanger (IHX) transfers heat from the primary to the secondary system, and eight steam generators in parallel and two superheaters in parallel transfer heat from the secondary to the steam system.

The EBR-II plant contains the normal types of temperature, pressure, and flow rate sensors. In addition, for each of the loss-of-flow tests discussed here, a special instrumented driver fuel subassembly (XX08) was placed in the reactor core. This subassembly contains 58 driver fuel elements, two flowmeters, and 22 fast-response thermocouples.

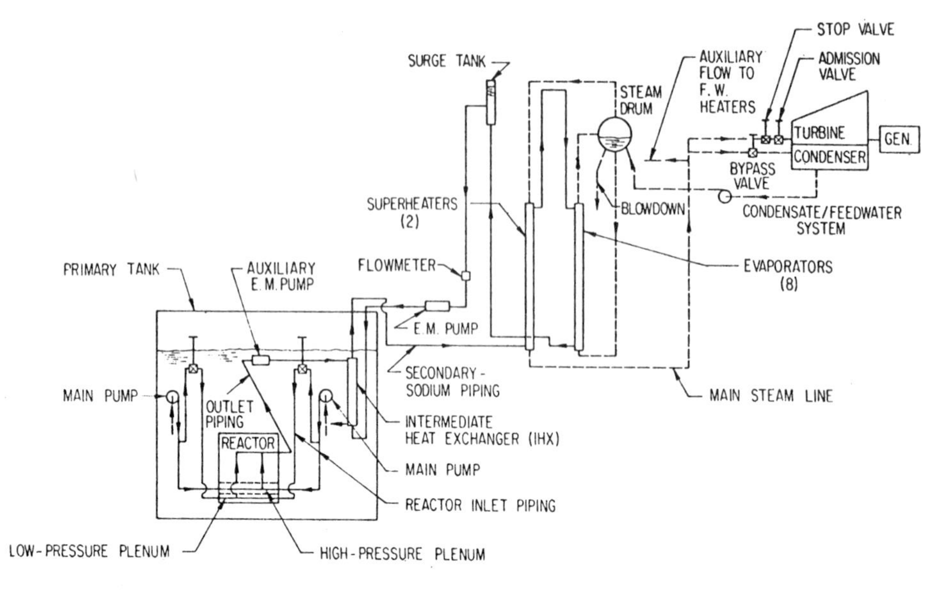

Fig. 1. Schematic of EBR-II Plant.

In one loss-of-forced flow test, the primary tank temperature was reduced to 351°C, the reactor was brought to 27.4% of rated power at 29.2% of rated flow, the auxiliary pump was turned off and the secondary pump trip bypassed. After 192 min of operation the primary pumps were turned off. Taking this point as t = 0, the reactor scrammed automatically at 1.8 s on a low flow signal, pump number 1 dropped to zero speed at 36.5 s; and pump number 2 went to zero speed at 42.5 s. Power to the secondary and feedwater pumps was maintained throughout this test.

Measured results and corresponding NATDEMO[4] predictions for the reactor and XX08 flow rates are shown in Figs. 2 and 3.[5] Measured and calculated coolant temperature at the top of the core in a driver subassembly are compared in Fig. 4. In all cases, agreement between measured and predicted results is good.

A second similar test was conducted, except that the initial power and flow rate were each 35% of their nominal values, and the secondary pump was tripped 17 s after the primary pumps were tripped. Comparisons of measured and calculated results showed good agreement and are shown in Figs. 5 and 6.[4]

On January 10, 1978, EBR-II was involved in a loss of site power while operating at rated power and flow. This resulted in concurrent scram, loss of primary, secondary, and feedwater flow, and trip-out of the turbine. The primary auxiliary pump remained on. Key plant data were automatically logged from about 20 s prior to the transient through the course of it. NATDEMO was again used to predict the response of the plant to this upset condition. Representative measured and calculated results are shown in Figs. 7 and 8; agreement is good.[6]

These results have been encouraging in that they have confirmed the adequacy of convective flow for post-transient cooling and are providing the data by which major safety codes can be updated, eventually allowing reliable extrapolation to

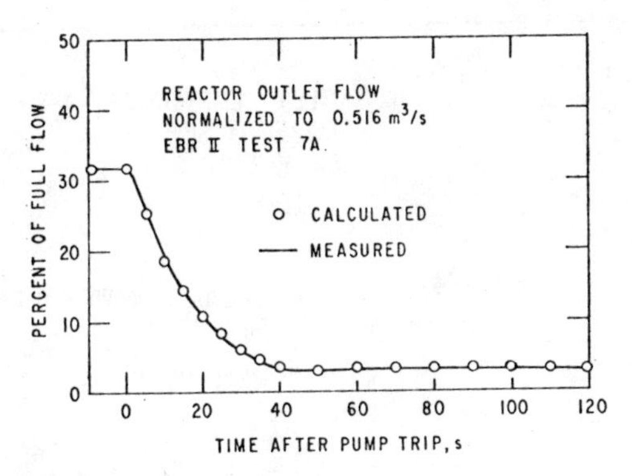

Fig. 2. Measured and Calculated Coolant Flow Rate at the Reactor Outlet (Total Primary Flow Rate).

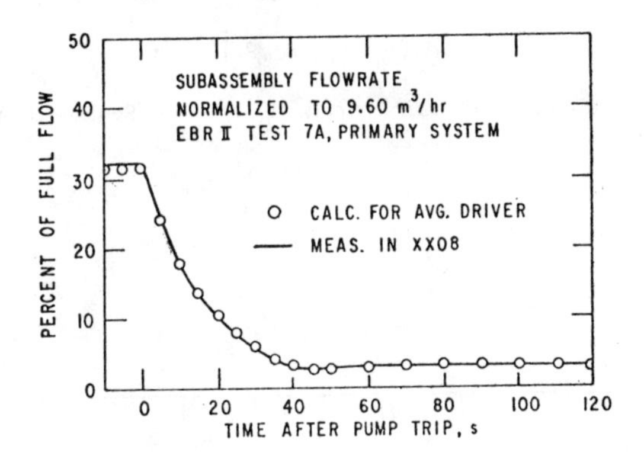

Fig. 3. Measured and Calculated Coolant Flow Rate in Instrumented Driver Subassembly XX08.

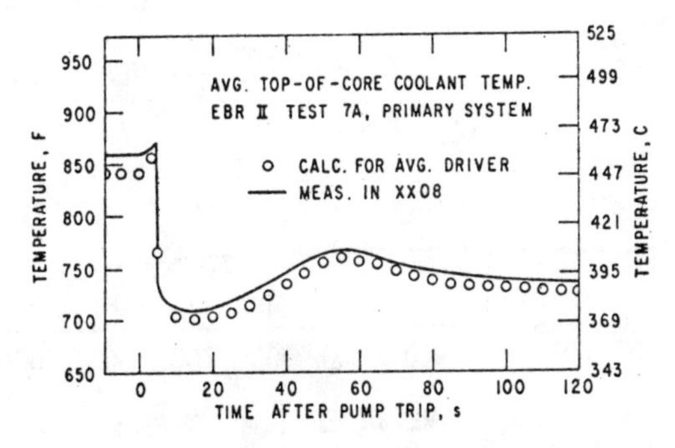

Fig. 4. Measured and Calculated Temperature at Core Top in Driver Subassemblies.

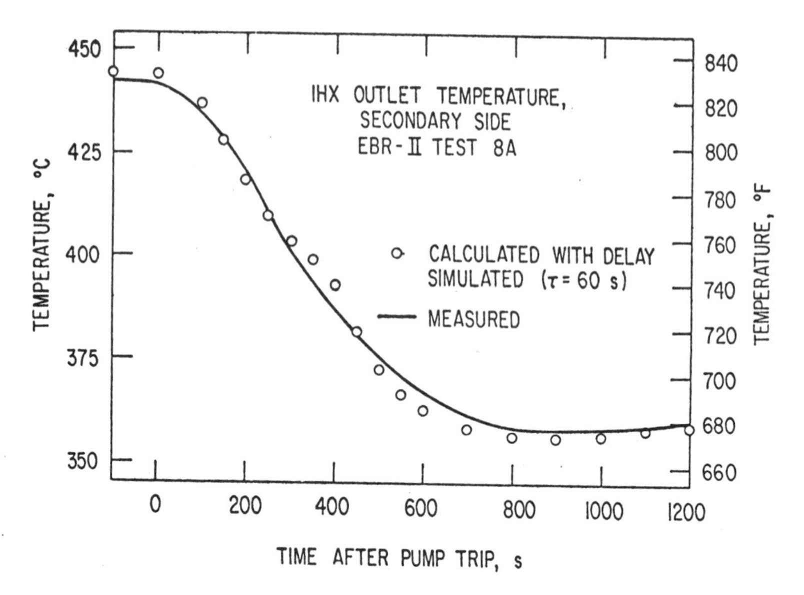

Fig. 5. Measured and Calculated IHX Outlet Sodium Temperature on Secondary Side.

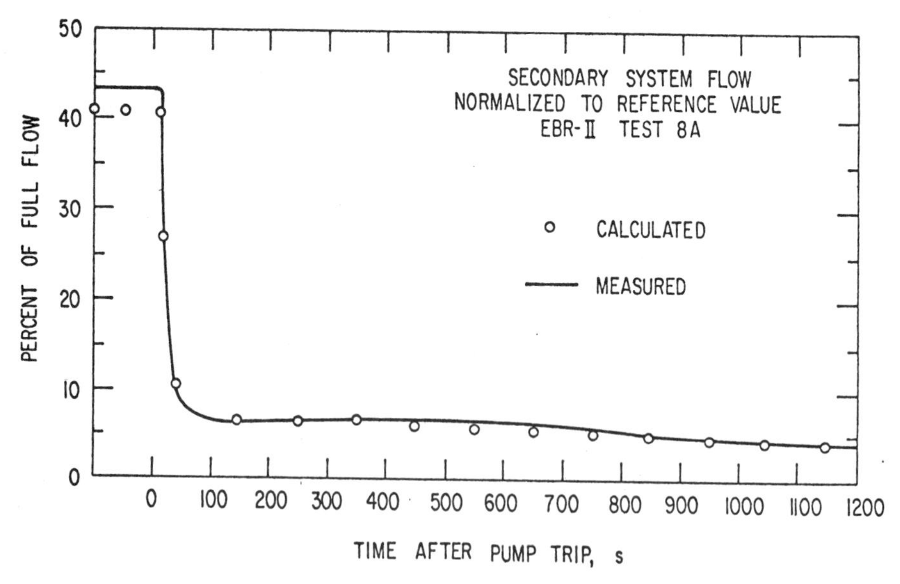

Fig. 6. Measured and Calculated Response of Secondary Sodium Flow Rate.

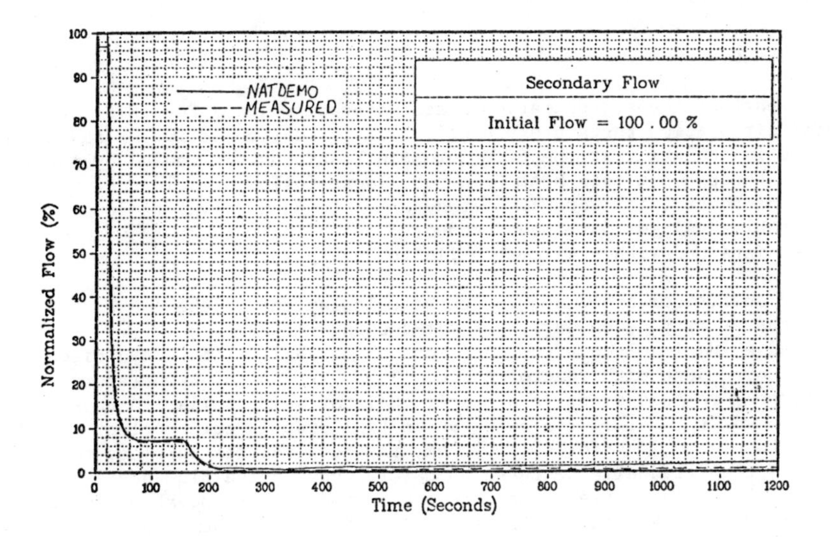

Fig. 7. Measured and NATDEMO-calculated Secondary-sodium Flow.

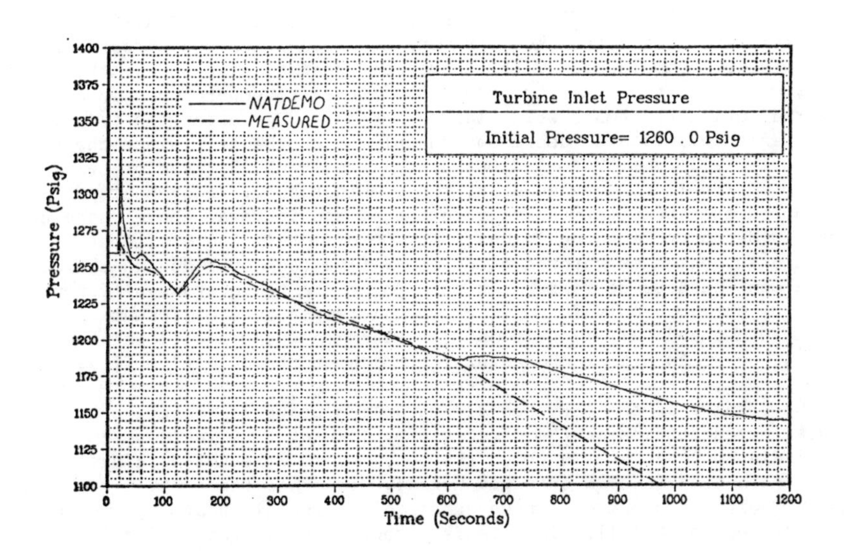

Fig. 8. Measured and NATDEMO-calculated Turbine Inlet Pressure. Conversion factor: 1 psi = 6.895 kPa.

the case of most interest, i.e., loss of flow with failure to scram for existing and proposed LMFBR designs.

2.3 FFTF Work

As part of the pre-operational test program planned for FFTF, certain natural circulation tests are planned. The primary objective of these tests is to directly confirm that the FFTF core can be adequately cooled by natural circulation through the main heat transport loops following a loss of on-site electrical power (with scram). Although the tests have yet to be conducted, considerable analyses have been conducted which indicate that core temperatures can be maintained within allowable limits during such events.

The analytical tool used for these calculations is the IANUS code[7] which models the primary, secondary, and air-cooled heat dumps of FFTF. The reference case studied[8] involved a complete loss of capability of forced circulation cooling coincident with reactor scram. The core coolant temperatures and flow rate respond to this event in a manner qualitatively similar to those predicted and measured in EBR-II. Following an initial drop in core coolant temperatures due to the overcooling following scram, the temperatures increase to a peak value and then gradually decrease as natural convective flow develops. The peak hot channel coolant outlet temperature was predicted to be about 800°C, occurring 110 s after the start of the transient and 20 s after the complete stoppage of the pumps. The minimum reactor flow rate was determined to be about 2.7% of the initial value.

As a result of these calculations, along with extensive supportive work establishing probable margins, modeling or parameter uncertainties, etc., considerable confidence has been generated in the capability of the main heat transport loops in FFTF to provide adequate natural convective decay heat removal. Final confirmation will be obtained from the start-up plant testing program.

2.4 PFR Work

A number of loss-of-flow tests have been conducted at PFR.[9] The most meaningful tests conducted thus far were those in which the effects of a site-power loss were simulated. With the reactor operating at 50 MWt the pumps and reactor were simultaneously tripped. Coolant exit temperatures were carefully monitored during the tests which lasted for 350-600 s after the trips. As expected, exit temperatures decreased sharply after the trip since the power level immediately fell from 50 to 1-2 MWt as the result of the reactor trip. However, as the pump impellers coasted down, the combination of decay heat and decreased flow caused exit temperatures to turn around and increase above those under normal 50 MWt operating conditions. At approximately 350 s after the trip exit temperatures peaked as natural circulatory flow took over and as the decay heating continued to decrease. At this point the temperature differential across the core was only 70°F. (The normal temperature differential across the core at 50 MWt and full flow is 30°F.) Plans call for additional tests at higher levels of power to better characterize the onset of convective flow.

The results of the PFR thermal-hydraulics program prompted a series of computer-based studies on the effects of a loss-of-flow incident in which the reactor did not trip and in which the pony motors were not working.[10] The PFR is characterized by a strongly negative power coefficient. This means that under LOF without scram conditions the coolant rapidly heats up and, as it does, the power begins to fall. This tends strongly to mitigate the effects of the loss of flow.

They concluded that sodium boiling can be avoided for a loss of pumping power (without reactor scram) at the beginning of a reactor run, but may occur if the loss of flow were to occur near the end of a run (with higher decay heat levels). In the latter case, sodium boiling would occur near the top of the core where the sodium void reactivity coefficient is negative, and would have the initial effect of further reducing reactor power. Whether boiling would progress toward the core center depends upon a number of factors, including the effect on coolant flow rate through the core. It is an area of investigation to be pursued further (for example, boiling would augment natural convection but may retard "residual" forced flow with pump coastdown because of increased flow impedance). Tests conducted at DFR have established that mixed oxide elements can survive boiling (particularly at the low heat-flux conditions at the time of boiling), so the issue becomes one of determining whether boiling could progress to the point where net positive reactivity insertion could occur. If not, then the PFR can be shown to carry considerable inherent capabilities to survive a loss-of-flow event without scram.

The information generated by these and other tests (and analyses) should allow design principles to be developed by which inherent protection against loss-of-flow events is provided.

3. INADVERTENT REACTIVITY INSERTION

During a reactivity-induced transient, the temperature of fuel, cladding, and coolant change throughout the reactor. As the power level increases, these temperatures also increase. The rate at which reactor power increases is dependent on the rate of reactivity insertion, initial power level and the feedback reactivity of the system being considered. When a reactivity-related trip parameter exceeds a predetermined level and/or rate, the transient is terminated when the shutdown system operates.

In designing a shutdown system, the first task is to identify individual reactor faults, classify them according to probability of occurrence, and design a system to accommodate each fault within limits appropriate to the probability of the event (at all times ensuring safety). An example of that classification is given in Table 1. It may be seen that as a general rule, the most common events are the less severe; they are of most interest for fatigue analysis of plant components. Generally, only the unlikely events challenge the shutdown system design.

Fuel-pin design requirements and transient acceptance criteria for the shutdown system at the FFTF are shown in Table 2 (from Ref. [12]). It can be seen from this table that as the severity of the event increases and the probability decreases, the allowable damage (acceptance criteria) becomes less restrictive. Categorizing events in this way permits a logical shutdown system design to be developed.

Once a shutdown system has been designed, it is of interest to analyze its response to a wide range of reactivity insertion rates. Generally, these analyses demonstrate significant capability beyond that actually required. For example, analysis of the FFTF shutdown system established its capability to protect against reactivity insertion rates of up to $3.0/s, well above that possible from control rod motion alone. Similar results were obtained from analysis of the EBR-II as discussed briefly below.

Table 1. Examples of Duty Cycle and Off-normal Events Assumed for Design Studies

	Transient	Number of Events Over Plant Lifetime
Normal Operation	Startup from refueling temperature	223
	Loop out-of-service (N-1 operation)	80
	Stepload increase or decrease of 10% of full power	1000 up 1000 down
Anticipated Events (upset)	Uncontrolled rod insertion	10
	Uncontrolled rod withdrawal	10
	Reactor startup with an excessive step power change	40
	Loss of power to one primary pump	5 per loop
	Four loop natural circulation	5
Unlikely Events (emergency)	Primary pumps mechanical failure	4
	Uncontrolled rod withdrawal from 100% power	1 cycle
	Uncontrolled rod withdrawal from startup to trip point with delayed manual trip	1 cycle

Table 2 (from Ref. [12])

Fuel Pin Design Requirement and Transient Acceptance Criteria

Category	Type Event	Requirement	Examples	Acceptance Criteria*
Normal Operation	Steady state	Maintain cladding integrity to design lifetime	Steady state operation	0.2% cladding strain
	Operational transients		Startup and shutdown	
Upset	Operational incident	Maintain cladding integrity to design lifetime	Scram; loss of power to one pump; reactivity insertion ∿3¢/s	0.3% cladding strain
Emergency	Minor incident	Maintain cladding integrity	Loss of all electrical power; reactivity insertion ∿10¢/s; continuous flow reduction	0.7% cladding strain
Faulted	Major incident	Maintain coolable geometry	Design earthquake Reactivity insertion <3$/s	

*Total strain.

3.1 Reactor TOP Analysis

For purposes of comparison, the response of EBR-II and FFTF to insertion rates associated with control rod motion at full reactor power are given in Table 3. below.

Table 3. Comparison of Power Change Associated with Reactivity Insertion for EBR-II and FFTF

Reactivity Insertion Rate (¢/s)	Total Power Change (%)	Average Rate-of-Change in Reactor Power	
		FFTF (%/s)	EBR-II (%/s)
3	25	3.0	4.2
10	25	11.0	12.5

EBR-II is seen to respond slightly faster. It also has a more restrictive shutdown criteria because of the formation of a clad-fuel eutectic at 715°C. It is, therefore, appropriate to consider the shutdown response at EBR-II as a limiting case.

Reactivity Insertion at Startup. Calculations were performed for a typical core at EBR-II with the nominal-feedback-reactivity model to establish response of the system as a function of reactivity-insertion rate at startup.[13] Of particular interest was comparing the time from initiation of a transient to that for a given reactor period, a given power level, and peak temperature of the inside surface of driver-fuel cladding. The calculations were made for unbounded linear ramp insertion rates* of 0.001-10 ¢/s. Figure 9 plots time to a 17 s reactor period, to 115% power level peak and to driver fuel inside-cladding temperatures of 715 and 816°C for ramp rates in that range. As expected, the first parameter reached is a 17 s period, and the margin between that parameter and the temperature limits is quite large. For ramp rates in excess of about 0.05 $/s, a 17-s period is reached within a few milliseconds. The margin between 115% power and the temperature limits is relatively small, although there is adequate margin to accomplish reactor trip over the range of reactivity insertion rates up to about 1.5 $/s. This rate is well above that possible with control rod insertion alone. Step insertions of up to $1.0 can also be tolerated

Subassembly-outlet-temperature trips, because of their slower inherent response, are capable of effective protection only for low insertion rates at startup.

Reactivity Insertion at Power. The first difference to note in comparing response of a reactor at power to one at startup is that, below ramp rates of about 0.03 $/s, a trip on reactor period is *not* reached before inside-cladding temperature limits are reached (see Fig. 10). This results because the strong reactivity feedback at power limits the rate of power rise. Short period is therefore not an effective parameter upon which to trip at power (at EBR-II, period trips are in fact bypassed at power).

*Reactivity inserted continuously at a particular ramp rate until a particular temperature limit is reached. A finite or limited ramp insertion is one in which reactivity is inserted at a particular ramp rate and stops when a specific amount of reactivity has been inserted.

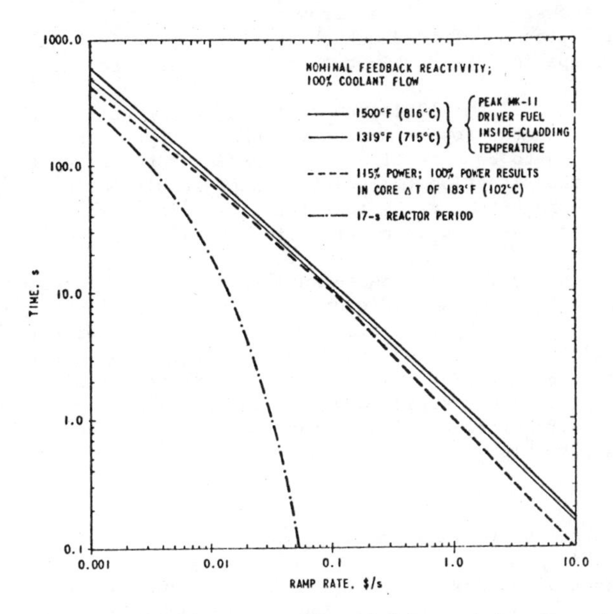

Fig. 9. Response of EBR-II to Unbounded Reactivity Insertion: Core Initially Just Critical.

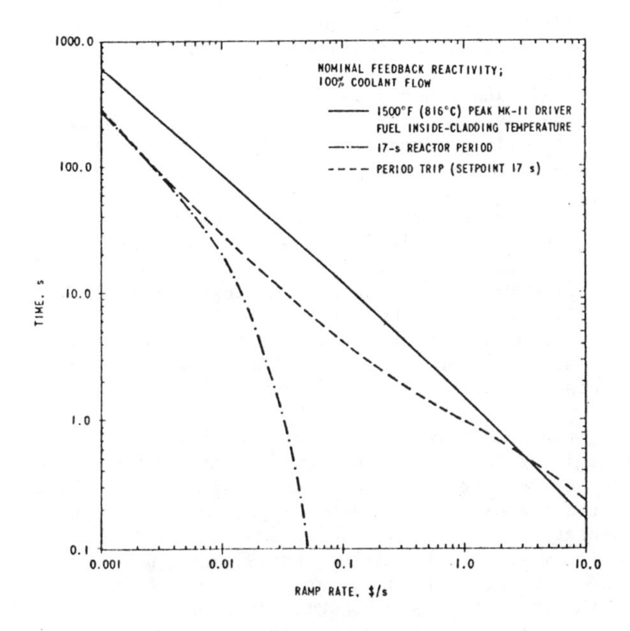

10. Response of EBR-II and Period-trip Circuitry to Unbounded Reactivity Insertion: Core Initially Just Critical.

There is, however, significant time available for trip on power level. This results because of the closer proximity of the trip point level to initial power level and because of feedback that is effective in reducing the rate of power rise. Trips on subassembly outlet temperature are also effective.

Shown in Figs. 11 and 12 is the response of trip circuitry associated with reactor period, reactor power level, and subassembly outlet temperature. Power-level trip is seen to be effective for insertion rates of up to about 3.0 $/s. In contrast to the situation for startup, subassembly-outlet-temperature trips are also seen to be effective against reactivity insertion at power, providing effective response for ramp rates of up to about 0.3 $/s.

A general conclusion from these analyses and others conducted for other fast reactors is that there is significant design margin available in shutdown systems for coping with reactivity insertion events. The question then becomes one of providing sufficient redundancy and diversity to ensure that the shutdown system operates when required. For this reason, current research is focused on mechanical aspects of shutdown systems to ensure their reliability. Included are systems that do not require external signals for shutdown, but are actuated by their inherent response to high core temperatures.

3.2 Self-actuated Shutdown Systems

Self-actuated shutdown systems span a wide range of devices, all intended to shut the reactor down without requiring active mechanical systems external to the reactor "core." In the U.S., several concepts are being developed.

The first concept utilizes "Curie-point magnet-release mechanisms to trigger gravity-driven insertion of the absorber"[13] into the core. The objective is to design a system that will respond to high sodium outlet temperature (on the order of 570°C) with sufficient reliability and response time to ensure protection for identified LOF and TOP events. To reduce insertion time, another concept utilizes a "bistable fluidic valve to hydraulically insert the absorber."

Results of preliminary analysis and testing for these concepts are encouraging. Response times on the order of 2 s appear easily obtainable, which is sufficient to protect against identified reactivity insertion events, including a 60¢ step insertion from a seismic event.

To provide inherent shutdown for higher insertion rates (possible only if the systems discussed previously fail), suggestions have been made for modifictions in fuel pin designs that will provide for fuel melting and movement away from core centerline *without* breach of cladding. Generally, these involve annular fuel pin designs. This concept is not new (see U.S. patent #3,932,217 filed November 1973), but will require testing to demonstrate feasibility. Such tests are now being designed in the U.S., using the TREAT reactor to simulate these severe transients.

Generally, the objective is to develop a fuel-element design in which fuel melting will begin at the axial core center, implying a low conductivity fuel so that peak temperatures are not significantly affected by the axial temperature distribution of the coolant. For slow ramps, molten fuel would flow down the void in the center of the element, under the influence of gravity, to a region of lower reactivity worth. For higher ramps, internal motion would conceivably be accelerated by increased fission gas pressure. While such concepts pose difficulties, they represent a worthwhile attempt to develop fuel pin designs that limit the potential for entry into an energetic HCDA by keeping molten

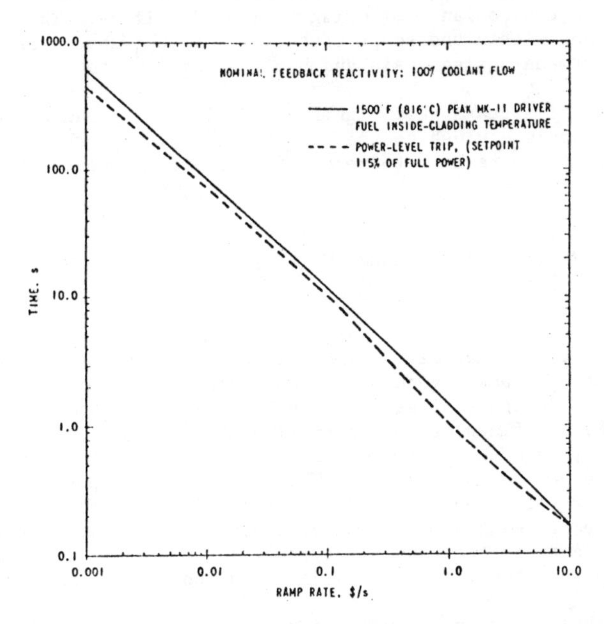

Fig. 11. Response of EBR-II and Power-level-trip Circuitry to Unbounded Reactivity Insertion: Core Initially Just Critical.

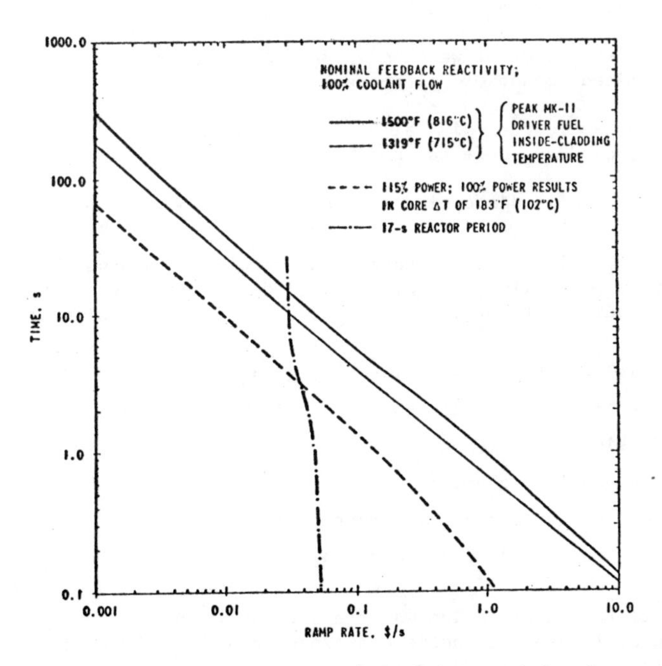

Fig. 12. Response of EBR-II to Unbounded Reactivity Insertion: Core Initially at Full Power (62.5 MWt).

fuel out of contact with sodium, as opposed to dealing solely with the transport of molten fuel in the coolant following fuel-clad failure. The discussion of such events is not within the scope of this paper, but should be recognized as the interface between design basis and hypothetical events. (See Chapter 14.)

In summary, it appears that significant reliability can be achieved for shutdown system designs close to the point of considering them to represent inherent response of the reactor to reactivity insertion events.

4. LOCAL DAMAGE WITHIN A SUBASSEMBLY (LOCAL FAULTS)*

4.1 Background

A local fault is generally defined as any condition which results in a localized mismatch of power generation and coolant flow beyond the bounds considered in normal design evaluations. This power-flow mismatch may either be a local effect which influences only a small number of fuel pins within the pin bundle, or it may be a subassembly-wide condition under which the performance of every pin in the bundle is affected. The basic concern about local faults is the potential for the propagation of fuel-pin failure which might make continued operation unsafe or lead to a loss of coolable geometry within the subassembly. The ultimate concern is the possibility that the damage will propagate to other subassemblies before the situation is detected and the reactor shut down.

Based on their origins or initiating events, the overall local-fault issues can be grouped into the following four categories:

1. Debris in the primary system, which may become lodged in the fuel subassembly and cause flow blockages.

2. Enrichment error which may cause local overpower and release a limited amount of molten fuel.

3. Stochastic fuel pin failures such as fuel pin end-plug weld defects and cladding defects, and

4. Fuel bundle distortion such as fuel pin bowing and hexcan buckling.

All of these local faults have been postulated to have the potential to lead to pin-to-pin failure propagation, either due to overheating or due to mechanical loading, or both, through one or more of the following mechanisms:

1. Development of insulating blockages,
2. Development of heat-generating blockages,
3. Fission gas release from fuel pins,
4. Fuel-coolant chemical reactions,
5. Molten fuel-coolant interaction (MFCI),
6. Pin bowing and hexcan buckling.

*The material for this section was prepared in large part by R. M. Fryer, S. A. Chang, and J. I. Sackett and is contained in a draft Program Plan for Local Fault Testing in EBR-II, March 1980.

The first in-reactor attempt in the U.S. to address local fault issues began with the Fuel-element Failure Propagation Loop (FEFPL) Project, now called the Sodium Loop Safety Facility (SLSF) Program. It had high priority in the safety program; in 1973, the Atomic Energy Commission listed 9 priority R&D issues with "fuel-failure thresholds" and FEFP ranked 2 and 3 after "criteria, codes, and standards." However, while the FEFPL (or SLSF) was being built, out-of-pile research, and more operating experience with failed fuel (French and Soviet) began to dismiss aspects of FEFP in the early 1970s. Progress was striking. A blockage or fuel-failure test, scheduled to be the first SLSF test, was postponed while analyses and out-of-pile experimental results here and abroad indicated that FEFP was not only of less consequence than once believed, but also perhaps entirely benign. No mechanisms by which rapid FEFP can occur for steady operation has been identified; only slow FEFP now appears possible.

Has slow FEFP ever been observed? Possibly, in Manufacture-Franco Belge-au-Bouche-Sodium (MFBS)-6, Mol 7B, EBR-II Run-Beyond-Cladding-Breach (RBCB)-1, BOR-60, and DRF, for examples. However, all have resulted in localized (self-limiting) damage. To support this conclusion, credible local faults and a history of FEFP, are set forth in Tables 4 and 5. A point of these tables is that local faults have occurred and have also been found to not jeopardize safety.

Table 4. Possible LMFBR Local Faults and Their Consequences*

Local Fault (Initiator)	Consequence (Phenomena)	Operational Status
Cladding defect	Fuel failure	Observed
Cladding swelling	Fuel failure	Observed
Local overenrichment	Fuel failure (with possible fuel melting)	Observed (not observed)
Inert coolant-channel blockage	Fuel failure	Observed
S/A inlet blockage	S/A-to-S/A failure propagation	Observed
Loose spacer wire	Fuel failure (Localized hot spot)	Observed
Broken spacer wire	Benign	Observed
Pin bowing	Local hot spots, fuel failure	Observed
Pin distortion	Local hot spots, fuel failure	Observed
Excess Na oxygen	Local blockages	Observed
Fuel swelling	Local blockages	Observed

*Tables 5 and 6 excerpted from D. K. Warriner, "The LMFBR Fuel-Design Environment for Endurance Testing, Primarily of Oxide Fuel Elements with Local Faults," accepted for presentation at the Nuclear Engineering Conference, ASME Century-2 Emerging Technology Conference, San Francisco, August 18-21, 1980.

Table 5. Possible Local-fault Consequences and End State(s)

Consequences	Further Consequences and End State(s)	Operational Status
Fuel failure	Severe breach, loss of fuel, damage of adjacent pin, $Na_3(U,Pu)O_4$ formation, blockage formation	Observed
Molten fuel ejection	e.g., mild FCI,* mechanical failure of other pins, and overheating to saturation if flow halted long enough	Not observed (to be simulated in an SLSF test)
Local heat-generating blockages	Fuel failure, or, much less likely, molten fuel ejection	Either not observed or not reported (to be simulated in an SLSF test)
S/A-to-S/A propagation	Whole-core involvement; meltdown beyond initiated S/As	Not observed

*FCI = Fuel-Coolant Interaction.

4.2 Current Status

The LMFBR safety technology on local faults has been developed from extensive efforts made to study their consequences by all nations devoted to the LMFBR development, and with the support of a vast amount of experimental data and many years of operational experience. However, the emphasis in the past was placed on mixed oxide driver fuel with CW 316 stainless-steel cladding, primarily operating under steady-state conditions. Potential failure propagation mechanisms that have been addressed include insulating blockages, heat-generating blockages, fission-gas release, fuel-coolant chemical interaction, molten-fuel coolant interaction, and mechanical interaction between pins (see Refs. [15] through [24]). All have been found to be either of extremely low probability or to lead to only slow propagative failure, if any.

Another encouraging result, from run-beyond-cladding-breach (RBCB) testing, is increasing confirmation that local faults can be detected, and protective action taken, before significant damage occurs. The most promising systems detect delayed neutrons, emitted by precursors which are in turn released in significant quantity when fuel is in direct contact with the sodium coolant. Also, recent improvements in thermocouple design, regarding both reliability and response time, has extended their range of application for detecting local flow disturbance in subassemblies.

4.3 Reactor Operational Safety Concerns

As discussed above, the current safety technology status on local faults is encouraging, with the potential for rapid FEFP shown to be extremely low for steady-state operation. The only remaining issues are concerned with the impact of upset and emergency transients on degraded elements, if degraded elements are to be tolerated during operation. These issues are of concern to designers insofar as they dictate design and operating philosophy. They are also important to the safety analyst for establishment of operating limits. Considered below is the operating strategy that influences the objectives of both.

For the purpose of maintaining high availability and reaching the goal of economic and safe nuclear power plant operation, the operational strategy is: (1) to operate the reactor with all fuel subassemblies to their designed goal burnups, (2) if this is impossible because of a system malfunction or a local fault in a fuel subassembly, to continue the operation to the next scheduled refueling time, (3) if this is not permissible because of the severity of the problem, to operate the reactor as long as possible to maximize the fuel utilization and to shut down for remedial action when the problem reaches a predetermined operating limit and, (4) to shut down the reactor immediately when a severe problem occurs.

To support this operational strategy, tests are being undertaken to address each condition. In particular, if breached elements are to remain in-core, they must be shown to be capable of surviving the same plant upsets appropriate for unbreached fuel elements. It is expected that these tests will provide the final data needed to demonstrate that local faults are indeed of no consequence to safety, even when breached elements are operated in-core.

5. FAILURES IN THE SECONDARY SODIUM OR STEAM SYSTEM

5.1 Failure of Feedwater Pumps

The loss of the feedwater pump in most plants would be immediately reflected by a decrease in steam drum water level. Operator response under these conditions normally requires a manual switchover to an auxiliary pump. If no action is taken, the steam drum would eventually drain and the evaporators and superheaters would begin "drying out." With steam generation no longer available as a heat-removal mechanism, the temperature of the secondary coolant would increase. Heat transfer between the primary and secondary coolants would decrease and, as a consequence, the temperature of the inlet sodium to the reactor would increase. The increase in inlet sodium temperature would shut the reactor down because of negative reactivity feedback, without requiring shutdown system action. The reactor core would not be endangered. Whether the resulting primary sodium temperatures (typically intermediate between reactor inlet and outlet temperature) would damage primary system components is a question that could only be answered for specific designs, but in principle damage should be minor.

The effects of total feedwater loss and system "dryout" on the turbine-generator would also be benign. Under normal operating conditions the turbine is controlled by steam header pressure. As header pressure decreases a pressure regulation system reduces the generator load. When all load eventually vanishes the turbine-generator will trip on a reverse power signal and the turbine stop valve will close.

The effects of dryout on the evaporators and superheaters are less easily identified. Typically, the inlet sodium side of the evaporators and superheaters would decrease in temperature whereas the temperatures of the respective outlet sides would increase. If permitted to continue the sodium sides and eventually the steam and water sides of the superheaters and evaporators, respectively, would tend to approach isothermal conditions at a temperature dictated by the core outlet coolant.

A similar situation would exist in the hot and cold legs of the secondary sodium system. Temperatures in the hot and cold legs would decrease and increase respectively. Stresses in the cold sides of the evaporators, superheaters, and secondary sodium loop would develop, but whether these would be sufficiently severe to cause mechanical damage is not known and is, in any event, beyond the scope of this discussion. The important conclusions are as follows. In the event of a situation in which feedwater supply is lost and in which no remedial action is taken, nothing of an adverse nuclear nature will happen. The reactor will shut itself down.

5.2 Failure of the Secondary Sodium Pumps

A logical extension of the hypothetical case described above is one in which the secondary sodium pump fails and, contrary to the explicit instructions of emergency procedures, no remedial action is taken. Except for a difference in time base, the consequences of such an event in the primary side will be essentially similar to those described immediately above. Assuming that feedwater supply continues, convective currents will be established in the secondary loop. Some heat, less than 10% of that normally produced, will flow from the primary system to the secondary system and will dissipate via the production of steam. The remaining heat from the still operating reactor will heat the bulk primary sodium. The attendant temperature increase at the reactor inlet, acting through the negative temperature coefficient, will cause reactivity decreases which, in turn, will shut the reactor down.

The results of calculations for this event at EBR-II show that when the power level falls to zero (reactor just critical) the inlet and outlet temperatures will equilibrate in the range, 760-810°F. Since the outlet temperature under normal operating conditions is 883°F an actual decrease in outlet temperature would take place. Clearly, then, the reactor portion of the plant would self-protect if all pumped secondary flow is lost and if no remedial action (a reactor shutdown) is taken.

5.3 Decay Heat Removal

An important area of concern in the operation of all nuclear power plants is the ability of plant systems to remove decay heat from the core immediately after plant shutdown. Here, fast breeders enjoy specific advantages. Such advantages are discussed below.

- Absence of Depressurization Phenomena

Aside from the dynamic pressure needed to overcome impedance effects in the core, pipes, heat exchanger, etc., fast reactors operate under nearly atmospheric pressure. Coolant flashing from depressurization, a phenomenon possible in pressurized water reactors, simply cannot occur.

- Reliable Decay Heat Removal

Four modes of decay heat removal are normally available. These, in the order of their priorities, are: forced cooling by the primary pump, forced cooling by the auxiliary pump, natural circulation of coolant through the core, and the dissipation of primary system heat to the atmosphere by shutdown coolers, normally operating by natural convective flow.

- Large Sodium Inventory

The core of an LMFBR cannot be uncovered without catastrophic failure of the low pressure primary vessel and its guard vessels. The complete submersion of the core under massive quantities of sodium assures that coolant will always be present during shutdown conditions.

It is concluded that the loss of heat removal from the secondary sodium or steam systems causes reactor response which is both predictable and benign. Significant advantages exist because of the low pressure of the sodium coolant, the large margin to sodium boiling, and reliable cooling of the core by natural convection of sodium.

6. SUMMARY AND CONCLUSIONS

1. LMFBRs can be designed to have features that tend to make themselves inherently immune to the consequences of component failure compounded by operator error. The cores of such systems will tend to self-protect under upset conditions, even if remedial action is not taken.

2. Tests have been conducted in PFR, EBR-II and others to measure the adequacy of natural convective flow for core cooling. The results of these tests have confirmed the existence of convective flow at rates high enough to cool the core immediately following a reactor trip upon LOF.

3. The results also suggest that a properly designed system can undergo a LOF without scram (hypothetical) without causing serious fuel problems.

4. The use of self-actuated shutdown systems are attractive possibilities as means for improving the reliability of plant-protection systems. In a real sense, these systems can be considered to represent "inherent" response of reactor cores to high temperature.

5. Loss of cooling in the secondary sodium or steam systems with no remedial action would lead to an increase in sodium inlet temperature at the reactor core. The reactor would shut itself down through reactivity feedbacks, with an equilibrium temperature that could be accommodated by primary system structures.

6. The INFCE conclusions stated earlier, namely that LMFBRs represent an inprovement in safety over conventional power reactor systems, appears to be supportable at least in response in anticipated upset.

REFERENCES

1. Nucleonics Week, August 30, 1979, Vol. 20, No. 35.

2. Balloffet, Y., _et al._, "Calculations of the Loss of Flow Accident in Large LMFBR: Influence of Core Parameters," Proc. of the Int. Mtg. on Fast Reactor Safety Tech., Seattle, Washington, August 19-23, 1979, pp.635-644.

3. Freslon, H., et al., "Analysis of the Dynamic Behavior of the PHENIX and SUPER PHENIX Reactors During Certain Accident Sequences," Proc. of the Int. Mtg. on Fast Reactor Safety Tech., August 19-23, 1979, pp. 1617-1626.

4. Mohr, D. and Feldman, E. E., "A Dynamic Simulation of the EBR-II Plant During Natural Convection with the NATDEMO Code," Specialists' Mtg. on Decay Heat Removal and Natural Convection in FBRs, Brookhaven National Laboratory, Upton, Long Island, New York, February 28-29, 1980.

5. Singer, R. M., et al., "Response of EBR-II to a Complete Loss of Primary Forced Flow During Power Operation," Specialists' Mtg. on Decay Heat Removal and Natural Convection in FBRs, Brookhaven National Laboratory, Upton, Long Island, New York, February 28-29, 1980.

6. Mohr, D. and Fedlman, E. E., "NATDEMO Code Analysis of EBR-II Plant," Specialists' Mtg. on Decay Heat Removal and Natural Convection in FBRs, Brookhaven National Laboratory, Upton, Long Island, New York, February 28-29, 1980.

7. Additon, S. L., McCall, S. L., and Wolfe, T. B., "Simulation of the Overall FFTF Plant Performance," HEDL-TC-556, March 1976.

8. Additon, S. L. and Purziah, E. A., "Natural Circulation in FFTF, A Loop Type LMFBR," Symp. on the Thermal and Hydraulic Aspects of Nuclear Reactor Safety, Vol. II, LMFBRs Ed., O. C. Jones, Jr. and S. G. Bankoff, Publsh. ASME, May 1977.

9. Anderson, R., et al., "Observation on Coolant Flow Patterns in the PFR Primary Circuit During Natural Circulation Experiments," Specialists' Mtg. on Decay Heat Removal and Natural Convection in PFRs, Brookhaven National Laboratory, February 28-29, 1980.

10. Dawson, C. W., "A Theoretical Analysis of the Establishment of Natural Circulation in the Dounreay Fast Reactor," Specialists' Mtg. on Decay Heat Removal and Natural Convection in PFRs, Brookhaven National Laboratory, February 28-29, 1980.

11. Ford, D. J. and Webster, E. B., "Reactor Physics Experience on the Prototype Fast Reactor," Int. Symp. on Fast Reactor Physics, September 24-28, 1979, Aix-en provence.

12. Yee, A. K., Baars, R. E., and Stepnewski, D. D., "Responde of FFTF Core to Protected Reactivity Addition Transients," Proc. of the Int. Mtg. on Fast Reactor Safety Tech., Seattle, Washington, August 19-23, 1979.

13. Dean, E. D. and Sackett, J. I., "Response of EBR-II to Reactivity Insertion, ANL-75-41.

14. U.S. Department of Energy, Fast Reactor Safety Program Progress Report, ANL/TMC 80-1, December 1979.

15. Warinner, D. K. and Cho, D. H., "Status and Needs of Local Fault Accommodation in LMFBRs," Proc. Int. Mtg. on Fast Reactor Safety Tech., Seattle, Washington, August 19-23, 1979, pp. 444-452.

16. Han, J. T. and Fontana, M. H., "Blockages in LMFBR Fuel Assemblies," Symp. on the Thermal and Hydraulic Aspects of Nuclear Reactor Safety, Vol. 2, Liquid Metal Fast Breeder Reactors, ASME, New York, 1977, pp. 51-121

17. Smith, D. C. G., et al., "DFR Special Experiments," Design, Construction, and Operating Experience of Demonstration LMFBRs, April 10-14, 1978, Bologna, Italy, pp. 249-262, Int. Atomic Energy Agency, Vienna, 1978.

18. Warinner, D. K., "Test Requirements for Sodium Loop Safety Facility In-Reactor Experiment T6 (P4)," ANL/RAS 77-12, April 1977. (Availability: U.S. DOE Technical Information Center.)

19. Kramer, W., et al., "In-pile Experiments "MOL-7C" Related to Pin-to-Pin Failure Propagation," Proc. Int. Mtg. on Fast Reactor Safety Tech., Seattle, Washington, August 19-23, 1979, pp. 473-482.

20. "Status of LMFBR Safety Technology, 1. Fission Gas Release in Reactor Safety," Final Draft, CSNI Report No. 40, Nuclear Safety Division, OECD Nuclear Energy Agency, Paris, France, November 1979.

21. "FFTF Final Safety Analysis Report, Vol. 7, Appendix C., Local Fuel Failure Events," HEDL-TI-75001, Vol. 7, p. C.10.4, December 1975. (Availability: U.S. DOE Technical Information Center.)

22. "Final Report on SLSF In-pile Experiment P1," ANL/RAS 77-29, August 1977. (Availability: U.S. DOE Technical Information Center.)

23. Crawford, R. M., et al., "Studies on LMFBR Subassembly Boundary Integrity," ANL-75-27, May 1975.

24. Marr, W. W., et al., "Analytical Investigation of Certain Aspects of LMFBR Failure Propagation," ANL-76-19, February 1976.

25. Johnson, T. R., Pavlik, J. R., and Baker, L. Jr., "Postaccident Heat Removal: Large-scale Molten-fuel-sodium Interaction Experiments," ANL-75-12, February 1975.

6 SINGLE- AND TWO-PHASE FLOW

by
J.G. Collier
UKAEA, Harwell, UK

The basic concepts of single-phase and two-phase flow in the core and primary circuit of a nuclear reactor are reviewed. The purpose of this chapter is to provide sufficient background information of a fundamental nature to allow the student to understand material in later chapters which relates to two-phase flow. Attention is particularly directed at boiling and two-phase flow aspects because these states often occur in design basis accidents. Methods of establishing the flow pattern, pressure drop, void fraction, critical flow and transient response in pipes and components are briefly considered as an introduction to the later Chapters where the application of these methods in accident sequences is discussed in more detail.

1. INTRODUCTION

In Chapter 3 we discussed the central role that thermohydraulic considerations play in determining the behaviour of a nuclear reactor. We also established that the integrity of a nuclear fuel element over a range of plant conditions, was basically determined by metallurgical considerations directly associated with its temperature and that this, in turn, was determined from a balance between the heat generation, internally within the fuel, and the heat transfer processes taking the heat away from the fuel element surface to the coolant.

The following Chapters describe in some detail the events which occur during design basis accidents in light water- and liquid metal-cooled reactors. However, before launching into such detail there is a need to consider some of the fundamental aspects of fluid dynamics and heat transfer in the different types of nuclear reactor. This Chapter and that which follows therefore discuss the various flow and heat transfer processes which can take place in a fuel channel and in the primary circuit of a nuclear reactor. Because two-phase liquid-vapour states often occur in the design basis accidents in water- and liquid metal-cooled reactors and because these states result in complex thermohydraulic regimes a considerable amount of attention is directed at boiling and two-phase flow aspects.

2. SINGLE PHASE FLOW

The single phase flow of liquids and gases in pipes is sufficiently well understood for engineering design purposes. It is usually possible to characterise the flow in terms of a few dimensionless groups such as Reynolds number, Mach number and friction factor. The single phase frictional pressure gradient in a channel is given, for example, by

$$-\left(\frac{dp_F}{dz}\right)_{LO} = \frac{2\, f_{LO}\, G^2 v_L}{D} \tag{1}$$

where the subscript L refers to the liquid phase and the subscript LO refers to the total flow as liquid. Well tried expressions are available to evaluate the Fanning friction factor f_{LO} for fully developed laminar or turbulent flow in pipes with varying degrees of wall roughness. For other geometries, such as annuli or rod bundles, use can be made of the hydraulic equivalent diameter (D_e) concept for approximate values or of experimentally determined fraction factors if greater accuracy is necessary. Single phase flows may be affected by the presence of heating or cooling at the pipe wall changing the velocity profile and inducing buoyancy effects, by obstructions such as grids or spacers and by the fact that the flow is not fully developed. Fully-developed flow in any component of a nuclear reactor circuit is unlikely because of the relative proximity of the various major components and of the complexities of the pipework. Further details on the single-phase flows in channels and rod bundles are available in a recent publication(1).

3. TWO-PHASE FLOW

The onset of boiling within the core of the reactor and of two-phase flow in the reactor pipework introduces considerable further complications. In two-phase flows there are not only twice the number of usual properties (such as density, viscosity etc.) but additional properties (surface tension, latent heat, vapour quality) associated with the phase change. The number of independent dimensionless groups increases by one for every additional property variable and so we are left with ten or more dimensionless groupings to cope with if we wish to "model" a two-phase flow.

A further difficulty is the wide variety of topological configurations the gas and the liquid can adopt. In its simplest form this particular problem can be considered as the specification of a flow pattern. However, many analytical models for two-phase flow also require information about the rates of momentum, heat and mass transfer across the various interfaces between gas and liquid.

A third major difficulty is concerned with non-equilibrium effects. As with single phase flow we must consider not only departures from hydrodynamic equilibrium (i.e. not "fully developed" flow) but, in addition, for two-phase flows we must also take account of departures from thermodynamic equilibrium (i.e. gas and liquid phases at different temperatures). Finally, two-phase flows often occur under transient conditions and it is necessary to ask the question "Are the local instantaneous properties of the flow during a transient identical to those appropriate for the equivalent fully-developed steady state flow?" This is what is usually assumed but, in general, the flow pattern, void fraction and friction factor will differ from the equivalent steady flow values. To what extent they actually differ is being studied at the present time.

4. TWO-PHASE FLOW PATTERNS

The flow patterns encountered in vertical co-current gas-liquid up-flow are well known, viz.

(a) <u>bubbly flow</u> in which the gas or vapour phase is distributed as discrete bubbles in a continuous liquid phase

(b) <u>slug flow</u> in which the gas or vapour bubbles are approximately the diameter of the tube and the liquid is contained in liquid slugs which separate successive gas bubbles

(c) <u>churn flow</u> which is formed by the breakdown of the large vapour bubbles in slug flow and results in an oscillatory or time varying flow with the gas or vapour passing in a more or less chaotic manner through the liquid

(d) <u>annular flow</u> in which a liquid film forms on the tube wall with a continuous central gas or vapour core

(e) <u>wispy-annular flow</u> in which a relatively thick liquid film aerated by gas bubbles is formed on the tube wall, together with a considerable amount of liquid entrained in a central gas or vapour core.

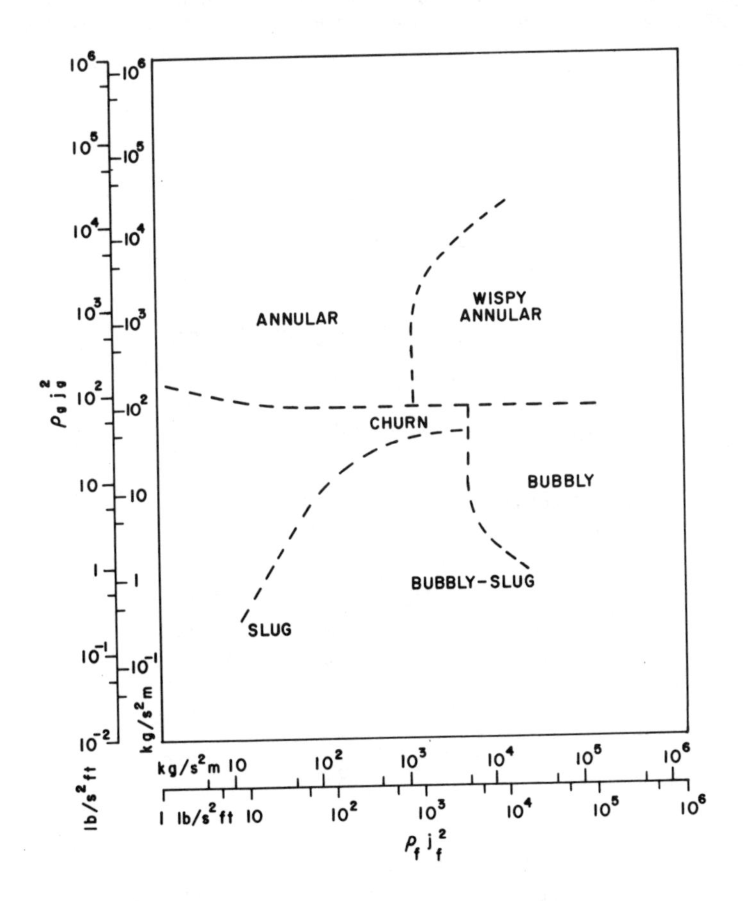

Fig.6.1 Flow Pattern Map for Vertical Flow (Hewitt and Roberts[2])

The identification of which flow pattern prevails under any particular flow condition is carried out by reference to a flow pattern map. Figure 6.1 shows a flow pattern map[2] obtained from observations on low-pressure air-water and high pressure steam-water flow in small vertical tubes. The axes represent the superficial momentum fluxes of the water ($\rho_L j_L^2$) and steam ($\rho_G j_G^2$).

The flow patterns observed in co-current two-phase flow in horizontal and inclined pipes are complicated by asymmetry of the phases resulting from the influence of gravity. A flow pattern map for horizontal flow which has been widely used, particularly in the petrochemical industry is that of Baker[3]. In 1974 Mandhane et al[4] published a new map based on a large number of flow pattern observations in horizontal pipes in the 13-50 mm diameter range. More recently, Taitel and Dukler[5] have proposed a theoretically based flow pattern chart. In Figure 6.2 the predictions obtained from the Taitel/Dukler chart are compared with the map offered by Mandhane et al for the situation of an air-water mixture at 1 bar pressure and 25°C flowing in a 25mm i.d. pipe. The agreement is very satisfactory. The Taitel/Dukler chart is, however, to be preferred because it takes into account the different influence of both pipe diameter and fluid properties on each of the flow pattern transitions. Finally it is noted that this analysis has been extended to cover transient flows[6].

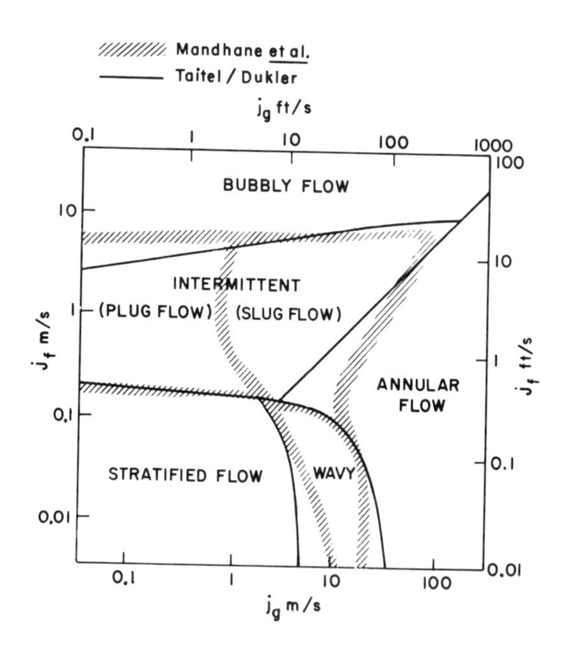

Fig.6.2 Comparison of Taitel/Dukler[5] prediction with Mandhane[4] Map (25mm dia. horizontal pipe)

5. TWO-PHASE FLOW MODELS

Various analytical models of two-phase gas-liquid flow have been published. These can be identified under the following four headings,

Homogeneous flow where the flow is assumed to behave as an equivalent single phase flow with pseudo properties.

Separated flow where the phases are considered to be artificially segregated. This introduces a requirement to model the interaction between the phases.

Drift flux model which is a particular separated flow model in which attention is focussed on the relative motion between the phases.

Flow pattern models where the phases are considered to be arranged in one of three or four definite prescribed geometries (based on observations of the topology of the phases).

The local two-phase pressure gradient is made up of three separate components; a frictional term, an accelerational term and a static head term, viz

$$-\left(\frac{dp}{dz}\right)_{TP} = -\left(\frac{dp}{dz}F\right) - \left(\frac{dp}{dz}a\right) - \left(\frac{dp}{dz}z\right) \quad (2)$$

In general the evaluation of the pressure drop across a channel involves integrating the local pressure gradient along the channel length.

Homogeneous model[16]

The basic premises upon which the model is based are the assumptions of

(a) equal vapour and liquid linear velocities

(b) the attainment of thermodynamic equilibrium between the phases

(c) the use of a suitably defined single phase friction factor for two-phase flow.

These assumptions lead to the following expressions for the three pressure gradient components:

$$-\left(\frac{dp}{dz}F\right) = \frac{2\, f_{TP}\, G^2 \bar{v}}{D} \quad (3)$$

$$-\left(\frac{dp}{dz}a\right) = G^2 \frac{d(\bar{v})}{dz} \quad (4)$$

$$-\left(\frac{dp}{dz}z\right) = \frac{g \sin\theta}{\bar{v}} \quad (5)$$

where θ is the angle of inclination of the pipe to the horizontal and $\bar{v}$ is the homogeneous fluid specific volume defined as the total volumetric flow divided by the total mass flow

$$\bar{v} = xv_G + (1-x)\, v_L \tag{6}$$

A number of different approaches have been made to the definition of the two-phase friction factor f_{TP}. Basically either the liquid phase viscosity μ_L or some two-phase mean viscosity $\bar{\mu}$ can be used to evaluate the Reynolds number.

Separated flow

The basic premises upon which the separated flow model is based are the assumptions of:

(a) constant but not necessarily equal velocities for the vapour and liquid phases

(b) the attainment of thermodynamic equilibrium between the phases

(c) the use of empirical correlations or simplified concepts to relate the two-phase friction multiplier (ϕ_{LO}^2) and the void fraction (α) to the independent variables of the flow.

Using the separated flow model the three components of the pressure gradient are respectively given by

$$-\left(\frac{dp}{dz}\,F\right) = -\left(\frac{dp}{dz}\,F\right)_{LO} \phi_{LO}^2 \tag{7}$$

where ϕ_{LO}^2 is known as the two-phase frictional multiplier and $-(dp/dz)_{LO}$ is the frictional pressure gradient calculated from the Fanning equation (eq.(1)) for the total flow (water and steam) assumed to flow as water

$$-\left(\frac{dp}{dz}\,a\right) = G^2 \frac{d}{dz}\left[\frac{x^2 v_G}{\alpha} + \frac{(1-x)^2\, v_L}{(1-\alpha)}\right] \tag{8}$$

where G is the total mass velocity, x is the thermodynamic steam quality, v_G and v_L are the respective steam and water specific volumes and α is the fractional cross sectional area occupied by steam (the void fraction)

$$-\left(\frac{dp}{dz}\,z\right) = g \sin\theta \left[\frac{\alpha}{v_G} + \frac{(1-\alpha)}{v_L}\right] \tag{9}$$

where g is the acceleration due to gravity and θ is the angle of inclination of the pipe to the horizontal.

Expressions are required for ϕ_{LO}^2 and α and these are discussed below.

Drift flux model

This model was initially developed by Zuber[33] and Wallis[34] and has been extended by Ishii[31,32] and others since. It allows consideration of counter current flows and the main assumptions are

(a) the shear stress at the channel wall is ignored

(b) the flow is basically one-dimensional.

TABLE 6.1

TWO-PHASE PRESSURE DROP CORRELATIONS FOR STEAM-WATER MIXTURES

	CORRELATION
Homogeneous[16]	$\phi_{LO}^2 = \left[1 + x\left(\frac{\rho_L-\rho_G}{\rho_G}\right)\right]\left[1 + x\left(\frac{\mu_L-\mu_G}{\mu_G}\right)\right]^{-1/4}$
Baroczy[11]	$\phi_{LO}^2 = \Omega\phi_{LO}^2\,(G = 1356\ kg/m^2s)$ where $\phi_{LO\,(G=1356)}^2 = f([\mu_L/\mu_G)^{0.2}\ (\rho_G/\rho_L)],x)$; $\Omega = f([(\mu_L/\mu_G)^{0.2}\ (\rho_G/\rho_L)],x)$
Chisholm[12]	$\phi_{LO}^2 = 1 + (\Gamma^2 - 1)\left[Bx^{(2-n)/2}(1-x)^{(2-n)/2} + x^{2-n}\right]$ where $\Gamma = [(dp/dz)_{GO}/(dp/dz)_{LO}]^{1/2}$ and $B = f(G,\Gamma)$
CISE[13]	$\left(\frac{dp}{dz}\right)_{TP} = \left[\frac{K\ G^n\ \overline{v}^{0.86}\ \sigma^{0.4}}{D^{1.2}}\right]$ where $\overline{v} = [x/\rho_G + (1-x)/\rho_L]$; $K = f$ (geometry); $n = f$ (geometry)
Martinelli-Nelson[14]	$\phi_{LO}^2 = fn(p,x)$
Smith-Macbeth[15]	$\phi_{LO}^2 = \left[\left\{\left(e(1-x) + \left(\frac{\rho_L}{\rho_G}\right)x\right)\left(e(1-x) + x\right)\right\}^{1/2} + (1-e)\ (1-x)\right]^2$ where $e = 0.4$

TABLE 6.2

TWO-PHASE PRESSURE DROP DATA FOR STEAM-WATER MIXTURES

Data Bank	ESDU[9]			Friedel[7]			Idsinga[8]			Harwell[10]		
Flow Direction	Upflow, downflow and horizontal			Upflow only			Upflow and horizontal			Upflow and horizontal		
Correlation	n	e	σ	n	e	σ	n	e	σ	n	e	σ
Homogeneous(16)	1709	-13.0	34.2	2705	-19.9	42.0	2238	-26.0	22.8	4313	-23.1	34.6
Baroczy(11)	1447	4.2	30.5	2705	-11.6	36.7	2238	- 8.8	29.7	4313	- 2.2	30.8
Chisholm(12)	1536	19.0	36.0	2705	- 3.8	36.0	2238	0.5	40.5	4313	13.9	34.4
CISE(13)	-	-	-	2705	16.3	28.0	2225	22.6	28.9	-	-	-
Martinelli-Nelson(14)	1422	16.3	36.6	-	-	-	2238	47.8	43.7	-	-	-
Smith-Macbeth(15)	-	-	-	-	-	-	-	-	-	4313	-16.6	24.1

n = number of data points analysed

e = mean error (%) = $(\Delta p_{cal} - \Delta p_{exp}) \times 100/\Delta p_{exp}$

σ = standard deviation of errors about the mean (%)

TABLE 6.3

VOID FRACTION CORRELATIONS FOR STEAM-WATER MIXTURES

	CORRELATION
Lockhart & Martinelli[22]	$\alpha = f(X)$ where $X = \left[\left(\frac{dp}{dz}\right)_L \Big/ \left(\frac{dp}{dz}\right)_G\right]^{\frac{1}{2}}$
Hughmark[23]	$\alpha = K$ where $K = f\,[Re, Fr, (1-\beta)]$
Smith[19]	$S = e + (1-e)\sqrt{\frac{\rho_L/\rho_G + e\,(1/x - 1)}{1 + e\,(1/x - 1)}}$ where $e = 0.4$
CISE[20]	$S = f(G, D, \rho_L, \rho_G, \mu_L, \sigma, \beta) + 1$
Chisholm[21]	$S = \left[x\left(\frac{\rho_L}{\rho_G}\right) + (1-x)\right]^{\frac{1}{2}}$
Thom[24]	$S = f\left(\frac{\rho_L}{\rho_G}\right)$
Bankoff-Jones[25]	$S = \left[\frac{1-\alpha}{A - \alpha + (1-A)\alpha^B}\right]$ where $A, B = f(p)$
Bryce[18]	$S = \left[\frac{1-\alpha}{A - \alpha + (1-A)\alpha^B}\right]$ where $A = f(p, G, X, \rho_G, \rho_L)$, $B = f(p, \rho_G, \rho_L)$

TABLE 6.4

VOID FRACTION DATA FOR STEAM-WATER MIXTURES

	Analysis of mean density						Analysis of slip ratio					
Data Bank	Friedel[7]			Bryce[18]			ESDU[17]			Bryce[18]		
Correlation	n	e	σ	n	e	σ	n	e	σ	n	e	σ
Lockhart & Martinelli[22]	-	-	-	-	-	-	598	-57.6	50.3	-	-	-
Hughmark[23]	484	-10.8	33.0	-	-	-	598	- 9.1	29.2	-	-	-
Smith[19]	484	0.5	26.8	639	8.6	31.5	-	-	-	639	18.0	77.8
CISE[20]	484	9.3	35.0	639	-1.4	22.7	598	-23.7	27.2	639	- 1.2	68.6
Chisholm[21]	484	- 0.4	26.0	-	-	-	598	-14.5	30.8	-	-	-
Thom[24]	484	7.4	36.5	639	43.3	61.7	-	-	-	639	132.6	200.0
Bankoff-Jones[25]	-	-	-	639	9.32	31.6	-	-	-	639	34.5	137.6
Bryce[18]	-	-	-	639	0.1	20.7	-	-	-	639	6.1	86.9

n = number of data points analysed
e = mean error (%) = (cal-exp)x100/exp
σ = standard deviation of error about mean (%)

The basic continuity equation of the drift flux model can be written as

$$\bar{\alpha} = \frac{j_G}{C_o\,(j_L + j_G) + \bar{u}_{Gj}} \tag{10}$$

In this equation j_G and j_L are the volumetric fluxes (or sometimes known as the "superficial velocities") of the phases given by the volumetric flowrate of the phase divided by the cross sectional flow area; $\bar{\alpha}$ is the area average void fraction; C_o is a distribution parameter correcting the basic one-dimensional theory to allow for concentration and velocity profiles across the channel and $\bar{u}_{Gj}$ is a *weighted mean drift velocity*.

Recommended values for both C_o and $\bar{u}_{Gj}$ are given by Ishii[32] according to the flow pattern and pressure range.

6. EVALUATION OF PRESSURE GRADIENT AND VOID FRACTION

To evaluate the local pressure gradient, expressions are required for the functions ϕ^2_{LO} and α. A very large number of correlations have been proposed for these functions and a summary of the better known correlations is given in Table 6.1. Recently a number of workers[7,8,9,10] have carried out systematic comparisons between these various correlations and data banks containing large numbers of experimental pressure drop measurements for steam-water mixtures. A summary of these various comparisons is given in Table 6.2. It can be seen that the various studies agree that the most accurate correlations for ϕ^2_{LO} are those of Baroczy[11], Chisholm[12] and CISE[13] but in each case the standard deviation of errors about the mean is 30-35%.

To evaluate the changes in momentum (or kinetic energy) and also the mean density of a two-phase flow, it is necessary to be able to establish the local void fraction or fraction of the flow cross section occupied by steam. Once again a large number of correlations have been proposed for the evaluation of void fraction (α). Sometimes, the correlation is expressed in terms of the slip ratio (S) which is related to the void fraction by the identity

$$S = \left(\frac{x}{1-x}\right)\left(\frac{\rho_L}{\rho_G}\right)\left(\frac{1-\alpha}{\alpha}\right) \tag{11}$$

A summary of the better known correlations for S or α is given in Table 6.3. Similarly, systematic comparisons[7,17,18] have been carried out between these various void fraction correlations and data banks containing large numbers of experimental measurements of either void fraction or fluid density measurements for steam-water mixtures. A summary of these comparisons is given in Table 6.4. It can be seen that the various studies agree that the most accurate correlations for void fraction are those of Smith[19], CISE[20] and Chisholm[21] with the latter having the added advantage of great simplicity. Again the standard deviation of the error on mean density is 20-30%.

These various two-phase flow correlations for pressure drop and various may be moderately satisfactory for steady state conditions over a limited range of variables but as pointed out by both Ishii[28] and Gardner[29] they may be quite inadequate for rapid transients. It is known that in single phase flows transient effects on wall shear do exist[30].

7. PRIMARY CIRCUIT COMPONENTS[26]

An important aspect of the design of any primary reactor circuit is an understanding of the behaviour of the various components and features present in the circuit. Such features include amongst others, bends, valves, manifolds, expansions, contractions and pumps.

Sudden enlargements

For sudden enlargements the following equation is recommended for the calculation of the static pressure change across the feature

$$\Delta p = \frac{G_1^2 \sigma (1-\sigma)}{\rho_L} \left[\frac{(1-x)^2}{(1-\alpha)} + \left(\frac{\rho_L}{\rho_G}\right) \frac{x^2}{\alpha} \right] \tag{12}$$

where G_1 is the mass velocity based on the upstream pipe area and σ is the ratio of upstream to downstream cross sectional areas. The void fraction α is based on that for pipe flow in the upstream section. For high mass velocities ($G > 2000$ kg/m^{2}s) the homogeneous model is acceptable.

Sudden contractions

For a sudden contraction the homogeneous model is satisfactory and the corresponding equation for the static pressure change is

$$\Delta p = \frac{G_2^2}{2\rho_L} \left[\left(\frac{1}{C_c} - 1\right)^2 + \left(1 - \frac{1}{\sigma^2}\right) \right] \left[1 + \left(\frac{\rho_L}{\rho_G}\right) x \right] \tag{13}$$

In this equation G_2 is the mass velocity based on the downstream pipe area and σ is again the ratio of the upstream to downstream cross sectional areas. C_c is the coefficient of contraction which is a function of σ. Transient behaviour at discontinuities have been studied by Weisman[27].

Bends

Data on the pressure change around circular arc bends for single phase flow is given in a number of handbooks. The loss coefficient is a function of Reynolds number, the ratio R/D where R is the radius of the bend and D is the pipe diameter and the deflection angle θ. The loss coefficient is also affected by the length of the outlet section. For two-phase flow, the pressure drop around the bend is up to 2.5 times higher than predicted using single phase pressure loss measurements and multiplying by an experimentally evaluated straight pipe friction multiplier. Chisholm[38] has presented a correlation for two-phase flow around bends.

Tees

Single-phase flow behaviour at a single sharp-edged tee junction has also been examined in many handbooks. Empirical equations devised by Gardel[39] have been found to be a satisfactory basis for calculating single phase loss coefficients for situations where the flow either divides or combines at a tee.

Manifolds and headers

A simple manifold may typically consist of a series of tee functions uniformed spaced along a blind-ended tube or header. The design of such headers or manifolds requires that both the momentum changes across the successive tee junctions and the friction pressure loss in the interconnecting pipework be taken into account. The arrangement of the inlet and exit from a pipework system can greatly influence the uniformity of flow. Figure 6.3 shows the experimentally observed flow distributions for two different piping arrangements. The flow distibution is particularly unfavourable in the case of asymmetric flows as shown. Uniform flow distributions are obtained by keeping the total branch area small compared with the manifold area and the head loss in the branch large compared with the inlet velocity head in the manifold.

For two phase flow at a tee junction there is a dramatic separation effect. The gas phase, having little inertia, tends to be separated and flows into the branch stream whilst the liquid travels past the tee. In a manifold system the separation effect at the individual tee junctions tend to accomulate giving a concentration of liquid towards the downstream end of the manifold as shown in Figure 6.4.

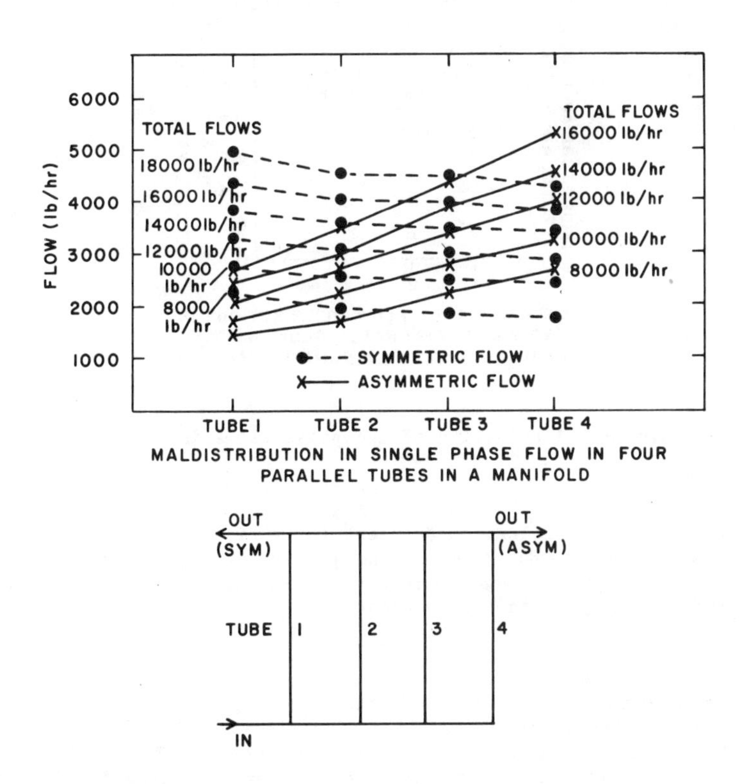

Fig.6.3 Maldistribution in Single Phase Flow in a Four Tube Pipework System

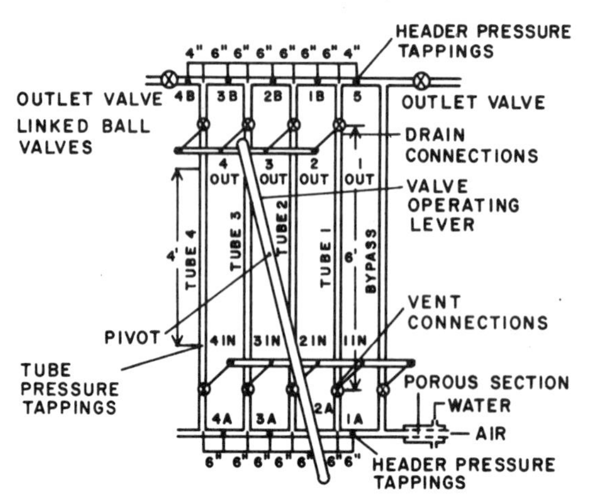

METHOD OF MEASURING VOID FRACTION SIMULTANEOUSLY IN FOUR PARALLEL TUBES CARRYING TWO PHASE FLOW

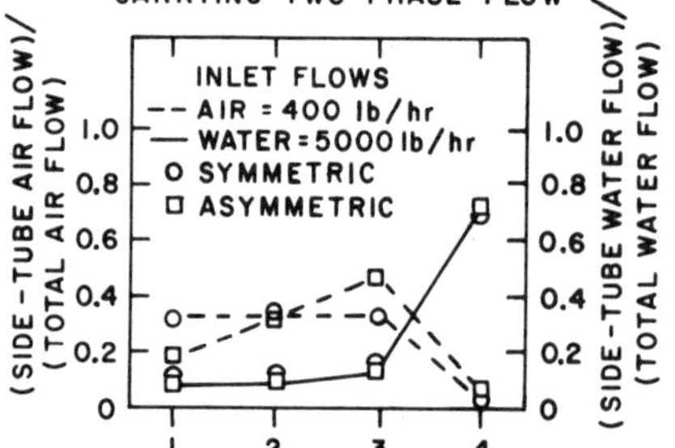

AIR AND WATER FLOW DISTRIBUTIONS IN TWO PHASE AIR-WATER FLOW IN FOUR PARALLEL TUBES IN A COMMON MANIFOLD

Fig.6.4 Maldistribution of Two-phase Flows in Manifolds

Valves

Published data on the loss coefficient for various types of valve can only be used to give a rough indication of the pressure drop across a valve. The influence of a valve may extend to thirty or more diameters. Two-phase flow through valves has been considered by Chisholm[40].

Pumps

The thermohydraulic behaviour of the primary system of a water-cooled reactor, during a transient or a loss-of-coolant accident is considerably influenced by the characteristics of the primary pumps. The condition of the pump (electrics lost, locked rotor etc.) in the damaged circuit influences its resistance and therefore the flow distribution to the breach. The behaviour of the pumps in the intact circuits likewise affect the ciculation in this part of the system and through the core itself.

The hydraulic performance of a centrifugal pump, as characterised by static head (H) and torque (T) is determined by the angular rotational speed (ω) volumetric flow (Q) and fluid density (ρ). Considering first just single phase liquid flow, if the head, volumetric flow and speed are all positive for normal pump operation then the normal head flow characteristic occupies just one quadrant of a four quadrant diagram. All the unusual head/flow characteristics occupy the remaining three quadrants for a positive rotation and all four quadrants for a negative rotation. Various sets of complete pump characteristics have been reported for specific pumps. This information is presented in terms of so-called "homologous curves". These are dimensionless extensions of the four quadrant pump characteristics in which the parameters H (head), T (torque), Q (volumetric flow) and ω (impeller speed) are made dimensionless by dividing by the appropriate "rated" (subscript R - point of maximum efficiency) parameters. Separate curves are used for head and torque.

Analysis of pump behaviour under two-phase flow conditions is usually based on pump inlet fluid conditions. The single phase homologous curves for a pump may be used to describe pump performance during two-phase conditions provided allowance is made for the degration of both head and torque which occur for two-phase flow. The degree of degration for both head and torque is a function of the upstream void fraction. Figure 6.5 shows the shape of these degration curves(41). Loss of performance occurs rapidly at void fractions above 18-20% with the maximum degration of head occurring for void fractions of 60-80%. A limited comparison(42) between steady state and transient two-phase behaviour for the MOD-1 pump indicates similar behaviour in the normal operating region suggesting that inertial effects in the pump impeller may be small.

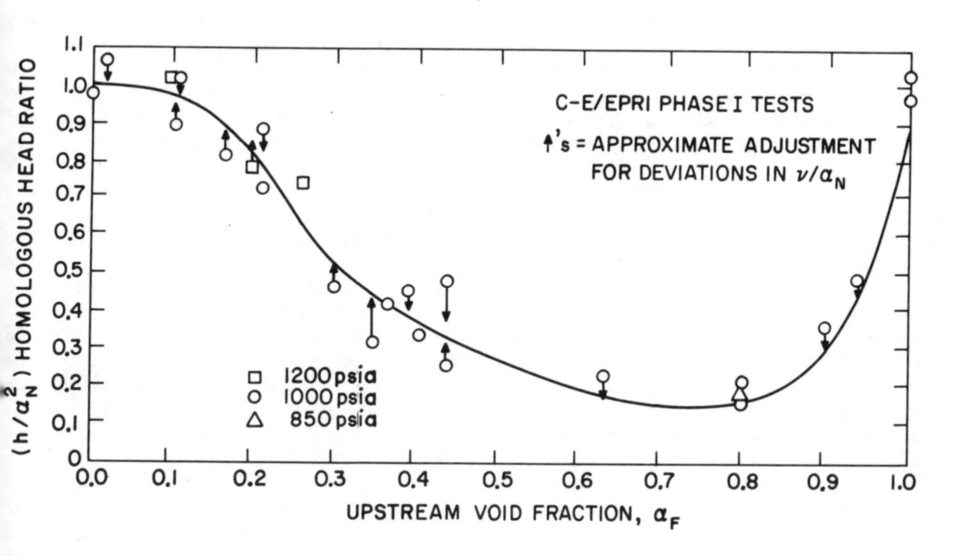

Fig.6.5(a) Effect of Void Fraction on Homologous Head Ratio for Rated Speed and near-rated Flow(41)

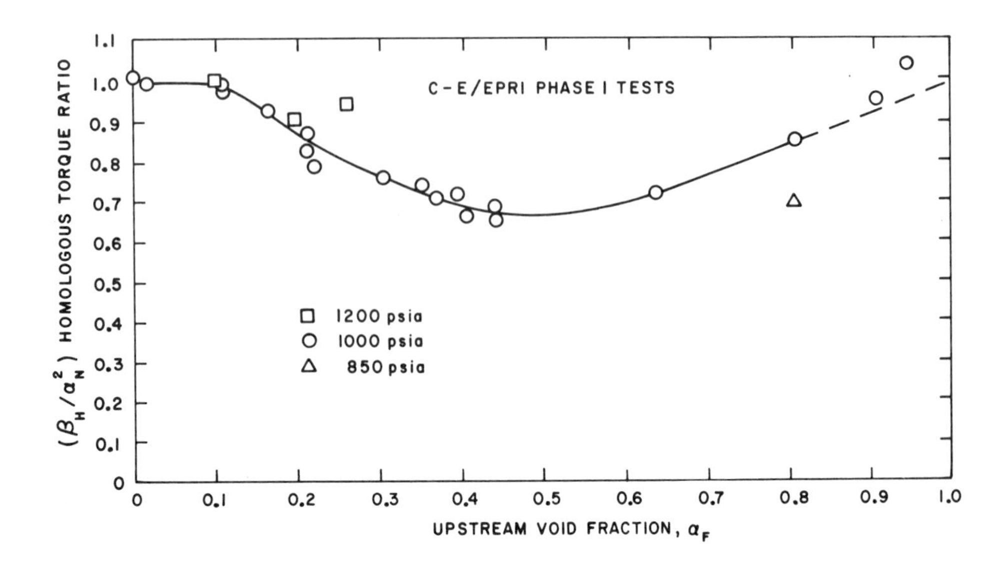

Fig.6.5(b) Effect of Void Fraction on Homologous Torque Ratio for Rated Speed and near-rated Flow[41]

8. COUNTERCURRENT FLOW

So far we have considered situations where the liquid and vapour phases flow concurrently. There are, however, a number of situations where countercurrent flow can exist in a nuclear reactor circuit. Examples are the flow pattern in the downcomer annulus surrounding the core of a PWR during injection of emergency core cooling, the cooling of BWR fuel rod bundles at low flows and the motion of molten cladding materials during a whole core accident in an LMFBR.

Countercurrent flow cannot exist over a wide range of flowrates. Consider the situation shown in Figure 6.6. Liquid is injected into a vertical tube through a porous wall and falls as a liquid film to the base of the tube where it is extracted again through a porous wall. Initially there is no gas flow but gas now enters the bottom of the tube and flows upwards over the liquid film. The various flow transitions which occur as the gas flow is increased are shown in the diagram. Initially, as the gas flowrate increases, nothing happens; the liquid film still runs down. However, a point is reached where large waves are formed on the liquid film and are carried upwards by the gas phase. Liquid is carried above the injection point and the liquid now flows both up and down. This initial transition is called the *flooding point*. If the gas velocity is increased further all the injected liquid is carried upwards in the form of an annular climbing film. If the process is now reversed and the gas phase progressively reduced from this high value there comes a point where the liquid phase, in addition to flowing upwards in a climbing film, will start to creep downwards again. This point is termed the *flow reversal* point.

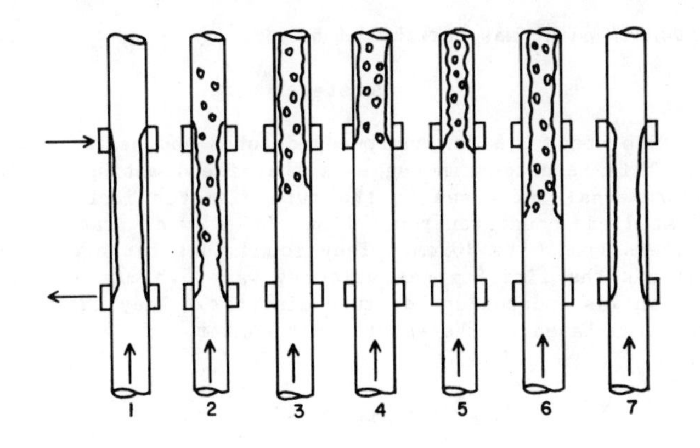

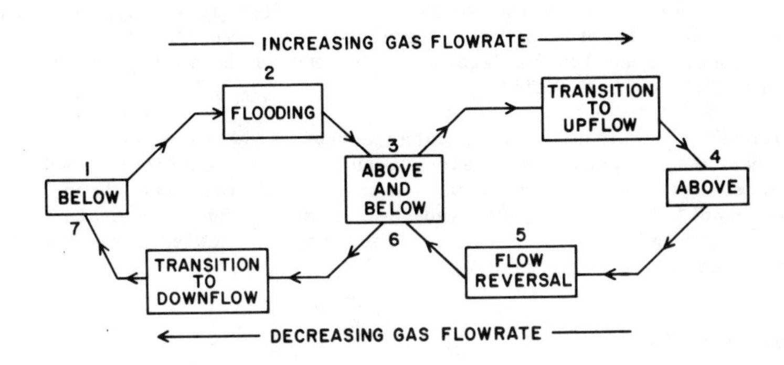

Fig.6.6 Flooding and Flow Reversal in Counter-Current Flow

Wallis(34) has provided empirical correlations of the *flooding point* in terms of non-dimensional superficial velocities.

$$j_G^* = j_G \rho_G^{0.5} \, [gD\,(\rho_L - \rho_G)]^{-0.5} \tag{14}$$

$$j_L^* = j_L \rho_L^{0.5} \, [gD\,(\rho_L - \rho_G)]^{-0.5} \tag{15}$$

The correlation can be expressed as

$$j_G^{*\frac{1}{2}} + j_L^{*\frac{1}{2}} = C \tag{16}$$

The value of the constant C varies significantly with the entry conditions for the liquid and gas. Typically a value of 0.88 is used for round-edged tubes and 0.725 for sharp-edged tubes.

The *flow reversal* point was correlated by

$$j_G^* \approx \text{constant} \tag{17}$$

where the constant was about 0.5 for sharp-edged tubes and nearer 0.9 for smooth entry situations. This relationship suggests that the limiting gas velocity at the point of flow reversal increases as the tube diameter increases. Pushkina and Sorokin[43] studied countercurrent flow in a wide range of vertical tubes having diameters from 6 to 309mm. They found that for air/water flows at atmospheric conditions the limiting gas velocity was 15.8 m/s and, contrary to the Wallis prediction was independent of tube diameter. They therefore proposed an alternative relation based on the Kutateladze number, Ku

$$Ku = j_G \rho_G^{\frac{1}{2}} \left[g\sigma \left(\rho_L - \rho_G \right) \right]^{-\frac{1}{4}} = 3.2 \tag{18}$$

Recently Richter and Lovel[44] have confirmed the validity of this equation. These various approaches to the scaling of the flow reversal point have caused difficulties in interpreting the results from experiments carried out at different geometric scales to determine the end of bypassing of the ECCS fluid during a loss of coolant accident in a PWR.

Although most studies of countercurrent flow have been concerned with single channels, there are situations during accident sequences when countercurrent flow can occur between a number of parallel channels. Piggott and Ackermann[45] have developed a simple but accurate method of establishing the multiple channel behaviour from a knowledge of the single tube characteristics.

9. CRITICAL FLOW

Critical or choking flows can occur during the depressurisation of a reactor circuit. For single phase compressible fluids the critical mass velocity can be readily obtained from the energy balance. This leads to

$$G^2_{crit} = -\left(\frac{\partial p}{\partial v} \right)_s \tag{19}$$

The derivative is taken at constant entropy consistent with the usual assumption of no friction and thermodynamic reversibility. In two-phase flow[46] a maximum or critical flow is also observed. The *homogeneous model* can be utilised to establish this critical flow but comparison with experimental values shows that the model underpredicts the flowrate for short pipes and for conditions where the fluid is a saturated or subcooled liquid. The use of a separated or slip flow model[47,48] improves the situation somehat but the underprediction near the saturated liquid state remains. It is clear that thermodynamic equilibrium is not being maintained under these conditions and that departures from equilibrium must be taken into account. Empirical correlations have been proposed, such as that by Henry and Fauske[49] and semitheretical models[50] have been developed which allow for the processes of bubble nucleation and bubble growth in the flashing fluid. These same models also allow a prediction of acoustic wave propagation in vapour-liquid mixtures as a function of frequency.

10. DYNAMIC ASPECTS

Increasingly, attention is being turned to the dynamic aspects of single phase and two-phase flows and the onset of hydrodynamic instabilities. All boiling and condensing systems are susceptible to spatial and temporal instabilities of various kinds.

System noise - fluctuating pressure and momentum fluxes linked to pumps, compressors, to vortex shedding or aeroelastic excitation, or in two-phase flow to particular flow patterns such as slug flow.

Excursive instability - "vapour locking" in a single channel or parallel channel system where a maximum occurs in the pressure drop/flow characteristic.

Chugging instability - caused by liquid superheating and ejecting fluid from channel ("geysering"). This affects mainly liquid metal-cooled reactor systems.

Oscillating instability - "density wave" instability with feedback and interaction between the various pressure drop components (caused by the time lag introduced by the density head term due to the finite speed of propagation of a "density wave").

The propagation of a density (voidage) disturbance is governed by the equation

$$\frac{\partial \alpha}{\partial t} + V_w \frac{\partial \alpha}{\partial z} = N \quad (20)$$

In this equation V_w is the velocity of the density wave and N is the rate of generation of vapour. The density wave velocity is given by

$$V_w = \left(\frac{\partial j_G}{\partial \alpha}\right)_j \quad (21)$$

The application of the drift flux model to periodic and transient two-phase flows has been developed by Zuber[35], Hancox and Nicoll[36] and their associates[37].

11. CONCLUDING REMARKS

In this Chapter we have attempted to comment upon the various fluid dynamic conditions which can occur within nuclear reactors concentrating, in particular, upon the complexities of the two-phase flow regime. In general, analysis of the flows inside a reactor during an accident transient is based on the assumption that the instantaneous local hydrodynamic condition is similar to that for the equivalent fully developed steady state flow condition. It has however been shown that this assumption is incorrect because of (a) *flow history effects*; i.e. the flow is not generally "fully developed" in a reactor circuit, (b) *lack of thermodynamic equilibrium*; this is particularly important in flashing flows during a depressurisation and during the reflood phase of the accident and lastly (c) *transient effects* on the velocity profile and upon wall shear.

These effects are being studied in relation to the particular flow patterns occurring and phenomenological models are being developed which recognise these deviations from the assumptions usually made in hydrodynamic computer codes.

REFERENCES

1. Kakac, S. and Spalding, D.B. (editors). "Turbulent Forced Convection in Channels and Bundles - Theory and applications to Heat Exchangers and Nuclear Reactors" Volumes 1 and 2, published by Hemisphere Publishing Corporation (1979).

2. Hewitt, G.F. and Roberts, D.N. "Studies of two-phase flow patterns by simultaneous X-ray and flash photography", AERE-M2159 HMSO (1969).

3. Baker, O. "Design of pipe lines for simultaneous flow of oil and gas", Oil and Gas Journal, 26 July (1954).

4. Mandhane, J.M., Gregory, G.A. and Aziz, K. "A flow pattern map for gas-liquid flow in horizontal pipes", Int. J. Multi-phase Flow, Vol.1 pp.537-553 (1974).

5. Taitel, Y. and Dukler, A.E. "A model for predicting flow regime transitions in horizontal and near horizontal gas-liquid flow", A.I.Ch.E. J. Vol.22, pp.47-55 (1976).

6. Dukler, A.E. "Modelling Two Phase Flow and Heat Transfer", Keynote paper KS-11 presented at 6th Int. Heat Transfer Conf., Toronto, Canada, August 1978.

7. Friedel, L. "Mean void fraction and pressure drop: Comparison of some correlations with experimental data", Paper A7, European Two-Phase Flow Group Meeting, Grenoble, 6-9th June 1977.

8. Idsinga, W. "An assessment of two-phase pressure drop correlations for steam-water systems", M.Sc. Thesis, M.I.T., May 1975.

9. "The frictional component of pressure gradient for two-phase gas or vapour/liquid flow through straight pipes", Engineering Sciences Data Unit (ESDU) published September 1976, London.

10. Ward, J.A. Private communication (1977).

11. Baroczy, C.J. "A systematic correlation of two-phase pressure drop", Chem. Engng. Prog. Symp. Series 62, No.64, pp.323 (1966).

12. Chisholm, D. "Pressure gradients due to friction during flow of evaporating two-phase mixtures in smooth tubes and channels", Int. J. Heat Mass Transfer Vol.16, pp.347-358 (1973).

13. Lombardi, C. and Peddrochi, E. "A pressure drop correlation in two-phase flow", Energia Nucleare, Vol.19, No.2.

14. Martinelli, R.C. and Nelson, D.B. "Prediction of pressure drop during forced circulation boiling of water", Trans. ASME 70, p.695 (1948).

15. Macbeth, R.V. Private communication quoted in paper by Brittain, I. and Fayers, F.J. "A review of U.K. developments in thermal-hydraulic methods for loss of coolant accidents", Paper presented at CSNI Meeting on Transient Two-Phase Flow, Toronto, August 3rd-4th 1976.

16. Collier, J.G. "Convective Boiling and Condensation", (2 ed.) published by McGraw Hill Book Co. Ltd. (UK), 1980.

17. "The gravitational component of pressure gradient for two-phase gas or vapour/liquid flow through straight pipes", Engineering Sciences Data Unit (ESDU) published 1977, London.

18. Bryce, W.M. "A new flow dependent slip correlation which gives hyperbolic steam-water mixture flow equations", AEEW-R1099 (1977).

19. Smith, S.L. "Void fractions in two-phase flow. A correlation based on an equal velocity head model", Proc. Inst. Mech. Engng. Vol.184, Pt.1, No.36, pp.647-664 (1969-70).

20. Premoli, A., Di Francesco, D., Prima, A. "An empirical correlation for evaluating two-phase mixture density under adiabatic conditions", European Two-Phase Flow Group Meeting, Paper B9, Milan, June 1970.

21. Chisholm, D. "Research note: void fraction during two-phase flow", J. Mech. Engng. Sci. Vol.15, No.3, pp.235-236 (1973).

22. Lockhart, R.W. and Martinelli, R.C. "Proposed correlation of data for isothermal two-phase two component flow in pipes", Chemical Engineering Progress Vol.45, No.1, pp.39-48 (1949).

23. Hughmark, G.A. "Hold-up in gas-liquid flow", Chemical Engineering Progress, Vol.58, No.4, pp.62-65 (1962).

24. Thom, J.R.S. "Prediction of pressure drop during forced circulation boiling of water", Int. J. Heat Mass Transfer, Vol.7, pp.709-724, (1964).

25. Jones, A.B. "Hydrodynamic stability of a boiling channel", KAPL-2170 (1961).

26. Collier, J.G. "Single-Phase and Two-Phase Flow behaviour in Primary Circuit Components", pp.313-355, Vol.1 "Two Phase Flows and Heat Transfer" Editors S. Kakac and F. Mayinger, published by Hemisphere Publishing Corp. (1977).

27. Weisman, J., Husain, A. and Harshe, B. "Two-phase pressure drop across abrupt area changes and restrictions", Two-phase Transport and Reactor Safety, Vol.4, pp.1281-1316, Veziroglu, T.N. and Kakac, S., Editors, published by Hemisphere Publishing Corp. (1978).

28. Ishii, M. "Two-Phase Flows and Heat Transfer", Vol.III, p.1461, Editors S. Kakac and F. Mayinger, published by Hemisphere Publishing Corp. (1977).

29. Gardner, G.G. Private communication (1979).

30. Koshkin et al. "Unsteady Heat Transfer", Izdatel'stvo Mechinostroenie, Moscow (1973).

31. Ishii, M. "Thermo-fluid Dynamic Theory of Two-phase Flow", Eyrolles, Paris (1975).

32. Ishii, M. "One dimensional drift flux model and constitutive equations for relative motion between phases in various two-phase flow regimes", ANL-77-47 (1977).

33. Zuber, N. and Findlay, J. "Average volumetric concentration in two phase flow systems", Trans. ASME J. Heat Transfer 87, 453, (1965).

34. Wallis, G.B. "One dimensional two-phase flow", published by McGraw Hill (1969).

35. Zuber, N. and Staub, F.W. "The propagation and the wave form of the vapour volumetric concentration in boiling forced convection system under oscillating conditions", Int. J. Heat Mass Transfer 9, pp.871-895 (1966).

36. Hancox, W.T. and Nicoll, W.B. "A general technique for the prediction of void distribution in non-steady two-phase forced convection", Int. J. Heat Mass Transfer 14, pp.1377-1394 (1971).

37. Inayatullah, G. and Nicoll, W.B. "Application of the drift flux formulation to the prediction of steady periodic and transient two-phase flows", Vol.1, "Two Phase Flows and Heat Transfer", editors S. Kakac and F. Mayinger, published by Hemisphere Publishing Corp. (1977).

38. Chisholm, D. "Pressure losses in bends and tees during steam-water flow", NEL Report No.318 (1967).

39. Gardel, A. "Pressure drops in flows through T-shaped fittings", Bull. Techn. Suisse Rom. 83, 9, pp.123-130 and 10, pp.143-148, April/May (1957).

40. Chisholm, D. and Sutherland, L.A. "Prediction of pressure gradients in pipeline systems during two phase flow", Paper 4 presented at Symp. on Fluid Mechanics and Measurements in Two-Phase Flow Systems, Leeds, 24-25 September, 1969.

41. Kreps, D.A. and Kennedy, W.G. "C-E/EPRI Two-Phase Primary Pump Performance Programme", Proc. of Topical Meeting on Thermal Reactor Safety, Sun Valley, Idaho, July 31-Aug. 4 (1977).

42. Olson, D.J. "Single and Two-phase performance characteristics of the MOD-1 semiscale pump under steady state and transient fluid conditions", ANCR-1165, Oct. 1974.

43. Pushkina, O.L. and Sorokin, Y.L. "Breakdown of liquid film motion in vertical tubes", Heat Transfer - Soviet Research 1 (5), 56-64, (1969).

44. Richter, H.J. and Lovell, T.W. "The effect of scale on two-phase countercurrent flow flooding in vertical tubes", Final report on NRC contract No. AT(49-24) 0329, Thayer School of Engng., Dartmouth College, Hanover, New Hampshire (1977).

45. Piggott, B.D.G. and Ackermann, M.C. "A study of countercurrent flow and flooding in parallel channels", paper presented at the 1980 ICHMT International Seminar on Nuclear Reactor Safety Heat Transfer, Dubrovnik (1980).

46. Saha, P. "A review of two-phase steam-water critical flow models with emphasis on thermal non-equilibrium", NUREG-CR-0417, Brookhaven Nat. Lab. (1978).

47. Levy, S. "Prediction of two-phase critical flow rate", J. of Heat Transfer, Trans. ASME Series C, Vol.87, 53-58, (1965).

48. Fauske, H. "Contribution to the Theory of Two-Phase one Component Critical Flow", ANL-6633 (1962).

49. Henry, R.E., Fauske, H.K. and McComos, S.T. "Two-Phase critical flow at low qualities", Pt.I and II, Nucl. Sci. and Engng. Vol.41, pp.79-91, 92-98 (1970).

50. Ardron, K.H., Duffey, R.B. and Hall, P.C. "Studies of the physical models used in transient two-phase flow analysis", Conf. on Heat and Fluid Flow in Water Reactor Safety, I. Mech. E., Manchester 13-15th September, 1977.

7 SINGLE- AND TWO-PHASE HEAT TRANSFER

by
J.G. Collier
UKAEA, Harwell, UK

The various regions of single phase and two-phase heat transfer encountered in nuclear reactors are identified with the aid of a three dimensional diagram—the "heat transfer (or boiling) surface". The boundaries between the various regions are discussed and correlations given to allow the appropriate heat transfer rates to be established. This chapter thus serves as an introduction to later chapters providing the student with sufficient fundamental background to allow the descriptions of nuclear system heat transfer behavior to be easily understood.

1. INTRODUCTION

The heat transfer processes within a nuclear reactor may be discussed, firstly, in terms of the thermodynamic phase state of the coolant and secondly, in terms of the method of circulation of the coolant. With liquid-cooled reactors it is necessary to consider liquid, two-phase and vapour states since limiting fault conditions are conceivable where such phase states will be present in the reactor. In the particular case of water-cooled reactors even supercritical phase states are possible. With gas-cooled reactors, the situation is a good deal simpler in that only the gaseous state needs to be considered. Three modes of circulation of the coolant can occur; forced circulation, natural circulation and natural convection. Forced circulation of the coolant is the normal mode of cooling a reactor and is provided by either pumps or gas circulators, the exception being some early designs of natural circulation boiling water reactor. The coolant may also circulate as a result of buoyancy forces acting on regions where the temperature of the fluid is higher and its density lower than the average value. If this circulation is via the external reactor circuit it is referred to as "natural circulation". Natural circulation will only occur if the heat sink (the steam generator) is located at a higher elevation than the heat source (the reactor core) and if the circuit is properly designed with natural circulation in mind. In the case of liquid-cooled reactors, natural circulation can be prevented by the unexpected presence of vapour or gas locks in the reactor pipework.

If the circulation caused by buoyancy forces is restricted to the reactor vessel it is usually referred to as "natural convection". Partial heat sinks may exist in the reactor vessel itself; for example, the graphite moderator in a gas-cooled reactor. These partial heat sinks are however usually insufficient to balance the heat source and the reactor will then increase in temperature until a new equilibrium state is achieved.

In this Chapter we will concentrate on the heat transfer regions present in a vertical heated channel as a liquid is evaporated into vapour. Simple correlations will be given for the estimation of parameters of design interest such as the critical heat flux and the heat transfer coefficient in terms of physical properties, mass flow, cross sectional geometry and local values of vapour quality and heat flux. Because the majority of the world's reactors are cooled by light (ordinary) water we will concentrate upon this particular coolant.

2. THERMO-HYDRAULIC ANALYSIS METHODS[1,2]

In order to establish the behaviour of a fuel assembly under a variety of operating conditions, including the length of time in the reactor, its location within the core, control rod positions etc. mathematical models are established[1,2,3,4,5,6] in which the fuel assembly is considered to be sub-divided into a number of parallel interacting flow sub-channels between the fuel rods. The equations of mass, momentum and energy conservation are solved to give radial and axial variations in fluid enthalpy and mass velocity. The components of the fuel rod (pellet, gas gap, clad) are modelled by a series of radial nodes at a number of axial levels. Heat transfer within the individual fuel rods is treated by a two-dimensional heat conduction equation whilst for the flow sub-channels, interchange of mass, heat and momentum is allowed been neighbouring sub-channels, the amounts being governed by the conservation equations. Once the *local* enthalpy (i) and the *local* mass velocity (G) at each sub-channel node has been established, suitable correlations are used to define the heat transfer regime. Basically this involves a comparison of the *local* heat flux (ϕ) through the fuel rod cladding at any time instant with the critical heat flux (ϕ_{crit}) for these same fluid-dynamic conditions. If ϕ is greater than ϕ_{crit} then overheating of the fuel can occur. The extent of this overheating can be established using the appropriate correlation for the heat transfer coefficient. From this analysis the local temperature of the fuel cladding (T_W) is established

$$T_W = \text{fn}\,(\phi,\ i,\ G,\ \text{system pressure, system geometry}) \qquad (1)$$

The fuel clad surface temperature is thus a function of at least five variables, some of which are themselves interdependent. To simplify further considerations, assume that the mass velocity, system pressure and geometry all remain constant. Then the clad temperature is only a function of the local heat flux and local enthalpy. In a three dimensional diagram where clad temperature, local heat flux and local enthalpy are the three axes, this relationship can be presented as a surface - the so-called "heat transfer (or boiling) surface". We will discuss what such a surface actually looks like in the following sections. Note that this "heat transfer surface" is not unique but is itself a function of the variables we did not consider, namely mass velocity, system pressure and geometry. Note also that in establishing the respective correlations for the dependant variables such as critical heat flux and heat transfer coefficient the *local* values of fluid enthalpy, velocity, surface heat flux etc. are used. But single phase flow, to some extent, and two-phase liquid-vapour flow, to a much greater extent, is demonstrably dependent upon the flow history. Hence there is a need to try to model the behaviour in a more realistic manner than is possible by considering only point conditions and we will return briefly to that topic towards the end of this Chapter.

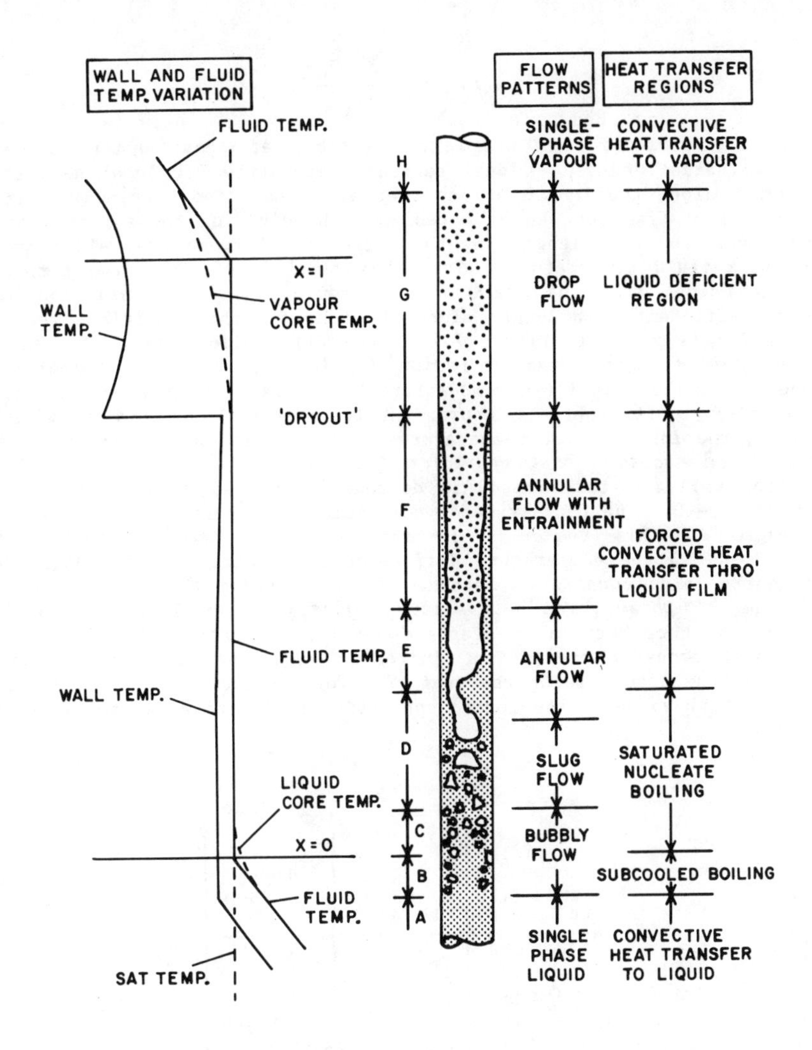

Fig.7.1 Regions of Heat Transfer in Convective Boiling

3. HEAT TRANSFER REGIONS IN A VERTICAL TUBE[7]

The heat transfer regions encountered when a liquid is evaporated in vertical heated tube are shown in Figure 7.1. Boiling cannot occur until the tube wall temperature exceeds the saturation temperature. The amount by which the wall temperature exceeds the saturation temperature is known as the "degree of superheat" ΔT_{SAT} and the difference between the saturation and the local bulk fluid temperature is known as the "degree of subcooling" ΔT_{SUB}. In the two-phase region where net vapour generation occurs, the variable characterising the heat transfer mechanism is the thermodynamic vapour "quality" (x). At any distance, z along the tube this is given by:

$$x(z) = \frac{i(z) - i_f}{i_{fg}} \tag{2}$$

It is useful at this stage to describe, at least qualitatively, the progressive variation of the local surface temperature (or local heat transfer coefficient) along the length of the tube as evaporation proceeds. The local heat transfer coefficient can be established by dividing the surface heat flux (constant over the tube length) by the difference between the wall temperature and the bulk fluid temperature. Typical variations of these two temperatures with length along the tube are shown in Figure 7.1. The variation of heat transfer coefficient with length along the tube for the conditions represented in Figure 7.1 is given in Figure 7.2 (curve (i), solid line). In the *single phase convective heat transfer region* (region A) the wall temperature is displaced above the bulk fluid temperature by a relatively constant amount, (the heat transfer coefficient is approximately constant) and is modified only slightly by the influence of temperature on the liquid physical properties. In the *sub-cooled nucleate boiling region* (region B) the temperature difference between the wall and the bulk fluid decreases linearly with length up to the point where x = 0. The heat transfer coefficient, therefore, increases linearly with length in this region. In the *saturated nucleate boiling region* (regions C and D) the temperature difference and, therefore, the heat transfer coefficient remains constant. Because of the reducing thickness of the water film in the *two-phase forced convective region* (regions E and F) the difference in temperature between the surface and the saturation temperature reduces and the heat transfer coefficient increases with increasing length or steam quality. At the dryout point the heat transfer coefficient is suddenly reduced from a very high value in the forced convective region to a value near to that

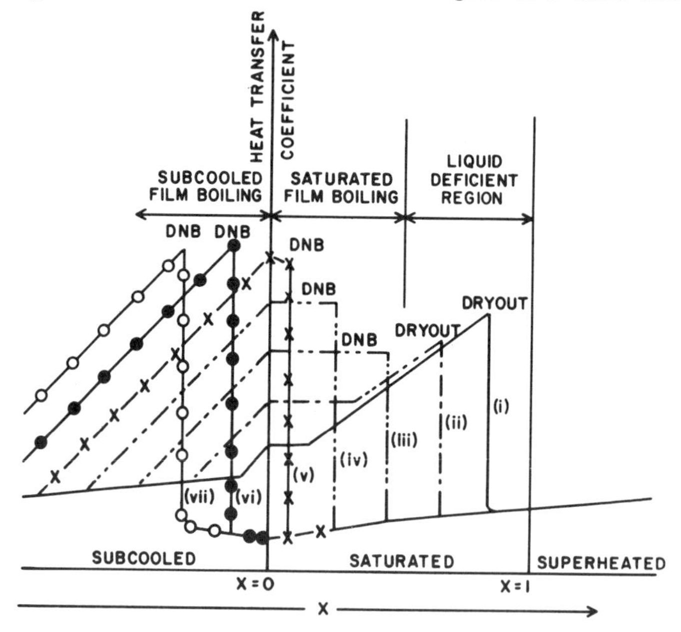

Fig.7.2 Variation of Heat Transfer Coefficient with Quality with increasing Heat Flux as Parameter

expected for heat transfer by forced convection to dry saturated steam. As the quality increases through the *liquid deficient region* (region G) so the steam velocity increases and the difference in temperature between the surface and the saturation value decreases with the corresponding rise in heat transfer coefficient. Finally, in the single-phase steam region ($x > 1$) the wall temperature is once again displaced by a constant amount above the bulk fluid temperature and the heat transfer coefficient levels out to that corresponding to convective heat transfer to a single phase steam flow.

The above comments have been restricted to the case where a relatively low heat flux is supplied to the walls of the tube. The effect of progressively increasing the surface heat flux whilst keeping the inlet flow-rate constant, will now be considered with reference to Figures 7.2 and 7.3. Figure 7.2 shows the heat transfer coefficient plotted against steam quality with increasing heat flux as parameter (curves (i)-(vii)). Figure 7.3 shows the various regions of two-phase heat transfer in forced convective boiling on a three-dimensional diagram with heat flux, steam quality and temperature as co-ordinates - "the heat transfer surface". Curve (i) relates to the conditions shown in Figure 7.1

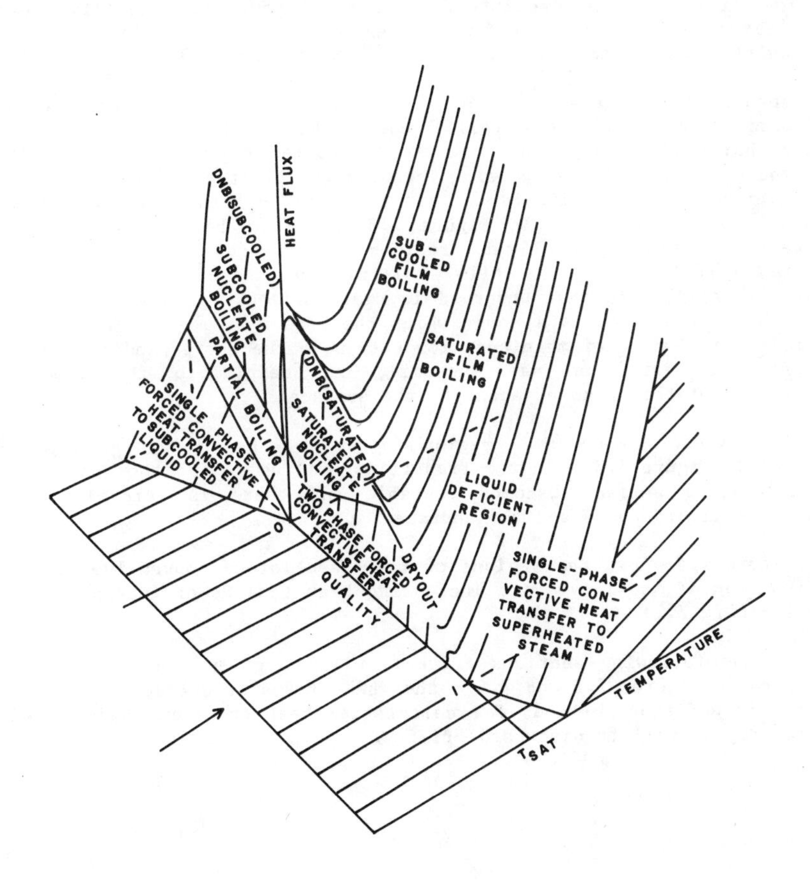

Fig.7.3 The Heat Transfer (or Boiling) Surface

for a low heat flux being supplied to the walls of a tube. The temperature pattern shown in Figure 7.1 will be recognised as the projection in plan view (temperature-quality co-ordinates) of curve (i). Curve (ii) shows the influence of increasing the heat flux. Sub-cooled boiling is initiated sooner, the heat transfer coefficient in the nucleate boiling region is higher but is unaffected in the two-phase forced convective region. Dryout occurs at a lower steam quality. Curve (iii) shows the influence of a further increase in heat flux. Again, sub-cooled boiling is initiated earlier and the heat transfer is again higher in the nucleate boiling region. As the steam quality increases, before the two-phase forced convective region is initiated, and while bubble nucleation is still occurring, an abrupt deterioration in the cooling process takes place. This transition is essentially similar to the critical heat flux phenomenon in saturated pool boiling and will be termed *departure from nucleate boiling* (DNB).

The mechanism of heat transfer under conditions where the critical heat flux (DNB or dryout) has been exceeded is dependent on whether the initial condition was the process of "boiling" (i.e. bubble nucleation in the sub-cooled or low steam quality regions) or the process of "evaporation" (i.e. evaporation at the water film-steam core interface in the higher steam quality areas). In the latter case, the liquid deficient region is initiated; in the former case the resulting mechanism is one of *film boiling* (Figure 7.3).

Returning to Figures 7.2 and 7.3, it can be seen that further increases in heat flux (curves (vi) and (vii)) cause the condition of "departure from nucleate boiling" (DNB) to occur in the sub-cooled region with the whole of the saturated or "quality" region being occupied by, firstly, "film boiling" and, in the latter stages, the "liquid deficient region" - both relatively inefficient modes of heat transfer. In Figure 7.3 the film boiling surface has been arbitrarily divided into two regions: *sub-cooled film boiling* and *saturated film boiling*. Film boiling in forced convective flow is essentially similar to that observed in pool boiling. An insulating steam film covers the heating surface through which the heat must pass. The heat transfer coefficient is orders of magnitude lower than in the corresponding region before the critical heat flux was exceeded, due mainly to the lower thermal conductivity of the steam adjacent to the surface.

A *transition boiling* region occupies the reverse slope (partly obscured) in Figure 7.3. In the transition boiling region the vapour film next to the heating surface becomes unstable and there is intermittent contact between the liquid phase and the surface.

Figure 7.4 shows the regions of two-phase forced convective heat transfer as a function of "quality" with increasing heat flux as ordinates (an elevation view of Figure 7.2).

In the following sections criteria will be given whereby the boundaries delineated in Figures 7.3 and 7.4 - the "heat transfer surface" - can be established. In addition, methods for calculating heat transfer rates in each heat transfer region will be discussed briefly.

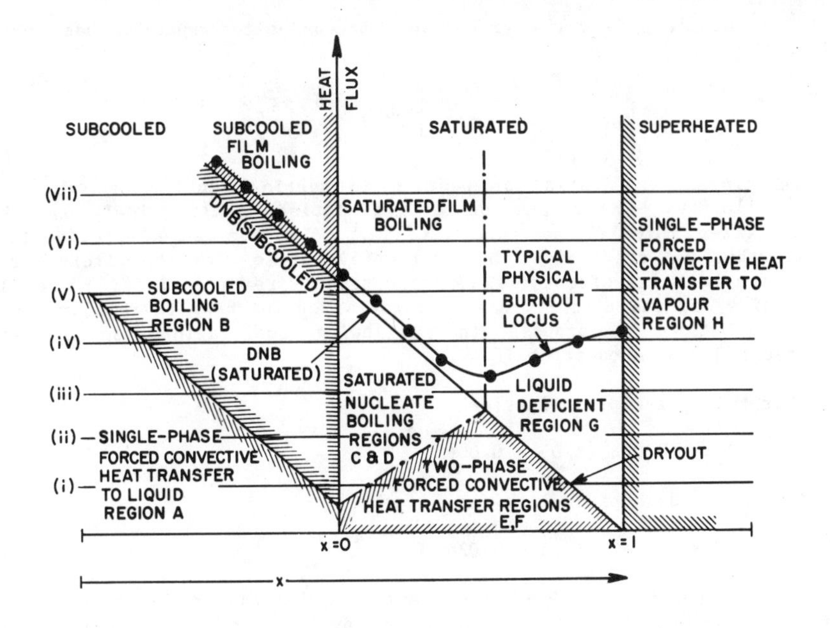

Fig.7.4 Regions of Two-Phase Forced Convective Heat Transfer as a Function of Quality with increasing Heat Flux as ordinate

4. SINGLE-PHASE LIQUID HEAT TRANSFER

The tube surface temperature in region A, convective heat transfer to single-phase liquid, is given by

$$T_W = T_f(z) + \Delta T_f \tag{3}$$

and

$$T_f = \phi / h_{fo} \tag{4}$$

where ΔT_f is the temperature difference between the tube inside surface and the mean bulk liquid temperature at a length z from the tube inlet, h_{fo} is the heat transfer coefficient to single-phase liquid under forced convection. The liquid in the channel may be in laminar or turbulent flow. For laminar flow a variety of theoretical relationships are available, depending on the boundary conditions, i.e. constant surface heat flux or surface temperature, developing velocity profile or fully-developed flow. The following empirical equation based on experimental data takes into account the effect of varying physical properties across the flow stream and the influence of natural convection.

$$\left[\frac{h_{fo}D}{k_L}\right] = 0.17\left[\frac{GD}{\mu_L}\right]^{0.33}\left[\frac{c_p\mu}{k}\right]_L^{0.43}\left[\frac{Pr_L}{Pr_W}\right]^{0.25}\left[\frac{D^3\rho_L^{\,2}g\beta\Delta T}{\mu_L^{\,2}}\right]^{0.1} \tag{5}$$

This relationship is valid for heating in vertical upflow or cooling in vertical downflow for $z/D > 50$ and $\{GD/\mu_L\} < 2000$.

For turbulent flow the well-known Dittus-Boelter equation has been found satisfactory

$$\left[\frac{h_{fo}D}{k_L}\right] = 0.023 \left[\frac{GD}{\mu_L}\right]^{0.8} \left[\frac{c_p\mu}{k}\right]_L^{1/3} \tag{6}$$

This relationship is valid for heating in vertical upflow or $z/D > 50$ and $\{GD/\mu_L\} > 10{,}000$. The Dittus-Boelter correlation is also widely used for the flow of water parallel to tube bundles. The available experimental data indicate that the coefficient in equation(6) varies with the pitch-to-diameter ratio. The heat transfer coefficients obtained from equation (6) are slightly conservative for the pitch-to-diameter ratios of interest to reactor designers. Tong and Weisman[1] propose the following correlations to replace the coefficient 0.023 in equation (6).

For triangular-pitch lattices

$$C = 0.026\ (S/D) - 0.006 \tag{7}$$

and for square pitch lattices

$$C = 0.042\ (S/D) - 0.024 \tag{8}$$

where S = tube pitch and D = tube diameter. Variations in the heat transfer coefficient around the periphery of a rod have been found to be negligible for S/D ratios > 1.2.

5. THE ONSET OF SUBCOOLED NUCLEATE BOILING

Consider the variations of temperature of the tube inner surface at a point, z, from the inlet as the heat flux is steadily increased at a given inlet subcooling and mass velocity. Figure 7.5 shows the relationship in a qualitative form. Three regions are shown as the single-phase (sub-saturation) region AB, the "partial boiling" region BCDE and the fully-developed subcooled boiling region EF. Figure 7.5 will be easily recognised as a part-section in the subcooled region through the surface illustrated in Figure 7.3.

As the heat flux is increased at constant subcooling, the relationship between the surface temperature, T_W, and the heat flux will follow the line ABD' until the first bubbles nucleate. A higher degree of superheat is necessary to initiate the first bubble nucleation sites at a given heat flux then indicated by the curve ABCDE. When nucleation first occurs the surface temperature drops from D' to D and, for further increases in heat flux, follows the line DEF. The criterion for the onset of boiling can crudely be established as the intersection of line ABD' and the fully-developed boiling curve C'' EF.

A more refined treatment of the onset of nucleation can be derived by considering the temperature profile in the region adjacent to the heated wall. This treatment was originally proposed by Hsu for pool boiling and has been used to predict both the onset of subcooled boiling and the suppression of saturated nucleate boiling. Consider Figure 7.6; if the liquid is at a uniform temperature, T_G, then bubble nuclei of radius r^* will grow if this temperature exceeds that given by

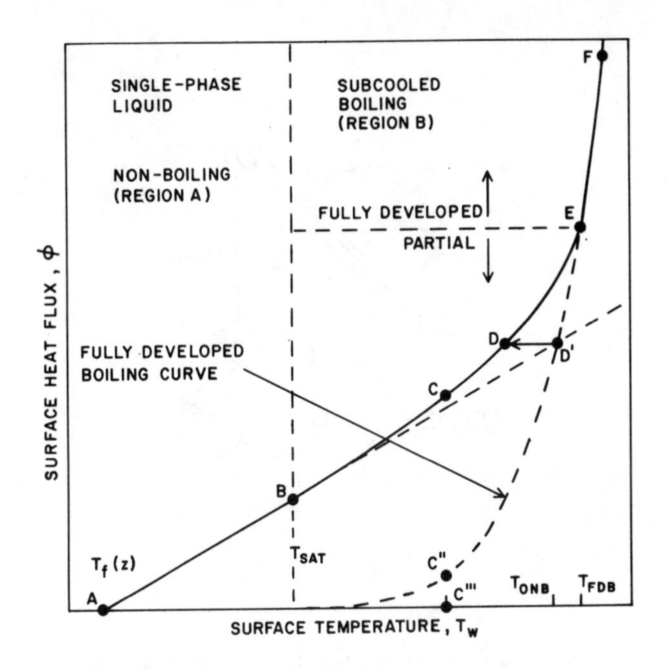

Fig.7.5 The Forced Convection Boiling Curve

$$(T_G - T_{SAT}) = \frac{R\ T_{SAT}\ T_G}{i_{fg}\ M} \ln \left[1 + \frac{2\sigma}{p\ r^*}\right] \tag{9}$$

In a heated system, there is a temperature gradient away from the wall which is approximated in the linear form

$$T_L(y) = T_W - \left(\frac{\phi y}{k}\right) \tag{10}$$

The postulate of Hsu is that the bubble nuclei on cavities in the heated wall will only grow if the lowest temperature on the bubble surface (i.e. that furthest away from the wall) is greater than T_G. Allowing for the distortion of the liquid temperature profile by the bubble by plotting T_G against nr^* and T_L against y on the same ordinates (Figure 7.6) it will be seen that only that range of bubbles for which T_L is greater than T_G (i.e. radii, r_{MIN} to r_{MAX}) can grow. If, over the whole field, T_L is less than T_G, then no nuclei will grow. When equation (10) is just tangent to equation (9) then nuclei of a critical radius, r_{CRIT} will grow. Thus, for nucleation of the critical nucleus (n = 1)

$$T_L(y) = T_G \text{ and } \frac{dT_L(y)}{dy} = \frac{dT_G}{dr} \tag{11}$$

Analytical solution of the equations was carried out by Davis and Anderson[8] who obtained the following expressions

$$\phi_{ONB} = \frac{k_L}{4B} (\Delta T_{SAT})^2_{ONB} \tag{12}$$

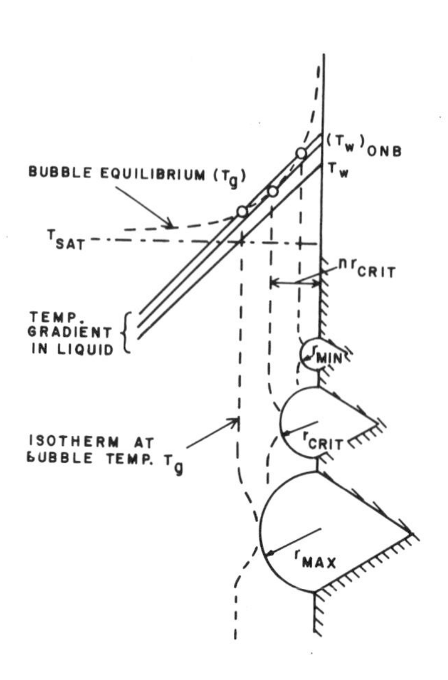

Fig.7.6 Onset of Nucleation in Subcooled Boiling

$$\text{and } r_{CRIT} = \sqrt{\frac{Bk_L}{\phi}} \tag{13}$$

$$\text{where } B = \left[\frac{2\sigma T_{SAT}(v_G - v_L)}{i_{fg}}\right] \tag{14}$$

The above treatments can only predict the onset of nucleate boiling accurately if there is a sufficiently wide range of "active" cavity sizes available. At low fluxes and low pressures the values of r_{CRIT} predicted from equation (13) may be so large that no "active" sites of this size are present on the heating surface. In this case an estimate of the largest "active" cavity size available on the heating surface must be made. Reasonable agreement with experimental data for water was found when a maximum "active" cavity size of 1 μm radius was used.

For the subcooled boiling region, equation (12) must be solved simultaneously with the heat transfer equation

$$\phi = h_{fo}(\Delta T_{SAT} + \Delta T_{SUB}(z)) \tag{15}$$

to give the heat flux ϕ_{ONB} and $(\Delta T_{SAT})_{ONB}$ required for the onset of boiling. The boundary between region A and region B shown in Figure 7.4 can be derived from such a simultaneous solution. The treatment may also be extended to cover the influence of dissolved gases upon the onset of nucleation.

6. SUBCOOLED NUCLEATE BOILING

Once boiling has been initiated, only comparatively few nucleation sites are operating initially so that a proportion of the heat will be transferred by normal single-phase processes between patches of bubbles. As the surface temperature increases, so the number of bubble sites also increases and the area for single-phase heat transfer decreases. Finally, the whole surface is covered by bubble sites, boiling is "fully-developed" and the single-phase component reduces to zero. In the "fully-developed" boiling region, velocity and subcooling, both of which have a strong influence on single-phase heat transfer, have little or no effect on surface temperature as observed experimentally.

Jens and Lottes(9) summarised experiments on subcooled boiling of water flowing upwards in vertical electrically-heated stainless steel or nickel tubes, having inside diameters ranging from 3.63 to 5.74 mm. System pressures ranged from 7 bars to 172 bars, water temperatures from 115°C to 340°C, mass velocities from 11 to 1.05×10^4 kg/m^2s, and heat fluxes up to 12.5 MW/m^2. These data were correlated by a dimensional equation valid for water only.

$$\Delta T_{SAT} = 25\phi^{0.25} e^{-p/62} \qquad (16)$$

where p is the absolute pressure in bar, ΔT_{SAT} is in °C and ϕ is in MW/m^2.

More recently, Thom(10) reported that the values of ΔT_{SAT} estimated from equation (16) were consistently low over the range of his experiments. A modified equation of the same form and valid only for water was suggested

$$\Delta T_{SAT} = 22.65\,\phi^{0.5} e^{-p/87} \qquad (17)$$

in SI units. This equation is recommended for calculational purposes. A number of studies of the basic mechanisms of subcooled boiling have been carried out but no generally accepted theory has so far emerged.

7. SATURATED NUCLEATE BOILING REGION

Here, the mechanism of heat transfer is essentially identical to that in the subcooled region. A thin layer of liquid near to the heated surface is superheated to a sufficient degree to allow nucleation. Just as with all other characterising variables, such as coolant temperature, void fraction, pressure gradient, etc., the heat transfer coefficient and heater surface temperature variation is smooth and continuous through the thermodynamic boundary (x = 0) marking the onset of saturated boiling. The methods and equations used to correlate experimental data in the subcooled region remain valid for this region with the provision that $T_f(z) = T_{SAT}$. Just as the heat transfer mechanism in the subcooled region is independent of the subcooling and, to a large degree, the mass velocity, so it may be inferred that the heat transfer process in this region is independent of the steam quality x(z), and the mass velocity. This, indeed, is found to be experimentally correct for the case of "fully-developed nucleate boiling". Thus, because the bulk temperature is constant in this region the heat transfer coefficient is also constant since ΔT_{SAT} is fixed.

8. SUPPRESSION OF SATURATED NUCLEATE BOILING

To maintain nucleate boiling on the surface, it is necessary that the wall temperatures exceeds the critical value for a specified heat flux. If the wall superheat is less than that given by equation (13) for the imposed surface heat flux, then nucleation does not take place; the value of ΔT_{SAT} for comparison with this equation is calculated from the ratio (ϕ/h_{TP}), where h_{TP} is the two-phase heat transfer coefficient in the absence of nucleation.

$$\phi_{ONB} = \frac{4B}{k_L} h_{TP}^{2} \qquad (18)$$

Equation (18) represents an equation for the boundary between the saturated nucleate boiling and the two-phase forced convective regions (Figure 7.4). This relationship is only valid on the basis of a complete range of "active" cavities on the heating surface. As with subcooled boiling, there is a similar "partial boiling" region between fully-developed saturated nucleate boiling and the two-phase forced convective region. In this transition region, both the forced convective and nucleate boiling mechanisms are significant.

Direct observations of nucleation in annular flow have been reported by Hewitt et al[11]. The results qualitatively confirmed the expected trends. However, the values of ϕ_{ONB} calculated from equation (18) are very much lower than those measured. This discrepancy is consistent with the findings of Davis and Anderson mentioned above. Equation (18) predicts that nucleation will persist at wall superheat down to 3°C but it is very unlikely that the large sizes of cavity required to be "active" down to this superheat were present on the stainless steel heating surface.

9. THE TWO-PHASE FORCED CONVECTION REGION

The two-phase forced convective region is most likely to be associated with the annular flow pattern. Heat is transferred by conduction or convection through the water film and steam is generated continuously at the interface. Extremely high heat transfer coefficients are possible in this region; values can be so high as to make accurate assessment difficult. Typical figures for water of up to 200 kW/m^2 °C have been reported.

Many workers have correlated their experimental results of heat transfer rates in the two-phase forced convective region in the form

$$\frac{h_{TP}}{h_f} = \text{fn}\left(\frac{1}{X_{tt}}\right) \qquad (19)$$

where h_f is the value of the single-phase liquid heat transfer coefficient based on the liquid component flow, and X_{tt} is the Matinelli parameter.

A number of relationships of the form of equation (19) have been proposed and in some cases these have been extended to cover the saturated nucleate boiling region also. However, these correlations do have a high mean error (± 30%) and a more satisfactory correlation has been proposed by Chen[12]. The correlation covers both the "saturated nucleate boiling region" and the "two-phase forced convective region". Both mechanisms occur to some degree over the entire range of the correlation and that the contributions made by the two mechanisms are additive.

$$h_{TP} = h_{NcB} + h_c \tag{20}$$

where

$$h_{NcB} = 0.00122 \left[\frac{k_L^{0.79} \, c_{pL}^{0.45} \, \rho_L^{0.49}}{\sigma^{0.5} \, \mu_L^{0.29} \, i_{fg}^{0.24} \, \rho_G^{0.24}} \right] \Delta T_{SAT}^{0.24} \, \Delta p^{0.75} (S) \tag{21}$$

$$\text{and } h_c = 0.023 \left[\frac{G(1-x)D}{\mu_L} \right]^{0.8} \left[\frac{\mu c_p}{k} \right]_L^{0.4} \left(\frac{k_L}{D} \right) (F) \tag{22}$$

The parameter S, the nucleate boiling suppression factor, and the parameter F, are graphical functions which may be approximated by the following curve-fits for S[21].

$$S = \begin{bmatrix} [1 + 0.12 \, (Re'_{TP})^{1.14}]^{-1} & , \quad Re'_{TP} < 32.5 \\ [1 + 0.42 \, (Re'_{TP})^{0.78}]^{-1} & , \quad 32.5 \leqslant Re'_{TP} < 70 \\ 0.1 & , \quad Re'_{TP} \geqslant 70 \end{bmatrix} \tag{23}$$

and for F

$$F = \begin{bmatrix} 1.0 & , \quad \frac{1}{X_{tt}} \leqslant 0.10 \\ 2.35 \, (\frac{1}{X_{tt}} + 0.213)^{0.736} & , \quad \frac{1}{X_{tt}} > 0.10 \end{bmatrix} \tag{24}$$

where

$$\frac{1}{X_{tt}} = \left(\frac{x}{1-x} \right)^{0.9} \left(\frac{\rho_L}{\rho_G} \right)^{0.5} \left(\frac{\mu_G}{\mu_L} \right)^{0.1} \tag{25}$$

and

$$Re'_{TP} = \left(\frac{G(1 - x)D}{\mu_L} \right) F^{1.25} \times 10^{-4} \tag{26}$$

10. THE CRITICAL HEAT FLUX CONDITION

The heat flux in the nucleate boiling regions cannot be increased indefinitely. At some critical value sufficient liquid is unable to reach the heating surface to cool it due to the rate at which vapour is leaving th surface and an abrupt deteriortion in the cooling process takes place - "departure from nucleate boiling" (DNB). Likewise, at some critical value of the "quality", complete evaporation of the liquid film may occur - "dryout". The term "critical heat flux" will be used to cover both these separate and distinct processes. A distinction should also be made between the precise point at which a more or less rapid surface temperature rise from a temperature close to the saturation value is initiated and the point at which failure of the heating surface due to rupture or melting occurs - "physical burnout". The term "burnout heat flux", which also suggests this latter condition, has been abused by many, who use it to note the heat flux at which the rapid deterioration of the cooling processes occurs. This ambiguity can be illustrated by reference to Figures 7.3 and 7.4. Low heat transfer coefficients coupled with the relatively high heat flux values required to initiate film boiling in the subcooled or low quality regions results in extremely high temperature differences at the critical heat flux condition. Failure of the heating surface usually occurs and thus the heat flux to initiate DNB is often identical with that to cause "physical burnout". However, this is not the case in the liquid deficient region where higher heat transfer coefficients and lower critical heat fluxes

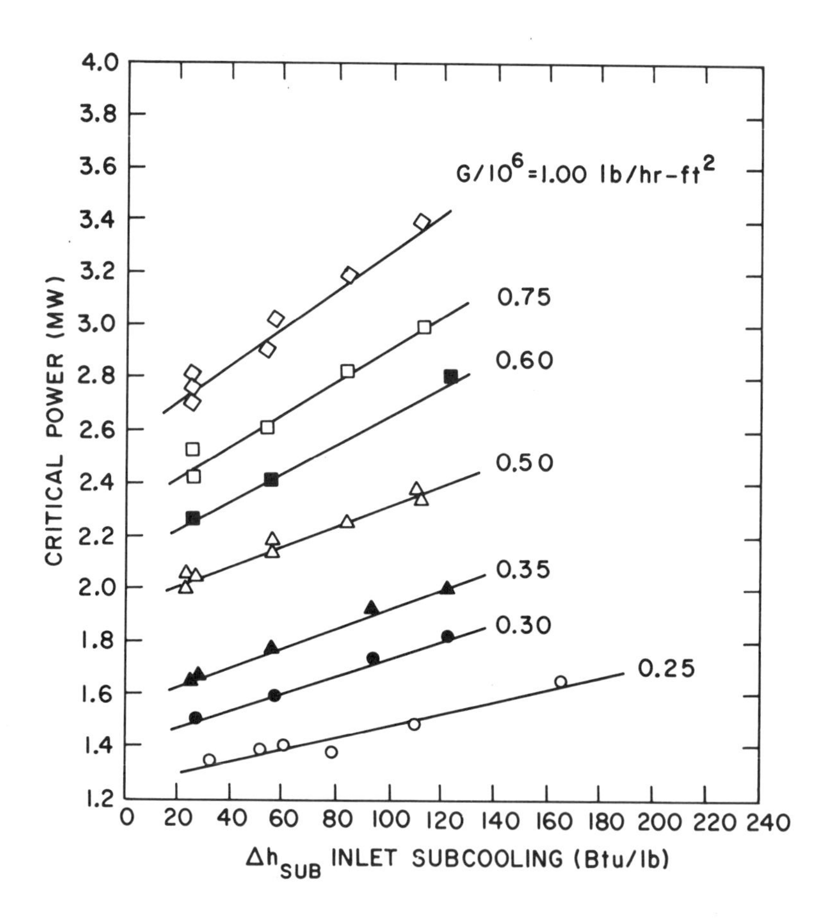

Fig.7.7 Critical power versus inlet subcooling, 16 rod x 12ft. cosine, uniform peaking, 6.9 MPa, various flowrates (Lahey and Moody[2])

cause only modest temperature excursions at "dryout". In this region the physical burnout locus denotes a particular isotherm representing the failure criterion for the chosen heating surface.

Experiments to determine the critical heat flux for water in vertical uniformly-heated and non-uniformly heated round tubes, annuli and rod clusters have been carried out in many countries over the past 30 years or so. Of particular importance, in relation to the behaviour of water reactors, are the tests carried out on full-scale electrically-heated replicas of PWR and BWR fuel elements. Typical of such tests are those undertaken at the General Electric ATLAS facility at San Jose in the USA. This facility has tested full-scale 7 x 7 and 8 x 8 fuel rod arrays. The results of such tests are proprietary information but Figure 7.7 shows data for a 4 x 4 array 3.66 m in length cosine heat flux profile plotted as critical power versus inlet subcooling with mass velocity as parameter. Similar tests have been undertaken on both 15 x 15 and 17 x 17 rod arrays by the vendors of PWRs.

Experimental CHF data are usually summarised in the form of correlations in which the critical heat flux is expressed as a function of various independent and, sometimes, dependent variables. Table 7.1 summarises a selection of the better-known correlations, particularly those in use in the UK. A complete compilation would run to well over 100 such correlations. Each correlation is usually optimised within a well-defined range of variables and extrapolation outside this range is not recommended. Correlations relating to complex geometries such as rod bundles may, in some cases, be based on the average flow properties across the complete flow cross section or be related to the conditions within a particular sub-channel. Correlations of this latter type are the W-3 correlation, the WSC correlations of Bowring[23] and the

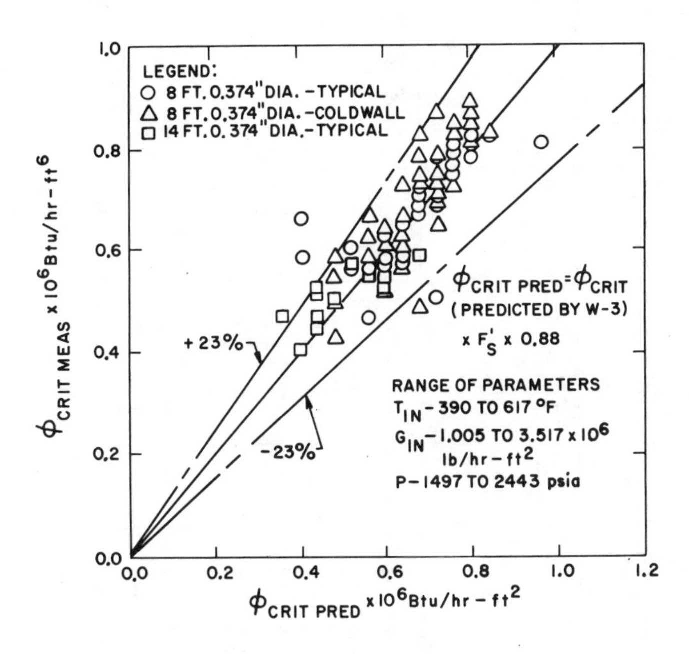

Fig.7.8 Critical Heat Flux data for 17 x 17 rod array (Westinghouse)

TABLE 7.1

SELECTION OF CRITICAL HEAT FLUX CORRELATIONS

AUTHOR	GEOMETRY	CORRELATION
Macbeth[13,14] Bowring[15]	Round Tubes Rod Bundles	$\phi_{CRIT} = \dfrac{A + CDG(\Delta i_{SUB})_i/4}{1 + Cz}$ where $A = fn(G,D,p)$ and $C = fn(G,D,p)$
CISE[16]	Round Tubes	$x_{CRIT} = a \left[\dfrac{z_{SAT}}{z_{SAT} + b}\right]$ where $a = fn(G,p)$, $b = fn(G,D,p)$
Barnett[17]	Annuli, Rod Bundles	$\phi_{CRIT} = \dfrac{A + B(\Delta i_{SUB})_i}{C + z}$ where $A = fn(G,D_h,D_e,p)$ $B = fn(G,D_h)$ $C = fn(G,D_e)$
Hewitt[18]	Round Tubes, Annuli, Rod Bundles	$x_{CRIT}\left[k_1(G)k_1(p)\right] = fn\left[k_2(G)k_2(D)z_{SAT}\right]$ where $k_1(G)$, $k_1(p)$, $k_2(G)$ and $k_2(D)$ are multiplying factors (graphical functions of G, D, and p)
Biasi et al[19]	Round Tubes	$\phi_{CRIT} = \dfrac{1.883 \times 10^3}{D^n G^{1/6}}\left[\dfrac{f(p)}{G^{1/6}} - x(z)\right]$ for low quality $\phi_{CRIT} = \dfrac{3.78 \times 10^3\, h(p)}{D^n G^{0.6}}[1 - x(z)]$ cgs units for high quality $n = 0.4$ for $D \geq 1$ cm $n = 0.6$ for $D < 1$ cm f(D) and h(p) are empirical functions of pressure
Westinghouse W-3[1,4,20]	PWR Rod Bundle Subchannel	$\phi_{CRIT} \times 10^{-6} = [(2.022-0.0004302p) + (0.1722 - 0.0000984p)\, e^{(18.177 - 0.004129p)x}]$ $\times [(0.1484 - 1.596x + 0.1729\|x\|)G \times 10^{-6} + 1.037] \times [1.157 - 0.869x] \times [0.2664 + 0.8357\, e^{-3.151D_e}] \times [0.8258 + 0.000794 (\Delta i_{SUB})_i]$ with corrections (i) for non-uniform heat flux, F (ii) for cold-wall, CWF (iii) for spacers, F'
GEXL[22]	BWR Rod bundles	$x_{CRIT} = fn(z_{SAT}, G, p, z, D_h,$ peaking factor)
Bowring[23]	BWR, PWR Rod bundle Subchannel	WSC-1, WSC-2
Babcock & Wilcox[24]	PWR Rod bundle Subchannel	$\phi_{CRIT} = a(b - c\, i_{fg}\, x)$ where $a = fn(G,D_e,p)$ $b = fn(G,p)$ $c = fn(G)$
Becker[25]	Round Tubes Annuli Rod Clusters	$\phi = G^{-0.5} f_1[x,p]\, f_2(D)$ where $f_1[x,p]$ and $f_2(D)$ are functions presented graphically. Additional corrections are added to account for heated perimeter effect.

B+W2 correlation. Typically, such correlations, when used with an appropriate sub-channel code, can predict critical heat fluxes in PWR and BWR rod clusters with an RMS error of 5-10%. Table 7.2, reproduced from reference (4), summarises the performance of these various correlations in relation to PWR and BWR conditions respectively. Figure 7.8 shows the predictions from the W-3 correlation compared with experimental measurements for a 17 x 17 PWR array.

11. HEAT TRANSFER BEYOND THE CRITICAL HEAT FLUX

There is some uncertainty about the heat transfer rates beyond the critical heat flux condition at low steam qualities and under subcooled conditions. This is because with water, at least, it is impossible to study such situations experimentally except under transient conditions. With water in the pressure range up to 100 bar, the critical heat flux values are in excess of 3 MW/m^2 for subcooled boiling, while typical values for the film boiling heat transfer coefficients in this region are 150-500 W/m^2 °C. Any attempt to pass through the critical heat flux under steady state conditions would produce heater surface temperatures in excess of 2000°C and would cause failure of most common metal surfaces - physical burnout.

12. TRANSITION BOILING

The use of experimental techniques[26,27,28] where the surface temperature rather than the surface heat flux is the controlling variable, has established quite definitely the existence of a "transition boiling" region in forced convection conditions as well as for pool boiling. A comprehensive review of the published information on transition boiling under forced convection conditions has been prepared by Groeneveld and Fung[29]. The Groeneveld and Fung review lists those studies carried out prior to 1976. Since that date, in response to the urgent demands of the nuclear industry, further studies have been initiated.

Various attempts have been made to produce correlations for the transition boiling region. Probably the most useful currently available is that by Tong and Young[30] which is given in terms of the heat flux in the transition boiling region (ϕ_{TB}) for a given surface temperature (T_W)

$$\phi_{TB} = \phi_{NB} \exp \left[- 0.001 \frac{x^{2/3}}{(dx/dz)} \left(\frac{\Delta T}{55.5} \right)^{(1 + 0.0029\, \Delta T)} \right] \qquad (27)$$

where ϕ_{NB} is the nucleate boiling heat flux (presumably equated with the critical heat flux) in W/m^2 and ΔT is the temperature difference ($T_W - T_{SAT}$) in °C. A transition boiling correlation having a wider range of application may be developed if the heat flux and wall temperature difference at the points of maximum and minimum on the "heat transfer surface" can be predicted with confidence. the present state of the art allows an accurate prediction of ϕ_{CRIT} and also ΔT_{CRIT} but the conditions at the minimum point are still subject to a large degree of uncertainty. The experimental data suggest that the conditions at the minimum point are a complex function of mass quality, flowrate and heat transfer surface properties[31].

TABLE 7.2

PERFORMANCE OF CRITICAL HEAT FLUX CORRELATIONS

(taken from reference (4))

(i) For PWR conditions

Correlation (used in conjunction with HAMBO[5])	47in. long ferrule grid 9 rod cluster		72in. long ferrule grid 9 rod cluster		84in. long mixing vane 25 rod cluster	
	e	σ	e	σ	e	σ
W-3	-7.2	8.8	-1.9	5.6	-12.4	8.3
WSC-2	6.7	6.7	0.3	5.7	2.3	9.8
B & W-2	5.2	8.4	-2.1	2.4	-6.2	4.4

e = mean error $(\phi_{CAL} - \phi_{EXP}) \times 100/\phi_{EXP}$

σ = std. deviation of errors about the mean (%)

(ii) For BWR conditions

Analysis	Correlation	Bundles with uniform heat flux e(%)	Bundles with non-uniform heat flux e(%)	Long bundles e(%)
Bundle average	Barnett[17]	9.86	9.36	21.33 12.79
Bundle average	Macbeth[14]	12.48	16.73	12.96 10.5
COBRA subchannel	Becker[25]	7.67	13.34	4.7 5.2
CISE subchannel	Gaspari[16]	7.9	9.68	14.6 8.7
CISE subchannel	Hewitt[18]	8.1	10.3	4.5 4.17

At low qualities and mass flow rates the flow regime would appear to be an "inverted annular" one with liquid in the centre and a thin vapour film adjacent to the heating surface. The vapour-liquid interface is not smooth, but irregular. These irregularities occur at random locations, but appear to retain their identity to some degree as they pass up the tube with velocities of the same order as that of the liquid core. The vapour in the film adjacent to the heating surface would appear to travel at a higher velocity.

Various analytical models have been used to predict film boiling heat transfer coefficients. A selection of these correlations is given in Table 7.3. The evidence at the present time is that laminar film boiling with a smooth interface only occurs over relatively short distances (7-10 cms) downstream of the dryout or "rewet" front. At longer distances the coefficient becomes independent of distance and takes on a value considerably (approximately a factor of 2) higher than the laminar solution would indicate. Figure 7.9 shows some experimental results for film boiling inside a vertical heated tube for water at low pressure. These data were taken using a transient technique[32] and show a considerable effect of flow direction (upflow or downflow) for saturated film boiling. This effect is absent for subcooled film boiling.

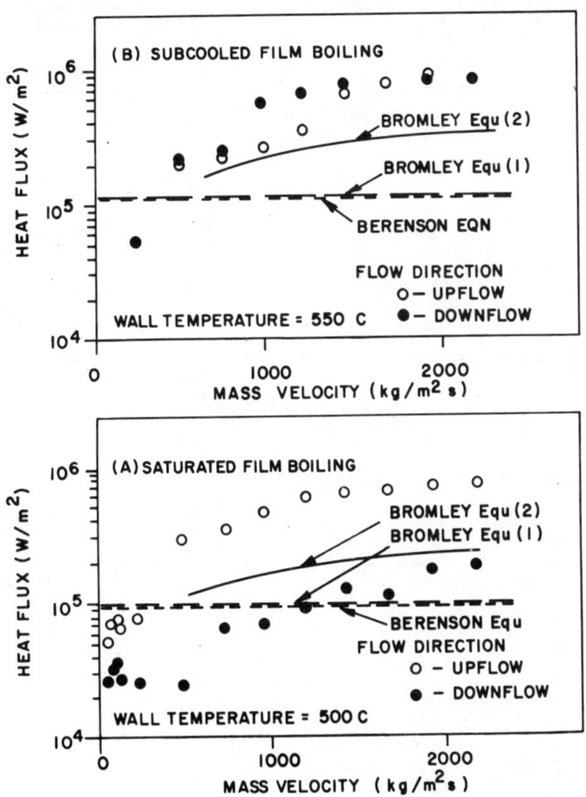

Fig.7.9 Comparison of experimental film boiling data for water[32] with correlations

TABLE 7.3

FILM BOILING HEAT TRANSFER CORRELATIONS

Geometry	Correlation
Flat plate, vertical, laminar flow	$\left[\frac{\bar{h}(z)z}{k_G}\right] = C\left[\frac{z^3 g\rho_G(\rho_L-\rho_G)i'_{fg}}{k_G\,\mu_G\,\Delta T}\right]^{1/4}$ C = 0.943 for zero interfacial shear stress (τ_i=0) C = 0.667 for zero interfacial velocity (u_i=0)
Flat plate, vertical, turbulent flow	$\left[\frac{h(z)z}{k_G}\right] = 0.056\,Re_G^{0.2}\left[\left(\frac{c_p\mu}{k}\right)_G \frac{z^3 g\rho_G(\rho_L-\rho_G)}{\mu_G^2}\right]^{1/3}$
Flat plate, horizontal (Berenson)	$h = 0.425\left[\frac{k_G^3 g\rho_G(\rho_L-\rho_G)i'_{fg}}{\mu_G\,\Delta T\,\frac{\lambda c}{2\pi}}\right]^{1/4}$ where $(\frac{\lambda c}{2\pi})$ is the characteristic bubble spacing = $\left[\frac{\sigma}{g(\rho_L-\rho_G)}\right]^{1/2}$
Cylinder, external surface, vertical (Bailey)	$h = 0.40\left[\frac{k_G^3 g\rho_G(\rho_L-\rho_G)i'_{fg}}{\mu_G\,\Delta T\,r}\right]^{1/4}$ r = cylinder (fuel rod) radius
Cylinder, external surface, horizontal, stagnant (Bromley)	$h = 0.62\left[\frac{k_G^3 g\rho_G(\rho_L-\rho_G)i'_{fg}}{\mu_G\,\Delta T\,D}\right]^{1/4}$ Bromley EQN (1)
Cylinder, external surface, horizontal, flowing (Bromley)	$h = 2.7\left[\frac{u\,k_G\,\rho_G\,i'_{fg}}{D\,\Delta T}\right]^{1/2}$ Bromley EQN (2)

14. LIQUID DEFICIENT REGION

Heat transfer rates in the liquid deficient region are bounded by two limiting situations, viz.

(1) complete departure from equilibrium. The rate of heat transfer from the steam phase to the entrained water droplets is so slow that their presence is simply ignored and the steam temperature $T_G(z)$ downstream of the

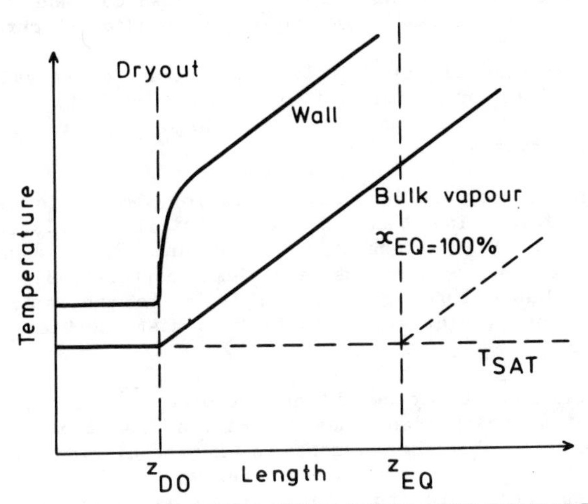

(a) COMPLETE LACK OF THERMODYNAMIC EQUILIBRIUM.

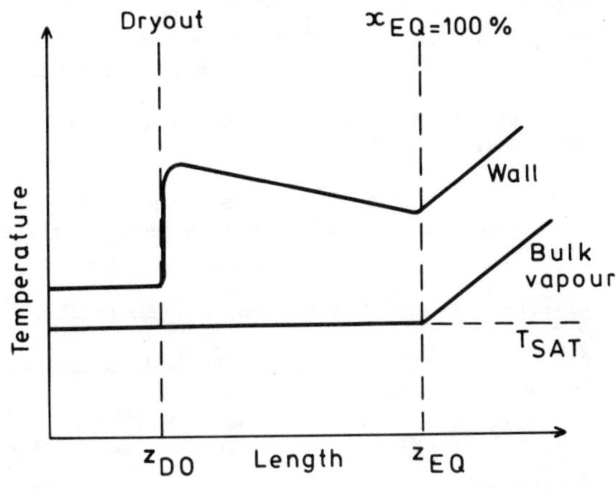

(b) COMPLETE THERMODYNAMIC EQUILIBRIUM.

Fig.7.10 Limiting Conditions for Post-dryout Heat Transfer

dryout point is calculated on the basis that all the heat added to the fluid goes into superheating the steam. The wall temperature $T_W(z)$ is calculated using a conventional single-phase heat transfer correlation (Figure 7.10).

(2) complete thermodynamic equilibrium. The rate of heat transfer fromthe steam phase to the entrained water droplets is so fast that the steam temperature $T_G(z)$ remains at the saturation temperature until the energy balance indicates all the droplets have evaporated. The wall temperature $T_W(z)$ is again calculated using a conventional single-phase heat transfer correlation, this time with allowance made for the increasing steam velocity resulting from droplet evaporation (Figure 7.10).

It is known that liquid deficient heat transfer behaviour tends towards situation (1) at low pressures and low velocities, whilst at high pressure (approaching the critical condition) and high flow rates (> 3000 kg/m^{2}s) situation (2) pertains.

A selection of correlations and models for the liquid deficient region is given in Table 7.4. The best of the empirical correlations are those of Groeneveld(33) and Slaughterback(34). Figure 7.11 shows heat transfer coefficients predicted by various empirical correlations as a function of pressure. The Slaughterback correlation is to be preferred at low pressures because it does not predict artificially high heat transfer coefficients under these conditions.

The correlation of Groeneveld and Delorme(35) is representative of the type of correlation which recognises departures from thermodynamic equilibrium and produces considerable improvement in the prediction of wall temperature compared with the empirical correlation. However, such correlations assume that the extent of the departure from thermodynamic equilibrium is characterised by the "local" conditions, i.e. is unaffected by what has happened preceding the point being considered. This obviously not the case.

A comprehensive model of heat transfer in the post-dryout region must take into account the various paths by which heat is transferred from the surface to the bulk vapour phase. Six separate mechanisms can be identified.

(i) heat transfer from the surface to water droplets which impact the wall ("wet" collisions)

(ii) heat transfer from the surface to liquid droplets which enter the thermal boundary layer but which do not "wet" the surface ("dry" collisions)

(iii) convective heat transfer from the surface to the bulk steam

(iv) convective heat transfer from the bulk steam to suspended liquid droplets in the steam core

(v) radiation heat transfer from the surface to the liquid droplets

(vi) radiation heat transfer from the surface to the bulk steam.

One of the first phenomenological models proposed was that of Bennett et al(36) which is a one-dimensional model starting from known equilibrium conditions at the dryout point. This model, however, did not consider

TABLE 7.4

LIQUID DEFICIENT REGION HEAT TRANSFER CORRELATIONS

Author	Correlation	σ(%)
Groeneveld[33] Slaughterback[34] (empirical)	$Nu_G = a\left[Re_G\left\{x + \frac{\rho_G}{\rho_L}(1-x)\right\}\right]^b Pr_{G,w}{}^c Y^d$ where $Y = 1 - 0.1\left(\frac{\rho_L}{\rho_G} - 1\right)^{0.4}(1-x)^{0.4}$ and a,b,c and d are indices	11.5%
Mattson et al[38] (empirical)	$h = C\exp(-0.5\sqrt{\Delta T}) + a\,Re_G{}^b\,Pr_{G,w}{}^c\,D_e{}^f\,k_G{}^h x^j$ where C, a,b,c,f,h and j are indices	17.28%
Groeneveld and Delorme[35] (taking into account non-equilibrium)	$x(z) - x^*(z) = \exp(-\tan\psi)$ where $\psi = fn(Re_{TP}, p, x_e)$ and $x(z) - x^*(z)$ is the difference between the thermodynamic "equilibrium" quality and the "true" steam quality	6.7%
Bennett et al[36]	Simultaneous solution of differential equations for (i) the change of "true" steam quality ($x^*(z)$) with respect to length (z) (ii) the change of vapour temperature (T_G) with respect to length (z) (iii) the acceleration of liquid droplets (iv) the evaporation of liquid droplets	-
Iloeje et al[26]	$\phi = \phi_{dc} + \phi_{dw} + \phi_c$ Three step model taking into account (a) heat transfer from surface to liquid droplets which hit wall (ϕ_{dc}) (b) heat transfer from surface to liquid droplets which enter boundary layer but which do not "wet" wall (ϕ_{dw}) (c) convective heat transfer to bulk steam (ϕ_c)	-

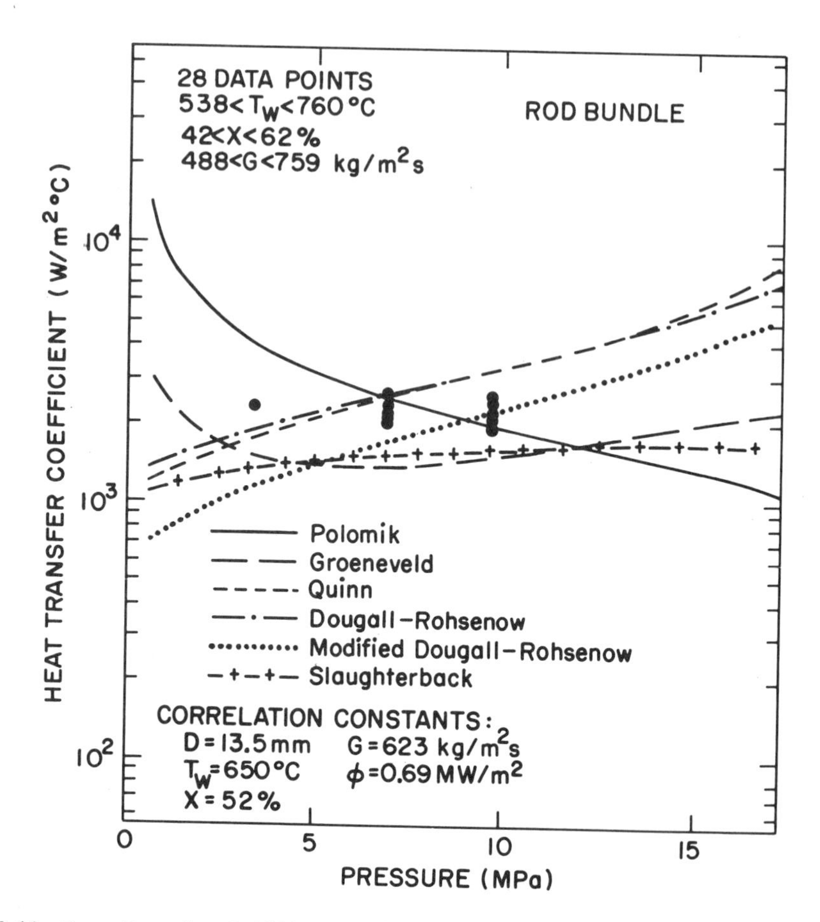

Fig.7-11 Heat Transfer Coefficients in Post-dryout Region as a Function of pressure (Slaughterback[34])

mechanisms (i) and (ii). However, good agreement with experimental wall temperature profiles was seen. More recently, Iloeje et al[26] have proposed a three step model which takes account of mechanisms (i), (ii) and (iii).

15. SINGLE-PHASE VAPOUR HEAT TRANSFER

Based on experimental data in tubes and annuli, the following correlation of Bishop et al[37] is recommended for steam at moderately high pressures and Reynolds numbers:

$$\left(\frac{hD}{k_G}\right)_f = 0.0073 \left(\frac{DG}{\mu}\right)_f^{0.886} \left(\frac{c_p \mu}{k}\right)_f^{0.61} \left[1 + \frac{2.76}{z/D}\right] \tag{28}$$

where subscript f refers to film conditions evaluated at $T = (T_W+T_b)/2$.

16. NON-EQUILIBRIUM EFFECTS

Many of the difficulties which arise from the use of correlations involving just local parameters are concerned with non-equilibrium effects. It is necessary to consider both departures from hydrodynamic equilibrium (i.e. not "fully developed" flow) and from thermodynamic equilibrium (i.e. vapour and liquid phases at different temperatures). By way of an example, consider an isothermal air-water flow in which all the water is injected at the channel entrance close to the wall and the air flows down the centre of the channel. After a finite distance the flow will have rearranged itself such that the disposition of liquid and gas does not change further with distance. Under boiling conditions, however, such "fully developed" flows can never be achieved because of the progressively increasing vapour flow with distance along the heated channel. However, analytical methods have now been proposed in order to calculate such hydrodynamically developing two-phase flows[39,40,41]. Turning to departures from thermodynamic equilibrium, these can become apparent - as compared with the "one-dimensional equilibrium model" used in most sub-channel codes - as a result of non-uniformities of temperature across the section, transient boundary layers around bubbles or droplets and kinetic theory limits on evaporation and condensation rates. Thus, vapour bubbles can exist in sub-cooled liquids and liquid droplets in superheated vapours. Departures from equilibrium where the two-phases exist at different mean temperatures can also occur in post-dryout heat transfer, in flashing and critical flows and in accident situations such as blowdown, reflooding, quenching and fuel-coolant interactions.

17. TRANSIENT EFFECTS

It is usual to assume that the heat transfer regime at any particular instant of time during a transient corresponds to that arrived at by inserting the local instantaneous hydrodynamic parameters into the appropriate heat transfer correlations. This can only be an approximation. For example it can be established theoretically that the rates of momentum, heat and mass transfer which occur during accelerating single phase laminar and turbulent flows differ appreciably from the local instantaneous rates arrived at from applying the steady state correlations.

Similar considerations apply to the onset of the critical heat flux. Most computer codes assume that the CHF condition is reached based on the use of the local instantaneous conditions in a particular steady state critical heat flux correlation. In general this approach gives a satisfactory prediction[42] but there are occasions when the onset of the critical condition is delayed well after the point where it is predicted to occur. Henry and Leung[43] have shown that this phenomenon is associated with the occurrence of fully developed subcooled boiling on the surface prior to the initiation of the transient. The existence of nucleation sites on the heat transfer surface allows extended cooling during the flow reversal in the core following a large LOCA in a PWR. Those surfaces which are mainly cooled by single phase forced convection during normal operating conditions are unable to activate the nucleation sites for boiling until the heat transfer surface has risen in temperature to a level which essentially satisfies the criterion for stable film boiling.

18. CONCLUDING REMARKS

In this Chapter we have tried to identify the various heat transfer regimes which occur during single phase and two-phase flows in water-cooled reactors. The concept of a complete "heat transfer surface" has been developed. The pressure upon which this "heat transfer surface" is based is that of the heat transfer rates can be described by the instantaneous local hydrodynamic conditions. Thus it is assumed that the heat flux calculation can be undertaken on the basis of the local conditions at the point being considered at the given instant of time. It has been shown that this is not strictly true because of (a) flow history effect, i.e. developing flowstates upstream of the point being considered, (b) lack of thermodynamic equilibrium, i.e. unequal temperatures for water and steam, and (c) transient effects upon wall shear and heat transfer rates as well as upon the critical heat flux. Models are being developed to incorporate some of these effects into the thermohydraulic analysis of water-cooled reactors.

REFERENCES

(1) Tong, L.S. and Weisman, J. "Thermal Analysis of Pressurised Water Reactors", American Nuclear Society (1970).

(2) Lahey, R.T. and Moody, F.J. "The thermalhydraulics of a Boiling water Reactor", American Nuclear Society (1977).

(3) Bjornard, T.A. and Griffith, P. "PWR Blowdown Heat Transfer", Symp. on the Thermal and Hydraulic Aspects of Nuclear Reactor Safety, Atlanta Georgia, Nov.27 - Dec.2 (1977), Vol.1, Light Water Reactors, pp.17-41, ASME.

(4) Weisman, J. and Bowring, R.W. "Methods for detailed thermal and hydraulic analysis of water-cooled reactors". Nucl. Science and Engineering 57, 255-276 (1975).

(5) Bowring, R.W. "HAMBO - a computer program for the subchannel analysis of the hydraulic and burnout characteristics of rod clusters." Pt.I General description AEEW-R524 (1967), Pt.II The equations AEEW-R582 (1968).

(6) Rowe, D.S. "COBRA - a digital computer program for thermal hydraulic subchannel analysis of rod bundle nuclear fuel elements". BNWL-371 (1967), BNWL-1229 (1970), BNWL-B-82 (1972) and BNWL-1695 (1973).

(7) Collier, J.G. "Convective Boiling and Condensation" (2 ed) published by McGraw Hill Book Co. (UK) Ltd., (1980).

(8) Davis, E.J. and Anderson, G.H. "The Incipience of Nucleate Boiling in Forced Convection Flow". AIChE Journal 12(4), 774-780 (July 1966).

(9) Jens, W.H. and Lottes, P.A. "Analysis of heat transfer burnout, pressure drop and density data for high pressure water". ANL-4627 (May 1951).

(10) Thom, J.R.S., Walker, W.M., Fallon, T.A. and Reising, G.F.S. "Boiling in subcooled water during flow up heated tubes or annuli". Paper 6 presented at the Symposium on Boiling Heat Transfer in Steam Generating Units and Heat Exchangers held in Manchester, 15-16 September 1965 by Inst. of Mech. Eng. (London).

(11) Hewitt, G.F., Kearsey, H.A., Lacey, P.M.C. and Pulling, D.J. "Burnout and nucleation in climbing film flow". AERE-R4374 (1963).

(12) Chen, J.C. "A correlation for boiling heat transfer to saturated fluids in convective flow". Paper presented to 6th National Heat Transfer Conference, Boston, 11-14 August 1963, ASME preprint 63-HT-34.

(13) Thompson, B. and Macbeth, R.V. "Boiling water heat transfer - burnout in uniformly heated round tubes: a compilation of world data with accurate correlations". AEEW-R356 (1964).

(14) Macbeth, R.V. "Burnout analysis - Part 5: Examination of published world data for rod bundles". AEEW-R358 (1964).

(15) Bowring, R.W. "A simple but accurate round tube uniform heat flux dryout correlation over the range 0.7 - 17 MN/m^2 (100-2500 psia)" AEEW-R789 (1972).

(16) Bertoletti, S. et al. "Heat Transfer Crisis with Steam-Water Mixtures". Energia Nucleare 12(3), March 1965. See also Gaspari, G.P. et al. "Some considerations on CHF in rod clusters in annular dispersed vertical upward two-phase flow". 4th Int. Conf. Heat Transfer, Paris (1970).

(17) Barnett, P.G. "A correlation of burnout data for uniformly heated annuli and its use for predicting burnout in uniformly heated rod bundles". AEEW-R463 (1966).

(18) Hewitt, G.F. and Kearsey, H.A. "Correlation of critical heat flux for vertical flow of water in uniformly heated tubes, annuli and rod bundles". AERE-R5590 (1970).

(19) Biasi, L. et al. "Studies on Burnout" Part 3, Energia Nucleare 14(9) 530-536 (1967).

(20) Tong, L.S. Nuc. Science and Engng. 33, 7 (1968).

(21) Butterworth, D. Private Communication (1978).

(22) GETAB - "General Electric BWR Thermal Analysis Basis Data, correlation and design application". NEDO-10958 (1973).

(23) Bowring, R.W. "WSC-2: A subchannel dryout correlation for water-cooled clusters over the pressure range 3.4 - 15.9 MPa (500-2300 psia)". AEEW-R983 (May 1979).

(24) Gellerstedt, J.S. et al. "Correlation of critical heat flux in a bundle cooled by pressurised water". Proc. Symp. Two-Phase Flow and Heat Transfer in Rod Bundles, p.63, ASME Winter Meeting, Los Angeles (1969).

(25) Becker, K.M. "A burnout correlation for flow of boiling water in vertical rod bundles". AE276 AB Atomenergi (1967).

(26) Iloeje, O.C., Plummer, D.N., Rohsenow, W.M. and Griffith, P. "A study of wall rewet and heat transfer in dispersed vertical flow". MIT Dept. of Mech. Engng. Report 72718-92 (September 1974).

(27) Plummer, D.N., Iloeje, O.C., Rohsenow, W.M., Griffith, P. and Ganic, E. "Post critical heat transfer to flowing liquid in a vertical tube". MIT Dept. of Mech. Engng. Report 72718-91 (September 1974).

(28) Groeneveld, D.C. "Effect of a heat flux spike on the downstream dryout behaviour". J. of Heat Transfer, pp.121-125 (May 1974).

(29) Groeneveld, D.C. and Fung, K.K. "Forced Convection Transition Boiling - a review of literature and comparison of prediction methods". AECL-5543 (1976).

(30) Tong, L.S. and Young, J.D. "A phenomenological transition and film boiling heat transfer correlation". Paper B 39 Vol.3, Proc. of 5th Int. Heat Transfer Conference, Tokyo, September 1974.

(31) Iloeje, O.C., Plummer, D.N., Rohensow, W.M. and Griffith, P. "An investigation of the collapse and surface rewet in film boiling in forced vertical flow". J. of Heat Transfer, pp.166-172, (May 1975).

(32) Ralph, J.C., Sanderson, S. and Ward, J.A. "Post dryout heat transfer under low flow and low quality conditions". Symp. on Thermal and Hydraulic Aspects of Nuclear Safety, 1977 ASME Winter Meeting, Atlanta, Georgia, Nov.27 - Dec.2 (1977).

(33) Groeneveld, D.C. "Post-dryout heat transfer at reactor operating conditions". AECL-4513, Nat. Topical Meeting on Water Reactor Safety, ANS Salt Lake City, Utah, March 26-28 (1973).

(34) Slaughterback, D.C. et al. "Statistical regression analysis of experimental data for flow film boiling heat transfer". ASME-AIChE Heat Transfer Conference, Atlanta, August 1973.

(35) Groeneveld, D.C. and Delorme, G.G.J. "Prediction of thermal non-equilibrium in the post-dryout region". Nuc. Engng. and Design 36, 17-36, (1976).

(36) Bennett, A.W., Hewitt, G.F., Kearsey, H.A. and Keeys, R.K.F. "Heat transfer to steam-water mixtures flowing in uniformly heated tubes in which the critical heat flux has been exceeded". Thermodynamics and Fluid Mechanics Convention, Bristol, 278-29 March 1968, I. Mech. E. See also AERE-R5373 (1967).

(37) Bishop, A.A. et al. "Forced convection heat transfer to superheated steam at high pressure and high Prandtl numbers". ASME Paper 65-WA/HT-35 (1965).

(38) Mattson, R.J. et al. "Regression analysis of post CHF - flow boiling data". Paper B3.8 Vol.4, Proc. of 5th Int. Heat Transfer Conference, Tokyo, September (1974).

(39) Whalley, P.B., Hutchinson, P. and Hewitt, G.F. "The calculation of critical heat flux in forced convection boiling". Paper B6.11, 5th Int. Heat Transfer Conference, Tokyo, Japan (1974).

(40) Whalley, P.B., Hutchinson, P. and Hewitt, G.F. "Prediction of annular flow parameters for transient conditions and for complex geometries". European Two-Phase Flow Group Meeting, Haifa, Israel, June 1975.

(41) Whalley, P.B. "The calculation of dryout in a rod bundle". AERE-R8319. European Two-Phase Flow Group Meeting, Erlangen, Germany, June 1976 (Paper A9).

(42) Leung, J.C.M. "Critical heat flux under transient conditions - a literature survey". NUREG/CR-0056 (ANL-78-39) June 1978.

(43) Henry, R.E. and Leung, J.C.M. "A mechanism for transient critical heat flux". ANS Topical Meeting on Thermal Reactor Safety, Sun Valley ID (August 1977).

8 NUCLEAR SYSTEM SAFETY MODELING

by
R.T. Lahey, Jr.
RPI, Troy, NY, USA

This chapter will be concerned with the formulation of the conservation equations of flowing two-phase system. In particular, the problem of interest here will be modeling the coolant in nuclear reactor systems during normal and off-design operating conditions.

1. BASIC CONCEPTS

The local volumetric vapor (i.e., void) fraction is defined as,

$$\alpha = \iiint_{V_v} dV / \iiint_{V} dV = V_v/(V_\ell + V_v) \simeq A_v/A_{x-s} \tag{1}$$

It is important to realize that void fraction is a time-averaged quantity. That is, it is the time-average of the phase indicator function (X_k),

$$\alpha(\underline{x},t) = \overline{X_v(\underline{x},t)} \stackrel{\Delta}{=} \frac{1}{T}\int_{t-T/2}^{t+T/2} X_v(\underline{x},t')dt' \tag{2a}$$

where,

$$X_k(\underline{x},t) \stackrel{\Delta}{=} \begin{cases} 1, \text{ if } \underline{x} \text{ is in phase-k at time-t} \\ 0, \text{ otherwise} \end{cases} \tag{2b}$$

and, T is the time interval over which averaging is to be accomplished.

Care must be taken in choosing T, since if it is too large, the transient information remaining in the resultant averaged signal may be severely limited. Indeed, time-averaging can be thought of as low pass filtering. The larger the T, the lower the effective break frequency, and thus less the frequency content of the filtered signal.

We shall not go into an extensive discussion of the various methods of time-averaging here. Those readers interested in a more in-depth treatment of the subject are referred to the work of Ishii [1]. For our purposes here, let us assume that all quantities of interest are appropriately [1] time-averaged.

Let us now consider spatial-averaging. In general, one can perform chordal, cross-sectional and volumetric spatial-averaging of the variables of interest. For our purposes here, we need only consider cross-sectional averaging, which for an arbitrary variable, ξ, is given by,

$$\langle \xi \rangle = \frac{1}{A_{x-s}} \iint_{A_{x-s}} \xi dA \tag{3a}$$

$$\langle \xi_\ell \rangle_\ell = \frac{1}{(1 - \langle \alpha \rangle) A_{x-s}} \iint_{A_{x-s}} \xi_\ell (1-\alpha) dA = \frac{\langle (1-\alpha) \xi_\ell \rangle}{\langle 1-\alpha \rangle} \tag{3b}$$

$$\langle \xi_v \rangle_v = \frac{1}{\langle \alpha \rangle A_{x-s}} \iint_{A_{x-s}} \xi \, \alpha dA = \frac{\langle \alpha \xi_v \rangle}{\langle \alpha \rangle} \tag{3c}$$

Since we will deal only with one-dimensional conservation equations, these definitions are sufficient. Those readers interested in a more in-depth treatment of spatial-averaging techniques are referred to the work of Delhaye and Achard [2].

2. CONSERVATION EQUATIONS

Let us now consider the one-dimensional phasic conservation equations of two-phase flow. These equations form the basis for the so-called two-fluid model which is in wide use throughout the nuclear industry.

2.1 Mass Conservation

The principle of conservation of mass can be written as,

$$\left\{ \begin{array}{l} \text{Mass} \\ \text{outflow} \\ \text{rate from} \\ \text{the control volume} \end{array} \right\} - \left\{ \begin{array}{l} \text{Mass} \\ \text{inflow} \\ \text{rate to the} \\ \text{control volume} \end{array} \right\} + \left\{ \begin{array}{l} \text{Mass} \\ \text{storage} \\ \text{rate in the} \\ \text{control volume} \end{array} \right\} = 0 \tag{4}$$

For simplicity, the control volume shown in Fig. 1 is drawn for the annular flow regime; however, the results which we shall derive are valid for any flow regime.

If we now apply Eq. (4) to the liquid phase in the control volume shown in Fig. 1, we obtain:

$$\Gamma A_{x-s} \Delta z + \rho_\ell (1 - \langle \alpha \rangle) \langle u_\ell \rangle_\ell A_{x-s} + \frac{\partial}{\partial z} \left[\rho_\ell (1 - \langle \alpha \rangle) \langle u_\ell \rangle_\ell A_{x-s} \right] \Delta z$$

$$- \rho_\ell (1 - \langle \alpha \rangle) \langle u_\ell \rangle_\ell A_{x-s} + \frac{\partial}{\partial t} \left[\rho_\ell A_{x-s} (1 - \langle \alpha \rangle) \Delta z \right] = 0 \tag{5}$$

or,

$$\frac{\partial}{\partial t} \left[\rho_\ell (1 - \langle \alpha \rangle) A_{x-s} \right] + \frac{\partial}{\partial z} \left[\rho_\ell (1 - \langle \alpha \rangle) \langle u_\ell \rangle_\ell A_{x-s} \right] = - \Gamma A_{x-s} \tag{6}$$

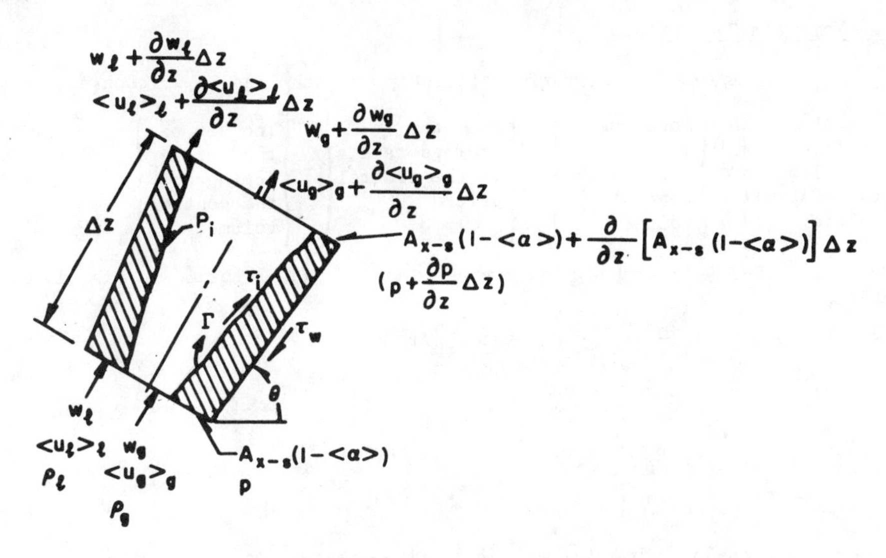

Fig. 1 Control Volume for Separated Flow (Each Phase)

Similarly, if we apply Eq. (4) to the vapor phase in the control volume shown in Fig. 1, we obtain,

$$\frac{\partial}{\partial t}\left[\rho <\alpha>) A_{x-s}\right] + \frac{\partial}{\partial z}\left[\rho_v <\alpha><u_v>_v A_{x-s}\right] = \Gamma A_{x-s} \tag{7}$$

The term Γ, on the right-hand side of Eqs. (6) and (7), is the volumetric vapor generation term. This interfacial mass transfer rate must be constituted before these equations can be integrated.

It is interesting to note that if Eqs. (6) and (7) are added, the Γ term cancels out, and we obtain the well-known one-dimensional continuity equation of a two-phase mixture,

$$\frac{\partial}{\partial t}\left[<\rho> A_{x-s}\right] + \frac{\partial}{\partial z}\left[G\ A_{x-s}\right] = 0 \tag{8}$$

where, the mass flux (G) is given by,

$$G = \rho_\ell (1 - <\alpha>)<u_\ell>_\ell + \rho_v <\alpha><u_v>_v \tag{9}$$

and the mixture density by,

$$<\rho> = \rho_\ell (1 - <\alpha>) + \rho_v <\alpha> \tag{10}$$

2.2 Momentum Conservation

The principle of conservation of linear momentum can be written as,

$$\left\{\begin{array}{l}\text{Momentum}\\ \text{outflow}\\ \text{rate from}\\ \text{the control}\\ \text{volume}\end{array}\right\} - \left\{\begin{array}{l}\text{Momentum}\\ \text{inflow}\\ \text{rate to the}\\ \text{control}\\ \text{volume}\end{array}\right\} + \left\{\begin{array}{l}\text{Momentum}\\ \text{storage rate}\\ \text{in the control}\\ \text{volume}\end{array}\right\} = \left\{\begin{array}{l}\text{The sum of}\\ \text{the forces}\\ \text{acting on}\\ \text{the control}\\ \text{volume}\end{array}\right\} \tag{11}$$

If we apply Eq. (11) to the liquid phase in the control volume in Fig. 1, we obtain,

$$\begin{aligned}
&\rho_\ell A_{x-s} \langle (1-\alpha) u_\ell^2 \rangle + \frac{\partial}{\partial z}\left[\rho_\ell A_{x-s} \langle (1-\alpha) u_\ell^2 \rangle\right]\Delta z \\
&+ \Gamma U_i A_{x-s} \Delta z - \rho_\ell A_{x-s} \langle (1-\alpha) u_\ell^2 \rangle \\
&+ \frac{\partial}{\partial t}\left[\rho_\ell (1 - \langle\alpha\rangle) A_{x-s} \langle u_\ell \rangle_\ell \Delta z\right] = -\frac{\partial}{\partial z}\left[p\,(1 - \langle\alpha\rangle) A_{x-s}\right]\Delta z \\
&+ p_i \frac{\partial[(1 - \langle\alpha\rangle) A_{x-s}]}{\partial z} \Delta z - g \rho_\ell (1 - \langle\alpha\rangle) A_{x-s} \Delta z \sin\theta \\
&- \tau_{w_\ell} P_{f_\ell} \Delta z + \tau_i P_i \Delta z
\end{aligned} \tag{12}$$

where,

U_i = axial velocity at the vapor/liquid interface

p_i = interfacial pressure

τ_i = interfacial shear

P_i = interfacial perimeter

τ_{w_ℓ} = wall shear (on liquid phase)

It is frequently assumed that $p_i = p_\ell = p_v = p$. If we make this assumption and separate the average of the product into the product of the averages by defining the distribution coefficient, C_ℓ,

$$C_\ell = \frac{\langle (1-\alpha) u_\ell^2 \rangle}{\langle 1-\alpha \rangle \langle u_\ell \rangle_\ell^2} \tag{13}$$

then Eq. (12) may be rewritten as

$$- (1 - \langle\alpha\rangle) \frac{\partial p}{\partial z} - g\, \rho_\ell (1 - \langle\alpha\rangle) \sin\theta - \frac{\tau_{w_\ell} P_{f_\ell}}{A_{x-s}}$$

$$+ \frac{\tau_i P_i}{A_{x-s}} = \frac{\partial}{\partial t} [\rho_\ell (1 - \langle\alpha\rangle) \langle u_\ell \rangle_\ell]$$

$$+ \frac{C_\ell}{A_{x-s}} \frac{\partial}{\partial z} [\rho_\ell A_{x-s} (1 - \langle\alpha\rangle) \langle u_\ell \rangle_\ell^2] + \rho_\ell (1 - \langle\alpha\rangle) \langle u_\ell \rangle_\ell^2 \frac{\partial}{\partial z} C_\ell$$

$$+ \Gamma U_i \tag{14}$$

If we now apply Eq. (11) to the vapor phase in Fig. 1, we obtain,

$$- \langle\alpha\rangle \frac{\partial p}{\partial z} - g \rho_v \langle\alpha\rangle \sin\theta - \frac{\tau_i P_i}{A_{x-s}} - \frac{\tau_{w_v} P_{f_v}}{A_{x-s}}$$

$$= \frac{\partial}{\partial t} [\rho_v \langle\alpha\rangle \langle u_v \rangle_v] + \frac{C_v}{A_{x-s}} \frac{\partial}{\partial z} [\rho_v A_{x-s} \langle\alpha\rangle \langle u_v \rangle_v^2]$$

$$+ \rho_v \langle\alpha\rangle \langle u_v \rangle_v^2 \frac{\partial}{\partial z} C_v - \Gamma U_i \tag{15}$$

where, we have made all the same assumptions as before, and have defined the vapor phase distribution coefficient as,

$$C_v = \frac{\langle \alpha u_v^2 \rangle}{\langle\alpha\rangle \langle u_v \rangle_v^2} \tag{16}$$

It is interesting to note that if Eq. (15) is added to Eq. (14), we obtain the one-dimensional momentum equation of the two-phase mixture,

$$- \frac{\partial p}{\partial z} - g \langle\rho\rangle \sin\theta - \frac{\tau_w P_f}{A_{x-s}} = \frac{\partial}{\partial t} G + \frac{1}{A_{x-s}} \frac{\partial}{\partial z} [\rho_\ell A_{x-s} C_\ell (1 - \langle\alpha\rangle) \langle u_\ell \rangle_\ell^2$$

$$+ \rho_v A_{x-s} C_v \langle\alpha\rangle \langle u_v \rangle_v^2] \tag{17}$$

In the mixture momentum equation, all unknown interfacial transfer terms have cancelled out. In contrast, in order to obtain closure for the phasic momentum equations, we must constitute explicit expressions for the interfacial transfer terms: Γ, U_i and τ_i, the wall shear, τ_{w_k}, and the phasic distribution coefficients, C_k. Before discussing these transfer laws, let us first consider the phasic energy equations.

2.3 Energy Conservation

The principle of conservation of energy (i.e., the First Law of Thermodynamics) can be written as,

$$\begin{Bmatrix} \text{Outflow} \\ \text{rate of} \\ \text{energy from} \\ \text{the control} \\ \text{volume} \end{Bmatrix} - \begin{Bmatrix} \text{Inflow} \\ \text{rate of} \\ \text{energy to} \\ \text{the control} \\ \text{volume} \end{Bmatrix} + \begin{Bmatrix} \text{Storage rate} \\ \text{of energy} \\ \text{in the} \\ \text{control} \\ \text{volume} \end{Bmatrix} = 0 \tag{18}$$

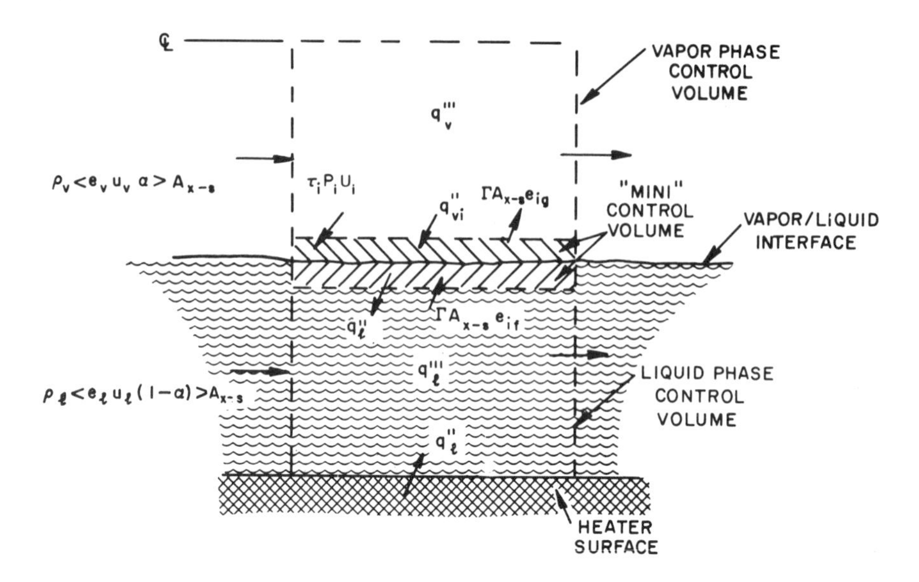

Fig. 2 Control Volumes for Energy Balance

It is convenient to consider the three control volumes, shown in Fig.2, one at a time. Let us first consider the application of Eq. (18) to the liquid phase in Fig. 2,

$$\rho_\ell <e_\ell u_\ell (1-\alpha)> A_{x-s} + \frac{\partial}{\partial z}[\rho_\ell <e_\ell u_\ell (1-\alpha)> A_{x-s}]\Delta z$$

$$+ p_i \frac{\partial < 1-\alpha>}{\partial t} A_{x-s} \Delta z^* + \Gamma e_{f_i} A_{x-s} \Delta z$$

$$- \rho_\ell < e_\ell u_\ell (1-\alpha)> A_{x-s} - q_\ell'' P_{H_\ell} \Delta z - q_\ell''' A_{x-s} < 1 - \alpha> \Delta z - q_{i\ell}'' P_i \Delta z$$

$$+ \frac{\partial}{\partial t}[\rho_\ell < e_\ell - p_\ell/\rho_\ell)(1-\alpha)> A_{x-s} \Delta z] = 0 \tag{19}$$

where, the total convective energy of phase-k (e_k) is the sum of the specific enthalpy, kinetic energy and potential energy:

---

* This term can be recognized as the pdV work term.

$$e_k = h_k + \frac{u_k^2}{2} + g\,z\,\sin\theta \qquad (20)$$

Assuming saturation conditions at the vapor/liquid interface, the appropriate energy outflow term is,

$$e_{f_i} = h_f + \frac{U_i^2}{2} + g\,z\,\sin\theta \qquad (21)$$

where the liquid phase subscript "ℓ" has been replaced by "f" to denote saturation.

Following the work of Yadigaroglu and Lahey [3], we can separate the average of the product into the product of the averages by defining a "total energy" distribution coefficient as,

$$K_\ell = \frac{<e_\ell u_\ell (1-\alpha)>}{<e_\ell>_\ell <u_\ell>_\ell <1-\alpha>} \qquad (22)$$

where,

$$<e_k>_k = <h_k>_k + \frac{1}{2} <u_k^2>_k + gz\,\sin\theta \qquad (23)$$

Thus, assuming $p_i = p_\ell = p$, and a time-invariant cross-sectional area (A_{x-s}), Eqs. (19) and (22) yield,

$$\frac{\partial}{\partial t}[\rho_\ell <e_\ell>_\ell (1 - <\alpha>)A_{x-s}] - (1 - <\alpha>)A_{x-s}\frac{\partial p}{\partial t} + K_\ell \frac{\partial}{\partial z}[\rho_\ell <e_\ell>_\ell <u_\ell>_\ell (1-<\alpha>)A_{x-s}]$$

$$+ \rho_\ell <e_\ell>_\ell <u_\ell>_\ell (1 - <\alpha>)A_{x-s}\frac{\partial K_\ell}{\partial z} = q_\ell'' P_{H_\ell} + q_\ell''' A_{x-s}(1 - <\alpha>)$$

$$+ q_{i_\ell}'' P_i - \Gamma\, e_{f_i} A_{x-s} \qquad (24)$$

Let us now apply Eq. (18) to the vapor phase in Fig. 2,

$$\rho_v <e_v u_v \alpha> A_{x-s} + \frac{\partial}{\partial z}[\rho_v <e_v u_v \alpha> A_{x-s}]\Delta z + p_i \frac{\partial <\alpha>}{\partial t}\Delta z\, A_{x-s}$$

$$+ \tau_i P_i U_R\,\Delta z + q_{vi}'' P_i \Delta z - \Gamma\, e_{g_i} A_{x-s}\,\Delta z$$

$$- \rho_v <e_v u_v \alpha> A_{x-s} - q_v''' A_{x-s} <\alpha> \Delta z - q_v'' P_{H_v}\,\Delta z$$

$$+ \frac{\partial}{\partial t}[\rho_v <(e_v - p_v/\rho_v)\alpha> A_{x-s}\Delta z] = 0 \qquad (25)$$

If we proceed as before, assume $p_i=p_v=p$, and define,

$$K_v = \frac{\langle e_v u_v \alpha \rangle}{\langle e_v \rangle_v \langle u_v \rangle_v \langle \alpha \rangle} \tag{26}$$

Eqs. (25) and (26) yield,

$$\frac{\partial}{\partial t}[\rho_v \langle e_v \rangle_v \langle \alpha \rangle A_{x-s}] - \langle \alpha \rangle A_{x-s} \frac{\partial p}{\partial t} + K_v \frac{\partial}{\partial z}[\rho_v \langle e_v \rangle_v \langle u_v \rangle_v \langle \alpha \rangle A_{x-s}]$$

$$+ \rho_v \langle e_v \rangle_v \langle u_v \rangle_v \langle \alpha \rangle A_{x-s} \frac{\partial K_v}{\partial z} = q_v''' A_{x-s} \langle \alpha \rangle$$

$$+ q_v'' P_{H_v} + \Gamma e_{g_i} A_{x-s} - \tau_i P_i U_R - q_{vi}'' P_i \tag{27}$$

where, noting the subscript "g" denotes saturated vapor,

$$e_{g_i} = h_g + \frac{U_i^2}{2} + gz \sin\theta \tag{28}$$

We must determine relationships between the interfacial transfer parameters: Γ, e_{g_i}, e_{f_i}, q_i'', τ_i and U_R. To do this, let us consider the "mini" control volume shown at the interface in Fig. 2. This control volume is presumed to be so small that we can neglect all temporal (i.e., storage) terms. If we apply Eq. (18) to this control volume,

$$\Gamma A_{x-s} e_{g_i} \Delta z + q_{i\ell}'' P_i \Delta z - \Gamma A_{x-s} e_{f_i} \Delta z - q_{vi}'' P_i \Delta z - \tau_i P_i U_R \Delta z = 0 \tag{29}$$

Equation (29) is the so-called "jump condition" [1] for interfacial energy transfer. It can be used to define the volumetric vapor generation rate (Γ),

$$\Gamma = [(q_{vi}'' - q_{i\ell}'')P_i + \tau_i P_i U_R] \,/\, [(e_{g_i} - e_{f_i})A_{x-s}] \tag{30}$$

For most cases of practical concern, we can neglect the interfacial shear work and the kinetic and potential energies in Eq. (30). Thus, we can write,

$$\Gamma \simeq (q_{vi}'' - q_{i\ell}'')P_i / h_{fg} A_{x-s} \tag{31}$$

It is interesting to note that the mixture energy equation can be obtained by combining Eqs. (22) and (24) and Eqs. (26) and (27), and adding the resultant equations to yield, after combination with Eq. (29),

$$\frac{\partial}{\partial t}[\rho_\ell (1 - \langle \alpha \rangle)(\langle e_\ell \rangle_\ell - p/\rho_\ell) + \rho_v \langle \alpha \rangle (\langle e_v \rangle_v - p/\rho_v)]$$

$$+ \frac{1}{A_{x-s}} \frac{\partial}{\partial z}[\rho_\ell K_\ell \langle e_\ell \rangle_\ell \langle u_\ell \rangle_\ell (1 - \langle \alpha \rangle) A_{x-s} + \rho_v K_v \langle e_v \rangle_v \langle u_v \rangle_v \langle \alpha \rangle A_{x-s}]$$

$$= \frac{q'' P_H}{A_{x-s}} + q''' \tag{32}$$

3. CONSTITUTIVE EQUATIONS

The analysis of non-equilibrium two-phase flow requires that one specify the functional form of the constitutive conditions in terms of system geometry and the variables of interest. For six equations, the two-fluid model presented in Eqs. (6), (7), (14), (15), (24) and (27), let us choose the following six variables as our primary unknowns:

$<\alpha>$, $<u_\ell>_\ell$, $<u_v>_v$, $<h_\ell>_\ell$, $<h_v>_v$, p.

To obtain closure, we must now specify expressions for the following 25 parameters which appear in the 6 conservation equations:

Γ, U_R, $<e_\ell>_\ell$, $<e_v>_v$, e_{g_i}, e_{f_i}, τ_{w_ℓ}, τ_{w_v}, τ_i, P_{f_ℓ}, P_{f_v}, P_{H_ℓ}, P_{H_v}, q''_ℓ, q'''_ℓ,

q''_v, q'''_v, $q''_{i\ell}$, q''_{vi}, U_i, P_i.

A number of these parameters have already been specified. Specifically, Γ is defined in terms of the other parameters in Eq. (30), and the relative velocity is given by, $U_R = <u_v>_v - <u_\ell>_\ell$. Moreover, the definition of $<e_\ell>_\ell$ and $<e_v>_v$ given in Eqs. (23) relates these parameters to the primary unknowns, $<h_\ell>_\ell$, $<h_v>_v$, $<u_\ell>_\ell$ and $<u_v>_v$. Similarly, Eqs. (21) and (28) relate e_{f_i} and e_{g_i}.

Expressions for the distribution coefficients C_ℓ, C_v, K_ℓ and K_v are given in Eqs. (13), (16), (22) and (26), respectively. However, these expressions can only be evaluated if one knows the appropriate lateral distributions. One can approximate these distributions with power-law profile techniques [4]; however, in practice, these parameters are frequently set to unity.

The wall shears, τ_{w_ℓ}, τ_{w_v} are normally evaluated in the form:

$$\tau_{w_k} = \frac{1}{8} f_k \rho_k <u_k>_k^2 \phi_k^2 \tag{33}$$

where f_k is the Darcy-Weisbach factor of phase-k ($f_k \simeq 0.02$), and ϕ_k^2 is an empirical two-phase multiplier. The generalized interfacial shear can be evaluated as,

$$\tau_i = \frac{C_D \rho_c}{2} [<u_v>_v - <u_\ell>_\ell]^2 + F_{vm} \frac{A_{x-s}}{P_i} \tag{34}$$

where F_{vm} is the volumetric virtual mass force, and where ρ_c is the density of the continuous phase (e.g., the liquid phase for bubbly flow and the vapor phase for dispersed annular flow). Let us first consider the form of the interfacial drag coefficient (C_D). This parameter depends on the flow regime under consideration. For example, if we have churn-turbulent bubbly flow, then Hench [5] recommends,

$$C_D = \frac{1.098}{4} <R_b>(1 - <\alpha>)^3 = 0.2745 <R_b>(1 - <\alpha>)^3 \tag{35}$$

where $<R_b>$ is the mean bubble size (in cm). In contrast, if we have annular flow, Wallis [6] recommends,

$$C_D = 0.005\ [1 + 75(1 - \langle \alpha \rangle)] \tag{36}$$

The appropriate (i.e., objective) form of the virtual mass force is given by [7],

$$F_{vm} = \rho_c\, C_{vm} \langle \alpha \rangle \left[\frac{D_v[\langle u_v \rangle_v - \langle u_\ell \rangle_\ell]}{Dt} + (\langle u_v \rangle_v - \langle u_\ell \rangle_\ell) \left[(\lambda - 2) \frac{\partial \langle u_v \rangle_v}{\partial z} + (1 - \lambda) \frac{\partial \langle u_\ell \rangle_\ell}{\partial z} \right] \right] \tag{37}$$

where, the material derivative is defined as,

$$\frac{D_v(\ )}{Dt} = \frac{\partial (\ )}{\partial t} + \langle u_v \rangle_v \frac{\partial (\ )}{\partial z} \tag{38}$$

and the virtual volume coefficient C_{vm} and parameter λ are flow regime dependent. For example [7], for low quality bubbly flow, $C_{vm} = \frac{1}{2}$ and $\lambda = 2.0$, while for high quality droplet flow, $C_{vm} = \frac{1}{2} \frac{\rho_v}{\rho_\ell}$ and $\lambda = 0$. In general, however, these parameters must be determined experimentally.

Let us now consider the frictional and heated perimeters of each phase (P_{f_k}, P_{H_k}). The specification of the appropriate constitutive relationships is a state-of-the-art problem; however, in some cases these parameters may be determined directly from the geometry and the flow regime. For instance, for ideal annular flow: $P_H = P_{H_\ell}$, $P_f = P_{f_\ell}$ and $P_{H_v} = P_{f_v} = 0$. In contrast, for stratified two-phase flow, the perimeters must be apportioned corresponding to the surface in contact with each phase, and are thus proportional to void fraction ($\langle \alpha \rangle$). One way to partition the perimeters for stratified horizontal flow is with the use of flow regime maps. Perhaps the best map is that due to Taitel and Dukler [8], which gives the liquid depth as a function of the Martinelli parameter (X). It should be clear, however, that much more work is needed to synthesize truly general constitutive laws for these parameters.

The heat flux to each phase, q_ℓ'' and q_v'', are normally specified in terms of an appropriate phasic convective heat transfer coefficient (H_k). That is,

$$q_k'' = H_k(T_w - \langle T_k \rangle_k) \tag{39}$$

where T_w is the temperature of the heated wall and the phasic mean temperature ($\langle T_k \rangle$) is determined from an equation of state of the form,

$$\langle T_k \rangle_k = f(p, \langle h_k \rangle_k) \tag{40}$$

Naturally, if one of the phases does not wet the heated surface, then its heat flux is zero.

The volumetric generation terms, q_k''', frequently ignored, but should be included if gamma or neutron heating is appreciable. In this case, the appropriately averaged volumetric heating must come from a separate physics calculation

The interfacial heat fluxes, q''_{ik}, are also frequently modelled in terms of various appropriate interfacial heat transfer coefficients. That is, for the case shown in Fig. 2 (in which we can assume the interface is at saturation temperature),

$$q''_{i\ell} = H_{i\ell} (T_{sat} - \langle T_\ell \rangle_\ell) \tag{41a}$$

$$q''_{vi} = H_{vi} (\langle T_v \rangle_v - T_{sat}) \tag{41b}$$

The interfacial heat transfer coefficient will depend on the flow regime in question. For instance, in high quality dispersed droplet two-phase flow Andersen et al [9] recommends,

$$H_{i\ell} = \frac{2}{3} \pi^2 \frac{\kappa_\ell}{D_d} \tag{42a}$$

$$H_{vi} = \frac{\kappa_v}{D_d} \left\{ 3.2 + 0.75 \left[\frac{\rho_v D_d U_R}{\mu_v} \right]^{0.5} Pr_v^{0.33} \right\} \tag{42b}$$

It is interesting to note that a large interfacial heat transfer coefficient implies that the phasic temperatures will be essentially saturation temperature, and thus the effect of Thermodynamic non-equilibrium can be easily investigated by varying the choice of H_{ki}.

The last two parameters are perhaps the most difficult to constitute. The velocity at the interface, U_i, has been a source of controversy for many years. Some investigators argue that, since it is the liquid phase which is evaporating, U_i should be the mean liquid phase velocity, $\langle u_\ell \rangle_\ell$. Wallis [6] has argued that the analysis of entropy production implies that we should use the average of the two phasic velocities,

$$U_i = \frac{1}{2} [\langle u_v \rangle_v + \langle u_\ell \rangle_\ell]$$

Indeed, this choice of U_i leads to symmetry in the conservation equations for evaporation and condensation.

The basis of the controversy can be seen in Fig. 3. For annular flow, we expect that the interfacial velocity will be somewhere between the 1-D phasic velocities. It has been previously suggested [10] that a general constitutive model is given by,

$$U_i = \eta \langle u_\ell \rangle_\ell + (1-\eta) \langle u_v \rangle_v$$

where η is an arbitrary parameter. If we choose η=1.0, Eq. (43) reduces to the classical assumption; while for $\eta = \frac{1}{2}$, we obtain the Wallis model.

The final parameter of interest, the interfacial perimeter (P_i), is of fundamental importance in interfacial momentum and heat transfer. Since this parameter always appears in the conservation equations in the form, P_i/A_{x-s}, we can equally well work in terms of this grouping, which is known as the interfacial area density. Unfortunately, these fundamental parameters are difficult to measure and model. Extensive experimental programs have been conducted [11, 12] in which measurements of the flow regime dependent interfacial area density were attempted. Unfortunately these programs have not produced unambiguous

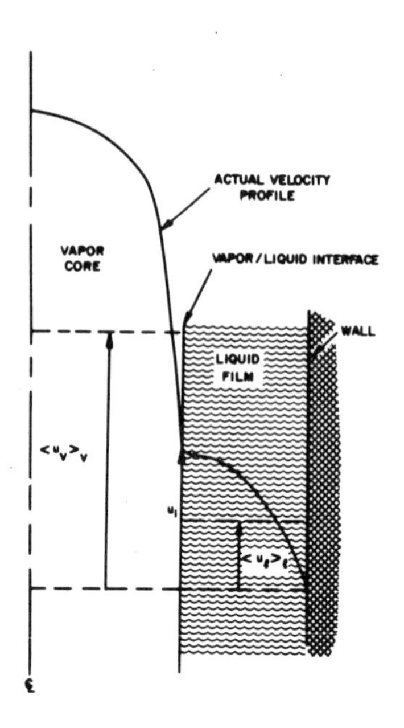

Fig. 3 Representation of Interfacial Velocity in Idealized Annular Flow

results and models.

For certain idealized flow regimes, one can work out the interfacial area density from geometric considerations. For instance, ideal annular flow in a tube (of radius R) has an interfacial area density of,

$$P_i/A_{x-s} = 2\pi r_i/\pi R^2 = 2\sqrt{<\alpha>/R} \tag{44}$$

while for uniform bubbly flow,

$$P_i/A_{x-s} = \frac{\pi D_b^2 <\alpha>}{\frac{1}{6}\pi D_b^3} = \frac{6<\alpha>}{D_b} \tag{44b}$$

Although these theoretical relationships are helpful in indicating trends, they are not recommended for practical calculations. The current state-of-the-art is to treat the interfacial area density (P_i/A_{x-s}) as a "knob" which is empirically adjusted to obtain agreement with separate-effects data. Indeed, it is common practice to "tune" many of the constitutive models in the two-fluid model.

4. CLASSIFICATION OF THE SYSTEM

Let us assume that we have now constituted expressions for the 25 parameters which appear in the two-fluid model, and thus we have achieved closure. It is important to realize that the choice of the functional form of these constitutive equations has an important effect on the classification, and numerical stability, of the resultant set of equations. Thus, let us now consider the mathematical implications of the two-fluid model discussed herein. The six conservation equations, given in Eqs. (6), (7), (14), (15), (24), and (27), are partial differential equations. In contrast, with the exception of the virtual mass force, given in Eqs. (34) and (37), the constitutive equations are algebraic (at least the way we have modelled them).

The system of one-dimensional conservation and constitutive equations can be written in matrix form as,

$$\underline{\underline{A}}\frac{\partial \underline{u}}{\partial t} + \underline{\underline{B}}\frac{\partial \underline{u}}{\partial z} = \underline{c} \tag{45}$$

where, neglecting kinetic and potential energy terms,

$$\underline{u} = (\langle\alpha\rangle, \langle u_\ell\rangle_\ell \ \langle u_v\rangle_v, \langle h_\ell\rangle_\ell, \langle h_v\rangle_v, p) \tag{46}$$

and, $\underline{\underline{A}}$ and $\underline{\underline{B}}$ are square matrices. As an example, the continuity equation for the vapor phase, Eq. (7), can be expressed as*,

$$\rho_v \frac{\partial\langle\alpha\rangle}{\partial t} + \langle\alpha\rangle \left[\frac{\partial \rho_v}{\partial p}\bigg|_{\langle h_v\rangle_v} \frac{\partial p}{\partial t} + \frac{\partial \rho_v}{\partial \langle h_v\rangle_v}\bigg|_p \frac{\partial\langle h_v\rangle_v}{\partial t} \right]$$

$$+ \rho_v \langle u_v\rangle_v \frac{\partial\langle\alpha\rangle}{\partial z} + \rho_v\langle\alpha\rangle \frac{\partial\langle u_v\rangle_v}{\partial z} + \frac{1}{A_{x-s}} \rho_v \langle\alpha\rangle\langle u_v\rangle_v \frac{\partial A_{x-s}}{\partial z}$$

$$+ \langle u_v\rangle_v \langle\alpha\rangle \left[\frac{\partial \rho_v}{\partial p}\bigg|_{\langle h_v\rangle_v} \frac{\partial p}{\partial z} + \frac{\partial \rho_v}{\partial \langle h_v\rangle_v}\bigg|_p \frac{\partial\langle h_v\rangle_v}{\partial z} \right] = \Gamma \tag{47}$$

Thus, if this is the second equation in the matrix formulation, the appropriate matrix elements are:

$$\left.\begin{aligned} a_{21} &= \rho_v \\ a_{22} &= a_{23} = a_{24} = 0 \\ a_{25} &= \langle\alpha\rangle \frac{\partial \rho_v}{\partial \langle h_v\rangle_v}\bigg|_p \\ a_{26} &= \langle\alpha\rangle \frac{\partial \rho_v}{\partial p}\bigg|_{\langle h_v\rangle_v} \end{aligned}\right\} \tag{48a}$$

* Assuming the flow area, A_{x-x}, is time-independent.

$$
\left.
\begin{aligned}
b_{21} &= \rho_v \langle u_v \rangle_v \\
b_{22} &= 0 \\
b_{23} &= \rho_v \langle \alpha \rangle \\
b_{24} &= 0 \\
b_{25} &= \langle u_v \rangle_v \langle \alpha \rangle \left(\frac{\partial \rho_v}{\partial \langle h_v \rangle_v} \right) \Bigg|_p \\
b_{26} &= \langle u_v \rangle_v \langle \alpha \rangle \left(\frac{\partial \rho_v}{\partial p} \right) \Bigg|_{\langle h_v \rangle_v}
\end{aligned}
\right\} \tag{48b}
$$

$$
c_2 = \Gamma - \frac{1}{A_{x-s}} \rho_v \langle \alpha \rangle \langle u_v \rangle_v \frac{\partial A_{x-s}}{\partial z} \tag{48c}
$$

An expansion of the other five conservation equations yields corresponding matrix elements.

To determine the mathematical nature of the system of equations, given in Eq. (45), we must evaluate the eigenvalues (ν_i). This is accomplished by calculating [13],

$$
\det [\underline{\underline{A}}\nu - \underline{\underline{B}}] = 0 \tag{49}
$$

If the six eigenvalues (ν_i) are real and distinct, the mathematical system is hyperbolic. If we have complex conjugate eigenvalues, the system is elliptic. Finally, if all eigenvalues are real, but some vanish or are equal, then the system is parabolic.

Physically, one would not expect a system of equations describing two-phase flow to have complex eigenvalues. Indeed, it is well known [13] that an elliptic system is completely specified by the boundary conditions. Since the boundaries for our one-dimensional problem are both space and time, complex eigenvalues imply that events in the future influence current phenomena. Clearly, this cannot be the case. Moreover, from laboratory experience we know that a well-posed problem should propagate information (i.e., waves) at well defined rates (i.e., along characteristics). Hence, if our problem is properly formulated, we should expect only real eigenvalues. The fact that some models don't yield real eigenvalues is due to the fact that some essential physical phenomena has been left out of the conservation or constitutive equations. It should be noted, however, that the only phenomena which can change the eigenvalues are those which contain space and time derivatives. If, for instance, we chose to include a model for the so-called Basset force (i.e., the transient development of the wall and interfacial shear) in the form of a first order relaxation process*, the elements in the matrix A would change and thus the

* $\frac{d\tau}{dt} + \tau/T = C$

eigenvalues would change. On the other hand, if we varied the drag coefficient (C_D) in Eq. (34), the eigenvalues would be uneffected.

The current state-of-the-art is such that the constitutive equations in most two-fluid models lead to complex eigenvalues. This frequently results in numerical difficulties, since complex eigenvalues imply exponential growth of perturbations.

Two basic appraoches have evolved to overcome these numerical problems. The approach used in the TRAC code is to introduce so-called numerical viscosity to damp out any instabilities. While this approach allows one to numerically integrate the ill-posed system, it can lead to considerable numerical diffusion and loss of spatial resolution.

The other appraoch is to try to introduce differential terms into the constitutive relations which make the eigenvalues real. This can be in the form of arbitrary relationships, such as proposed by Bouré [15], or in terms of additional physics, such as the virtual mass force, which is not always included, but has been considered herein in Eqs. (34) and (37).

The reader should realize that all the discussion so far has been concerned with one-dimensional two-fluid models. It should be clear that even for these relatively simple cases, considerably more R&D work is needed before the mathematical systems can be considered to be general, accurate and well-posed. For the transient analysis of three-dimensional two-phase flows, such as is done in TRAC vessel/plenum modules [14], our understanding of constitutive modelling is even more primative.

5. MODELLING OF LWR SAFETY PHENOMENA

Since one of the prime purposes of 1-D two fluid models is the safety analysis of current generation Light Water Nuclear Reactors (LWRs), it is of interest to consider what modifications, and additions, are needed to the generic six-equation model just discussed. In particular, we will be concerned with modelling the relevant Emergency Core Cooling (ECC) phenomena during the design basis Loss-of-Coolant-Accident (LOCA).

One mode of heat transfer which we have neglected in our discussion so far is radiation heat transfer. This mode can become quite significant during the LOCA reflood phase, in which the fuel rods may be at elevated temperatures. The heat radiated from the rods is absorbed, and re-emitted from the superheated vapor and saturated liquid droplets, and is exchanged between the rods and structural components. For such situations, the heat flux terms in Eqs. (39) must be supplemented by a radiation heat flux, $\underline{q}''_R$, where, making classical gray body assumptions [10],

$$q''_{j_R} = \Lambda_{ij}\,\Omega_i - Q_{ij}\,S_i \qquad (1 \le j \le N^*) \tag{50}$$

and,

$$\Lambda_{ij} = \frac{\varepsilon_i}{(1-\varepsilon_i)}\,[\delta_{ij} - X^{-1}_{ij}] \tag{51a}$$

---

* N is the number of surfaces under consideration.

$$Q_{ij} = \frac{\varepsilon_i}{(1-\varepsilon_i)} X_{ij}^{-1} \tag{51b}$$

$$X_{ij} = [\delta_{ij} - (1-\varepsilon_i)F_{i-j} (1-\varepsilon_v-\varepsilon_\ell+2\varepsilon_v\varepsilon_\ell)]/\varepsilon_i \tag{51c}$$

$$\Omega_i = \sigma T_i^4 \tag{51d}$$

$$S_i = \frac{(1-\varepsilon_i)}{\varepsilon_i} [\varepsilon_v \sigma < T_v >_v^4 (1-\varepsilon_\ell)$$

$$+ \varepsilon_\ell \sigma < T_\ell >_\ell^4 (1-\varepsilon_v)] \tag{51e}$$

$$\varepsilon_k = 1 - \exp(-a_k L_M) \tag{51f}$$

$$L_M = D_H \tag{51g}$$

The radiation heat transfer network is shown in Fig. 4. More refined thermal networks are possible [9]; however, this model does include the relevant physics involved.

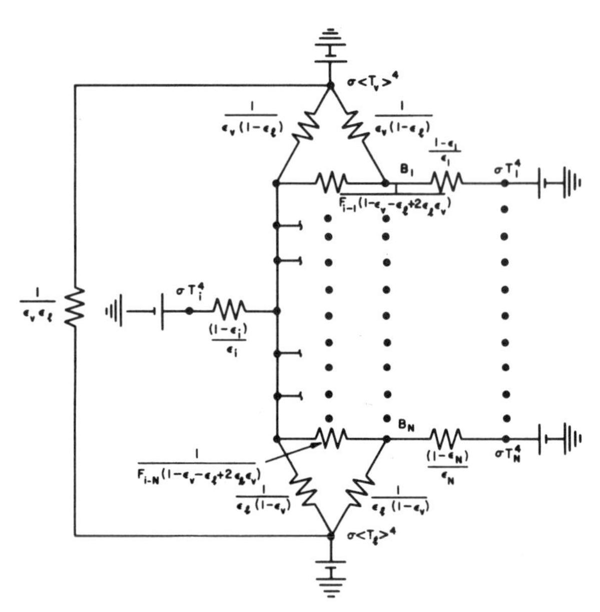

Fig. 4 Equivalent Electrical Analog for a Homogeneous Mixture of a Gray Gas and Liquid Droplets in a Gray Enclosure (Rod Bundle)

It is interesting to note that the net radiant energy absorbed by the vapor and liquid phase, respectively, are [10],

$$q'_{R_v} = \sum_{j=1}^{N} \sum_{i=1}^{N} P_{H_i} F_{i-j} \varepsilon_v (1-\varepsilon_\ell)[B_i - \sigma <T_v>_v^4]$$

$$+ \sum_{i=1}^{N} P_{H_i} \varepsilon_v \varepsilon_\ell (\sigma T_{SAT}^4 - \sigma <T_v>_v^4) \quad (52)$$

$$q'_{R_\ell} = \sum_{j=1}^{N} \sum_{i=1}^{N} P_{H_i} F_{i-j} \varepsilon_\ell (1-\varepsilon_v)[B_i - \sigma T_{SAT}^4]$$

$$+ \sum_{i=1}^{N} P_{H_i} \varepsilon_v \varepsilon_\ell (\sigma <T_v>_v^4 - \sigma T_{SAT}^4) \quad (53)$$

where, the radiosity vector ($\underline{B}$) is given by,

$$B_j = \psi_{ij}[\Omega_i + S_i] \quad (54)$$

The radiation heat transfer (per unit length) to the liquid, q'_{R_ℓ}, must also be included in volumetric vapor generation term, Γ; thus, Eq. (30) becomes,

$$\Gamma = \frac{[(q''_{vi} - q''_{i\ell})P_i + \tau_i P_i U_R + q'_{R_\ell}]}{(e_{g_i} - e_{f_i})A_{x-s}} \quad (55)$$

The reader should realize that in order to perform realistic ECC analysis, a heat transfer package must be specified for the convective heat transfer coefficient, H_k, and, for BWRs, a falling film rewet model. These heat transfer models are not the subject of this chapter, but have been covered in other chapters.

In addition, a transient conduction model is needed for the nuclear fuel and structural components. A typical radial conduction model [10] can be written in the form,

$$\underline{T}^{(n+1)} = \underline{\underline{C}}\, \underline{T}^{(n)} + \underline{Q}^{(n)} \quad (56)$$

where, $T_i^{(n+1)}$ is the temperature of (radial) node-i, at time $t=t_o+(n+1)\Delta t$, and Q_i is the vector which contains the nodal internal generation and the convective and radiation boundary conditions. As shown schematically in Fig. 5, for $1 \le i \le m-1$, the internal generation is normally determined from decay heat curves, such as given in Fig. 6.

For high temperatures, one must consider the heat generation due to the exothermic zirc/water reaction,

$$Zr + 2H_2O \rightarrow ZrO_2 + 2H_2 + \Delta \quad (57)$$

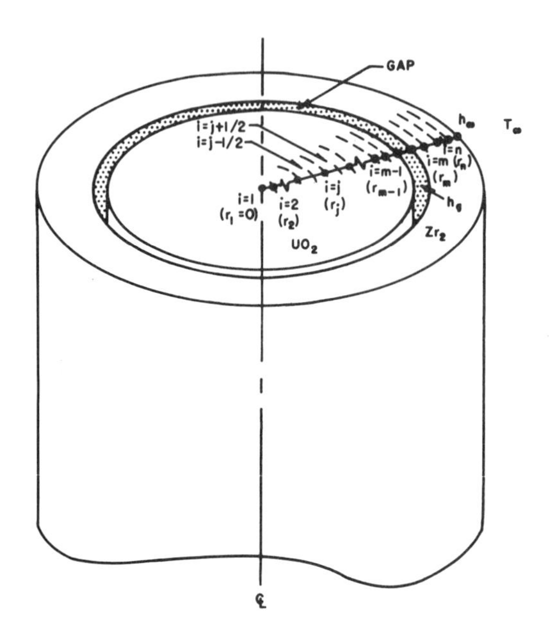

Fig. 5 Spatial Grid for Transient Conduction Analysis of Uranium Fuel Element

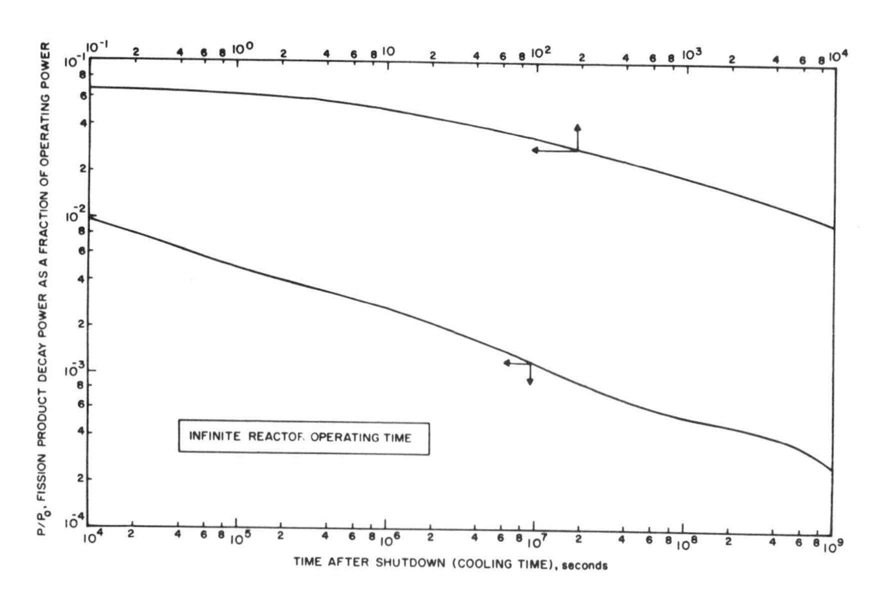

Fig. 6 Decay Heat Curve

This chemical reaction may generate heat in the outer (i=n) nodes and, if the cladding is breached, in the inner (i=m) nodes. At high cladding temperatures (T_w>1255K), this effect must be considered. The Baker-Just model [16] is frequently used in licensing calculations of the heat generated (Δ) by the zirc/water reaction, although this model tends to be conservative, since it does not take into account the constraint on the reaction rate due to steam diffusion limitations.

A typical BWR ECC situation in a fuel rod bundle is shown in Fig. 7. The type of thermal-hydraulic detail that must be modelled by the two-fluid LWR safety code is clearly shown. Figure 8 shows how the heat transfer package, rod conduction model and two-fluid thermal-hydraulics are integrated for the calculation of peak clad temperature. Typical results for a hypothetical BWR/6 LOCA are shown in Fig. 9.

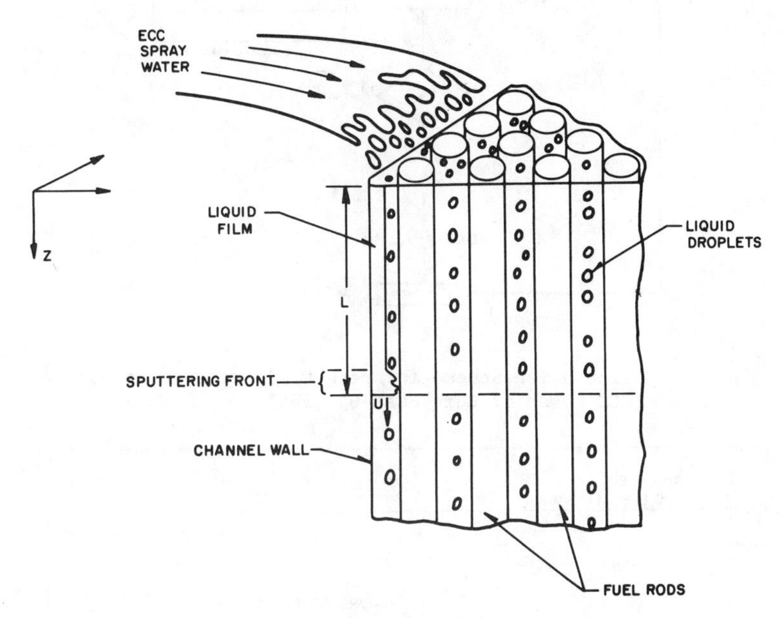

Fig. 7 A Segment of a BWR Fuel Rod Bundle During ECCS Spray Cooling

6. DRIFT-FLUX TECHNIQUES

The above discussion of one-dimensional two-fluid models has shown the importance of the phasic interaction terms. Unfortunately, the functional form of many of these terms are currently not well understood. Thus, alternate approaches have been developed in which empirical correlations are substituted for many of the phasic interaction terms. These approaches include the use of phasic slip correlations and the specification of lateral distribution effects through the use of drift-flux relations [4].

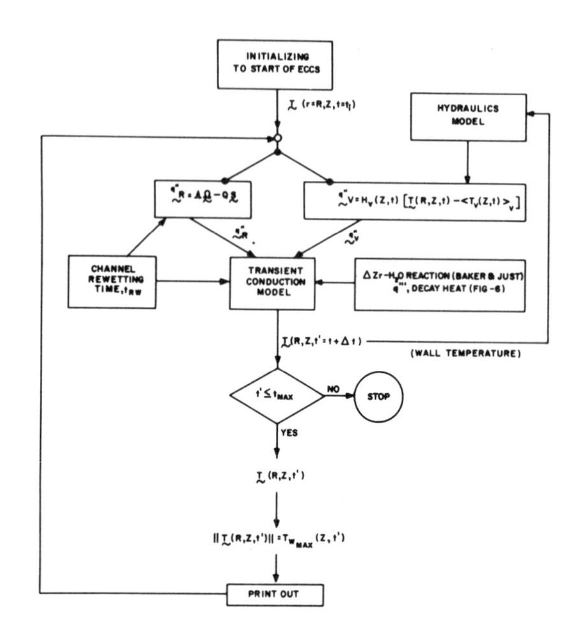

Fig. 8 Calculation Scheme for Peak Clad Temperature During ECCS Phase of Core Heat-Up (BWR)

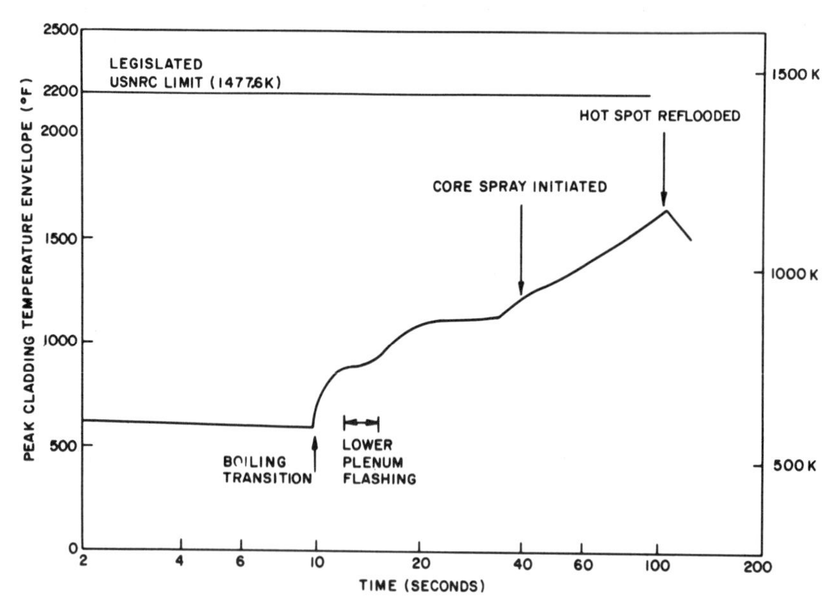

Fig. 9 Typical BWR/6 Peak Cladding Temperature Following a Design Basis Accident

All drift-flux models involve the use of the drift-flux parameters, C_o and V_{gj}, to specify the interfacial momentum transfer. The physical significance of these parameters can be understood through consideration of the following derivation. If we start with the identity,

$$u_g = j + (u_g - j) \tag{58}$$

where, the total volumetric flux is given by,

$$j \triangleq j_g + j_f \tag{59}$$

and recall the definition of the superficial local velocity of the vapor phase (j_g)

$$j_g = \alpha \, u_g \tag{60}$$

then, Eq. (58) can be written,

$$j_g = \alpha \, j + \alpha(u_g - j)$$

Averaging across the cross-sectional flow area,

$$\langle j_g \rangle = C_o \langle j \rangle \langle \alpha \rangle + V_{gj} \langle \alpha \rangle \tag{61}$$

where,

$$C_o \triangleq \langle \alpha j \rangle / \langle \alpha \rangle \langle j \rangle \tag{62a}$$

$$V_{gj} \triangleq \langle (u_g - j)\alpha \rangle / \langle \alpha \rangle \tag{62b}$$

Obviously these "drift-flux" parameters have been introduced so that we can separate the average of the product of two variables into the product of the averages.

It is informative to rewrite Eq. (61) in the following form,

$$\langle u_g \rangle_g = \langle j_g \rangle / \langle \alpha \rangle = C_o \langle j \rangle + V_{gj} \tag{63}$$

In this form we see that the void concentration parameter, C_o, quantifies the lateral distribution of the vapor phase, and the drift velocity, V_{gj}, specifies the degree of local slip. If we plot Eq. (63) in Fig. 10, we see that C_o and V_{gj} are the slope and intercept, respectively. Since an experimentalist normally measures $\langle j \rangle$, $\langle j_g \rangle$ and $\langle \alpha \rangle$ (and thus knows $\langle u_g \rangle_g$), we can readily determine the drift-flux parameters from data. Indeed, as shown in Fig. 11, if the data are plotted three-dimensionally, we see that Eq. (63) would envelope the data. It is interesting to note that the projection of this envelope is the so-called flooding curve; thus if we choose the void dependence of C_o and V_{gj} correctly, we will automatically predict the counter-current flow limitation (CCFL).

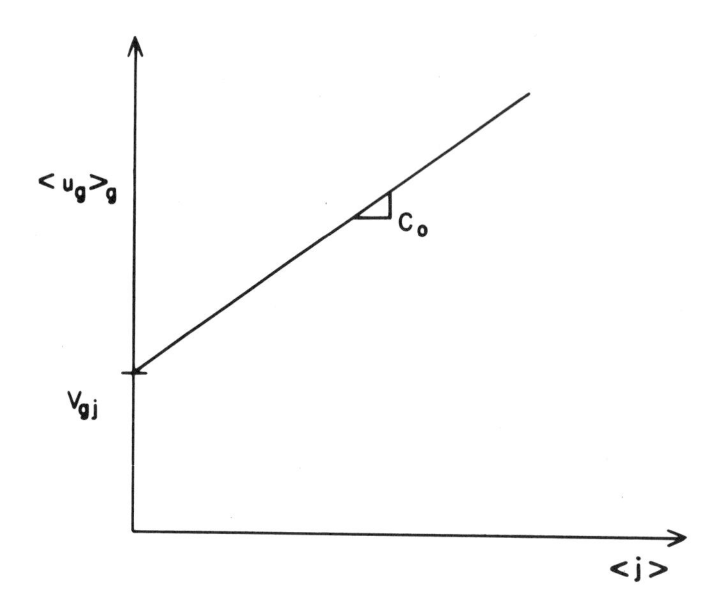

Fig. 10 Drift-Flux Parameters

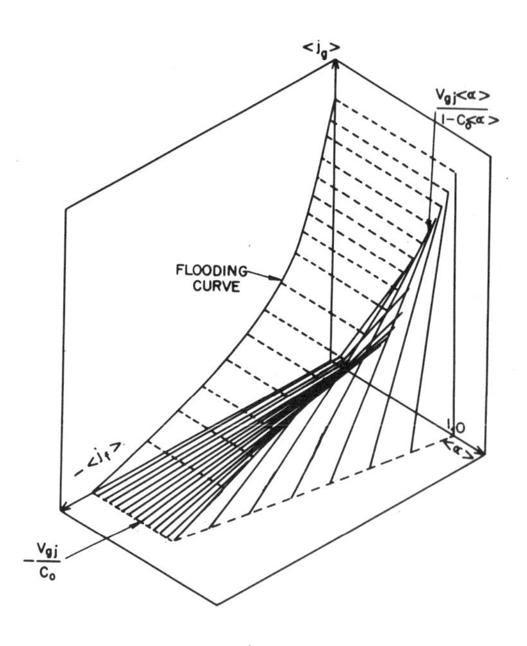

Fig. 11 Flooding Curve and the 3-D Surface of the Drift-Flux Model

It is interesting to rewrite Eq. (61) as,

$$\langle \alpha \rangle = \frac{\langle j_g \rangle}{C_o[\langle j_f \rangle + \langle j_g \rangle] + V_{gj}} \tag{64}$$

This can be factored into,

$$\langle j_g \rangle [1 - C_o \langle \alpha \rangle] = C_o \langle \alpha \rangle \langle j_f \rangle + \langle \alpha \rangle V_{gj} \tag{65}$$

Dividing through by $\langle \alpha \rangle (1 - \langle \alpha \rangle)$,

$$\langle u_g \rangle_g \frac{[1 - C_o \langle \alpha \rangle]}{[1 - \langle \alpha \rangle]} = C_o \langle u_f \rangle_f + \frac{V_{gj}}{(1 - \langle \alpha \rangle)} \tag{66}$$

which can be written as,

$$S = \frac{\langle u_g \rangle_g}{\langle u_f \rangle_f} = \frac{[1 - \langle \alpha \rangle]}{[1/C_o - \langle \alpha \rangle]} + \frac{V_{gj}(1 - \langle \alpha \rangle)}{(1 - C_o \langle \alpha \rangle) \langle j_f \rangle_f} \tag{67}$$

Clearly the slip ratio (S) is specified by the drift-flux parameters. The local slip is quantified by V_{gj} and the global slip by C_o. Thus, once we have specified these parameters, the interfacial drag has been implicitly specified.

Various drift-flux models have been developed for the investigation of reactor safety. These models can involve only the three (mixture) conservation equations, or can involve four or five conservation equations. RELAP/5 [17] is an example of a five equation drift-flux model. In this code, the phasic continuity and energy equations are used in conjunction with a mixture momentum equation.

The simplest drift-flux model, which still allows for the calculation of phasic slip and non-thermodynamic equilibrium, is the four-equation drift-flux model. The essence of such models will now be described.

The Eulerian forms of the mixture conservation equations (8), (17) and (32) can be rewritten in the following (Lagrangian) forms,

$$\frac{\partial}{\partial t}[\rho_\ell(1 - \langle \alpha \rangle) + \rho_v \langle \alpha \rangle] + \frac{\partial}{\partial z}[\rho_\ell \langle j_\ell \rangle$$

$$+ \rho_v \langle j_v \rangle] + \frac{1}{A_{x-s}}[\rho_\ell \langle j_\ell \rangle + \rho_v \langle j_v \rangle] \frac{\partial A_{x-s}}{\partial z} = 0 \tag{68}$$

$$- \frac{\partial p}{\partial z} - g[\rho_\ell(1 - \langle \alpha \rangle) + \rho_v \langle \alpha \rangle] - \frac{\tau_w P_f}{A_{x-s}} - \Gamma \left[\frac{\langle j_v \rangle}{\langle \alpha \rangle} - \frac{\langle j_\ell \rangle}{(1 - \langle \alpha \rangle)}\right]$$

$$= \rho_\ell (1 - <\alpha>) \frac{\partial}{\partial t} \left(\frac{<j_\ell>}{(1 - <\alpha>)} \right) + \rho_\ell <j_\ell> \frac{\partial}{\partial z} \left(C_\ell \frac{<j_\ell>}{1 - <\alpha>} \right)$$

$$+ \rho_v <\alpha> \frac{\partial}{\partial t} \left(\frac{<j_v>}{<\alpha>} \right) + \rho_v <j_v> \frac{\partial}{\partial z} \left(C_v \frac{<j_v>}{<\alpha>} \right) \tag{69}$$

and neglecting kinetic and potential energy terms,

$$(<h_v>_v - <h_\ell>_\ell)\ \Gamma = \frac{q''P_H}{A_{x-s}} + q''' + \frac{\partial p}{\partial t} - \rho_v <\alpha> \frac{\partial <h_v>_v}{\partial t}$$

$$- \rho_v <j_v> \frac{\partial}{\partial z} (K_v <h_v>_v) - \rho_\ell (1 - <\alpha>) \frac{\partial <h_\ell>_\ell}{\partial t}$$

$$- \rho_\ell <j_\ell> \frac{\partial}{\partial z} (K_\ell <h_\ell>_\ell) \tag{70}$$

Obviously, Eq. (68) is the mixture continuity equation, (69) the mixture momentum equation and (70) the mixture energy equation. The fourth independent conservation equation is the vapor phase continuity equation, which can be written as,

$$\frac{\partial <\alpha>}{\partial t} + \frac{\partial <j_v>}{\partial z} = \frac{\Gamma}{\rho_v} - \frac{<\alpha>}{\rho_v} \frac{\partial \rho_v}{\partial t}$$

$$- \frac{<j_v>}{\rho_v} \frac{\partial \rho_v}{\partial z} - \frac{j_v}{A_{x-s}} \frac{\partial A_{x-s}}{\partial z} \tag{71}$$

Closure of this system of equations is provided by the drift-flux expression, Eq. (61). While the choice of independent variables is arbitrary, it is convenient to choose.

$$\underline{u} = (<h>, p, <j>, <\alpha>)^T \tag{72}$$

where,

$$<h> \triangleq <h_v>_v <\alpha> + <h_\ell>_\ell (1 - <\alpha>) \tag{73}$$

Once we determine $<j>$ and $<\alpha>$, Eq. (61) yields,

$$<j_v> = <\alpha> C_o <j> + <\alpha> V_{gj} \tag{74a}$$

and,

$$<j_\ell> = <j>(1 - <\alpha>C_o) - <\alpha> V_{gj} \tag{74b}$$

The basic solution strategy is best seen by combining Eq. (74a) and the vapor phase continuity equation, (71),

$$\frac{\partial <\alpha>}{\partial t} + \frac{\partial}{\partial z} [<\alpha>C_o<j> + <\alpha>V_{gj}] = \frac{\Gamma}{\rho_v} - \frac{<\alpha>}{\rho_v A_{x-s}} \frac{D_v}{Dt} (\rho_v A_{x-s})$$

Assuming C_o is flow regime dependent, i.e., $C_o = C_o(<\alpha>)$, and expanding,

$$\frac{\partial\langle\alpha\rangle}{\partial t} + \left\{ \langle j\rangle \left(C_o + \langle\alpha\rangle \frac{dC_o}{d\langle\alpha\rangle}\right) + V_{gj} + \langle\alpha\rangle \frac{dV_{gj}}{d\langle\alpha\rangle} \right\} \frac{\partial\langle\alpha\rangle}{\partial z}$$

$$= \frac{\Gamma}{\rho_v} - C_o\langle\alpha\rangle \frac{\partial\langle j\rangle}{\partial z} - \frac{\langle\alpha\rangle}{\rho_v A_{x-s}} \frac{D_v}{Dt} (\rho_v A_{x-s}) \tag{75}$$

Equation (75) can be written more compactly as,

$$\frac{D_K\langle\alpha\rangle}{Dt} \triangleq \frac{\partial\langle\alpha\rangle}{\partial t} + C_K \frac{\partial\langle\alpha\rangle}{\partial z} = \frac{\Gamma}{\rho_v} - C_o\langle\alpha\rangle \frac{\partial\langle j\rangle}{\partial z} - \frac{\langle\alpha\rangle}{\rho_v A_{x-s}} \frac{D_v}{Dt} (\rho_v A_{x-s}) \tag{76}$$

where, the so-called kinematic wave velocity is given by,

$$C_K \triangleq \langle j\rangle \left(C_o + \langle\alpha\rangle \frac{dC_o}{d\langle\alpha\rangle}\right) + V_{gj} + \langle\alpha\rangle \frac{dV_{gj}}{d\langle\alpha\rangle} \tag{77}$$

This is the celerity at which discontinuities in void fraction propagate.

Equations (77), (70), (69) and (68) are essentially the set of equations which are used in MAYU-4 [18]. These equations are in Lagrangian form and are thus conveniently integrated along the characteristics. As in two-fluid models, one must take care that the eigenvalues of the overall drift-flux model are all real, or the numerical scheme utilized may become unstable. Fortunately, reasonable choices of C_o and V_{gj} normally lead to well-posed systems.

It should be obvious that slip is provided through the drift-flux parameters. The means of calculating liquid subcooling and vapor superheat is less obvious and thus will be explained here. The mixture energy equation, Eq. (70), defines the volumetric vapor generation rate (Γ). For bulk boiling conditions, it becomes,

$$\Gamma = \frac{1}{h_{fg}} \left[\frac{q''P_H}{A_{x-s}} + \frac{\partial p}{\partial t} \right.$$

$$- \rho\langle\alpha\rangle \frac{\partial h_g}{\partial t} - \rho_g\langle j_g\rangle \frac{\partial h_g}{\partial z}$$

$$\left. - \rho_f(1 - \langle\alpha\rangle) \frac{\partial h_f}{\partial t} - \rho_f\langle j_f\rangle \frac{\partial h_f}{\partial z} \right] \tag{78}$$

For subcooled boiling, the volumetric vapor generation can be given by [19],

$$\Gamma = \frac{1}{h_{fg}} \left\{ q''P_H\left[1 - \frac{(h_f - \langle h_\ell\rangle_\ell)}{(h_f - \langle h_{\ell_d}\rangle_\ell)}\right] - H_o \frac{h_{fg}}{v_{fg}} \langle\alpha\rangle(T_{SAT} - \langle T_\ell\rangle_\ell) \right.$$

$$\left. + \langle\alpha\rangle \frac{\partial p}{\partial t} - \rho_g\langle\alpha\rangle \frac{\partial\langle h_g\rangle_g}{\partial t} - \rho_g\langle j_g\rangle \frac{\partial\langle h_g\rangle_g}{\partial z} \right\} \tag{79}$$

where the liquid subcooling at the initiation of subcooled boiling is given by [20],

$$h_f - \langle h_{\ell_d} \rangle_\ell = \begin{cases} 0.0022\, q'' D_H c_{p_\ell} / \kappa_\ell , & Pe < 70{,}000 \\ 154\, q''/G, & Pe > 70{,}000 \end{cases} \quad (80a)$$

where, the Peclet number (Pe) is defined as,

$$Pe = \frac{G D_H c_{p_\ell}}{\kappa_\ell} \quad (80b)$$

For heat transfer situations beyond the boiling transition point, in which superheated vapor is expected, the following expression [21] can be used for Γ,

$$\Gamma = \frac{1}{h_{fg}} \left\{ \frac{K_1 \kappa_v}{D_H^2} (1 - \langle\alpha\rangle)(\langle T_v \rangle_v - T_{SAT}) \right.$$

$$\left. - \rho_f (1 - \langle\alpha\rangle) \frac{\partial h_f}{\partial t} - \rho_f j_f \frac{\partial h_f}{\partial z} + (1 - \langle\alpha\rangle) \frac{\partial p}{\partial t} \right. \quad (81)$$

where,

$$K_1 = \left\{ 12 + 5.508 \left[\frac{\rho_v \left(\frac{\langle j_v \rangle}{\langle\alpha\rangle} - \frac{\langle j_f \rangle}{(1 - \langle\alpha\rangle)} \right) \delta}{\mu_v} \right]^{0.55} \left(\frac{c_{p_v} \mu_v}{\kappa_v} \right)^{0.33} \right\} \left(\frac{D_H}{\delta} \right)^2 \quad (81a)$$

and,

$$(\delta / D_H) = 1.47 \left[\frac{\rho_v \langle j_{v_{B.T.}} \rangle^2 \sqrt{\frac{\sigma}{g(\rho_f - \rho_v)}}}{\langle\alpha\rangle} \right]^{-0.675} \left(\frac{\langle j_f \rangle}{\langle j_{f_{B.T.}} \rangle} \right)^{1/3} \quad (81b)$$

It is important for the reader to realize that for cases of non-thermodynamic equilibrium, we have been able to calculate the non-equilibrium temperature of one phase by assuming the other is at saturation. In the subcooled boiling case we assumed the vapor was at saturation temperature while for the beyond B.T. case, we assumed the residual liquid drops were saturated. While these assumptions are considered valid for the thermal-hydraulic situations described here, they may not hold for other situations of interest. For the evaluation of cases in which both phases are in non-equilibrium, one must use either a five equation drift-flux model or a six-equation, two-fluid model.

7. MIXTURE MODELS

In many cases of practical significance, the two-phase system is essentially in thermodynamic equilibrium. Moreover, for some cases, the system may be well modelled by a homogeneous mixture (i.e., one which has no slip). For such cases, it is convenient to just use the three mixture conservation equations. One form of the mixture equations is given in Eqs. (8), (17) and (32); however, there are many other equivalent forms.

An extensive discussion of the equivalent forms of the mixture conservation equations has been given previously [10], and thus will not be repeated here.

It is of interest, however, to point out that one compact form of the drift-flux mixture momentum equation is (10),

$$-\frac{\partial p}{\partial z} - g\langle\rho\rangle - \frac{\tau_w P_f}{A_{x-s}} - \frac{1}{A_{x-s}}\frac{\partial}{\partial z}\left[A_{x-s}\frac{\langle\alpha\rangle}{(1-\langle\alpha\rangle)}\frac{\rho_f\rho_g}{\langle\rho\rangle}(V'_{gj})^2\right]$$

$$= \frac{1}{g_c}\left[\frac{\partial G}{\partial t} + \frac{1}{A_{x-s}}\frac{\partial}{\partial z}\left(\frac{G^2 A_{x-s}}{\langle\rho\rangle}\right)\right] \quad (82)$$

where,

$$V'_{gj} = \langle u_g\rangle_g - \langle j\rangle = V_{gj} + (C_o - 1)\langle j\rangle \quad (83)$$

and, for simplicity, the assumption has been made that $C_\ell = C_v = 1.0$.

It should be noted that the drift-flux term appears as a spatial gradient and represents the net momentum flux of the two-phase mixture with respect to the center-of-mass velocity, $G/\langle\rho\rangle$. This term is really a spatial acceleration type term; however, it can be regarded as an additional volumetric force in the same sense that the Reynolds stress term of single-phase turbulence is considered a force.

From Eq. (67), we see that $C_o = 1.0$ and $V_{gj} = 0.0$ imply homogeneous (S=1.0) flow. Thus, for homogeneous flow, the only change in the mixture momentum equation, Eq. (82), is the vanishing of the gradient term which represents the drift-flux induced force. That is, the only difference between the homogeneous and slip flow mixture momentum equations is the drift-flux gradient term, which is an exact differential and thus does not change the complexity of the numerical integration scheme.

Numerous digital computer codes, which are based on mixture conservation equations, have been written for the safety analysis of nuclear reactor systems. One of the best known in RELAP/4 [22] which is based on homogeneous mixture equations. Codes of this type have been the "work-horses" of the industry, but are currently being replaced by more detailed codes such as TRAC. Due, however, to the relatively long-running time required by both of these codes, it can be anticipated that drift-flux techniques, and analyses, will play a more important role in the future, particularly for the safety analysis of such quasi-steady phenomena as small-break LOCA.

While the two-phase equations used in the analysis of reactor safety can be recast into various equivalent forms, the basic structure, and the problems concerning each formulation, has now been given. Subsequent chapters will describe the use of such equations for reactor safety evaluations.

NOMENCLATURE

a_k = Plank mean absorption coefficient of phase-k.

A_{x-s} = Cross sectional flow area.

B_i = Radiosity vector.

C_D = Interfacial drag coefficient.

C_k = Distribution coefficient of phase-k.

C_K = Kinematic wave velocity.

C_o = Void concentration parameter.

c_{p_k} = Specific heat of phase-k.

C_{vm} = Virtual volume coefficient.

D_H = Hydraulic diameter.

D_b = Bubble diameter.

D_d = Droplet diameter.

e_k = Total convected energy of phase-k.

e_{k_i} = Total convected energy at the interface of phase-k.

$\langle e_k \rangle_k$ = One-dimensional average of total convected energy of phase-k.

f_k = Darcy-Weisbach friction factor of phase-k.

F_{v-m} = Volumetric virtual mass force.

F_{i-j} = View factor.

G = Mass flux.

g = Acceleration of gravity.

H_o = Condensation coefficient.

H_{ik} = Interfacial heat transfer coefficient.

H_k = Wall heat transfer coefficient of phase-k.

h_k = Enthalpy of phase-k.

$\langle h_k \rangle_k$ = One-dimensional averaged enthalpy of phase-k.

h_{fg} = Latent heat of vaporization.

j = Volumetric flux of two-phase mixture.

$\langle j \rangle$ = One-dimensional averaged volumetric flux of two-phase mixture.

j_k = Volumetric flux of phase-k.

$\langle j_k \rangle$ = One-dimensional averaged volumetric flux of phase-k.

K_k = Total (convected) energy distribution coefficient of phase-k.

L_m = Mean beam length.

p = Static pressure.

p_i = Interfacial pressure.

Pe = Peclet number.

P_f = Total friction perimeter.

P_{f_k} = Friction perimeter of phase-k.

P_i = Interfacial perimeter.

Pr_k = Prandtl number of phase-k.

$\underline{Q}$ = Nodal internal generation and surface heat transfer vector.

q'_{r_k} = Radiation heat transfer (per unit axial length) to phase-k.

q'' = Total wall heat flux.

q''_k = Wall heat flux to phase-k.

q''_{ik} = Interfacial heat flux to phase-k.

q''' = Volumetric internal generation of the two-phase mixture.

q'''_k = Volumetric internal generation of phase-k.

R = Radius of tube.

r_i = Radial position of interface.

S = $\langle u_v \rangle_v / \langle u_\ell \rangle_\ell$, Slip ratio

T = Averaging time internal.

T_w = Wall temperature.

$\langle T_k \rangle_k$ = One-dimensional averaged temperature of phase-k.

t = Time.

$\underline{T}^{(n)}$ = Temperature vector (at n^{th} time step)

T_{SAT} = Saturation temperature.

U_R = $\langle u_v \rangle_v - \langle u_\ell \rangle_\ell$, Relative velocity.

U_i = Interfacial velocity (in axial direction).

u_k = Velocity of phase-k.

$\langle u_k \rangle_k$ = One-dimensional averaged velocity of phase-k.

V_{gj} = Local drift-velocity.

V'_{gj} = $\langle u_g \rangle_g - \langle j \rangle$

V = Volume.

X = Martinelli parameter.

X_k = Phase indicator function.

z = Axial position.

GREEK SYMBOLS

α = Local void fraction.

$\langle \alpha \rangle$ = One-dimensional averaged void fraction.

Γ = Volumetric vaporization rate.

ρ_k = Density of phase-k.

$\langle \rho \rangle$ = $\rho_\ell(1-\langle \alpha \rangle) + \rho_v \langle \alpha \rangle$, density of two-phase mixture.

δ = Droplet diameter.

λ = An arbitray parameter in virtual mass acceleration.

κ_k = Thermal conductivity of phase-k.

μ_k = Viscosity of phase-k.

τ_w = Total wall shear.

τ_{w_k} = Wall shear on phase-k.

τ_i = Interfacial shear.

θ = Orientation (angle) from horizontal.

η = Weighting parameter in interfacial velocity model.

σ = Stefan-Boltzmann constant.

ε_i = Emissivity of surface-i.

ϕ_k^2 = Two-phase multiplier for phase-k.

SUBSCRIPTS

c	=	Continuous phase
v	=	Vapor
g	=	saturated vapor
ℓ	=	Liquid
f	=	Saturated liquid

REFERENCES

[1] Ishii,M., "Thermo-Fluid Dynamic Theory of Two-Phase Flow", Eyrolles, 1975.

[2] Delhaye, J.M. and Achard, J.L., "On the Averaging Operators Introduced in Two-Phase Flow Modeling", proceedings of the CSNI Specialists Meeting, Vol.-I, Toronto, 1976.

[3] Yadigoraglu, G. and Lahey, R.T., Jr., "On the Various Forms of the Conservation Equations in Two-Phase Flow", Int. J. Multiphase Flow, Vol. 2, pp 477-494, 1976.

[4] Zuber, N. and Findlay, J.A., "Average Volumetric Concentration in Two-Phase Flow Systems", J. Heat Transfer, 1965

[5] Hench, J.E. and Johnston, J.P., "Two-dimensional diffuser performance with subsonic, two-phase air/water flow", APED - 5477, 1968.

[6] Wallis, G.B., "One-Dimensional Two-Phase Flow", McGraw Hill Book Company, 1969.

[7] Drew, D., Cheng, L., Lahey, R.T., Jr., "The Analysis of Virtual Mass Effects in Two-Phase Flow", Int. J. Multiphase Flow, Vol. 5, 1979.

[8] Taitel, T. and Dukler, A.E., "A Model for Predicting Flow Regime Transitions in Horizontal and Near Horizontal Gas-Liquid Flow" A.I. Ch.E Journal, Vol-22, 1976.

[9] Andersen, J.G.M., Andersen, P.S., Olsen, A. and Miettinen, J., "NORCOOL A Model for Analysis of a BWR under LOCA conditions", NORHAV-D-29, 1976.

[10] Lahey, R.T., Jr., and Moody, F.J., "The Thermal-Hydraulics of a Boiling Water Nuclear Reactor", ANS Monograph, 1977.

[11] Khachadour, A.M., "Interfacial Area Measurement Using a Radioisotope Technique", M.S. Thesis, McMaster University (Canada), 1975.

[12] Veteau, J.M., "Mesure des Aires Interfaciales Dans Les Ecoulements Diphasiques", CEA-R-5005, 1979.

[13] Garabedian, P.R., "Partial Differential Equations", John Wiley & Sons Inc., 1964.

[14] LASL Staff, "TRAC-P1A: An Advanced Best Estimate Computer Program for PWR LOCA Analysis", LA-7777-MS (NUREG/CR-0665), 1979.

[15] Bouré, J., "On a Unified Presentation of the Non-Equilibrium Two-Phase Flows," ASME Symposium Volume, Nonequilibrium Two-Phase Flows, 1975.

[16] Baker, L. and Just, L.C., "Studies of Metal-Water Reactions at High Temperatures", ANL-6548, 1962.

[17] Ransom, V. H. et al, "RELAP/5 MOD-0 Code Description," CDAP-TR-057, 1979.

[18] Punches, W.C., "MAYU-04, A Method to Evaluate Transient Thermal-Hydraulic Conditions in Rod Bundles", GEAP-23517, 1977.

[19] Lahey, R.T., Jr., "A Mechanistic Subcooled Boiling Model" paper FB-4, Proceedings of Sixth Int. Heat Transfer Conf., 1978.

[20] Saha, P. and Zuber, N., "Point of Net Vapor Generation and Vapor Void Fraction in Subcooled Boiling", Proceedings of Fifth Int. Heat Transfer Conf., Vol-IV, 1974.

[21] Saha, P., Shiralkar, B.S., Dix, G.E., "A Post-Dryout Heat Transfer Model Based on Actual Vapor Generation in the Dispersed Droplet Region", ASME Preprint 77-HT-80, 1977.

[22] Katsma, K. R. et al, "RELAP/4 MOD-4 A Computer Program for the Transient Thermal-Hydraulic Analysis of Nuclear Reactors and Systems - User's Manual," ANCR-NUREG 1335, 1979.

PART 3

DESIGN BASIS ACCIDENT: LIGHT WATER REACTORS

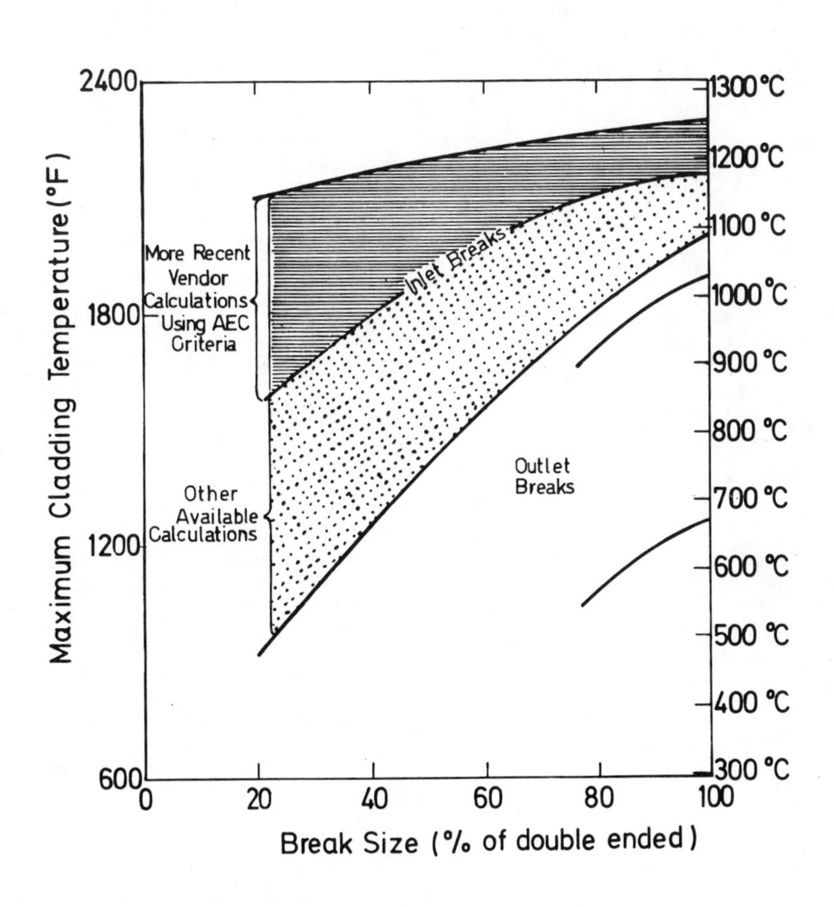

9 LOSS OF COOLANT ACCIDENT

by
W.L. Riebold
CEC – JRC, Ispra, Italy

A summary of the essential design features of both the main cooling system and the emergency core cooling system (ECCS) of the two light-water reactor (LWR) types precedes the main sections of this paper. Following the introductory material provided in Chapter 2, details are given to allow the student to understand specific behavior which is included in following chapters.

The verbal-pictorial description of the loss-of-coolant accident (LOCA) covers the whole transient period from the start of depressurization caused by a rupture within the reactor cooling system, through to complete refilling of the core. Distinction is made between large break and small break LOCAs as well as between LOCA in PWRs and BWRs, and consideration of the differences between a liquid line break and a steam line break is given.

The paper finishes with a brief consideration of a LOCA without an effective ECCS, including an introduction to possible collapse of the core and pressure vessel melt through which will be supplemented by more detailed developments in Chapters 17 and 18. Concluding remarks include a summary of the ECCS acceptance criteria and a list of the critical LOCA phenomena.

1. INTRODUCTION

The loss-of-coolant accident (LOCA) in light-water reactors (LWR) is the hypothetical Design Basis Accident: it represents the postulated, most severe but highly unlikely accident conditions against which the reactor safety systems have to be designed. It is for this reason that, in order to demonstrate the safety of reactor systems the efforts of the reactor safety research programs in various countries have for several years concentrated on the LOCA. Their particular aim is to understand the whole thermohydraulic processes associated with the LOCA in order to explore the safety margins contained in the present licensing requirements and thus provide the basis for a quantification of the existing conservatism.

A verbal-pictorial description of the thermohydraulic scenario associated with the LOCA is the main subject of the present paper. A brief introduction deals with the initiating events and the classification of LOCAs. In view of some fundamental differences with respect to the physically significant thermohydraulic phenomena occurring during such a transient, the course of

LOCAs is described separately for large break and small break accidents. Again, LOCAs in pressurized-water reactors (PWR) and in boiling-water reactors (BWR) are considered separately due to the differences in the design features of both the main cooling system and the emergency core cooling system (ECCS) of the two types of light-water reactors. To provide a reference basis for the main section of the paper, and to ease the understanding of the function of the main cooling system and the emergency core cooling system (ECCS) as well as their interaction with each other during a LOCA transient, these systems for both reactor types, PWRs and BWRs, are summarized at the beginning of the paper.

For the sake of completeness, a brief consideration of the main events of a LOCA without an effective ECCS has been included.

The paper concludes with a summary of the relevant ECCS acceptance criteria and of the critical, physical phenomena occurring during the various phases of a LOCA.

Material for this paper has been drawn from a large variety of sources and in particular from ref. [1], [2] and [4].

2. NUCLEAR STEAM SUPPLY SYSTEMS WITH LIGHT-WATER REACTORS

This section is intended as both a refresher for those already familiar to some extent with nuclear power reactors, and a reference for the subsequent chapters dealing with the course of a loss-of-coolant accident (LOCA) and related safety issues.

A short description will be given of the main design features of light-water reactors (LWR) in common use in commercial nuclear steam supply systems (NSSS). This description is subdivided into two parts: the first one (2.1) deals with the main cooling system, and the subsequent one (2.2) with the emergency core cooling system (ECCS).

2.1 The LWR Main Cooling System

There are two basic types of light-water reactor: the pressurized-water reactor (PWR) and the boiling-water reactor (BWR).

The principal difference between the two reactor systems is related to the design concept of transferring energy from the reactor fuel to the turbine:

- PWRs employ a "dual system" consisting of a high-pressure "primary cooling system" (PCS) for the heat removal from the reactor core, and an intermediate-pressure "secondary cooling system" (SCS) for steam generation and delivery to the turbine. The two systems are coupled together through the steam generators which provide the intermediate heat exchange process. PWRs are therefore called "indirect cycle" systems. A simpli-

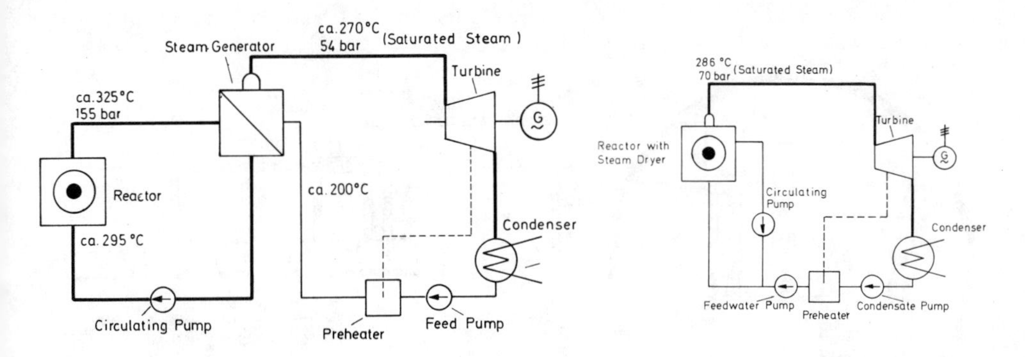

Fig. 1: Simplified Flow Diagram of a PWR Nuclear Power Plant with « dual » or « indirect cycle » system [2]

Fig. 2: Simplified Flow Diagram of BWR with « direct cycle » system and forced circulation [2]

fied flow diagram of a PWR nuclear power plant is shown in Fig. 1.

- BWRs employ a "single system" of intermediate pressure, where the energy supplied from the hot fuel to the reactor coolant is transported directly (without an intermediate heat exchange process), as steam, to the turbine. BWRs are therefore called "direct cycle" systems. A simplified flow diagram of a BWR nuclear power plant is shown in Fig. 2.

Since the working fluid of the reactor is normally weakly contaminated with radioactivity, the principal difference between the two reactor systems is the isolation of the reactor working fluid from the turbine-generator steam supply, as indicated in Fig. 3 and 4 which show schematically the basic components of PWR and BWR power generating systems.

Today the basic lay-out of each of the two reactor types is largely standardized. However, reactors of the same type but from different manufacturers show differences in the detailed design of individual components and subsystems which give rise to interesting comparisons. In the following sections consideration will be limited to the Kraftwerk Union (KWU) designs of both PWRs and BWRs, and to the Westinghouse (W) and General Electric (GE) designs for PWRs and BWRs respectively.

The pressurized-water reactor (PWR) cooling system. The high-pressure primary cooling system of PWRs at present in operation is made up of 2, 3 or 4 parallel loops depending on the power of the reactor plant and on the reactor manufacturer. Fig. 3 shows a schematic representation of such a loop; it comprises a steam generator (SG) (3), a main coolant circulation pump (MCP) (2) and the associated main coolant piping, and is connected to the inlet and outlet nozzles of the reactor pressure vessel (RPV) (1). A flow diagram of a nuclear power plant with a 4-loop PWR is shown in Fig. 5. A pressurizer, (4) in Fig. 3 and 5, provided with electrical heaters and a spray and pressure relief system, is connected through a surge line with the hot leg of one of the primary loops; its purpose is to maintain the primary coolant at a constant

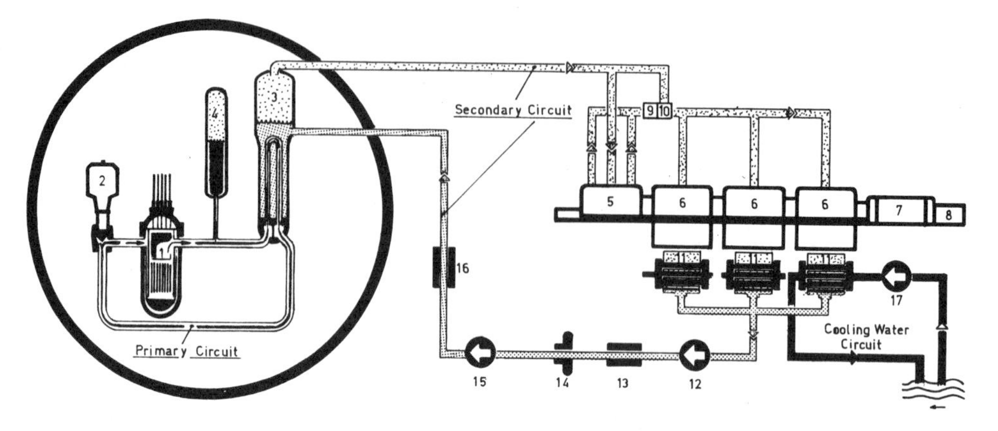

Fig. 3: Schematic Idealization of a PWR Power Generating System [10]

Primary Cooling Circuit
1 Reactor Pressure Vessel
2 Coolant Circulation Pump
3 Steam Generator
4 Pressurizer

Secondary Cooling Circuit
5 Turbine, high-pressure part
6 Turbine, low-pressure part
7 Generator
8 Exciter Unit
9 Water Separator
10 Superheater
11 Condensers
12 Main Condensate Pump
13 Low-Pressure Preheater
14 Feedwater Storage Tank
15 Main Feedwater Pump
16 High-Pressure Preheater

Cooling Water Circuit
17 Main Cooling Water Pump.

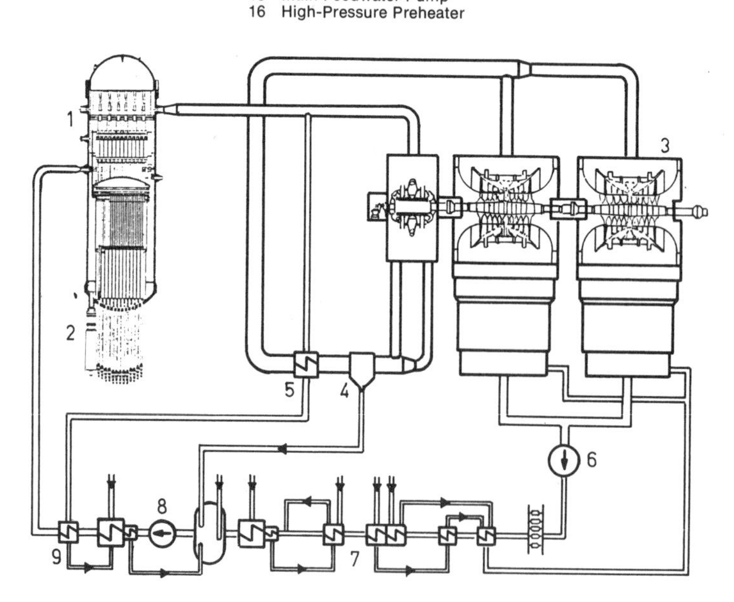

Fig. 4: Schematic Idealization of a BWR Power Generating System [14]

1 Reactor Pressure Vessel
2 Axial Coolant Circulation Pump
3 Turbine with Condenser
4 Water Separator
5 Intermediate Superheater
6 Condensate Pump
7 Low-Pressure Preheater
8 Feedwater Pump
9 High-Pressure Preheater

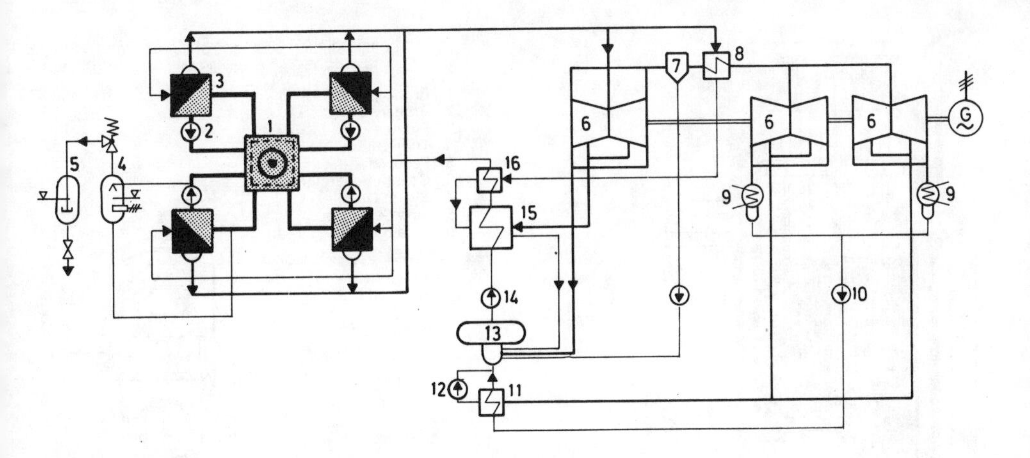

Fig. 5: Flow Diagram of a 4-Loop PWR Nuclear Power Plant [11]

1 Reactor
2 Coolant Circulation Pump
3 Steam Generator
4 Pressurizer
5 Quench Tank
6 Turbine-Generator
7 Water Separator
8 Intermediate Superheater
9 Condenser
10 Main Condensate Pump
11 Low-Pressure Preheater
12 Subsidiary Condensate Pump
13 Feedwater Storage Tank
14 Main Feedwater Pump
15 High-Pressure Preheater
16 Condensate Cooler

operating pressure of about 160 bar. Reactor coolant water at this pressure is circulated by large coolant pumps through the primary loops of the reactor and the steam generators and is heated up within the reactor core to a temperature of about 325°C which is low enough to prevent the water from bulk boiling. Within the steam generators this high-pressure reactor coolant water circulates inside straight or U-shaped, vertical heating tubes. The outer surfaces of these tubes are in contact with the secondary cooling system water which circulates through the steam generator secondary side at a considerably lower pressure and temperature, of about 54 bar and 270°C respectively. Heat transferred within the steam generators from the hot, high-pressure, primary water inside the tubes, to the secondary water outside the tubes, causes the latter to boil and produce steam for the turbines. Before leaving the steam generators, the steam passes through steam dryers arranged above the free surface of the secondary water within the steam drum of the steam generators.

The high-pressure envelope of a PWR is thus represented by the reactor pressure vessel, the steam generator heating tubes and inlet and outlet plena, the circulation pump housing, the pressurizer and the connecting pipe system. It is the integrity of this envelope which prevents a loss-of-coolant accident. The whole high-pressure envelope of a PWR is located within the containment.

A longitudinal section of the PWR pressure vessel and its internals is shown in Fig. 6, which also indicates the coolant water flow path. The core of a 1300 MWe reactor is made up of about 200 fuel elements; each element consists typically of 256 fuel rods assembled to a bundle of square cross section

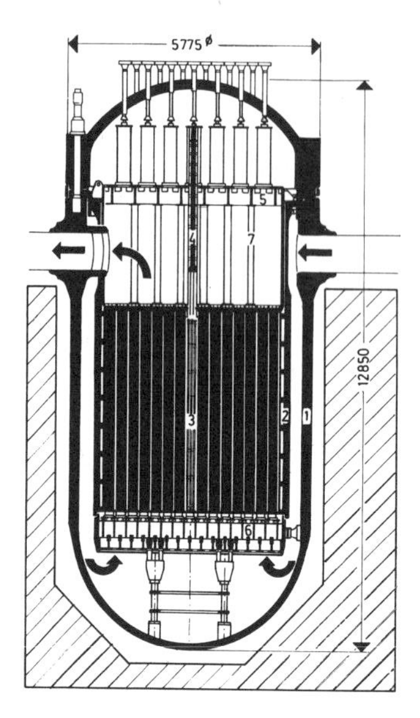

Fig. 6: Reactor Pressure Vessel and Internals of a 1300 MWe KWU PWR [10]

1 Reactor Pressure Vessel (RPV)
2 Core Barrel with Thermal Shield
3 Fuel Element
4 Control Rod
5 Upper Support Grid
6 Lower Support Grid
7 Support

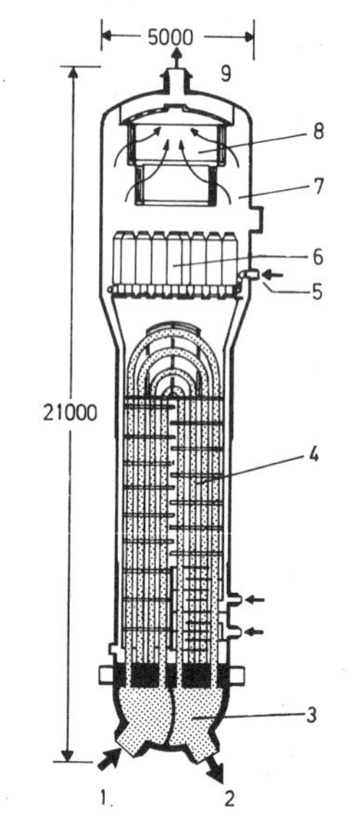

Fig.7: Longitudinal Section of a U-tube Steam Generator (KWU PWR) [10]

1 Primary Inlet
2 Primary Outlet
3 Plenum
4 Tube Bundle
5 Feedwater Inlet
6 Coarse Separator
7 Steam Drum
8 Fine Separator
9 Steam Outlet

measuring about 225 mm along each side. PWR fuel assemblies are open arrays allowing radial mixing of the primary fluid as it passes through the reactor core.

Fig. 7 shows a longitudinal section of a U-tube steam generator. A schematic flow diagram of the main heat removal system (secondary cooling system) of a PWR nuclear power plant is given in Fig. 8.

<u>The boiling-water reactor (BWR) cooling system.</u> It can be seen from Fig. 9, and indeed from certain of the previous diagrams, that the main heat removal system of a BWR is fundamentally identical to that of a PWR, except that the steam is taken directly from the reactor pressure vessel and the feedwater is returned there; the steam pressure and temperature of about 70 bar and 286 ^{o}C respectively are essentially the same.

Hence, the major parts of a BWR nuclear steam supply system are the reactor pressure vessel and the equipment inside the vessel, which are schematically presented in Fig. 10a and 10b for the KWU and GE designs.

In both cases the coolant water enters the reactor core from below and boils inside the core (1), Fig. 10a, 10b. A steam water mixture of about 28% steam quality (about 66% steam void fraction) leaves the core at the top and passes through steam separators at the top of the core shroud. Steam then

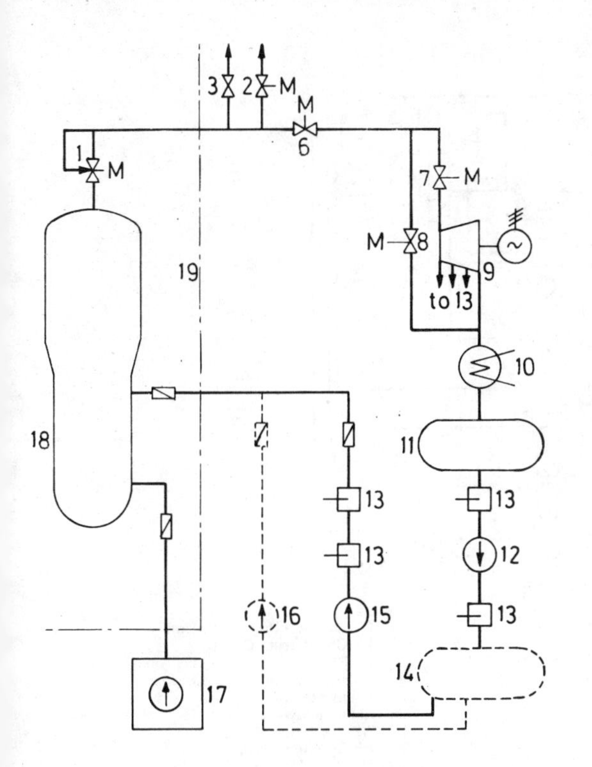

Fig. 8: Schematic Flow Diagram of the Main Heat Removal System (Secondary Cooling System) of a KWU PWR [2]

1 Combined Armature (Shut-off, Safety-Bypass Valve, in some KWU Plants)
2 Relief Valve
3 Safety Valve
6 Quick Shut-Down Armature
7 Turbine Shut-off and Control Valve
8 Turbine Bypass
9 Turbine
10 Condenser
11 Condensate Container
12 Condensate Pump
13 Preheater
14 Feedwater Storage Tank
15 Main Feedwater Pumps
16 Start-up and Run-down Pump
17 Auxiliary Feedwater System
18 Steam Generator
19 Safety Containment

flows through steam dryers (3, 4, 5, 6) to remove residual water, and finally leaves the pressure vessel with about 0.2% moisture (about 1% water volume fraction) before flowing through parallel main steam lines (7, 8) to the turbine generator. Water returning from the turbine condensers (6, 7) enters the annular downcomer between core shroud and pressure vessel and there mixes with water flowing downwards from the steam separators. Circulation pumps (5, 3) at the bottom of the downcomer - 8 axial pumps in KWU-BWRs, 8 jet pumps in GE-BWRs - drive the water flow through the bottom of the reactor vessel and the region of the control rod guide tubes, back to the core. In the case of the GE design, water is withdrawn from the annular downcomer by two motor-driven recirculation pumps outside the pressure vessel and is returned to provide the driving flow for the jet pumps (see Fig. 10b).

The reactor core is made up of about 780 fuel elements; each element consists of typically 64 fuel rods assembled to a bundle of square cross section measuring about 134 mm along each side. Each bundle is installed in a metal channel which is open at the top and the bottom end, allowing only vertical coolant flow through the assembly. The closed sides of BWR fuel arrays prevent lateral flow of coolant between adjacent assemblies in the reactor core.

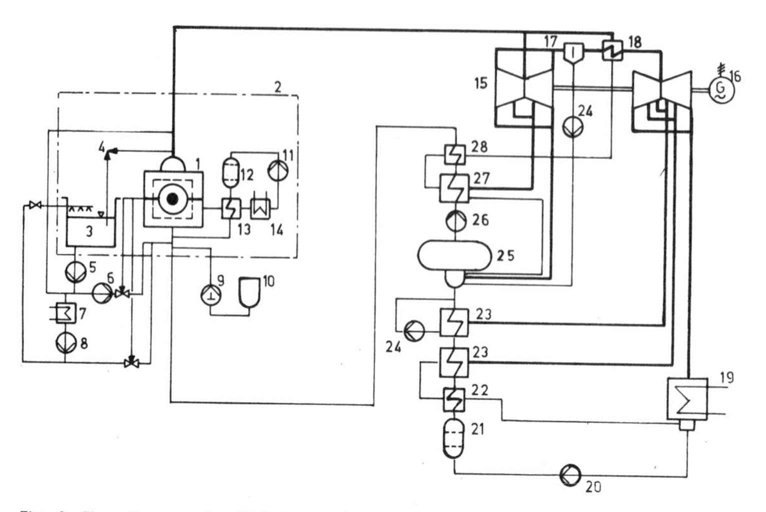

Fig. 9: Flow Diagram of a BWR Nuclear Power Plant [12]

1 Reactor
2 Containment
3 Condensation Pool
4 Safety and Relief Valves
5 Pre-Pump
6 HP Pump
7 Cooler
8 LP Pump
9 Poisoning Pump
10 Poisoning Solution Pool
11 Coolant Cleanup Pump
12 Cleanup Filter
13 Regenerative Heat Exchanger
14 Cleanup Cooler
15 Turbine
16 Generator
17 Water Separator
18 Intermediate Superheater
19 Condenser
20 Main Condensate Pump
21 Condensate Upgrading Unit
22 LP Condensate Cooler
23 LP Preheater
24 Subsidiary Condensate Pump
25 Feedwater Tank
26 Main Feedwater Pump
27 HP Preheater
28 HP Condensate Cooler

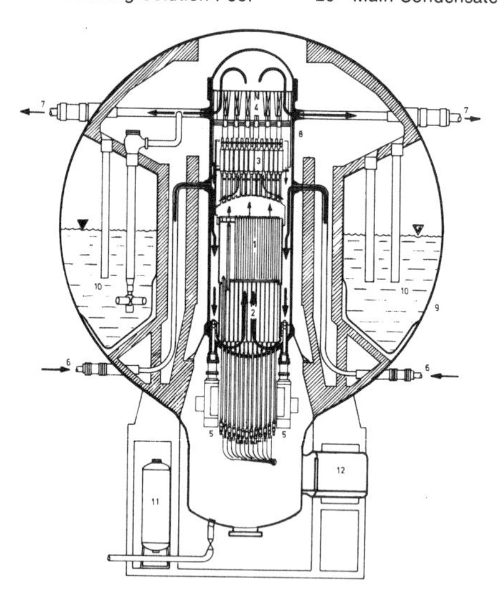

Fig. 10 a: Schematic Diagram of a KWU — BWR Nuclear Steam Supply System [13]

1 Fuel Elements
2 Control Rods
3 Steam-Water Separator
4 Steam Dryer
5 (Axial) Circulation Pumps
6 Feedwater Lines
7 Steam Lines
8 Reactor Pressure Vessel
9 Containment
10 Condensation/Pressure Suppression Pool
11 Reactor Scram System
12 Air Lock

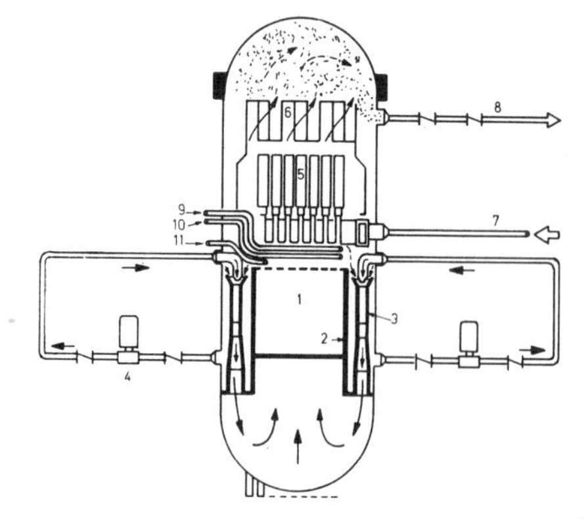

Fig. 10 b: Schematic Diagram of a GE - BWR Nuclear Steam Supply System [1]

1 Core
2 Shroud
3 Jet Pump
4 Recirculation Pump
5 Steam Separators
6 Steam Dryers
7 Make-up from Condenser
8 Main Steam Flow to Turbine
9 High Pressure Core Spray (1)
10 Low Pressure Core Spray (1)
11 Low Pressure Coolant Injection (3)

Table I: Typical operational parameters for a 1200 MWe PWR and BWR (KWU design) [10],[14]

Parameters		PWR	BWR
Total Plant			
Thermal power rating	MW_{th}	3752	3840
Electrical power output	MWe	1240	1249
Power plant efficiency (netto)	%	33,2	32,5
NSSS			
Reactor coolant flow	t/h	72000	51480
Reactor coolant outlet pressure	bar	155	70,6
Reactor coolant outlet temperature	°C	323	286
Steam flow to turbine	t/h	7150	7000
Steam pressure at turbine inlet	bar	51,7	67
Steam temperature at turbine inlet	°C	266	283
Feedwater final temperature	°C	210	215
Reactor core			
Average heat flux	W/cm^2	61	50,5
Average fuel power	kW/kg U	36,7	26
Average power density	kW/dm^3	92,3	56
Fuel element number		193	784
Fuel rod number per element		236	63
Active fuel rod length	mm	3900	3760
Fuel rod outside diameter	mm	10,75	12,5
Fuel		UO_2	UO_2
Average fuel enrichment	%w.	2,48	2,0
Fuel weight	t	101,7	147
Reactor pressure vessel			
Inner diameter	mm	5000	6620
Cylinder wall thickness	mm	243 + 5	163 + 8
Total height	mm	13250	22676
Total weight	t	530	785
Pumps			
Number/type		4/centrifugal pumps	8/intern. ax. pumps
Pump pressure head	m liqu. column	93	31,1
Pumping capacity	t/h	18000	8731 m^3/h
Pump speed	min^{-1}	1500	1838
Driving power	kW	8550	1030
Containment			
Sphere diameter or cylinder diameter	m	56	29
Design pressure	bar	5,7	4
Steam Power Plant			
Turbine type/number		HP/1 LP/3	high-pressure condensation turbine/1
Turbine speed	min^{-1}	1500	1500
Condenser cooling water flow	m^3/h	210000	158000 t/h
Condenser pressure	bar	0,044	0,0863

Due to the additional equipment required inside the vessel for coolant water circulation, steam water separation and steam drying, a BWR reactor pressure vessel has about three times the volume of a typical PWR pressure vessel.

Typical operational parameters for PWRs and BWRs are listed in Table 1.

2.2 The LWR Emergency Core Cooling System (ECCS)

Each nuclear power plant has many different safety features installed which are provided to compensate for any kind of departure from normal reactor operating conditions, irrespective of the cause. Additional backup safety devices and systems are provided to protect against damage to the plant or harm to operators and to the general public.

The final objective of all these safety features, indeed the main aim in LWR safety, is - in case of an accident - to prevent the release of radioactive material in excess of prescribed limits, by maintaining the integrity of adequate barriers.

The emergency core cooling system (ECCS) is the most important of the safety features installed in a nuclear reactor to achieve this goal. Its main purpose is to maintain the reactor core in a "coolable geometry" in the event of a loss-of-coolant accident (LOCA) caused by a break anywhere in the main cooling system, of any size from a small leak up to a double-ended rupture of a main coolant pipe (PWR) or of a recirculation line (GE-BWR). To cope with this requirement, the ECCS has

- to provide sufficient core cooling by removing both residual stored energy and radioactive decay heat from the core at a rate such as to limit the maximum fuel cladding temperature to a value less than the prescribed limit of 1200°C and thereby to prevent the cladding from losing structural integrity;
- to have sufficient capacity, diversity, reliability, and redundancy to cool the core under all accident conditions presently postulated [3].

In view of this objective, the tasks of the safety systems installed on both LWR types are fundamentally the same, and therefore similar solutions can be used. However, the differences between the main cooling systems are significant, and must be taken into account in the design of the ECCSs. In view of these differences, the systems for the two LWR types are considered separately in the following two chapters.

The PWR emergency core cooling system (ECCS). To cope with the requirements outlined before, the PWR emergency core cooling (ECC) and residual heat removal (RHR) systems, as shown schematically in Fig. 11, consist of several independent subsystems: these include high- and low-pressure, active and passive systems. Each subsystem is characterized by redundancy of equipment and flow path.

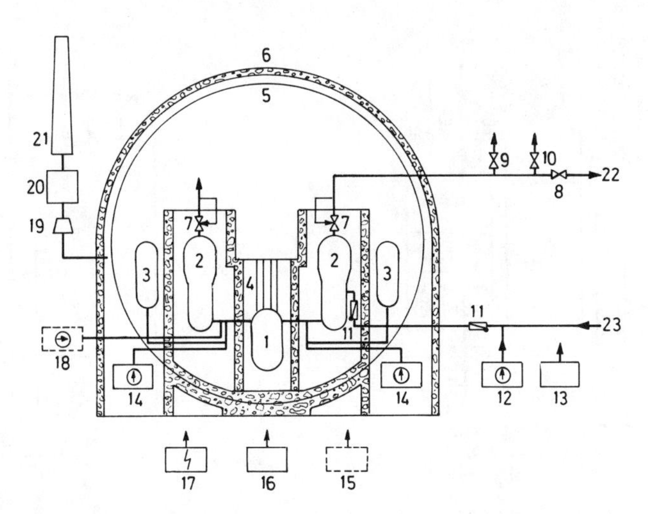

Fig. 11: Schematic Representation of Safety Related Subsystems of a PWR NSSS [2]

1 Reactor Pressure Vessel (RPV)
2 Steam Generator (SG)
3 Accumulator (ECCS)
4 Fuel handling pool (no safety system)
5 Safety Containment
6 Concrete Containment
7 Combined Safety & Shut-off Valve Unit
8 Shut-off Valve Unit
9 Safety Valve
10 Relief Valve
11 Check Valve
12 Auxiliary Feedwater System
13 Emergency System
14 Emergency Coolg. & Resid. Heat Removal (Long Term Cooling) System
15 Control System (no safety system)
16 Reactor Protection System
17 Power and Emergency Supply System
18 Volume Control System (only partially a safety system)
19 Blower
20 Filter
21 Chimney
22 to Turbine
23 from Main Feedwater Pumps. The PCS Safety Valves are located on Pressurizer not represented here.

The flow diagrams of the low-pressure part of the emergency cooling and residual heat removal system for a KWU and Westinghouse PWR, are shown in Fig. 12 and 13. In both cases we distinguish between the accumulator injection system and the low-pressure injection (LPIS) or recirculation system.

The accumulator injection system consists of large tanks, typically one (W) or two (KWU) for each primary loop, containing cold borated water under nitrogen gas at a pressure of about 30 bar (KWU) or 45 bar (W). These tanks are connected through check valves and piping to the main coolant pipes. Westinghouse use cold leg injection (between pump and RPV) only; KWU inject in both cold and hot (between RPV and SG) legs. In case of a large break LOCA, with a rapid decrease of the primary system pressure, these tanks discharge their water automatically into the primary system as soon as the primary system pressure has decreased below the accumulator pressure and opened the check valves; there is no need for activation of pumps, valves or other equipment: the accumulator injection system is thus a completely passive system. The volume of the individual tanks, about 35 m^3, is identical for both accumulator systems; the KWU system thus has both a higher redundancy, and twice the total water volume.

The low-pressure injection system (LPIS) is an active system containing low head, high delivery pumps ((10) in Fig. 12, (2) in Fig. 13) and several

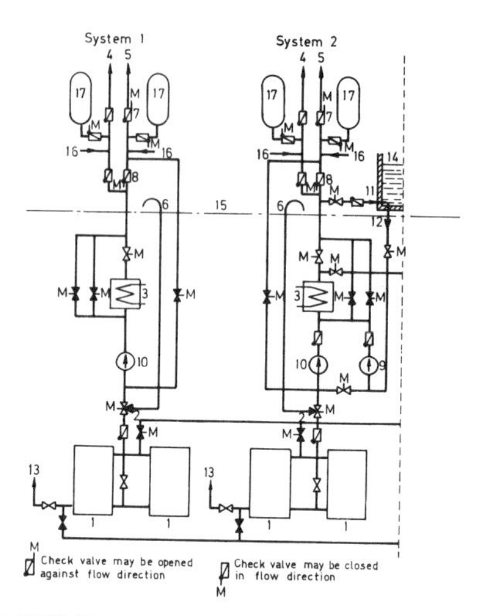

Fig. 12: KWU Emergency Cooling and Residual Heat Removal System (simplified).
One half of two mirror symmetrical systems of two subsystems each [2]

1 Reflooding Water Storage Tank
2 Three-way Valves
3 Heat Exchanger
4 Injection into Cold Leg
5 Injection into Hot Leg
6 Sump Suction Location
7 Check Valve, to open against flow direction
8 Check Valve, to close in flow direction
9 Fuel Storage Pool Water Pump
10 Residual Heat Removal Pump
11 Inlet of Fuel Storage Pool Water
12 Discharge of Fuel Storage Pool Water
13 Suction Line for HP (Safety) Injection Pumps
14 Fuel Storage Pool
15 Containment
16 Injection from HPIS
17 Accumulators

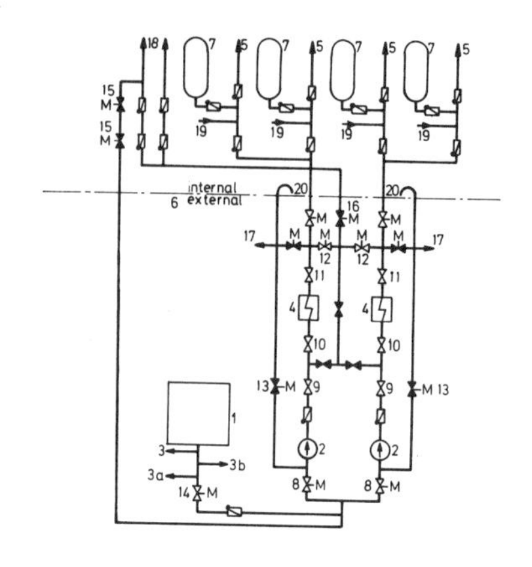

Fig. 13: Westinghouse Emergency Cooling and Residual Heat Removal System, Accumulators with Lower Pressure System Simplified Scheme [2]

1 Reflood Water Storage Tank (only one)
2 (Low Pressure) Residual Heat Removal Pump
3-3.b Discharge for Make-up Pumps, HP Injection Pumps, Safety Containment Spray Pumps
4 Heat Exchanger
5 To the 4 Cold Legs
6 Safety Containment
7 Accumulator
8-14 Valves
15 Valves for Suction from Hot Leg Nr. 4
16 Valve for Injection into Hot Legs Nr. 2 & 4
17 Suction Pipes for HP System
18 Injection into Hot Legs Nr. 2 & 4
19 Injection from HP System
20 Containment Sump Suction Locations

valves which have to be activated. There are three different operation phases of this system:

- the injection phase: in case of a loss-of-coolant accident due to a large or intermediate size break within the primary system, the LPIS starts operation when the primary system pressure has decreased to about 10 bar; water from the reflood water storage tank (1), (one in a W plant, one double tank for each subsystem in a KWU plant) is injected into the primary system to refill the reactor pressure vessel after the accumulators have emptied.
- the recirculation phase: when the storage tank(s) have emptied, the LPIS pump suction side is switched to the containment sump ((6) in Fig. 12, (20) in Fig. 13) where water from the break has accumulated; the LPIS continues to operate in this mode for as long as heat removal is required.

- the normal residual heat removal phase: in case of a normal reactor run-down, primary coolant water is withdrawn from the hot legs (one only in W plants ((18) in Fig. 13), four in KWU plants ((5) in Fig. 12)), cooled down and returned to the cold legs.

The flow diagrams of the high-pressure part of the ECC systems for KWU and Westinghouse PWRs are shown in Fig. 14 and 15. This is normally used to compensate for primary coolant losses in the case of small leaks within the primary system; it is again an active system and starts operating when the primary system pressure falls below about 110 bar.

Both HPISs consist essentially of high head, low delivery pumps ((2) in Fig. 12, (1) in Fig. 15), sometimes called safety injection pumps, which take water from the reflood water storage tank(s) ((1) in Fig. 12 and 13) and feed it into the injection lines. On Westinghouse plants cold leg injection is used; on KWU plants hot leg injection is normally used but in case of a hot leg break, injection will be performed into the cold leg of the broken loop. The Westing-

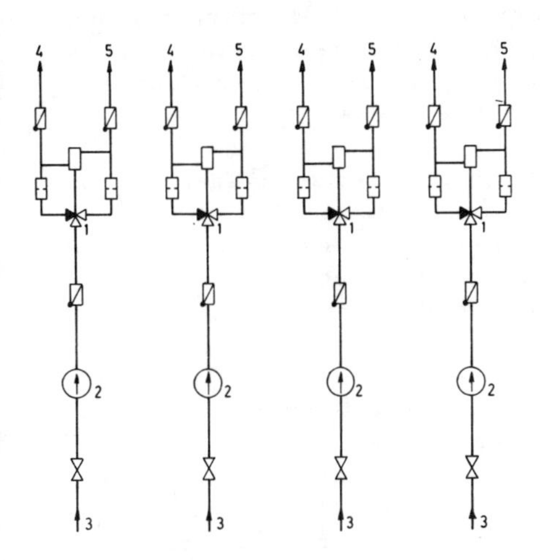

Fig. 14: KWU HP (Safety) Injection System [2]

1 Three-way Valve, controlled by pressure difference in both injection lines. Injection normally into hot legs except one hot leg is broken
2 HP (Safety) Injection Pumps
3 Suction Lines from Reflooding Water Storage Tanks (1 in Fig. 12)
4 Injection into Cold Leg (4 in Fig. 12)
5 Injection into Hot Legs (5 in Fig. 12)

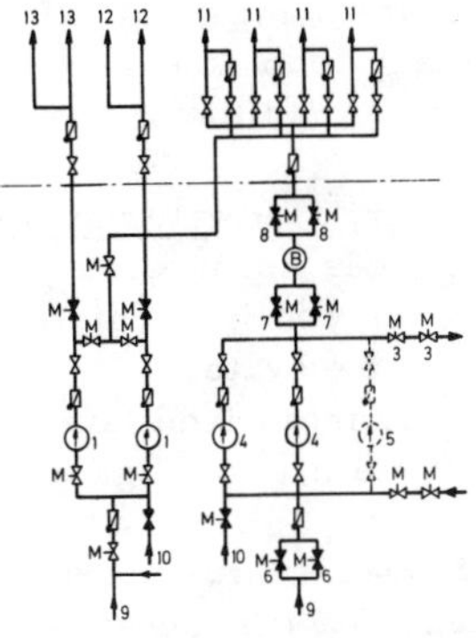

Fig. 15: Westinghouse HP Safety Injection System (During the injection phase the HPIS injects only into cold legs) [2]

1 HP Injection Pumps
2 Valves in Suction Line of Volume Control System (closed in emergency cooling case)
3 Valves in Pressure Line of Volume Control System (closed in emergency cooling case)
4 Make-up Pumps of Volume Control System (Centrifugal pumps)
5 Make-up Pumps of Volume Control System (Displacement pump)
6 Valves in Suction Line (open in emergency case)
7-8 Valves in Pressure Line to Borating System (open in emergency case)
9 Suction Lines from Reflooding Water Storage Tank (1 in Fig. 13)
10 Suction Line from LP Pumps (17 in Fig. 13)
11 Injection into Cold Legs (19 in Fig. 13)
12 Injection into Hot Legs Nr. 2 & 4 (18 in Fig. 13)
13 Injection into Hot Legs Nr. 1 & 3
B = Boric acid tank

house HPIS is also used in the recirculation phase. The make-up or volume control system in W plants can be used in addition to the HPIS after operating the appropriate valves. However, the make-up pumps' head corresponds to the normal coolant operation pressure so that there is the possibility of over-feeding in case of a small leak LOCA.

Besides the differences already mentioned between the KWU and the Westinghouse ECCS design features, there is a fundamental difference which should be stressed. In the Westinghouse plants individual components (pumps, valves, relays, etc.) are duplicated or triplicated, but remain interconnected with each other within the whole system: e.g. Westinghouse plants use two interconnected HPIS together with the make-up or volume control system. In contrast, in KWU plants there are several identical and self-sufficient subsystems which are as a whole redundant to each other: for example the high-pressure injection system (HPIS) consists of four completely independent subsystems or branches.

Both active ECC systems need power from normal or emergency supply sources to operate pumps and valves. These systems are energized automatically by primary loop coolant pressure sensing and level sensing switches which connect power to the appropriate driving systems.

The auxiliary feedwater system (AFS) is a further safety feature of PWRs which, although it is not normally considered under the heading of ECCS, is also an emergency cooling system. The AFS - in both Westinghouse and KWU plants - supplies feedwater to the steam generators (secondary side) in the event of failure of main feedwater pumps or other essential parts of the main heat removal system, see Fig. 8.

The AFS operation is started in all accident events leading to a reactor scram and, hence, a turbine trip and a shut-off of the main feedwater pumps. It is only really needed in the case of small breaks, where continued heat removal from the primary cooling system via the steam generators is indispensible and is achieved by means of secondary loop steam release through the vapour relief and safety valves, (2) and (3) in Fig. 8.

Figs. 16 and 17 reveal again the fundamental difference between the interconnected redundancy in Westinghouse plants and the completely independent redundancy of KWU plants.

The BWR emergency core cooling system (ECCS). BWRs, like PWRs, have an emergency core cooling system consisting of several independent subsystems, designed to provide sufficient core cooling in case of an unplanned depressurization or a loss of coolant from the reactor. Figs. 18 and 19 show a schematic representation of a KWU and a GE BWR emergency cooling and residual heat removal system. In both cases we have essentially two subsystems, a high-pressure and a low-pressure one; all systems are activated upon receiving a signal of either high drywell pressure or low reactor vessel water level.

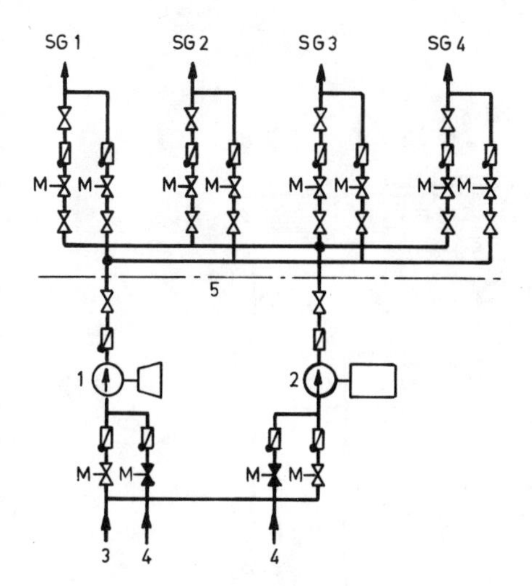

Fig. 16: Westinghouse Auxiliary Feedwater System [2]

1 Auxiliary Feedwater Pump: Turbine driven
2 Auxiliary Feedwater Pump: Diesel driven
3 from Condensate Tank
4 from Service Water System
5 Safety Containment

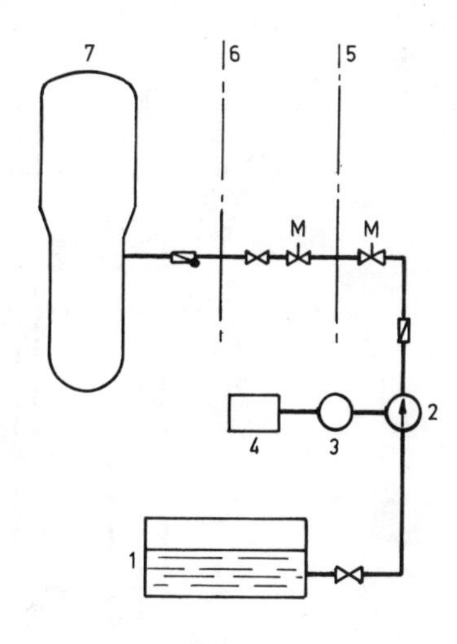

Fig. 17: One of four KWU Auxiliary Feedwater Systems [2]

1 Deionate Tank
2 Auxiliary Feedwater Pump
3 Generator
4 Diesel Engine
5 Concrete Containment
6 Safety Containment
7 Steam Generator

The high-pressure core spray system (HPCS) - one in GE plants, three in KWU plants (in more recent KWU plants it is referred to as residual heat removal system) - is aimed at providing adequate core cooling in the event of small leaks or failure of feedwater pumps or other minor water supply systems. Coolant water is taken from the pressure suppression pool (GE and KWU plants) and/or the condensate storage tank (only GE plants), and is fed to a spray nozzle array installed above the reactor core.

The automatic depressurization system (ADS) is actuated - on reactor water level signal "low" and/or on containment pressure signal "high" - independently if the feedwater pumps and the HPCS system are not capable of maintaining the prescribed reactor water level; by discharging steam through pressure relief valves into the pressure suppression pool, the reactor pressure is reduced in order to allow the low-pressure emergency core cooling systems to start operation and provide sufficient cooling.

In case of high drywell pressure caused by a rupture in the steam line or in the feedwater line, steam or steam-water mixture is passed through vent pipes to the pressure suppression pool for condensation. In addition, KWU plants use a drywell containment spray system supplied with water from one low-pressure ECC system pump, to reduce the drywell pressure.

The low-pressure core spray system (LPCS, in GE plants only), and

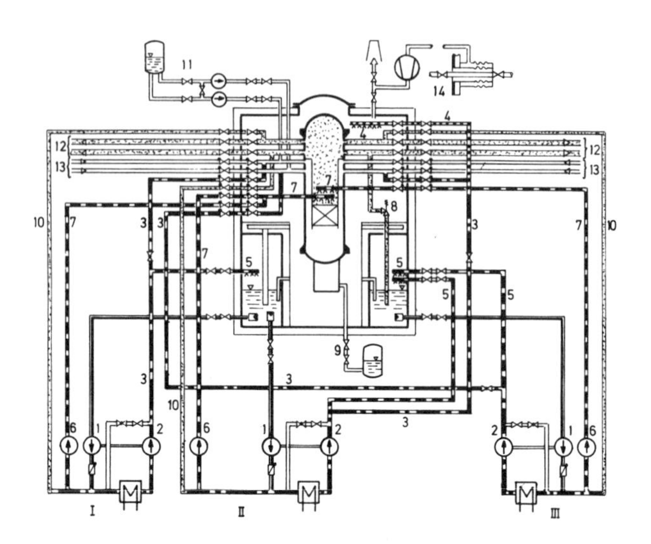

Fig. 18: Emergency Cooling &Residual Heat Removal System of a KWU BWR [14]

1 Initial Stage } LP Pump
2 LP Stage
3 LP Coolant Injection Lines
4 Containment Spray System
5 Condensation Pool Spray System
6 HP Pump
7 HP Core Spray System
8 Automatic Depressurization System Relief Valves
9 Reactor Scram System
10 Suction Line for Normal Residual Heat Removal System
11 Poisoning System
12 Steam Lines
13 Feedwater Lines
14 Penetration Suction System
10 + 2 +3 Residual Heat Removal Circuit

Fig. 19: Emergency cooling and Residual Heat Removal System of a GE BWR [2]

1 HP Pump
2 Condensate Tank
3 Suction Line for HP System
4 HP Spray Distributor
5 HP Shut-off Armature
6 HP Test Line
7 Suction Line for LP Spray System
8 LP Pump
9 LP Spray Distributor
10 Heat Exchanger
11 Suction Location for LP Systems
12 Circulation Loop

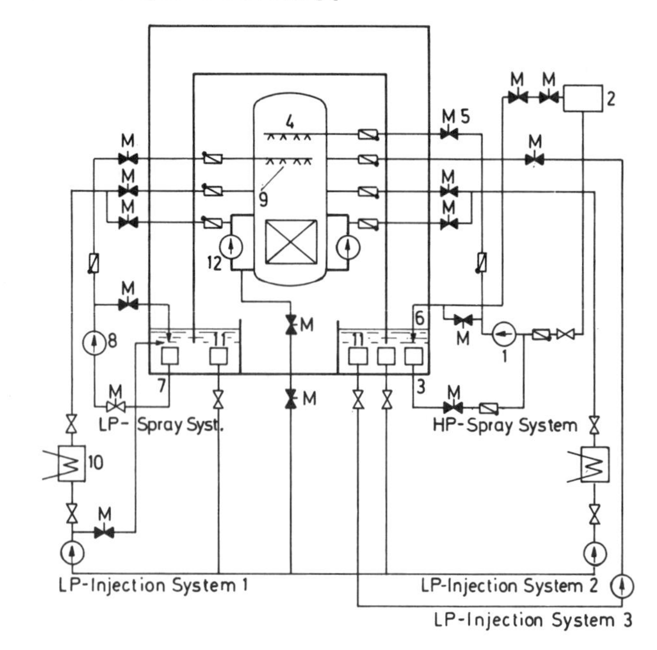

the three low-pressure coolant injection (LPCI) systems (GE and KWU plants) are activated on low reactor water level or high drywell pressure and start operation when the reactor pressure is low enough.

The LPCS (GE, one system) system pump provides spray cooling to the fuel assemblies by taking water from the pressure suppression pool and feeding a separate, low pressure spray header installed in the reactor vessel above the core.

For emergency cooling, the three LPCI systems (KWU and GE plants) pump water from the pressure suppression pool to the reactor vessel (GE plants) or into three of four feedwater lines (KWU). The same systems are also used for residual heat removal in case of a normal reactor run-down. For this purpose the pumps (all 3 in KWU plants, 2 only in GE plants) take water from three steam lines (KWU plants) or from one recirculation line (GE), pass it through heat exchangers, and return it to three feedwater lines (KWU) or to two recirculation lines (GE).

Finally, the pressure suppression pool water may be cooled down by an appropriate spray system supplied with water by the LPCI system pumps.

The LPCI system is capable of protecting the core following even a large break, by reflooding the core and maintaining a sufficient water level.

In the case of a steam line break outside the containment, major steam losses from the reactor are prevented by closing isolation valves placed in each steam line on both sides of the containment penetration. In addition, flow restrictors installed in the reactor vessel outlet nozzles limit the steam losses during the period before these valves are fully closed, or if they should fail.

3. THE LOSS-OF-COOLANT ACCIDENT

3.1 Initiating Events

There are several events which may cause a departure from normal operating conditions in a LWR. Such events may be distinguished from each other according to three basic criteria:

- their nature or origin:
  - events from outside: chemical explosions, earthquakes, air crash, sabotage
  - events from inside: operational transient, failures within the main cooling system, operator errors
- their frequency, ranging:
  - from frequent occurrences yielding little or no fuel integrity problem
  - to extreme situations with a low likelihood of occurrence, having the greatest risk of moderate fuel damage

- their consequences:

  . events which allow return to normal operating conditions and the continuation of reactor operation, after compensation by the normal control systems: operational transients

  . events requiring the intervention of installed safety systems to bring them under control, which, however, cause a temporary shut-down of the reactor, which allows time for checks and repairs if necessary: upset conditions

  . accidents, whose consequences may not be held within the admissible limits by the installed safety systems: emergency conditions

  . limiting fault conditions.

Two distinct types of event can cause an imbalance between heat produced and heat removed leading to a potential risk of inadmissible high fuel temperatures:

- transients, where either the heat production is above, or the heat reval is below the nominal value
- a loss of coolant, where core cooling fluid is escaping from a leak or break within the primary cooling system.

This type of transient is usually dealt with under the heading of "anticipated transients without SCRAM" (ATWS), or in the less important (with regard to safety) category, "anticipated transients with SCRAM". Included are events such as loss-of-coolant flow, partial loss of heat sink, loss-of-coolant pressure etc., which are normally expected to leave the primary cooling system undamaged and may therefore be considered of less importance in the context of safety analysis.

A loss of coolant caused by a major rupture within the primary cooling system is one of several "limiting fault condition occurrences". These are faults which are not expected to occur but are postulated because their consequences could include the risk of a major release of radioactive material. The loss-of-coolant accident (LOCA)caused by the double-ended break of a main coolant inlet pipe (cold leg between pump and RPV) of the primary cooling system of a PWR, or of the recirculation line in a GE BWR, is the most drastic such fault against which the safety systems have to be designed; it is therefore referred to as the design basis accident (DBA) determining the specifications for the limiting, safety related design of the reactor plant, particularly of the emergency core cooling system (ECCS).

3.2 Classification of LOCAs

The severity of a LWR LOCA caused by a rupture within the primary cooling system and, hence, its possible consequences, is essentially determined by the rupture characteristics in terms of break location and break size.

Within the primary cooling system we have to distinguish between fracture of the reactor pressure vessel and breaks within the piping system.

The most serious accident by far, the catastrophic rupture of the reactor pressure vessel below the core level, would leave no defense against core melt-down and the consequent release of a substantial amount of radioactive material, because cooling water could not be contained near the core. The ECCSs are not designed to cope with such an accident, because such a failure of the reactor pressure vessel is considered a virtually inconceivable event. In fact, thorough evaluations have yielded a probability of occurrence of leaks or breaks in the reactor pressure vessel which is several orders of magnitude smaller than that of leaks or breaks in the pipework [2].

In fact, piping breaks within the primary cooling system have a considerably higher probability of occurrence due to the relatively large number of pipes and the wide range of diameters, extending between about 20 and 800 mm.

Fig. 20 gives a simplified representation of the various possible break sizes and locations within the primary cooling system of a 1300 MWe KWU PWR.

As shown, possible break sizes range from the largest one, a double-ended or guillotine rupture of a main coolant pipe, which is referred to as a 2 x 100% [1)] break, down to an offset shear break of a single steam generator

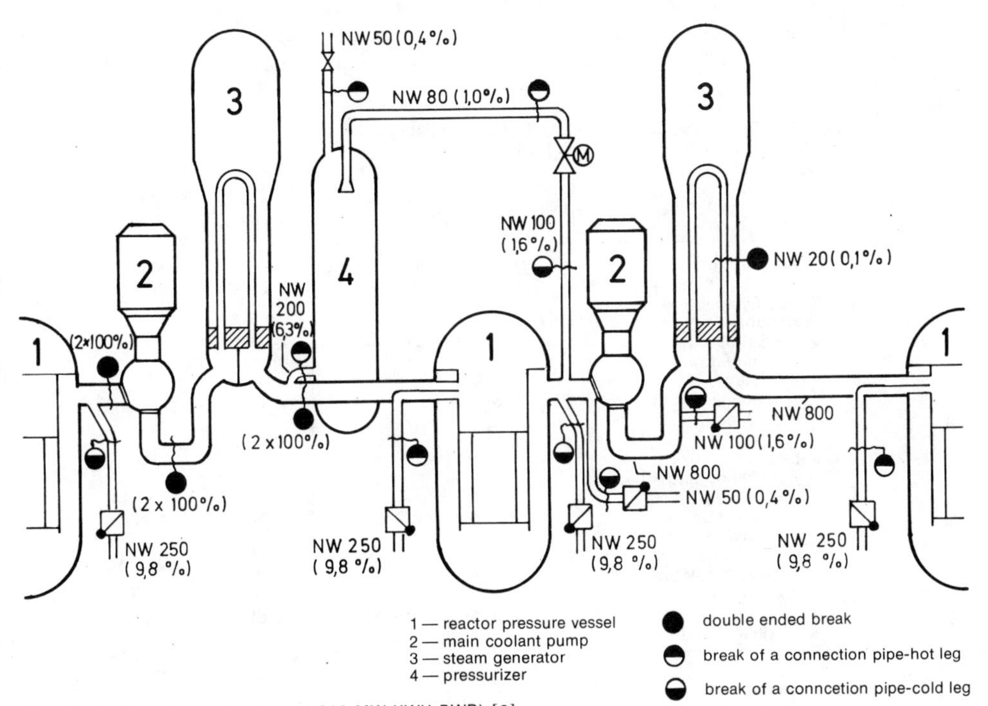

Fig. 20: Break sizes and locations (1300 MW KWU PWR) [8]

1) percentages refer to the main coolant pipe flow cross section.

tube of about 0.1%. Connection line breaks are usually considered to be single-ended ruptures with sizes ranging up to about 10%. Instrumentation lines - not shown in Fig. 20 - with diameters of about 20 mm would represent the smallest pipe break size of about 0.05%.

In addition, for the case of a rupture of the main coolant piping, distinction is made in Fig. 20, between cold leg ruptures, between pump and RPV or between SG and pump, and hot leg ruptures, between RPV and SG. A connection line rupture may also be considered as a cold or hot leg rupture, depending on where the line joins the main coolant piping.

Within this wide rupture size spectrum, a distinction can be made between small, intermediate and large LOCAs according to break size. Such a distinction was introduced for the first time in the Reactor Safety Study of the USAEC [7]; the resulting classification of LOCAs is based on the number and type of safety systems called upon to prevent fuel damage and provide sufficient long-term cooling. Table 2 gives a classification of LOCAs [2] together with their estimated average probability of occurrence [9], and, for the case of a cold leg break LOCA in a KWU PWR with 4 accumulators and combined cold and hot leg injection, the minimum number of functional safety systems required for effective emergency core cooling.

The quantitative values for the break size area and percentage should not be considered fixed; they certainly vary between plants from different manu-

Table 2: LOCA Classification, Safety System Functioning [2] [1)]
Probability of Occurrence [9]

	Large Break		Intermediate Break	Small Break
- Break Size (cm^2)	> 1000	400 — 1000	80 — 400	< 80
%	> 25	10 — 25	2 — 10	< 2
- Safety System Functions				
• HPIS (hot leg)	—	—	2 of 4	2 of 4
• Accumulators				
- hot leg injection	3 of 4	3 of 4	2 of 4	—
- cold leg injection	2 of 4	2 of 4	2 of 4	—
• LPIS				
- for reflooding				
• hot leg injection	2 of 4	2 of 4	2 of 4	2 of 4
• cold leg injection	1 of 4	1 of 4	1 of 4	1 of 4
- for recirculation				
• hot leg injection	2 of 4	2 of 4	2 of 4	2 of 4
• Auxiliary Feed-water Injection	—	—	—	1 of 4 2 of 4
- Average Probability of Occurrence /9/	10^{-4}/a	$3 \cdot 10^{-4}$/a	$8 \cdot 10^{-4}$/a	$3 \cdot 10^{-3}$/a

1) *for a 1300 MWe KWU PWR*

facturer and may do so even between various plants from the same manufacturer.

Another, more generally valid classification of LOCAs can be made using the occurrence of physically significant phenomena as criteria for the distinction between small, intermediate and large break LOCAs. Such phenomena are phase separation, additional energy rejection, and mixture level behaviour within the primary circuit. Such a classification is given in Table 3.

Table 3: Phenomenological LOCA Classification

Phenomenon	Break Size		
	Large	Intermediate	Small
- Phase Separation before ECC Water Injection	no	yes	yes
- Additional Heat Rejection in SG Required for Depressurization below LPIS Initiation Point	no	no	yes

4. THE COURSE OF LOCAs

Since a LOCA causes a system depressurization, both the course of a LOCA and the depressurization or transient time are strongly affected by the break size, and to some extent also by the break location. The reactor system response to a depressurization accident, and the ECCS performance required is therefore different for small and large break LOCAs, and to some extent different also for PWRs and BWRs.

In view of these differences, in the following sections large and small break LOCAs, and PWR and BWR LOCAs, are considered separately.

For both LWR types, the most severe demands on ECCS performance are imposed by the hypothetical, sudden severance of a main coolant pipe which is referred to as the design basis accident. In this context, most attention has been given to the double-ended (2 x 100%) or guillotine break of a large diameter cold leg or inlet pipe for PWRs, or recirculation pump inlet (suction) pipe for BWRs.

In order to evaluate ECCS performance during a LOCA, conservative initial conditions are assumed, namely reactor operation at 102% of full power when the postulated break occurs. Fuel cladding temperature in a PWR at this time will be close to the coolant temperature, about 320°C; the average temperature of the hottest fuel pellet will be well above 1100°C, and the peak temperature on the pellet axis greater than 2200°C. The excess heat content of the fuel at this average temperature is called the "stored heat"; it is to some extent determined by the thermal resistance of the gap between the fuel and the cladding. Both stored heat and gap thermal resistance strongly affect the temperature history of the cladding during the early period of a LOCA.

A further conservative assumption is a heat generation rate from radioactive fission product decay which is 20% higher than the proposed ANS[1] standard.

The course of a large break LOCA with an effective ECCS will first be described for a PWR, then for a BWR plant. LOCAs without an effective ECCS will be dealt with briefly thereafter.

The relationship between significant LOCA phenomena and ECCS acceptance criteria will be briefly outlined in the subsequent section.

4.1 The Large Break LOCA in PWRs

The sequence of events occurring in a large break LOCA for the case of an effective ECCS, can be divided into four sequential periods: blowdown, refill, reflood and long-term cooling. This sequence of events together with the relevant thermohydraulic phenomena involved will be considered with the aid of Figs. 21, 23 and 24. The curves presented are generalized results from several calculations for two different pipe break locations ("inlet" or "cold leg" break, and "outlet" or "hot leg" break) and characterize the behaviour of the most significant quantities during a LOCA [1].

Figs. 23 and 24 give a snapshot-type representation of the expected coolant flow behaviour within the primary cooling system of a PWR at several sequential instants during the LOCA transient for both cold leg (Figs. 23 a-f) and hot leg (Figs. 24 a-f) break location.

The total duration of such a LOCA, from pipe break to the start of long-term cooling, may extend to about 2 to 5 minutes. The blowdown period, characterized by a rapid depressurization and expulsion of coolant from the primary system, occupies the first 20 to 30 seconds (30-60 seconds in a BWR) only. The peak cladding temperatures during reflooding normally occur at about 40-120 seconds.

The blowdown period.

(a) Subcooled depressurization, Figs. 23a and 24a: Immediately following the hypothetical severance of a large coolant pipe, primary system water is expelled through the break into the containment, causing the system pressure to drop in a few tens of milliseconds down to the highest local fluid saturation pressure, due to subcooled depressurization (1)*). This violent press-

---

[1]) American Nuclear Society (ANS)

*) Numbers in brackets refer to the corresponding numbers in Fig. 21.

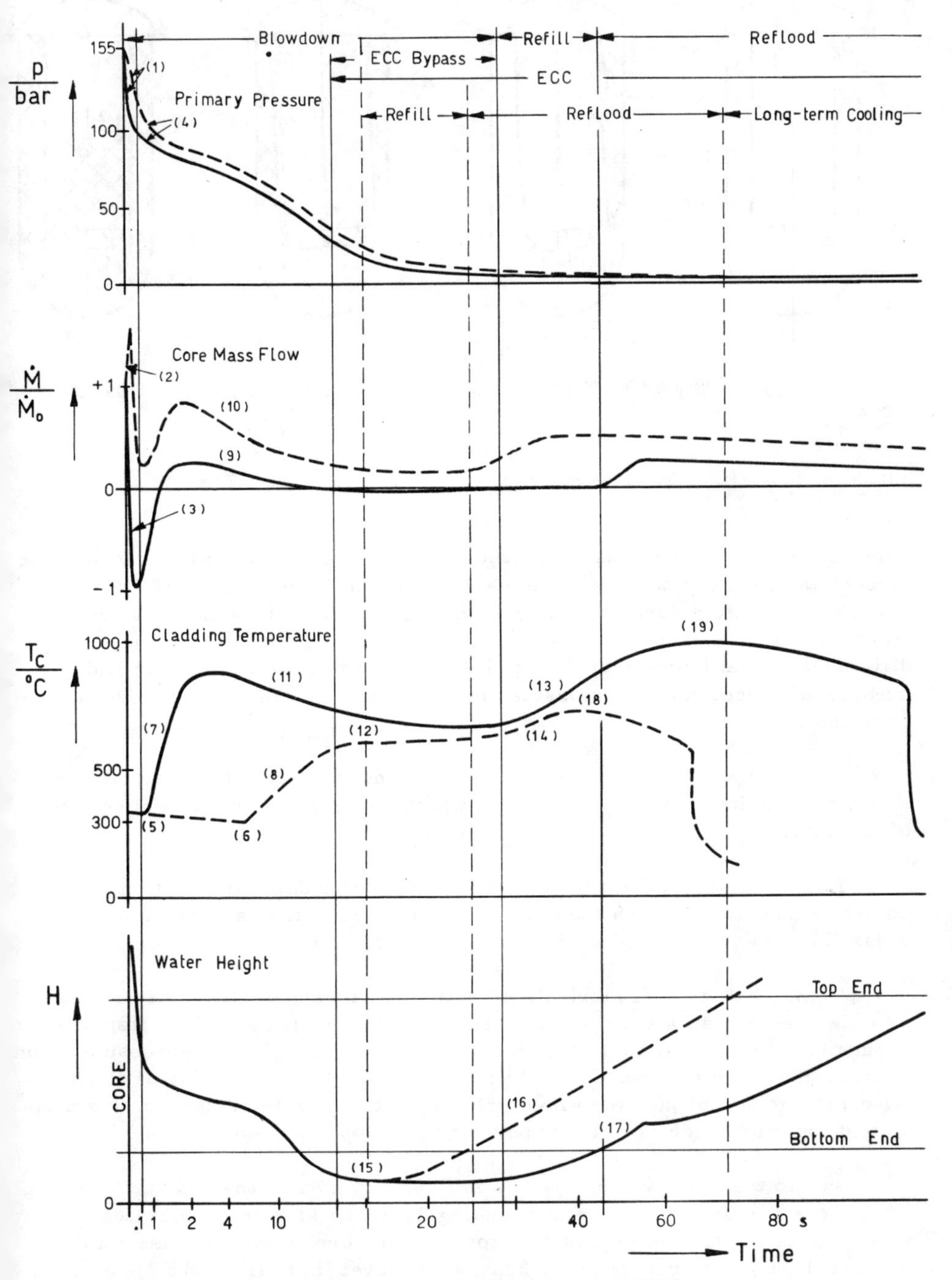

Fig. 21: Generalized Loss - of - Coolant Behaviour for Large Pipe Breaks in a PWR [1]

——— Inlet Break
– – – – Outlet Break

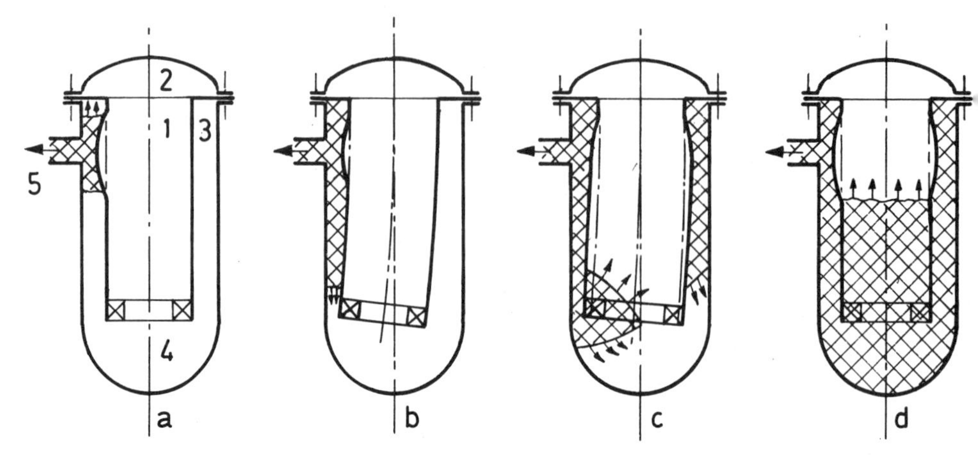

Fig. 22: Depressurization Wave and Resulting Deformation of the Core Barrel after a Large Cold Leg Break [2]

1 Core Barrel
2 RPV Head
3 Downcomer
4 Lower Plenum
5 Break Location close to Inlet Nozzle ; shadowed areas: zones of reduced pressure

release - blowdown - is characterized by a depressurization wave propagating through the primary cooling system and the reactor pressure vessel where it causes a dynamic deformation of the core barrel. This is impressively illustrated in Fig. 22, taken from ref. [2]. The highest possible local pressure difference is that between system pressure and saturation pressure; it decreases with decreasing break size and increasing distance of break location from the RPV.

At the rupture position, the establishment of critical flow velocity determines the maximum break mass flow, which governs the subsequent course of the LOCA.

During this very first blowdown, or subcooled depressurization phase, coolant water flow through the core is accelerated in the case of an outlet pipe break (2), and decelerated in the case of an inlet pipe break (3).

(b) Saturated depressurization: After the coolant pressure has reached or fallen below the local saturation pressure, boiling (flashing) starts; this occurs less than 100 ms into the transient and as a result the depressurization process continues at a considerably reduced rate (4). The boiling or flashing front propagates through the whole primary cooling system, starting from the hottest location which is in the upper core and upper plenum region.

Due to the negative reactivity coefficient in LWRs, the resulting voiding in the core region, and the corresponding decrease in water moderator density, causes the fission process to stop and the core power generation to fall down to the fission product decay heat power level (initially about 7% of the

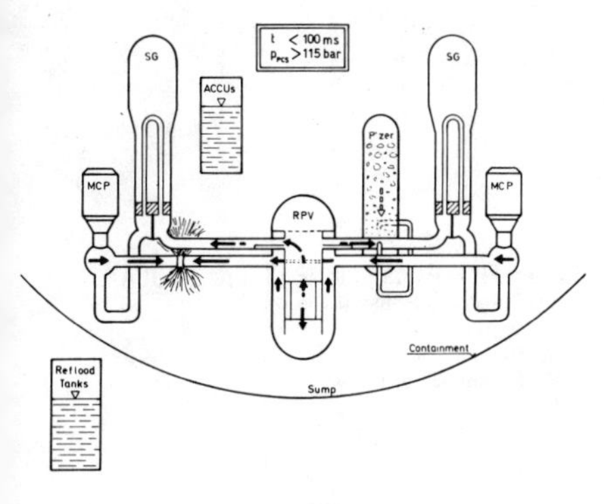

Fig. 23: Coolant Flow Paths in a DE-CL Break LOCA in PWRs
a) Subcooled Blowdown Phase
Water Flow
Water-Steam (low quality) Flow

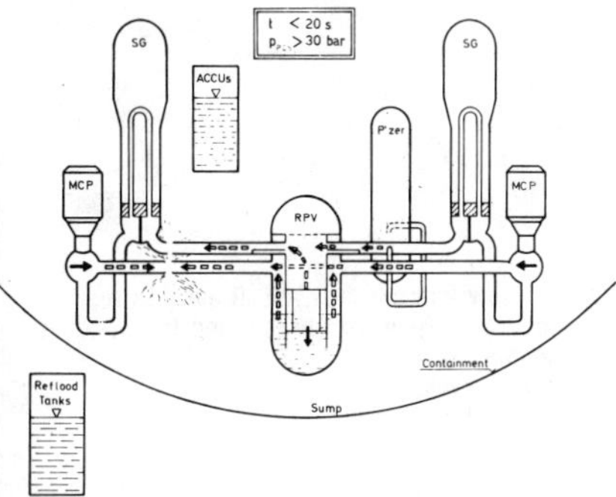

Fig. 23: Coolant Flow Paths in a DE-CL Break LOCA
b) Saturated Blowdown Phase
Steam-Water Flow

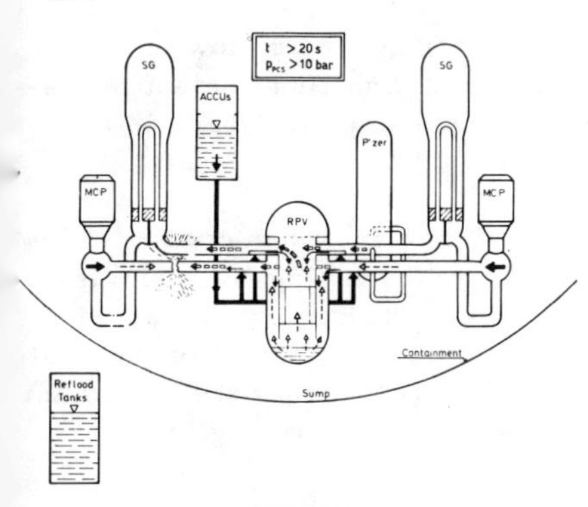

Fig. 23: Coolant Flow Paths in a DE-CL Break LOCA
c) ECC Bypass Phase (Accu's only)
Water Flow
Steam-Water Flow
Steam Flow

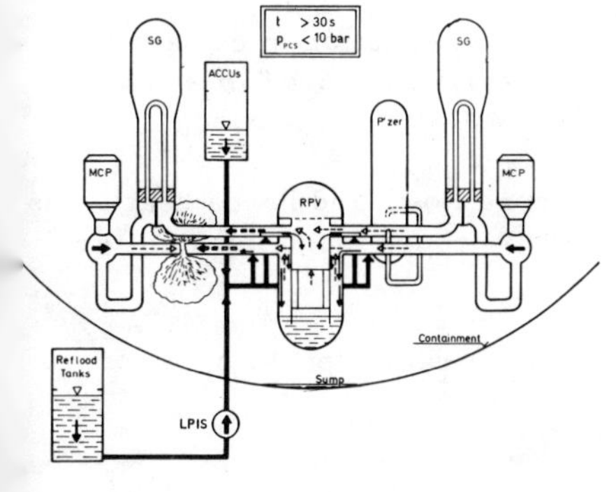

Fig. 23: Coolant Flow Paths in a DE-CL Break LOCA
d) ECC Refill Phase (Accu's + LPIS)
Water Flow
Steam Flow
Steam-Water Flow

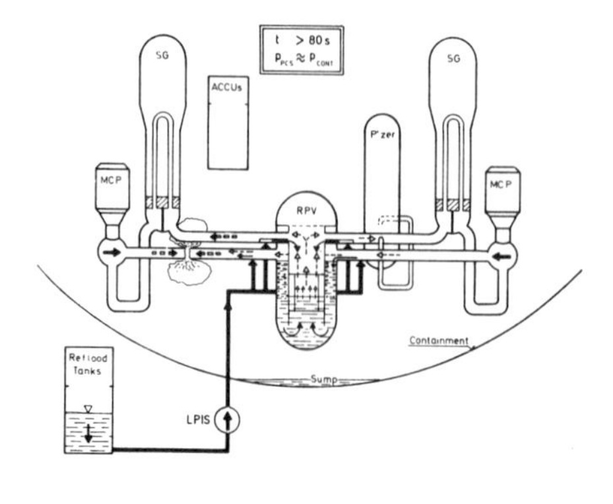

Fig. 23: Coolant Flow Paths in a DE-CL Break LOCA
e) ECC Reflood Phase (LPIS only)
Water Flow
Steam Flow
Steam-Water Flow

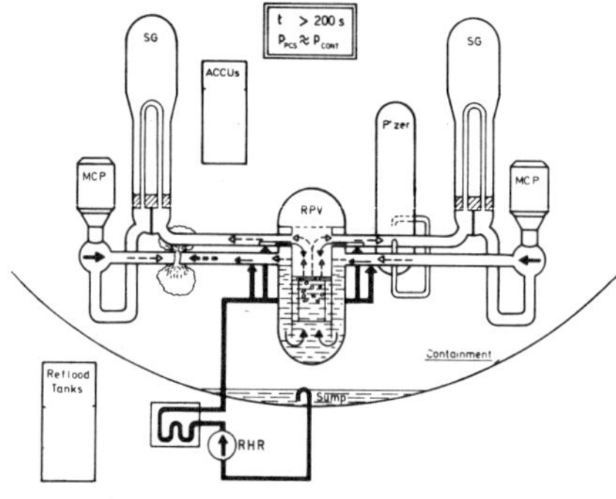

Fig. 23: Coolant Flow Paths in a DE-CL Break LOCA
f) Residual Heat Removal Phase (Long-term Cooling)
Water Flow
Steam Flow
Steam-Water Flow

rated power). The reactor scram, which is usually triggered by low pressurizer pressure ($\lesssim$ 145 bar) and water level signal, and high containment pressure signal, occurs later, at about 0.4 s into the LOCA transient, and is in principle not required in case of large LOCAs in LWRs.

(c) Boiling transition, Figs. 23b and 24b: At the time when coolant voiding starts within the core, coolant flow conditions are changing from single-phase to two-phase flow. This, together with reduced coolant pressure and reduced coolant flow through the core causes a drastic degradation of the cooling conditions for the fuel rods: the critical heat flux falls below the maximum heat flux, boiling transition (DNB) occurs, and the fuel rod cooling breaks down.

In "inlet break" conditions, due to substantial reduction in coolant flow, stagnation and flow reversal, DNB occurs very early, at about 0.5 - 0.8 s into the LOCA transient (5).

In "outlet break" conditions, substantial core flow, and hence heat trans fer, is preserved for an extended period, so DNB occurs much later, after several seconds (6).

(d) Stored heat redistribution, first clad temperature peak: As a conse-

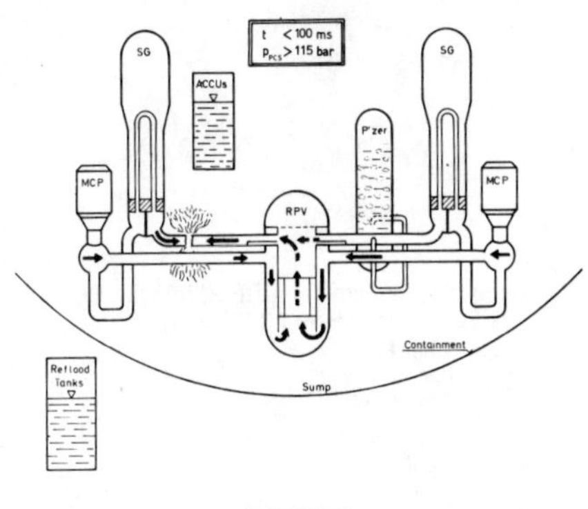

Fig. 24: Coolant Flow Paths in a DE-HL Break LOCA in PWRs

a) Subcooled Blowdown Phase

Water Flow

Water-Steam (low quality) Flow

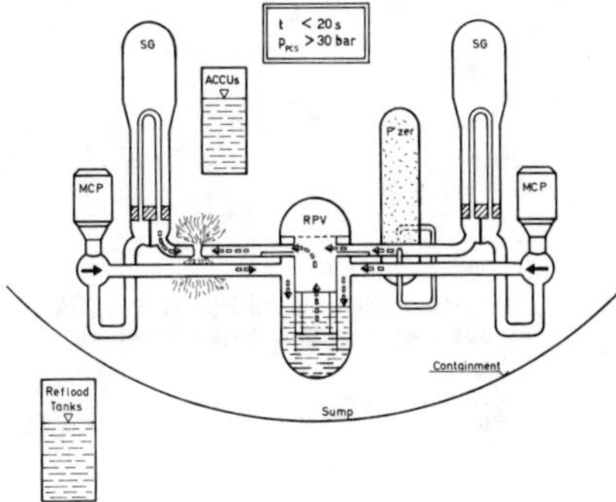

Fig. 24: Coolant Flow Paths in a DE-HL Break LOCA

b) Saturated Blowdown Phase

Steam-Water Flow

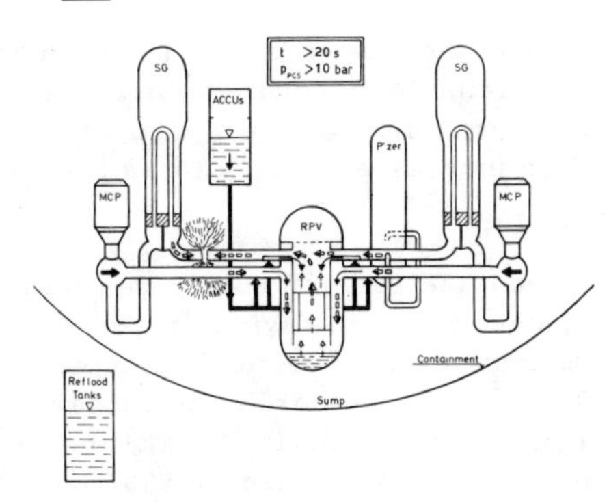

Fig. 24: Coolant Flow Paths in a DE-HL Break LOCA

c) ECC Refill Phase (Accu's only)

Water Flow

Steam-Water Flow

Steam Flow

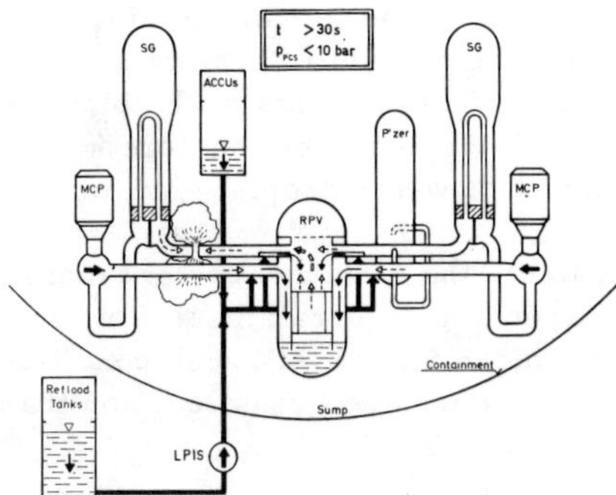

Fig. 24: Coolant Flow Paths in a DE-HL Break LOCA

d) ECC Refill Phase (Accu's + LPIS)

Water Flow

Steam-Water Flow

Steam Flow

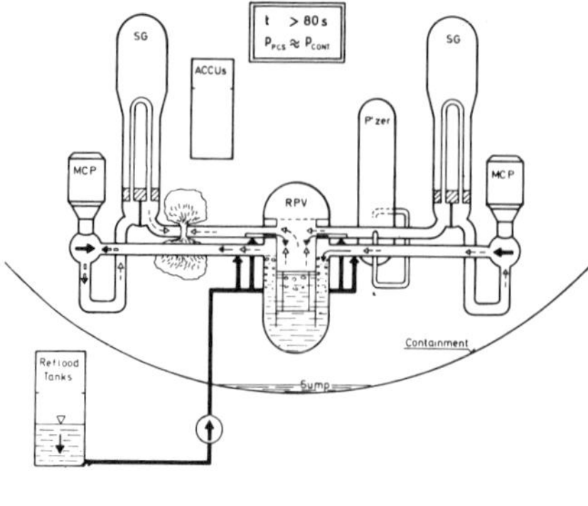

Fig. 24: Coolant Flow Paths in a DE-HL Break LOCA
e) ECC Reflood Phase (LPIS only)
← Water Flow
←-- Steam Flow
←▫▫ Steam-Water Flow

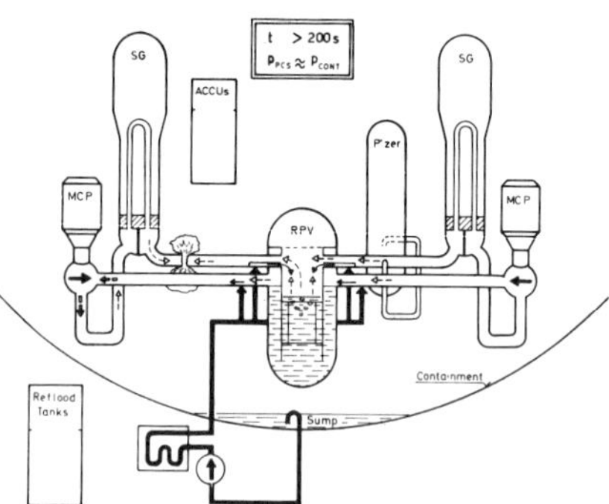

Fig. 24: Coolant Flow Paths in a DE-HL Break LOCA
f) Residual Heat Removal Phase (Long-term Cooling)
← Water Flow
←-- Steam Flow
←▫▫ Steam-Water Flow

quence of the sudden loss of heat removal from the fuel rods, the large amount of stored heat within the fuel redistributes leading to an equalization of the internal temperature distribution (the temperature on the axis drops and the outer surface temperature of the fuel pellets increases). This causes the cladding temperature to start rising sharply.

If there were no heat removal at all from the fuel rod during this initial blowdown period, and neglecting the decay heat generation within the fuel, the cladding temperature would increase to its maximum theoretical value which is the average fuel temperature, about 1100-1200°C. According to conservative calculations for the worst case of a double-ended cold leg break, the real maximum cladding temperature to be expected will barely exceed 900°C. This value is limited predominantly by three effects: (a) the thermal resistance of the gap between fuel pellets and cladding, (b) the amount of heat removed by water rushing through the core during the very first instants of the blowdown period, and (c) the amount of heat subsequently removed by the core steam flow, which is affected strongly by the amount of water droplet entrainment.

Several seconds into the LOCA transient, the effective coolant mass flow through the core, which is strongly depending on break mass flow and loop components behaviour, is significantly higher in outlet break conditions due to the much lower flow resistance between core and break location than under inlet break conditions (9, 10).

The difference in the total amount of heat removed during this initial LOCA period for the two break conditions, resulting from the difference in core mass flow, is reflected clearly in the difference in cladding temperature behaviour both with respect to the slope of the temperature increase (7, 8) and the maximum temperature reached.

(e) Persisting heat sources and cooling deterioration: In addition to the stored heat, heat from two further sources has to be removed during this initial LOCA transient: these are the fission product decay heat, and, possibly, for cladding temperatures of 980°C and higher, the heat generated by the chemical reaction between zircaloy and steam to form hydrogen and zirconium dioxide.

The fission product decay heat produced within the first minute of a large LOCA (7% of rated power at reactor shut-down, 5% at 10 s, 2% after 10 minutes) may amount to roughly the same order of magnitude as the stored heat released during the same period.

At temperatures of about 1100°C or higher, the heat released from the metal-water reaction during a period of one minute, may again amount to the same order of magnitude as the decay heat.

Therefore, following fuel rod temperature equalization due to stored heat redistribution, the further behaviour of the cladding temperature is essentially governed by the imbalance between decay heat produced and heat transferred to the coolant, leading to a termination of the cladding temperature rise (12) under outlet break conditions or even to a slight temperature decrease (11) under inlet break conditions. However, since the cooling conditions continue to deteriorate, the cladding temperature finally rises due to adiabatic fission product decay heating (13, 14).

For temperatures above about 650°C, the cladding is expected to start ballooning as a result of the combination of decreased mechanical strength (due to temperature increase) and the increasing differential pressure between internal gas pressure (increasing with temperature) and decreasing external system pressure. Above about 750°C this swelling phenomenon may become significant, with the risk of a cladding rupture.

The continuing expulsion of coolant from the primary system through the break into the containment causes a continuing depressurization of the primary system and an accompanying decrease of water inventory; finally the water level within the reactor vessel will fall below the lower end of the core (15).

(f) ECC phase: Accumulators, Figs. 23c and 24c: When the primary system pressure has dropped below the nitrogen pressure within the accumulators of the emergency core cooling system (see Fig. 11-13), cold water as auxiliary or emergency coolant is discharged from the accumulators through automatically opening check valves and appropriate injection lines into the primary system. Thereby, the emergency core cooling phase is started with the aim of replacing the coolant lost through the break. This occurs after about 10-15 s into the LOCA transient depending on both the system depressurization rate and the accumulator pressure which is about 45 bar in Westinghouse plants, and about 27 bar in KWU plants.

In Westinghouse plants, the auxiliary coolant injection is performed into all cold legs or inlet pipes close to the RPV and, in more recent plants only, also into the upper plenum through the so-called upper head injection (UHI) device.

In all KWU plants, accumulators inject into all cold legs or inlet pipes, and into all hot legs or outlet pipes, both injection locations being placed close to the RPV.

(g) Bypass phase, Fig. 23c: Since at this time in the LOCA transient the system pressure is still high with respect to the containment pressure, there is still a substantial break mass flow.

In hot leg or outlet break conditions with continuing upward flow through the core, the auxiliary coolant injected into the cold legs will penetrate the downcomer annulus without being impeded, will reach and refill the lower plenum, and finally cause the water level to enter into the core region for the subsequent reflooding of the core (16).

In cold leg or inlet break conditions, lower plenum refilling is considerably delayed, essentially by two effects:

- steam water counter-current flow in the downcomer annulus: steam escaping from the core during reversed core flow, together with steam from continued lower plenum water boiloff flows upwards through the downcomer and impedes emergency coolant water from the cold leg injection from penetrating the downcomer; this effect is further increased by emergency coolant flashing due to stored heat release from the hot RPV walls.
- accumulator ECC bypass: the major part of the emergency coolant injected into the cold legs is entrained by the steam flow from the intact loop cold legs around the upper downcomer annulus to the break, and so carried directly out of the break without penetrating the downcomer.

Hence, in cold leg break conditions, it has to be assumed that all auxiliary coolant injected into the cold legs during this initial ECCS operation period bypasses the lower plenum ("Bypass phase") and leaves the primary system directly through the breaks, thereby considerably delaying the lower plenum refill (17).

(h) End of blowdown (end of bypass): The blowdown phase is usually considered to have terminated when pressure equalization between primary system and containment, at about 2-4 bar, has been achieved and break mass flow has become negligible, which is the case in either break location at about 30-40 s into the LOCA transient.

Under inlet break conditions, it is only after steam flow from the intact loop cold legs to the break has become negligible, that gravity forces begin to overcome the entrainment forces and emergency water starts to penetrate the downcomer and to refill the pressure vessel.

(i) LPIS activation, Figs. 23d and 24d: After about 30 s for the case where emergency power is required, or when the system pressure has dropped below about 10 bars (KWU plants), the low pressure injection system (LPIS) starts operating. For a short period auxiliary coolant water is supplied from both the accumulators and the LPIS, until the accumulators have emptied. The LPIS continues to inject water for as long as is required, taking water from the reflood tanks and, later on, from the containment sump.

The high-pressure injection system (HPIS) is not needed in large break LOCAs; firstly, because the pressure decreases so rapidly that both accumulators and LPIS are very quickly activated, secondly because it would not make a significant contribution due to the low pump delivery. Besides that, for the case where emergency power is required, the delay for both HPIS and LPIS is determined by the start-up time of the emergency power system.

The refill period. The refill period starts when auxiliary coolant water reaches the pressure vessel lower plenum for the first time and causes the water level to start rising again; refill is terminated when the water level has reached the bottom end of the core.

Adiabatic core heat-up, Figs. 23d and 24d: During the whole period from before accumulator injection initiation, up to the end of refill, the reactor core is essentially uncovered and the fuel rods are uncooled except by thermal radiation and by the small natural convection current in the steam-filled core. Due to the decay heat release, the core temperatures rise adiabatically during this period (13, 14) at a rate of about 8-12°C/s for PWRs and of about 5-7°C/s for BWRs. If they started rising from about 800°C, they would reach values above 1100°C after about 30-50 s, when the zircaloy-steam reaction becomes a significant additional energy source. Therefore, the lower plenum water level at the end of blowdown, and termination of lower plenum refill are the two critical parameters determining the maximum fuel cladding temperatures which can possibly be reached during this refill period, which is the period with the poorest core cooling of the whole LOCA.

The reflood period, Figs. 23e and 24e. The reflood period begins when the water level in the reactor pressure vessel has reached the bottom end of, and starts to rise up the core.

(a) Second peak clad temperature: As the emergency coolant water enters the core it is heated up and starts boiling. At about 0.5 m above the bottom end of the core, due to the very hot cladding surface, the boiling process becomes very vigorous and causes a rapid flow of steam upwards through the core. This flow entrains a substantial amount of water droplets which provide an initial cooling of the hotter core regions. This cooling effect becomes increasingly effective as the water level rises, progressively reducing the rate of rise of cladding temperature and finally reversing the rise in temperature of the hot spot at about 60-80 s into the LOCA transient.

(b) Quenching: When the cladding temperature has again sufficiently decreased (to about 350-550°C) the auxiliary coolant water finally rewets the cladding surface and causes a sharp temperature drop due to substantially higher cooling rates (quenching). This quench front rises into the core from the bottom end (in case of cold leg injection only) or propagates from both ends into the core (in case of cold and hot leg or upper head injection). When the whole core has quenched and the water level has finally risen above the top end of the core, the reflooding period is considered to be terminated; this occurs at 1-2 minutes into the LOCA transient.

At the beginning of the LOCA, the reactor's power-producing fission process is shut-down by loss of the coolant-moderator and/or by SCRAM. During the reflooding period measures are required to ensure that reinitiation of the fission process cannot occur. Such measures are provided by the SCRAM system and by injection of neutron absorbing poisons (boron) along with the auxiliary coolant water.

(c) Steam binding: The course of core reflooding as described above, may however, in certain cases be adversely affected. The velocity with which the water level can rise in the lower plenum (during refill) and the core (during reflood) is determined, and hence limited, by the balance between its driving force, and the flow resistance which the steam encounters between core and break location. Since the driving force, represented by the water level difference between downcomer and core, is limited, the steam flow resistance becomes important and gives rise to the so-called "steam binding" problem.

In outlet break conditions the steam flow resistance is relatively low and steam can easily escape from the reactor core.

In case of inlet breaks, however, the steam has to overcome the resistance of the hot leg pipework, steam generators and pumps before reaching the break.

This flow resistance is further increased by additional energy transfer from the secondary system fluid to the steam flow as it passes through the steam generators; this causes a considerable increase in volume, by evaporation of entrained water and superheating of steam. An additional steam flow resistance may result from water which has collected within the U-tube between steam generator and pump.

On the other hand, the limited driving force, having a maximum value of about 0.5 bar in steady-state flooding conditions, may be further reduced by two effects, (a) a decrease in the downcomer fluid density caused by evaporation of water due to stored heat release from hot RPV walls and (b) a decrease in the effective downcomer head if typical manometer tube oscillations between downcomer and core occur.

This steam binding effect is even more pronounced in the case of a pipe break between steam generator and pump [2]. Thus, steam binding reduces the reflooding rate and thereby reduces heat transfer between fuel rods and coolant fluid causing an extension of the reflooding period which is accompanied by an extended adiabatic cladding temperature increase, a higher second peak temperature, and a delayed temperature turnaround.

(d) Steam generator U-tube rupture: The steam binding problem may be further exacerbated if rupture of a few U-tubes (one to ten) of a steam generator occurs during a large break LOCA. The dumping of secondary steam into the depressurized primary system acts as a further volume and energy source, increases thereby the reflood backpressure and reduces finally the reflood rate. (According to an estimate reported in [1], the rupture of one (!) U-tube (equivalent of about 6 cm^2 break size) could result in a reduction of the reflood rate of about 17%).

(e) Gap resistance: The relative heights of the first and second cladding temperature peaks depend on the thermal resistance of the gap between fuel pellets and cladding: low gap resistance causes the stored heat release to the cladding to occur during the initial LOCA period leading to a higher first peak temperature, and vice-versa for a high gap resistance.

(f) Metal-water reaction: A higher second peak cladding temperature increases the potential for significant zircaloy cladding oxidation due to the metal-water reaction which becomes increasingly important above about 1000°C. Significant cladding oxidation results in cladding embrittlement: this may be sufficient to cause cladding failure during the subsequent quenching process (by ECC water) which induces very large stresses. The possible release of fuel pellets and cladding fragments may impair the coolability of the core within the hotter core regions during the subsequent residual heat removal period.

(g) Design measures against steam binding: From the foregoing, it is clear that measures to reduce the potential for steam binding may be essential in order to maintain the core in a coolable geometry. Such measures are:

- vent valves in Babcock & Wilcox plants, installed in the upper core barrel wall for allowing vapour discharge directly from the upper plenum into the upper downcomer annulus and through the cold leg to the break, thus by-passing the whole of the loop;
- hot leg injection (KWU plants), or upper head injection (Westinghouse plants), provides water to the upper plenum to condensate vapour and thus reduce the reflood backpressure.

(h) Experimental reflood results: Recent experimental reflood investigations with combined cold and hot leg injection performed with the PKL test facility, have shown a significant reduction in cladding temperature turnaround time, see Fig. 25.

The effect of the reflooding rate on cladding temperature rise, turnaround time and quench time is shown in Fig. 26 which illustrates experimen-

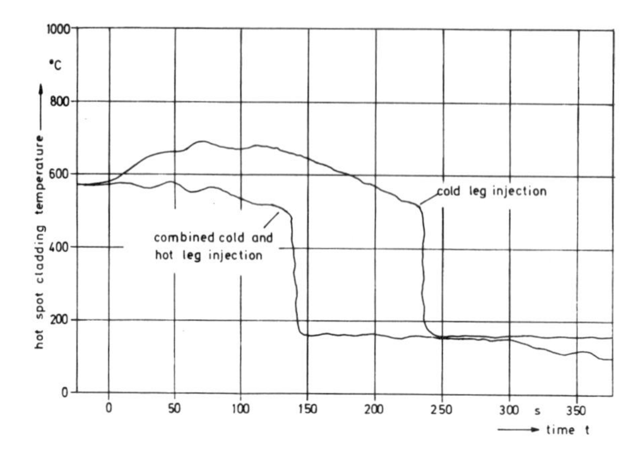

Fig. 25: Comparison of Cold Leg and Combined Cold and Hot Leg Injection [15]

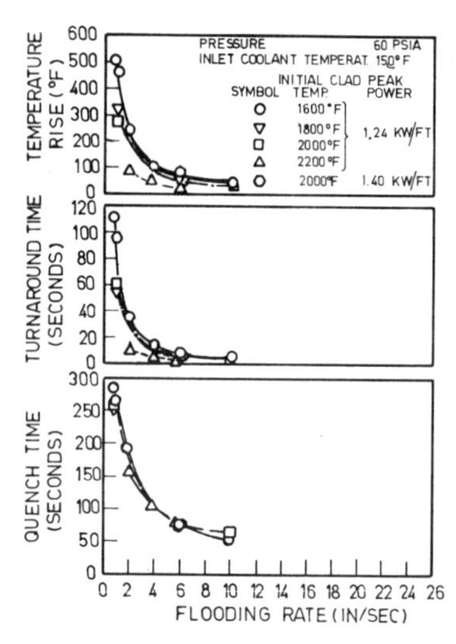

Fig. 26: Summary of PWR-FLECHT results as a function of flooding rate: effect on temperature rise, turnaround time, and quench time [1]

tal results from PWR-FLECHT*)-tests.

(i) Second peak clad temperature (PCT) vs. break size: Fig. 27 shows the calculated effect of break size and location on the maximum cladding temperature (second peak). There are several design basis accident (DBA) conditions which influence the cooling system behaviour during a PWR-LOCA and thus determine the final requirements for the ECC system design. The strong influence of the large inlet break is evident.

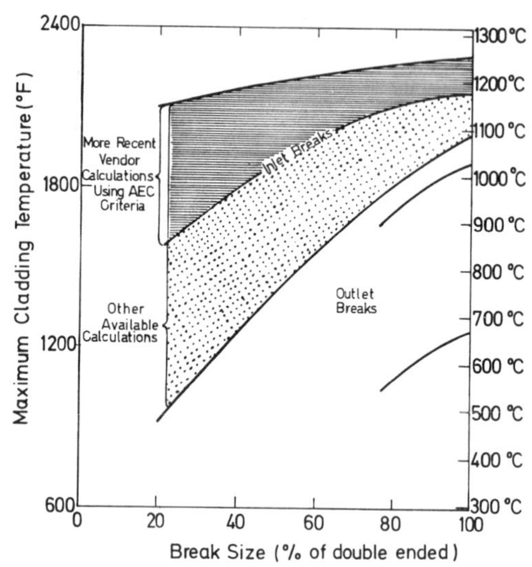

Fig. 27: Generalized comparison of maximum cladding temperature for various primary system pipe break conditions in a PWR. [1]

*) FLECHT = Full Length Emergency Cooling Heat Transfer. Experiments performed at Westinghouse USA for PWR and BWR geometries and conditions.

Long-term cooling period, Figs. 23f and 24f. After termination of the reflooding period, the low-pressure injection system continues to operate. When the reflood water storage tanks have emptied, the LPIS pump suction is switched to the containment sump, where most of the auxiliary coolant water so far supplied to the reactor has finally collected, after escaping from the primary system as steam and being condensed in the containment. The cooling maintained during this period is to ensure the long-term dissipation of the decay heat which, 30 days after reactor shut-down, still amounts to about 5 MW for a 3800 MWth PWR.

4.2 The Large Break LOCA in BWRs

The sequence of events occurring in large break BWR LOCAs is essentially similar to that in PWR LOCAs. It will be considered, together with the relevant thermohydraulic phenomena involved, by the aid of Fig. 28 which is taken from ref. [1]. The curves presented there are generalized results holding for either a postulated liquid circulation line (GE BWRs only) break or for a postulated steam line break (GE and KWU BWRs).

During normal reactor operation a significant amount of the cooling system fluid is at saturation conditions, and only a small part of it is at slightly subcooled conditions. Therefore, in either break condition, there is only a limited subcooled depressurization following an instantaneous severance of a large coolant pipe.

Liquid line break.

(a) Liquid break flow period: An offset shear break in the pump suction line of one of the two recirculation loops (GE) represents a double-ended rupture; it is the worst pipe break in a GE BWR and is therefore the design basis accident (DBA) for the ECCS. In case of such an accident, the core mass flow drops to about one-half of the nominal value (1)*); pump coast down in the intact recirculation loop allows the four associated jet pumps to continue to supply coolant water to the core (2) and are assisted by a strong natural circulation flow.

During the initial LOCA period the depressurization process (3) is rather slow because (a) the critical break discharge flow is water flow and hence limited, (b) a large part (about 40%) of the total reactor vessel volume is occupied by steam, and (c) the steam line shut-off valves close within a few seconds isolating the RPV from the main heat sink, so that the system coolant can escape from the break only.

For either break location, both GE and KWU BWR plants experience a similar core flow behaviour to that of PWRs in the case of an outlet pipe break: the depressurizing coolant maintains an upward flow through the core.

---

*) numbers in brackets refer to the corresponding numbers in Fig. 28.

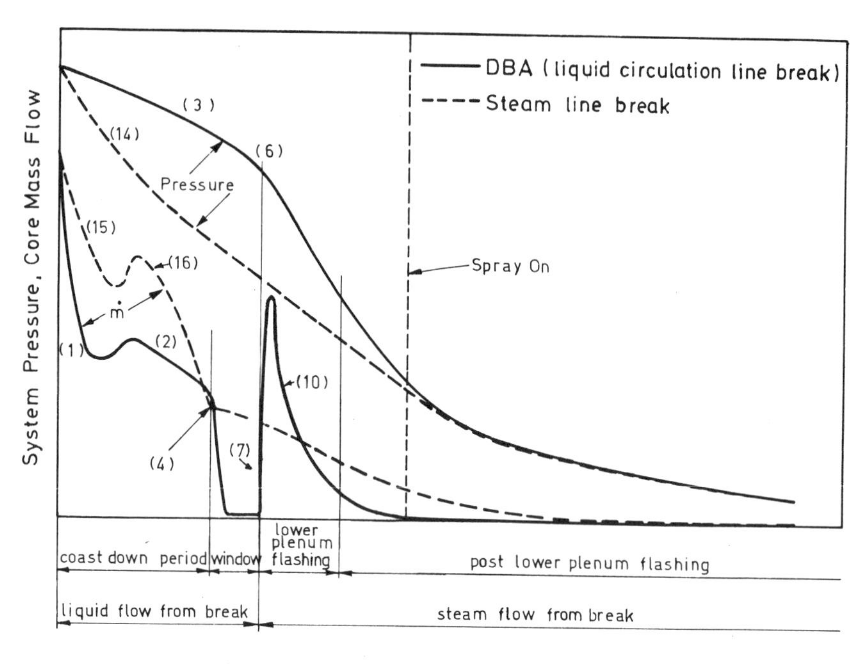

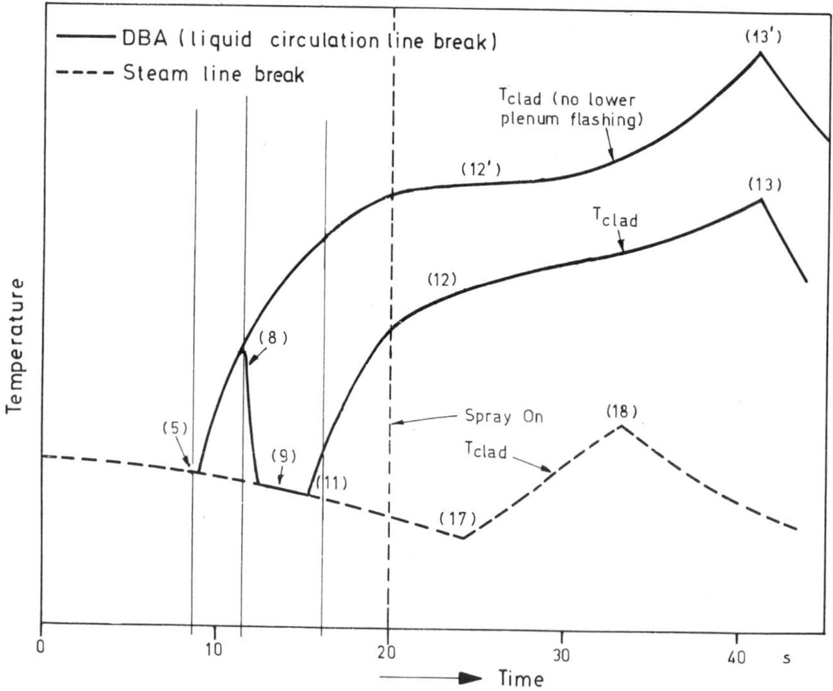

Fig. 28: Generalized loss-of-coolant behaviour for large pipe breaks in a BWR [1]

(b) "Window": After the liquid level in the reactor vessel has dropped below the jet pump inlet, the intact recirculation loop pump cavitates, causing the core mass flow to drop rapidly (4) to nearly zero and the critical heat flux (CHF) to fall below the maximum heat flux. At the same time (about 5-10 s into the LOCA transient) DNB occurs (5) and the cladding temperature starts rising instantaneously due to stored energy redistribution within the fuel rods (c.f. sectn. 4.1 (d)).

The time period between the start of recirculation pump cavitation (water level below jet pump inlet) and the start of lower plenum flashing (water level below recirculation line outlet) is referred to as the "window" period during which the core mass flow and hence the core cooling is nearly zero.

(c) Steam break flow period: When the falling liquid level in the downcomer has uncovered the recirculation line outlet, the break discharge flow changes to steam and thereby causes the depressurization rate to increase substantially (6). The system pressure reaches the saturation pressure of the residual water within the lower plenum, which starts flashing violently ("lower plenum flashing", end of "window" period) and forces a two-phase mixture flow through the core (7). The resulting increased core mass flow leads to sufficiently improved cooling conditions to cause the cladding temperature rise to terminate (8) and to return to about fluid saturation temperature (9).

The flashing process, however, rapidly decreases with decreasing water inventory in the lower plenum; therefore the core mass flow is again reduced (10) and, because depressurization has continued, once again critical heat flux is exceeded causing a second DNB and cladding temperature rise (11).

After the stored energy redistribution within the fuel has finally been completed, and the temperature within the fuel rod has equalized, further cladding temperature rise is determined by the decay heat release.

(d) ECCS period: About 20 s after break has occurred, the high pressure core spray (HPCS) system becomes effective and top spray flow is developed at a system pressure below about 20 bars. Heat removal from the fuel rods is then increased by both a radiation sink due to spray water wetting the fuel canister walls, and a steam convective flow resulting from spray water evaporation. The resulting improved cooling conditions reduce the cladding temperature rise (12) until the reactor vessel lower plenum is filled with water accumulated from the spray system and from the low pressure coolant injection (LPCI) system which meanwhile has also started to operate. Core flooding is then initiated and develops in a way similar to that in PWRs, finally causing the cladding temperature to turn around (13).

(e) CCFL: The steam binding effect inhibiting the reflood rate does not occur for either break location in a BWR, because of relatively low frictional pressure drop in the steam flow path to the break.

However, a delay in the core reflooding process, similar to that in PWRs due to the steam binding effect, may also occur in BWRs due to the countercurrent steam and water flow. The counter-current flow limited (CCFL) flooding situation may be caused by the steam flow velocity upwards through the upper tie plate being so high that spray water is prevented from running down into the fuel bundles. It thus accumulates in the upper plenum until a breakthrough occurs into some of the lower powered bundles which finally allow the spray water to reach the lower plenum with however a corresponding delay for refilling and core reflooding.

(f) Without lower plenum flashing: If lower plenum flashing, and the effective core cooling (8) thereby induced does not occur, the cladding temperature rise from the early DNB (5) continues (12') and reaches considerably higher values (13'). As a consequence of the higher temperature level additional energy is produced by the induced metal-water reaction causing a significantly increased rate of temperature rise prior to the event of flooding (13').

Steam line break. In case of a steam line break the depressurization rate is considerably higher (14)[1]) because break flow is only steam, from a higher location in the vessel. Therefore the liquid coolant fraction in the reactor remains high, the recirculation loop pumps are in coast-down operation and the jet pumps maintain a significant core mass flow (15).

Continuing depressurization will finally cause the recirculation pump to cavitate and, therefore, the core mass flow to decrease (16). However, both the flashing process and natural circulation occurring will be strong enough to provide a reasonable core flow and give rise to sufficient steam cooling to remove the major part of the stored heat and maintain the cladding temperature near the fluid saturation temperature.

Cooling conditions, nevertheless, continue to deteriorate and boiling transition (DNB) occurs (17), causing the cladding temperature to rise at a rate commensurate with the remaining stored heat and the decay heat release. Because the fluid loss up to now has been mainly steam and steam-water mixture, the residual system coolant inventory is still relatively high (the lower plenum remains filled), and considerably less emergency coolant water is necessary to reflood the core than in case of recirculation line break; hence, cladding temperature turnaround occurs at a considerably earlier time and lower temperature (18).

Metal-water reaction and cladding embrittlement are much the same as those of a PWR, see sectn. 4.1. The general fuel cladding behaviour, however, is different according to the smaller fuel rod internal pressure.

Fig. 29 shows the calculated effect of break size and location on the maximum cladding temperature in a BWR.

---

1) in GE-BWRs; in KWU-BWRs the steam line break size is smaller than a liquid line break size.

4.3 The Small Break LOCA

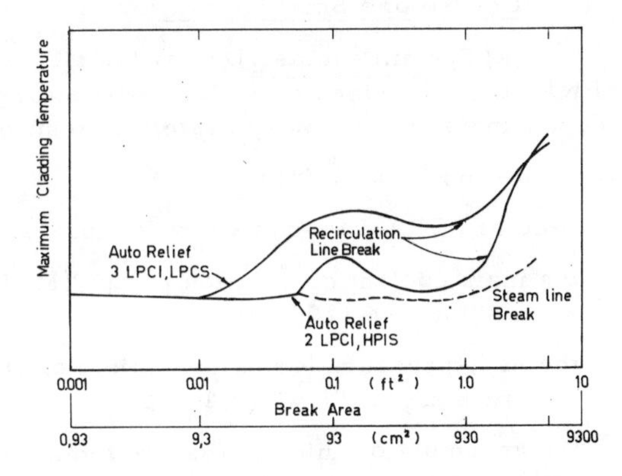

Fig. 29: Generalized comparison of maximum cladding temperature for various pipe break conditions in a BWR (from General Electric Co.) [1]

<u>General characterization.</u> As shown in Table 2, break sizes of 2% (of a main coolant pipe flow cross section) and less are usually considered as small breaks.

Although there are no firmly established boundaries as yet between small and intermediate or large breaks, a further criterion other than absolute or relative break size, and number and type of safety systems called upon (see Table 2), which might be used for the determination of the upper limit of what is "small", is the occurrence of physically significant phenomena or events which govern the course of the LOCA transient.

One such event would certainly be the need for the main heat removal system (or secondary cooling system) of a PWR to reject additional heat from the primary cooling system in order to depressurize it to the LPIS activation pressure.

Physically significant phenomena would be e.g.: (a) the mixture level behaviour in the reactor core, (b) the time dependence of the heat transport direction within the steam generators, (c) flow regime conditions, to mention only the most important ones, see Table 3.

The overall risk contribution is determined by these physically significant phenomena and by the average probability of occurrence of a certain break size, see Table 2. The last depends on both the pipe diameter spectrum of a reactor main cooling system (an example is shown in Fig. 30), and on empirical results for pipe rupture probability as function of pipe diameter; this pipe diameter spectrum may vary even for plants of the same manufacturer.

The course of a small break LOCA might be significantly affected also by the excitation criteria for the preparation and initiation of emergency cooling systems.

The following considerations are of a still more general character; they will essentially deal with the small break LOCA course in PWRs. The Three Mile Island reactor 2 accident of March 28, 1979, near Harrisburg, Pennsylvania, USA, is not described separately in this context. However, reference will be made to some extent and where appropriate.

Course of a Small Break LOCA.

(a) System events: During the LOCA transient, in the case of a KWU - PWR, the following actions of the reactor protection system are called upon depending on the primary system pressure [8]:

on the primary side:

- reactor SCRAM is excited by low pressurizer signal ($\leq$ 145 bar)
- the main coolant circulation pumps (MCP) are tripped on system pressure signal "low" ($\leq$ 110 bar)
- the high-pressure (or safety) injection (HPI) pumps are started on system pressure signal "low" ($\leq$ 110 bar)
- the accumulator injection is started automatically when the system pressure has fallen below 27 bar
- the low pressure injection (LPI) or residual heat removal pumps start to deliver at a system pressure below 10 bar

on the secondary side:

- turbine trip is caused on low pressurizer pressure signal ($\leq$ 145 bar)
- in the emergency power case, the turbine bypass ((8) in Fig. 8) remains closed thereby excluding the main steam bypass to the condensers

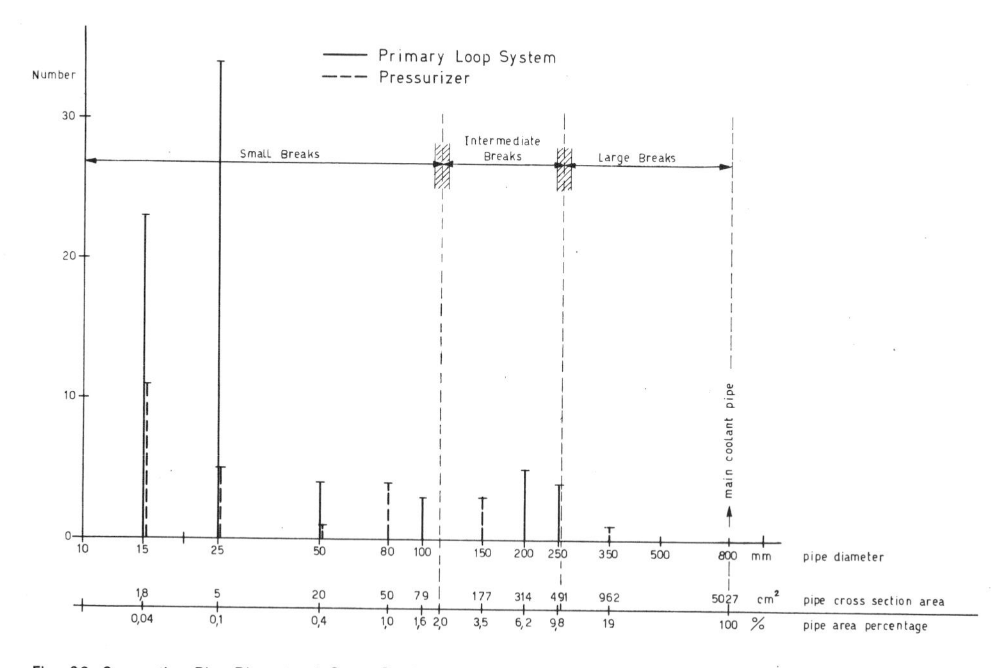

Fig. 30: Connection Pipe Diameter / Cross Section / Percentage Spectrum of a PWR

- the secondary side has to be run down with a temperature gradient of 100 K/h, which is initiated on low pressurizer pressure signal (≤ 145 bar) and high containment pressure signal (> 30 mbar).

The emergency cooling criteria demand that 2 out of the following 3 conditions be met [8]:

- primary cooling system pressure ≤ 110 bar
- pressurizer level < 2.28 m
- differential pressure between plant and operation building, and atmosphere > 30 mbar.

These conditions cause the following actions:

- the HPI pumps "on"
- the main coolant pumps "off"
- containment building "isolation" from environment.

(b) General thermohydraulics: In a small-break accident, the rate of the primary system pressure decrease due to the loss of coolant from the break depends primarily on the break size, as shown in Fig. 31. In addition, the break location (main coolant pipe, reactor pressure vessel, connecting pipes, pressurizer valves) and, in large horizontal pipes, the break orientation (top, bottom or side) determines whether the break flow will be water, water-steam mixture, or steam only, and this affects the depressurization rate which in-

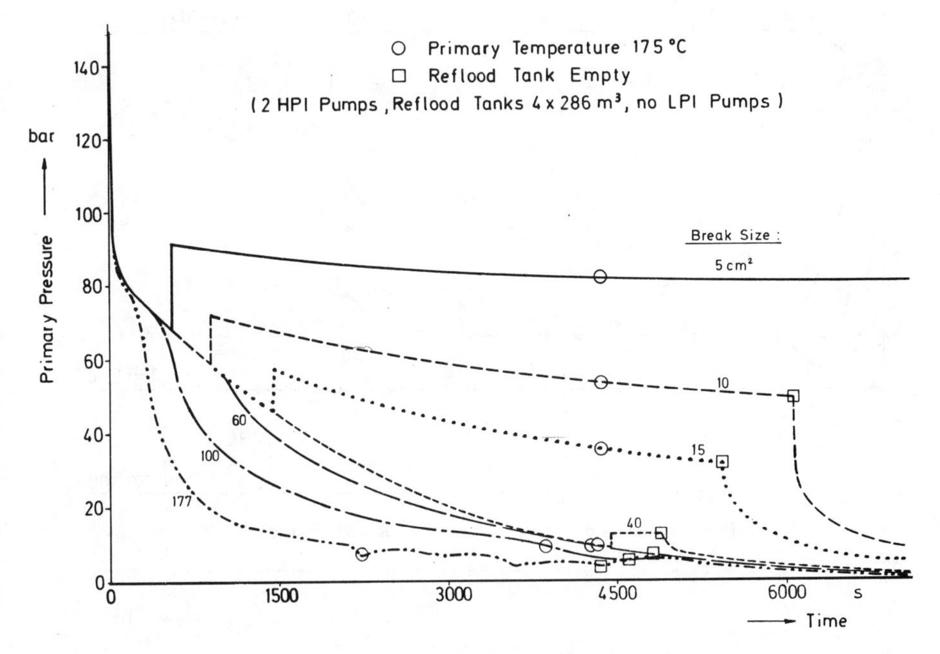

Fig. 31: Primary Pressure vs. Time for Small Break LOCAs in a PWR [8]

creases with increasing flow quality upstream the break.

After the high pressure injection system (HPIS) has started operating, cold borated water from the reflood tanks (see Figs. 12 to 15) is fed into the primary system hot legs (KWU plants) or cold legs (Westinghouse plants). This provides compensation for the primary coolant loss and removes primary system energy (fuel stored energy, fission product decay heat, system internal energy) in addition to that removed by energy rejection through the break and through the steam generators. Finally it increases the primary system depressurization rate in order that the low pressure subsystems of the ECCS may start operation as soon as possible.

Core cooling and primary system heat removal will be strongly influenced by the extent to which a substantial coolant circulation is maintained, initially by continued pump operation and, after pump shut-off, by natural circulation.

During this natural circulation period of a small break LOCA which is essentially the whole remaining LOCA period after MCP shut-off, the gravitational heads representing the driving force for the coolant mass flow rate are determined by (1) the formation of mixture levels due to phase separation between water and vapour, see Fig. 32, (2) the distribution of the residual

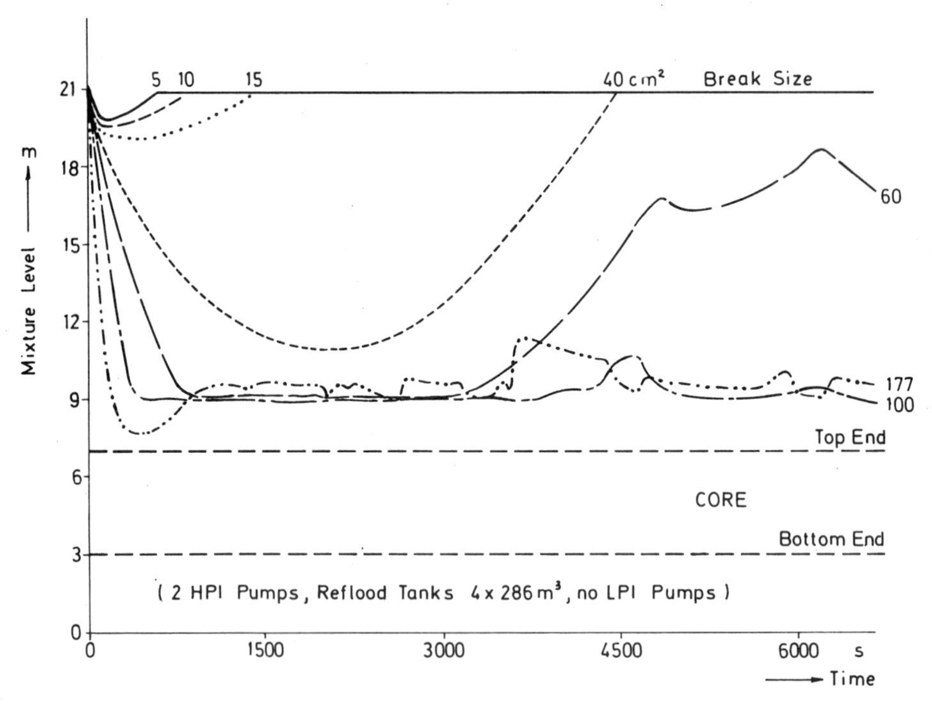

Fig. 32: Mixture Level vs. Time for Small Break LOCAs in a PWR [8]

coolant mass inventories within the various regions of the primary circuits, and (3) the relative elevation of those regions. Under such conditions water-vapour counter-current flow may even occur in some regions. This is the case especially within the hot leg pipes where primary vapour condenses within the riser leg of the steam generator U-tubes, and saturated water can flow back from the steam generator into the hot leg pipes (reflux condenser).

Temperature equalization in the fuel rods will be achieved at around coolant (saturation) temperature. DNB is unlikely to occur throughout the whole LOCA transient provided the core remains covered by liquid coolant.

The steam generated on the steam generator secondary sides is released through the turbine bypass valve and condensers. In the case of loss of site power, however, the condensers cannot operate as heat sinks; the steam then has to be released by blowing down the secondary side through the appropriate valves (1, 2, 3 in Fig. 8), which requires the operation of the auxiliary feedwater system (AFS), see Figs. 16 and 17.

In the case of a very small break LOCA (less than about 1% break size), the primary system energy rejection through the break which is assisted by the HPIS cold emergency coolant water, is insufficient to depressurize the system rapidly enough. Additional energy then has to be rejected through the steam generators by blowing down the secondary side as described before. Under these conditions the vapour relief valves are controlled so as to yield a secondary side temperature gradient of 100 K/h (KWU plants).

A rapid depressurization of the primary system must be aimed at in order to minimize the primary fluid loss and thus leave the core in safer conditions.

After the primary system pressure has decreased to below the accumulators pressure (27 bars in KWU plants, 45 bars in Westinghouse plants), emergency coolant water discharge from the accumulators combines with that from the HPIS.

Finally, at a system pressure below about 10 bars, the residual heat removal system (RHR), see Figs. 12 and 13, starts operating and takes over the heat removal from the primary system, which finally consists of the fission product decay heat only, for as long a time period as required.

The total time duration of a small break LOCA depends essentially on break size and location and may range from several minutes to several hours. The LOCA is considered terminated when the system reaches pressure equalization with the containment pressure, or when the water level in the reactor has again risen above the top end of the core if it had ever been below during the LOCA, or when it again reaches the outlet nozzle elevation.

In BWRs, relief valves on the steam lines are operated by the automatic depressurization system (ADS) to reduce the system pressure in the reactor vessel rapidly in case the high-pressure core spray (HPCS) system should fail to maintain the water level above the core top end.

(c) Primary system pressure and mixture level behaviour: Brief consideration will be given to the Figs. 31 and 32. A comparison of these figures, showing the primary pressure and the mixture level history as function of several break sizes and for specific ECCS function conditions, reveals that there is a substantial difference in the behaviour of both quantities for break sizes less than about 40 cm^2 (ca. 1%):

- for break sizes less than 1%, the mixture level decrease within the primary system is limited to above the RPV outlet nozzle elevation and to a time period which decreases with decreasing break size; emergency coolant water fed-in by the HPI pumps is sufficient to compensate the coolant loss from the break and to refill the primary system; the system pressure, which so far had decreased following the saturation line, at this point experiences a step increase to a level which is determined by the HPI pumps' head and delivery characteristics; the pressure finally falls down again to the saturation value when the reflood water storage tanks have emptied (squares in Fig. 31);
- for break sizes larger than 1%, the mixture level drops rapidly down to the RPV outlet nozzle elevation and, for break sizes larger than about 100 cm^2 or 2.5%, even further down, into the upper plenum, with the potential for core uncovery with increasing break size; the primary pressure decreases throughout the LOCA period following the saturation line; the HPI pump delivery is not sufficient to compensate for the coolant loss from the break;
- the circles in Fig. 31 indicate the instant in time when the primary coolant temperature has decreased to below the saturation value corresponding to the LPIS initiation pressure (about 10 bars); at this instant the LPIS will start operation if the system pressure is low enough.

In Westinghouse plants, in addition to the HPIS pumps the make-up pumps of the volume control system ((4) in Fig. 15) can also be used for providing high-pressure injection of emergency coolant water. However, since the pressure head of these pumps is sufficient for the normal primary system operation pressure, their use gives rise to the potential for over-feeding the primary system.

Certain break locations, especially those on the top of the pressurizer, can give rise to unusual liquid coolant distributions within the primary system and hence cause liquid level indications which can mislead operators into manual intervention in the ECCS operation, as occurred during the TMI-2 accident.

In particular, should the pressurizer relief valve stick open, the pressurizer may completely fill with water within the first few minutes, causing the HPIS to be switched off with the aim of preventing the pressurizer and the primary system from going solid. Since, in spite of the indications to the contrary there may in fact still be a liquid-vapour interface within the primary system, perhaps even within the RPV, the redistribution of the liquid mass inventory after HPIS pumps have shut down can result in a core uncovery. A similar effect can occur when the main coolant pumps have been kept in operation to assist coolant circulation, and are shut down at a later instant.

Finally, natural circulation within the primary system may be significantly affected by the coolant vapour itself and even more, by incondensable gases. These could result either from fission gas release or from metal-water reaction within the core, should this have occurred due to a temporary uncovering of the core as happened during the TMI-2 accident.

These are only a few of the more general considerations influencing the thermohydraulic phenomena occurring during the course of a small break LOCA in PWRs.

A more detailed and systematic description requires more experimental and theoretical work. Up to now only a rather limited amount has been performed, compared with that done in connection with large break LOCAs.

5. LOCA WITHOUT EFFECTIVE "ECCS"

A large double-ended break LOCA without effective ECCS would inevitably lead to a core melt-down. A probable sequence of events describing the core history would be the following [1]:

Blowdown. The primary cooling system behaviour during the early blowdown period is similar to that described in section 4.1 (under the heading: (a) "The blowdown period"); it starts to differ only from about 15 s onwards when ECCS normally starts operation.

The whole blowdown period would be completed after about 20 s. At this time, it is estimated that roughly one half of the stored energy in the fuel rods will have been removed by the coolant escaping from the break; about 10% of the original water inventory in the reactor vessel will remain, and the core temperatures might reach 800 to 900°C.

Core heat-up by decay plus chemical heating. After blowdown, decay heat production will cause the core temperature to rise at a rate of about 11°C/s for a PWR or 7°C/s for a BWR. If starting from 900°C, temperatures higher than 1100°C could be reached within about 30-50 s causing the zircaloy-steam reaction to become a significant additional energy source.

As cladding temperatures increase above 1100°C the zircaloy-steam reaction rate becomes dependent on the availability of steam. The cladding melting point, 1850°C, could be reached after a reaction time period of about one minute.

Core collapse and collection in the reactor vessel. Assuming that all the decay heat is retained and that 50% of the cladding undergoes the metal-water reaction, the energy released could be enough to raise the average core temperature up to the melting point of UO_2. In the most pessimistic case this would take at least 10 minutes from the time of the break, and most probably up to 60 minutes, resulting in the collection of a major portion of the core, in a molten state, within the lower pressure vessel head.

Pressure vessel melt-through and collection at the bottom of containment. If it is assumed that 50-80% of the fuel has collected in the lower pressure vessel and that this represents that part of the core with the greatest decay heating, a melt-through of the pressure vessel could be expected between 30 and 120 minutes after pipe break.

The molten core material, together with the molten portion of the vessel bottom, will interact with the water pool, which is expected to have formed at the containment bottom, and may cause a steam explosion. Two serious effects could occur as consequences of such an explosion: (a) the scattering of molten core material to higher containment regions with the potential there to cause a localized melt-through, and (b) the subdivision of molten core material: this increases the reaction rate with water and hence the hydrogen production with the potential risk for a containment failure due to overpressure.

Behaviour of molten core mass at the containment bottom. The molten core and vessel material having a weight of about 100-200 tons would displace the water in the bottom of the containment and come into direct contact with the concrete. This would be heated up and disintegrate; the concrete residue would finally float on top of the molten core. Containment vessel penetration by the molten core material may occur anything from a few hours to a few days after pipe break.

6. THE ECCS ACCEPTANCE CRITERIA AND CRITICAL LOCA PHENOMENA

6.1 The ECCS Acceptance Criteria

There are five main acceptance criteria which are related to the calculated consequences of the most serious LOCA with functioning ECCS.

(1) The calculated peak cladding temperatures may not exceed 1200°C

(2) To limit oxidation-induced cladding embrittlement, the maximum cladding oxidation calculated for the hottest fuel rod shall not exceed 17% of the cladding thickness

(3) The maximum hydrogen generation from the zirconium cladding reaction with water or steam calculated for the entire reactor must not exceed 1% of the totally possible amount

(4) All core changes calculated should leave the core in a coolable geometry

(5) With effective ECCS, the long-term core cooling system must be capable of removing sufficient energy from the primary system for whatever time period is required to maintain the calculated core temperatures at acceptably low levels.

6.2 The Critical LOCA Phenomena

The critical phenomena occurring during a LOCA are those which significantly affect the course of a LOCA. For safety analysis purposes they have to be modelled so as to yield conservative calculation results and satisfy the ECCS acceptance criteria.

Those phenomena will be listed briefly and separately for the three main LOCA periods [1], [2].

(A) <u>Blowdown period.</u>

(1) Initial Conditions:
- 102% core power
- nominal reactor operation conditions

(2) Energy Sources:
- high stored energy due to high gap resistance
- fission product decay heat = ANS + 20%
- reactor internals heat transfer (piping, vessel wall etc.)
- PWR primary-to-secondary heat transfer
- metal-water reaction

(3) Depressurization wave:
- propagation through the primary cooling system (PCS)

(4) Primary System Coolant Behaviour:

- Break Flow:
  - break size
  - discharge flow model
- Break Location:
  - affects flow stagnation point within PCS
- Frictional Pressure Drop:
  - two-phase friction multiplier
- Pump Characteristics:
  - two-phase pump model
- ECC Fluid "bypass":
  - end-of-bypass definition = begin of refill
- Core Flow Distribution:
  - cross flow induced by fuel rod swelling blockage
- Lower Plenum Fluid Entrainment:
  - residual fluid in lower plenum after blowdown
- Two-Phase Flow Models:
  - relative velocity
  - constitutive relationship liquid/vapour

(5) Core Thermal Behaviour:

- CHF and DNB:
  - steady-state and transient CHF models
- Post-CHF Heat Transfer:
  - no return to nucleate boiling to be assumed
- Fuel Rod Behaviour:
  - swelling and rupture of clad
  - gap resistance
- Core Geometry Changes:
  - fuel rod bowing

- inter-rod contact

(B) Refill Period.

(1) Energy Sources: . See Blowdown Period

(2) ECC/Primary System Coolant Behaviour:

- Break Flow Discharge Rate: . transition from choked to unchoked flow
- Lower Plenum Refill Rate: . ECC penetration of downcomer
- Steam/ECCS Fluid Interaction: . steam flow in unbroken loops during ECC injection

(3) Core Thermal Behaviour:

- Post CHF Heat Transfer: . see Blowdown Period
- Convection Eddy Cooling (PWR): . hot rods induce convection eddies in steam-filled core

(C) Reflood Period.

(1) Energy Sources: . See Blowdown Period

(2) Structural Behaviour:

- Thermal Shocks:
  - . cladding brittle failure
  - . temperature limit 1200°C
  - . 17% equiv. clad oxidation limit
- Oscillatory Loads: . manometer-like osciallations between core and downcomer annulus
- Core Geometry Changes: . coolable geometry required

(3) ECC/Primary System Coolant Behaviour:

- Break Flow Discharge Rate: . unchoked steam flow
- Oscillatory Flow: . see under (C) (2) above
- Reflood Rate:
- Droplet "carry over" through Core: . fluid flow at core outlet plane vs. flow at core inlet plane
- Steam Binding:
  - . pump Δp (locked impeller)
  - . steam generator superheat
  - . steam generator tube leak

- Core Flow Distribution: . blockage and radial flow

(4) Core Thermal Behaviour:

- PWR Reflood Heat Transfer:
  . oscillatory flow (no benefits allowed)
  . low reflood rate restrictions (steam heat transfer only)
  . containment pressure influence (lowest pressure value for heat transfer correlations)

- BWR Core Spray/Reflood Heat Transfer: .

A detailed consideration of the most significant of these phenomena will be given in other chapters of this course.

SUMMARY

Two facts should be pointed out characterizing an important aspect of the verbal-pictorial description of the course of a loss-of-coolant accident (LOCA) in light-water reactors (LWRs), the primary objective of this chapter:

- the subdivision into the two essential parts for the large break LOCA and for the small break LOCA
- the difference in detail in which these two parts have been dealt with, and which reflects, to more or less the same extent, the present knowledge on these two types of LOCA.

The much more detailed description of the course of a large break LOCA is a direct consequence of the fact that in the past almost exclusive emphasis in the licensing process has been put on the "design basis accident" concept. Therefore, much more knowledge about the large break LOCA has become available from major research efforts concentrated on the experimental and analytical investigation of the various phenomena characterizing and controlling the course of this type of accident. Indeed, these phenomena are now well identified, the essential mechanisms are sufficiently understood to allow a satisfactory consideration not only in "evaluation model" (EM) calculations but also in "best estimate" (BE) calculations.

There have, however, remained uncertainties in essentially two areas: (1) the non-equilibrium phenomena, whose effect can however, at least for the purpose of licensing considerations, be sufficiently taken into account by appropriate conservative assumptions; (2) the multi-dimensional flow behaviour in the core and upper plenum region during the late LOCA period (refill, reflood). In both areas, more experimental information and increased modelling effort is required with a view to achieving satisfactory prediction capabilities of large system computer codes with respect also to these effects on the course of a large break LOCA. Furthermore, much analytical effort is necessary in the near future in order finally to make use of the considerable amount of results which has become available during the last few years, from pertinent experimental research conducted in the area of separate effects as well as of integral system effects.

The clearly less detailed description of the course of a small break LOCA reflects the fact that in the past this accident concept has received much less emphasis in the licensing process. However, whilst the research effort in this area had already increased before the TMI-2 reactor accident at Harrisburg, USA, such effort has been considerably amplified after this accident, and not only in the USA. Indeed, the small break LOCA is a possible type of accident for which experimental and analytical research is still required in areas like water-vapour phase separation, mixture level behaviour and distribution, influence of noncondensable gases on heat transfer and natural circulation, severe fuel damage, etc., to mention only the most important ones. Such research work is at present already under way, and first results are just now becoming available.

Finally it must be pointed out, that because of the serious potential consequences of a major release of radioactivity, and in view of existing safety-related technological opportunities, there should be a continuing major effort to improve light-water safety as well as to understand and mitigate the consequences of possible accidents.

ACKNOWLEDGEMENT

The author owes Mr. Tom Fortescue a debt of particular gratitude for the assistance he has given with correcting the English.

REFERENCES

1. Report to the American Physical Society by the Study Group on Light Water Reactor Safety. 1975. Review of Modern Phys., Vol. 27, Suppl. No. 1.

2. Smidt, D. 1979. Reaktor-Sicherheitstechnik. Springer Verlag, Berlin-Heidelberg-New York.

3. Thermal and hydraulic aspects of nuclear reactor safety. Vol. 1: Light-water reactors. Editors: O.C. Jones and S.G. Bankoff. The Am. Soc. of Mech. Eng., 1977.

4. Riebold, W. 1977. Introduction to the problems related to the loss-of-coolant accident. Ispra Courses on "Thermohydraulic problems related to reactor safety", part B: LWR related thermohydraulic problems. Commission of the European Communities, Joint Research Centre - Ispra Establishment, Italy, September 29-30.

5. Case, E.G. 1974. Analysis of a sudden major loss of coolant accompanied by serious failure of emergency core cooling. Nuclear Safety, Vol. 15, No. 3, May-June.

6. USAEC-WASH-1250. 1973. The safety of nuclear power reactors (light-

water cooled) and related facilities.

7. USAEC-WASH-1400, NUREG 75/014. 1975. Reactor safety study. An assessment of accident risks in US commercial nuclear power plants. Appendix IX: Safety design rationale for nuclear power plants.

8. Weisshäupl, H.A. ECCS features and LOCA behaviour of the KWU 1300 MW PWR. Working paper presented at the 14th meeting of the ad-hoc specialist working group on the LOBI programme, Part B., 1980.

9. Die deutsche Risikostudie. Kurzfassung. Der Bundesminister für Forschung und Technologie, Gesellschaft für Reaktorsicherheit., 1979.

10. Kernkraftwerk Biblis. KWU-Broschüre 10952 17615. Best. Nr. KWU-166a, Jan. 1976.

11. The 1300 MW standard nuclear power station with pressurized water reactor. KWU-Broschüre 10893 97520. Order No. KWU 336-101. Sept. 1975.

12. Braun, W., Hundt, D. and Steinert, Ch. 1975. Leichtwasserreaktoren in Deutschland. ETZ - A 96. Nr. Jan. 1975, pp. 24-29.

13. Kernkraftwerk Brunsbüttel. KWU-Broschüre 10770 11745. Best. Nr. 287. Nov. 1974.

14. Kernkraftwerk Gundremmingen Block B und C. KWU-Broschüre 10851 97515. Best. Nr. 338. Sept. 1975.

15. Hein, D., Maginger, F. and Winkler, F. 1977. The influence of loop resistance on refilling and reflooding in the PKL-tests. Paper presented at the water reactor safety research information meeting, Gaithersburg (Md.), USA, Nov. 7-11.

10 BLOWDOWN PHASE

by
W.L. Riebold
CEC – JRC, Ispra, Italy

M. Reocreux
CEA-DSN, Fontenay-aux-Roses, France

O.C. Jones, Jr.
BNL-AUI, Upton, NY, USA

This Chapter describes in detail the sequence of events which occur during the first phase of a design basis accident in a light water reactor—the blowdown phase. In the main part of this chapter the most significant thermohydraulic phenomena occurring during the high-pressure or blowdown phase of a loss-of-coolant accident (LOCA) in lightwater reactors (LWR) are considered under three subheadings: a general, physical description, the modelling problems, and a discussion of particular aspects. Specific aspects of the blowdown phase which are to be covered include the pressure wave and flashing boundary propagation, critical flow, blowdown heat transfer, and two-phase pump behavior. Since details of many of the thermohydraulic behavior including correlations are included elsewhere in this book, they are not duplicated herein. Rather, existing correlations currently used are identified through appropriate references. New work on an area of importance to the entire DBA sequence, thermal nonequilibrium, is summarized in an appendix. The goal of this chapter, then, is to provide an overview of the rapidly evolving state-of-the-art of describing the thermal hydraulic behavior of an LWR during blowdown.

1. INTRODUCTION

The objective of this chapter is to consider the various physical phenomena occurring within the main cooling system of light-water reactors (LWR) during the high-pressure or blowdown phase of a loss-of-coolant accident (LOCA). There is a rather large variety of thermodynamic, hydrodynamic, thermal or mechanical phenomena which may occur simultaneously or sequentially during a LOCA transient. They are dealt with under a few appropriate headings which at the same time represent the most important of these phenomena affecting or even controlling the course of the transient. However, a clear separation is not generally possible, since several such phenomena can occur simultaneously, affecting each other to a not unimportant extent. This is especially true for the case of nonequilibrium effects which may significantly influence the flashing process, the two-phase flow, the critical flow and the heat transfer process; another very prominent example is the dependence on each other of two-phase flow and heat transfer mechanisms.

The sequential arrangement of the various sections reflects as far as possible the chronological occurrence of the respective phenomenon on the one hand and allows, again as far as possible, a logical evolution in the treatment of the various phenomena on the other.

The individual sections are in most cases structured in such a way, that a general description of the phenomenon precedes the consideration of modelling aspects. It was intentional that in the subheadings on modelling, no correlations for the calculation of the respective quantities have been given. The reason for that is that in nearly all cases no generally valid or sufficiently verified correlation can be given. In most cases the validity ranges of different correlations are different depending on the data from which they have been derived. Accordingly, for each application case and range an appropriate and careful choice has to be made. An exhaustive description of the correlations for each single heading would have by far exceeded the limits of the present chapter. Many of the correlations currently used are identical with those described elsewhere in this book. For instance the fundamentals are described in Chapters 6 and 7 whereas post-dryout heat transfer is detailed in Chapter 12. Therefore, preference was given to a brief outline of the state-of-the-art, and explicit reference was made to pertinent literature where a sufficiently complete representation of the specific subject is to be found. Each heading is concluded, where appropriate, by a brief discussion directed at a validation of the state-of-the-art.

Exception from this general line is made with respect to the thermal non-equilibrium phenomenon which in terms of the nonequilibrium vapor generation rate plays a predominant role at several instants during a LOCA transient and in several regions within the primary cooling system of a reactor. It is of primary importance in flashing flows and post-dryout heat transfer. Recent analytical work on flashing nonequilibrium and post-dryout has therefore been described more in detail in Appendix A to this paper.

2. DEPRESSURIZATION WAVE PROPAGATION

2.1 Description

The opening of a break within the primary cooling system of a LWR initiates a loss-of-coolant accident and puts the high-pressure coolant fluid instantaneously into communication with the low-pressure containment atmosphere. The resultant depressurization wave propagates through the primary loop system and the pressure vessel. During the first milliseconds this propagation occurs in a single-phase liquid flow. Due to the small compressibility of the liquid the propagation velocity depends also on the elasticity of the piping and structures. The wave propagation through the pressure vessel is especially complicated, since complex wave reflections and interactions with the internal structures cause a 3-dimensional behavior. This is shown in exaggeration in Fig. 1 representing the dynamic deformation of the core barrel and the resultant relation between the waves in the downcomer and in the core. Disregarding the pressure wave reflections, the maximum possible local loads are given by the difference between operation pressure and saturation pressure. Therefore in the case of a boiling water reactor (BWR) there is practically no such load occurring.

The forces on the internal structure caused by the pressure waves are the highest during the initial phase of the accident. These forces decrease with decreasing break size and with increasing distance of the break location from the internal structure under consideration. The aim of the quantitative analysis of these dynamic loads is to determine their potential for any structural damage and especially for any damage to the core integrity which could degrade the coolability of the core for the remaining LOCA period.

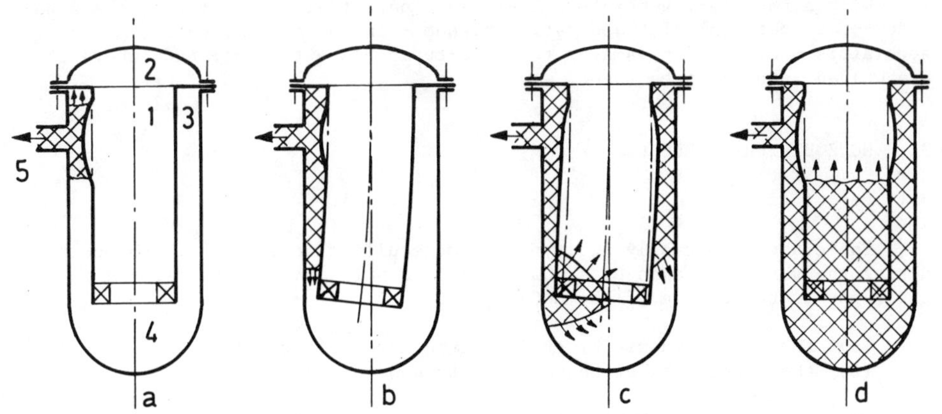

Figure 1. Depressurization wave and resulting deformation of the core barrel after a large cold leg break.[2]

1 Core Barrel
2 RPV Head
3 Downcomer
4 Lower Plenum
5 Break Location close to Inlet Nozzle; shadowed areas: zones of reduced pressure

2.2 Modelling

The analytical description of the wave propagation in single-phase liquid flow does not raise any physical difficulties. The elasticity of piping and structures can be taken into account. Numerical difficulties may arise from the 2- or 3-dimensional calculations required when the coupling between fluid and structure deformation has to be considered.

The deformation shown in Fig. 1 can not be represented by a one-dimensional modelling of the phenomena. By means of an appropriate nodalisation, however, it is possible to obtain a de-facto two-dimensional structure which allows the calculation of the pressure wave propagation within the projection of the developed annular downcomer space.[2] This method is used in codes like DAPSY,[3] PWINCAD,[4] WHAMMOD, and LECK.[5] The time-dependent pressure distribution load on the pressure vessel internals may be calculated assuming stiff walls as fluid envelopes. The mechanical stresses and deformations are then calculated separately by a structural mechanics code, e.g. STRUDL II. In this case, the continuous fluid-structure interaction is not taken into account, and the assumption of stiff walls is considered to yield conservative results since structural elasticity causes a reduction of pressure peaks.

2.3 Discussion

Besides the pressure vessel region and the coolant piping, there are also the main coolant circulation pumps and the steam generators, where the depressurization wave may affect the integrity of the component.

Within the coolant pump housing the depressurization wave may cause an axial displacement of the impeller such as to make it touch and, especially in case of the broken loop pump with the potential for impeller overspeed in cold leg break conditions, eat into the pump housing.

Within the steam generators there is a potential for rupturing the separator plate between inlet and outlet plenum - in the case of U-tube steam generators - or causing damages to the U-tubes close to their ends where they are welded into the tube sheet.

3. NONEQUILIBRIUM EFFECTS

3.1 Description

There are essentially two types of nonequilibrium effects which occur during the various stages of a LOCA transient which may influence, and in some cases govern the course of the LOCA:

(a) the thermodynamic or phase change nonequilibrium;
(b) the mechanical or phase velocity nonequilibrium.

The mechanical nonequilibrium where phase velocities differ persists during the whole two-phase flow period of a LOCA. Its influence on the course of a LOCA varies with both the region within the primary cooling system where it occurs, and the time into the transient; it will be dealt with implicitly in Sections 5 and 6.

Thermal nonequilibrium during phase change can be important in essentially four areas of different relevance for a LOCA:[6]

(a) rapid depressurization (flashing) as it may occur in the early stage of a LOCA;

(b) post-dryout heat transfer as it may exist during the blowdown and the refill/reflood period of a LOCA;

(c) condensation due to the contact of subcooled water and hot steam occurring during the emergency core cooling period (water injection, spray cooling);

(d) subcooled boiling during quasi-steady state and transient periods of a LOCA.

The second and the fourth area will be implicitly dealt with relative to current practice in Section 7; the third area is discussed in Chapters 11 and 12. Recent developments regarding the general framework of relaxation theory is covered in Appendix A with particular attention to flashing and post-dryout.

The first area, the phase change nonequilibrium in rapid depressurization conditions, is of particular interest with regard to two aspects:

- the early system pressure behavior;

- the critical flow through pipes discharging two-phase mixtures to the surroundings.

In both cases, the fluid is exposed to a strong pressure decrease and, hence, a departure from equilibrium occurs where the phases can not follow the changes from one equilibrium state to another. The required exchange of matter by phase change across interfaces is limited by the ability of the fluid to transfer heat to or from generally pre-existing interfaces (heterogeneous nucleation), and by the amount of interfaces available for this transfer.

Therefore, to take into consideration the problem of nonequilibrium phase change under such depressurization conditions requires the availability of constitutive relations describing the actual rates of liquid-vapor mass exchange, i.e. the volumetric rate of vapor generation.

3.2 Modelling

This vapor generation process is essentially governed by the following parameters:

(a) the time delay for bubble nucleation;

(b) the nucleation rate;

(c) the number of bubble nuclei per unit mass (or volume) of liquids;

(d) the bubble growth;

(e) the density,

and may be additionally affected by other parameters like the presence of dissolved gases and particulates.

Models developed for predicting critical flow in most cases do not describe the mechanisms of vapor generation during rapid depressurization or flashing. The vaporization rate under the latter conditions can be calculated directly by several correlations,[6] which, however, result from different models and assumptions.

A more recent model has been developed by Wolfert.[7] He proposed two correlations for the vaporization rate. One results from the application of a heat transfer coefficient between translating bubbles and the liquid, by allowing a relative velocity between bubbles and liquid, and yields a vaporization rate proportional to the first power of liquid superheating. The other one results from the application of the bubble growth law of Plesset and Zwick,[8] considers the case with no relative velocity, and leads to a vaporization rate proportional to the second power of liquid superheating. Bubble density must be specified as a parameter. A comparison of both correlations shows that at low vapor void fraction and high liquid superheating the correlation for zero relative velocity dominates; for high vapor void fraction the vaporization rate is determined primarily by the other correlation implying a relative velocity. The net vapor generation rate per unit volume is then obtained by simply adding both terms. No justification is given for this model, but from a comparison with experimental data of Friz et al.,[9] good agreement has been obtained for a bubble density of 10^9 per m^3, see Fig. 2. In that experiment a fixed volume of subcooled water at high pressure was suddenly expanded to a slightly larger volume, and the pressure-time history was recorded.

Fig. 2 shows clearly the drastic initial pressure drop below the liquid saturation pressure due to thermal nonequilibrium effects during a sudden expansion (pressure undershoot), and the subsequent pressure recovery to saturation pressure.

The initial undershoot is caused by a flashing delay of the hot liquid and leads to an increasing liquid superheat and, hence, thermal nonequilibrium. This superheat is the driving force for both the activation of bubble nucleation sites, and the bubble growth. The induced vaporization process results in

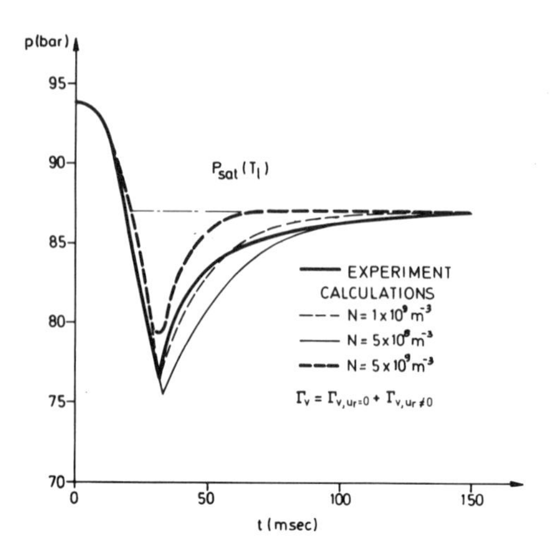

Figure 2. Comparison of the predictions of Wolfert[7] with the experimental measurements of Friz, et al.[9]

the reduction of thermal nonequilibrium and a strong fluid expansion which causes the pressure to return to the saturation value.

Recent analytical work on flashing nonequilibrium has been undertaken emphasizing the mechanistic behavior governing the early stages from inception up to approximately 20% voids. The general development for nonequilibrium phase change shows the process to be intrinsically an initial value relaxation problem (Appendix A), requiring the modelling of three factors: flashing inception which is the flashing analog of boiling inception; interfacial area density; interfacial heat flux. The former problem has been solved for ideal fluids without dissolved gas or particulates. The area density is very simply modelled but still requires specification of a nucleation or bubble density. The latter problem has been approximated as a conduction controlled heat flux in terms of the bubble age for a given local superheat, similar to Wolfert's model, but accounting for depressurization effects which can be significant. Accurate prediction of early void growth once inception nucleation density has been specified has been demonstrated.

4. FLASHING BOUNDARY PROPAGATION

4.1 Description

During the phase of depressurization wave propagation, which lasts about 50 ms, the fluid pressure within the primary cooling system, due to subcooled depressurization, drops very rapidly down to the saturation pressure corresponding to the fluid temperature. The decompression rate within the broken loops varies with the fluid temperature. It amounts to about 10,000 bar/s in the cold leg or RPV inlet region, and to about 1,000 bar/s in the hot leg or RPV outlet region. Within the intact loops, this decompression rate is more or less the same everywhere and amounts to about 500 bar/s; hence, it appears to be imposed by that of the hot region of the cooling system.

Pressure stabilization in the broken loop at the respective saturation pressure, and in the intact loops at the saturation pressure corresponding to the highest fluid temperatures, results from the flashing process occurring which induces a strong expansion of the fluid and thereby limits the rate of further pressure decrease during the following saturated depressurization phase to less than about 20 bar/s.

Typical initial pressure undershoot shown in Fig. 2 is usually observed as a consequence of thermal nonequilibrium in all single volume small scale blowdown experiments like CANON,[10] Edward's pipe,[11] and large scale vessel blowdown experiments like Battelle-Frankfurt[12] or Marviken.[13] In loop system blowdown experiments like Semiscale, LOFT and LOBI, a certain initial pressure undershoot can still be observed in the region close to the break location; but it is less pronounced and less clear, which may be due to several reasons like wave reflection, higher system elasticity, larger number of nucleation sites, etc., or simply to the balance between decompression rate and turbulent effects outlined in the appendix. This has yet to be determined.

After the flashing has started in the hottest parts of the cooling system, the pressure continues to decrease to a much smaller extent and reaches gradually the saturation pressure of the regions at lower temperature. This depressurization process is accompanied by the propagation of a flashing boundary, which starts within the upper plenum and continues into the core to reach the lower plenum, and into the hot legs of the loops to penetrate the steam generators and to finally reach the cold legs and the downcomer. This flashing propagation occurs also with a certain delay without, however, inducing a pressure undershoot as it was observed in the very beginning of the transient when the first bubbles appeared in the loop system. This is due to the increased compressibility of the vaporizing and strongly expanding fluid. After about 3 to 5 seconds into the LOCA transient, the whole primary cooling system has flashed.

The flashing process, especially that in large volumes of the primary cooling system such as the lower and upper plena, causes flow reversal in certain loop regions depending on the break location, and affects the position and/or the displacement of the flow stagnation point. These phenomena are particularly important when they occur within the core and cause a drastic degradation of the heat transfer conditions and consequently a strong increase of the cladding temperature. It is for this reason that the flashing process, its propagation and the possible flashing delay are of primary importance since they may significantly affect the prediction of the accident sequences.

4.2 Modelling

A correct description of the previously mentioned phenomena needs, of course, nonequilibrium models. Since nearly all codes presently available such as RELAP, etc., which are used also for licensing calculations, are equilibrium codes, they are completely unable to describe phenomena like flashing delay, liquid superheat, and superheat resolution by vaporization.

Because of the lack in existing codes, one of the primary goals of advanced codes is to take into account these nonequilibrium effects. For this purpose continuity equations are used for each phase where the mass transfer appears explicitly and the fluid nonequilibrium results from appropriate mass transfer laws. To correctly describe the flashing process, such laws have to consider:

(a) the nucleation process as the predominant effect in the pressure undershoot and the flashing delay phenomena;

(b) the fully established flashing under nonequilibrium conditions which governs the pressure recovery and the dynamics of the flashing propagation.

Essentially three approaches are used to determine the respective mass transfer laws:

(a) the analytical approach: the derivation of the laws is based exclusively on theoretical considerations; for example, starting from a given bubble population the evolution of the bubble size is described by theoretical conduction - convection calculations around the bubble. Such an approach appears to be very general and easy to extend to all scale experiments. However, there are still parameters which need to be adjusted like number and distribution of bubble nuclei, and for which experimental data in various flows and at different scale are still lacking. Furthermore, this approach has still to rely on simplifying assumptions - like neglection of bubble interactions - which limit the application range of the model;

(b) the global approach: this is the correlation type approach, which in its principle is very simple, but extrapolation to other types of experiments may be very doubtful;

(c) the mixed approach: this approach tries to combine the two preceding ones: the general form of the laws resulting from theoretical considerations are adjusted globally on experiments.

4.3 Discussion

All approaches mentioned above have been used before, the first and third ones predominating. Examples of the first one are the development of Ardron[14] and that of Wolfert[7] in the DRUFAN code. The mass transfer laws developed by Bauer et al.[15] are used in the third type of approach.

In spite of the strong efforts made in all these developments to obtain a purely physical description, there remains the need of model qualification on experimental data. Especially the description of the nucleation process calls for further development since it depends on many parameters difficult to quantify like water quality, impurities, wall effects, etc.

5. TWO-PHASE FLOW

5.1 Description of the Main Features

In the preceding sections it has been shown how two-phase flow starts at the beginning of the accident. Since two-phase flow will persist during the whole accident, its proper modelling is essential for the description and the understanding of all phenomena occurring during the course of an accident. Fundamental descriptions are given in Chapters 6-8.

The description of a two-phase mixture requires the prediction of both the thermal and the dynamic state and relative amounts of each phase.

The state of the phases is generally referred to as an equilibrium state which has to be defined for both the thermal and the dynamic phenomena. The mixture is considered to be in thermal equilibrium with both phases at saturation conditions. Such equilibrium state can be achieved in the case of very slow transients. Dynamic or mechanical equilibrium is given when both phases are at the same velocity. This equilibrium again can be reached only in very slow transients, and in flow configurations where interphase friction is strongly coupling the two phases. When both equilibrium states are given, the two-phase flow is referred to as homogeneous (equal velocities) equilibrium (both phases at saturation) flow.

In reality, these equilibrium states are never achieved, and the departure from equilibrium may become important as the fluid particles are submitted to strong transients.

As discussed already relative to the flashing propagation process, the thermal nonequilibrium can be due to the depressurization process which causes the fluid pressure to decrease below the saturation pressure corresponding to the temperature of the fluid. The induced vaporization process then tends to resolve the nonequilibrium by decreasing both the fluid temperature and the pressure undershoot. Thermal nonequilibrium occurs also in the case of strong heat release from the wall which superheats the fluid close to the wall. This superheated fluid tends to go back to thermal equilibrium by vaporizing or inducing vaporization in the core of the flow.

Mechanical nonequilibrium is given by unequal phase velocities. These are caused by the difference of phase densities combined with gravity forces in the case of low velocity flow, and with acceleration forces in the case of high velocity flow, or simply due to weak coupling between phases as may occur in annular or mist flows. This inequality of velocities may be taken into account as the velocity (slip) ratio, as the velocity difference, or as a drift velocity. The friction forces between the two phases tend to decrease this unequality of the velocities and hence to resolve the mechanical nonequilibrium.

In both cases of mechanical and thermal nonequilibrium there are driving forces built-up which tend to re-establish equilibrium. These driving forces are in actuality expressed as the gradients which drive the transfer of heat or momentum between the phases: temperature gradient for thermal effects and velocity gradient for momentum transfer. In the former case the temperature gradients may be accentuated by increasing the pressure change rates or heat fluxes. In the latter case, gradients may be caused by gravitational or other acceleration effects. Both may be affected by changes in flow regime which alters the coupling between the phases. In both cases, the driving forces induce exchange processes between the two phases which tend to make both phases approach their equilibrium state. Therefore, the main problems in describing two-phase flows arise from the need to determine those transfer processes not only between the phases but also between the fluid and the flow boundaries like pipe walls, fuel cladding, steam-generator tubes. Furthermore, and similar to single-phase flows, two-phase flow also may show a one-, two- or three-dimensional behavior which has to be described in the most correct way as well. However, going to 2-D or 3-D calculations implies computing models which are increasingly complicated and difficult to handle.

There is one further flow feature to be mentioned which, especially in reactor cooling systems, may be of importance and which is given by the degree of flow establishment. Flow establishment is strongly depending on the relative length of the flow path expressed by the length to diameter (L/D) ratio. Since in a reactor cooling system, apart from the core and the steam generators, the straight parts of the coolant ducts (pipes, flow channels) have almost always an L/D ratio of less than 10, flow will almost never be established. This raises additional problems for the flow modelling which generally is verified analytically as well as globally on experiments where usually more or less fully established flow conditions are achieved. At present, none of the available flow models is able to completely take into account this effect of non-established flow.

5.2 Modelling

A short review will be given of the existing models. A detailed developmental history and description of computer models for system safety analysis is given in Chapter 13, and will not be repeated herein. A brief outline, however, should prove useful herein for continuity purposes and is included as follows. For all phases of a LOCA (blowdown, refill, reflood) the same types of two-phase flow models are used. Special features of those models are the transfer laws which take into account the various ranges of the thermohydraulic parameters (pressure, velocity, density, etc.), of the flow patterns (bubble, slug, annular, stratified flow, etc.), and of the fluid to wall interactions (heat transfer, radiation, friction, etc.), encountered during an accident.

Models used in codes of the first generation. The most wide-spread and best-known code of the first generation type is the RELAP code. In this type of code, the basic two-phase flow model is the homogeneous equilibrium model and, hence, the two phases are assumed to be at saturation and equal velocity. The whole cooling system is represented by chains of volumes which are connected to each other by junctions.

The mass and energy balances are described within such volumes or nodes, whereas the flow is determined in the junctions by using the momentum equations. This approach, in fact, consists in a volume averaging procedure of the local mass, energy and momentum equations, with a staggered volume for the momentum. For this averaging procedure, volume averaged parameters and area averaged parameters (at volume boundaries) have to be related to each other. Those relations are generally equality or arithmetic averages.

To some extent, mechanical nonequilibrium effects can be taken into account by appropriately extending this basic model. In RELAP for example this is achieved by introducing a bubble rise model and a slip model. The bubble rise model allows one to calculate a mixture level on the basis of a void fraction profile within the mixture. The slip model uses a correlation for determining the slip at the junction which is taken into account for the mass and energy balance within the volume. However, no mechanical nonequilibrium effect is considered in the momentum equation.

Models used in advanced codes. Models used in advanced codes should be the most physical ones. With a view to this goal, they have to describe all essential features of two-phase flows; i.e. they have to include thermal and mechanical nonequilibrium effects, and they have to use a 0-D, 1-D, 2-D or even 3-D description of individual phenomena when these show a 0-D, 1-D or 2-D or 3-D behavior.

However, many fundamental questions arise from writing such models, such as the significance of time and spatial averaging required to obtain a practical set of equations. A rational derivation results in a number of more or less questionable assumptions needed, (see Boure,[16] Reocreux,[17] Boure[18]).

The most important phenomenon which physically governs the thermal nonequilibrium, is the heat transfer process. Therefore, one of the primary tasks for developing advanced code models is to determine the respective transfer laws, which represent simultaneously the main differences between the various models.

For describing the thermal nonequilibrium, rates of mass and energy transfer are to be taken into account. This implies that the continuity and the energy equation for each individual phase has to be written. However, sometimes only one energy equation is used which then needs an additional assumption for the other phase, as for example that vapor is at saturation. The form of energy and mass transfer laws used in the models determines the differences between those models. Some examples for the way to derive these laws have been discussed in the section on flashing propagation.

There are two types of models for describing mechanical nonequilibrium:

(a) the drift models consider mechanical nonequilibrium by algebraic correlations[19] which also take profile effects into account;

(b) the two fluid models express the momentum transfer explicitly using the momentum equation of each individual phase.[20,21]

5.3 Discussion

It is evident that the models used in codes like RELAP cannot predict correctly, if at all, phenomena including large nonequilibrium effects. Those codes meanwhile have been used so extensively that one knows "how to apply them." Therefore they may be considered as practical tools giving satisfactory results for the blowdown phase of a LOCA. Their applicability for predicting the refill and reflood phase is, however, much reduced.

The advanced code models should eventually represent an improvement and a considerable step ahead. The two-fluid model e.g. is considered to be more appropriate for describing the dynamics of the phase velocity difference variations than is the drift flux model. On the other hand the two-fluid model which uses separate field equations for each phase must also include relationships which express the coupling interactions at interfaces. These relationships are virtually unknown at present and must thus be approximated in an "ad hoc" manner. The drift flux model is instead based generally on a description of volume- or area-averaged behavior of a mixture through which one phase migrates. Relationships required for closure are thus in terms of global quantities which have been intensively studied and for which well defined correlations exist in number. The physical knowledge thereby accumulated in these correlations, is certainly of great advantage when links are made between drift flux and two-fluid models like in [22] and [23].

In fact, momentum transfer laws as well as energy and mass transfer laws still require much verification against experiments in order to cover the whole range of parameters involved. Intensive effort is required in this area especially for determining the interaction laws required to couple the phases for closure in the two-fluid model.

6. CRITICAL FLOW

6.1 Description

The concept of critical flow which is also referred to as "choked" or "sonic flow" is well known. Let us consider two communicating volumes each containing the same fluid at the same conditions. Keeping constant the fluid conditions in one volume, and decreasing the pressure in the other one, the flow rate between the two volumes increases with the pressure difference until it reaches a maximum value at a certain pressure ratio between the two volumes.

A further decrease of the downstream pressure has no further effect on the flow rate. This maximum flow rate is called the critical flow rate. The flow rate independence from downstream conditions is due to the fact that somewhere in the flow duct some average fluid velocity is equal to the sound velocity and, hence, sound (pressure) waves become stationary and do not propagate further upstream. This location in the flow duct is called the critical cross-section. In a single-phase compressible flow this situation is referred to as that of sonic flow conditions where the flow Mach number (ratio of fluid velocity to sound velocity) is equal to one.

In loss-of-coolant accidents, the pressure difference between the primary cooling system and the containment remains very high during the whole blowdown phase, and even during a part of the refill phase. In fact, for the whole time period where the primary pressure is higher than about 5 bar, the coolant flow at the break location is submitted to the above mentioned critical pressure difference or sonic flow conditions and, hence, is called critical.

Critical or choked flow can also occur in locations upstream of the break location provided the flow velocities at that location are high enough (about 50 - 100 m/s). This can be the case for instance in the coolant pumps, (circulation pump in PWRs, jet pumps or axial pumps in BWRs).

As in single-phase or gas flow, the choked flow is mainly governed by two effects:

(a) the phenomena occurring in the critical section (critical conditions);

(b) the flow evolution upstream the critical section, leading to the critical conditions.

In single-phase or gas flow these phenomena are quite well known and predicted, whereas in two-phase flow their understanding and prediction raises still many problems.

To determine the fluid state in the critical section, conditions have to be elaborated and verified by flow parameters which show that the flow is indeed critical. For this purpose, in single-phase flow essentially two equivalent conditions are used:

(a) the first one requires a maximum for the specific mass velocity G and is expressed by $(\partial G/\partial p)_S = 0$. This requirement cannot easily be "extrapolated" to two-phase flows since they are defined by seven parameters (p_G, p_L, V_G, V_L, S_G, S_L) instead of only three parameters p, V, S in case of a single-phase flow;

(b) the second condition requires that the fluid velocity be equal to the sound velocity. An "extrapolation" of this requirement to two-phase flows implies that a representative average velocity be defined since the individual phase velocities are generally unequal.

The usual way to handle the problem of critical two-phase flow conditions is described in Section 6.3.

Since upstream of the critical section the fluid is subjected to strong acceleration, the evolution of the flow parameters to reach critical conditions experiences very steep gradients. Under these conditions, the thermal and mechanical nonequilibrium effects become very important and represent the main problems in critical two-phase flow calculations. Many critical flow experiments have shown the importance of these effects,[24,17] which are difficult to predict since the fluid particles are submitted to such strong transients that the use of nonequilibrium relationship obtained from experiments with established flow conditions is questionable.

6.2 Effect of Critical Flow on the Evolution of a LOCA

Critical flow mainly affects the "maximum possible flow." At the break location, critical flow conditions determine the maximum break mass flow rate and are thereby limiting both the loss of coolant from, and the depressurization rate of the primary cooling system. The depressurization rate is the governing quantity and, hence, of particular importance for the course of a LOCA since it determines the point in time when the various safety and emergency cooling systems (high pressure injection system, accumulators, low pressure injection system) start operating.

Changes in the critical flow rate cause considerable changes in the maximum cladding temperature. This has been shown in sensitivity studies performed by varying the break area in PWR LOCA prediction calculations.

Another, generally less considered effect of critical flow is that of the coolant flow distribution within the two branches of the broken loop upstream of the two ends of the break. According to the fundamental definition of critical flow, the critical flow rate is determined by the fluid conditions upstream of the critical section. In case of a double-ended cold leg break, the thermodynamic conditions of the fluid in the pipe section close to the reactor downcomer are different from those in the pipe section close to the coolant pump and steam generator. Therefore, the critical flow rates and, hence, the break flow rates are different at each end of the break. This difference in the break flow rate causes different coolant mass flow rates within the respective upstream piping and finally determines the location of the flow stagnation point, since this location is given by equal pressure drop along either flow path to the break. On the other hand, the location of the stagnation point strongly affects the core mass flow and, hence, the core cooling. It is finally for the last reason that a correct determination and discrimination of the critical or break mass flow is of primary importance.

Critical flow may, however, occur also at other locations within the primary loop system, especially within the impeller of the broken loop pump. Then again the coolant flow distribution within the broken loop and also within the unbroken loops may be changed and thereby also the location of the flow stagnation point, which finally affects again the cladding temperature. In this special case, the prediction of critical two-phase flow is particularly difficult since 2-D and 3-D effects must also be taken into account.

A further effect of critical flow is that of the two-phase jet forces on the loop and containment structure. The total jet or mass force is given by the momentum flux at the critical section, and will be distributed over the whole two-phase jet according to the shape of this jet. This spatial distribution as well as the total amount of the force have to be known in order to determine the potential for any damage to the loop or containment structures, especially within the first seconds of the accident when the momentum flux is the highest.

6.3 Models Used in the Codes

Analytical models. Only the characteristic features of the main models usually applied in codes will be briefly outlined here. For more details reference has to be made to pertinent literature e.g. [25,26].

(a) Homogeneous equilibrium model (HEM)[27]

In this model the two-phase mixture is assumed to be in thermal equilibrium ($T_G = T_L = T_{SAT}$) and in mechanical equilibrium ($V_G = V_L = V_M$). The description of the two-phase mixture in this case is equivalent to the description of a single-phase fluid where the fluid properties are obtained from appropriate equations of state. For the critical flow calculations the single-phase theory can be applied.

The HEM model is very often used in nodal codes like RELAP. It is a self-consistent model for predicting the critical pressure and critical mass flux for given upstream stagnation conditions which may be either saturation conditions for all vapor qualities, or subcooled water or superheated vapor conditions.

The critical mass flow rate obtained is too small due to the assumption of infinite rates of heat, mass and momentum exchange between the phases. This underprediction becomes significant for the flow through very short nozzles and tubes.[25]

(b) Homogeneous frozen model

In the "homogeneous frozen model" mechanical equilibrium is assumed, and, in contrast to the HEM model case, the evolution of the two-phase mixture close to the critical section is assumed to be so fast that practically no phase change can occur. This causes the introduction of the additional condition $dx/dp = 0$ into the homogeneous critical conditions.

This model is generally applied to the case of short pipes or nozzles where the fluid transit time is very small. It is useful for small vapor qualities in the bubble flow region, where the influence of different phase velocities is small compared to the effect of thermal nonequilibrium between the phases. In this region, the model yields an overprediction of the critical flow rate due to disregard of heat and mass transfer between the phases.[25]

(c) Moody and Fauske models[28,29,30]

Both models assume thermal equilibrium but allow mechanical nonequilibrium. To determine the slip ratio as the additional variable to be taken into account, a further critical flow condition has to be aded to the usual one $\partial G/\partial p = 0$, which yields the slip ratio in the critical section:

(1) in case of the Fauske model, maximizing the flow momentum results in the slip ratio correlation;

(2) in case of the Moody model, maximizing the energy results in the slip ratio correlation.

The Moody model has been widely applied in nodal codes. It predicts the critical pressure and critical flow rate as functions of the upstream stagnation conditions in the saturated region. The assumption of thermal equilibrium is partially compensated for by an overprediction of the slip velocity.[25]

(d) Henry and Henry-Fauske models[24,31]

These models assume thermal nonequilibrium and use only one critical condition, $(\partial G/\partial p)_{H_0} = 0$, where H_0 is the stagnation enthalpy. The thermal nonequilibrium is taken into account by introducing a highly empirical relation between the actual quality and the equilibrium quality into the flow rate correlation. This makes it suited for application in the region of low vapor quality as well as for the transition from initially subcooled liquid.[25]

(e) Models related to fluid flow models

In all models described so far, the critical flow modelling has been performed independently from the general fluid flow modelling. A consequence of this fact is that the assumptions used for calculating the fluid flow evolution may be completely different from those used for calculating the critical flow. Such a situation is given for instance when a critical flow model considering slip between the phases, like the Moody model, is used together with a homogeneous two-phase flow model. In reality, however, and on a rational basis, too, the choked flow properties are part of the general fluid flow properties. Reocreux[17] has shown that in all 1-D two-phase flow models which result in a system of first order partial differential equations, critical flow is obtained when the determinant of the matrix of $\partial/\partial z$ derivatives is zero, and when simultaneously a certain coherence condition is satisfied which yields the location of the critical section.

These conditions are rigorous, and by using them each flow model incorporates implicitly its own choked flow results. Critical flow is then obtained by describing the flow evolution down to the critical section where the previously outlined conditions are satisfied. By this procedure nonequilibrium effects may be taken into account by appropriate and constant transfer laws.

The most advanced model of this kind is the two-fluid model where thermal and mechanical nonequilibrium is described by the mass, energy and momentum transfer between the phases. Reocreux[17] has shown that in this model these transfer processes have to be expressed by algebraic terms mainly describing the fluid behavior in established flow conditions, and by derivative terms describing the dynamic interactions between the phases. These interactions directly affect the critical conditions since the respective terms are contained in the determinant.

In other types of modelling, some of these interactions are implicitly taken into account. In the drift flux model for instance they are obtained from the drift relation for the phase velocities.

Practical application of critical flow models. The first category of models (HEM, frozen model, Moody, Henry-Fauske) is mainly used in the first generation nodal codes of which RELAP is the most prominent representative. In these codes, the flow rate calculated by the general two-phase flow model (mostly the homogeneous equilibrium model) is compared at each time step with the flow rate obtained from the critical flow model. If the calculated flow rate is smaller than the critical flow rate, the flow is considered subcritical and the calculation is continued. In the opposite case the flow is declared critical, and the critical value is taken as the flow rate. Since the general two-phase flow model may be different from the critical flow model with respect to the assumptions used, some inconsistencies may appear, which are generally erased by a recommended "proper" nodalization close to the break location.

In these models, a "contraction" coefficient is often used which is claimed to take into account some "vena contracta" effect at the break location. However, since these "contraction" coefficients are not the same for all models, it appears more likely to consider them as a measure for adjustment purposes. For the Moody model for instance a coefficient of 0.6 to 0.8 is recommended to match experimental data, whereas the HEM model requires a coefficient of 1.0.

In the second category of models (related to fluid models), the critical condition (determinant = 0) represents a singularity of the model. Two procedures are used:

(a) the first one is similar to that used in nodal codes, i.e. when critical conditions are reached, the critical value of flow rates is used, and this imposed value figures as boundary condition substituting the usual condition on the exit pressure;

(b) the second procedure (used e.g. in TRAC code) keeps the boundary condition on the exit pressure. The choked flow behavior is obtained from the numerical behavior of the system. This procedure in some cases results in inconsistencies, particularly at the location of the critical section (e.g. in nozzles).

In the case of the second category of models a difficulty[21] may arise from the fact that the singular condition (determinant = 0) for the discretized system is different from the one for the analytical system before discretization. In such a case, special care has to be taken in the numerical analysis.

6.4 Discussion

The need for a detailed knowledge of critical flow mechanisms is sometimes put in question when its application to prediction calculations for a reactor cooling system is considered. Since sensitivity studies on the break area are considered necessary, it is felt that these studies might cover implicitly uncertainties on choked flow effects and particularly on effects of the contraction coefficient. This is only partially true for two reasons:

(a) the possible influence of different coolant flow distribution in the loop regions upstream of the two break ends;

(b) the real flow rate is needed for the purpose of comparison between system calculations and results of integral experiments.

Although the verification of the models on critical flow experiments[26] will not be discussed, it can be pointed out that in case of the first category of models, the calculation results generally agree well with experimental data. Some rules have been established to obtain "reasonable" results. For large break experiments it is generally agreed that

(a) the Moody model, applying a contraction coefficient of 0.6 to 0.8, can fit experimental data when the upstream conditions are two-phase flow;

(b) the combination of the Henry-Fauske model, followed by HEM at a transition quality of 0.02 and with a contraction coefficient of 1.0 can give reasonable results in case of subcooled and two-phase upstream conditions.

In the case of small break experiments the application of these models may give rise to some difficulties. For the LOFT test L3-1, or the Semiscale test S02-6 for example, the calculated depressurization rate is too small, and a contraction coefficient greater than one would be required to fit the experimental data.

Results from the second category of models are very dependent on the physical realism of the transfer laws; some sets of transfer laws yield reasonable results. But in those cases the critical flow cannot be dissociated from the fluid flow evolution model for which the evaluation of the transfer laws needs to be further improved.

Finally, it should be mentioned that choking of initially subcooled flows has recently been found highly dependent on the actual inception for the start of phase change.[39,41] In the case of critical flows through restrictions, properly describing the flashing inception superheat and location was entirely sufficient to enable accurate predictions of all data to be made within ±5% using single-phase theory. None of the models described above have any inclusion of inception nonequilibrium (or nucleation delay). Undoubtedly, then, some of the correlation factors they have included are of a compensatory nature due to the incorrect or incomplete physical descriptions implied by this, as well as other, omissions.

7. BLOWDOWN HEAT TRANSFER

7.1 General Remarks

Heat removal from an LWR core during a loss-of-coolant accident currently represents the most important task in the field of reactor safety. Predicting the rate of heat removal is an important and difficult problem, since heat flux and local fluid conditions are directly dependent on each other and need to be simultaneously predicted.

The most important quantity to be determined in the final analysis is the fuel cladding temperature which results from the balance between

(a) heat supplied from the fuel;

(b) heat removed from the fuel rod by the coolant flowing through the core.

During the blowdown phase, the heat supply from the fuel is initially constituted mainly by the energy redistribution leading to a temperature equalization, and is later due to the decay heat production. Hence, stored heat and decay heat production are two essential input quantities to the heat transfer problem during blowdown.

Heat removal from the fuel rods depends directly on the evolution of the coolant fluid conditions within the core, which, in turn, result from the overall system response to depressurization and blowdown. One important parameter in this context is the location of the flow stagnation point, which is not a fixed one, and is sensitive to several effects like flashing propagation, pumps behavior, choked flow distribution on each side of a double-ended break, etc. If this flow stagnation point is travelling through the core, fluid flow becomes nearly zero, heat transfer becomes negligible and, hence, cladding temperature increases.

A correct prediction of core heat transfer during blowdown requires therefore a correct prediction of fluid flow conditions in the core region during blowdown. The heat transfer mechanisms occurring in two-phase flow conditions are qualitatively well known. The various heat transfer regimes are represented in Fig. 3, where the surface heat flux is plotted as function of the difference between wall and fluid saturation temperature, and have been discussed in detail in Chapter 7. During the blowdown phase of a LOCA, all these heat transfer regimes may occur and persist for different time periods according to the fluid flow conditions which may cover a rather wide variation range including flow decrease and reversal, upwards and downwards flow, low and high quality flow at high and low pressure and mass flow rates. The various possible heat transfer regimes have been extensively reviewed and reported in [34] and [35]; they will be briefly described in the following subsection, together with the modelling methods.

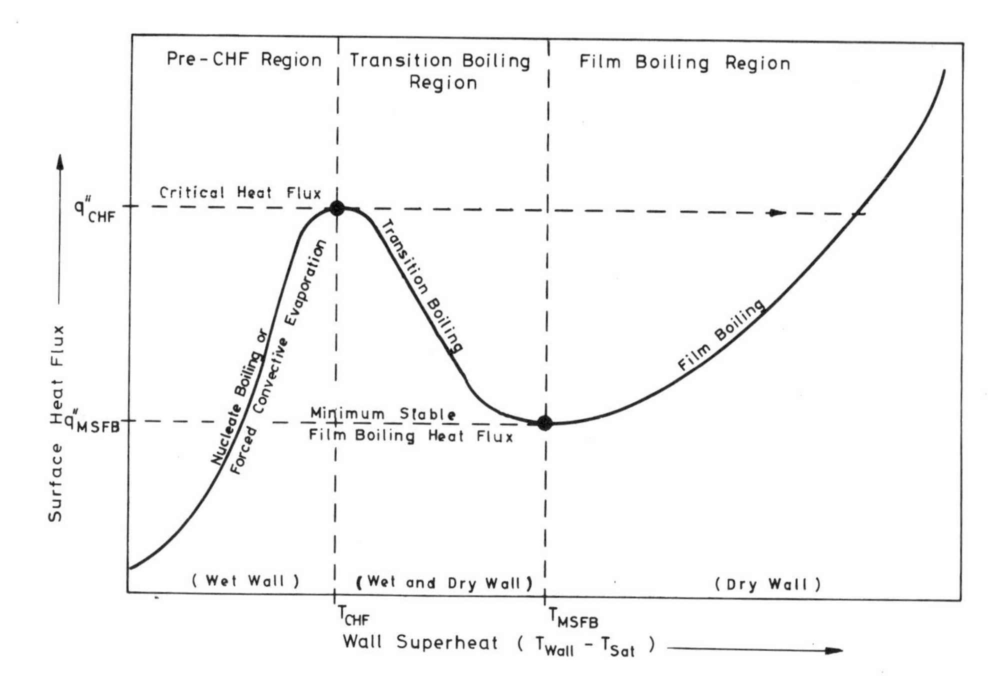

Figure 3. Generalized boiling curve.

7.2 Modelling

Modelling of heat transfer during blowdown requires, as mentioned before:

(a) the precise determination of the fluid flow conditions;

(b) the elaboration of a heat transfer package containing the respective correlations.

An extensive review and summary of the currently recommended heat transfer correlations to be used for the various heat transfer regimes is given in [34] and [35].

(a) Convective heat transfer to single-phase fluid

These heat transfer conditions prevail during the steady-state normal operation of PWRs, and they persist also during the initial part of the subcooled depressurization phase of a LOCA, hence, only a very short period of time.

For modelling the convective heat transfer to single-phase fluid, the Sieder-Tate correlation is applied which holds for fully developed turbulent flow (subcooled liquid or superheated vapor).

(b) Pre-CHF boiling heat transfer

In this case heat transfer may take place to subcooled liquid as well as to saturated two-phase mixture.

Subcooled nucleate boiling can occur as soon as the wall temperature exceeds the minimum temperature required for bubble nucleation. Such a situation exists during the normal steady-state PWR operation (in the hotter parts of the core) as well as during the subcooled depressurization phase of a LOCA, again a very short period. The subcooled nucleate boiling heat transfer can be calculated by the extended Chen correlation which compares satisfactorily with experimental data.

During the saturated depressurization phase, which immediately follows the subcooled depressurization phase and then persists up to the end of blowdown, almost all heat transfer mechanisms described below are occurring. One of these is the heat transfer to saturated two-phase mixtures, where essentially two mechanisms are possible:

(1) saturated nucleate boiling in the low quality region;

(2) forced convective evaporation in the high quality high flow region.

For determining the boiling heat transfer to saturated two-phase mixtures, the Chen correlation is recommended which covers both the low and high quality region, and which for low flows approaches automatically the well-known Forster-Zuber relation for pool boiling.

(c) Transient critical heat flux (CHF)

Critical heat flux refers to a reduced heat transfer condition which may result from two different mechanisms. One of them is "departure from nucleate boiling" (DNB) and occurs when the nucleation rate has become so high that the vapor rushing away from the wall surface prevents the liquid from reaching the wall. In pool boiling or low flow boiling conditions, the liquid inflow is limited by the Holmhotz instability occurring in unstable water-vapor counterflow conditions. In forced convective boiling conditions, DNB may be controlled by an agglomeration of bubble nucleation sites for high subcooling or by bubble clouding for slight subcooling and low quality.[35] The other mechanism resulting in CHF is "dryout" which usually occurs in the annular flow regime when the flow rate of the liquid film at the wall approaches zero. CHF under DNB conditions generally limits the steady-state power level of PWRs. This is because the steepness of the operating line on the heat flux-enthalpy plane due to large mass fluxes and high inlet subcoolings results in CHF at low quality or bulk subcooled conditions. For the BWR, however, the inlet subcooling is lower due to lower pressure and the mass fluxes are also lowered in order to achieve safe boiling conditions. The shallow operating line thus results in an intermediate or high quality CHF limit representative of annular flow dryout. In transient blowdown conditions, however, either mechanism may be encountered in either reactor.

For determining the transient critical heat flux, the Biasi correlation may be used for the high flow region. It can be easily evaluated on the basis of local conditions and has a wide range of applicability. Indeed, it is valid over the whole pressure range of interest and may be used for downflow and upflow. In the low flow region, the Zuber-Lienhard-Dhir correlation may indeed be used. This is essentially the Zuber relationship which is physically based on pool boiling CHF and derived for horizontal plate geometry. Lienhard and Dhir extended its use to vertical rod bundle geometry.

One important task for predicting CHF is the determination of the respective wall temperature at which CHF occurs. This wall temperature (T_{CHF}) may be calculated from the Thom correlation for subcooled nucleate boiling, or the Chen correlation, once the heat flux at CHF is known.

(d) Forced convective transition boiling

This heat transfer region is bounded by two firm limits, one of which is the CHF point forming the low temperature boundary and the othere is the minimum stable film boiling point forming the high temperature boundary. According to Fig. 3, the coordinates of these two points are (T_{CHF}, q''_{CHF}) and (T_{MSFB}, q''_{MSFB}). The transition boiling heat transfer mechanism is a combination of both nucleate boiling (wet wall) and film boiling (dry wall) heat transfer. Both mechanisms are unstable and may exist at any given location on a heating surface for any given fraction of time.

During transition boiling, and in the high quality region, most of the heat is transferred by droplet-wall interaction, and is dependent on droplet size, droplet impact velocity and angle, and on surface roughness. The same variables also influence the minimum (stable film boiling) heat flux q''_{MSFB} and the corresponding surface temperature T_{MSFB}.

For the transition boiling heat transfer a large number of correlations have been proposed, see [35], which may be divided into three groups:

(1) the correlations like those of Ramu, or Mattson or Tong, which contain boiling and convective components, and which are usually claimed to hold in both the transition boiling and the film boiling region;

(2) the phenomenological correlations like those of Iloeje, or Tong and Young, are based on a physical model for heat transfer in the transition boiling region;

(3) the empirical correlations like those of Ellion, or Berenson, or Mcdonough which all have a simple form and cannot be extrapolated outside the range of data from which they have been derived.

(e) Minimum stable film boiling point

Two different considerations have been applied so far for determining the minimum stable film boiling point. One involves the problem of determining the minimum wall temperature required to support stable film boiling, see [34], and the other consideration involves the mechanisms responsible for initiating transition boiling or partial rewetting, see [35].

According to the first approach, there are two mechanisms responsible for determining T_{MSFB} in pool boiling, namely the classic film instability mechanism and the phenomenon of homogeneous nucleation.

In the second approach, two theories are proposed. According to the first one, this minimum temperature is a thermodynamic property of the fluid, namely the maximum liquid temperature, hence, a function of pressure, and prevails in fast transients. According to the second theory, rewetting begins due to hydrodynamic instabilities depending on velocities, densities, and viscosities of both phases and on surface tension, which are the controlling factors in low flow and low pressure conditions.

The minimum stable film boiling temperature (T_{MSFB}) can be calculated from a correlation derived by Henry using the equation of the instantaneous contact temperature of two initially isothermal bodies together with the assumption that this temperature be equal to the homogeneous nucleation temperature of water.

The rewetting front propagation is primarily governed by axial heat conduction within the region close to the rewetting front, and is discussed in detail in Chapter 12.

(f) Film boiling or dry-wall heat transfer to two-phase mixture

Within this heat transfer region the heated surface is cooled by radiation, forced convection to the vapor and by interaction of the liquid and the heated surface. The vapor can be significantly superheated even in the presence of liquid droplets; therefore, thermal nonequilibrium can be an important effect in this region.

As in the case of early nonequilibrium flashing, post dryout has also been shown to be an initial value process in which the following three factors must be considered (Appendix A):

(1) inception of the nonequilibrium process taken at the dryout point;

(2) interfacial area density which was modelled as due to uniform spherical droplets;

(3) interfacial heat flux calculated by standard means applicable to convection to rigid spheres.

Heat transfer from the wall to the superheated vapor may be calculated assuming the vapor flows alone in the duct at its actual velocity. Recent application of this methodology has shown that the nonequilibrium quality may be predicted within 2-3% and wall temperatures for all data examined within 8-12°C RMS in spite of a three-times pressure extrapolation to 207 bars from the data base.

The term film boiling is actually a term used to represent several different situations where different flow regimes may exist. These three flow regimes are picture in Fig. 4 and include the following situations.

(1) In the liquid deficient regime, encountered usually at high void fractions (greater than 80%), a dispersed spray of droplets is

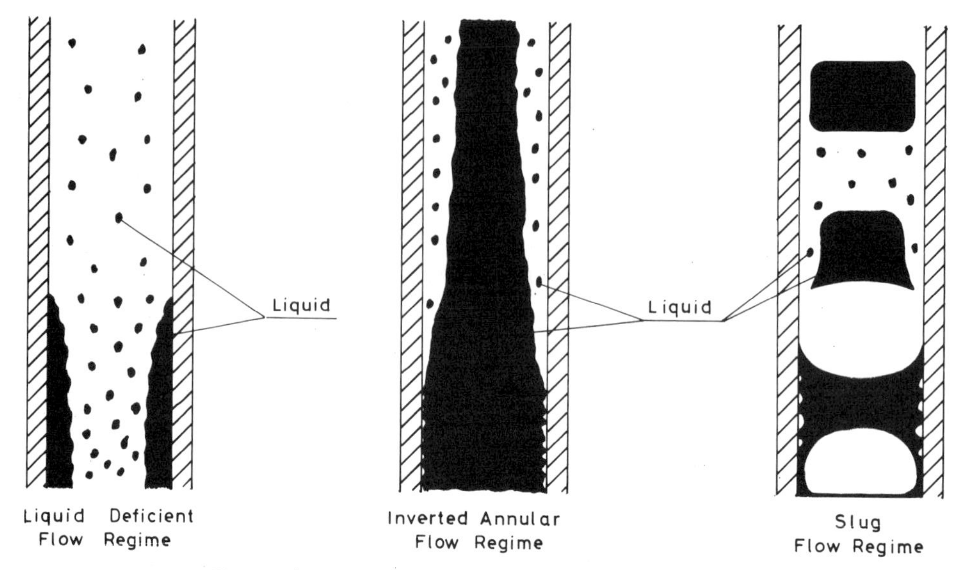

Figure 4. Flow regimes during film boiling.

entrained by the vapor. The vapor temperature depends on the wall-vapor and the vapor-droplet heat exchange. For heated surface temperatures less than the minimum temperature, some wetting of the wall and appreciable droplet vaporization may occur. For wall temperatures higher than the minimum temperatures, only dry collisions and, hence, little heat transfer to small droplets can take place.

(2) In the inverted annular flow regime, encountered usually at low void fractions (less than 30%), a continuous liquid core which may contain entrained vapor bubbles, is surrounded by a vapor annulus which may contain entrained droplets. Heat transfer occurs from the wall to the vapor and from the vapor to the wavy liquid core. The latter takes place by forced convective evaporation and is, hence, more efficient than the former single-phase convection.

(3) The slug flow film boiling regime represents a transition between the above two cases; it is usually occurring at low flows and at void fractions which are too high for inverted annular film boiling and too low for dispersed flow film boiling. Several theories have been proposed for the transition of inverted annular flow into slug flow.[35]

For the whole range of dry-wall heat transfer to two-phase mixture (film boiling), the following correlations have found acceptance.[38]

(1) the Groeneveld correlation, together with the Miropol'skij correlation holds for the high mass velocity range (greater than 800 kg/m^2 sec) and contains a two-phase Reynolds number term that induces the calculation of the vapor velocity using the homogeneous void fraction model;

(2) the modified Groeneveld correlation incorporating the drift flux void fraction is to be used for pressures above 14 bars;

(3) for pressure below 14 bars, a modified Dittus-Boelter correlation is proposed which, for high mass fluxes where the drift flux and the homogeneous models predict equal void fractions, reduces to the well-known Dougall-Rohsenow correlation for dispersed flow film boiling. An advantage of the Dittus-Boelter correlation is that it accounts for the influence of mass flux and flow direction on the heat transfer (through the void fraction);

(4) the modified Bromley film boiling equation is proposed for the low flow region.

A much more detailed distinction with respect to flow and heat transfer regimes has been proposed by Groeneveld and Gardiner,[35] by considering the inverted annular and slug flow film boiling region separately from the liquid deficient heat transfer region.

7.3 Discussion

There are still several shortcomings generally characterizing the state-of-the-art in the prediction of the various blowdown heat transfer mechanisms. To mention only the most important and common ones: there is the use of tube correlations to describe rod bundle behavior; the inadequate description of post-CHF heat transfer for very low flow rates, for which even tube data are lacking; downflow post-CHF heat transfer data are virtually nonexistent; nearly all post-CHF correlations are based on steady-state data while frequently these correlations are applied to transient conditions.

On the other hand, prediction methods based on steady-state tube data seem to overestimate the cladding temperature of fuel rods. This effect is mainly due to the fact that the grid spacers of fuel bundles which are not contained in the tubes, cause an improvement of downstream heat transfer which may prevent high vapor superheats and result in a reduction of the cladding temperature.

Heat transfer correlations established from experiments are generally lacking due to the fact that the experimental fluid conditions which could not be directly measured have been determined with homogeneous equilibrium models. Therefore, heat transfer in two-phase flow conditions is usually referred to the fluid saturation temperature, and globally to the mixture instead of distinguishing between the two phases. This is generally insufficient for the case of the two fluid models used in advanced codes. In addition, since the volumetric flow rate is an important physical parameter in the heat transfer mechanism, the use of a slip model instead of a homogeneous model may strongly affect the heat transfer calculation in the case of different volumetric flow rate of the two phases. Hence, for the sake of physical consistency, each heat transfer package should be used together with those flow models which have been employed for its determination. Consequently, the use of advanced flow models requires appropriate modifications and new assessment of the heat transfer models.

Finally, the physical consistency of some models must be examined in any given situation. For instance, for the case of film boiling post-dryout heat transfer, it is now known that this process is one of a relaxation phenomena with a forcing function. As such it is an initial value problem. It is easily shown that the rod temperatures obtained are dependent on the location and fluid state at the onset of CHF, as well as on the path followed by the post-CHF fluid. A local conditions correlation (such as any of those described above) can not properly account for these effects, and rather is an attempt to correlate not only the post-dryout heat transfer mechanisms but also the critical heat flux. It is no wonder that even the best of these produce RMS errors which may exceed 100°C in prediction of wall temperatures.

8. TWO-PHASE PUMP BEHAVIOR

8.1 Description

The behavior of the coolant circulation pumps is characterized mainly by the three parameters:

(a) the flow rate;

(b) the pressure difference between inlet and outlet, referred to as the pump head;

(c) the rotation speed.

During the blowdown phase of a LOCA, these parameters may take a wide range of possible negative and positive values.

In the context of loss-of-coolant accident considerations, the behavior of the broken loop circulation pump is of primary interest. Generally it is assumed that at the beginning of the accident, there is a loss of electrical power supply so that the pumps are rotating only on their inertia. In the case of a cold-leg break, the initial positive coolant mass flow in the pump of the broken loop will increase due to the break flow, and thereby the rotation speed of the pump impeller will be increased too. It is essentially dependent on the pump design whether under these conditions the impeller will eat into the housing and hence be blocked, or whether the impeller may reach a sufficient overspeed to explode.

In case of a hot-leg break, the broken loop pump impeller will ultimately experience flow reversal and may successively encounter one or more of the following operation states:

(a) negative flow at positve impeller speed causing the impeller speed to decrease;

(b) an impeller blockage at zero speed depending upon the pump design;

(c) additional system faults in some cases or simply negative impeller speeds with negative flow depending on system design.

Hence, during a loss of coolant accident, the broken loop pump may operate in all three quadrants of its operational-characteristics diagram, and furthermore also under two-phase flow conditions. An essential task in the framework of LOCA analysis is, therefore, the prediction of the pump behavior under such quite abnormal operation conditions. In addition, as already mentioned in Section 6, critical flow can occur within the pump especially in case of speed increase. This can occur due to the high tangential velocities which might be reached at the impeller outlet.

8.2 Effect of Pump Behavior on the Course of a LOCA

The importance of the location of the flow stagnation point on the course of a LOCA has already been pointed out before. This location depends on the flow pressure drop distribution within the primary cooling loops. Since the coolant pumps represent an important singularity in this pressure drop distribution, their effect on the LOCA course is evident. This holds particularly for the pump within the broken loop. A correct prediction of the core cooling during the blowdown period of a LOCA depends strongly on a correct prediction of the flow distribution and thus requires therefore a good knowledge of the pump behavior.

8.3 Modelling

The important models usually applied are point models yielding pump head and torque by applying two-phase multipliers to the single-phase pump characteristics. The impeller rotation speed is obtained by solving the mechanical equation of motion for the impeller, including the applied torques.

These models need both the single-phase pump characteristics in all relevant quadrants of the pump operation diagram, and the two-phase multipliers from pump experiments under two-phase flow conditions. Furthermore, they depend strongly on the pump geometry. Therefore, even if pump experiments are performed at different scales, the extrapolation of the results to full size pumps remains questionable. This is due to the fact that similarity factors for small scale pump experiments are difficult to be established depending also on which phenomenon shall be kept unchanged. For example, the pump speed cannot be chosen such that the fluid velocities at the impeller inlet and outlet are the same for the small scale pump as for the full size pump. On the other hand, for investigating critical flow in pumps, the pump rotation speed cannot be kept unchanged with respect to the full size pump speed, because in that case the small size pump would never experience critical flow due to the smaller fluid velocities. If the pump speed is increased to allow critical flow, distortions will occur at the impeller inlet and with respect to the fluid transit time.

With a view to solving this extrapolation problem, attempts have been made towards more sophisticated models. One such attempt[33] results in an axial model calculating the transient evolution of a fluid particle along its path through the pump. In this case, the pump is represented by a pipe with variable cross section. In regions of different cross section, sudden flow direction changes of the fluid are taken into account. The flow momentum transfer in the impeller is modelled by momentum transfer at the wall of the corresponding part of the pipe. The flow evolution in this idealized representation of the pump is calculated by an axial two-phase flow model allowing one to consider both mechanical and thermal nonequilibrium.

Such a model developed for single-phase flow has shown satisfactory results for the extrapolation between pumps of different geometry and different scale. In two-phase flow conditions, encouraging results have been obtained for choked flow prediction, but further improvement and verification is still required.

8.4 Discussion

Among the variety of problems arising from the pump behavior prediction under LOCA or transient two-phase flow conditions there is the particular problem of scaling and extrapolation. The real phenomena occurring in a pump under those conditions are 2-D and 3-D and involve the interaction with rotating parts. Since the present state of knowledge does not allow one to perform such 2-D or 3-D calculations, all problems of modelling must include scaling and extrapolation difficulties as well.

In the point models, the global approach using two-phase multipliers hides all kinds of effects especially geometrical effects. In the extrapolation procedure these models are applicable only to one type of pump, because the two-phase multipliers vary with pump type. For an axial model, the geometrical dimensions are computational input parameters, and therefore it is expected that geometrical differences may appropriately be taken into account. Comparison of the models against different types of small scale experiments will then increase the confidence in extrapolation. Nevertheless, this concept can not be expected to completely solve the problem, since the 1-D approximation is a rather severe one with regard to real 2-D and 3-D phenomena and behavior.

In view of the large influence of the pump behavior on the course of a LOCA, pump modelling still needs important improvements.

9. SUMMARY

Among the large variety of thermohydraulic phenomena occurring during a loss-of-coolant accident (LOCA) in light-water reactors (LWR), the most important ones have been dealt with which significantly affect or control the course of such a LOCA.

The aim in this chapter was to supply an overview of the state-of-the-art of both the present knowledge of the various physical mechanisms, and the problems arising from the modelling of those mechanisms.

The following general conclusions may be drawn:

(a) In spite of the very extensive investigations already performed, more experimental and analytical effort is still needed for the necessary improvement of both a thorough understanding and a sufficiently detailed description of the various phenomena and mechanisms. This effort is also needed to generate the basis for a better assessment of existing correlations and modelling methods. This need holds particularly for the heat transfer mechanisms in the transition and subcooled film boiling regime where virtually no data are available for low pressure, low flow and low quality conditions. Furthermore, the existing data are mostly obtained from simple geometry experiments (tubes, annuli) and steady-state conditions.

(b) More space is dedicated to the analytical description of the thermal nonequilibrium phenomenon in flashing flow and in post-dryout heat transfer, by concentrating especially on the nonequilibrium vapor generation rates.

(c) Attention has to be drawn to the need of combining both heat transfer models and critical two-phase flow models in each case only with the appropriate and physically compatible two-phase flow models.

(d) A definite trend can be stated which is away from evaluation models (EM), which yield conservative results, towards best estimate models (BE) leading to more realistic results.

APPENDIX A. NONEQUILIBRIUM CONSIDERATIONS

The difficulties with the early stages of the blowdown phase stem to a great extent from our present inability to accurately specify the nonequilibrium aspects of the flow field, both single-phase and two-phase. In the single phase region, large liquid superheats can develop caused by the decompression that the hot liquid undergoes due to both system depressurization and convective effects. These superheats control to a large measure the vaporization rates once flashing begins. Both mechanical and thermal nonequilibrium effects have been recognized to be important in the two-phase regions. Mechanical nonequilibrium results from differences between the velocity of the two phases requiring us to take a nonhomogeneous viewpoint of the flow field. The complicacies which arise in the field equations and boundary and interfacial conditions are discussed in Chapter 8. The problem of predicting the difference in phase velocities, discussed in Chapter 6, has a strong interactive effect in determining the pressure gradients and hence the critical discharge rates. Similarly, the pressure gradients determine the Lagrangian decompression rates of the fluid which governs the rate of departure from thermal equilibrium, and thus have a strong interactive effect on the nonequilibrium vapor generation rates. Only the latter shall be discussed herein.

In addition to the nonequilibrium conditions encountered during the early flashing stages of blowdown, the conditions encountered in the reactor core once the critical heat flux condition has been exceeded are controlled by strong nonequilibrium behavior characterized by a superheated vapor and a saturated liquid. While on the surface this situation appears substantially different from the flashing case, it will be shown that both are governed by the same inhomogeneous relaxation equation. The differences between the two are largely due to the predominant mode of heat transfer controlling the interphase exchange, and the interactions with the surrounding solid surface.

It is the purpose of this appendix to summarize recent activity undertaken to accurately describe the case of thermal nonequilibrium of interest to nuclear safety specialists. The general developments shall first be described followed by specifics relating to first the flashing case then the post-dryout situation. A summary section will recapitulate the major conclusions of this appendix.

A.1 General Development

The general way in which the problem of nonequilibrium phase change enters into questions of reactor safety is through the need to provide constitutive relations describing the actual rates of liquid-vapor mass exchange during hypothetical accident sequences. For example, the continuity equation for phase-k may be written for the constant area case as

$$\frac{\partial}{\partial t}[\alpha_k\rho_k] + \hat{\nabla}\cdot[\alpha_k\rho_k\hat{v}_k] = \Gamma_k \tag{A.1}$$

where α_k, ρ_k, and $\hat{v}_k$ are volume fraction, density, and velocity respectively for phase-k. Equation (A.1) shows that mass conservation for phase-k depends heavily on the formulation used for the phase-k volumetric rate of vapor generation (mass per unit time per unit volume) Γ_k. Under conditions approximating thermodynamic equilibrium, the equilibrium path generation rates, Γ_{ek}, are easily specified. However, under more complex situations, such as when cold water flows into the lower plenum of a water reactor and contacts superheated steam, the determination of the mass transfer rates may be exceedingly difficult. It has only been under the most simple circumstances of quasi-steady state that any reasonable progress has been made to date in the following cases: subcooled boiling, post dryout, and to some extent flashing.

In all cases of interest to nuclear safety, we are concerned with situations where the thermal nonequilibrium between phases is important whereas each phase taken locally by itself can be considered to be in thermodynamic equilibrium. Thus, thermodynamic properties can always be specified in terms of one or two independent properties (usually pressure and temperature or saturation pressure or temperature), and liquid and vapor are at their mutual saturation condition along surfaces of contact. In addition, most conditions occur with one phase at saturation, and the other either subcooled or superheated, net phase change occurring in the direction the bulk temperature difference would indicate. Subcooled boiling appears to be the predominant exception. The nonequilibrium encountered is thus associated with the limitations in thermal transfer rates and the inability of the fluid to follow successively between equilibrium states. The rate of change of equilibrium states, when compared with the relaxation processes involved, therefore, sets up a relaxation potential for the nonequilibrium processes. In such circumstances, energy transfer occurs generally due to thermal differences which may be represented by the bulk phase-k temperature differences. Such cases will be the only ones considered herein.

A general summary of nonequilibrium considerations for water reactor systems was presented by Jones and Saha[36] during the 1977 Winter ASME Meetings Symposium on the Thermal and Hydraulic Aspects of Nuclear Reactor Safety. Subsequent work has emphasized flashing and condensation. From a physical viewpoint the evaporative mass flux at an interface results from heat transferred to the interface $\hat{q}_k''\cdot\hat{n}$, stretching of the interface resulting in energy storage $\sigma\hat{\nabla}_s\cdot\hat{u}_s$, and viscous heating due to interfacial shear $(\bar{\bar{\tau}}\cdot\hat{n})\,(\hat{u}_k-\hat{u}_s)$, and may be expressed by

$$G_{\ell v} = \frac{\sum_{k=\ell,v} \hat{q}_k''\cdot\hat{n}_k + \sigma\hat{\nabla}_s\cdot\hat{u}_s - \sum_{k=\ell,v} (\bar{\bar{\tau}}\cdot\hat{n}_k)\cdot(\hat{u}_k-\hat{u}_s)}{\Delta i_{fg} + \tfrac{1}{2}\hat{u}_r\cdot\sum_{k=\ell,v} (\hat{u}_k-\hat{u}_s)} \tag{A.2}$$

where $\hat{u}_s$ is the velocity of the interfacial surface and $\hat{\nabla}_s$ the surface gradient operator. For nuclear systems the surface tension and viscous effects are generally negligible. In addition, the mechanical energy exchange due to relative velocities $\hat{u}_r$ are less than 5% of the latent heat Δi_{fg}, even at critical flow rates, so that a simplified expression becomes

$$G_{\ell v} = \frac{1}{\Delta i_{fg}} \sum_{k=\ell,v} \hat{q}_k'' \cdot \hat{n}_k \ . \tag{A.3}$$

We thus see that at an interface separating a liquid and a vapor phase, evaporation can only occur if there is a net quantity of energy supplied to that interface. From a mechanistic viewpoint, then, our thermodynamic concepts indicate that evaporation can occur only when there is a temperature difference between the two phases--i.e., only nonequilibrium phase change is possible.

The evaporative mass flux may be integrated over all interfacial area A_i in an infinitesimal volume of length dz in the streamwise direction. An expression for Γ_v is thus obtained as

$$\Gamma_v = \frac{1}{A\Delta i_{fg}} \int_{\xi_i} \sum_{k=\ell,v} \hat{q}_k'' \cdot \hat{n}_k \frac{dA}{dz} \ . \tag{A.4}$$

In this case, ξ_i is the whole interfacial perimeter determined by the intersection of the plane of area A and the liquid vapor interfaces. The interfacial area density dA/dz is thus seen to be equally important in determining the evaporation rates. More shall be discussed regarding this point in subsequent sections. From a mechanistic viewpoint, however, the difficulty in determining Γ_v stems directly from the difficulties involved in predicting interfacial heat flux and area densities. To date this has been accomplished only in a semiempirical manner and then only under quasi-static conditions. Finally, Equation (A.4) shows us that for a given value of Γ_v, since the heat flux is directly proportional to the phase temperature difference, only as the interfacial area density gets very large can the temperatures approach each other--i.e., can the system approach thermal equilibrium.

Quasi-Static Nonequilibrium Relaxation. If the temporal effects can be ignored relative to the spatial gradient effects, Equation (A.1) may be written as

$$\hat{\nabla}\cdot(\alpha_p\rho_k\hat{v}_k) = \hat{\nabla}\cdot(\hat{G}_k) = \Gamma_k. \tag{A.5}$$

The term $\alpha_k\rho_k\hat{v}_k = \hat{G}_k$, the phase-k mass flux, which has a magnitude of Gx_k, x being the flowing phase-k quality defined by $x_k = G_k/G$. If the liquid and vapor flow in the same z-direction then (A.5) may be written as

$$\frac{d}{dz}(Gx_k) = \Gamma_k \tag{A.6}$$

for the quasi-one dimensional situation. Since G is constant in the quasi-static case (constant area) we may write (A.6) for the nonequilibrium case

$$\frac{dx_k}{dz} = \frac{\Gamma_k}{G} \tag{A.7}$$

and the limiting equilibrium case

$$\frac{dx_{ek}}{dz} = \frac{\Gamma_{ek}}{G}, \qquad x_k \to x_{ek}\ . \tag{A.8}$$

Since we are interested in the departure from equilibrium, $Q_k = x_{ek} - x_k$, we may utilize (A.7) and (A.8) to obtain in dimensionless form

i.e.;

$$\frac{dQ_k}{dx_{ek}} + N_{rk}Q_k = 1 \tag{A.9}$$

which is very similar in form to the inhomogeneous relaxation equation of Debye[37,38] for dielectric relaxation. Note that we have recognized that the nonequilibrium phase-k generation rate Γ_k may be expressed in terms of the equilibrium value as

$$\Gamma_k = \Gamma_{ek}N_{rk}Q_k \tag{A.10}$$

In this form, Equation (A.9) shows that the quantity Q_k, the nonequilibrium, behaves in a relaxation manner with a forcing function. The inhomogeneity in (A.9) forces the nonequilibrium to increase whereas the relaxation number N_{rk} determines the rate at which the nonequilibrium relaxes back to zero (x approaches x_e). The solution to (A.9) for constant N_{rk} and for initial conditions x_{eko} and Q_{ko} is

$$(Q_k - \frac{1}{N_{rk}}) = (Q_{ko} - \frac{1}{N_{rk}})\ e^{-N_{rk}(x_{ek}-x_{eko})} \qquad (A.11)$$

showing the exponential nature expected in relaxation processes.

The utilization of Equation (A.9) and the relaxation concept rests on our ability to determine the relaxation parameter N_{rk}. This determination has advanced much more in post-dryout work than in flashing. In either case, however, the process would be similar. Theoretical calculations of N_{rk} could be compared with those experimentally determined and a correlation thus obtained. This has not yet been done for flashing flows.

A.2 Flashing Flows

Nonequilibrium methods for flashing flows were summarized in Reference [39] and shall not be repeated herein. Only a summary of the development and recent work shall be presented.

Direct integration of (A.5) in the one dimensional case yields for the vapor phase

$$\alpha\rho_v v_v = \int_{z_o}^{z} \Gamma_v(\xi)d\xi \ . \qquad (A.12)$$

z_o is the location of flashing inception where $\alpha_o \sim 0$, and which together with (A.4) shows the following parameters must be adequately defined:

a) point of flashing inception
b) interfacial area density
c) net interfacial heat flux.

These shall be discussed in order.

Flashing Inception

This work has been summarized by Jones[40] for straight pipes and by Abuaf, Jones, and Wu[41] for restricted flows. The superheat at flashing inception may be expressed through a combination of static[42] and convective[40] decompressive effects while accounting for the effects of nozzle convergence[41] on turbulent pressure fluctuations.

The flashing inception superheat has been correlated in terms of the underpressure at flashing inception Δp_{Fi}. This underpressure is the difference in pressure between the saturation value and the inception value, $\Delta p_{Fi} \equiv p_{sat} - p_{Fi}$, given by

$$\Delta p^+_{Fi} \equiv \frac{\Delta p_{Fi}}{\Delta p^\circ_{Fio}} = \sqrt{1+13.25(\Sigma'_o+\Delta\Sigma')^{0.8}} - \frac{27}{2}\left(\frac{\overline{u'^2}}{U^2}\right)\Phi(\Delta\Sigma')^{2/3}. \qquad (A.13)$$

Δp°_{Fio} is the limiting <u>static</u> expansion undershoot at vanishing expansion rates ($\Delta\Sigma'=0$, $\Sigma'_o \to 0$), where $\Sigma'_o = \partial p/\partial t$ and $\Delta\Sigma' = v\partial p/\partial z$. In this correlation the turbulent fluctuation intensity is given by

$$\sqrt{\frac{\overline{u'^2}}{U^2}} = 0.072 \frac{A}{A_o} \qquad A \leq A_o \tag{A.14}$$

whereas the parameter Φ is given as

$$\Phi = \frac{\psi}{\Delta p_{Fio}^{\circ}} \tag{A.15}$$

with

$$\psi = \begin{cases} (2d\sqrt{\rho_\ell}/f)^{2/3}, \text{ friction dominated} \\ (A\sqrt{\rho_\ell}/[dA/dz])^{2/3}, \text{ acceleration dominated.} \end{cases} \tag{A.16}$$

The limiting undershoot is specified as[7]

$$\Delta p_{Fio}^{\circ} = 0.258 \frac{\sigma^{3/2}}{\sqrt{kT_c}} \frac{T_r^{13.76}}{(1-\frac{\rho_g}{\rho_f})} \tag{A.17}$$

in terms of the critical and reduced initial fluid temperatures, T_c and T_r, the surface tension σ, Boltzman's constant k, and the densities of the saturated liquid and vapor respectively, ρ_f and ρ_g. This Equation (A.13) has been tested over the range of temperatures $0.515 \leq T_r \leq 0.935$ and static decompression rates $0.004 \leq \Sigma_o' \leq 1.803$ Matm/sec,[42] for water only. The turbulence effects have only been tested against two low pressure steady flow systems in straight pipes ($\Sigma_o'=0$) and several nozzles where $A/A_o \leq 0.25$. It is known that the correlation predicts static values, Δp_{Fio}°, twice those observed at very low expansion rates, but in these cases the superheats observed are generally less than 2-4°C. The expected error is $\sim$10% of what Δp_{Fi} would be with $\Delta\Sigma'=0$ within the range specified above--i.e., at superheats up to values in excess of 50°C.

The two cases of straight pipes and of area contractions are discussed below. In the general sense the behavior is seen to be described as follows.

In a physical plane of Δp_{Fi} versus Σ', the static flashing inception ($\Delta\Sigma'=0$) would appear as a family of curves beginning at constant values, Δp_{Fio}°, with vanishingly small expansion rates. Then, Δp_{Fi} increases slowly assymptotic to $\Sigma_o'^{0.4}$ at high expansion rates. These are shown as the lighter lines at fixed temperature in Figure A.1. For practical purposes, Δp_{Fio}° is limited between $\sim$0.2 bar at 100°C and $\sim$9.5 bar at $\sim$300°C after which it decreases again to zero due to vanishing surface tension as the critical point is approached. Less than 15 percent increase in inception undershoot is noticed for expansion rates less than 10 kbar/s while almost a tenfold increase is predicted at 10 Mbar/s expansion rates (beyond the correlation range).

When the convective expansion rate effects are taken into account, four parameters must be considered: Δp_{Fio}°, Σ_o', ψ, and, $\Delta\Sigma'$. In this case a curve starting at the static coordinate values of (Δp_{Fio}, Σ_o') departs by increasingly larger amounts from the static curve for increasing <u>total</u> expansion rates depending on the parameter ψ. These are the two typical families of lines shown as the dark curves in Figure A.1 departing the 175°C static decompression curve at values of 0.001 and 0.1 Mbar/s for Σ_o'. Whether this curve first increases, or decreases monotonically, with increasing Σ' depends on the initial

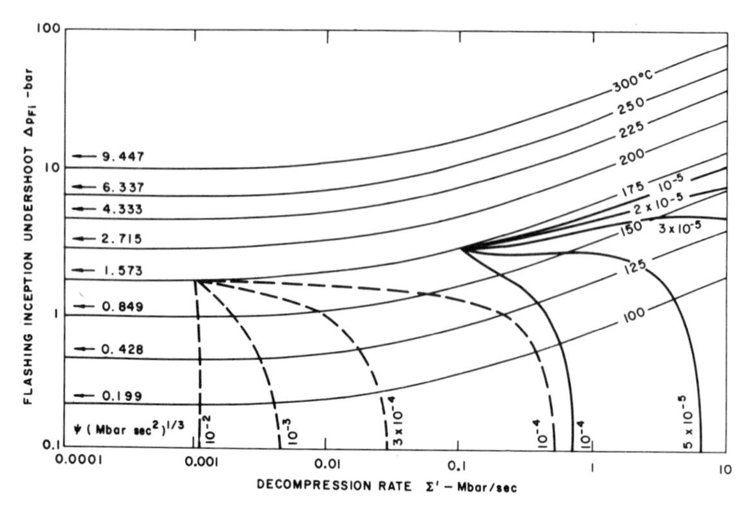

Figure A.1. Physical combination of the static flashing inception correlation of Alamgir and Lienhard[42] with the flowing turbulence effects described in light lines: static decompression effects only. Dark lines: turbulence effects included. (BNL Neg. 4-367-80)

slope at the departure point. If two given rates of static decompression are sufficiently small and dominated by large values of convective expansion rates such that a given total value of Σ' is large with respect to the two static values, curves of the same value of ψ will tend to coalesce. This is seen for the two curves in Figure A.1 where $\psi=10^{-4}$ (Mbar-s^2)$^{1/3}$. Indeed, for Σ' greater than approximately 0.1, the two different curves for $\psi=5\times10^{-5}$ are virtually indistinguishable.

The dimensionless plane suggested by Equation (A.13) reduces the complexity by one dimension while still keeping the general pictorial behavior of the phenomenon unchanged. By plotting the dimensionless undershoot, Δp_{Fi}^+, as a function of the convective expansion rate, $\Delta\Sigma'$, (Figure A.2), the essence of the physical behavior is maintained while providing a much simplified picture. It is noted that the expected range of ψ for friction-caused expansions is 10^{-4} to 10^{-2} (Mbar-s^2)$^{1/3}$ while those of accelerated cases is one to two orders of magnitude smaller for the various nozzles encountered in the critical flow literature. The normal range for Φ is then seen to be a six-order range between 0.1 and 10^5 (s/Mbar)$^{2/3}$. In the case of the Reference [43] experiments, ψ is near 0.004 (Mbar-s^2)$^{1/3}$ while, based on the extrapolated values of Δp_{Fio}, Φ is near 20,000 (s/Mbar)$^{2/3}$.

As seen in Figure A.2, at a given value of Σ_o' the curves become assymptotic at small Σ' to the ratio of the static value of undershoot to that for vanishing expansion rates at a given temperature. For increasing convective expansions, the undershoot may or may not first increase, depending on Σ_o' and Φ, but then decreases rapidly to zero, indicating the disappearance of any significant amounts of nonequilibrium.

Straight Pipes. In straight pipes, the flashing inception criteria have been tested on the data of Reocreux[43] and Seynhaeve, et al.[44] These data are shown in Figure A.3. The quantity Δp_{Fio} is the product of $\Delta p_{Fio}{}^\circ$ and the

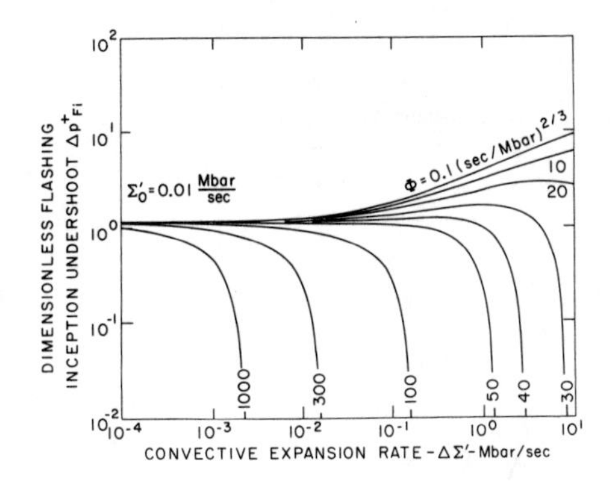

(a) 0.01 Mbar/s
(BNL Neg. 4-368-80)

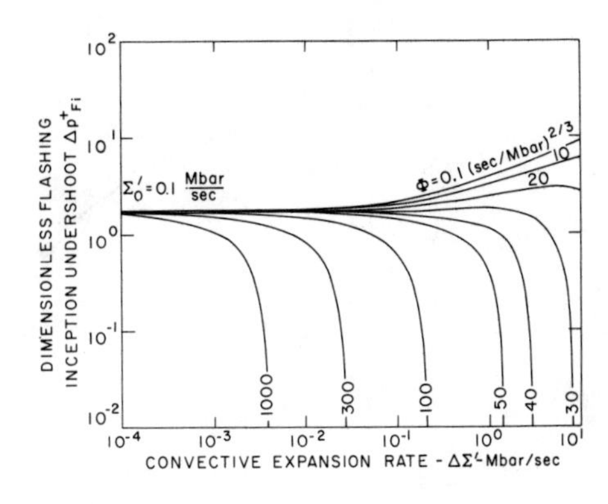

(b) 0.1 Mbar/s
(BNL Neg. 4-370-80)

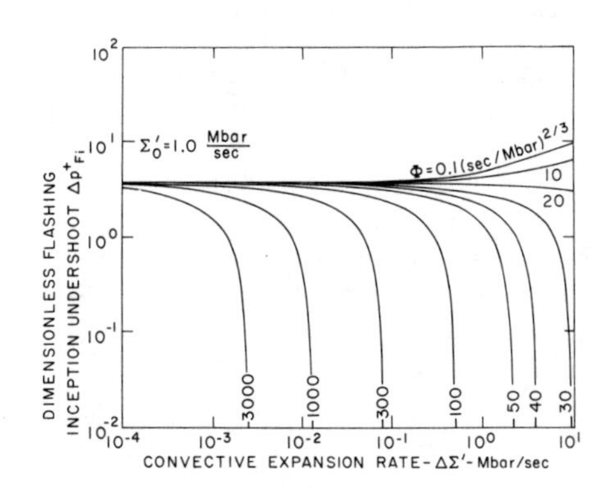

(c) 1.0 Mbar/s
(BNL Neg. 4-369-80)

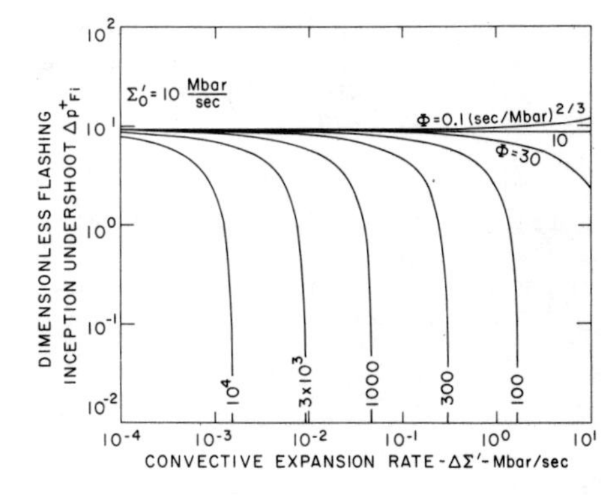

(d) 10 Mbar/s
(BNL Neg. 4-371-80)

Figure A.2. Dimensionless flashing inception correlation combining Alamgir and Lienhard[42] with the turbulence effects developed in [40].

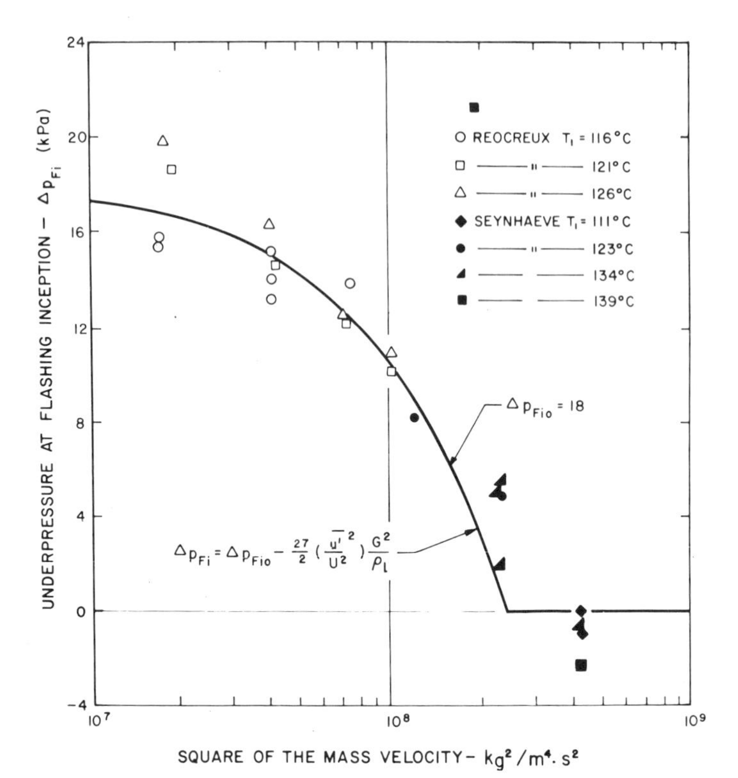

Figure A.3. Comparison of the flashing inception data of Reocreux[42] and of Seynhaeve, et al.[44] with the theory developed herein using the approximate static flashing over-expansion value of 18 kPa for the computation. (BNL Neg. 3-238-79)

radical shown in A.13, this product representing the correlation of Alamgir and Lienhard[42] if $\Delta\Sigma'=0$. Note that a limiting value of Equation (A.13) is vanishing superheat represented by the horizontal line to the right of $G^2 \sim 2.3 \times 10^8$ $kg^2/m^4\text{-sec}^2$. Thus, the correlation becomes

$$\Delta p_{Fi}^* \equiv \frac{\Delta p_{Fi}}{\Delta p_{Fio}} = \text{Max} \begin{cases} 0 \\ 1 - 27(\frac{\overline{u}'^2}{U^2})\text{Fi} \end{cases} \tag{A.18}$$

where Fi is the flashing index given by

$$\text{Fi} = \frac{G^2}{2\rho_\ell \Delta p_{Fio}} \tag{A.19}$$

and

$$\Delta p_{Fio} = \Delta p_{Fio}^\circ \sqrt{1+13.25\Sigma_o'^{0.8}} \ . \tag{A.20}$$

The flashing index is the reciprocal of the well known cavitation index used to predict cavitation onset. The correlation seems well confirmed by these few data available.

Nozzles, Orifices, and Other Contractions. In nozzles, the effect of the contraction seems to be to suppress turbulence substantially.[45] The expression given by (A.14) is only an approximation when compared against Uberoi's data.[45] It has no significance in cases where the turbulence is not fully developed at the start of the convergence but may apply in the form

$$\sqrt{\frac{\overline{u'^2}}{U^2}} = \sqrt{\left(\frac{\overline{u'^2}}{U^2}\right)_o} \quad \frac{A}{A_o} \qquad \text{(A.21)}$$

if the inlet or initial turbulent intensity can be specified a priori or otherwise determined. While Equation (A.14) or alternatively (A.21) has been tested for the limiting cases of $A=A_o$ and $A^2/A_o^2<0.0625$, it has not been tested in the range between where convergence rates may be extremely slight. Note that in the latter case the turbulence effects have been suppressed by almost 94%, leaving the expansion rate dominated superheats (first term only in Equation A.13) depressed by only a small fraction of a degree centigrade for most low expansion rate data. Even for the data of Brown[46] the remaining superheat out of a total of $\sim$50°C would still be 47-49°C. The scatter in the data would mask the trends which may otherwise be predicted on the basis of turbulent considerations. Thus, Equation (A.14) or (A.21) serve only to link in a convenient way two limiting values, the region between which remains largely untested.

The utilization of (A.13) toward predicting the onset of flashing inception in nozzles is shown in Figure A.4 for two nearly identical temperatures. In one case the initial inlet subcooling was ($\sim$50°C) (Figure A.4a) while in the other case the initial subcooling was negligible. By using the local geometry, the local expansion rates can be easily calculated if the flow rates are known. By plotting the local pressure as measured on $p_{sat}-p$ vs Σ' coordinates, these local values may be compared with Equation (A.13), in this case with the second term having the turbulence effects neglected. What is observed is that in all cases examined flashing inception occurs virtually at the throat.

The implication of the preceeding remarks is obvious. Critical flashing flows in nozzles occurs with flashing onset at the throat at a well defined pressure, with single phase flow upstream. Such cases can usually be calculated utilizing single phase concepts, and such is indeed the case here. Figure A.5 summarizes the data of Brown,[46] Zimmer, et al.,[47] Simoneau[48] (liquid nitrogen), and Sozzi and Sutherland.[49] It should be noted that all these data are predicted with a discharge coefficient of 0.94±0.04 in good agreement with the expected values for single phase nozzles. Each nozzle separately can be correlated with a slightly different discharge coefficient usually with a smaller error except those of Reference [49] which may in some cases have had inlet voids in nozzle 1.

The only other data examined were those of Powell[50] which have defied analysis to date. These data, with subcooled inlet conditions, covered a wide range of inlet pressures (2800-17000 kPa) and inlet temperatures (203-288°C). The comparison of the measured critical mass fluxes at the throat with the calculated values (solid lines) are presented in Figure A.6 for various inlet pressures and inlet temperatures. The calculated critical mass fluxes are generally within ±5 percent of the experimentally measured ones for the entire range of inlet pressures and temperatures reported.

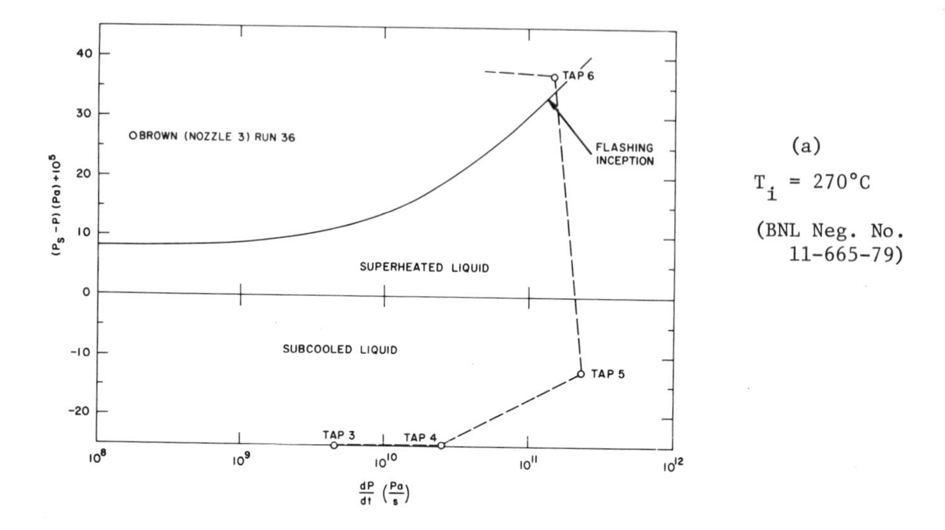

(a)

$T_i = 270°C$

(BNL Neg. No. 11-665-79)

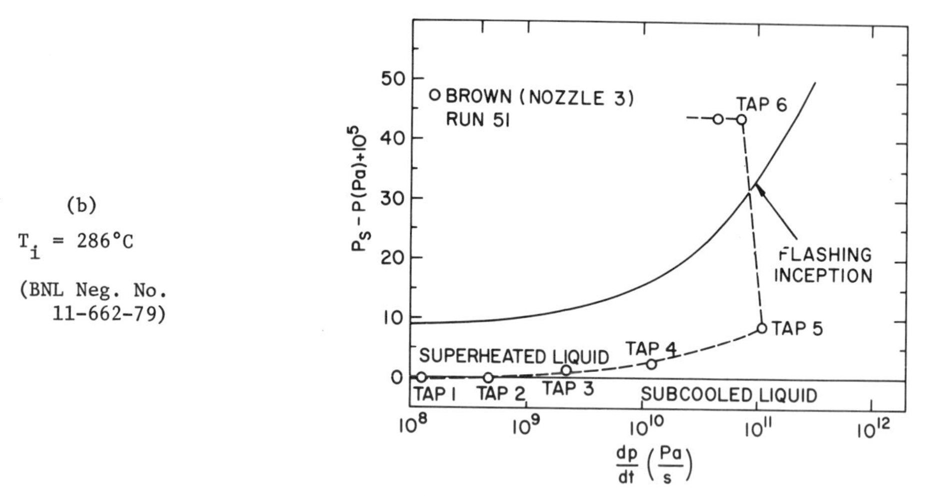

(b)

$T_i = 286°C$

(BNL Neg. No. 11-662-79)

Figure A.4. Comparison of the flashing inception predicted by Alamgir and Lienhard[42] (solid line) with the locus of the liquid depressurization history (circles connected by dashed line) in Brown's nozzle.[46]

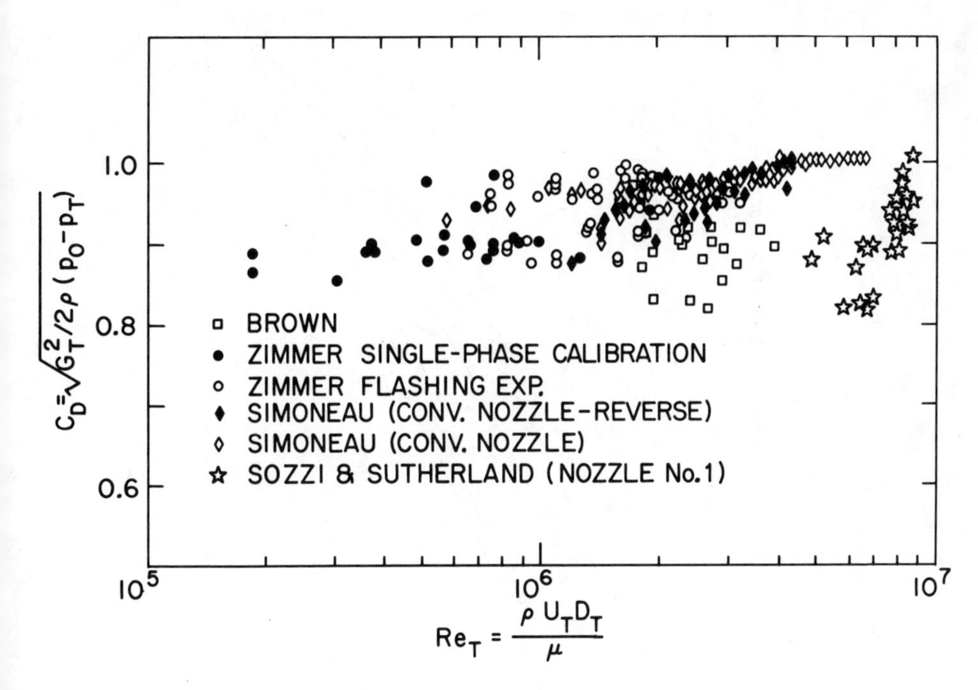

Figure A.5. Variation of the discharge coefficient with the Reynolds number for data in a converging-diverging nozzle with subcooled liquid inlet conditions. (BNL Neg. 3-12-80)

Interfacial Area Density and Heat Flux

A reformulation of (A.4) taking into account a population of bubbles (the low void condition) having been nucleated at different points along a stream tube yields

$$\Gamma_v(z) = \frac{\pi}{A\Delta i_{fg}} \int_{z_o}^{z} q_i''(z,z')\delta(z,z')\frac{dN}{dz'}dz' . \tag{A.22}$$

The nucleation site density, dN/dz, results in bubbles in dV = Adz having different Lagrangian histories, resulting in different interfacial heat fluxes and bubble sizes. Many workers in the past have attempted to utilize descriptions of bubble growth in a uniformly superheated, constant pressure liquid to define the size and heat flux required for Equation (A.22). These estimates were shown to be highly inaccurate by Jones and Zuber [51] who developed expressions for heat flux and size of bubbles growing in variable pressure fields yielding

$$q_i'' = \frac{k_\ell F_o}{\sqrt{\pi a_\ell t}} + \frac{k_\ell}{\sqrt{\pi a_\ell}} \int_o^t \frac{F'(\eta)}{\sqrt{t-\eta}} d\eta . \tag{A.23}$$

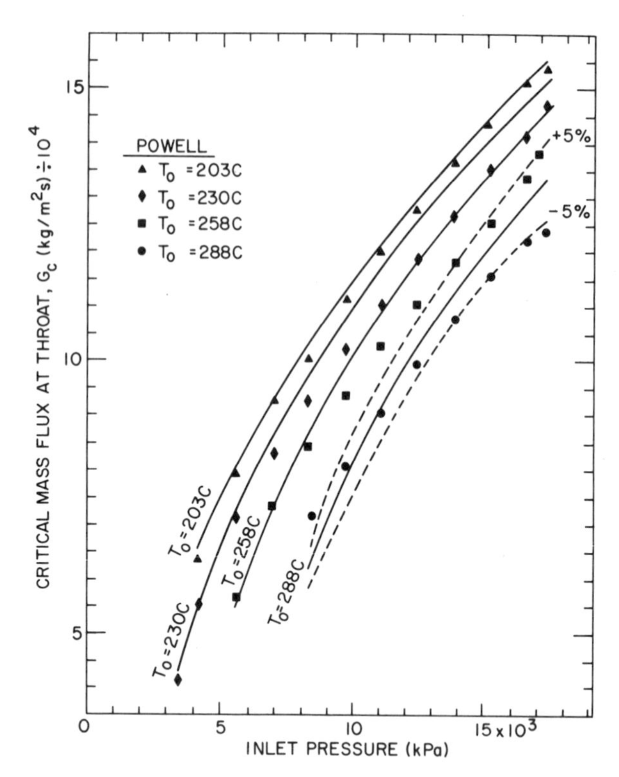

Figure A.6. Comparison of critical throat mass fluxes measured by Powell[50] with those calculated by the present method for different nozzle inlet pressures and temperatures. (BNL Neg. 3-97-80)

$F'(t)$ is the time rate of change of saturation temperature due to pressure changes. The bubble size in terms of the nucleation size R_o is given by

$$\frac{\delta}{2} \equiv R(t) = \left(\frac{\rho_{vo}}{\rho_v}\right)^{2/3} \Bigg\{ R_o + \ldots \tag{A.24}$$

$$\ldots + \frac{K_s k_t \Delta T_s}{\rho_{vo} \Delta i_{fg} \sqrt{\pi a_\ell}} \left[2\sqrt{t} + \frac{2}{3}\left(\frac{\Delta i_{fg}}{RT_o} - 1\right) \int_o^T \left(\frac{T_o}{T} - 1\right) \eta^{-\frac{1}{2}} \, d\eta \right] \ldots$$

$$\ldots + \frac{K_s k_t}{\rho_{vo} \Delta i_{fg} \sqrt{\pi a_\ell}} \left[\int_o^t \int_o^\eta \frac{F'(\xi)}{(\eta-\xi)^{\frac{1}{2}}} \, d\xi d\eta + \ldots \right.$$

$$\left. \ldots + \frac{2}{3}\left(\frac{\Delta i_{fg}}{RT_o} - 1\right) \int_o^t \left(\frac{T_o}{T} - 1\right) \int_o^\eta \frac{F'(\xi)}{(\eta-\xi)^{\frac{1}{2}}} \, d\xi d\eta \right] \Bigg\}$$

From a mechanistic viewpoint, reasonably accurate results were obtained in comparison with single bubble data and might even be obtained in the case of flashing if the population could be accurately identified. There is reason to expect that the latter could be accomplished due to the autocatalytic nature of decompressive bubble growth.[51] From a practical standpoint, the triple integration along bubble trajectories to obtain Γ_v from these equations does not seem reasonable.

Since bubbles growing in a Lagrangian field having decompression occurring as t^n will have bubble radius growth as $t^{n+1/2}$, the resulting void fraction growth from these bubbles will be as $t^{3(n+1/2)}$ a very strong function of time.[51] Thus, bubbles first nucleated in a decompressing population will strongly dominate the vapor source term. It seems reasonable, then, to initially treat the nucleation rate as a delta function leading, therefore, to uniformly sized, identical history bubbles at a given cross section. As a consequence, the bubble population density, interfacial area density, size and void fraction may all be related through

$$\frac{1}{A}\int_{\xi_i} \frac{dA}{dz} = \frac{6\alpha}{\delta} = \pi N_b \delta^2 \quad . \tag{A.25}$$

Similarly, since the bubbles are treated as identical in a given cross section, the net interfacial heat flux will also be identical at any location. For short decompression times especially appropos of critical flow situations, the heat flux is assumed to be described approximately through the well known expression, modified to utilize local instantaneous temperatures,

$$\sum_{k=1,v} \vec{q}''_k \cdot \vec{n}_k \sim K_s \frac{k_\ell}{\sqrt{\pi a_\ell t}} (T_\ell - T_s) \quad \text{as } t \to o \tag{A.26}$$

where K_s is the sphericity correction factor of $\sqrt{3}$, (Plesset and Zwick[52]), or $\pi/2$, (Forster and Zuber[53]). That this is indeed true has been previously demonstrated.[39]

The equation obtained for the vapor source term is thus a combination of (A.25) and (A.26) together with (A.4) yielding

$$\Gamma_v = K_s \, [36\pi N_b]^{1/3} \, \alpha^{2/3} \, \frac{k_\ell (T_\ell - T_s)}{\Delta i_{fg} \sqrt{\pi a_\ell t}} \quad . \tag{A.27}$$

Once the superheat at flashing inception can be determined through the use of (A.13), the inception bubble size can be determined through Laplace's equation

$$R_o = \frac{2\sigma}{\Delta p_{Fi}} \quad . \tag{A.28}$$

By knowing the initial population density, the initial void fraction can be determined. Work is currently in progress to model the population density which may be more successful than past efforts given the new information on inception superheats.

Lacking a model for population density, (A.27) may be rewritten as

$$\Gamma_v = C_\Gamma \frac{k_\ell}{\sqrt{a_\ell t}} \frac{(T_\ell - T_s)}{\Delta i_{fg}} \alpha^{2/3} \tag{A.29}$$

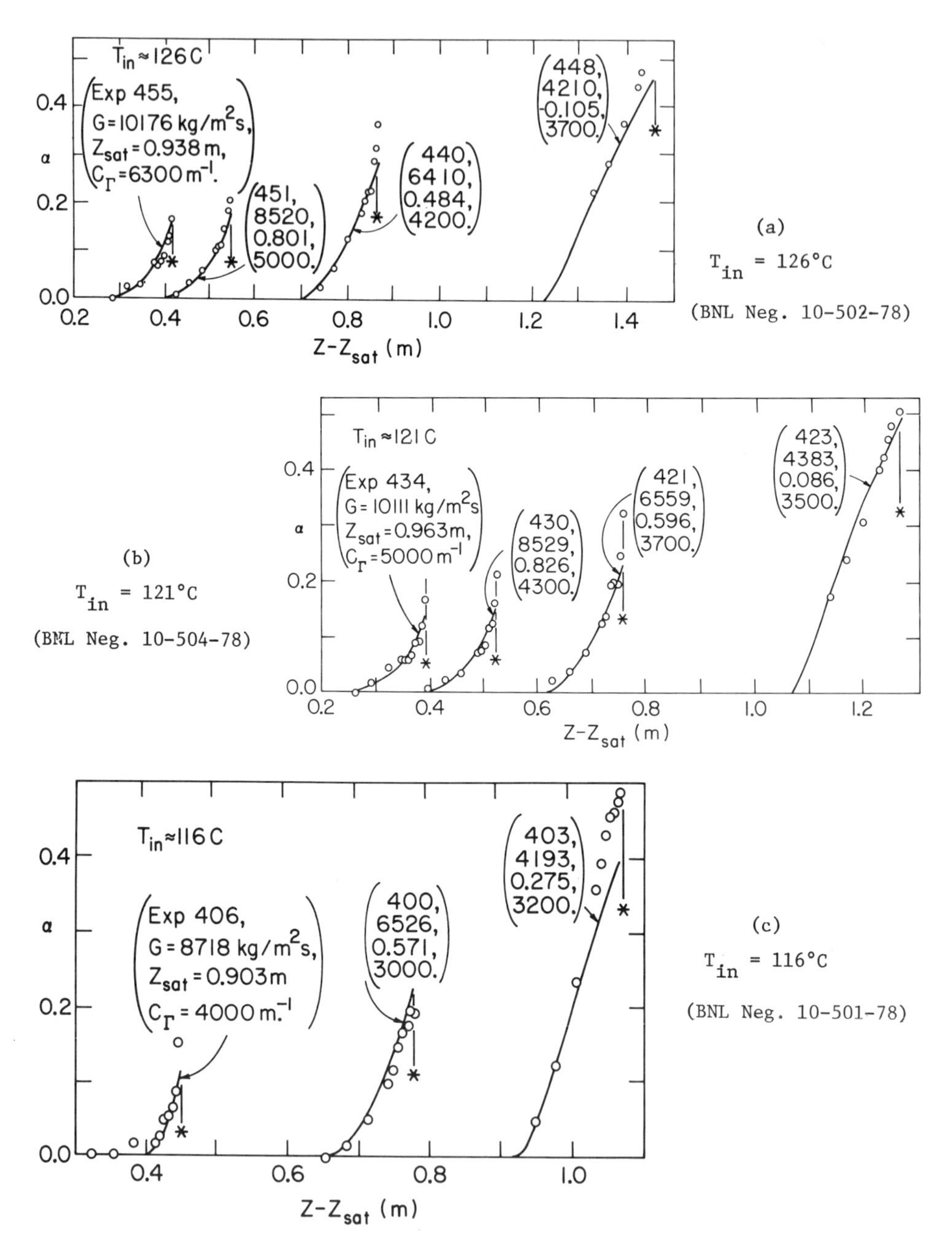

Figure A.7. Comparisons of the model of Wu, et al.[54] with the data of Reocreux.[43]

where t is obviously the Lagrangian lifetime of the voids nucleated at the inception point. By utilizing the energy equation together with an appropriate expression for slip, Wu, et al.[54] developed comparisons of (A.29) with the data of Reocreux[43] as shown in Figure A.7. At this time, there is some reason to expect that the values of C_Γ may apply to other geometries under similar circumstances, indicating a possibility of obtaining the desired correlation for N_b. The agreement in Figure A.7 indicates that this methodology may prove adequate for prediction of nonequilibrium vapor growth in flashing flows up to the range of 15-20% void fractions representing the limit of bubbly flow. For higher void fractions, equilibrium methods may indeed prove adequate.

A.3 Post Dryout Heat Transfer

Aside from subcooled boiling discussed in Chapter 7 and flashing nonequilibrium discussed in the last section, heat transfer in the post-dryout regime represents a third heat transfer regime where nonequilibrium effects are encountered and in this case are very important in properly defining the thermal behavior of the fuel elements.

The elements of the traditional viewpoint of post-dryout heat transfer have been described in Chapter 7 and some models introduced. Furthermore, the most commonly considered correlations to predict the heat transfer rates in reactor systems are summarized in Table 1 of Chapter 12. What is characteristic in virtually all these correlations is that the steam is treated as being in thermal equilibrium with the liquid--it is not. Similarly, virtually all correlations presented up through the mid 1970's were point correlations treating post-dryout as a local phenomena--it is not. Since the path that the nonequilibrium takes is dependent on the CHF starting point, it is easily seen that point correlations are really attempts to correlate simultaneously the critical quality and the integral effects of post-dryout heat transfer as well as the local heat transfer conditions. Thus, it is not reasonable to expect local empirical formulations to predict such conditions accurately and by and large they don't. Even in the best cases, the expected errors may be well over 100°C.

In the late 1960's, however, the existence of the nonequilibrium was not only recognized but workers simultaneously in the U.S. and the U.K. developed the first models to predict the qualitative behavior. Thus, both Bennett, et al.[55,56] and Forslund and Rohsenow[57,58] developed the two-step concept where heat was transferred directly from the dry fuel element surface to superheated steam by forced convection and then, also by convection, heat transfer occurred between the steam and the liquid droplets which were saturated, thus causing vaporization. Their models were unwieldy, however, and generalized correlations were not developed.

In 1977 at the National Heat Transfer Conference in Salt Lake City, two papers were published which simplified the method and presented definite correlations, both of which used similar techniques. The paper by Saha, Shiralkar, and Dix[59] utilized the dimensional form of Equation (A.9) to develop methods for predicting behavior in water systems only. Jones and Zuber[60] utilized Equation (A.9) directly and developed a general correlation based on only two sets of widely different data: the 1.7-bar nitrogen data of Forslund and Rohsenow[57] and the 69-bar water data of Bennett, et al.[55] Detailed theoretical foundations were presented in Reference (60).

The key to developing the correlations lay in the use of Equations (A.9), (A.10), and (A.4) implicitly while recognizing that post-dryout is an initial value phenomena similar to flashing. While flashing begins at the inception point where upstream flow is single phase, post-dryout behavior begins at the dryout point where conditions are taken to be nearly at equilibrium. At this critical heat flux (CHF) location, the onset of the phenomena is assumed to occur simultaneously with the evolution of uniformly sized droplets. Whatever the mechanism, the assumption that they could be treated as a delta-function in space-time was implicit. In such cases, the interfacial area density is given by

$$\frac{1}{A}\int_{\xi_i} \frac{dA}{dz} = \frac{6(1-\alpha)}{\delta} = \pi N_d \delta^2 \tag{A.30}$$

whereas the interfacial heat transfer was taken as

$$\sum_{k=\ell,v} \hat{q}_k'' \cdot \hat{n}_k = h(T_v - T_{sat}) \quad . \tag{A.31}$$

So that the vapor source term becomes (Equation A.4)

$$\Gamma_v = \frac{6h(T_v - T_{sat})(1-\alpha)}{\delta \, \Delta i_{fg}} \quad . \tag{A.32}$$

The difficulty in (A.32) is the determination of h, α, T_v, and δ simultaneously. Saha used the Heineman heat transfer correlation[61] for vapor-to-wall heat transfer applicable for steam only. He then correlated the droplet size at dryout to obtain

$$\delta_{DO} + 1.47\left[\frac{\rho_g J_{g,DO}^2 \sqrt{\sigma/g\Delta\rho}}{\sigma}\right]^{-0.675} \quad . \tag{A.33}$$

The local droplet size was then related to that at dryout given by (A.33) by assuming the droplet number flux remains constant and integrating (A.9). Results were quite good as indicated by comparisons shown in the paper.

Jones and Zuber,[60] on the other hand, obtained a correlation for the observed superheat relaxation numbers $N_{Sr} \equiv N_v = N_{rk}$ in terms of the calculated value S given by

$$N_{Sr} = \begin{cases} 14.3 \; Bo \left(\dfrac{S}{\sqrt{p_r}}\right)^2 & \dfrac{S}{\sqrt{p_r}} \le 0.22 \\[2ex] 1.23 \; Bo \left(\dfrac{S}{\sqrt{p_r}}\right)^{3/8} & \dfrac{S}{\sqrt{p_r}} > 0.22 \quad . \end{cases} \tag{A.34}$$

In this case Bo was the Boussinesque number

$$Bo = \frac{G}{\rho_\ell \sqrt{g\delta_{DO}}} \quad , \tag{A.35}$$

the reduced pressure $p_r = p/p_c$ where p_c is the critical pressure of the fluid, and the parameter S was the calculated relaxation number

$$S = \frac{3}{2}\left(\frac{N_d \pi}{6}\right)^{2/3} \frac{k_v D \Delta i_{fg}}{C_{pu} q_w'' x} Nu_\delta \; (1-\alpha)^{1/3} \quad . \tag{A.36}$$

To obtain values for S, simultaneous compatibility must be obtained between expressions for initial droplet size: $$\delta_c = \left(\frac{27}{4}\right)^{1/6}\left(\frac{C_D \sigma We_c}{\alpha g \Delta\rho}\right)^{1/2} \tag{A.37}$$

drag coefficient: $$C_D = \frac{24}{Re_\delta}(1+0.1Re_\delta)^{0.75} \tag{A.38}$$

droplet evaporation: $$\delta = \delta_c\left(\frac{1-\alpha}{1-\alpha_o}\right)^{1/3} \tag{A.39}$$

slip ratio: $$\frac{v_v}{v_\ell} = \left(1 - \frac{\alpha\rho_v v_v}{Gx}\right)^{-1} \tag{A.40}$$

droplet Nusselt No.: $$Nu_\delta = 2 + 0.74Re_\delta^{1/2} Pr_v^{1/3} . \tag{A.41}$$

The droplet Nusselt number is $Nu_\delta = h\delta/k_v$ while the droplet Reynolds number is given by $Re_\delta = \rho_v \delta v_r/\mu_v$ and the relative velocity is obtained by balancing drag and buoyancy

$$v_r = \sqrt{\frac{\alpha\delta g\Delta\rho}{3C_D\rho_v}} . \tag{A.42}$$

The critical Weber number We_c was taken as 7.5 although little effect was noticed in the range 1-20.

In obtaining the correlation for the superheat relaxation number given by (A.34), the experimental data were differentiated to yield

$$N_{Sr} = \frac{x_i - x_{i+1}}{x_{e_{i+1}} - x_{e_i}} \cdot \frac{1}{\overline{x}_e - \overline{x}} \tag{A.43}$$

where $\overline{x}_e$ and $\overline{x}$ represent the averaged values between the ith and the (i+1) location on a test section.

The correlation data base and root mean square error in computed values of actual quality are shown in Table A.1. The overall standard duration in quality prediction for the 671 nitrogen data points of Forslund and Rohsenow[57] was 0.035 whereas for the 1084 water data points of Bennett, et al.[55] the standard duration was 0.020 for an average overall 1755 points of 0.027 testifying to the smoothing effects of the correlation.

The advantage of the procedure described is that the correlation first of all smoothed the differentiated data (Figure A.8) yielding an integration parameter which then provided further smoothing to the predictions. The wall temperatures both for the data base and for all other data tested were all predicted with a mean square error of 8-12°C depending on the conditions over the range of wall temperatures between 600°C and 1200°C. Typical of the comparisons observed for the data base are those shown in Figure A.9, and for other data in Figure A.10. Even when the correlation was extended by a factor of 3 over the range of its data base the comparisons are seen to be quite good (Figure A.11).

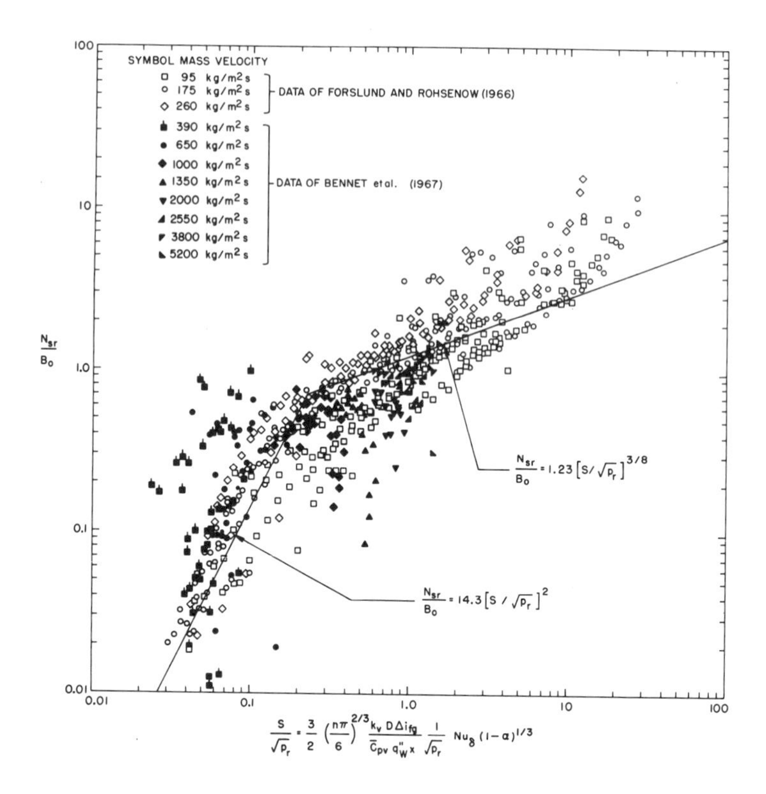

Figure A.8. Post-dryout correlation of the integration parameter of Jones and Zuber.[60] (BNL Neg. 2-36-77)

A.4 Summary

The current status of nonequilibrium phase exchange has been summarized. Several important features apparently common to all situations of interest in the field of nuclear safety heat transfer include the following

1. If the quantity Q_k, the nonequilibrium, is the difference between equilibrium and actual flowing quality of phase k $x_{ek}-x_e$, then Q_k is described by the inhomogeneous relaxation equation

$$\frac{dQ_k}{dx_e} + N_{rk}Q_k = 1 \quad .$$

2. All proper descriptions of Q_k require specification of starting (inception) criteria, Q_{ko} and x_{eo}.

3. The phase-k relaxation parameter N_{rk} which is the relative phase change rate per unit nonequilibrium is not a function of Q_k but may be a function of x_k,--i.e.: a local variable.

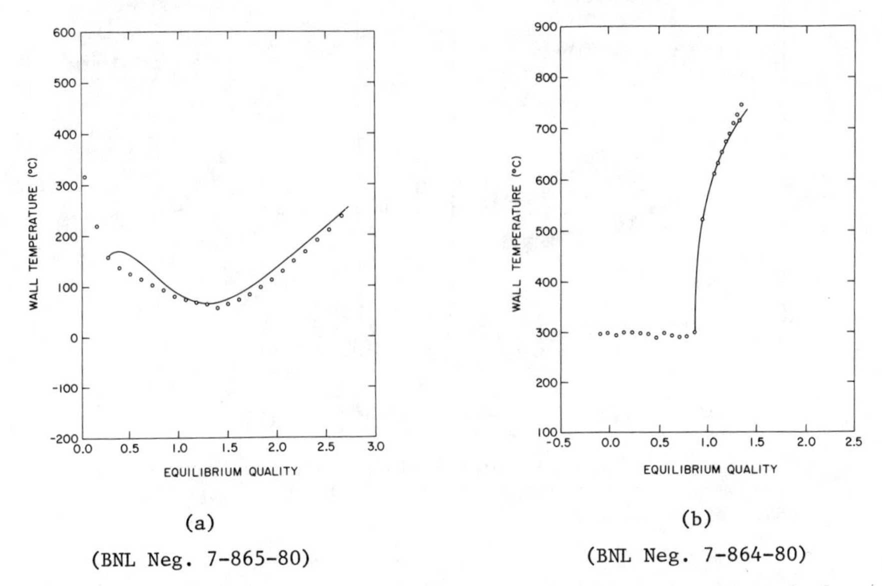

(a)

(BNL Neg. 7-865-80)

(b)

(BNL Neg. 7-864-80)

Figure A.9. Comparison of the predictions of Jones and Zuber[60] with data of (a) Forslund and Rohsenow,[57] and (b) Bennett, et al.[55]

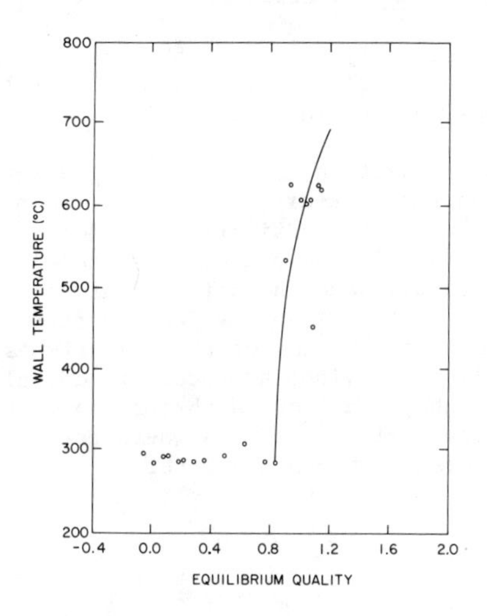

Figure A.10. Comparisons of the Jones and Zuber correlation[60] with nondata base data of Jansen and Kervinen.[62] (BNL Neg. 10-863-80)

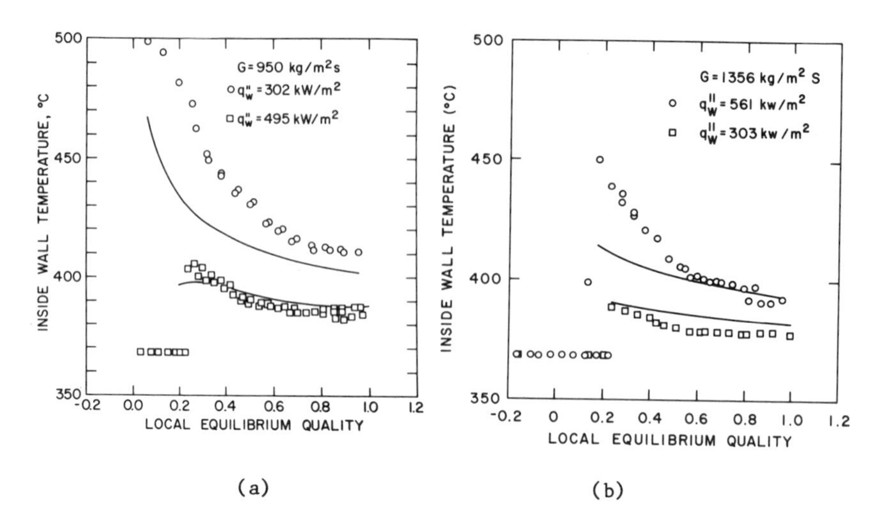

Figure A.11. Comparison of calculated wall temperatures with those measured by Swenson, et al.[63] at 207 bar (p_r=0.94). (BNL Neg. 7-747-77)

4. Proper specification of the relaxation parameter requires local determination of

   (a) Interfacial area density
   (b) Net interfacial heat transfer rates.

5. The relaxation process is an initial value, path dependent process and hence Q_k is not a locally described variable.

Based on the above, it is easily seen that the prediction of local wall temperatures in post-dryout, and local void fraction in flashing cannot be expected to be accurately accomplished by local criteria.

Application of the relaxation methods developed were described for both the flashing process and the post-dryout process. Both seemingly different situations were shown to be quite similar--the major differences being in the specification of the nonequilibrium phase, interfacial area density, interfacial heat transfer mechanisms, and inception point--as shown in Table A.2. While application of the method is not as far advanced for the flashing case, early results are encouraging in view of the comparisons provided with data, especially the explanations provided and accurate calculations of critical discharge of initially subcooled liquids through nozzles and orifices. Application to the existing post-dryout data where possible shows accurate prediction of wall temperatures possible within 8-12°C.

Table A.1

Summary of the Root-Mean-Square Deviation in the Actual Quality Calculated Compared with the Data Base Data of Forslund and Rohsenow[57] and of Bennett, et al.[55]

$0.05 \leq Pr \leq 0.31$
$95 \leq G \leq 5200$ kg/m^2--s
$5.79 \leq D \leq 12.37$ mm
$16 \leq q_w'' \leq 1836$ kw/M^2
$121 \leq L \leq 553$ cm
$5 \leq z_c \leq 553$ cm
$0.13 \leq x_e \leq 3.2$

	Mass Velocity Range kg/m-s	Number of Runs	Number of Points	Mean Square Deviation
Forslund & Rohsenow (1966)	95	13	346	0.0401
	175	13	222	0.0230
	260	12	186	0.0316
	390	16	149	0.0256
	650	16	156	0.0253
Bennett et al. (1967)	1000	12	124	0.0229
	1350	17	171	0.0254
	1970	16	131	0.0211
	2550	16	140	0.0140
	3850	14	157	0.0084
	5200	6	61	0.0077

Table A.2

Comparison of Flashing and Post-Dryout Nonequilibrium Considerations

Quantity	Flashing $k=\ell$	Post-Dryout $k=v$
Fluid at Saturation	vapor	liquid
Nonequilibrium Phase	liquid superheat	vapor superheat
Inception Criteria	$x_o \sim o$ initial superheat	$x_o = x_c$ critical quality
Flow Pattern Assumed	bubbly	droplet
Interfacial Area Density	$\frac{6\alpha}{\delta}$	$\frac{6(1-\alpha)}{\delta}$
Interfacial Heat Transfer Mechanism	transient conduction to bubbles	droplet convection
Interfacial Heat Flux	$\frac{K_s k_\ell (T_\ell - T_s)}{\sqrt{\pi a_\ell t}}$	$h(T_v - T_s)$
Major Forcing Function	Lagrangian decompression	wall heating

Nomenclature

English

a	Thermal diffusivity
A	Area
Bo	Boussinesque number
C	Coefficient
C_D	Drag coefficient or Discharge coefficient
C_p	Specific heat
D	Tube diameter
Fi	Flashing index
F_o	Initial superheat
F'	dT_s/dt
G	Mass velocity
h	Heat transfer coefficient
i	Enthalpy
k	Thermal conductivity
K_s	Sphericity factor
L	Length
n	Unit normal
N	Droplet number density
N_{rk}	Relaxation number for phase-k.
N_{sr}	Superheat relaxation number
Nu	Nusselt number
p	Pressure
P	Perimeter
Pr	Prandtl number
q"	Heat flux
R	Bubble radius
Re	Reynolds number
t	Time
T	Temperature
u'	Velocity Fluctuation
U	Local average velocity
v	Velocity
V	Volume
We	Weber number
x	Quality
z	Axial location

Greek

α	Void fraction
Γ	Phase generation rate
δ	Droplet or bubble diameter
Δi_{fg}	Latent heat
μ	Viscosity
ρ	Density
ϕ	Generalized flux
Φ	Parameter (Equation A.15)
ψ	Parameter (Equation A.16)
σ	Surface tension
Σ_o'	Static expansion rate
$\Delta\Sigma'$	Convective expansion rate
τ	Shear stress
ξ_i	Interface perimeter

Subscripts

c	Critical heat flux or choked condition
δ	Droplet
e	Equilibrium
f	Saturated liquid
Fi	Flashing inception
Fio	Flashing inception without turbulence
g	Saturated vapor
hom	Homogeneous
i	Interface or inlet
k	Phase-k
ℓ	Liquid
o	Initial or stagnation
r	Relative or reduced or relaxation
s	Saturated value
T	Throat
v	Vapor
w	Wall

REFERENCES

[1] Reocreux, M., "The Blowdown, Refill and Reflood Phase During a LOCA," LWRS/5/80 Ispra Courses 1980 on Thermohydraulic Problems Related to LWR Safety, May 19-13, 1980.

[2] Smidt, D., Reaktor-Sicherheitstechnik, Springer-Verlag, Berlin - Heidelberg - New York, 1979.

[3] Grillenberger, T., "The Computer Code DAPSY for the Calculation of Pressure Wave Propagation in the Primary Coolant System of LWR," Report MRR-I-66 (1976).

[4] Pana, P., and Müller, M., Nuclear Engng. Design, 36 (1976), pp. 183-190.

[5] Hughes, G.A., "Kurzbeschreibung des digitalen Rechenprogrammes LECK," KWU-Report Nr. 33/74.

[6] Jones, O.C., and Bankoff, S.G., Ed., Thermal and Hydraulic Aspects of Nuclear Reactor Safety, Vol. 1: Light-Water Reactors, American Society of Mechanical Engineers, 1977.

[7] Wolfert, K., "The Simulation of Blowdown Processes with Consideration of Thermodynamic Nonequilibrium Phenomena," paper presented at the OECD/NEA Specialists' Meeting on Transient Two-Phase Flow, Toronto, Canada, August 1976.

[8] Plessset, M.S., and Zwick, S.A., "The Growth of Vapor Bubbles in Superheated Liquids," Journal of Applied Physics, Vol. 25, 1954, pp. 493-500.

[9] Friz, G., Riebold, W., and Schulze, W., "Studies on Thermodynamic Nonequilibrium in Flashing Water," paper presented at the OECD/NEA Specialists' Meeting on Transient Two-Phase Flow, Toronto, Canada, August 1976.

[10] Rousseau, J.C., and Riegel, B., "Blowdown Experiment on CANON; Interpretation by the BERTHA Code," European Two-Phase Flow Meeting, Haifa, 1975.

[11] Edwards, A.R., and O'Brien, T.P., "Studies of Phenomena Connected with the Depressurization of Water Reactors," J. Brit. Nucl. Energy Soc., 9, p. 125 (1970).

[12] Technischer Bericht BF-RS 16 B-40-4-4: SWR- and DWR-DE-Versuchsprogramm, Battelle, May 1979.

[13] Ericson, L., and Hall, D.G., "The Marviken Critical Flow Tests - A Description and Early Results," ENS-ANS International Topical Meeting on Nuclear Power Reactor Safety, Brussels, October 16-19, 1978.

[14] Ardron, K.H., "A two-fluid model for critical vapor-liquid flow," Int. J. Multiphase Flow, Vol. 4, pp. 323-337, 1978.

[15] Bauer, E.G., Houdayer, G.R., and Sureau, H.M., "A Nonequilibrium Axial Flow Model and Application to Loss-of-Coolant Accident Analysis," Transient Two-Phase Flow Proceedings of the CSNI Specialists Meeting, Toronto 1976, Vol. 1, pp. 429-457.

[16] Boure, J.A., and Reocreux, M.L., "General Equations of Two-Phase Flows - Applications to Critical Flows and to Non-Steady Flows," Fourth All-Union Heat Transfer Conference, Minsk 1972.

[17] Reocreux, M., "Contribution to the Study of Critical Flow Rates in Two-Phase Water-Vapor Flow," PhD Thesis, University of Grenoble, 1974, Translated as NUREG TR-0002, 1977.

[18] Boure, J.A., "Mathematic Modelling and the Two-Phase Constitutive Equations," European Two-Phase Flow Group Meeting, Haifa, 1975.

[19] Zuber, N., and Findlay, J.A., "Average Volumetric Concentrations in Two-Phase Flow Systems," Trans. ASME J. Heat Transfer, 87C, p. 453, 1965.

[20] Delhaye, J.M., "Equations Fondamentales des Ecoulements Diphasiques," CEA R-3429, 1968.

[21] Houdayer, G., Pinet, B., and Vigneron, M., Hexeco-code, "A Six Equation Model," 2nd Transient Two-Phase Flow CSNI Specialist Meeting, Paris, 1978.

[22] Ishii, M., and Chowla, T.C., "Two-Fluid Model and Momentum Interaction Between Phases," Washington Information Meeting, 1979.

[23] Rousseau, J.C., and Toth, J., Private communication, 1979.

[24] Henry, R.E., A Study of One- and Two-Component, Two-Phase Critical Flows at Low Qualities," ANL 7430, March 1968.

[25] Städtke, H., "Critical Flow Models for Application in Blowdown Computer Codes," THRS/B10/77 Ispra Courses on Thermohydraulic Problems Related to Reactor Safety, Part B: LWR Related Thermohydraulic Problems, September 19-30, 1977.

[26] Städtke, H., "Modelling of Break Mass Flow for Loss-of-Coolant Accidents," LWRS/80/10 Ispra Courses, Thermohydraulic Problems Related to LWR Safety, Ispra, 19-23 May, 1980.

[27] Wallis, G.B., One-dimensional two-phase flow, Ch. 2 MacGraw Hill Book Company, 1969.

[28] Moody, F.J., "Maximum Flow Rate of a Single-Component, Two-Phase Mixture," Trans. ASME, J. Heat Transfer, 87C, pp. 134-142, 1965.

[29] Moody, F.J., "Maximum two-phase vessel blowdown for pipes," Trans. ASME, J. Heat Transfer, 88C, pp. 285-295, 1966.

[30] Fauske, H.K., "Contribution to the Theory of Two-Phase One-Component Critical Flow," ANL 6633, October 1962.

[31] Henry, R.E., and Fauske, H.K.," The Two-Phase Critical Flow of One-Component Mixture in Nozzles, Orifices and Short Tubes," Trans. ASME, J. Heat Transfer, 93C, pp. 179-187, 1971.

[32] Latrobe, A., and Grand, D., "On Some Difficulties in Numerical Modelling Transient Two-Phase Flow," CSNI Specialist Meeting, Toronto 1976.

[33] Reocreux, M., Sureau, H., Thibaudeau, J., Chabrillac, M., Courtaud, M., and Gomolinski, M., "French Thermal-Hydraulic Studies for the Development of Advanced Safety Codes for PWR," ENS-ANS International Tropical Meeting on Nuclear Power Reactor Safety, Brussels, October 16-19, 1978.

[34] Bjornard, T.A., and Griffith, P., "PWR Blowdown Heat Transfer," In [6], pp. 17-41.

[35] Groeneveld, D.C., and Gardiner, S.R.M., "Post-CHF Heat Transfer Under Forced Convective Conditions," In [6], pp. 43-73.

[36] Jones, O.C., Jr., and Saha, P., "Nonequilibrium Aspects of Water Reactor Safety," in Thermal and Hydraulic Aspects of Nuclear Reactor Safety, Vol. 1: Light Water Reactors, O.C. Jones, Jr., and S.G. Bankoff Eds., ASME, New York, 1977.

[37] Debye, P., "Polar Molecules," Chemical Catalogue, New York, 1929, Trans. Faraday Soc., 30, 1934, p. 679.

[38] Higasi, K., "Dielectric Relaxation and Molecular Structure," Res. Inst. Appl. Elec., Hokkaido Univ., Sapporo, Japan, 1961.

[39] Jones, O.C., Jr., "Inception and Development of Voids in Flashing Liquids," Proc. U.S.-Japan Seminar or Two-Phase Flow Dynamics, August 1-3, 1979, to be published by Hemisphere Press, Washington.

[40] Jones, O.C., Jr., "Flashing Inception in Flowing Liquids," Trans. ASME, J. Heat Trans., 102C, pp. 439-444, 1980.

[41] Abuaf, N., Jones, O.C., Jr., and Wu, B.J.C., "Critical Flashing Flows in Nozzles with Subcooled Inlet Conditions," in Polyphase Flow and Transport Technology, R.A. Bajura, Ed., Amer. Soc. Mech. Eng., pp. 65-73, 1980.

[42] Alamgir, Md., and Lienhard, J.H., "Correlation of Pressure Undershoot during Hot-Water Depressurization," TRANS. ASME, J. Heat Transfer, 1980, in press.

[43] Reocreux, M., "Contributions a l'Etude das Debits Critiques en Econlement Diphasique Eau-Vapeur," Ph.D. Thesis, Universite Scientifique et Medicale de Grenoble, France, 1974.

[44] Seynhaeve, J.M., Giot, M.M., and Fritte, A.A., "Nonequilibrium Effects on Critical Flow Rates at Low Qualities," presented at the Specialists Meeting on transient Two-Phase Flow, Toronto, August 3-4, 1976.

[45] Uberoi, M.S., "Effect of Wind Tunnel Contraction on Free-Stream Turbulence," J. Aeron. Sci., 23, 8, pp. 754-764, 1956.

[46] Brown, R.A., "Flashing Expansion of Water Through a Converging-Diverging Nozzle," MS Thesis, University of California, Berkeley, USAEC Report UCRL-6665-T, 1961.

[47] Zimmer, G.A., Wu, B.J.C., Leonhard, W.J., Abuaf, N., and Jones, O.C., Jr., "Pressure and Void Distributions in a Converging-Diverging Nozzle with Nonequilibrium Water Vapor Generation," BNL-NUREG-26003, 1979.

[48] Simoneau, R.J., NASA-Lewis Research Center, personal communications, 1979.

[49] Sozzi, G.C., and Sutherland, W.A., "Critical Flow of Saturated and Subcooled Water at High Pressure," General Electric Company Report NEDO-13418, 1975.

[50] Powell, A.W., "Flow of Subcooled Water Through Nozzles," Westinghouse Electric Corporation, WAPD-PT-(V)-90, April 1961.

[51] Jones, O.C., Jr., and Zuber, N., "Bubble Growth in Variable Pressure Fields," Trans. ASME, J. Heat Trans., 100C, pp. 453-459, 1978.

[52] Plesset, M.S., and Zwick, S.A., "The Growth of Vapor Bubbles in Superheated Liquids," J. Appl. Phys., 25, pp. 493-500, 1954.

[53] Forster, H.K., and Zuber, N., "Growth of a Vapor Bubble in a Superheated Liquid," J. Appl. Phys., 25, pp. 474-478, 1954.

[54] Wu, B.J.C., Saha, P., Abuaf, N., and Jones, O.C., Jr., "A One-Dimensional Model of Vapor Generation in Steady Flashing Flow," Interim Milestone Report, BNL-NUREG-25709, October, 1978.

[55] Bennett, A.W., Hewitt, G.F., Kearsey, H.A., and Keeys, R.F.K., "Heat Transfer to Steam-Water Mixture Flowing in Uniformly Heated Tubes in which the Critical Heat Flux has been Exceeded," AERE-R 5373, October, 1967.

[56] Bennett, A.W., et al., "Heat Transfer to Steam-Water Mixture Flowing in Uniformly Heated Tubes in which the Critical Heat Flux has been Exceeded," Paper #27 presented at the Thermodynamics and Fluid Mechanics Convention, Bristol, March, 1968.

[57] Forslund, R.P., and Rohsenow, W.M., "Thermal Non-Equilibrium in Dispersed Flow Film Boiling in a Vertical Tube," MIT Report 75312-44, November, 1966.

[58] Forslund, R.P., and Rohsenow, W.M., "Dispersed Flow Film Boiling," Trans., ASME Ser. C. J. Heat Trans., 90, pp. 399-407, 1968.

[59] Saha, P., Shiralkar, B.S., and Dix, G.E., "A Post-Dryout Heat Transfer Model Based on Actual Vapor Generation Rate in Dispersed Droplet Regime," ASME Paper 77-HT-80, presented at the 17th National Heat Transfer Conference, Salt Lake City, Utah, August 14-17, 1977.

[60] Jones, O.C., Jr., and Zuber, N., "Post-CHF Heat Transfer: A Non-Equilibrium Relaxation Model," ASME Paper 77-HT-79, presented at the 17th National Heat Transfer Conference, Salt Lake City, Utah, August 14-17, 1977.

[61] Heineman, J.B., "An Experimental Investigation of Heat Transfer to Superheated Steam in Round and Rectangular Ducts," ANL-6213, 1960.

[62] Janssen, E., and Kervinen, J.A., "Film Boiling and Rewetting," General Electric Co. Report NEDO-20975, 1975.

[63] Swenson, H.S., Carver, J.R., and Szoeke, G., "The Effects of Nucleate Boiling vs. Film Boiling on Heat Transfer in Power Boiler Tubes," J. Eng. Power, Transaction ASME, 84, 1965, pp. 365-371.

11 EMERGENCY COOLING WATER INJECTION

by

F. Mayinger

Univ. Hannover, Hannover, FRG

This chapter describes the second major phase of a loss of coolant accident in a light water reactor—the ECC injection phase. Specific aspects of emergency cooling systems are described and the thermalhydraulic phenomena which occur during this phase are delineated.

1. INTRODUCTION

Licensing rules /1/ and safety deliberations for nuclear reactors require to take in account accidents from inside as well as from outside. Impacts from outside may be for example earthquakes, gas cloud explosions, aircraft crash, or sabotage. Accidents from inside may be

loss of all electrical power supply

loss of flow due to pump failure, or

loss of coolant (LOCA) due to a break in the primary system.

For the last mentioned occurance an emergency core cooling system is needed and for the other two accidents from inside an emergency feed water system for pressurized water reactors acting on the secondary side and a controlled venting of the steam generators usually must be put in operation.

In this paper only the emergency cooling system acting in case of a loss of coolant accident shall be discussed. The deliberations will be concentrated to pressurized water reactors mainly because there the physical phenomena are more manifold and hydrodynamically more complicated.

The largest break in the primary system of a pressurized water reactor to be taken in account for emergency core cooling would be the guillotine rupture of one of the main coolant pipes, having a diameter up to 0,8 m in a 1300 MW_{el} nuclear power station. However, much more probable is a small leak, for example a rupture of an instrumentation nozzle or the opening of a safety valve in the primary system. The emergency core cooling system (ECCS) has to overcome the whole spectrum of these break possibilities and therefore it has to be able to cool the core and to transport the decay heat in a most efficient way which means for large breaks it has to feed in a large amount of water as soon as possible at low pressure, and for small leaks lower quantities, however, at high pressure.

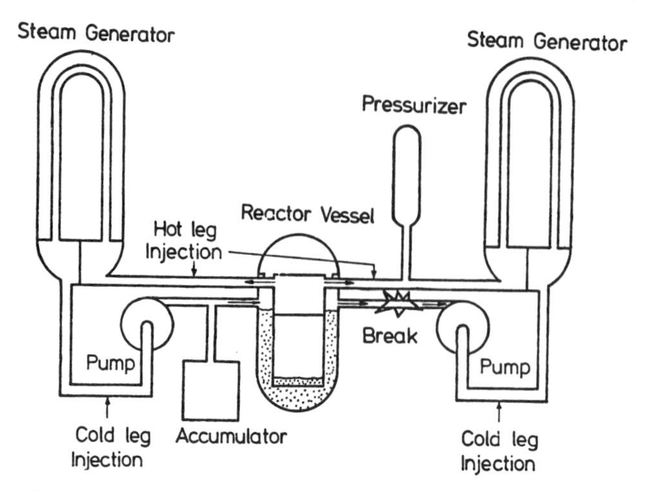

Fig. 1. Flow pathes during Blowdown and Reflood of a PWR

2. SEQUENCES OF THE LOSS OF COOLANT ACCIDENT

The hydrodynamic behaviour during the sequences of a loss of coolant accident is strongly dependent on the size and the location of the break. Generally we distinguish between the hot legs and the cold legs in the primary loops /2/ of a pressurized water reactor. The hot legs - as shown in fig. 1 - represent the flow path from the pressure vessel to the steam generator and the cold leg incorporates the pump surge line from the steam generator and the connecting line between pump and pressure vessel. During steady state conditions the flow is always upwards in the core, going from the cold leg to the hot leg in the primary system of a pressurized water reactor (PWR).

In the case of a large break there is a very sudden pressure

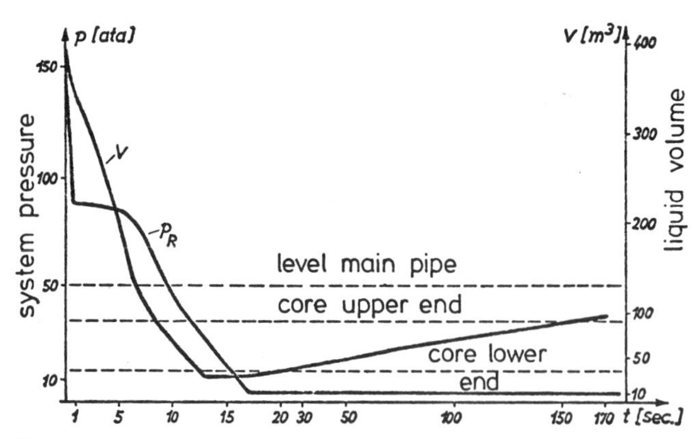

Fig. 2. System pressure and liquid volume in the core during blow down

decrease, as shown in fig. 2 in the primary system until the originally subcooled water reaches its saturation pressure. This pressure decrease creates strong shok waves in the primary system, which cause strong forces on the core structure.

The then following depressurization has a much slower gradient with respect to time and produces a very violent flashing of the water. Afterwards a two-phase flow of high velocity at the beginning is created and directed to the leak location. So in case of a cold leg leak the originally upwards orientated flow is reversed, which implies a short stagnant period with low cooling capability in the core. The pressure in the primary system finally reaches equilibrium with the atmosphere in the containment and this period of the loss of coolant accident, which is called blowdown, is finished.

Parallel to the pressure decrease also the inventory of the coolant in the primary system becomes smaller due to mass flow out of the break. Long before pressure equilibrium between primary system and containment is reached, however, the emergency core cooling system becomes activ and feeds in new cooling water into the pressure vessel. Therefore after a short period, in which there is no or few water in the core, the water level is rising again. The hydrodynamic phenomena following the blowdown are subdivided in a refill and a reflood period.

During the refill period, water is provided to the pressure vessel, however, does not yet reach in its liquid state, the hot rods of the core due to different physical reasons. So for example by injecting via the cold leg the water has to flow through the downcomer and to fill up the lower plenum before reaching the lower grid plate - see fig. 3 -. During this period already vapour is produced by the heat stored in the walls and in the structure of the primary system, which flows through the core and has a slight cooling action. This evaporation at the hot walls, however, may even prevent the flowing down of the emergency coolant in the downcomer, completely or partially for a short time as demonstrated

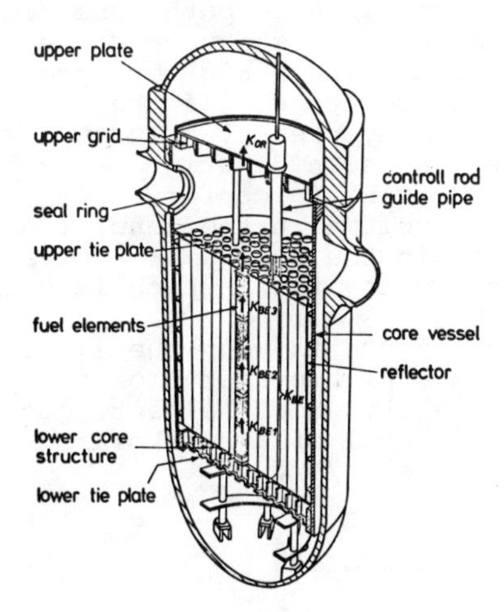

Fig. 3. Structure of the reactor pressure vessel

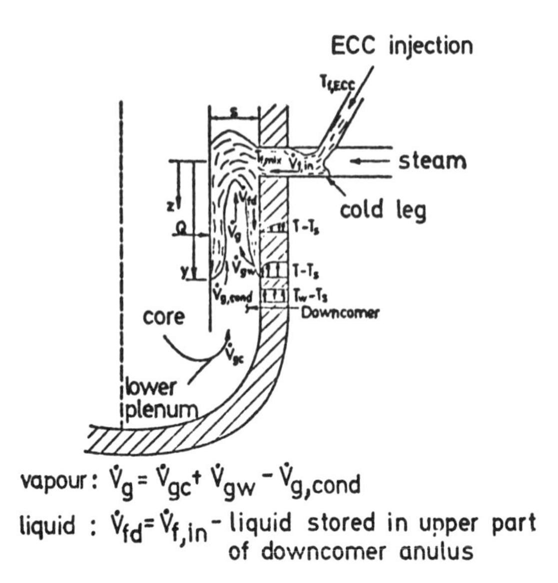

Fig. 4. Downcomerflow with cold leg injection

in fig. 4. Due to bouancy forces and pressure drop the vapour produced at the hot walls may flow against the injected water which again may lead locally and temporarily to flooding conditions as wellknown in apparatus - like bubble columns or pebble beds - in chemical engineering plants. This holdup, however, is hydrodynamically very unstable, especially also due to condensation effects at the water-vapour interface and so chugging may occur until stable counter current flow in the downcomer is reached /3/.

When the water reaches the lower end of the fuel elements a very strong evaporation starts and the vapour produced, has to flow to the leak which in case of a break location in the cold leg, may be a very complicated path (see fig. 1) through the steam generator and the pump. This may be connected with a high flow resistance and such needing a high overpressure in the upper plenum of the pressure vessel. If the resistance for this steam flow is too high it may prevent a further rising of the water level in the core and such deteriorate the cooling action considerably. The only driving force for rising the water level in the core is the water column in the annular downcomer, which means that overpressures in the upper plenum of a few hundred millibar may block the cooling flow, which is called steam binding.

The reflooding period starts when the liquid level is rising in the core which causes a high evaporation and produces a two-phase mixture of high quality which helps to cool and to quench the core long before the level of a non-boiling liquid would fill up the system completely /4/.

3. EMERGENCY CORE COOLING SYSTEMS

There are different philosophies how to cool the core in a most efficient way and how to guarantee that the ECC-water reaches the fuel elements as soon as possible and that it is not blocked

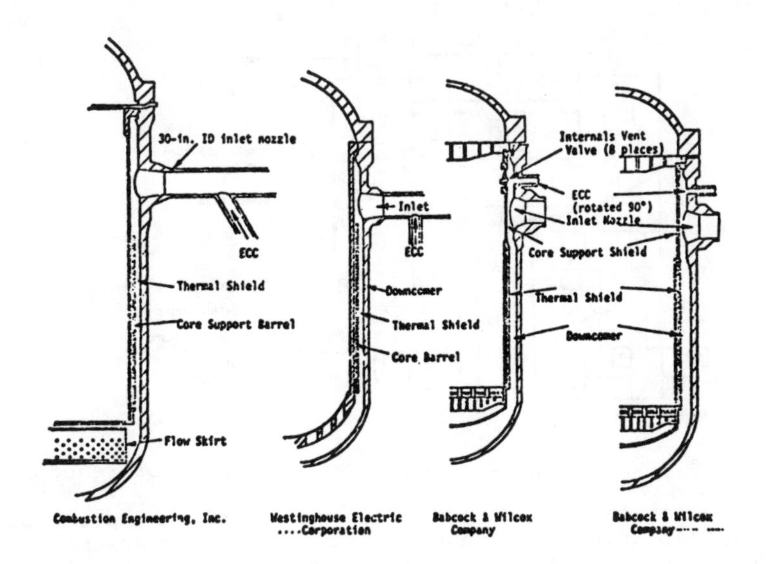

Fig. 5. PWR Configurations for Various Manufacturers Designs

by steam binding. Emergency core cooling water can be injected at different positions at and around the pressure vessel as shown in fig. 5. From a hydrodynamic point of view one of the best solutions certainly would be to inject the water immediately into the lower plenum. This, however, would not only mean that the design of the pressure vessel could not be optimal from the mechanical point of view, because additional nozzles increase the probability for breaks. But it could also endanger the long term cooling of the core if an ECC-nozzle would break and the water instead of filling up the core flows out immediately again. Even an ECC-water injecting pipe brought down through the downcomer, may deteriorate the reflooding of the core by acting like a syphon and sucking out the water. Most vendors inject the ECC-water through the cold legs only, as shown in fig. 5. To overcome the steam binding problem one vendor - see also fig. 5 - provides the core support shield above the core with internal vent valves which open automatically at a slight overpressure in the upper plenum and allow to equalize the pressure between downcomer and upper plenum. So steam binding can be avoided and the cooling of the core can be made more effectively.

A German vendor /5/ is injecting the ECC-water in the cold leg as well as in the hot leg, for two reasons, namely

to avoid the refilling period at least partially by allowing the ECC-water to reach the core directly by falling down through the upper tie-plate and

to reduce the steam binding effect by condensing part of the steam at the surface of the hot leg injected cold water.

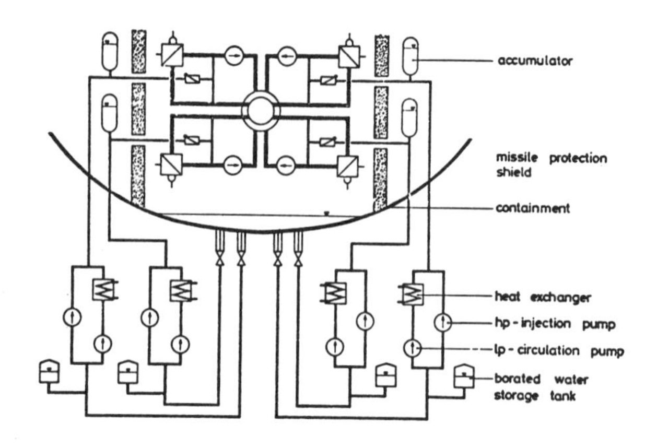

Fig. 6. Core cooling system for a pressurized water reactor

In the strategy of nuclear safety measures the principal of redundancy /4/ is used, because it has also to be assumed that at the time of the accident a component of the ECC-system fails or is under inspection. Therefore most of the pressurized water reactors consists of 4 identical subsystems as shown in a simplified way for a German 4-loop plant in fig. 6. Each subsystem consists of a high pressure pump and a low pressure pump which get their water at least for the first half hour from a borated water storage tank and which are connected to both sides of one loop, i.e. to the cold and the hot leg. For the very first time the cooling water, however, is not provided by the pumps but by accumulators which are connected via self opening valves with the primary loops. These accumulators are filled up to 2/3 with water which is pressurized by nitrogen. The nitrogen pressure varies between approximately 40 and 25 bar, depending from the hydrodynamic assumptions made in ECC-calculations. If one starts from the conservative assumption that at the end of blowdown the lower plenum is empty then the lower pressure has some benefits because a later action of the accumulator would spare water for refilling and prevent it to the blown out from the lower plenum. Other hydrodynamic assumptions may argue in a different way.

In the German licensing procedure it has to be assumed that in case of an accident one ECC-subsystem fails a second one is under inspection and the third one partially feeds to the break. This means that only one ECC-subsystem would be fully available for cooling the core which is very conservative with respect to the ECCS efficiency.

In plants with only three primary loops - the 900 MW-class - the 4th subsystem has to be connected in a special way with controlling valves to the loops.

When the storage tank filled with borated water is empty, the emergency core cooling pumps take their water from the containment which in the meantime was collected there. This water, however, is preheated and therefore for long term cooling a heat sink

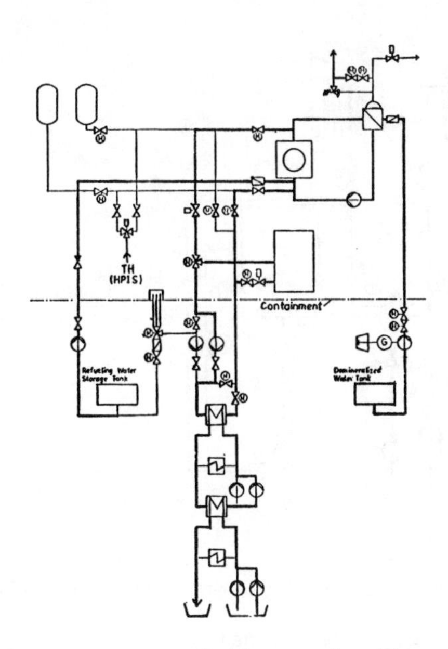

Fig. 7. Principle measures of the protection against external and internal accidents

in the form of a heat exchanger is needed from now on. This heat exchanger is connected with a cooling chain on its secondary side with a river as shown in fig. 7. This cooling chain is designed in two mechanically separated circuits to avoid that any radioactive particles are carried to the river.

On the right hand side of fig. 7 the emergency feed water system is shown which has to become active in case of the loss of all power supply systems. A pump, driven by a Diesel generator injects water to the secondary side of each steam generator in such an amount that the water level there stays constant at a given value while a release valve allows to blow out steam from the secondary side and so reducing the temperature. By this the steam generator can act as a heat sink for the primary system of the reactor and the heat from the core is transported to the steam generator either by liquid natural convection or by vapour flow. In the latter case the steam generator acts on its primary side as a reflux condenser. This heat sink behaviour of the steam generator is also needed in case of a small leak in a primary system, where the mass flow out of the leak is to small to transport enough energy to cool the core. Pressurized water reactors, manufactures by German vendors therefore must have an automatic blowdown system for the secondary side of the steam generators reducing the temperature there by 100 K per hour.

In reality the valve system connecting the ECC-pumps, the heat exchanger, the containment and the water reservoir with the primary loops and the core are much more complicated than shown in fig. 6. A more detailed sketch of the flow paths and the flow diagram for one ECC-subsystem is given in fig. 8. The ECC design

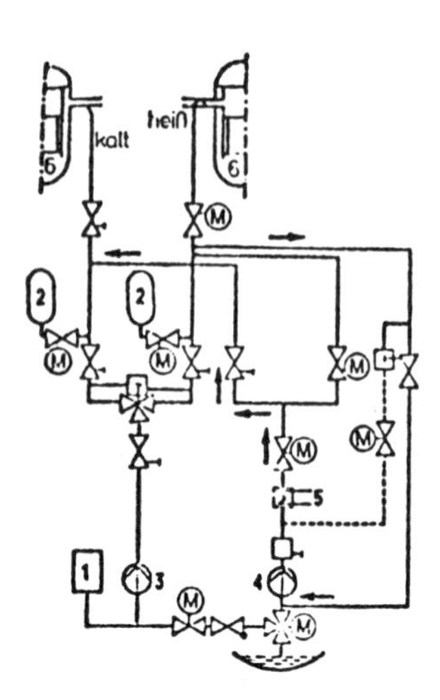

Fig. 8. Emergency core cooling system for KWU-PWR's (1 of 4 subsystems)
1 Store tank for borated water
2 Accumulator
3 High pressure injection pump
4 Post heat removal pump
5 Heat exchanger - post heat removal
6 Reactor pressure vessel

shown there for a PWR of a German vendor, has two accumulators, each connected to the hot- or cold leg respectively instead of one as in older concepts.

Hot leg injection must not be mixed up with the concept of upper head injection as used in US vendors design for long term cooling. The upper head injection has smaller mass flow rates and becomes later active than the hot leg concept. Boiling water reactors have a spray cooling system at the top of the core which injects water into the upper plenum. One reason why such an arrangement is needed for boiling water reactors is the assumption that one or several control rod nozzles at the bottom of the pressure vessel may break and such a flooding of the core with injection of ECC-water from below only, may be difficult. Sometimes also an incore spray system was under discussion which, however, never was manufactured. This incore spray system would have the benefit that the water immediately reaches the fuel elements. Its design due to the narrow core structure, however, is difficult and it cannot be excluded that the very tiny spray nozzles may be blocked under certain conditions.

4. FLUID DYNAMIC AND THERMODYNAMIC PHENOMENA DURING EMERGENCY CORE COOLING

When the ECC-water injection starts, the walls of the pressure vessel are still hot and therefore in the downcomer - as mentioned before - counter current flow may arise with water going

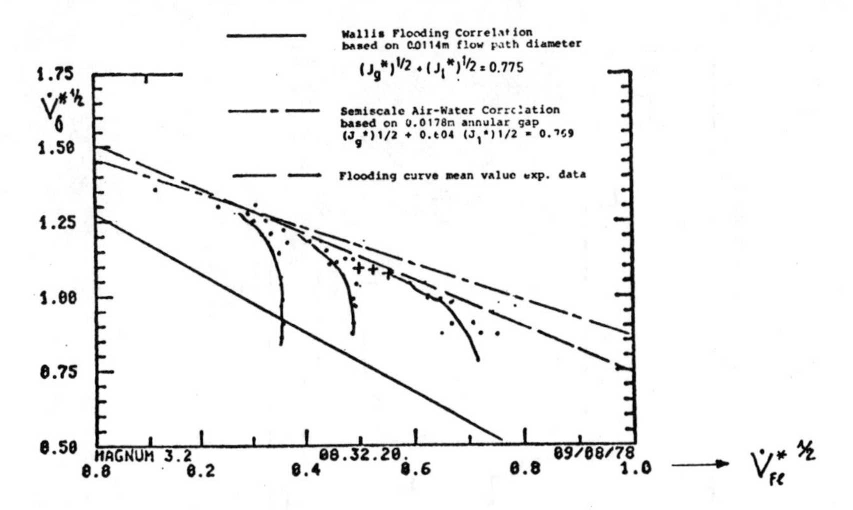

Fig. 9. Countercurrent flow of vapour and gas by Mohr

down and steam flowing up. This counter current flow may be blocked when either the steam flow becomes to high or the water flow to low. This problem of counter current flow and the stability criteria for flooding were under investigation theoretically and experimentally several times. Wallis /6, 7/ presented a flooding correlation based on 0,0114 m flow path diameter which gives an early flooding compared to experiments for example performed in the semiscale test rig /8/, as shown in fig. 9 /9/. One has to be aware of the fact that flooding is besides the volumetric flow rates of vapour and liquid strongly dependent on the geometrical conditions. In a narrow annulus flooding is much more likely than in a wide and open area.

The evaporation process during the refilling period that is when the water penetrates the annular downcomer and fills up the lower plenum, droplets are produced and are carried over into and through the core. These droplets give a significant support to the cooling of the core. Analysing the diameter of these droplets one finds two families of size as demonstrated in fig. 10 /10/. Lee gave a good explanation for the droplet forming mechanism which makes clear why these two classes of size exist. The droplets in the micron range are produced when a vapour bubble penetrating the liquid surface destroys the thin liquid membran formed above it. By disruption of this membran a large number of very tiny droplets are formed. The rising bubble leaves a trough in the liquid behind it which is filled up by inward radial dynamic mass flow. This again produces a water jet directed upward and disintegrating into one or several droplets. This phenomenon produces the large droplets in the millimeter range.

The droplets carried over through the core during refilling and mainly during reflooding may form a water reservoir above the fuel element end-boxes and above the upper tie plate. This water reservoir formation is much stronger with hot leg injection of ECC-water. In this case, however, the water pool in

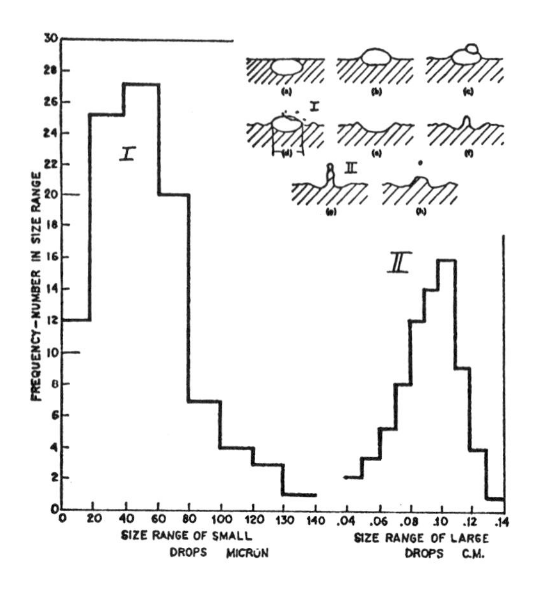

Fig. 1o. Droplet size distribution at carry over of the free surface by Newitt - 1 cm above the liquid surface

the upper plenum is subcooled. The holdup of the water in and above the upper fuel element end-boxes is a function of the steam flow and of the gravity forces. Possible steam water interactions at the upper core support plate during reflood are shown in fig.11.

With saturated water above the upper core support plate, liquid fallback is possible as long as the gravity force is higher than the stagnation pressure of the vapour resulting from the kinetic energy of the vapour flow. With increasing steam flow the liquid becomes suspended in the upper plenum and a froth layer is formed there. Finally when the vapour flow still is enlarged gas tubes may form as long as the liquid layer is not to thick.

With hot leg injection the amount of liquid being present above the upper core support plate, consists mainly of cold ECC-water, i.e. it is under subcooled conditions. This thermodynamic non-equilibrium between steam flowing up and water in the upper plenum changes the interactions between both phases strongly. Steam can condens when touching the water surface or when penetrating through the froth layer. This condensation is combined with a strong local and temporal depressurization effect which accelerates the liquid downwards through the fuel element end-boxes into the core region and such improving the cooling and quenching efficiency.

The mentioned steam binding effect is not only a function of the amount of steam flowing in the hot leg but also strongly dependent on the liquid carried along by the steam, i.e. the two-phase quality in the hot leg. Therefore the realistic correlations for predicting the ECC-behaviour need reliable information

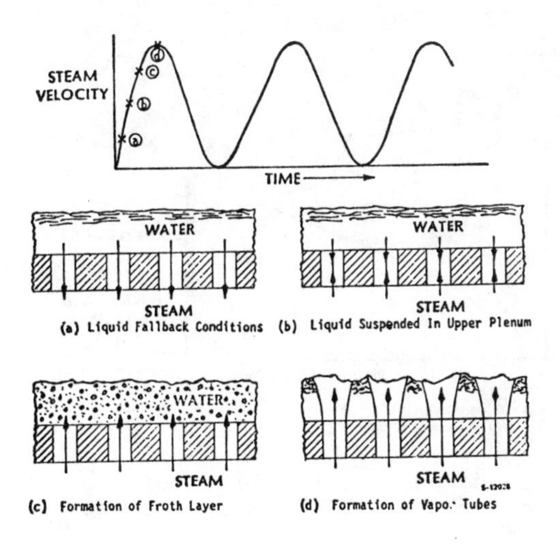

Fig. 11. Possible steam-water interactions at upper core support plate during reflood

about the amount of liquid carried through the hot leg to the steam generator /11, 12/. The flow path in the upper plenum and the droplet trajectories, however, are very complicated and depend strongly on the design of the upper plenum structure and the amount of water being present there. The control rod shrouds are covered with a liquid layer from which by vapour cross flow as shown in fig. 12, droplets are formed travelling to the main coolant nozzle. Another mechanism for forming droplets can be observed if the structure is hot enough to evaporate the liquid partially and to form bubbles in the liquid layer. Then the same droplet formation takes place as already mentioned in connection with the refilling discussion and demonstrated in fig. 10.

The droplets carried over through the hot leg and the steam generator will be collected in the pump bow and may form a liquid seal there, preventing the vapour produced in the core to flow to

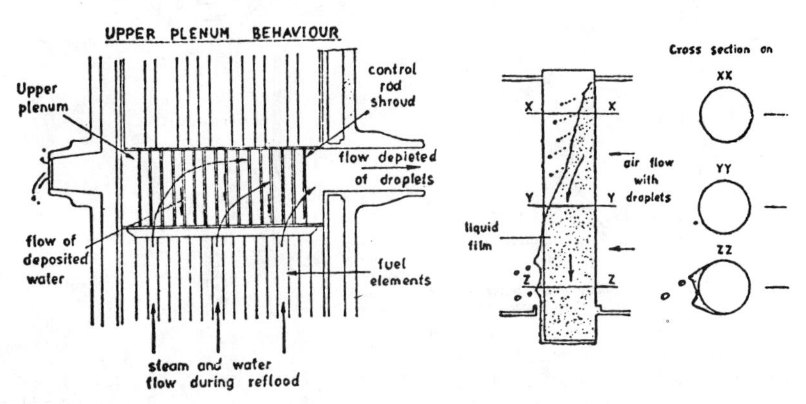

Fig. 12. Flow pathes in the upper plenum

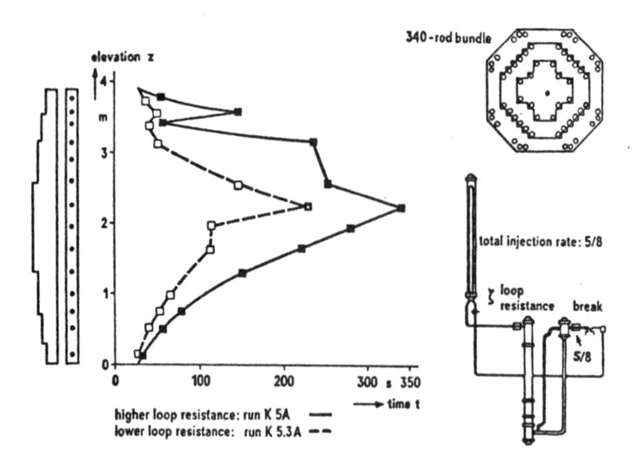

Fig. 13. Comparison of quenchfront progression with cold leg injection influence of loop resistance

the leak in case of a cold leg break. So the loop behaviour may have a strong influence into the cooling efficiency during the reflood phase. Experiments /13/ performed in a test rig imitating the primary loops of a pressurized water reactor and having a electrically heated core of 340 rods, demonstrated this loop influence as shown in fig. 13 where the longitudinal progress of the rewetting front on a rod in the centerline of the core is plotted versus the time. Also with cold leg injection the condensation front - also called quench front - penetrates the core from the bottom and from the top. The progress of the quench front from the top downwards is due to liquid fallback from the upper plenum out of a liquid layer which was formed there by droplet carry over. With high loop resistance, however, it takes much longer until the quench front reaches the middle level of the core, i.e. the rod is quenched completely than with low loop resistance.

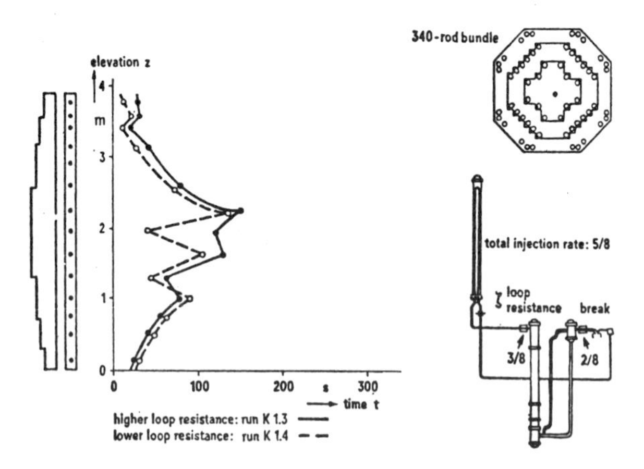

Fig. 14. Comparison of quenchfront progression with combined cold and hot leg injection Influence of loop resistance

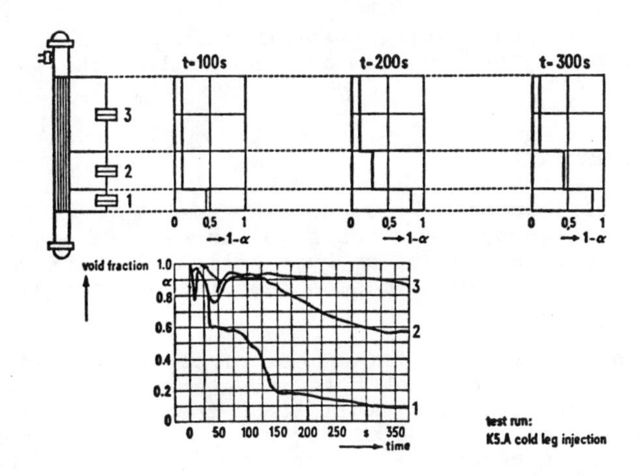

Fig. 15. PKL-Axial coolant distribution

The influence of the loop resistance becomes much smaller if in addition to cold leg injection also ECC-water is injected through the hot leg /15/ which effect can be seen from fig. 14. The vapour produced in the core is now to an large expend condensed in the upper plenum and therefore there is no or less need for good flow conditions through the hot leg.

The quenching of the rods along their complete length is finished before the core is fully filled up with water. Comparing fig. 15 where the void fraction in the core after distinct time intervals is shown, with fig. 13 presenting the quench front progress, measured in the same test, demonstrates that after 300 s the core is approximately only half full with water whereas the rods are almost completely quenched. Hot leg injection here again causes some changes in the reflooding behaviour, as shown in fig. 16. Due to the fact that a part of the ECC-water comes from the top the difference in water level between the downcomer and the core is not the only responsible force for filling up

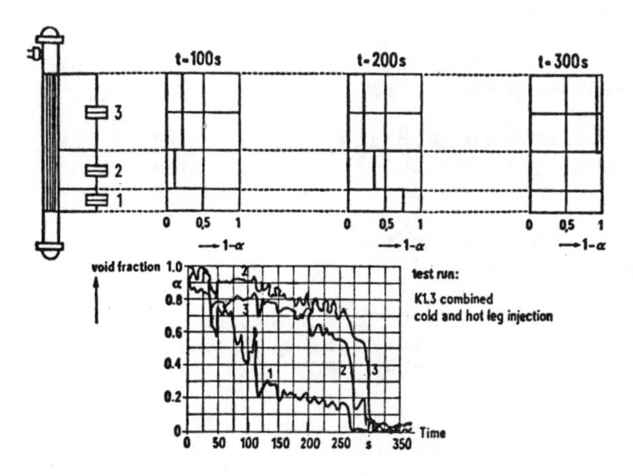

Fig. 16. PKL-Axial coolant distribution

the core. The water penetrating through the upper grid plate downward immediately reaches the core and is added to the water inventory there.

5. CONCLUDING REMARKS

Doubtless all emergency core cooling systems used in a somewhat different design by the vendors guarantee a safe and efficient cooling of the core in case of an hypothetical loss of coolant accident. Each of the special design features has certain advantages and disadvantages as well.

The highest probability for a loss of coolant accident, however, is not with a large break but with a small leak anywhere in the primary system. With very small leaks the high pressure injection system - see left side of fig. 7 - may be efficient enough to keep the pressure in the primary loops above the level where the emergency core cooling system - also the high pressure pump - can became activ. But also with acting high pressure emergency pumps the energy transport through the leak may be much smaller than the decay heat produced in the core. Therefore another heat think - namely the steam generator - is needed. An automatic control of this heat transportation mechanism seems to be very important to avoid operators mistakes.

Licensing rules require a different grade of redundancy of the emergency cooling systems . Closely connected with redundancy deliberations is the question up to what extent the different ECC-subsystems may be mashed together or whether a complete separation of each subsystem - electrically and hydraulically - has to be required. Possibilities for interconnection of subsystems may improve the efficiency of the whole ECC-system because a failure in one part can perhaps be overcome by using another subsystem. On the other hand, however, the probability for a failure is much higher if a defect in one component could influence more than one subsystem.

REFERENCES

1. RSK Leitlinien für Druckwasserreaktoren, available from GRS Köln, Glockengasse 1, FRG

2. Smidt, D., Reaktor-Sicherheitstechnik, Sicherheitssysteme und Störfallanalyse für Leichtwasserreaktoren und schnelle Brüter, Springer-Verlag Berlin-Heidelberg-New York, 1979

3. Wallis, G. B., One-dimensional Two-phase Flow, McGraw-Hill Book Company 1969, Kap. 4, S. 89

4. Mayinger, F., Entwicklungen und Kenntnisstand bei der Notkühlanalyse von Leichtwasserreaktoren, Atomkernenergie, Verlag Karl Thiemig, München, Bd. 32, 1978, Lfg. 4, S. 220 - 228

5. Mayinger, F., Winkler, F., Hein, D., The Efficiency of Combined Cold and Hot Leg Injection, Trans.Amer.Nucl.Soc. 26, 1977, S. 4o8-4o9

6. Block, J.A., Wallis, G.B., Effect of Hot Walls on Flow in a Simulated PWR Downcomer during LOCA, Creare Inc. Report TN-188, May 1974

7. Wallis, G.B., An Analysis of Cold Leg ECC Flow Oscillations, Creare Inc. TN-196, Aug. 1974

8. Semiscale Mod-1 Program and System Description for Reflood Heat Transfer Tests, ANCR-NUREG-1305, May 1976

9. Mohr, C.M., Lopez, D.A., EG & G. Idaho Falls: Quick Look Report on supplemental air-water upper plenum tests 1978

10. Newitt, D.M., Liquid Entrainment, 1. The mechanism of drop formation from gas or vapour bubbles, Trans.Inst.Chem.Eng. Vol. 32 (1964)

11. Viecenz, H.J., Blasenaufstieg und Phasenseparation in Behältern bei Dampfeinleitung und Druckentlastung, Dissertation der Universität Hannover, 198o

12. Langner, H., Untersuchungen des Entrainment-Verhaltens in stationären und transienten zweiphasigen Ringströmungen, Dissertation der Universität Hannover, 1978.

13. Technischer Bericht zu Forschungsvorhaben RS 36-RS 36/1 Notkühlprogramm, Niederdruckversuche, DWR-Wiederauffüllversuche, KWU Erlangen, RE 23/060/76, Dez. 1976

12 REFLOOD AND REWET HEAT TRANSFER

by

F. Mayinger

Univ. Hannover, Hannover, FRG

The final phase of the design basis accident in a water reactor is described in this chapter. Both the flow phenomena and heat transfer phenomena which occur during reflooding of the core and rewetting of the fuel rods are detailed, including the influence of various parameters relating to the reactor conditions.

1. INTRODUCTION

Licensing calculations usually assume that after the end of blowdown following a loss of coolant accident with a large or medium sized break the fuel rods of the core are unwetted with cladding temperatures far above the saturation conditions of the primary coolant. The emergency core cooling water injected by the accumulators and the pumps cannot wet immediately the core but a precooling procedure by vapour and two-phase mixtures has to take place before rewetting and by this a sudden and large temperature decrease can occur. There are good reasons for drawing the conclusion that these assumptions concerning the thermal impact onto the fuel rods are too conservative and that a large part of the rods - at least the low and medium powered ones - will not be exposed to high cladding temperatures due to the fact, that the emergency cooling water from the accumulators is effective early enough i.e. before a very large temperature rise in the cladding, due to decay heat production and heat conduction from the pellets can take place. In the following, however, a high temperature arise will be imputed to describe all fluid dynamic and thermodynamic effects.

The quenching front along each rod in the core can progress from the bottom and from the top, because the precooling process carried over enough water to the upper plenum which may fall or flow back again.

Discussing the reflood and rewet heat transfer we have to take in account all flow phenomena and heat transfer mechanisms being possible under these complicated thermohydraulic conditions.

2. FLOW PHENOMENA

Discussing the heat transfer during reflood we have to distinguish between wetted and unwetted portions of the core. The heat transfer mechanism in the unwetted region is more complicated and shows a higher variety of possibilities than in the wetted one. However, even in the wetted region we have to distinguish between a part where nucleate boiling prevails and another where

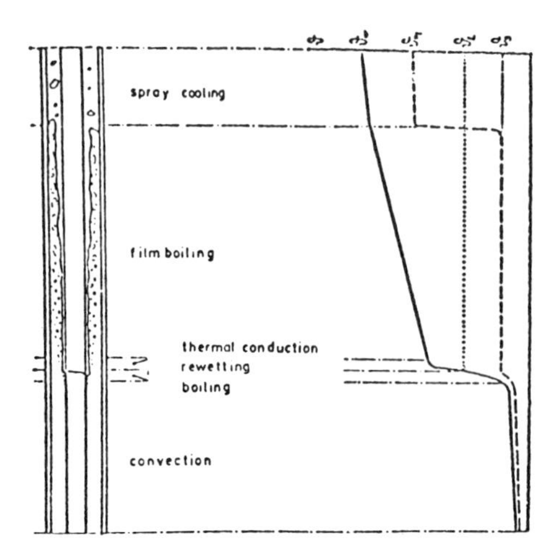

Fig. 1. Heat-transfer areas; surface and fluid temperatures

heat is transported by liquid convection only, as long as flooding from the bottom is concerned. In the unwetted region we can distinguish three heat transfer conditions, namely film boiling, spray cooling, and transition boiling. All three conditions are generally called post dryout heat transfer /1/.

With bottom flooding the liquid level may be far above the rewetting boundary at the cladding of the rod, as demonstrated in fig.1.

Water flowing down from the top in the form of a falling film, can only wet the rod as long as the cladding temperature is not too far above its saturation temperature. At the border to the high temperature region which in this case is downwards from the

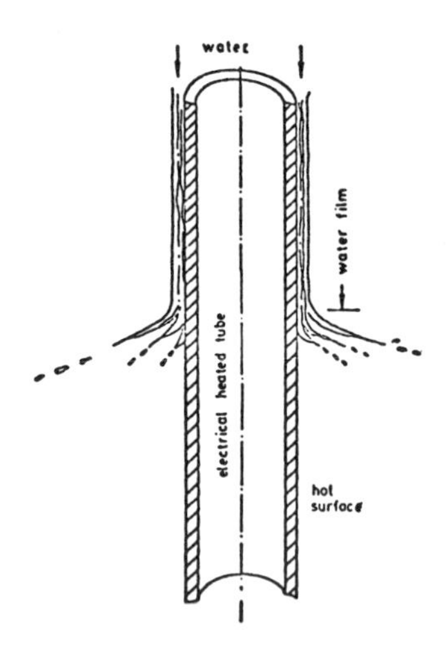

Fig. 2. Spraying

wetted part sputtering occurs as shown in fig. 2. These sputtering phenomenon may be due to a violent local evaporation and momentum effects by bouyancy induced vapour upflow. If the steam production in the core is very high, strong counter current flow effects may exist like in fig. 3, and finally the water downflow may be blocked by the shear stress and momentum transfer from the upflowing steam. This flooding condition is usually not very stable and so chugging occurs, a flow phenomenon, where periodically water is ejected upward out of the core and is falling back again into it.

The simplest way to consider the temperature change in the cladding with respect to time and elevation is to derive the heat flux to the coolant from the boiling curve given by Nukiyama /2/. Under flooding conditions at the bottom of the core - zone I in fig. 4 - heat is transported by liquid convection which in upward direction (zone II) changes into nucleate boiling. The heat flux reaches a peak just beneath the wetting front and before boiling crisis conditions. In the wetting area (zone III) transition boiling may exist and upwards from this position film boiling (zone IV) and finally spray cooling with fog flow (zone V) is responsible for cooling the rod. At the wetting front and in the transition boiling region, the rod undergoes temporarily and locally a sudden temperature drop, as shown on the right hand side of fig. 4 and in fig. 5. Not always spray cooling is preceded by film boiling. This is dependent on the flooding velocity and the temperature of the cladding.

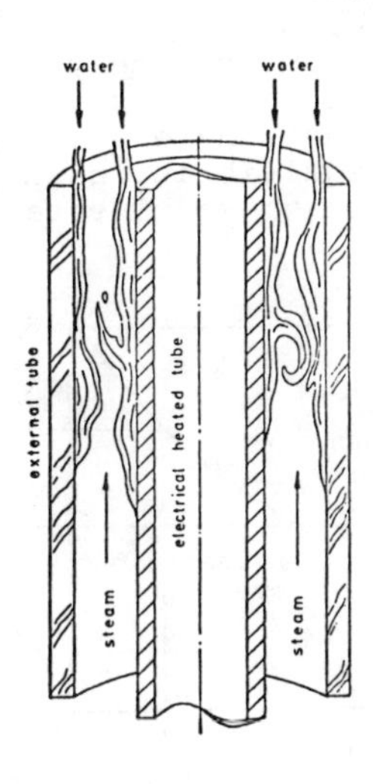

Fig. 3. Flooding

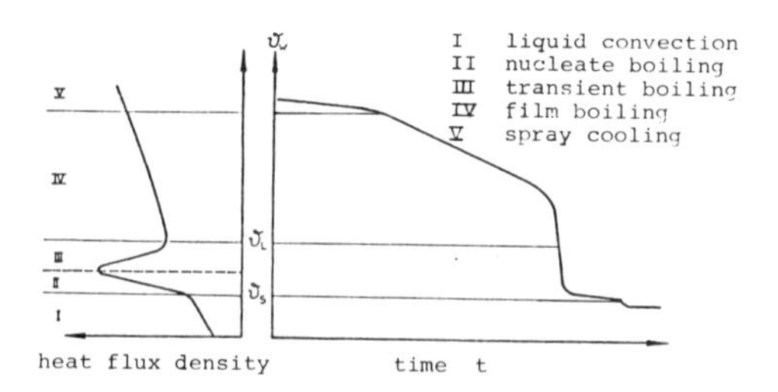

Fig. 4. Coordination of the boiling curve given by Nukiyama at the progress of the temperature under flooding conditions

3. HEAT TRANSFER IN THE UNWETTED REGION

3.1 SPRAY COOLING AND DISPERSED FLOW HEAT TRANSFER

In dispersed flow small droplets are embedded and suspended in vapour flowing through the cooling channels formed by the fuel rod array. The droplets can not wet the wall, however, have the ability to cool down the boundary layer near the heated wall to an large extend. Former theories for predicting the heat transfer efficiency of spray cooling therefore started from the wellknown equations for single phase vapour flow and added a correction term for considering the cooling improvement due to the droplets by lowering the boundary layer temperature /3, 4/. Examples for heat transfer correlations based on these assumptions are given in table 1.

An important simplification in these correlations is the assumption that vapour and droplets are in thermodynamic equilibrium i.e. that they have approximately the same temperature. In reality, however, there is large dis-equilibrium between vapour and the droplets as measurements have shown and as demonstrated in fig. 6. Equations based on equilibrium assumptions predict a much lower cladding temperature than experimentally observed.

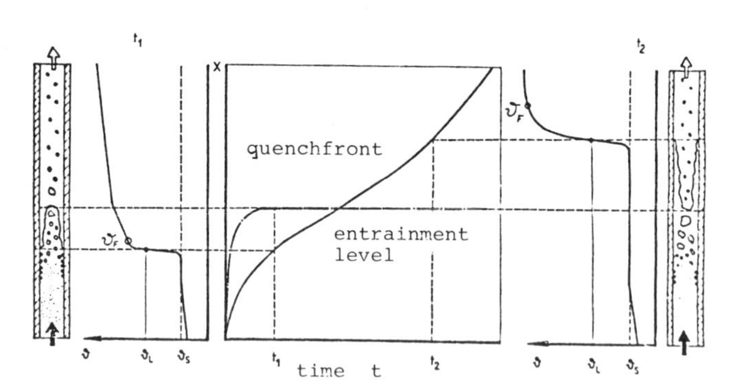

Fig. 5. Quenching with bottom flooding

Equations and references (ref. see reference list in /1/)	Range of applicability:				Comments
	p [bar]	$\dot{m} \left[\frac{kg}{m^2 s}\right]$	$\dot{x}$	geometry	
Miropolski (1963): $\alpha = 0{,}023 \cdot \frac{\lambda_V}{d_h} Re^{0,8} \cdot Pr^{0,8} \cdot Y$ /67/ $Y = 1 - 0{,}1 \cdot (\frac{\rho_V}{\rho_L} - 1)^{0,4} (1-\dot{x})^{0,4}$	40 ... 220	700...2000	0,06...1,0	tubes	
Dougall-Rohsenow (1963) $\alpha = 0{,}023 \cdot \frac{\lambda_V}{d_h} Re^{0,8} \cdot Pr^{0,4}$ $Re = \frac{\dot{m}\, d_h}{\eta^*} \cdot (\frac{\rho_V}{\rho_L}(1-\dot{x}) + \dot{x})$	< 35	1660...3650	< 0,5	tubes	
Groeneveld (1959) $\alpha = 3{,}27 \cdot 10^{-3} \frac{\lambda_V}{d_h} Re_V^{0,9} Pr^{1,33} \cdot Y^{-1,5}$ /23/	34 ... 215	700...5300	0,1 ... 0,9	tubes and annulus	Y see Miropolski (1963)
Groeneveld (1971) - Regressionsanalyse $Nu = a\,(Re_V \cdot (\dot{x} + \frac{\rho_V}{\rho_L}(1-\dot{x})))^b\, Pr_w^c\, Y^d\, q^e\, \frac{D_{heat}}{D_{hyd}}$	68 ... 215	700...5300	0,1 ... 0,9	tubes and annuli	Coefficents and exponents see ref./20/present paper
Bishop (1964) /58/ $\alpha = 0{,}0193\, Re^{0,8} \cdot Pr^{1,23} \cdot (\frac{\rho_V}{\rho_L})^{0,068} \cdot (\frac{\rho_V}{\rho_L}(1-\varepsilon)+\varepsilon)^{0,18}$	165 ... 215	1350...3400	0,1 ... 1,0	tubes	
Herkenrath, Mörk-Morkenst. (1969) /56; 57/ $Nu = 0{,}06 \cdot \{(Re_F \cdot (\dot{x} + \frac{\rho_V}{\rho_L}(1-\dot{x})) \cdot \frac{\rho_W}{\rho_V} \cdot Pr_W\}^{0,8} \cdot \ldots \cdot (\frac{\dot{m}}{1000})^{0,4} \cdot (\frac{p}{p_{CRIT}})^{2,7}$	140... 220	750...4100	0,1 ... 1,0	tubes	$\dot{m}$ in kg/m²s
Cumo, Brevi /54/ (1969) $\alpha = 8{,}9 \cdot 10^{-3} \cdot \frac{\lambda_V}{d_h} \cdot Re^{0,84} \cdot Pr^{1,33} \cdot \left(\frac{1-\dot{x}_{Do}}{\dot{x}-\dot{x}_{Do}}\right)^{0,124}$	50	500...3000	0,4 ... 1,0	tubes	Two-step model: includes droplet-wall heat-transfer
Polomik (1961,1961,1967): /43/ $Nu = 1{,}36 \cdot 10^{-3} \cdot Re_f^{0,853} Pr_f^{0,33} (\dot{x}/1-\dot{x})^{0,147} (\rho_V/\rho_L)^{0,66}$; $Nu = 0{,}0039 \cdot (Re_V \cdot (\dot{x} + \rho_V/\rho_L \cdot (1-\dot{x})))^{0,9}$; $Nu = 0{,}023 \cdot Re_f^{0,292} \cdot Pr_f^{0,33} \cdot (\dot{x}/1-\dot{x})^{0,01} \cdot \dot{q}^{0,47}$ $\dot{q}$ (BTU)	55 ... 100	1000...2450	0,4 ... 0,7	annuli	Coefficients and exponents obtained from least error analysis using high pressure data
Lee (1970): /53/ $T_W - T_{SAT} = 1{,}915 \left[\frac{\dot{q}}{\dot{m}\,(\dot{x} + \frac{1-\dot{x}}{4,15})}\right]$	140 ... 180	1000...4000	0,3 ... 0,75	tubes	
Tong (1964): /70/ $Nu = 0{,}005 \left(\frac{D\, \bar{w}\, \rho_v}{\eta_v}\right)^{0,8} \cdot Pr_W^{0,5}$	>138	>700	> 0 1	tubes	equation should be used at low quantity subcooled conditions
Polomik (1967): /48/ $Nu = 0\,00115\,(Re_V)^{0,9}\, Pr_V^{0,3} (\frac{T_W}{T_{SAT}} - 1)^{-0,15}$	40... 100	700...2700	0,2 ... 1,0	2 rod bundles	
Collier (1972): /79/ $\dot{q} \left[\frac{D^{0,2}}{(\dot{m} \cdot \dot{x})^{0,8}}\right] = \frac{1}{389} (T_W - T_{SAT})^m$ $m = 1{,}284 - 0{,}00312 \cdot \dot{m}$	69	> 1000	0,15...1,0	tubes and annuli	$T_W - T_{SAT} < 200\ ^\circ C$
Bishop (1965): /65/ $Nu = 0{,}055 \left[Re_W \cdot \frac{\rho_{V,W}}{\rho_V} (\dot{x} + \frac{\rho_V}{\rho_L}(1-\dot{x}))\right]^{0,82} \cdot Pr_W^{0,96} \cdot \ldots \cdot (\frac{\rho_V}{\rho_L})^{0,34} \cdot (1 + \frac{26,9}{L/D})$	165... 215	1350...3400	0,1 1,0	tubes	

Table 1. Post-Dryout Correlations

Assumptions with frozen flow and maximum possible disequilibrium give better agreement, however, overpredict the cladding temperature.

Newer models developed for example by Ganic /5/ or by Iloeje /6/ take in account thermodynamic non-equilibrium and treat in detail the hydrodynamic and thermodynamic interactions at the phase boundaries.

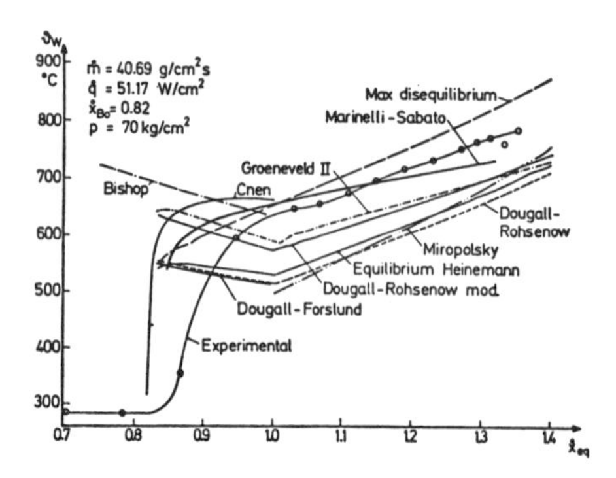

Fig. 6. Comparison of experimental Post-Dryout data with correlations assuming thermodynamic equilibrium

Because the temperature of the steam is not constant across the flow area, the droplets which enter the superheated boundary layer next to the wall, without actually wetting the wall, are more likely to evaporate than droplets in the less superheated steam core. Finally, at high wall temperatures the heat radiation to the steam and the droplets has to be taken into account. Thus the following effects and possible heat transport paths from the wall to the fluid are to be found:

a) Heat transfer from the wall to the liquid drops which reach the thermal boundary layer without wetting the wall: so called dry collisions.

b) Heat transfer from the wall to the droplets which temporarily wet the wall, so called wet collisions.

c) Convective heat transfer from the wall to the steam.

d) Convective heat transfer from the steam to the droplets entrained in the vapour core.

e) Radiation heat transfer from the wall to the liquid droplets.

f) Radiation heat transfer from the wall to the steam.

Short-term wetting of the wall by liquid droplets with sufficiently high kinetic energy can only occur at moderate excess temperatures, in which case Iloeje /6/ assumes that, as shown in fig. 7, steam bubbles are formed on the contact surface of the droplet which then disperse the liquid into smaller droplets. These are then catapulted away by the expansion effect of the developing steam, and a thin liquid film is left on the wall which is later evaporated.

3.2 FILMBOILING

With film boiling the hot surface is separated by vapour film

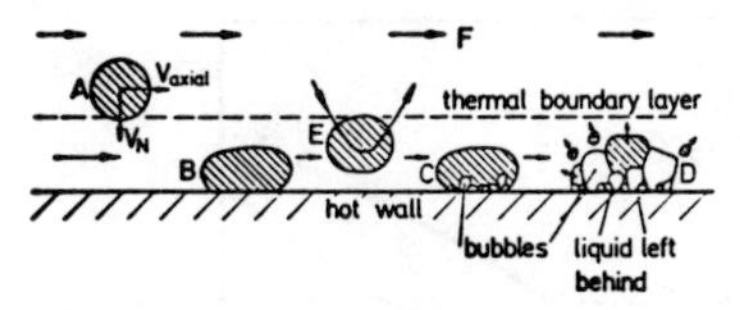

Dispersed flow heat transfer process

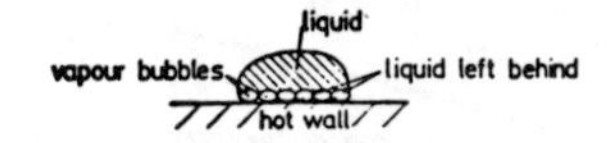

Idealized bubble geometry at end of bubble growth - square array

Fig. 7. Dispersed flow heat transfer process

from the liquid. The simplest way to correlate the heat transfer through the vapour film, starts from the assumption that the vapour film flow is laminar with a linear temperature distribution. If in addition the shear stress between the vapour and the water is neglected, one gets - as shown by Collier /7/ for a vertical wall

$$\frac{h(z)\,z}{k_V} = \sqrt[4]{\frac{z^3 \cdot g \cdot \varrho_V \cdot (\varrho_L - \varrho_V) \cdot h_{fg}}{4 \cdot k_V \cdot \eta_V \cdot (\vartheta_W - \vartheta_S)}} \tag{1}$$

Equation (1) contains the temperature difference $\Delta\vartheta = (\vartheta_W - \vartheta_S)$ between the wall and the saturation temperature of the liquid which means that an iterative solution for the heat transfer coefficient h is necessary. In addition the position where the unwetted region starts from (z = 0) has to be known.

For horizontally orientated cylinders and natural convective flow, Bromley /8/ gave the equation

$$h = 0{,}62 \cdot \sqrt[4]{\frac{k_V \cdot g \cdot \varrho_V \cdot (\varrho_L - \varrho_V) \cdot h_{fg}}{\eta_V \cdot \Delta\vartheta \cdot D_{Zyl}}} \tag{2}$$

The situation changes considerably if the liquid surrounding the vapour film is subcooled. The heat transfer then may be strongly increased due to condensation effects at the vapour-liquid interface.

Theoretical models describing the heat transfer with film boiling in subcooled liquids were presented by Sparrow and Cess /9/, Nishikawa and Ito /10/ and Lauer /11/. There are large differences in the heat transfer coefficient predicted by these theories as shown in fig. 8, where also some experimental data measured by Lauer are given. The equation by Lauer is quite simple

$$\dot{q}''(z) = \left(\frac{k_V}{\delta_V} + h_{rad}\right)(\vartheta_W - \vartheta_S) \tag{3}$$

Lauer takes in account a heat transfer coefficient h_{Str} due to

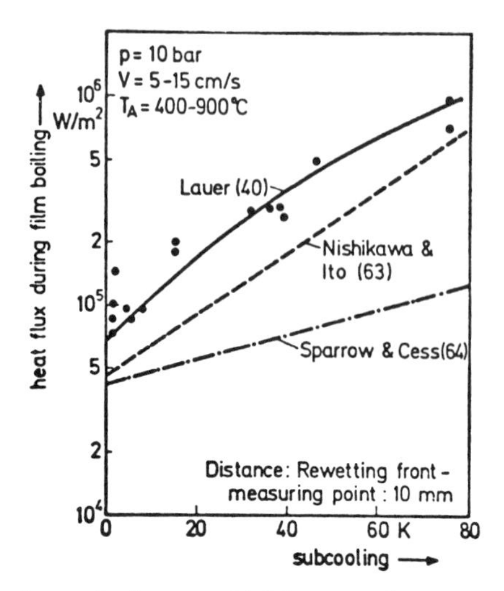

Fig. 8. Comparison between different heat transfer models concerning film pool boiling with subcooled liquid

radiation and calculates the local thickness δ_V of the vapour film using an energy balance over the vapour mass flow rate produced.

3.3 TRANSITION BOILING

With transition boiling the situation is much more confused than with film boiling and dispersed flow cooling because due to the difficult experimental conditions and the complicated thermo-hydraulic conditions only a few measurements exist in the literature. The heat transport process with transition boiling gives heat flux quantities between film boiling and nucleate boiling. Liquid may temporarily wet the surface, however, still cladding-vapour heat transport prevails. A comparison of data measured by Groeneveld /12/ and a correlation by Hsu /13/ shows poor agreement as seen from fig. 9.

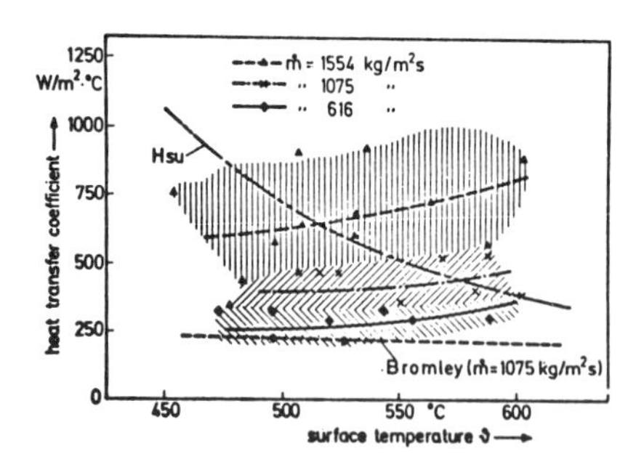

Fig. 9. Transient film boiling heat transfer coefficients versus surface temperature

Especially for predicting the fuel rod behaviour under high pressure emergency core cooling - for example with small leaks - experimental and theoretical research in transition boiling is needed.

4. REWETTING PROCESS

4.1 LEIDENFROST PHENOMENON

The Leidenfrost phenomenon already known since approximately 200 years, is still not fully understood from the thermodynamic point of view. Leidenfrost temperatures measured in water (see fig. 10) and in freon (see fig. 11) give values which lay 100 K (water) respectively 40 K (freon) above the saturation temperature at low pressures. At high pressures the Leidenfrost temperature is more closely to the saturation temperature.

The Leidenfrost temperature can be strongly influenced by the roughness of the surface to be wetted and the flow conditions near the wetting point. With water a sudden increase in the wall superheating - i.e. in the Leidenfrost temperature from 250 K at 1 bar to 300 K at 5 bar - was observed by several experimentators, which is very important for emergency core cooling with large breaks and low containment pressure.

Similar as for transition boiling subcooling of the liquid - i.e. of the emergency core cooling water - has a high influence on the Leidenfrost temperature. Fig. 12 shows that with increasing subcooling, rewetting can be observed at higher temperatures and even at 1 bar pressure and only 25 K subcooling, which corresponds to a temperature of approximately 75°C in the ECC water, the cladding of the fuel element can be quenched even at temperatures as high as 500°C.

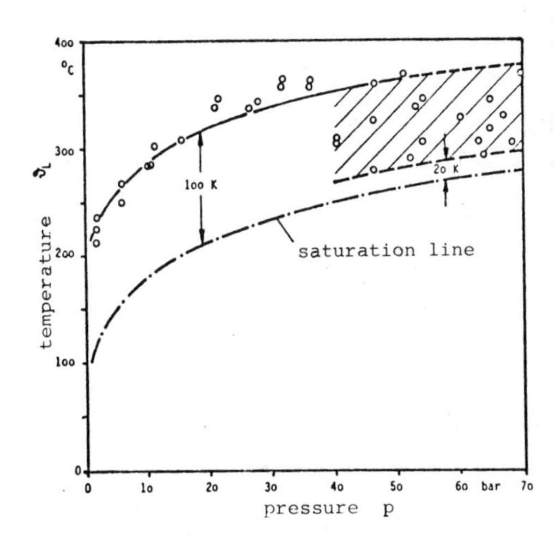

ig. 10. Leidenfrost temperature as a function of pressure
Fluid: water

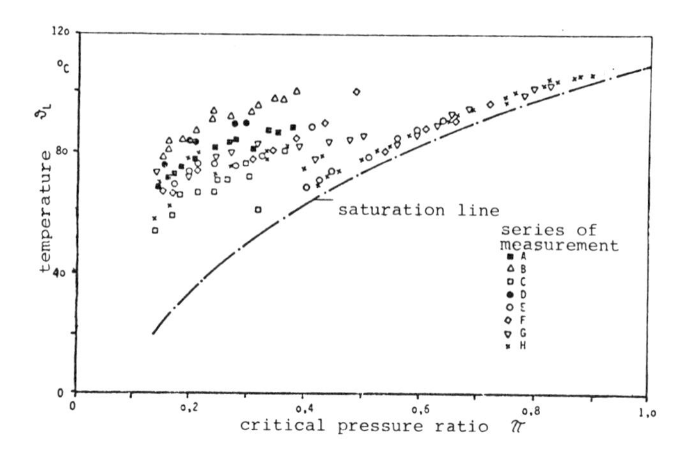

Fig. 11. Leidenfrost temperature as a function of critical pressure ratio; Fluid: freon R 12

4.2 HEAT CONDUCTION IN THE CLADDING

A energy balance around the border between the filmboiling or spray cooling region and the nucleate boiling region, shows that axial heat conduction in the material of the cladding and the velocity of the progress of the wetting front play an important role. For a first rough estimation a sudden change in the heat transfer coefficient from nucleate boiling to film boiling

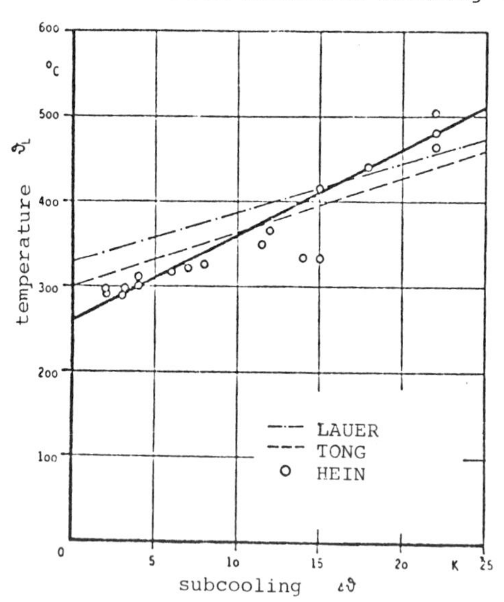

Fig. 12. Leidenfrost temperature as a function of subcooling of the liquid at 1 bar

or spray cooling can be assumed at the border between the wetted and unwetted region. This sudden change in the heat transfer coefficient may be assumed to travel with constant velocity along the rod. For various moving velocities u of the sudden change in the heat transfer coefficient different temperature-courses are caused at the wetting front as shown in fig. 13. With higher travelling velocities of the wetting front the temperature in the wall at the point where the sudden jump in the heat transfer coefficient is assumed increases. Taking in account that the re-wetting and by this the jump in the heat transfer coefficient is fixed to a certain temperature - Leidenfrost temperature - (x = 0 in fig. 13) than one can draw the conclusion that for a given jump in the heat transfer coefficients for nucleate boiling and film boiling and for a given axial heat conduction only a single velocity u can exist, fulfilling all these conditions. This velocity u corresponds to the wetting front progress velocity, derived from Yamanouchi /15/ and other authors from an energy balance

$$u = \frac{1}{\varrho \cdot c} \cdot \sqrt{\frac{h \cdot k}{\delta}} \cdot \sqrt{\frac{(\vartheta_0 - \vartheta_S)^2}{(\vartheta_W - \vartheta_S)(\vartheta_W - \vartheta_0)}} \tag{4}$$

Yamanouchi proposes to take for the heat transfer coefficient h in the wetted region a constant value of $h = 5.10^4$ W/m²K and to use for the temperature difference $(\vartheta_O - \vartheta_S)$ the value of 100 K.

With this travelling velocity we can write an energy balance for the cladding around the wetting front which reads in a two dimensional form

$$\frac{\partial^2 \vartheta}{\partial r^2} + \frac{1}{r} \cdot \frac{\partial \vartheta}{\partial r} + \frac{\partial^2 \vartheta}{\partial x^2} - \frac{\varrho \cdot c}{k} \cdot u \cdot \frac{\partial \vartheta}{\partial x} + \frac{q'''}{k} = 0 \tag{5}$$

which also includes a heat source density q''' taking in account the decay heat by radioactive products. There are various solutions of equation (5) in the literature /15, 16/.

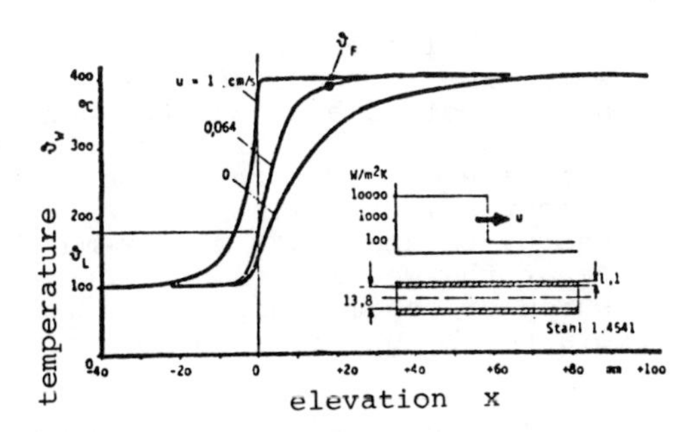

Fig. 13. Shape of the temperature curve in the region of the moving wetting front

Literaturstelle	verwendete experimentelle Werte	Wärmeübergangskoeffizienten W/m²K	ϑ_L °C	Bemerkungen
Yamanouchi /Y1/ Semeria und Martinet /S9/	Yamanouchi /Y1/	$\alpha_2 = 2\cdot10^5 - 10^6$ $\alpha_3 = 0$	150	α_2 α_3
Sun et al. /S8/	Yamanoucni /Y1/ Duffey und Porthouse /D4/	$\alpha_2 = 1{,}7\cdot10^4$ $\alpha_3 = \frac{\alpha_2}{N} e^{-0{,}05x}$	260	Vorkühlung berücksichtigt α_2 α_3
Chun und Chon /C7/	Case et al /C10/	$\alpha_2 = 2{,}56\cdot10^4$ $\alpha_3 = 170$ $\alpha_4 = 0$	260	Länge des Nebelkühlbereiches und des Wassermitrisses zur Berechnung von α_3 α_2 α_3 α_4
Thompson /T4/	Bennett et al /B18/	$\alpha_1, \alpha_2 \approx 7\cdot10^6$ (Maximalwert) $\alpha_3 = 0$	$\vartheta_S + 100$	Druck 7 - 70 bar $\alpha_2 - R\cdot\Delta\vartheta_S^3$ α_3
Elias und Yadigaroglu /E6/	Duffey und Porthouse /D4/	$\alpha_1 = 170$ $\alpha_2, \alpha_3, \alpha_4$ = Approximation der Siedekurve	260	α_1 α_2 α_3 α_i
Andreoni /A1/	Andreoni /A1/	α_1 mit Jens-Lottes α_2 aus Benetzungsversuchen α_3 aus Experiment	$\vartheta_S + 200$	α_1 α_2 α_3

Table 2a. Heat conduction models

4.3 REWETTING MODELS

For describing the reflood and rewet heat transfer, the physical conditions in the immediate neighbourhood of the quenching front have to be wellknown. One reason for this is, that in this region large quantities of heat are destored from the wall producing vapour. For solving the heat balance and heat conduction equation (5) different assumptions were made in the literature with regard to the course of the heat transfer coefficients at the quenching front. A survey of these assumptions was made by Elias and Yadigaroglu /17/ and is presented in table 2a and b. The references shown in these tables can be found in the thesis of Hein /14/. All these rewetting theories are based on energy balances and differ mainly in assuming the heat transfer course at the wetting front. Future research has to clarify how the measured or assumed courses in the heat transfer coefficients can be explained by physical phenomena.

5. REACTOR CONDITIONS - INFLUENCE OF DIFFERENT PARAMETERS

Under reactor conditions there is a close interaction between the reflood process in the core and the flow effects in the loops - unbroken and broken - of the primary system. The flooding velocity

Literaturstelle	verwendete experimentelle Werte	Wärmeübergangskoeffizienten W/m²K	ϑ_L °C	Bemerkungen
Duffey und Porthouse /D5/	Yamanouchi /Y1/ Duffey et al. /D5/ Andreoni /A1/ Martini /M1/ Thompson /T2/ Campanile /C9/	$\alpha_2 = 10^4 - 2\cdot 10^6$ $\alpha_3 = 0$	190 - 250	α_2 α_3
Coney /C5/	Bennett /B12/	$\alpha_2 = 0{,}95 - 1{,}3\cdot 10^6$	$\vartheta_S + 68$	Druck: 7 - 70 bar α_2 α_3 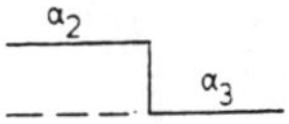
Blair /B10/	Thompson /T2/	$\alpha_2 = 1{,}7\cdot 10^4$ $\alpha_3 = 0$	260	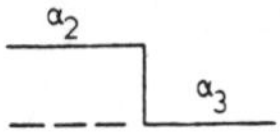 Zylindergeometrie α_2 α_3
Thompson /T4/	Bennett /B18/	$\alpha_1, \alpha_2 = 4 - 8\cdot 10^5$ (Maximalwert) $\alpha_3 = 0$	$\vartheta_S + 100$	numerische Lösung α_3 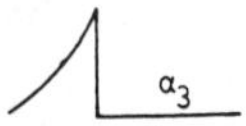
Tien und Yao /T3/				Wiener-Hopf-Verfahren 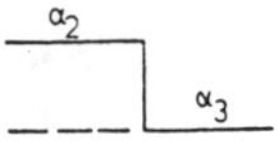 α_2 α_3
Dua und Tien /D2/	Duffey /D4/ Yamanouchi /Y1/	$\alpha_2 = 1{,}7\cdot 10^4$ $q''_3 = \frac{q_0}{N}\cdot e^{-ax}$	260	Wiener-Hopf-Verfahren mit Vorkühlung 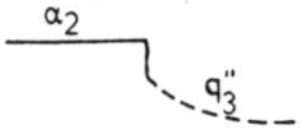 α_2 q''_3

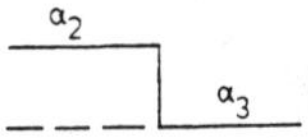

Table 2b. Wetting models

i.e. the rise of the collapsed water level, depends strongly from the pressure drop in the intact loops. Increasing flooding velocity improves the heat transfer and accelerates the progress of the wetting front.

On principal the temporal and local temperature behaviour during reflood is shown in fig. 14. The initial conditions of the experiment were arranged in such a way, that the wall of the test channel in full length was slowly heated up to a temperature of 600°C in a stagnant steam atmosphere and then with starting the flooding the decay heat flux of 3 W/cm² was imposed. Due to the low rising velocity of the water the heat transport mechanism was too weak at the beginning of the flooding to carry away the full decay heat and therefore a temperature increase in the upperpart of the test channel can be observed. Only the lower half is cooled good enough and undergoes an early quenching. In this special example with the very low flooding velocity, it takes almost 500 s until the heat transported by spray cooling is equal to the heat generated by fission product decay. The then following cooling down by dispersed flow or by film boiling goes on until the Leidenfrost temperature is locally reached.

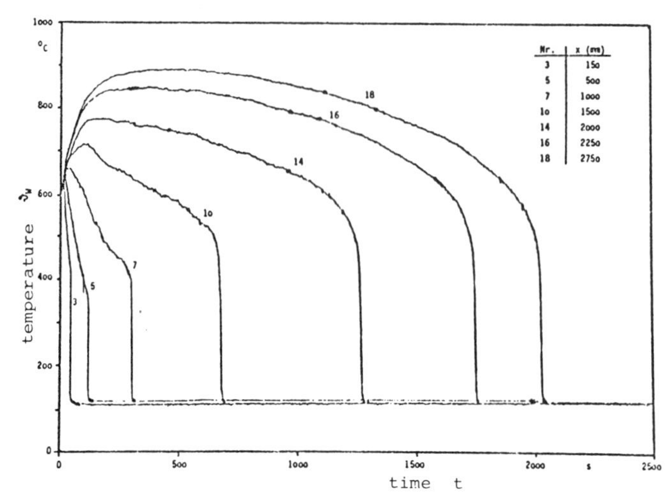

Fig. 14. Local wall temperature under flooding conditions (p = 4,5 bar; w = 2 cm/s, $\Delta\vartheta_E$ = 35 K; q" = 3 W/cm²)

The heat transfer coefficients corresponding to these test conditions are shown in fig. 15 for different heights in the test channel. The figure demonstrates that the spray cooling or film boiling heat transfer coefficient is not strongly dependent from the local position under these low flooding rates. With higher water velocities, however, the quality of the vapour water mixture along the channel increases faster and then the heat transfer coefficients are considerably different in various channel heights.

From these heat transfer conditions which again are a function of several thermo- and fluiddynamic parameters, the velocity of the progress of the quenching front is governed. How the flooding velocity influences this progress of the wetting front is shown in fig.1 for similar initial test conditions like in the fig. 14 and 15, however, with a higher subcooling of the ECC-water. In this special test arrangement, quenching from the top occured only at low flood-

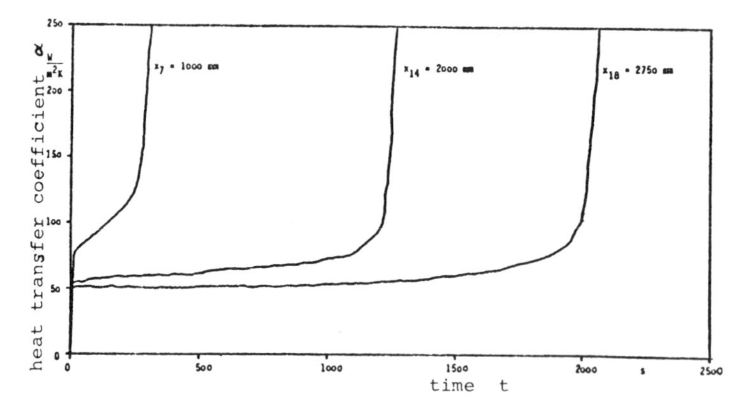

Fig. 15. Local heat transfer coefficient under flooding conditions (p = 4,5 bar; w= 2 cm/s, $\Delta\vartheta_E$ = 35 K; q" = 3 W/cm²)

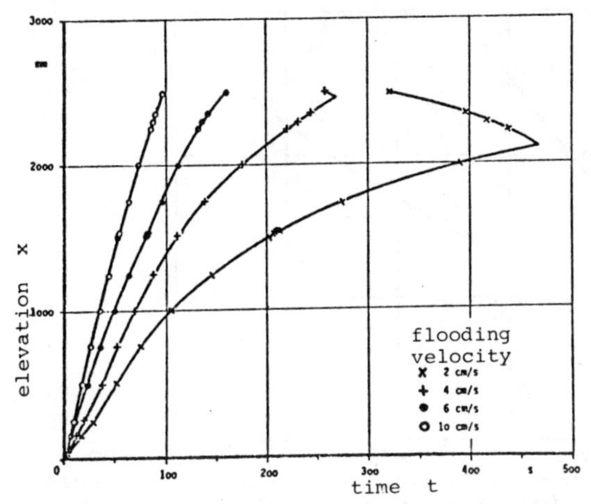

Fig. 16. The influence of the flooding velocity on the progress of the wetting front
(p = 4,5 bar; $\Delta\vartheta_E$ = 75 K; q" = 3 W/cm²)

ing velocities because it takes some time until enough water is carried over to the upper plenum for reflux cooling.

This reflux cooling effect by water fall back is also depending from the grade of subcooling of the ECC-water. Due to condensation effects higher subcooled water carried over to the upper plenum can fall back earlier and by this support the quenching from the top. Saturated water flowing back has poorer wetting abilities. However, the subcooling of the ECC-water also influences the progress of the quench front from the bottom strongly as shown in fig. 17.

With increasing pressure the density of the vapour becomes higher and by this its cooling ability in the unwetted region. This effect can easily and immediately be seen from the wellknown Colborn heat transfer equation, where the Nusselt-number is a function of the Reynolds-number in the 0,75 th power. In addition with in-

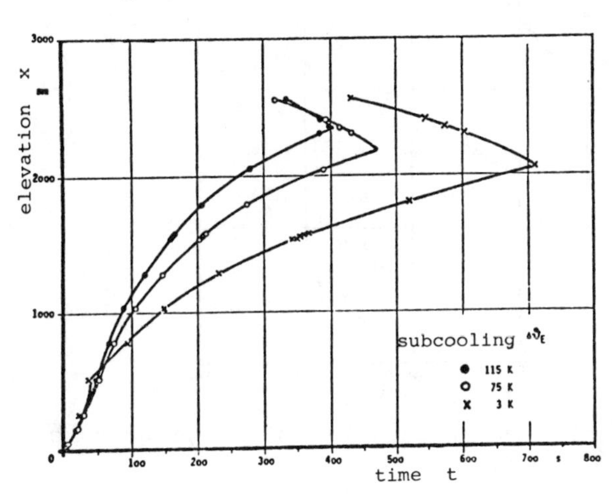

Fig. 17. The influence of the subcooling of the ECCS-water on the progress of the wetting front
(p=4,5 bar, w=2cm/s)

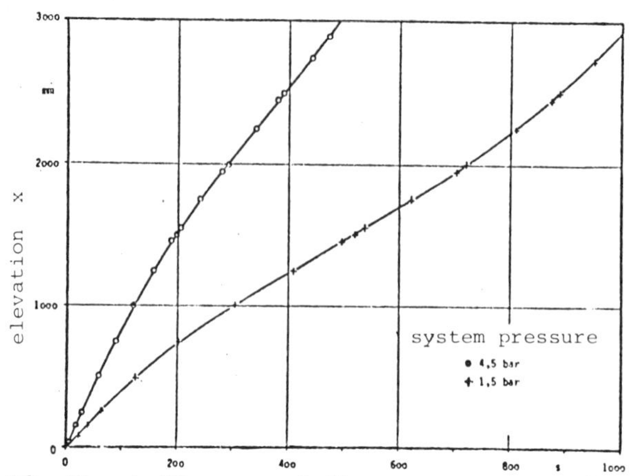

Fig. 18. The influence of the pressure on the progress of the wetting front (w = 6 cm/s, $\Delta\vartheta_E$ = 3 K, q" = 3 W/cm²)

creasing vapour density the carry over of water droplets is enlarged which also improves the cooling. These are the main reasons, why the progress of the wetting is strongly affected by the pressure as demonstrated in fig. 18 where only wetting from the bottom was researched.

It can be easily understood that the amount of decay heat produced in the core has an influence into the wetting progress too. At given heat transfer conditions the energy balance becomes worse at higher decay heat production and the rods are heated up to a larger extent. Which results in a later quenching as shown in fig. 19.

A critical comparison of the information given in the fig. 15 and 16 rise the question, why the velocity of the wetting front is so strongly dependent from the ECC-water flow rate although the heat transfer coefficient is almost not influenced by the rising

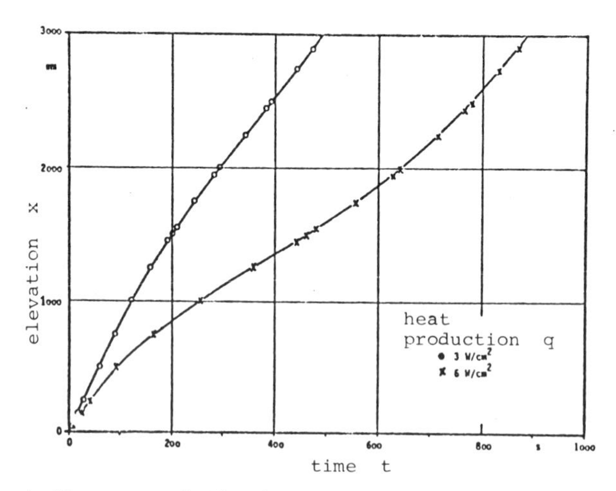

Fig. 19. The influence of the heat production on the progress of the wetting front
(w = 6 cm/s, $\Delta\vartheta_E$ = 3 K, p = 4,5 bar)

velocity of the water. One reason for this is the superheating of the vapour in the core during the flooding process, which is much higher at lower flow rates than at larger ones. There is a large thermodynamic dis-equilibrium and temperature difference between the wall and the vapour as demonstrated in fig. 20. The heat flux density from the wall to the cooling fluid results from the product of heat transfer coefficient and the temperature difference between the wall and the fluid. Higher flooding rates increase the quality of the dispersed flow and by this reduce the superheating of the steam.

The interfacial heat transfer between the liquid and the vapour in the dispersed flow region is affected by the difference in velocity between the water droplets and the vapour. The slip between the phases is a strong function of the quality under flooding conditions, as shown in fig. 21. There are two reasons for this, in high quality flow only small droplets exist in the vapour which due to strong drag forces are carried away quickly. So steam and water droplets have almost the same velocity. In the low quality region the vapour is bubbling through a liquid pool of very low water velocities. So even with moderate vapour velocities the slip ratio reaches high values due to the almost stagnant liquid.

The parameter studies presented up to now were performed at smooth heated walls without any geometrical disturbances like spacers being present along the cladding of the fuel rods in a reactor. These spacers - having no decay heat source and low heat capacity - rewet comparatively early and cause a local cooling of the rod. So the spacers can form new wetting nuclei at their position in the core and can influence the progress of the wetting front considerably as shown in fig. 22, where the wetting behaviour with and without spacers is compared.

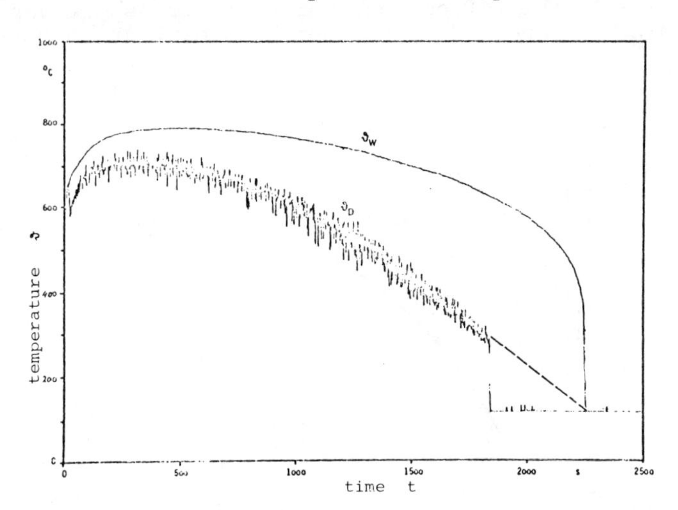

Fig. 2o. The progress of the wall- and steam temperature in the region of the test tube exit (p = 1,5 bar, w = 2 cm/s, $\Delta\vartheta_E$ = 35 K, q" = 3 W/cm²)

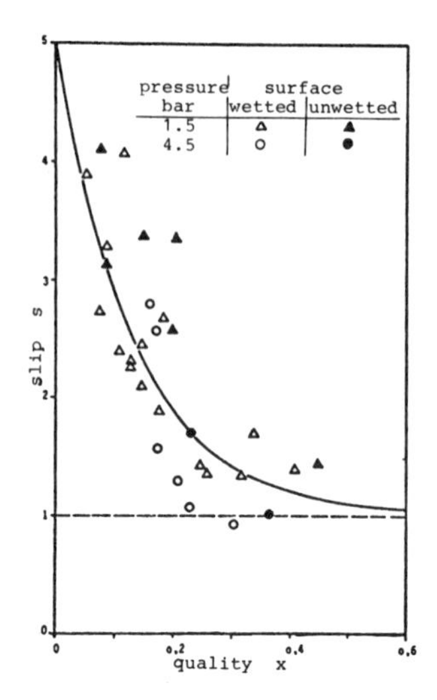

Fig. 21. The influence of the quality on the slip between the phases under flooding conditions

6. CONCLUDING REMARKS

The fundamental aspects of thermodynamic and fluiddynamic phenomena influencing the quenching process, discussed here, were mainly restricted to a one-dimensional flow behaviour in the core or in one of its subchannels. In reality cross flow and mixing may occur in the core during reflooding, which can transport vapour and liquid from lower powered zones quenching earlier to hotter

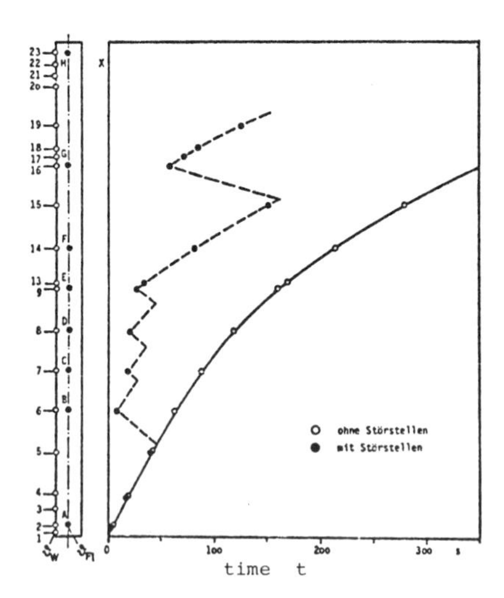

Fig. 22. The influence of new wetting nucleus on the progress of the wetting front

channels and so influencing the heat transfer considerably. The reflux cooling from the top is depending from the water collected in the upper plenum by carry over effects, however, also by the stagnation pressure of the vapour flow, holding the liquid level there.

In a real reactor there are never fully constant fluiddynamic conditions over the whole cross section of the pressure vessel in the upper plenum and at the upper tie-plate. Therefore in the regions of lower vapour flow rate or higher water levels, the liquid in the upper plenum can fall back and consequently the vapour will give place this downward penetrating liquid by concentrating into neighboured flow areas. This causes strong mixing effects.

The water level in the upper plenum, however, is also dependent from the carry over of droplets into the large pipelines of the hot leg and from there to the break opening. This droplet trajectories are of three-dimensional nature. Their behaviour can only be tested in large scale experiments.

The amount of water stored in the upper plenum and the quality in the pipes of the primary loops carrying the mixture to the break influence steam binding strongly and by this the rising velocity of the ECC-water level.

All these effects can only be tested in large scale experiments, representing the geometrical conditions in their full size. New experimental activities within an international contract researching the large volumetric and three-dimensional effects are underway /19/. Within these activities the thermodynamic and hydrodynamic behaviour in the upper endboxes at the tie-plate and in the upper plenum of a full sized pressure vessel of a PWR as well as the refill and reflood behaviour in test rigs with a core of 2000 rods will be researched.

The theoretical prediction of the cooling behaviour during the refill and reflood phase can only be performed by very complicated computer programs, which are based on the wellknown conservation equations and contain a large number of constitutive equations describing a variety of physical phenomena. Usually these constitutive equations have up to now a very empirical character and a lot of fundamental research is needed in the near future to improve the knowledge about fluiddynamic and thermodynamic effects for getting theoretically well based correlations. To perform these experiments improved and advanced measuring techniques are needed, which have to be developed too.

REFERENCES

1. Collier, J.G., Post-Dryout Heat Transfer - A review of current position, Proceedings of the "NATO Advanced Study Institute on Two-Phase Flow and Heat Transfer", 1976, Istanbul, Turkey, Hemisphere Publishing, New York

2. Nukiyama, S., Maximum and minimum values of heat transmitted from metal to boiling water under atmospheric pressure, J.Soc. mech. Engrs., Japan 37, 53 - 54 and 367 - 374 (1934)

3. Hein, D., Köhler, W., The role of Thermal Non-equilibrium in Post-Dryout Heat Transfer paper presented at the "European Two-phase Flow group meeting 1976", Grenoble (France)

4. Bailey, N.A., Dryout and Post-Dryout Heat Transfer at Low Flow in a Single Tube Test Section, AEEW-Report No.1068, 1977

5. Ganić, E.N., and Rohsenow, W.M., Int. Journal of Heat and Mass Transfer 20 (1977) 855

6. Iloeje, O.C., Plummer, D.N., Rohsenow, W.M., and Griffith, P., A study of wall rewet and heat transfer in dispersed vertical flow, MIT Dept. of Mech. Engng., Report 72718-92, Sept. 1974

7. Collier, J.G., Post-dryout heat transfer - A review of the current position, European Two-Phase Flow Group Meeting 1975, Paper F 5, Haifa, Juni 1975

8. Bromley, L.A., Heat transfer in stable film boiling, Chem. Eng. Progr. 46 (5), 221 - 227 (1950)

9. Sparrow, E.M., R.D. Cess, The effect of subcooled liquid on laminar film boiling, J. Heat Transfer 84C, 55 - 62 (1962)

10. Nishikawa, K., Ito, T., Two-phase boundary layer treatment of free convective film boiling, Int. J. Heat Mass Transfer 9, S. 103 - 115 (1966)

11. Lauer, H., Untersuchung des Wärmeübergangs und der Wiederbenetzung beim Abkühlen heißer Metallkörper, EUR 5702.d, 1976

12. Groeneveld, D.C., and Young, J.M., Film Boiling and Rewetting Heat Transfer during Bottom Flooding of a hot Tube, Proceedings sixth international Heat Transfer Conference 1978,

13. Hsu, Y.Y., Tentative Correlations of Reflood Heat Transfer, LOCA-research Highlights (April 1 - June 30), 1975

14. Hein, D., Modellvorstellungen zum Wiederbenetzen durch Fluten, Dissertation Universität Hannover, 198o

15. Yamanouchi, A., Effect of core spray cooling in transient state after loss of coolant accident, J. Nucl. Sci. Techn. 5 (11), 547 - 558 (1968)

16. Andersen, J.G.M., Hansen, P., Two-dimensional heat conduction in rewetting phenomenon, Report NORHAV-D-6, Danish Atomic Energy Commission Research Establishment, Risø, Denmark (1974)

17. Elias, E., Yadiagaroglu, G., Rewetting and liquid entrainment during reflooding - state of the art, EPRI NP-435, May 1977

18. RS 36 - Abschlußbericht

19. Mayinger, F., GRS-Fachgespräche, May 198o

13 LWR SYSTEM SAFETY ANALYSIS*

by
J.F. Jackson and D.R. Liles
LASL
Los Alamos, NM, USA
V.H. Ransom and L.J. Ybarrondo
E.G.&G. Idaho Inc
Idaho FAlls, ID, USA

In this final chapter relating to the design basis accident of a light water reactor, methods currently used or being developed for analysis are described. A brief review of the evolution of light-water reactor safety analysis codes is presented. Included is a summary comparison of the technical capabilities of major system codes. Three recent codes are described in more detail to serve as examples of currently used techniques. Example comparisons between calculated results using these codes and experimental data are given. Finally, a brief evaluation of current code capability and future development trends is presented.

1. INTRODUCTION

This paper discusses the evolution of light-water reactor (LWR) accident analysis techniques and describes three available computer codes to illustrate current capability. Emphasis is given to the Loss-of-Coolant Accident (LOCA), although the analysis methods discussed can also be used for other postulated accidents. To further limit the scope, the discussion will be restricted to systems-analysis codes, i.e., those that describe the overall thermal-hydraulic behavior of the entire primary, and in some cases, secondary system during an accident.

Most of the analytical development effort over the past 14 years has been devoted to the large-break LOCA. The accident at Three-Mile Island, however, has resulted in an increased priority on techniques that can deal with the much longer transients that ensue from smaller leaks. Although most of the material in this paper will concentrate on the traditional large-break LOCA analysis methods, their applicability to small break accidents will also be discussed where appropriate.

Much of the earlier work also focused on development of licensing, or "evaluation model" (EM) codes. These codes embody a number of agreed upon "conservatisms" in the modeling to conform to established licensing rules. Recently, there has been more emphasis on developing "best-estimate" codes that try to model the system behavior as accurately as possible. Such codes are much more amenable to experimental assessment and can serve to evaluate the safety margins inherent in EM models. Emphasis in this paper is on best-estimate codes.

*Work performed under the auspices of the U.S. Nuclear Regulatory Commission.

In the first section, we trace the development of LWR accident codes from 1966 through the present time. This will include a discussion of the role of analysis in LWR nuclear safety research, followed by a review of some of the important physical phenomena that have been identified, and the associated technical issues that have required resolution. Finally to illustrate the breadth of the effort and the substantial progress that has been achieved, a chart summarizing the historical evolution of safety code capabilities is presented.

The next three sections are devoted to brief descriptions of three recent accident-analysis system codes developed or under development in the United States. This will include selected separate effects and integral systems test data comparisons. The last section briefly summarizes current capabilities and anticipated development activities.

The three system code descriptions given in this chapter include brief discussions of the numerical methods involved. More detailed information on the numerical methods is presented in Appendix A at the end of Chapter 19, where some of the pertinent basic numerical concepts are reviewed and several buzz terms used in the code descriptions are defined. The appendix at the end of this chapter and Appendix B after Chapter 19 present examples of the finite difference equations and solution strategies used in current reactor safety codes.

1.1 The Role of Analysis in LWR Safety

Analysis has played a unique role in nuclear reactor safety for two main reasons. First, the full-scale demonstration experiments (or actual events) that are normally available to evaluate the accident behavior of industrial products (e.g., automobiles and aircraft) are not available nor practical to obtain in the case of nuclear power plants. This is because the diversity of reactor system designs and the numerous potential events to be considered make the required large number of full-scale experiments prohibitively expensive. Consequently, a greater than usual responsibility has been placed on the reactor safety analyst to be rigorous and accurate in the developing and testing of analysis tools.

A second, and perhaps related, reason stems from the philosophy that has evolved in the United States of making extensive use of analysis as an investigative, design, and evaluation tool. Let us briefly examine this philosophy. First, nuclear plants are designed to be clearly safe in normal operation and incorporate substantial allowance for off-normal operation and system/component failures. Second, analyses are used to determine those system/component failures that could affect safety so that appropriate preventative actions may be taken. Third, it is still presumed that some of the system/component failures will occur and that the provided safety features (with margin and redundancy) will keep the nuclear plant safe in spite of such failures. Finally, some of the safety features themselves are assumed to fail during the accidents they are designed to mitigate so the consequences can be analyzed. This process of repeated investigative analysis is used to identify possible weaknesses of nuclear safety systems so they can be rectified. The result is to reduce the credibility of severe nuclear accidents to an acceptably low level. Thus, nuclear-system safety analysis has been and will continue to be a very important element of LWR nuclear safety.

1.2 Scope of Analysis Development

The development of analytical methods for reactor safety analysis has been one of the most comprehensive analytical efforts undertaken in the United States. This effort has involved several national laboratories, industrial firms, government agencies, and many universities working cooperatively. The completeness and accuracy with which LWR transient behavior under LOCA conditions can be modeled has improved steadily. The analytical tools are continuing to be improved and tested to achieve even greater accuracy, predictive reliability, and economy.

The development of these methods has been and continues to be a very challenging task. The physical phenomena that can exist under postulated accident conditions have required the development of new analytical models and associated numerical solution methods. The large number of components and the complexity of accident phenomena have necessitated innovative application of even the most sophisticated modern computers to achieve the desired results in practical computation times.

The requirement to model two-phase flow conditions has either directly or indirectly accounted for the greatest part of the technical development effort. Under LOCA conditions, nonhomogeneous (relative motion between the phases), nonequilibrium (temperature difference between the phases), and in some cases, multidimensional flow effects can be important. New hydrodynamic models had to be developed to account for these phenomena. The presence of two-phase flow also influences the performance of pumps; the flow through valves, orifices, and postulated breaks; and convective heat transfer mechanisms. The ability to model with accuracy the discharge rate from an assumed system leak is particularly important since it determines the rate of coolant loss and system depressurization.

Another important area is that of heat transfer. Under design conditions, the heat transfer process in a reactor core is in the well characterized subcooled and nucleate boiling regimes. However under postulated accident conditions, the heat transfer extends into the transition and film boiling regimes. During the emergency coolant reflood phase, the maximum core temperature is dependent upon the details of the film boiling process and in particular, on the transition back to nucleate boiling associated with quenching the hot fuel rods. Axial heat conduction along the very steep temperature gradients near the quench front is also very important.

Substantial progress has been made in understanding basic two-phase thermal-hydraulic phenomena and in their quantification with empirical correlations. These phenomena and their characterization have been the topics of several other chapters. The main purpose of systems-analysis computer codes is to synthesize this knowledge into a consistent framework of conservation relations so that they can be applied to practical reactor safety problems. The advantage of computer modeling is that it allows one to treat the complexity inherent in reactor accident behavior. Advances in computer analysis techniques have had to go hand-in-hand with advances in phenomenological modeling. More efficient and reliable numerical solution strategies have had to be developed. Issues such as convergence, accuracy, stability, and economy have also had to be addressed.

For example, one issue of wide discussion has been the formulation of a macroscopic model for two-phase flow.[1] Attempts to formulate the macroscopic Eulerian-type equations for a nonhomogeneous two-phase mixture have resulted in systems of differential equations that have complex characteristics (sometimes referred to as being ill-posed).[2-4] This may imply an unstable character for solutions to initial-boundary value problems. Several sets of equations have been proposed, even some that have real roots, but there is no single set that has universal acceptance. Most agree that the difficulty is a result of the inability to describe accurately the differential character of all the fluid interactions[5] and the inability to characterize the covariant terms that arise in the integral averaging process. Even though this issue lacks complete resolution, it has not prevented the development of successful numerical models for the flow of two-phase fluids.[6-8] The reason for this is that the imperfection in the differential models primarily affects the short-wave length behavior of the solutions. Generally, these effects are at shorter wave lengths than can be resolved numerically for practical mesh spacings. Thus, the ill-posed issue is of more academic than practical importance as far as the accurate simulation of LWR systems is concerned.

Stable solution behavior is achieved through the damping or numerical dissipation inherent in the schemes used to solve the differential equations. This numerical dissipation is the result of implicitness, use of donored-flux terms, and inherent viscosity associated with the difference operations. The net result is that the shortest wave-length components of the initial data decay and a well-posed numerical initial-boundary-value problem is obtained. The complexity of most models makes analytical investigation of stability impractical. Stability has been achieved by use of methods proven to be stable for simpler problems and then investigated by numerical experimentation with representative test problems.

The accuracy of analytical models and associated numerical schemes has many facets, i.e., accuracy of the physical description, fluid properties, empirical correlations, and numerical discretization. When calculated results are compared to data, all of these inaccuracies are combined. Careful study is required to separate the sources of inaccuracy. Experience with a particular method gained through application to many separate effects and integral system experiments is probably the best and usual assessment technique. The quest for accuracy is sometimes at odds with the need for economy in terms of required computer time. The use of multidimensional and complex system representations can result in very large systems of equations that must be solved with attendant large computational times. The balance between detail of representation and economy is one that can vary, depending upon the end use of the results. If system component interaction is of interest, then the entire system must be represented even if a compromise in detail is required. If, on the other hand, the phenomena of interest are local or of short duration, then a more detailed representation can be used.

A related issue is the trade-off between simple, and often highly empirical, models and more complex models that are rooted more strongly in fundamental principles. Although the more empirical models are often more economical, they may not extrapolate to new (and untested) regimes as reliably as the more fundamental models.

The extensive range of operation of LWR system components under postulated LOCA conditions places an additional burden on the modeler. Small perturbation theory of linear models is too restrictive to be of use under such conditions since many components and the physical phenomena exhibit highly nonlinear behavior over the range of interest. Thus, each system component model needs to be very general and capable of operation over a wide range. As an example, a pump model must be capable of representing the performance for both positive and negative flow, positive and negative head, forward and reverse rotation, and fluid conditions ranging from subcooled liquid to all vapor. Such comprehensive representation is frequently made difficult by a lack of data covering the range of potential operation.

2. EVOLUTION OF ANALYTICAL METHODS

In the past 14 years, significant progress has been made in all areas of nuclear safety research and development. In particular, the LWR system codes used for safety analysis have improved substantially. The purpose of this section is to summarize the evolution of this improvement.

2.1 Historical Perspective

The year 1966 is a reasonable point of reference from which to measure progress because on October 27, 1966, Mr. H. L. Price, Director of Regulation, U.S. Atomic Energy Commission (AEC) appointed a task force to conduct a review of power reactor emergency core-cooling systems and core protection.[9] Mr. Price's letter of appointment stated, "Because of the increasing size and complexity of nuclear power plants, the AEC regulatory staff and Advisory Committee on Reactor Safeguards (ACRS) have become increasingly interested in the adequacy of emergency core cooling systems and the phenomena associated with core meltdown . . ."

There are four principal driving forces that have contributed to the continual evolution and improvement of nuclear safety system codes in the United States:

1. The appointment and subsequent report of the task force mentioned above,[9]

2. The emergency core-cooling (ECC) hearings,[10]

3. The research and development conducted in accordance with the Water Reactor Safety Program Plans,[11] and

4. The philosophy of nuclear safety design and evaluation that has evolved in the United States.

The codes used by the pressurized-water reactor (PWR) and boiling-water reactor (BWR) vendors have evolved somewhat separately because of the different geometry and transient behavior of the two reactor systems. In fact, the BWR codes used for LOCA analysis have remained relatively constant in form and content although they have been influenced by the substantial changes in the PWR system-analysis codes. The main emphasis in the following discussions will be devoted to the PWR system-analysis code evolution unless otherwise indicated.

FLASH[12] is the genesis of the reactor system codes used for large-break PWR LOCA analysis. It was developed in the U.S. Naval program. The following statements from page vi of Ref. 12 will help illustrate the state of this type of analysis in 1966 and the progress achieved in 14 years.

> "In previous treatments of the loss-of-coolant accident, the primary system has been represented by a single volume filled with steam and water. In some analyses, the steam and water phases have been assumed to be completely separated and in others, to be completely homogeneous. Core cooling has been assumed to be essentially perfect

until the water inventory fell below some preassigned critical value, after which core cooling has been assumed to be essentially zero. In using these treatments, results were found to depend critically upon the a-priori assumptions concerning the separated or homogeneous state of the coolant and on the value assumed for the critical water inventory. In general, it has been impossible to justify any particular set of assumptions on technical grounds."

The authors go on to say,

"It was to avoid the necessity for making these a-priori assumptions that FLASH was developed. FLASH divides the primary system into three volumes, each of which contains both a homogeneous mixture and a separated steam phase. The degree of separation is calculated continuously. The explicit core-cooling calculations avoid the need for any assumptions concerning water inventories."

The authors were also quite realistic about their achievement and were prophetic about where improvements could be made.

"The model used in FLASH represents a considerable simplification of the actual system geometry. On the other hand, the FLASH model attempts to account for the behavior of every component of the primary system during a loss-of-coolant accident. At present, data on the performance of many of these components under the extreme off-design conditions which prevail during a loss-of-coolant accident are unavailable. As this information becomes available, it can be factored in the existing structure of FLASH. For the present, however, FLASH provides a considerable extension of our ability to calculate what might happen in the primary system during a loss-of-coolant accident."

Table I illustrates the chronological development of the principal PWR oriented codes from FLASH and FLASH-2.[13] The RELAP(SE) series[7,14-19] and TRAC[8] have all been sponsored by the U.S. Nuclear Regulatory Commission (USNRC) or its predecessor the Atomic Energy Commission. All the PWR vendor codes have developed in a manner similar in substance to the USNRC-sponsored codes; therefore, they will not be separately addressed for purposes of brevity.

It is important to note that RELAP5/MOD0[12] and TRAC P1A[13] are offset in Table I to illustrate that they represent a quantum step forward in technical capability, flexibility of use, user convenience, level of experimental assessment, and potential economy of operation. RELP5/MOD0 and TRAC P1A were developed because it was clear that the RELAP series up to RELAP4 could not achieve the technical capability, flexibility of use, user convenience, and economy of operation required for best-estimate nuclear safety calculations.

2.2 Technical Evolution of System Codes

Table II illustrates the technical evolution of the computer codes listed in their chronogogical order of development in Table I. Most of the categories used to classify the capabilities of the codes were previously used in Refs. 20 and 21. These categories are intended to be representative of the significant progress achieved by each code and are not intended to be complete in the absolute sense of listing every improvement each code represented. The most significant advance(s) offered by each code is highlighted by the accented rectangle(s). It is clear that TRAC and RELAP5 are quite superior technically and mechanistically to the other codes.

TABLE I

CHRONOLOGICAL EVOLUTION OF LWR LARGE-BREAK-LOCA SYSTEM CODES

	Computer Code Name	Date
A.	Homogeneous and Equilibrium Hydrodynamics Equation Base:	
	1. FLASH	May 1966
	2. RELAPSE	September 1966
	3. FLASH-2	April 1967
	4. RELAP 2	March 1968
	5. RELAP 3	June 1970
	6. RELAP4/MOD3	October 1975
	7. RELAP4/MOD5	September 1976
	8. RELAP4/MOD6	January 1978
	9. RELAP4/MOD7	March 1980
B.	Nonhomogeneous and Nonequilibrium Hydrodynamics Equation Base	
	1. TRAC (PlA and BDO)	PlA - March 1979 BDO - February 1980
	2. RELAP 5	May 1979

It is worthwhile reflecting on the continuing incessant drive for technical excellence, completeness, and precision illustrated by Table II. In the ongoing development of LWR technology, plant designs have become more sophisticated, power densities have become higher to improve economy, and available reactor plant sites have become less favorable. At the same time, people have become more concerned about the quality of their environment. These factors, in addition to the four mentioned earlier, generated increasing needs for improved plant integrity, reliability, and assurance of safety system performance. These increasing needs placed further demands and responsibilities on analysts for measureability in design and safety assessment techniques and rigor in their application. The basic principles are recognized. Special emphasis was given to determining the important LOCA physical phenomena, translating the LOCA phenomena into equations, solving the equations numerically, molding the equations in computations, evaluating the relative conservatism and realism of various assumptions, and testing the resultant system computer codes for completeness and precision using data from component and systems experiments.

TABLE 2

TECHNICAL COMPARISON OF LWR LARGE LOCA SYSTEM CODES

CODE CAPABILITIES	FLASH	RELAPSE	FLASH-2	RELAP2	FLASH-4	RELAP3	RELAP4/MOD3	RELAP4/MOD5	RELAP4/MOD7	RELAP4/MOD7	TRAC(P1A/BDO	RELAP5
A. Reactor System Representation												
• PWR	Yes	Yes	Yes	Yes	Yes	Yes	Yes	Yes	Yes	Yes	Yes	Yes
• BWR	No	No	No	Yes	No	Yes	Yes	Yes	Yes	Yes	Yes	No
• Control Volumes	3	3	20	3	20	20	100	75	75	Dyn.Sto. Comp.Sto.Ltd.	Dyn.Sto. Comp.Sto.Ltd.	Dyn. Sto. Comp.Sto.Ltd.
• One/Multi-dimensional	Quasi 2 Dim.	Quasi 1 Dim.	Quasi 1 Dim.	Quasi 1 Dim.	Quasi 1 Dim.	Quasi 1 Dim.	1 Dim.	1 Dim.	1 Dim.	1 Dim.	Multi. Dim.	1 Dim.
• ECC System - LPI	Fill Table	Fill Table	Fill Table	Fill Table	Fill Table	Press. Dep.Fill	Press. Dep. Fill	Time/Press. Dep. Fill	Time/Press. Dep. Fill	Time/Press. Dep. Fill	Time/Press. Dep. Fill	Fill Table
- HPI	Hot Vol Only	Fill Table	Fill Table	Fill Table	Fill Table	Press. Dep.Fill	Press. Dep. Fill	Time/Press Dep. Fill	Time/Press. Dep. Fill	Time/Press. Dep. Fill	Time/Press. Dep. Fill	Fill Table
-Accumulator	No	Fill Table	Fill Table	Fill Table	Fill Table	Press. Dep.Fill	N_2 Driven	Control Vol. w/Air	Control Vol. w/Polytropic Air	Control Vol. w/Polytropic Air	Active Control Vol. w/Air	Active Control Vol.w/Air
• Secondary System	Const. h	Flow Dep. Const. h	Improved Linear Models	Flow Dep. Const. h	Improved Linear Models	Flow Dep. Const. h	Yes.w/Min. Control Logic	Yes.w/Min. Control Logic	Yes.Added Natural Conv. HT to MOD5	Yes.Added Natural Conv. HT to MOD5	Adequate Pri./ Sec.Model-Steam Gen. Only	Same as TRAC(P1A/ BDQ)
• Trip Logic	No	No	No	No	No	Yes	Yes	Yes	And/or Logic Logic Added	And/or Logic Logic Added	Complex Trip Logic Allowed	Complex Trip LogicAllowed
• Check Valves	No	No	No	No	No	Yes	Yes	Yes	Yes	Yes	Yes	Yes
• Motor Act.Val.	No	No	No	No	No	No	Yes	Yes	Yes	Yes	Tabular	Yes
B. Hydro Model												
• Homogeneous & Equil.	Yes	Yes	Yes	Yes	Yes	Yes	Yes	Slip Added	Slip Added	Slip Added	No	Inc. as Option
• Momentum Flux	No	No	No	No	No	No	Yes	Yes	Yes	Yes	Yes	Yes
• Nonhomo. & Nonequil.	No	No	No	No	No	No	No	No	No	Explicit Non-equil.ECC Mix	Two Fluid, Six Equations	2 Fluid, Five Equations
• Numerics	Explicit	Explicit	Explicit	Explicit	Implicit	Explicit	Implicit	Implicit	Implicit	Implicit	Semi-orFully-Implicit	Semi-Implicit
• Arbitrary Network	No	No	Yes	No	Yes	Yes	Yes	Yes	Yes	Yes	Yes	Yes

TABLE 2

TECHNICAL COMPARISON OF LWR LARGE LOCA SYSTEM CODES

CODE CAPABILITIES	FLASH	RELAPSE	FLASH-2	RELAP2	FLASH-4	RELAP3	RELAP4/MOD3	RELAP4/MOD5	RELAP4/MOD6	RELAP4/MOD7	TRAC(P1A/BD0)	RELAP5
C. Model Improvements												
• Fuel - Type	Plate	Plate	Plate	Plate	Plate	Pin/ Plate	Pin/Plate	Pin/Plate	Pin/Plate	Pin/Plate	Pin	Pin/Plate
- Average/Hot Channel	Yes Very Simple	Yes Very Simple	Yes Very Simple	Yes Very Simple	Yes Very Simple	Yes	Yes	Yes	Yes	Yes	Yes, 3D Multi-Rod	Yes, Multi-Rod
- Gap	No	No	No	No	No	No	Yes	Yes Ross & Stoute Used	Yes	Yes Advanced Base on FRAP	Yes	No
- Thermal Properties	No	No	No	No	No	Yes	Yes	Yes	Yes	Yes, From MATPRO	Yes	No
- Conduction	No	No	No	No	No	Yes	Yes	Yes	Yes	Yes	Yes	Yes
• Metal-Water Reaction	No	No	Yes	No	Yes	No	Yes	Yes	Yes, Added Cathcart-Paul	Yes	Yes	No
• Reactor Kinetics	Very Limited	Point Kinetics	Improved FLASH	Point Kinetics	Very Limited	Point Kinetics	Point Kinetics	Point Kinetics	Point Kinetics	Point Kinetics	Point Kinetics	Point Kinetics
• Phase Separation	Constant Vel. Steam	Constant Vel. Steam	Constant Vel. Steam	Variable Bubble Rise	Constant Vel. Steam	Constant Vel. Steam	Variable Bubble Rise	Variable Bubble Rise	Variable Bubble Rise	Variable Bubble Rise	Uses Two-Fluid Model	Uses Two-Fluid Model
• Integral Blow-down Reflood Calculation	No	No	No	No	No	No	No	Reflood Model Must Renod. by Hand	Improved Re-flood Must Renod. by Hand	Automatic Renod.	Automatic Renod.	Yes but not fully Developed
• Metal Heat Conduction	No	No	Yes Lumped	No	Yes Lumped	No	Yes Distributed	Yes Distributed	Yes Distributed	Yes Distributed	Yes Lumped Dist. Pipes	Yes Distributed
• Water/Steam Properties	Limited Iter. Table Lookup	Limited Iter. Table Lookup	Limited Iter. Table Lookup	Limited Iter. Table Lookup w/Mem.	Limited Iter. Table Lookup Sat.Line	Limited Iter. Table Lookup w/Mem.	Extensive Iter. Table Lookup w/Memory	Extensive Iter. Table Lookup w/Memory	Extensive Iter. Table Lookup w/Memory	Extensive Iter. Table Lookup w/Memory	Correlations	Extensive Direct Table Lookup w/Memory
• Heat Transfer Correlations	Conv. Nucleate & Film Boiling Only. Surface Temp. Dictated Transition Between Correlations					7 Modes HT +Transition on Quality, Press., Mass Flux	Improved	Continued Improvement	Improved Correl. & Logic-Diff. Correl. for Blowdown/ Reflood	Improved Correl. & Logic-Diff. Correl. for Blowdown/ Reflood	1 Set for Entire Trans. Phases Treated Separately	Same as RELAP4/MOD6 Blowdown Faster Logic

TABLE 2

TECHNICAL COMPARISON OF LWR LARGE LOCA SYSTEM CODES

CODE CAPABILITIES	FLASH	RELAPSE	FLASH-2	RELAP2	FLASH-4	RELAP3	RELAP4/MOD3	RELAP4/MOD5	RELAP4/MOD6	RELAP4/MOD7	TRAC(P1A/BD0)	RELAP5
C. Model Improvements (Continued)												
• For/Rev. Loss Coeff.	No	No	No	No	No	No	Yes	Yes	Yes	Yes	No	Yes
• Pump Representation	Time Tables	Time Tables	Time Tables	Time Tables	Time Tables	Time Tables	Homologous Pump Curves	Homologous Pump Curves	Homologous Pump Curves	Homologous Pump Curves	Homologous Pump Curves	Homologous Pump Curves
• Choked Flow Model	Yes	Yes	Yes	Yes	Yes	Yes	Several Options	Several Options	Several Options	Several Options	Num. Choking Fine Mesh	2 Fluid Bound. Cond.
D. User Convenience												
• Plots	No	No	No	Yes	Yes	Yes	Yes	Yes	Yes	Yes	Yes	Yes
• Restart	No	No	No	No	No	Yes	Yes	Yes	Yes	Yes+Renod.	Yes+Renod. on Restart	Yes
• Automatic Steady-State	No	No	No	No	No	No	No	No	No	Yes, PWR Only	Yes	No
• Input - Free Format	No	No	Yes	No	No	No	Yes	Yes	Yes	Yes	No	Yes
- Input Diagnostics	Limited	Limited	Limited	Limited	Limited	Limited	Processes Beyond 1st Error	Same as RELAP4/MOD3	Same as RELAP4/MOD3	Same as RELAP4/MOD3	Limited	Excellent process all input
E. Computers Used												
	CDC 6600	IBM 7044	CDC 6600	IBM 7044 CDC 6600 Univac	CDC 6600	IBM 7044 CDC 6600 CDC 7600	IBM360 CDC6600 CDC7600	IBM360 CDC7600	IBM360 CDC7600	CDC7600	CDC7600	CDC7600

3. RELAP4/MOD6 DESCRIPTION AND EXAMPLE COMPUTATIONS

The RELAP4[17-19] computer code was developed to describe the thermal-hydraulic behavior of LWRs subjected to postulated transients such as a loss-of-coolant, pump failure, or nuclear power excursion. It can also analyze the behavior of part of a system, provided the appropriate thermal-hydraulic boundary conditions are supplied. It calculates the interrelated effects of coolant thermal-hydraulics, system heat transfer, and core neutronics. Because the program was developed to solve a large variety of problems, the user must specify the applicable program options and the system to be analyzed.

3.1 Program Status

RELAP4/MOD7[19] is the most recent version of the RELAP4 code to be released for general use. At this time, RELAP4/MOD6[18] is probably the most extensively used version of the code and most of the discussion herein refers to this version. Where appropriate, improvements that are available in MOD7 will be described.

RELAP4/MOD7 is the culmination of an extensive development effort. This series of codes is based on a homogeneous equilibrium fluid model (HEM) to which many refinements have been added to give a partial account for nonhomogeneous and nonequilibrium effects. The advanced codes, TRAC and RELAP5, are based on more fundamental approaches for modeling nonhomogeneous and nonequilibrium two-phase fluid flow, and in this respect, they represent significant departures from the RELAP4 efforts.

In spite of the limitation of the HEM assumption, these codes have served a very useful function and have provided the nuclear industry with a powerful analytical capability. This capability has been utilized extensively in the design of safety systems and has played a key role in the power reactor licensing process. In fact, the RELAP4 code is still the basic analysis tool for demonstrating that the licensing requirements can be met by a particular plant design. The shift of this function to the advanced codes will occur as experience with, and confidence in, these codes is established.

Those versions of RELAP4 up to and including RELAP4/MOD5[17] were intended primarily as blowdown and refill codes, i. e., they were designed to calculate system phenomena from initial operating conditions to the time of pipe rupture, through system decompression, and up to the initiation of core recovery with emergency core coolant. In the RELAP4/MOD6[18] version, the calculational capabilities were extended from blowdown and refill through core reflood for PWR systems. Finally, RELAP4/MOD7[19] includes improved user conveniences and modeling improvements that permit a continuous or integral calculation of the blowdown and reflood phases of a LOCA.

The evolution of RELAP4 has passed through many cycles of model revision and addition to extend its applicability to situations where the basic assumptions were inadequate. This process led to the production of models to account for nonhomogeneous and nonequilibrium effects. These models are not completely general and, consequently, require considerable knowledge on the part of the user to produce correct results. For these reasons, and in view of the progress of the advanced codes, the RELAP4/MOD7 version of the code is to be the last of this series.

3.2 Model Description

The RELAP4/MOD6 program consists of program controls, fluid dynamics models, heat transfer models, and a reactor kinetics model, all coupled by a numerical solution scheme that advances in time. Each of these parts is summarized in the following pragraphs.

Program Controls. The program input features are used to specify the problem dimensions and constants, time-step size, trip controls for reactor-system transient behavior, and output. Controls are also provided for restarting a problem and producing a plotting tape. There are three basic options that are selected by input--Standard RELAP4, RELAP4-EM, and RELAP CONTAINMENT.

Hydrodynamic Model. The basic modeling philosophy embodied in the RELAP4 code is one in which the system to be modeled is divided into a number of subcontrol volumes that are connected by junctions or flow paths. Mass and energy are conserved in each control volume and an approximate momentum equation is used to calculate the flow at each junction. As RELAP4 has evolved, numerous specialized models have been developed to account for phenomena such as phase separation, thermal nonequilibrium, heat transfer effects, pumps, valves, multiple stream mixing, etc. The user must specify through the program controls which of these models is to be used in a particular problem. Such modeling decisions do influence the results, and care must be taken that the models are not misapplied. The model variations are too numerous to describe in this limited discussion, so the interested reader is referred to the users manuals.[17-19]

The RELAP4 hydrodynamic model is based on the assumption that the flow process is essentially one-dimensional so that area-averaged properties can be represented as functions of one space variable and time. In addition, the basic model assumes a homogeneous and equilibrium mixture exists at each point in the system. The mass, energy, and flow equations are integrated over a fixed control volume to obtain integrated stream-tube differential relations.

The HEM model includes only the mixture mass conservation equation. The basic mass-dependent variable is the fluid total mass or the density in each control volume. The mass fluxes at each junction connected to a control volume are defined by means of a donor formulation, i.e., the fluid properties of the source are used to compute the mass flux. A Wilson bubble rise model[22] can be selected by the user to approximate nonhomogeneous effects in vertical control volumes, and a slip model is available for approximation of nonhomogenous effects in horizontal control volumes. Both of these models use emperical constants specified by the users.

Like the mass equation, the HEM model only includes the mixture total energy equation. The mixture internal energy in the fluid control volume is the fundamental dependent variable and is expressed in terms of the junction energy flux and fluid total enthalpy. Here again, a donor formulation is used to establish the junction energy properties, although an "enthalpy transport" model can be specified to give a partial account for nonhomogenous and non-equilibrium effects. The enthalpy transport model consists of a quasi-steady approximation to the distribution/energy source terms so that the junction or "edge" energies differ from the volume average values in a manner dependent upon the process. This model can be used to approximate the nonequilibrium effects downstream of emergency core coolant (ECC) injection points and to approximate the energy gradients present in the reflood process.

Four basic forms of the fluid flow equation have evolved and are included in RELAP4.[17] Each form is based on a particular set of assumptions. The user must choose the form most appropriate for a particular junction. The four basic forms are: Form 1--Compressible Single-Stream Flow with Momentum Flux, Form 2--Compressible Two-Stream Flow with 1-D Momentum Mixing, Form 3--Incompressible Single-Stream Flow without Momentum Flux, and Form 4--Compressible Single-Stream Integral Momentum Equation.

The choice of the flow equation form depends upon the purpose and detail of the desired calculation. Forms 1 and 2 include a 1-D momentum flux term. These are applicable when the control volumes represent a 1-D stream tube. Form 2 should be used only when two streams can combine and exchange momenta on a 1-D basis. Form 3 provides an alternate to the compressible flow equation with the momentum flux term for modeling multidimensional geometries. An alternate form of the momentum equation developed by Zuber[23] is obtained by using a different control volume approach yielding the compressible integral momentum equation (Form 4).

Heat Transfer Model. The transfer of thermal energy between the fluid and the boundaries is modeled by a combination of transient conduction and convective heat transfer correlations. The thermal interactions that are modeled in this way include reactor fuel pin to fluid, steam generator primary fluid to wall to secondary fluid, and vessel/piping system stored energy to fluid. Models also exist for internal heat generation in the wall or fluid due to electrical or gamma heating. The transient conduction is calculated using a Crank-Nickolson finite difference technique for the 1-D transient heat conduction equation.[24] Slab, cylindrical, or spherical geometry can be represented. The geometry and conditions of the heat conductor are specified by the user.

The convective heat transfer at fluid boundary interfaces is the boundary condition for the transient conduction solution and is the source or sink of thermal energy to the fluid. The code uses convective heat transfer correlations to calculate the critical heat flux (CHF), pre-CHF heat transfer, and post-CHF heat transfer. The basic approach used in RELAP4/MOD6 is to construct a heat transfer surface for the wall heat flux as a function of the wall superheat and fluid quality. This heat flux surface is constructed from a variety of correlations for different ranges of the independent variables. In general, it is necessary to represent a wide range of conditions from subcooled liquid forced convection to two-phase film boiling. The details of this subject are discussed in other chapters. The users manual for a particular code version of interest should be consulted for specific information on the correlations used.

Component Models. The hydrodynamic and heat transfer models are quite general and can be applied to any thermal-fluid system (within limits established by the basic assumptions). However, there are several models that are specific to certain components such as pumps, jet pumps, fuel rods, valves, controls, etc. These are briefly described in the following.

Both the quasi-steady hydrodynamic performance and the transient operation of a centrifugal pump are modeled. The quasi-steady performance is modeled using emperically established homologous curves that relate the centrifugal pump similarity parameters for single-phase operation. From these curves any one of the parameters (head, volumetric flow, or speed) can be established from the remaining two. Pump performance under two-phase conditions is modeled using a head degradation parameter, which is a function of the pumped fluid void fraction. The transient mechanical operation of the pump is modeled by applying the angular acceleration relation for the pump and motor. The motor power is variable to enable pump trip and coastdown to be simulated.

For jet pumps, the momentum exchange between the drive flow and the pumped fluid is modeled using a special form of the momentum equation that includes the mixing effect of multiple streams at different velocities. Discontinuities that occur upon flow reversal are smoothed.

The fuel model consists of a space independent model for the fission- and radioactive-decay energy generation processes. The model includes reactivity feedback effects from the fuel temperature, water density, and water temperatures. The kinetics equations are solved using a numerical method similar to the IREKIN[25] code. The thermal energy generated in the fuel is transferred to the coolant by means of conduction through the ceramic fuel pellet, across the interface/gap between the fuel and the clad, and finally, across the clad. The conduction through the fuel pellet and the clad can be accurately characterized, but the conduction across the fuel/clad interface requires greater detail. The gap dimension varies with fuel and clad temperature and even when the fuel and clad are in contact, there remains a significant resistance. A dynamic fuel model is included for establishing the gap resistance due to change in the gap dimensions and change in pressure of the gas within the gap. Several other phenomena such as axial fuel/clad expansion, fuel/clad swelling, and metal-water reaction are also considered in the fuel model.[26]

On/off and check-valve models are included. The on/off valve is activated by a logical test on one or a combination of system variables such as time, pressure, temperature, etc. The check-valve model can include the effect of fluid forces and the inertia of the popet.

Control functions can be simulated by means of logical trips using time or any system parameters. The action that can be taken includes reactor scram, open/close valves, motors on/off, and even some change to the models that are employed for a period of operation.

Solution Method. The basic numerical scheme used in RELAP4/MOD6 and MOD7 is essentially the same as the original scheme developed for the FLASH4 code.[27] The approach consists of expressing the pressure in each control volume in terms of the corresponding mass and energy that exist in the volume at the end of a finite time interval. The expressions thus obtained for the pressures are in terms of the junction mass and energy fluxes so that substitution of these expressions into the momentum or flow equations at the appropriate junctions results in a system of coupled linear difference equations for the junctions flows. The coupling for a system of consecutive control volumes is such that a tridiagonal matrix of linear equations in the new time junction flows is obtained.

When branches or cross connections are present, the solution matrix is no longer tridiagonal. Thus the general solution scheme for the system of equations consists of a reduction algorithm by which the tridiagonal portions of the matrix are reduced directly followed by inversion of the reduced matrix. The matrix inversion can be accomplished by direct or iterative methods. The RELAP4/MOD6 solution scheme uses a direct matrix solver for small systems, less than 14 volumes, and an iterative scheme for large systems. When the iterative solution scheme is used, the timestep must be small enough to obtain a diagonally dominant matrix.

Once the solution for the junction flows is obtained, the remaining variables such as control volume mass, energy, pressure, etc., are obtained by back substitution into the respective conservation equations and use of the constitutive relations (equation-of-state, wall friction, heat transfer, etc.).

Programming. The RELAP4 program is written in FORTRAN IV, and the MOD5 and MOD6 versions are operable on both the IBM-360 and CDC-7600 computers; while the MOD7 version is only operable on the CDC-7600 and -176 series computers. The running time of a RELAP4/MOD6 loss-of-coolant problem can vary from minutes to hours, depending primarily upon the number of fluid volumes used, the coolant break size, and the number of heat conduction nodes used throughout a given system representation. For LWR models ranging from 15 to 40 fluid volumes, running time through refill may range from 15 to 60 minutes or more on the CDC-7600 and from 2 to 8 hours or more on the IBM-360/75.

3.3 RELAP4 Example Calculations

The RELAP4/MOD6 code has been used extensively in integral system simulations for some time, and for this reason, we will present results for one integral experiment. The example is a recent Semiscale Mod-3 test, S-07-6, which was included in the assessment effort on RELAP4/MOD6.[28] This experiment has also been modeled using RELAP5 and thus serves as a basis for comparison of the relative performance of the codes.

Mod-3 designates the latest major hardware modifications in the Semiscale Test Facility. Whereas early Semiscale testing was directed toward Loss-of-Fluid Test Facility (LOFT) counterpart and blowdown heat transfer experiments, Mod-3 was designed to model LOCA behavior in PWRs more easily.[29] The Mod-3 system differs from the previously operated Mod-1 system in three important aspects. First, Mod-3 has a new vessel that contains a full-length (3.66-m) heated core, has a full-length upper plenum and upper head with internal structures representative of those in a full-sized PWR with upper-head injection (UHI), and has an external downcomer. Second, an active pump and an active steam generator have been added to the broken loop. Third, the break simulation has the capability to represent communicative breaks of various sizes. Fig. 1 is an isometric sketch of the Mod-3 facility showing the most important features.

Test S-07-6 was the first integral blowdown and reflood experiment to be performed in the Mod-3 system. The test was a 200% cold-leg break with cold-leg ECC injection. A complete set of initial and operating conditions for this test is given in Ref. 30.

RELAP4/MOD6 Model for Semiscale Mod-3 Test S-07-6. Two RELAP4/MOD6 models were used to predict the behavior of Test S-07-6.[28] A blowdown model was used to calculate the transient response from the time of the simulated piping break until the end of the lower-plenum refill. A separate reflood model was used to model the reflood period through rod quench.

The model nodalization diagram used in the analysis of the blowdown and refill phases is shown in Fig. 2. The model includes 52 control volumes and 67 junctions. One control volume is used to represent the lower plenum, one the core mixer box, and two sets of 5 volumes for the hot and average channels in the core. The inlet annulus, guide tube, support tubes, and the upper head are each represented by one volume. The downcomer and upper plenum are each represented by three volumes.

A total of 50 heat slabs was used to represent heat conducting solids in contact with the coolant in the core, downcomer, steam generators, vessel, and piping. The high-power rods in the core are represented by 12 axially-stacked heat slabs, and the low-power rods are similarly represented by 5 heat slabs.

The heat transfer correlations used in the calculation are highly influential in determining the core thermal and hydraulic response. In this pretest prediction, the set of RELAP4 heat transfer correlations designated HTS2 was used. The heat transfer correlations as they are applied in specific regions of the boiling curve are tabulated in Table III.

The vertical-slip option was used in all downcomer, core-, guide- and support-tube junctions. The bubble-rise option was used in the following locations: (1) intact- and broken-loop accumulators, (2) the pressurizer, (3) the intact- and broken-loop steam-generator secondaries, (4) the pressure suppression tank, and (5) the upper head. The bubble-rise model was used in the upper and lower plenum, to be consistent with the use of the slip model in the core.

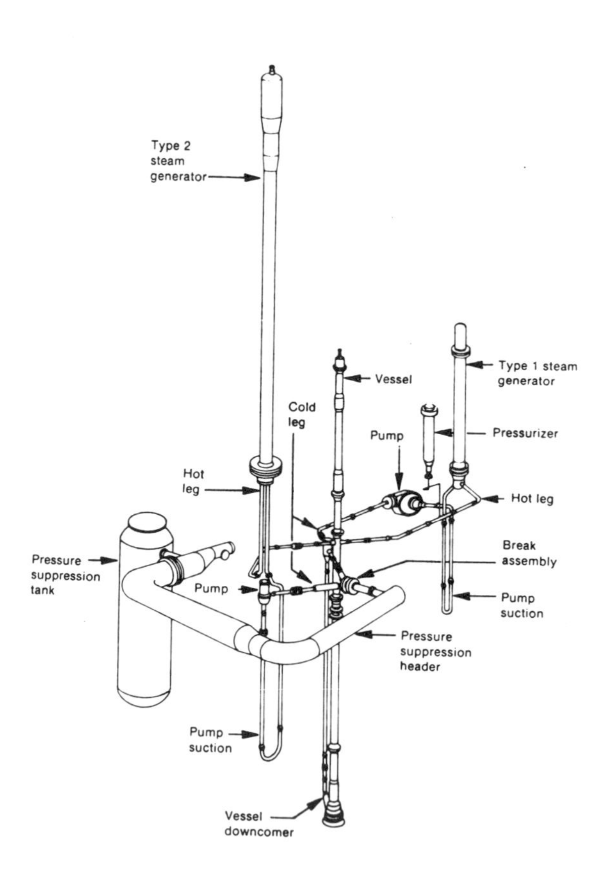

Fig. 1. Isometric sketch of the Semiscale Mod-3 Facility.

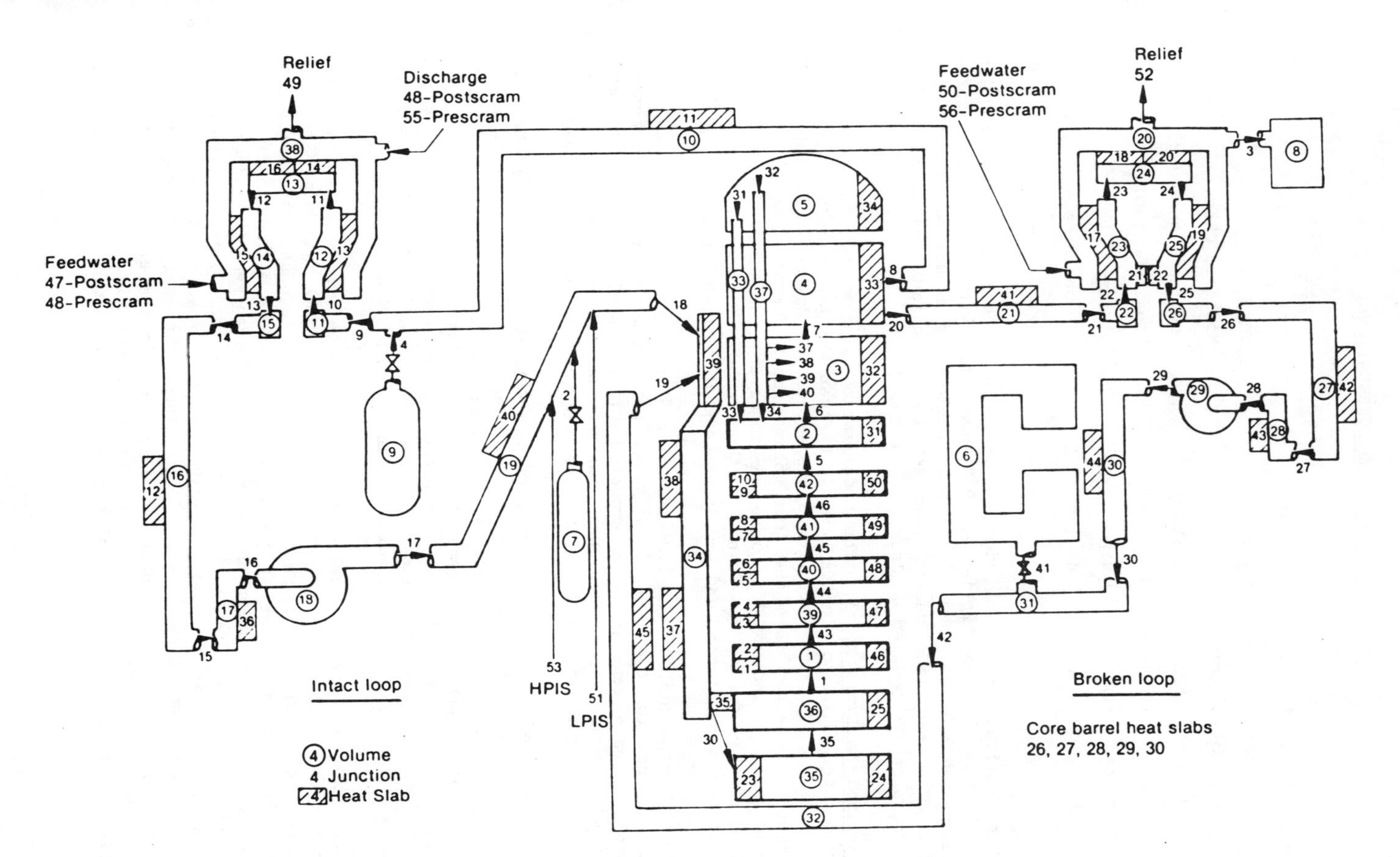

Fig. 2. RELAP4 blowdown nodalization, Mod-3 Test S-07-6.

TABLE III

RELAP4/MOD6 HEAT-TRANSFER CORRELATIONS, TEST S-07-6

Region	Correlation
Subcooled forced convection	Dittus-Boelter
Saturated nucleate boiling	Chen
Subcooled nucleate boiling	Modified Chen
High-flow transition boiling	Modified Ton-Young
High-flow film boiling	Condie-Bengston
Forced convection to vapor	Dittus-Boelter
Low flow, low void fraction	Hsu and Bromley-Pomeranz

The critical-flow model used was the Henry-Fauske/Homogeneous Equilibrium model (HF/HEM). A multiplier of 0.84 was used with the HEM for saturated blowdown and 1.0 was used with the subcooled and saturated HF critical-flow model. The transition quality was set to the default value of 0.02. The break nozzles are modeled as Junctions 26 and 27.

Reflood Analysis Model. The nodalization diagram for the model used to analyze the reflood portion is similar to that for the blowdown portion, shown in Fig. 2, but with the following differences. The downcomer is modeled by two fluid volumes rather than the four volumes used during blowdown. The pressurizer volume was discarded to reduce computer running time since the pressurizer empties during blowdown. Further nodalization changes include fewer volumes in piping, plena, and the steam-generator primary side.

The incompressible momentum equation form that excludes momentum-flux terms was used for the reflood analysis, i.e., kinetic terms were assumed to be small. Phase separation was modeled in the upper plenum and in the two downcomer volumes. For the upper plenum, the Wilson bubble-rise model was used. Complete phase separation was assumed for the downcomer volumes.

Initial conditions for the reflood analysis were taken from the blowdown analysis at the calculated end of lower plenum refill. Because the nodalization was different for the two analyses, an attempt was made to preserve fluid quality in regions that were lumped together. Heater rod temperatures were reinitialized at current fluid temperatures. ECC injection was specified by extrapolating calculated ECC behavior from blowdown through accumulator emptying and continuing the low-pressure injection.

Principal performance evaluators for the blowdown transient are system pressure for hydraulic processes and hot-channel cladding temperature histories for thermal response. An important diagnostic indicator is the density fluid in the lower plenum. Figure 3 shows a comparison of system pressure between calculation and experiment. The agreement is adequate until about 13 s into the transient. After that the calculated depressurization rate is greater than the measured rate and the experimental end of blowdown lags the calculated time by about 15 s. Thus, the analysis shows the end of the refill period to be at about 45 s, whereas in the experiment, this event occurs after 50 s. The cladding temperature history in the hot channel (Fig. 4) indicates an underprediction of the maximum temperature by as much as 75 K.

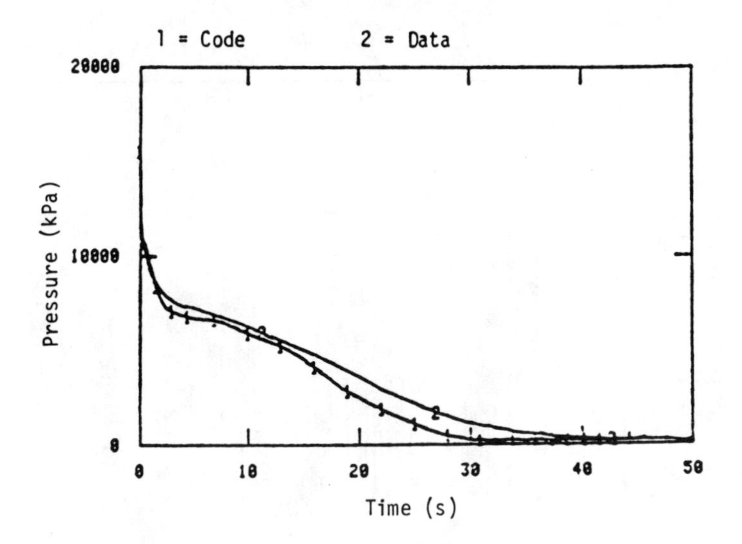

Fig. 3. System pressure history for Test S-07-6 blowdown.

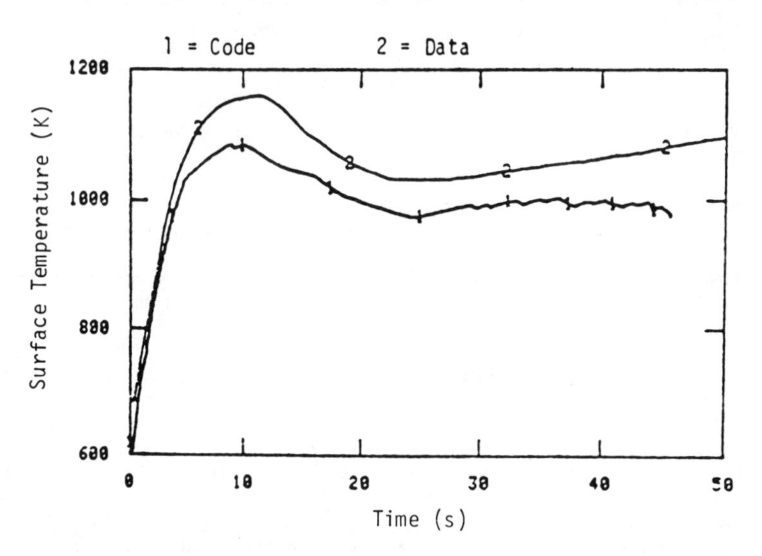

Fig. 4. Cladding surface temperature at the 1.84-m elevation in the hot channel for Test S-07-6 blow down.

Reflood Calculations and Data Comparisons. The reflood phase was considered to start at 45 s into the transient. This was the calculated time-to-lower-plenum refill. During reflood, the maximum cladding temperature, time to temperature turnaround, and time-to-rod quench are normally considered to be important performance evaluators. Parameters such as the downcomer liquid level, core liquid level, and core inlet density are considered diagnostic indicators.

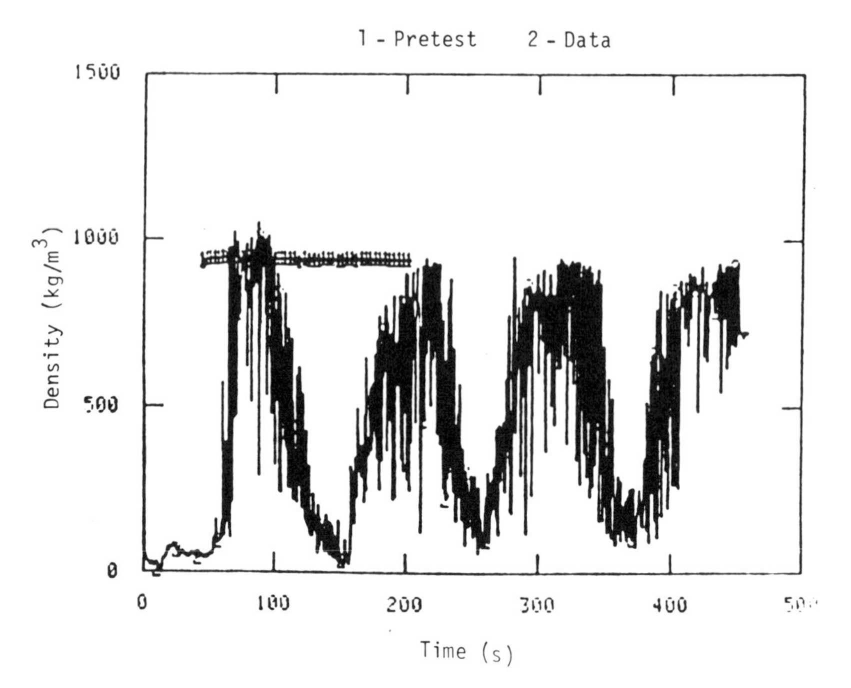

Fig. 5. Calculated and measured core inlet density, Test S-07-6 reflood.

Figure 5 shows the core inlet density as calculated and measured. The measurement shows a voiding of the core at about 100 s, followed by a long-period oscillation in the inlet flow. The calculation indicates that the fluid at the core inlet remains a dense liquid. The observed downcomer voiding also followed the pattern of core inlet density, whereas the calculation shows the downcomer to remain liquid-filled.

The base-case code-data comparisons demonstrated a need for incorporating mechanisms of liquid voiding in the modeling of the Semiscale Mod-3 downcomer. A primary contributor to this voiding behavior was determined, on posttest review, to be extensive vapor generation attributable to unanticipated heat transfer from the downcomer wall to the fluid. An additional study was made, incorporating wall heat transfer in the analysis and providing some facility for voiding downcomer liquid by changing to a Wilson bubble-rise model in the downcomer volumes. The results of the study indicated a tendency to improve but failed to provide acceptable code-data agreement.

Temperature histories are shown in Fig. 6 for the cladding at a location approximately at core midplane. The measured temperature is compared with the results of both the base-case analysis and the revised analysis. When voiding first occurs at about 100 s, both analysis and measurement (Fig. 6) show a tendency for the temperature to decrease as the core outlet flow also decreases. This decrease is followed by a slight temperature rise in the revised calculation and a major rise in the measured temperature--the difference being attributable to the failure of the calculation to sustain the voiding characteristic.

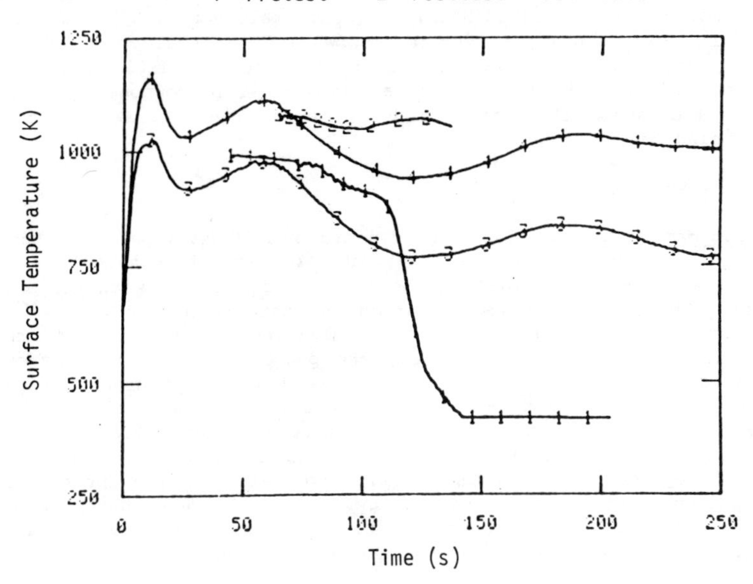

Fig. 6. Cladding surface temperature at the 1.84-m elevation in the hot channel for Test S-07-6 with additional calculation.

4. RELAP5 DESCRIPTION AND EXAMPLE COMPUTATIONS

The RELAP5 development objective[8] is an economical and user-convenient code for system transient simulation of LWR LOCA and non-LOCA transients. RELAP5 is an advanced, 1-D fast-running system analysis code. It is a completely new code based on a nonhomogeneous, nonequlibrium hydrodynamic model and features top-down structural design, with the significant programming elements coupled in modular fashion. To a great extent, the development of RELAP5 has been influenced by the experience gained through the development and usage of the RELAP4 series of codes. This is evident in the emphasis placed on the convenience with which both the developer and the user can interface with the code.

The RELAP5 code includes the thermal-hydraulic and mechanical models used to describe the processes that occur during transient operation and postulated accidents in an LWR. Component process models are included for pipes, branches, abrupt flow area changes, pumps, accumulator, valves, plant trips, heat transfer, neutronics, and choked flow. These, as well as other models, have been integrated into a versatile system code framework.

4.1 Program Status

The RELAP5 code is now operational, has been tested on hypothetical problems as well as actual experimental systems, and is in use at the Idaho National Engineering Laboratory for pre and posttest predictions of the LOFT, Semiscale and PBF experiments. The first version, RELAP5/MOD0, is available from the

National Energy Software Center at Argonne National Laboratory. A code description and user's manual are also available. All the discussion and example computations presented herein refer to this version that was developed for modeling the blowdown portion of an LWR LOCA. Development of RELAP5 is continuing, and a new version will be completed during 1980 that includes an accumulator model, point neutronics, a noncondensible component of the vapor phase, small-break stratification models, improved heat transfer models, and faster running capability,

4.2 Model Description

Hydrodynamic Model. The hydrodynamic model developed for the RELAP5 code[31-33] includes the important physics of the two-phase-flow process, while incorporating any simplifying assumptions consistent with the end use of the model. The principal simplification is that one of the phases exists at the saturation state. Generally, it is sufficient to specify that the least massive phase be at saturation, i.e., the phase that is either appearing or disappearing. The specification of one phase temperature greatly reduces the amount of constitutive information that must be provided relative to interphase and overall energy transfer. All interphase energy transfer mechanisms are implicitly lumped in the vapor mass generation model. Thus, a single correlation replaces the need for constitutive relations for interphase energy transfer, distribution of external energy transfer between phases, and distribution of energy transfer between sensible heat and heat of vaporization. In addition, only a single overall energy equation is required.

The two-fluid nonequilibrium hydrodynamic model includes options for simpler hydrodynamic models. Included are a homogeneous flow model and/or a thermal equilibrium model. The two-fluid or homogeneous flow models can be used with either the nonequilibrium or equilibrium thermal models, i.e., four combinations. The primary reason for inclusion of the homogeneous/equilibrium option is to permit the code to be compared to existing HEM code results such as RELAP4 for the purpose of checkout and development.

Field Equations. The basic field equations[34] for the two-fluid nonequilibrium model consist of the two phasic continuity equations, the two phasic momentum equations, and the mixture total energy equation--a total of five equations. The equations are employed in stream-tube differential form with time and one space dimension as independent variables and in terms of dependent variables, which are time- and volume-averaged quantities. The phasic mass conservation equations are summed and differenced to obtain a mixture continuity equation and an equation for the temporal variation of the mixture quality.

The phasic momentum equations are also used as a sum and difference. The sum equation is obtained by direct summation of the phasic momentum equations with the interface conditions substituted where appropriate. The difference of the phasic momentum equations is obtained by first dividing the vapor and liquid phasic momentum equations by the respective product of phasic void fraction and density and, subsequently, subtracting. Here again, the interface momentum conditions are employed.

The mixture total energy equation is obtained by summing the phasic energy equations. This mixture equation is transformed into the equivalent thermal energy equation by using the momentum equations to obtain a mechanical energy equation, which is subsequently subtracted from the total energy equation. Here again the interface conditions are employed to simplify the resulting energy equation. The reason for selecting the thermal energy equation rather than the total energy equation is that the development of the numerical scheme is simplified. The thermal energy equation does not involve time derivatives of the kinetic energy and thus fewer new time variables will appear in the approximate finite difference equations.

State Relations. The dependent variables that appear as temporal and/or spatial derivatives in the five field equations are density, pressure, static quality, mixture internal energy, and the two phasic velocities. The phasic properties also appear in the spatial derivatives and as coefficients of the derivatives. To obtain a determinant system, the state relationship must be employed wherein density and the phasic properties are expressed as functions of the pressure, static quality, and mixture internal energy. The state of the system is established from this information and from the specification that one of the phases exists at the prevailing saturation condition.

The state of each phase is established by specification of the pressure and phasic internal energy (only the pressure is needed to specify the state of the saturated phase). For the case of subcooled liquid or superheated vapor, these states are established using tabular equilibrium data as a function of the pressure and phasic internal energy. For the pseudo states of superheated liquid and subcooled steam, the properties are extrapolated along isobars using property derivatives evaluated at the corresponding saturation state.

In addition to the state properties, derivatives of the mixture density with respect to the pressure, static quality, and mixture internal energy are required in the numerical solution scheme. These derivatives can be expressed in terms of the isothermal compressibility and the isobaric coefficient of thermal expansion, both of which are available from the state properties data.

Constitutive Relations. A primary feature of the RELAP5 hydrodynamic model is that only two basic interphase constitutive relations are required, i.e., interphase mass transfer and interphase drag. The specification that one phase exists at local saturation conditions replaces the need for energy transfer and partitioning constitutive relations, both between phases and between each phase and the wall. The only heat transfer correlation required is the overall wall-to-fluid correlation. The RELAP4/MOD6[18] convective heat transfer correlations are used for this purpose. The remaining required constitutive relation is for wall friction. Here again, an existing two-phase multiplier correlation has been adapted.[18]

In summary, four constitutive relations are required by the hydrodynamic model--the vapor generation rate, the interphase drag, the wall friction, and the wall heat transfer. These relations are primarily empirical in nature as opposed to the field equations that characterize the fluid dynamic behavior. However, the ability of any numerical hydrodynamic model to agree with or predict physical phenomena with accuracy will depend heavily on the accuracy of the constitutive relations.

The vapor generation rate is the result of several mechanisms such as interphase energy transfer rate, the energy partitioning between phase change and sensible heat, interphase surface area, nucleation site density, turbulence level, etc. In RELAP5, all of these separate but interacting mechanisms are modeled by a single dimensionless correlation. This vapor generation model was developed by merging the results of three independent and widely varying investigations. The three approaches are (1) a mechanistic model by Jones and Saha[35] based on interphase energy exchange, (2) an empirical dimensional correlation by Houdayer, et al.,[36] from the Moby Dick data, and (3) the results of a dimensional analysis to establish the dimensionless groups and functional form of the vapor generation rate. The last of these efforts was completed as a part of the RELAP5 project[3] in order to establish the scale dependence of the vapor generation function.

Interphase drag consists of two parts--the dynamic drag due to the virtual mass acceleration and the steady drag arising from viscous shear between phases. The dynamic drag has been included because of the effect it has on the sound speed and hence, the choking criterion. The dynamic drag is calculated based on the induced mass of a spherical bubble (or drop) in a mixture of vapor bubbles (or liquid drops) and liquid (or vapor). The steady drag depends on the flow regime and the relative phase velocity.

The flow regime map used in RELAP5 is a simplified Bennett map[37] for vertical flow and is similar to the one used by TRAC.[8] The flow regimes are classified into the general categories of dispersed, separated, churn turbulent, and transitional flow.

Constitutive relations for the steady drag are formulated for the separated and dispersed flows based on semimechanistic models. The drag in the transition regimes is calculated by linear interpolation on the reciprocal values of the separated or dispersed-flow drag coefficients defined at the boundaries of the particular transition region. This yields a continuous variation in the calculated relative velocity. The calculation of the drag due to the virtual mass effect is based on an objective and symmetric formulation of the relative acceleration proposed by Lahey.[38] This formulation involves spatial and temporal derivatives of the phase velocities with a correlation for the virtual mass coefficient of Zuber's.[39]

The wall friction force terms only include wall shear effects. Form losses due to abrupt area change are calculated using mechanistic form-loss models. Other form losses due to elbows or complicated flow passage geometry are modeled by specified energy loss coefficients. Wall shear losses in piping systems are usually small compared to form losses, thus a relatively simple approach that yields an accurate steady-state frictional pressure drop is employed.

The HTFS modification of the Baroczy two-phase friction multiplier correlation[40] was used with the Colebrook correlation for the single phase friction factor including wall roughness effects. Both laminar and turbulent flow regimes are included. The two-fluid hydrodynamic model requires that the wall friction force be partitioned between the liquid and vapor phases. The method used in RELAP5 is based on void fraction partitioning of the friction force. The phasic friction components are normalized so that the sum of the phasic frictional forces agrees with that derived from the two-phase multiplier approach.

Heat Transfer Correlations. The wall heat transfer correlations used in RELAP5/MOD0 are adaptations of the blowdown heat transfer package from RELAP4/MOD6.[18] In adapting the RELAP4 package, the correlations were converted to scientific notation units and only the Condie-Bengston correlations were retained for use in the transition and film boiling regions. In addition, the procedure for applying correlations was modified to eliminate the need for iteration and to allow the same procedure to be usable for both steady-state and transient calculations.

Special Process Models. Special models are used in RELAP5 for those processes that have small relaxation times or are so complex in nature that they must be modeled by quasi-steady empirical models. Break flow, internal choking, abrupt area change, and branching are examples of processes having short relaxation times compared to component transport times. The hydrodynamic performance of pumps and valves are examples of processes that are too complex to be modeled from first principles, so empirical correlations are used. The use of quasi-steady models for break flow and flow at abrupt area changes results in considerable computer time savings since it eliminates the need for fine nodalization at such points.

A break flow model[41,42] is included for calculation of the mass discharge from the system at such points as a pipe break or a nozzle in the case of scaled experiments like Semiscale or LOFT. Generally, the flow at such breaks is choked until the system pressure nears the containment pressure. The RELAP5 break flow model is used to predict the flow at such system discharge points and is also used to predict and calculate choked flow at internal points in the system. The model is based on characteristic theory in which a criterion is developed for the conditions under which propagation of pressure signals upstream just ceases. This theory applies to all two-phase conditions. Additional theoretical considerations have been employed to extend the break flow model to conditions of subcooled liquid flow that flashes at the point of mass discharge.

The general reactor system contains piping networks that consist of many sudden area changes and orifices. In order to apply more efficiently the hydrodynamic model to such systems, analytical models for these components have been developed.[43] The RELAP5 abrupt-area-change model is based on the Bourda-Carnot[44] formulation for a sudden enlargement and standard pipe flow relations, including vena-contracta effect, for sudden contractions and/or orifices. Quasi-steady continuity and momentum balances are employed at points of abrupt area change. The numerical implementation of these balances is such that the hydrodynamic losses are independent of the upstream and the downstream nodalization. In effect, the quasi-steady balances are employed as jump conditions that couple fluid components having abrupt change in cross-sectional area. This coupling process is achieved without change to the basic linear semi-implicit numerical time-advancement scheme.

In order to model flow in interconnected piping networks, it is necessary to model the two-phase fluid process at tees, wyes, and plenums. A general description of the two-phase flow process is complicated by the possibility of phase separation effects.[45] However, there are many situations where wye or plenum branching is adequate for both flow merging and division. Typical situations are parallel flow paths through the reactor core, jet pump flow mixing sections, and any branch from a vessel of large cross section (in this case the fluid momentum is small and it is entirely permissible to neglect the momentum convective terms). For branching situations where phase separation effects due to momentum and/or body force effects are important, a branching algorithm has been developed[7] in which the parallel or wye branching model is used to map the 2-D situation onto the 1-D space of the fluid model.

The RELAP5 pump model is a straightforward conversion of the RELAP4 centrifugal pump model.[17] The pump is interfaced to the unequal velocity hydrodynamic model of RELAP5 quite simply by assuming that the head developed by the pump is similar to a body force term that appears only in the mixture momentum equation. The pump dissipation term for the thermal energy equation is computed from the total pump power (given by torque times rotational speed) minus the rate of fluid reversible energy addition.

Numerical Methods. The RELAP5 numerical solution scheme[34] is based on replacing the system of differential equations with a system of finite difference equations, which are partially implicit in time. In all cases, the implicit terms are formulated to produce a linear time advancement matrix, which is solved by direct inversion using a sparse matrix algorithm. An additional feature of the scheme is that the implicitness has been selected such that the five field equations can be reduced to a single difference equation per fluid control volume or mesh cell in terms of the hydrodynamic pressure. Thus, only an NxN system of difference equations must be solved simultaneously at each timestep. (N is the total number of control volumes used to simulate the fluid system.)

The difference equations are based on the concept of a control volume or "mesh cell" in which mass and energy are conserved by equating accumulation to rate of influx through the cell boundaries. This results in defining mass and energy volume average properties and requiring knowledge of velocities at the volume inlets and outlets (junctions). The junction velocities are conveniently defined through use of momentum control volumes that are centered on the mass and energy cell inlets and outlets. This produces a numerical scheme having a staggered spatial mesh. The scaler properties (pressure, energy, and quality) of the flow are defined at cell centers and vector quantities (velocities) are defined at the cell junctions. The resulting 1-D spatial noding is illustrated in Fig. 7. The term cell is used throughout the discussion to mean an increment in the spatial variable corresponding to the mass and energy control volume.

The mass and energy difference equations for each cell are obtained by integrating the stream-tube formulations for the mass and energy equations with respect to the spatial variable, x, from the junction at x_j to x_{j+1}. The momentum equations, on the other hand, are integrated with respect to the spatial variable from cell center to adjoining cell center (x_K to x_L) as seen in Fig. 7. In all cases, the correlation coefficents for averaged products are taken as unity so that averaged products are replaced directly with products of averages.

Several general guidelines were followed in developing the overall numerical scheme. These guidelines are summarized below.

1. Mass and energy inventories are very important quantities in water reactor safety analysis and as such the numerical scheme should be consistent and conservative in these quantities (a greater degree of approximation for momentum effects was considered acceptable). Both mass and energy are convected from the same cell and each is evaluated at the same time level (i.e., mass density is evaluated at old time level so energy density is also evaluated at old time).

2. In order to achieve fast execution speed, implicit evaluation is used only for those terms necessary for numerical stability, elimination of the wave propagation time-step limit, and those phenomena known to have small time constants. Thus, implicit evaluation is used for the velocity in mass and energy transport terms, the pressure gradient in the momentum equations, and the interphase mass and momentum exchange terms.

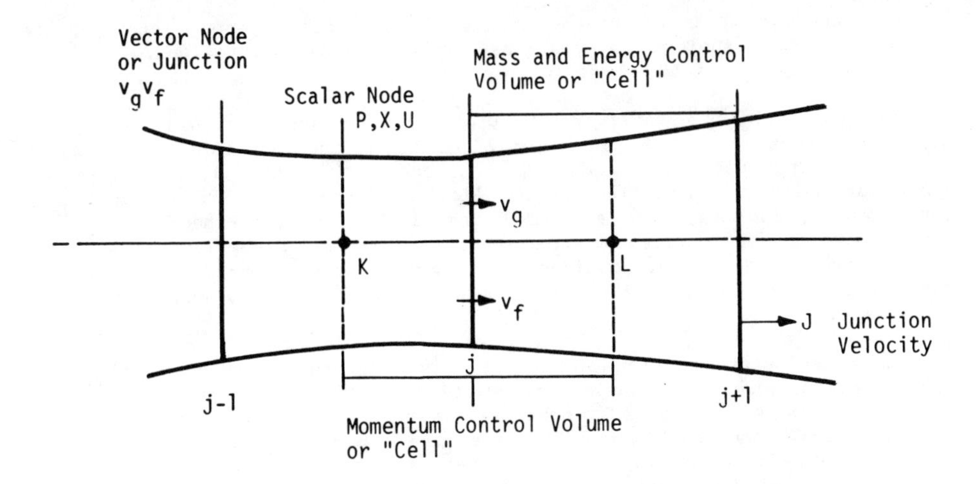

Fig. 7. Difference equation nodalization schematic.

3. To further enhance computing speed, the time level evalutions were selected so that the resulting implicit terms are linear in the new time variables. Where it was necessary to retain nonlinearities, two term Taylor series expansions about old time values were used to obtain a formulation linear in the new time variables (higher order terms were neglected). High computing speed is achieved by eliminating the need to solve large systems of nonlinear equations iteratively.

A well-posed numerical problem is obtained as the result of several factors. These include the selective use of implicitness (evaluation of spatial gradient terms at the new time), donor formulations for the mass and energy flux terms and use of a "donor-like" formulation for the momentum flux terms. The well-posed final numerical scheme (as well as its accuracy) has been demonstrated by extensive numerical testing during development.

System Code Development. The primary emphasis in the system code design of RELAP5 has been to achieve an economical code. Attention has been focused on reducing computer time per mesh point per advancement. Timestep control algorithms have been included to minimize the number of advancements. Dynamic storage has been used to keep the computer memory requirements to a minimum.

The user's time in setting up, debugging, and interpreting results is also significant. User conveniences significantly reduce overall simulation costs. The system code includes many modeling disciplines such as hydrodynamics, pumps, valve actions, heat transfer, and neutronics. Because the details of the system must be described to the program, the requirement for a large amount of data cannot be avoided. Thus, user-oriented input, extensive error checking, and several forms of printed and plotted output are provided.

Experience gained from the development and use of RELAP4 has shown that to achieve true economy, the code structure must provide for ease of addition and modification. Care has been taken in RELAP5 to structure data files and program organization in a modular fashion in order to achieve this goal.

The RELAP5 code is organized into four basic parts: input, steady-state initialization, transient calculations, and the output functions. Each of these parts is summarized in the following.

The code contains extensive input processing routines designed to help the user find input errors and in a small number of checkout runs to obtain an error-free input data deck. The input processor is designed to process all input for every job submitted and to list the errors. In this respect the input routine is similar to a FORTRAN complier. The error-checking routines find impossible or conflicting data specifications and misapplications of the various models. The input routines also process program control data for such functions as major/minor edits, writing of restart records, and creation of plot files.

The steady-state portion of the code is intended to produce the initial conditions for starting transient calculations. This capability is currently incomplete, and the input and generalized restart features are used to provide this function. The initial conditions can be input, or a transient calculation can be made to achieve a steady state and then the generalized restart feature is used to modify the configuration and initiate the transient. Transient hydrodynamic, heat transfer, and neutronics calculations are performed in the transient portion of the code. Other functions are time-step regulations and trip logic calculations.

The output portion of RELAP5 provides both major and minor output edits at specified intervals, prepares restart records, and generates plot files for graphical output. RELAP5 has an internal plotting feature for graphical output, and can also be used with any external plotting package. Internal diagnostic edits are provided whenever the code fails due to water property errors, which are generally symptomatic of an unrealistic modeling condition.

4.3 RELAP5 Example Calculations

The RELAP5 code has been used to model several separate-effects experiments and some limited system experiments. The separate effects experiments that have been modeled include the Edwards 3-inch[46] and 8-inch pipe blowdowns; the Edwards Phase II two-pipe blowdown; the Moby Dick Run 447; the General Electric one-foot vessel level swell; the Marviken III Tests 4,[42] 22, and 24; Semiscale Tests Mod-2 S-01-4a, Mod-2 S-06-2, and Mod-3 S-07-6;[47] and the LOFT Tests L3-0, L3-1, and L3-2. These tests have been used for developmental assessment. In all cases the performance achieved using the code has been good. The LOFT system test simulations are the most recent applications of the code and good agreement with data was achieved while requiring a CPU time less than real time for the L3-2 experiment prepediction.

Three representative applications of the code that will be summarized in the following discussion are the Marviken III Test 4, the Semiscale Mod-3 Test S-07-6, and the LOFT LOCE L3-2 small-break test.

Marviken III Test 4. The RELAP5 code was used to simulate the Marviken III Test 4[42] to evaluate the code's ability to predict the hydrodynamic behavior of a large-scale blowdown test. Simulation of the test allows evaluation of the choked flow model under conditions that are comparable to those expected in an LWR during a postulated LOCA.

The purpose of the Marviken III Test 4* was to establish choked flow rate data for a large scale nozzle (500 mm in diameter) with subcooled and low-quality water conditions at the nozzle inlet. A schematic of the pressure vessel, discharge pipe, and test nozzle is shown in Fig. 8. The pressure vessel was initially filled with water to an elevation of 16.8 m above the discharge pipe inlet. The steam dome above the water level was saturated at 4.94 MPa. The water level was at nearly saturation conditions for about 6 m below. The water was subcooled by about 30 K below the saturated fluid after a small transition zone. The initial temperature profile is also shown in Fig. 8.

The nodalization used for the numerical simulation is also illustrated in Fig. 8. A nearly uniform cell length of about 1 m was used everywhere. No special nodalization was used in the nozzle region. This was possible because RELAP5 includes an analytical choking criterion that is applied at the throat of the nozzle.

The calculated blowdown transient was simulated by opening the discharge pipe outlet to the ambient pressure. The measured data consisted of pressures, differential pressures, temperatures, and mass discharge rates inferred from pitot-static pressure data. Corresponding values were calculated and comparisons are shown in Figs. 9 and 10. Figure 9 shows the pressure history at the vessel top. Except for an initial nonequilibrium undershoot (at about 3 s after rupture), the depressurization process was essentially in equilibrium. The calculated blowdown rate was in agreement with the system blowdown rate, and since the depressurization rate was controlled by the break mass flow, the fact that the pressure profiles were in agreement demonstrates that the discharge flow was modeled accurately.

A comparison of the calculated and inferred discharge flow rates is shown in Fig. 10. The clear transition from subcooled to two-phase critical flow is shown both in the test data and in the calculations at about 17 to 20 s after rupture. This was reflected in the numerical calculations as the code automatically switched from the subcooled choked-flow criterion to the two-phase criterion.[42]

Comparison of the RELAP5 calculations with the Marviken III Test 4 results provides a good evaluation of the ability of a two-phase thermal-hydraulic model to predict mass discharge rates correctly under choked-flow conditions at large scale.

<u>Semiscale Mod-3 Test S-07-6</u>. The Semiscale Mod-3 system[29] and a summary of the Test S-07-6[30] are given as a part of the RELAP4 modeling of this same test. The application of RELAP5 to this test[47] is a good example of integral system behavior prediction capability and also provides an example of the benefits obtained by the use of an advanced hydrodynamic model. In particular, the RELAP5 results agree much better with data than the RELAP4 results. Test S-07-6 response was quite different from previous Semiscale experiments and was characterized by several periods in which refill of the downcomer and partial reflooding of the core was followed by a rapid reduction in both the downcomer and core liquid inventories (mass depletion).

---

*This test is one in a series of tests performed as a multinational project at the Marviken Power Station by A. B. Atomenegri Sweden. The test results are reported by L. Ericson, et al., in "Interim Report Results from Test 4," MXC-204, May 1979.

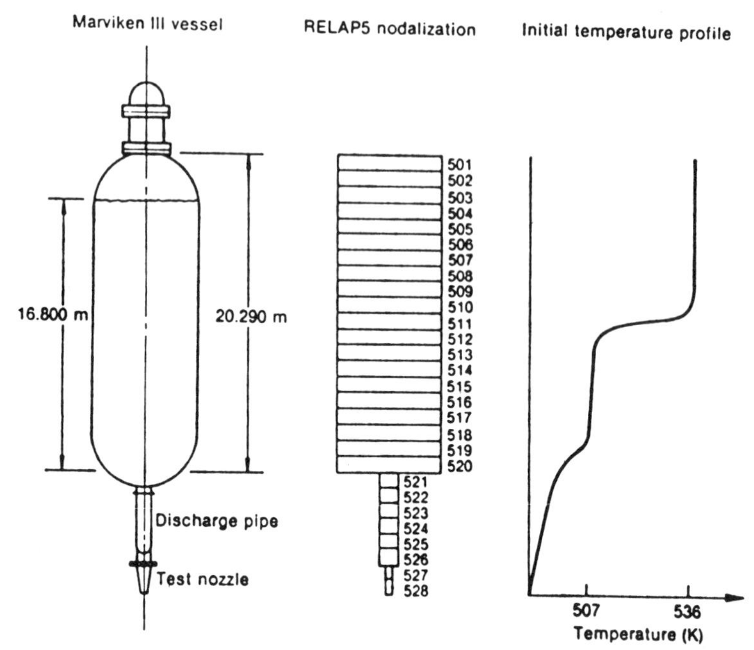

Fig. 8. Marviken III Test 4 vessel schematic, RELAP5 nodalization, and initial temperature profile.

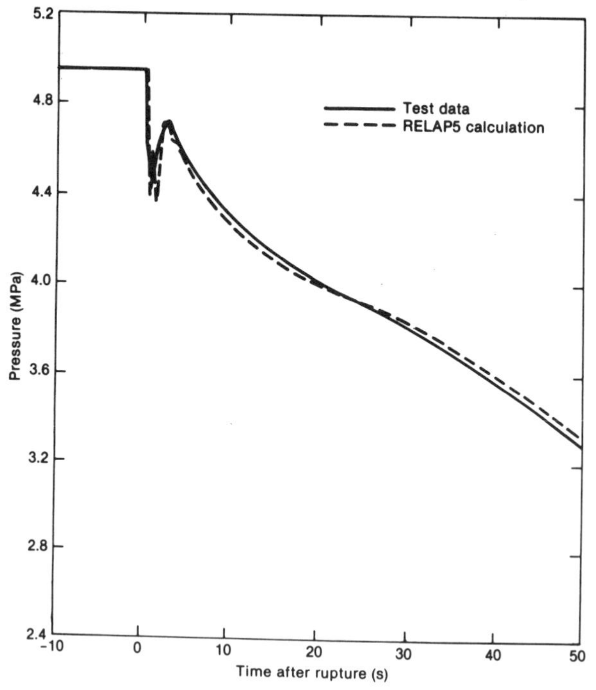

Fig. 9. Calculated and measured pressure at vessel top (Cell 501 in RELAP5 nodalization).

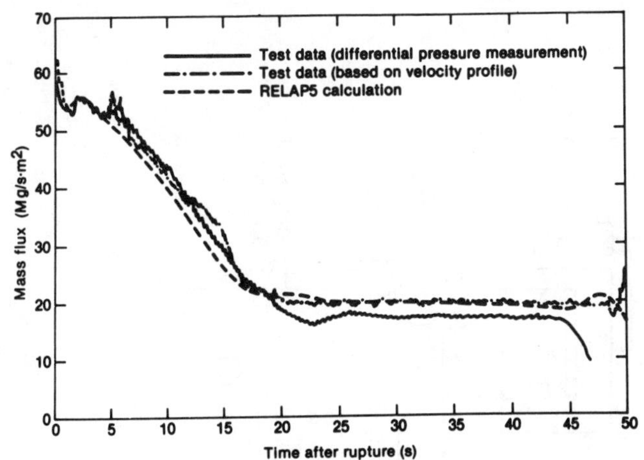

Fig. 10. Calculated and measured mass flux at nozzle inlet (Cell 526 in RELAP5 nodalization).

The flow oscillations resulted in core temperature oscillations, and complete quenching of the core did not occur until 500 s after rupture.

The RELAP5 model of the Semiscale Mod-3 system is divided into control volumes connected by junctions. The code uses component oriented modeling so that large sections of the system can be identified as a component. The component can then be subdivided to obtain the needed detail. A schematic of the model for the Semiscale Mod-3 system is shown in Fig. 11. This nodalization diagram can be compared to the isometric drawing of the Semiscale system shown in Fig. 2. A total of 133 control volumes interconnected by 143 flow junctions were used. A total of 109 conduction heat structures (shown as shaded areas in Fig. 11) were used to represent heat transfer from pipes and structural parts of the system. Twelve heat structures (one for each power step) were used to represent low-power heater rods.

The model was initialized by running the calculation at prescribed initial conditions until steady state was reached. The transient calculation was then made from the initiation of rupture through blowdown and reflood to 200 s. The core power, pump coastdown, and ECC rates were taken from the experimental data.

The break flow rate controls the rate at which the system empties, the depressurization rate, and the core flow behavior. The transition of break flow to two-phase flow was calculated to occur at 2 s, while the test data indicated two-phase choked flow at 3 s. The calculated break flow rate in the two-phase flow region was slightly higher than the measured data indicated. The slightly higher calculated total break flow rate resulted in slightly more rapid depressurization of the system. Figure 12 shows the calculated pressure at the vessel upper plenum. The calculated pressure was higher than shown by the test data before 5 s and lower thereafter.

In spite of slight discrepancies, the RELAP5 calculation of the blowdown behavior of the Mod-3 system was very good. This was reflected in the core heater temperature. Figure 13 shows the rod temperature in the hot channel at 184 cm above the bottom of the rod. The calculated temperature was taken at

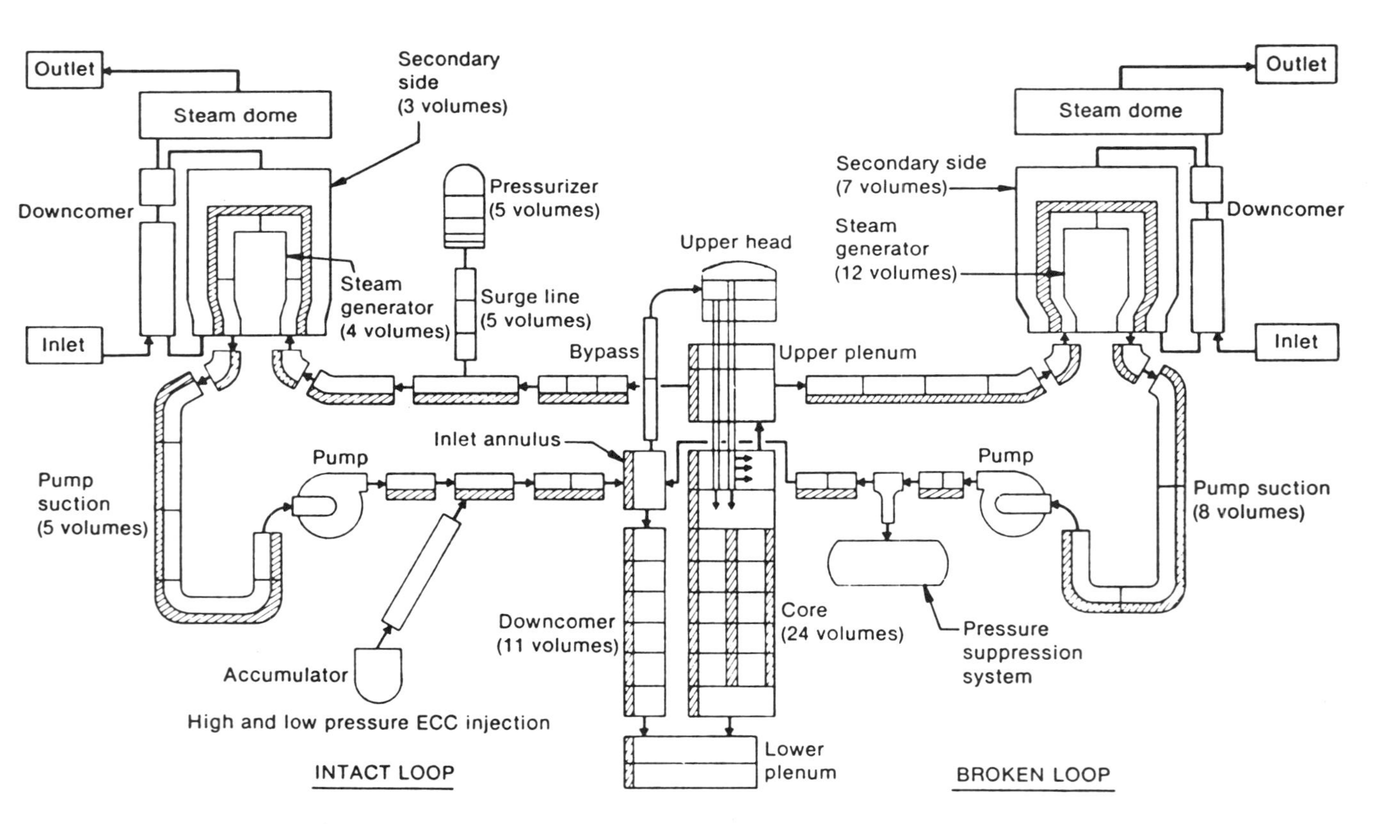

Fig. 11. RELAP5 nodalization for Semiscale Mod-3 system.

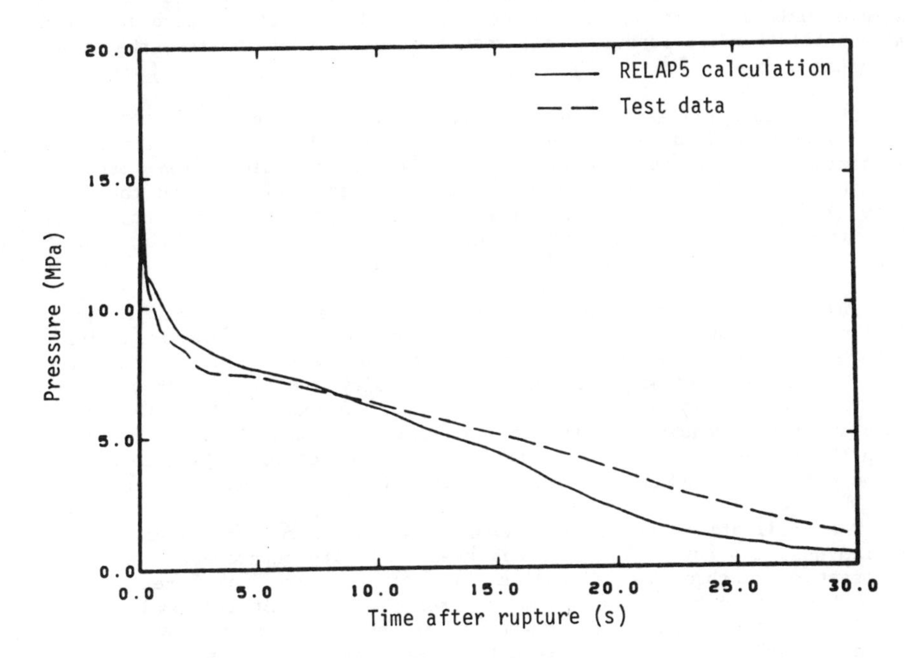

Fig. 12. Calculated and measured pressure in the upper region.

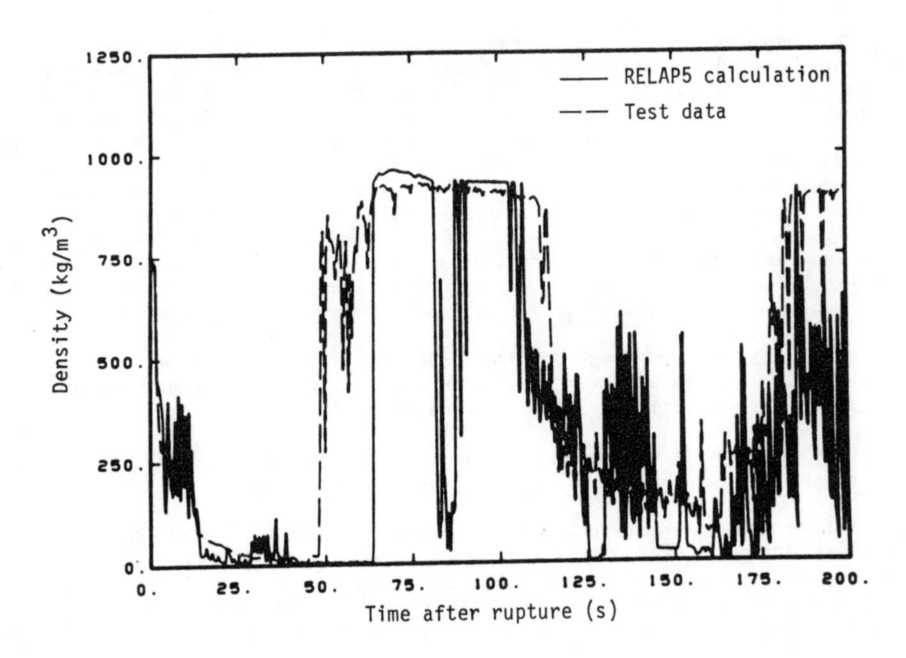

Fig. 13. Calculated and measured fluid density at the middle of downcomer.

slightly under the surface (80-K difference existed at the initial steady-state conditions). Calculated temperatures reached a peak of 1150 K, which agreed with the test data. The decrease in heater rod temperature beginning at about 12 s after rupture was a result of water draining from the upper head into the core.

The RELAP5 simulation of the phenomena associated with ECC injection, refill, and reflood were equally encouraging. The test was simulated from initiation of pipe rupture through reflood in one calculation without renodalization. At 19 s, the system pressure reached 4.14 MPa, and the ECC water from the accumulator began to flow into the system. During the accumulator ECC injection period, the difference in the calculated vapor and liquid temperature clearly indicated thermal nonequilibrium existed.

One of the most interesting aspects of Test S-07-6 was the multiple filling and emptying of the downcomer and core as mentioned earlier. The downcomer depletion behavior and the effect on the core thermal response during reflood was reflected in the fluid density. Figure 13 shows calculated and measured densities at the center of the downcomer. Both the calculated values for the density and the periodic mass depletion behavior compared well with the measured data. In the calculation, the ECC water penetrated into the downcomer at 65 s, while the test data showed this to occur at 50 s.

The oscillations in the downcomer mass flow were controlled by the time period when the subcooled water was present in the downcomer. When the fluid temperature reached the saturation temperature, vapor was generated and the hydrostatic head in the downcomer decreased. Some of the coolant was then expelled from the top of the downcomer. When the downcomer hydrostatic head decreased sufficiently as a result of mass depletion, the ECC water could again flow into the downcomer. The heating, expelling, and refilling process was repeated periodically.

The measured heater rod surface temperature rose and decreased as water left and entered the core. The calculation also showed the oscillation in the heater rod temperature. Figure 14 shows the calculated and measured rod temperatures at the hot and average channels near the axial peak power zone. The calculated maximum and minimum temperatures compared well with the test data before 100 s. The calculated frequency of the temperature oscillation was close to the measured frequency after 100 s. The calculated temperature of the hot channel gave a lower value while the temperature of the average channel was too high. The average of the two temperatures fell within the measured data. Two channels were used to model the core and no cross-flow was allowed between the two channels. This core model appears to be the cause for the discrepancy between the calculated and measured rod temperatures.

This analysis confirmed that Semiscale Mod-3 Test S-07-6 was modeled well by RELAP5 during the blowdown period and that it gives reasonable quantitative results for the refill and reflood periods of the test.

LOFT TEST L3-2. The results presented here represent the first time that the RELAP5 code was used for a formal pretest prediction.

A LOFT Facility description and a summary of the LOCE L3-2 key events are included in Ref. 48. The nodalization used in the RELAP5 calculation was similar to the nodalization used for the RELAP4 blowdown calculation of LOFT LOCE L2-3.[48] In areas where significant elevation differences exist, the RELAP5 nodalization was increased to define steep density gradients. The

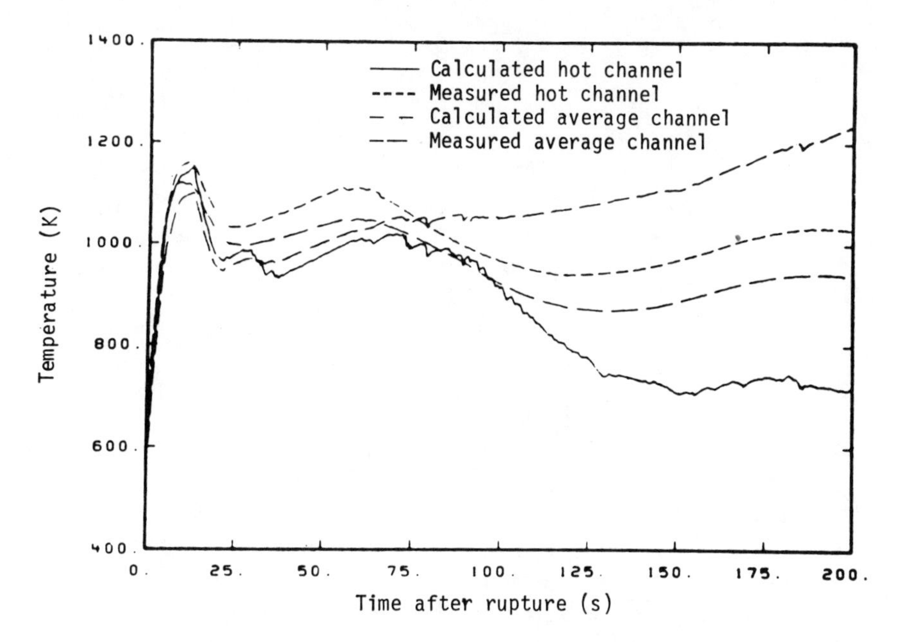

Fig. 14. Calculated and measured temperature of heater rod at the hot and averaged channels near axial power peak.

RELAP5 nodalization also includes simulation of the potential bypass flow path between the reactor vessel inlet annulus and upper plenum. The nodalization scheme is shown in Fig. 15.

The liquid separator and mist extractor of the steam generator secondary system are modeled by modifying the donor formulation of the convective terms at the separator junction (Component 10). The steam flow control valve is assumed to have a linear area change with stem position and a zero-inertia constant speed driver. The RELAP5/MOD0 valve subroutine required modification to model this type of valve. The sophisticated trip logic in RELAP5 allows simulation of the valve controller. The steam generator outflow is connected to the air-cooled condenser (Component 16) where the pressure is specified. The feed flow is input as a function of time.

The ECCS System is represented by Components 168, 500, and 505 (see Fig. 15). LOFT Accumulator A, Component 168, is modeled using the RELAP5 accumulator model. The LPIS and HPIS pump models, Components 505 and 500, respectively, required modification to the time-dependent junction subroutine in RELAP5 so that the flow provided by these components could be specified as a function of downstream pressure. The orifice at the break plane is modeled by a valve having an open area equal to the area of the drilled break orifice.

Heat conduction between the primary and secondary sides of the steam generator is through heat Structure 5-2, the steam generator tubes. The reactor pressure vessel, filler blocks, core filler, upper and lower core support structures, and core also were modeled using heat conductors. The system was modeled with no heat loss to the surroundings.

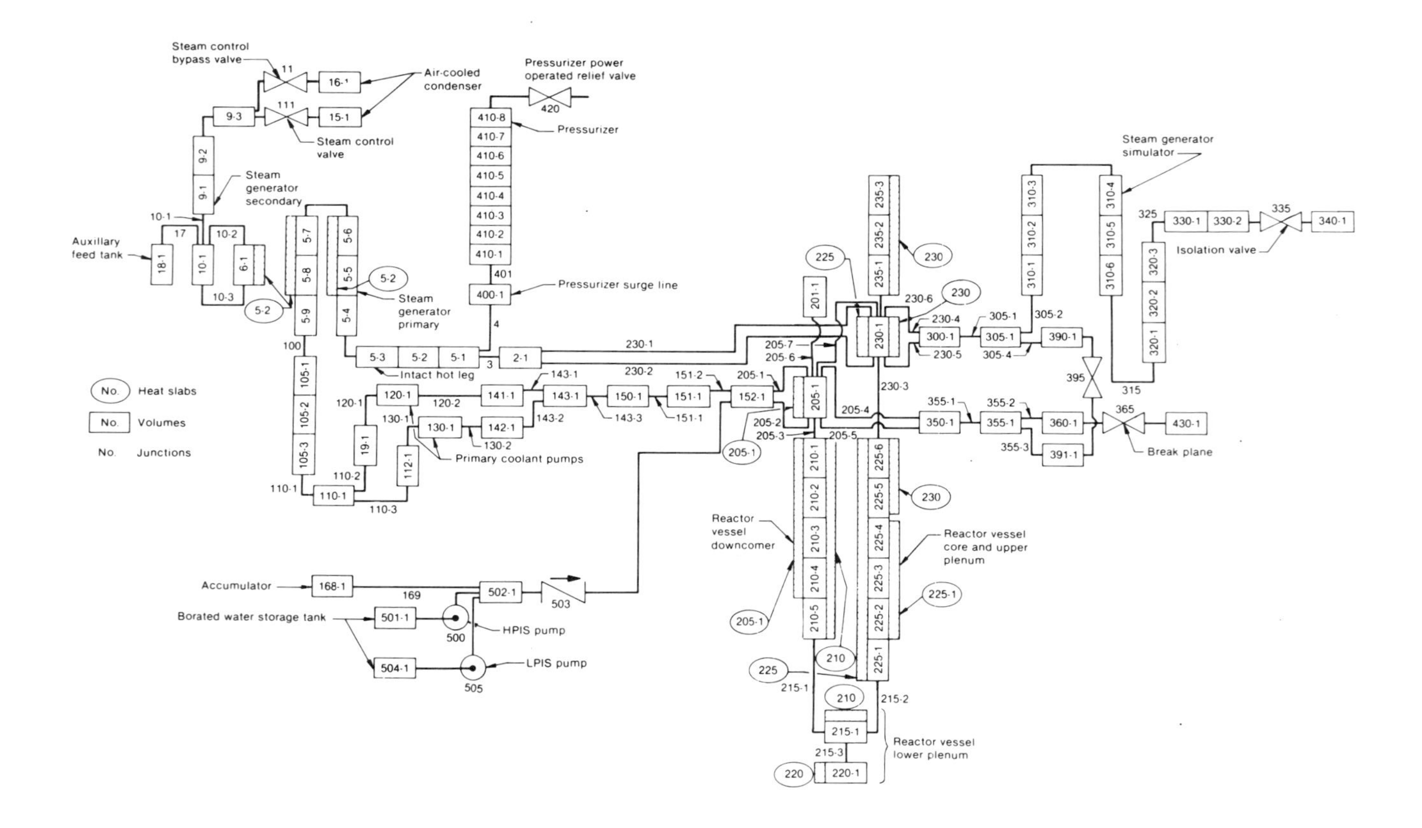

Fig. 15. LOFT RELAP5 model schematic diagram.

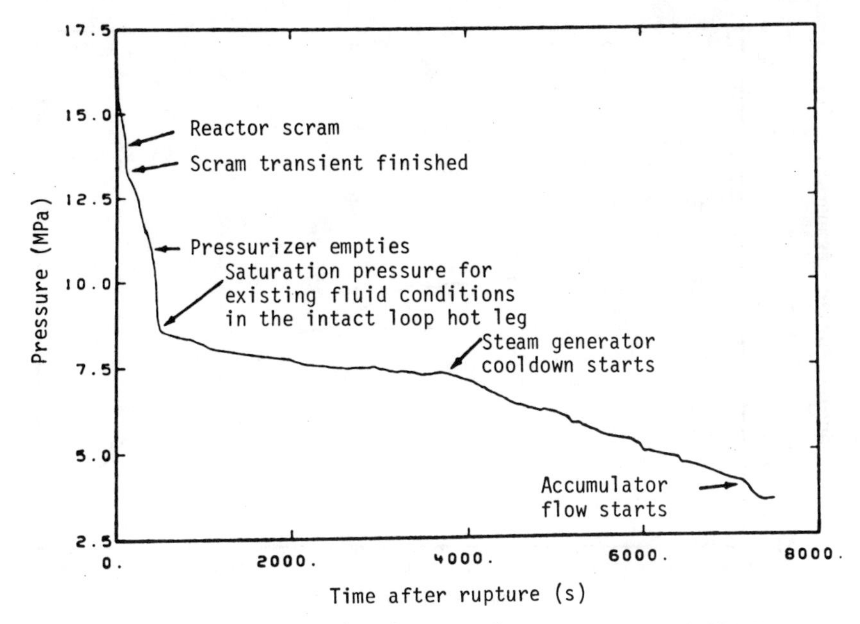

Fig. 16. RELAP5 predicted upper plenum pressure response.

The transient was initiated by opening the cold leg quick-opening blowdown valve. For the first 94 s after experiment initiation, the pressure is predicted to decrease 1.41 MPa from the initial value, causing a reactor scram to initiate as shown in Fig. 16. During the next 13 s, the steam generator removes more energy from the primary system than the reactor core adds, resulting in a net stored energy loss in the primary loop. The resulting density increase in the primary coolant places a further demand on the pressurizer, resulting in the high depressurization rate after 94 s. At 107 s, the steam flow control valve shuts completely, mitigating the rapid pressure decline in the primary system. At 127 s, HPIS is initiated by low pressure in the hot leg; but at about 400 s, the pressurizer empties and the primary system rapidly approaches the saturation pressure corresponding to the fluid temperature in the intact loop hot leg. The steam flow control valve is predicted to start opening at about 150 s. A small flow in the intact loop carries thermal waves, generated by the valve opening, throughout the system. The steam valve is predicted to stay closed in the period between 1000 and 2000 s and to start opening again at about 2000 s, reducing pressure to about 7.2 MPa. At this point, HPIS flow is about equal to break flow.

After 1 h, steam is removed from the steam generator by opening the steam flow control bypass valve in such a manner to cause cooling in the steam generator secondary side of 44.4 K per hour. This energy removal causes cooling in the primary loop of 42.5 K per hour. The calculations, therefore, indicate that the steam generator cooling will be effective in the primary system. After 1.1 h of cooldown, the auxiliary feed pump is turned on to fill the steam generator secondary side. The addition of this cold water causes a cooling rate in the steam generator greater than 44.4 K per hour. The steam flow control bypass valve is, therefore, shut whenever the cooldown exceeds 44.4 K per hour. The liquid level in the reactor vessel upper plenum was not

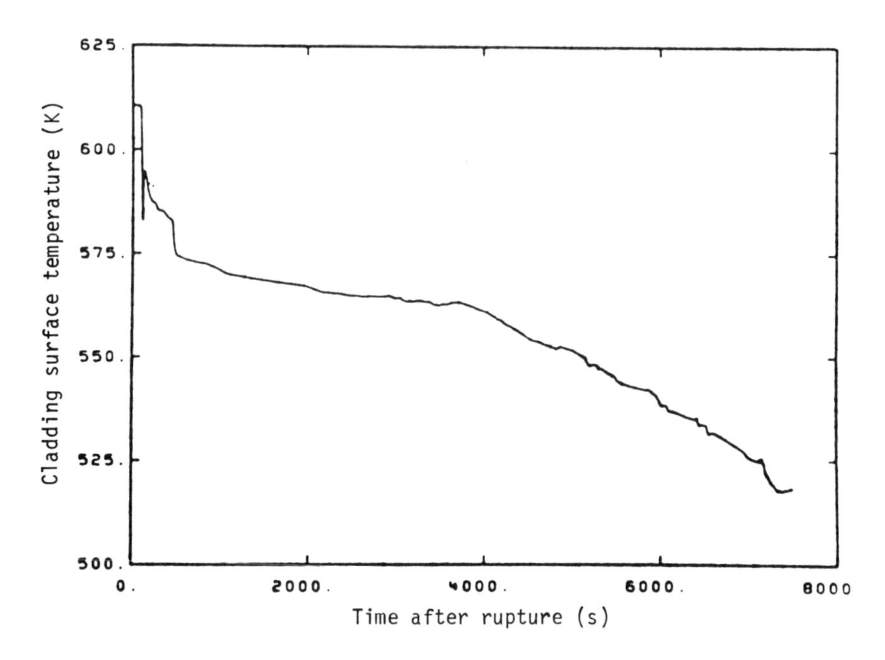

Fig. 17. RELAP5 predicted cladding surface temperature.

predicted to drop below the top of the active core, thus the cladding surface temperature response shown in Fig. 17 is calculated to be benign.

The calculations shown here were run faster than the simulated time, i.e., less than 7500 s of CDC-7600 CPU seconds were required to simulate system behavior to 7500 s. The faster-than-real time calculational speed achieved in this application was a milestone in the RELAP5 code development. The achievement of real time computational capability suggests the possibility of future applications such as system simulators, on-line diagnostic computation, and system control.

5. TRAC DESCRIPTION AND EXAMPLE COMPUTATIONS

This section presents a summary overview of the Transient Reactor Analysis Code (TRAC) and a few comparisons between TRAC calculations and experimental data. Detailed descriptions of TRAC are given in Refs. 8 and 49, while a summary of several experimental comparisons is given in Ref. 50.

5.1 Goals and Development Guidelines

A key goal of the TRAC development effort is to provide an advanced best-estimate LWR systems code that can credibly predict the accident behavior of LWRs. The desired predictive credibility is to be established through the careful assessment of code calculations against a sufficiently broad range of pertinent experimental data.

To accomplish this goal, the following guidelines for the development of the initial versions of TRAC were adopted.

1. Eliminate user-selected modeling options and parameter variations (or "tuning dials") to the degree possible. If numerous combinations of modeling options are used in assessing a code against various experiments, it is difficult to know what options might be appropriate for a new situation where no direct experimental data exist (e.g., an accident in an actual reactor). The goal of the TRAC assessment effort is to predict adequately a broad range of experiments with no user tuning from one test to the next.

2. Model important physical phenomena in as fundamental a way as is practical. Basic modeling should generally extrapolate to new situations with more reliability than highly empirical approaches. Such basic modeling also tends to provide more detail on the thermal-hydraulic behavior of a system.

3. Provide sufficient flexibility to allow modeling of all major LWR designs and pertinent experimental configurations.

Versions of TRAC developed according to the above guidelines are referred to as "detailed" versions.

An additional goal of the TRAC effort is to provide fast running code versions that can be used for such applications as parametric studies, scoping calculations, licensing applications, and very long transients (e.g., small breaks). Some of the major guidelines being followed in the development of the fast running versions are as follows.

1. Use less detailed (and usually more empirical) modeling to achieve short running times.

2. Keep as much in common as possible between the fast running and detailed versions to minimize the amount of needed experimental assessment.

3. Calibrate the fast running versions against the carefully assessed detailed versions for specific applications.

5.2 Development Status

The initial versions of TRAC were detailed versions designed primarily to analyze large-break LOCAs in PWRs. The first version, TRAC-P1, was released by The Los Alamos Scientific Laboratory (LASL) on a limited basis in March 1978. An improved version, TRAC-P1A,[8] was released through the National Energy Software Center in March 1979. A further refined and improved version, called TRAC-PD2,* is scheduled for release in the spring of 1980.

The initial fast running version, TRAC-PF1, is currently under development at LASL. The experimental assessment process will start in the spring of 1980, with its public release planned for late 1980. The development of BWR

*All future versions of TRAC will be designated as TRAC-xyz, where x=P for PWR versions and = B for BWR versions; y=D for detailed versions and = F for fast running versions; and z is a version identification number.

versions is being carried out at the Idaho National Engineering Laboratory (INEL). The initial BWR version, TRAC-BD0, was completed in February 1980. The first BWR release version, TRAC-BD1, is under development.

TRAC-P1A will be used as a reference version for this paper. In some cases, the modeling in TRAC-PD2 (as well as calculated results) will be referred to, however.

5.3 Model Description

Some of the important modeling characteristics of TRAC-P1A are summarized in the following section. These characteristics typically reflect the state of the art in the various areas and were incorporated in pursuit of the goals and guidelines outlined above for detailed versions of TRAC.

Multidimensional Fluid Dynamics. Although the flow within the ex-vessel components is treated in one dimension, a full 3-D (r,θ,z) flow calculation can be used within the reactor vessel. This is done to allow an accurate calculation of the complex multidimensional flow patterns inside the reactor vessel that can play an important role in determining accident behavior. For example, phenomena such as ECC downcomer penetration during blowdown, multi-dimensional plenum and core flow effects, and upper plenum de-entrainment and fallback during reflood can be treated directly.

The flow can be blocked across specified boundaries within a vessel to allow modeling of internal structures such as the downcomer. Flow restrictions can also be specified as appropriate to model structures like core support plates. One-dimensional components can be connected to any vessel mesh cell face (including interior mesh cells) to model the appropriate loop connections. Each of these features is illustrated in Fig. 18 where a simplified 3-D vessel noding is illustrated.

The 3-D hydrodynamics treatment in the vessel will reduce to 2-D (x-y) or even 1-D geometry when this is appropriate.

Nonhomogeneous, Nonequilibrium Modeling. A full two-fluid (six-equation) hydrodynamics approach is used to describe the steam-water flow within the reactor vessel, thereby allowing such important phenomena as countercurrent flow to be treated explicitly. The flow in the 1-D loop components is described using a five-equation drift-flux model, which differs from the standard four-equation drift-flux approach by the addition of a separate vapor energy equation. Thus, it is not necessary to make any assumptions regarding the temperature of either phase. This provides a consistent nonequilibrium thermodynamic treatment in both the vessel and loop components and permits more accurate modeling of the fluid dynamics through a direct treatment of flashing and condensation effects.

Flow-Regime-Dependent Constitutive Equation Package. The basic field equations must be supplemented by a number of so-called constitutive equations to obtain closure. These equations describe the transfer of mass, momentum, and energy between the steam-water phases as well as the interaction of these phases with the system structure. Because the nature of these interactions is strongly dependent on the flow topology, a flow-regime-dependent constitutive equation package has been incorporated into the code. The ability of TRAC to successfully meet its goal of eliminating user-selected modeling options

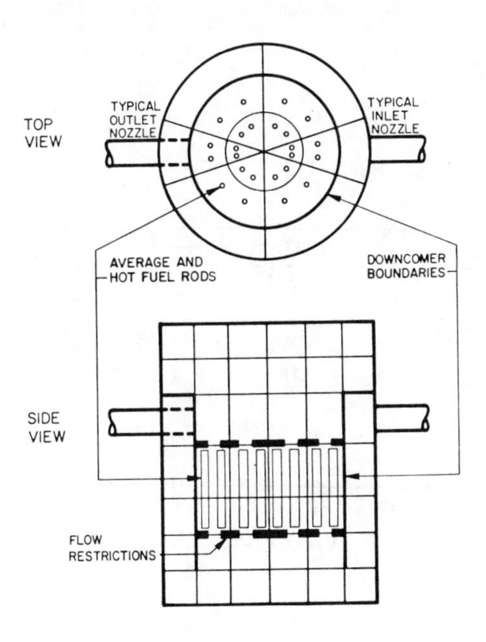

Fig. 18. Illustration of TRAC 3-D configuration.

hinges on the ability of this package to recognize flow regimes adequately and to supply appropriate correlations. The flow regimes currently considered are bubbly, slug, and annular (or annular mist) with appropriate transition regions.

In the case of the five-equation drift-flux model used in 1-D components, the interphase slip correlation is also flow-regime dependent. A flow-regime map has also been incorporated into TRAC for this purpose. This is shown in Fig. 19 to serve as an illustration of the form of these maps. As can be seen, the flow-regime selection is made on the basis of void (steam) fraction and the magnitude of the overall mass flux.

The details of the constitutive equation package are beyond the scope of this chapter; however, many of the phenomena and correlations are discussed in other chapters. The specific correlations used in TRAC-P1A are given in Ref. 8. Although the constitutive relations in TRAC will be improved in the future, assessment calculations performed to date indicate that a fairly wide range of conditions can be adequately treated with the current package.

Comprehensive Heat Transfer. Heat transfer models in TRAC include conduction models to calculate temperature fields in structural materials and fuel rods, and convection models to provide heat transfer between structure and coolant. Heat transfer to the two-phase fluid is calculated using a generalized boiling curve constructed from a library of heat transfer correlations based on local surface and fluid conditions. The heat transfer regimes and correlations used for this purpose[8] are summarized in Table IV.

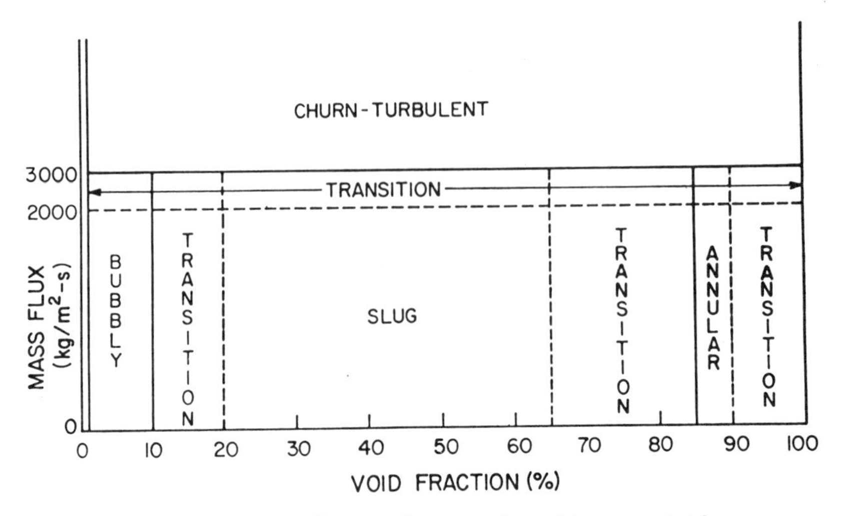

Fig. 19. TRAC flow regime map for slip correlations.

TABLE IV

HEAT TRANSFER CORRELATIONS IN TRAC-P1A

Regime	Correlation
Forced convection to single-phase liquid	laminar flow : constant Nusselt number turbulent flow : Dittus-Boelter
Nucleate boiling and forced convection vaporization	Chen
Critical heat flux	low flow : Zuber pool boiling high flow : Biasi
Transition boiling	log-log interpolation
Minimum stable film boiling	low pressure : Henry-Berenson high pressure : homogeneous nucleation
Film boiling	modified Bromley Dougall-Rohsenow
Forced convection to single-phase vapor	free convection : McAdams turbulent flow : Dittus-Boelter
Forced convection to two-phase mixture	laminar flow : constant Nusselt number turbulent flow : Dittus-Boelter
Horizontal film condensation	Chato
Vertical film condensation	Nusselt theory
Turbulent film condensation	Carpenter and Colburn

Conduction models are used to calculate temperature fields in 1-D (cylindrical) pipe walls, lumped-parameter slabs, and 1-D (cylindrical) fuel rod geometries. Pipe wall conduction is used in the components outside the vessel, whereas the slab and fuel rod conduction models are used in the vessel module. The fuel rod conduction analysis accounts for gap conductivity changes, metal-water reaction, and quenching phenomena. A fine-mesh axial renoding capability is available for fuel rods to allow more detailed modeling of reflood heat transfer and tracking of quench fronts due to bottom flooding and falling films.

In TRAC-P1A, quench fronts are advanced using an empirical velocity correlation. Experience with this approach has indicated that it is difficult to model low flooding rate experimental data accurately. A new reflood model has been incorporated into TRAC-PD2 that explicitly accounts for axial heat conduction near the front.

Each fluid mesh cell in the core region can contain an arbitrary number of fuel rods for the purpose of fluid dynamics calculations. However, heat transfer calculations are only performed on one average rod and one hot rod in each core mesh cell as shown in Fig. 18. The average rod represents the average of the ensemble of rods in the mesh cell, and its thermal calculation couples directly to the fluid dynamics. A spatial power peaking factor and local fluid conditions in the mesh cell are used in the hot rod calculation, but this calculation does not feed back to the hydrodynamics. The total core power level is determined from either a table lookup or from the solution of the point-reactor kinetics equations, including decay heat (6 delayed neutron groups and 11 decay heat groups). The spatial power distribution is specified by separate radial and axial power shapes in the core plus a radial distribution in the fuel rod.

Component and Functional Modularity. TRAC is completely modular by component. The component modules are assembled through input data to model virtually any PWR design or experimental configuration. This gives TRAC great versatility in the possible range of applications. It also allows component modules to be improved, modified, or added without disturbing the remainder of the code. Modules are available to model accumulators, pipes, pressurizers, pumps, steam generators, tees, valves, and vessels with associated internals.

TRAC is also modular by function. This means that the major aspects of the calculations are performed in separate modules. For example, the basic 1-D hydrodynamics solution algorithm, the wall temperature field solution algorithm, heat transfer coefficient selection, and other functions are performed in separate sets of routines that are accessed by all component modules. This type of modularity allows the code to be readily upgraded as improved correlations and experimental information become available.

5.4 Numerical Methods

A summary of the basic numerical methods in TRAC is given in this section. A more detailed description of the finite difference equations and solution strategy is presented in Appendix A.

The system of field and constitutive equations is solved using a staggered differencing scheme[51,52] on an Eulerian mesh. In this approach, the velocities are located at the mesh cell surfaces, while volume properties such as pressure, temperature, energy, and density are located at the mesh cell centers. A semi-implicit time differencing is normally used, with donor cell averaging employed to produce stability. When the semi-implicit approach is used, a standard Courant stability criterion must be observed.

The 1-D flow equations are written in two separate finite difference forms. One form is the semi-implicit, staggered difference approach mentioned above.[53] The second form is an unconditionally stable fully implicit approach.[54] The latter form is used in 1-D components where very high flow velocities are expected locally (such as near a break during the blowdown phase of a LOCA). In such cases, the fluid velocity Courant condition would necessitate very small time-step sizes in a semi-implicit formulation. A fully-implicit component can then be substituted. Thus, TRAC allows the user to blend semi- and fully-implicit formulations in the same calculation to improve computing efficiency. The actual finite-difference equations used in TRAC are too lengthy to reproduce here (especially the 3-D equations). The 1-D drift-flux equations are given in Appendix A, while the others can be found in Ref. 8.

Iterative methods are generally used to solve the finite-difference equations. Each timestep in the transient calculation consists of several passes through all the components in the system. These passes, whose purpose is to converge to the solution of the nonlinear finite-difference equations, are called outer iterations. If the outer iteration process fails to converge, the integration time-step size is reduced and the timestep is repeated.

The solution procedure during an outer iteration begins with a linearization of the equations for each 1-D component. This results in a block tridiagonal system in which linear variations in pressure and other independent variables (vapor fraction, liquid temperature, and vapor temperature) are solved in terms of variations in the junction velocities for the component. If there are no vessels in the calculation, these linearized equations are combined with the linearized junction momentum equations to obtain a closed linear system for the junction velocity variations. This system is solved by direct methods and a back substitution is made to update the remaining independent variables. Therefore, there is no inner iteration process involved for 1-D components.

When one or more vessels are present, the variations in the 1-D component junction velocities are solved in terms of the pressure variations at the vessel junctions. These equations are combined with the remaining linearized equations in the vessel to provide a closed set of linear equations. Because the matrix is usually too large for direct inversion, this set of linear equations is solved by Gauss-Seidel iteration (an option that allows for a direct inversion in the case of relatively small problems has recently been made available). When this vessel inner iteration process has converged, back substitution through the 1-D components again completes the solution of the full linear system. A single pass through this procedure provides the solution for the linearized finite difference equations. Subsequent passes for the same timestep result in a Newton-Raphson iteration scheme with quadratic convergence on the nonlinear difference equations.

A steady-state capability is also included in TRAC to provide time-independent solutions. These may be of interest in their own right or can serve as initial conditions for subsequent transient calculations. Two types of calculations are available within the steady-state capability: (1) a Generalized steady-state calculation and (2) a PWR initialization calculation.

The first is used to find steady-state conditions for a system of arbitrary configuration. The second is applicable to PWR systems and is used to adjust certain loop parameters to match a set of user-specified flow conditions.

Both calculations utilize the transient fluid dynamics and heat transfer routines to search for steady-state conditions. The search is terminated when the normalized rates of change of fluid and thermal variables are reduced below a user-specified criterion throughout the system. For a given problem, computer running times for steady-state calculations are generally much smaller than those for transient calculations.

All TRAC versions to date have been developed on CDC-7600 computers. TRAC-PD2 is currently being converted for use on the CRAY-1. It is anticipated that future release versions will also be converted for use on IBM computers. Fast running versions will additionally be available for use on DEC/VAX machines. Computer running time is highly problem dependent. It is a function of the total mesh cells in the problem and the maximum allowable time-step size. The total run time for a given transient can be estimated from a unit run time of 2 to 3 ms per mesh cell per timestep on a CDC-7600 with an average time-step size of 5 ms.

5.5 TRAC Example Calculations

A major part of the TRAC development effort involves the comparison of calculations with experimental data. This experimental assessment process proceeds in two phases. The first phase, called developmental assessment, is an integral part of the code development effort. It consists of numerous posttest analyses of experiments covering all aspects of LOCA phenomenology and serves as an aid to model development and evaluation. A code version is not released until it has adequately analyzed a predetermined set of experiments. Data comparisons from nine experiments were formally documented as part of the release of TRAC-P1A.[50] These experiments are listed in Table V to illustrate the scope of the developmental assessment process. The assessment set to be documented with the release of TRAC-PD2 is considerably larger.

Following the release of a given version of TRAC, an independent assessment phase is initiated. This phase emphasizes blind pretest predictions to establish predictive capability. The independent assessment to date has emphasized LOFT experiments but is expanding to include other facilities such as LOBI and the Japanese Cylindrical Core Test Facility.

Detailed comparisons of TRAC calculations and experimental data have been published elsewhere; thus, only a brief summary of two test comparisons from Ref. 50, followed by a more recent LOFT comparison, will be presented here. The first is an analysis of some countercurrent flow experiments performed at Creare, Inc., to study ECC downcomer bypass phenomena. This serves to illustrate the use of separate-effects tests in the model development process. The other two will be analyses of a semiscale and a LOFT experiment to illustrate the role of systems data comparisons.

Creare Countercurrent Flow Experiments. The Creare countercurrent flow experiments investigated the effects on ECC penetration to the lower plenum of countercurrent steam flow rate, downcomer wall superheat, and ECC subcooling. The basic component of the Creare test facility is a 1/15-scale (linear dimension), multiloop, cylindrical model of a PWR downcomer region. A detailed description of this facility and its operation is given in Ref. 55. The configuration used in the tests analyzed here is the so-called "base-line"

TABLE IV

TRAC-P1A DEVELOPMENTAL ASSESSMENT ANALYSES

No.	Experiment	Thermal - Hydraulic Effects
1	Edwards Horizontal Pipe Blowdown (Standard Problem 1)	Separate effects, 1-D critical flow, phase change, slip, wall friction
2	CISE Upheated Pipe Blowdown (Test 4)	Same as 1 plus pipe wall heat transfer, flow area changes, and gravitational effects
3	CISE Heated Pipe Blowdown (Test R)	Same as 2 plus critical heat flux (CHF)
4	Marviken Full-Scale Vessel (Test 4)	Same as 1 plus full-scale effects
5	Semiscale 1-1/2 Loop Isothermal Blowdown (Test 1011, Standard Problem 2)	Synergistic and systems effects 1-D flow, phase change, slip wall friction, critical nozzle flow
6	Semiscale Mod-1 Heated Loop Blowdown (Test S-02-8, Standard Problem 5)	Same as 5 plus 3-D vessel model with rod heat transfer including nucleate boiling, DNB, and post-DNB
7	Creare Countercurrent Flow Experiments	Separate effects, countercurrent flow, interfacial drag and heat transfer, condensation
8	FLECHT Forced Flooding Tests	Separate effects, reflood heat transfer, quench front propagation, liquid entrainment and carryover
9	Nonnuclear LOFT Blowdown with Cold Leg Injection (Test L1-4, Standard Problem 7)	Integral effects during blowdown and refill, scale midway between Semi-scale and full-scale PWR

configuration having a 0.0127-m (0.5-in.) downcomer gap and a "deep plenum" geometry. The vessel has four cold legs oriented 90° to each other. Three of these legs are assumed to be "intact" and are connected to ECC injection lines. A single "broken" leg connects to the pressure suppression tank.

The test procedure is as follows. A constant steam flow rate through the vessel is established. The steam enters at the top of the vessel, flows down the center of the vessel into the lower plenum, up the downcomer, and out the broken cold leg. After reaching a steady steam flow rate, water is injected simultaneously into the three intact cold legs at a constant preset flow rate. After a short transient period, the plenum normally begins to fill. The test is run until the lower plenum is full or until the filling rate can be determined. A complete penetration, or flooding, curve is composed of a set of tests at a given liquid injection rate and liquid temperature with the steam flow rate varied over a range such that water delivery ranges from complete delivery to complete bypass.

The TRAC model of the Creare vessel is shown in Fig. 20. The 3-D vessel module used 112 computational cells with the mesh lines indicated in the figure.

The calculational procedure paralleled the Creare experimental procedure. A steady-state calculation was performed to establish a constant reverse steam flow and lower plenum pressure. This steady-state calculation was run until J^*_{gc} (the dimensionless reverse core steam flow rate) reached a constant value. This normally took about 3 s of simulation time. The transient calculation was then started with the initiation of ECC injection into the three intact cold legs. Results for two ECC injection rates and levels of subcooling are compared in Fig. 21. The low subcooling cases injected 30 gpm of ECC water at 212°F, while the high subcooling cases injected 60 gpm at 150°F. The reactor scale injection flow rate is 60 gpm. The system pressure ranged from 1 to 3 atm.

The basis for selecting these two penetration curves was to separate the basic phenomena determining whether ECC bypass or delivery will occur. These phenomena are interfacial momentum and energy exchange between the liquid and the steam. For a low subcooling case, the only effect that can produce bypass is the interfacial drag between the steam and the liquid. The calculated penetration curve for this case gives an appraisal of the constitutive relationship describing interfacial momentum exchange. Moreover since the calculations cover the range of complete bypass to complete dumping, different flow regimes exist in the downcomer at the bypass point than at the complete delivery point. In the high subcooling case, the interfacial heat transfer becomes significant in determining the quantity of liquid delivered. As can be seen, the TRAC calculations agreed very well with both of the experimental penetration curves. Comparisons such as these indicate that complex multi-dimensional phenomena, such as ECC bypass, can indeed be modeled with the rather fundamental modeling approach taken in the detailed versions of TRAC.

Semiscale Test S-02-8. Test S-02-8 was a simulation of a PWR double-ended cold-leg break performed in the semiscale Mod-1 facility.[56] In the Mod-1 configuration, nuclear heating is simulated with 40 electrically heated rods. The TRAC model of this experiment is shown in Fig. 22. This figure illustrates how the various TRAC components are connected to simulate a PWR type configuration. The calculation involved nearly every TRAC component module. The model had a total of 111 mesh cells in the 1-D loop components and 152 cells in the 3-D vessel component.

The initial steady-state conditions calculated with TRAC for use at the start of blowdown agreed very well with the experimental data. Some of the key transient results are compared in Figs. 23-25. The calculated and measured lower plenum pressures agree quite well as can be seen in Fig. 23. This indicates that TRAC did a reasonable job of analyzing overall system performance. A comparison of the cladding temperatures in the highest power region is shown in Fig. 24. The calculated results are compared with a band of measured values that encompassed all the cladding thermocouples in the lower half of the highest power step in Semiscale. With the exception of a slightly advanced departure from nucleate boiling, TRAC does an excellent job of predicting the cladding temperature response in this high power zone.

The final comparison presented is the hot-leg break mass flow rate shown in Fig. 25. The small rise in the flow rate at about 12 s is due to a slug of higher density fluid coming from the intact hot leg. The TRAC results show a small rise at this time, but underpredict the magnitude. In general, however, break flow rates were predicted quite well without the use of any empirical break flow model.

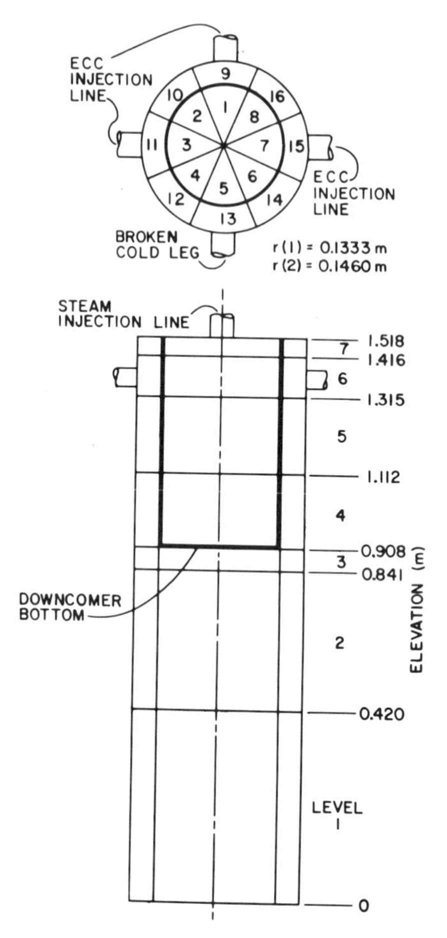

Fig. 20. TRAC noding for Creare 1/15-scale vessel.

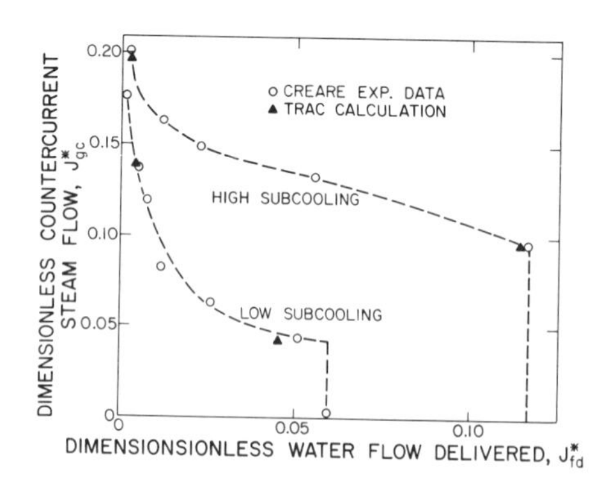

Fig. 21. Flooding curves for Creare experiments.

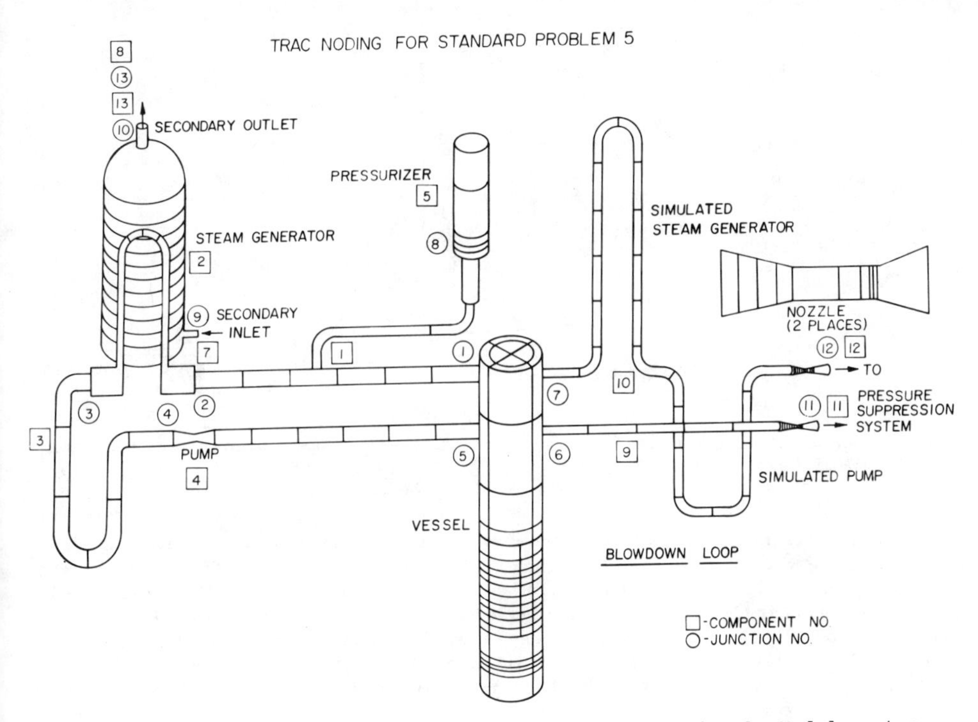

Fig. 22. TRAC noding and component schematic for Semiscale Mod-1 system.

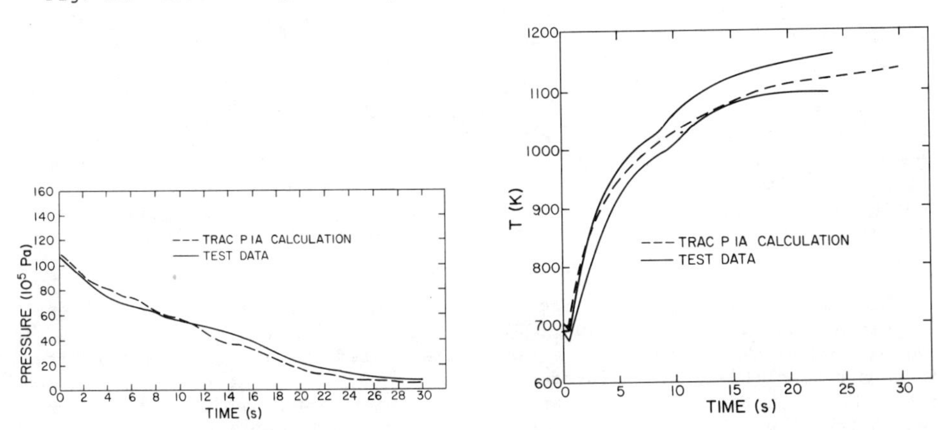

Fig. 23. Lower plenum pressure for Semiscale Test S-02-8.

Fig. 24. Cladding temperature in high power zone for Semiscale Test S-02-8.

<u>LOFT Test L2-3</u>. Because LOFT is the only available nuclear-heated integral-effects test facility, several of the recent LOFT tests have been extensively analyzed with TRAC. This has included both blind pretest predictions and extensive posttest analyses. As an example, a few key results from Test L2-3 will be compared with the TRAC pretest predictions.

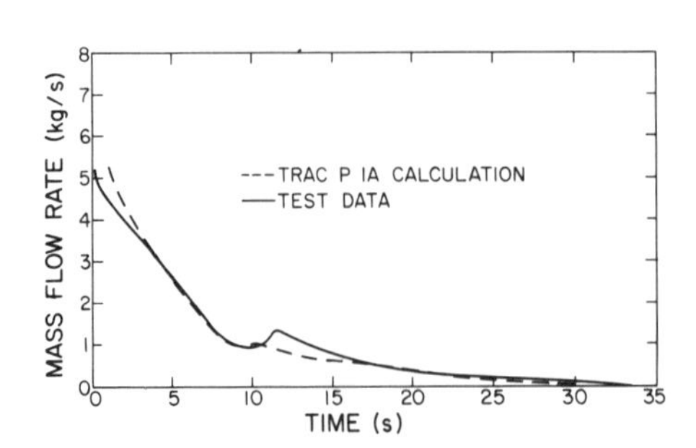

Fig. 25. Hot-leg break mass flow rate for Semiscale Test S-02-8.

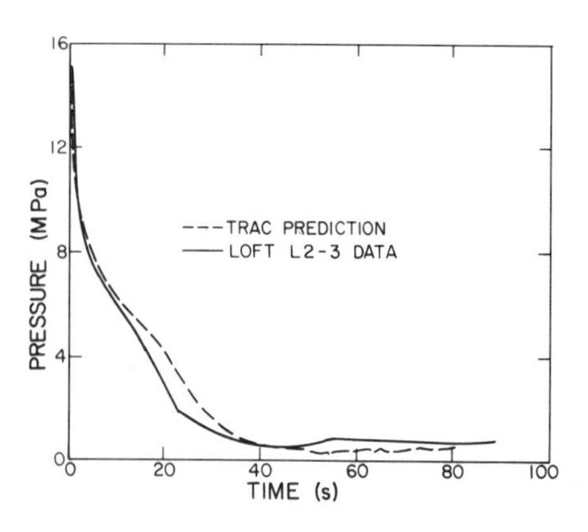

Fig. 26. Upper plenum pressure for LOFT Test L2-3.

Test L2-3 simulated a double-ended cold leg break and was conducted from an initial power of 37 Mwt.[57] The TRAC model consisted of 27 components with a total of 322 fluid mesh cells. There were a total of 12 axial levels in the vessel, including 5 axial levels within the core region. A total of 192 fluid cells were used within the vessel, including 60 within the core itself. The reflood fine mesh was initiated 10 s after accumulator injection started. There were 5 uniform fine-mesh intervals for each axial level, giving a total of 25 fine cells.

The initial system thermal and hydraulic conditions for the pretest calculation were obtained using the steady-state option. Good agreement was obtained between the calculated and measured conditions.

Pretest transient results are compared with the data[50] in Figs. 26-29. The upper plenum pressure is shown in Fig. 26. The calculation depressurized somewhat more slowly than the data and resulted in delayed ECC injection. Figure 27 shows the broken loop cold-leg mass flow rate. The early underprediction of the break flow is consistent with the overprediction of system pressure. In general, however, these results are in quite good agreement.

One of the more interesting (and unexpected) results from the early nuclear-heated blowdown tests in LOFT was the observed early rewet behavior in the high-power core region. The TRAC pretest predictions for an earlier test in this series (L2-2) had predicted rewet to occur in the lower power regions, but did not predict early rewetting in the high power region. It was suspected that the minimum film boiling correlation in TRAC-P1A might be at fault because it was based on low pressure data. The minimum film boiling correlation of Iloeje[59] was subsequently tried and found to give much better results. Because of this, two blind pretest predictions of Test L2-3 were performed. One used the standard release version of TRAC-P1A, while the second had the Iloeje correlation implemented. Temperatures from these two predictions are compared with data in Figs. 28 and 29.

The temperatures in a low-power peripheral region are compared in Fig. 29. As can be seen, both of the TRAC pretest predictions agreed quite well

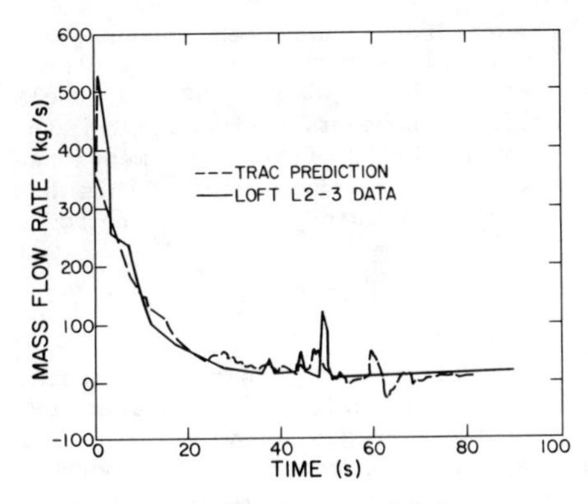

Fig. 27. Broken loop cold-leg mass flow rate for LOFT Test L2-3.

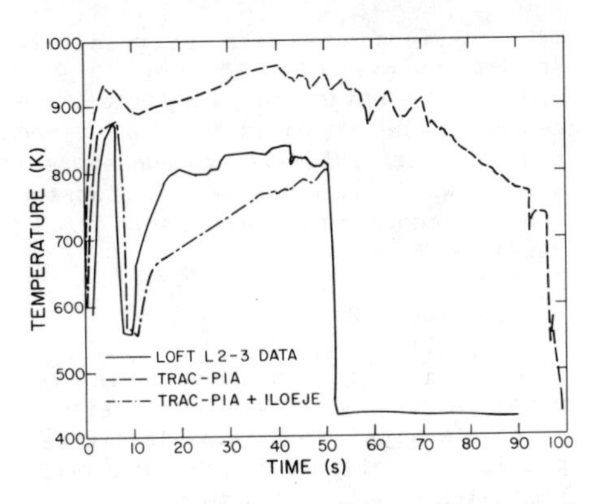

Fig. 28. TRAC pretest prediction of cladding temperature at core midplane for LOFT Test L2-3.

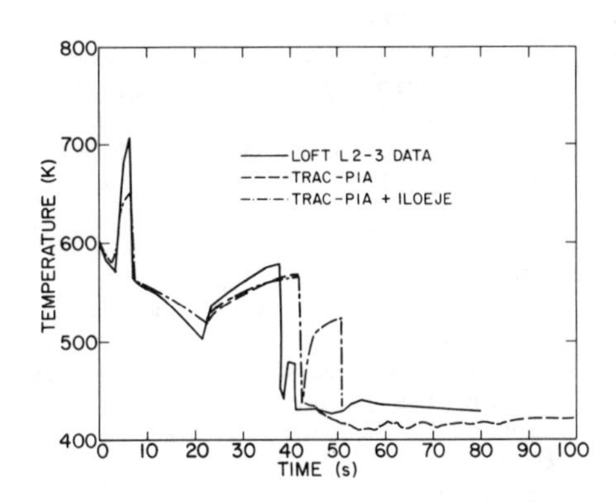

Fig. 29. TRAC pretest predictions of cladding temperature at outer core periphery for LOFT Test L2-3

with the data. In the higher power region near the center of the core, however, standard TRAC-P1A did not predict the early rewet, while the version with the Iloeje correlation did (see Fig. 28). These results again indicate that an improved film boiling correlation is necessary to calculate the rewet phenomena in LOFT. To decide on an appropriate correlation to be implemented in future release versions of TRAC, additional data from other facilities are currently being examined.

Difficulties in accurately predicting the dryout behavior in parts of the core were also observed. These appear to be related to the underprediction of the broken cold-leg flow during the first 5 s of the transient. Underprediction of the cold-leg break flow was mainly because vapor generation model in TRAC-P1A does not account for the effect of delayed nucleation. This problem was not evident in the broken loop hot leg because of the higher flow resistance and temperature. In general, the calculated behavior in the intact loop was qualitatively good.

Experience to date in predicting LOFT experiments has been encouraging. In particular, TRAC has done a good job of predicting the overall thermal-hydraulic behavior of the large-break LOCA tests. Areas where specific models needed to be improved have also been identified, however. For example, the tendency of TRAC-P1A to underpredict subcooled break flows, as mentioned above, was also observed in subsequent small-break test analyses. This has led to modeling improvements that will be incorporated into future code versions.

6. CONCLUSIONS

The material presented in this chapter illustrates that substantial progress has been made in meeting the need for reliable and efficient LWR safety analysis codes. The recent best-estimate codes provide increased predictive reliability through much more comprehensive modeling of important thermal-hydraulic phenomena and much more extensive and methodical assessment against experimental data. In addition, the replacement of user-controlled options and tuning dials with more fundamental modeling is enhancing the predictive credibility of these new codes.

As indicated earlier, much of the code development and assessment effort to date has been mainly directed toward the large-break LOCA. The blowdown and refill phases of these accidents can be well characterized with the available system codes, and steady progress is being made in modeling the reflood phase. Perhaps the most difficult issue remaining with regard to large-break LOCAs is establishing the ability of the codes to extrapolate to full-size PWR behavior.

Although much of the code development to date is also applicable to small break LOCAs and other postulated transients, some new modeling features are required. In addition, the codes must be very fast running to allow analysis of very long transients. These features are currently being emphasized in the development of RELAP5 and the fast running versions of the TRAC code. Considerable assessment remains to be done in this area.

The authors believe that the accurate and economical performance already achieved will be significantly improved. These codes can be used to examine a broad range of postulated accident conditions to assist in implementing improved designs and procedures for accident prevention and investigation. Additionally, codes with faster-than-real-time running capability could potentially be used in on-line accident diagnostic systems that could allow operators to respond more effectively to off-normal conditions.

Finally, code development will continue to meet increasing needs for improved plant integrity and reliability, and assurance of safety system performance. These needs in the nuclear area are stimulated by our country's need for a stable, economical, and safe supply of energy. For the analyst and experimentalist in nuclear safety, these increasing needs mean greater demands and responsibilities for measurability in design and safety assessment techniques and rigor in their application.

APPENDIX A - REVIEW OF SOME NUMERICAL TECHNIQUES USED IN LWR SAFETY CODES

A.1 INTRODUCTION

Any reasonable set of equations that could be used to describe the thermal hydraulics of a nuclear reactor under accident conditions would be far too complex to allow analytic solutions. The increasingly sophisticated digital computers that are available for scientific purposes do allow numerical simulations of the fluid mechanics and heat transfer, however. Two fundamental difficulties with any numerical procedure are: (a) approximations must be made that may not always preserve the exact character of the original equations and (b) insights into the solutions (such as unstable regions, peak variable values, etc.) are not attained as readily as with analytical solutions. Despite these difficulties, numerical modeling can provide information on the response of the whole system that cannot otherwise be obtained. This Appendix covers some of the aspects of hydrodynamic modeling of large reactor systems using finite difference schemes.

A.2 FINITE DIFFERENCE EQUATIONS

Consider the following set of hydrodynamics equations

$$\frac{\partial \rho}{\partial t} + \frac{\partial}{\partial x} \rho V = 0 \qquad \text{(A-1)}$$

$$\rho \frac{\partial V}{\partial t} + \rho V \frac{\partial V}{\partial x} + \frac{\partial P}{\partial x} = 0 \qquad \text{(A-2)}$$

$$\rho = \rho \ (P) \quad . \qquad \text{(A-3)}$$

Here, ρ is the density, V the velocity, and P the pressure.

We wish to approximate the partial differential equations in such a way that a digital computer can integrate the equations in space and time with a consistent technique (i.e., one which in the limit as $\Delta t \rightarrow 0$ and $\Delta x \rightarrow 0$ returns a solution of the original differential equations).

Although there are a vast number of procedures that could be used, all of the current major reactor system codes (TRAC, RELAP5, and COBRA TF) employ the same basic difference scheme and solution technique. This particular scheme is relatively easy to code, is stable for moderate time-step sizes, and is robust and reliable.

The spacial differencing uses a staggered mesh[A-1]--staggered because the momentum equations are written over volumes half a mesh cell up or downstream from the volumes over which the scalar field equations are provided. The heavy lines in the following figure indicate a typical 1-D computational mesh, while the dotted lines show where the momentum equations will be written

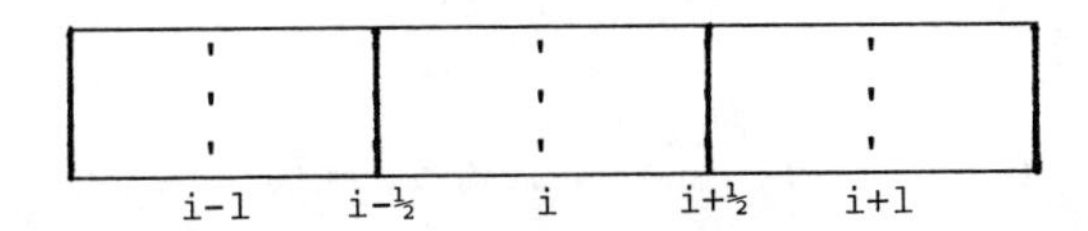

The subscript i indicates a mesh cell center. All the thermodynamic variables, i.e., ρ and p, are located at the center of the scalar cells, while the velocities are located at the center of the momemtun cells (scalar cell edges).

Reference A-2 by Liles and Reed provides a more complete description of the solution procedure that is now delineated. A finite difference approximation to Eq. (A-1) becomes:

$$\frac{\rho_i^{n+1} - \rho_i^n}{\Delta t} = \frac{\left(\rho^n V^{n+1}\right)_{i-\frac{1}{2}} - \left(\rho^n V^{n+1}\right)_{i+\frac{1}{2}}}{\Delta x} , \qquad \text{(A-4)}$$

where the superscript n implies the old time quantities and n+1 implies the new time quantities. The momentum equation, Eq. (A-2), becomes

$$\frac{V_{i+\frac{1}{2}}^{n+1} - V_{i+\frac{1}{2}}^{n}}{\Delta t} = -V_{i+\frac{1}{2}}^{n} \frac{\left(V_{i+1}^{n} - V_{i}^{n}\right)}{\Delta x} - \frac{1}{\rho_{i+\frac{1}{2}}} \frac{\left(P_{i+1}^{n+1} - P_{i}^{n+1}\right)}{\Delta x} . \qquad \text{(A-5)}$$

The time levels for the convective terms are chosen to allow the wave speed Courant limit $\frac{(V+c)\Delta t}{\Delta x}$ to be violated (see Ref. A-3), but not the material Courant limit, $\frac{V\Delta t}{\Delta x} < 1$. In these expressions c is the speed of sound and V is the material velocity. It should be noted that Eq. (A-5) can be rewritten as:

$$V_{i+\frac{1}{2}}^{n+1} = \tilde{V}_{i+\frac{1}{2}} - \frac{\Delta t}{\rho_{i+\frac{1}{2}} \Delta x} \left(\delta P_{i+1} - \delta P_{i}\right) , \qquad \text{(A-6)}$$

where $$\tilde{V}_{i+\frac{1}{2}} = V_{i+\frac{1}{2}}^{n} - V_{i+\frac{1}{2}} \frac{\Delta t}{\Delta x} \left(V_{i+1}^{n} - V_{i}^{n}\right) - \frac{\Delta t}{\rho_{i+\frac{1}{2}}} \frac{\left(P_{i+1}^{n} - P_{i}^{n}\right)}{\Delta x} \qquad \text{(A-7)}$$

and $$\delta P_i = P_i^{n+1} - P_i^n , \qquad \text{(A-8)}$$

which is the pressure change during the timestep.

We shall next linearize the equation of state

$$d\rho = \frac{\partial \rho}{\partial P} dP \qquad \text{(A-9)}$$

and write a first order Taylor series expansion for ρ_i^{n+1}, i.e.,

$$\rho_i^{n+1} = \rho_i^n + \left(\frac{\partial \rho_i}{\partial P}\right)^n dP_i . \qquad \text{(A-10)}$$

Combining (A-4), (A-6), (A-9), and (A-10), we obtain (one more momentum equation must be written for $V_{i-\frac{1}{2}}^{n+1}$)

$$\left.\frac{\partial\rho}{\partial P}\right)_i^n \delta P_i = \frac{\delta t}{\delta x}\left[\rho_{i-\frac{1}{2}}^n\left(\tilde{V}_{i-\frac{1}{2}} + \frac{\Delta t}{\Delta x}\frac{1}{\rho_{i-\frac{1}{2}}^n}\left(\delta P_{i-1} - \delta P_i\right)\right)\right.$$

$$\left. - \rho_{i+\frac{1}{2}}^n\left(\tilde{V}_{i+\frac{1}{2}} + \frac{\Delta t}{\Delta x}\frac{1}{\rho_{i+\frac{1}{2}}^n}\left(\delta P_i - \delta P_{i+1}\right)\right)\right] . \qquad \text{(A-11)}$$

It should be noted that in order to complete the finite difference scheme, variable values are required at locations where they are not defined. Auxiliary equations are needed to obtain closure. To obtain stability, donor-cell averages are used.

$$\rho_{i+\frac{1}{2}} = \rho_i , \qquad \text{if } V_{i+\frac{1}{2}} \geq 0$$

$$\rho_{i+\frac{1}{2}} = \rho_{i+1} , \qquad \text{if } V_{i+\frac{1}{2}} < 0 . \qquad \text{(A-12)}$$

It should be noted that pressures in the momentum equation cannot be donor celled. If they were, waves could not always propagate in all directions, and stagnation regions could be uncoupled from the remainder of the flow field. One of the great virtues of the staggered difference scheme is that close coupling of the pressure gradient term occurs naturally with the velocity and results in diagonally dominant matrices (if $\frac{V\Delta t}{\Delta x} < 1$) .

A.3 SOLUTION STRATEGY

This provides us with a tridiagonal matrix in pressure to solve. Direct elimination or a Gauss-Seidel iteration can be used. The total calculational sequence occurs as follows: (a) a first pass over the mesh for Eq. (A-7) provides "trial" new time velocities, (b) Eq. (A-11) is solved for pressure with all the old-time quantities known either from initialization or the previous timestep, (c) the thermodynamic equation of state (Eq. (A-3)) is solved for the new densities. This finishes one timestep.

A.4 1-D DRIFT-FLUX EQUATIONS IN TRAC

This same technique can be used for the more complicated two-phase flow equations in reactor safety codes. A minimum model that describes adequately both thermal and velocity nonequilibrium is the five equation drift-flux model. This approach is used in the 1-D components in TRAC. The following text develops these equations and describes how they are solved.

Field Equations. The differential field equations[A-4] for the two-phase, five-equation drift-flux model are given below. The subscripts g and ℓ refer to the gas and liquid phase, and m denotes mixture quantities. α is the volume fraction of the vapor.

Mixture Mass Equation

$$\frac{\partial}{\partial t} \rho_m + \frac{\partial}{\partial x} (\rho_m V_m) = 0 \tag{A-13}$$

Vapor Mass Equation

$$\frac{\partial}{\partial t} (\alpha\rho_g) + \frac{\partial}{\partial x} (\alpha\rho_g V_m) + \frac{\partial}{\partial x} \left[\frac{\alpha\rho_g (1-\alpha) \rho_\ell V_r}{\rho_m}\right] = \Gamma \tag{A-14}$$

Mixture Equation of Motion

$$\frac{\partial}{\partial t} V_m + V_m \frac{\partial}{\partial x} V_m + \frac{1}{\rho_m} \frac{\partial}{\partial x} \left[\frac{\alpha\rho_g (1-\alpha) \rho_\ell V_r^2}{\rho_m}\right] = - \frac{1}{\rho_m} \frac{\partial p}{\partial x} - K V_m |V_m| + g \tag{A-15}$$

Vapor Thermal Energy Equation

$$\frac{\partial}{\partial t} (\alpha\rho_g e_g) + \frac{\partial}{\partial x} (\alpha\rho_g V_m e_g) + \frac{\partial}{\partial x} \left[\frac{\alpha\rho_g (1-\alpha) \rho_\ell V_r e_g}{\rho_m}\right] + p \frac{\partial}{\partial x} (\alpha V_m)$$

$$+ p \frac{\partial}{\partial x} \left[\frac{\alpha (1-\alpha) \rho_\ell}{\rho_m} V_r\right] = q_{wg} + q_{ig} - p \frac{\partial \alpha}{\partial t} + \Gamma h_{sg} \tag{A-16}$$

Mixture Thermal Energy Equation

$$\frac{\partial}{\partial t} (\rho_m e_m) + \frac{\partial}{\partial x} (\rho_m e_m V_m) + \frac{\partial}{\partial x} \left[\frac{(1-\alpha) \rho_\ell \alpha\rho_g (e_g - e_\ell)}{\rho_m} V_r\right] + p \frac{\partial V_m}{\partial x}$$

$$+ p \frac{\partial}{\partial x} \left[\frac{\alpha (1-\alpha) (\rho_\ell - \rho_g)}{\rho_m} V_r\right] = q_{wg} + q_{w\ell} \quad , \tag{A-17}$$

where

$$\rho_m = \alpha\rho_g + (1-\alpha)\rho_\ell \quad , \tag{A-18}$$

$$V_m = \frac{\alpha\rho_g V_g + (1-\alpha)\,\rho_\ell\, V_\ell}{\rho_m} \quad , \tag{A-19}$$

and

$$V_r = V_g - V_\ell \ . \tag{A-20}$$

The expression for e_m is the same as Eq. (A-19) with V replaced by e. In addition to the thermodynamic relations that are required for closure, specifications for the relative velocity (V_r), the interfacial heat transfer (q_{ig}), the phase change rate (Γ), the wall shear coefficient (K), and the wall heat transfers (q_{wg} and $q_{w\ell}$) are required. The correlations used for these quantities will not be discussed. Gamma can be evaluated from a simple thermal energy jump relation

$$\Gamma = \frac{-q_{ig} - q_{i\ell}}{h_{sg} - h_{s\ell}} \quad , \tag{A-21}$$

where

$$q_{ig} = h_{ig}A_i \; (T_s - T_g)/vol \tag{A-22}$$

and

$$q_{i\ell} = h_{i\ell}A_i \; (T_s - T_\ell)/vol \tag{A-23}$$

The quantities h_{ig}, $h_{i\ell}$, and A_i are evaluated using a complicated estimate of the flow regime interfacial heat transfers.

Wall heat transfer terms assume the form:

$$q_{wg} = h_{wg}A_{wg}(T_w - T_g)/vol \tag{A-24}$$

and

$$q_{w\ell} = h_{w\ell}A_{w\ell}(T_w - T_\ell)/vol \tag{A-25}$$

Finite Difference Equations. The 1-D flow equations in TRAC have been written in two separate finite difference forms. The first form of the difference equations is semi-implicit and has a time step size stability limit of the form

$$\Delta t < \left|\frac{\Delta x}{V}\right| \quad ,$$

where Δx is the mesh spacing and V the fluid velocity. In blowdown problems, this timestep is usually prohibitively small due to the high velocities near the break. To alleviate this problem, a set of unconditionally stable, fully-implicit difference equations was written for use in pipes where the fluid velocities are expected to be high. Only the first semi-implicit set will be considered.

The equations are solved for 1-D pipes using the staggered difference scheme on the Eulerian mesh. State variables such as pressure, internal energy, and void fraction are obtained at the center of the mesh cells, which have length Δx_j, and the mean and relative velocities are obtained at the cell boundaries. Because of this staggered difference scheme, it is necessary to form spatial averages of various quantities to obtain the finite difference form of the divergence operators. To produce stability in the partially implicit method, a donor-cell average is used of the form,

$$
\begin{aligned}
\langle YV \rangle_{j+\frac{1}{2}} &= Y_j V_{j+\frac{1}{2}} && \text{for } V_{j+\frac{1}{2}} \geq 0 \\
&= Y_{j+1} V_{j+\frac{1}{2}} && \text{for } V_{j+\frac{1}{2}} < 0 \quad ,
\end{aligned}
\tag{A-26}
$$

where Y is any state variable or combination of state variables. An integer subscript indicates that a quantity is evaluated at mesh cell center and a half integer denotes that it is obtained at a cell boundary. With this notation, the finite difference divergence operator is

$$
\nabla_j (YV) = \{A_{j+\frac{1}{2}} \langle YV \rangle_{j+\frac{1}{2}} - A_{j-\frac{1}{2}} \langle YV \rangle_{j-\frac{1}{2}}\}/\text{vol}_j \quad ,
\tag{A-27}
$$

where A is the cross-section area, and vol_j is the volume of the jth cell. Slight variations of these donor-cell terms appear in the velocity equation of motion. Donor-cell averages are of the form

$$
\begin{aligned}
\langle YV_r^2 \rangle_j &= Y_j V_{r,j-\frac{1}{2}}^2 && \text{for } V_{r,j-\frac{1}{2}} \geq 0 \\
&= Y_j V_{r,j+\frac{1}{2}}^2 && \text{for } V_{r,j+\frac{1}{2}} < 0
\end{aligned}
\tag{A-28}
$$

and the donor-cell of the term $V_m \nabla V_m$ is

$$
\begin{aligned}
V_{m,j+\frac{1}{2}} \nabla_{j+\frac{1}{2}} V_m &= V_{m,j+\frac{1}{2}} (V_{m,j+\frac{1}{2}} - V_{m,j-\frac{1}{2}})/\Delta x_j && \text{for } V_{m,j+\frac{1}{2}} \geq 0 \\
&= V_{m,j+\frac{1}{2}} (V_{m,j+3/2} - V_{m,j+\frac{1}{2}})/\Delta x_{j+1} && \text{for } V_{m,j+\frac{1}{2}} < 0 \quad .
\end{aligned}
\tag{A-29}
$$

Given the preceding notation, the finite difference equations for the partially implicit method are:

Mixture Mass Equation

$$(\rho_m^{n+1} - \rho_m^n)_j/\Delta t + \nabla_j(\rho_m^n V_m^{n+1}) = 0 \quad , \qquad \text{(A-30)}$$

Vapor Mass Equation

$$(\alpha^{n+1}\rho_g^{n+1} - \alpha^n\rho_g^n)_j/\Delta t + \nabla_j(\alpha^n\rho_g^n V_m^{n+1}) + \nabla_j(\rho_f^n V_r^n) = \Gamma \quad , \qquad \text{(A-31)}$$

Mixture Energy Equation

$$(\rho_m^{n+1}e_m^{n+1} - \rho_m^n e_m^n)_j/\Delta t + \nabla_j(\rho_m^n e_m^n V_m^{n+1}) + \nabla_j\left[\rho_f^n\,(e_g^n - e_\ell^n)V_r^n\right]$$

$$= - p_j^{n+1}\nabla_j\left\{V_m^{n+1} + \left[\rho_f^n\left(\frac{1}{\rho_g^n} - \frac{1}{\rho_\ell^n}\right)V_r^n\right]\right\} + q_{j,wg} + q_{j,w\ell} \quad , \qquad \text{(A-32)}$$

Vapor Energy Equation

$$\left[(\alpha\rho_g e_g)^{n+1} - (\alpha\rho_g e_g)^n\right]_j/\Delta t + \nabla_j(\alpha^n\rho_g^n e_g^n V_m^{n+1}) + \nabla_j(\rho_f^n e_g^n V_r^n) + p^{n+1}\nabla_j(\alpha^n V_m^{n+1})$$

$$+ p^n\nabla_j(\rho_f^n V_r^n/\rho_g^n) = -p_j^{n+1}(\alpha_j^{n+1} - \alpha_j^n)/\Delta t + (q_{wg} + q_{ig} + \Gamma h_{sg}^{n+1})_j \quad , \qquad \text{(A-33)}$$

Mixture Equation of Motion

$$(V_m^{n+1} - V_m^n)_{j+\frac{1}{2}}/\Delta t + V_{m,j+\frac{1}{2}}^n\,\nabla_{j+\frac{1}{2}}\,V_m^n$$

$$= - \left\{(p_{j+1}^{n+1} - p_j^{n+1})/\overline{\Delta x}_{j+\frac{1}{2}} + \nabla_{j+\frac{1}{2}}(\rho_f V_r^2)^n\right\}/\bar{\rho}_{m,j+\frac{1}{2}}^n + g^n - K^n V_m^{n+1}|V_m^n| \quad , \qquad \text{(A-34)}$$

where

$$\overline{\Delta x}_{j+\frac{1}{2}} = (\Delta x_j + \Delta x_{j+1})/2 \quad , \tag{A-35}$$

$$\rho_f^n = \frac{\alpha^n(1-\alpha^n)\ \rho_g^n \rho_\ell^n}{\rho_m^n} \quad , \tag{A-36}$$

and

$$\begin{aligned} \bar{\rho}_{m,j+\frac{1}{2}}^{n} &= \rho_{m,j}^{n} \qquad \text{for } V_{j+\frac{1}{2}} \geq 0 \\ &= \rho_{m,j+1}^{n} \qquad \text{for } V_{j+\frac{1}{2}} < 0 \quad . \end{aligned} \tag{A-37}$$

If the appropriate caloric and equations-of-state are inserted and a first-order Taylor series is used to evaluate the new time state quantities, we obtain block tridiagonal matrices in the variables α, P, T_L, T_V. These can then be solved and passes made through the thermodynamics to obtain new densities and energies for both phases. This basic numerical procedure extends to the two fluid model and may be also used in multiple spacial dimensions (although the resulting arrays are no longer tridiagonal).

This Appendix will not address a wide range of numerical problems such as numerical diffusion, numerical viscoscity, and formal trunciation accuracy. Reference A-3 contains discussions of some of these other important points.

A.5 REFERENCES

A-1. F. H. Harlow and A. R. Amsden, J. of Comp. Physics 18, p 40 (1975).

A-2. D. R. Liles and W. H. Reed, J. of Comp. Physics 26, p 390 (1978).

A-3. P. J. Roache, Computational Fluid Dynamics, Hermosa Publishers, Albuquerque, New Mexico (1976).

A-4. M. Ishii, "Thermo-Fluid Dynamic Theory of Two-Phase Flow," Eyrolles, Paris (1975).

REFERENCES

1. M. Ishii, Thermo-Fluid Dynamic Theory of Two-Phase Flow, Collection de la Direction des Etudes et Recherches d'Electricite de France, 22, xxix, p 248 (1975).

2. D. Gidaspow (Chairman), "Modeling of Two-Phase Flow," Proceedings of Round Table Discussion RT-1-2 at the Fifth International Heat Transfer Conference, Tokyo, Japan, Sepetember 3-7, 1974, also in Heat Trans., 3, 1974.

3. P. S. Anderson, P. Astrup, L. Eget, O. Rathmann, "Numerical Experience With the Two-Fluid Model, RISQUE," Proceedings of NAS Water Reactor Safety Meeting, July 31 - August 4, 1977.

4. R. Jackson, "The Present Status of Fluid Mechanical Theories of Fluidization," Chemical Engineering Progress Symposium Series, Number 105, Vol. 66, 1970.

5. J. D. Ramshaw and J. A. Trapp, "Characteristics, Stability and Short-Wave Length Phenomena in Two-Phase Flow Equation Systems," Aerojet Nuclear Company Report, ANCR-1272 (1976).

6. F. H. Harlow and A. R. Amsden, "Flow of Interpenetrating Material Phases," Journal of Computational Physics, 18 (1975) pp 440-464.

7. V. H. Ransom, et al., "RELAP5 Code Development," CDAP-TR-057 (VOL. I) (May 1979).

8. "TRAC-P1A - An Advanced Best-Estimate Computer Program for PWR LOCA Analysis," NUREG/CR-0665 (May 1979).

9. Emergency Core Cooling, Report of Advisory Task Force on Power Reactor Emergency Cooling, USAEC, W. K. Ergen, Chairman (1967).

10. "Acceptance Criteria for Emergency Core Cooling Systems for Light-Water Cooled Nuclear Power Reactors Pursuant to the AEC's Notice of Rule-Making Hearing 50-1," November 26, 1971, Concluded August 25, 1972, Concluding Statement of Position of the Regulatory Staff, Docket No. RM-501-1 (April 16, 1973).

11. "Water Reactor Safety Research Program, A Description of Current and Planned Research," NUREG-0006 (February 1979).

12. S. G. Margolis, J. A. Redfield, "FLASH: A Program for Digital Simulation of the Loss-of-Coolant Accident," WAPD-TM-534 (May 1966).

13. J. A. Redfield, J. H. Murphy, V. C. Davis, "FLASH-2: A FORTRAN/IV Program for the Digital Simulation of a Multinode Reactor Plant During Loss-of-Coolant," WAPD-TM-666 (April 1967).

14. K. V. Moore, L. C. Richardson, J. W. Sielinsky, "RELAPSE-I - A Digital Program for Reactor Blowdown and Power Excursion Analysis," PTR-803 (September 1966).

15. K. V. Moore, W. H. Rettig, "RELAP2 -- A Digital Program for Reactor Blowdown and Power Excursion Analysis," IDO-17263 (March 1968).

16. W. H. Rettig, et al., "RELAP3 -- A Computer Program for Reactor Blowdown Analysis," IN-1321 (June 1970).

17. K. R. Katsma, G. L. Singer, T. R. Charlton, et al., RELAP4/MOD5 - A Computer Program for Transient Thermal-Hydraulic Analysis of Nuclear Reactors and Related Systems, User's Manual, Volumes I-III, ANCR-NUREG-1335. Volume I - RELAP4/MOD5 Descriptions, Volume II - Program Implementation, Volume III - Checkout Applications (September 1976).

18. S. R. Fisher, et al., "RELAP4-MOD6 - A Computer Program for Transient Thermal-Hydraulic Analysis of Nuclear Reactors and Related Systems," CDAP-TR-003 (January 1978).

19. S. R. Behling, "RELAP4/MOD7 - A Best-Estimate Computer Program to Calculate Thermal and Hydraulic Phenomena in a Nuclear Reactor of Related Systems," to be published.

20. L. J. Ybarrondo, C. W. Solbrig, H. S. Isbin, "The 'Calculated' Loss-of-Coolant Accident: A Review," AIChE Monograph Series, No. 7, (1972).

21. C. W. Solbrig, L. J. Ybarrondo, and R. J. Wagner, "Idaho Nuclear Code Automation: A Standardized and Modularized Code Structure," Proceedings of Conference on New Developments in Reactor Mathematics and Applications, CONF-710302, (Vol. 1) (March 29-31, 1971).

22. J. F. Wilson, R. J. Grenda, and J. F. Patterson, "The Velocity of Rising Steam in a Bubbling Two-Phase Mixture," Transactions of the American Nuclear Society (May 1962).

23. "Evaluation of LOCA Hydrodynamics," Regulatory Staff Technical Review, USAEC (November 1974).

24. R. J. Wagner, "HEAT 1--A One-Dimensional Time-Dependent or Steady-State Heat Conduction Code for the IBM-650," IDO-16867 (April 1963).

25. R. J. Wagner, "IREKIN - Program for the Numerical Solution of the Reactor Kinetics Equations," IDO-17114 (January 1966).

26. J. A. Dearien, L. J. Siefken, and M. P. Bohn, "FRAP-T3 - A Computer Code for the Transient Analysis of Oxide Fuel Rods," TFBP-TR-194 (August 1977).

27. T. A. Porsching, J. H. Murphy, J. A. Redfield, and V. C. Davis, "Flash-4 a Fully Implicit FORTRAN IV Program for the Digital Simulation of Transients in a Reactor Plant," WAPD-TM-840 (March 1969).

28. "Assessment of the RELAP4/MOD6 Thermal-Hydraulic Transient Code for PWR Experimental Applications - Addendum - Analysis Completed and Reported in FY 1979," EG G-CAAP-5022 (February 1980).

29. M. L. Patton, "Semiscale Mod-3 Test Program and System Descriptions," TREE-NUREG-1212 (July 1978).

30. V. Esparza, K. E. Sacket, and K. Stanger, "Experiment Data Report for Semiscale MOD-3 Integral Blowdown and Reflood Heat Transfer Test S-07-6 (Baseline Test Series)," NUREG/CR-0467, TREE-1226, (January 1979).

31. J. A. Trapp and V. H. Ransom, "RELAP5 Hydrodynamic Model: Progress Summary - Field Equations," SRD-126-76, EG G Report (June 1976).

32. V. H. Ransom and J. A. Trapp, "RELAP5 Hydrodynamic Model: Progress Summary PILOT Code," PG-R-76-013, EG G Report (December 1976).

33. V. H. Ransom, et al., "RELAP5 Code Development and Results," Presented at the Fifth Water Reactor Safety Research Information Meeting (November 7-11, 1977).

34. V. H. Ransom, and J. A. Trapp, "RELAP5 Progress Summary, PILOT Code Hydrodynamic Model and Numerical Scheme," Idaho National Engineering Laboratory report No. CD-AP-TR-005 (January 1978).

35. O. C. Jones, Jr. and P. Saha, "Volumetric Vapor Generation in Nonequilibrium, Two-Phase Flows," Notes prepared for Advanced Code Review Group Meeting, Water Reactor Safety Research Division, U.S. Nuclear Regulatory Commission, Washington, D.C. 20555, (June 2, 1977).

36. G. Houdayer, et al., "Modeling of Two-Phase Flow with Thermal and Mechanical Nonequilibrium," paper presented at the Fifth Water Reactor Safety Research Information Meeting, Washington, D.C., (November 7-11, 1977).

37. A. W. Bennett, et al., "Flow Visualization Study of Boiling Water at High Pressure," AERE-R 4874 (1965).

38. R. T. Lahey, Jr., "RPI Two-Phase Flow Modeling Program," presented at Fifth Water Reactor Safety Research Information Meeting, (November 7-11, 1977).

39. N. Zuber, "On the Dispersed Two-Phase Flow in the Laminar Flow Region," Chemical Engineering Science, Vol. 19, (1964) pp 897-917.

40. K. T. Claxton, J. G. Collier, and A. J. Ward, "H.T.F.S. Correlation for Two-Phase Pressure Drop and Void Friction in Tubes," AERE-R7162 (1972).

41. V. H. Ransom and J. A. Trapp, "RELAP5 Progress Summary Analytic Choking Criterion for Two-Phase Flow," EG G Report, CDAP-TR-013 (1978).

42. V. H. Ransom and J. A. Trapp, "The RELAP5 Choked Flow Model and Application to a Large Scale Flow Test," presented at ASME Heat Transfer Division, Nuclear Reactor Thermal-Hydraulic 1980 Topical Meeting, Saratoga, New York.

43. J. A. Trapp and V. H. Ransom, "RELAP5 Hydrodynamic Model Progress Summary Abrupt Area Changes and Parallel Branching," PG-R-77-92, EG G Report (November 1977).

44. J. K. Vennard, Elementary Fluid Mechanics 4th Edition, John Wiley and Sons, 1965.

45. J. G. Collier, "Advanced Study Institute on Two-Phase Flows and Heat Transfer, ASI Proceedings, Istanbul, Turkey (August 1976).

46. K. E. Carlson, V. H. Ransom, and R. J. Wagner, "The Application of RELAP5 to a PIPE Blowdown Experiment," presented at ASME Heat Transfer Division, Nuclear Reactor Thermal-Hydraulic 1980 Topical Meeting, Saratoga, New York.

47. H. H. Kuo, V. H. Ransom, and D. M. Snider, "Calculated Thermal Hydraulic Response for Semiscale Mod-3 Test S-07-6 Using RELAP5--A New LWR System Analysis Code," Presented at the ASME Heat Transfer Division Nuclear Reactor Thermal-Hydraulic 1980 Topical Meeting, Saratoga, New York.

48. E. J. Kee, et al., "Best-Estimate Prediction for LOFT Nuclear Experiment L3-2," EG G-LOFF-5089 (February 1980).

49. "TRAC-Pl: An Advanced Best-Estimate Computer Program for PWR LOCA Analysis. Vol. I: Methods, Models, User Information, and Programming Details," Los Alamos Scientific Laboratory report LA-7279-MS, Vol. I (NUREG/CR-0063) (June 1978).

50. J. C. Vigil and K. A. Williams (Compilers), "TRAC-PlA Developmental Assessment Calculations," Los Alamos Scientific Laboratory report LA-8056-MS (NUREG/CR-1059) (October 1979).

51. F. H. Harlow and A. R. Amsden, "A Numerical Fluid Dynamics Calculation Method for All Flow Speeds," J. Comp. Physics 8, 197 (1971).

52. F. H. Harlow and A. R. Amsden, "KACHINA: An Eulerian Computer Program for Multifield Fluid Flows," Los Alamos Scientific Laboratory report LA-5680 (1975).

53. D. R. Liles and W. H. Reed, "A Semi-Implicit Method for Two-Phase Fluid Dynamics," accepted for publication in J. of Comp. Physics.

54. J. H. Mahaffy and D. R. Liles, "Application of Implicit Numerical Methods to Problems in Two-Phase Flow," Los Alamos Scientific Laboratory report LA-7770-MS (1979).

55. C. J. Crowley, J. A. Block, and C. N. Cary, "Downcomer Effects in a 1/15-Scale PWR Geometry - Experimental Data Report," Creare, Inc. report NUREG-0281 (May 1977).

56. L. J. Ball, D. J. Hanson, K. A. Dietz, and D. J. Olson, "Semiscale MOD-1 Test S-02-8 (Blowdown Heat Transfer Test)," Aerojet Nuclear Company report ANCR-NUREG-1238 (August 1976).

57. D. L. Reeder, "LOFT System and Test Description (5.5-ft Nuclear Core 1 LOCES)," EG G Idaho, Inc. report TREE-1208 (NUREG/CR-0247) (July 1978).

58. P. G. Prassinos, B. M. Galusha, and D. B. Engleman, "Experiment Data Report for LOFT Power Ascension Experiment L2-3," Idaho National Engineering Laboratory report TREE-1326 (NUREG/CR-0792) (July 1979).

59. O. C. Iloeje, D. N. Plummer, W. M. Rohsenow, and P. Griffith, "An Investigation of the Collapse and Surface Rewet in Film Boiling in Forced Vertical Flow," J. of Heat Transfer, pp 166-172 (May 1975).

PART 4

DESIGN BASIS ACCIDENT: LIQUID METAL FAST BREEDER REACTORS

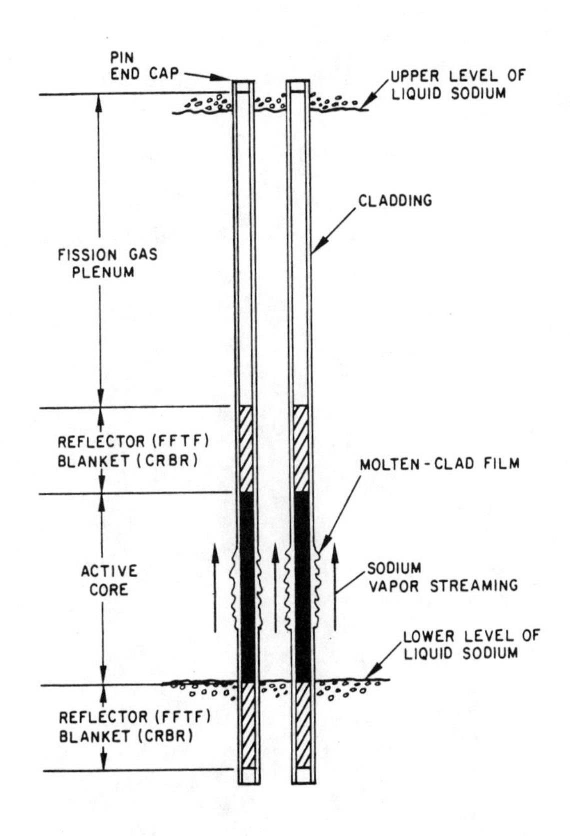

14 CORE DISRUPTIVE ACCIDENTS

by

H.K. Fauske*

ANL, Argonne, Il, USA

This chapter begins the next sequence of the text—a summary of the DBA in liquid metal fast breeder reactors. The initiators for the core disruptive accident are reviewed followed by considerations relating to failure propagation. Emphasis is on the mechanistic approach to analyzing the sequence of events from initiation, through transition, to termination.

1. INTRODUCTION

In this chapter we put major emphasis on low-probability (but potentially large-consequence) accidents -- e.g., core-disruptive accidents (CDAs) -- because the issue of significant risk from fast reactors, as from thermal reactors, becomes only of concern in the case of a CDA capable of breaching the containment. The potential causes of containment failures under CDA conditions include:

a. Rapid failure due to CDA energetics, and

b. Relatively slower failures due to CDA fuel debris causing long-term pressurization.

In discussing CDAs we will stress the balanced approach aimed at reducing both the probability and consequences of a CDA.[1]

2. CDA INITIATORS AND CHANGING TRENDS

Three classes of possible initiating conditions have been identified:

(1) Those resulting in a reactivity insertion at a rate so great that the reactor plant-protection system would be unable to respond in time,

(2) Malfunctions or faults within the design basis of the reactor plant-protection system, but in combination with failure of the plant-protection system [these can be broken down into fuel-failure propagation, whole-core loss-of-flow (LOF), and transient overpower (TOP)], and

(3) Malfunctions or faults leading to interruption of heat-removal capability even with shutdown, such as loss-of-heat-sink.

*Present Address: Fauske and Associates, Incorporated, 627 Executive Drive, Willowbrook, Illinois 60521.

From a probabilistic point of view, it has become customary in the U.S. to consider only the second and third class of initiators because the first class (including control-rod ejection, gas-bubble intake, and failure of core-support structure or core-restraint system) can be discounted on the grounds that they can be effectively precluded by design.[2] CDAs are therefore considered to have three general initiators:

1. Fuel-failure propagation,
2. Loss of plant-protection systems (unprotected transients), and
3. Loss of heat sink.

However, it is becoming increasingly desirable to confirm intrinsic reactor plant and systems capabilities to remove and/or terminate these three CDA initiators with only limited damage. These intrinsic capabilities include:

a. Margins in performance of fuel, coolant, and structures,
b. Margins in system performance (degraded-mode component operation),
c. Inherent accident-limiting processes, and/or
d. Self-actuated safety systems that do not depend on external power supply or operator interaction.

2.1 Fuel-failure Propagation

With the introduction of multiple subassembly inlets, subassembly disruption need only be considered as a hypothetical event, because extensive efforts have been unable to identify any sequences of events that could result in disruption.[3] The principal reasons for this conclusion are: 1) local faults (including fission-gas release, molten-fuel release, and localized boiling) have been shown to be isolated events (no rapid propagation), and 2) slow blockage propagation, if it should occur at all, appears self-limiting and can be detected. We conclude that fuel-failure propagation no longer remains a potential candidate as a CDA initiator.

2.2 Loss of Plant-protection Systems

Despite the use of two independent and diverse plant-protection shutdown systems (electromechanical devices), it has become common practice to consider anticipated transients without scram (ATWS) such as the LOF, and to proceed with a mechanistic evaluation of such events (see Section 3). For the LOF case,* current technology applied to current U.S. design practice suggests that major core disruption would be unavoidable.[1]

However, recent developments indicate that *practical* and inherently safe shutdown devices can be incorporated as part of the reactor core design so as to ensure shutdown for all anticipated transients well before the sodium boiling temperature and/or fuel failure limits have been reached.[5,6] The most promising technique to date appears to be intrinsic actuation of the absorber material at elevated temperature by electromagnetic means. The introduction of such systems should largely eliminate the need for detailed mechanistic considerations of ATWS accidents.

*We note that the current French design with its potential negative feedback characteristics may avoid sodium boiling altogether.[4]

2.3 Loss of Heat Sink

Inherent decay-heat removal capability exists within the primary coolant boundary of current LMFBR designs. This is because 1) the system can be designed such that a break in the primary coolant boundary does not result in a loss of cooling to the reactor core (in contrast to the case of an LWR), and 2) even if single-phase natural circulation is not established, sodium boiling in the core is an acceptable means for removing decay heat* -- i.e., fuel melting would be initiated only if the coolant level drops below the core, which is inherently prevented by design.

This inherent capability to remove decay heat from the core, together with current developments that include an independent (dedicated) direct auxiliary shutdown heat removal system,[9,10] should assure the necessary reliability of decay-heat removal.** Heat removal from the primary system again can be provided by natural circulation, while practical considerations are likely to necessitate forced circulation on the ambient side. For this type of accident (assuming failure of the normal heat transport systems), considerable time is available to ensure proper functioning of active components. The power is reduced rapidly to decay-heat levels following scram, and the subsequent heatup therefore occurs over hours or days, as compared to seconds for the unprotected accidents.

In summary, with the probabilistic viewpoint it appears that the second and third classes of CDA initiators identified above can also be discounted on the grounds that they can be effectively mitigated by appropriate design features. This implies that detailed mechanistic considerations beyond early termination become unnecessary. Obviously, this has serious implications relative to future R&D needs as well as to the need for additional safety test facilities.

3. MECHANISTIC APPROACH TO CDA

In the absence of mitigating design features, the assessment of CDA consequences in the U.S. has largely been based on detailed mechanistic analyses of the unprotected TOP and LOF accident. That work is briefly summarized below, with major emphasis on the unprotected-LOF accident. This accident is believed to provide a reasonable envelope for possible power-to-flow mismatches, including the TOP accident.

3.1 Unprotected TOP

The principal areas of interest for this accident class, relative to the potential for whole-core involvement, include: (a) fuel-failure phenomenology, (b) pin-failure incoherencies (in space and time), and (c) postfailure fuel motion.[11]

---

*For subcooled plenum conditions (below and above the core), analysis[7] and simulation experiments[8] indicate that 5% of nominal power can be safely removed without resulting in dryout and overheating in the absence of forced-flow conditions. However, the duration of sodium boiling would be short because the large driving head provided by boiling would drive the system into single-phase natural circulation.

**This feature also has important implications relative to long-term in-vessel accommodation of fuel debris following postulated core disruption from unprotected accidents, and will be discussed in Section 4.

In the first area, available analytical and experimental data generally support axial failure location well above the core midplane [this is definitely the case where the reactivity rate is limited to the maximum speed of the control rod drive (several cents per second)], suggesting that there is sufficient negative reactivity feedback for neutronic shutdown by the fuel moving axially within the fuel pin to the failure location. Fuel sweepout in the coolant channels, while desirable, apparently is not necessary for shutdown.

The key remaining concern about limiting core damage involves the potential for fuel plugging and its effect on coolability, which relates to the second and third areas of interest. Evidence is available from TREAT fuel-pin disruption tests with (U,Pu) oxide fuel that plugging can occur[12] (although these have generally been carried out with less than prototypic hydraulic conditions and fuel sample sizes). However, arguments based on intrasubassembly temperature incoherencies, and hence incoherencies of fuel-pin failure, might prevent complete subassembly blockages.

3.2 Unprotected LOF

This accident has been studied extensively during the last two decades. These studies have produced numerous reports dealing specifically with the initiating, transition, and termination phases of the accident.[13-15]

Initiating Phase. The potential for rapid reactivity insertion associated with the voiding process was first connected to the possibility of large superheating prior to boiling inception. This concern was justified at the time in view of early laboratory experiments and the propensity for alkali metals to wet stainless-steel surfaces. However, with the elimination of this concern,* the possibility of augmented reactivity addition due to accelerated sodium voiding from fuel-coolant interaction (FCI) emerged. However, in this case no experimental data base existed for the concern, and, in fact, for such to occur, rather unphysical assumptions must be postulated -- such as pin-hole failure near the midplane with fuel moving in the opposite direction to the liquid sodium. This picture is contrary to experimental evidence in which the fuel has clearly been shown to be dispersive for such power conditions.[12] This, together with the fact that the reactivity worth of the fuel motion is more than enough to cancel the reactivity worth associated with sodium voiding, leads one to conclude that this concern is not justified. Ongoing and planned experiments in the current TREAT facility are expected to completely eliminate this concern. In the absence of any significant fuel vapor pressures being developed during the initiating phase, a gradual transition to a largely disrupted core is predicted.[17]

Transition Phase. The major concern here is the potential for an energetic recriticality event if the core should become "bottled-up" or if there is a pressure-driven recompaction associated with a fuel-coolant interaction.[17,18]

The oxide fuel-steel mixture would generally tend to be dispersive even at decay-heat power level,[1,17,19] and would therefore tend to seek colder regions. The dimensions of the fuel assembly structures above and below the active core region are such that the propensity for fuel freezing and plugging is quite

*The possibility of significant superheating was eliminated by recognizing the important effect of inert gases present in the real reactor system.[16] This result limits the net ramp rate from sodium voiding in a typical large homogeneous core to less than 15 $/s.[14,15]

high.[20] This has therefore led to concern over a "bottled-up" core[14,17] and associated concern about dynamic effects with adverse fuel motions.[21] In the latter computations, the majority of the fuel is assumed to remain within the active core region.

In this regard, it is important to notice that extensive experiments with reactor materials clearly illustrate that substantial penetration of fuel into the subassembly structure around the active core region is possible prior to complete plugging.[22] In fact, a major fraction of the fuel could be initially displaced in this manner. Subsequent remelting and dilution by blanket material into the core region would render an *energetic* recriticality extremely unlikely. It is anticipated that recriticality events from dynamic effects will be limited to rather modest ramp rates (10-25 $/s).[23]

Furthermore, the subassembly hexcan (steel wall) subjected to molten core material on one side and cooled by liquid sodium on the other side has also shown an inability to contain molten fuel.[24] Control- and safety-rod channels would therefore appear to be natural escape paths for molten fuel as it develops because these channels are large enough to render freezing and complete plugging highly unlikely. Further analysis and experimental validation is desirable in this area, because demonstration of early fuel removal capabilities will remove the remaining concern relative to the potential of energetic recriticality events. As discussed further below in connection with pressure-driven recompaction, the presence of liquid sodium will not interfere with the necessary fuel transport.

Two geometric configurations have been of concern relative to the pressure-driven recriticality problem: 1) ejection of molten fuel from the active core region into the upper (or lower) subassembly structure (blanket and fission-product plenum) that might contain liquid sodium, and 2) injection of molten fuel directly into the upper (or lower) sodium plenum following unplugging of a postulated "bottled-up" core.

In the case of oxide fuel, the first concern was addressed by the Upper Plenum Injection (UPI) experiments.[25] These tests, which were carried out in a "constrained" geometry, showed no potential for sustained pressure events capable of leading to flow reversal and recompaction of fuel. This result also suggests that the presence of sodium will not inhibit fuel removal in channels where fuel freezing is not possible.

The second concern has been addressed by numerous other applicable tests carried out in "unconstrained" geometry. Those tests again show the absence of sustained pressure events consistent with the Interface Temperature Spontaneous Nucleation criterion.[26]

Termination Phase. Based on the above discussion, the fuel is most likely to be removed from the core in a rather benign fashion, and the main problem is one of demonstrating core-debris accommodation. Unrealistic physical processes must be postulated to achieve highly energetic prompt-burst conditions leading to a true hydrodynamic disassembly of the reactor core. Such calculations are, however, useful in establishing pressure-volume relationships to be used to upgrade any weak links in the design established principally from functional requirements as well as assessing safety margins provided by the resulting design. In this regard, current design practices offer a substantial and adequate safety margin. Work potential corresponding to several hundred MJ is readily accommodated by the primary coolant system boundary.

We note that, in the absence of significant sodium entrapment, the relatively cold clad material (especially in blanket and fission plenum regions) can be expected to substantially affect the fuel-vapor expansion work potential from a postulated disassembly event.[27,28] For the FFTF and CRBRP (with fission plenum at the top of the fuel pin), an otherwise-postulated energetic CDA event might be reduced to an essentially benign event, if early fuel freezing and plugging can be ruled out.

Finally, the possibility of sodium entrainment and work augmentation must be considered as the postulated hot fuel bubble expands beyond the core region into the upper sodium plenum. Here, early considerations developed for the FFTF and later applied to CRBRP included various optimum ways for sodium entrainment during fuel-bubble expansion up to slug impact. These calculations resulted in an augmentation factor of, at most, 2.[29] Compared to the large and rather arbitrary initial ramp rate associated with the CDA event, this factor cannot be considered significant. Later work has argued that sodium entrainment will lead to a reduction in work potential, perhaps by a factor of 2.[30] In any case, we note that such mitigation, principally due to fuel-vapor condensation, is not likely to be substantial, as compared to a one-component system where vapor could be visualized to condense rather rapidly.[31] It is concluded that little work augmentation beyond that provided by the fuel vapor process is likely, and that the work potential from the latter can be readily contained with current design.[1,2]

4. END-OF-SPECTRUM CDA CONSIDERATIONS

Although current developments indicate that the ATWS accidents eventually can be eliminated as a source for CDAs, historic trends nevertheless suggest that unprotected reactivity accidents will continue to receive attention. From a catastrophe viewpoint, a large spectrum of events can be foreseen (such as failure of core support structure, a large gas bubble entering the core, a localized perturbation leading to inward compaction of the core, etc.), resulting in rapid reactivity insertion that might be as high as 100 $/s with the core still largely full of liquid sodium. This viewpoint, as compared to the non-catastrophic or probabilistic viewpoint, requires nonmechanistic end-of-spectrum estimates of consequences in order to quantify or envelope the possible associated risk.

4.1 Energetics

The fuel-vapor potential associated with reactor disassembly from ramp rates as large as 100 $/s is not significant for the sodium-in case.[32] The problem here is rather the potential for an energetic fuel-coolant interaction (FCI), because theoretically this process can provide at least a factor of ten increase in work potential above that possible by fuel vapor expansion alone. The FCI concern seems well justified because: 1) of previous experience with non-LMFBR systems, such as SL-1, BORAX-1, and SPERT-1[33-35] which clearly demonstrated that rather energetic vapor explosions are possible, and 2) the combination of LMFBR reactor core design and the nature of the end-of-spectrum reactivity accidents lead to a *coarsely* mixed region of fuel and sodium to begin with, a condition frequently suggested to be capable of leading to a sustained propagating energetic vapor explosion or detonation.[37] It is therefore not surprising that the FCI or vapor explosion problem has emerged as *the* key LMFBR safety issue.*

---

*In view of the importance of this issue, a separate subsection in this report is devoted to a discussion of vapor explosions.

Large-scale explosive sodium-vaporization events on a millisecond time scale are unlikely under these conditions.* This position is supported by the TREAT meltdown experiments, particularly, the S-series test results.[12] The principal findings from these tests can be summarized as follows: For times up to the stopping of the piston, the principal working fluid clearly appeared to be fuel vapor. After the piston stopped, sodium vaporization events took place. Similar results are noted with the more recent ACPR tests sponsored by NRC.[38] We note that sodium vaporization events following stopping of the piston in these autoclave experiments should not be unexpected, because the sodium can no longer escape from the hot fuel. This condition would appear nonprototypic relative to HCDA conditions as noted further below.

The possibility of augmenting the fuel-vapor pressure with extensive sodium vaporization on a time scale of tens of milliseconds must also be considered.

Again, large-scale sodium vaporization is unlikely because:

- Initial mixing upon fuel failure is largely prevented, because the sodium must move before the hot fuel can enter the coolant channel. (Note there is typically less than 1% volume change of the liquid sodium due to pressure change resulting from fuel failure.) This is consistent with interpretation of the above in-pile experiments.

- The space emptied by the expulsion of liquid sodium is immediately occupied by the expanding fuel mass.** This process appears to occur both during the acoustic as well as the inertial phase associated with sodium expulsion. We note that the inertial pressure gradient developed in the liquid sodium slug is such that it is unlikely that the fuel ejects through breached cladding into the liquid sodium slug appreciably ahead of the fuel-sodium interface.

- The presence of solid cladding, a necessary requirement for the "sodium-in" case, largely prevents fuel entrainment as the liquid sodium is being expelled as a result of the expanding fuel mass.

- During the expulsion of liquid sodium, an insignificant quantity of liquid sodium is left behind on the fuel-pin structure (fuel blanket and fission plenum regions), because the expanding fuel mass is heavier than liquid sodium. We note that a heavier fluid pushed by a lighter fluid (e.g., gas-liquid system) would leave behind a substantial film of the heavier fluid ($\sim$0.15 liquid fraction).[39] This

---

*These conditions (coarse mixing when fuel and coolant are separated by solid cladding material) are in complete contrast to that visualized for the SPERT-1, BORAX-1, and SL-1 nuclear transients that ultimately led to energetic steam explosions. The presence of aluminum cladding (with time constant equal to or less than the nuclear period) and water coolant (favoring film boiling) leads to ideal conditions (molten fuel and molten cladding sufficiently *finely* dispersed in the coolant and only separated by thin vapor blankets) for producing a propagating steam explosion. This picture is consistent with the observation that the relatively fine fragmentation and intermixing necessary for large-scale vapor explosion appears to be largely a prepropagation phenomenon.

**This largely eliminates the concern about autocatalytic effects, because the reactivity loss associated with fuel expansion is more than enough to compensate for reactivity gain due to sodium expulsion, as indicated earlier.

largely explains the absence of any sustained pressurization events in the Upper Plenum Injection experiments,[25] which simulated fuel expansion into a prototypic fuel-pin-subassembly structure containing liquid sodium.

Therefore, it would appear difficult to trap a significant amount of sodium as fuel expands following the power surge associated with an arbitrary reactivity insertion event. This is to be compared to 1) the in-pile experiments noted above, where sodium is trapped as a result of stopping the piston, i.e., resulting in a constant-volume system with considerable void space and room for mixing, and 2) the most recent CORRECT-II experiments,[40] where bare molten fuel is flooded with liquid sodium in the presence of substantial external constraint, i.e., again ensuring confined boiling and vaporization of liquid sodium resulting in pressurization events on a time scale of tens of milliseconds.

4.2 Vapor Explosion

Analogous to chemical explosives in which energy-rich fuels and oxygen-rich compounds must be uniformly and finely intermixed, the hot and cold liquids must be *finely* intermixed in order to obtain the necessary energy transfer on a time scale consistent with explosive behavior. In the detonation theory,[36] this process is postulated to occur during the propagation stage. The necessary mixing prior to propagation, which is referred to as coarse premixing, where a characteristic dimension of the order 1 cm is frequently referred to is of same dimension as that present in current LMFBR core design. Therefore, if a suitable trigger mechanism is available, it has been argued that it may be difficult to rule out a large-scale detonating vapor explosion for the LMFBR case.

However, experimental observations clearly suggest that fine-scale fragmentation and mixing must occur during the dwell period prior to the triggering and propagation stages. Some examples are:

- In large-scale tests with aluminum/water, the mixture has a mean thickness of $\sim$500 μm and this is achieved after a dwell time of the order 1/2 s.[41] This is confirmed by cases in which no interaction is triggered; the form of the solidified metal provides a fine "honeycomb" structure.

- To explain the measured vapor production prior to the explosive propagation event in large-scale experiments such as Freon-oil,[42] a mean thickness as small as $\sim$100 μm is suggested.[43]

- Propagating large-scale explosive events with systems such as Al-H_2O, Sn-H_2O, Freon-H_2O, etc., can readily be explained by a particle size in the range of 100-1000 μm.

- The mode of heat transfer is sufficiently rapid even in the presence of a thin vapor film.[44]

- No experimental evidence is available for shock velocity fragmentation (as implied by detonation theory) taking part in observed vapor explosions.[45] In numerous experiments, the metal is found to be in a cokey form with high surface area after the interaction. Small-scale experiments, where hydrodynamic and vapor collapse fragmentation modes could clearly be ruled out, produced similar results;[46] i.e., is consistent with the honeycomb-like structure.

Because fragmentation and intermixing are suggested to be largely pre-propagation phenomena, the following requirements must be satisfied:

- Mechanical disturbances must be generated so as to produce the necessary fragmentation; i.e., numerous physical contacts between the hot and cold fluid are necessary.

- Sufficient void space must be available so as to allow adequate intermixing.

- Minimum energy transport is required during the fragmentation and intermixing step so as to prevent premature interaction; i.e., film-boiling-like behavior must be assured.

These seemingly imposing requirements can only be satisfied if the instantaneous interface temperature between the hot and the cold liquid exceeds the spontaneous nucleation temperature.[29,47] A close-packed set of critical-size bubbles will nucleate, grow, and coalesce, resulting in vapor blanketing and rapid cut-off of heat transfer. Also, if the ambient pressure is sufficiently low, the short-lived nucleation process will produce adequate localized pressure to ensure fragmentation, but insufficient to collapse the void space, hence ensuring both intermixing and minimum energy transfer.

The above observations are in complete accord with the well-defined experimentally observed threshold effects, including:

- Lower-temperature onset of vapor explosions,[48-50] and
- A pressure cut-off of vapor explosions.[51-53]

The significance of the Interface Temperature, Spontaneous Nucleation criterion for onset of large-scale vapor explosions relates directly to the LMFBR UO_2/Na case. For this system, the contact temperature is well below the spontaneous nucleation temperature. We conclude, therefore, that large-scale vapor explosions are rigorously excluded for the conditions of an LMFBR core meltdown accident with oxide fuel. This includes the sodium-in case as well.

In well over a hundred tests involving molten UO_2 and liquid sodium under a variety of contact modes, including numerous in-pile tests, no large-scale vapor explosions or energetic fuel-coolant interactions have ever been observed.

In summary, based on available analysis and experiments, we therefore conclude that, even for the end-of-spectrum CDA conditions such as considered above, the anticipated energetics would be benign. Rather pessimistic assumptions in direct conflict with experimental data and analysis must be made in order to violate the primary coolant system boundary based on current design practices.

4.3 Core Debris Accommodation

In assessing means for long-term fuel-debris accommodation following a hypothetical core disruption, as discussed above, the following must be considered:

a. The inherent capability of the LMFBR to remove decay heat within the primary system, including provision for a direct and independent emergency path from the primary system,

b. The highly unlikely event of significant energetics so as to violate the primary system and its dedicated decay-heat removal system, and

c. The possibility of fuel-debris-bed coolability either by simple conduction and/or by sodium boiling.

By assuring adequate surfaces so as to make the latter phenomenon effective, attractive arguments for indefinite fuel-debris coolability by passive means can be made within the primary coolant system boundary.

A simple illustration of an adequate surface area is the installation of a flat plate in the lower part of the reactor vessel. For current loop designs, such a surface would be capable of removing decay heats of up to 4% of full power, based on current fuel-debris-bed coolability technology.*[54] This ability to provide in-vessel coolability of an essentially completely disrupted core is unique to the LMFBR concept, and eliminates the many concerns about invoking ex-vessel coolability arguments. For the LMFBR, these include chemical reaction of sodium with concrete in the event of a liner failure, the heating of concrete and the resultant release of water vapor from the concrete, the generation of hydrogen as a result of sodium-water or sodium-concrete reactions in an inert atmosphere, and possibly the interaction of molten fuel with concrete.

However, even in this case there is substantial evidence becoming available that, in the absence of an engineered core catcher of any kind (referred to as the "core-on-the-floor" concept), the inherent response of the reactor structures and the use of mitigating features (including provision for active containment cooling, thereby ensuring condensation and refluxing of sodium) enable the fuel debris to be cooled by sodium evaporation and to be accommodated for an indefinite period without violating the final containment boundary. In summary, both in-vessel and ex-vessel accommodation appear to be viable options for the long-term cooling of fuel debris.

5. CONCLUDING REMARKS

Recent developments as they relate to practical use of intrinsic reactor plant and system capabilities to assure reactor shutdown and decay-heat removal for all anticipated transients appear very promising and deserve increased attention. [See Chapter 5 for detailed discussions). The successful demonstration of such inherent capabilities will undoubtedly reduce detailed mechanistic consideration of core-disruptive accidents, as well as significantly affecting future safety R&D needs, including the need for new safety test facilities.

Despite this promising evolution, historic trends suggest that unprotected accidents in connection with rather arbitrary reactivity events will continue to receive attention. Therefore, an increased emphasis on nonmechanistic end-of-spectrum estimates of consequences is likely to emerge in order to quantify or envelope the possible associated risk. Here, efforts to date show that

---

*For the French Super Phenix design, a tray system is provided in the lower tank so as to accommodate a major fraction of the fuel in the event of a postulated CDA.[5]

highly energetic events from postulated core-disruptive events appear very unlikely. Rather pessimistic assumptions in direct conflict with experimental data and analysis must be made in order to violate the primary coolant system boundary based on current design practices.

This observation, together with the LMFBR's inherent capability of removing decay heat, suggest that indefinite coolability of the disrupted fuel can be accomplished within the primary coolant system boundary. It follows that the risk associated with LMFBR deployment is negligibly small, even if a CDA is postulated to occur.

REFERENCES

1. Fauske, H. K., "The Role of Core Disruptive Accidents in Design and Licensing of LMFBRs," Nuclear Safety, Vol. 17, No. 5, September-October 1976.

2. "Final Safety Analysis Report, Fast Flux Test Facility," October 1976.

3. Fauske, H. K., Grolmes, M. A., and Chan, S. H., "Assessment of Fuel Failure Propagation in LMFBRs," Proc. Fast Reactor Safety and Related Physics, CONF-761001, Chicago, Illinois, October 5-8, 1976.

4. Freslon, H., et al., "Analyses of the Dynamic Behavior of the Phenix and Super-Phenix Reactors During Some Accident Sequences," Proc. Int. Mtg. on Fast Reactor Safety Technology, Seattle, Washington, August 19-23, 1979.

5. Villeneue, J., et al., "Safety-Related Design and Development of Components for the Creys-Malville Plant," Proc. Intl. Mtg. on Fast Reactor Safety Technology, Seattle, Washington, August 19-23, 1979.

6. Dupen, C. F. G., et al., "Self-Actuated Shutdown Systems for Commercial LMFBRs," Proc. Int. Mtg. on Fast Reactor Safety Technology, Seattle, Washington, August 19-23, 1979.

7. Dunn, F. E., "Fuel Pin Coolability in Low Power Voiding," Trans. Am. Nucl. Soc., Vol. 28, 1978.

8. Hinkle, W. D., "Water Tests for Determining Postvoiding Behavior of the LMFBRs," MIT-EL 76-005, June 1976.

9. Vossebrecker, H. and Kellner, A., "Inherent Safety Characteristics of Loop-type LMFBRs," Proc. Int. Mtg. on Fast Reactor Safety Technology, Seattle, Washington, August 19-23, 1979.

10. Smith, R. D., "Design and Development of Systems and Components for Safety," Proc. Int. Mtg. on Fast Reactor Safety Technology, Seattle, Washington, August 19-23, 1979.

11. Wilburn, N. P., et al., "An Updated Assessment on the Unprotected Transient Overpower Accident in the FTR," HEDL-TI 75285, October 1977.

12. Dickerman, C. E., et al., "Status and Summary of TREAT In-pile Experiments on LMFBR Response to Hypothetical Core-disruptive Accidents," Symp. on the Thermal and Hydraulic Aspects of Nuclear Reactor Safety, Vol. 2: Liquid Metal Fast Breeder Reactors, edited by O. C. Jones, Jr. and S. G. Bankoff, ASME publication, 1977.

13. Stevenson, M. G., et al., "Current Status and Experimental Basis of the SAS LMFBR Accident Analysis Code System," Proc. Fast Reactor Safety Mtg., April 2-4, 1974, Beverly Hills, California, CONF-740401.

14. Jackson, J. F., et al., "Trends in LMFBR Hypothetical-accident Analysis," Proc. Fast Reactor Safety Mtg., April 2-4, 1974, Beverly Hills, California, CONF-740401.

15. Fauske, H. K., "Summary on Accident Energetics, Including Coolant Dynamics, Cladding and Fuel Relocation, and Molten-Fuel-Coolant Interaction," Proc. Intl. Mtg. on Fast Reactor Safety and Related Physics, October 5-8, 1976, Chicago, Illinois.

16. Fauske, H. K., "Nucleation of Liquid Sodium in Fast Reactors," J. Reactor Technology, Vol. 15, No. 4, Winter 1972-73.

17. Ostensen, R. W., et al., "The Transition Phase in LMFBR Hypothetical Accidents," Proc. Int. Mtg. on Fast Reactor Safety and Related Physics, CONF-761001, p. 895, 1976.

18. Boudreau, J. E. and Jackson, J. F., "Recriticality Considerations in LMFBR Accidents," Proc. Fast Reactor Safety Mtg., April 2-4, 1974, Beverly Hills, California, CONF-740401-P3.

19. Fauske, H. K., "Assessment of Accident Energetics in LMFBR Core-disruptive Accidents," Nuclear Eng. & Design, Vol. 42, No. 19, 1977.

20. Ostensen, R. W., et al., "Fuel Flow and Freezing in the Upper Subassembly Structure Following an LMFBR Disassembly," Trans. Am. Nucl. Soc., Vol. 18, 1974.

21. Bohl, W. R., "Some Recriticality Studies with SIMMER-II," Proc. Int. Mtg. on Fast Reactor Safety Technology, Seattle, Washington, August 19-23, 1979.

22. Epstein, M., et al., "Analytical and Experimental Studies of Transient Fuel Freezing," Proc. Int. Mtg. on Fast Reactor Safety and Related Physics, October 1976, Chicago, Illinois, Vol. IV, CONF-761001.

23. Theofanous, T. G., "Multiphase Transients with Coolant and Core Materials in LMFBR Core Disruptive Accident Energetics Evaluation," NUREG/CR-0224, July 1978.

24. Hakim, S. G. and Kennedy, J. M., "Development of Transition-phase Code," ANL-RDP-70, April 1978.

25. Henry, R. E., et al., "Experiments on Pressure-driven Fuel Compaction with Reactor Materials," Proc. Fast Reactor Safety and Related Physics, Chicago, Illinois, October 5-8, 1976, CONF-761001.

26. Fauske, H. K., "Some Aspects of Liquid-liquid Heat Transfer and Explosive Boiling," Proc. ANS Topical Meeting on Fast Reactor Safety, Beverly Hills, California, CONF-740401-P2, p. 992, 1974.

27. Theofanous, T. G. and Fauske, H. K., "An Energy Dissipation Mechanism Due to the Cladding of the Fission-gas Plenum During an HCDA," ANL/RAS 72-31, September 1972.

28. Bell, C. R. and Boudreau, J. E., "Application of SIMMER-I to the Post-disassembly Fluid Dynamic Behavior within an LMFBR Reactor Vessel," Proc. Fast Reactor Safety and Related Physics, CONF-761001, Chicago, Illinois, October 5-8, 1976.

29. Cho, D. H., Epstein, M., and Fauske, H. K., "Work Potential Resulting from a Voided-core Disassembly," Trans. Am. Nucl. Soc., Vol. 18, No. 220, June 1974.

30. Corradini, M. L., "Heat Transfer and Fluid Flow Aspects of Fuel-Coolant Interactions," COO-2781-12TR, September 1978.

31. Theofanous, T. G., personal communication, October 1979.

32. Nicholson, R. B. and Jackson, J. F., "A Sensitivity Study for Fast Reactor Disassembly Calculations," ANL-7952, January 1974.

33. Miller, R. W., Sola, A., and McCardell, R. K., "Report of the SPERT-1 Destructive Test Program on an Aluminum, Plate-type, Water Moderated Reactor," IDO-16883, 1964.

34. Deitrich, J. R., "Experimental Investigation of the Self-limitation of Power During Reactivity Transients in a Subcooled Water-Moderated Reactor-BORAX-1 Experiments, 1954," AECD-3668, 1965.

35. SL-1 Project, "Final Report of SL-1 Recovery Operations," IDO-19311.

36. Board, S. J., Hall, R. W., and Hall, R. S., Nature, 254, 319, 1975.

37. Board, S. J. and Caldarola, L., "Fuel-Coolant Interactions in Fast Reactors," in Thermal and Hydraulic Aspects of Nuclear Reactor Safety," O. C. Jones, Jr. and S. G. Bankoff, eds., 2, 195, ASME, New York, 1977.

38. Schmidt, T. R., "LMFBR Prompt-Burst Excursion (PBE) Experiments in the Annular Core Pulse Reactor (ACPR)," Proc. Fast Reactor Safety and Related Physics, CONF-761001, Chicago, Illinois, October 5-8, 1976.

39. Grolmes, M. A. and Fauske, H. K., "Liquid Film Thickness and Mechanism for Dryout during LMFBR Loss-of-flow Accidents," Trans. Am. Nucl. Soc., Vol. 14, No. 743, 1971.

40. Amblard, M. and Jacobs, H., "Fuel-Coolant Interactions - The CORRECT-II UO_2-Na Experiment," Proc. Int. Mtg. on Fast Reactor Safety Technology, Seattle, Washington, August 19-23, 1979.

41. Briggs, A. J., "Experimental Studies of Thermal Interactions at AEE Winfrith," Proc. Third Specialist Meeting on Sodium Fuel Interactions in Fast Reactors, p. 595, Tokyo, Japan, March 22-26, 1976. Also unpublished communication, A. J. Briggs and G. J. Vaughn, 1979.

42. Anderson, R. P. and Armstrong, D. R., "R-22 Vapor Explosions," ASME Symp. on the Thermal and Hydraulic Aspects of Nuclear Reactor Safety, Winter Annual Meeting, Atlanta, Georgia, November 27 - December 2, 1977.

43. Fauske, H. K., unpublished notes, June 1977.

44. Hall, W. B., paper presented at FCI Mtg., Argonne National Laboratory, SNIDOC(77), 173, Nuclear Eng. Agency, OECD, Paris, 1977.

45. Fishlock, T. P., "Calculations on Propagating Vapor Explosions for the Aluminum/Water and UO_2/Sodium Systems," Fourth CSNI Specialist Meeting on Fuel-Coolant Interactions in Nuclear Reactor Safety, CSNI Report 37, 1979.

46. Baines, M. and Board, S. J., "Transient Contact and Boiling Phenomena on Solid and Liquid Surfaces at High Temperatures," CEGB Report RD/B/N 4075, 1978.

47. Fauske, H. K., "The Role of Nucleation in Vapor Explosions," Trans. Am. Nucl. Soc., Vol. 51, No. 813, 1972. See also H. K. Fauske, Nucl. Sci. Eng., Vol. 51, p. 95, 1973.

48. Henry, R. E., et al., "Large-scale Vapor Explosions," Proc. of the Fast Reactor Safety Mtg., Beverly Hills, California, April 1974, CONF-740401-P2.

49. Board, S. J. and Hall, R. W., "Recent Advances in Understanding Large-scale Vapor Explosions," Proc. of the Third Specialist Meeting on Sodium/Fuel Interaction in Fast Reactors, PNC N251 76-21 (Vols. 1 and 2), Tokyo, Japan, March 22-26, 1976.

50. Dullforce, T. A., Reynolds, J. A., and Peckover, R. S., "Interface Temperature Criteria and the Spontaneous Triggering of Small-scale Fuel-Coolant Interactions," CLM-P517, Culham Laboratory, UKAEA.

51. Henry, R. E., Fauske, H. K., and McUmber, L. M., "Vapor Explosion Experimnts with Simulant Fluids," Proc. Int. Mtg. Fast Reactor Safety and Related Physics, Chicago, Illinois, October 5-8, 1976.

52. Hohman, H., Kottowski, H. M., and Henry, R. E., "The Effect of Pressure on $NaCl-H_2O$ Explosions," Fourth CSNI Specialist Meeting on Fuel-Coolant Interactions in Nuclear Reactor Safety, CSNI Report 37, 1979.

53. Buxton, L. D., Nelson, L. S., and Benedick, W. B., "Steam Explosion Triggering and Efficiency Studies," Fourth CSNI Specialist Meeting on Fuel-Coolant Interactions in Nuclear Reactor Safety, CSNI Report 37, 1979.

54. Baker, L. Jr., "Core Debris Behavior and Interactions with Concrete," Nuclear Engineering and Design, Vol. 42, 1977.

15 ACCIDENT INITIATION AND CLAD RELOCATION

by

M. Epstein

ANL, Argonne, IL, USA

This chapter summarizes the concerns which relate to initiation of a core disruptive accident. Specific emphasis is placed on a mechanistic approach through analysis of separate effects to specify the initial conditions leading to the onset of clad motion followed by molten clad dynamics.

1. INTRODUCTION

Two years ago the author had the honor of writing a short monograph on heat transfer topics related to liquid-metal fast breeder reactor (LMFBR) safety technology [1]. The present lecture on the same subject is developed from this review article. As in the earlier survey, we attempt to follow the accident sequence resulting from a loss-of-flow (LOF) accident in a small LMFBR as a basis for identifying the major heat-transfer problems and the current state of the efforts to solve them. The lecture is divided into two chapters. The specific topics of importance to the initiation of the LOF accident, namely sodium boiling and clad relocation, are discussed in this chapter. The next chapter is devoted to phenomena presently thought to be important to the meltdown phase of the accident. For example, the topics of boiling flow regimes for fuel-steel mixtures and the freezing of such mixtures in the cold pin structure above and below the core region are subsections of the next chapter.

While the organizational approach is similar to that found in Ref. [1], the discussion that follows is considerably more detailed. Since it would take many chapters, if not volumes, to do justice to all work on the LOF accident, the scope of this lecture has been limited to presenting only the most fundamental studies of the two-phase flow and heat-transfer phenomena occurring in the LOF accident sequence. Some of these phenomenological studies have provided the basis for the development of mathematical models which have been incorporated into accident analysis codes. The nature of the lecture is largely determined by the fact that the phenomena are described in terms of simple, one-dimensional physical pictures. Presentations are purposely made restrictive; it is felt that this will help the reader to grasp the underlying ideas. Accordingly, we discuss neither numerical computations that deal with several interrelated or multi-dimensional phenomena, nor computer codes that attempt to describe the progression of the accident from beginning to end. Complex in-pile or out-of-pile experiments are introduced only when they are needed to put the more basic studies in perspective. Topics outside the domain of heat transfer, such as neutronics for molten fuel regions and the physical chemistry of fission gas release, are also excluded except when they have special relevance for completing the description of particular phenomena.

It should also be mentioned that in the present lecture a detailed account of who did what in fast reactor safety heat-transfer research has been avoided. Most of the review is based on research performed here at Argonne National

Laboratory. Other workers in fast reactor safety research will need no reminding that the names which feature most in an author's list of references are not necessarily those of the most important contributors to the subject. The author's excuse for referring so often to "Argonne work" stems from a desire to present as unified and coherent an account of the heat transfer phenomena as possible, and the picture would have become blurred if sufficient space were taken to do justice to other workers or organizations whose ideas or formulations, equally valid though they may be, differ from those contained herein.

Although some knowledge of boiling, melting and freezing processes is presupposed, a section in each chapter presents selected aspects of these subjects. Their incompleteness is recognized.

2. THE LOSS-OF-FLOW ACCIDENT AND THE TRANSITION PHASE

An analysis of the initiating phase of an unprotected flow coastdown accident in a small reactor like the Fast Flux Test Facility Reactor (FFTF) has been reported in Ref. [2]. The accident involves assuming that the coolant pumps lose power and that the plant protective system fails to scram the reactor. The analysis indicates that the reactor subassemblies successively progress through the stages of coolant (liquid sodium) boiling and voiding followed by melting of the cladding and fuel. Individual subassemblies begin to lose their identity as they become engulfed by a growing region of molten fuel and steel. At this point, the accident enters what is termed the transition phase [2]. The pressures generated within the molten region are too low to cause a massive dispersal of the molten fuel from the core region; however, sufficient local axial expansion of the molten region takes place to lower the reactivity and maintain the reactor in a subcritical condition. Thus the molten fuel-steel region is continuously heated by fission and decay processes, but there is no energetic excursion that leads to a permanent disassembly of the core. (Should a hydrodynamic disassembly occur, the transition phase would be over.)

Extended fuel dispersal may be inhibited by partial or total channel plugging in the relatively cold "unheated" regions above and below the active core zone owing to solidification of either molten clad or fuel. Continued decay heating causes further core melting and boilup of the fuel-steel mixture. The mixture should "fill" the available volume and, if passages leading from the core are blocked, rapid melting attack on the upper and lower structures will begin, ultimately allowing paths to develop for the removal of molten core material. This transition phase boil-out process should result in little or no damage to the reactor vessel. Clearly, the transition phase can be divided into three study areas: coolant motion, clad motion, and fuel motion. In this chapter and in the next chapter, we will present the results of basic analytical and experimental research on specific phenomena related to the motion of these materials during the transition phase.

Before leaving this section, it is worthwhile to mention that the transition phase may arise in a variety of hypothetical accident situations. As mentioned above, one example is a flow coastdown accident without protective scram. Another one would involve a transient overpower accident where the reactor core is partially voided without introducing high reactivity insertion rates. Almost complete flow blockage within a single subassembly could also lead to a situation similar to the transition phase, although the scale would be much smaller. Generally, the molten fuel-steel region may be any size from a single subassembly up to the whole core.

3. TWO-PHASE FLOW REGIMES

Almost all the phenomena predicted to take place under LMFBR loss-of-flow accident conditions involve the flow of vapor-liquid mixtures. The two-phase flow may be the consequence of sodium boiling within the intact reactor core shortly after the inception of flow decay, or it may be the result of vaporization of molten steel droplets embedded within the growing region of fuel melt that forms following the ejection of the sodium coolant from the reactor core. In either case the movement of reactor core material (sodium, steel clad, and fuel) is influenced by the prevailing two-phase flow pattern. Since the intact LMFBR core is not in its most reactive configuration, there is always the possibility that such material motion can lead to prompt-critical reactivity excursions and hydrodynamic disassembly of the reactor core. Thus the neutronic stability of the reactor during accident conditions is essentially dependent on the maintenance of a particular two-phase flow pattern. It is therefore desirable to make brief mention of the relevant flow patterns and the mechanism that controls the transition from one flow pattern to another.

In our discussion of flow regimes we will ignore the vapor-liquid phase change process, as some of the most important features of boiling two-phase mixtures are solely the results of fluid dynamic effects. The flow regimes that may occur in boiling fuel-steel mixtures can be illustrated by the process of injecting gas through a porous surface at the bottom of a stagnant pool of liquid (see Fig. 1). Successively, three distinct modes of behavior have been observed as the gas flow rate is increased [3,4,5]:

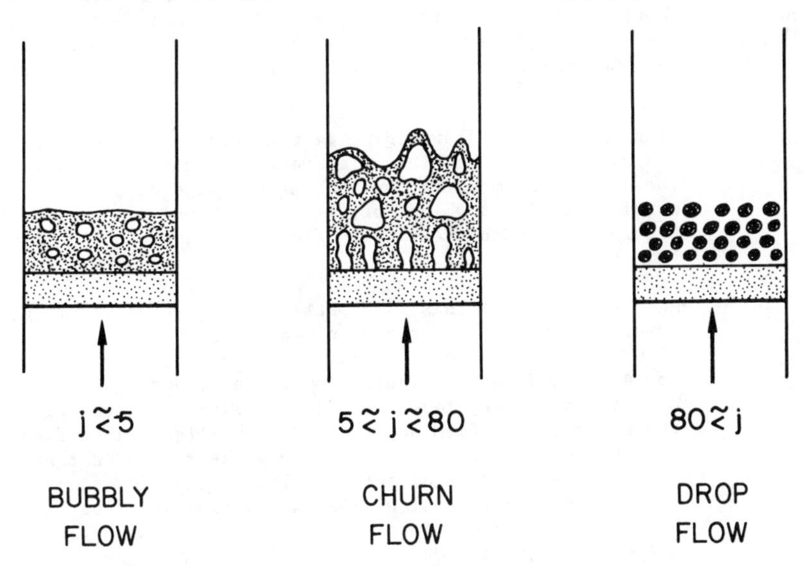

Fig. 1. Two-phase flow regimes formed by passing fluid through a porous plate at the base of a vertical container of large diameter. The volumetric flux j (cm/sec) refers to air-water mixture.

(i) Bubbly flow: Individual pores in the plate surface operate independently as bubble sources, producing a rising column of bubbles. Increases in the gas flow rate results in an increased population of bubbles in the pool. In shallow liquid pools, the bubble population in the pool can increase to the point where the gas volume (void) fraction approaches 60 percent.

(ii) Churn flow: The dispersed bubbles become so closely packed that they eventually agglomerate into large bubbles in order to accommodate the increased gas flow rate. With further increases in gas flow, the flow pattern becomes unsteady and chaotic. Large gas slugs rise through the pool in this regime resulting in regions of localized unsteady recirculation and intermittent contractions or chugging of the pool. The churn pattern is somewhat similar to the slug pattern observed when gas-liquid mixtures flow in a vertical pipe (see below); it is, however, much more chaotic and disordered.

(iii) Droplet flow: At very large gas flow rates, the liquid phase breaks up into droplets which are supported by the gas stream. Note that the transition from churn to drop flow is characterized by the liquid changing from the continuous phase to the discontinuous phase.

Unlike boiling or bubbling in stagnant liquid pools, two-phase flow processes in pipes or channels usually involve net motion of both the gas (or vapor) and liquid phases. Flow regimes associated with upward two-phase flow in a vertical pipe or channel are of particular interest to transition phase analysis, specifically in the area of sodium boiling and subsequent molten clad motion. The flow patterns described above for gas injection in liquid pools have also been observed in vertical channel flows. Two additional flow patterns have been identified for two-phase flow in long vertical channels [6,7] (see Fig. 2):

(iv) Slug flow: This flow regime is characterized by a series of individual large bubbles which have a diameter almost equal to the channel diameter. The bubbles are separated by a slug of continuous liquid which "bridges" the channel. Between the bubble and the channel liquid flows downwards in the form of a thin film.

(v) Annular flow: In this flow pattern the gas (or vapor) moves upward in the center or core of the channel while a liquid film flows downward along the wall (contercurrent flow). However, for a fixed downward liquid flow there is a critical gas velocity at which very large waves of amplitude several times the thickness of the film appear on the gas-liquid interface. This results in a large increase in the pressure gradient and the liquid flow reverses direction. This transition from countercurrent to cocurrent flow is known as flooding (see Figure 2). Flooding is of great importance both to sodium boiling and molten clad relocation.

It is important to note that quite often two or more flow patterns will be present simultaneously. For example, in annular flow the gas core usually contains a significant number of entrained droplets so that the flow regime may be best described as annular drop flow.

The physical process that can lead to flow regime transitions such as bubbly to churn, churn to droplet dispersed, or slug to annular has been clearly identified by Kutateladze [8]. The flow transition mechanism emerges from the well-known fact that wind blowing over a liquid surface generates

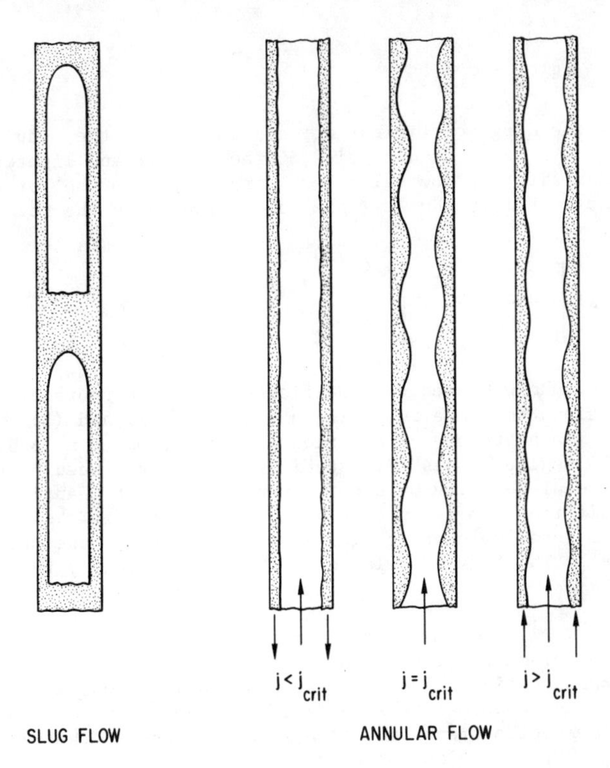

Fig. 2. The slug and annular flow regimes in a vertical pipe; showing the transition from countercurrent to cocurrent annular flow.

surface waves. The flow of a gas or vapor relative to the surface of the liquid in gas-liquid mixtures may therefore be expected to generate surface waves by the same type of instability phenomenon. A change in flow regime should occur when the ratio of amplitude to wavelength for these waves becomes too large for the crests to remain an integral part of the liquid structure. Based on wind-wave theory, the maximum particle dimension, λ, for the discontinuous phase within the two-phase mixture is given by

$$\lambda \sim \frac{\sigma}{\rho_c U^2} \tag{1}$$

where σ is the liquid surface tension, ρ_c is the density of the continuous phase and U is the local velocity of the lighter phase relative to the heavier phase. In writing equation (1), it was assumed that viscous effects are absent in the two fluids. Criterion (1) is often referred to as the Weber number criterion for the breakup of drops or the deformation of bubbles in a fluid stream, or as the Kelvin-Helmholtz wavelength of (capillary wave) disturbances that result in the destruction of fluid jets or films.

In steady state the relative velocity U is expressed in terms of the balance between buoyancy and drag forces:

$$C_D \frac{\rho_c U^2}{2} \sim \lambda g(\rho_H - \rho_L) \quad (2)$$

which defines the drag coefficient C_D, and where g is the gravitational constant and ρ_H and ρ_L are the densities of the heavier and lighter fluid, respectively. Substituting the value for λ from (1) into the force balance condition, the stability (Kutateladze) condition takes the form

$$U_{crit} = K' \left[\frac{(\rho_H - \rho_L)g\sigma}{\rho_c^2} \right]^{1/4} \quad (3)$$

where K' is an unknown numerical coefficient that incorporates C_D and many of the approximations that are implicit in equations (1) and (2). The critical velocity U_{crit} as expressed by equation (3) is the maximum local velocity at which one flow regime "loses" its stability and is replaced by another flow regime. The local velocity U is not a very convenient quantity to work with since it is difficult to measure in any experimental rig. Often it is possible to measure the total volumetric flux j of the lighter phase relative to the heavier phase.† The flux is related to the local velocity by

$$j = \alpha U \quad (4)$$

where α is the volumetric concentration of the lighter phase averaged over the whole flow cross section. In two-phase gas-liquid flows, α is termed the void fraction and j is known as the superficial gas velocity. Assuming that a particular flow regime transition always takes place at the same value of α, we can replace U in equation (3) with j:

$$j_{crit} = K \left[\frac{(\rho_H - \rho_L)g\sigma}{\rho_c^2} \right]^{1/4} \quad (5)$$

where the product $K'\alpha$ is abbreviated as K.

From the derivation, it is clear that equation (5) represents an extremely rough result. As might be expected this criterion for the onset of flow regime change is not universally valid and will not apply in certain specific two-phase flow situations. Moreover, numerous competing criterion exist for predicting flow regime transitions, and a completely general synthesis, as suggested by criterion (5), is yet to be achieved. It may, however, be remarked that a rather large fraction of the available data is in good agreement with this simple theory (see below). Criterion (5) will prove valuable in providing insight into trends pertaining to coolant dynamics, and cladding and fuel relocation phenomena.

The stability constant K takes on various values depending on the flow regime transition under consideration. The values of K for a number of flow regime transitions of importance to transition phase processes are given in Table 1 along with the appropriate literature citations. The flow regimes are illustrated in Figs. 1 and 2. Table 1 was first suggested by Fauske [12,13]

---

† The volumetric flux is the total volume fluid flow rate divided by the flow cross sectional area.

for use in fast reactor safety applications. A complete table of K values is given in Ref. [8].

Table 1. Values of the Stability Parameter K

Transition	K and ρ_c	Reference
Formation of churn flow from bubbly flow	$\sim 0.3;\ \rho_c = \rho_H$	[6]
Onset of droplet dispersion at the surface of a liquid pool or the fluidization of a heavier liquid by a lighter liquid.	$\sim 0.2;\ \rho_c = \rho_L$	[5,9]
Flooding (reversal of direction) of falling liquid films in an upward gas flow when channel diameter is not important (see Section 9).	$\sim 3.0;\ \rho_c = \rho_L$	[10,11]

Several of the overall properties of boiling fuel-steel mixtures can be predicted if one can describe the upward volumetric gas flux j for a given flow pattern. As with flow regime transitions, much of what is known about the superficial velocity j comes from studies of the steady injection of gases through the bottom of vertical liquid columns or pools (as illustrated in Fig. 1). A wide variety of gas injection data is well correlated by an empirical equation of the simple form:

$$j = v_\infty \alpha^m (1 - \alpha)^n \tag{6}$$

where α is the gas void fraction, v_∞ is a characteristic gas velocity which is proportional to the terminal velocity of a single particle (droplets or bubbles) in an infinite fluid medium, and m and n are phenomenological quantities to be determined from experiment.

When gas is bubbled through a stagnant liquid at low flow rates, Wallis [6] recommends the relation

$$j = 1.2 \left[\frac{(\rho_H - \rho_L) g \sigma}{\rho_H^2} \right]^{1/4} \alpha(1 - \alpha) \tag{7}$$

The bubbly flow pattern is usually found to break down at approximately α = 0.4. Substituting this value into equation (7) we recover criterion (5) for the onset of churn flow from bubbly flow (see also Table 1). In other words, equation (7) represents the superficial gas velocity over the whole range of possible void fraction values 0-0.4, while criterion (5) represents only the departure from the bubbly flow regime at α = 0.4. It should be mentioned that these conclusions are based mostly on experience with gas introduced into tap water through a porous plate. As pointed out by Wallis [4], the void fraction range over which the bubbly flow pattern is stable is particularly sensitive to impurities and the method in which bubbles are produced. The addition of certain surface active agents will allow bubbly flow to persist up to practically 100 percent void fraction in the form of a stable foam. On the other hand, the transition from bubbly to churn flow at about 10 percent void fraction in distilled water has been reported by Wallis [4]. Transition to churn flow also occurs at low void fractions in tap water when the gas is introduced through large orifices [3].

According to Zuber and Hench [3], the superficial velocity of the gas in the churn regime is well represented by

$$j = 1.53 \left[\frac{(\rho_H - \rho_L) g \sigma}{\rho_H^{\ 2}} \right]^{1/4} \quad \frac{\alpha}{1 - \alpha} \tag{8}$$

The transition from the bubbly flow regime to the churn regime is not a sudden and dramatic instability. Instead there exists a transitional, developing flow pattern in which churn flow gradually replaces bubbly flow as the gas flow rate is increased [4]. It is not possible to conduct repeatable measurements of j versus α in this transitional regime and therefore it is indescribable analytically. Unfortunately, the transitional flow pattern may persist over a large range of gas flow rates. This is discussed in more detail in the next chapter.

Wallis [6] suggested the following expression for the superficial gas velocity necessary to sustain a droplet cloud:

$$j = 1.4 \left[\frac{(\rho_H - \rho_L) g \sigma}{\rho_L^{\ 2}} \right]^{1/4} \alpha^2 \tag{9}$$

He maintains that a change in the flow pattern might be expected at the value of α below which the droplets pack together. For spherical droplets this occurs at $\alpha \simeq 0.4$. Substituting this value into equation (9) we recover the minimum superficial velocity for fluidization as given by equation (5) and Table 1. It should be cautioned, however, that there is no direct experimental evidence in support of equation (9) for suspension of drops in gases; it is based exclusively on data taken in liquid-liquid systems.

It is interesting to note that by introducing an experimentally supported two-phase viscosity into equation (2), Ishii and Zuber [14] correctly predicted the superficial velocity-void fraction relationships given above for bubbly, churn and droplet flows.

4. INCIPIENT BOILING SUPERHEAT

For the postulated LOF accident in the FFTF, boiling of the liquid sodium coolant within the volume heated fuel region (active core region) of the reactor begins approximately 10 to 15 sec after the inception of flow decay. The early sodium void growth is largely dictated by the temperature at which the process of boiling can start. Thus the first problem area pertinent to the LOF accident and, therefore, pertinent to the transition phase is the degree of liquid superheat required to initiate boiling in liquid sodium. In laboratory experiments superheats as high as 500°C have been observed in liquid sodium and potassium metals. However, it has been convincingly demonstrated that incipient-boiling superheats in fast reactors will be very low, owing to the effects of inert gas trapped in wall cavities or circulating in the form of bubbles.

In boiling of any liquid on an engineering surface bubbles arise, as a rule, from certain preferred cavities, cracks, or scratches on the surface. Presumably, this is due to their ability to trap vapor or gas and thus promote boiling. The degree of liquid wetting can be of decisive importance to the effectiveness of a cavity in trapping and maintaining gas over a long period of time. The degree of wetting can be ascertained from a measurement of the contact angle θ. If this angle is measured from the surface of the cavity to the interface between the liquid and gas through the liquid phase, then the wetting is more complete the smaller θ. Complete wetting corresponds to $\theta = 0$, complete non-wetting to $\theta = 180°$.

When a system is filled with a liquid, the liquid may or may not penetrate to the bottom of a representative cavity. A cavity that completely fills with liquid cannot serve as a nucleation site. Bankoff [15] developed a simple criterion for the wetting requirements for a cavity to retain gas during the advancement of a liquid front across the cavity. This criterion should provide the conditions for a cavity or scratch to be a stable nucleation site after filling a metal (reactor) loop with liquid sodium. We consider the advancement of a semi-infinite region of liquid across a scratch

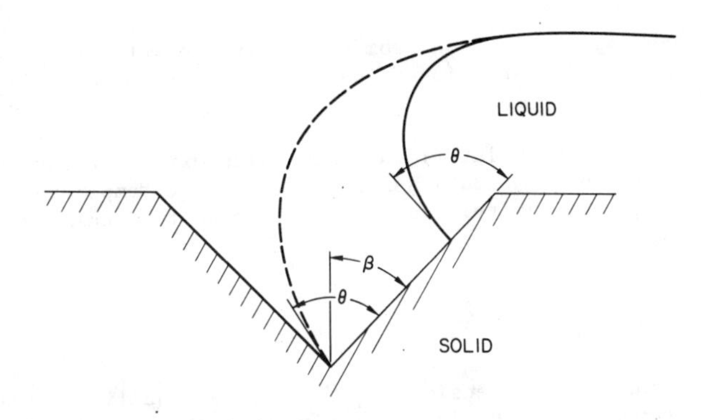

Fig. 3. Condition for the entrapment of gas in the advance of a liquid flow across a V-shaped groove. From Bankoff [15].

with a wedge-shaped cross-section (see Fig. 3) such that the angle of inclination with the vertical is β. The geometric conditions for gas entrapment is readily seen to be

$$\theta > 2\beta \tag{10}$$

This result implies that scratches on well-wetted surfaces will never serve as nucleation sites for boiling since $\theta \sim 0°$. The ability of liquid sodium to wet metal surfaces at temperatures above 300°C [16] means that all the usual V-shaped scratches found on metal surfaces will be filled with liquid sodium, suggesting that a very large liquid superheat will be required to initiate boiling. This, however, is plainly contrary to observations of sodium boiling in experimental loops.

Holtz [17], in an attempt to reconcile the apparent contradiction just cited, devised a simple model for estimating the incipient-boiling superheat for sodium in contact with a metal surface. Initially, during the fill stage, the cavity surfaces are assumed to be covered with oxide, and hence poorly wetted. The sodium therefore, will trap gas and will tend not to penetrate the cavities. Figure 4 represents the sequence of events postulated by Holtz.

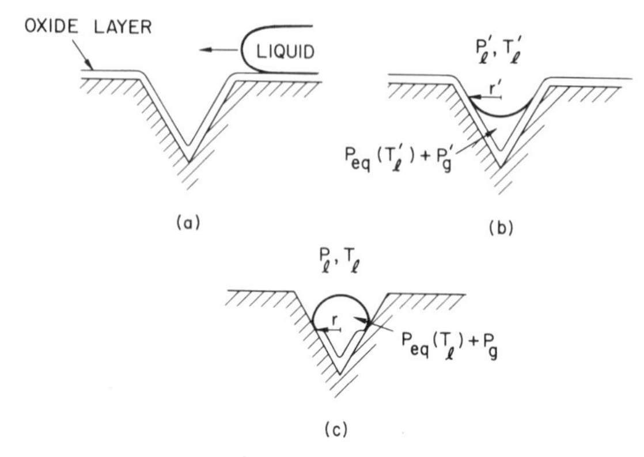

Fig. 4. Transition from nonwetting to wetting via the dissolution of metal oxide. From Holtz [17].

Figure 4a shows the cavity interiors coated with oxide. Figure 4b shows a non-wetting condition ($\theta \sim 180°$) as liquid sodium at temperature T'_ℓ and pressure P'_ℓ enters the cavity. In this fill stage the meniscus equilibrium condition is

$$P'_\ell - P_{eq}(T'_\ell) - P'_g = \frac{2\sigma(T'_\ell)\cos\beta}{r'} \tag{11}$$

where $\sigma(T'_\ell)$ is the surface tension, $P_{eq}(T'_\ell)$ is the equilibrium vapor pressure at the liquid temperature, P'_g is the pressure exerted by the inert gas and r' is the radius of the cavity at the base of the meniscus. As the temperature is raised and/or the pressure is lowered the oxide layer is continuously dissolved by the liquid sodium until complete wetting is achieved ($\theta \sim 0°$) at the inception of boiling (see Figure 4c). At the boiling temperature, T_ℓ, and pressure, P_ℓ, the equilibrium conditions for the meniscus is now

$$P_g + P_{eq}(T_\ell) - P_\ell = \frac{2\sigma(T_\ell)\cos\beta}{r} \tag{12}$$

Assuming that the meniscus does not wander during the switch from non-wetting to wetting, $r' = r$, and we obtain

$$P_{eq}(T_\ell) = P_\ell - P_g + \frac{\sigma(T_\ell)}{\sigma(T'_\ell)}\left[P'_\ell - P_{eq}(T'_\ell) - P'_g\right] \tag{13}$$

Using the sodium vapor pressure curve to connect P_{eq} with T_ℓ we can calculate the liquid temperature T_ℓ required to initiate boiling. Equation (13) reveals the importance of knowing the pressure, temperature, and gas content conditions when the system is filled or during operating conditions prior to the heating transient. For this reason Holtz's model has been denoted the pressure-temperature history model.

Bankoff [18] and Deane and Rohsenow [19] pointed out that it is not necessary to assume that active surface cavities start out non-wetted but change to wetted at incipient boiling. Cavities with sharp-edged entry regions, as diagrammed in Fig. 5, can act as nucleation sites regardless of the

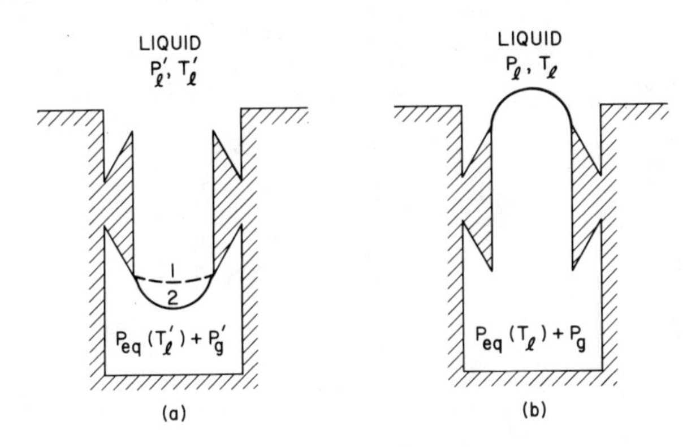

Fig. 5. Sharp-edged cavity. From Bankoff [18].

degree of wetting of the surface. When liquid penetrates downward into the cavity, the gas-liquid interface at the lower edge assumes different consecutive shapes of which two are indicated in Fig. 5a. The radius of curvature of the earliest of these is very large so that the pressure difference across the meniscus is small. The radius of curvature has a minimum value when the bubble is a hemisphere. Curve 2 in Fig. 5a corresponds to this state of maximum capillary pressure. Thus the cavity with the smallest neck radius such that the meniscus forms a hemisphere at the lower edge should function as a nucleation site. Upon heating the system, the increasing vapor pressure in the cavity must overcome the same maximum capillary pressure to expel the vapor past the upper sharp edge (Fig. 5b). The resulting equation is then identical to equation (13) given by Holtz.

Henry and Singer [20] and Logan et. al. [21] have presented some experimental measurements of incipient boiling superheat of sodium flowing in a heated channel. In the work reported in [20], the system pressure was held constant at the liquid boiling pressure, i.e., $P_\ell = P'_\ell$, and the activations of wall cavities were accomplished by increasing the temperature from T'_ℓ up to T_ℓ at boiling inception. In the experiments described in [21] the boiling event was initiated by simultaneously increasing the temperature from T'_ℓ to T_ℓ and decreasing the pressure from P'_ℓ to P_ℓ. Henry [22] analyzed the data of [20] and [21] by assuming that a complete loss of history occurs after boiling is terminated and that the conditions governing the next boiling event are dictated by the operating (pressure and temperature) conditions between each run. He further assumed sufficiently clean surfaces and flowing sodium so that an insignificant amount of gas phase is present. Thus the Holtz formula (13) reduces to

$$P_{eq}(T_\ell) = P_\ell + \frac{\sigma(T_\ell)}{\sigma(T'_\ell)}\left[P'_\ell - P_{eq}(T'_\ell)\right] \qquad (14)$$

The data are compared with equation (14) in Figs. 6 and 7, and the agreement

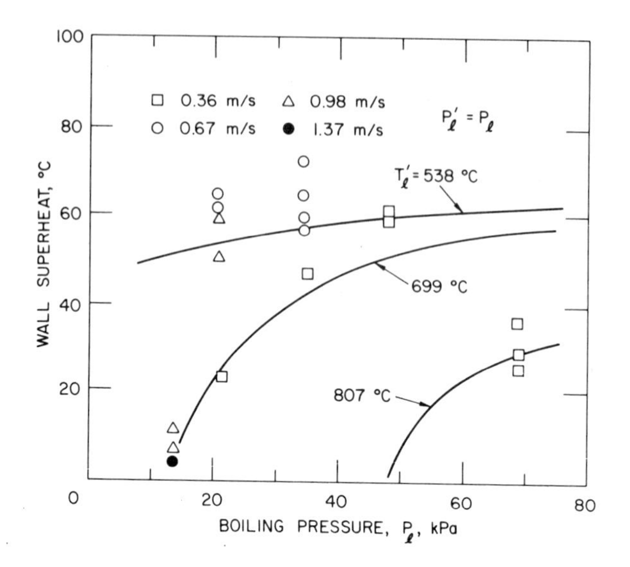

Fig. 6. Comparison of pressure-temperature history model, equation (14), with convective sodium superheat data of [20]. From Henry [22].

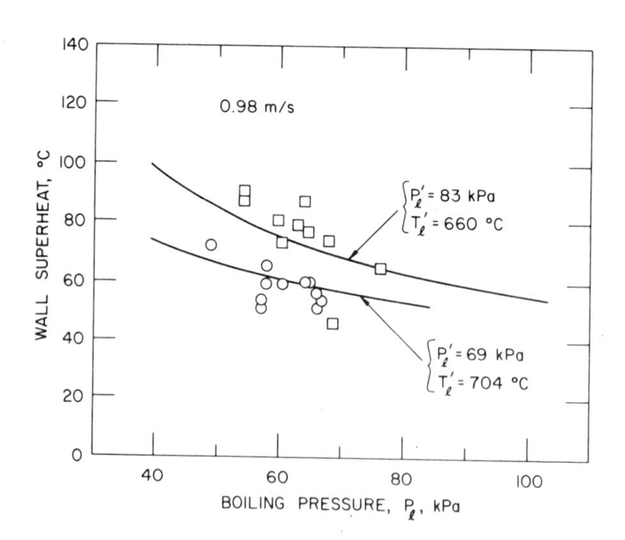

Fig. 7. Comparison of pressure-temperature history model, equation (14), with convective sodium superheat data of [21]. From Henry [22].

is good over the sodium flow velocity range investigated. Thus the memory effects are adequately described by Holtz's model.

The pressure-temperature history model was exploited by Holtz, Fauske and Eggen [23] to predict incipient boiling superheats occurring during various postulated LMFBR accidents. During normal reactor operation, liquid sodium passes through the core and enters an upper plenum region containing warm sodium with a free surface in contact with argon gas. Since the solubility of argon in sodium increases with increasing temperature, the sodium is likely to be saturated with dissolved argon at the plenum temperature. The liquid sodium then flows from the plenum into the heat exchanger where it is cooled and therefore supersaturated with argon. Here the dissolved argon may precipitate from the sodium in the form of minute gas bubbles. These bubbles tend to re-dissolve in the sodium as it is heated in flowing back through the reactor core. Depending on the initial size of the bubbles, they may disappear in the core region. In order to estimate incipient-boiling superheats Holtz, Fauske and Eggen [23] ignored the possibility of gas bubble formation and assumed that the inert argon gas remains in the liquid sodium solution. The sodium superheats required for incipient boiling in reactors of the pool design were calculated for both total subassembly flow-blockage and power-excursion type accidents. In the pool design, the maximum superheat required to initiate boiling was estimated to be 48°C for the flow blockage and 23°C for the power excursion type accident. Results of the calculations for the loop system for both accident conditions indicate that boiling from surface cavities will occur at essentially zero superheat.

As mentioned above, gas bubbles in the flowing sodium as well as surface cavities which remain voided over a long period of time will be present in real reactor systems. If the cavity radius falls within the range of the gas bubble radius distribution the cavities may ultimately achieve equilibrium with the gas bubbles,† making it very difficult to determine the source of boiling, as bubbles or cavities may act as nucleation sites. In either case, the magnitude of the incipient-superheat temperature should be approximately the same.

Liquid sodium superheat experiments were performed by France et. al. [24] in a forced convection loop. The test section was designed to simulate a single reactor fuel element. Sodium flowed vertically upward through an annulus formed by a cylindrical heater surrounded by a steel tube and entered a plenum equipped with an argon cover gas. In all tests, boiling was initiated in a manner similar to a loss-of-flow accident situation. Test series were conducted utilizing a plenum temperature representative of pool type reactors, ∿360°C, and a higher plenum temperature of about 470°C as maintained in a LMFBR loop type system. Superheat data taken under pool type LMFBR conditions compared well with the Holtz pressure-temperature history model. The experiments clearly demonstrated that in a loop system operating at steady-state for a period of time sufficient to saturate the sodium with argon gas the superheat at the onset of boiling is zero. Of course, boiling initiation under the condition of zero superheat is most desirable in an LMFBR because it minimizes the potential for high voiding rates.

† A steady state situation is established in which the rate of transport of gas from a cavity to large bubbles is identically equal to the rate of transport of gas from small bubbles to the cavity.

5. FLOW REGIME FOR BOILING SODIUM

As mentioned earlier, in order to describe the sodium coolant motion during LOF accident conditions a knowledge of the flow regime is required. Fauske [13] predicted the flow patterns for boiling sodium in the narrow fast-reactor channels under accident conditions by fairly straightforward application of the two-phase flow concepts outlined in Section 3. While this process is generally of a highly transient nature, a quasi-steady state calculation should be sufficient to indicate the flow patterns that must be considered in a more complex transient analysis. From a simple energy balance written for the boiling sodium in the reactor channel, the superficial sodium-vapor velocity j at the edge of the boiling zone of instantaneous length Z is given by

$$j = \frac{\dot{p}Z}{A\rho_v h_{v_\ell}} \tag{15}$$

where $\dot{p}$ is the reactor power density in Watt/m, A is the flow cross-sectional area, ρ_v is the vapor density, and h_{v_ℓ} is the latent heat of evaporation. It is assumed here that all the power generated is removed by sodium evaporation. By eliminating j between equations (5) and (15) we obtain the power density-boiling length criterion that can be used to delineate the various flow regime boundaries that can occur in channel boiling:

$$\dot{p}Z = A\rho_v h_{v_\ell} K \left[\frac{(\rho_H - \rho_L)g\sigma}{\rho_c^2}\right]^{1/4} \tag{16}$$

The stability constant K is obtained from Table 1 of Section 3.

The flow regime transitions have been conveniently displayed by Fauske [13] in the form shown in Fig. 8 for a nominal power density $\dot{p}_o$ = 33.0 kW/m and a total active core height Z_o = 0.92 m. Fauske postulated that the onset of drop fluidization, which represents the breakdown of the churn-bubbly flow pattern in liquid pools or columns ($K \sim 0.3$; $\rho_c = \rho_H$), corresponds to the flow regime transition from bubbly-churn flow to slug flow in the reactor channel. This transition is represented by the lower curve in Figure 8. The transition from slug to annular flow is reasoned to occur when the superficial vapor velocity exceeds the critical flooding velocity ($K \sim 3.0$; $\rho_c = \rho_L$). It is clear from Fig. 8 that liquid sodium boiling in LMFBR channel geometry is dominated by the slug and annular flow regimes rather than the bubbly-churn flow regime encountered in high pressure light-water reactors.

It is also seen from Fig. 8 that the analysis of sodium voiding should be divided into two parts: (1) the early voiding period requiring slug flow analysis and (2) the final stages of voiding which is best described by the annular flow pattern.

6. SLUG FLOW: THE CYLINDRICAL BUBBLE MODEL

In the analysis of slug flow it is assumed that the overall flow pattern can be predicted if the growth of a single slug bubble can be described. This is accomplished by considering a control volume containing one bubble and part of the liquid slugs on each side of it.

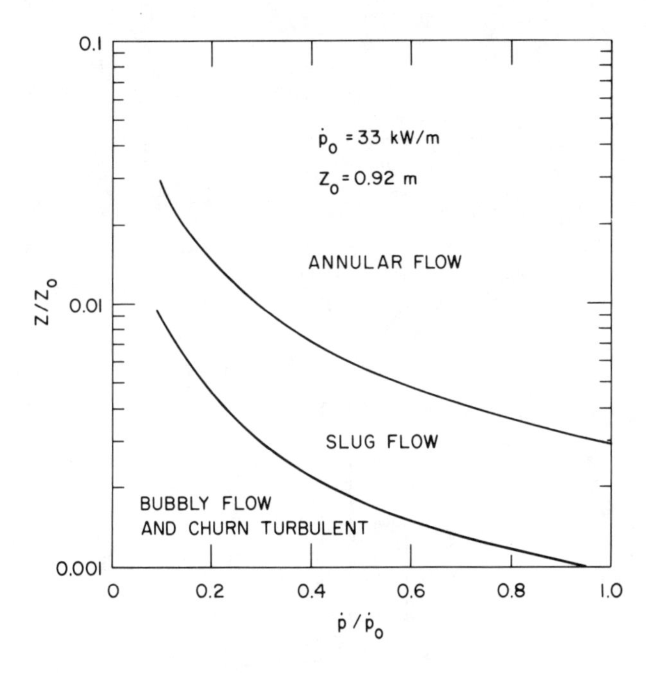

Fig. 8. Flow regime map for sodium boiling in a channel; LMFBR conditions. From Fauske [13].

One of the most successful models for the growth of a slug bubble was developed by Ford, Fauske and Bankoff [25]. The model was first applied by these workers to the configuration shown in Fig. 9, which is a schematic of the test section used in their experiment. The test section consisted of a constant-diameter tube capped by a variable diameter plenum. A rapid depressurization technique was used to initiate slug bubble growth in superheated liquid Freon-113 within the tube. The plenum and upper portion of the tube were provided with a cooling jacket, while the lower part of the tube was enclosed within a heating jacket. This arrangement simulates the continuous heat generation in the active core region and the relatively cold axial blanket zone of a fast reactor. During the experiment, the vapor bubble was observed to act like a piston, "pushing" the liquid out of the tube into the plenum as it expands (see Fig. 9), leaving behind a residual liquid Freon film on the tube wall.

The analysis of Ref. [25] is based on the following important assumptions: (1) the vapor bubble growth begins as an infinitesimally small strip or disc of vapor occupying the entire channel (see Fig. 10) with the exception of any film remaining on the tube wall; (2) the liquid film occupies only a small fraction of the cross sectional area of the tube; (3) radial temperature gradients in the upper and lower liquid slugs are neglected; (4) the vapor bubble pressure and temperature are considered to be spatially uniform and the vapor in the bubble is assumed to be in equilibrium with the liquid at the bubble surface; and (5) the thermal boundary layer thickness in the film is small compared with the film thickness and there is negligible drainage within the film.

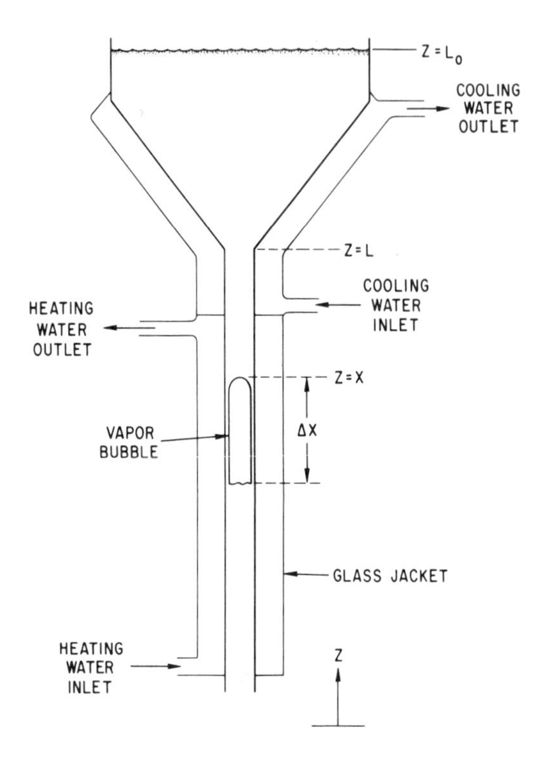

Fig. 9. Schematic diagram of the experimental apparatus for cylindrical bubble growth and liquid slug expulsion. From Ford, Fauske, and Bankoff [25].

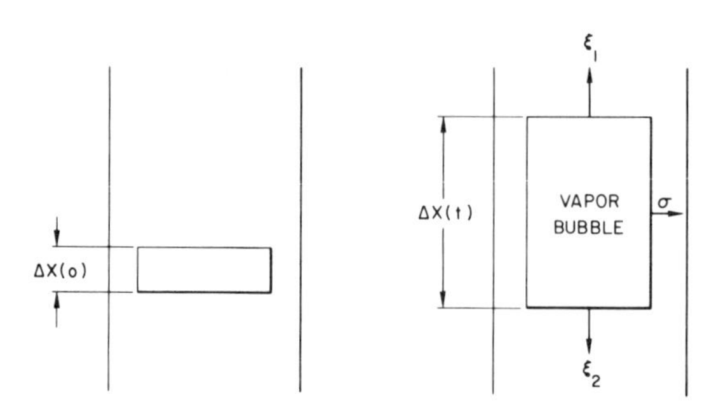

Fig. 10. Model for cylindrical bubble growth. From Ford, Fauske and Bankoff [25].

The one-dimensional equation of motion for the velocity v of the liquid slug within the tube can be integrated once to give

$$\frac{dv}{dt} = \frac{(P_v - P_o)}{\rho_\ell (L - x)} + \frac{v^2}{2(L - x)} - \frac{g(L_o - x)}{L - x} - \frac{2fv^2}{D} - g \tag{17}$$

where P_v is the bubble vapor pressure, P_o is the plenum pressure, f is the single phase coefficient of friction, D is the tube diameter, ρ_ℓ is the liquid density, g is the gravitational constant, x is the location of the upper bubble interface, L is the length of the circular tube, L_o is the location of the liquid surface in the plenum, and t is the time. The quantities x, L and L_o are measured from the bottom of the tube. In arriving at the above equation, the frictionless Bernoulli equation was used to eliminate the pressure at z = L in favor of the height of the plenum. A macroscopic energy balance for the vapor bubble shown in Fig. 9 yields

$$\rho_v h_{v\ell} A \frac{d\Delta x}{dt} + A\Delta x \frac{d}{dt} (\rho_v h_v) = kA \frac{\partial T}{\partial \xi_1} (o,t) + kA \frac{\partial T}{\partial \xi_2} (o,t)$$

$$+ k \int_{A_f} \frac{\partial T}{\partial \sigma} (o,t,t') \, dt + A\Delta x \frac{dP_v}{dt} \tag{18}$$

where ρ_v is the vapor density, $h_{v\ell}$ is the heat of vaporization, A is the cross-sectional area of the tube, Δx is the instantaneous axial bubble length, T is the temperature within the liquid slugs or the liquid film, k is the liquid thermal conductivity, h_v is the enthalpy of the vapor, A_f is the area of the liquid film in contact with the bubble, and ξ_1, ξ_2, and σ are the distances in the upper and lower liquid slugs and the liquid film, respectively, measured from the vapor-liquid interface. In the third term on the right-hand side of equation (18), t' is the time of arrival of the vapor bubble at location z = x. Because of the axial motion of the bubble, fresh liquid-film surface area is continuously exposed to the vapor; therefore, the local temperature gradient at the film surface is a function of both t and t'. This behavior is quite different from spherical bubble growth where the temperature gradient in the liquid does not depend on the voiding history.

The temperature gradients at the upper and lower liquid-vapor interfaces are found by solving the heat-conduction equation. For the upper slug this has the form

$$\alpha \frac{\partial^2 T}{\partial \xi_1^2} (\xi_1,t) + \frac{Q(\xi_1,t)}{\rho_\ell C_p} = \frac{\partial T}{\partial \xi_1} (\xi_1,t) \tag{19}$$

subject to the conditions

$$T(\xi_1,0) = F(\xi), \quad T(o,t) = \phi(t), \quad T(\infty,t) < \infty \tag{20}$$

where α and C_p are the liquid thermal diffusivity and heat capacity, respectively, $Q(\xi_1,t)$ represents the "volumetric" heat input to the liquid slug from the wall, $F(\xi)$ is the axial temperature profile in the tube at the onset of bubble formation and $\phi(t)$ is the equilibrium bubble vapor temperature determined by the pressure within the bubble. Equation (19) can be easily solved by the Laplace transformation method [26]. The result for the temperature gradient is:

$$\frac{\partial T}{\partial \xi_1}(o,t) = -\int_o^t \frac{1}{\sqrt{\pi\alpha(t-\tau)}} \frac{d\phi(\tau)}{d\tau} d\tau$$

$$+ \int_o^\infty \int_o^t \frac{\xi_1}{2k\sqrt{\pi\alpha(t-\tau)^3}} \exp\left[\frac{-\xi_1^2}{4\alpha(t-\tau)}\right] Q(\xi_1,t)d\tau d\xi_1$$

$$+ \int_o^\infty \frac{F(\xi)}{2\alpha} \frac{\xi_1}{\sqrt{\pi\alpha t^3}} \exp\left[\frac{-\xi_1^2}{4\alpha t}\right] d\xi_1 - \frac{\phi(0)}{\sqrt{\pi\alpha t}} \qquad (21)$$

The heat conduction equation and corresponding initial and boundary conditions for the liquid film separating the vapor bubble from the tube wall are

$$\alpha \frac{\partial^2 T}{\partial \sigma^2}(\sigma,t,t') = \frac{\partial T}{\partial t}(\sigma,t,t') \quad \text{for } t > t' \qquad (22)$$

$$T(\sigma,t',t') = T_w; \quad T(\infty,t,t') = T_w; \quad T(o,t,t') = \phi(t) \qquad (23)$$

Again, using the Laplace transformation method, we find for $t > t'$

$$\frac{\partial T}{\partial \sigma}(o,t,t') = \frac{T_w - \phi(t')}{\sqrt{\pi\alpha(t-t')}} - \int_{t'}^t \frac{1}{\sqrt{\pi\alpha(t-t'-\tau)}} \frac{d\phi(\tau)}{d\tau} d\tau \qquad (24)$$

Subject to assumptions (1) - (5), equations (17), (18), (21) and (24) describe the evolution of a cylindrical bubble in a vertical channel. An iterative scheme was developed to solve this system and is reported in [25]. Briefly, the method of solution consisted of estimating the temperature, $\phi(t)$, at the bubble surface which determines the bubble pressure. Equation (17) is then integrated over one time-step to determine the position of the upper surface of the bubble. This initial estimate is systematically refined so that the resulting trial solution satisfies the energy equation (18) as well as the momentum equation (17).

Figure 11 shows the comparison between the predicted and experimental bubble growth for a liquid superheat of 33°C. It is seen that the model predicts the liquid expulsion process very well. The two other solutions exhibited in Fig. 11 correspond to the cases of inertial-controlled bubble growth and heat-conduction controlled growth. It is apparent that these limiting cases predict bubble expulsion rates that are considerably larger than indicated by experiment. Note that an appreciable departure from the familiar square-root growth law for spherical bubbles is obtained with cylindrical bubble growth. The continuous exposure of fresh liquid surface in cylindrical growth results in growth rates that increase with time.

The cylindrical bubble model presented above has been incorporated into the SAS accident analysis code to describe the early stages of sodium voiding [27]. Similar models have been used to predict sodium coolant displacement due to rapid fission gas release [28] or molten fuel-coolant interactions following fuel pin failures [29,30]. However, the kinetics of fission gas

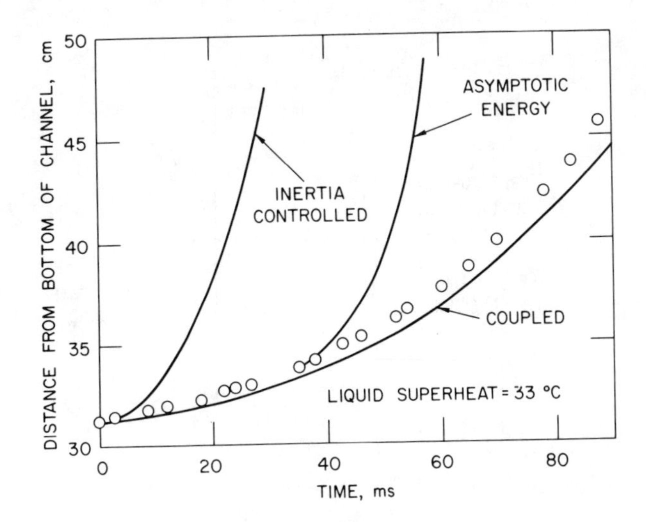

Fig. 11. Distance of upper liquid-vapor interface from bottom of tube versus time: comparison of theory with experiment. From Ford, Fauske and Bankoff [25].

release and sodium vapor production in these cases is so poorly understood that these analyses are highly parametric in nature.

7. LIQUID SODIUM FILM THICKNESS

As noted in the previous section, one of the important characteristics of the cylindrical-bubble slug expulsion is the liquid film left behind on the heated walls. While realistic sodium voiding rates can be calculated with only limited knowledge of the magnitude of the film thickness, no reliable estimate of the occurrence of film dryout nor clad melting can be made.

To determine the residual film thickness, Fauske, Ford and Grolmes [31] conducted an experimental study of the liquid film thickness during cylindrical bubble growth in glass tubes. Liquid Freon 11 or 113 was uniformly superheated and cylindrical bubble growth was initiated at the base of the tube, as shown in Fig. 12. In these experiments, the free liquid surface (slug) displacement was photographed simultaneously with the vapor bubble displacement. Since the liquid film drainage is small, mass continuity requires that

$$VA = A_v V_v \tag{25}$$

where A and A_v are the cross sectional areas of the tube and vapor bubble, and V and V_v are the upward velocities of the liquid slug and the vapor bubble, respectively. We can calculate the void fraction α from the definition

$$\alpha \equiv \frac{A_v}{A} = \frac{V}{V_v} \tag{26}$$

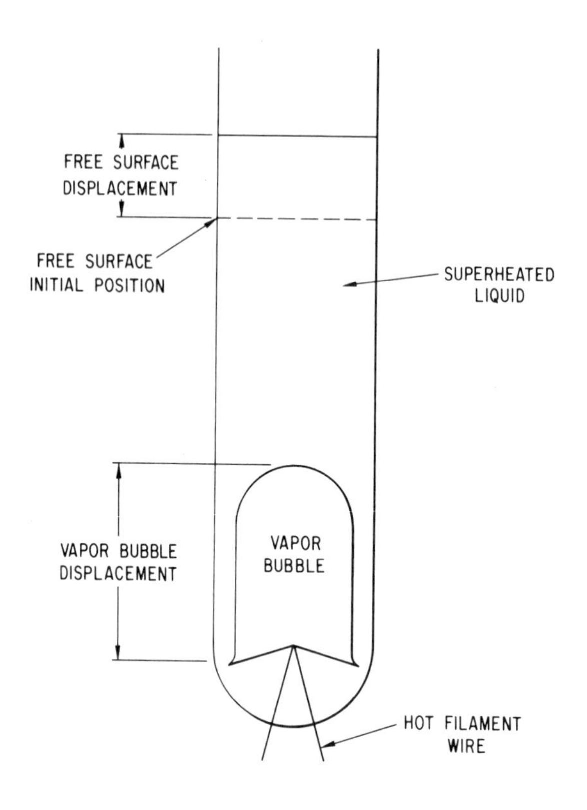

Fig. 12. Schematic diagram of the experimental apparatus for slug expulsion of a superheated liquid. From Fauske, Ford and Grolmes [31].

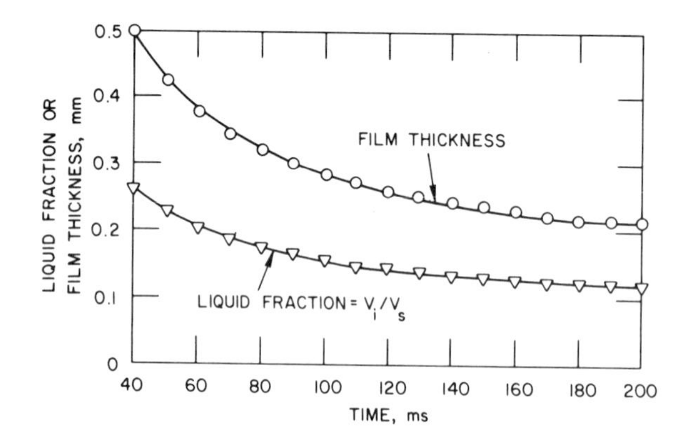

Fig. 13. Film-thickness versus time measurement for slug expulsion of superheated liquid Freon-113. From Fauske, Ford and Grolmes [31].

The velocities V and V_v were determined from the measured displacement data. Figure 13 shows typical liquid fraction-and film thickness-time histories. It should be noted that over the range of superheats investigated in this study (5 to 50°C), the liquid fraction $(1 - \alpha)$ approached an asymptotic value of approximately 0.15. Fauske et. al. [31] suggested that this is because the flow in the liquid ahead of the bubble is turbulent and that the bubble velocity is determined by the (maximum) velocity in the turbulent core. On the other hand, the liquid slug velocity represents the average velocity. Therefore, the ratio V/V_v in equation (26), if assumed to be the ratio of the average-to-maximum for turbulent flow, takes on a constant value independent of flow velocity.

8. ANNULAR FLOW REGIME FOR BOILING SODIUM

During a flow coastdown accident, the flow pattern immediately following the onset of sodium boiling is best described as a multiple-bubble slug flow. That is, during early sodium voiding, there are a number of separate vapor bubbles and liquid slugs in the coolant channel (see Fig. 2). This flow pattern rapidly changes into bidirectional cocurrent annular flow in which a single, long "bubble" occupies the flow channel, ejecting coolant in both the upward and downward directions from the active core region. The initial value of the void fraction, α, required to start the solution of the equations for two-phase annular flow, is determined by the thickness of the liquid film left behind during the coolant expulsion, as discussed in Section 7.

While the pressure within the slug-flow bubble is taken as uniform (Section 6), an axial pressure gradient will develop within the expanding annular-flow bubble as the vapor velocity within the bubble increases. The axial pressure gradient not only affects the motion of the liquid slugs but the behavior of the liquid film as well. As discussed in Section 5, the transition to the annular flow regime occurs when the sodium vapor velocity exceeds the critical flooding velocity, and, therefore, large waves appear on the film which give rise to net axial transport of liquid within the film in the direction of vapor flow. An approximate estimate of the film motion can be obtained by writing the appropriate momentum equations for the liquid film velocity V_ℓ and the vapor flow velocity V_v. Assuming quasi-steady incompressible flow and neglecting gravity, these are

$$\frac{dP}{dz} + \frac{4}{D}\tau_w = 0 \qquad (27)$$

for both film and vapor and

$$\sqrt{\alpha}\,\frac{dP}{dz} + \frac{4}{D}\tau_i = 0 \qquad (28)$$

for the vapor, where dP/dz is the pressure gradient over the vapor length z, D is the hydraulic diamter of the channel, and τ_w and τ_i are the shear stresses at the wall and at the vapor-liquid interface, respectively. The shear stress at the wall should be about the same as it is for single-phase all-liquid flow:

$$\tau_w = \frac{1}{2} f \rho_\ell V_\ell^{\,2} \qquad (29)$$

where ρ_ℓ is the liquid density and the friction factor $f \simeq 0.005$. Wallis [6] developed an empirical correlation for τ_i which accounts for the "wavy" nature of cocurrent, upward annular flow. For thin liquid films in channels this is

$$\tau_i = \frac{1}{2} f\rho_v V_v^2[1 + 75(1 - \alpha)] \qquad (30)$$

where ρ_v is the vapor density.

Eliminating dP/dz between equations (27) and (28) yields

$$\frac{V_\ell}{V_v} = \sqrt{\frac{\rho_v}{\rho_\ell}} \frac{[1 + 75\ (1 - \alpha)]^{1/2}}{\alpha^{1/4}} \qquad (31)$$

A simple mass balance on the liquid film over the vapor length z results in an expression for the rate of decrease of the film thickness, or, more conveniently, the rate of increase of the void fraction due to axial film transport:

$$\frac{d\alpha}{dt} = \frac{V_\ell(1 - \alpha)}{z} \qquad (32)$$

Invoking energy balance (15) and equation (31) leads to the differential equation:

$$\frac{d\alpha}{dt} = \frac{[1 + 75(1 - \alpha)]^{1/2}(1 - \alpha)}{\alpha^{5/4}} \frac{\dot{P}_o}{Ah_{v\ell}\sqrt{\rho_v\rho_\ell}} \qquad (33)$$

Since the void fraction varies between 0.85 and 1.0, we may, as a first approximation, set $\alpha^{5/4} \simeq 1.0$ in equation (33). Integrating the result from the initial liquid fraction $(1 - \alpha) = 0.15$ at time $t = 0$ to the time at which the liquid fraction is ten times less than this initial value yields

$$t_{DRY} = \frac{1.2\ Ah_{v\ell}\sqrt{\rho_v\rho_\ell}}{\dot{P}_o} \qquad (34)$$

The quantity t_{DRY} may be regarded as the time span between the onset of voiding and the appearance of local dryout spots on the clad surface. The estimated dryout time for typical LMFBR conditions is approximately 30 msec. This estimate of clad dryout, based on axial film transport, was first reported by Grolmes and Fauske [32]. Following film dryout, cladding temperatures rise rapidly and melting starts within about 1 second under nominal reactor conditions.

9. MOLTEN CLAD DYNAMICS

Recapitulating, boiling in LMFBR channel geometry is dominated by the slug and annular flow regimes rather than the bubbly flow regime encountered in high-pressure light-water reactors. As a result, sodium coolant is ejected in both the upward and downward directions from the active core region, leaving behind a residual liquid sodium film on the steel clad fuel pins. Rapid removal of this film by sodium vapor shear forces leads to clad dryout followed by clad melting. Meanwhile, intermittent contact between the lower liquid sodium level and the bottom of the heated fuel section (chugging) results in sodium evaporation and upward streaming of sodium vapor. The vapor passes through the active core zone and condenses in the vicinity of the upper liquid sodium level, now located well above the active core region at the top of the fuel pins (see Fig. 14). Flow of the liquid sodium back into the active core

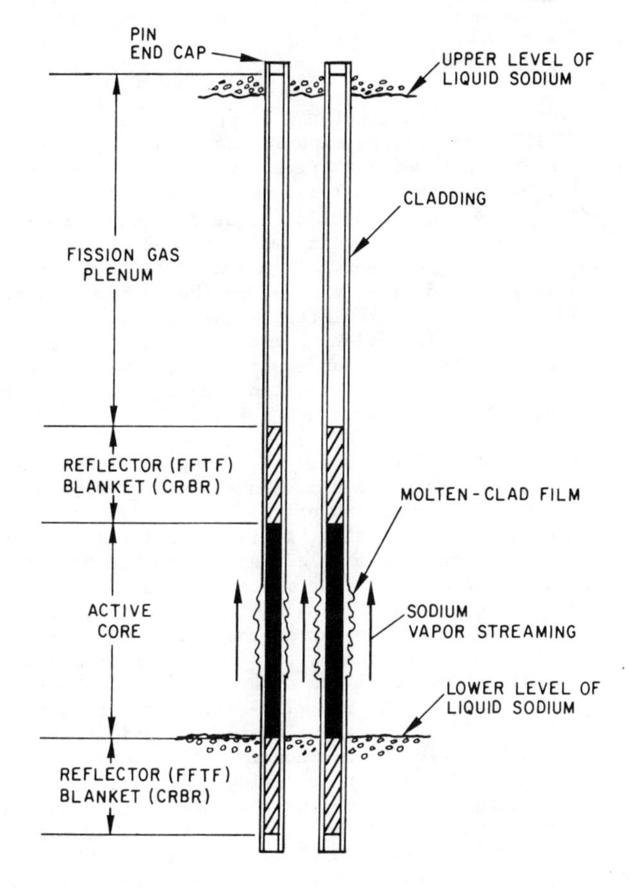

Fig. 14. Schematic of LMFBR subassembly channel at the inception of molten clad motion.

zone (rewetting) will not take place as long as the rate of evaporation exceeds the rate of condensation, which is generally the case in whole-core LOF accidents.† The hydrostatic pressure head between the upper and lower liquid sodium levels (in the liquid sodium external to the voided region) dictates a sodium vapor velocity of approximately 80 m/sec. Using the Kutateladze correlation for flooding, equation (5) with K = 3 and $\rho_c = \rho_L$, the required

† The voiding characteristics described above are based on calculations performed with the SAS accident analysis computer code [27], which incorporates the physical picture for film dryout presented in the previous section together with the cylindrical bubble growth model discussed in Section 6. It is to be noted that the SAS calculated voiding behavior compares well with both out-of-pile and in-pile experiments designed to simulate the unprotected loss-of-flow accident [33,34].

velocity to bring about clad flooding is estimated to be about the same as the available sodium vapor velocity. Another semi-empirical correlation that is widely utilized to predict flooding is the correlation of Wallis [6]:

$$j_{crit} = \sqrt{\frac{gD(\rho_\ell - \rho_g)}{\rho_g}} \tag{35}$$

where D is the hydraulic diameter of the channel. The data obtained by Wallis and Makkenchery [11] from flooding experiments with air and water in vertical tubes indicate that the Kutateladze criterion (5) is appropriate for large-diameter channels, but for the small channel dimensions of interest in LMFBR applications, flooding estimates should be based on equation (35). Using equation (35), we predict clad flooding when the sodium vapor velocity exceeds 20 m/sec. Thus it is expected that molten clad will begin to move in the upward direction as a liquid film, similar to the sodium film motion discussed in Section 8. The problem is formulated by asking what the ensuing clad motion is. Knowledge of this motion is important to assessment of (i) reactor reactivity feedback effects, (ii) subsequent molten fuel motion, and (iii) partial or total channel plugging if molten clad reaches the unheated regions above and below the active core zone and solidifies.

The first attempt to analyze molten cladding motion was made by Bohl and Heames [35]. Their model is a one-dimensional annular flow model and is designed to fit within the framework of the SAS accident analysis code. To fulfill this purpose, the molden clad is divided into discrete axial segments which can combine with other molten cladding segments. Due consideration is given to the existence of waves on the molten cladding surface, and the Wallis [11] film-gas interaction model of annular two-phase flow, equation (30), is used to calculate the frictional shear at the sodium vapor-molten clad interface. A more sophisticated version of this model was presented in [36]. A one-dimensional annular flow model was also developed by Ishii, Chen, and Grolmes [37]. Again the shear at the film-sodium interface is represented by equation (30). The results using these models indicate that sustained clad motion in the upward direction is unlikely. The cladding which melts first moves upward under the influence of streaming sodium vapor, as illustrated in Fig. 15a. This cladding material begins to freeze above the heated region, resulting in a larger hydraulic resistance to both sodium vapor flow and further upward penetration of molten cladding. Consequently, downward flow of additional melting clad occurs. Complete (frozen) clad plugs are predicted below the heated fuel region. These results are in qualitative agreement with loss-of-flow experiments conducted in the TREAT reactor [38,39]. Post-test analysis showed that most of the clad material drained to form a substantial plug at the lower end of the fuel section, and a much smaller upper blockage was observed above the fuel section.

Cladding motion experiments using a single-pin, simulated cladding-relocation apparatus were carried out by Henry et. al. [40]. The objective of these experiments was to examine the applicability of available flooding correlations, e.g., equation (35), which are based on air-water data, to the high-density, high surface-tension molten cladding. As shown in Fig. 16, Wood's metal "clad" was cast into an undercut section in a copper tube and then machined down to the outer diameter of the copper tubing. Steam was forced to flow into the top of the copper tubing, then through the section coated with Wood's metal, and finally out through the bottom of the copper tubing. In this manner the upper end of the Wood's metal section melted first, as would be expected under actual core meltdown conditions. Argon flowed vertically upward through an annular region between the 20 cm-long

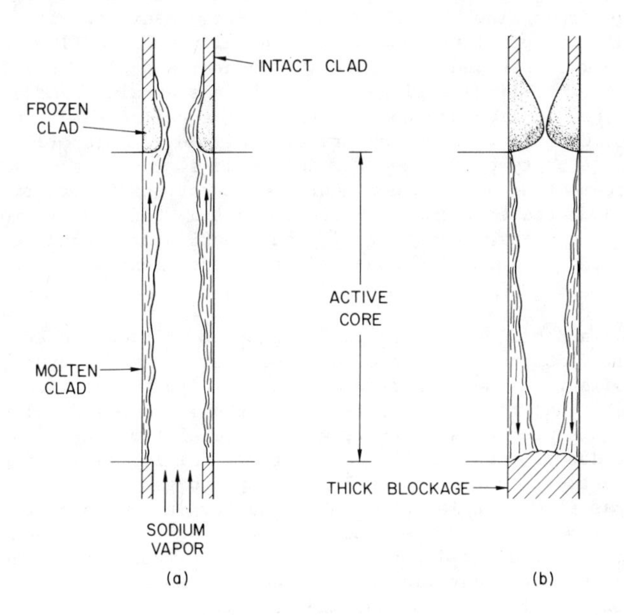

Fig. 15. Clad flooding and deflooding process: (a) upward clad motion (flooding); (b) downward clad motion (deflooding).

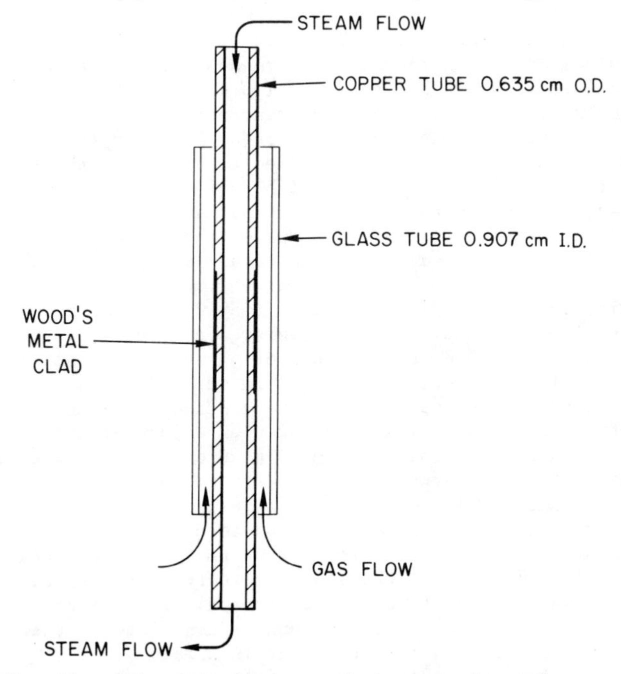

Fig. 16. Schematic diagram of the experimental test section for simulated clad motion in a cylindrical annulus. From Henry et al. [40].

section of 0.038-cm Wood's metal cladding (inner wall) and a glass tube (outer wall). A large argon flow bypass was provided to insure constant inlet pressure throughout an experiment. At a flow velocity of about 10.0 m/sec, the molten cladding was observed to bridge the channel and attempted to flow upward. However, the liquid bridge across the gas annulus decreased the gas flow rate and the cladding drained. When the gas velocity was increased to 18 m/sec, upward cocurrent flow was observed. Using expression (35), we predict a critical gas velocity of 9 m/sec for this Wood's metal-argon system, which is in good agreement with the experimental observations. More recently, Henry et. al. [41] conducted experiments using argon and Wood's metal in a ten-pin bundle arranged in a triangular array. The flooding data obtained with these multi-pin experiments were also found to be in agreement with the predicted gas velocities.

Theofanous and his coworkers, in a series of critical papers [42-45], provided detailed observations of clad relocation dynamics beyond the stage of incipient flooding. These workers were unhappy with the application of steady-state two-phase annular flow concepts to the problem of clad relocation. They remarked that this problem differs from traditional flooding studies which involve the sudden reversal in flow of a steady falling film along the inner surface of a vertical tube or channel in an upward gas flow. In conventional studies, the gas flow rate is increased gradually until flooding followed by upward liquid film motion is observed. Theofanous et. al. argued that different behavior might be anticipated for molten cladding, however, since in this case an initially stationary liquid film suddenly appears (due to melting) in a gas flow that far exceeds the flooding velocity. Moreover, the liquid film is not constantly fed into the system from some external source so that upward clad relocation causes film thinning and consequently any initial irregularities or waves on the surface of the film will tend to decay.

The experimental studies of Theofanous et. al. [42-45] were conducted in a vertical channel having a rectangular cross section. In their early studies [42,43], a uniform liquid film flow (ethyl alcohol) down one side of the channel was established first. A countercurrent flow of air was then established, and the flow rate was increased gradually while the film flow behavior was observed. This experiment was modified to allow for the effects of an initially stationary melt film with a high surface tension like that of stainless steel [44]. Wood's metal alloy was used as the cladding simulant. The metal was cast as a solid film on one side of the channel ("semi-annular" geometry). After an upward air flow was established, the metal film was rapidly melted by irradiating the backside of the channel wall with an infrared lamp bank. The most recent investigation [45] was performed with a melt film on both sides of the channel ("fully-annular" flow, as shown in Fig. 17). This work included detailed measurements of (1) molten metal film relocation velocities, (2) film thicknesses during film depletion and (3) pressure drops over the melt zone. These quantities were continuously monitored throughout the transient. Each channel wall consisted of a 15 cm-long, 0.038 cm-thick Wood's metal cladding simulant cast on to a 46 cm-long, thin steel plate. The transient liquid film measurements were based on the electrical resistance of the melt film-channel wall composite. Six voltage taps were connected to the channel wall over the solid film region and to a point 6 cm above the film for the film thickness measurements. Four pressure taps penetrated the narrow test section sidewall to allow transient pressure drop measurements over both the film region and the length of the test section. An important point established in these runs was the need for good wetting between the melt film and the downstream substrate in order to maintain a film flow. Untreated test section surfaces resulted in local metal film accumulation and entrainment of the film as it advanced upwards [44]. The wetting behavior of the system was

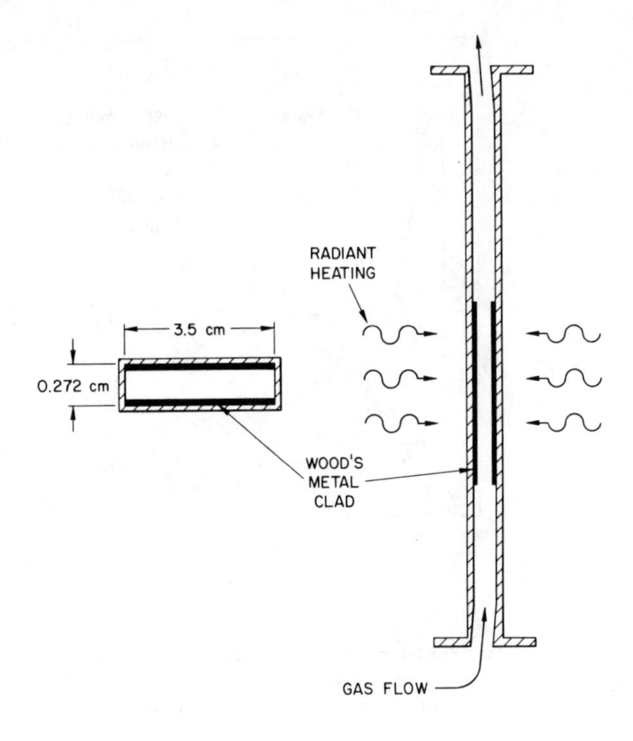

Fig. 17. Schematic diagram of the experimental test section for simulated clad motion in a large aspect channel. From Marrotte and Theofanous [45].

improved by coating the test section walls with a thin layer of Wood's metal (∿0.0005 cm). In some of the runs, one of the film-coated test plates was removed and replaced with an acrylic window in order to "capture" the relocation process with high-speed photography (semi-annular runs). Refreezing of the molten metal film did not occur since the heating lamps were left on during the transient.

Once again, Wallis' steady-state correlation (35) for the onset of flooding (flow reversal) was found to be accurate even under conditions of transient flow of a high-surface tension melt film. The final disposition of the film mass N_m (normalized by the initial solid film mass) is shown in Fig. 18 as a function of the dimensionless superficial gas velocity j/j_{crit}, with j_{crit} given by equation (35). It is seen that the conditions $j > j_{crit}$ corresponds to upward clad relocation in the form of a melt film, and that the fraction of the mass remaining in the original area of the solid film peaks at the flooding condition $j = j_{crit}$. Since the film thickness was continuously measured during the transient, the frictional pressure drop as a function of void

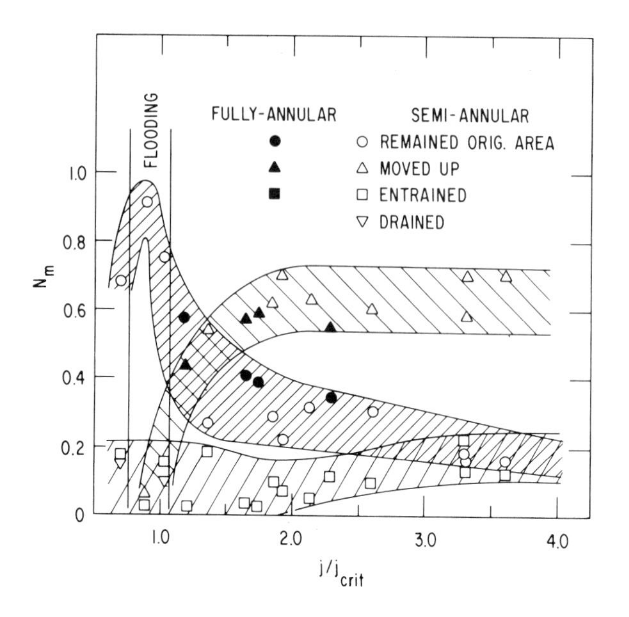

Fig. 18. Postmortem distribution of clad film mass (normalized) versus j/j_{crit} for all runs. From Marrotte and Theofanous [45].

fraction could be determined and compared with that given by steady-state cocurrent annular flow theory; namely

$$\frac{dP}{dz} = -\frac{4}{D} \cdot \frac{1 + 75(1 - \alpha)}{\sqrt{\alpha}} \cdot \frac{1}{2} f\rho_v V_v^2 \tag{36}$$

Equation (36) is obtained from equations (28) and (30). It is convenient to make (36) dimensionless by forming the ratio

$$\phi \equiv \frac{dP/dz}{(dP/dz)_{\alpha=1}} \tag{37}$$

where $(dP/dz)_{\alpha=1}$ is the frictional pressure gradient for all-gas flow through the same channel with the same mass flow rate as the two-phase flow.† For the slit geometry of interest here, it is readily demonstrated that

$$\phi = \frac{1 + 75(1 - \alpha)}{\alpha^3} \tag{38}$$

Figure 19 shows a comparison between equation (38) and the pressure drops observed by Marrotte and Theofanous [45] under sustained upward clad relocation conditions. The measured frictional pressure drops fall well below

† This ratio between the frictional pressure gradients is usually known as the "two-phase multiplier".

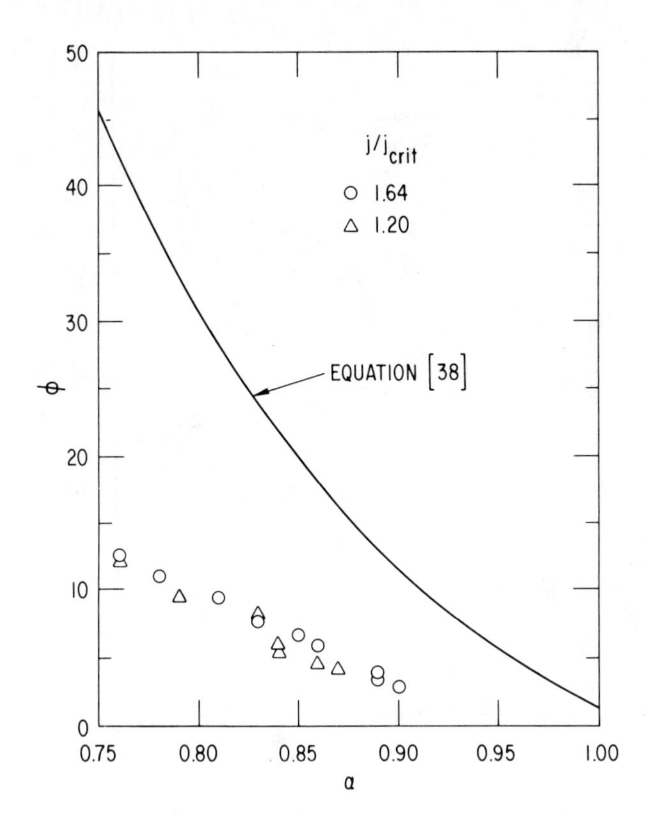

Fig. 19. Two-phase multiplier versus void fraction. From Marrotte and Theofanous [45].

those based on equation (36). The reduced pressure drops appear to be quite consistent with the small surface waves revealed by the high-speed photography. Apparently there is no mechanism to produce film accumulation and large waves. The film "stretches" as it advances so that surface disturbances which are primarily responsible for friction are bounded and a relatively smooth annular flow persists. Alternatively, damping of the waves may be due to the high surface tension of the melt film. A series of steady-state experiments with high-surface tension films in upward cocurrent flow with gas could well determine the effect of surface tension on interfacial friction. The film front was observed to advance in a jerk-like motion on a stop and go basis. This behavior was attributed to the small-amplitude wave motion superposed on the film; that is, the relocation velocity of the cladding is controlled by waves that build up on the film and then decay as they move over the clean surface in the downstream region. Figure 20 shows both the wave velocity and the film relocation velocity plotted against j/j_{crit}. The fact that the two velocities exhibit the same general trend lends support to the notion that the advancement of the film front is controlled by the wave motion.

Based on these observations, Marrotte and Theofanous [45] concluded that under reactor loss-of-flow accident conditions sustained upward clad relocation velocities should not exceed 60 cm/sec. This is well below the value predicted by the one-dimensional SAS clad relocation model [35] of several

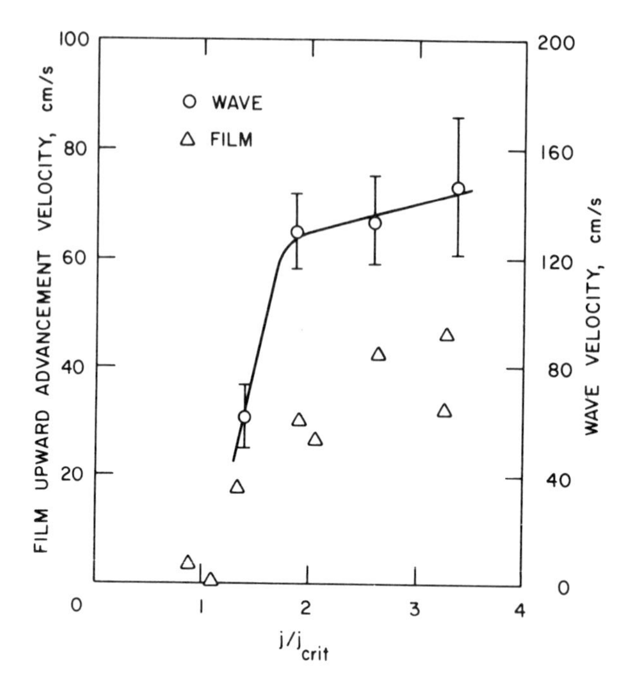

Fig. 20. Film upward advancement velocity and wave velocity versus j/j_{crit}. From Marrotte and Theofanous [45].

hundreds of cm/sec and somewhat below the value of about 90 cm/sec obtained with the more recent models developed by Bohl [36] and Ishii, Chen, and Grolmes [37].

10. MOLTEN CLAD SOLIDIFICATION

Compared with clad motion, there has been relatively little interest in clad freezing as an experimental research topic. The only experimental study to date on freezing of an annular flow is that reported by Tsinteris [46]. His apparatus is illustrated schematically in Fig. 21. The heart of the apparatus was a two-phase mixing section, which consisted of two concentric tubes. Wood's metal flowed from a tank under pressure into the annular region between the tubes. A hypodermic needle placed between the tank and the mixing section served as a control valve by which a constant metal flow rate was obtained. Nitrogen gas from a pressurized bottle was fed into the inner tube. The two fluids emerged from the top of the mixing section in the form of an annular flow and entered a freeze section made out of thick-walled stainless-steel tubing cooled by either dry ice or liquid nitrogen. The inside diameter of the freeze tube was 0.55 cm. Each run was started by first establishing the gas flow through the inner tube. The liquid metal was then allowed to flow out of the tank and into the mixing section. The annular two-phase flow entered the freeze tube where it solidified. At the end of each experiment the penetration distance into the cold tube was measured and the shape of the frozen crust was noted.

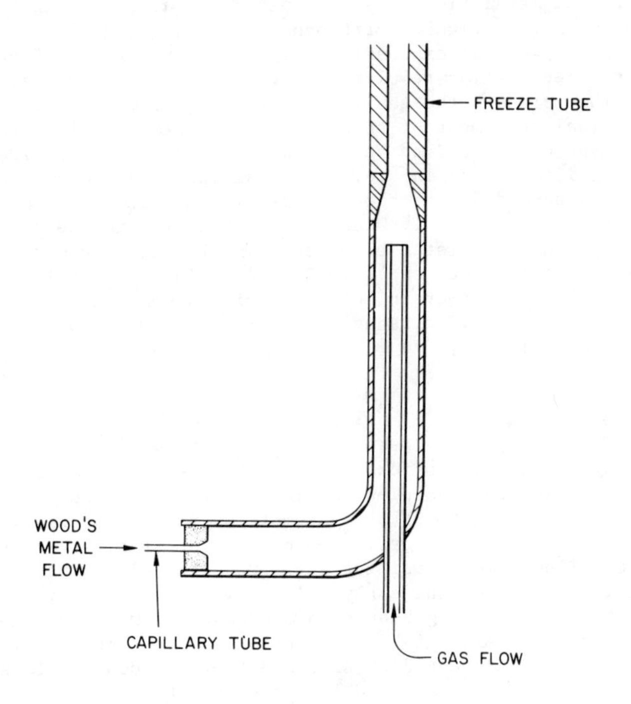

Fig. 21. Schematic diagram of the test section for the solidification of an annular two-phase flow. From Tsinteris [46].

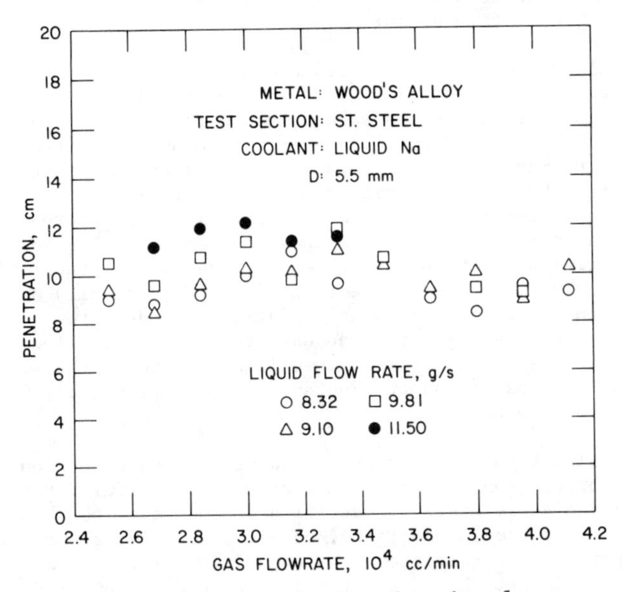

Fig. 22. Measured penetration lengths of an annular two-phase flow before freezing.

Some of the Wood's metal penetration data as a function of the gas flow velocity with the metal flow rate as a parameter are shown in Fig. 22. The gas flow rates in this figure correspond to 1.5 j_{crit} up to 2.38 j_{crit} for flooding based on Wallis' expression (35). Over the range of velocities indicated, the penetration results are seen to be somewhat insensitive to both the liquid flow rate and the gas velocity. No attempt was made by Tsinteris at a mathematical treatment of this two-phase solidification process. The frozen specimens were found to be hollow, with the crust profiles looking very much like that illustrated in Fig. 15b. Tsinteris concluded that the absence of a complete plug was due to the self-regulating nature of solidification in an upward annular flow. Early Wood's metal freezing in the cold tube results in a larger hydraulic resistance to both gas flow and further upward penetration of molten metal. When the flooding criterion $j_{crit} > 1$ is violated, the liquid flow will begin to drain leaving behind a central cavity in the frozen region. Recall that similar behavior is predicted with the one-dimensional clad motion models [36,37].

11. SUMMARY

It appears that the applicability of the pressure-temperature history model for estimating the incipient sodium superheat has been firmly established. Application of the model to LMFBR systems results in essentially zero superheats for conditions of interest. Two-phase flow stability criteria indicate that transient liquid-metal boiling in narrow channels is best characterized by the annular flow pattern; the early void growth being described by slug flow. The growth of a typical slug bubble in a channel is well described by the cylindrical bubble model, which takes full account of growth limitations due to liquid slug inertia and heat diffusion and thermal boundary layer variation in time and space. Experimental data indicate that the slug bubbles occupy the whole channel cross section except for a liquid film left on the channel walls. Because of the high reactor power density the flow of vapor and liquid film is predicted to be cocurrent due to the rather strong shear stress generated at the wavy interface between the film and the vapor core. The induced film motion leads to rapid film dryout.

Sodium boiling has been discussed above as if it occurred in a single tube or channel. The movement of fluid in the radial direction from the boiling to the nonboiling interconnected subchannels within a real reactor subassembly has been ignored. The magnitude of this two-dimensional effect is largely determined by the magnitude of the radial temperature gradient within the subassembly and the shape of the axial temperature profile downstream of the active fuel zone at the inception of sodium boiling. If temperature conditions provide for a relatively small two-phase boiling zone, the liquid sodium can flow around the central boiling channels for sometime without the rapid sodium flow decay and expulsion described in Section 8. It is generally agreed, however, that a one-dimensional single-channel treatment is adequate for describing subassembly voiding resulting from an unprotected loss-of-flow accident. For other aspects of the problem of sodium boiling including the work accomplished in Europe, the reviews of Han and Fontana [47], Peppler [48], Costa [49], and Kottowski [50] may be helpful.

On the basis of the experimental evidence cited in Section 9, the steady-state Wallis correlation for flooding inception is useful for predicting the onset of upward clad relocation under transient conditions. Under sustained flooding (cocurrent flow) conditions, however, an order-of-magnitude lower pressure drop than that predicted by standard correlations was observed. This difference has been attributed to transient wave development phenomena at the molten clad-gas interface. As with sodium boiling, clad motion in a reactor subassembly is not necessarily one dimensional. Radial melting incoherencies occur since fuel pins located away from the center of the subassembly experience an extended time to clad melting. This departure from one-dimensional melting behavior and its effect on clad motion has been addressed by Fauske [51], Henry et al. [52], Ishii and Chen [53], and Marrotte and Theofanous [45].

ACKNOWLEDGMENTS

This work was performed under the auspices of the U.S. Department of Energy. The author wishes to thank Kathy Rank for an excellent job of typing the manuscript.

REFERENCES

1. Epstein, M., "Melting, Boiling and Freezing: The Transition Phase in Fast Reactor Safety Analysis," in Symposium on the Thermal and Hydraulic Aspects of Nuclear Reactor Safety, 2, ASME Press, New York, 1977, pp. 171-193.

2. Jackson, J. F., Stevenson, M. G., Marchaterre, J. F., Sevy, R. H., Avery, R., and Ott, K. O., "Trends in LMFBR Hypothetical-Accident Analysis," Proc. Fast Reactor Safety Meeting, USAEC Report CONF-74041-P3, 1974, pp. 1241-1264.

3. Zuber, N., and Hench, J., "Steady State and Transient Void Fraction of Bubbling Systems and Their Operating Limits, Part I, Steady State Operation," General Electric Report 62GL100, July, 1962.

4. Wallis, G. B., "Some Hydrodynamic Aspects of Two-Phase Flow and Boiling," Paper No. 38, Intern. Heat Transfer Conf., Boulder, Colo., ASME, 1961.

5. Kutateladze, S. S. and Moskvicheva, V. N., "Hydrodynamics of a Two-Component Layer as Related to the Theory of Crises in the Process of Boiling," Soviet Physics: Technical Physics, 4, 1960, pp. 1037-1040.

6. Wallis, G. B., One-Dimensional Two-Phase Flow, McGraw-Hill, N.Y., 1969.

7. Hewitt, G. F. and Hall-Taylor, N. S., Annular Two-Phase Flow, Pergamon Press, 1970; see also Collier, J. G., Convective Boiling and Condensation, McGraw-Hill, 1972.

8. Kutateladze, S. S., "Elements of the Hydrodynamics of Gas-Liquid Systems," Fluid Mechanics - Soviet Research, 1, 1972, pp. 29-50.

9. Wallis, G. B., "Two-Phase Flow Aspects of Pool Boiling from a Horizontal Surface," Paper No. 3, Two-Phase Fluid Flow Symposium, Institution of Mechanical Engineers, London, 1961.

10. Pushkina, O. L. and Sorokin, Yu. L., "Breakdown of Liquid Film Motion in Vertical Tubes," Heat Transfer - Soviet Research, 1, 1969, pp. 56-64.

11. Wallis, G. B. and Makkenchery, S., "The Hanging Film Phenomenon in Vertical Annular Two-Phase Flow," Journal of Fluids Engineering, 96, 1974, pp. 297-298.

12. Fauske, H. K., "Boiling Flow Regime Maps in LMFBR HCDA Analysis," Trans. Am. Nucl. Soc., 22, 1975, pp. 385-386.

13. Fauske, H. K., "Flow Regime Map for Sodium Boiling," presented at the Sixth Liquid Metal Working Group Meeting, Risley, England, October, 1975.

14. Ishii, M., and Zuber, N., "Drag Coefficient and Relative Velocity in Bubbly, Droplet or Particulate Flows, " AIChE Journal, 25, 1979, pp. 843-855.

15. Bankoff, S. G., "Entrapment of Gas in the Spreading of Liquid over a Rough Surface," AIChE Journal, 4, 1958, pp. 24-26.

16. Todd, J. J. and Turner, S., "The Surface Tension of Liquid Sodium and its Wetting Behavior on Nickel and Stainless Steels," TRG Report 1459(R), UK.A.E.A., Risley, England, 1968.

17. Holtz, R. E., "The Effect of the Pressure-Temperature History upon Incipient-Boiling Superheats in Liquid Metals," ANL-7184, 1966.

18. Bankoff, S. G., Simulation in Boiling Heat Transfer," in Cocurrent Gas-Liquid Flow, E. Rhodes and D. S. Scott eds., Plenum Press, NY, 1969, pp. 283-301.

19. Deane, C. W. and Rohsenow, W. M., "Mechanism and Behavior of Nucleate Boiling Heat Transfer to the Alkali Liquid Metals," Report No. DSR 76303-65, MIT, Cambridge, Mass., October, 1969.

20. Henry, R. E. and Singer, R. M., "Forced Convection Sodium Superheat," Trans. Am. Nucl. Soc., 14, 1971, pp. 723-724.

21. Logan, D., Baroszy, C. J., Landoni, J. A., and Morewitz, H. A., "Studies of Boiling Initiation for Sodium Flowing in a Heated Channel," Report No. AI-AEC-12767, Atomics International, 1969.

22. Henry, R. E., "Incipient-Boiling Superheat in a Convective Sodium System," Trans. Am. Nucl. Soc., 15, 1972, pp. 410-411.

23. Holtz, R. E., Fauske, H. K., and Eggen, D. T., "The Prediction of Incipient Boiling Superheats in Liquid Metal Cooled Reactor Systems," Nucl. Eng. and Design, 16, 1971, pp. 253-265.

24. France, D. M., Carlson, R. D., Rohde, P. P., and Charmoli, G. T., "Experimental Determination of Sodium Superheat Employing LMFBR Simulation Parameters," J. of Heat Transfer, 96, 1974, pp. 359-364.

25. Ford, W. D., Fauske, H. K., and Bankoff, S. G., "The Slug Expulsion of Freon-113 by Rapid Depressurization of a Vertical Tube," Int. J. Heat Mass Transfer, 14, 1971, pp. 133-139.

26. Carlsaw, H. S. and Jaeger, J. C., Conduction of Heat in Solids, 2nd edn., Oxford University Press, Oxford, Chapter XII, 1959.

27. Stevenson, M. G., Bohl, W. R., Dunn, F. E., Heames, T. J., Hoppner, G., and Smith, L. L., "Current Status and Experimental Basis of the SAS LMFBR Accident Analysis Code System," Proc. Fast Reactor Safety Meeting, USAEC Report CONF-740401-P3, 1974, pp. 1303-1317.

28. Chawla, T. C. and Hoglund, B. M., "A Study of Coolant Transients During a Rapid Fission Gas Release in a Fast Reactor Subassembly," Nucl. Sci. and Engng., 44, 1971, pp. 320-?.

29. Cronenberg, A. W., Fauske, H. K., and Eggen, D. T., "Analysis of the Coolant Behavior Following Fuel Failure and Molten-Fuel-Sodium Interaction in a Fast Nuclear Reactor," Nucl. Sci. Engng., 50, 1973, pp. 53-62.

30. Wider, H. U., Jackson, J. F., Smith, L. L., and Eggen, D. T., "An Improved Analysis of Fuel Motion During an Overpower Excursion," Proc. Fast Reactor Safety Meeting, USAEC Report CONF-740401-P3, 1974, pp. 1541-1555.

31. Fauske, H. K., Ford, W. D., and Grolmes, M. A., "Liquid Film Thickness for Slug Explusion," Trans. Am. Nucl. Soc., 13, 1970, pp. 646-647.

32. Grolmes, M. A. and Fauske, H. K., "Liquid Film Thicknesses and Mechanism for Dryout During LMFBR Loss-of-Flow Accidents," Trans. Am. Nucl. Soc., 14, 1979, pp. 743-744.

33. Henry. E. R., Singer, R. M., Lambert, G. A., McUmber, L. M., Quinn, D. J., Spleha, E. A., Erickson, E. G., Jeans, W. C., and Parker, N. E., "Sodium Expulsion Tests for Seven-Pin Geometry," Proc. Fast Reactor Safety Meeting, USAEC Report CONF-740401-P3, 1974, pp. 1188-1201.

34. Grolmes, M. A., Holtz, R. E., Spencer, B. W., Miller, C. E., and Kramer, N. A., "R-Series Loss-of-Flow Safety Experiments in TREAT," Proc. Fast Reactor Safety Meeting, USAEC Report CONF-740401-P3, 1974, pp. 279-302.

35. Bohl, W. R. and Heames, T. J., "A Clad Motion Model for LMFBR Loss-of-Flow Accident Analysis," Trans. Am. Nucl. Soc., 16, 1973, p. 358.

36. Bohl, W. R., "CLAP: A Cladding Action Program for LMFBR HCDA Analysis," Trans. Am. Nucl. Soc., 23, 1976, pp. 348-349.

37. Ishii, M., Chen, W. L., and Grolmes, M. A., "Molten Clad Motion Model for Fast Reactor Loss-of-Flow Accidents," Nucl. Sci. Eng., 60, 1976, pp. 435-451.

38. Deitrich, L. W., Barts, E. W., DeVolpi, A., Dickerman, C. E., Eberhart, J. G., Carter, J. C., Fischer, A. K., Murphy, W. F., and Stanford, G. S., "Fuel Dynamics Experiments Supporting FTR Loss-of-Flow Analysis," Proc. Fast Reactor Safety Meeting, USAEC Report CONF-74041-P1, 1974, pp. 239-253.

39. Dickerman, C. E., Barts, E. W., DeVolpi, A., Holtz, R. E., Neimark, L. A., and Rothman, A. B., "Recent Results from TREAT Tests on Fuel, Cladding, and Coolant Motion," Trans. Am. Nucl. Soc., 20, 1975, pp. 534-535.

40. Henry, R. E., Jeans, W. C., Quinn, D. J., and Spleha, E. A., "Cladding Relocation Experiments," Trans. Am. Nucl. Soc., 18, 1974, pp. 209-210.

41. Henry, R. E., Jeans, W. C., Quinn, D. J., and Spleha, E. A., "Cladding Relocation Experiments," Proc. Intl. Mtg. on Fast Reactor Safety and Related Physics, USERDA Report CONF-761001-Vol. IV, 1976, pp. 1691-1695.

42. DiMonte, M., and Theofanous, T. C., "Cladding Relocation Dynamics: Incoherency Effects," Trans. Am. Nucl. Soc., 22, 1975, pp. 405-406.

43. Theofanous, T. C., DiMonte, M., and Patel, P. D., "Incoherency Effects in Clad Relocation Dynamics for LMFBR CDA Analyses," Nucl. Eng. and Des., 36, 1976, pp. 59-67.

44. Theofanous, T. G., Prather, W., Chen, M., Spies, T. P., and Lois, L., "Clad Relocation Dynamics--The Physics and Accident Evolution Implications," Proc. Intl. Mtg. on Fast Reactor Safety and Related Physics, USERDA Report CONF-761001-Vol. IV, 1976, pp. 1697-1706.

45. Marrotte, G. N. and Theofanous, T. G., "Fundamental Aspects of Molten Clad Relocation," paper presented at the Specialists" Workshop on Predictive Analysis of Material Dynamics in LMFBR Safety Experiments, Los Alamos, NM, March 13-15, 1979; see also Marrotte, G. N., M.S. Thesis Purdue Univ., May 1978.

46. Tsinteris, A., "Solidification of Liquid Metals Under Forced Flow Conditions with Applications to the LMFBR Loss of Flow Accident," M.S. Thesis, Purdue University, December 1975; see also Theofanous, T. G., NUREG Report CR-0224, June 1978.

47. Han, J. T. and Fontana, M. H., "Blockages in LMFBR Fuel Assemblies," in Symposium on the Thermal and Hydraulic Aspects of Nuclear Reactor Safety, 2, ASME Press, New York, 1977, pp. 51-121.

48. Peppler, W., "Sodium Boiling in Fast Reactors: A State-of-the-Art Review," in Symposium on the Thermal and Hydraulic Aspects of Nuclear Reactor Safety, 2, ASME Press, New York, 1977, pp. 123-153.

49. Costa, J., "Contribution to the Study of Sodium Boiling During Slow Pump Coastdown in LMFBR Subassemblies," in Symposium on the Thermal and Hydraulic Aspects of Nuclear Reactor Safety, 2, ASME Press, New York, 1977, pp. 155-169.

50. Kottowski, H., "Sodium Boiling," (published elsewhere in this text).

51. Fauske, H. K., "Some Comments on Cladding and Early Fuel Relocation in LMFBR Core Disruptive Accidents," Trans. Am. Nucl. Soc., 21, 1975, pp. 322-323.

52. Henry, R. E., Jenas, W. C., Quinn, D. J., and Spleha, E. A., "28-Pin Cladding Relocation Experiments," Trans. Am. Nucl. Soc., 27, 1977, pp. 498-500.

53. Ishii, M., and Chen, W. L., "Evaluation of Incoherency Effect on Cladding Motion in R-series Test," Trans. Am. Nucl. Soc., 28, 1978, pp. 442-443.

16 FUEL MOTION

by

M. Epstein

ANL, Argonne, IL, USA

This chapter is concerned with the motion of the reactor fuel following the inception of fuel melting in the loss-of-flow accident sequence. Some phenomena connected with the hydrodynamics of boiling fuel-steel mixtures within the confines of the active core zone are considered. The significance of steel melting and fuel crust formation at the boundaries of such mixtures is discussed in some detail. Fuel motion outside the active core region involves the freezing of fuel in the channels that surround the active core zone. The current understanding of this subject is also presented in some detail.

1. BOILING FLOW REGIME MAP FOR FUEL-STEEL MIXTURES

In the LOF accident sequence, fuel melting will follow clad melting. The time interval between the complete melting of cladding and that of fuel decreases with increasing power level; in fact, at a sufficiently large power level fuel melting may precede the onset of clad melting. Clad motion studies based on the one-dimensional single-channel models (see Chapter 15, Section 9) indicate that, in general, the melting fuel begins to lose its geometry when a significant portion of molten cladding is still present in the heated fuel region. In fact, it has been reasoned that practically all of the molten cladding will remain within the active core zone owing to two-dimensional effects. The nonuniform radial cladding melting pattern that must develop in a reactor subassembly following sodium voiding (melting incoherency) led Fauske [1] to conclude that the molten cladding is, essentially, suspended by sodium vapor streaming with the net result of little upward or downward molten clad motion. He suggested that the large increase in the pressure gradient in the region of melting clad in a flooded condition leads to vapor flow diversion to outer subchannels where the cladding is not yet molten.* This bypass vapor flow will allow the molten cladding to quickly unflood and to begin to drain. However, maximum vapor flow will quickly be reestablished in the region of molten cladding, and the flooding process will begin again. Thus, it was concluded that insignificant net molten cladding motion is to be expected. Experimental verification of this possibility requires a rod bundle geometry having the appropriate length scale and a sufficient number of subchannels to simulate gas flow diversion. Accordingly, a 28-pin clad relocation experiment has been performed, using Wood's metal to simulate the cladding material [2]. The results are in agreement with the conclusion of Ref. [1].

---

* Clad melting will begin in the central part of the hottest subassemblies.

With relatively little upward or downward movement of the clad, molten fuel-steel interfaces will be established in the core region. Fuel disruption due to release of entrapped fission gas from the fuel melt is likely to lead to swelling and frothing and consequent mixing with the molten cladding. The boiling point of steel is very close to the melting point of fuel so that the molten fuel-steel mixture behaves as a saturated liquid subjected to volumetric heat generation [3]. Thus heat transfer from the fuel to the entrained steel will result in rapid steel vaporization and further dispersal and boilup of the mixture.† A detailed quantitative description of this process (sometimes referred to as the "transition to the transition phase") including the effects of fuel axial thermal expansion, buckling of fuel pellet stacks, and fission gas release seems to be too complicated to be practical. However, if we neglect these important early contributions to fuel motion, the subsequent dispersal and boilup of fuel by steel vaporization can be illustrated by fairly straightforward application of two-phase flow theory.

Fauske [5] extended the two-phase flow stability criterion suggested by Kutateladze (see Chapter 15, Section 3) for predicting changes in flow regimes to a boiling system with internal heat generation:

$$j_{crit} = K\left[\frac{(\rho_H - \rho_L)g\sigma}{\rho_c^{\,2}}\right]^{1/4} \tag{1}$$

where j_{crit} is the critical superficial velocity of the lighter fluid phase, ρ_c is the density of the continuous phase, σ is the surface tension of the heavy fluid, and ρ_H and ρ_L are the densities of the heavier and lighter fluid, respectively. The dimensionless stability parameter K takes on various values depending on the flow regime transition under consideration (see Table 1, Chapter 15).

Now in a volume-heated boiling pool, the velocity j at height z above the bottom of the pool is given by

$$j = \frac{\dot{Q}(1 - \bar{\alpha})z}{\rho_L h_{fg}} \tag{2}$$

where $\dot{Q}$ is the volumetric heat generation rate (neutron and decay heating in the reactor system) that is converted to latent heat of evaporation, h_{fg}, and $\bar{\alpha}$ is the average vapor void fraction for that portion of the pool that lies below height z. By eliminating j between equations (1) and (2), Fauske was able to construct the pertinent flow regime boundaries. The flow regime map for a fuel-steel volume-heated pool appears in Fig. 1. Fauske also constructed a map in which the flow regimes are related to required criticality dimensions for molten fuel (see Fig. 2). A most important conclusion to be derived from Fauske's work is that a criticality configuration is only possible below approximately 1 percent of nominal reactor power (Fig. 2). In addition, we note from Fig. 1 that the fluidization limit for the droplet regime is exceeded for a very small fraction of nominal power and, therefore, Fauske concluded that a boiling fuel-steel mixture in the reactor core region must first exist in the form of molten fuel-steel drops suspended in steel vapor. In the years immediately following Fauske's analysis of boiling flow regimes, the relevance

† Entrained steel particles in the fuel has been observed during postmortem examinations of the TREAT meltdown experiments [4].

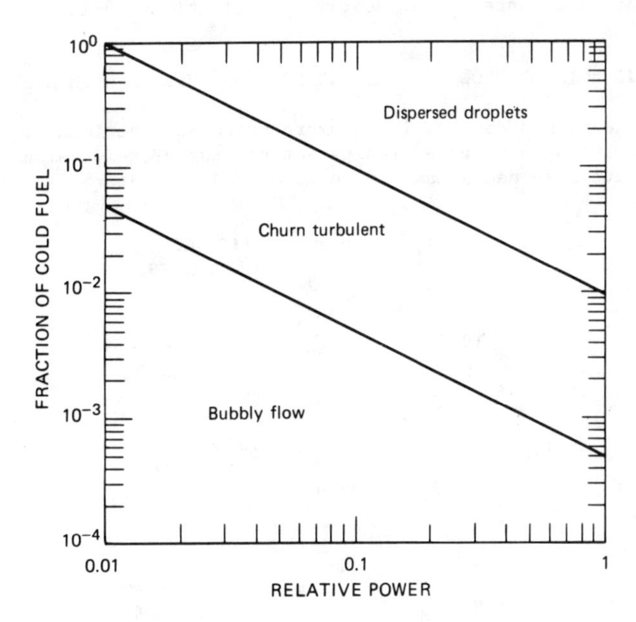

Fig. 1. Flow regimes in a boiling fuel-steel pool. From Fauske [5].

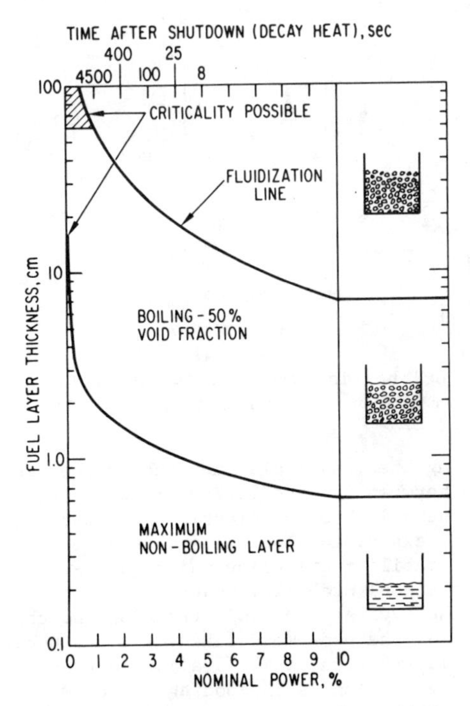

Fig. 2. Flow regime map indicating required conditions for fuel criticality. From Fauske [5].

of the Kutateladze correlation to flow regime transitions served as the basis for a series of fundamental experimental investigations [6-10].

2. VOLUMETRIC BOILING FLOW REGIMES AND POOL BOILUP; EXPERIMENTAL STUDIES

Farahat, Henry and Santori [6] have reported an experimental study carried out in a microwave oven using a transparent boiling vessel, as shown in Fig. 3. The microwave oven had a maximum power of 600 W. Water was used as the

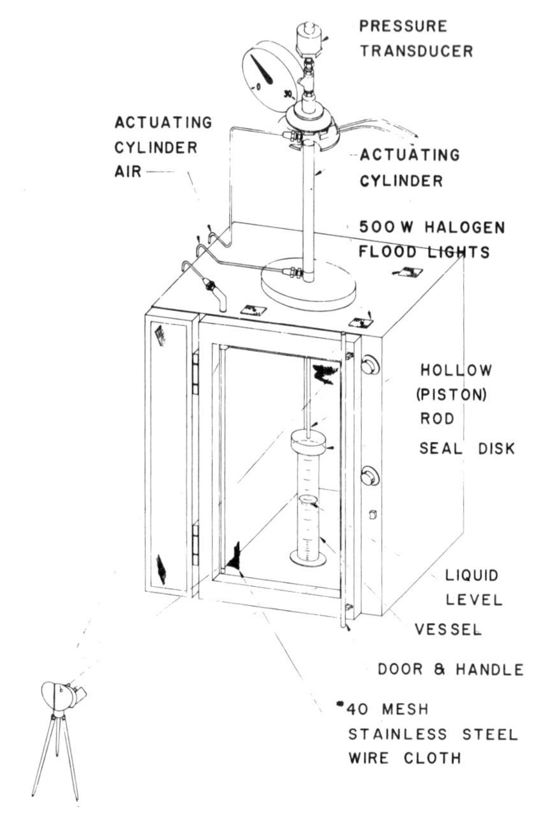

Fig. 3. Pool boiling with volumetric heating using a microwave oven. From Farahat, Henry and Santori [6].

test fluid. By varying the vessel diameter and/or the liquid height, volumetric heating rates between 1.2 and 20.2 W/cm^3 were obtained. The water was seeded with small, neutrally buoyant plastic spheres to insure ample bulk nucleation sites. The experimental results were based on two studies conducted both (i) to test the stability criterion under quasi-steady boiling conditions and (ii) to examine the pool stability under bottled up conditions. In part (i) measurements of the average pool void fraction and the superficial vapor velocities at the top of the pool were taken. At a superficial vapor velocity of $\sim$5 cm/sec and a void fraction of $\sim$0.75 a marked transition from normal bubbly flow was observed, which is in good agreement with equation 1. However, the flow regime which the authors observed at superficial velocities above 5 cm/sec was not the churn-turbulent flow regime as might be expected from examination of Fig. 2. Instead, the void fractions dramatically increased to values approaching 95 percent, with the flow pattern adopting a cellular vapor-liquid appearance typical of a foam regime. In part (ii), in order to simulate the (freeze) plugging phenomena as boiling fuel attempts to move out of the core region (see Section 6), the water was allowed to boil up to the top of the vessel at which time the vessel was sealed with a plug. The pool behavior was monitored by high-speed movies and pressure instrumentation. Immediately following the plugging of the vessel, the pool was observed to pressurize, yet the pool maintained its boiled up state. These observations were interpreted as evidence that small axial (and radial) heat loss to the glass container boundaries is capable of sustaining a boiled up state, i.e., heat losses promote boilup.

Ginsberg, Jones and Chen [7] conducted experimental investigations of two-phase flow patterns in volume-heated boiling pools using a joule resistance heating technique. Their experiments were performed with a zinc sulfate-demineralized water electrolyte solution in a rectangular column open at the top (see Fig. 4). The column, constructed of clear plastic, was 120 cm high

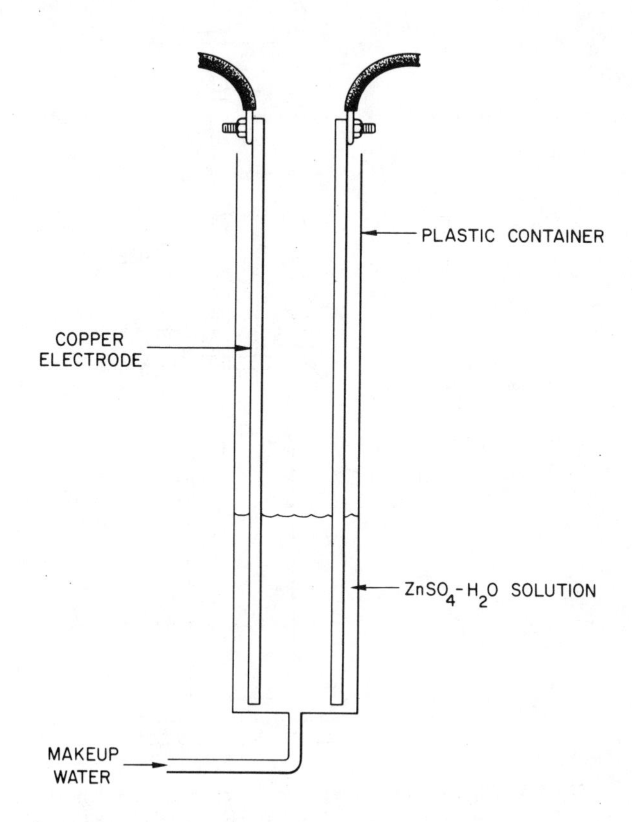

Fig. 4. Pool boiling with volumetric heating using the Joule resistance technique. From Ginsberg, Jones and Chen [7].

and 8.9 x 6.4 cm in cross section. Two opposing vertical copper electrodes were placed into the pool and traversed the height of the column. Internal heating within the pool, sufficient to cause boiling, was obtained by passing an electrical current between the electrodes. The water was allowed to boil off freely from the top of the pool while the liquid level was kept fixed with makeup water. Photographic observations of the flow patterns were made together with measurements of the average void fraction and the applied power. Ginsberg et. al. [7] concluded from their observations that criterion (1) for flow regime transitions does not apply to the volume-heated system studied. A bubbly flow regime was always observed for superficial vapor velocities up to approximately 5 times the critical velocity given by equation (1). Moreover, a churn flow regime was observed at superficial vapor velocities well above the critical velocity for the transition to dispersed droplet flow represented by equation (1).

Orth et. al. [8] conducted experiments involving a gas-water analogue in which capillary tubes were used to obtain a continuous, volumetric addition of gas bubbles to a pool of water. The apparatus is shown in Fig. 5. The test

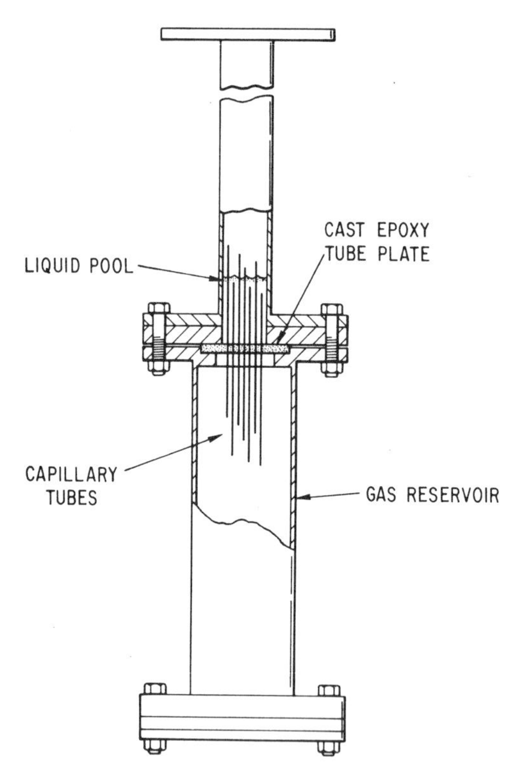

Fig. 5. Schematic diagram of volumetric gas injection experiment. From Orth et. al. [8].

section was a square (5.1 x 5.1 cm) Lucite column. An epoxy plate separated the column from a high-pressure gas reservoir. Stainless-steel capillary tubes 20 cm long, 0.15 mm ID, were cast into and penetrated the epoxy plate. The tubes were regularly spaced in the plane normal to the test section and arranged such that the upper ends of the tubes were vertically staggered over a volume of the test section of height 10 cm. Tests were conducted with epoxy plates having either 49 tubes or 621 tubes. Tests were also conducted with an epoxy plate in which the tops of the capillary tubes were flush with the upper plate surface. This allowed the bubbling behavior with volumetric gas injection to be compared with the results obtained with the injection of gas through a "perforated" plate at the bottom of the liquid column. Tap-water was poured into the column to various initial heights ranging from 1/2 to 6 times the height of the capillary tube matrix, i.e., from 5 to 60 cm. During the experiment, nitrogen gas was fed steadily into the liquid column through the capillary tubes. A static pressure probe that was free to move vertically through the pool was employed to obtain local void fractions as a function of distance above the bottom of the pool. As is usual with this system (see Chapter 15, Section 3), the ideal bubbly flow pattern, in which bubbles rise steadily and uniformly, was gradually transformed into churn flow as the gas flow was increased from zero. A "compact bubbly flow regime" appeared to represent a transition region between ideal bubbly flow and churn flow. In this regime, the column showed a close packing of bubbles, and void fractions as high as 65 percent were measured for the shorter initial column heights. The flow patterns produced with the injection of gas through the perforated plate were observed to be similar to those obtained with volumetric bubbling; in fact, for superficial gas velocities above the transition from bubbly flow (see below), the void fraction versus height data for the volumetric and surface injection cases were practically identical. This suggests that the change in flow pattern to churn flow tends to produce recirculation in the column which obliterates the effects of volumetric injection. In agreement with previous studies [6,7], the transition to droplet-dispersed flow was never observed.

Perhaps, in retrospect, the absence of the transition to droplet dispersed flow when j exceeds the value given by equation (1) with $K \sim 0.2$ and $\rho_c = \rho_L$ $(= \rho_v)$ is not too surprising. A careful examination of the experimental support for this equation [11,12] (see also Chapter 15, Section 3) reveals that it may not apply to boiling or bubbling in deep liquid pools. Wallis [11] reports a stability constant $K = 0.12$ which he obtained in experiments involving the blowing of gas through a very shallow pool of liquid. He associated the appearance of a fine mist above the pool with the onset of fluidization.† Kutateladze [12] "bubbled" water into pools of mercury and pools of carbon tetrachloride and considered dispersed flow to begin when the mercury or carbon tetrachloride broke up into drops, i.e., became the discontinuous phase. The work reported in [6-8] seems to indicate that these techniques are not applicable to steam- or gas-water pools. It has been suggested [7] that Dukler and Taitel's correlation [13] for the onset of drop annular flow, namely

$$j_{crit} \simeq 1.5 \left[\frac{(\rho_\ell - \rho_v) g \sigma}{\rho_v^{\,2}} \right]^{1/4} \qquad (3)$$

would be more appropriate for predicting the onset of fluidization in liquid pools. Based on equation (3), a superficial steam velocity of ∿1000 cm/sec is required to levitate water droplets. No experiments have so far been carried out at sufficiently elevated power levels by which such speculation might be

† Accordingly, the condition $K = 0.12$ might be interpreted as a "misting" condition rather than as a condition for fluidization.

checked. Note that equation (3) is equivalent to equation (1) with K set equal to 1.5 and $\rho_c = \rho_v$. It should be pointed out that the reported absence of the drop flow regime could well be due to the relatively narrow channels employed in the above-referenced studies [6-8]. The liquid has a tendency to stick to the walls of the channels so that a drop flow might never be obtained in such geometries, whereas fluidization might readily be achieved in a wide pool.

Figure 6 shows a graphical summary of the results reported in [6-8] in terms of the ranges of observed superficial velocities versus pool-averaged void fractions. The superficial velocity j in Fig. 6 is normalized by the critical velocity for the breakdown of bubbly flow, equation (1) with K = 0.3 and ρ_c taken as the liquid density ρ_ℓ:

$$j_{crit} = 0.3 \left[\frac{(\rho_\ell - \rho_v) g \sigma}{\rho_\ell^2} \right]^{1/4} \tag{4}$$

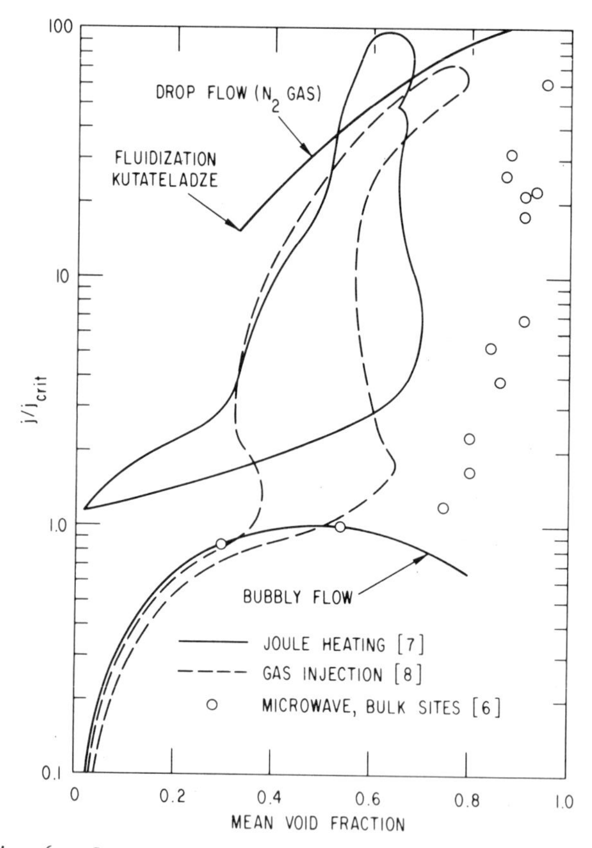

Fig. 6. Superficial velocity versus average void fraction for several methods of vapor production.

The figure clearly illustrates the way in which completely different results can be obtained for the same values of the superficial velocity depending on the method of vapor production. The semi-empirical j versus α relations for two "ideal" flow regimes, bubbly and particle (drop) flow, are also included in Fig. 6 [see equations (7) and (9) in Chapter 15]. The results obtained in the microwave oven with bulk nucleation sites in the liquid [6] and the results obtained with volumetric gas injection [8] follow the ideal bubbly flow curve at low superficial velocities. The departure of the data from the bubbly flow curve at approximately $j = 1.0$ is in good agreement with equation (4). Interestingly enough, while the transition to drop flow was never visually observed, the data obtained with gas injection asymptotically approach the ideal drop flow curve at high superficial velocities. It is seen from Fig. 6 that the ideal bubbly flow pattern never developed in the joule-heating experiments [7]. Evidently the bubbly flow pattern that persists up to relatively high vapor velocities, in the Ginsberg et. al. experiments [7], is some "non-ideal" bubbly flow pattern.

It seems, at present, that all the data in Fig. 6 that lie above the bubbly flow curve, except perhaps, the data of Farahat et. al. [6], represent a transition region between ideal bubbly flow and some yet undetermined flow regime (most likely drop-annular flow). The different quantitative results simply reflect the different modes of vapor production, the purity of the liquids, and the initial column heights investigated. This transition region should not be confused with the analytically tractable, churn flow regime observed when gas-liquid mixtures flow upwards in long pipes [13,14]. It is recommended that a new term be introduced in order to differentiate it from true churn flow.

The persistence of foaming in the boiling experiments of Farahat et. al. [6] is quite interesting. Presumably these workers satisfied the two necessary requirements for the production of a "dynamic foam" [21]. First, a sufficient number of bulk nucleation sites was provided so that the vapor formed as numerous small bubbles instead of several large ones. Second, the approach to steady-state boiling was rapid and therefore, the rate of rise of the pool surface exceeded the rate of bubble rise (relative to the liquid) and consequent agglomeration, thereby "locking" the bubbles into a foam regime. An alternative and, perhaps, more reasonable explanation for foaming in these experiments has been provided by Ginsberg et. al. [7]: "Chemical and particulate present in liquids may enhance foam development and stability [15]. It is therefore possible that the particulate additives, with its associated red water-soluble coloring agent, enhanced the stability of bubbles against coalescence and hence, led to a foam regime."

Foam, whatever its cause, can provide the mechanism by which large quantities of molten fuel can be rapidly removed from the core region to a permanently subcritical state. This was clearly demonstrated in experiments conducted by Henry, Smith, and Weber [10], which were designed to simulate the sudden "venting" of pressurized core material from the active core region (see Chapter 15, Section 2). Again, these experiments were carried out in a microwave oven with a transparent Lucite vessel and water as the test fluid (see Fig. 3). The test vessel was sealed at the top with a Lucite cap penetrated by a vent tube. The venting process was initiated by opening a ball valve which was installed in the vent tube, as shown in Fig. 7. Each run was started by turning the oven on and internally heating the water until the pool boiled up to occupy the total volume of the test vessel. At this time, the ball valve in the vent line was closed and the pool was allowed to pressurize to a pressure level of approximately 220 kPa ($\sim$2.2 atm). Then, the ball valve was

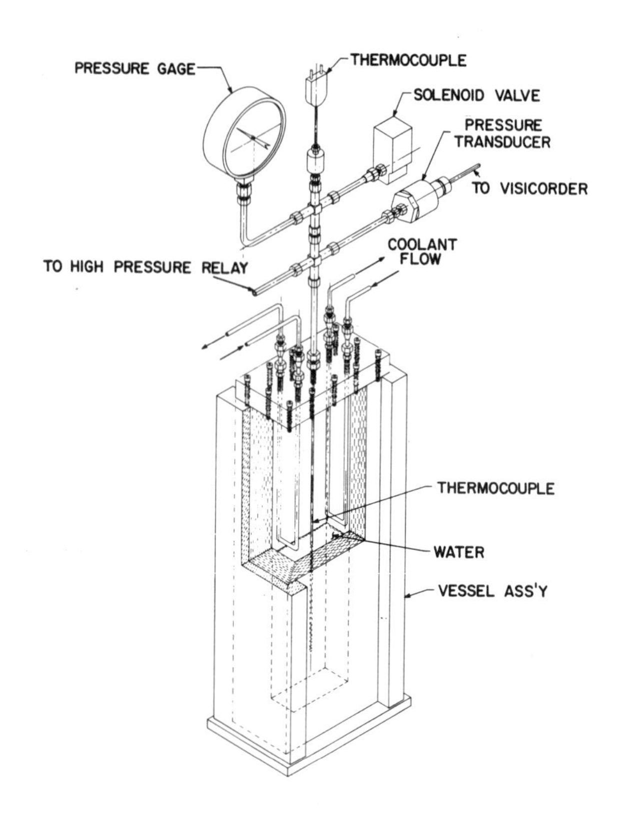

Fig. 7. Experimental test vessel for depressurization of an internally heated pool. From Henry, Smith and Weber [10].

open, and depending on the conditions desired, the power was either left on or turned off. The pool was allowed to depressurize back to atmospheric pressure. The venting flow patterns were changed by the addition of solid particles to the liquid or by using an aqueous solution. The techniques used were: (1) seeding the liquid with the same neutrally buoyant plastic spheres employed by Farahat et. al. [6], (2) placing steel spheres at the bottom of the vessel, or (3) adding small quantities of a liquid foaming agent to the water pool together with the steel spheres.

The effects of flow pattern on the amount of liquid removed from the test vessel is clearly illustrated in Fig. 8. The measured fraction of water remaining in the vessel at the end of the venting process is plotted against the initial liquid fraction. We note that the amount of residual liquid is essentially independent of the applied internal heating for the power levels achieved in the experiment ($\lesssim 3$ W/cm^3). The remaining liquid fraction is also somewhat insensitive to the initial liquid fraction. The high speed motion pictures that accompanied the experiment revealed that in the presence of bulk nucleation sites (plastic spheres) the foam flow pattern prevailed from the initiation of venting until the process was complete. Obviously, the foam

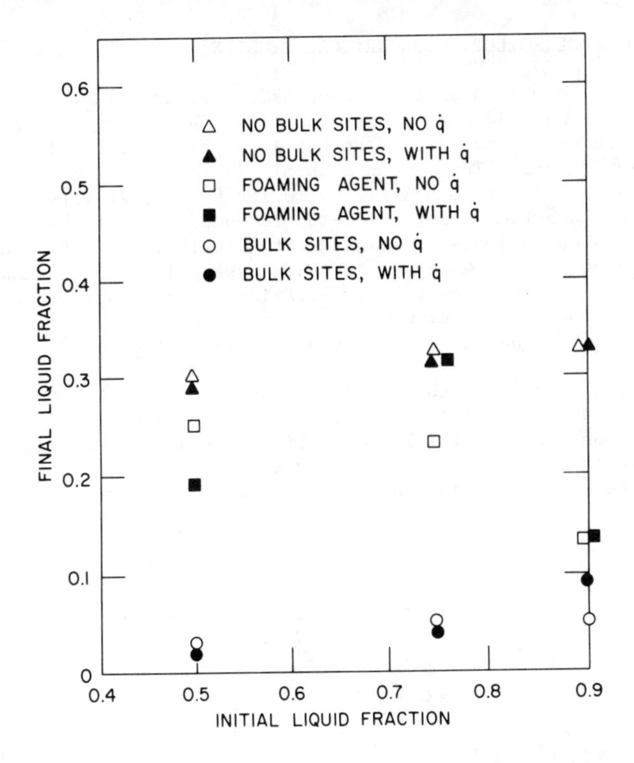

Fig. 8. Final liquid fraction as a function of initial liquid fraction; effects of boiling flow regime. From Henry, Smith and Weber [10].

flow regime results in minimum vapor drift (minimum slip) and hence represents the maximum boilup condition. With stainless steel balls a churn-bubbly flow pattern was evident during most of the transient and minimum boilup was observed. The addition of a foaming agent to the water with the steel balls increased the foaming within the system and stabilized the foam for a greater length of time; however, the foam was late in getting started and significant vapor drift still occurred at the onset of venting.

It is worth noting that Koontz [9] constructed an experimental facility to study volumetric boiling in a two-component system. Water was placed in a test vessel and heated volumetrically using a microwave oven. Liquid hexane (or heptane) entered the test section from below through a perforated plate. The dispersed hexane phase boils (evaporates) while rising through the continuous water phase. The flow regime observed by Koontz appears to be the same bubbly-churn "transitional" regime reported in [7,8]. One important concern that cannot be addressed in single-component volume-heated boiling studies is the possibility that the volatile steel component will separate from the heavier molten UO_2. It is hoped that the two-component volume-heated pool reported by Koontz [9] will yield valuable information in this area. A study of stratification and mixing within boiling immiscible mixtures using simulant materials and a gas-liquid analogue has recently been completed [16].

3. TRANSIENT POOL BOILUP: THEORETICAL STUDIES

A simple one-dimensional theoretical study of transient volume-heated boiling pools was carried out by Epstein [17]. The analysis is based on the single main assumption that at any height in a transient boiling pool there are no inertia effects. Thus, when a bubble is formed, it is assumed instantaneously to attain its terminal rise velocity so that gravity is balanced against the forces between the liquid-bubble components of the bubbly mixture. Moreover, pressure gradients associated with liquid inertia and superheat in the liquid are neglected. Subject to these assumptions the boilup process was determined entirely from a continuity equation (without use of momentum and energy equations) in combination with a phenomenological "flux function" law [e.g., equation (6), Chapter 15] for the relative motion of vapor and liquid.

The physical model and the coordinate system are shown schematically in Fig. 9. A liquid pool of initial height H_o is contained in a vertical tube open to the atmosphere. Initially, the liquid is at its saturation temperature and there is no boiling. Then the volumetric heat generation rate (power input) is raised from zero to $\dot{Q}$ and bulk boiling commences.

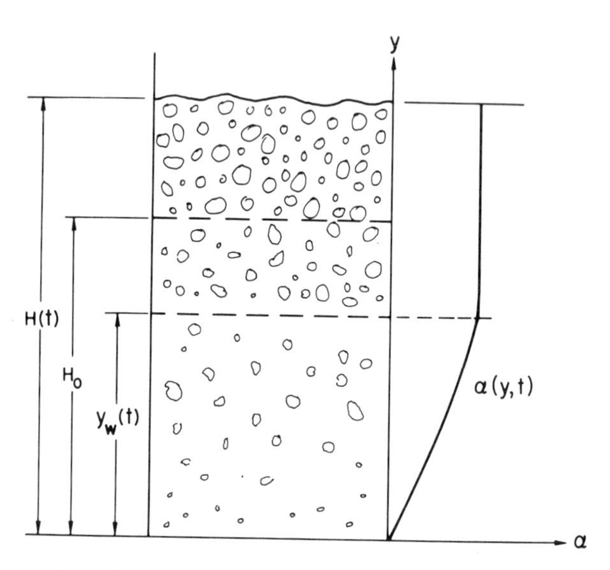

Fig. 9. Transient growth of a boiling pool; configuration and notation.

The vapor phase satisfies the continuity equation

$$\frac{\partial \alpha}{\partial t} + \frac{\partial \alpha u_v}{\partial y} = \frac{\dot{Q}}{h_{v\ell}\rho_v} (1 - \alpha)^m \tag{5}$$

where u_v is the local upward vapor velocity. Similarly, from liquid continuity, we have

$$-\frac{\partial \alpha}{\partial t} + \frac{\partial (1 - \alpha)u_\ell}{\partial y} = -\frac{\dot{Q}}{h_{v\ell}\rho_\ell} (1 - \alpha)^m \tag{6}$$

where u_ℓ is the local liquid velocity. The vapor velocity is related to the liquid velocity by the relation

$$u_v = u_\ell + v_\infty(1 - \alpha)^{n-1} \tag{7}$$

In the above equations $\dot{Q}$ is the volumetric heat generation rate that causes liquid evaporation, $h_{v\ell}$ is the latent heat of vaporization, ρ_v and ρ_ℓ are the densities of the vapor and liquid, α is the vapor void fraction, v_∞ is a characteristic vapor velocity [see Chapter 15, equation (6)], t is the time, and y is the vertical coordinate measured from the bottom of the pool. Equation (7) follows from the empirical law governing the volumetric vapor flux. The index n is a phenomenological quantity to be determined from experiment; n = 2 for bubbly flow and n = 0 for churn flow [see Chapter 15, equations (7) and (8)]. The factor $(1 - \alpha)^m$ results from assuming that the power input to the liquid depends upon the void fraction. In the case of nuclear fission heating, for example, the heat generation rate is proportional to the local density of the pool which leads to m = 1. This is also the case when the liquid is heated by the dissipation of electrical energy. On the other hand, an experiment on pool boilup may consist of suddenly introducing gas uniformly throughout a liquid pool by a group of orifices - a system in which m = 0 best represents the production of void.

Adding equations (5) and (6) gives the following partial differential equation with the time derivatives removed:

$$\frac{\partial \alpha u_v}{\partial y} + \frac{\partial (1 - \alpha) u_\ell}{\partial y} = \frac{\dot{Q}}{h_{v\ell}\rho_v}(1 - \alpha)^m \tag{8}$$

where we have assumed $\rho_v/\rho_\ell << 1$. A solution of the equation, obeying the boundary conditions that $u_\ell = 0$ and $\alpha = 0$ at the bottom of the pool (y = 0), yields

$$\alpha u_v + (1 - \alpha)u_\ell = \frac{\dot{Q}}{h_{v\ell}\rho_v}\int_0^y (1 - \alpha)^m \, dy \tag{9}$$

Eliminating u_ℓ between equations (7) and (9) leads to the following relation between vapor velocity and void fraction:

$$u_v = v_\infty(1 - \alpha)^n + \frac{\dot{Q}}{h_{v\ell}\rho_v}\int_0^y (1 - \alpha)^m \, dy \tag{10}$$

Finally, substituting equation (10) for u_v into the vapor continuity equation (5), we obtain the equation governing the void fraction distribution:

$$\frac{\partial \alpha}{\partial t} + \left[v_\infty(1 + n)(1 - \alpha)^n - v_\infty n(1 - \alpha)^{n-1} + \frac{\dot{Q}}{h_{v\ell}\rho_v}\int_0^y (1 - \alpha)^m \, dy \right] \frac{\partial \alpha}{\partial y}$$

$$= \frac{\dot{Q}}{h_{v\ell}\rho_v}(1 - \alpha)^{m+1} \tag{11}$$

It is convenient at this time to introduce the following dimensionless independent variables:

$$\tau \equiv \frac{\dot{Q}}{h_{v\ell}\rho_v} t \tag{12}$$

$$Z \equiv \frac{\dot{Q}}{v_\infty h_{v\ell}\rho_v} y \tag{13}$$

With this choice of notation, equation (11) transforms to:

$$\frac{\partial \alpha}{\partial \tau} + \left[(1 + n)(1 - \alpha)^n - n(1 - \alpha)^{n-1} + \int_0^Z (1 - \alpha)^m dZ \right] \frac{\partial \alpha}{dZ}$$

$$= (1 - \alpha)^{m+1} \tag{14}$$

A closed-form solution of equation (14) can be obtained by the method of characteristics. The solution is equivalent to the solution of the following pair of ordinary differential equations:

$$d\tau = \frac{dZ}{(1 + n)(1 - \alpha)^n - n(1 - \alpha)^{n+1} + \int_0^Z (1 - \alpha)^m dZ} = \frac{d\alpha}{(1 - \alpha)^{m+1}} \tag{15}$$

subject to the conditions $\alpha = 0$ at $Z = 0$, and $\alpha = 0$ at $\tau = 0$ for all Z. Solving the right-hand equation for $dZ/d\alpha$ and differentiating the result with respect to α, we obtain the second-order differential equation

$$\frac{d}{d\alpha}\left[(1 - \alpha)^{m+2} \frac{dZ}{d\alpha} \right] = n(n - 1)(1 - \alpha)^{n-1} - n(n + 1)(1 - \alpha)^n \tag{16}$$

where the integral term has been removed. The void fraction can also be related to the independent variable τ by equating the first and third terms in equation (15):

$$(1 - \alpha)^{-(m+1)} \, d\alpha = d\tau \tag{17}$$

The above two equations can be readily integrated and the results allow us to obtain the limiting characteristic, Z_w:

$$Z_w = \frac{n - 1}{n - m - 1}\left[(1 + m\tau)^{(1+m-n)/m} - 1 \right]$$

$$- \frac{n}{n - m}\left[(1 + m\tau)^{(m-n)/m} - 1 \right] \tag{18}$$

The region below this limiting characteristic, $Z < Z_w$, corresponds to the steady-state regime, while the region above, $Z > Z_w$, is the transient domain. This limiting characteristic or "continuity wave" thus gives the time required to achieve steady-state boiling conditions at any position Z. From the solutions of equations (16) and (17), we may determine the void fraction distribution during the boilup transient; i.e.,

$$\frac{n-1}{n-m-1}\left[(1-\alpha)^{n-m-1}-1\right]-\frac{n}{n-m}\left[(1-\alpha)^{n-m}-1\right]=Z; \quad Z < Z_w \quad (19)$$

and

$$\alpha = 1 - \frac{1}{(1+m\tau)^{1/m}} \quad ; \quad Z > Z_w \quad (20)$$

It should be mentioned that in the solution of equation (16) we used the condition $dZ/d\alpha = 1$ when $\alpha = 0$ which follows from the right-hand equation (15).

Clearly, the height of the pool, H(t), must satisfy the integral equation

$$H(t) = H_o + \int_o^{H(t)} \alpha dy \quad (21)$$

where H_o is the pool height at time t = 0. The pool height is conveniently described in terms of the dimensionless instantaneous pool height $h(\tau)$:

$$h \equiv \frac{\dot{Q}H}{v_\infty h_{v\ell}\rho_v} \quad (22)$$

Thus, Eq. 19 transforms to

$$h(\tau) = h_o + \int_o^{h(\tau)} \alpha dZ \quad (23)$$

where h_o is the dimensionless initial pool height given by

$$h_o \equiv \frac{\dot{Q}H_o}{v_\infty h_{v\ell}\rho_v} \quad (24)$$

The pool height follows immediately from equations (18) - (20) and (23) and is

$$h(\tau) = h_o(1+m\tau)^{1/m} + \frac{n}{n-m+1}(1+m\tau)^{(m-n)/m}$$
$$- \frac{n-1}{n-m}(1+m\tau)^{(m-n+1)/m}$$
$$+ \frac{m-1}{(n-m)(n-m+1)}(1+m\tau)^{1/m} + Z_w(\tau) \quad (25)$$

In the limit as $m \to 0$, $(1+m\tau)^{1/m} \to \exp(\tau)$ so that

$$h(\tau) = h_o - \frac{1}{n(n+1)} \exp(\tau) + \frac{1}{n} \exp[(1-n)\tau] - \frac{1}{n+1} \exp(-n\tau) \qquad (26)$$

Equation (26) shows that the pool height is an exponential function of time when the vapor production rate does not depend on the pool density. Typical pool height evolution is shown in Fig. 10, for the dimensionless initial pool height $h_o = 1$, the flow regime index $n = 0,1$, and the void production index $m = 0,1$. Examination of this result reveals that pool growth ultimately ends when $n = 0$ but continues indefinitely for $n = 1$. This result is discussed below. One notes that the effect of nonzero m is to retard the pool growth. This is expected since less void is produced when the vapor production rate is pool density-dependent.

A typical history of a boiling pool with zero initial void fraction is as follows: As boiling begins, a bubbly mixture appears throughout the liquid column in which the vapor void fraction is uniform, except at the very bottom of the pool where a region of nonuniform void fraction is formed (see Fig. 9). The location of the discontinuity between these regions, $Z_w(\tau)$, is given by equation (18). To accommodate the vapor generated by the boiling process, the surface of the pool begins to rise. The position of the pool surface in time is given by equation (25). When the changes in void fraction propagated from the bottom overtake the rising surface, i.e., $Z_w = h$, the transient boiling period is postulated to be complete. For the $n = 1.0$ case exhibited in Fig. 10, the limiting characteristic or wave that propagates the steady-state void fraction profile upward from the bottom of the pool cannot catch up

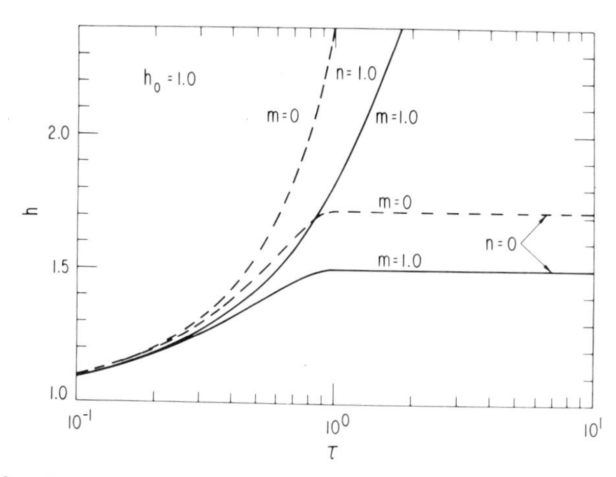

Fig. 10. Predicted dimensionless pool height-time relation; dependence on flow regime index n, for constant power ($m = 0$) and for pool-density dependent power ($m = 1$). From Epstein [17].

with the rising surface of the pool. The pool will continue to expand. There can either be a change in flow regime (i.e., a sudden decrease in the value of the index n) or a rejection of excess liquid material at the top of the pool container.

Using the model presented above, it was demonstrated by Epstein [17] that it takes 1 sec for a nuclear fuel pool to boilup in the churn regime (n = 0) to twice its initial height at a power level equivalent to only 0.25 percent of nominal power.

Several detailed analytical studies followed this preliminary work. In two companion papers, Condiff and Epstein [18,19] reformulated the model of Ref. [17] in rather general terms and noted that variations in the form of the vapor flux function can account for qualitatively different predictions for the boilup behavior. In the first paper [18], attention was confined to equation construction and a rigorous treatment of simple pool boilup (foam-free pool). The second paper was concerned with the phenomena of formation, decay, and interaction of vapor void discontinuities with foaming. Condiff, Epstein and Grolmes [20] utilized the theory presented in the second paper [19] to illustrate the qualitative features of transient boilup with foam formation for a special choice of flux function. Figure 11 shows a time (τ)

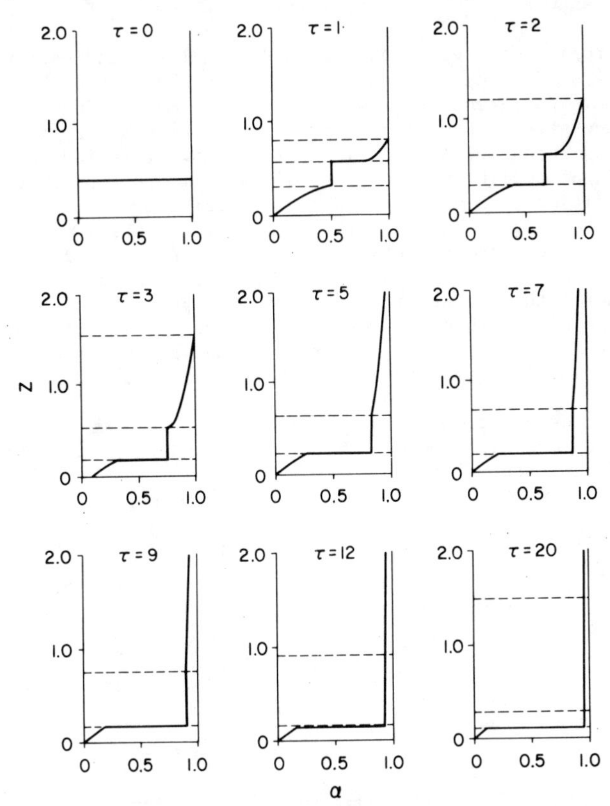

Fig. 11. Time sequence of void fraction profiles during transient volumetric pool boiling. From Condiff, Epstein and Grolmes [20].

sequence of void fraction profiles for an initially saturated liquid pool of depth $h_o = 0.4$. The initial pool top (dark line) serves as a discontinuity of vapor void fraction which propagates upward. The jump in void fraction across it to values less than unity corresponds to formation of a foam which rises like lather above the propagating pool top discontinuity. Just below the discontinuity, there is a transient region in which the vapor void fraction is uniform. The void value within this region increases with increasing time. Below this transient region a steady-state region of nonuniform void fraction is formed. The foaming pool top discontinuity diminishes in strength and disappears at $\tau = 2$. Prior to this disappearance at time $\tau = 1$, a second discontinuity begins to form between the transient and steady state regions and moves downward in time. For times $\tau > 9.58$ foam formation begins again, this time above the lower discontinuity. The transient section is forced to lift above this foam and acts as a widening buffer zone between separate regions of foam. The qualitative similarity between the overall flow structure predicted in [20] and an actual bubble-foam flow pattern is noted in [21]. It should be mentioned that boilup behavior is sensitive to the initial pool depth. Other boilup possibilities are discussed in [20].

4. SIMULTANEOUS MELTING AND SOLIDIFICATION

The problem of predicting molten UO_2 fuel solidification rates arises in several areas of study of the transition phase of a hypothetical accident sequence in a fast reactor. Knowledge of the magnitudes of fuel solidification rates at the steel boundaries of a boiling fuel pool are necessary for a complete understanding of pool stability, as well as for bounding the rate of propagation of boiling regions beyond the incident subassemblies (see Section 5). In addition, the upward dispersal of boiling core material involves the transient freezing of fuel-steel mixtures in the steel channels above the active core zone. A distinctive and most important characteristic of fuel freezing on solid steel is that the surface of the underlying steel will melt upon contact with the fuel if the initial temperature of the steel wall is sufficiently high. In other words, the UO_2-steel interface temperature falls between the fusion temperatures for these substances, resulting in solidification in the initially molten UO_2 and melting in the initially solid steel.

A basic description of the problem is provided by Fig. 12. We consider the infinite composite region formed by the sudden contact of material 1 with material 2 along the plane at $x = 0$. Material 1 in the half-space $x > 0$ is

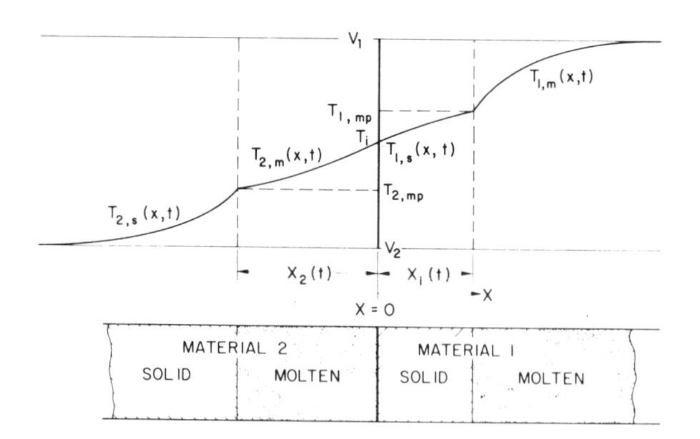

Fig. 12. Schematic of physical configuration and instantaneous temperature profile following sudden contacting of two semi-infinite regions.

initially molten at temperature $V_1 > T_{mp,1}$ and material 2 in the half-space $x < 0$ is initially solid at temperature $V_2 < T_{mp,2}$. At time zero a solidification front appears at the plane of contact $x = 0$ and moves in the positive x-direction into material 1; similarly, a melting front appears at $x = 0$ and moves in the negative x-direction into material 2. At time t the temperature distribution appears as in Fig. 12, where $X_1(t)$ and $X_2(t)$ represent the thickness of solidified material 1 and the thickness of melted material 2, respectively. In the analysis that follows, it is assumed that very intimate thermal contact is maintained at $x = 0$ and that the change of volume upon solidification or melting can be neglected. In addition, all physical properties are considered constant.

The temperature profile in each of the four regions shown in Fig. 12 must obey a partial differential equation of the form

$$\frac{\partial^2 T}{\partial x^2} = \frac{1}{\alpha}\frac{\partial T}{\partial t} \tag{27}$$

where α is the thermal diffusivity of the material. At the solidification and melting fronts we have the boundary conditions

$$T_{1,s}(X_1,t) = T_{1,m}(X_1,t) = T_{1,mp} \tag{28}$$

$$T_{2,s}(-X_2,t) = T_{2,m}(-X_2,t) = T_{2,mp} \tag{29}$$

$$K_{1,s}\frac{\partial T_{1,s}}{\partial x} - K_{1,m}\frac{\partial T_{1,m}}{\partial x} = h_{\ell s,1}\rho_1 \frac{dX_1}{dt} \quad \text{at} \quad x = X_1(t) \tag{30}$$

$$K_{2,s}\frac{\partial T_{2,s}}{\partial x} - K_{2,m}\frac{\partial T_{2,m}}{\partial x} = -h_{\ell s,2}\rho_2 \frac{dX_2}{dt} \quad \text{at} \quad x = -X_2(t) \tag{31}$$

where the last two conditions represent the liberation and absorption of latent heat of fusion, $h_{\ell s}$, at these fronts, K is the thermal conductivity and ρ is the density. The boundary conditions to be satisfied at $x = 0$ are

$$T_{1,s}(0,t) = T_{2,m}(0,t) \tag{32}$$

and

$$K_{1,s}\left[\frac{\partial T_{1,s}}{\partial x}\right]_{x=0} = K_{2,m}\left[\frac{\partial T_{2,m}}{\partial x}\right]_{x=0} . \tag{33}$$

Finally, the temperature profiles are subject to the following boundedness conditions as $|x|$ goes to infinity:

$$T_{1,m}(\infty,t) = V_1, \quad T_{2,s}(-\infty,t) = V_2 \tag{34}$$

The above equations pose the classical Neumann problem [22] for which the solution is [23]:

$$\frac{T_{1,s} - T_{1,mp}}{T_{1,mp} - T_{2,mp}} = - \frac{\sigma \left\{ \text{erf}(\chi_1) - \text{erf}\left[\frac{x}{2(\alpha_{1,s} t)^{1/2}} \right] \right\}}{\text{erf}(\chi_2) + \sigma\, \text{erf}(\chi_1)} \tag{35}$$

for the solidified region $0 < x < X_1(t)$ in material 1;

$$\frac{T_{1,m} - V_1}{V_1 - T_{1,mp}} = - \frac{\text{erfc}\left[\frac{x}{2(\alpha_{1,m} t)^{1/2}} \right]}{\text{erfc}\,(\beta_1 \chi_1)} \tag{36}$$

for the molten region $X_1(t) < x < \infty$ in material 1; where

$$\sigma = \left(\frac{\alpha_{1,s}}{\alpha_{2,m}} \right)^{1/2} \frac{K_{2,m}}{K_{1,s}} \tag{37}$$

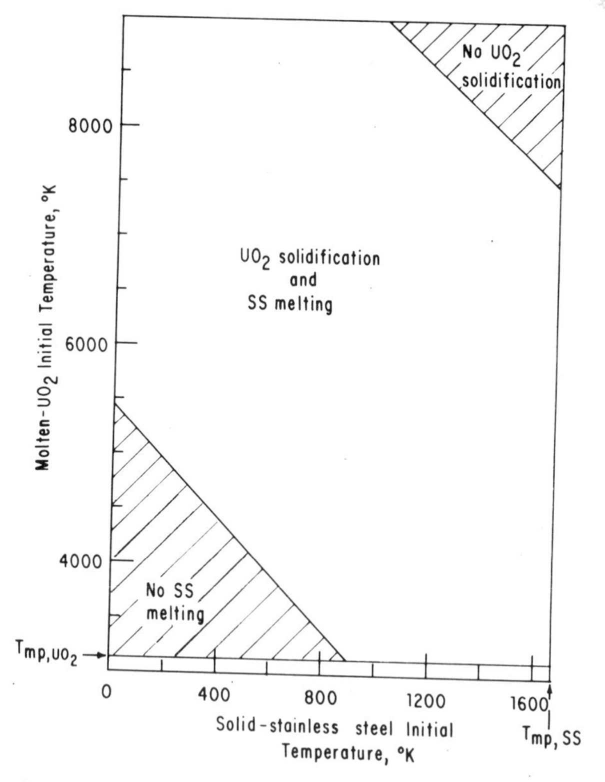

Fig. 13. Initial-temperature map for initially molten UO_2 contacting initially solid stainless steel. From Epstein [23].

$$\beta_1 = \left(\frac{\alpha_{1,s}}{\alpha_{1,m}}\right)^{1/2} \qquad (38)$$

Similar equations are obtained for the temperature profile in material 2 [23]. The "growth" constants χ_1 and χ_2 are determined from the interface condition (33) subject to the unique requirement that $X_1(t)$ and $X_2(t)$ satisfy

$$X_1(t) = 2\chi_1(\alpha_{1,s}t)^{1/2} \qquad (39)$$

$$X_2(t) = 2\chi_2(\alpha_{2,m}t)^{1/2} \quad , \qquad (40)$$

It is this last requirement, dictating that the interface temperature at the plane of contact remain fixed at the value

$$\frac{T_i - T_{2,mp}}{T_{1,mp} - T_{2,mp}} = \frac{\text{erf}(\chi_2)}{\text{erf}(\chi_2) + \sigma\ \text{erf}(\chi_1)} \qquad (41)$$

which is characteristic of the Neumann problem and its similarity solution.

The above formulation was applied to the problem of initially molten UO_2 in contact with initially solid stainless steel. Figure 13 is an initial temperature map for the UO_2-steel system, showing the region (unshaded) in which simultaneous UO_2 solidification and steel melting occur. The high values given for the initial molten UO_2 temperatures in Fig. 13 have been intended to allow for molten UO_2-solid steel contacting in the presence of high pressures. The especially high values that appear in Fig. 13 are not expected to occur in practice but serve to illustrate how high one must heat molten UO_2 before it fails to solidify upon contact with solid stainless steel. We note that if the initial temperature of the stainless steel exceeds ∿700°C, we can expect simultaneous steel melting and fuel solidification to occur.

5. STABILITY OF A FUEL CRUST

In the previous section we have supposed that simultaneous fuel crust growth and steel melting take place in a quiescent medium; in fact this can seldom be so. In practically all fast reactor safety applications, the melting solid steel is immersed in a flow of molten UO_2. The steel melting rate in this circumstance depends on the behavior of the solid UO_2 layer which forms when the flowing fuel comes into contact with the solid steel surface and then "floats" on the steel melt. Since the frozen fuel layer does not have a solid surface to "stick" to, mechanical processes for crust removal may prevent the fuel crust from insulating the melting steel surface. In considering heat exchange between the molten fuel stream and steel structure, the mechanical stability of the growing fuel crust is a most important concern. In the absence of the solid fuel layer, the melting steel is exposed to the hot fuel (at ∿3000°C) and the "driving force" for heat losses is the difference between the fuel temperature and the steel melting temperature, which can be of the order of 1600°C. On the other hand, if heat is transferred from the fuel flow to the steel melt front through a layer of frozen fuel, which isolates the melting steel from the fuel flow, the relevant driving temperature is reduced by about an order of magnitude to ∿150°C.

A description of the behavior of a fuel crust growing between an internally-heated molten UO_2 pool and its lower solid steel boundary is very important to transition phase analysis. If the formation of a solid fuel crust on the lower steel boundary is inhibited by the rapid growth of the intervening lighter steel melt layer, the steel melt will rise up through the heavier molten fuel region in the form of steel droplets (Taylor instability). This situation represents an effective large volume heat sink which can prevent pool boilup. It seems likely that the fate of the crust depends on a "race" between the crust growth process and the rate of development of buoyancy forces due to the presence of the underlying steel melt. Thus, slow crust growth should lead to a thin, weak crust which becomes laterally unstable through sidewise buckling, whereas rapid crust freezing will lead to a stable crust. A theoretical treatment of the mechanical stability of the crust was reported by Epstein [24]. This work parallels that of Rayleigh and Kelvin for the generation of waves at the plane interface between two different fluids but with both surface tension at the crust surface and elastic forces within the crust acting to stabilize the motion.

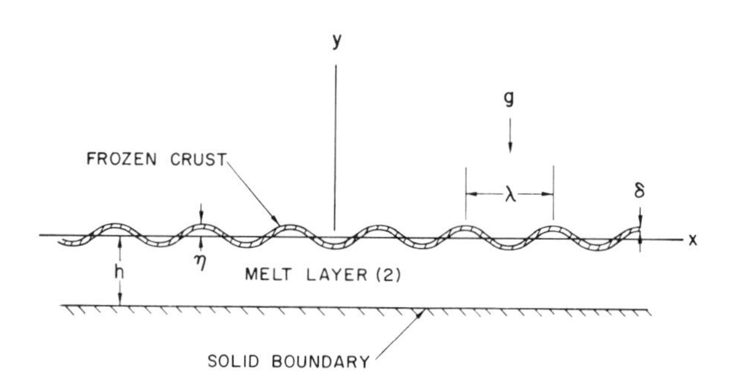

Fig. 14. Schematic diagram of frozen layer stability model.

Figure 14 represents a schematic illustration of a growing, submerged frozen crust. The upper fluid (material 1) of density ρ_1, occupies the space $y > 0$. The lower fluid layer (material 2) of density ρ_2 occupies the space $0 < y < -h$. A thin frozen layer of material 1 of instantaneous thickness δ separates the two fluids. Here we are concerned with crust stability in a gravity field only, when the upper fluid is more dense than the lower fluid ($\rho_1 > \rho_2$).

The frozen layer is deflected and set in two-dimensional motion according to the infinitesimal disturbance

$$\eta = \eta_0 e^{ikx+nt} \tag{42}$$

where k is the wave number, $2\pi/k$ is the wavelength λ of the disturbance, η_0 is the initial amplitude and n is the growth constant of the disturbance (complex frequency). The linearized hydrodynamic equation for each of the two fluid regions shown in Fig. 14 are of the form

$$\frac{\partial u}{\partial x} + \frac{\partial v}{\partial y} = 0 \quad (43)$$

$$\frac{\partial u}{\partial t} = - \frac{1}{\rho} \frac{\partial P}{\partial x} \quad (44)$$

$$\frac{\partial v}{\partial t} = - \frac{1}{\rho} \frac{\partial P}{\partial y} - g \quad (45)$$

and are connected through the following compatibility (matching) conditions at the frozen crust "interface" located at $y = \eta(x,t)$:

$$D \frac{\partial^4 \eta}{\partial x^4} - 2\sigma \frac{\partial^2 \eta}{\partial x^2} = P_2 - P_1 \quad (46)$$

$$v_1(x,\eta,t) = \frac{\partial \eta}{\partial t} \quad (47)$$

$$v_2(x,\eta,t) = \frac{\partial \eta}{\partial t} \quad (48)$$

where u and v are the small-disturbance velocities in the x- and y-directions, respectively, and P is the liquid pressure. Equation (46) expresses the condition of continuity of pressure across the interfacial crust. The first two terms in equation (46) represent the crust elasticity and the surface tension forces at the upper and lower crust surfaces, respectively, which act to stabilize the crust. Here D is the crust stiffness given by

$$D = \frac{E\delta^3}{12(1 - \varepsilon^2)} \quad (49)$$

and σ is the avearge surface free energy (tension) for the upper and lower crust surfaces; i.e., $\sigma = (\sigma_1 + \sigma_2)/2$. Equations (47) and (48) state that the velocity of either fluid at the crust surface is equal to the velocity of the crust itself. Finally, equations (43) - (45) are subject to the following conditions at $y \to \infty$ and $y = -h$:

$$v_1(x,\infty,t) = 0, \qquad v_2(x,-h,t) = 0 \quad (50)$$

Equations (43) - (45) have solutions of the form [25]

$$u = - \frac{\partial \phi}{\partial x}, \qquad v = - \frac{\partial \phi}{\partial y} \quad (51)$$

$$P = \rho \frac{\partial \phi}{\partial t} - \rho g y \quad (52)$$

where the velocity potential ϕ must satisfy

$$\frac{\partial^2 \phi}{\partial x^2} + \frac{\partial^2 \phi}{\partial y^2} = 0 \quad (53)$$

and the conditions

$$\frac{\partial \phi_1}{\partial y}(x,\infty,t) = 0, \qquad \frac{\partial \phi_2}{\partial y}(x,-h,t) = 0 \tag{54}$$

For the upper fluid, we assume the solution

$$\phi_1(x,y,t) = A_1 e^{-ky} e^{ikx+nt} \tag{55}$$

and for the lower fluid,

$$\phi_2(x,y,t) = A_2 \frac{\cosh[k(y + h)]}{\sinh (kh)} e^{ikx+nt} \tag{56}$$

where A_1 and A_2 are unknown constants. Note that equations (55) and (56) are consistent with boundary conditions (54). The pressure condition across the crust, equation (46), yields [see equations (42), (51), and (52)]

$$D\eta_0 k^4 + 2\sigma\eta_0 k^2 = \rho_2 n A_2 \coth (kh) - \rho_1 n A_1 + (\rho_1 - \rho_2) g\eta_0 \tag{57}$$

The kinematic conditions (47) and (48) give

$$A_1 k = \eta_0 n, \qquad -A_2 k = \eta_0 n \tag{58}$$

Eliminating A_1 and A_2 between equations (57) and (58) results in the desired expression for the growth rate of the disturbance:

$$n^2 = \frac{g(\rho_1 - \rho_2)k - Dk^5 - 2\sigma k^3}{\rho_1 + \rho_2 \coth (kh)} \tag{59}$$

We see from equation (59) that the amplitude of the initial disturbance grows only when $n > 0$, i.e., when

$$g(\rho_1 - \rho_2)k - (Dk^5 + 2\sigma k^3) > 0 \tag{60}$$

or crust motion is stable for wave numbers larger than a "cutoff wave number," k_c, given by

$$k_c = \left[\sqrt{\left(\frac{\sigma}{D}\right)^2 + \frac{(\rho_1 - \rho_2)g}{D}} - \frac{\sigma}{D}\right]^{1/2} \tag{61}$$

When the term $(\rho_1 - \rho_2)gD/\sigma^2 << 1$, crust elasticity limitations to the instability will be negligible. In such cases the crust behaves like a membrane with only the presence of surface energy to remove the instability.

The amplitude of the disturbance grows most rapidly when the wave number k in equation (59) is such that n, or n^2, is a maximum; i.e., $dn^2/dk = 0$. At this "most probable wave number," k_p, we have

$$g(\rho_1 - \rho_2) - (5Dk^4 + 6\sigma k^2) = \frac{\rho_2 h[1 - \coth^2(kh)]}{\rho_1 + \rho_2 \coth (kh)} [g(\rho_1 - \rho_2)k - (Dk^5 + 2\sigma k^3)] \tag{62}$$

In the general case, this equation is complex and cannot be solved analytically. We confine our analysis here to the two limiting cases: shallow lower liquid layers and deep lower liquid layers. When kh is small; i.e., the depth of the lower liquid is small compared with the wavelength of the disturbance, we find coth (kh) = 1/(kh), and equation (62) simplifies to

$$3Dk^4 + 4\sigma k^2 - g(\rho_1 - \rho_2) = 0 \tag{63}$$

from which the most probable wave number is obtained:

$$k_p = \left[\sqrt{\left(\frac{2}{3}\frac{\sigma}{D}\right)^2 + \frac{(\rho_1 - \rho_2)g}{3D}} - \frac{2}{3}\frac{\sigma}{D}\right]^{1/2} \quad ; \quad k_p h << 1 \tag{64}$$

If in equation (62) kh >> 1; i.e., the depth of the melt layer is great compared with the wavelength, we find

$$k_p = \left[\sqrt{\left(\frac{3\sigma}{5D}\right)^2 + \frac{(\rho_1 - \rho_2)g}{5D}} - \frac{3\sigma}{5D}\right]^{1/2} \quad ; \quad k_p h >> 1 \tag{65}$$

The growth time for an unstable crust wave or the breakup period of the crust is of the order of 1/n [26]. That is, during a time interval

$$\tau = \frac{1}{n(k_p)} \tag{66}$$

the amplitude of the most probable wavelength increases e times (see equation (42)). Again we consider the limiting case kh << 1, and from equations (59), (64), and (66) we obtain

$$\tau = \left[\frac{3\rho_2}{2hk_p^2[g(\rho_1 - \rho_2) - \sigma k_p^2]}\right]^{1/2} \quad ; \quad k_p h << 1 \tag{67}$$

where k_p is given by equation (64). If on the other hand we make kh large, we find

$$\tau = \left[\frac{5(\rho_1 + \rho_2)}{4k_p[g(\rho_1 - \rho_2) - \sigma k_p^2]}\right]^{1/2} \quad ; \quad k_p h >> 1 \tag{68}$$

where k_p is given by equation (65). The time, τ, it takes for the instability to develop as a function of the UO_2 fuel crust thickness is shown in Fig. 15 for selected values of the thickness of the lower steel melt layer. These values should be compared with the crust growth time. Conduction theory predicts the familiar square root crust growth law (see Section 4):

$$\delta = 2\chi\sqrt{\alpha t} \tag{69}$$

where χ, the "growth constant" is a function of the physical properties and initial temperatures of the upper and lower fluid materials. Clearly, the characteristic time, τ_f, for crust growth is

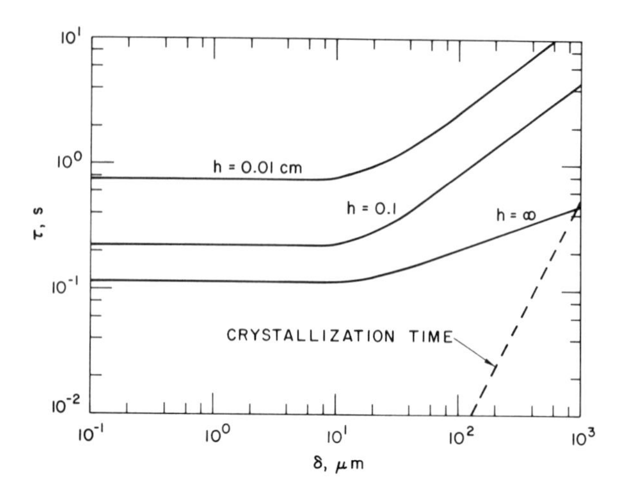

Fig. 15. Time for unstable wave growth versus UO_2 frozen layer thickness (for steel melt layers of infinite and finite depth) in a gravity field. From Epstein [24].

$$\tau_f = \frac{\delta^2}{4\chi^2\alpha} \tag{70}$$

This time is also plotted in Fig. 15 (dashed curve). The time for UO_2 crust growth on melting steel τ_f is completely negligible on the scale of the time for the growth of the most unstable wave for crust thicknesses below about 0.03 cm.

The two characteristic times, τ and τ_f are seen to be comparable when the effects of surface tension may be neglected (see Fig. 15). In this limit of large δ, we find the simple asymptotic behavior (see equations (49), (64), (65), (67), and (68))

$$\tau = \left[\frac{3\rho_2}{2gh(\rho_1 - \rho_2)}\right]^{1/2}\left[\frac{3E\delta^3}{12(1 - \varepsilon^2)g(\rho_1 - \rho_2)}\right]^{1/4} \tag{71}$$

for $k_p h \ll 1$ and

$$\tau = \left[\frac{5}{4g}\,\frac{\rho_1 + \rho_2}{\rho_1 - \rho_2}\right]^{1/2}\left[\frac{5E\delta^3}{12(1 - \varepsilon^2)g(\rho_1 - \rho_2)}\right]^{1/8} \tag{72}$$

for $k_p h \gg 1$. The foregoing results are not directly applicable to a crust growing between two different fluid materials. Clearly, the variation of crust thickness with time plays a role in the instability phenomena. We should expect, nevertheless, that the assumption

$$\tau = \tau_f \tag{73}$$

should furnish an approximation of the right order of magnitude for the crust thickness at the onset of instability.

Using equation (70) in equations (71) and (72) to obtain δ and substituting the results in equation (61), yields the approximations for the cutoff wavelength at the onset of crust instability:

$$\lambda_c = 7.11\left[\frac{\chi^4\alpha^2\rho_2}{hg(\rho_1 - \rho_2)}\right]^{3/10}\left[\frac{E}{(1-\varepsilon^2)g(\rho_1-\rho_2)}\right]^{2/5}; \quad k_c h \ll 1 \tag{74}$$

$$\lambda_c = 6.41\left[\frac{\chi^4\alpha^2}{g}\,\frac{\rho_1+\rho_2}{\rho_1-\rho_2}\right]^{3/13}\left[\frac{E}{(1-\varepsilon^2)g(\rho_1-\rho_2)}\right]^{4/13}; \quad k_c h \gg 1 \tag{75}$$

For a crust growing on a melting solid we have from conduction theory (see Section 4)

$$h = 2\chi_m\sqrt{\alpha_m t} \tag{76}$$

where χ_m is the melting constant and, like the growth constant χ, is strictly a function of the physical properties and initial temperatures of the upper and lower materials. In this case the crust breakup criterion is $\tau = \tau_f = \tau_m$ where $\tau_m = h^2/(4\chi_m^{\ 2}\alpha_m)$ is the characteristic time for melting. In most cases of interest the characteristic times are comparable when $k_p h \ll 1$. Thus, from equations (61), (69), and (71) we have the cutoff wavelength for the instability of a crust growing on a melting solid, viz

$$\lambda_c = 5.61\left[\frac{\chi^5\alpha^2\rho_2}{\chi_m(\alpha_m/\alpha)^{1/2}g(\rho_1-\rho_2)}\right]^{3/14}\left[\frac{E}{(1-\varepsilon^2)g(\rho_1-\rho_2)}\right]^{5/14} \tag{77}$$

The cutoff wavelength in equations (74), (75), and (77) can be interpreted as the critical wavelength for crust breakup. If the upper and lower materials are included between two parallel vertical walls, this imposes an upper limit to the admissible wavelength. Equation (77) predicts that UO_2 crust instability on melting steel must manifest itself by the formation of crust waves that exceed $\sim$2.0 m in length. These waves are sufficiently long compared with molten UO_2 pool (reactor core) geometry to ensure the stability of the frozen UO_2 crust.

A series of simple experiments were performed by Epstein [24] in which a relatively hot, heavy liquid was suddenly poured over a column of lighter cold material (liquid or solid). The lower liquid was placed in a glass test vessel. If a lower material in solid form was desired, the liquid was frozen in place at the bottom of the test vessel. The upper liquid was first placed in a pouring vessel. The test was initiated by tipping the pouring vessel, allowing liquid to impact and spread over the surface of the lower material. The liquid was poured rapidly enough to ensure that the rate of accumulation of the upper liquid (layer) exceeded the rate of phase conversion at the lower material-upper liquid interface. The initial temperature of the upper liquid was just above its melting point so that forced convection heat transfer was negligible over the duration of the experiment.

The experimental observations of the mechanical stability of the crust growing into the upper liquid along with crust stability predictions are reported in Table 1. In experiment 1, liquid Freon Fluorocarbon 112A ($\sim$50°C) was poured over water at $\sim$20°C in a 7-cm diam vessel. The Freon 112A was observed to freeze upon contact with the surface of the water forming a wavy but stable crust which prevented the water (ρ_2 = 1g cm^{-3}) from rising through the denser liquid Freon (ρ_1 = 1.6g cm^{-3}). In this case a prediction of marginal (or neutral) stability was obtained since the calculated cutoff wavelength is approximately equal to the vessel diameter. In experiment 2, a very stable, smooth lead crust (ρ_1 = 11.7g cm^{-3}) grew above melting gallium (ρ_2 = 6.0g cm^{-3}) in agreement with the theory. Experiments 3-5, involving an ice crust growing on cold liquid octane, indicated a critical vessel diameter below which the crust is stable, as predicted by equation (75). For a fixed vessel diameter, ice crust breakup was observed on a column of liquid octane whereas stable crust growth was observed on melting octane (compare experiment 3 with experiment 6) in agreement with theory. For ice growth on melting octane we predict ice breakup when the vessel diameter exceeds $\sim$16 cm. This was not determined experimentally (see experiments 6-8). Instead, smooth, stable ice covers were observed to grow on melting octane in vessels as large as 30 cm in diameter. It may be that in situations where large cutoff wavelengths (say $\gtrsim$ 10 cm) are required, natural sources of small disturbances are insufficient to start the instability.

Epstein et. al. [27] conducted experimental investigations of the behavior of a freeze layer on a melting surface in a stagnation-region flow geometry. Their experimental apparatus is shown in Fig. 16. Water from a large constant-temperature reservoir is passed through a nozzle. The water jet is directed upward against the lower end of a meltable rod, having a diameter about twice that of the nozzle. The selection of the diameters of the nozzle and the impingement surface was made to ensure that a flat melting surface was maintained, so that melting rate measurements were indeed restricted to the stagnation region. The meltable rod is attached to a slide so that its vertical position is adjustable through the action of a high-tension cable. The movement of the cable was controlled with a variable speed electric motor. The jet is deflected by the melting impingement surface thereby setting up a flow of liquid normal to the jet that spreads radially outwards and forms a free "fan" jet a short distance from the edge of the melting surface. A thin plate, which serves as a melt-front finder (or indicator), is located a fixed distance above the base of the apparatus by means of a support pin. During a run, the melting rod was "fed" into the jet at a rate such that the indicator plate remained submerged in the thin fan jet deflected by the melting surface. This measure proved to be fully successful in achieving a constant separation distance between the nozzle and the melting surface. The rod melting rates were determined with the aid of instrumentation which converts the downward

Table 1. Results of Crust-stability Experiments

Expt	Material Pair (Upper) (Lower)	Melting Point, °C	Specific Gravity	Physical State and Temp, °C	Vessel Diameter d, cm	Cutoff Wavelength λ_c, cm	Crust-stability Theory	Observation[a]
1	Freon-112A H_2O	40 0	1.6 1.0	Liquid, 50 Liquid, 20	7	6[b]	Marginal $\lambda_c \simeq d$	Stable
2	Lead Gallium	327 30	11.70 6.0	Liquid, 335 Solid, -20	7	280	Stable $\lambda_c >> d$	Stable
3	H_2O C_8H_{18}	0 -57	1.0 0.7	Liquid, 2 Liquid, -30	7	3	Unstable $\lambda_c < d$	Unstable
4	"	"	"	Liquid, 2 Liquid, -55	7	6	Marginal $\lambda_c \simeq d$	Unstable
5	"	"	"	"	3	6	Stable $\lambda_c > d$	Stable[c]
6	"	"	"	Liquid, 2 Solid, -78	7	16	Stable $\lambda_c \simeq d$	Stable
7	"	"	"	"	15	16	Marginal $\lambda_c \simeq d$	Stable
8	"	"	"	"	30	16	Unstable $\lambda_c < d$	Stable

[a]The crust was pronounced stable if it prevented most of the lower liquid or melt from rising to the surface of the upper liquid. Occasionally a small amount of lower liquid material was observed to escape from beneath the crust through small openings that formed between the vessel wall and the crust edge.

[b]The elastic constants of solid Freon Fluorocarbon 112A were taken to be comparable to those of other ethylene-type molecules for which such properties are available; in particular E = 400 MPa and ε = 0.4 for polytetrafluorethylene were chosen.

[c]A relatively thick continuous ice layer containing trapped octane droplets was observed.

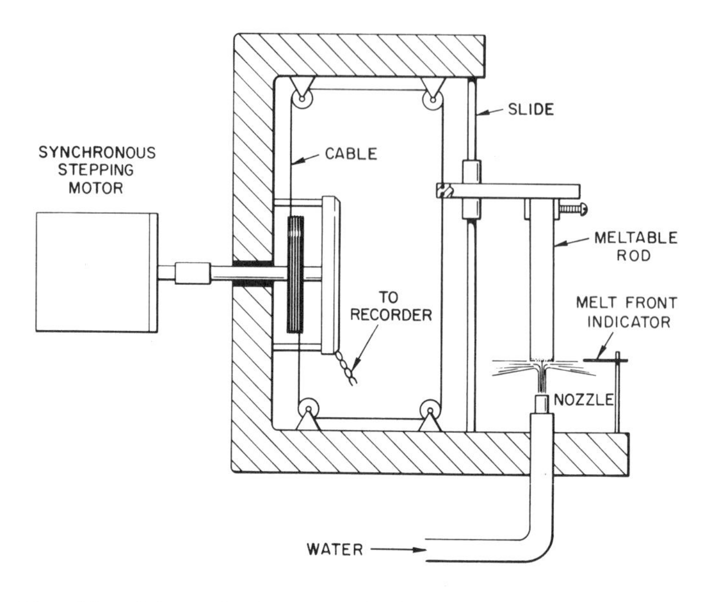

Fig. 16. Schematic diagram of the experimental apparatus for jet solidification. From Epstein et. al. [27].

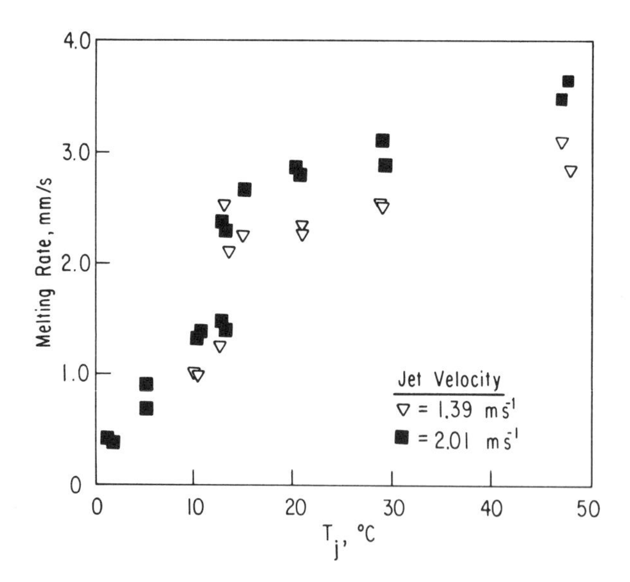

Fig. 17. Melting rate of octane rod as a function of jet temperature; jet velocity as a parameter. From Epstein et. al. [27].

rod displacement history into an electric signal that can be read and recorded on an electronic strip-chart recorder. The test specimens were cylindrical rods of frozen octane (m.p. -56.6°C) and frozen mercury (m.p. -38.9°C) about 15 cm high and 0.95 cm in diameter.

For a given jet velocity, two sharply defined melting regimes for the water-jet/ octane-rod system were observed by Epstein et. al. [27]. This is illustrated in Fig. 17, which shows a typical plot of melting velocity of an octane rod against water jet temperature. The melting rate data fall on two different curves depending on whether the jet temperature is less than or greater than 13°C. For jet temperatures in the range $0 < T_j < 13°C$, the melting velocity increases rapidly with increasing jet temperature, whereas for $T_j > 13°C$ the melting velocity is relatively insensitive to the jet temperature. This behavior is believed to be directly related to the formation of a protective ice crust on the octane melt layer when the jet temperature falls below 13°C, as manifest by the step change in rod melting rate. If heat is transferred from the liquid jet to the octane melt front through a layer of ice, which isolates the melting octane from the water jet, the heat flux is proportional to the difference between the jet temperature and the ice melting temperature. Therefore, we expect the melting rate to fall off rapidly as the jet temperature decreases, and tend to zero when $T_j \rightarrow 0°C$. This behavior is clearly seen in Fig. 17. On the other hand, beyond the point of incipient crust formation where $T_j > 13°C$, the thermally protective ice crust is absent. The melting octane is directly exposed to the hot water jet, and the relevant temperature difference for heat convection is the jet temperature minus the octane melting temperature (-56.5°C). Thus, the convective heat flux can only vary by at most 50 percent over the jet temperature range 14 - 50°C.

Experiments carried out with the water-jet/octane-rod system at increased jet velocities of 2.80 and 3.16 m s^{-1} also revealed the step-change behavior shown in Fig. 17 when T_j was about 13°C. This threshold was reproducible to within ± 1.0°C. The jet temperature required for a step change in melting rate appears to be insensitive to jet velocity. Thus, this step change is <u>not</u> likely due to a change in flow pattern such as a transition from laminar to turbulent flow in the impingement region. Again, the observed discontinuity in the melting rate is consistent with the formation of a protective ice crust on melting octane when $T_j < 13°C$. A laminar-axisymmetric jet flow model was developed by Epstein et. al. to describe melting heat transfer in the presence of jet solidification. This crust/melt-layer model predicts the onset of crust formation to be independent of jet velocity. Moreover, calculations indicated that, for water jet impingement on melting octane, incipient ice crust formation is inevitable when the jet temperature drops below 15°C; this prediction is in excellent agreement with the temperature at which the step change in melting rate was observed.

The melting rate data obtained by Epstein et. al. [27] for water jet impingement on melting mercury are shown in Fig. 18. Qualitatively, the melting behavior is similar to that of octane at low jet temperatures insofar as the melting velocity decreases with decreasing T_j, approaching zero when $T_j \rightarrow 0°C$, illustrating once again the thermally protective nature of the ice crust. The one obvious distinction between the melting rate data for the octane rod and for the mercury rod is that, over the jet temperature range investigated, the melting mercury rod does not appear to lose its protective ice cover. The results of calculations using the jet solidification model showed that incipient crust formation will occur at a water jet temperature equal to 117°C; consequently, for the range of jet temperatures studied in [27] an ice crust will always be present on melting mercury. This explains the absence of a temperature threshold or melting rate discontinuity in Fig.

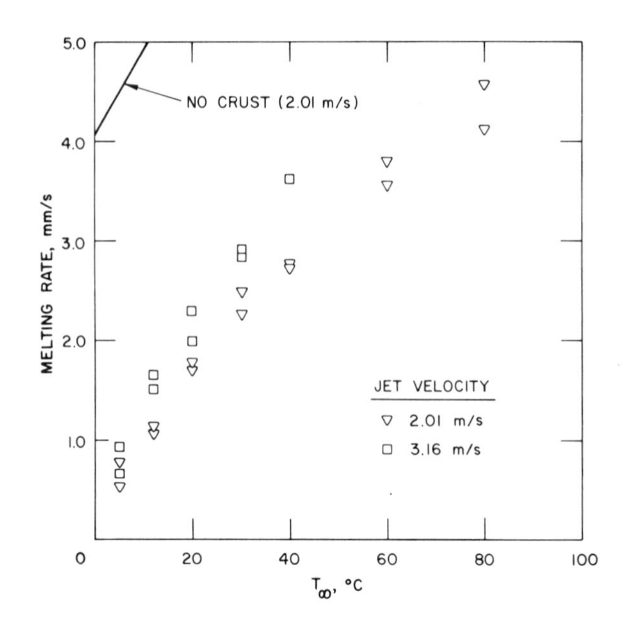

Fig. 18. Melting rate of mercury rod as a function of jet temperature; jet velocity as a parameter. From Epstein et. al. [27].

18. Also shown in Fig. 18, is the result of calculations of the melting rate based on a two-material axisymmetric flow model that accurately predicts melting rates in situations in which the impinging liquid does not freeze on the melting solid [28]. This no-crust melting theory is seen to overestimate the mercury melting rate by about one order of magnitude for $0 < T_j < 10°C$; it is found to overestimate the melting rate by a factor of 4.0 for T_j as high as 80°C.

As a practical matter, if the flowing material has a large potential for solidification on the wall material, as in the above-referenced experiments, crust removal leading to exposed wall-melt layers is difficult to attain. Even if crust breakup and removal take place, it is doubtful that molten-stream/wall-melt interfaces would appear for any significant length of time. Clearly, if the time for flow recrystallization is negligible compared to a time characterizing the rapidity of mechanical crust removal, the distinction between mechanically stable and unstable crusts loses all its significance. Under such conditions, crust growth would be expected to influence the wall-melting rate strongly. This is the case for UO_2 crusting on melting steel.

Suppose that a portion of the solid steel surface is suddenly exposed to molten fuel as a result of the turbulent moving fuel stream, sweeping away a section of protective fuel crust and underlying steel melt. Fuel will then quickly fill this clearing and contact the solid steel surface. Let us estimate the time necessary to grow a new protective fuel crust. If perfect thermal contact is achieved, an interface temperature of ∿1600°C at the surface of fuel-steel separation will be established instantaneously (see Section 4).

In the opposite case of poor thermal contact, the steel surface will not melt and the fuel crust is stable.

As the temperature of molten fuel is lowered, the kinetics of the formation and breakage of solid fuel clusters (embryos) dictate that spontaneous nucleation will occur when the fuel temperature is near ∿2100°C [29]. (A good rule of thumb for predicing the lowest temperature to which a pure liquid could be supercooled is $\sim 0.8\ T_{mp}$, where T_{mp} is the melting point in degrees Kelvin.) In other words, whether crystallization begins at an interface or within the melt, the molten fuel cannot be expected to remain liquid below this temperature. In fact, the process of homogeneous nucleation represents the maximum supercooling that can be tolerated by pure fuel. The presence of the fuel-steel interface will most likely encourage nucleation, thereby lowering the tolerance of the fuel melt to supercooling.

The fact that the interface temperature between molten fuel and steel falls well below the spontaneous nucleation temperature indicates that fuel crystallization cannot proceed until a sufficiently thick thermal boundary layer, δ, is developed to "uncover" crystalline embryos of the critical size r_{crit}, namely,

$$r_{crit} \simeq 2.5 \frac{\sigma}{h_{s\ell}} \tag{78}$$

where $h_{s\ell}$ is the latent heat of fusion (per cubic centimeter) and σ is the interfacial tension [30]. Conduction theory predicts the familiar square-root growth law for the thermal layer:

$$\delta = \sqrt{\pi \alpha t}\ . \tag{79}$$

The waiting time for fuel crystallization is obtained by invoking the equality $\delta \simeq 2r_{crit}$, or

$$t \simeq \frac{8}{\alpha} \left(\frac{\sigma}{h_{s\ell}} \right)^2 . \tag{80}$$

For molten UO_2 with a thermal diffusivity $\alpha = 0.007\ cm^2\ s^{-1}$, $\sigma \simeq 500\ g\ s^{-2}$, and $h_{s\ell} = 2.8 \times 10^{-10}\ g\ cm^{-1}\ s^{-2}$, we get

$$t \simeq 3.6 \times 10^{-13}\ s\ .$$

It thus turns out that the time to crystallization is negligibly small. This must be considered as a rather crude approximation, since for very short times, the Fourier conduction formulation (79) may not be valid. The calculation does, however, demonstrate that UO_2 crystal growth at the UO_2-steel interface must take place practically instantly.

Of course, the steel melt layer will not be protected from the fuel flow until individual crystalline embryos combine to form a frozen layer in thermodynamic equilibrium with the fuel stream (at the fuel solidification front). Under this condition, when conduction-controlled crust growth prevails, the temperature drop on the fuel side of the growing crust is limited to the difference between the free stream temperature and the fuel melting temperature. Early in the crystallization process, there is no conduction restriction to crust growth. Instead, freezing is controlled by the rate at which molecules cross the liquid-solid interface, that is, by the molecular reordering process. During this period, the fuel temperature at the solidification front, T, rises to the fuel melting temperature T_{mp}, while the solidification front propagates into the UO_2 melt at a velocity $\dot{R}$ which can be described by the exponential law [29]

$$\dot{R}(\text{cm s}^{-1}) = 0.7884 \times 10^{-4}\phi[\exp(-1.5\phi^{-1}) - 19.4 \exp(4.47\phi^{-1})], \qquad (81)$$

where

$$\phi \equiv \frac{T}{T_{mp}} .$$

At the spontaneous nucleation temperature of ∿2100°C established for UO_2, the solidification velocity is predicted to be

$$\dot{R} \simeq 508 \text{ cm s}^{-1} . \qquad (82)$$

For conduction-limited crust growth, the velocity of solidification is usually expressed by

$$\dot{R} = \lambda\sqrt{\alpha/t} , \qquad (83)$$

where the growth constant $\lambda \simeq 0.9$ for a UO_2 layer growing on melting steel (see Section 4). The time at which kinetically controlled UO_2 solidification gives way to conduction-limited growth can be obtained approximately by equating equations (82) and (83) and solving for t:

$$t \simeq \frac{\lambda^2\alpha}{\dot{R}^2} = 2.2 \times 10^{-8} \text{ s} . \qquad (84)$$

The time needed for establishing conduction-controlled UO_2 crust growth on melting steel is indeed very short. Let us compare this time with the time it takes turbulence eddies to remove sections of crust from the melting wall. The turbulent or eddy velocity u_{turb} in the wall layer is small compared to that of the free stream, U_∞. In terms of an order of magnitude (see Ref. [26], Chapter I), we have for flow past a flat plate

$$u_{turb} \simeq 0.19 \frac{U_\infty}{Re^{0.1}} , \qquad (85)$$

where Re is the flow Reynolds number based on the length scale in the direction of flow. For a UO_2 flow of $\sim 10^3$ cm s^{-1} and a length scale of ∿10 cm, $Re \simeq 10^6$. According to equation (85), eddy velocities, which cause crust motion away from the wall, are of the order of magnitude of

$$u_{turb} \simeq 45.0 \text{ cm s}^{-1} .$$

Over the period of time it takes to grow a protective crust, of the order of 10^{-8} s [see equation (84)], the crust displacement due to turbulence is only 10^{-6} cm, a distance equivalent to about 1/10 the crust thickness. Clearly, there is insufficient time to remove the crust, considering the short period during which fuel crystallization will "cement" growing cracks or "repair" existing gaps between crust sections. This result indicates that predictions of steel melting rates must incorporate the insulating effects of a solid fuel layer.

6. FREEZING OF FLOWING FUEL IN A STEEL CHANNEL

As mentioned in Chapter 15, Section 2, the sequence of events for the postulated LOF accident includes the upward dispersal of boiling core material (molten fuel and molten steel) out of the active core region. The extent of penetration of the molten material into the largely intact pin structure above the core is a major concern, as fuel freezing can lead to a temporarily "bottled-up" core condition.

There is extensive literature on the freezing of liquids in channels, and a brief review of the investigations conducted during the last decade is presented in [31]. All of this work is limited to (i) studies of the shape of the frozen layer and the freezing section pressure drop under steady-state conditions or (ii) the transient growth of the frozen layer for the case where a fully developed channel flow is suddenly disturbed by the introduction of a uniform and steady sub-freezing temperature along a certain length of the channel well. The problem of importance to fast nuclear reactor safety analysis is the transient freezing of molten core material as it penetrates into an initially empty channel. This problem differs from the work mentioned above in that the freezing length increases with time. As demonstrated in Section 4, another distinctive characteristic of fuel freezing is that the inside surface of the steel channel wall will melt upon contact with the penetrating liquid fuel if the initial temperature of the channel wall is sufficiently high (i.e., $\stackrel{\sim}{>}$ 700°C). The work on freezing of an advancing flow in the absence of melting channel walls will be presented first.

The first study of freezing of a penetrating tube flow was performed by Cheung and Baker [32]. They conducted a series of tests in which various liquids were allowed to flow under gravity into long copper tubes cooled by liquid nitrogen or by a dry ice-acetone bath. Different constant driving heads in the liquid reservoir were used, and the penetration distances were measured. An empirical correlation of the transient freezing data was obtained. The authors concluded that their experimental results were in good qualitative agreement with the classical conduction freezing model; i.e., freezing is governed by transient heat conduction through a growing frozen layer at the channel wall (conduction model). Madejski [33] obtained a simple solution to the penetration and freezing of a liquid at its fusion temperature in a cold channel. His heat conduction model is based on the assumption that the pressure drop over the instantaneous freezing length is the same as in channels of constant cross-section. The penetration of a liquid at its freezing temperature into a tube was treated theoretically and experimentally by Epstein, Yim and Cheung [31]. We present here in some detail the model for transient freezing developed by these workers since this represents one of the more complete models for liquid penetration and freezing in an initially empty tube.

We consider the transient flow and freezing system shown in Fig. 19 (a schematic diagram of the apparatus used in Refs. [31,32]). The reservoir (or tank) contains a liquid at its freezing temperature T_{mp} (saturated liquid) having density ρ and kinematic viscosity ν. At time $t = 0$, the liquid is allowed to drain out at the bottom of the tank and enter a cold tube at temperature $T_w (T_w << T_{mp})$. The problem is formulated by asking what the flow

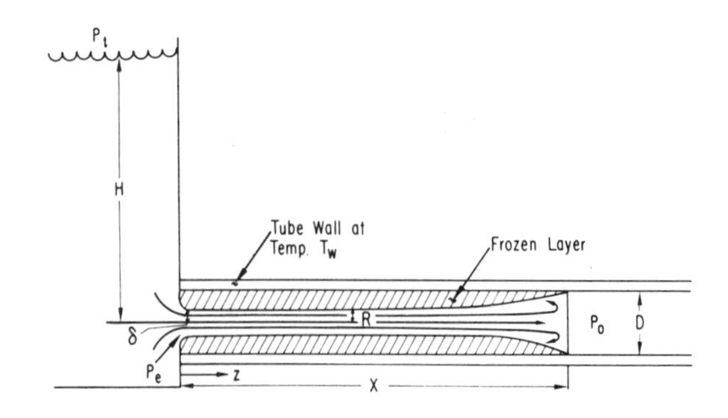

Fig. 19. Penetration of liquid limited by crust growth, showing instantaneous frozen layer profile, and nomenclature.

penetration length X(t) is at the moment solidification is complete in the inlet tube region. We make the assumptions that (1) the temperature at the liquid-solid interface is equal to the equilibrium solidification temperature; (2) all physical properties are constant and the densities of the liquid and frozen deposit are taken to be the same; (3) the liquid-injection pressure or tank pressure, P_t, is sufficiently large so that most of the liquid displacement occurs under turbulent flow conditions; (4) liquid flow inertia and entrance losses are small relative to the frictional pressure drop; (5) the liquid penetration or flow rate is insensitive to details of the crust shape; and (6) the tube wall is maintained at constant temperature T_w.

Assumptions 1-5 have been discussed in some detail in References [31,34]. Criteria for the validity of assumptions 3 and 4 are presented in [31]. Assumption 5 is verified theoretically in [34].

According to the model, the liquid shear at the solid-liquid interface is well-described locally by (assumption 3)

$$\tau = \frac{1}{2} f \rho u^2 \tag{86}$$

where the friction factor, f, is obtained from the Blasius formula:

$$f = 0.0791 \left(\frac{2uR}{\nu} \right)^{-1/4} \tag{87}$$

In the above expressions, a slug-flow axial velocity profile u = u(z,t) is assumed and ρ is the density of the liquid and R(z,t) is the radial interface location. The applicability of equations (86) and (87) in the presence of transient crust growth is discussed in Reference [31]. Briefly, we require the time necessary for the development of the turbulent velocity profile to be small compared with the time it takes for solidification to be complete at the channel entrance. In almost all cases of practical interest, this requirement is met.

The momentum balance between pressure and frictional forces over the instantaneous freezing length, X(t), takes the form

$$\frac{2}{\rho}\int_0^X \frac{\tau}{R}\,dz = \frac{\Delta P}{\rho} \tag{88}$$

We assume a square-root function for the axial variation of the crust shape (assumption 5):

$$\frac{R}{R_0} = 1 - \left(1 - \frac{\delta}{R_0}\right)\left(1 - \frac{z}{X}\right)^{1/2} \tag{89}$$

where $\delta(t)$ is the instantaneous radius of the tube at the inlet location $z = 0$ (see Fig. 19). Equation (89) satisfies the conditions $R = \delta$ at $z = 0$ and $R = R_0$ at $z = X$, where R_0 is the initial radius of the tube. An overall mass balance on the incompressible advancing flow between z and X(t) leads to the following relation between the local axial velocity u and the velocity of the leading edge, dX/dt:

$$uR^2 = \frac{dX}{dt} R_0^2 \tag{90}$$

To complete the system of equations it remains to specify $\delta(t)$. Clearly, the crust growth behavior in the inlet region is equivalent to the problem of freezing a saturated liquid inside a cylinder. The problem of the inward freezing of a circular cylinder has been treated in numerous papers (see, e.g., [35,36]). Stefan [36] utilized an approximate collocation method which takes full account of the movement of the phase conversion front and transient heat conduction within the frozen layer. For freezing a saturated liquid inside a cylinder this technique yields the inverted solution.

$$2(\delta/R_0)^2 \ln(\delta/R_0) - (\delta/R_0)^2 + 1$$

$$= \frac{4\alpha_s}{R_0^2}\left[[1 + 2c_s(T_{mp} - T_w)/h_{\ell s}]^{1/2} - 1\right]t \tag{91}$$

A comparison of equation (91) with numerical results [35] shows that the error is less than ∿20.0 percent. In equation (91) c_s and α_s are the heat capacity and thermal diffusivity of the frozen layer, respectively, and $h_{\ell s}$ is the latent heat of fusion.

Substituting equations (86), (87) and (89) into equation (88), performing the indicated integration, the equation of motion for the penetrating liquid becomes

$$0.1582\ Re^{-1/4}\left(\frac{dX/dt}{\sqrt{2\Delta P/\rho}}\right)^{7/4} \cdot \frac{X}{R} F(\delta/R_0) = 1.0 \tag{92}$$

where

$$F(\delta/R_0) = \frac{2}{(1 - \delta/R_0)^2}\left|\,[(\delta/R_0)^{-m+1} - 1]/(m - 1)\right.$$

$$\left. - [(\delta/R_0)^{-m+2} - 1]/(m - 2)\,\right| \tag{93}$$

In the above expressions the index m = 19/4 and Re is defined as the Reynolds number based on the total pressure drop $\Delta P = P_t - P_o + 2gH$ (see Fig. 19):

$$Re = \frac{(2\Delta P/\rho)^{1/2} D}{\nu} \tag{94}$$

The final liquid penetration length, X_p, before the tube freezes shut is found by eliminating t in equation (92) in favor of the instantaneous radius, $\delta(t)$, at the tube inlet via equation (91) and then integrating the result between the limits X = 0 when $\delta = R_0$ and $X = X_p$ when $\delta = 0$ to obtain

$$\frac{X_p}{D} = 0.155\ Re^{8/11} \left[\frac{Pr(\alpha_\ell/\alpha_s)}{B}\right]^{7/11} \tag{95}$$

where

$$B \equiv [1 + 2c_s(T_{mp} - T_w)/h_{\ell s}] \tag{96}$$

To test the validity of the model, some experimental studies were carried out in a long cold tube [31]. The cold tube consisted of a copper coil immersed in a bath of liquid coolant. When the tube was cooled to the bath temperature, a valve which separated the coil and tank was opened to initiate the flow of liquid. The valve was located in a short connecting tube (not shown in Fig. 19) which separated the tank and freezing coil. The tank containing the saturated liquid was pressurized up to ∿ 1 atm so that the liquid flowed into the cold tube under the influence of both gravity and a pressure gradient. The tank was constructed of glass, with a convergent glass nozzle in the wall near the bottom which fed the connecting tube. The liquid penetration before freezing was obtained by measuring the difference in weights of the copper coil before and after penetration. The principal working fluid used in this study was Freon 112A which melts at 40.5°C. The penetration data for Freon 112A compares well with equation (95), as shown in Fig. 20. Experiments were

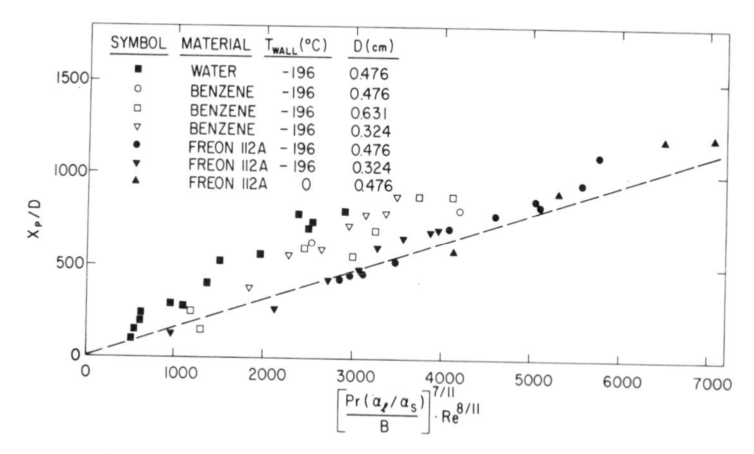

Fig. 20. Comparison of measured penetration lengths with equation (95) (dashed). From Epstein, Yim, and Cheung [31].

also carried out with both water and benzene flows. The predicted penetration lengths fall below the experimental data, due to the difficulty in maintaining a constant tube-wall temperature with these materials. The Biot number, or index of the relative resistance to heat transfer in the boiling liquid nitrogen coolant and in the growing frozen layer, is ~ 0.09 for Freon 112A. This is to be compared with Biot numbers of 0.4 and 1.6 for benzene and water flows, respectively. Good agreement between theory and experiment was obtained for all freezing materials by replacing equation (95) with a more elaborate solidification model that accounts for resistance to heat transfer in the liquid coolant [37].

The foregoing analysis has been recently extended by Epstein, Stachyra, and Lambert [38] to consider the amount of liquid displaced into or through a rod bundle before freezing shut. When the supply pressure drop ΔP is sufficiently low, the rod bundle freezes shut before the liquid can advance to the bundle exit at $z = L$. In this case the final liquid penetration length was found to be given by

$$\frac{X_p}{D_h} = 0.085\left(\frac{\nu}{\chi^2 \alpha_s}\right)^{7/11}\left(\frac{\Delta P D_h^2}{\rho \nu^2}\right)^{4/11} \tag{97}$$

where, again, χ is the growth constant for crust growth on a cold wall (see Section 4), and D_h is the hydraulic diameter of the rod bundle. At high injection pressures, an amount of liquid mass m_p will emerge from the rod-bundle exit before the inlet region is closed by the solidified layers. The theory reveals that

$$\frac{m_p}{\rho L C D_h} = 3.33 \times 10^{-3}\left(\frac{D_h \nu}{\chi^2 \alpha_s L}\right)\left(\frac{\Delta P D_h}{\rho \nu^2 L}\right)^{4/7} - 0.159 \tag{98}$$

where C is the wetted perimeter of the rod bundle.

It should be recognized that the theory reviewed here will not apply when the liquid entering the tube is at a temperature higher than the freezing point (superheated). In this case complete occlusion of the tube with frozen material ultimately occurs at some location between the inlet and the leading edge of the liquid flow. Calculation of the penetration distance in this most general situation is quite difficult but would constitute a useful next step.

Ostensen and Jackson [39] were the first to consider the fuel penetration behavior in the presence of a melting steel channel wall. These workers expressed the opinion that the rate of freezing of fuel in steel channels is controlled by turbulent heat transport to the steel wall. It was postulated by these authors that since the steel surface is molten, a solid fuel layer would not have an underlying surface to cling to. Moreover, the region just behind the leading edge of the flow, where freezing is expected to occur first, appears as a "slush" and freezing is complete when the latent heat of fusion is "removed" from the slush by further (turbulent) heat loss to the steel channel wall ("bulk-freezing model").

Figure 21 describes the situation in which molten fuel flowing in a channel is being frozen in a bulk manner just behind the fuel front. The energy given

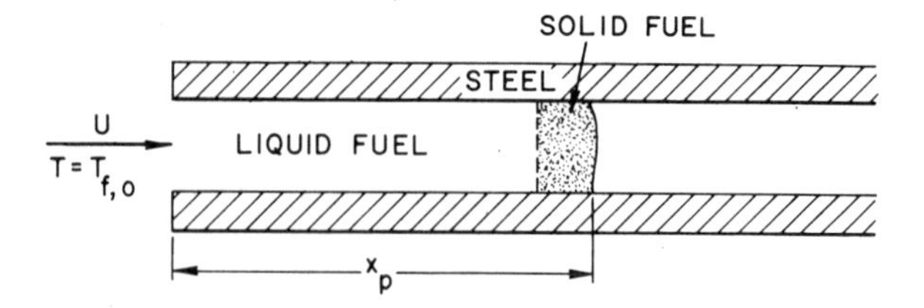

Fig. 21. Penetration of molten fuel limited by bulk solidification.

up by an element of molten fuel of hydraulic diameter D, length Δx, initial temperature $T_{f,0}$, and velocity U, upon cooling to the solid state, is

$$\frac{\pi}{4} D^2 \rho_f (\Delta x) [c_f (T_{f,0} - T_{f,mp}) + h_{\ell s,f}] \ .$$

where the subscript f refers to properties of the fuel. If we set this equal to the total heat transferred by turbulent convection over the fuel element penetration distance X_p, namely,

$$\pi D (\Delta x) h (T_{f,0} - T_w) X_p / U \ ,$$

we obtain

$$X_p = \frac{1}{4} \frac{UD\rho_f c_f}{h} \frac{h_{\ell s,f}/c_f + (T_{f,0} - T_{f,mp})}{T_{f,0} - T_w} \ , \tag{99}$$

where h is the turbulent heat-transfer coefficient and T_w is the temperature of the channel wall. In deriving equation (99), we have neglected the fact that changes in the mean temperature of the molten fuel element occur as the element proceeds along the channel. It can be shown that equation (99) is a valid approximation providing $(T_{f,0} - T_{f,mp})/(T_{f,0} - T_w) << 1$. Typically, this ratio is ~ 0.25. Assuming that turbulent heat transport within the channel is well represented by the Reynolds' analogy, which provides a linear relationship between heat-transfer coefficient and the frictional shear stress τ, we have†

$$h = c_f \frac{\tau}{U} = c_f \frac{\frac{1}{2} f \rho_f U^2}{U} = \frac{f}{2} \rho_f c_f U \ , \tag{100}$$

where f is the dimensionless coefficient of friction ($f \sim 0.005$). In using equation (100), we have ignored the complex "tumbling" flow pattern that must exist in the vicinity of the fuel front. Combining equations (99) and (100) leads to the following simple expression for X_p:

† This is an excellent approximation for molten UO_2 since its Prandtl number is close to unity.

$$X_p = \frac{1}{2}\frac{D}{f}\frac{h_{\ell s,f}/c_f + (T_{f,0} - T_{f,mp})}{T_{f,0} - T_w} . \tag{101}$$

Note that the bulk freezing model indicates that the penetration length is proportional to the channel diameter and independent of the supply pressure, whereas in the conduction model, X_p is approximately proportional to the square of the hydraulic diameter and the cube root of the pressure drop.

It is of interest to compare both the conduction model and the bulk freezing model with experiment. No experimental data are available on the transient freezing of pure molten ceramic fuel in turbulent pipe flow through a melting steel channel. A series of tests in which a chemically produced mixture of high-pressure nitrogen gas, 3200°C molten uranium dioxide and metallic molybdenum was injected into a steel rod bundle is reported in references [40-42]. In the thermite tests of Ref. 40, molten UO_2 at $\sim$50 atm pressure was injected into a seven-pin subassembly structure with hollow steel pins having a 0.5-cm-diameter flow path. At the conclusion of the experiments the major fraction of the frozen thermite debris ("clinker") was located at approximately 30 cm from the pin-bundle inlet. Behind the clinker, an empty subassembly was observed indicating a severe ablation effect. The remaining frozen thermite debris was found to be spread out over the intact pin-bundle structure ahead of the clinker zone. Taking T_w equal to the melting temperature of steel ($\sim$ 1400°C), we get X_p = 24 cm with the bulk freezing model [equation (101)]. This is to be compared with the conduction model, equation (97), which predicts X_p = 320 cm. Clearly, the conduction theory on which equation (97) is based does not explain fuel freezing under the conditions that existed in these experiments. On the other hand, the bulk freezing model is in good agreement with the experiments.

While the experimental results appeared to support Ostensen and Jackson's hypothesis, it was not clearly stated by the authors as to exactly how bulk freezing takes place. One would expect the appearance of some type of interfacial site from which freezing could begin. Epstein et. al. [43] concluded that the posttest observation of an ablated entrance region within the steel pin bundle is compatible with a bulk-freezing mechanism. These workers proposed that if conditions in the fuel flow are such to prevent crust growth then the steel wall melting can become severe. The relatively cold melting steel rapidly mixes with the fuel by turbulent diffusion and causes the fuel to freeze in a bulk manner; i.e., bulk freezing takes place when the fuel crust growth process at the channel wall is transformed to a steel ablation process (ablation-induced freezing model). Figure 22 illustrates the postulated sequence of events following molten fuel penetration into a reactor subassembly tube bundle. The suggested mechanism appears to be consistent with the rough, shiny, frozen debris observed at the conclusion of the thermite freezing experiments, where the shiny portions would be caused by the presence of entrained steel. This steel ablation-induced freezing concept was used to obtain a simple expression for molten fuel penetration into steel channels. The expression is similar in form to the equation for fuel penetration based on the Ostensen-Jackson bulk freezing model, differing only by a factor of $\sim$ 2.

In a subsequent experimental study [41], Henry et. al. replaced the hollow steel pins with prototypic fast reactor steel clad pins. The first 15 cm of each pin was solid steel (Inconel reflector) followed by 120 cm of hollow cladding, and the last 2.5 cm was again solid steel (pin end cap). In all the tests, the initial temperature of the rod bundle was 900°C. Two

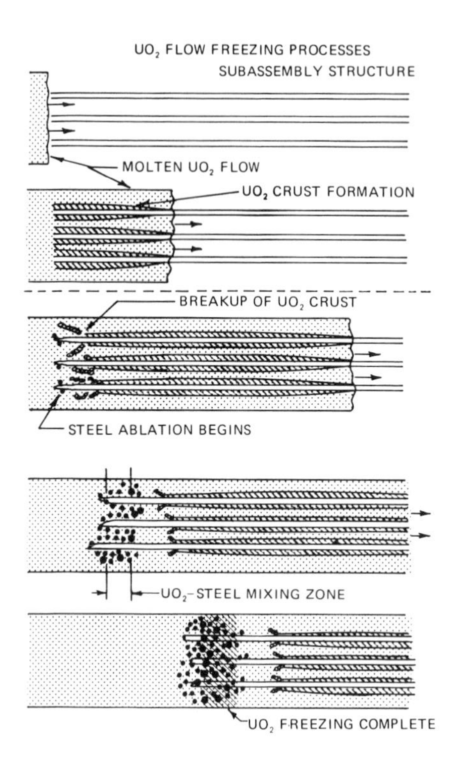

Fig. 22. Steel-ablation-induced freezing in a reactor subassembly tube bundle. From Epstein et. al. [43].

different injectors were used in this experiment in order to assess the influence of thermite mass on penetration behavior. The molten thermite penetration results using a 0.5 kg thermite load were similar to those of the earlier work [40] and thus showed little influence of pin geometry on the penetration process. Appreciable differences were observed, however, when the thermite injector inventory was increased to 2.0 kg. The flow front exceeded the previous 30-cm penetration length and came to rest in the vicinity of the pin end cap, a distance of ∿ 135 cm from the pin bundle inlet. As in previous tests, a severe melting attack of the steel pins in the reflector region by the penetrating thermite mixture was in evidence. Inspection of the frozen front revealed solid steel plugs containing only trace amounts of thermite fuel. This must be compared with the rough, clinker-like debris observed at the conclusion of the 0.5 kg injections. In addition, as much as 300g of molten steel emerged from the opposite end of the 7-pin test bundle. The authors concluded that the observed penetration behavior is clearly incompatible with a bulk freezing mechanism. It was postulated that, while steel ablation rapidly leads to fuel freezing in a bulk manner, bulk freezing does not provide significant containment capability. Instead, frozen steel is required to stop the flow and fuel penetration is conduction controlled in the steel component. They proposed that a complete blockage of pin bundle channels is preceded by the following three stages: (i) ablation and extrainment of the steel channel wall by the flowing thermite mixture in the pin bundle inlet (reflector) region; (ii) melted steel becomes the continuous phase in the region behind the flow front; and (iii) the penetrating molten steel-thermite mixture is cooled below the temperature that is necessary to sustain ablation, and conduction-limited steel crust growth begins on the channel walls. Complete occlusion of the channels, however, can only occur in the pin end cap region which "contains" enough heat capacity to freeze the flowing steel during the injection period. Using a simple conduction model, the authors demonstrated that the short penetration distances for the 0.5 kg injections were not caused by fuel freezing but were simply due to the limited quantity of material employed.

It is interesting to note that the penetration distances obtained in these seven-pin tests with inconel reflector geometry and a hydraulic diameter D_h = 0.3 cm agree reasonably well with the simple conduction model. The application of equation (97) to the flowing fuel results in X_p = 130 cm. In addition to the experiments with rod-bundle geometry, Henry et. al. [41] conducted a series of experiments in which 0.5 kg of thermite reaction products was injected into thick-walled steel tubes. A flow blockage did not occur in these experiments owing to the relatively short tube lengths employed (∿ 40.0 cm). The observed crust thickness left behind in the test section at the completion of each run as a function of tube temperature was found to be consistent with heat-conduction controlled crust growth.

The results of the most recent set of thermite injection tests have been reported in [42]. The test pins were 150 cm in length in which the first 36 cm at the injector end contained a stack of depleted UO_2 pellets (blanket region). Once again, two injector sizes were employed, capable of delivering 0.5 and 2.0 kg of molten reaction products into the pin bundles. In one of these tests, 2.0 kg was injected with the steel initial temperature at 900°C. The leading edge of the molten material traveled 140 cm, in agreement with the tests performed in [41] under the same conditions of temperature and thermite load. In two other tests the bundle temperature was set at 300°C, well below the initial steel temperature corresponding to incipient steel ablation upon contact with molten UO_2. Both tests used a 2.0 kg thermite load, but only about 25 percent of this mass actually entered the bundles. The leading edge of the fuel came to rest 40 cm from the bundle inlet. This reduction in penetration length, from 130 cm to 40 cm, as the bundle temperature was decreased from 900°C to 300°C cannot be explained by the conduction model. Moreover, using the bulk freezing model, we predict a penetration length of 17 cm, which is over a factor of 2 less than the observed value.

The only experimental study of a freezing liquid flowing through a melting channel under well-defined flow conditions is that reported by Yim et. al. [44]. In these experiments, hot Freon 112A (mp 40.5°C) was injected into an ice pipe. A schematic diagram of their experimental apparatus is shown in Fig. 23. Two solenoid valves separated a 150 cm-long, thick-walled ice pipe from a Freon reservoir. The reservoir was connected at the top to a large tank pressurized with nitrogen gas to maintain a constant Freon injection

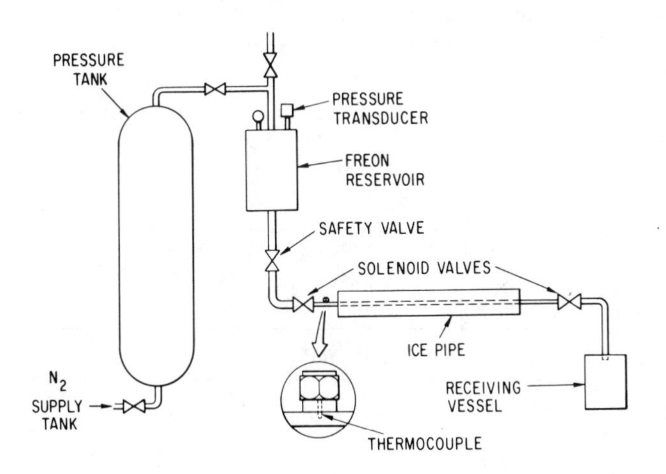

Fig. 23. Experimental apparatus for the solidification of Freon 112A in a melting ice pipe, From Yim, et. al. [44].

pressure. Each run was started by activating the two solenoid valves simultaneously for a preset duration and thereby allowing the hot Freon to enter the ice pipe. The Freon was allowed to flow through the ice pipe and into a receiving vessel for a period of 6-11 s depending on the injection pressure, at which time the solenoid valves were closed simultaneously and the Freon flow shutoff. At the end of each run the Freon and the melted ice collected in the receiving vessel were separated and their individual masses recorded. The shape of the melted ice channel was determined from the frozen Freon "casting" that remained in the ice pipe (i.e., trapped between the two solenoid valves). The Freon crust behavior during a run was visible through the ice pipe wall.

At high injection pressures and temperatures, the inner ice wall appeared to be partially covered by a random array of Freon crust "spots". The spots seemed to be slightly elongated in the direction of flow. The crust spots were not permanent, but disappeared and sometimes were replaced by a new spot at the same site. The departure or disappearance of crust spots was attributed to a combination of crust melting (see below) and crust mobility owing to the underlying layer of melted ice. At lower pressures and/or temperatures, sections of Freon crust were clearly seen to move or slide along the ice pipe wall in the direction of flow with a jerking motion. At times, the crust sections were observed to accumulate in the exit region of the ice pipe right in front of the exit tube forming a Freon "crust jam". For some runs performed at low initial Freon temperatures, tiny solid Freon particles were seen flowing into the receiving vessel. This Freon "hail" was assumed to be caused by turbulent mixing between the relatively cold melted ice and the flowing Freon. Perhaps some of the melted ice in the form of water droplets is rapidly entrained by the bulk Freon flow, thus providing evidence for ablation-induced freezing. However, this process did not lead to a flow blockage.

How does mixing between the flowing Freon and the melted ice take place in the presence of an intervening Freon crust? To explain this Yim et. al. [44] presented a model in which the ice melting process is controlled by the growth and remelting of the Freon crust sections on the ice pipe wall. When Freon flow commences a Freon crust is postulated to build up on the melting ice wall, as sketched in Fig. 24. The growth of the Freon crust comes to a

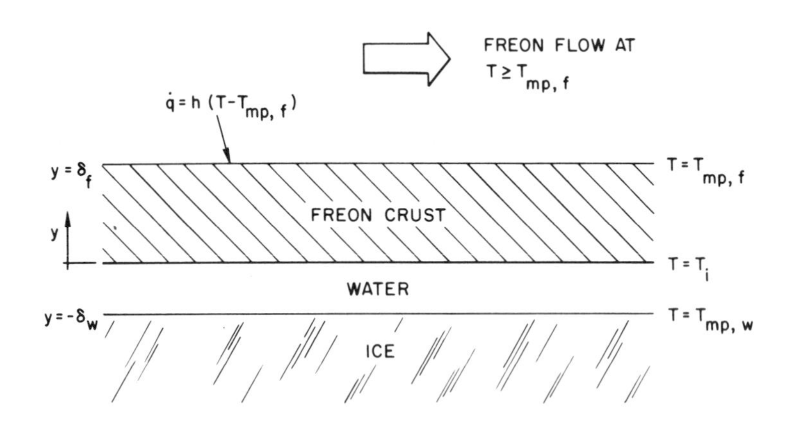

Fig. 24. Postulated simultaneous ice wall melting and Freon crust formation. From Yim et. al. [44].

stop when the conduction heat flux into the melting ice layer balances convection from the Freon flow. Then the frozen Freon layer will begin to melt; it ultimately disappears when melting is complete. At this instant the exposed water layer is removed (entrained) by the highly turbulent Freon flow and the process of Freon crust growth and decay begins again. This implies that, except for the short period of time during which the melted ice is removed from the wall, the Freon crust protects the melting ice surface from direct contact with the hot flowing Freon melt. If the time interval between initial growth to complete removal of the crust is small in comparison with the Freon injection period, ice melting can be visualized as a process in which ice is continuously being removed (ablated) at a rate proportional to the difference between the local Freon flow temperature and the Freon fusion temperature, $T - T_{mp,f}$.

Agreement was obtained between the theory and measurements of the amount of ice melted as shown in Fig. 25. The variation of the mass of melted ice

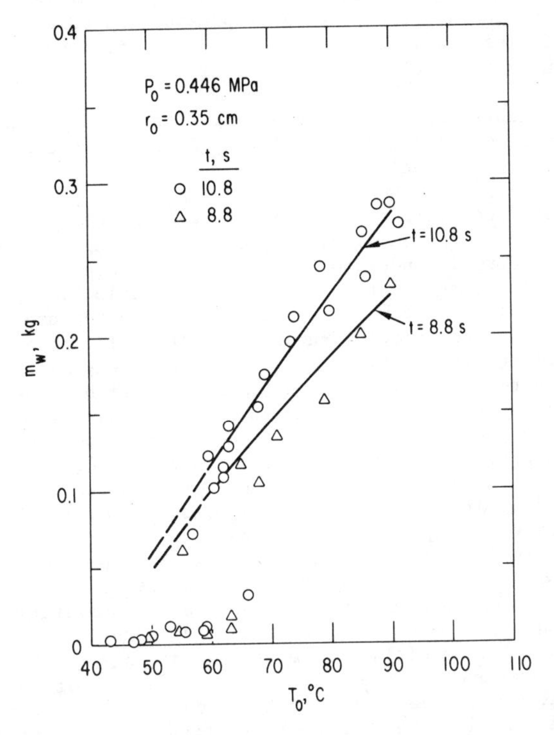

Fig. 25. Mass of melted ice collected vs. injection temperature; comparison between theory and experiment. From Yim et. al. [44].

collected, m_w, with initial Freon flow temperature, T_o, is shown in this figure. The sudden drop in m_w as T_o falls below $\sim 70°C$ was attributed to relatively rapid accumulation of Freon crust in the ice pipe exit region.

7. SUMMARY

The results of simulation experiments of boiling fuel-steel mixtures (pools) failed to support the existence of the droplet-dispersed flow at low equivalent power levels, which is the flow regime predicted by the Kutateladze stability criterion. Pool-average void fraction data by various investigators indicate that the mode of vapor production, the purity of the liquid, and the initial pool height all play a role in pool boilup behavior. As a result, the observed "churn" flow regime cannot be defined in quantitative terms; that is, a fully developed two-phase flow condition is nowhere achieved within the shallow pools of interest. While these complications preclude accurate predictions of boilup of fuel-steel pools, there is a consensus of opinion among investigators that, even for power levels as low as $\sim$1 percent, intense pool boilup should occur in fuel-steel pools with the average pool void fraction exceeding 60 percent.

The effects of two-phase flow pattern on the amount of molten fuel that can be removed from the active core region was clearly illustrated in experiments designed to simulate the venting of pressurized core material through relief paths. It was concluded that the foam flow pattern provides the most efficient mechanism for fuel removal.

All the available experimental and analytical evidence to date indicate that predictions of steel melting by flowing fuel must account for the insulating effect of a growing fuel crust.

Compared with the present knowledge of freezing of low-melting simulant materials in tubes and rod bundles, as yet the freezing of thermite mixtures in rod-bundle geometry is understood poorly. The conduction model is reasonably successful with simple materials injected into initially empty channels with non-melting walls. The model was even found to represent reasonably well the behavior of hot Freon flowing through a melting ice pipe. Moreover, the freezing of thermite mixtures in steel tubes can be explained in terms of simple heat-conduction theory. However, difficulties have been encountered with the freezing of thermite mixtures in rod bundles. At this stage all we can say is that the observed penetration lengths fall between the predictions based on the conduction model and the bulk freezing model. The need for better predictions is evident from the fact that the two models lead to penetration lengths that differ by more than a factor of 5.0. Of course, the very idea of there being a single dominant mechanism for flow blockage in this complicated freezing process involving a multi-phase/multi-component mixture may be an over-simplification. The thermite injection studies themselves thus far have been of an exploratory and often primarily qualitative nature. While thermite releases offer an attractive means of generating molten fuel, it is difficult to control the initial or release conditions of the thermite mixture. Another uncertainty involves the flow regime. The thermite discharge is best described as a high-void fraction, droplet dispersed flow, which probably becomes more dense during the course of the transient. Knowledge of the physical properties, specifically thermal conductivity, of the thermite reactant product mixture is as yet almost totally lacking. A trend toward careful control of experimental conditions and toward measurement of the penetration behavior of pure molten UO_2 would be most desirable.

ACKNOWLEDGMENTS

This work was performed under the auspices of the U.S. Department of Energy. The author wishes to thank Kathy Rank for an excellent job of typing the manuscript.

REFERENCES

1. Fauske, H. K., "Some Comments on Cladding and Early Fuel Relocation in LMFBR Core Disruptive Accidents," Trans Am. Nucl. Soc., 21, 1975, pp. 322-323.

2. Henry, R. E., Jeans, W. C., Quinn, D. J., and Spleha, E. A., "28-Pin Cladding Relocation Experiments," Trans. Am. Nucl. Soc., 27, 1977, pp. 498-500.

3. Ostensen, R. W. and Jackson, J. F., "Dynamic Behavior of a Partially Molten LMFBR Core," Trans. Am. Nucl. Soc., Vol. 17, 1974, pp. 220.

4. Dickerman, C. E., Barts, E. W., DeVolpi, A., Holtz, R. E., Neimark, L. A., and Rothman, A. B., "Recent Results from TREAT Tests on Fuel, Cladding, and Coolant Motion," Trans. Am. Nucl. Soc., Vol. 20, 1975, pp. 534-535.

5. Fauske, H. K., "Boiling Flow Regime Maps in LMFBR HCDA Analysis," Trans. Am. Nucl. Soc., Vol. 22, 1975, pp. 385-386.

6. Farahat, M., Henry, R. E., and Santori, J., "Fuel Dispersal Experiments with Simulant Fluids," Proc. Intl. Mtg. on Fast Reactor Safety and Related Physics, USERDA Report CONF-761001-Vol. IV, 1976, pp. 1707-1714.

7. Ginsberg, T., Jones, O. C., and Chen, J. C., "Flow Behavior of Volume-Heated Boiling Pools: Implications with Respect to Transition Phase Accident Conditions," Nuclear Technology, 46, 1979, pp. 391-398.

8. Orth, K. W., Epstein, M., Linehan, J. H., Lambert, G. A., and Stachyra, L. J., "Hydrodynamic Aspects of Volumetric Boiling," Trans. Am. Nucl. Soc., 33, pp. 545-546; see also Orth, K. W., M.S. Thesis Marquette University, in preparation.

9. Koontz, F. A., "Volumetric Boiling - A Fundamental Study of the Phenomena Pertaining to LMFBR Safety," M.S. Thesis Purdue Univ., August, 1977; see also Theofanous, T. G., NUREG Report CR-0224, June 1978.

10. Henry, R. E., Smith, J. L., and Weber, J. T., "Depressurization of Internally Heated Boiling Pools," ASME/AIChE 18th National Heat Transfer Conference, San Diego, CA, August, 1979, paper no. 79-HT-101.

11. Wallis, G. B., "Two-Phase Flow Aspects of Pool Boiling from a Horizontal Surface," paper no. 3, Two-Phase Fluid Flow Symposium, Institution of Mechanical Engineers, London, 1962.

12. Kutateladze, S. S. and Moshvichevz, V. N., "Hydrodynamics of a Two-Component Layer as Related to the Theory of Crises in the Process of Boiling," Soviet Physics: Technical Physics, 4, 1960, pp. 1037-1040.

13. Dukler, A. E. and Taitel, Y., "Flow Regime Transitions for Upward Gas-Liquid Flow: A Preliminary Approach Through Physical Modeling," NUREG Report 0162, January, 1977.

14. Zuber, N. and Findlay, J. A., "Average Volumetric Concentration in Two-Phase Flow Systems," J. of Heat Transfer, 87, 1964, pp. 453-468.

15. Bikerman, J. J., Foams, Springer, Verlag, New York, Chapter 5, 1973.

16. Epstein, M., Petrie, D. J., Linehan, J. H., Lambert, G. A., and Cho. D. H., "Hydrodynamic Aspects of Stratification and Mixing Within Boiling Immiscible Mixtures," to be presented at the annual meeting of the American Nuclear Society, Las Vegas, June, 1980.

17. Epstein, M., "Transient Behavior of a Volume-Heated Boiling Pool," ASME Winter Meeting, Houston, Texas, Dec. 1975, paper no. 75-WA/HT-31.

18. Condiff, D. W., and Epstein, M., "Transient Volumetric Pool Boiling - I Convex Flux Relations," Chem. Engr. Sci., Vol. 31, 1976, pp. 1139-1148.

19. Condiff, D. W., and Epstein, M., "Transient Volumetric Pool Boiling - II Non-Convex Flux Relations," Chem. Eng. Sci., Vol. 31, 1976, pp. 1149-1161.

20. Condiff, D. W., Epstein, M., and Grolmes, M. A., "Transient Volumetric Pool Boiling with Foaming," Chem. Eng. Progr. Symposium Ser., 73, 1977, pp. 86-96.

21. Epstein, M., "Melting, Boiling and Freezing: The Transition Phase in Fast Reactor Safety Analysis," in Symposium on the Thermal and Hydraulic Aspects of Nuclear Reactor Safety, 2, ASME Press, New York, 1977, pp. 171-193.

22. Carslaw, H. S., and Jaeger, J. C., Conduction of Heat in Solids, 2nd edn. Oxford University Press, Oxford, Chapter XI, 1959.

23. Epstein, M., "Heat Conduction in the UO_2-Cladding Composite Body with Simultaneous Solidification and Melting," Nucl. Sci. Eng., 51, 1973, pp. 84-87.

24. Epstein, M., "Stability of a Submerged Frozen Crust," J. of Heat Transfer, 99, 1977, pp. 527-537.

25. Lamb, H., Hydrodynamics, Dover, 1932, pp. 455-462.

26. Levich, V. G., Physicochemical Hydrodynamics, Prentice-Hall, Chapter XI, 1962.

27. Epstein, M., Swedish, M. J., Linehan, J. H., Lambert, G. A., Hauser, G. M., and Stachyra, L. J., "Simultaneous Melting and Freezing in the Impingement Region of a Liquid Jet," AIChE Journal, in press, 1980.

28. Swedish, M. J., Epstein, M., Linehan, J. H., Lambert, G. A., Hauser, G. M., and Stachyra, L. J., "Surface Ablation in the Impingement Region of a Liquid Jet," AIChE Journal, 25, 1979, pp. 630-638.

29. Cronenberg, A. W., and Fauske, H. K., "UO_2 Solidification Phenomena Associated with Rapid Cooling in Liquid Sodium," J. Nuclear Materials, 52, 1974, pp. 24-32.

30. Frenkel, J., Kinetic Theory of Liquids, Dover, Chapter VII, 1952.

31. Epstein, M., Yim, A., and Cheung, F. B., "Freezing-Controlled Penetration of a Saturated Liquid into a Cold Tube," Trans. ASME J. Heat Transfer, Vol. 99, 1977, pp. 233-238.

32. Cheung, F. B., and Baker, L., Jr., "Transient Freezing of Liquids in Tube Flow," Nucl. Sci. Eng., Vol. 60, 1976, pp. 1-9.

33. Madejski, J., "Solidification in Flow Through Channels and into Cavities," Int. J. Heat Mass Trans., Vol. 19, 1976, pp. 1351-1356.

34. Epstein, M., and Hauser, G. M., "Freezing of an Advancing Tube Flow," Journal of Heat Transfer, Trans. ASME, Series C, Vol. 99, 1977, pp. 687-689.

35. Tao, L. C., "Generalized Numerical Solutions of Freezing a Saturated Liquid in Cylinders and Spheres," AIChE Journal, Vol. 13, 1967, pp. 165-169.

36. Stephan, K., "Influence of Heat Transfer on Melting and Solidification in Forced Flow., International Journal of Heat and Mass Transfer, Vol. 13, 1969, pp. 199-214.

37. Epstein, M., and Hauser, C. M., "Solidification of a Liquid Penetrating into a Convectively Cooled Tube," Letters in Heat and Mass Transfer, Vol. 5. 1978, pp. 19-28.

38. Epstein, M., Stachyra, L. J., and Lambert, G. A., "Transient Solidification in Flow into a Rod Bundle," J. Heat Transfer, in press, 1980.

39. Ostensen, R. W., and Jackson, J. F., "Extended Fuel Motion Study," in Reactor Development Program Progress Report ANL-RDP-18, Argonne National Laboratory, July 1973, pp. 7.4-7.7.

40. Ostensen, R. W., Henry, R. E., Jackson, J. F., Goldfuss, G. T., Gunther, W. H., and Parker, N. E., "Fuel Flow and Freezing in the Upper Subassembly Structure Following an LMFBR Disassembly," Trans. Am. Nucl. Soc., Vol. 18, 1974, pp. 214-215.

41. Epstein, M., Henry, R. E., Grolmes, M. A., Fauske, H. K., Goldfuss, G. T., Quinn, D. J., and Roth, R. L., "Analytical and Experimental Studies of Transient Fuel Freezing," in Proceedings of the International Meeting on Fast Reactor Safety and Related Physics, Vol. IV, USERDA Conf. No. 761001, Chicago, October 1976, pp. 1788-1798.

42. Spencer, B. W., Roth, R. L., Goldfuss, G. T., and Henry, R. E., "Results of Fuel Freezing Tests with Simulated CRBR-Type Fuel Pins," Trans. Am. Nucl. Soc., Vol. 30, 1978, p. 446.

43. Epstein, M., Grolmes, M. A., Henry, R. E., and Fauske, H. K., "Transient Freezing of A Flowing Ceramic Fuel in a Steel Channel," Nucl. Sci. Eng., Vol. 61, 1976, pp. 310-323.

44. Yim, A., Epstein, M., Bankoff, S. G., Lambert, G. A., and Hauser, G. M., "Freezing-Melting Heat Transfer in a Tube Flow," Int. J. Heat Mass Transfer, 21, 1978, pp. 1185-1198.

17 IN-VESSEL POST ACCIDENT HEAT REMOVAL

by

J. Costa

GENG, Grenoble, France

Once the core disruptive accident has occurred, attention must be given to the removal of heat from a disrupted geometry—the subject of this chapter. After a presentation of the In-Vessel PAHR problem characteristics and the various paths through which the Decay Heat is removed, the different physical phenomena encountered are described: Debris formation, dispersion and settling, debris bed behaviour and cooling, molten pool behaviour and cooling, boiling pool, natural convection in sodium.

It is concluded that considerable progress has been made over the past ten years but that prototypic experiments are still needed to improve our understanding of the long term behaviour of a molten pool of fuel (and structure materials) immersed in liquid sodium (in particular the stability of the solid crust).

1. INTRODUCTION

In recent years, a large amount of research has been devoted to Post Accident Heat Removal (PAHR) in Liquid Metal Cooled Fast Breeder Reactors (LMFBR). Four PAHR "information exchange" meetings were organized, the first two by Sandia Laboratories at Albuquerque in 1974 and 1975 (Coats, 1975 ; Coats, 1977) the third by Argonne National Laboratory in 1977 (Baker and Bingle, 1978) and the fourth by the ISPRA establishment (Coen and Holtbecker, 1979). Together with the proceedings of the SEATTLE Conference, 1979, these meetings provide the basic material for this review.

PAHR is one of the many aspects of the LMFBR's safety. As opposed to the other safety problems, dealing with initiating events, melt down, or disassembly phases, where characteristic times are short - of the order of the millisecond or second or event minute - PAHR deals with long term phenomena, time scales of the order of days or even month. On this time scale some physico-chemical reactions such as those between sodium and core debris may be significant.

The PAHR initial situations are not very well defined, they result from any type of accident in which there is melting of the fuel : subassembly accident, core disruptive accident, core melt down accident...

Neutronic shut down is supposed to have been achieved and the only heat to remove is the decay heat. Indicative values of the decay heat are given in figure 1. This comes from some delayed fissions and absorption of α, β and γ emitted by fission products, actinides and activated reactor materials (structure and coolant). Soon after the shut down the decay heat in the fuel is at a level of 5 to 6% of the nominal power.

The reactor is supposed to have Decay Heat Removal Systems (DHRS) capable of removing the decay heat in case of failure of normal heat removal systems (secondary circuits) (Grand, 1980).

For safety, PAHR means the evaluation of long term consequences of accidents for the public and the plant itself : "potential risks" and "residual risks" once special design features have been recommended to designers (core catchers for example).

For designers PAHR means a demonstration of the performances of the safety systems introduced In Vessel (SUPERPHENIX) or Ex Vessel (SNR 300) to reduce the potential risks.

In both cases a detailed comprehension of all the phenomena involved is needed (recriticality, physico-chemical phenomena, heat transfer phenomena...).

In this lecture devoted to In-Vessel PAHR problems only debris formation, debris dispersion and settling, debris bed behaviour and cooling, molten layer behaviour, boiling pool, and natural convection in sodium will be treated. Ex Vessel PAHR problems are treated in an other lecture (Peckover, 1980).

2. GENERAL CONSIDERATIONS

A typical In-Vessel PAHR situation for a pool type LMFBR following a whole core accident is shown in figure 2. The total decay heat (DH) comes from the activated structural materials q_1, the activated sodium q_2 and from the fuel and fission products q_3. The normal decay heat removal systems (DHRS) which are designed to remove this decay heat without exceeding the maximum permissible temperature on the reactor vessel are supposed to be in operation. After the whole core accident, (disassembly or melt down) the core is no longer in its normal coolable configuration and the decay heat sources are spread in the reactor vessel.

The fraction q_1 from the stainless steel activation is supposed to heat directly the sodium of the primary circuit (Na2).

- The fraction q_2 coming from the activated sodium (^{24}Na) is distributed uniformly in the sodium.
- The remaining fraction q_3 (the most important) is distributed, for example, either : in the cover gas plenum (fraction q_{31} coming from the gaseous or volatile fission products ; in the sodium (fraction q_{32} coming from the fine fuel debris and fission products in suspension) ; on various structures of the main vessel (fraction q_{33} of the fuel expelled from the core during the hypothetical core disruptive accident) ; is still in the core (fraction q_{34} in the case of a partial melt down accident) ; or on the core catcher (fraction q_{35}) after melting of the core supporting structures (Cassoudesalle and Lerigoleur, 1978).

The fractions q_1, q_2, q_{31}, q_{32}, q_{33} and q_{34} are supposed to heat directly the sodium of the primary vessel while the fraction q_{35} is divided into two

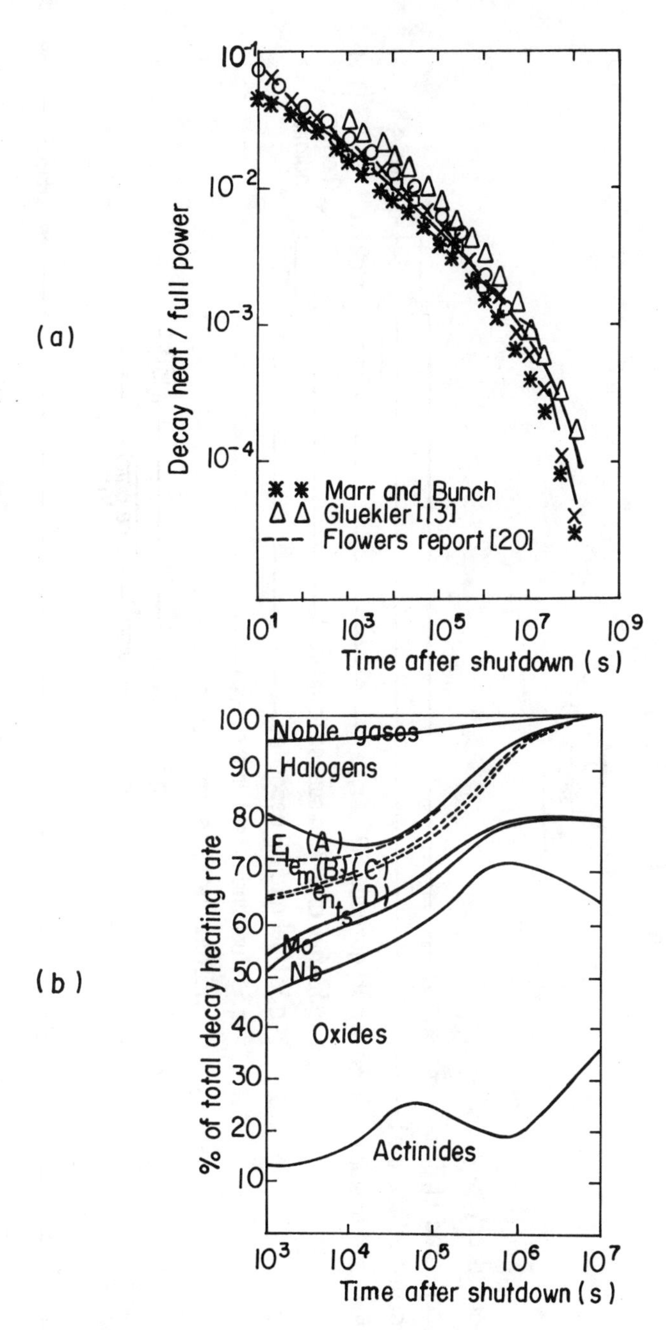

FIG. 1 Decay heat (from TURLAND and PECKOVER, 1977)
a) total decay heat.
b) cumulative contribution for the decay heating rate.

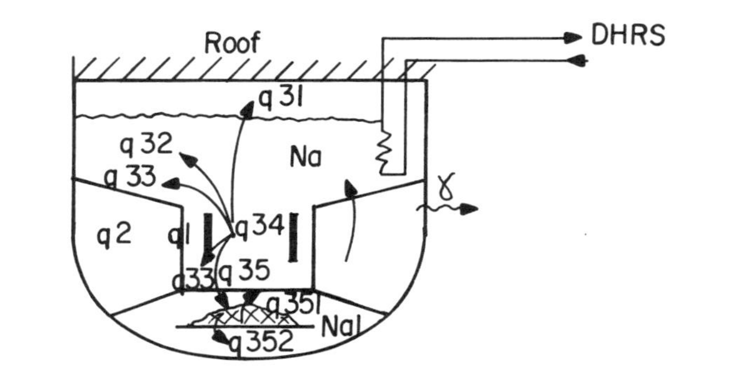

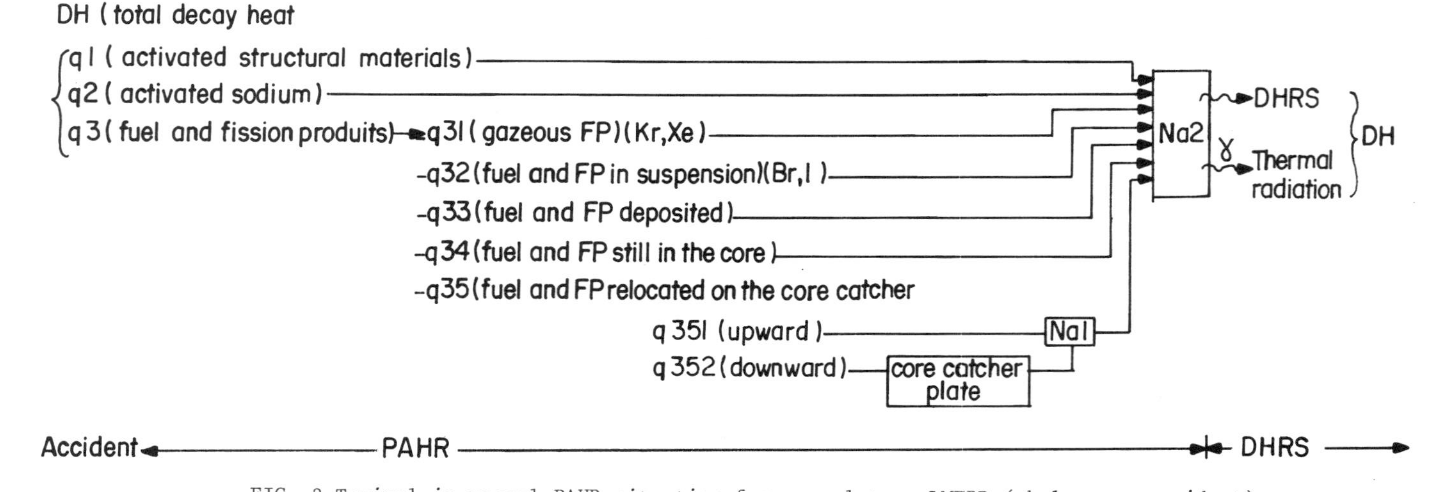

FIG. 2 Typical in vessel PAHR situation for a pool type LMFBR (whole core accident).

parts : q_{351} upward and q_{352} downward through the core catcher plate, both heating the sodium Nal at the bottom of the vessel around the core catcher. q_{35} is then transferred to the DHRS, the emergency intermediate heat exchangers through the thermal resistance of various cavities filled with sodium.

When the fuel is supposed to melt through the main vessel the situation is an Ex-Vessel PAHR situation.

Any type of accident can lead to a PAHR situation ; the different types generally considered are :

- Hypothetical core disruptive accidents (HCDA). They can result in a large amount of fuel expelled upward ; for SNR 300 for instance, (Vossebrecker and Grönefeld, 1978) almost the total fuel inventory is swept out in the upper plenum. The main vessel, the emergency intermediate heat exchanger and the core catcher have to resist to the mechanical loads of the disassembly phase.
- Core melt down accidents. These are less severe on the mechanical side but can lead to a larger amount of molten fuel, which is more difficult to control than the dispersed fuel in the case of a disruptive accident.

Besides those whole core accidents which are recognized to be very unlikely the local subassembly accident involving one or more subassemblies, which is less improbable, has to be considered (the Fermi accident).

Donati et al., 1978, have studied the various plugging situations, leading to melting of the fuel in a subassembly.

In the subassembly accident as in the whole core accident not all the decay heat remains dissipated in the subassembly-equivalent fractions of decay heat, q_{31} (gaseous), q_{32} and q_{33} escape upward while a fraction q_{35} might escape downward. The question is to know whether the fraction q_{34} that remains in the subassembly can be removed by the inter-subassembly sodium without dammaging the subassembly wrapper or if a propagation mecanism can take place.

3. DEBRIS FORMATION

Core debris resulting from a core melt down accident contains fuel, structural materials (steel) and control rod materials, each with their different physical properties. Their configurations cover the wide spectrum between fine particles of a few microns in diameter to large lumps of molten material through embrittled fuel pellets.

Fine fuel particles result from a contact between molten fuel with liquid coolant. Several fragmentation mechanisms have been envisaged in the framework of studies on Fuel Coolant Interaction (FCI) ; the most important are (Cronenberg and Grolmes, 1975) :

- thermal fragmentation,
- hydrodynamic fragmentation,
- gas release fragmentation.

Thermal fragmentation results from thermal stresses in the solidified crust formed upon contact with the coolant ; the debris has sharp angles because it is formed in the solid state. This fragmentation process is very common and explains the fine debris obtained in Molten Fuel Contact experiments in which no energetics or violent effects have been observed.

Hydrodynamic fragmentation results from vapor film collapse or large acceleration differences between the heavy material (fuel or structure material) and the light one (sodium). Formed by this process, while they are still in a liquid phase, the fragments should have rounded shapes.

In practice it is very difficult to separate those two mechanisms in molten fuel coolant contact experiments in which, although-non energetic, some rapid collapse or vapor formation is obtained.

Structural material not directly in contact with the fuel might form larger sized debris because it would not be completely molten.

The gas release fragmentation results from a increase in the solubility with the temperature - upon the negative thermal shock, the dissolved gas tends to escape from the fuel.

Typical particle size distributions are given in figure 3.

In general the size distribution is log normal with 50% of the mass fragmented in particles of a size less than 1 mm. The smallest particles have a size of the order of 1 μm. The size of the largest is not very well defined, sizes of the order of cm have been obtained in the FRAG 6 experiment (Chu et al., 1979). The maximum size depends upon the size of the jet of liquid fuel, the depth of the coolant pool and its temperature.

All those results have been obtained with an excess in mass of sodium. When this is not the case large amounts of molten fuel can remain non-fragmented. That possibility has to be taken into account in the analysis of the accident scenario : formation of pools in a subassembly - on the diagrid - on the core catcher...

Residual coolant flow and radiation in some peripheral zone of a subassembly can provide sufficient cooling to prevent a complete melting of the pellets - the cladding material being completely molten and removed or not. Thus piles of pellets can remain in the core. That possibility too has to be considered in accident scenario, as it is an intermediate stage between the normal core configuration and the completely molten fuel.

The influence of the physical properties of the molten material on the shape of the fragments can be observed since a melt of iron and alumina was tested. In the FRAG 1 experiment, a mild interaction, the fragments are all single phase (rounded shape for iron, jagged outline for alumina) while for FRAG 2 experiments, a more violent interaction, the fragments are intermixed (Chu et al., 1979).

Those experimental results cannot be directly extrapolated to a reactor situation. What is the influence of the fission products ? They can be either solid, liquid or gas. They can increase the thermal diffusivity of the fuel thus increase the violence of interaction, but also the gas may prevent a direct contact with the coolant.

4. CORE DEBRIS DISPERSION AND SETTLING

Core debris may remain in the core, or be expelled upward, or settle downward.

Core debris can be expelled upward by energetic interactions, nuclear

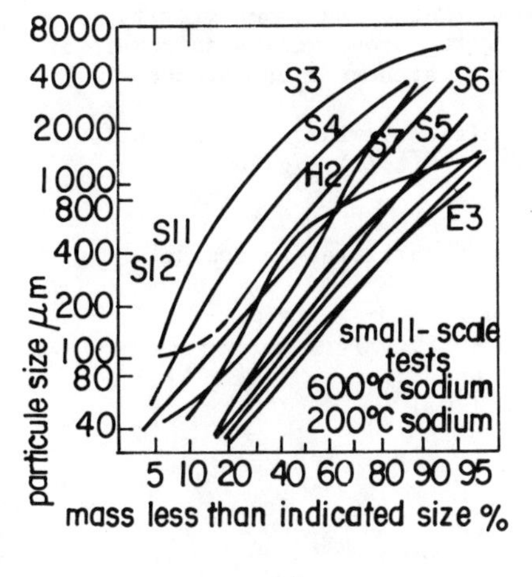

a)

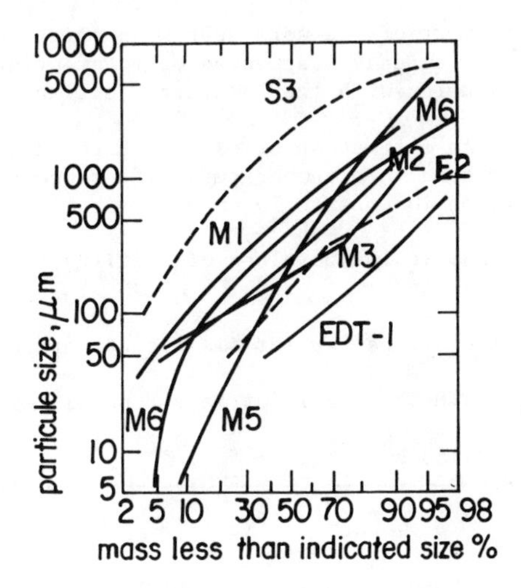

b)

c)

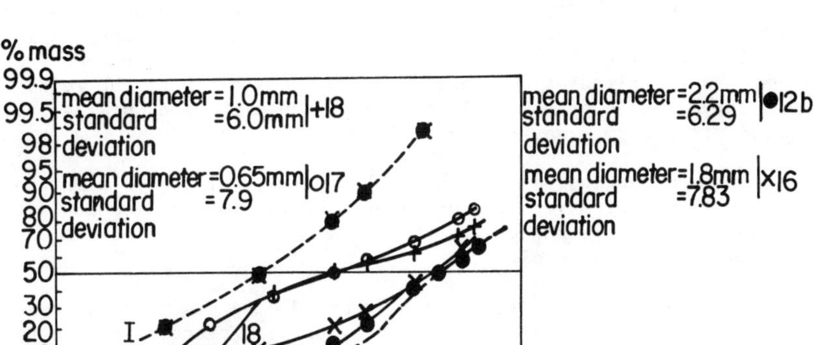

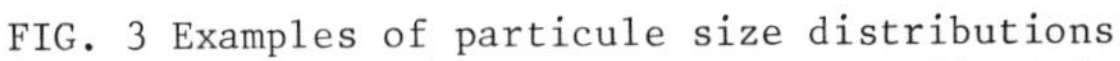

FIG. 3 Examples of particule size distributions
a) from treat tests and from smell scale laboratory experiments.
b) from larger scale out-of-pile experiments.
c) from CENG experiments.
a and b from GLUEKLER and BAKER, 1977.
c from AMBLARD and BERTHOUD, 1979.

excursion or entrainment by a residual flow of coolant : Those debris should be of relatively small size and, once in the upper plenum, part will remain in suspension in the sodium and part will settle on internals in zones at very low coolant velocity.

Part of the core debris expelled downward or dropping under gravity, either already fragmented or fragmenting during the drop, will also remain in suspension with the remaining settling on a core catcher or on internals.

An estimation of natural circulation velocities in large sodium pools submitted to a temperature difference - $V \sim \sqrt{g\ \beta\ \Delta T\ L}$ (1) - gives values between 10 cm/s to 50 cm/s.

Approximate values of settling velocities of particles in sodium are given by the Stokes law for small size particles (up to 25 μm) $v = \frac{g(\rho_f - \rho_{Na})\ d^2}{18\ \mu_{Na}}$ (2)

or by the Bird law for larger particles (up to 1 mm)

$$v = \frac{0,153\ g^{0,71}\ (\rho_f - \rho_{Na})^{0,71}\ d^{1,14}}{\rho_{Na}^{0,29}\ \mu_{Na}^{0,43}} \quad (3)$$

Those formulae are approximate because they do not take into account the decay heat of the particles or the agitation of the sodium. In any case it seems that particles of sizes below 50 μm will be entrained by the sodium movement and settle in zones where sodium is stagnant. Not many studies have been devoted to the formation of the debris layer or a heap. One can imagine that in case of a rapid release of a large amount of fragmented material in the downward direction, the drop of material will entrain some sodium and create some recirculating flows above the core catcher. The resulting effect might be a spread of the small debris on the plate and outside while the large debris might form a heap (figure 4).

As a consequence of the possibility that fuel debris might deposit in zones of low sodium velocity designers should avoid hollow zones were fuel debris might settle and create thermal stresses or even melting of major components of the vessel.

What has not been swept out or dropped can remain in the core if a sufficient residual cooling is provided by natural circulation. Up to now not many studies have been devoted to this topic.

5. DEBRIS BED BEHAVIOUR AND COOLING

Once a debris bed is formed by settling of core debris on a supporting structure (fuel and structure materials) it might stay as it is if sodium provides enough cooling or change progressively into a molten pool if sodium boils and the central zone dries out.

The characteristics of a debris bed are its composition, the distribution of particule size, the height, the specific power, the porosity,... Because large particles have a larger falling velocity, they will deposit first. Thus the particles size distribution is not uniform in a debris bed. Rivard and Coats, 1979, have proposed a debris bed formation model for a log-normal distribution of particles.

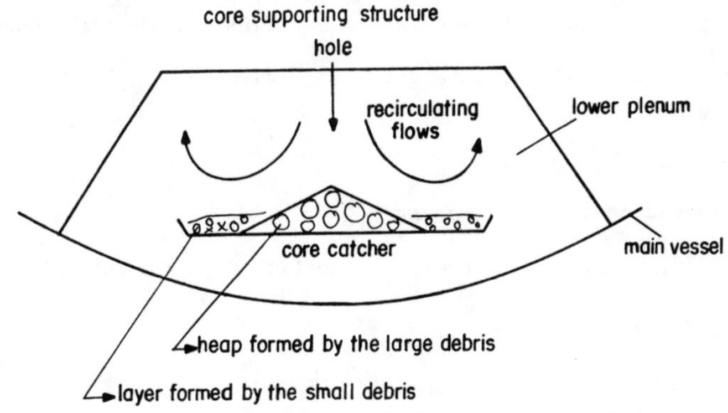

FIG. 4 Deposition of debris on core catcher.

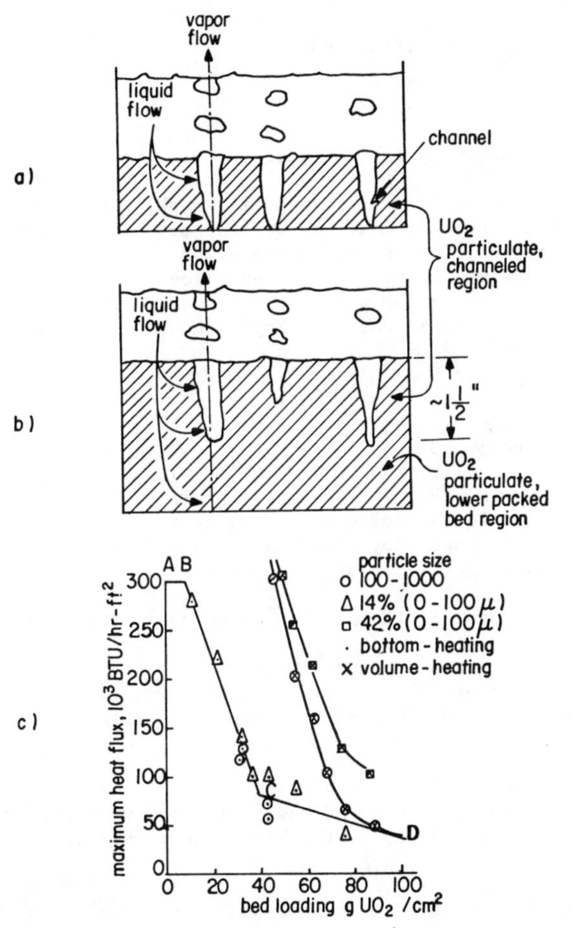

FIG. 5 Boiling debris bed cooling mechanisms
a) shallow bed
b) deep bed
c) corresponding D.O. heat flux

}from GABOR et al., 1974.

Simple conduction calculations can be made in such a system when the characteristics of an equivalent homogeneous layer can be estimated : equivalent density, specific heat, specific power density, thermal conductivity.

The porosity is the first characteristic to evaluate

$$\varepsilon = \frac{V - V_d}{V_d} \qquad (4)$$

it can be measured easily on samples.

Lerigoleur, 1978, proposed a correlation to fit ANL and SANDIA data :

$$\varepsilon = 0{,}57 - 1.33\ 10^{-4}\ C \qquad (5)$$

From it the mean density can be obtained

$$\rho = \varepsilon\ \rho\ Na + (1 - \varepsilon)\ \rho_f \qquad (6)$$

and also the effective specific decay heat

$$q_B = \varepsilon\ q \qquad (7)$$

The equivalent conductivity is more difficult to evaluate.

Theoretical evaluations can be made in coupling parallel and serie conduction in the heterogeneous mixture. But as soon as the temperature difference between the central zone and the upper boundary exceeds a maximum value estimated from a maximum internal Rayleigh number, natural convection develops in the sodium thus leading to an apparent increase of the thermal conductivity. Hardee and Nilson, 1977 make use of a Rayleigh number defined by Buretta,1972 :

$$Rab = \frac{\rho\ g\ \beta\ L\ K}{\mu a} \left(\frac{q\ L^2}{k_B}\right) \qquad (8)$$

Whose critical value is 33.

Nijsing and Swalm, 1978 made use of correlations developed by Kulacki and Goldstein, 1972 for liquid layers to determine the equivalent conductivity of the debris bed in the upper layer and made conduction calculations to evaluate the temperature distribution in debris beds.

Upward heat removal is further increased when sodium boiling takes place in the bed.

From observation of laboratory experiments two distinct boiling situations can be distinguished : shallow beds and deep beds (figure 5) Gabor et al., 1974.

The vapor formed in the central layer of the debris bed escapes through chimneys formed in the debris - the bed is shallow when the chimney height is as high as the boiling bed layer, the bed is deep when the chimney is not so high.

Models have been developed by Dhir and Catton, 1976, to predict debris bed dry-out - their ideas are that the maximum heat flux is limited by the critical velocity of the vapor in the exit of the chimney in a shallow bed and by the maximum downflow of liquid in the deep bed.

Many experiments have been performed with simulant materials or real materials but with simulant heating - bottom heated or internal heating by electrical effects. Most of these were devoted to saturated boiling but some of them were devoted to the influence of sodium subcooling, Dhir and Catton, 1979, or particle size distribution, Dhir, 1979.

We will only mention here the new In-Pile SANDIA results (Rivard, 1979). In these 3 experiments, D1, D2, D3, particles of UO_2, 100 - 1000 μm immersed in sodium were fission heated in a vessel of 102 mm diameter insulated on the periphery and bottom. Data are compared to previously published work. They do not differ very much from Gabor et al., 1974 data while they do differ from the Dhir and Catton, 1975 correlation. The Hardee and Nilson, 1977, formulation produces qualitative agreement with the D series data. Observations during the post dry out regime indicate that the dry out zone is slowly expanding.

Lipinski and Rivard, 1979, developed new one dimensional deep bed models for incipient dry out and have used them to make predictions of the next D-4 experiment.

Incomplete self-leveling of a not initially flat debris bed has been observed in laboratory experiments. More realistic experiments (composition of the debris bed and sodium environment) should be made to come to a more precise conclusion.

6. MOLTEN POOL BEHAVIOUR AND COOLING

A molten pool can be formed directly by the fall of a large amount of molten materials on the catcher plate or the bottom of the vessel, assuming that no extensive fragmentation occurs during the fall. Mechanical damage that could be produced by the impact have to be taken into consideration in this case.

The molten pool is made of liquid fuel and liquid steel. If they are initially mixed, with time the fuel, which is more dense, should sediment while molten steel and some liquid fission products should occupy the upper part of the layer.

At its bottom the molten pool should be separated from the supporting plate by a layer of solid debris of UO_2, as well as at the top because of the contact with sodium.

If the molten pool results from the evolution of a debris bed after dry out, the liquid layer is surrounded by upper and lower dry out beds of solid debris.

The evolution with time of such a molten layer raises several questions : as to the :

- migration of fission products (liquid, volatiles, solids)
- stability of the upper solid crust
- internal boiling
- downward heat flux

Most of the decay heat comes from the fission products, thus it is important to know where they are located. In most of the current calculations they are assumed to be distributed uniformely. But if they are more dense in the upper region the downward heat flux will be reduced.

Because of thermal stresses cracks will form in the upper solid crust, thus we ask if a stable situation exists or is there a progressive destruction of the layer by penetration of sodium through the cracks, vaporisation more or less violent, expulsion of debris, formation of a new crust, formation of cracks, penetration of sodium, ?...

Internal boiling depends upon the maximum temperature reached in the pool, thus what is the stability of the upper layer of a boiling pool ?

For the designers, the problem consists of demonstrating that the supporting plate is not damaged by this molten layer. The answer depends upon the physical properties of the plate (refractory material or steel) and upon the downward heat flux.

Many studies have been devoted to those questions, including simple conduction calculations, studies on the internal natural convection in various conditions...

From a simple conduction calculation the maximum thickness of a liquid phase layer should be rather small (5 to 6 cm for usual values of post accident situations) ; for thicker layers the maximum temperature in the central region exceeds the saturation temperature of the fuel.

In fact in the upper part of the layer, as in a debris bed, the fluid is not in a stable condition and natural convection takes place. The resulting enhancement of heat transfer, which can be taken into account in the calculations by an effective conductivity several times higher than normal flattens the temperature profile and a much thicker layer is predicted.

Kulacki and Goldstein, 1972 and Jahn and Reineke, 1974, were among the first to study experimentally internally heated fluid layers while numerical and theoretical approaches were developing, Peckover and Hutchinson, 1973, Suo Antilla and Catton, 1975, Jahn and Reineke, 1974, Biasi et al., 1977. Table 8 and figure 7 both in Gluekler and Baker, 1977, summarize the results. Heat transfer in steady convection are expressed by correlations of the form $Nu = C\ Ra^m$. For instance for an infinite horizontal layer with isothermal surfaces, Kulacki and Goldstein (1972) proposed the following correlations :

$$Nu_u = 0.329\ Ra'^{0.236} \quad \text{(in the upward direction)} \qquad (9)$$

$$Nu_d = 1.428\ Ra'^{0.094} \quad \text{(in the downward direction)} \qquad (10)$$

measured for $Pr \simeq 6$ and for Ra' in the range $1280 - 2.4\ 10^7$. Further studies by Kulacki et al., 1977 covered particular aspects such as a layer bounded below by a portion of a sphere, the transient response of a layer to a step change in the Rayleigh number, and a vertical cylinder as a new geometry. Fieg, 1978 also performed studies in cylinders. New studies are taking into account the two fluid layers (Reineke et al., 1978) and Kulacki et al., 1978).

Results from out of pile experiments with real fuel (up to 20 kg) done by ANL (Stein et al., 1978) but without sodium do not agree with usual predictions. The downward heat flux is approximately a factor of two greater.

The explanation given by ANL, which has yet to be demonstrated, is that this is due to the influence of internal radiation at that high temperature level. Three recent in-pile experiments have been obtained at Sandia (Plein et al., 1979), done with a small amount of fuel debris (without sodium) - 834 g for MP1 and MP2 and 628 g for MP3. Post-experiment observations show a void in the center surrounded by a thick solid crust. Possible molten fuel/coolant contacts depend upon the mechanical strength and stability of that crust. The third experiment, a UO_2/steel experiment demontrated that a dried out UO_2 debris bed could melt a steel substructure if not sufficiently cooled.

Special mention should also be made of a molten pool formed inside a subassembly.

The same stability problems exist but mainly sideward instead of downward. Is the side wall protected by a tight solid crust of fuel or can the molten fuel attack the wrapper through cracks ? Practically no studies have been done in this particular geometry.

7. BOILING POOLS BEHAVIOUR AND HEAT TRANSFER

If the thickness of the molten layer is large enough, the temperature in the central zone might exceed the saturation temperature of the fuel and boiling occurs.

At a lower temperature boiling of steel might also occur if steel is still present in the molten layer.

The local vapor production can be expressed in terms of the specific heat generation q and the void fraction :

$$\frac{dm_v}{dy} = \frac{q(1 - \alpha)}{h_{fg}} \qquad (11)$$

introducing the superficial velocity and using the Kutateladze number :

$$K = \frac{u_v \sqrt{\rho_c}}{\sqrt[4]{\sigma g(\rho_H - \rho_\ell)}} \qquad (12)$$

a first idea of the flow pattern can be found. At low power densities the bubbly flow is the most probable.

The rising of bubbles in the pool has several effects : transport of heat by latent heat and enhancement of liquid recirculating flows.

A consequence of this is a further increase of the equivalent thermal conductivity as compared to the molten pool.

Because of vapour formation the solid crust on top of the layer is submitted to deformations and will probably break, leading to to possible reentries of sodium and interactions with the liquid fuel. Thus, the existance of a boiling pool is questionable. Up to now no realistic experiment has been performed to validate this hypothesis.

Nervertheless, assuming a boiling pool exists, one of the main questions is the downward heat transfer coefficient. What is the influence of boiling on it ?

Experiments with injection of gas through the bottom of the layer indicate an enhancement of several times the downward heat flux but they are not relevant. The destruction of the thermal sublayer by the gas injection is representative of bubbling of gas coming from a concrete support, for instance, but not representative of internal boiling.

No prototypic experiments have been done in this field because it is very difficult to heat to boiling oxide fuel.

Simulant experiments (salt water) have been performed by ANL (Hesson and Gunther, 1972 - Stein et al., 1974) in rectangular cavities aiming at a simulation of a molten layer. Downward heat flux significantly larger than values calculated by conduction from a stagnant layer have been obtained. Agitation created by the bubbles is responsible of this increase in heat transfer.

More recent studies on two phase flow pattern and heat transfer in boiling were done at BNL (Ginsberg et al., 1978) in relation with the transition phase; in this case the geometry is a vertical cylinder and thus more relevant to a PAHR situation after a subassembly accident.

Extrapolations of salt water heat transfer results to UO_2 pool conditions were done by Kazimi, 1977. It is concluded that sideward heat fluxes will be higher than both the upward and downward flux and that the magnitude of the heat fluxes will depend upon the potential for fuel vapor to escape or condense at the upper surface of the pool.

8. NATURAL CONVECTION IN SODIUM AROUND THE CORE CATCHER

Although it is expected that adequate protection of the public against accidents should be achieved by high reliability of protective systems, minor changes to the design or low expensive additional components might improve the in-vessel debris containment capabilities. Their configuration should eliminate any risk of recriticality and protect the main vessel from any melt-through. Catcher plates or sacrificial layers might be used. In this lecture only catcher plate thermohydraulic problems will be presented.

The mechanical resistance of the core catcher plate depends upon its mechanical properties and upon its temperature level and gradients (Kayser and Lerigoleur, 1978).

Starting from the temperature of the heat sink or the temperature of the sodium around the Decay Heat Removal Systems, the mean temperature of the sodium around the core catcher results from the thermal resistance through the sodium of the primary circuit (large cavities of sodium). The temperature of the catcher plate results in a further thermal resistance between the sodium and the plate. Heat transfer correlations given by Glueckler and Baker, 1977 are not specific to liquid metals characterized by a very low Prandtl number.

Experiments have been carried out by Amblard et al., 1978 to investigate both phenomena. Experiments were performed with water as a simulant of sodium.

Basic studies were carried out in a simplified geometry : a closed cavity cooled by the sides and either heated from the top and adiabatic at the bottom or heated from the bottom, the top being adiabatic. This configuration is typical of in sodium PAHR problems.

For downward heat flux the results compare rather well with the theoretical prediction of Fujii et al., 1973 in turbulent regime ($GrPr^2$ from to 10^{11} to 10^{13}) given for an horizontal plate facing downward in an infinite medium and correlated by $Nu = 0,52(GrPr^2)^{0,17}$ (13) while for laminar regime in closed cavities heat transfer values 50% higher than those obtained in an infinite medium have been measured.

It is thus concluded that the boundaries do not influence very much the heat transfer coefficient by natural convection in turbulent regime. This coefficient is only few times higher than the one obtained by conduction.

For the upward heat flux available sodium results in the literature, subbotin et al., 1960, Kutateladze, 1959 and Mac Donald and Connolly, 1960, cover a limited range of parameter ($GrPr^2$ 10^3 - 10^7) while the range of interest for an LMFBR is (10^{11} - 10^{13}). Extrapolation over such a wide range cannot be considered reliable, thus sodium tests should be done in this field.

In simplified geometries computer codes solving the coupled mass, momentum and energy equations for single phase flows can be used-for instance the Notung code (Vossebrecker and Grönefeld, 1978).

Recently, Davies and Sheriff (1980) presented results of a study of natural convection to sodium from a flat plate.

For a horizontal downward facing surface in sodium, the following experimental correlation was obtained :

$$Nu = 1.19\ (0.522 - 0,0437Pr - 0.06733\ Pr^2)\ (Gr^{*}_{a}\ Pr^2)^{1/6} \qquad (14)$$

These experimental results (for $Gr^{*}_{a}\ Pr^2$ in the range 10^5 to 10^6) are of the order of 20% higher than the Fujii and al., (1973) theory.

For a heating plate inclined at 15°

$$Nu_x = \frac{0.90(Gr^{*}_{x}\ \cos 75°.Pr^2)^{1/5}}{(4 + 9\ Pr^{1/2} + 10\ Pr)^{1/5}} \qquad (15)$$

These experimental results (for $Gr^{*}_{x}\ Pr^2$ in the range 10^1 10 10^7) are of the order of 10% lower than theory.

9. CONCLUSIONS

A brief review of the main physical phenomena encountered in the analysis of In-Vessel Post Accident Heat Removal Problems has been presented.

A large amount of work has been carried out on this topic over the past decade.

Analytical studies, experiments with simulant materials, some out-of-pile experiments with real materials and few in-pile experiments with real materials have been performed.

A rather good understanding of most of the phenomena is available but key questions still remain, dealing with the long term behaviour of a molten fuel pool or boiling pool surrounded by solid crusts and immersed in a large pool of liquid sodium. Can a solid and tight crust exist ? What happens if it breaks ? In this field there is a need for prototypic experiments.

In PAHR problems it is very difficult to be deterministic since most of the boundary conditions are not very well defined and because of the coupling between the heat source and the heat transfer area which changes with time. Thus "judgment" plays a very important role in evaluating not only the extreme situations taken into account in safety analysis and design studies, but also in evaluating the mitigating effects.

NOMENCLATURE

a = thermal diffusivity

C = Bed Loading (kg/m^2)

C_K = Kozeni constant = 4.94

d = diameter

D = diameter

Gr^*_a = modified Grashof number based on the half width of the heating plate

Gr^*_x = modified Grashof number based on the distance x from the leading edge of the heating plate

h_{fg} = latent heat of vaporisation

k = thermal conductivity

K = permeability = $\frac{1}{36}\frac{D\rho}{C_K}\frac{\varepsilon^3}{(1-\varepsilon)^2}$

L = characteristic length

Nu = Nusselt number

Nu_x = local Nusselt number

Pr = Prandtl number

q = fraction of Decay Heat (see fig. 2)

Ra = Rayleigh number (Ra_b : defined by BURETTA)

Ra' = Rayleigh number based on the volumetric heat source

v = velocity

V = volume

v_v = superficial velocity

β = expansion coefficient

ΔT = temperature difference

ε = porosity

μ = dynamic viscosity

ν = cinematic viscosity

ρ = specific mass

σ = surface tension

Subscript :

B = bed

C = continuous phase

d = debris

f = fuel

h = heavy phase

ℓ = light phase

Na = sodium

p = particle

REFERENCES

Amblard, M. and Berthoud, G., 1979. "UO_2-Na Interactions - The CORECT II Experiment". Proceedings of the 4th CSNI Specialist Meeting on Fuel Coolant Interaction in Nuclear Reactor Safety, Bornemouth, UK, April 2-5, 1979, Tattersall, R.B., UKAEA, AEE Winfrith, Ed., 1979, CSNI report n°37, Vol. 2, pp. 508-532.

Amblard, M., Martin, R. and Tenchine, D., 1978. "Heat Transfer by Natural Convection around the Core Catcher Plate". Proceedings of the Fourth Post-Accident Heat Removal (PAHR) information Exchange Meeting, Varese, Italy, October 10-12, 1978. European Appl. Res. Rept. - Nucl. Sci. Technol., Vol. 1, N°6, 1979, pp. 130-143, Harwood Academic Publishers.

Baker, L., Jr. and Bingle, D.J., Eds, 1978. "Proceedings of the Third Post-Accident Heat Removal "Information Exchange" Meeting", ANL, November 2-4, 1977, report ANL-78-10, 447 p..

Biasi, L., Castellano, L. and Holtbecker, M., 1977. "Molten Pool Theoretical Studies". Proceedings of the Third Post-Accident Heat Removal (PAHR) "Information Exchange" Meeting, ANL, November 2-4, 1977, report ANL-78-10, pp. 122-129.

Buretta, R., 1972. "Thermal Convection in a Fluid Filled Porous Layer with Uniform Internal Heat Somes". Ph. D. Dissertation, Univ. of Minnesota, 1972.

Cassoudesalle, G. and Lerigoleur, C.. "Thermal Calculations to Investigate the Integrity of Melt Through Time of the Fuel Support Grid for a Postulated Core Melt-Down". Proceedings of the Fourth Post-Accident Heat Removal (PAHR) Information Exchange Meeting, Varese, Italy, October 10-12, 1978. European Appl. Res. Rept. Nucl. Sci. Technol., Vol. 1, N°6, 1979, pp. 220-229. Harwood Academic Publishers.

Chu, T.Y., Beattie, A.G., Drotning, W.D. and Powers, D.A., 1979. "Medium-Scale Melt-Sodium Fragmentation Experiments". Proceedings of the Internation ANS/ENS Meeting on Fast Reactor Safety Technology, Seattle, August 19-23, 1979, pp. 742-757.

Coats, R.L., Ed., 1975, "Proceedings of the First Annual Post-Accident Heat Removal Information Exchange", Albuquerque, October 8-9, 1974, SAND, 75-0497.

Coats , R.L., Ed., 1977. "Proceeding of the second Annual Post-Accident Heat Removal Information Exchange", Albuquerque, November 13-14, 1975, SAND 76-9008, 364 p..

Coen, V. and Holtbecker, H., Eds, 1979. "Post-Accident Heat Removal - Proceedings of the Fourth Post-Accident Heat Removal (PAHR) Information Exchange Meeting", Varese, Italy, October 10-12, 1978. European Appl. Res. Rept. - Nucl. Sci. Technol., Vol. 1, N°6, 1979, pp. 1315-1712, Harwood Academic Publishers.

Cronenberg, A.W. and Grolmes, M.A. , 1975. "Fragmentation Modeling Relative to the Break up of Molten UO_2 in Sodium". Nuclear Safety, Vol. 16, N°6, November-December 1975, pp. 683-700.

Davies, N.W. and Sheriff, N., 1980 "An Experimental Investigation of Natural Convection Heat Transfer from Downward Facing Surfaces in Sodium for Fast Reactor Internal Core Catchers" International Seminar of the ICHMT, Dubrovnik, Yu, Sept., 1-5, 1980- To be published.

Dhir, V.K., 1979. "Dry Out Heat Fluxes in Debris Beds Containing Particles of Different Size Distributions". Proceedings of the International ANS/ENS Meeting on Fast Reactor Safety Technology, Seattle, August, 19-23, 1979, pp. 770-780.

Dhir, V.K. and Catton, I., 1975. "Prediction of Dry Out Heat Fluxes in Beds of Volumetrically Heated Particles". Proc. Int. ANS/ENS Meeting on Fast Reactor Safety and Related Physics, Chicago, October 5-8, 1975, Vol. 4, pp. 2026-2036.

Dhir, V.K. and Catton, I., 1979. "Natural Convective Cooling of Debris Bed". Proceedings of the Fourth Post-Accident Heat Removal (PAHR) Information Exchange Meeting, Varese, Italy, October 10-12, 1978. European Appl. Res. Rept. - Nucl. Sci. Technol., Vol. 1, N°6, 1979, pp. 34-54, Harwood Academic Publishers.

Donati, A., Garcin, J. and Fortunato, M., 1978. "Build-up to a Molten Fuel Pool inside a Plugged Subassembly". Proceedings of the Fourth Post-Accident Heat Removal (PAHR) Information Exchange Meeting, Varese, Italy, October 10-12, 1978. European Appl. Res. Rept. - Nucl. Sci. Technol., Vol. 1, N°6, 1979, pp. 172-181, Harwood Academic Publishers.

Fieg, G., 1978. "Heat Transfer Measurements of Internally Heated Liquids in Cylindrical convection cells". Proceedings of the Fourth Post-Accident Heat Removal (PAHR) Information Exchange Meeting, Varese, Italy, October, 10-12, 1978, European Appl. Res. Rept. - Nucl. Sci. Technol., Vol. 1, N°6, 1979, pp. 144-153, Harwood Academic Publishers.

Fujii, T., Honda, H. and Marioka, I.. "A Theoretical Study of Natural Convection Heat Transfer from Downward Facing Horizontal Surfaces with Uniform Heat Flux". Int. J. Heat Mass Transfer, Vol. 16, 1973, p. 611.

Gabor, J.D., Sowa, E.S. Baker, L., Jr. and Cassulo, J.C., 1974. "Studies and Experiments on Heat Removal From Fuel Debris in Sodium". Proceedings of ANS Fast Reactor Safety Meeting, Beverly Hills, April 1974, Conf. 740401, pp. 823-844.

Ginsberg, T., Jones, D.C. and Chen, J.C., 1978. "Volume-Heated Boiling Pool Flow Behaviour and Application to Transition Phase Accident Conditions". Proceedings of ENS/ANS Int. Topical Meeting on Nuclear Power Reactor Safety, Brussels, October 16-19, 1978, pp. 1295-1310.

Gluekler, E.L. and Baker, L., Jr., 1977. "Post-Accident Heat Removal in LMFBR's." Thermal and Hydraulic Aspects of Nuclear Reactor Safety, Vol. 2, Liquid Metal Fast Breeder Reactors, ASME Publication, Jone, O.C. and Bankoff, S.G., Eds, pp. 285-324.

Grand, D., 1980. "Natural Convection Cooling". Summer School of the ICHMT, Dubrovnik, Yu., August 25-29, 1980. To be published.

Hardee, H.C. and Nilson, R.H., 1977. "Natural Convection in Porous Media with Heat Generation". SAND 760433, December 1977.

Hesson, J.C. and Gunther, W.H., 1972. "Heat Transfer from Boiling Pools with Internal Heat Generation". Trans. Amer. Nucl. Soc., 15, 1972, pp. 864-865.

Jahn, M. and Reineke, H.H., 1974."Free Convection Heat Transfer with Internal Heat Sources, Calculations and Measurements". Paper NC 2-8, 5th Int. Heat Transfer Conf., Tokyo, 1974.

Kayser, G. and Lerigoleur, C., 1979. "An Internal Core Catcher for a Pool LMFBR and Connected Studies". International ANS/ENS Meeting on Fast Reactor Safety Technology, Seattle, August 19-23, 1979, pp. 781-791.

Kazimi, M., 1977. "The Potential for Fuel Boiling under Decay Heat Conditions". Proceeding of the Third Post-Accident Heat Removal Information Exchange, ANL, November 2-4, 1977, ANL-78-10, Baker, K., Jr. and Bingle, D.J., Eds, pp. 377-383.

Kulacki, F.A., Emara, A.A, Korpela, S.A., Lambha, N.K. and Min, J.H., 1977. "Natural Convection with Internal Energy Sources : Some Recent Experimental and Numerical Results with Post-Accident Heat Removal Applications". Proceedings of the Third Post-Accident Heat Removal "Information Exchange", ANL, November 2-4, 1977, ANL report n°78-10, pp. 101-113.

Kulacki, F.A. and Goldstein, R.J., 1972. "Thermal Convection in a Horizontal Fluid Layer with Uniform Volumetric Energy Sources". J. Fluid Mech., Vol. 55, 1972, p. 271.

Kulacki, F.A., Min, J.H., Nguyen, A.T. and Keymani, M., 1978. "Steady and Transient Natural Convection in Single and Multi-fluid Layers with Heat Generation". Proceedings of the Fourth Post-Accident Heat Removal (PAHR) Information Exc nge Meeting, Varese, Italy, October 10-12, 1978. European Appl. Res. Rept. - Nucl. Sci. Technol., Vol. 1, N°6, 1979, pp. 98-113, Harwood Academic Publishers.

Kutateladze, S.S., 1959, "Liquid Metal Heat Transfer Media", Chapman and Hall Ltd., London.

Lerigoleur, C., 1978. "Particle Bed Dry-Out in Sodium". Proceedings of the Fourth Post-Accident Heat Removal (PAHR) Information Exchange Meeting, Varese, Italy, October 10-12, 1978. European Appl. Res. Rept. - Nucl. Sci. Technol., Vol. 1, N°6, 1979, pp. 67-84, Harwood Academic Publishers

Lipinski, R.J. and Rivard, J.B., 1979. "Debris Bed Heat Removal Models : Boiling and Dry-out with Top and Bottom Cooling". Proceedings of International ANS/ENS Meeting on Fast Reactor Safety Technology, Seattle, August 19-23, 1979, pp. 758-769.

Mac Donald, J.S. and Connolly, T.J., 1960. "Invertigation of Natural Convection Heat Transfer in Liquid Sodium". Nuclear Science and Engineering, Vol. 8, p. 363.

Nijsing, R. and Schwalm, D., 1978. "Asymptotic Temperature Distribution and Final Composition of a Particle Bed with a Molten Fuel Layer". Proceedings of the Fourth Post-Accident Heat Removal (PAHR) Information Exchange Meeting, Varese, Italy, October 10-12, 1978. European Appl. Res. Rept. - Nucl. Sci. Technol., Vol. 1, N°6, 1979, pp. 55-66, Harwood Academic Publishers.

Peckover, R.S., 1980. "Ex-Vessel PAHR Problems". Summer School of the ICHMT, Dubrovnik, Yu., August 25-29, 1980. To be published.

Peckover, R.S., and Hutchinson, I.H., 1973. UKAEA, Culham, report n° CLM-R-123, 1973.

Plein, M.G., Lipinski, R.J., Carlson, G.A. and Varela, D.W., 1979. "Summary of the First Three In-Core PAHR Molten Pool Experiments". Proceedings of the International ANS/ENS Meeting on Fast Reactor Technology, Seattle, August 19-23, 1979, pp. 356-369.

Reineke, H.M., Schramm, R. and Steinberner, U., 1978. "Heat Transfer from a Stratified Two-Fluid System with Internal Heat Sources". Proceedings of the Fourth Post-Accident Heat Removal (PAHR) Information Exchange Meeting, Varese, Italy, October 10-12, 1978. European Appl. Res. Rept. - Nucl. Sci. Technol., Vol. 1, N°6, 1979, pp. 87-97, Harwood Academic Publishers.

Rivard, J.B., 1979. "Cooling of Particulate Debris Bed : Analysis of the Initial D Series Experiments". Proceedings of the Fourth Post-Accident Heat Removal (PAHR) Information Exchange Meeting, Varese, Italy, October 10-12, 1978. European Appl. Res. Rept. - Nucl. Sci. Technol., Vol. 1, N°6, 1979, pp. 23-33, Harwood Academic Publishers.

Rivard, J.B. and Coats, R.L., 1979. "Post-Accident Heat Removal : an overview of some in-vessel safety considerations". Proceedings of the International ANS/ENS Meeting on Fast Reactor Safety Technology, Seattle, August 19-23, 1979, pp. 709-720.

Stein, R.P., Baker, L., Jr., Gunther, W.H. and Cook, C.. "Heat Transfer from Internally Heated Molten UO_2 Pools". Proceedings of the Fourth Post-Accident Heat Removal (PAHR) Information Exchange Meeting, Varese, Italy, October 10-12, 1978. European Appl. Res. Rept. - Nucl. Sci. Technol., Vol. 1, N°6, 1979, pp. 156-163, Harwood Academic Publishers.

Stein, R.P., Hesson, J.C. and Gunther, W.H., 1974. "Studies of Heat Removal From Heat Generating Boiling Pools". ANS Fast Reactor Safety Conference, Beverley Hills, April 2-4, 1974. CONF-74041, pp. 865-880.

Subbotin, V.I., Sorokin, D.N., Ovechkin, D.M. and Kudryavtse, A.P., 1972. "Heat Transfer in Boiling Metals by Natural Convection". Russian Translation, Programs for Scientific Translations, Jerusalem.

Suo Antilla, A.T. and Catton, I., 1975. "The Effect of a Stabilizing Temperature Gradient on Heat Transfer from a Molten Fuel Layer with Volumetric Heating". Trans. ASME, 74 WA/HT-45.

Turland, B.D. and Peckover, R.S., 1977. "The Distribution of Fission Product Decay Heat". Proceedings of the Third Post-Accident Heat Removal"Information Exchange", ANL, November 2-4, 1977, report ANL-78-10, pp. 17-25, Baker, L., Jr. and Bingle, D.J., Eds, 1978.

Vossebrecker, H. and Grönefeld, G., 1978. "Retention of Dispersed Fuel within the Reactor Vessel of the SNR 300". Post-Accident Heat Removal. Proceedings of the Fourth Post-Accident Heat Removal (PAHR) Information Exchange Meeting, Varese, Italy, October 10-12, 1978. European Appl. Res. Rept. - Nucl. Sci. Technol., Vol. 1, n°6, 1979, pp. 388-398, Harwood Academic Publishers.

18 EX-VESSEL POST ACCIDENT HEAT REMOVAL

by

R. Peckover
UKAEA, Abingdon, UK

Once the molten core material has breached the reactor vessel, consideration must be given to interaction of the material with the containment structure surrounding the vessel. In this chapter a detailed description of these interactions is provided.

1. INTRODUCTION

1.1 Generalities

The adequacy of the containment of fast reactors has traditionally been evaluated by analysing the response of the containment to a spectrum of core disruptive accidents. Reactors are carefully designed to minimize the potential for serious accidents and incorporate systems which will terminate all identifiable accident initiators without serious core damage. If the reactor containment is not breached then there is no hazard to the public. It is clearly wise not only to take all reasonable precautions to ensure that the containment is not penetrated but also to examine the consequences of penetration to assess whether it is worth installing additional features to mitigate the effects of such penetrations, bearing in mind the low probabilities of the events under consideration.

Essentially an initially secure containment can be penetrated in only two ways: either by mechanical damage, or by melt-through. The ability of the containment structure to withstand large transient pressures, missile damage,earthquakes and over-pressurization on a longer timescale are discussed in other chapters. Melt-through of the containment requires large localized heat fluxes to be incident on the containment envelope, and this in practice can only occur if core debris, including the heat generating fission products, can be brought close to the containment while at elevated temperatures. Clearly the key feature is the ability of the liquids and structures within and around the containment to remove the heat generated within the core debris without melting through the containment, and it is for this reason that the analysis of the relatively unpressurised movement and cooling of core debris after it has emerged from the original core matrix goes by the name of Post Accident Heat Removal studies (or PAHR). Within the overall containment envelope is the primary vessel, which apart from pipe penetrations, provides an inner or primary containment vessel, within which the core material normally resides. It is convenient to subdivide PAHR studies into two parts: In-vessel and Ex-vessel. It is clearly desirable to arrange if at all possible that core debris remains within the vessel. An examination of the situations which may occur and review of the basic knowledge required to analyse the behaviour of core debris in-vessel is the subject of chapter 17 by J.Costa. This chapter will be concerned with the interaction of core debris and sodium with materials outside the primary vessel. These obviously must include water and concrete; but other materials which may have been placed there because of their beneficial properties in the event of melt-through can also be considered. Ultimately one is concerned with the prevention of the escape of radioactive fission products, either into the atmosphere as aerosols, or into the ground water.

There can be no doubt that below-vessel containments can be designed which will ensure that the core debris does not escape into the ground. Apart from questions of cost, there may be penalties to pay in terms of undesirable side-effects. In particular the interaction between ex-vessel material and core debris and sodium may under appropriate circumstances result in the generation of large quantities of vapour, particularly H_2O and CO_2; the accumulation of these in the vapour blanket within the primary vessel, or within the secondary containment building above the primary vessel might lead to unacceptable levels of pressurization and aerosol emission. Moreover if the gases are reduced to H_2 and CO , these can contribute to the risk of fire within the containment building. Thus although it is necessary to concentrate on the detailed local fluid dynamics and heat transfer (since it is only at relatively localized 'hot-spots' that melt-through can occur), yet the global implications for the containment as a whole must be examined in assessing what precautions to take to control the movement of core-debris ex-vessel.

An additional complication for PAHR studies lies in the fact that it is concerned with what happens after the debris emerges from the core matrix, and the form and quantity and direction of emergence of the core debris depend on the previous history of the core meltdown for which there exists, inevitably, a wide envelope of possible realizations. This means that the viewpoint in PAHR studies is different from that of reactor safety studies of earlier parts of the accident sequence, in which the initial conditions and geometry may be considered to be well defined and the object is to find the envelope and the most probable form of the subsequent course of events. Crudely speaking in PAHR studies it is the final states which can be best defined, and the object is to decide which parts of the envelope of initial conditions evolve towards desirable final states, and indeed to show that only desirable final states are accessible. In fact this blurring of the perspective does have an important simplifying effect, since it means that limiting models which bound the range of behaviour become more important than detailed mechanistic descriptions.

1.2 Concepts of Containment

If core debris succeeds in melting its way through the bottom of the reactor vessel, it will in some of the current designs fall into the reactor cavity, which is likely to include a steel liner which is designed to prevent contact of the underlying concrete or brick with sodium especially in the event of a small leak. Liner penetration by molten core debris (especially in the form of a localized jet) is quite possible, so that in this event the substrate material would be contacted by core debris and by sodium. If this is felt to be undesirable then there exist various concepts to assure post accident fuel containment each with various advantages and disadvantages. These can be classified as

(1) the cooled crucible concept. Here an auxiliary cooling system is provided to cool the walls of the cavity and so inhibit melt through. This would certainly need an additional inner liner of some ceramic material to minimize the thermal shock and erosion of the liner when the debris arrives.

(2) sacrificial barrier. This is similar to the crucible concept except that a quite massive sacrificial liner is provided. In an extreme form the barrier can become a sacrificial bed filling the reactor cavity up to the base of the reactor vessel and enveloping it. The melted liner material should be mutually soluble with the fuel so that the heat source is diluted and considerable delay time occurs before the pool approached the cooled wall. Magnesia and Alumina are likely candidate materials for a sacrificial barrier.

(3) stable barrier concept. In this concept, a refractory material is used which is capable of containing the core debris even at high temperatures. Graphite is a promising material for this application.

(4) catch tray concept. A system of trays is arranged so as to intercept the downward motion of the core debris, and to cool it under sodium. This concept is particularly promising in-vessel for pool type reactors and has been discussed in chapter 17.

An alternative to these engineered containment concepts is to rely on the inherent retention capability of the concrete basemat beneath the reactor cavity. If one can show that the core debris will only partially penetrate the basemat, then this last choice is obviously the simplest and cheapest.

To evaluate the inherent retention capability of the basemat and to compare the advantages and disadvantages of the various concepts just sketched a wide range of physical and chemical phenomena need to be adequately understood.

These include

- (i) the fragmentation of core debris
- (ii) the response of liners
- (iii) interactions of sodium with concrete
- (iv) interactions of fuel and steel with concrete
- (v) sodium water-interactions
- (vi) interactions of potential barrier materials with core debris and sodium
- (vii) heat transfer from molten pools
- (viii) growth of molten pools by melting into concrete or sacrificial materials.

These topics will be considered in the following sections. Finally the implications for ex-vessel core debris retention will be considered. A select bibliography is attached.

2. CHARACTERIZATION OF CORE DEBRIS

2.1. Physical Form

In a hypothetical core disruptive accident (HCDA), significant quantities of core material - oxide fuel, and steel from clad and wrappers and core structure - may be ejected from the original core region. Downward relocation of fuel and steel may occur either by axial melting through the lower assembly structure or by streaming through the coolant channels. Upward relocation depends on the extent that the mixture of core debris can be swept through the axial length of the pin bundle. The quantities ejected (if any) depend on several factors including the degree of overheating in the core and whether fuel-coolant interactions have occurred. The dynamics of molten clad, and the freezing of flowing fuel in a steel channel have been reviewed by Epstein [35]. There exists considerable potential for cooling the fuel material in the original core region [11], but it is difficult to place any useful limits on the amount of debris which might emerge.

There is a wide range of evidence that when molten UO_2 and steel are quenched in sodium, the core debris is broken up into fragements mainly in the 10μ to 1000μ size range. There is substantial scatter between different experiments (see [15], fig.3 in this volume). Since the conditions of accident initiation can vary considerably, this envelope of possible distributions must be accepted as a datum for PAHR.

The majority of particulate will settle onto any horizontal or near horizontal surfaces, though perhaps 10-20% of the particulate may be swept into the primary piping [109]. The initial 'support' for a particulate bed may fail, in which case the debris will fall to a lower level. Its arrival on a new 'support' could be accompanied by both mechanical and thermal shock. If the debris falls as a concentrated stream of material some erosion of the impacted surface may occur during this initial transient. Even if a particulate bed overheats and forms a molten pool, it is likely when the supports fails that the molten stream of falling debris will be refragmented by the sodium [102]. Debris bed behaviour and cooling is considered by Costa [15] in chapter 17 in this volume; see also [84] and [13] and references therein.

While within the primary vessel the chemical state of the core debris remains reasonably well defined. There is some possibility that oxide fuel particulate may be attacked by sodium above 500°C,resulting in the formation of sodium pluto-uranate. If this does occur it may modify the particle size distribution and the cooling capabilities of a sodium-logged particulate bed. and thus the likelihood of support melt-through (see §3).

If the primary vessel and any steel cavity liner have been penetrated, the debris could contact concrete or some other materials placed to control the movement of core debris. Chemical reactions between sodium, the constituents of core debris and the concrete or other substrate material can then result in significant changes in the form of the core debris. These are discussed in §4 and §5.

2.2. Decay Heat Source Distribution

The internal heating in an agglomeration of core debris is due to the presence of fission products (which term is taken here to include the actinides). The level of decay heat depends on (a) the elapse time since shutdown (b) the mean burn-up in the reactor at shutdown and (c) the fraction of the fission products present. The fall-off of decay heat levels with time is discussed in [114]; it scales with operating power of the reactor on stream (see also [15], fig 1a in this volume). The ratio (f) of decay heat to operating power is $\sim$ 7% after one second.Using adiabatic assumptions, Rivard [98] calculates that even if a particulate bed were formed at this early time, it would take a minimum of 16s for sodium (initially at 733K) to boil, 33s for steel to melt and 88s for UO_2 to melt when the corresponding values of f are $\sim$ 5%, 4.5% and 3.7%. Since a typical fast reactor power level is roughly 150kW/kg of oxide for the core as a whole, it follows that 5kW/kg provides an upper bound to the power levels of interest.The time for most of the debris to settle is of order a few tens of seconds (because of its high density relative to sodium); this therefore is less important than the time for a bed to heat up, though the extent to which the particulate has cooled during settling makes this also rather imprecise,(further discussion of settling is given in §4 of [15]).

It is believed that the primary vessel is unlikely to be penetrated in less than half an hour, so that for ex-vessel PAHR, the timescales of interest are long, - days and months - during which period the total rate of decay heat generation has fallen off considerably. At one day,f $\sim \frac{1}{2}$%, and after one year, f $<$ 0.1%.

It is of some interest to know the integrated decay heat to be dissipated. For a 3 GWt reactor, the heat content of the core and surrounding structure is around 20 GJ, which corresponds to a few hundred seconds of full reactor power. The decay heat generated within the first three years after shutdown, assuming typical burn-up times, is $\sim$ 100TJ, which corresponds to roughly 10 hours of full reactor power [86]. This indicates that when considering the evolution of debris pools, the initial temperature of the debris is not an important parameter.

In analysing the behaviour of core debris, in which the decay heat is a primary driving term, it is necessary to consider the locations of the responsible fission products. In the early stages between 20 and 30% of the heat is produced by volatile elements[114,11]which may either be dissolved in the sodium or present in the cover gas; this distributed heat source, although having important radiological consequences, is of little concern when examining debris behaviour. The contribution from volatiles falls to less than 10% of the total heat after 10 days, as the lower boiling point isotopes tend also to have shorter half-lives. The remainder of the fission products can be divided into an oxide group which will remain with the fuel and a metallic group [61,62]which are insoluble in the fuel and tend to form a separate alloy within the fuel matrix; after fuel melting,it is likely that this alloy will separate out. The alloy is soluble in molten steel (arising from clad etc) and so this heat source, which at times up to 10 days accounts for 20-30% of the total, is likely to be located with the steel debris from cladding and internal structure, given time and opportunity. The balance of the decay heat (50-70% at times of interest) is produced by elements which form oxides in the fuel,including transuranic elements (accounting for between 10 and 30% of the total heat). These oxides are soluble in molten fuel and thus this heat source is likely to remain with the oxide fuel, unless very high temperatures are reached when a greater portion of the fission products may volatilize. The contribution of the different fission products to the total decay heat is discussed in [9] and [114]. (see also [15], fig 1(b) in this volume). Roughly speaking the decay heat in the oxide phase is at least twice that in the metallic phase.

3. RESPONSE OF A LINER TO HOT DEBRIS

3.1. Melt-through

The cavity below the primary vessel is likely to have a steel liner to ensure that small leaks of sodium do not contact the concrete structure of the cavity. After the failure of the primary vessel, molten debris can penetrate into the cavity, and sodium will also enter. Experiments [102] in which sodium and molten UO_2 have been poured into a crucible indicate that the core debris is likely to be re-fragmented during its fall into the cavity, and that core debris is likely to be distributed over the cavity floor. If one ignores for the moment non-uniformities in the particulate bed distribution, then an important question is to ascertain what heat fluxes from the debris to the plate are necessary for the plate to melt-through over a wide area. This basic problem, the melting through of a steel plate, is also of central relevance to earlier stages in the accident sequence. The various catch tray devices proposed as in-vessel debris barriers and the bottom of the reactor vessel itself may be considered as steel plates, so that it is appropriate to consider here the melt-through of a steel plate for heat transfer coefficients beneath the plate varying from high values(if flowing sodium is present) to rather low values (when the plate is in contact with a low conductivity concrete). Simple scoping calculations can be made if it is assumed that the heat flux density Φ applied to the top surface of the plate is constant. Let the plate be of thickness a, with thermal conductivity k, specific heat c and melting point T_m . Let the material beneath the plate be at a temperature T_a, with a heat transfer coefficient of h (which for sodium might typically be $\sim$ 2kW/(m^2K)). The plate will melt through completely if

$$\Phi > h.\Theta_m \quad (3.1)$$

where $\Theta_m \equiv T_m - T_a$; whereas no melting will occur if

$$\Phi < h\Theta_m/(1+ \frac{ha}{k}) \quad (3.2)$$

If $\Theta_m \sim 10^3$K and $h \sim 2.10^3$ W/(m^{2}K), corresponding to sodium at 450°C, and the plate is steel, 20mm thick, of conductivity 20W/(mK), then fluxes less than 0.6 MW/m^2 will give no melting, and fluxes greater than 2MW/m^2 will give complete melt-through. Note that irrespective of h, there must be some melting if $\Phi > k\Theta_m/a$ (which is approximately 1MW/m^2 for the parameter values above).

If the plate is backed by a poor conductor (e.g. a reactor cavity liner in contact with concrete) then the time for the plate to melt completely, t_c, is given approximately by

$$t_c = \rho c \Theta_m (1+S) a/\Phi \qquad (3.3)$$

where ρ is the density of the plate, $\mathcal{L}$ is the latent heat of fusion of the plate and S is the Stefan number $\equiv \mathcal{L}/c\Theta_m$. For a 20mm steel plate, $t_c = (130/\Phi_m)$ seconds (where Φ_m is measured in MW/m^2). Thus if $\Phi_m = 2$, $t_c \approx 60$ seconds.

Calculation of the time for melt through for finite heat transfer coefficients beneath the plate requires a numerical solution. The times for melting of a steel plate are given in [78]. For example a flux of 1MW/(m^2) applied to a 20mm steel plate with $h \sim 2$kW/(m^{2}K) will result in 0.3 of the plate thickness melting; almost all of this will occur within about 4 minutes. Fig 1 indicates the extent of melting of a 2.5 cm stainless steeel slab backed by coolant at 500°C, as the heat flux applied to the top surface, and the heat transfer coefficient to the material beneath the plate, are varied. For concrete beneath the plate, the effective heat transfer coefficient is rather small so that the time to melt-through for a given applied flux can be approximated by (3.3).

If debris is heaped in one place, much higher downward heat fluxes are likely to occur locally, giving a hot spot on the plate. Such an area is the place where melt-through is most likely to occur. To help to ameliorate this one might consider placing a thin skin of a highly conducting metal on the upper surface of the steel plate [6]For an internal core catcher or the reactor vessel bottom the choice of available materials is rather restricted since it is undesirable to add to the inventory of materials already inside the primary vessel; for the reactor cavity, which would not be subject to a sodium environment under normal operation, the constraints are more relaxed. Turland [116] has examined the effect of such a highly conducting skin, and found significant attentuation of the local heat flux density at the hot spot, for debris accumulation of width up to 10 times the skin thickness. This however affords little extra protection against broader accumulations. An alternative approach is to place poorly conducting but highly refractory material on the top surface of the plate so that higher heat fluxes are required for melting to occur; the interaction between core debris and candidate refractories is discussed in §8.

3.2 Creep Rupture

The circumstances under which a steel plate would melt-through was considered in the previous section. It is also possible for a steel plate supporting hot debris to be so weakened by the thermal stresses imposed on it that it fails mechanically. Dalle Donne et al [34] have calculated the maximum admissible heat flux through a plate, on the assumption that thermal stresses up to twice the yield strength are acceptable, where the yield strength $\sigma_{0.2}$ corresponds to 0.2% strain. On this basis, the maximum acceptable, temperature difference across the plate is given by

$$\Delta T = 2\sigma_{0.2}(1-\nu)/(\alpha E) \qquad (3.4)$$

where ν is Poisson's ratio (=0.33), α is the coefficient of linear thermal expansion and E is Young's Modulus. For steel $\alpha = 1.1 \times 10^{-5} K^{-1}$, and $E = 2.1 \times 10^{10}$ kg/m^2. If $\sigma_{0.2}$ is taken to be 5×10^7 Kg/m^2, then $\Delta T = 300K$. For a steel plate of thickness 20mm, this gives a maximum acceptable flux of ≈ 0.35 MW/m^2.

If creep is considered, then the fact that the yield stress is temperature and time dependent must be taken into account. Kayser and Le Rigoleur [56]consider the formation of a 'plastic hinge' if the plate is rigidly supported at a horizontal spacing of s. Taking into account a 10% safety factor, the maximum load,P,is then given (in kg/m^2) by

$$P = 1.5\left(\frac{e}{s}\right)^2 . \sigma_{0.2} \tag{3.5}$$

where $\sigma_{0.2}$ is now a function of the time and temperature history of the plate and e is the thickness of that part of the plate which retains its strength. As an example, for a flux of $0.2 MW/m^2$ with s taken as 400mm, [65] gives e to be 16mm and P to be 7200 kg/m^2. More generally they present limiting loads as a function of downward heat flux after various elapse times. For a given heat flux,the limiting load after 100 hours is a factor of 4 smaller than that after 1 hour.

For a liner which is closely positioned against a layer of concrete, the additional support provided by the concrete may make a certain yielding of the liner acceptable. More serious may be buckling induced in the liner by the thermal stresses if the liner is tied back into the concrete at regular intervals. If flaws develop in the plate, this may allow ingress of sodium behind the liner, and the initiation of chemical reactions between sodium and water within the concrete.

An analysis by Gluekler et al [56] of the structural integrity of the bottom of the reactor vessel when loaded with core debris suggests that vessel failure would occur as a result of loss of strength of the steel prior to melting, and that the failure would involve a substantial area of the vessel. This indicates that a large quantity of sodium and core debris would descend rather suddenly into the reactor cavity beneath. To guard against the transient erosion and thermal shock that the impact would produce in the cavity liner, a refractory inner liner appears to be required.

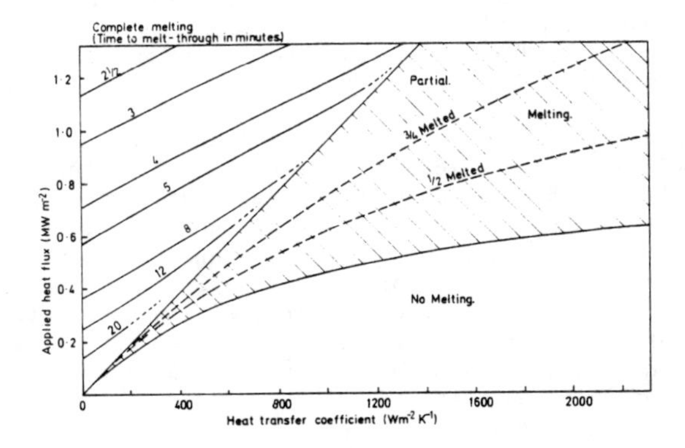

Fig.1. The melting of a 2.5 cm steel slab as the applied heat flux, and the heat transfer coefficient to coolant (at 500^oC), are varied.

4. SODIUM WATER INTERACTIONS AND HYDROGEN GENERATION

4.1. Hydrogen Production

Sodium and water can react exothermically under appropriate conditions to generate hydrogen. Since this can contribute to the risk of fire within the containment, water is carefully excluded from the sodium circuits under normal reactor operation. However the concrete of external reactor structures contains water within the cement and to some extent within the aggregate material. If core debris penetrates the reactor vessel, and any further steel liner present in the reactor cavity, it may open up a pathway which enables sodium to come into contact with water within the concrete.

The following set of reactions involve the direct combination of sodium with water, resulting in hydrogen production [32]

$$Na + H_2O = NaOH + \tfrac{1}{2}H_2 \qquad \Delta H(1100K) = -\ 36.5\ kcal \qquad (4.1)$$

$$2Na + H_2O = Na_2O + H_2 \qquad \Delta H(1100K) = -\ 39.2\ kcal \qquad (4.2)$$

$$NaOH + Na = Na_2O + \tfrac{1}{2}H_2 \qquad \Delta H(1100K) = -\ 2.7\ kcal \qquad (4.3)$$

All three reactions are exothermic, though the heat release in reaction (4.3) is negligible. Other reactions between sodium,water and the constituents of concrete such as silica are indicated in §5.

Water is present in concrete as evaporable water and chemically bound water. Evaporable water forms part of the cement and usually constitutes 2-3 wt% of concrete. This water is driven off from the concrete at temperatures in the range 20-200°C. Chemically bound water is present for example in calcareous aggregate in the form of hydroxides such as $Ca(OH)_2$ and $Ca_3(Al(OH)_6)_2$ which decompose at temperatures within the range 200-600° C.

The kinetics of the loss of water from concrete may be empirically described by first-order rate equations of the form

$$\frac{d\alpha}{dt} = (1-\alpha)\exp(a-(b/T)) \qquad (4.4)$$

where α is the weight fraction, t is the time in minutes, and T is the absolute temperature. Powers [89] finds experimental data can be approximately represented by a=14 b=5.5 x 10^3 for evaporable water, and a=28 b=20.5 x 10^3 for chemically bound water.

Since one mole of water contains one mole of molecular hydrogen,the maximum weight of hydrogen released cannot exceed (1/9)th of the weight of water available. Equivalently 55 kg-moles of molecular hydrogen is the maximum that can be generated from 1 tonne of water. In common concretes,water can constitute as much as 7wt%. Thus the weight of hydrogen generated can be up to 1% of the weight of concrete involved. For any naturally occurring substrate,a similar limit can be calculated once the rock composition is specified. If the quantity of concrete or rock from which water will be lost can be estimated a priori,then these simple considerations enable a crude but hard upper limit to be placed on the quantity of hydrogen generated.

To carry out accurate calculations of the quantity of the hydrogen produced involves solving the evolutionary equations which describe the growth of a meltpool, the propogation of thermal fronts into the substrate and the upward flow of the gaseous decomposition products (H_2O and CO_2) into the pool. Some of this is discussed later. Basically water is released by heating the concrete. If a high thermal flux is incident on the face of a concrete slab, then one may consider there to be 3 regions - a layer of molten concrete, a layer of decomposing concrete, and a layer of virgin concrete. Within the layer of decomposing concrete the fraction of the water content remaining will vary from zero at temperatures in excess of 600°C to effectively unity at 30-40°C, adjacent to the virgin concrete layer. The quantity of concrete from which water has been lost is then the volume of the molten concrete plus the appropriately weighted fraction of the decomposing concrete. A detailed model, and associated computer code, has been developed by Knight and Beck [68] which enables temperature profiles and water release rates from decomposing concrete to be predicted.

4.2. Effect on Containment

It has already been mentioned that the interaction of ex-vessel material with core debris and sodium may lead to pressurization of the containment, involving the risk of aerosol emission. To assess the response of the containment a number of scoping studies have been carried out; to illustrate the general lines of such analysis the application of the HEDL computer code called CACECO [80, 81] is discussed.

The CACECO code model deals with the various subdivisions of the reactor buildings - reactor vessel, reactor cavity, reactor containment building (RCB)- in a lumped manner. It can predict the pressure - temperature response of up to four cells containing both liquid and gas phases. In general it employs homogeneous models and neglects any kinetics and diffusion effects. The atmosphere in each cell is homogeneous, having the same temperature at all locations. Energy added to the atmosphere is assumed to be transmitted instanteously, as are the pressure effects. Chemical reactions are not diffusion controlled and go to completion instanteously regardless of dilution until one reactant is completely depleted. The HEDL version of CACECO assumes thermal equilibrium in each cell; this forces the atmosphere and pool in each cell to be at the same temperature. This obviously overpredicts heat transfer rates,and need not be conservative. A version of CACECO used at BNL to examine the containment response of FFTF removes this restriction [53]. The cell volumes are assumed to be in contact with appropriate heat structures, and mass transfer between cells is represented by a vent pathway model. The cross-sectional area of such vents can be prescribed as can the times when such inter-cell venting is allowed. This corresponds to the postulated failure of certain seals, which can be activated by appropriate 'blow-out' conditions.

Fig.2 shows the containment geometry used by [53]for FFTF. For ex-vessel analysis they used only 3 cells, combining the reactor vessel and cavity into a single cell. To deal with the attack on sub-cavity materials, it is necessary to couple in codes developed to analyse melt-concrete interactions, such as GROWS [73] or INTER [77]. The philosophy of codes modelling the evolution of melt pools is discussed in §7. The key variables which are the end-products of CACECO type calculations are (i) the pressure (ii) the temperature and (iii) the hydrogen concentrations within the RCB as functions of time; the times taken for the over pressure to reach some prescribed safety limit and for the hydrogen concentration to reach the flammability limit (normally taken to be 4%) are important, since decisions may then need to be made over controlled venting. The results of global analyses of this type are only as good as the detailed physics which is modelled. Nevertheless parametric analysis enables sensitive parts of the model to be identified; for these more detailed investigation is desirable.

Various conclusions can be drawn from the calculations of [53] for FFTF:

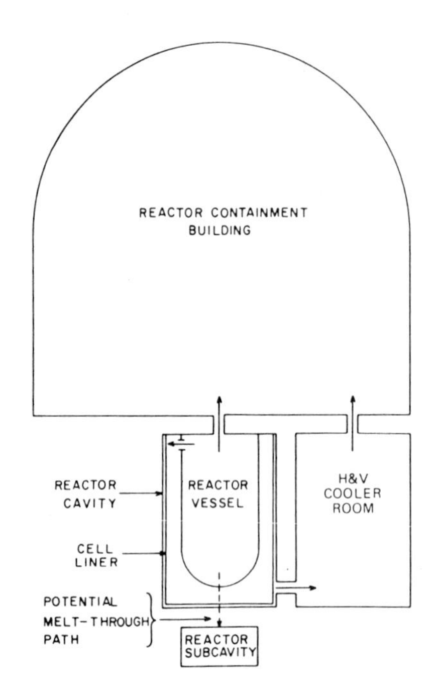

Fig.2. Diagram of the subdivisions of a reactor used in the CACECO code. [53]

(i) the time for hydrogen and pressure build-up in the RCB is almost an order of magnitude less if the cavity liner fails than if it does not. (ii) melting attack on subcavity concrete can lead to more rapid hydrogen evolution, but the RCB pressure build-up may be rather slower (iii) combustion of sodium vapour in the RCB is the largest contributor to the transient. This implies that as far as the integrity of the RCB is concerned, those parameters that facilitate sodium vaporization are the controlling factors in the accident sequence.

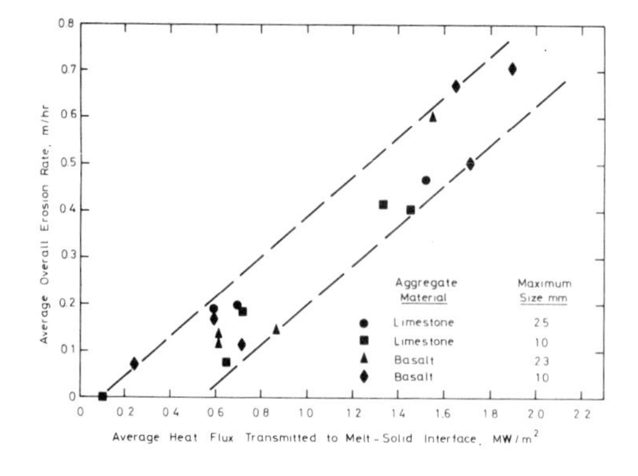

Fig.3. Overall Erosion Rate vs net heat flux to concrete melting front.[75]

5. INTERACTIONS WITH CONCRETE

5.1. Thermal Attack

When hot sodium and/or core debris contact concrete, heat is conducted into the concrete and a complex sequence of chemical reactions can occur. The chemical reactions are considered in later sections, but it is to be remarked here that many of these are exothermic, and so contribute to the thermal energy available for heating the concrete. The conduction of heat into the concrete leads to dehydration (see §4) and gas release. Besides the release of water vapour, the decomposition of the aggregate can lead to the evolution of CO_2. For example calcite ($CaCO_3$) and dolomite ($MgCa(CO_3)_2$) which are found in limestone concrete decompose to calcia (CaO) and magnesia (MgO) with the release of CO_2. This decomposition occurs in the temperature range 600-1000°C.

For typical concretes with a basaltic aggregate, the weight loss due to the evolution of CO_2 is less than 2wt%; for concretes with limestone aggregate however the decarboxylation of calcium carbonate in the aggregate and cementitious particles may involve more than 22% weight loss. As Knight and Beck remark [68], this increases the porosity of the concrete substantially - an effect which must be incorporated in any model of the thermal decomposition of limestone-based concrete. The kinetics of the loss of CO_2 from concrete may also be represented by an equation of the form (4.4). For CO_2, Powers [89] reported $a=16\pm10$, $b=(19\pm7)\times10^3$ using a least squares fit to experimental data. The volume of gas evolved from limestone concrete is rather large, and its rapid and vigorous production will violently agitate the overlying melt. This stirring action must be taken into account in modelling the evolution of melt-pools (see §7).

A number of experiments have been performed [74,24,76] in which concrete has been transiently heated. For small samples (<0.2m) without constraint, no spalling or fragmentation has been observed. For larger sections, spalling confined to within a few centimetres of the heated surface has been observed.

Analysis of the ANL results [24] by Baker [19] shows that the temperature dependence of the effective thermal diffusivity (and conductivity) must be taken into account in order to represent experimental results satisfactorily. Moreover the effective values appear to be modified by the heat transfer due to the migration of water vapour and CO_2. In the steady state, the thermal conductivity of many concretes can be represented by

$$k = 2.25(1-[T/2000]) \quad W/mK \qquad (5.1)$$

where T is in Kelvin [19]. The effective values are rather high at low temperatures, and significantly lower at higher temperatures. This may be explained in terms of the movement of water away from the heated surface at low temperatures and the flow of steam and CO_2 through the more porous concrete towards the heated surface at higher temperatures.

If the thermal flux incident on a slab of concrete is sufficiently high then not only will the concrete decompose, it will also melt. Melting for both basaltic and limestone concrete occurs first at $\sim$ 1100°C in the cement; complete liquefaction depends on aggregate type and occurs at 1400±50°C. The latent heat associated with concrete is $\sim \frac{1}{2}$MJ/kg and is small compared with the calorific input required to raise concrete to its liquidus temperature. At Sandia, concrete has been heated in two different ways by radiant heating and by a plasmajet. Muir [75] finds that the overall erosion rate is linearly proportional to the net heat flux reaching the melting interface. A flux of $1MW/m^2$ corresponds to approximately 0.25m/hour; the erosion of 1 cm^3 of concrete requires $\sim$ 10kJ. The lack of any systematic dependence on the type of concrete (fig.3) implies that it is the cement rather than the aggregate which is the main determinant of the rate of frontal advance.

Heating concrete can also result in the formation of cracks since large temperature gradients will produce substantial differential thermal expansion between different locations in the concrete and between aggregate and cement. Severe cracking of the concrete crucible is reported in [16] for some sodium-concrete tests. Steel re-inforcements provide a further source of thermal stress. Failure modes of heated concrete are considered in [58]. The importance of cracking is not clear. In some circumstances sodium has penetrated down the cracks and the increased surface area has presumably enhanced the reaction rate. However the vigorous evolution of water vapour and CO_2 up the cracks may prevent significant ingress of sodium. The role of cracks in the substrate in the growth of debris melt pools is considered in §7.

5.2. Sodium-Concrete Interactions

After the failure of the liner in the reactor cavity, it is expected that sodium as well as core debris will pass through the penetration and that the main reaction initially will be between sodium and concrete. Sodium attacks bare concrete quite vigorously. Experimentally observed penetration rates vary from 12mm/hr for basaltic concrete and Tennessee limestone concrete to 25mm/hr for high density magnetite concrete.

The sequence of chemical reactions involved in the attack of sodium on the concrete is complex and poorly understood. Clearly the water evolving from the concrete is important, and sodium-water reactions have already been discussed in §4.1; these reactions are major sources of heating during the interaction.

The other gas which evolves from concrete as it decomposes thermally is CO_2. This can react with sodium oxide or with sodium as follows:

$$Na_2O + CO_2 \rightarrow Na_2CO_3 \quad (5.2)$$

$$4Na + CO_2 \rightarrow 2Na_2O + C \quad (5.3)$$

$$4Na + 3CO_2 \rightarrow 2Na_2CO_3 + C \quad (5.4)$$

$$Na + 2CO_2 \rightarrow Na_2CO_3 + CO \quad (5.5)$$

Reaction (5.2) is favoured when sodium oxide is available in the melt. Reactions (5.3) and (5.4) are thermodynamically favoured over reactions such as (5.5) which would evolve CO. Baker [19] reports elemental carbon in the reactant mixtures at the end of some experiments. These results imply that while there is an excess of sodium present above the concrete surface, only negligible quantities of CO_2 or CO can emerge into the containment atmosphere.

Concrete aggregates are mixtures of refractory oxides. The main constituents are silica, calcia, alumina, magnesia and ferrite. The most reactive of the concrete constituents is silica (SO_2) or silica-bearing minerals such as basalt. Sodium silicate is very stable, and can be formed by the reactions

$$Na + 3SiO_2 \rightarrow 2Na_2SiO_3 + Si \quad (5.6)$$

$$Na_2O + SiO_2 \rightarrow Na_2SiO_3 \quad (5.7)$$

Magnetite (Fe_3O_4) is expected to react with sodium to generate FeO:

$$Fe_3O_4 + 2Na \rightarrow Na_2O + 3FeO \quad (5.8)$$

FeO may itself be reduced by

$$FeO + 2Na \rightarrow Na_2O + Fe \quad (5.9)$$

While liquid sodium is present, the temperature is limited to the boiling point of sodium (1156K). This is below the melting points of calcia and magnesia and neither should react with sodium. The accumulation of reaction products at the sodium/concrete surface may be expected to hinder contact between sodium and the reactive constituents of concrete, and penetration would be limited thereby. This however is currently a source of debate [53].

Further discussion of chemical reactions between sodium and concrete may be found in reviews by Baker and Gluekler [9,19], in papers presented at the Seattle Conference [16,32], and in earlier collections [2-5, 10].

5.3 Steel-Concrete Interactions

While liquid sodium is locally present, no reactions are expected between (solid) steel and concrete. Once a particulate bed has dried out, the temperature may rise sufficiently for the steel component to melt and it can be expected to flow down and envelope the still solid fuel particulate (note however that experiments at JRC Ispra [79] suggest that this may not immediately occur at steel melting, but may be delayed until ~2250°C, the temperature at which steel wets oxide fuel).

When molten steel has been poured into concrete crucibles, the melt layer was vigorously agitated by the passage of gases released from the concrete. Both H_2O and CO_2 are reduced and essentially only H_2 and CO emerge from the melt layer [24,75]. The steel is oxidized to a mixture of nickel, chromium and iron oxides which form a single phase layer overlaying the molten steel. This layer also dissolves the oxide constituents of concrete - CaO, SiO_2, and MgO.

5.4 Fuel-Concrete Interactions

It is to be expected that $(UPu)O_2$ fuel will dissolve in a molten mixture oxides arising from the interaction of steel with concrete. Experiments have been carried out at ANL with UO_2 microspheres and molten Columbia River basalt [31]. These indicate that the solubility of UO_2 in basalt is 6wt% at 1350°C, 10wt% at 1560°C and 50wt% at 2200°C. Capp et al [29] heated basalt with a UO_2 pellet at temperatures ranging from 1600 to 2500°C. At the higher temperatures there was apparently total solution of the UO_2 combined with the production of a hard-glass like melt. Detailed analysis showed the separation of the silica/ UO_2 phase from the alumina.

Some small-scale tests of the attack of molten UO_2 on concrete have been performed [24]. Molten UO_2 was found to penetrate concrete at rates ranging from 0.003 to 0.04mm/s depending on the power input. The solidified pools appeared to be an opaque glass having a density of 3000-4000kg/m^3, and analysis showed it to consist of solutions of UO_2 and the oxides of concrete.

6 HEAT TRANSFER FROM MOLTEN POOLS

6.1. Simple Internally Heated Pools

Pools of molten debris may form in-vessel either from the slumping of molten material directly from the core matrix, or from the overheating of a particulate bed. Such pools may occur on catcher plates, or on the bottom of the vessel itself. In-vessel molten pool behaviour is discussed in [15 , §6]. Similar pools can form in the reactor cavity after vessel failure. Molten pools of core debris mixed with concrete or other 'substrate' material need also to be considered in the event that vessel bottom and cavity liner are penetrated. The melting of a molten pool into its supporting structure both laterally and vertically downwards depends on the distribution of heat fluxes from the pool to the melting front over its surface. In this section, §6, we shall suppose the pool dimensions to be invariant, and consider the spatial distribution of heat rejection from such a pool. The growth of a molten pool is considered in §7.

Consider initially a horizontal layer of material of infinite extent, with thermal conductivity K, uniform volumetric heat source density H (in watts/m^3) and of depth L. Let the upper surface be at temperature T_1 and the lower surface at $T_o \equiv T_1 + \Delta T$. If convection is not occurring in the layer, then the temperature distribution is simply a parabolic conduction profile. The heat flux through a surface can be expressed non-dimensionally as a Nusselt number Nu by

$$Nu = \phi L/k(T_{max}-T_i) \qquad (6.1)$$

where ϕ is the heat flux density through the surface, T_{max} is the maximum temperature in the layer, and T_i is the temperature of the surface (T_o or T_1 as appropriate). It is convenient to define $A \equiv k\Delta T/HL^2$, so that A=0 when $T_1 = T_o$ and $A=\infty$ when H=0. If the subscripts u and d denote up and down, then in this simple case

$$Nu_u=4/(1+2A) \; ; \; Nu_d=4/(1-2A) \qquad (6.2)$$

The layer may be considered as consisting of two sublayers separated by the temperature maximum (see fig 4). The lower sublayer of thickness $L_d = L(\frac{1}{2}-A)$ is stable against convection; the upper sublayer of thickness $L_u = L(\frac{1}{2}+A)$ is potentially unstable. A modified Nusselt number Nu' may be defined, based not on the complete layer depth, but on the sublayer adjacent to the relevant surface. In this case we obtain the simple result that $Nu_u' = Nu_d' = 2$.

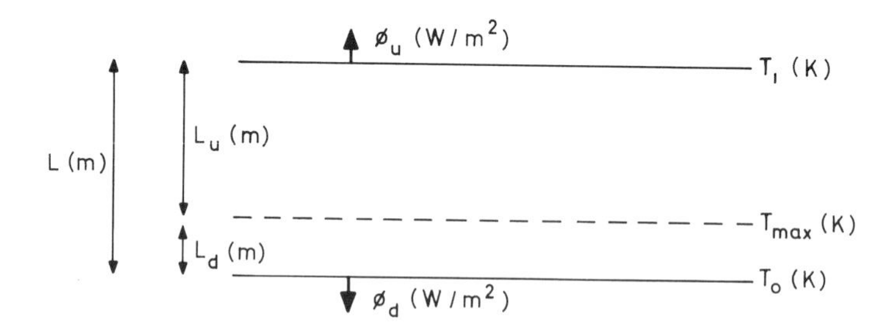

Fig.4. Division of a heat generating horizontal layer of fluid into two sublayers of thickness L_u and L_d.

When temperature gradients in the layer are sufficiently steep, convection occurs. The relevant non-dimensional parameters are the Rayleigh number Ra defined by

$$Ra = \frac{\beta g L^3 \Delta T}{\nu \kappa} \qquad (6.3)$$

and a modified Rayleigh number Ra_H defined by $Ra_H \equiv Ra/A$. Ra is sometimes called the external Rayleigh number since it depends on the applied temperature difference; Ra_H is then the internal Rayleigh number since the internal heat sources here provide the de-stabilizing potential. A number of experimental studies for internally heated layers have been carried out [44-46,63,69-72,92-3, 107]. For an adiabatic bottom, and isothermal upper boundary, it is found that

$$Nu_u = 0.4\ Ra_H^{0.226} \qquad (6.4)$$

represents the data upto $Ra_H \approx 10^9$. For isothermal surfaces with $T_1=T_o$, the fluxes can be correlated by

$$\left.\begin{aligned} Nu_u &= 0.3\ Ra_H^{0.236} \\ Nu_d &= 1.4\ Ra_H^{0.095} \end{aligned}\right\} \qquad (6.5)$$

Examination of the convective patterns indicates that a convecting layer may be conveniently considered as composed of two sublayers - a lower sublayer which is effectively stagnant, and an upper convecting layer. The lower sublayer has a downward facing parabolic profile; essentially all the heat generated in the lower sublayer is conducted downwards. The division between upper and lower sublayers corresponds to the horizontally averaged temperature maximum and so can be considered as an insulating membrane. If modified Nusselt numbers are based on the sub-layer depths, then $Nu_d' \approx 2$, and the correlation for the upper sub-layer is well represented by (6.4) with Nu_u and Ra_H replaced by Nu_u' and Ra_H'. Baker, Faw and Kulacki [26] have applied this idea to the case when the layer surfaces are unequal, and obtain the following approximate correlation:

$$(1-\eta)^{0.870}/(\eta^2+2A) = 0.173\ Ra_H^{0.226} \qquad (6.6)$$

for the fraction η of heat conducted downwards (η is defined as $Nu_d/(Nu_d+Nu_u)$) Note that although Ra_H is instrinsically positive, Ra may take either sign; here Ra is >0 when the lower surface is at the higher temperature. For a liquid layer sandwiched between solid crusts of the same substance, so that $T_1=T_o$, the fraction of heat downwards decreases with increasing Ra_H, asymptotically as $\sim Ra_H^{-1/9}$. If the thermal resistance above the layer is rather high (for example a vapour cavity is present) then a larger fraction may be expected to go downwards.

The form of convection is one in which a broad region filling the centre of a cell is rising and being heated. At the upper boundary the liquid deposits its heat and the now cooler liquid returns downwards in plumes which becomes narrower as convection becomes stronger. This is true also for turbulent convection though the locations of the cold falling plumes are continually changing.

For a laterally bounded layer with the sidewalls cooled, the basic behaviour is not modified [100]. Cool descending plumes form on the sidewalls and may be considered to be extensions of the boundary layers on the upper surface. If the pool aspect ratio differs little from the aspect ratio of the cellular structure which would occur in the infinite layer, then the correlations would be modified only to the extent that conduction through the sidewalls must be subtracted off.

Examination of the data of Mayinger et al reported in [18] indicates that, for a right cylinder of radius r and depth L, the fraction of heat conducted through the sidewalls can be approximated to first order by $1/(1+[r/L])$. The results of Fieg [45] are consistent with this. If the system is sufficiently broad, multiple cells can form; Reineke [95] finds in a hemisphere that flow is downwards in the centre, upwards at some intermediate radius, and downward at the edges. Reineke has also obtained the local Nusselt number on the hemisphere surface as a function of angle (Θ) from the vertical. If ϕ_d and ϕ_s are typical flux densities downwards and sidewards then $\phi(\Theta)=\phi_s \sin^2\Theta+\phi_d \cos^2\Theta$ is a reasonable representation of the data. (see also [52])

Experiments have been carried out at ANL [104] in which up to 20 kg of UO_2 has been electrically heated. The results are not in agreement with the above correlations, and significantly larger downward heat fluxes are reported. Various explanations have been put forward; one possibility appears to be that the different electric resistances of solid and liquid UO_2 result in larger flows of electric current through the solid crusts than initially calculated [118].

For debris pools formed above concrete, there will, as earlier sections have indicated, be a considerable bubbling of vapour through the pool. The passage of bubbles agitates the pool in a fashion similar to that of bulk boiling and the ratio of heat rejected laterally to that rejected upwards is likely to be affected in the same way. (Boiling heat transfer from self-heated pools is discussed in §7 of [15]). An important difference however lies in the fact that gas evolved from the lower boundary must pass through the lower sublayer, whereas for internal boiling the lower sublayer is relatively free of vapour. The influence of the gases released from the concrete on the dynamics of a molten pool and on the heat transfer to the boundaries can be simulated by injecting gas through the lower surface of the pool. Small scale laboratory experiments of this kind have been carried out in several places.[27,43,46]. A review of the literature and an assessment of its relevance to ex-vessel debris pools has been carried out by Blottner [28].

The superficial gas velocity v_s at the surface has a significant influence on the type of bubble flow that occurs. For water, Wallis [119] found 3 regimes when air was blown into water:- (i) laminar bubbly flow (v_s<0.06m/s); (ii) foam flow (0.06<v_s<0.76 m/s) and (iii) blanketed or film flow (v_s>0.76m/s). The critical velocity v_1 for transition from regime (i) to (ii) is given according to Wallis, by

$$v_1 = 0.38\ (\sigma/\rho_\ell \mathcal{A})^{\frac{1}{2}} \qquad (6.7)$$

where σ is the surface tension, ρ_ℓ is the liquid density, and $\mathcal{A}$ is the Laplace constant defined by $\mathcal{A}^2 = \sigma/\{g(\rho_\ell-\rho_g)\}$. Here ρ_g is the gas density and g the gravitational constant. The critical velocity v_2 for transition from regime (ii to (iii) is given according to Kutateladze by

$$v_2 = K_s(\sigma/\rho_g \mathcal{A})^{\frac{1}{2}} \qquad (6.8)$$

where the parameter K_s is a constant of order 10^{-1}. Blottner has estimated the critical velocities for water vapour bubbling through molten UO_2, molten steel and slag; typical superficial velocities of the gas released from concrete melts [90] fall in the foam flow regime for all three pool materials if the ablation rate of concrete is taken to be $\sim 10^{-4}$m/s consistent with [75]. The slag is representative of the mainly oxidic layer formed from molten concrete, oxidised steel, dissolved UO_2 and reaction products with sodium; its second critical velocity v_2 is much lower than that for molten steel or UO_2, being perhaps twice the superficial velocities from experimental concrete melts.

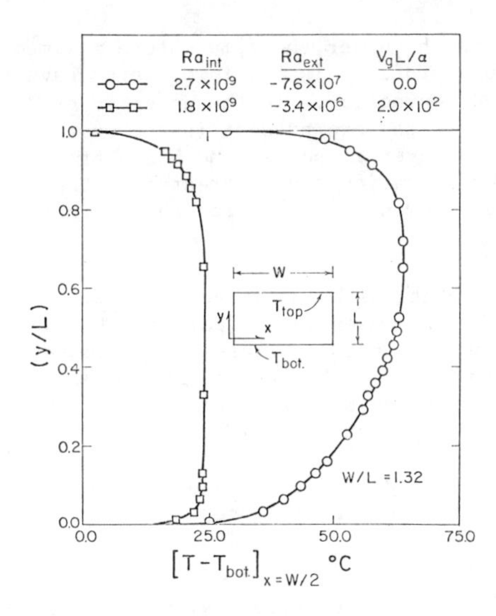

Fig.5 Vertical temperature distribution along the pool centre-line with and without gas injection - the gas flattens the profile [43]

The experimental investigations show that pool maximum temperatures are considerably reduced by gas injection compared with natural convection (see fig.5); the pool is well stirred by the gas, with a nearly isothermal temperature distribution. Natural convection effects become negligible relative to the effects of gas injection even at rather small superficial velocities, in as much as the Nusselt number is independent of the internal Rayleigh number. For v_s in the range 10^{-5} m/s to 4×10^{-3} m/s, [43] find the heat transfer coefficient for downward heat transfer h_d can be correlated by

$$h_d = (k/W).a\,(R_f)^b \qquad (6.9)$$

where k is the thermal conductivity, W is the pool width, and R_f is the Reynolds-Froude number $\equiv (v_s{}^3/\nu g)$ where ν is the kinematic viscosity of the pool fluid. For an isothermal upper boundary a=200, b=0.5 whereas for a free upper boundary a=230, b=0.07. The sidewall heat transfer coefficient, h_s, is comparable with h_d. [27] find similar qualitative behaviour in a study with v_s ranging up to 5×10^{-2} m/s but the dependence of h_d on v_s for a free boundary surface is more marked: $h_d \propto v_s^{0.4}$. It is to be remarked that neither study finds any significant dependence on the depth of the pool except at very low gas injection rates. Further studies in the foam flow regime are clearly required.

6.2. Multi-layer pools

The metallic steel phase is found to form a distinct layer in some melting experiments. An oxide phase rich in fuel is denser than steel and would lie beneath such a metallic phase; however a slag phase in which the constituents of concrete and oxidized steel have strongly diluted the fuel is lighter than steel and so would float above the steel. Simulation experiments have been carried out [46,71,97] in which mass transfer has been ignored, and heat sources have been present in either or both of two immiscible layers arranged with the lighter layer above. In these circumstances the coupling between the two layers is essentially only through the thermal boundary condition at their mutual interface. Heat transfer through the layers can be satisfactorily represented by combining correlations of the general type(6.6). If significant heat sources are present in the lower layer but not in the upper, the temperature maximum occurs in the lower layer and temperature profiles in the lower layer have the form characteristic of internally heated convection; in the upper layer Bénard convection occurs. If significant heat sources are present in the upper layer but not in the lower layer, then the lower layer is cooler and has a stably stratified conduction profile; this configuration can be treated in Baker-Faw-Kulacki framework [26] when appropriate allowance is made for the thermal resistance of the lower layer.

Werle has carried out two-layer experiments with gas injection [120] in the case where the heat sources are in the lower denser layer (water). The upper layer was silicon oil. The passage of the gas bubbles causes an irregular wavy movement of the interface. At moderate values of the superficial velocity v_s, small water droplets are carried (along with the bubbles) into the oil layer where they coalesce and descend as large droplets to the interface. With increasing v_s a mixing layer develops at the interface; above a critical value of v_s (4 mm/s for this system), transport of oil into the water is also observed and the system begins to form a more or less homogeneous mixture. The upward heat transfer can be represented by

$$Nu_u = C(v_s).Ra^{1/3} \qquad (6.10)$$

where the parameter C increases with superficial gas velocity. Similar effects may be expected in accident conditions when the fuel-rich oxide phase lies beneath a steel phase.

6.3. Effects of Mass Transfer

Convection within a molten debris pool will be effected by the fact that as it grows, lighter material will be absorbed through the melting front. This melt layer material has additional buoyancy and so the capacity to drive more intense convection. Farhadieh[40,41] has carried out experiments to investigate the mechanism controlling the penetration of a hot liquid pool into a solid with which it is miscible. The liquid pool, simulant of oxide core debris, consisted of an aqueous solution of KI and NaBr, with densities in the range $996 < \rho_L < 1700$ kg/m^3. The solid, 'sacrificial' material, was polyethylene glycol with density of 1095 kg/m^3 at melting. The liquid/solid interface was horizontal, with hot liquid above. The ratio b of bulk pool to melt layer density is an important parameter. If $b < 1$ the melted solid remains below the bulk liquid and the melting rate follows pure conduction; if $b > 1$ four flow regimes can be distinguished. For $1 < b < 1.09$, the melt layer is not thermally convecting, the flow is controlled by material convection with the flow pattern consisting of needle-like streamers protruding through the conductive layer. At $b \sim 1.09$ strong thermal convection in the melt layer sets in and the heat transfer coefficient increases sharply. For $1.09 < b < 1.25$ the melt layer is strongly convecting with a flattish temperature profile. For $b > 1.25$ the melt layer is washed away into the pool as rapidly as it is formed by the vigorous turbulence generated and the heat transfer coefficient is further increased (see fig.6).

For $b > 1$, the liquid in the pool in the vicinity of the melting front will be lighter than the bulk pool material, due to the higher concentration of the liberated bad material there. The actual density gradient will be the weighted sum of the temperature and concentration profiles within the pool. Qualitatively it is clear that the larger the density difference between the pool and the melted bed material, the more vigorous the convection, and the greater the enhancement of the downward heat transfer. Thermohaline convection has been intensively studied in an oceanographic context (see [117], and [54], and references therein) and the different regimes mapped out. In the reactor context the density difference between melted bed material (e.g. concrete, MgO,Al_2O_3) and the pool is likely to be large initially; as at this stage concentration driven convection will enhance the rate of melt front propogation at later times when the oxide phase has been well diluted. The concentration gradients may be rather small, and the bottom of the pool more quiescent.

The effect of mass transfer on heat transfer at a melting vertical side wall has also been considered [11,66,110,41]. For a fixed cooled vertical sidewall, a thermal boundary layer in which convection is downward will exist adjacent the wall; the same will be true for a vertical melting front advancing into a solid of the same chemical constitution as the pool (e.g. a hot steel pool adjacent to a steel plate). On the other hand the addition of lighter material at the interface may retard or even reverse the direction of flow at the boundaries.

The calculation of appropriate (time-dependent) heat transfer coefficients for heat transfer to different positions on the melting front in a growing melt pool is thus complex. The instantaneous size of the pool, the rate of melt incorporation, and the rate of gas injection all influence the rate of heat transfer and these processes interact with each other dynamically.

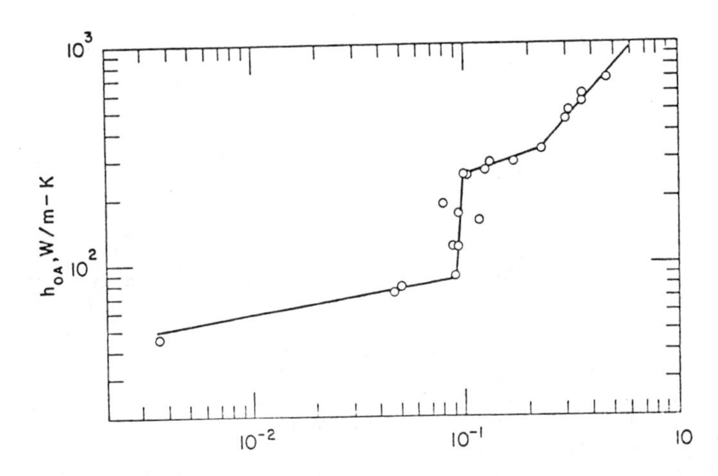

Fig.6 Overall heat transfer coefficient as a function of b - 1 where b is the ratio of bulk pool to melt layer density.[41]

7. EVOLUTION OF MOLTEN POOLS

7.1. Miscible Pools

The modelling of the penetration of core debris into a substrate of concrete, rock, or specially chosen barrier materials is complicated by the chemical reactions between the materials, and particularly by the range of time scales over which species diffusion and dissolution processes take place. Core debris itself can be thought of as composed of two distinct components:

(i) an oxide fuel phase with the oxide fission products; and
(ii) a metallic (steel-based) phase with the metallic fission products.

The constituents of a substrate can combine with the oxide fuel to form a single phase, given time. If the debris pool is rather turbulent, the two phases may be so intermingled, as in an emulsion, that the pool may be considered to be homogeneous and effectively completely miscible with the substrate. Oxidation of the metallic phase by the passage of H_2O and CO_2 can also turn the pool into a single oxide phase. In this section we shall consider the debris pool to be miscible with the substrate and discuss its evolution. In §7.2 the penetration of debris with a significant metallic phase is discussed.

If then it is postulated that a meltpool has formed beneath the reactor vessel in the solid substrate with which the molten debris is miscible, the energy balance is

$$\underset{(1)}{-F_{up}} + \underset{(2)}{Q(t)} = \underset{(3)}{\int_V \frac{\partial}{\partial t}\left[\rho_\ell c_\ell T\right] dV'} + \underset{(4)}{\rho L^* \int_\Omega \underline{u}.d\underline{S}} - \underset{(5)}{k\int_\Omega \nabla T.d\underline{S}} \qquad (7.1)$$

(see fig.7). In the pool ρ_ℓ and c_ℓ are the density and specific heat, and $T(\underline{r},t)$ is the temperature. The terms in the equation are as follows (1) the heat transfer upwards through the top surface of the pool; (2) the decay heat source $Q(t)$ which is the integral of the heat source density within the pool (3) the change in pool temperature (4) the advance of the melting front and (5) thermal conduction into the bed. The pool volume is V and its lower surface is denoted by Ω. If the solid does not evolve gas on heating, and has a well-defined melting point, we may identify ρ, L^* and k as the density, latent heat of fusion and the thermal conductivity of the substrate. For concrete the picture is more complicated. The aggregate and cementious parts of the concrete melt at somewhat different temperatures, and CO_2 and water vapour are released at different stages as the thermal front propagates into the concrete. Essentially the penetrating front consists of a number of phase change fronts propagating in close succession. Nevertheless the form of the energy balance equation can be retained if the front is considered to be thin, provided ρL^* is identified as a modified latent heat per unit volume which includes the heat required to release the water vapour and CO_2 as well as melting the concrete,and where $k\nabla T$ is the effective flux of heat away from the front.

The quantities of interest are the shape and size of the pool when it reaches its maximum extent and the time it takes to do so. Hence it is the location of the penetrating front as a function of time which is required. The local velocity of advance $\underline{u}$ of the (multiple)front is determined by a generalized Stefan condition

$$\underline{F}_1(\Omega) - \underline{F}_2(\Omega) = (\rho L^*)\underline{u} \qquad (7.2)$$

where $\underline{F}_2$ is the heat flux density conducted away from the front into the sub-

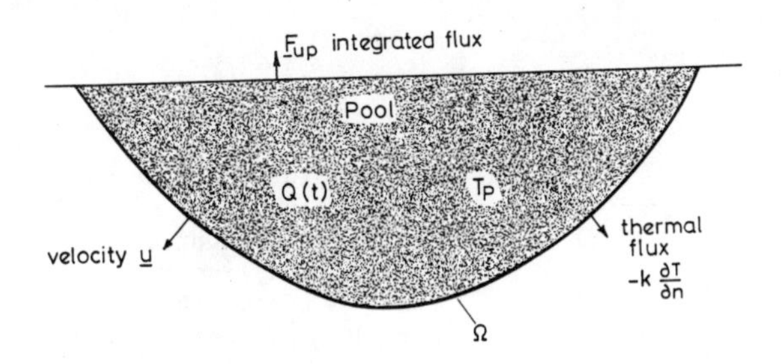

Fig.7. A schematic melt-pool, of bulk temperature T_p, containing distributed heat sources giving a spatially integrated heat source $Q(t)$. A total thermal flux F_{up} is transferred upwards. The lower surface of the pool Ω is advancing at a velocity $\underline{u}(\underline{r},t)$.

strate and $\underline{F}_1$ is the heat reaching the front from the pool. The distribution of the heat flux density $\underline{F}_1$ over the propagation front is the main factor determining the shape and size of the melt pool. If no vapour is evolved the local heat transfer coefficient at positions on the front is determined by the convection generated by the density gradients within the pool, arising from the transfer of lighter material into the pool through the melting front, and from thermal buoyancy effects. If vapour is generated, the convection patterns can be substantially modified by the passage of vapour bubbles; moreover vapour films separating concrete from the pool can provide an additional thermal resistance whose magnitude varies with position on the front [17].

Four temporal stages of pool growth can be distinguished (though the time of transition from one stage to another is rather arbitrary):

(i) an early period when the pool temperature can change rapidly; the movement of the melting front, though relatively rapid, is limited by the thermal resistance of the boundary layer of the pool. During this stage term (3) in eq. (7.1) is important; for subsequent stages,when the pool is close to its melting point,term (3) can be neglected.

(ii) a period of 'fast' melting front advance where the thermal boundary layer ahead of the front is steep, and heat conducted into the cold substrate preheats it locally to the bed melting point.Term (5) in eq.(7.1)is $\approx \rho c \Delta T \int \underline{u}.d\underline{S}$ where c is the specific heat of the solid substrate and ΔT is the difference between the melting temperature and the ambient temperature in the substrate far from the pool.

(iii) a period of conduction dominated growth: the thermal conduction front increases in thickness and the melting front slows. At the end of this period, the maximum pool size is reached.

(iv) a final period of pool shrinkage as the decay heating function continues to decrease, and heat is conducted away through the substrate.

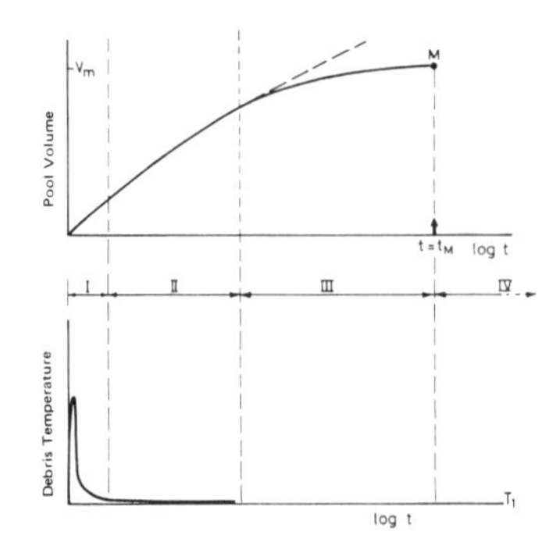

Fig.8. The evolution of the volume and mean temperature of a miscible pool, showing the four temporal phases. The dashed line - - - indicates pool growth if the thermal boundary layer in the substrate remained thin. T_1 is the melting temperature of the substrate.[83].

In general the pool will attain its maximum melted volume, its maximum depth and maximum width at somewhat different times; the time of maximum volume is chosen here as the instant separating stages 3 and 4. Fig.8 shows schematically the evolution of pool volume and mean pool temperature through the four stages.

The shape of the molten pool depends on the distribution of the heat flux density F_l over the melting front surface. If F_d is the downward flux density at the bottom of the pool and F_s is the lateral flux density at the side of the pool, then F, will vary smoothly as a function of the angle θ between the normal to the front and the vertical. The form for a non-boiling pool without concentration gradients or gas evolution is given in §6.1. The ratio (F_s/F_d) may be used as a measure of the tendency for the pool to grow laterally; it is a function of several parameters including the difference between the densities of the debris (ρ_2) and the substrate (ρ_1). This is strikingly illustrated in some simulation experiments involving water-soluble waxes in which the initial

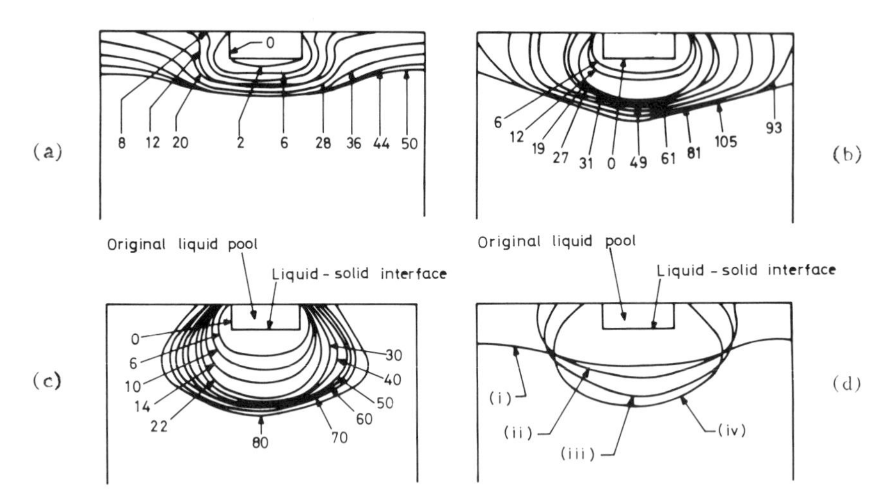

Fig.9. Advance of aqueous solutions into a water-soluble wax bed [40]. The successive positions of the melting front are labelled by the time in minutes in (a-c). Pool shapes after 50 min. are compared in (d). Bed density (ρ_1)=1.15 tonne/m^3; initial density (ρ_2)of pool(in tonne/m^3): 1.5[(a) and (i)]; 1.7[(b) and (ii)]; 2.0[(iii)]; 2.3[(c) and (iv)].

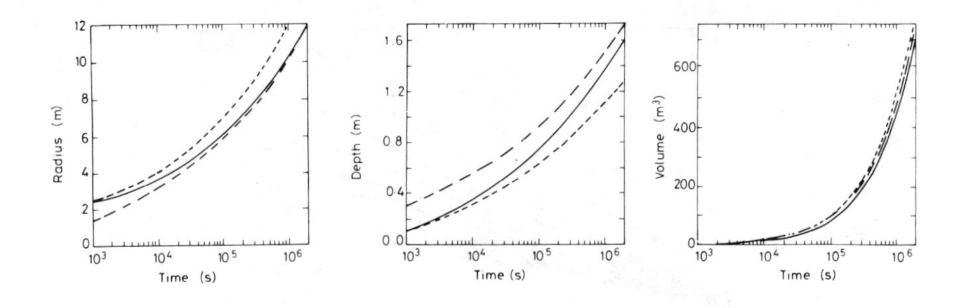

Fig.10 Effects on the radius, depth and volume of a miscible melt pool produced by varying the initial depth D_o and the lateral heat transfer coefficient ($\propto\alpha$). The initial debris volume was $1.8m^3$; $Q_o = 3GW$.
---- $D_o = 0.1m$, $\alpha = 1.0$ ———— $D_o = 0.1m$, $\alpha = 0.6$
— — — $D_o = 0.3m$, $\alpha = 0.6$ [86]

value of ρ_2 was varied by changing the composition of the electrolyte simulating the molten debris (see fig. 9) When $\rho_2-\rho_1$ was initially small, pool growth was mainly lateral. For larger values of $\rho_2-\rho_1$, initial pool growth was mainly downwards; only after sufficient dilution of the pool by molten substrate material did strong lateral growth occur.[40]

Although the shape of the pool is sensitive to the distribution of $\underline{F}_1(\Omega)$, the volume of the pool is much less sensitive. Fig.10 shows the results of a sensitivity study [86] in which α (which is proportional to F_s) was varied by almost a factor of 2. Although the radii and depths of the pool were markedly different in the two cases, the pool volumes were not (note that the plots are log-linear so that the melting front is actually decelerating). The volume of the pool at any time was also reasonably insensitive to the initial distribution of the debris (provided it is compact) - the factor of 3 change in the initial depth of debris corresponds to a change in aspect ratio by a factor of 5 for a fixed volume of core debris.

Thus in order to calculate the maximum volume of the debris pool as it is diluted by molten substrate, and the time taken to reach this maximum, any reasonable distribution of heat flux to the melting front may be assumed and a good approximate result attained. The most convenient assumption is that the debris pool grows as a downward facing hemisphere with F_1 a function of time but not of position on the melting front. The insensitivity at later times to the initial debris distribution allows this also to be taken in hemispherical form. Fig.11 illustrates the growth of such a hemispherical pool for debris melting into a substrate with the thermophysical properties of basalt [86]. The transition from stage 2 to stage 3 occurs in this case after $\sim 10^6$ secs, as the increase thereafter of the thermal boundary layer ahead of the melt front indicates. The maximum size of the pool is strongly influenced by the decay heat available, which in turn depends on the mean irradiation time t_q of the fuel within the reactor at shutdown; for fresh fuel the decay heat function falls off rather rapidly. A convenient empirical form for the power of the decay heat Q_1 supplied to the melt-front as a whole is [83]

$$Q_1(t) = \lambda\, Q_o\left\{(t_f/t)^{\frac{1}{4}} - (t_f/[t+t_q])^{\frac{1}{4}}\right\} \qquad (7.3)$$

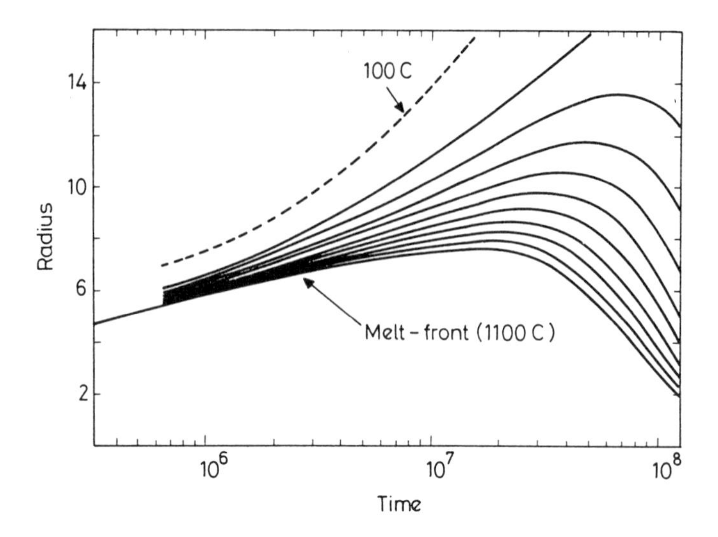

Fig.11 The growth of a hemispherical melt-pool calculated using the ISOTHM computer code [86]. The isotherms shown are equally spaced in temperature; the dashed isotherm is essentially at ambient temperature.

where t is the time elapsed since shutdown, $t_f \approx 1.5 \times 10^{-4}$ sec for a fast reactor and Q_o is the nominal power of the reactor on stream. The multiplier λ can be considered as the product of two factors λ_1 and λ_2 where λ_1 is the fraction of the core debris which reaches the substrate and λ_2 is the fraction of the decay heat generated in the pool which is not transported upwards to the overlying coolant. In terms of the quantities in eq.(7.1)

$$Q_1 = \lambda_2 Q = Q - F_{up} \tag{7.4}$$

It is convenient to introduce a length scale b defined by $b = Q_o/2\pi k(\Delta T)$ in which k is the thermal conductivity of the bed. This is the radius of the hemisphere whose surface would, in equilibrium, be able to conduct away the full reactor power Q_o. Note that b is not a 'typical' length scale; it is much larger than the maximum pool radius (for current reactors $b \sim 10$ km, $r_m \sim 10$m). A time scale t_b for the system can be defined by

$$t_b = \left\{ (2\pi b^3/3)\ \rho\ (L + (1+\gamma)c\ \Delta T) \right\} / \lambda Q_o \tag{7.5}$$

The term in curly brackets in (7.5) is the sum of the enthalpy in a pool of radius b and the enthalpy gained by the surrounding substrate (measured above ambient). γ is the ratio of sensible heat in the substrate to that in the pool; detailed studies [83] show that when the pool reaches its maximum size, $\gamma \approx 4$. For infinite burn-up, the maximum volume of the pool is given by [85]

$$V_\infty = b^3 (t_f/t_b)^{\frac{1}{2}} \tag{7.6}$$

and the time to reach maximum size is approximately t_∞ where

$$t_\infty = 0.2 t_b . (t_f/t_b)^{1/3} \tag{7.7}$$

For finite burn-up the maximum pool volume V_m is given by

$$V_m = V_\infty \quad \text{if } t_q \geq t_\infty$$
$$= V_\infty (t_q/t_\infty)^{\frac{3}{4}} \quad \text{if } t_q \leq t_\infty \qquad (7.8)$$

The time to maximum pool volume is $\sim t_\infty$ for high burn-up ($t_q \geq t_\infty$) and $\sim t_\infty.(t_q/t_\infty)^{2/5}$ for low burn-up ($t_q < t_\infty$). The variation of the maximum pool size with irradiation time is shown in fig. 12 for an alumina substrate. It compares the formula (7.8) - the straight line segments-with detailed calculations using the ISOTHM code [86].

If it assumed rather pessimistically that one third of the core debris escaped through the bottom of the primary vessel, and realistically that 80% of the heat in the pool is transported upwards to the overlying sodium then λ=0.06. In practice λ will almost always be very much smaller than this,because of debris dispersal within the reactor vessel and the presence of structures within the vessel which act as barriers to the downward penetration of core debris. For λ = 0.06 for a 3GWt reactor, V_∞ = 75m^3 (corresponding to a hemisphere of radius r_∞ of 3.3m) and t_∞ = 5 x 10^6 secs. For 240-day burn-up, t_q = 2.1x10^7 sec, so that $t_q/t_\infty \approx 4$. Thus $V_m \approx V_\infty$ in this case.

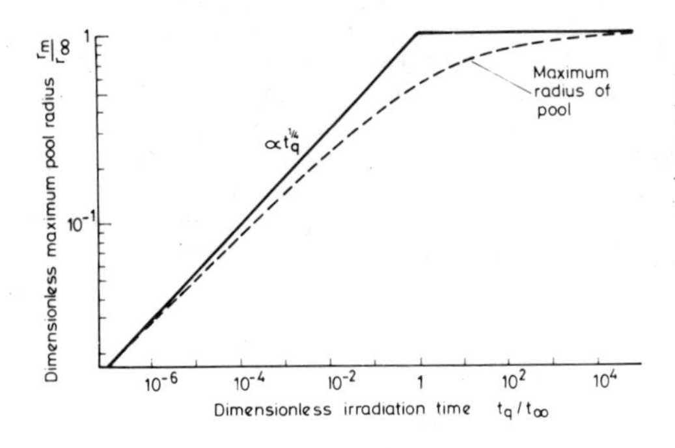

Fig.12. Maximum radius r_m for a hemispherical miscible melt-pool as a function of irradiation time t_q. —— (approximation);- - -(ISOTHM). r_∞ and t_∞ are maximum radius and time to achieve it for infinite burn up [85].

As a first approach to following the evolution of a debris pool both in shape and size, one can assume that the pool has a shape that can be specified by two parameters which change with time. In the INTER code [77] it assumed that the pool is a spherical segment, specified by the radius of curvature R and the pool depth h. In GROWS-II [21]the pool is assumed to have horizontal upper and lower surfaces and sidewalls circular in vertical cross-section; the pool is thus specified by the overall radius of the pool R, and the radius of curvature R_o through a sidewall. Right cylinders of depth D and radius R were assumed in PAMPUR1 [86]; in PAMPUR2 [86] hemispheroids were assumed,specified by the lengths of the major and minor axes. The advantage of restricting the pool shape to a single class of similar solids lies in the fact that the flux to the melt-front surface can then be represented easily as some function of the horizontal and vertical flux densities F_s and F_d which can be investigated experimentally in crucibles of fixed size; in this way the data on heat transfer from molten pools discussed in §6 can be incorporated and a coupling between the dynamics inside the pool and outside the pool established. The results from GROWS-II and INTER [21,91] both support the point that the volume of

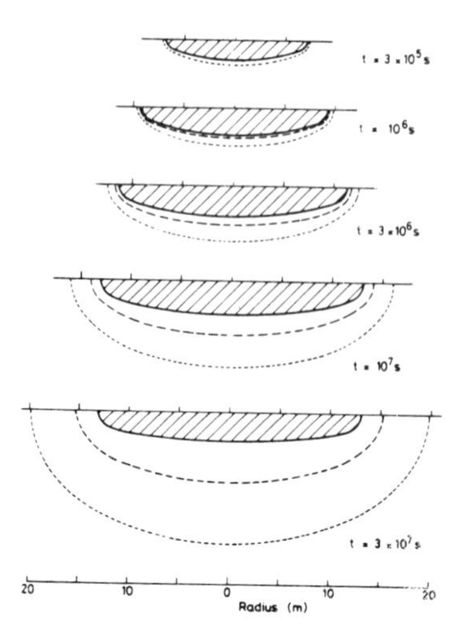

Fig.13. Growth of a miscible melt-pool containing debris from a 1GWe core calculated using the CONIMM computer code [111]which allows the shape of the melt-pool to evolve freely. The pool is shown hatched; also shown are the 600°C isotherm(— — —) and the 160°C isotherm(----).

material melted at any given time is relatively insensitive to the initial shape and aspect ratio of the pool. When sufficient data has been accumulated on heat and mass transfer within pools of a wide range of shapes, these two-parameter representations of the melting front can be dispensed with, and the pool dynamics can be coupled with the heat transfer in the substrate through a freely evolving/ melt-front. In this context, one method particularly well suited to handling the heat transfer in the substrate is the isotherm migration method (IMM)[38] which follows the location of particular isotherms including the melting isotherm. Fig.13 shows the evolution in the shape of a melt-pool (calculated using a 2D generalization of the IMM[111]) for a 1GWe core using a simplified heat transfer model for the pool interior, but including upward heat transfer to overlying sodium.

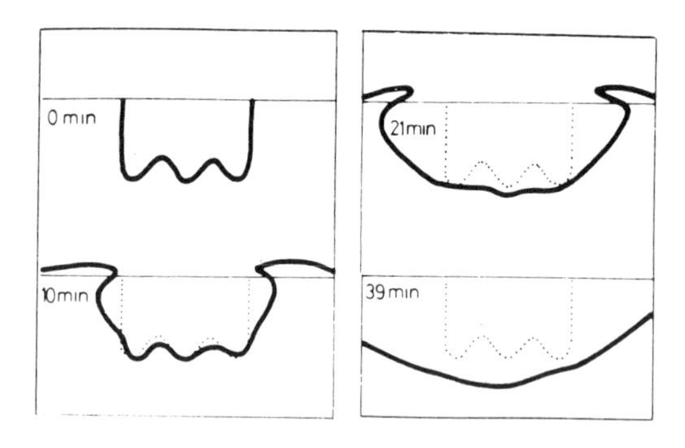

Fig.14. Decay of an initially imposed sinusoidal perturbation to the melt-front in a simulation experiment. The melt-front only is shown, together with its initial position(....).After 40 min the melt-front is smooth and its perturbed history lost.[115].

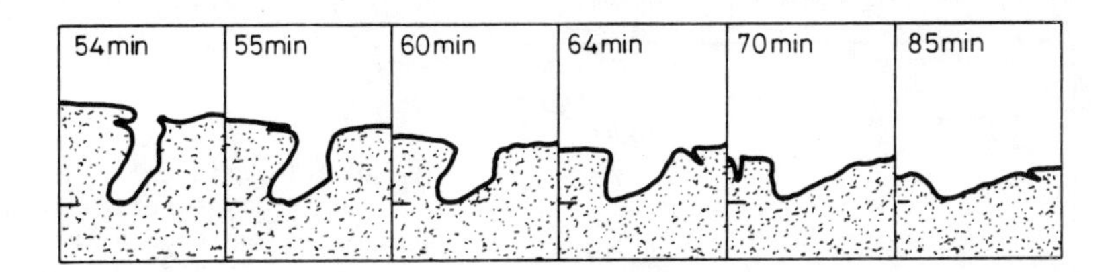

Fig.15 Laboratory simulation using a substrate of cracked wax. The advance of the melting front in the vicinity a sub-critical fissure is shown. [115].

In modelling the growth of a melt-pool it is customary to assume that the advancing melt-front is a plane or simple curved surface. Penetration times based on this assumption would be overestimated if the pool grew in a less regular fashion. Possible mechanisms for producing irregular growth are:-

(i) that small perturbations to the interface grow
(ii) that a cracked substrate allows enhanced penetration

Simulant experiments with joule heated electrolyte and a wax bed show that perturbations of the interface do not grow - they decay [115]; see fig.14. This result is supported by a linear analysis for homogeneous materials[113]. For the advance of a melt-front when gas is being evolved and chemical reactions are occurring between substrate and pool material, the situation in the vicinity of the advancing (multiple) front is complex and much analysis remains to be done. So far there is no experimental evidence for the spontaneous development of deep corrugations at the melt-front, so that models assuming simple forms for the melt-front appear to be valid.

If the substrate contains cracks and fissures then pool liquid can leak into these fissures. For a small fissure, the faster local advance of the solid/liquid interface is only temporary since the additional surface area of the fissure enables heat to be conducted more rapidly from the melt in the fissure to the substrate than from the main pool. The main melt-front thus catches up and subsumes this local penetration [115];see fig.15. On the other hand if the fissure is wide enough, the penetration of heat-generating molten debris into it will result in sufficient heat being conducted to the sidewalls of the fissure for it to be enlarged, thus allowing further penetration [87]. The critical crack width depends on the density of decay heat in the diluted melt-pool, and this in turn depends on the intrinsic fall-off in the decay heat level, and the extent to which the debris has been diluted by substrate material. Fig.16 shows the critical radius for a cylindrical crack entered by debris t_0 secs after shutdown, based on reasonable assumptions for miscible melt-pool growth. Fissures are likely to be present in the substrate as a result of thermal stresses induced by the advancing melt-front but will probably be too thin to significantly perturb the general advance of the melt-front. On the other hand if a sacrificial bed of ceramic blocks is placed under the reactor vessel, it is clear that the blocks must be placed sufficiently close together for the gaps between to be less than the critical dimension. Assuming that debris does not emerge from the reactor vessel in less than ½ hr. after shutdown, gaps less than 10mm should be sub-critical. The penetration of molten fuel through the interstices of the reactor support structure is of course a problem closely related to that of fissure penetration.

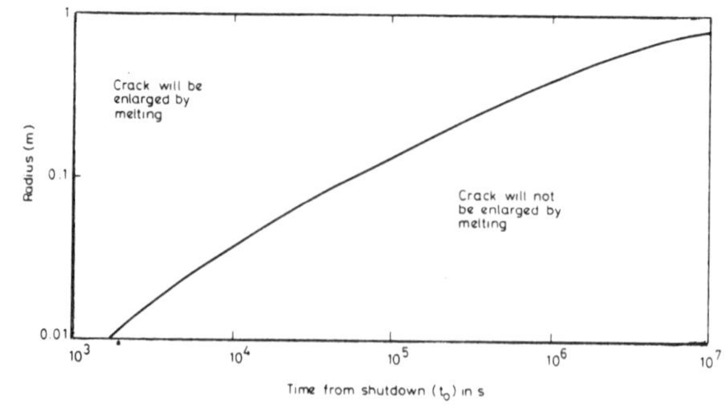

Fig.16. Critical radius for a cyclindrical crack entered by molten but diluted debris, t_o sec. after shutdown.

7.2 Metallic Phase Penetration.

The fact that a debris pool may consist of several distinct layers has already been discussed in §6.2. If it is assumed that the pool has not been homogenized by violent agitation induced by the passage of gas released from the heating and decomposition of the substrate, then it may be considered to consist of up to three layers. The lowest layer (I), is rich in oxide fuel, and so is heavier than steel; it is being diluted by substrate material with which it is miscible. The middle layer (II) is metallic, mainly steel and containing the metallic fission product alloys. The upper layer (III) is again oxidic, lighter than steel, consisting of melted substrate material, laced with some oxide fuel, and oxides from the oxidation of steel if H_2O and CO_2 are released from the substrate. All three layers will contain fission product heating to some extent. If the substrate does not release gas (e.g. if it were an alumina sacrificial barrier) then initially the upper layer will be absent and oxide fuel will sit as a slurry in the bottom of the steel pool. Some of the melted substrate will dissolve some of the oxide fuel and so will form a layer I; however much of the substrate material will be sufficiently buoyant to pass in the form of bubbles up through the pool and as a liquid film up around the sides of the pool past the metallic layer. After some time, all of the oxide fuel will have been dissolved in melted substrate material and the resulting phase will be much lighter (by a factor of ~ 2) than the steel phase and so will migrate to layer III. At this stage layer I may be ignored, and the configuration can be simplified conceptually to a hot metallic layer sitting on a colder substrate with which it is immiscible and which melts as a result of heat transferred to it. The melted substrate material is lighter than the metallic phase and Rayleigh-Taylor instability at the interface between melt and metallic phase will occur. Simulation experiments involving the melting of benzene under water [42,121] illustrate this effect vividly. The melted material is detached from the liquid substrate film in a regular pattern of liquid droplets, at a spacing dictated by the Rayleigh-Taylor wavelength. These drops pass through the metallic layer entraining some of it and induce a counter-flow which transports hotter material down towards the interface; this enhances the heat transfer from the metallic layer to the sacrificial bed. Substantial downward heat fluxes result, and the formula for these given in [121] gives reasonable agreement with the velocity of melt-front advance for simulant experiments. The melting of a steel support by a superposed molten layer of oxide fuel (whose melting point lies below the boiling point of steel) can also be described in this framework.

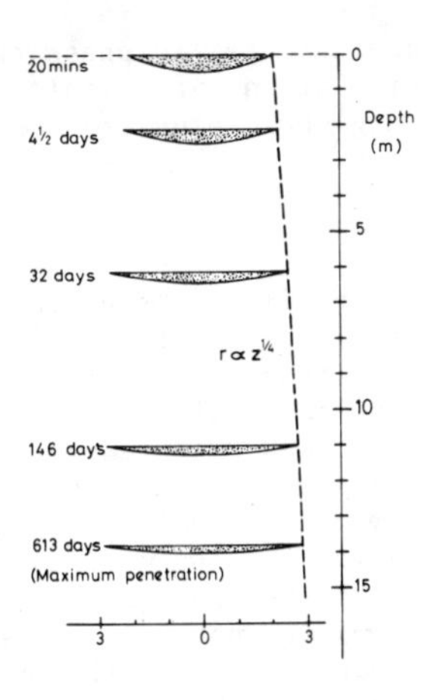

Fig.17. Descent of lens-shaped immiscible pool (volume $3m^3$) into a substrate of conductivity 2W/(mk) when λQ_o in eq.(7.3) is 100MW [85].

If the substrate releases gas, then a vapour layer may be conceived as lying between the substrate and the metallic phase. Again Rayleigh-Taylor instabilities can be expected at the gas/metal interface. Dhir,Catton et al [30,37,122]applied Taylor instability theory to this situation, obtained good agreement with simulation experiments involving the sublimation of dry ice, and proposed that their heat transfer data could be used to predict the penetration of metallic core melt into concrete. This concept has been incorporated into the WECHSL code [94] and the agreement between computer predictions and thermite tests in a concrete crucible is quite satisfactory, though the tests were over rather a short period.

Since the metallic phase is not diluted by the substrate, the heat flux density to the melting front is not attenuated by the increase in pool surface area which results from dilution. Consequently a heat generating metallic pool would penetrate downwards much further than would a miscible pool.

The penetration into the substrate by metallic phase core debris may be considered in a fashion analogous to that for miscible pools in §7.1, if it is assumed that the steel-based phase settles at the bottom of the main pool as a shallow lens-shaped sub-pool, immiscible with the substrate material. In a simplified approach taken in [85] the metallic phase sub-pool is assumed to be of constant volume; to allow its shape to evolve it is taken to be a spherical segment specified by its radius of curvature R which increases with time.Equation (7.3) can be used with Q_1 now meaning the decay heat power supplied to the lower surface of the metallic meltpool. Fig.17 shows the descent of such a lens-shaped pool when λQ_o is taken to be 100MW. Note that in this model the pool grows slowly in width and becomes shallower as it descends. This illustrative calculation assumes that the heat flux to the melting front is evenly distributed; however the same framework is applicable whatever the distribution of heat flux, since the mass transfer of melted substrate material whether around the pool edges or through it in the form of Taylor bubbles simply influences the convective motions in the pool and hence $\underline{F}_1(\Omega)$. Scoping studies indicate that the penetration depth is strongly dependent on the fraction of fission products which migrate to the metallic phase and on the thermal conductivity of the substrate [85].

Under some circumstances a crust of solid metallic phase may form on the underside of the metallic pool, and compel melted substrate to flow upwards around the side of the metallic pool. In this case the metallic phase shape becomes fixed and it will descend as a globule. Simulation experiments and a simple model for the downward penetration of such a globule are also presented in [85].

7.3. Further Comments

The analysis of the growth of real melt-pools into miscible substrates requires (i) good analytical tools for representing the movement of melting fronts and (ii) a good understanding of how the heat transfer from self-heated pools is effected by some or all of (a) mass transfer into the pool, (b) internal radiation, (c) gas agitation and (d) boiling. Both these areas required considerable development and the situation now is much more satisfactory than it was a decade ago. Of the experimental work on heat transfer from molten pools discussed in §6 , almost all has taken place within the past ten years. The modelling of the growth of molten pools would have been very crude with the size and speed of computers available then, for moving boundary problems are not well suited to coarse finite difference meshes. Putting together both facets of the problem in a realistic fashion is now feasible, as the calculations by Reineke et al [96] illustrate. Nevertheless the available data for heat transfer from internally heated pools is still incomplete and uncertain [25] concerning the effects of density differences arising from mass transfer into the pool and the bubbling of gas through the pool - both having significant effects on the rate of penetration into structural materials. For pools containing two immiscible phases, all the single phase modelling difficulties are present besides the need to represent the coupling between the two phases properly. As far as substrate materials are concerned those that release gases on heating (such as concrete or basalt) complicate the pool dynamics, but this is partially compensated for, if the gases are oxidizing, by the oxidation of the steel phase so that only a single liquid phase is present, at least after sufficient time. Conversely, two discrete liquid phases are present in the melt pool formed in a substrate such as magnesia from which gas is not released.

8. OTHER BARRIER MATERIALS

The behaviour of core debris ex-vessel depends on the constitution of the materials it encounters. Heating of concrete results in the evolution of water vapour and CO_2 (§5.1) which may pressurize the secondary containment even if sodium were absent (e.g.in a GCFR). These gases are reduced to H_2 and CO on passage through sodium (§5.2) and can lead to exothermic reactions within the secondary containment building (§4.2). Sodium also reacts with some of the constituents of common concretes (§5.2) producing additional heat which contributes to the further thermal attack on adjacent concrete. An obvious question thus arises as to whether there are alternative materials which might be used as the top layer of the substrate beneath the reactor vessel and which on balance have more desirable characteristics than concrete. Containment concepts which could make use of alternative materials have been listed in §1.2; a comparison of them from an engineering point of view is presented in §9. This section concentrates on the basic physics and chemistry of alternative materials which have been proposed.

The following characteristics are desirable for molten debris ex-vessel: (i) it should be contained in as small an envelope as possible (ii) its temperature should be as low as possible (iii) it should not make aerosol emission from the secondary containment building significantly more likely. With real materials these aims are to some extent mutually incompatible, and the optimum design must involve compromise between them.

The debris may be restricted to a small envelope by a refractory material with which it does not interact chemically and whose melting point exceeds any temperature the debris may reach (i.e. it is effectively crucibled by a stable barrier material). Candidate materials are those with high melting points; a number of those which melt above 2000°C have been listed by Gluekler [55],(see also [8]). These are mainly refractory oxides and carbides; elements such as carbon and tungsten may also be considered. Refractory oxides are considered in §8.1; carbon and carbides in §8.2. A somewhat larger,but still relatively small envelope, could be achieved if the debris were diluted by sufficient sacrificial material with a high latent heat of melting for the pool temperature to be limited to be below the melting point of candidate crucible materials. This would result in a small, high temperature pool. Magnesia and alumina are possible crucible materials.

To keep the debris relatively cool, there are two alternatives. Either the debris can be mixed with a good conductor (such as sodium) so that the decay heat generated is rejected rapidly, or it can be diluted with sufficient quantity of a relatively low melting point material for the heat generated to be absorbed as latent heat. This latter alternative results in a relatively large low temperature debris pool, where the increase in pool surface area allows enhanced heat transfer to the substrate. Borax and related borosilicates might be used in this fashion (§8.3); refractory oxides may also be appropriate.

The generation of gases which would not be mainly dissolved in the sodium pool would result in the additional pressurization of the secondary containment. Thus heat and gas evolving reactions between the substrate material and core debris or sodium are undesirable. Candidate materials must also be considered from this point of view.

8.1 Refractory oxides

Four obvious candidate oxides are magnesia (MgO), alumina (Al_2O_3), zirconia (ZrO_2) and silica (SiO_2) which has a significantly lower melting point than the others. A factor in their favour is that chemical reactions with stainless steel and fuel debris will occur at most to a minor extent. These materials are used for lining crucibles in steel making so it is to be expected that they would not react with stainless steel- experiments with MgO and steel reported by Stein et al [103] and Swanson et al[108] confirm this.

Silica is attractive because it is widely available and cheap, its use however would appear to be precluded because it reacts strongly with sodium(see §5.2 and also [47]).

Magnesia has a high specific heat (1.3kJ/(kg.K)), a high melting point ($\sim 2800^oC$) and a high latent heat of melting ($\sim$1.8MJ/kg). It is known that MgO and UO_2 are miscible in the liquid state and form liquid solutions with a eutectic composition of about 50 mol % MgO in UO_2. Eutectic formation was reported in experiments performed at Aerospace [108]; the eutectic temperature (in this context effectively the melting point at the MgO/UO_2 interface) was found to be in the range 2200-2300^oC [103]. Despite the large thermal gradients through the magnesia, cracking due to thermal shock was restricted to the top 5mm. There is some evidence of UO_2 diffusing into solid magnesia ahead of the melt front. Magnesia is a relatively light (3.5 tonne/M^3) so that there is some possibility of a two-layer pool developing - the upper layer rich in MgO, the lower layer close to the entectic composition.

Alumina has a specific heat of 1.1 kJ/(kg.K), a melting point of $\sim$ 2000^oC and a latent heat of melting of $\sim$ 1.1 MJ/kg, all somewhat lower than those of magnesia. However these are partially compensated for by a higher density (4 tonne/m^3). The pure alumina - UO_2 system has been examined by Capp et al [29] whose experiments indicated that the eutectic temperature was below 1950^oC and that the eutectic composition was UO_2 - 83 mol % Al_2O_3. This implies that larger pool volumes would be needed to attain the eutectic composition for alumina than for magnesia. Capp reports two zones in the melt: the upper layer was a liquid containing UO_2 and Al_2O_3 which was in equilibrium with undissolved UO_2 which had sunk to the bottom of the crucible. Experiments on alumina brick (86% Al_2O_3, 8% SiO_2) with UO_2 at 2050^oC showed that the UO_2 had diffused throughout the sample, and mixing appeared to be better with admixture of silica than without it; in the samples after resolidification, a marked vertical concentration gradient in UO_2, with the concentration a factor of 10 higher at the bottom than at the top.

Zirconia has a lower specific heat (0.6kJ/(kg.k)) and latent heat ($\sim$0.7MJ/kg) than either magnesia or alumina. Its density however is significantly higher (5.7 tonne/m^3)and it does have a high melting point ($\sim 2760^oC$.).

The compatibility of these oxides has been thoroughly reviewed by Fink[47] Pure magnesia is compatible with sodium, as is alumina. However binder material greatly affects the compatibility, which decreases as the percentage of silica in the material increases. For alumina, Fink concludes that if the silica content is greater than 8%, alumina samples are not stable in high-temperature or boiling sodium. For magnesia, Harklase (98% purity MgO) performed well in tests with sodium close to its boiling point. For zirconia, experiments at ANL reported by Fink [47] indicate that except in a high-fired high purity form, zirconia is not compatible with sodium.

Magnesia, alumina, and zirconia are all available commercially at an economic cost, but zirconia would appear on compatibility grounds to be inferior to the other two. Both magnesia and alumina are comparable to concrete as sacrificial heat absorbing materials: magnesia requires $\sim$ 17GJ/m^3 to raise it to the eutectic temperature and melt it, and alumina requires $\sim$13GJ/m^3. Muir's erosion rates for concrete [75] correspond to $14GJ/m^3$. They are superior in as much as when dry they do not react with sodium or lead to gas evolution on heating. Maximum pool volumes in both alumina and magnesia beds are calculated in [86] using the hemispherical shape assumption.

If the aim is not to dilute the debris, but to provide an effective and stable barrier against it, then expensive materials may be considered since only relatively small quantities of these might be required. Both thoria(ThO_2) and depleted UO_2 have high melting points and their high densities will ensure they do not float above the core debris. A liner of thickness 80mm of ThO_2 or UO_2 has been considered for the SNR-300 by INTERATOM [64]. Thoria was found to stand up to thermal shock more satisfactorily than urania. Fink [47] reports that thoria is completely compatible with high-temperature and boiling sodium, in contrast to the known formation of sodium uranate (and sodium pluto uranate) under some conditions. Knowledge of the high temperature compatibility of thoria with sodium and core debris is important since it is one of the few materials which might be used in PAHR experiments to crucible such mixtures of real reactor materials. So far experiments in which UO_2 and stainless steel have been held in crucibles of thoria at temperatures approaching the melting point of UO_2 for tens of minutes have been encouraging.

8.2 Graphite

Carbon-based compounds are of interest because several have high melting points and so might be suitable as barrier materials. Graphite moderators are used in gas cooled reactors, and there exists the possibility that carbide fuel might be introduced at some stage into LMFBRs.

Graphite is a cheap material, with a well established technology for its processing and production. It has a good specific heat, and remains solid up to approximately the melting point of UO_2. It is resistant to sodium attack, although alkali metal atoms are known to diffuse into graphite, leading to internal strains. Over the course of a few hours carbon rods remain intact when exposed to boiling sodium, but after a month there is serious deterioration[47]. Graphite was identified by Gluekler [55] in his preliminary selection as a candidate material; it could perhaps be used outside the primary vessel either as a liner for a crucible beneath the primary vessel designed to accommodate sodium and molten core debris or as a barrier material per se. It has been further studied by Peehs et al [88]with a view to its use in SNR-300 as a crucible liner. Tests on the compatibility between UO_2-pellets, UO_2/stainless steel mixtures and graphite were carried out and the pellets were completely converted to carbides at $2400^{\circ}C$, when liquid uranium carbide phases occur. In the presence of stainless steel the liquefaction temperature was lowered to $2000^{\circ}C$. The conversion rates as indicated by the CO generation were high, with the interactions completed within 15 min. Fink et al [48] found that interaction between graphite and UO_2 alone resulted in UC formation between $1600^{\circ}C$ and $1830^{\circ}C$, and the formation of both UC and UC_2 at higher temperatures. With added stainless steel, they found that reactions were inhibited below $1950^{\circ}C$ in accord with Peehs' results. Above this temperature, the reactions

$$\left.\begin{aligned} Fe + UO_2 + 4C &\rightarrow FeUC_2 + 2CO \\ UO_2 + 3C &\rightarrow UC + 2CO \\ UO_2 + 4C &\rightarrow UC_2 + 2CO \end{aligned}\right\} \quad (8.1)$$

take place. The presence of steel greatly enhanced the rate of CO evolution. It is suggested [49] that $FeUC_2$ which melts at or below $1950^{\circ}C$ acts as a carrier for carbon, thereby accelerating the UO_2-graphite reaction rate. It would appear that any molten material in which graphite is soluble may fulfill this role and thus increase the rate of CO production. Graphite has been proposed as a sacrificial material [59] on the basis that $FeUC_2$ would form a well-diluted melt that stays liquid to temperatures as low 1000°C. In fact the melting point is likely to be rather higher than this and the CO generated unacceptable for the integrity of the secondary containment building.

Some aspects of the interactions in a system containing UO_2, stainless steel, sodium hydroxide and sodium carbonate are also discussed by Fink [48]and Baker [19]; the chemistry is complex.

8.3 Borax

Dalle Donne et al [33-35] have proposed that borax would be a suitable material to dilute the core debris in a core catcher for a GCFR. The basis of the design is that solid borax ($Na_2B_4O_7$) should be placed in a number of thin steel cubical boxes stacked in up to seven layers. The oxide fuel dissolves in liquid borax upto 50 wt% UO_2. The dissolution process is controlled by the successive melting of the steel boxes so that liquid borax is already at the melting point of steel, where a solution of 40 wt% of UO_2 in borax may be achieved in 10-15 minutes. This leads to a relatively low temperature pool of well diluted debris. The object of the steel partitions is to prevent the debris simply falling to the bottom of the borax pool. By requiring the steel to melt, the borax temperature is kept sufficiently high for a satisfactory rate of dissolution of the UO_2 in the borax. The idea appears promising for GCFRs, but there are some possible sources of difficulty. Borax boils at 1575°C, and borax vapour blanketting of hot fuel debris might interrupt the dissolution process. Also since the box walls are thin, the boxes may need to be vented to avoid pressure disequilibria. Vented boxes allow the ingress of moisture - the hygroscopic reaction of borax is well known. Moreover vents might allow borax vapour to be transported within the primary system - and boron is a neutronic poison.

Recently Dalle Donne et al [36] have proposed that a variant of this proposal might be suitable for a LMFBR, with sodium metaborate ($2NaBO_2$) used in place of borax. Sodium metaborate is compatible with sodium, and experiments reported in [36] indicate that sodium metaborate dissolves UO_2 about 10 to 15 times faster than borax does. Provided the metaborate bed is ex-vessel it would appear to have the advantages of the GCFR version. The vapour blanketting is more severe since the metaborate boils at 1434°C. Moreover this might make it difficult to use steel for the box material. On the other hand the steel separating partitions might fail mechanically at a temperature well below the steel melting point, as a result of thermal stresses induced by the weight of hot debris above it. For an LMFBR such a metaborate bed must obviously be ex-vessel.

Neither borax nor sodium metaborate are expensive materials. Moreover the presence of substantial quantities of boron in the debris pool should ensure that secondary criticality in the molten debris was excluded. This proposal deserves further investigation, for knowledge of the thermophysical properties of sodium metaborate, both alone and containing UO_2 in solution is relatively sparse.

9. DISCUSSION AND CONCLUSION

9.1. Comparison of concepts.

The various types of concept which have been proposed to assure post accident fuel containment were briefly categorized in §1.2. In this section some comments are offered on the strengths and weaknesses of these ideas in the light of the physical and chemical phenomena considered in §2-§8.

A reactor in normal operation is provided with a wealth of instrumentation to ensure that any significant deviations from normal operation can be rapidly sensed and the reactor shutdown. It is conceivable but unlikely that fault conditions could develop in which some of the core becomes molten despite the safety features incorporated into the design; the most likely outcome of such partial melting is simply that the fuel and steel will remain within the general envelope of the core matrix in a modified geometry and will be coolable there by the normal decay heat removal systems. It is conceivable but even more unlikely that fault conditions could develop which would result in significant quantities of the core material emerging into the primary vessel. There is a considerable body of evidence that in such circumstances the hot core debris is likely to fragment and settle in the form of debris beds on the wide variety of structures which exist within the primary vessel. Such debris beds are coolable under sodium for quite a wide range of conditions. (see [15]); indeed it is believed that they may be coolable in all credible configurations, particularly if some design of in-vessel core debris tray is incorporated beneath the core to shield the bottom of the primary vessel. Nevertheless it is not currently possible to assert without qualification that there are no circumstances under which core debris will melt-through the primary vessel. It is in this context, of providing additional assurance of safety in the extremely unlikely circumstances of the penetration of core debris into the cavity beneath the reactor vessel, that the requirements for ex-vessel post accident heat removal and fuel constraint should be considered.

The obvious approach is to explore to what extent the reactor building and substructure would be capable of containing the debris once it has emerged from the primary vessel. The potential for accommodating post-accident decay heat,for a limited period time, by the subvessel concrete and other reactor building struc tures is term the 'inherent retention capability'. Beneath the reactor vessel is the reactor cavity,lined, in the simplest version, with stainless steel to separate the underlying concrete from sodium in the case of a small leak in the primary vessel during normal operation. When flooded with sodium after a melt-through this may be considered as a downward extension of the primary envelope. It has the advantages of in-vessel cooling - large quantities of sodium are present as a temporary heat sink, the surviving heat rejection mechanisms for DHR below the reduced level of the sodium pool are available, and the chemical compatibilities of all materials are well known. Its weaknesses are that the steel liner is vulnerable to small flaws leading to sodium/concrete reactions, and that it is uncertain whether the liner would be able to withstand sudden contact with a stream of hot debris released on vessel melt-through. Nevertheless the intrinsic decay heat level decreases with time so some benefit must accrue from the need to penetrate the steel liner before contacting the concrete.

The concrete basemat provides a considerable capability for diluting and restraining the core debris. As a sacrificial material it needs 10kJ to melt 1cc of concrete so that the molten pool which would form within the concrete would be of a restricted size. The water and CO_2 released from the concrete decomposition are available to oxidize the steel debris phase, which results ultimately in a single phase oxide pool, whose characteristics are more easily established than those of a two-phase oxide/metallic pool. The disadvantages appear to be that (i) reactions between sodium and water in the concrete are exothermic and this increases the quantity of heat to be removed before the debris can settle into a stable configuration (ii) the gas generation is large and leads to pressurization of the reactor containment building and (iii) the modelling of the molten pool behaviour is extremely complex because of the influence of mass transfer and chemical reactions on the pool growth: this makes it difficult to assign high levels of confidence to the predictions of computer models used to simulate this behaviour. There is also the possibility that the concrete structures will suffer a deterioration in strength, although this is probably unimportant if the reactor vessel is supported from above. Reliance on the inherent retention capability is the option chosen for the CRBR project.

Engineered ex-vessel containment provides an alternative to this approach. In the cooled crucible concept, the reactor cavity is modified to include a thin refractive liner as a stable barrier, and auxiliary cooling is provided to keep the inner surface of the crucible at an acceptably low temperature such that the steel structures retain their integrity (viz 600-800°C)[9] Such a crucible concept is shown in fig.6 of [19], and was recommended by Gluekler [6] as an attractive option. A disadvantage of this design is that auxiliary cooling is essential, and it must be so arranged that it is available immediately after a melt-through has occurred, if the integrity of the liner is to be assured. The liner could be graphite or thoria or urania. The first leads to the undesirable production of CO ; the other two are expensive. The stable barrier plus auxiliary cooling system is very attractive in principle for it limits the penetration and so can be compact, enabling a smaller overall containment envelope to be claimed. The SNR-300 core catcher [67] is of this kind (see fig.18 of [11]); it consists of a spreader (including a neutron absorber) and a barrier of depleted UO_2 or thoria over NaK cooling coils. An important uncertainty for such designs lies in the distribution of the debris. If the debris is accummulated more deeply at some position on the liner then a local hot spot will arise, and this may be sufficiently severe for the barrier material to become overheated and the cooling system beneath attacked by the debris. Moreover the arrival of the debris may produce unacceptable thermal and mechanical shock. Overall this type of concept has many valuable features.

If the crucible is given a massive liner, then the concept has become one of a sacrificial bed. Sacrificial beds may be passive, in which case they may need to be relatively large, or additional cooling may be provided to limit the size and temperature of the debris pool within a smaller bed (but see below). Zivi first suggested the use of a passive ceramic bed [11], such as depleted UO_2. This would be very expensive and would lead to a high temperature pool. Basalt has also been suggested, but this is really little different from the use of the concrete basemat as a sacrificial bed because exothermic reactors with sodium will still occur. The growth of melt pools into sacrificial beds of alumina or magnesia have been examined, [25,86] and provided the mass transfer within the pool can be represented properly, [9] the modelling of such melt pools should be in reasonable accord with reality. A passive sacrificial bed is shown in [55], fig 6-1. A sacrificial bed with auxiliary cooling is the concept chosen for the CRBR parallel design option (see fig.17 in [11]. In this concept a sacrificial bed 5m deep of magnesia blocks surmounted by a debris spreader is placed beneath the guard vessel. The bed is a right cylinder of 7m radius surrounded by a cooling system embedded in graphite, and finally the concrete containment vessel basemat. Such a device is expensive, and in view of the fact that the auxiliary cooling system cannot be inspected, it is difficult to be assured that the cooling circuits will function correctly on demand. The presence of sodium above the bed to act as a heat rejection pathway would help to limit the size of the melt-pool. One curious feature of a sacrificial bed with auxiliary cooling is the fact that auxiliary cooling only limits the size of the pool if the size of the bed has been chosen correctly. This is because the cooling system only takes away substantial heat fluxes if the pool boundary approaches close to the cooling system. If the bed volume enclosed by the cooling system is too small the melting front will advance and melt the cooling system since the heat fluxes conducted into the bed ahead of the melt front are too large for the cooling system to handle. If the bed volume enclosed by the cooling system is too large, then the pool size will be self limiting without significant influence from the The cooling system is helpful in the latter case in assisting the pool to cool and resolidify more rapidly than it would have done otherwise. The metallic phase of the debris poses a residual problem for refraction oxide beds, since with no oxidizing agent present the immiscible metallic phase containing the metallic fission products will penetrate much further downwards than would the pool boundary of a miscible pool. An ideal ex-vessel sacrificial bed concept might involve a multi-component bed containing an oxide to dilute the UO_2, sufficient water-bearing material to oxidize the steel but not to provide problems for the RCB, a neutron poison to give assurance against recriticality, a more refractory outer barrier to limit penetration to a known envelope, and some impermeable outer liner to keep out additional water during the lifetime of the reactor.

If magnesia is used as a sacrificial bed the melt-pool will have a temperature in excess of $2000^{\circ}C$. If liquid sodium overlays the pool, some disruption of the interface between the sodium and the melt crust could be envisaged as a result of violent boiling and intermittent sodium penetration. The magnitude of this effect is clearly a function of pool temperature. An alternative approach is to use a diluent with a low melting point; the metaborate bed proposed by Dalle Donne (see §8.3) is a concept in this category.

9.2. Concluding Remarks

At present in-vessel retention is being studied vigorously. The acceptability of the concrete basemat as the next line of defence depends on whether transients in the RCB can be handled satisfactorily. There seems considerable scope among the existing concepts to give a high level of assurance that in the event of a core melt-down the molten core debris can be cooled safely in a pre-determined envelope.

ACKNOWLEDGEMENTS

I am grateful to my colleague Dr.B.D.Turland for reading this manuscript critically and for many stimulating discussions on this topic. This work has been carried out under the auspices of the UK Safety and Reliability Directorate, Culcheth.

REFERENCES AND SELECT BIBLIOGRAPHY

A. General

[1] Thompson T J and Beckerley J G (1973) The Technology of Nuclear Reactor Safety, Volume 2: Reactor Materials and Engineering. 820 pp. M.I.T.Press.

[2] Proceedings of the Fast Reactor Safety Meeting, Beverly Hills, April,1974, USERDA Report CONF-740401.

[3] Core meltdown experimental review (1975) Sandia Report SAND74-0382.

[4] Coats R L (Editor) Proceedings of the Second Post-Accident Heat Removal Information Exchange Albuquerque. Nov.1975. Sandia Laboratories Report SAND76-9008.

[5] Proceedings of the ANS-ENS International Meeting on Fast Reactor Safety and Related Physics, Chicago Oct.1976, USERDA Report CONF-761001.

[6] Gluekler E L (1976) Ex-vessel Core Retention Concept for Early Sized LMFBR, General Electric Report GEAP-14121.

[7] Jones O C Jr. and Bankoff S G (1977) The Thermal and Hydraulic Aspects of Nuclear Safety.Vol.2: Liquid Metal Fast Breeder Reactors. American Soc. of Mech. Eng.New York.329 pp.

[8] Peckover R S (1977) Post Accident Heat Removal. Paper presented at the meeting on "Thermohydraulic Problems related to Reactor Safety", JRC Ispra, September 1977 [UKAEA Culham Lab.Reprint CLM-P518].

[9] Gluekler E L and Baker L Jr.(1977) Post-Accident Heat Removal in LMFBR'S. pp 285-324 in ref. [7].

[10] Baker L Jr and Bingle J D (Editors) Proceedings of The Third Post-Accident Heat Removal Information Exchange, ANL. Nov.1977, Argonne Nat.Lab.Report ANL-78-10.

[11] Kazimi M S and Chen J C (1978) A Condensed Review of the Technology of Post-Accident Heat Removal for the Liquid-Metal Fast Breeder Reactor.Nuclear Technology 38, 339-366.

[12] Coen V and Holtbecker H (Editors) Proceedings of the Fourth Post-Accident Heat Removal Information Exchange, Varese, Italy Oct.1978. European Applied Research Reports - Nucl·Sci.Tech. Vol.1 (1979) p 1315-1712.

[13] Joly C and Le Rigoleur C (1979) General and Particular Aspects of the Particulate Bed Behaviour in the PAHR Situation for LMFBR's. Lecture presented at the Conference on "Fluid Dynamics of Porous Media in Energy Applications" Von Karman Institute for Fluid Dynamics Feb.1979.

[14] Proceedings of International Meeting on Fast Reactor Technology,Seattle Aug.1979.

[15] Costa J (1980) In-Vessel Post Accident Heat Removal. Proceedings of the ICHMT Summer School, Dubrovnik Aug.25-29 1980. (this volume).

B. Specific.

[16] Acton R U, Sallach R A, Smaardyk J E and Kent L A (1979) Sodium interaction with Concrete and firebrick. Paper VI/D/6 in ref.[14].

[17] Alsmeyer H, Barleon L, Günther C, Müller U and Reimann M (1979) Models and model experiments on the interaction of core melts and underlying materials. European Appl.Res.Rept. - Nucl.Sci. Technol.6, 1516-1524.

[18] Baker L.Jr (1976) Analysis of heat transfer from molten-fuel layers. Argonne National Lab.report ANL-RDP-54 pp 8.5-8.9.

[19] Baker L,Jr (1977) Core Debris Behavior and Interactions with Concrete.Nucl. Eng. & Design 42, 137-150.

[20] Baker L Jr.,Chasanov M G, Gabor J D, Gunther W H, Cassulo J C, Bingle J D and Mansoori G A (1976) Heat removal from molten fuel pools. pp 2056-2065 in Ref.[5].

[21] Baker L Jr, Cheung F B and Bingle J D(1978) Core debris penetration into concrete pp 223-232 in [10].

[22] Baker L,Jn.,Cheung F B and Bingle J D(1979) Growth of internally-heated core debris pools into soluble and insoluble structures. European Appl.Res. Rept. - Nucl.Sci.Technol.6, 1525-1533.

[23] Baker L,Cheung F B and Bingle J D(1979)Transient heat transfer in concrete. European Appl.Res.Rept.-Nucl.Sci.Technol.6, 1595-1604.

[24] Baker L Jr.,Cheung F B, Chasanov M G, Sowa E S, Bingle J D, Staahl G, Pavlik J R and Holloway W(1976) Interactions of LMFBR core debris with concrete. pp 2105-2114 in ref.[5].

[25] Baker L Jr,Cheung F B, Farhadieh R, Stein R P, Gabor J D and Bingle J D(1979) Thermal interactions of a molten core debris pool with surrounding structural materials. Paper II/C/5 in ref[14].

[26] Baker L Jr.,Faw R E and Kulacki F A(1976) Post accident heat removal - Part I: Heat transfer within an internally heated, non boiling layer. Nucl.Sci.and Eng.61, 222-230.

[27] Bergholz R F and Bjorge R (1977) Bubble-induced heat transfer in a heat generating liquid layer. pp 361-368 in ref.[10].

[28] Blottner F G (1979) Hydrodynamics and heat transfer characteristics of liquid pools with bubble agitation Sandia Laboratories report SAND 79-1132.

[29] Capp P D, James J M, Kingsbury A and Smith D L(1979) Sacrificial material UO_2 interactions - experimental studies at AEE Winfrith.European Appl.Res.Rept. - Nucl.Sci.Technol.6, 1639-1649.

[30] Castle J N, Dhir V K and Catton I (1977) On the heat transfer from a pool of molten UO_2 to a concrete slab.Trans.Am.Nucl.Soc 26, 356-7.

[31] Chasanov M G et al(1974) Reactor safety and physical property studies - annual report July 1973-June 1974. Argonne Report ANL-8120.

[32] Colburn R P, Muhlestein L D, Hasseberger J A, Mahncke A J (1979) Sodium-concrete reactions. Paper VI/D/3 in ref. [14].

[33] Dalle Donne M,Dorner S, Schumacher G, Artnik J, Goetzmann C A and Rau P (1976) Post-accident heat removal for gas-cooled fast reactors.pp 2016-2025 in ref.[5].

[34] Dalle Donne M,Dorner S and Schumacher G(1978) Development Work for a Borax Internal Core-Catcher for a Gas-Cooled Fast Reactor.Nuclear Technology 39,138-154.

[35] Dalle Donne M,Dorner S and Schumacher G(1979) Borax as sacrificial material for an internal core catcher of a nuclear reactor.European Appl.Res.Rept.-Nucl. Sci.Technol.6, 1650-1667.

[36] Dalle Donne M, Dorner S, Fieg G, Schumacher G and Werle M (1979) Development work for fast reactor core-catchers on the basis of sodium borates,Paper II/C/6 in ref [14].

[37] Dhir V K, Castle J N and Catton I (1977) Role of Taylor instability on sublimation of a horizontal slab of dry ice. J.Heat Transfer 99,411-418.

[38] Dix R C and Cizek J (1970) The isotherm migration method for transient heat conduction analysis in "Heat Transfer 1970" edited by U Grigull and E Hahne. Elsevier.

[39] Epstein M(1977) Melting,Boiling and Freezing: The "Transition Phase" in Fast Reactor Safety Analysis,pp 171-193 in ref.[7].

[40]Farhadieh R, Baker L Jr. and Faw R E(1975) Studies of pool growth with simulant materials. p.271-278 in ref.[4].

[41] Farhadieh R and Chen M M(1977) Mechanisms controlling the downward and sideward penetration of a hot liquid pool into a solid. Nucl.Eng.Des.58,65-74.

[42] Faw R E and Baker L Jr (1976) Nucl.Sci.Eng.61,321.

[43] Felde D K, Musicki Z and Abdel-Khalik S I(1979) Growth of volumetrically-heated pools in miscible gas-releasing solid beds.European Appl.Res.Rept.-Nucl. Sci.Technol.6, 1567-1576.

[44] Fiedler H E and Wille R (1970) Turbulente freie konvection in einer horizontalen flüssigkeitsschicht mit volumen-wärmequelle.Proc.4th Int.Heat. Transf.Conf.paper NMC4.

[45] Fieg G(1979) Heat transfer measurements for internally heated liquids in cylindrical convection cells.European Appl.Res.Rept.-Nucl.Sci.Technol.6,1458-1467.

[46] Fieg G (1976) Experimental investigations of heat transfer characteristics in liquid layers with internal heat sources. pp 2047-2055 in ref.[5].

[47] Fink J K (1978) A Review of Studies on the compatibility of sodium with refractory ceramics.ANL Report.

[48] Fink J K, Heiberger J J and Leibowitz L (1978) Interaction of UO_2 with reactor materials pp 415-423 in ref [10].

[49] Fink J K, Heiberger J J and Leibowitz L (1979) Studies of the role of molten materials in interactions with UO_2 and graphite. Paper VI/D/7 in ref.[14].

[50] Fischer J,Schilb J D and Chasanov M G(1973) Investigation of the distribution of fission products among molten fuel and reactor phases.J.Nucl.Materials 48, 233-240.

[51] Gabor J D, Baker L Jr. and Cassulo J C(1975) Heat removal from heat generating pools. Ref.[4] above p.133-150.

[52] Gabor J D, Baker L, Cassulo J C, Erskine D J and Warner J G (1980) Heat transfer to curved surfaces from heat generating pools. J.Heat Transfer 102, 519-524.

[53] Gasser R D and Pratt W T(1980) Containment response to postulated core meltdown accidents in FFTF. Nuclear Technology 47, 282-307.

[54] Gershuni G Z and Zhukhovitsky E M(1976) Convective Stability of incompressible fluids 330 pp Israel Program for Scientific Translations,Jerusalem.

[55] Gluekler EL(1975) Status of post-accident retention concepts and models. General Electric Report GEAP-14048.

[56] Gluekler EL,Clever RM and Rumble E T(1977)Investigations of the Structural Integrity of a Reactor Vessel for a Postulated Core Meltdown Accident.Proc.Fourth International Conference on Structural Mechanics in Reactor Technology.Paper G10-E.

[57] Gluekler E L and Dayan A(1976) Considerations of the third line of assurance - Post-accident heat removal and core retention in containment. Ref. pp 1995-2004 in ref.[5].

[58] Gluekler E L and Dayan A(1978) Concrete Failure Modes at Elevated Temperatures, p.233-242 in [10].

[59] Goetzmann C A(1976) Chemical interaction possibilities between core and core catcher materials. pp 251-254 in [4].

[60] Hilliard R K (1975) Sodium-concrete reactions, linear response and sodium fire extinguishment.p.297-322 in ref.[4].

[61] Hodkin D J, Potter P E and Hills R F(1979) The constitution of a molten oxide fast reactor core. European Appl.Res.Rept.-Nucl.Sci.Technol.6,1605-1613.

[62] Hodkin D J and Potter P E(1980) On the chemical constitution of a molten oxide core of a fast breeder reactor. Rev.int.hautes Tempér,Refract.17,70-81.

[63] Jahn M and Reineke H H(1974) Free convection heat transfer with internal heat sources.Proc.5th Int.Trans.Conf.Tokyo. Paper NC2.8.

[64] Jung J, te Hessen E, and Mollerfeld H(1979) Material investigations for verification of the core catcher concept for SNR-300. European Appl.Res. Rept.-Nucl.Sci.Technol.6, 1614-1623.

[65] Kayser G and Le Rigoleur C(1979) The Main Problems of an Internal Core Catcher for a Pool-Type Commercial LMFBR. European Appl.Res.Rept.-Nucl.Sci. Technol.6,1694-1701.

[66] Kazimi M S, Tsai S S and Gasser R D(1977) Post accident fuel relocation and heat removal in the LMFBR. Brookhaven Nat.Lab.Report BNL-50570.

[67] Keintzel G and Rothfuss H (1978) Design of the core catcher system for the fast breeder reactor SNR 300. Kerntecknik 20,127-134.

[68] Knight R L and Beck J V(1979) Model and Computer Code for Energy and Mass Transport in decomposing concrete and related materials. Paper VI/D/D5 in [14].

[69] Kulacki F A and Emara A A(1975) High Rayleigh number convection in enclosed fluid layers with internal heat sources. US-NRC Report NUREG-75/065.

[70] Kulacki F A and Goldstein RJ(1972) Thermal convection in a horizontal fluid layer with uniform volumetric energy sources. J.Fluid Mech.55, 271.

[71] Kulacki F A, Min J H, Nguyen A T and Keyhani M(1979) Steady and transient natural convection in single- and multi-fluid layers with heat generation. European Appl.Res.Rept.-Nucl.Sci.Technol.6, 1412-1427.

[72] Kulacki F A and Nagle M E(1975) Natural convection in a horizontal fluid layer with volumetric heat sources. J.Heat.Trans.97C,204.

[73] Kumar R, Baker L Jr and Chasanov M G(1974) Ex-vessel Considerations in Post Accident Heat Removal.ANL Report ANL/RAS 74-29.

[74]Meachum S A(1976) The interactions of Tennessee limestone aggregate concrete with liquid sodium. Westinghouse Report WARD-D-0141.

[75] Muir J F(1977) Response of concrete to high heat fluxes. Trans.American Nucl.Soc.26, 399.

[76] Muir J F, Powers D A and Dahlgren D A(1976) Studies on molten fuel-concrete interactions.pp 2095-2104 in [5].

[77] Murfin W A(1977) A Preliminary Model for Core/Concrete Interaction. Sandia Laboratories Report SAND 77-0370

[78] Palentine J E and Turland B D (1980) The Melting of Horizontal Stainless-Steel Plates Subjected to Constant Flux Thermal Loads. Proceedings of the ICHMT Conference on "Nuclear Reactor Safety Heat Transfer"(edited by S.G.Bankoff) Hemisphere Press.

[79] Palinski R(1980) Private Communication.

[80] Peak R D(1977) User's Guide to CACECO Containment Analysis Code.Hanford EDL Report HEDL-TC-859.

[81] Peak R D, Simpson D E and Stepnewski D D(1977) Response of Liquid Metal Fast Breeder Reactor Containment to a hypothetical core meltdown accident. Nucl.Eng.Des.42,169.

[82] Peckover R S(1974). The Thermal Containment of Reactor Core Material After Shutdown.Proc.Fast Reactor Safety Meeting Los Angeles,April,1974: USAEC Report CONF-740401 p 802-822.

[83] Peckover R S (1978) Thermal Moving Boundary Problems Arising in Reactor Safety Problems. Paper presented at the Symposium on 'Free and Moving Boundary Problems in Heat Flow and Diffusion,' Durham(UKAEA Culham Reprint CLM-P562).

[84] Peckover R S(1980) The thermohydraulics of self-heated particulate beds. Proceedings of 9th meeting of the Liquid Metal Working Group, Rome June 1980 (edited by H M Kottowski) pp 789-812.

[85] Peckover, R S, Dullforce, T A and Turland, B D(1979) Models of Core Melt Behaviour after a postulated reactor vessel melt-through. Paper VI/D/9 in [14].

[86] Peckover R S, Turland B D and Whipple R T P (1977) On the growth of melting pools in sacrificial materials. pp 179-186 in [10].

[87] Peckover R S, Adlam J A and Turland B D(1979) On the effects of cracks within sacrificial bed material on the growth of molten pools. European Appl. Res.Rept. Nucl.Sci.Technol,6, 1624-1638.

[88] Peehs M,Hofer G. Friedrich H J and Heuvel H J (1976) Experimental investigations on the compatibility of an SNR-type corium with graphite. Ref.[5] above, pp 2077-2086.

[89] Powers D A(1977) Empirical models for the thermal decomposition of concrete and largescale melt/concrete interaction tests.Trans.Americal Nucl.Soc.26,400-401.

[90] Powers D A(1979) Influence of Gas-generation on melt/concrete interaction. International Symposium of Thermodynamics of Nuclear Materials,Feb.1979.

[91] Pyun J J, Gasser R D, Pratt W T and Bari R A(1978) Ex-vessel containment response to a core meltdown pp 327-334 in [10].

[92] Ralph J C, McGreevy R and Peckover R S(1976) Experiments in turbulent thermal convection driven by internal heat sources. pp 587-599 in 'Heat Transfer and Turbulent Buoyant Convection' (Spalding DB and Afgan N editors) Hemisphere/ McGraw-Hill.

[93] Ralph J C and Roberts D N (1974) Free convection heat transfer measurements in horizontal liquid layers with internal heat generation. UKAEA Report AERE-R7841.

[94] Reimann M and Murfin W B(1979) Calculations for the decompositon of concrete structures by a molten pool.European Appl.Res.Rept.-Nucl.Sci.Technol.6,1554-1566.

[95] Reineke H H(1975) Thermodynamic behaviour and heat transfer in the molten core.p.181-200 in ref.[4].

[96] Reineke H H, Rinkleff L and Schramm R(1978) Heat transfer between molten core material and concrete pp 114-121 in ref.[10].

[97] Reineke H H, Schramm R and Steinberner U(1979) Heat transfer from a stratified two-fluid system with internal heat sources. European Appl.Res. Rept.-Nucl.Sci.Technol.6, 1401-1411.

[98] Rivard J B (1978) Post-Accident Heat Removal: Debris-Bed Experiments D-2 and D-3. Sandia Report SAND78-1238.

[99] Schramm R and Reineke H H(1976) Thermohydraulics of the interactions between molten core material and reactor concrete. pp 2087-2094 in ref.[5].

[100] Smith W and Hammitt F G(1966) Natural Convection in a Rectangular Cavity with internal heat generation. Nucl.Sci.Eng.25, 328.

[101] Sowa E S, Gabor J D, Baker L Jr., Pavlik J R, Cassulo J C and Holloway W (1976) Studies of the formation and cooling of particulate fuel debris in sodium.pp2036-2046 in ref.[5].

[102]Sowa E S, Gabor J D, Pavlik J R and Cassulo J C(1977) Particulate bed-formation tests using thermite generated core debris. ANL Report RDP-61 pp.6.54-6.59

[103] Stein R P, Baker L, Gunther W H and Cook C(1978) Interaction of heat generating UO_2 with structural materials. pp 389-397 in ref.[10].

[104] Stein R P, Baker L,Jr., Gunther W H and Cook C(1979) Heat transfer from internally heated molten UO_2 pools. European Appl.Res.Rept.-Nucl.Sci.Technol. 6, 1468-1477.

[105] Stein R P, Hesson J C and Gunther W H(ANL) (1974) Studies of heat removal from heating generating boiling pools. pp 865-882. in ref.[2].

[106] Suo-Anttila A J, Catton I and Erdmann R C(1974) Boiling heat transfer from molten fuel layers. pp.845-864, in ref.[2].

[107] Suo-Anttila A J and Catton I (1975) Thermal convection experiments with internal heating.p.161-168 in ref.[4].

[108] Swanson D G, Zehms E.H, Aug C Y, van Paassen H L L, and Marchese A R(1978) Molten core debris interactions with core containment materials pp.408-414 in ref.[10].

[109] Tsai S S and Kazimi M S (1976) Particle-to-particle interaction effects on fuel debris redistribution in the LMFBR Outlet Plenum Trans.Am.Nucl.Soc.23,372.

[110] Tsai S S and Kazimi M S(1976) Horizontal heat transfer from molten fuel to an ex-vessel core retention bed. Trans.Am.Nucl.Soc. 24, 250.

[111] Turland B D(1979) A general formulation of the isotherm migration method for reactor accident analysis. Proc.5th Conf. on Structural Mechanics in Reactor Technology (Berlin), paper B3/5 North Holland.

[112] Turland B D and Peckover R S (1979) Melting front phenomena - Analysis and computation. European Appl. Res. Rept. - Nucl. Sci. Technol. 6,1499-1515.

[113] Turland B D and Peckover R S(1980) The stability of planar melting fronts in two-phase thermal Stefan problems. J.Inst.Maths Applics 25,1-15.

[114] Turland B D and Peckover R S (1978) The Distribution of Fission Product Decay Heat, pp17-25 in [10].

[115] Turland B D, Peckover R S and Dullforce T A (1977) On the stability of advancing melting fronts to spatial perturbations. Proc. Third Post Accident Heat Removal Conference, Argonne November 1977: Argonne Report ANL-78-10. pp 172-178.

[116] Turland B D (1978). The use of a Highly Conducting Liner in Reducing Local Heat Flux Density Concentrations. In 'Heat Transfer 1978' Vol.5, pp 161-165. Nat Research Council of Canada.

[117] Turner J S(1973) Buoyancy Effects in Fluids 367 pp. Cambridge University Press.

[118] Vossebrecker H (1980) Private Communication.

[119] Wallis G B (1962) Some hydrodynamics aspects of two-phase flow and boiling. Proc. Heat Transfer Conference, Boulder, Colorado, pp.319-340.

[120] Werle H (1979) Experimental investigation of heat between two horizontal liquid layers with gas injection. European Appl.Res.Rept.Nucl.Sci.Technol.6,1478-1485.

[121] Alsmyer H, Barleon L, Gunther Cl.,Muller U and Reimann M(1978) Some concepts on heat transfer from molten pools to underlying sacrificial beds. pp 161-171 in [10].

[122] Brinsfield W A, Toman W I, Dhir V K and Catton I(1979) Heat transfer and penetration from a heated pool to a substrate. Paper VI/D/8 in [14].

19 LMFBR SYSTEM SAFETY ANALYSIS

by

J.F. Jackson & M.G. Stevenson

LASL, Los Alamos, NM, USA

In order to adequately assess the safety characteristics of an LMFBR undergoing a design basis accident, a whole new class of computational tools capable of dealing with molten material dynamics needed to be developed. This chapter discusses the development of current state-of-the-art computational methods used in the United States for analysis of core meltdown accidents (Hypothetical Core Disruptive Accidents) in LMFBRs. The emphasis is on the phenomenological basis and numerical methods of the codes SAS, VENUS-II, SIMMER, and REXCO.

1. INTRODUCTION

The development of computational methods for Liquid-Metal Cooled Fast-Breeder Reactor (LMFBR) accident analysis has long been dominated by the difficult problem of core melt, disassembly, and possible primary system damage. Although other LMFBR analysis methods also are important in assuring public safety, this chapter will discuss only the computer methods used in the mechanistic analysis of LMFBR core meltdown, i.e., the Hypothetical Core Disruptive Accident (HCDA).

Over 20 years of development from the initial Bethe-Tait method[1] have provided increasingly complex computer codes that tax the fastest computers available. These codes synthesize the ever improving understanding of basic accident phenomena into a consistent analytical framework based on fundamental conservation laws. Experience has shown that many of these phenomena are complex and highly interactive (see Chapters 14-16). Furthermore, our understanding of them rests largely on small-scale experiments that often are performed with simulant materials. In many cases, direct experimental data are unobtainable. Thus, large mechanistic codes must be able to provide reliable and comprehensive predictions of full-scale accident behavior.

Because the primary sodium in an LMFBR operates at low (nearly atmospheric) pressure, system design can preclude a complete loss of coolant and core uncovering as the initiator of a whole-core melt. The only accident initiators that can lead to whole-core melt are (1) transients combined with failure to shut down (scram), (2) shutdown loss of decay heat removal, (3) core-wide failure propagation (such as initiated by local flow blockages), and (4) rapid and large reactivity insertions exceeding protective system response. Proper design can

---

Work performed under the auspices of the US Nuclear Regulatory Commission and the US Department of Energy.

virtually elminate all of these; but meltdowns initiated by at least one of these mechanisms are usually considered in licensing evaluations. In the US, the two transients that have received the most attention are the loss-of-flow and the transient overpower accidents.

Although many codes have been developed for HCDA analysis, we will concentrate on the phenomenological basis, numerical methods, and experimental verification of the major whole-core accident analysis codes or code series used in the US, i.e., SAS,[2-5] VENUS,[6,7] SIMMER,[8,9] and REXCO.[10] These or similar codes are being used in several countries for LMFBR HCDA analysis. Other US codes, such as MELT-III,[11] TWOPOOL,[12] and VENUS-III[13] provide important supplementary analysis capabilities, but will not be discussed in detail here.

It is extremely difficult to calculate in detail meltdown sequences from initiation through complete cessation of significant material motion in the primary system. Indeed, this goal has not yet been accomplished. However, the mechanistic methods noted above can now track sequences until neutronic events and the resulting energetics are largely over. For most cases, the primary remaining analysis problem is to assess the potential for melt-through of the vessel. Although the available methods provide a mechanistic framework for analyzing the energetics of accident sequences, there remain modeling limitations and uncertainties. There is a strong need for experimental verification, particularly of the modeling of the complex process of gradual core disruption. In addition, SAS and SIMMER are very long running. Their use in a production mode for many calculations is practical only on the largest computers.

2. CORE DISRUPTION ACCIDENT SEQUENCES

Before discussing the details of the codes noted above, it is useful to discuss their historical development as related to the phenomenology involved in whole-core-disruptive accident sequences. The details of core disruptive accident phenomenology are presented in Chapters 14-16. The purpose of this section is to review HCDAs from the viewpoint of mechanistic computer code development. In addition to the two-phase-flow phenomenology typical of Light-Water Reactor (LWR) accident analysis, LMFBR meltdown accident calculations must focus on reactivity events. Motion of any of the core materials (sodium, steel, fuel, and control) can have a significant reactivity effect. Fuel compaction, the concern of the original Bethe-Tait analysis,[1] has a potential for leading to severe excursions. Conversely, outward redistribution of fuel is the only way to achieve a permanently shutdown neutronic state unless control rods are inserted into the core during the accident sequence. The tight coupling of the complex material motions with reactivity effects demands considerable care in the modeling.

Until about 1973, HCDA sequences usually were analyzed in the four relatively unique phases, shown in Fig. 1 and discussed below.

2.1 Accident Initiation

This phase of a mechanistic accident analysis involves a calculation of the core neutronics and thermal behavior to the point of loss-of-subassembly geometry. Because most LMFBR designs use rigid subassembly ducts, intact subassemblies are coupled only through inlet and outlet thermal-hydraulic conditions and through reactor power. Subassemblies with similar power, flow, and irradiation conditions are generally, although not necessarily, lumped together in groups. All the subassemblies within a given group, or "channel," are

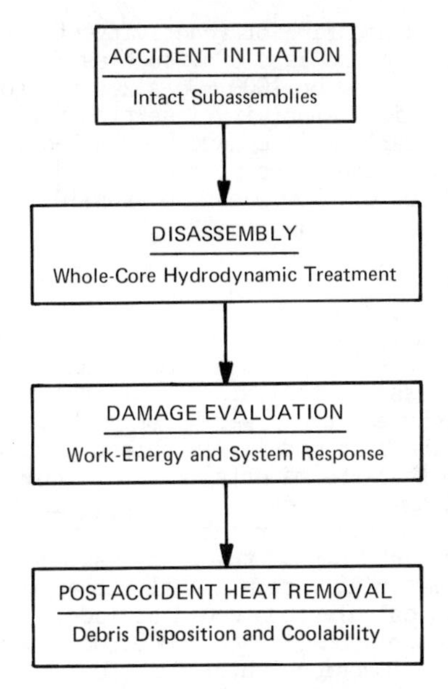

Fig. 1. Traditional approach to mechanistic accident analysis.

assumed to behave identically. Phenomena such as transient thermal-hydraulics, sodium boiling, fuel-pin mechanics and failure, cladding and fuel motion, and fuel-coolant interactions are treated with one-dimensional models. Figure 2 depicts this kind of multichannel core representation.[14] The multichannel SAS codes (SAS2A and its successors) were preceded in the US by multichannel codes such as FREADM[15] and TART.[16] These earlier codes included simple parametric models, particularly of sodium voiding, and were designed to allow a reasonably

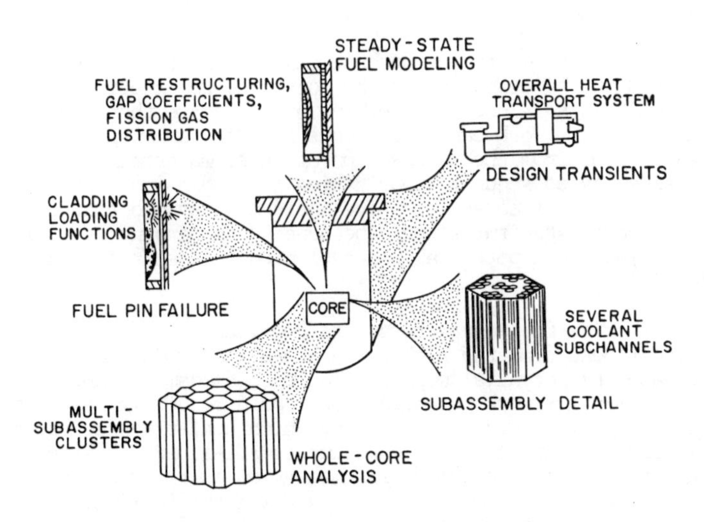

Fig. 2. Multichannel core representation in a one-dimensional model.

realistic but conservative accounting of reactivity effects for input to the core disassembly codes available in the late 60s, such as MARS.[17] At that time in the US, HCDA analysis was focused on 1000 MWe-sized reactors, and accident sequences were dominated by high reactivity insertion rates due to large, positive, sodium void worths. These early multichannel codes led to the recognition that a mechanistic treatment of noncoherence of sodium voiding among subassemblies of varying initial power and sodium flow was important.[18]

2.2 Disassembly Analysis

A "classical" disassembly typically begins when a prompt-critical neutronics excursion is induced. The resulting rapid heating and vaporization of the fuel produce relatively high pressures that disassemble the core, i.e., move the fuel outward rapidly, and thereby end the power burst. This disassembly process was first described by the Bethe-Tait model. In this approach, the core was treated as a homogeneous fluid so that the material motion during disassembly could be calculated using a hydrodynamic approach. The original Bethe-Tait calculations were done in spherical geometry. The reactor power was calculated using point kinetics with first-order perturbation theory to estimate the reactivity feedback associated with the material motion. Analytic approximations simplified the equations so that hand calculations could be made. The method was later codified in the Weak Explosion code.[19] At about the same time, methods using numerical solutions of one-dimensional hydrodynamics and neutron transport equations were developed, such as the AX-1 code.[20]

Variations and improvements[21] subsequently were made to the basic Bethe-Tait approach. Doppler feedback was included, improvements were made to the equation of state used to estimate the pressures, more accurate neutronics was implemented in some cases, and the capability for doing hydrodynamics calculations in two-dimensional (r,z) geometry was added. A number of disassembly computer codes were developed in the US as successive improvements were made, with VENUS-II[7] now being the most used of these classical disassembly codes.

2.3 Damage Evaluation

During or following reactor disassembly, the rate at which the thermal energy released in the transient can be converted to kinetic energy (work) must be established. This is needed to evaluate the damage that the excursion might cause to the system. Work can either come from the expansion of the core materials themselves or from the interaction of these materials with the sodium coolant causing its consequent vaporization and expansion. Early studies of the damage potential of core-disruptive accidents assumed that the work on the surrounding reactor structures would be done by the expanding fuel materials only. The maximum work potential was evaluated by isentropic expansion calculations.[22]

As larger oxide-fueled reactors were considered, it was recognized that the transfer of heat from the high-temperature fuel to the sodium could considerably increase the potential work available to do damage. Hicks and Menzies[23] calculated the maximum work potential that could result if heat transfer to the sodium were extremely rapid. Because very high heat-transfer rates had been observed in reactor accidents and experiments (SL-1, SPERT) attention turned to predicting these rates by assuming fragmentation of the fuel into sodium. Initially, this was done parametrically by varying the fuel particle size and mixing time and calculating the resulting heat-transfer rates as a basis for assessing damage potential. Early results of this type of calculation were reported by Cho, et al.[24] and Padilla.[25] A review of the historical development of work-energy conversion is available in Ref. 26.

Once the pressure source term is established, the response of the system can be analyzed. This usually is done with a hydrodynamic calculation of the pressure propagation coupled with an analysis of the structural response of the important system components. The REXCO series[10] of codes is widely used in the US for this purpose. The initial versions of REXCO used hydrodynamics methods similar to those already in use[27] at the Los Alamos Scientific Laboratory (LASL) and Lawrence Livermore Laboratory (LLL).

2.4 Postaccident Heat Removal

The final phase in HCDA analysis (excluding containment and consequence analysis) is an evaluation of postaccident heat removal (PAHR). The objective is to analyze the long-term decay heat removal from the fuel following a disruptive accident. This has mainly taken the form of defining possible post-disruptive fuel dispositions and of analyzing the subsequent coolability. The disposition of core materials and the required measures to assure cooling of the core debris are, of course, dependent on the reactor design under consideration.

Detailed mechanistic analysis methods are not yet in common use for this problem and we will not discuss this area further. Reviews of the phenomenology are available in Ref. 28.

2.5 Gradual Core Disruption and the Transition Phase

The relatively clear-cut separation into the phases discussed above has disappeared with more detailed analyses. As initiating-phase models improved, the resulting initial nuclear excursions became progressively milder, largely due to the more accurate representation of intersubassembly noncoherence. This noncoherence arises mainly from carefully accounting for the differences in power level and coolant flow rates among the subassemblies. The net result of this is that calculated accident sequences tend to proceed into a gradual meltdown of the core, instead of ending in a vigorous disassembly excursion. This does not mean that a disassembly excursion cannot be induced at some point in the accident, but merely that it is unlikely in the early stages of core meltdown. Another important reason for considering milder accident progressions was the concentration in the US on the Fast Flux Test Facility (FFTF) and the Clinch River Breeder Reactor (CRBR) reactors with positive sodium void worths much smaller than the 1000-MWe reactors emphasized in the middle-to-late 1960s.

As an example, a characterization of the CRBR unprotected loss-of-flow (ULOF) accident (assuming a preirradiated core) consists of (1) undercooling of the core; (2) sodium boiling and mild power increase leading both to cladding melting and relocation and to cladding failures in partially voided and non-voided subassemblies; (3) fuel swelling, slumping, and/or dispersal driven by fission gas and sodium vapor; and (4) melting of subassembly cans initiating a "transition phase" of gradual disruption. This latter phase is characterized by steel relocation and vaporization and fuel melting and motion until no core fuel subassembly structures remain intact. In this kind of accident sequence, reactivity insertion rates from sodium boiling are only a few $/s at the highest.[29] Similarly, cladding and fuel motion in intact subassemblies typically leads to at most a few $/s. Even in larger reactors with large positive sodium void reactivities, maximum positive voiding reactivity rates from normal boiling are given as only a few tens of $/s[30] by SAS3A calculations.

The definition of the transition phase led to the more generic accident progression diagram shown in Fig. 3. The progression from initiating phase to a classical disassembly is included as is the gradual core disruption of the transition phase.

In the transition phase, a subassembly successively progresses through coolant voiding and melting, generating pressures generally too low to cause a massive dispersal of molten fuel from the core region. The subassembly can walls melt quickly, and growing regions of molten fuel and steel begin to form in the hottest portions of the core. This stage of the accident is likely to be accompanied by a number of mild excursions induced by continued slumping in successively lagging subassemblies and by the re-entry of fuel that was temporarily dispersed by mild pressurizations. Recent SIMMER-II calculations[31] indicate that excursions occur with only a few tens of $/s for maximum reactivity insertion rates in the CRBR ULOF transition phase, but these increase as more subassemblies become involved.

The neutronic events in this sequence can be terminated by three mechanisms. First, the transition phase can end relatively benignly with gradual fuel removal

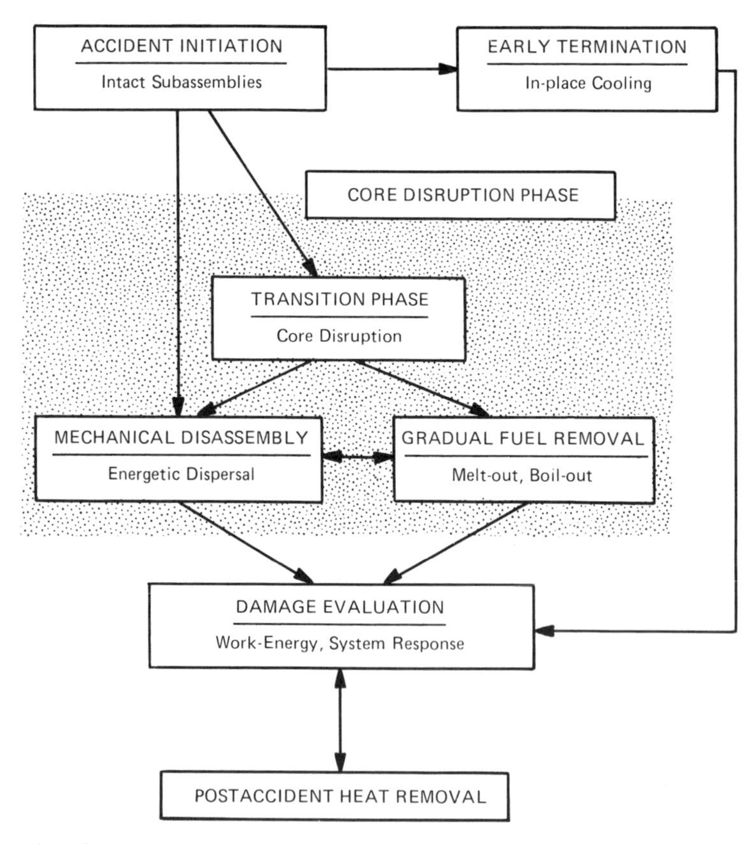

Fig. 3. Comprehensive accident progression diagram.

by mild expulsions, meltout processes, or both. Second, a bottled core blowdown can occur, in which a molten core completely contained by the surrounding structure gradually heats and pressurizes due to decay heat generation and/or mild neutronic bursts, then blows down (expands) following sudden opening (due to melt-through or mechanical failure) of the surrounding structure. Third, the rapid expansion following sufficiently energetic neutronic excursions can remove sufficient fuel from the core to leave the reactor permanently shut down.

A developing area of analysis has been the calculation of the extended motion or core expansion following either a disassembly-type excursion or a bottled-core blowdown. SIMMER-II calculations[32] of postdisassembly expansions have indicated considerably reduced thermal-to-kinetic-energy (work) conversion efficiencies compared to the more conventional isentropic expansion calculations. SIMMER-II can bridge the analysis areas between SAS and REXCO because of its ability to (1) follow directly from SAS initiating-phase calculations into a transition phase (possibly including several neutronic excursions and progressing to a completely disrupted core,[31] (2) follow the extended material motion after these excursions, and (3) determine pressure loadings on the system.

A different path to terminating a disruptive accident sequence before either a disassembly or a transition phase was first noted during the analysis[33] of unprotected transient overpower accidents (UTOP) in the FFTF and is indicated by the "early termination" path in Fig. 3.

A UTOP accident can be characterized by (1) pins failing before any sodium boiling, (2) the ejection of molten fuel and fission gas into flowing liquid sodium, (3) localized voiding by expanding fission gas and sodium vaporization, (4) dispersal out of the active core of some of the ejected fuel, and (5) some freezing of molten fuel but incomplete plugging of coolant subchannels. If the reactivity insertion driving the overpower ceases, then the fuel sweep-out can lead to a neutronically stable condition with in-place cooling of the remaining fuel debris and unfailed pins. Conversely, continuation of the driving reactivity or a lack of adequate in-place cooling could lead to whole-core involvement. If a transition phase developed, it would involve considerably different conditions than a ULOF because full coolant flow would continue, at least in undamaged channels, and all channels would experience the full-core pressure drop. Generally, transition phase behavior for a UTOP initiator has not been examined in any detail.

With the above as background, we now examine some of the specific codes. The following code descriptions include brief discussions of the numerical methods used to solve the equations. Some introductory information on numerical methods is presented in Appendix A. This includes defining some of the key terms used in the various code descriptions.

3. THE SAS ACCIDENT INITIATION CODES

The SAS codes represent a long development from the single-channel SAS1A version through SAS2A, SAS3A, SAS3D, and SAS4A (soon to be completed). This development, as for other LMFBR accident analysis methods, proceeded without the benefit of an adequate phenomenological data base. As might be expected, this resulted in some significant modeling deficiencies when compared to experimental data, although the successes[34,35] with some of the early models were gratifying given their ad hoc nature. The early models also have tended to lose credibility because of numerical deficiencies, particularly as their scope was extended into analyzing more slowly developing accident sequences. In the initial developments, to quote Waltar and Padilla,[14] "attention to mathematical solutions was normally only given to the degree necessary for obtaining a stable and accurate, but not necessarily efficient, numerical solution."

SAS1A[2] was a single-channel code that included point reactor kinetics; a combination of annular slip two-phase flow and single-bubble slug-boiling models;[36] an elastic-plastic cladding and elastic-fuel deformation model, DEFORM-I;[37] a transient fuel-pin heat-transfer model; and an internal switch to the MARS[17] disassembly module.

The sodium-boiling model was later completely rewritten as a multibubble slug-boiling model for the multichannel SAS2A version.[3] A new fuel-pin deformation module, DEFORM-II,[38] also was included in a version of SAS2A. SAS2A did not include the MARS disassembly module but provided punched output that could be used as input for initial conditions in the VENUS-II disassembly code.

The phenomenological structure of SAS3A is shown in Fig. 4. SAS3A[4] contained the first models, such as CLAZAS,[39] SAS/FCI,[40] and SLUMPY,[41] to allow complete calculations of cladding and fuel motions into the transition phase. SAS3D used the same models but restructured the code for more numerical efficiency and to allow up to 35 channels instead of the 10 allowed in SAS3A. SAS4A[5] is a new code with a new sodium-boiling module and with CLAP,[42] PLUTO2,[43] and LEVITATE[44] replacing the equivalent models of SAS3A and SAS3D. Here we will concentrate on the developments through SAS3A but will relate the newer models of the not yet released SAS4A to the earlier ones as appropriate.

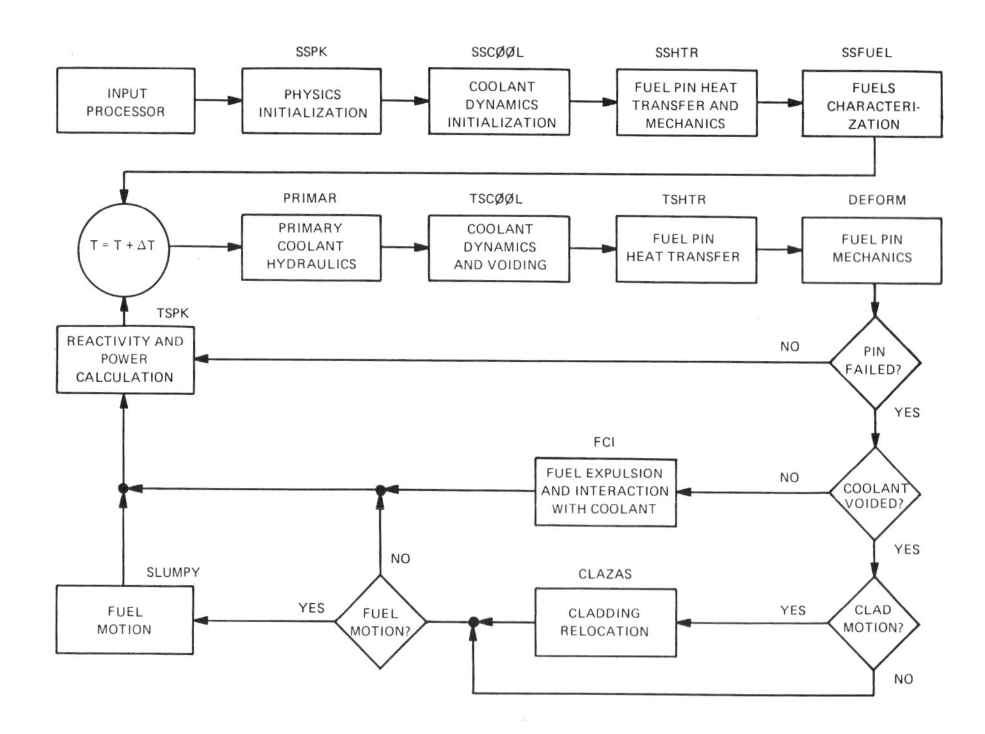

Fig. 4. SAS3A module relationships.

3.1 Sodium Boiling

The sodium-boiling model used in SAS3A and SAS3D began as a new version of the single-bubble slug-boiling model included as an option in SAS1A. The new version was similar to the models of Cronenberg, et al.[45] and of Schlechtendahl[46] and included an explicit calculation of vaporization of a liquid film remaining on pins after voiding, of condensation in colder regions, and of vaporization or condensation on the liquid interfaces. This single-bubble model was appropriate for situations involving high superheat.

However, early calculations with superheats lower than 70°C showed that the lower liquid slug heated well beyond saturation, indicating that additional bubbles should form. This model then was modified to allow a new vapor bubble to form in the liquid coolant at any point where the liquid temperature exceeded the local time-dependent saturation temperature by more than the original superheat criterion. It also was observed that, because of the high rate of vapor production and the resulting vapor velocities, large axial pressure gradients should exist in a large vapor bubble. For this reason the model was modified to account for the pressure gradients and the resulting multibubble model was included in the standard version of SAS2A and in SAS3A. Figure 5 indicates the conceptual basis of the model. It allows the formation and continued presence of a number of bubbles in a channel and follows the bubbles either as they move up a channel and collapse (or reach the exit) or as they become sustained and void the channel.

One-dimensional Eulerian equations for the upper and lower liquid slug (or the whole channel before boiling starts) are written in the form,

$$\frac{\partial \rho_\ell}{\partial t} + \frac{\partial G}{\partial z} = 0 \ , \tag{1}$$

$$\frac{\partial G}{\partial t} + \frac{\partial}{\partial z}\left(\frac{G^2}{\rho_\ell}\right) = \frac{\partial p}{\partial z} - \rho_\ell g - \left(\frac{\partial p}{\partial z}\right)_f \ , \tag{2}$$

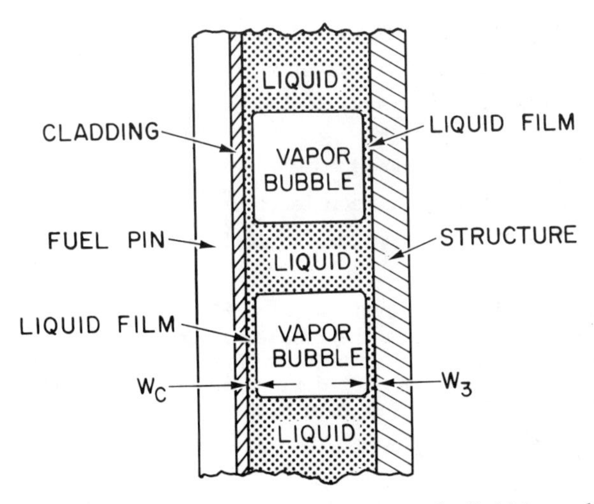

Fig. 5. Conceptualization of the SAS multibubble model.

where ρ_ℓ is the liquid density, G is the mass-flow rate, p is the pressure, g is the acceleration due to gravity, and $(\partial p/\partial z)_f$ is a friction pressure gradient. A lumped-parameter momentum equation for both upper and lower liquid slugs is obtained by integrating Eq. (2) over the length of the slug. The fixed grid used with the vapor region makes this something of a combined Eulerian-Lagrangian scheme.

For small bubbles, the pressure drop within the bubble is negligible and the bubble growth is determined by coupling the momentum equations for the liquid slugs with an energy and mass balance in the bubble, assuming saturation conditions and spatially uniform pressure and temperature. In the case of large bubbles, the voids can extend over regions with considerable axial surface temperature variation and the subsequent vaporization and condensation can cause high vapor velocities. This acceleration and the frictional effects are accounted for in an axially varying vapor pressure model. In this model, the vapor pressure, temperatures, and mass-flow rates are calculated at fixed nodes within the bubble and also at the lower and upper interfaces. This uses a simultaneous solution of the vapor cells momentum and continuity equations with the momentum equations for the liquid slugs setting the boundary conditions at the interfaces. Heat flow through the liquid-vapor interface is ignored in this model and the vapor velocity at the interface is assumed equal to the interface velocity. In addition, the pressures across this interface are considered continuous.

The SAS2A model accounted only for film dryout by vaporization, but for SAS3A a moving film treatment was added to account for film stripping by vapor shear. The frictional pressure drop in the vapor is increased due to the formation of waves or instabilities on the surfaces of the films. This is accounted for by using a vapor friction factor, f_δ, in the form

$$f_\delta = f_o \left[1 + \frac{300\, M_f (W_c + \gamma_2 W_3)}{D_h (1 + \gamma_2)} \right] , \tag{3}$$

where

f_o = single-phase friction factor,

W_c = liquid film thickness on cladding,

W_3 = liquid film thickness on structure,

$\gamma_2 = \dfrac{\text{surface area of structure}}{\text{surface area of cladding}}$,

D_h = hydraulic diameter, and

M_f = user-supplied multiplication factor.

The resulting large shear force on the liquid film tends to lead to early local dryouts.

To obtain numerical stability for reasonably large time steps, an implicit differencing scheme is used, with the resultant simultaneous set of equations solved by Gaussian elimination. Time-step control is used when vapor pressures in any cell increase by more than 10% during a cycle. The reduced time step, Δt, is calculated from the original time step, Δt_o and the fractional increase in pressure, $\Delta p/p$, by

$$\Delta t_n = 0.075 \quad t_o \frac{p}{|\Delta p|} . \tag{4}$$

The above procedure gives good results while allowing sufficiently large time-step sizes to render the calculations economically feasible. Overall, the SAS3A sodium boiling model has provided a reliable solution technique, particularly for loss-of-flow conditions, which has proven to stand well in comparison with numerous in-pile LOF experiments, such as in Refs. 47-49.

The basic aspects of the model are being retained for SAS4A, but refinements are being made, as discussed in Ref. 50, to extend its range of validity and to provide a more sophisticated solution technique. The new SAS4A model (1) accounts rigorously for the coexistence of sodium vapor and fission gas in voided regions; (2) allows heat transfer from dried-out cladding to the sodium vapor/fission gas mixture and superheating of the vapor; and (3) accounts for both spatial and temporal flow area changes, including those given by the cladding motion model. The numerical method used to solve the vapor conservation equations is a multicomponent generalization of the original Implicit Continuous Eulerian method.[51] As in the SAS3A moving film model, the vapor is coupled to the moving liquid sodium film on the cladding as long as the film remains. Both models essentially provide a two-velocity field (vapor and liquid) treatment for an annular flow regime. A comparison of the models for a calculation of a TREAT LOF test is given in Figs. 6 and 7. Despite variations in voiding profile oscillations, the overall behavior is very similar.

3.2 Cladding Relocation

In analyzing an unprotected loss-of-flow accident[52] for FFTF it was recognized that, under the expected relatively constant power conditions (the FFTF sodium void reactivity effect is small), cladding would melt some 2-3 s before fuel melting in the subassemblies that first voided. Thus, although there was no direct experimental evidence of such an effect, it was reasonable to assume that the molten cladding would relocate (possibly moving out of the heated region), freeze, and plug. The first supposition was that the cladding would drain downward and plug below the fuel region. However, following a suggestion by Fauske[53] it was recognized that upward sodium velocities could be high enough to cause flooding (wave formation), and the resulting high friction drag would lead to upward cladding motion similar to the sodium-film stripping effect included in the SAS voiding model. This could then lead to cladding freezing and an initial plug formation above the fuel region, followed by draining and plugging below the fuel region from cladding that melted later.

In addition to the mechanical effect on sodium voiding and fuel dispersal, cladding relocation can have a significant reactivity effect and a significant thermal effect in that removal of the cladding heat sink allows fuel temperatures to rise more rapidly. To analyze cladding relocation, the CLAZAS module[39] was developed for inclusion in SAS3A.

This model uses an explicit calculation of the motion of molten clad driven by the sodium-vapor pressure gradient and frictional drag. The physical picture is one of discrete clad segments that can combine by moving over other clad segments. Each segment moves under the forces of gravity, the channel pressure gradient, the frictional drag due to streaming sodium vapor, and the friction between moving clad and the fuel pin. Feedback effects are included to modify the coolant-channel sodium-vapor friction, the coolant channel hydraulic diameter, and the sodium-vapor flow area. Heat-transfer effects are calculated by including relationships describing the heat transfer from moving clad to stationary clad and from stationary fuel to moving clad, as well as a variable viscosity model for molten clad as a function of internal energy. Cladding

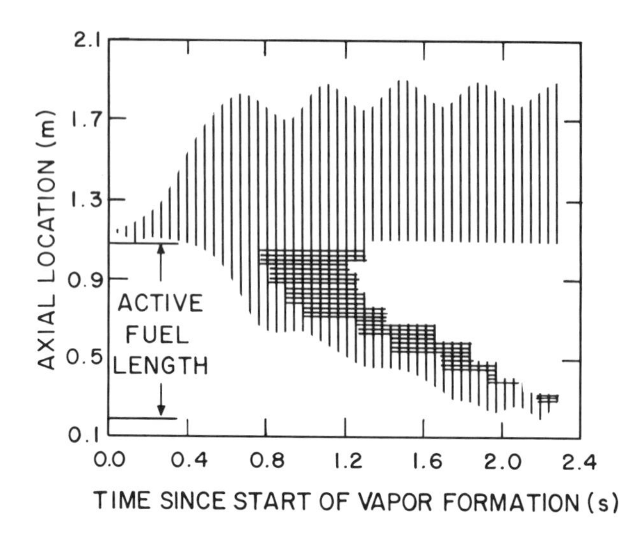

Fig. 6. SAS3D voiding, dryout, and clad melting pattern in a typical TREAT LOF test.

freezing is treated by increasing the cladding viscosity to a high value as the cladding internal energy decreases through the heat of fusion.

As an example, Fig. 8 shows an early CLAZAS calculation of the TREAT L2 experiment. This calculation was the first to indicate the suspected, but not previously observed, possibility of upper cladding blockage formation. It was a relatively successful example of ad hoc modeling, as the postmortem examinations did indeed indicate upper blockage formation.

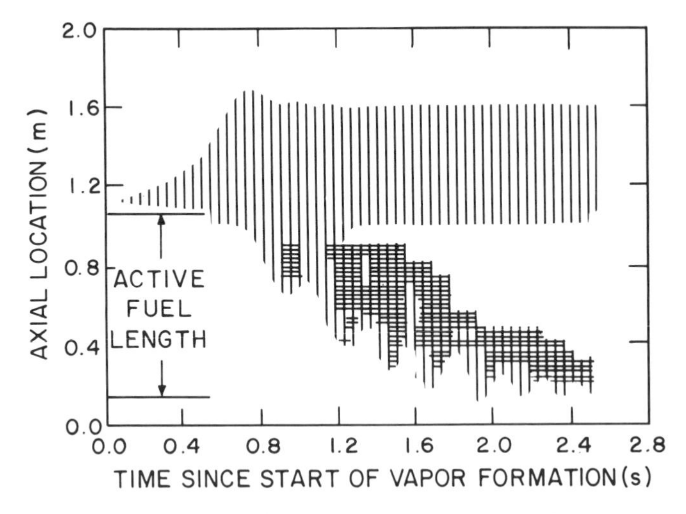

Fig. 7. SAS4A voiding, dryout, and clad melting pattern for the same problem shown in Fig. 6.

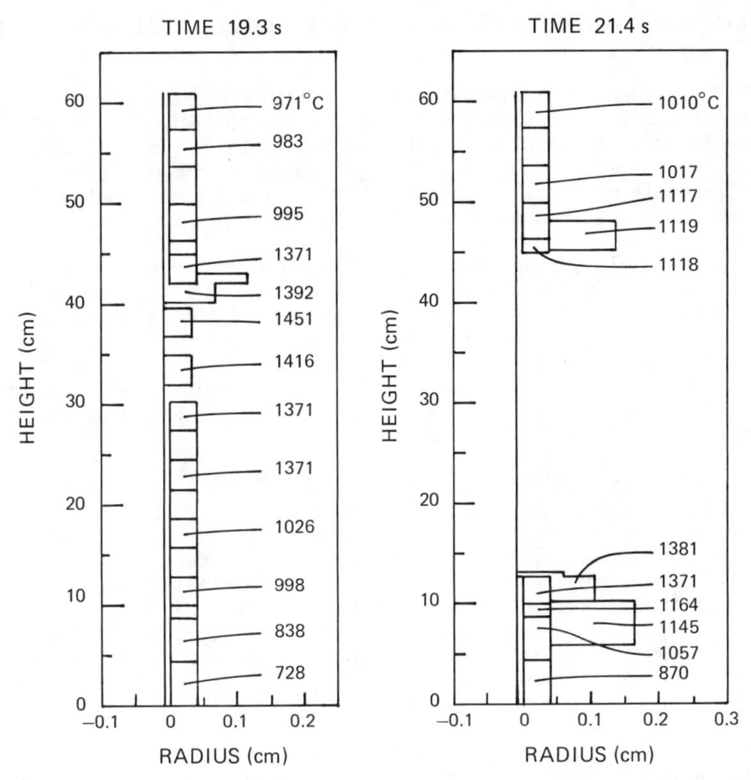

Fig. 8. CLAZAS calculation of L2 TREAT experiment showing cladding blockage at top and bottom of fuel column.

An improved model, CLAP,[42] was developed after CLAZAS and is being included in SAS4A. CLAP uses an Eulerian formulation rather than the Lagrangian-style cladding segments used in CLAZAS. Its predictions of smaller upper cladding blockages as compared to CLAZAS are in better agreement[54] with the observations from the more prototypic R-series TREAT LOF experiments.

3.3 Fuel Motion in Voided Subassemblies

A common and quite reasonable assumption in many fast reactor accident analyses has been that fuel, driven only by gravity, slumps as it melts. This was the basic assumption of the early Bethe-Tait analysis,[1] which assumed whole-core uniform slumping under gravity to provide a conservative bound for reactivity insertion rate. Gopinath and Dickerman[55] developed early slumping noncoherence arguments based on radial power distributions for EBR-II HCDA analyses. Because the mechanistic multichannel codes include this noncoherence effect directly, the emphasis during the last decade has been on including the several forces other than gravity that can act on fuel as it becomes mobile, either before or after initiation of melting. These forces include sodium vapor pressure gradients and shear, fission product pressures, and steel and fuel vapor pressures. The SLUMPY module[41] in SAS3A and SAS3D was the first mechanistic model to include all these effects, albeit still in a somewhat parametric fashion.

SLUMPY provides a one-dimensional compressible-hydrodynamics calculation and can be used either to supply detailed initial conditions to a VENUS-II or other two-dimensional disassembly calculation, or it can be used directly as a disassembly code within the limitation of one-dimensional motion (implying intact subassembly geometry). In most SLUMPY calculations, fuel motion is assumed to be initiated when melting begins in unrestructured fuel.

As the fuel melts, individual axial fuel segments join the "slumped" region treated in the SLUMPY calculation shown in Fig. 9. Unmelted fuel above the slumped region can fall into or be pushed out of this slumped region. The unmelted pin below the slumped region is assumed stationary. Both the upper and lower solid fuel segments restrict the area available for the slumped region. Axially limiting boundaries for the slumped region are dependent on user input; or if CLAZAS is used, the boundaries may be determined by the time-dependent positions of calculated clad blockages.

Figure 10 shows a SLUMPY calculation of the TREAT L4 experiment. This early calculation shows "eructations" of fuel similar to those indicated by TREAT hodoscope measurements. In this SLUMPY calculation, the imposed heat-transfer assumptions led to these eructations being induced by steel-vapor pressures. Whether this or other mechanisms such as fission gas release cause such observed fuel dispersal events is still a subject of much discussion (a good review of fission gas effects is given by Deitrich in Ref. 56). SLUMPY provides a framework for including several mechanisms, with the notable exceptions that treatment of massive fuel swelling prior to melting or a "chunk breakup" mode of fuel disruption is difficult.

One of the necessary aspects of HCDA hydrodynamic models is that continued energy generation by fission and subsequent material vaporization provides a continued internal driving pressure for the momentum equations. In addition

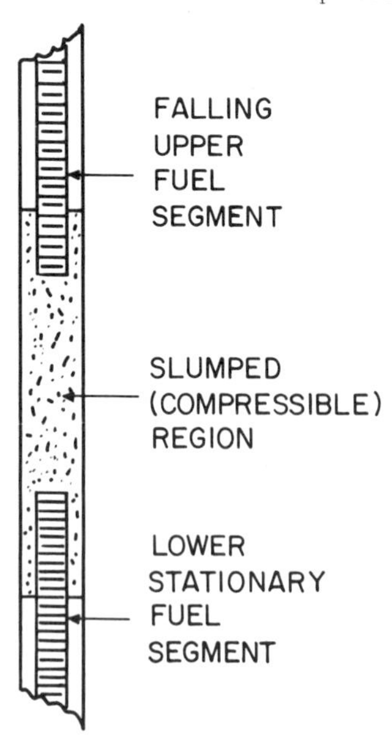

Fig. 9. Radially enlarged geometry showing fuel motion in a SLUMPY calculation.

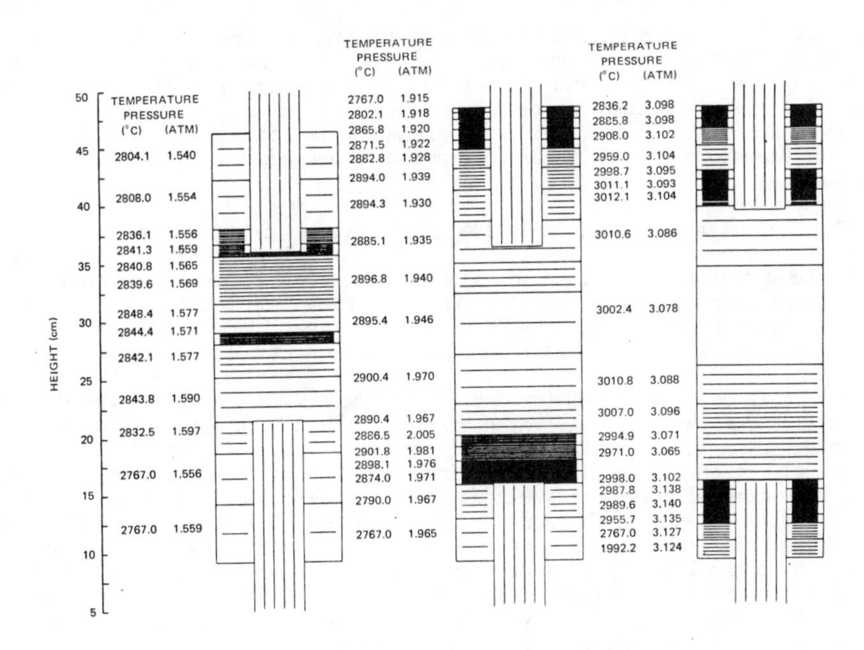

Fig. 10. SLUMPY calculation of L4 TREAT experiment showing eructation induced by steel-vapor pressure.

to solving the conservation equations for a multicomponent system, the rate of energy generation and its transfer among the several possible components must be explicitly accounted for, with a feedback to the pressure field in the momentum equation through the equation of state. The SLUMPY equation of state includes steel (iron) and fuel (UO_2) vapor pressure and uses fission gas or sodium vapor pressure as a background pressure when steel and fuel vapor pressures are low. Liquid compressibility is included when single-phase conditions are reached.

SLUMPY was the first multicomponent HCDA analysis model to account for both separate velocity and energy fields with momentum and heat transfer between the fields. Several assumptions concerning viscous and friction effects, drag forces in a suspension, and heat transfer to the structure were employed to provide a sufficient number of equations for numerical solution. Slip is allowed between fuel particles and fission gas, but two-phase steel/fuel-vapor mixtures are treated as homogeneous flow. For the first case, a noteworthy simplification is that the fission gas inertial term is neglected; that is, the fission gas momentum equation is treated in a quasi-steady-state fashion.

The initial solution technique used in SLUMPY was an Eulerian formulation with explicit time-differencing; but as is generally true for Eulerian methods (unless special techniques are used), definition of material interfaces proved difficult. Also, a severe numerical stability problem arose in regions of high-power density undergoing rapid fuel vapor expansions. For such reasons, the Eulerian approach was abandoned and an explicit Lagrangian approach with mesh rezoning was employed. Rezoning in SLUMPY takes place when an additional fuel node satisfies an initiation-of-motion criterion. If material rapidly vaporizes and the mesh cell grows, the pressure gradient is spread over a fairly large region. If rezoning then is performed using constant intervals, the pressure gradient at the top and bottom of the cell is increased substantially, causing rapid acceleration and compression of slower moving cells. This problem was resolved by performing the rezone so that cells have a certain minimum mass, instead of a minimum length. When the grid-mesh is rezoned, the field variables are transformed by a linear interpolation between the old and new mesh structure.

The minimum mass requirement also is significant in time-step control, which is based on a Courant condition.[57] When a cell with a small amount of vapor is compressed, the volume (or linear dimension in this one-dimensional formulation) is reduced substantially until single-phase conditions are reached. Thus, the denominator of the Courant number (the spatial mesh) is reduced significantly, requiring a compensating time-step reduction to ensure stability.

SLUMPY is being replaced for SAS4A with the LEVITATE model,[44] which provides a more consistently integrated treatment of fuel, steel, sodium, and fission gas dynamics in voided or nearly voided regions of channels. LEVITATE treats fuel motion inside disrupting pins and, as a pin becomes completely disrupted, the channel treatment includes three velocity fields (a liquid fuel and steel field, a solid fuel fragment or chunk field, and a mixed vapor field) in an Eulerian grid. The solution technique is explicit in time and is coupled integrally to the liquid sodium slug calculation in the channel. LEVITATE, through including various fuel breakup modes, several flow regimes, cladding and structure ablation, fuel-steel mixing, and fuel-steel mixture freezing, removes most of the modeling limitations of SLUMPY. Because of its similar explicit numerical treatment, however, it is not likely to decrease computing times significantly.

3.4 Fuel Pin Failures

Several possible modes of fuel pin failure (i.e., cladding failure) can occur during a disruptive accident. Some are relatively simple to predict, such as cladding rupture due to plenum fission gas failure or cladding melting following dryout during sodium boiling. However, the prediction of cladding rupture during an unprotected transient overpower accident is more difficult and has been the subject of considerable modeling and experimental effort. A central problem has been whether fuel-cladding relative thermal expansion or transient fission gas release from the fuel and internal pressurization is the major contributor to cladding damage and failure.

A straightforward analysis of relative thermal expansion, assuming elastic fuel and elastic-plastic cladding, was included in the DEFORM-I model of SAS1A. DEFORM-II in SAS2A, and as an option in SAS3A, provided a similar analysis but included an elastic-plastic fuel treatment and allowed treatment of a central void and transient fission gas release. However, it did not allow for radial cracking of the outer part of the fuel pellets and the resulting loss of strength, particularly important when high pressures from fission gas release, fuel expansion on melting, and fuel vapor pressure occur in the fuel central region. Also, a reliable high-temperature failure criterion based on plastic deformation was not available during the time SAS3A was developed.

For these reasons, a "burst pressure" failure was included in SAS/FCI. This assumed pin internal pressures could be transmitted directly to the cladding (with a 1/r loss) through radially cracked and strengthless fuel. Unlike the thermal expansion loading mode in which cladding strain relieves the loading, the loadings caused by transient fission gas pressurization are not sufficiently relieved by local cladding deformations. Thus, an ultimate strength failure criterion (the burst pressure) rather than a strain failure criterion was used in SAS/FCI.

Both DEFORM-II and the burst pressure criterion model in SAS/FCI have serious limitations, and several models have been developed that remove these limitations. One of the first of these is LAFM,[58] which uses a cladding life-fraction damage correlation to predict failure and has been verified by comparison to a large number of TREAT TOP experiments. LAFM calculations have indicated that depending on transient conditions, either thermal expansion, fission gas release, or a combination can be the cause of failure. A similar model, DEFORM-III,[59] has been developed for inclusion in SAS4A.

3.5 Fuel Motion/FCI in Unvoided Channels

Early analyses[60] of a mild overpower transient HCDA in FFTF assumed that if the fuel pins failed at the midplane, plenum fission gas release through the failure could cause rapid voiding, leading to a prompt-critical transient and core disassembly. This mechanism was recognized to be unrealistic, and later modeling concentrated on developing a more consistent scenario for the phenomena following fuel-pin failure. The FISFAX model[61] was used to evaluate a sodium-in prompt-critical transient in FFTF and was the first to include the effect of sodium-induced fuel-coolant interaction (FCI) dispersive fuel motion. This concept, along with more realistic estimates of fuel-failure locations, went into the models used in later FFTF analyses. These analyses indicated that a hydraulic fuel-sweepout mechanism could terminate mild reactivity transients well before prompt criticality was reached.[33] Calculations with the SAS/FCI module[40] in SAS3A also indicated early termination, although in both cases the modeling originally was developed to provide a realistic estimate of the reactivity insertion rates as initial conditions for core-disassembly calculations rather than to predict early accident termination.

SAS/FCI is essentially a "point" or lumped parameter model as opposed to the one-dimensional voiding, cladding motion, and fuel motion models in SAS3A. SAS/FCI integrates the effects of fuel/fission gas mixture expulsion into flowing liquid sodium, the thermal interaction between molten fuel and sodium (based on a Cho-Wright parametric model[24]), condensation of sodium vapor (noted to be important in an early analysis of Cronenberg[62]), sodium slug ejection, and the resulting sodium and fuel motion reactivity effects. Cladding failure time and location can be set by input criteria using fuel thermal conditions, although SAS/FCI also includes the internal pin pressure failure criterion noted above. Although the fundamental modeling employs a lumped-parameter treatment for temperature, pressure, and sodium density in the interaction zone, a pseudo-Lagrangian model is used to treat fuel motion. In this model, the fuel moves with a velocity found by linear interpolation of the interface velocities of the two constraining liquid sodium slugs. Fuel motion inside the pin is treated by assuming a uniform decrease in density. It can be easily envisioned that this overpredicts the positive reactivity effect of internal motion if the failure is at the core axial midplane. Conversely, negative reactivities are overpredicted with away-from-midplane failures.

The numerical method of SAS/FCI uses explicit time-differencing of a single ordinary differential equation describing the interaction-zone pressure. This single equation was derived analytically from the momentum equations that couple pin internal and coolant channel pressures and the FCI zone internal energies. Back substitution is used to obtain other dependent variables.

Although small time steps generally are required, the method still is quite efficient.

When pin failures and fuel-coolant interactions occur as calculated by SAS/FCI, resulting high pressures in the coolant channels are felt instantaneously in the inlet plenum (the liquid slugs are assumed incompressible). Rapid inlet pressurizations can be calculated, particularly when many subassemblies are in the channel undergoing the FCI process. This pressurization tends to increase flows in other channels and to delay heatup. Although it is unrealistic to include this coherence enforced by lumping many subassemblies together and assuming that all pins fail simultaneously, the possible importance of this effect led to the treatment of primary loop pressures and flow rates included in the PRIMAR-II module[63] of SAS3A, and more recently, PRIMAR-IV in SAS4A.[64]

A primary limitation of SAS/FCI has been its inability to treat fuel expulsion into partially voided regions of a channel, as is typical of some channels in an ULOF overpower condition (the LOF/TOP syndrome). This limitation and its unrealistic predictions of fuel motion reactivities have led to development of the one-dimensional codes PLUTO[65] (a Lagrangian code), PLUTO2[43] (Eulerian), and EPIC[66] (Eulerian, Particle-in-Cell). Both PLUTO2 and EPIC have been or are being coupled to a version of SAS (EPIC in SAS3D and PLUTO2 to be in SAS4A). Figure 11 shows a comparison of SAS/FCI and PLUTO2 calculations of the TREAT L8 experiment from Ref. 5.

These models provide the only available mechanistic treatment of what is termed as loss-of-flow-driven transient-overpower (LOF/TOP) conditions. If, in a loss-of-flow accident, voiding, cladding relocation, and/or fuel slumping do lead to a sufficiently high overpower (TOP) condition, pins in colder subassemblies can fail with fuel expulsion into liquid sodium. If large positive reactivities result following near midplane failures, autocatalytic behavior can be calculated, i.e., the high-power levels cause more rapid pin failures, inducing higher insertion rates, higher power levels, and even more rapid pin failures (SAS/FCI overpredicts this effect). PLUTO2 predicts[67] that such autocatalytic behavior is much less likely than given by SAS/FCI predictions.

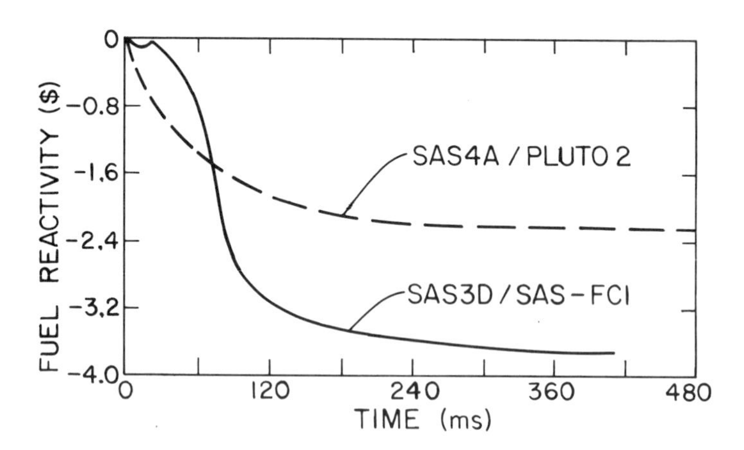

Fig. 11. Comparison of SAS/FCI and PLUTO2 postfailure fuel reactivity feedback calculations.

4. DISASSEMBLY MODELS -- VENUS-II

Numerous hydrodynamic disassembly models have been developed. They all share the basic approach, first proposed by Bethe and Tait,[1] of treating the core materials as a hydrodynamic fluid. Early models also assumed that the material densities remained constant during disassembly to facilitate analytical solutions of the equations.

The first model to provide a direct numerical solution to the multidimensional hydrodynamics equations was the VENUS[6] code developed at Argonne National Laboratory (ANL). This represented a significant advance in capability because the inclusion of an explicit calculation of the material density changes allowed the use of a more accurate density-dependent equation of state. The current version of this code, VENUS-II,[7] will be discussed as an example of the techniques used in the disassembly area.

The hydrodynamics equations are solved in r-z Lagrangian coordinates. Because the grid mesh deforms with the material, mass conservation is merely expressed by

$$\rho = \frac{\rho_o V_o}{V_t} , \tag{5}$$

where ρ and V_t are the density and total volume of the mesh cell and ρ_o and V_o are the initial values.

The radial and axial accelerations of each mesh point are calculated from the momentum equations

$$\ddot{r} = -\frac{1}{\rho}\frac{\partial p}{\partial r} \quad \text{and} \quad \ddot{z} = \frac{1}{\rho}\frac{\partial p}{\partial z} , \tag{6}$$

where r and z are the Eulerian coordinates of a given point in the Lagrangian grid mesh. The pressure, p, is the sum of the pressure obtained from the equation of state and the Von Neumann-Richtmyer viscous pressure.

The momentum equations are solved with an explicit, first-order, finite-difference technique previously used in hydrodynamics codes developed at LASL[27] and in the REXCO damage evaluation codes to be discussed later. In this approach, the positions, velocities, and accelerations are associated with the mesh points, while such quantities as the pressures, densities, and internal energies are associated with the mesh cell centers. A detailed development of the finite-difference equations and solution strategy is presented in Appendix B.

As with any explicit numerical calculation, the time-step size must be controlled properly to ensure a stable and accurate solution. The requirements on the time-step control are made especially stringent because of the density-dependent equation of state used. When single-phase conditions develop in a mesh cell, the resulting pressure is an extremely sensitive function of the internal energy and density.

The influence on the time-step size can be appreciated by considering a time step during which a cell being compressed passes from two-phase to single-phase conditions. As soon as the cell volume is decreased to the point where all the vapor space is eliminated, a large resisting pressure should start developing. Because finite time steps must be used, the usual situation is to reduce the volume somewhat beyond this point. This causes the compression to be overestimated. If the time steps are too large, the resulting pressure in the compressed cell can then be greatly overestimated.

The VENUS-II time-step size is controlled using the stability index

$$w = \frac{c^2}{A} \left(\frac{\Delta t}{1.2}\right)^2 + 4 \left|\frac{\Delta V}{V}\right| , \qquad (7)$$

where c is the speed of sound, A the mesh cell area, V the specific volume, and ΔV is the change in the specific volume during the time step. The time-step size, Δt, is adjusted to keep the maximum value of w for all mesh cells within empirically determined limits.

In general, the time-step size must be kept very small ($< 10\ \mu s$) whenever single-phase conditions are encountered. This does not seriously restrict the analysis of energetic excursions, as the time scales that must be covered are of only a few milliseconds. Fairly long running times are required for mild excursions that involve extended single-phase pressure conditions, however.

A significant limitation of a two-dimensional Lagrangian disassembly model, like VENUS-II, arises because it cannot easily calculate large material displacements. The model becomes inaccurate and eventually unstable when the distortions of the grid mesh grow too large. This restriction is not important in analyzing an initial energetic burst because the power excursion is normally terminated by a relatively small outward displacement of the fuel. Only a few centimeters of displacement occur before the power has decreased to insignificant levels. Although the Lagrangian calculations can be extended to larger displacements using rezoning techniques (this is done in some REXCO calculations), numerous other complications, such as the interaction of the expanding core materials with the surrounding structure, must then be considered.

5. NEUTRONICS

While far more effort has gone into refining numerical methods for steady-state neutronics analysis of fast reactors, neutron kinetics also has been an area of considerable interest. Most fast-reactor accident codes have used point kinetics (in particular, SAS3A and VENUS-II); but more recently, multidimensional kinetics methods are coming into wider use.

As is well known, when material density changes and the resulting flux shape and spectrum changes are small, first-order perturbation theory and point model kinetics are adequate. Generally, point kinetics is sufficiently accurate to predict neutronics behavior during initiating-phase events before substantial fuel motion. Similarly, in a classical core disassembly, large negative reactivities and neutronic shutdown result from relatively small outward motions and, again, perturbation theory and point kinetics are sufficiently accurate (as was assumed in the original Bethe-Tait analysis). Some space-dependent effects and differences in diffusion and transport theory can be found even in the two cases noted, but more important effects occur with the large fuel motions typical of large-scale fuel slumping or dispersal in the initiating phase and of the gross fuel motions occurring in the transition phase.

Two approaches have been taken in adding space-dependent kinetics to fast-reactor accident analysis codes. For one, direct finite-difference methods have been developed (e.g., one based on an explicit exponential extrapolation method to solve the two-dimensional multigroup time-dependent transport equations was included in SIMMER-I). The second approach began with the "improved quasistatic" method of Ott and Meneley[69] and provided the first two-dimensional time-dependent multigroup diffusion equation solution used for fast-reactor accident analysis. This method was used in the FX2 code,[70] which was coupled to the original VENUS code.[71] This particular combination was not very useful because the small displacements allowed by the VENUS Lagrangian treatment did not lead to significant space-dependent neutronics effects.

The quasistatic method factors the space-, energy-, and time-dependent flux (and angular-dependent in the case of transport theory) into a gradually developing time-dependent spatial flux shape and an amplitude function, that is,

$$\psi(x,E,t) = \phi(x,E,t)\ T(t) \ .$$

This form is substituted into the appropriate diffusion or transport equation, is weighted (generally with the stationary adjoint flux), integrated over all independent variables except time, and is normalized on ϕ, such that

$$\iint \frac{\phi^*(x,E,0)\ \phi(x,E,t)}{v}\, dxdE = \text{constant} \ .$$

An ordinary differential equation for T(t) results. This equation has the form of the usual point kinetics equations and specifies how the basic parameters can be calculated from the spatial flux shapes. The flux shapes can be calculated with conventional steady-state flux methods with some source terms added.

Because this method makes maximum use of existing static flux shape calculational methods and because it requires a minimum of full space-dependent flux calculations, the improved quasistatic method has been adopted for both SAS4A (as a two- or three-dimensional multigroup diffusion theory solution) and in SIMMER-II (as either a multigroup diffusion or S_n transport theory two-dimensional solution). In recent applications of SIMMER-II, both full space-dependent kinetics and a transport theory treatment have proved necessary for certain transition-phase problems.[72]

6. GENERAL CORE DISRUPTION AND EXTENDED MOTION -- SIMMER

The ultimate goal of the material motion analysis for a core-disruptive accident is to describe the fuel motion until a permanently subcritical and coolable configuration is attained. This provides initial conditions for damage evaluation and postaccident-heat-removal calculations. In many cases, this involves removing a sizable fraction of the fuel to a position well beyond the immediate core region. For accident sequences that do not proceed directly into a gross mechanical disassembly, this removal process may occur over many seconds and may involve complex interactions with the structure surrounding the core. Because such codes as SAS and VENUS-II were unable to address this problem properly, the development of several new models was instigated.

The limited Lagrangian methods led to the development of the extended material motion codes, such as TWOPOOL and VENUS-III at ANL and SIMMER at LASL. These are all based on the Implicit Continuous Eulerian (ICE) two-fluid method developed by Harlow and Amsden and used in the KACHINA code.[73] VENUS-III and TWOPOOL, with their emphasis on numerical efficiency, are very useful for a class of problems. Because SIMMER-II has a broader range of applicability achieved by including a structure field and all core materials, it will be the subject of the further discussion here.

The SIMMER (S_n, Implicit, Multifield, Multicomponent, Eulerian, Recriticality) development effort was begun in 1974 to provide a general mechanistic code framework to address all the disruptive accident analysis events occurring between those covered by the multichannel codes, such as SAS, and the damage assessment codes such as REXCO. One of the initial objectives was to provide a tool for resolving questions of recriticality[74] following core material expulsion after a mild disassembly. However, the SIMMER modeling framework was made sufficiently general to allow studies of transition phase development.

SIMMER-II is a coupled space- and time-dependent neutronics and multiphase, multicomponent Eulerian fluid dynamics computer code designed to calculate the two-dimensional (r,z) motion of LMFBR core materials during a core-disruptive accident. The neutronics calculation in SIMMER-II employs the quasistatic method to solve either multigroup diffusion or transport equations in two space dimensions (r,z) and in time. As an option, the point kinetics equations may be solved. The fluid dynamics calculation solves multicomponent, multifield coupled equations for mass, momentum, and energy conservation. (A field is a set of materials having the same velocity distribution.) In SIMMER-II all vapor components have one velocity and all liquid components have another velocity at a given space-time point. The two fields are momentum-coupled through drag terms. In addition to the two moving fields, a "structure field" is included to account for still-intact solid materials. Mobile particulate solid material also is allowed in the "liquid" field

Each velocity field contains several components, and each component is described by density and internal energy distributions. To account for different enrichment zones and different degrees of burnup, the fuel material in SIMMER-II can be separated into fertile and fissile components. For example, in an LMFBR fueled with mixed uranium and plutonium dioxide, the fertile component can be approximated as depleted uranium dioxide and the fissile component as plutonium dioxide. These two components are modeled by two distinct density distributions, but, because they are intimately mixed, only one energy distribution is necessary to determine their common local temperature. A control material component that usually is modeled as boron carbide (B_4C) is needed to treat the neutron absorption effect of this material and its thermal interactions with other materials in the reactor core. A fission gas component also is included, not only to provide pressure when it is released from the unrestructured fuel matrix to the vapor field, but also to account for the presence of noncondensibles in the phase transition models. Thus, SIMMER-II has six basic components: fertile fuel, fissile fuel, steel, sodium, control material, and fission gas. A separate accounting must be made of materials that melt and then freeze either as particulates in the liquid fuel or on still solid structural components (the can walls, for example).

Figure 12 summarizes the SIMMER computational and modeling framework. Within each fixed (Eulerian) grid mesh, considerable modeling detail is allowed through the multifield calculational technique; that is, liquid and vapor flow are treated with separate conservation equations and the structural field is used for solid elements (intact fuel, cladding and wire wraps, control rods, and subassembly can walls). Subassembly can walls, if intact, restrict radial motion between cells. In this way, the methodology can encompass a multichannel treatment as with an initiating-phase code such as SAS3A. Radial motion is allowed when can walls are calculated to fail as occurs in a transition phase.

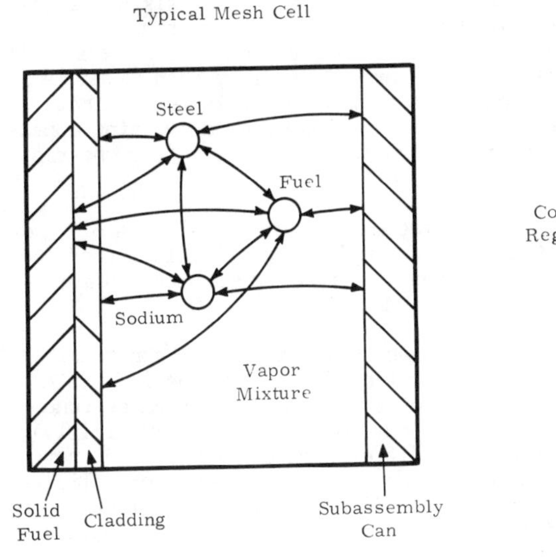

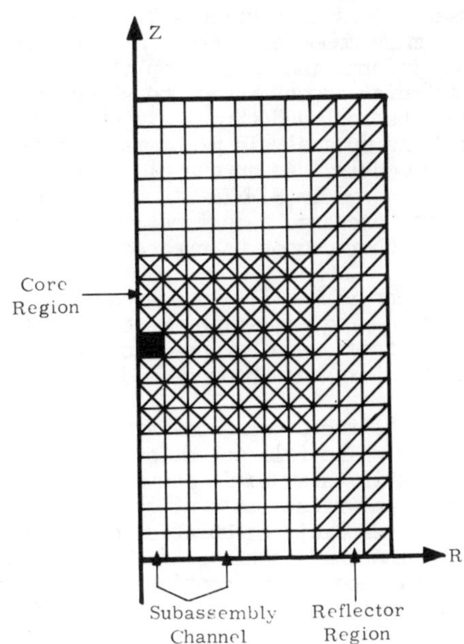

Arrows represent possible mass, energy, and momentum couplings.

Fig. 12. SIMMER modeling framework.

This modeling framework allows the inclusion of energy and momentum exchange between almost any phase (solid, liquid, or vapor) of any component (represented only as sodium, steel, and fuel in Fig. 12) and another phase of the same component or any phase of another component. This is indicated in Fig. 12 by the arrows. The liquid phase is represented here by dispersed droplets in the continuous vapor field, which is a mixture of component vapors. The modeling framework also includes mass exchange between all phases of most components (obvious exceptions are that solid sodium and solid or liquid fission gas are not included). Thus, structural components, including fuel, cladding and can wall steel, and control material can melt, begin to move, and eventually vaporize. Similarly, vapors can condense and liquids can freeze, all as dictated by heat-transfer processes and energy conservation.

The multicomponent field equations include numerous coefficients, generally as analytical functions, which model mass exchange, momentum exchange, and energy exchange among all the components. Usually, these are included in the form of generally accepted engineering correlations, such as heat transfer relations of the form

$$q_{ab} = h_{ab}A_{ab}(T_a - T_b) ,$$

where a and b could represent two phases of the same component (nonequilibrium is allowed), two components in the same phase (except that only one average vapor energy and temperature are used), or two components in two phases (such as liquid steel to solid intact fuel pellets). As is usual with such general relations, considerable care must go into the specific h, A, and ΔT quantities used. This is particularly true in the SIMMER framework, not only because of the many components treated, but also because these terms contain all the modeling information; for example, there is no standard geometry that fixes the interfacial areas. Even the structure field geometry and the resulting areas must be included through input parameters. This leads to some inconvenience in setting up analysis problems, but allows much more flexibility than fixed geometry models. As an example, a dispersed-droplet flow regime is the nominal regime used in SIMMER-II, but other flow regimes can be modeled by using appropriate exchange functions. The major limitation is that changes from one regime to another cannot be treated in a single calculation.

The SIMMER-II fluid dynamics numerical methods are based on those of the KACHINA program, which was the first to use the implicit multifield method. The methods were extended considerably to treat the many components and more complex exchange process models included in SIMMER-II.

The field equations are differenced in space using a partial donor cell method. The full set of spatially differenced time-dependent equations is solved in three basic steps: (1) calculation of the intra-cell exchange functions and rates (most of the intra-cell rate equations are solved implicitly), (2) an explicit calculation of the spatially differenced continuity and energy equations and portions of the momentum equations, and (3) a complete implicit calculation of the continuity and momentum equations (except that momentum convection is treated explicitly), and the pressure (from the equation of state).

For two-phase cells, an important part of the first step is an implicit calculation of the liquid droplet size, based on several criteria including a local Weber number. The implicit iterative solution scheme couples the Weber number criterion, the liquid vapor drag correlation, and truncated versions of the momentum equations. This implicit scheme stabilizes the droplet size calculation. The single-size (for each component in each mesh cell) droplet restriction provides only a crude approximation to the complex flow topologies likely to occur, but the droplet size does yield an interfacial area dependent on the relative velocity between liquid and vapor, which is important to both the energy and momentum exchange processes. The phase transition rates, also dependent on droplet size, are also calculated partially in an implicit manner, although not in the pressure iteration involved in the third step of the fluid-dynamics solution.

Step two, the explicit part of the fluid dynamics calculation, provides a relatively straightforward first solution of the continuity and energy equations. The third step is similar to the pressure iteration used in KACHINA although modified somewhat. An explicit step is taken on the momentum equations to calculate first-iteration advanced-time-step momentum fluxes. An iteration on the advanced-time pressure, density, and velocity distribution is made using the continuity and momentum equations and an equation of state (there are both analytical and tabular options included for the various equations of state needed). The momentum convection terms are included explicitly to avoid the full nonlinearity.

Numerous time-step controls exist in SIMMER-II, including those to account for rapid phase transitions. Because momentum convection is treated explicitly, a Courant limitation on material velocity (but not sound speed) results. That is, material cannot be convected more than one mesh in a single time step, but there is no similar restriction on pressure waves.

7. CONTAINMENT EVALUATION METHODS

The Reactor EXcursion COntainment (REXCO) codes were developed to analyze the response of LMFBR primary systems to core-disruptive accidents. An early version of REXCO-H[75] provided only a two-dimensional hydrodynamics analysis. A later version, REXCO-HEP,[10] also included the elastic and plastic deformations of the reactor structures and components.

The equations of motion are treated with the same two-dimensional Lagrangian formulation discussed earlier for the VENUS-II code. A main difference is in the equation of state, as the REXCO codes deal principally with sodium as a working fluid. In addition, an Eulerian formulation was developed to facilitate the calculation of larger degrees of material motion. This code, ICECO,[76] also is based on the ICE[51] formulation. The convective terms in the mass equation and the pressure gradients in the momentum equation are evaluated at advanced times. Elimination of the mass-flux between the mass and momentum equations, in conjunction with an equation of state relating density and pressure, reduces this set to a Poisson-type equation.

A newly developed technique[77] for adapting ICECO to handle moving boundaries (which is straightforward with the Lagrangian codes) treats the reactor vessel boundary as a locus of nodal points as in REXCO-HEP. Movement of these points may be the result, for example, of fluid impact that deforms the vessel boundary. The movement of this Lagrangian boundary then is mapped onto the fixed Eulerian grid, providing the regions at which the hydrodynamic boundary conditions must be satisfied.

8. VERIFICATION OF WHOLE-CORE ANALYSIS CODES

The many models discussed above have been tested against experimental data using several different approaches. The SAS code has not been tested in its full form; instead, the separate models were compared to data from integral experiments performed with a few (1, 3, or 7) test pins in the TREAT reactor or with 19 and 37 test pins in the Sodium Loop Safety Facility (SLSF) loop in the Engineering Test Reactor. Generally, the goal has been to achieve reasonable qualitative agreement with the test data, particularly regarding cladding and fuel motion. For the sodium boiling model, quantitative agreement of voiding rate has been achieved both for out-of-pile tests in the OPERA loop and in-pile R-Series 7-pin tests in TREAT.[47] Agreement with 19-pin LOF tests in the SLSF loop in the ETR is also good.[48,49]

No time-dependent quantitative data are available for cladding relocation. However, the main objective of cladding relocation models is to predict the timing and location of blockage formation. As noted earlier, a reasonably successful test was achieved with the original CLAZAS model when it predicted upper cladding blockages in the TREAT L2 test. However, the CLAZAS model tended to overpredict upper blockage in the longer pin R-Series experiments. The newer CLAP model to be released in SAS4A predicts smaller upper blockages, in better agreement with the tests.[5]

Verification of SLUMPY has been more difficult, as the only fuel motion data available are those from the fast neutron hodoscope at the TREAT reactor. SLUMPY calculations have been tuned to fit the relatively rapid eructations of fuel observed by the hodoscope in a number of TREAT tests. The first example of this was the comparison noted earlier of a SLUMPY calculation of the TREAT L4 experiment. In these calculations the cladding was assumed to remain on the fuel and was mixed with the fuel after motion was initiated. The heat-transfer assumptions were such that steel vapor pressures began moving fuel at about the time of the observed eructation. In this case, the final configuration of the fuel was qualitatively similar to that observed in posttest neutron radiographs.

Despite some successes in fitting experimental data, the large uncertainties in the fuel breakup, fission gas release, and heat-transfer parameters in SLUMPY do not allow a unique set of parameters for reliable extrapolation to other test conditions. The new LEVITATE model, which is much more mechanistic than SLUMPY, is being tested against TREAT experiments and may permit such extrapolations. Significant data base uncertainties still exist in defining timing and modes of fuel pellet breakup and of fission gas release, however.

SIMMER code verification has proceeded with a completely different approach. Because no reactor material tests are available on relevant SIMMER regimes, many simulant materials tests have been analyzed. As noted earlier, the SIMMER calculational formalism is different from that of codes like SAS. SIMMER solves the conservation equations at each mesh cell at every time step. There are no special subroutines for computing specific phenomena; for example, the equation set used to compute heat transfer from a fuel pin to liquid sodium in a channel also is used to compute liquid-fuel to liquid-sodium heat transfer. Only the exchange coefficients vary. Because no specialized subroutines compute such things as freezing, vaporization, cladding motion, fuel motion, or fuel-coolant interactions, SIMMER verification rests principally on proper representation of the exchange functions (heat transfer and drag coefficients, for example). These exchange functions are obtained from engineering correlations that are applicable only over specific limited ranges; thus SIMMER can best be verified only over those specific ranges. This conclusion has led to testing SIMMER only for specific accident problems.

The initial SIMMER verification effort has been on extended expansions resulting from assumed core disassemblies. This emphasis came about from the SIMMER results[78] indicating that the system kinetic energy following a prompt-critical burst in the CRBR would be reduced many fold from an ideal isentropic expansion. For example, in CRBR, a kinetic energy of about 100 MJ is associated with an isentropic expansion to the cover-gas volume following a voided-core disassembly driven by about a $75/s ramprate. SIMMER calculates a system kinetic energy at the cover-gas volume of about 5 MJ. This dramatic reduction in kinetic energy is caused by both fluid dynamics effects and rate-controlled phenomena.

SIMMER-II was used to analyze scale expansion experiments performed at the Stanford Research Institute (SRI)[79] to test its ability to calculate the hydrodynamics of bubble expansions. After some minor numerical method modifications, SIMMER-II predicted the experimental results with excellent accuracy.[80] Similar analyses of Purdue University OMEGA nitrogen bubble expansion experiments[81] and of COVA experiments[82] show good agreement except that shock fronts from high pressure (about 100 MPa and higher) expansions in the COVA experiments are smeared by the SIMMER Eulerian hydrodynamics formulation.

SIMMER also has been used to investigate rather fundamental fluid dynamics problems including single-phase shock propagation, entrance-length effects in laminar flow, and steady laminar flow. (These results have been documented in Ref. 83.) In general, the results of these basic physics calculations are excellent over the fluid dynamics regime from laminar to supersonic flow.

Based on all of these analyses, the SIMMER hydrodynamics formulation is felt to be verified[84] for the expansion-type problems, but verification of the rate effects in expansion problems is incomplete. Recently, SIMMER has been used to calculate several experiments having significant rate effects. These include Oak Ridge National Laboratory (ORNL) FAST/CRI experiments,[85] the SRI 1/30-scale flashing-source experiment,[80] Sandia prompt-burst excursion (PBE) experiments,[86] and the Purdue OMEGA flashing-source experiments.[87] In the last case, good agreement could be obtained only when a nucleation-site density model was added. In most other cases, the comparisons are sufficiently reasonable (given limited experimental data) that model changes are not justified.

Overall, SIMMER has been quite thoroughly tested for the bubble expansion problems. One remaining problem with the SIMMER modeling framework is that the Eulerian solution leads to "numerical" mixing. The mixing effect is dependent on mesh-cell size and is exaggerated when large mesh cells are used because the necessary averaging of quantities within the cells tends to equilibrate temperatures too rapidly (or at least nonmechanistically). This effect may not be significant in the small-scale experiment analyses because of the necessarily small mesh cells used, but it could be a factor in extrapolating "verified" rate effects models to full scale. Testing for other than expansion problems, such as those of the transition phase, has been very limited. One exception is that SIMMER was used to calculate,[88] relatively successfully, the TREAT R7 test for the full span of sodium boiling, cladding relocation, and fuel motion.

Verification of disassembly codes has been complicated by the fact that no disassembly transient experiments have been performed on LMFBRs. There have been, however, a limited number of disassembly tests performed on other reactor types. Although these reactors differed markedly from an LMFBR, the data obtained have provided a basis for experimental validation of LMFBR disassembly codes. The most applicable of these tests are the KIWI-TNT,[89] SNAPTRAN-2,[90] and SNAPTRAN-3[91] experiments.

The SNAPTRAN experiments were performed on SNAP 2/10A reactors. These reactors had small zirconium-hydride cores with fully enriched uranium. The SNAPTRAN-2 test was performed in the open atmosphere with no coolant present in the core. Rapid reactivity insertion was achieved by modifying the normal beryllium control-drum drive mechanisms to allow for extremely high-speed rotation. A total reactivity of $4.91 was inserted. In the SNAPTRAN-3 experiment, the beryllium reflector/control assembly was replaced by a neutron-absorbing binal sleeve. The core-sleeve assembly was then placed in a large tank of water. A prompt-critical excursion was initiated by the rapid removal of the sleeve, with the consequental increase in neutron reflection. A net reactivity of $3.60 was achieved upon sleeve removal. The core was full of NaK coolant during this test.

The KIWI-TNT test was the most energetic disassembly experiment available. It was performed on a modified KIWI-B-4E reactor. The core was fueled with fully enriched uranium-dicarbide beads dispersed in graphite assemblies. The excursion was initiated by the rapid rotation of beryllium control drums that resulted in a total reactivity insertion of $8.30. The test was performed with no coolant present in the core.

All three of these tests were analyzed with the VENUS-II code.[92] To apply the code to these reactors, modifications were made to the equation of state and fuel heat-capacity functions to account for the core materials not normally found in LMFBRs. All other aspects of the code, including hydrodynamics and point kinetics calculations, were used without modification.

The VENUS-II code accurately predicted the power history and total fission energy release in these three experiments. The most important calculated results, the total fission energy releases, are compared to the measured values in Table I. The calculated fission energies for the two SNAPTRAN tests fell within the uncertainty range of the experimental data, but the code overpredicted the KIWI-TNT experiment by 25%. Additional analysis of KIWI-TNT, however, showed this discrepancy could readily be explained by heat-transfer effects associated with the reactor's beaded fuel (a complication not present in an LMFBR).

Numerous parametric calculations also were performed that demonstrated the relative insensitivity of the VENUS-II results to reasonable uncertainties in the key parameters. In the SNAPTRAN experiments, the most sensitive parameters were those related to the temperature feedback effects that were dominant in these excursions. This is similar to the situation found in energetic LMFBR excursions where Doppler feedback is usually the dominant effect.

Barring unforeseen feedback effects, the code should be even more accurate when applied to energetic LMFBR transients.[93] This is because the key parameters (heat capacity, temperature coefficient, and equation of state) are relatively better known for an LMFBR. Thus, the good agreement observed in these experimental comparisons lends considerable confidence to VENUS-II LMFBR disassembly analyses.

The REXCO codes have received thorough testing of both hydrodynamics and structural methods by comparisons to FFTF scale model experiments performed at SRI[94] and to SNR-300 scale model experiments performed at Ispra.[95] Figure 13 shows REXCO- and ICECO-calculated plastic deformations for a thin vessel Ispra experiment. This indicates quite good agreement, and similarly good agreement was obtained in the other comparisons noted. REXCO- and ICECO-calculated hydrodynamics parameters (pressures, fluid impulses) both show good agreement, which is interesting because of the large difference in the numerical formulations, REXCO being an explicit Lagrangian code and ICECO being a semi-implicit Eulerian code.

Overall, experimental verification of HCDA analysis codes has proceeded slowly. Generally, hydrodynamics and structural dynamics models dealing with relatively clearly defined problems, such as sodium boiling and damage consequences (such as vessel deformation), have fared well in experimental comparisons. In more complex phenomenological situations, such as fuel pin disruption, combined fuel and cladding motion, and transition phase dynamics, treatment of complicated and changing geometries and multicomponent multiphase flows makes mechanistic modeling much more difficult. This, coupled with a lack of adequate data from in-reactor experiments and the near-impossibility of performing multicomponent simulant experiments with internal heat generation, means that

TABLE I

COMPARISON OF VENUS-II CALCULATIONS WITH EXPERIMENTAL DATA

Test	Experimental Energy Release (MW·s)	Calculated Energy Release (MW·s)
SNAPTRAN-2	38.0 - 47.3	47.0
SNAPTRAN-3	32.7 - 61.3	33.7
KIWI-TNT	8090.0 - 9889.0	12350.0

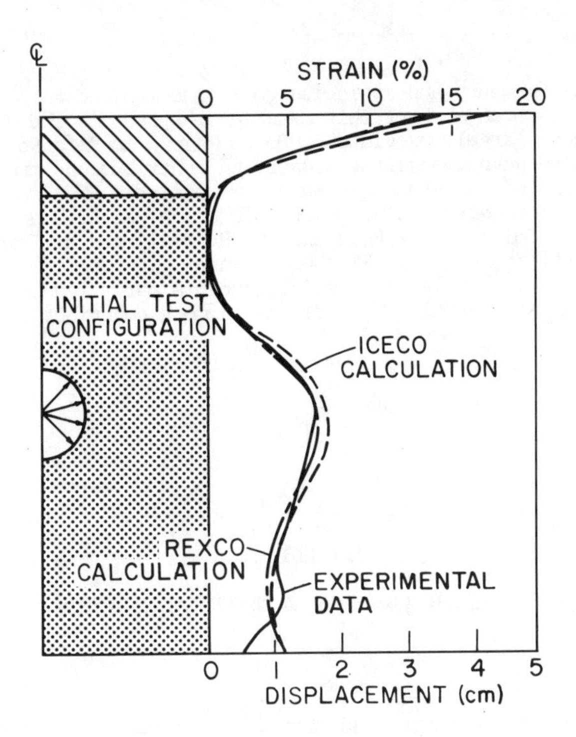

Fig. 13. REXCO/ICECO comparisons for deformations of a thin test vessel.

total verification is improbable for some time. However, most of the modeling is at least technically credible, and the codes still can be used to determine the sensitivity of important results to modeling uncertainties. Also, some conclusions on cladding or fuel motions based on specific analyses can be verified by integral reactivity measurements, such as in the SLSF loop, rather than by verifying the modeling itself. Of course, care must be taken in extrapolating such conclusions to the full-scale reactor case.

9. CONCLUSIONS

The combination of SAS3D or SAS4A, SIMMER-II, and REXCO or ICECO, as representing the state of the art in HCDA analysis techniques, generally can provide a reasonable treatment of complete HCDA sequences. Other codes, developed in the US or in other countries, also can fit in this combination of state-of-the-art codes; but SAS, SIMMER, and REXCO provide at least one set of detailed codes for straight-through calculations of most disruptive accident sequences. Much remains to be done in improving models and computational efficiencies and in verifying the codes, but defensible analyses are feasible.

ACKNOWLEDGEMENTS

Our acknowledgements and apologies go to the many co-authors on our previous papers from which much of this material was drawn. The descriptions of numerical methods, largely provided by D. P. Weber in Ref. 96, were particularly valuable. We also acknowledge the pioneering contributions made by key people in the development of the state-of-the-art mechanistic HCDA analysis techniques discussed here. Some of these are R. B. Nicholson, F. E. Dunn, T. J. Heames, W. R. Bohl, L. L. Smith, J. E. Boudreau, C. R. Bell, and Y. Chang. Most importantly, the efforts of G. J. Fischer in instilling the idea that mechanistic analysis is both worthwhile and possible and his early leadership in the development of analysis techniques, were an enormous contribution.

APPENDIX A

SOME BASIC CONCEPTS IN NUMERICAL ANALYSIS

A.1 INTRODUCTION

Several numerical concepts and methods are referred to in the reactor safety computer code descriptions in Chapters 13 and 19. This appendix provides introductory background material on some of the main concepts.

A.2 REVIEW OF BASIC CONCEPTS

Finite difference equations. Finite difference techniques are used to transform differential equations into coupled sets of algebraic equations that can be solved with the standard arithmetic and logical operations available in digital computers. In carrying this out, a network of mesh (or grid) points is established throughout the domain occupied by the independent variables. The variables and equation solution then are defined at the mesh points or at interpolated positions between them. The equations then are written for these mesh points with the differential operators approximated by difference expressions involving the variables at neighboring mesh points.

Eulerian vs Lagrangian coordinates. Fluid dynamics equations can be solved numerically in two different types of grid mesh coordinate systems.[97] In the Lagrangian approach, the grid mesh is assumed to move with the fluid. Thus, a given mesh cell contains the same fluid throughout the calculation. This approach is depicted in Fig. A.1 where we illustrate the deformation that might result from an applied central pressure. As long as the displacements are relatively small, this approach can work very well. As displacements become large, however, the mesh cells can become highly distorted. In fact, neighboring mesh points can even cross over one another, resulting in mesh cells with an "hourglass" shape. When these large distortions occur, the results typically are inaccurate. This difficulty can be circumvented somewhat by techniques that periodically regularize the grid. These technqiues are generally quite complicated and introduce errors into the calculation each time they are applied.

The use of Lagrangian coordinates can offer distinct advantages for certain types of small-displacement hydrodynamics calculations, however. For example, material interfaces can be readily retained. Also, the mathematical formulation is quite simple. This is illustrated in the development of the VENUS disassembly code finite difference equations presented in Appendix B of this chapter. The problem solved by VENUS is well suited to Lagrangian coordinates because core material displacements of only a few centimeters are typically required to terminate a prompt-critical neutronics excursion. The Lagrangian coordinate approach also was used in the early versions of the REXCO damage evaluation code. A later version of the code incorporated mesh cell renormalization to facilitate the analysis of larger displacements.

In the Eulerian approach, the grid mesh is fixed in space and the fluid is allowed to move from one mesh cell to another in response to pressure gradients. This is illustrated in Fig. A.2. This type of approach is much better suited to hydrodynamics problems where large material displacements are calculated. For example, it has been used exclusively in the LWR systems codes discussed in Chapter 13. All of the more recent LMFBR codes discussed in this chapter are also Eulerian.

The Eulerian approach does have some inherent problems. For example, interfaces are difficult to resolve precisely. There is also a tendency for artificial material diffusion. As material diffuses into a mesh cell, it is immediately homogenized throughout the cell (in basic Eulerian treatments). It

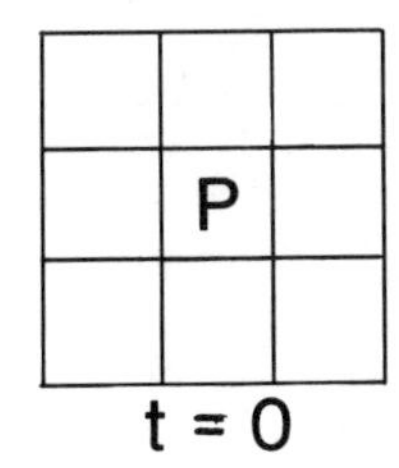

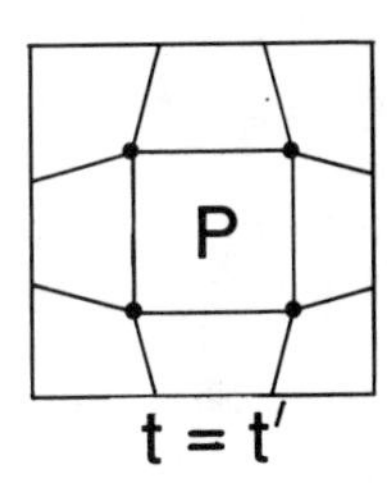

Fig. A-1. Illustration of Lagrangian coordinate approach.

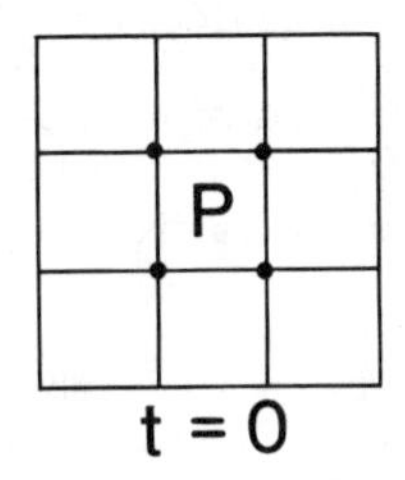

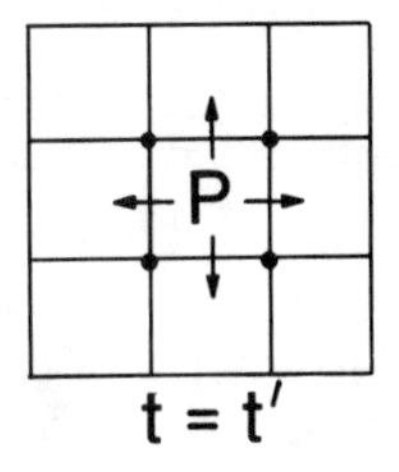

Fig. A-2. Illustration of Eulerian coordinate approach.

can then start diffusing through the other mesh cell boundaries during the next time step. This can result in small amounts of material diffusing through the grid much more rapidly than the material can physically move.

Some more complicated schemes have been developed that try to combine the advantages of both approaches. In one such scheme, the Particle-in-Cell (PIC) method,[98] the continuous fluid is replaced by a set of discrete Lagrangian mass points called particles. The motion of these particles through a fixed Eulerian mesh is calculated.

Explicit vs implicit. To illustrate the difference between explicit and implicit differencing schemes, consider the following parabolic equation[99]

$$\alpha \frac{\partial^2 u(x,t)}{\partial x^2} = \frac{\partial u(x,t)}{\partial t} \quad \alpha = \text{constant}. \qquad \text{(A-1)}$$

If we wish to solve for $u(x,t)$ numerically, we can superimpose a grid mesh with grid line indices of j and k in the t and x directions, respectively. This notation is illustrated in Fig. A.3. Let the solution at point (j,k) be denoted as $u_{j,k}$.

Suppose we have the solution up to t_j and we now wish to advance it to t_{j+1}. To do this we must put Eq. (A-1) into finite difference form. One way to do this is to start at point (i,j) and use the following central difference expression for $\partial^2 u/\partial x^2$:

$$\frac{\partial^2 u}{\partial x^2} \cong \left[\frac{u_{j,k+1} - u_{j,k}}{\Delta x} - \frac{u_{j,k} - u_{j,k-1}}{\Delta x}\right] \bullet \frac{1}{\Delta x} = \frac{u_{j,k+1} - 2u_{j,k} + u_{j,k-1}}{\Delta x^2} .$$

If we use a simple forward difference for the time derivative, we obtain

$$\frac{\partial u}{\partial t} \cong \frac{u_{j+1,k} - u_{j,k}}{\Delta t} .$$

Equation (A-1) then becomes

$$\frac{u_{i,k+1} - 2u_{j,k} + u_{j,k-1}}{(\Delta x)^2} = \frac{1}{\alpha}\left(\frac{u_{j+1,k} - u_{j,k}}{\Delta t}\right) . \qquad \text{(A-2)}$$

This expression involves the points in the grid mesh depicted in Fig. A.3.

All the quantities with subscript j are known; therefore the only unknown is $u_{j+1,k}$ (which is point D in Fig. A.3). Thus, Eq. (A-2) can be explicitly solved for $u_{j+1,k}$. This type of differencing approach is therefore referred to as explicit.

The main difficulty with explicit techniques is one of stability. For example, it can be shown that solutions obtained with Eq. (A-2) will only remain stable if

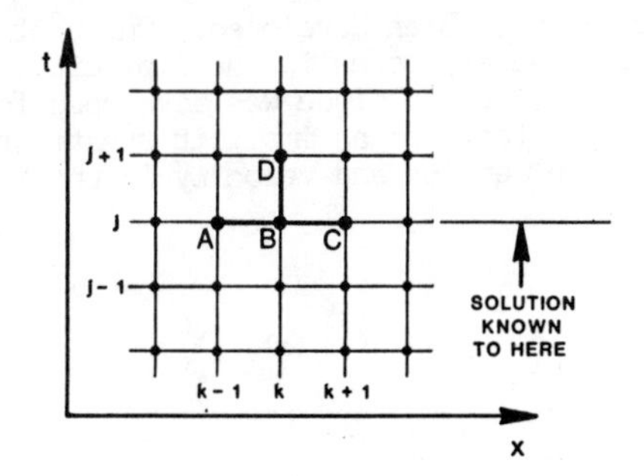

Fig. A-3. Grid mesh representing explicit differencing approach.

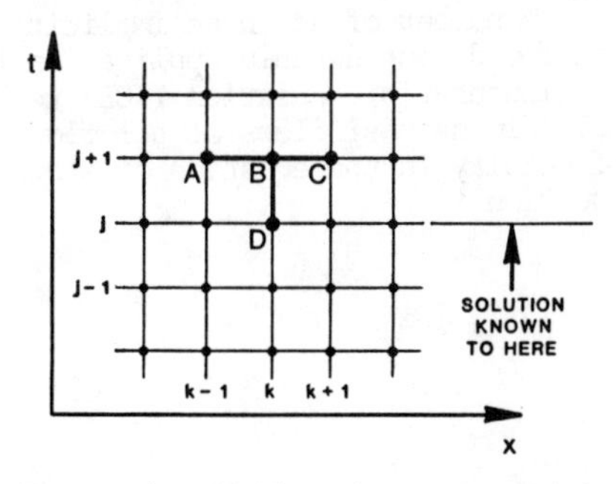

Fig. A-4. Grid mesh representing implicit differencing approach.

$$\Delta t < \frac{(\Delta x)^2}{2\alpha} \ .$$

Typically, this criterion forces the time-step size to be much smaller than would be required to control the truncation error. Thus, although explicit techniques are computationally simple, the restriction on time-step size only makes them practical for problems that cover short-time intervals. An explicit approach can be successfully used in reactor disassembly codes such as VENUS, because a typical disassembly calculation only covers a time interval of several milliseconds or less.

Now consider a new difference representation that is centered on point (j+1,k). If $\partial^2 u/\partial x^2$ is approximated again with a central difference expression, but $\partial u/\partial t$ is now represented by a backward difference expression, the result is

$$\frac{u_{j+1,k+1} - 2u_{j+1,k} + u_{j+1,k-1}}{(\Delta x)^2} = \frac{1}{\alpha} \frac{(u_{j+1,k} - u_{j,k})}{\Delta t} \ . \qquad \text{(A-3)}$$

The points involved in this difference scheme are shown in Fig. A.4. Now the values of u at points A, B, and C are all unknown. Although Eq. (A-3) cannot be solved explicitly for these unknowns, it can be written for all of the points k-1, 2, ..., n, resulting in n linear equations in the n unknowns $u_{j+1,k}$. These equations can then be solved with standard numerical matrix techniques. Such an approach is said to be implicit.

Although computationally more complex, implicit techniques can typically tolerate much larger time-step sizes than explicit techniques. In fact, most purely implicit formulations can be shown to be stable for any time-step size. In the semi-implicit approaches commonly used in reactor safety codes[100] (these usually use a donor-cell technique where some of the differencing depends on the direction of the flow), the time-step size must satisfy the so-called Courant condition,

$$\Delta t < \frac{\Delta x}{|v|\Delta t} \ ,$$

where v is the material velocity. This is much less restrictive than is obtained for explicit approaches.

A number of advanced implicit techniques have been developed. One that has found considerable application in reactor safety codes is the Implicit Continuous-fluid Eulerian (ICE) technique.[51] This technique was developed for multidimensional flows at all flow speeds. It features an implicit treatment of density in the equation of state and of the density and velocity in the mass equation.

APPENDIX B

HYDRODYNAMICS DIFFERENCE EQUATIONS AND SOLUTION STRATEGY FOR VENUS AND REXCO CODES

B.1 INTRODUCTION

This appendix gives a detailed example of the different equations and solution strategy used in a reactor safety computer code. The appendix in Chapter 13 discusses an example of an implicit-Eulerian technique, so we have chosen to develop an explicit-Lagrangian example in this appendix. For this purpose, we will discuss the numerical techniques used to solve the hydrodynamics equations in the original VENUS (disassembly) and REXCO (damage evaluation) codes. This method was previously developed and used at LASL.[98] The development prescribed here will follow that given in Ref. 101.

B.2 DIFFERENTIAL EQUATIONS

As the hydrodynamics equations are solved in Lagrangian coordinates, mass conservation is expressed merely by

$$\rho = \frac{\rho_o V_o}{V_t} , \tag{B-1}$$

where ρ and V_t are the density and total volume of a given mesh cell and ρ_o and V_o are the initial values. Equation (B-1) simply follows from the fact that each mesh cell contains the same material throughout the calculation.

The equations are solved in two-dimensional (r-z) geometry resulting in the following momentum equations:

$$\ddot{r} = -\frac{1}{\rho}\frac{\partial P}{\partial r} \quad \text{and} \quad \ddot{z} = \frac{1}{\rho}\frac{\partial P}{\partial z} . \tag{B-2}$$

Here, r and z are the Eulerian coordinates of a given point in the Lagrangian grid mesh, P is the pressure, and the dots refer to time differentiation. Thus, these equations give the acceleration of the Lagrangian grid points that result from appropriately averaged pressure gradients.

B.2 DIFFERENCE EQUATIONS

In developing the difference equations, the notation depicted in Fig. B.1 will be used. Because the Lagrangian grid moves with the fluid, it will become distorted as illustrated in the figure. The mesh points (the intersections of Lagrangian grid lines) are identified by a pair of integers (I,J). The mesh cells are identified by the integers of the lower left mesh point (in the undeformed Lagrangian meshes); for example, the mesh zone ODHA is identified as zone (I,J). Positions, velocities, and accelerations are assumed to be associated with the mesh points; they are identified by the mesh-point integers as subscripts. Specific volumes, densities and pressures, strains, and internal energies are assumed to be associated with the cells; they are identified by the mesh cell numbers as subscripts. The times t, $t + \Delta t$, and $t + \Delta t/2$ are identified by superscripts n, n + 1, and $n + \frac{1}{2}$. The latter also is used to identify the change in a quantity between t^n and t^{n+1}. Where appearing, the superscript 0 denotes the initial value of a quantity at the start of the problem.

In the Lagrangian formulation, mass conservation is automatically satisfied. However, a mass equation is needed for the computation of the new density ρ, which is used in the momentum equations and in the equation of state. This computation is performed as follows:

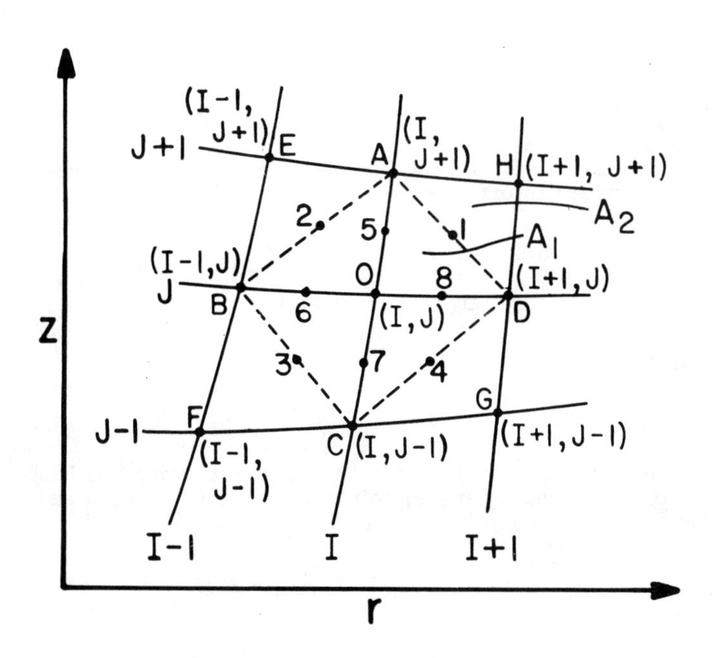

Fig. B-1. A typical mesh point (I,J) with the adjacent zones and points, illustrating the notation used in the finite-difference equations.

The area of the deformed mesh ODHA is the sum of two triangles AOD and AHD:

$$A_{I,J} = A_1 + A_2 \ ,$$

where

$$A_1 = \frac{1}{2}|\overline{OD} \times \overline{OA}| = \frac{1}{2}[(r_{I+1,J} - r_{I,J})(z_{I,J+1} - a_{I,J})$$

$$- (r_{I,J+1} - r_{I,J})(z_{I+1,J} - z_{I,J})] \ , \quad \text{(B-3)}$$

and

$$A_2 = \frac{1}{2}|\overline{HA} \times \overline{HD}| = \frac{1}{2}[(r_{I,J+1} - r_{I+1,J+1})(z_{I+1,J} - z_{I+1,J+1})$$

$$- (r_{I+1,J} - r_{I+1,J+1})(z_{I,J+1} - z_{I+1,J+1})] \ . \quad \text{(B-4)}$$

Substitution of Eqs. (B-1) and (B-2) for A_1 and A_2 gives

$$A_{I,J} = \frac{1}{2} (z_{I+1,J+1} - z_{I,J})(r_{I+1,J} - r_{I,J+1})$$

$$- (r_{I+1,J+1} - r_{I,J})(z_{I+1,J} - z_{I,J+1})] \ . \quad \text{(B-5)}$$

The volume V of a quadrilateral is calculated from

$$V_{I,J} = A_{I,J} 2\pi \bar{r}_{I,J} \ , \quad \text{(B-6)}$$

where $\bar{r}_{I,J}$ is the radius of the centroid of the area $A_{I,J}$. A reasonable approximation for moderate distortions is

$$\bar{r}_{I,J} = \frac{1}{4}(r_{I,J} + r_{I,J+1} + r_{I+1,J} + r_{I+1,J+1}) \ . \quad \text{(B-7)}$$

Thus, the finite difference form of Eq. (B-1) is just

$$\rho^n_{I,J} = \frac{(\rho_0 V_0)_{I,J}}{V^n_{I,J}} \ . \quad \text{(B-8)}$$

The momentum equations are put into finite-difference using the Midpoint Method. Writing a Taylor's expansion between points 0 and 5, 0 and 6, 0 and 7, and 0 and 8, yields four equations of the form

$$P_5 = P_0 + (z_5 - z_0)\frac{\partial P}{\partial z} + (r_5 - r_0)\frac{\partial P}{\partial r} + \frac{1}{2}(z_5 - z_0)^2 \frac{\partial^2 P}{\partial z^2}$$

$$+ (z_5 - z_0)(r_5 - r_0)\frac{\partial^2 P}{\partial z \partial r} + \frac{1}{2}(r_5 - r_0)^2 \frac{\partial^2 P}{\partial r^2} + \dots \ ,$$

where terms of third and higher order in $(z_5 - z_0)$ and $(r_5 - r_0)$ have been omitted. The system is overdetermined for $\partial P/\partial z$ and $\partial P/\partial r$. One method of removing the overdeterminacy is to solve first for $(P_5 - P_7)$ and $(P_6 - P_8)$, which gives

$$P_5 - P_7 = (z_5 - z_7)\frac{\partial P}{\partial z} + (r_5 - r_7)\frac{\partial P}{\partial r} + \frac{\partial^2 P}{\partial z^2}(z_5 - z_7)\delta_{z57} + \frac{\partial^2 P}{\partial r^2}(r_5 - r_7)\delta_{r57} + \frac{\partial^2 P}{\partial z \partial r}[(z_5 - z_7)\delta_{r57} + (r_5 - r_7)\delta_{z57}] + \ldots$$

and

$$P_6 - P_8 = (z_6 - z_8)\frac{\partial P}{\partial z} + (r_6 - r_8)\frac{\partial P}{\partial r} + \frac{\partial^2 P}{\partial z^2}(z_6 - z_8)\delta_{z68} + \frac{\partial^2 P}{\partial r^2}(r_6 - r_8)\delta_{r68} + \frac{\partial^2 P}{\partial z \partial r}[(z_6 - z_8)\delta_{r68} + (r_6 - r_8)\delta_{z68}] + \ldots ,$$

where

$$\delta_{z57} = \frac{1}{2}(z_5 + z_7) - z_0, \text{ etc.},$$

are measures of the asymmetry of the mesh. Second- and higher-order terms in the mesh size -- i.e., $(z_5 - z_7)$, $(r_5 - r_7)$, etc. -- are omitted.

The above equations are then solved for the gradients

$$\left.\begin{aligned} \frac{\partial P}{\partial z} &= \frac{1}{2A_3}[(P_5 - P_7)(r_6 - r_8) - (P_6 - P_8)(r_5 - r_7)] + R_z \\ &\text{and} \\ \frac{\partial P}{\partial r} &= \frac{-1}{2A_3}[(P_5 - P_7)(z_6 - z_8) - (P_6 - P_8)(z_5 - z_7)] + R_r \end{aligned}\right\} , \qquad \text{(B-9)}$$

where

$$A_3 = \frac{1}{2}[(z_5 - z_7)(r_6 - r_8) - (z_6 - z_8)(r_5 - r_7)]$$

represents the area of the quadrilateral 5678. The remaining terms, R_z and R_r, involve products of the mesh size and the δ's. If the mesh is nearly symmetric, these terms are negligible compared to those that have been retained.

On substituting Eq. (B-9) into the momentum equations, neglecting the remainder terms, and writing

$$P_5 = \frac{1}{2}(P_{I,J} + P_{I-1,J}), \quad P_6 = \frac{1}{2}(P_{I-1,J} + P_{I-1,J-1}),$$

$$P_7 = \frac{1}{2}(P_{I-1,J-1} + P_{I,J-1}), \quad P_8 = \frac{1}{2}(P_{I,J} + P_{I,J-1}),$$

$$r_5 = \frac{1}{2}(r_{I,J+1} + r_{I,J}), \quad r_6 = \frac{1}{2}(r_{I,J} + r_{I-1,J}),$$

$$r_7 = \frac{1}{2}(r_{I,J} + r_{I,J-1}), \quad r_8 = \frac{1}{2}(r_{I+1,J} + r_{I,J}),$$

$$z_5 = \frac{1}{2}(z_{I,J+1} + z_{I,J}), \quad z_6 = \frac{1}{2}(z_{I,J} + z_{I-1,J}),$$

$$z_7 = \frac{1}{2}(z_{I,J} + z_{I,J-1}), \text{ and } z_8 = \frac{1}{2}(z_{I,J} + z_{I+1,J}),$$

we obtain

$$\left.\begin{aligned} \ddot{r}_{I,J} &= \frac{-1}{(A\rho)_{I,J}}[(P_{I,J} - P_{I-1,J-1})(z_{I,J+1} - z_{I,J-1} \\ &\quad + z_{I-1,J} - z_{I+1,J}) - (P_{I-1,J} - P_{I,J-1}) \\ &\quad (z_{I,J+1} - z_{I,J-1} + z_{I+1,J} - z_{I-1,J})] \\ \text{and} \quad & \\ \ddot{z}_{I,J} &= \frac{1}{(A\rho)_{I,J}}[(P_{I,J} - P_{I-1,J-1})(r_{I,J+1} - r_{I,J-1} \\ &\quad + r_{I-1,J} - r_{I+1,J}) - (P_{I-1,J} - P_{I,J-1}) \\ &\quad (r_{I,J+1} - r_{I,J-1} + r_{I+1,J} - r_{I-1,J})] \end{aligned}\right\} , \qquad \text{(B-10)}$$

where

$$(A\rho)_{I,J} = A_{I,J}\rho_{I,J} + A_{I-1,J}\rho_{I-1,J} + A_{I-1,J-1}\rho_{I-1,J-1} + A_{I,J-1}\rho_{I,J-1} .$$

No time index is specified in Eq. (B-10), the implication being that all variables are prescribed for the same time.

B.4 SOLUTION PROCEDURE

It is assumed that all calculations are performed up through cycle $n(t = t^n)$, so that for all points and zones there are available the values of r^n, z^n, $\dot{r}^{n-\frac{1}{2}}$, $\dot{z}^{n-\frac{1}{2}}$, A^n, V^n, and p^n.

Next, the accelerations at time n are explicitly calculated from Eq. (B-10). These accelerations are then used to advance the velocities of all points to time $n + \frac{1}{2}$,

$$\dot{r}^{n+\frac{1}{2}} = \dot{r}^{n-\frac{1}{2}} + \ddot{r}^{n} \left[\tfrac{1}{2}(\Delta t^{n+\frac{1}{2}} + \Delta t^{n-\frac{1}{2}})\right]$$

and , (B-11)

$$\dot{z}^{n+\frac{1}{2}} = \dot{z}^{n-\frac{1}{2}} + \ddot{z}^{n} \left[\tfrac{1}{2}(\Delta t^{n+\frac{1}{2}} + \Delta t^{n-\frac{1}{2}})\right]$$

where

$$\Delta t^{n+\frac{1}{2}} = t^{n+1} - t^{n}$$

and

$$\Delta t^{n-\frac{1}{2}} = t^{n} - t^{n-1} \quad .$$

These velocities, in turn, are used to advance the coordinates of all points to time n + 1,

$$r^{n+1} = r^{n} + \dot{r}^{n+\frac{1}{2}} \, \Delta t^{n+\frac{1}{2}}$$

and . (B-12)

$$z^{n+1} = z^{n} + \dot{z}^{n+\frac{1}{2}} \, \Delta t^{n+\frac{1}{2}}$$

Once the coordinate positions have been advanced, the mesh-cell volumes are recalculated. The mesh-cell densities are updated using Eq. (B-8), and are used, in turn, to calculate the new mesh-cell pressures. This completes one time step in the hydrodynamics calculation.

REFERENCES

1. H. A. Bethe and J. H. Tait, "An Estimate of the Order of Magnitude of the Explosion When the Core of a Fast Reactor Collapses," UKAEA-RHM(56)/113 (1956).

2. D. R. MacFarlane, Ed., "SAS1A, A Computer Code for the Analysis of Fast Reactor Power and Flow Transients," Argonne National Laboratory report ANL-7607 (1970).

3. F. E. Dunn, G. J. Fischer, T. J. Heames, P. Pizzica, N. A. McNeal, W. R. Bohl, and S. M. Prastein, "The SAS2A LMFBR Accident Analysis Computer Code," Argonne National Laboratory report ANL-8138 (1975).

4. M. G. Stevenson, W. R. Bohl, F. E. Dunn, T. J. Heames, G. Hoppner, and L. L. Smith, "Current Status and Experimental Basis of the SAS LMFBR Accident Analysis Code System," Proc. ANS Conf. Fast Reactor Safety, Beverly Hills, California, 1974, USAEC-CONF 740401, p. 1303.

5. J. E. Cahalan, et al., "The Status and Experimental Basis of the SAS4A Accident Analysis Code System," Proc. Int. Mtg. on Fast Reactor Safety Technology, Seattle, 1979 (American Nuclear Society, LaGrange, Illinois, 1979), Vol. II, pp 603-614.

6. W. T. Sha and T. H. Hughes, "VENUS: A Two-dimensional Coupled Neutronics-Hydrodynamics Computer Program for Fast-Reactor Power Excursions," Argonne National Laboratory report ANL-7701 (October 1970).

7. J. F. Jackson and R. B. Nicholson, "VENUS-II: An LMFBR Disassembly Program," Argonne National Laboratory report ANL-7951 (1972).

8. C. R. Bell, et al., "SIMMER-I: An S_n, Implicit, Multifield, Multicomponent, Eulerian Recriticality Code for LMFBR Disrupted Core Analsyis," Los Alamos Scientific Laboratory report LA-NUREG-6467-MS (January 1977).

9. L. L. Smith, "SIMMER-II: A Computer Program for LMFBR Disrupted Core Analysis," Los Alamos Scientific Laboratory report LA-7515-M (October 1978).

10. Y. W. Chang and J. Gvildys, "REXCO-HEP: A Two-Dimensional Code for Calculating the Primary System Response in Fast Reactors," Argonne National Laboratory report ANL-75-19 (June 1975).

11. A. E. Waltar, et al., "MELT-III, A Neutronics, Thermal-Hydraulics Computer Program for Fast Reactor Safety Analysis," Hanford Engineering Development Laboratory report HEDL-TME 74-47 (December 1974).

12. J. J. Sienicki and P. B. Abramson, "The TWOPOOL Strategy and the Combined Compressible/Incompressible Flow Problem," Proc. of the Topical Meeting on Computational Methods in Nuclear Engineering, Williamsburg, Virginia, April 1979, pp. 1-69.

13. D. P. Weber, et al., "VENUS-III: An Eulerian Disassembly Code," Trans. Am. Nucl. Soc. 21, 219 (1975).

14. A. E. Waltar and A. Padilla, Jr., "Mathematical and Computational Techniques Employed in the Deterministic Approach to Liquid-Metal Fast Breeder Reactor Safety," Proc. Nat. Topical Mtg. of the ANS Mathematics and Computation Division, Tucson, 1977 (Am. Nuc. Society, LaGrange Park, Illinois 1977) Vol. 64, No. 2, pp. 418-451.

15. D. D. Freeman, "Coupling of Dynamics Calculations in the FREADM Code," Proc. Conf. Effective Use of Computers in the Nucl. Industry, Knoxville, Tennessee, April 1969, USAEC Report CONF 690401.

16. M. G. Stevenson and B. E. Bingham, "TART: An LMFBR Transient Analysis Project," in Proc. Conf. Effective Use of Computers in the Nucl. Industry, Knoxville, Tennessee, April 21-23, 1969, USAEC Report CONF 690401, pp. 16-29.

17. N. Hirakawa, "MARS, A Two-Dimensional Excursion Code," Atomic Power Development Associates report APDA-198 (June 1967).

18. M. G. Stevenson and J. H. Scott, "The Effect of Noncoherence in Fast Reactor Sodium Voiding Accidents," Trans. American Nuclear Society, San Francisco, California, December 1-4, 1969 (American Nuclear Society, Hinsdale, Illinois, 1970) Vol. 2 pp. 905-906.

19. J. W. Stephenson, Jr. and R. B. Nicholson, "Weak Explosion Program," ASTRA 417-G.0 (1961).

20. D. Okrent, et al., "AX-1, A Computing Program for Coupled Neutronics-Hydrodynamics Calculations on the IBM-704," Argonne National Laboratory report ANL-5977 (1959).

21. R. B. Nicholson, "Methods for Determining the Energy Release in Hypothetical Fast Reactor Meltdown Accidents," Nuc. Sci. Engr. 18, 207 (1964).

22. W. J. McCarthy and D. Okrent, "Fast Reactor Kinetics," The Technology of Nuclear Reactor Safety, Vol. 1, Chapter 10 (1964).

23. E. P. Hicks and D. C. Menzies, "Theoretical Studies on the Fast Reactor Maximum Accident," Proc. Conf. Safety Fuels and Core Design in Large Fast Power Reactors, October 11-14, 1965; also Argonne National Laboratory report ANL-7120 (1965).

24. D. H. Cho, W. L. Chen, and R. W. Wright, "Pressure Pulses and Mechanical Work from Molten Fuel-coolant Interactions: A Parametric Study," Trans. Am. Nucl. Soc. 14, 290 (1971).

25. A. Padilla, Jr., "Analysis of Mechanical Work Energy for LMFBR Maximum Accidents," Nuclear Technology 12, 348 (1971).

26. J. Marchaterre, et al., "Work-Energy Characterization for Core-Disruptive Accidents," Proc. Conf. on Fast Reactor Safety and Related Physics, Chicago, Illinois, October 1976. CONF-761001, Vol. III, p. 1121.

27. P. L. Browne and M. S. Hoyt, "HASTL -- A Numerical Calculation of Two-dimensional Lagrangian Hydrodynamics Utilizing the Concept of Space-dependent Time Steps," Los Alamos Scientific Laboratory report LA-3324 (1965).

28. L. Baker, Jr., "Core Debris Behavior and Interactions with Concrete," Nuclear Engineering and Design 42, No. 1, 137 (1977).

29. W. R. Bohl, J. E. Cahalan, and D. R. Ferguson, "An Analysis of the Unprotected Loss-of-Flow Accident in the Clinch River Breeder Reactors with an End-of-Equilibrium-Cycle Core," Argonne National Laboratory report ANL/RAS 77-15 (May 1977).

30. H. H. Hummel, Kalimullah, and P. A. Pizzica, "Physics and Pump Coast-down Calculations for a Model of a 4000 MWe Oxide-Fueled LMFBR," Argonne National Laboratory report ANL-76-77 (June 1976).

31. W. R. Bohl, "Some Recriticality Studies with SIMMER-II," Proc. Int. Mtg. on Fast Reactor Safety Technology, Seattle, 1979 (American Nuclear Society, LaGrange, Illinois, 1979) Vol. III, pp. 1415-1424.

32. C. R. Bell, J. E. Boudreau, J. H. Scott, and L. L. Smith, "Advances in the Mechanistic Assessment of Postdisassembly Energetics, Proc. Int. Mtg. on Fast Reactor Safety Technology, Seattle, 1979 (American Nuclear Society, LaGrange Park, Illinois, 1979) Vol. I, pp. 207-218.

33. A. E. Waltar, et al., "An Analysis of the Unprotected Transient Overpower Accident in the FTR," Hanford Engineering Development Laboratory report HEDL-TME 75-50 (June 1975).

34. W. R. Bohl and T. J. Heames, "CLAZAS: The SAS3A Clad Motion Model," Argonne National Laboratory report ANL/RAS 74-15 (August 1974).

35. W. R. Bohl, "SLUMPY: The SAS3A Fuel Motion Model for Loss-of-Flow," Argonne National Laboratory report ANL/RAS 74-18 (August 1974).

36. D. R. MacFarlane, N. A. McNeal, D. Meneley, and C. K. Sanathanan, "Theoretical Studies of Fast Reactors During Sodium Boiling Accidents," Proc. of Int. Conf. on Safety of Fast Reactors, Aix-en-Provence, France (1967).

37. C. K. Youngdahl, "Fuel Element Deformation Model for Fast Reactor Accident Safety Study Code," Nucl. Eng. Design 15, 149-186 (1971).

38. A. Watanabe, "The DEFORM-II Mathematical Analysis of Elastic, Viscous, and Plastic Deformation of a Reactor Fuel Pin," Argonne National Laboratory report ANL-8041 (1973).

39. Bohl, op. cit., p. 2.

40. L. L. Smith, J. R. Travis, M. G. Stevenson, F. E. Dunn, and G. J. Fischer, "SAS/FCI, A Fuel-Coolant Interaction Model for LMFBR Whole-Core Accident Analysis," Proc. ANS Topical Conf. on Mathematical Models and Computational Techniques for Analysis of Nuclear Systems, Ann Arbor, Michigan, 1973 (Am. Nuc. Society, LaGrange Park, Illinois, 1973) Vol. I.

41. W. R. Bohl and M. G. Stevenson, "A Fuel Motion Model for LMFBR Unprotected Loss-of-Flow Accident Analysis," Proc. ANS Topical Conf. on Mathematical Models and Computational Techniques for Analysis of Nuclear Systems, Ann Arbor, 1973 (American Nuclear Society, LaGrange Park, Illinois, 1973) Vol. II.

42. W. R. Bohl, "CLAP: A Cladding Action Program for LMFBR HDCA LOF Analysis," Transactions ANS, Toronto, 1976 (American Nuclear Society, Hinsdale, Illinois, 1976) Vol. 23, pp. 348-349.

43. H. U. Wider, A. M. Tentner, and P. A. Pizzica, "The PLUTO2 Overpower Excursion Code and a Comparison with EPIC," Proc. of the Inter. Mtg. Fast Reactor Safety, Seattle, August 1979 (Am. Nuc. Society, LaGrange Park, Illinois) Vol. I, pp. 121-130.

44. A. M. Tentner and H. U. Wider, "LEVITATE - A Mechanistic Model for the Analysis of Fuel and Cladding Dynamics Under LOF Conditions for SAS4A," Proc. Int. Mtg. Fast Reactor Safety Technology, Seattle, 1979 (American Nuclear Society, LaGrange Park, Illinois, 1979) Vol. IV, pp. 1998-2007.

45. A. W. Cronenberg, H. K. Fauske, D. G. Bankoff, and D. T. Eggen, "A Single Bubble Model for Sodium Expulsion from a Heated Channel," Nucl. Engr. and Design 16, 285 (1971).

46. E. G. Schlechtendahl, "Theoretical Investigations on Sodium Boiling in Fast Reactors," Nucl. Sci. Engr. 41, 99 (1970).

47. G. Hoppner and W. R. Bohl, "Analysis of Cladding and Fuel Motion in the R4 Experiment," Trans. Am. Nucl. Soc. 19, 239 (1974).

48. I. T. Hwang, et al., "Sodium Voiding Dynamics and Cladding Motion in a 37-Pin Fuel Assembly During an LOF Transient," Trans. Am. Nucl. Soc. 28, 445.

49. T. E. Kraft, et al., "Simulations of an Unprotected Loss-of-Flow Accident with a 37-Pin Bundle in the Sodium Loop Safety Facility," Proc. Int. Mtg. Fast Reactor Safety Technology, Seattle, 1979 (American Nuclear Society, LaGrange Park, Illinois) Vol. IV, pp. 1998-2007.

50. C. H. Bowers, L. L. Briggs, J. M. Kyser, and D. P. Weber, "An Improved Two-Component Sodium Voiding Model for the SAS4A Analysis Code," Proc. Int. Mtg. Fast Reactor Safety Technology, Seattle, 1979 (American Nuclear Society, LaGrange, Illinois, 1979) Vol. I, pp. 99-109.

51. F. H. Harlow and A. A. Amsden, "A Numerical Fluid Dynamics Calculation Method for all Flow Speeds," J. of Comp. Physics 8, 197 (1971).

52. W. R. Bohl, et al., "A Preliminary Study of the FFTF Hypothetical Flow Coastdown Accident," Argonne National Laboratory report ANL/RAS 71-39 (April 1972).

53. H. K. Fauske, Argonne National Laboratory, personal communication, July 1971.

54. Cahalan, op. cit., p. 609.

55. D. V. Gopinath and C. E. Dickerman, "Calculations of Coherence of Failure for Hypothetical Meltdown Accidents in an EBR-II-like Reactor," Argonne National Laboratory report ANL-6844 (1964).

56. L. W. Deitrich, "An Assessment of Early Fuel Dispersal in the Hypothetical Loss-of-Flow Accident," Proc. Int. Mtg. Fast Reactor Safety Technology, Seattle, 1979 (American Nuclear Society, LaGrange, Illinois, 1979) Vol. II, pp. 615-623.

57. R. D. Richtmeyer and K. W. Morton, Difference Methods for Initial-Value Problems (John Wiley and Sons, Inc., 1967).

58. Cahalan, op. cit., p. 609.

59. J. H. Scott and P. K. Mast, "Fuel Pin Failure Models and Fuel-Failure Thresholds for Core Disruptive Accident Analysis," Proc. Int. Mtg. on Fast Reactor Safety and Related Physics, Chicago, 1976, CONF-761001, Vol. III, pp. 1015-1026.

60. D. E. Simpson, A. E. Waltar, and A. Padilla, Jr., "Assessment of Magnitude and Uncertainties of Hypothetical Accidents for the FFTF," Hanford Engineering Development Laboratory report HEDL-TME 71-31 (1971).

61. P. G. Lorenzini and G. F. Flanagan, "An Evaluation of Fuel-Coolant Interaction During Disassembly of an LMFBR," Proc. Conf. New Developments in Reactor Mathematics and Applications, CONF-710302, Idaho Falls, Idaho (1971) p. 50.

62. A. W. Cronenberg, "A Thermodynamic Model for Molten UO_2-Na Interaction, Pertaining to Fast-Reactor Fuel-Failure Accidents," Argonne National Laboratory report ANL-7947 (1972).

63. F. E. Dunn, J. R. Travis, and L. L. Smith, "The PRIMAR-2 Primary Loop Module for the SAS3A Code," Argonne National Laboratory report ANL-RAS 76-5 (March 1976).

64. Cahalan, op. cit., p. 606.

65. H. U. Wider, et al., "An Improved Analysis of Fuel Motion During an Overpower Excursion," Fast Reactor Safety, Beverly Hills, 1974, CONF-740401-P3, p. 1541.

66. P. A. Pizzica and P. Abramson, "EPIC: A Computer Program for Fuel-Coolant Interactions," Int. Mtg. Fast Reactor Safety and Related Physics, Chicago, 1975, CONF-761001, Vol. III, pp. 979-987.

67. Wider, op. cit., p. 121.

68. J. VonNeumann and R. D. Richtmyer, "A Method for the Numerical Calculation of Hydrodynamic Shocks," J. Appl. Phys. 21, 3, 232 (1950).

69. K. O. Ott and D. A. Meneley, "Accuracy of the Quasistatic Treatment of Spatial Reactor Kinetics," Nucl. Sci. Eng. 36, 402 (1969).

70. D. A. Meneley, et al., "A Kinetic Model for Fast-Reactor Analysis in Two Dimensions," Dynamics of Nuclear Systems, (University of Arizona Press, Tucson, Arizona, 1972) pp. 483-500.

71. W. T. Sha and A. E. Waltar, "An Integrated Model for Analyzing Disruptive Accidents in Fast Reactors," Nucl. Sci. and Engr. 44, 135-156 (1971).

72. M. G. Stevenson, "Nuclear Reactor Safety Quarterly Progress Report, October 1-December 30, 1979," Los Alamos Scientific Laboratory report LA-8299-PR (to be published).

73. A. A. Amsden and F. H. Harlow, "KACHINA: An Eulerian Computer Program for Multifield Fluid Flows," Los Alamos Scientific Laboratory report LA-5680 (1974).

74. J. E. Boudreau and J. F. Jackson, "Recriticality Considerations in LMFBR Accidents," Proc. Fast Reactor Safety Mtg, Beverly Hills, April 2-4, 1974, CONF-740401-P3.

75. Y. Chang, J. Gvildys, and S. H. Fistedis, "Two-Dimensional Hydrodynamics Analysis for Primary Containment," Argonne National Laboratory report ANL-7498 (1969).

76. Chung-Yi Wang, "ICECO - An Implicit Eulerian Method for Calculating Fluid Transients in Fast-Reactor Containment," Argonne National Laboratory report ANL-75-81 (December 1975).

77. H. Y. Chu, et al., "A Generalized Eulerian Method in Reactor Containment Analysis and Its Comparison with Other Numerical Methods," Int. Mtg. on Fast Reactor Safety and Related Physics, Chicago, 1976, Vol. III, p. 1285.

78. C. R. Bell and J. E. Boudreau, "SIMMER-I Accident Consequence Calculations," Trans. ANS, 1977, TRANSAO 27-1-1028, Vol. 27, pp. 555-556.

79. R. J. Tobin and D. J. Cagliostro, "Effects of Vessel Structures on Simulated HCDA Bubble Expansions," Stanford Research Institute technical report 5 (November 1978).

80. A. J. Suo-Anttila, "Analysis of Postdisassembly Expansion Experiments," Proc. Int. Mtg. on Fast Reactor Safety Technology, Seattle, 1979 (Am. Nuc. Society, LaGrange Park, Illinois, 1979) Vol. IV, pp. 1848-1857.

81. M. G. Stevenson, "Nuclear Reactor Safety Quarterly Progress Report, October 1-December 30, 1979," Los Alamos Scientific Laboratory report LA-8299-PR (to be published).

82. P. E. Rexroth, Los Alamos Scientific Laboratory, private communication, 1978.

83. J. E. Boudreau, et al., "A Critical Assessment of SIMMER-II Models," Los Alamos Scientific Laboratory report (to be published).

84. J. H. Scott, "Overview and Status of the SIMMER Testing Program," Proc. Int. Mtg. on Fast Reactor Safety Technology, Seattle, 1979 (Am. Nuc. Society, LaGrange, Illinois, 1979) Vol. I, pp. 197-206.

85. A. J. Suo-Anttila, Los Alamos Scientific Laboratory, private communication, 1979.

86. J. L. Tomkins, J. T. Hitchcock, and M. F. Young, "Prompt Burst Energetics (PBE) Experiment Analyses Using the SIMMER-II Computer Code," Proc. Int. Mtg. Fast Reactor Safety Technology, Seattle, 1979 (Am. Nucl. Society, LaGrange, Illinois, 1979) Vol. II, pp. 1001-1010.

87. A. J. Suo-Antilla, Los Alamos Scientific Laboratory, private communication, 1980.

88. W. R. Bohl, "A SIMMER-II Analysis of the R-7 Treat Test," Los Alamos Scientific Laboratory report NUREG/CR-0760, LA-7763-MS (April 1979).

89. L. D. P. King, et al., "Description of the KIWI-TNT Excursion and Related Experiments," Los Alamos Scientific Laboratory report LA-3350-MS (1966).

90. J. F. Jackson, W. F. Rhoades, and L. I. Moss, "Analysis of SNAPTRAN-1 and -2 Reactor Kinetics Experiments," Atomics International report NAA-SR-11850 (1967).

91. W. E. Kessler, et al., "SNAPTRAN 2/10A-3 Destructive Test Results," Phillips Petroleum Co., Atomic Energy Division report IDO-17019 (January 1965).

92. T. F. Bott and J. F. Jackson, "Experimental Comparison Studies with the VENUS-II Disassembly Code," Proc. Conf. on Fast Reactor Safety and Related Physics, Chicago, 1976, CONF-761001, Vol. 3, p. 1134.

93. D. H. Barker, et al., "Improvement and Verification of Fast Reactor Safety Analysis Techniques: Progress Report," Department of Chemical Engineering, Brigham Young University report COO-2571-8 (1977).

94. S. H. Fistedis, et al., "Fast Reactor Containment Analysis, Recent Improvements, Applications, and Future Developments," Proc. Fast Reactor Safety Mtg., Beverly Hills, 1974, CONF-740401-P2, p. 763.

95. A. H. Marchertas, C. Y. Wang, and S. H. Fistedis, "A Comparison of ANL Containment Codes with SNR-300 Simulation Experiments," Int. Mtg. on Fast Reactor Methods, 1976, CONF-761001, Vol. III, p. 1324.

96. J. F. Jackson and D. P. Weber, "Hydrodynamics Methods in Fast Reactor Safety," Proc. Conf. on Computational Methods in Nuclear Engineering, Charleston, 1975, CONF-750413, Vol. I, pp. 27-52.

97. F. H. Harlow and A. A. Amsden, "Fluid Dynamics," Los Alamos Scientific Laboratory report LA-4700 (1971).

98. F. H. Harlow, "The Particle-in-Cell Method for Numerical Solution of Problems in Fluid Dynamics," Proc. Symposia in Applied Mathematics 15, 269 (1963).

99. R. W. Hornbeck, Numerical Methods (Quantum Publisher, Inc., New York, 1975).

100. Safety Code Development Group, "TRAC-P1A: An Advanced, Best-Estimate Computer Program for PWR LOCA Analysis," Los Alamos Scientific Laboratory report LA-7777-MS (May 1979).

101. P. L. Browne and M. S. Hoyt, "HASTL -- A Numerical Calculation of Two-dimensional Lagrangian Hydrodynamics Utilizing the Concept of Space-dependent Time Steps," Los Alamos Scientific Laboratory report LA-3324 (1965).

PART 5

SPECIAL TOPICS

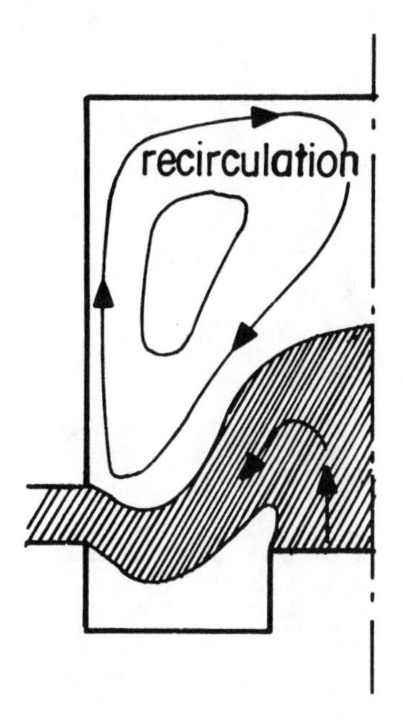

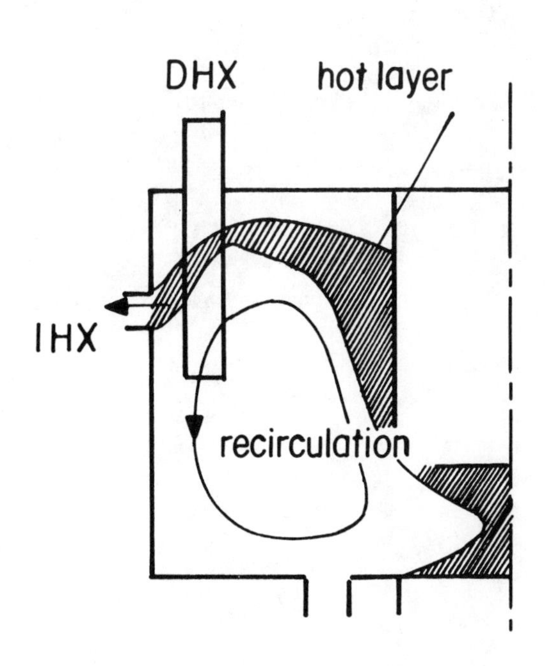

20 VAPOR EXPLOSIONS

by
S.G. Bankoff
Northwestern Univ., Evanston, IL, USA

Vapor explosions have come into prominence recently in connection with potential accidents in the nuclear reactor and the liquefied natural gas industries. Based on available evidence, the spontaneous nucleation (SN) theory for initiation and escalation appears to be valid. This implies that a large-scale explosion of molten UO_2 and highly subcooled sodium would be very difficult to achieve. The key uncertainty as to whether a supercritical steady-state detonation is theoretically possible is the fragmentation rate behind the shock. It is the purpose of this chapter to describe in detail the current thinking regarding vapor explosions. The questions currently unresolved are delineated, some relevant experiments are reviewed and calculations of specific importance given.

1. INTRODUCTION

If two nearly immiscible liquids are rapidly brought into intimate contact, and the temperature of one of the liquids is considerably above the normal boiling point of the second liquid, rapid vaporization will occur. In some cases, if the spontaneous nucleation temperature of the volatile liquid has been exceeded, explosive boiling, accompanied by the generation of shock waves, may ensue. If the shock waves have significant damage potential, the resulting explosion is termed a vapor explosion. The terms "physical explosion" and "thermal explosion" have also been used. It is thus distinguished from a chemical explosion, in which the energy release stems from chemical reactions. It is possible, of course, to have explosions in which both physical vapor production and chemical reactions contribute significantly, such as in the rapid mixing of aluminum well above its melting point with water. Vapor explosions are generally limited to liquid-liquid systems, in which a hot nonvolatile liquid is brought into contact with a colder volatile liquid. A characteristic feature is the rapid production of new interfacial area, which is necessary to obtain energy transfer on a time scale consistent with explosive behavior. If the freezing point of the hot liquid is above the cold liquid temperature, explosive boiling is always accompanied by extensive production of fine particulate material. On the other hand, extensive fragmentation may occur without shock formation, either because the mixing process is too slow, or because the initial constraints surrounding the mixing zone are too small to prevent rapid pressure relief.

Some of this material was presented at the Japan-U.S. Seminar on Two-Phase Flow Dynamics, Kobe, Japan (1979), sponsored by the Japan Society for the Promotion of Science and the U.S. National Science Foundation.

Because of the very short timescales involved, detailed mechanistic studies of vapor explosions are very difficult. Most experimenters have contented themselves with observing the results of bringing together, in various modes, a number of fuel-coolant combinations at different temperatures. Water is the most common coolant, because of its importance relative to industrial explosions in the metallurgical and paper industries, as well as to hypothetical accidents in light-water cooled nuclear reactors. In recent years, however, considerable effort has been expended in studying the interaction of liquid sodium with various LMFBR materials, such as molten UO_2 and steel, in a variety of configurations. Because of the high thermal conductivity and strong wetting ability of sodium, it is very difficult to maintain a continuous vapor film around the hot liquid drops, which results in markedly different behavior than with water as the coolant. The time scale for the fragmentation is 10^{-3} to 10^{-4} s, based upon a 0.01 m initial dispersal scale and a shock velocity through the liquid-vapor mixture $\sim 10^2$ m/s. Slower fragmentation and energy transfer may be relatively harmless, although each case must be separately and carefully analyzed. Less efficient fuel-coolant interactions may take place when initial mixing is poor, in which case a series of expansions and contractions may result in partial mixing and energy transfer in successive stages, as in shock tube experiments[3-7].

These broad categories of fuel-coolant interactions have been termed Category 1 and 2, respectively[54]. Category 1 corresponds to a "true explosion where neither the peak pressure nor its rise time ($<$ 1 ms) is affected by additional constraints other than those due to the fluid mixture itself". In Category 2 events, however, the pressure rise is determined by the degree of constraint provided by the system and the pressure rise time is comparatively long. Much attention has been focussed on Category 1, although Category 2 events seem much more probable in suggested LMFBR accident scenarios.

Vapor explosions may also be classified in terms of scale, since the basic mechanisms may be quite different. Small-scale explosions, involving one or a few drops of hot liquid, are inherently unsteady expansions, whereas large-scale explosions (involving say 20 kg or more of hot liquid) tend to be steady or quasi-steady detonations proceeding through a coarsely-premixed fuel-coolant-vapor region.

The following mechanistic classifications are therefore proposed in this paper:

1. Unsteady mixing, expansion and contraction
   a) Single cycle
   b) Multiple cycle

2. Steady or quasi-steady propagation of a shock front
   a) Subcritical pressures (and hence vapor) behind the shock
   b) Supercritical pressures behind the shock

All of the above types, with the exception of 2b, have been experimentally identified. There is currently considerable discussion as to whether an event of type 2b is possible, since this would have important consequences for reactor safety. Furthermore, a single interaction may involve both types 1 and 2. Thus, a large-scale interaction may begin by a sequence of local expansions and contractions, and then escalate into a full-fledged detonation wave.

Four stages of an efficient large-scale vapor explosion can be identified:

1. Coarse premixing of fuel and coolant without rapid energy transfer. This requires stable film boiling, when the two liquids initially come into contact.

2. Triggering event, consisting of collapse of the vapor blankets in some small region due to arrival of a pressure wave or cooling of the hot liquid.

3. Escalation of the fuel-coolant interaction by a cyclic mixing process, in which fragmentation and energy transfer occur simultaneously.

4. Propagation of a fully-developed detonation wave through the coarse fuel-coolant mixture.

Current interest centers around two theories:

1. The spontaneous nucleation (SN) hypothesis, due to Fauske[8,9], which gives necessary conditions for triggering and escalation of vapor explosions*.

2. Steady plane detonation theory applied to fuel-coolant interaction due to Board and Hall[10] (B-H theory).

These theories have sometimes been treated in the literature as being mutually exclusive. However, since they deal with entirely different phases of a large scale MFCI, it is clear that each can be considered quite independently of the other. We therefore focus this review on these two theories, together with modifications introduced by more recent evidence. In addition, some of the key concepts of rapid mixing and fragmentation will be discussed.

*We use the terms "fuel" and "coolant" from the nuclear safety area to represent generic terms for the hot and cold liquids, respectively.

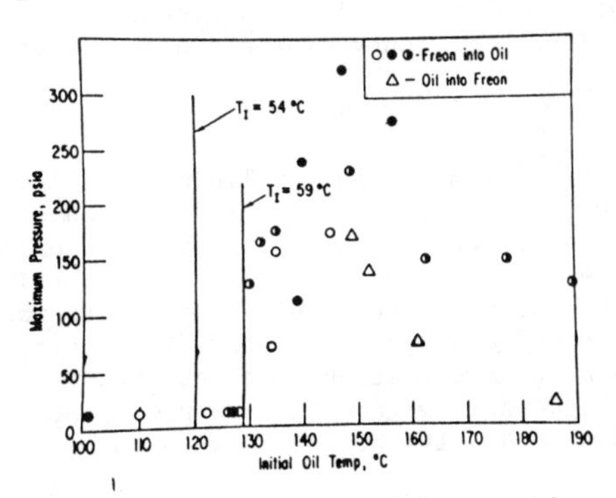

Fig. 1a Maximum measured interaction pressures for Freon-22-oil system[97]. The homogeneous nucl. temp. T_{hn} for Freon-22 lies in range 54-59°C.

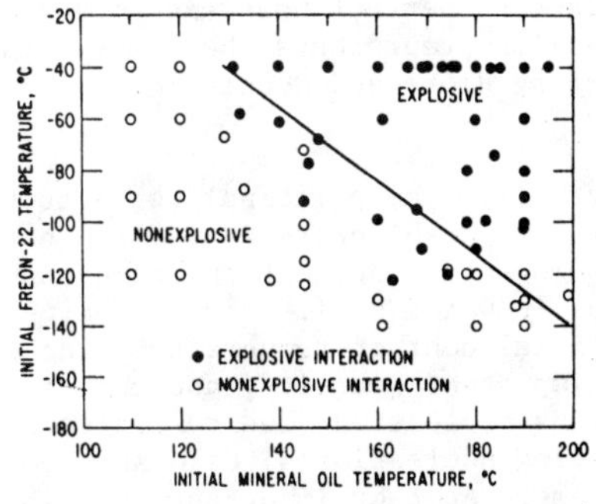

Fig. 1b Subcooled Freon-Mineral Oil Data. The diagonal lines represent the locus $T_{io} = T_{hn}$. The larger scatter is due to the uncertainty in Freon temp. at the instant of contact.

2. SN THEORY

It was hypothesized by Fauske[8,9] that the following conditions are necessary for a large-scale vapor explosion:**

A. Initial film boiling, which allows coarse premixing
B. Liquid-liquid contact (triggering event)
C. $T_{io} > T_{sn}$, where T_{io} is the interfacial initial-contact temperature and T_{sn} is the spontaneous nucleation temperature. T_{io} is calculated from the the conduction theory formula; $T_{sn} < T_{hn}$ (the homogeneous nucleation temperature)if there is a positive contact angle for a bubble growing at the interface.
D. Adequate physical and inertial constraints.

The controversial item is C. It is seen that C implies that A is satisfied. C has been modified by Bankoff to read:[2]

C' . Immediate local pressurization upon contact.

This takes into account that all large-scale explosions to date (Al/H_2O, $steel/H_2O$, $smelt/H_2O$, water/LNG, $lava/H_2O$) have proceeded under conditions where $T_{io} > T_{cr}$, the critical temperature of the coolant. Nucleation is then irrelevant The importance of this modification is that some earlier theories arbitrarily postulated a cut-off when $T_{io} = T_{cr}$. However, shock tube studies of destabilization of film boiling, both on a heated nickel tube[11] and on a molten liquid-metal surface contained in a heated crucible[7], have shown that the maximum heat transfer for the tube, as well as the maximum interaction pressure in the metal/water, after passage of the shock, occurs when $T_{io}/T_{cr} \cong 1.1-1.3$. Spontaneous nucleation is therefore sufficient, but not necessary, for C' to be satisfied. Since the local pressures are of order 10^2 bars, rapid mixing and fragmentation results.

Note that there is no contradiction between C and C', since spontaneous nucleation implies bubble growth times of order 10^{-6} s with densities $> 10^{12}$ cm^{-3}. On the other hand, the spontaneous nucleation requirement has also been interpreted[71] as implying a cutoff in heat transfer at the spontaneous nucleation temperature, so that the maximum pressure is the vapor pressure of the coolant at the fuel temperature. This is not necessarily true, as shown by the pressure-wave film boiling destabilization experiments[7,11] to be discussed below. Higher peak pressures can be achieved by mixing in confined geometries. The limit of mixing and perfect heat exchange at constant volume, followed by isentropic expansion, represents the theoretical maximum efficiency of the process, treated first by Hicks and Menzies[72].

**Actually, the principal concepts of initial film boiling, followed by collapse of the vapor blankets around the hot liquid and very rapid vapor formation due to homogeneous nucleation, were enunciated earlier in connection with LNG explosions in water[55-57]. Fauske sharpened the concepts by recognizing that the interfacial contact temperature, rather than the hot liquid bulk temperature was the important quantity, and that spontaneous nucleation, or vapor-free heterogeneous nucleation (due to statistical density fluctuations, rather than pre-existing nucleation sites) can occur, particularly with water, in poorly-wetting systems. Another important contribution was the distinction between large and small scale explosions.

2.1 Small Scale Studies

There is probably no real disagreement that extensive liquid-liquid contact with interfacial temperatures at or above the spontaneous nucleation temperature is a sufficient condition to produce at least a local explosion. For well-wetting liquid pairs $T_{sn} = T_{hn}$, and for small-scale experiments the requirement has been clearly established. Indeed, the best way to measure T_{hn}, (within 1^oK) is to determine the explosion temperature for a rising "coolant" drop in a non-isothermal "fuel" column[13,14]. With well-wetting liquids, such as saturated Freon-22, intermediate-scale pouring experiments exhibit similarly spectacular agreement with the criterion $T_{io} = T_{hn}$ as the temperature threshold for the initiation of a vapor explosion for saturated Freon[97]. Figs. 1a,b the subcooled scatter is somewhat larger, due to the uncertainty in the actual Freon bulk temperature at the time of contact, but the agreement is still very good. Of particular interest is the pressure cutoff observed. When the initial pressure was increased from 1 to 2 atm, no explosions were recorded over a range of initial fuel temperatures. This may be connected to the increased stability of film boiling, as in the "capture" theory of Henry and Fauske[98], or to slower bubble growth rates, leading to less vigorous mixing, or both. However, water drops rising in silicone oil explode at temperatures of 513-543^oK, well below $T_{hn} \simeq 583$ K, despite the presence of nucleate boiling from pre-existing sites and some miscibility between the two liquids[15]. A reasonable estimate for H_2O is thus $T_{sn} \sim 513^oK$, although lower figures have been suggested. This uncertainty has led to claims that T_{sn} is in fact quite arbitrary, so that Requirement C is then ambiguous. This is certainly not the case for well-wetting liquids, and a simple calculation shows that unrealistic contact angles ($\theta > 110^o$) are required to lower T_{sn} much more than 70^oK below T_{hn}.

It is not necessary to have long heating times and intimate liquid-liquid contact throughout the experiment in order for small drops to explode[73-76]. Small (~ 3 mm) drops of several organic liquids, including methanol, ethanol, and pentane, were allowed to fall 0-15 cm. onto the surface of a pool of hot silicone oil or glycerol. A stable liquid-liquid contact was obtained if the Weber number upon impact exceeded a critical value, which depended weakly (Fig. 2) on the surface temperature. In this case the droplet evaporated smoothly to dryness if $T_H < T_{hn}$, but exploded after a delay of 50-200 ms if $T_H > T_{hn}$. If the impact Weber number was too small, stable Leidenfrost boiling took place. These results led Fauske[9] to propose a fundamental distinction between small and large-scale explosions. Small-scale explosions may occur, since T_i begins to drift upwards when the thermal boundary thickness is of the same order as the drop dimensions. At the same time T_{sn} may increase because of improved wetting with time. A small explosion is obtained if $T_i = T_{sn}$ before the drop evaporates completely. This mechanism is inadequate, however, for propagation of a large-scale explosion, where a local explosion mist occur instantaneously upon contact in order to sustain a propagating detonation. The importance of this concept is that it precludes large-scale explosions in the UO_2/Na system, independent of any triggering mechanism, since $T_{io} \sim 1500$-1700 K, well below T_{hn} = 2100-2300 K. This has been vigorously disputed, both in terms of the validity of the SN theory[1,78] and the adequacy of the estimates of T_{io}, T_{sn} and T_{hn} fot this fuel-coolant pair[77]. We pursue here the SN evidence, and consider the UO_2/Na case later.

The difference between poor wetting and good wetting is clearly shown in the experiments of Board, Hall and Brown[79]. Dropping mineral oil into Freon-22 results in a sharp pressure peak threshold at $T_{io} = T_{hn}$, (Fig.3a), in agreement with the Henry, et al. experiments[20,97]. However, with Freon/22 water, the temperature threshold is not nearly as sharp, and significant interactions are obtained when $T_{io} < T_{hn}$ (Fig. 3b). However, as noted above $T_{sn} \simeq 513$ K for water in contact with hydrocarbons, as measured under ideal laboratory conditions[13,15]

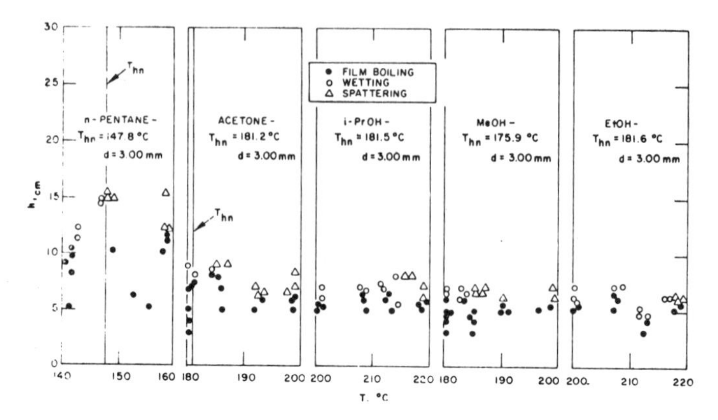

Fig. 2 Vaporization of several volatile liquids impinging on a silicone surface. (Waldram, Fauske and Bankoff[75])

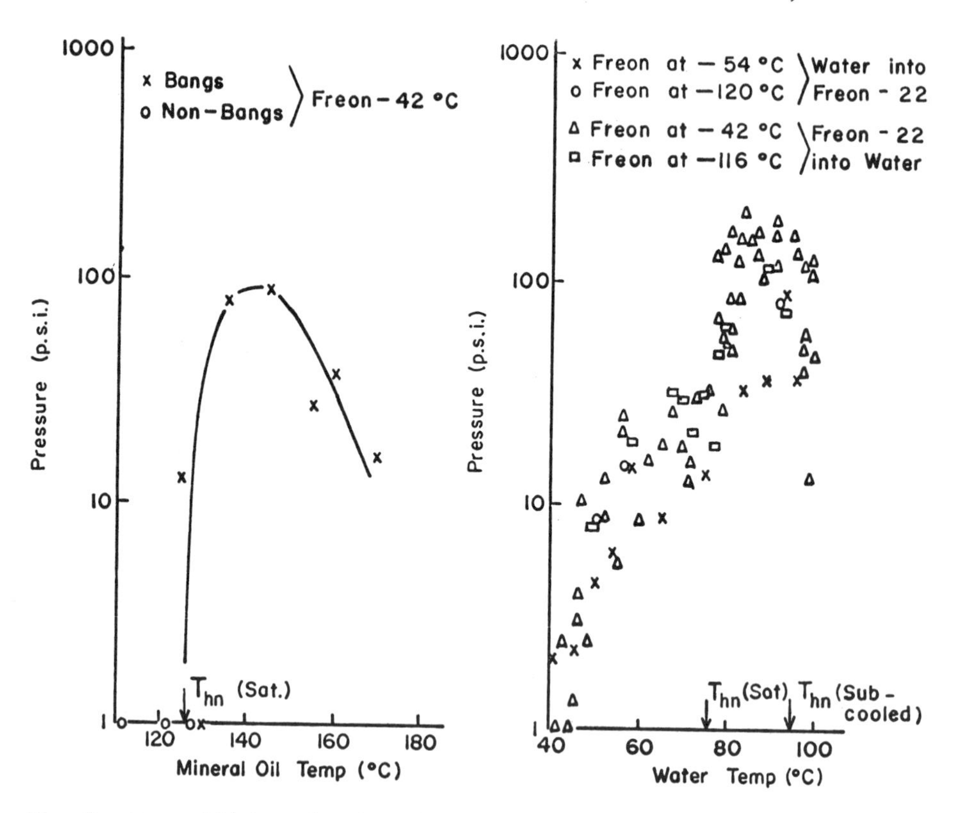

Fig. 3a Freon-22/mineral oil fuel-coolant interactions in the pouring mode. Note the excellent agreement with the homogeneous nucleation temperature interfacial criterion.

Fig. 3b Freon-22/water interactions in the pouring mode, using both saturated and subcooled Freon.

(quasi-static rising-drop experiments). This type of experiment has not been conducted with Freon-22/water (indeed, it would be very difficult because of freezing of the water), but if a similar decrease in the reduced temperature were experienced, it is seen that the data generally lie above $T_{io} = T_{sn}$. The reduced temperature is here appropriate, since a simple estimate of the limit-of-superheat is $T_{hn}/T_{cr} = 27/32$, based on the van der Waals equation of state[17]. It should be noted, however, that it is indeed difficult to suppress heterogeneous nucleation in a poorly-wetted aqueous system[18]. Skripov and Pavlov[16] found that rise times of ~ 0.04 ms were required to suppress heterogeneous nucleation on a pulsed platinum wire in water. The corresponding times for a number of organic liquids were ~ 1 ms. Wetting rates in the liquid UO_2/liquid Na system are not known, but it has been speculated that they are diffusion-limited, rather than reaction-rate-limited, and hence very fast[2].

2.2 Temperature-Interaction Zone

An important concept introduced by Dullforce, et al.[80] was the temperature interaction zone, outside of which interactions do not occur without external triggering. Early experiments by McCracken[81] showed that FCI's in water were very sensitive to initial conditions, so that great care is required in order to obtain reproducible results. The results of about 300 experiments in which 12 g. samples of molten tin were dropped into boiled distilled water are summarized in Fig. 4. The left-hand vertical boundary is that associated with the SN theory. The authors contest the SN theory on the basis of the large increase in contact area required for interaction on the time scale of the explosion, as well as the observed coolant cut-off temperature for interaction. Instead they propose a cyclic boiling model of bubble growth, collapse, jet penetration and vaporization. These objections are quite unclear, since none of the three circumstances discussed above is incompatible with the SN hypothesis. The large area increase simply indicates that fragmentation and heat transfer proceed together; the cutoff behavior is related to the stability of film boiling; and the cyclic boiling mechanism seems to be very reasonable, but perfectly compatible with SN. Indeed, later experiments of the same type from the same laboratory confirm the SN hypothesis (Fig. 5). With tin/water there is some ambiguity, since T_{sn} and the melting temperature, T_m, are quite close. However, they are quite different for Cerrotru (m.p. 412°K)/water and water/(Freon) R-22; and the spontaneous nucleation hypothesis is clearly confirmed. More recently, Tso and Tsien[83] have further subdivided the TIZ zone, taking into account the SN theory and the absence or presence of solidification.

2.3 LNG/Water Experiments

Enger and Hartman[21] reported results of an investigation of rapid phase transformation during liquefied natural gas spillage on water, which provided further insight into the nature of the triggering mechanism in vapor explosions. This work was motivated by a U.S. Bureau of Mines study of the hazards of marine transport of LNG, in which unexplained explosions occurred non-reproducibly in spill tests of LNG on water, both on a laboratory scale and on an artificial pond in quantities up to 475 liters. Several pairs of authors[64-66] suggested that the triggering mechanism was the rapid homogeneous nucleation of a thin layer of LNG in contact with water. An extensive program of spill studies with methane, ethane, propane, propylene, LNG, and mixtures thereof, primarily on water, in scales ranging from laboratory to tankcar-size spills, demonstrated conclusively that vapor explosions only occurred when the water temperature was in a narrow range, bounded from below by the homogeneous nucleation temperature of the hydrocarbon. For example, with propane a coherent ice layer formed, accompanied by nucleate boiling, when the water temperature was 0-52°C; vapor explosions were observed in the range 53-70°C; and film boiling, accompanied by

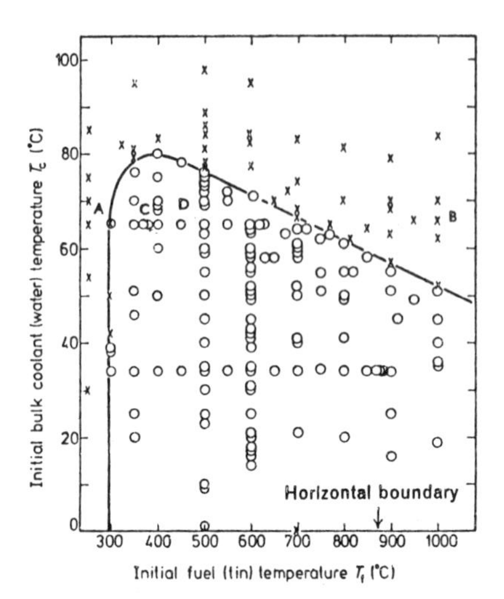

Fig. 4 Temperature interaction zone for 12 g of tin dropped through 3 cm into boiled distilled water: o indicates an interaction; x indicates no interaction.

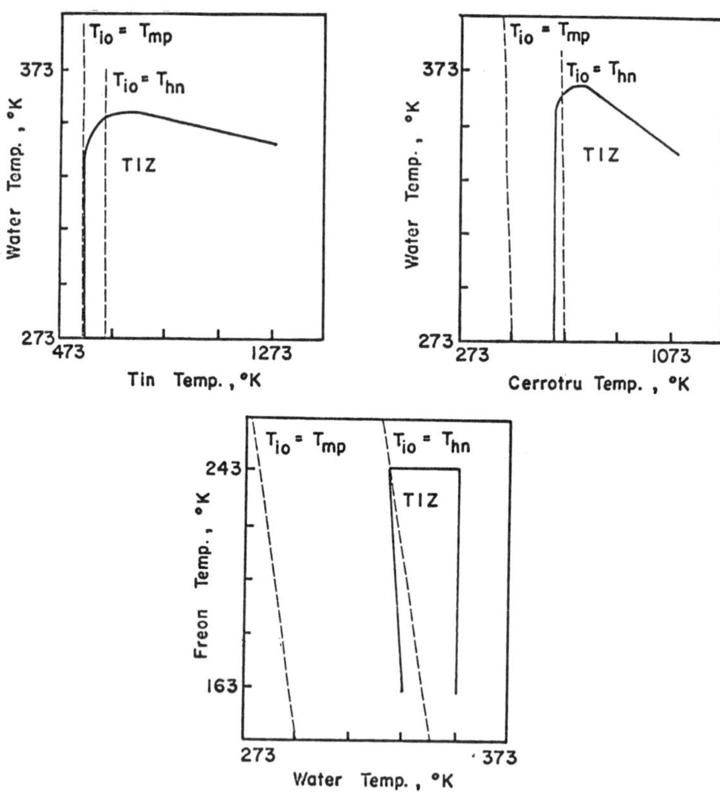

Fig. 5 Spontaneously-triggered temperature-interaction zone (TIZ) for several fuel-coolant pairs[21]. The left-hand boundary represents the lower temperature threshold for the vapor explosion to proceed, and is well represented by $T_{io} = T_{sn}$. The diagonal upper boundary represents a cutoff due to stable film boiling, and can be moved upwards by applying a pressure pulse[7].

rapid pops, in the range 71-82°C. No explosions were recorded with methane, ethane or nitrogen spills. In these cases ice formed during film boiling, followed eventually by nucleate boiling from the ice layer. As LNG "ages", its boiling point increases, due to differential distillation of methane away from the heavier components. For any particular mixture there was thus an explosive composition envelope, outside of which film boiling and ice formation occurred without an explosion. For example, only after 93% boil-off did spillage of commercial LNG in tank load quantities result in explosions. It was also possible to obtain delayed explosions with less-enriched mixtures in sufficient quantity by boil off from the hydrocarbon layer as it spread over the water surface, which could be predicted from a simple model of the spill spreading and vaporization rate. The spontaneous nucleation requirement was thus found consistently to be operative over a wide range of hydrocarbon mixtures and spill sizes.

The work on explosive ranges for hydrocarbon spills was extended by Porteous and Reid[67] to several cold liquids and hydrocarbon mixtures. The results are summarized in Table 1, and are in excellent agreement with the SN theory.

Table 1 EXPLOSIVE RANGES FOR HYDROCARBON SPILLS[67]

Hydrocarbon	Hot Liquid	Temperature of Hot Liquid - T_H °K	Explosive Range T_H/T_{hn}
ethane	methanol	306-331	1.14-1.23
ethane	methanol-water	296-304	1.10-1.13
propane	water	326-334	1.00-1.02
isobutane	ethylene glycol-water	374-379	1.04-1.05
n-butane	water	371	0.98
n-butane	ethylene glycol-water	387-398	1.03-1.06
propylene	water	317-346	0.97-1.06
isobutylene	ethylene glycol	379-408	1.03-1.10
ethane-propane	water	293	1.02-1.06
ethane-n-butane	water	278	1.00-1.01
		283	1.02-1.04
		293	1.05-1.08
		303	1.06-1.11
ethane-n-pentane	water	293	1.04-1.08
ethylene-n-butane	water	293	1.05-1.08
ethylene-n-butane	water	293	1.05-1.14
ethane-propane-n-butane	water	293	1.02-1.08
methane-ethane-propane-n-butane	--	293	1.08-1.13

One can note from these experiments, as with the earlier work, that freezing of the hotter liquid inhibits, but does not necessarily prevent vapor explosions. The process depends upon the competition between the rate of freezing after liquid-liquid contact is established locally, and the rate of crust breakup, due to pressure surges produced by vapor bubble growth and collapse, arrival of an external pressure shock wave, and/or local turbulent mixing.

Porteous and Reid also note that methane can be made to explode on water by high-velocity impact, although never by simple pouring. For this system $T_H/T_{hn} \sim 1.76$, so that the upper explosive limit depends strongly on the trigger energy, as well as the tendency to freeze rapidly. Similar observations were made by Anderson and Armstrong[68] in connection with liquid N_2 impacting on water. It is important to note that while the trigger energy affects the upper bound of the temperature-interaction zone, it does not affect the lower temperature bound, which is given by the SN theory. Wey, et al. also note that the lower temperature bound is not affected by the trigger energy[82].

2.4 Application to UO_2/Na FCI

The significance of this discussion relates mainly to the UO_2/Na case, for which $T_{io} < T_{hn}$. T_{hn} for Na = 2140-2400 K, based either on a hard-sphere equation-of-state thermodynamic limit-of superheat[19], or homogeneous nucleation theory[9]. The Fauske theory[9] hypothesizes that unless $T_{io} \geq T_{sn}$, a large-scale shock propagation cannot occur, since the delay time for local pressurization and fragmentation is too long. On the other hand, small-scale explosions, due to sodium drops entrapped in fuel which are then heated to T_{sn}, are not ruled out[22], nor are vigorous (but not explosive) interactions resulting from poor initial mixing in a confined geometry[23]. A comparison between T_{hn} and T_{io} for the UO_2/Na and the Al/H_2O systems[9] is shown in Fig.6. In contrast with the known explosivity of Al/H_2O, the UO_2/Na would be non-explosive upon sudden contact, according to this hypothesis. However, the interfacial contact temperature, T_i, would eventually rise to T_{hn} after a heating period whose length depends on the initial sodium temperature and the drop radius.

Several explosions of this type, with a delay of 50-200 ms, were recorded for a few grams of 400°C sodium injected into a crucible containing molten UO_2 by Armstrong, et al.[23]. Many other attempts to produce explosions with UO_2 and sodium have failed. These particular explosions are consistent with the above hypothesis, particularly since the sodium droplets might be trapped under a solid UO_2 crust.

An alternative hypothesis which has been advanced is that the delay time simply corresponded to the time for the vapor film to destabilize due to cooling of the hot liquid. However, subcooled film boiling of sodium, particularly from a low thermal conductivity surface, such as UO_2, is extremely unstable, as shown by quenching experiments[69], the Henry correlation[33] and unsteady-state conduction calculations under even the most conservative assumptions (no surface waves; uniform film thickness; saturated sodium)[70]. Thus, Hall calculates that for a stationary UO_2 sphere as large as 10^{-2} m radius, in stagnant sodium, T_∞ = 880°C and T_s = 1200°C (P_∞ = 9.4 atm), the maximum vapor film thickness is only 18 μm, which is comparable to the usual surface roughnesses.

Another set of experiments leading to large pressures (69 bars) has been recently reported by Amblard and Berthoud[23]. Sodium at 685°C was allowed to fall, in a shock-tube geometry, into a reaction chamber (7.76 liters) containing 4.9 kg molten UO_2 in a highly constrained geometry. The rise time was of the order of 10 ms, so that this interaction is more properly a Category 2, or Class 1b, interaction. Earlier tests with a larger volume chamber showed lower pressures, but several peaks, indicating repeated injection of coolant into the fuel, followed by vaporization, over-expansion and collapse. Similar results obtained by Sharon and Bankoff[7] in a shock tube geometry will be discussed later.

Finally, MacInnes, et al.[77] call attention to the effect of uncertainties in thermophysical properties of UO_2 on fuel-coolant interactions. It is suggested that the thermal conductivity of molten UO_2 may have been seriously underestimated, due to neglect of the electronic contribution to heat transport. No data on this point are available, nor is it clear that freezing, both before and upon, contact with liquid sodium will not occur, so that the relevant physical state is not the molten one. On the other hand, saturated sodium film

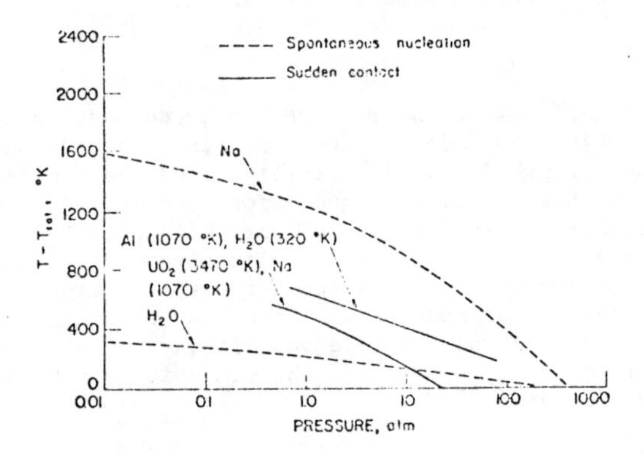

Fig. 6 Comparison between spontaneous nucleation temperatures and sudden contact temperatures for two different systems.[9]

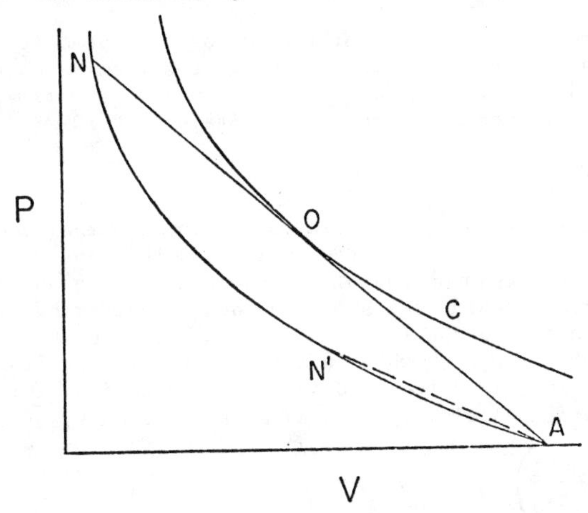

Fig. 7 Pressure-specific volume plot for thermal detonation wave. The curved line AN represents the shock adiabatic for isentropic compression from the initial state A to the von Neumann spike N. The curved line CO represents the equilibrium Hugoniot, where temperature and velocity equilibrium have been achieved. The straight line AON is the Rayleigh line, and the tangency condition at O implies that the mixture velocity is equal to the local sonic velocity relative to the shock fromt (Chapman-Jouguet condition).

For a dense dispersion, with unequal phase velocities, an extra coordinate must be added, such as $S + u_d/u_f$. The shock adiabat then becomes a surface in (P,V,S) space, and the system follows a curved path to the equilibrium state. The pressure is generally not a maximum immediately behind the shock front, but rises, as the fluid is slowed down by viscous drag from the drops. The von Neumann spike, N, is thus replaced by a pint on the shock adiabat surface, shown symbolically as N'.

boiling may be stable enough to allow mixing of the UO_2 and sodium on a coarse scale without significant heat transfer. This is a necessary precondition for propagation of a thermal detonation wave, to be discussed in the next section. Further experiments are needed to resolve this point.

3. B-H THEORY

Board, Hall and Hall[10] were the first to apply detonation theory to fuel-coolant interactions. The shock is treated as a plane, steady one-dimensional detonation proceeding through initially coarsely-mixed fuel, coolant liquid and coolant vapor. The shock collapses the vapor blankets, inducing a large relative velocity between the fuel drops and surrounding coolant liquid. At subcritical pressures boiling mechanisms are operative, but at supercritical pressures the drops can fragment only by boundary-layer stripping and/or Taylor instability. The latter mode is predicted to be dominant for Bond numbers, $Bo > 10^5$, based on local relative velocities[24]. The proportionality constant in the theoretical expression for the dimensionless breakup time was estimated from air-water data:[26-28]

$$T_b \equiv (\rho_d/\rho_f)^{\frac{1}{2}} (u_r t_b/r_d) \cong 44\, Bo^{-\frac{1}{4}} \tag{1}$$

$$Bo = \rho_f g r_d^2/\sigma = 3u_r^2\, r_d C_d/8\sigma \tag{2}$$

with $C_D \sim 2$. The time for velocity equilibrium was estimated from the time for acceleration of an isolated drop up to the free-stream velocity. Applying the single-phase (or homogeneous flow) conservation laws for mass, momentum and energy across the fragmentation zone, one obtains[58,85,86]

$$\tfrac{1}{2}(P_1+P_2)(V_1-V_2) = U_2-U_1 \tag{3}$$

where P,V and U are the pressure, specific volume and internal energy, respectively, of the mixture on a volumetric basis, and the subscripts 1 and 2 refer to the undisturbed region ahead of the shock and any plane behind the shock front. Immediately behind the shock, no heat transfer has yet taken place, and an equation of state $U_2(P_2,V_2)$ determines the shock adiabat, shown in Fig. 7 as the curved line AN ; while after temperature equilibration has taken place, the equation of state now determines the equilibrium Hugoniot (line CO). The propagation velocity of the shock is given by

$$u_1 = V_1 \left(\frac{P_2-P_1}{V_1-V_2}\right)^{\frac{1}{2}} \tag{4}$$

which implies that all downstream states lie on a straight line from the initial state A to a point on the shock adiabat, N. For a combustion shock in a gas with a single reaction proceeding at a finite rate, stability considerations dictate that the shock velocity be minimized, which implies that the line , which is tangent to the equilibrium Hugoniot, determines the shock velocity. The tangency point, O, is called the C-J (Chapman-Jouguet) point. The mixture is compressed by the shock front to the point, N, where the velocity relative to the shock is subsonic. This assumes that homogeneous flow, corresponding to equal velocities of the two liquids at every instant, prevails, so that the mixture internal energy is uniquely defined by the pressure and specific volume.* As fragmentation proceeds and the mixture expands, the velocity increases until, at the C-J point, it is sonic relative to the shock front. Taking the Chapman-Jouguet condition for the end of the fragmentation region, and solving the jump shock balance, supercritical pressures were predicted behind the shock (~ 10^2 MPa for tin/water and ~ 10^3 MPa for UO_2/Na). These calculations did not take into account the slowing-down of the coolant in a dense droplet dispersion, and more detailed analyses have therefore been performed.

3.1 Chapman-Jouguet Conditions

It is instructive, first of all, to examine the jump mass momentum, and energy balances across the shock front:

$$[\sum_i \alpha_i \rho_i u_i] = 0 \quad (5)$$

$$[\sum_i \alpha_i (P_i + \rho_i u_i^2)] = 0 \quad (6)$$

$$[\sum_i \alpha_i \rho_i u_i (h_i + \tfrac{1}{2} u_i^2)] = 0 \quad (7)$$

where the subscript i takes on the values cv, cl, and f, referring to coolant vapor, coolant liquid and fuel, respectively, and the square jump brackets refer to values of a point 1 before and a point 2 behind the shock. If velocity and pressure equality exists between the phases (homogeneous flow) at both points 1 and 2, Eqs. (5) and (6) reduce to those for classical single phase detonation theory[58]. Providing the mixture at point 2 has a well-defined and unique specific volume, temperature equilibrium is not necessary. In particular, a plane (called the Chapman-Jouguet or C-J plane) of local sonic velocity relative to the shock front must exist at the end of the fragmentation and velocity-equilibrium zone in order to prevent weakening of the shock by rarefaction waves from the far field. If the fragmentation process takes place by removing small debris particles, either by boundary layer stripping or by surface boiling mechanisms, the degree of fragmentation, E, at the C-J plane can be treated as a parameter. If now it is assumed that heat transfer from the unfragmented portions of the fuel drops can be neglected, and that the debris particles equilibrate in temperature and velocity instantaneously with the surrounding coolant, families of partial-fragmentation Hugoniot curves can be constructed, as in Fig. 8.[25] The pronounced knee on the curves for $E < 1$ marks the point at which vapor disappears as the pressure increases. From this simple model one sees immediately that the tangent from this particular initial point to the reaction adiabat for $E < 1$ will always intersect it close to the knee, and hence at subcritical pressures. Indeed, a number of experimental studies of tin-water propagating interactions have failed to yield supercritical pressures behind the shock front[78,87,88,89].

*In fact, there are large relative velocities between the unfragmented fuel drops and the coolant behind the shock. Hence the state of the system is no longer uniquely defined on a (P,V) diagram, but also depends on the slip velocity ratio, S. However, at the end of the reaction zone (C-J plane) the tangency condition can be applied, provided that velocity equilibrium has been attained. This poses the theoretical difficulty that the relative velocity becomes zero only at infinity, but in practice a close approximation may be obtained within a few centimeters or tens of centimeters. Likewise, supercritical detonations, if they exist, are never truly steady-state, but at best slowly-varying.

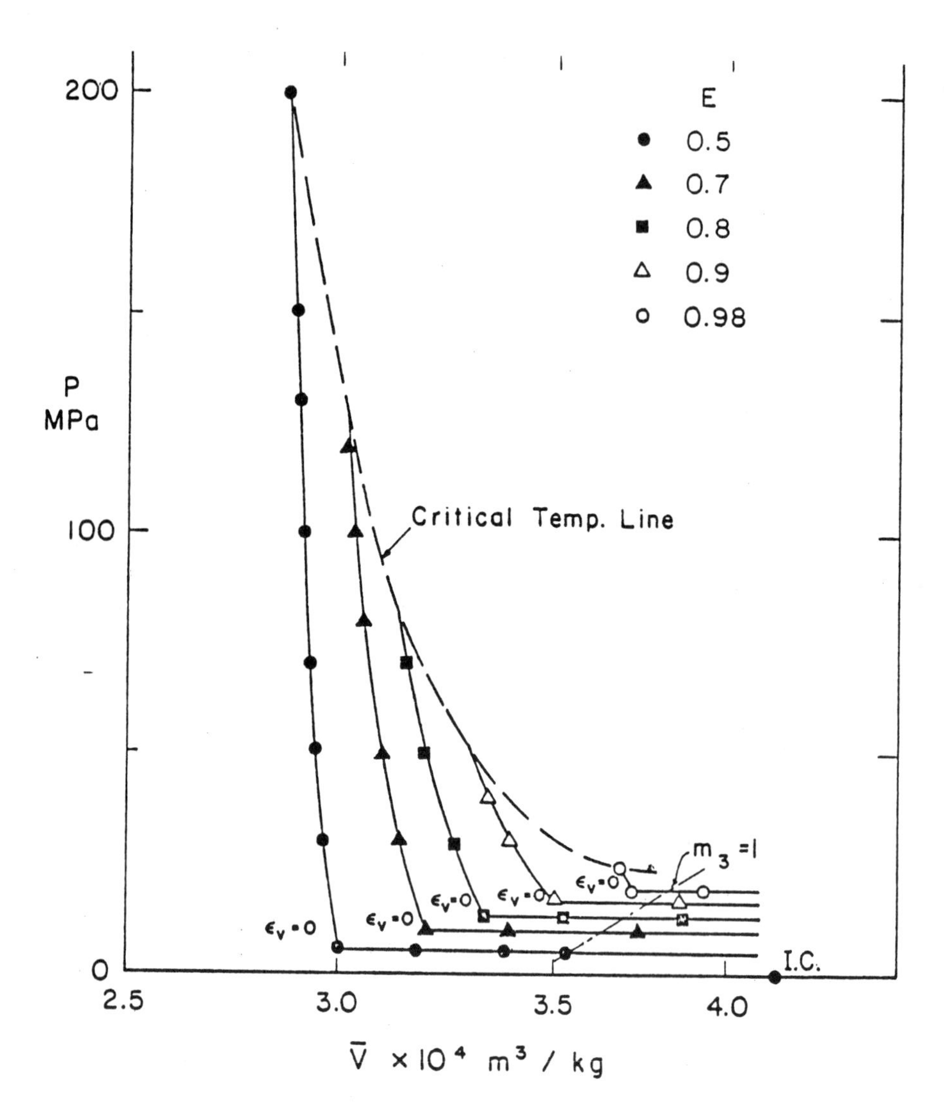

Fig. 8 Partial fragmentation Hugoniot curves for a tin-water mixture.[25] Initial conditions: Vol. fraction fuel drops, α_{d1} = 1/3 vol. fraction coolant vapor in coolant liquid/vapor mixture ϵ_{v1} = 0.5 (equal volume of coolant vapor and liquid). Mass fuel/mass coolant = 6.5. Parameter E is mass fraction of the fuel drops fragmented at the Chapman-Jouguet plane. m_3 = 1 represents the sonic velocity line calculated from a separated-flow formula.

For UO_2/Na the same effect exists, although less pronounced. For initially equal volumes of molten UO_2, liquid sodium and sodium vapor, Fig. 9 shows that only subcritical pressures can be achieved if E = 0.5 at the sonic velocity plane, but supercritical pressures are possible if E = 0.7. The points on the partial-fragmentation Hugoniots where the Mach number M_3 = 1 determined from the separated flow multiphase sonic velocity criterion [90,91,92] differ somewhat from the tangency condition points (homogeneous flow assumption), as might be expected.

The requirements for a stable C-J plane[58,59,90,92] have been analyzed. They demand that the expression for the pressure gradient, deduced from the separated-flow differential equations of the fragmentation zone, be indeterminate, corresponding to the simultaneous vanishing of the numerator and denominator. Physically this is necessary in order to match the time-independent fragmentation region to the time-dependent supersonic expansion zone. With the assumption that the fuel debris particles form a homogeneous fluid mixture with the coolant, the pressure gradient is given by[92]

$$\frac{dP}{dz} = \frac{\frac{3}{8} C_D \rho_f \frac{\alpha_d}{r_d} u_r^2 \left(\frac{1}{\rho_f u_f} - \frac{1}{\rho_d u_d}\right) - \Gamma_f\left(\frac{2}{\rho_f u_f} - \frac{u_d}{\rho_f u_f^2} + \frac{1}{\rho_d u_d}\right)}{\frac{\alpha_f}{\rho_f u_f^2}\left(1 - \frac{u_f^2}{C_f^2}\right) + \frac{\alpha_d}{\rho_d u_d^2}\left(1 - \frac{u_d^2}{C_d^2}\right)} \quad (8)$$

where the mass source function, Γ_f, is given by

$$\Gamma_f = \frac{\alpha_d \rho_d u_d}{1-E} \frac{dE}{dz} \quad (9)$$

At supercritical pressures, boiling mechanisms for fragmentation cannot be operative, and velocity equilibrium (u_r = 0) implies that Γ_f is also zero. The vanishing of the denominator of Eq. (8) is the well-known requirement for the attainment of choked flow in a separated-flow system. This may be termed a normal, or equilibrium, C-J plane, corresponding to zero entropy production. Small disturbances would be expected to damp out, due to irreversibilities, resulting in a stable condition. Strictly speaking, entropy production due to heat transfer does not cease until temperature equilibrium is achieved. Since there is a finite relative velocity cutoff for fragmentation at supercritical pressures, the requirement of zero entropy production implies that the C-J plane is located at a great distance behind the shock front. In practice, however, the fragmentation is essentially cut off in a relatively short effective zone length[92], as shown in Fig.10.

The conclusion that one can draw is that steady supercritical thermal detonations cannot exist, although in practice slowly-varying detonations are not ruled out. It is also possible that the numerator of Eq. (8) can be made to vanish at an isolated point, or points, with slip at supercritical pressures. However, in the absence of vapor there seems to be no obvious restoring force driving the perturbed system back to the singular C-J plane. On the other hand, at subcritical pressures a small increase in pressure tends to collapse the vapor layers, resulting in increased heat transfer and hence more vapor production. A stable C-J plane with u_f = 0 is then possible.

Further progress requires that an estimate be made of the reaction zone length for assumed initial conditions and drag coefficient as a function of dimensionless break-up time, T_b. The curves shown in Fig. 11 were obtained[92] by assuming the pressure and the degree of fragmentation at the C-J plane, calculating the C-J plane conditions from the component jump mass balances and

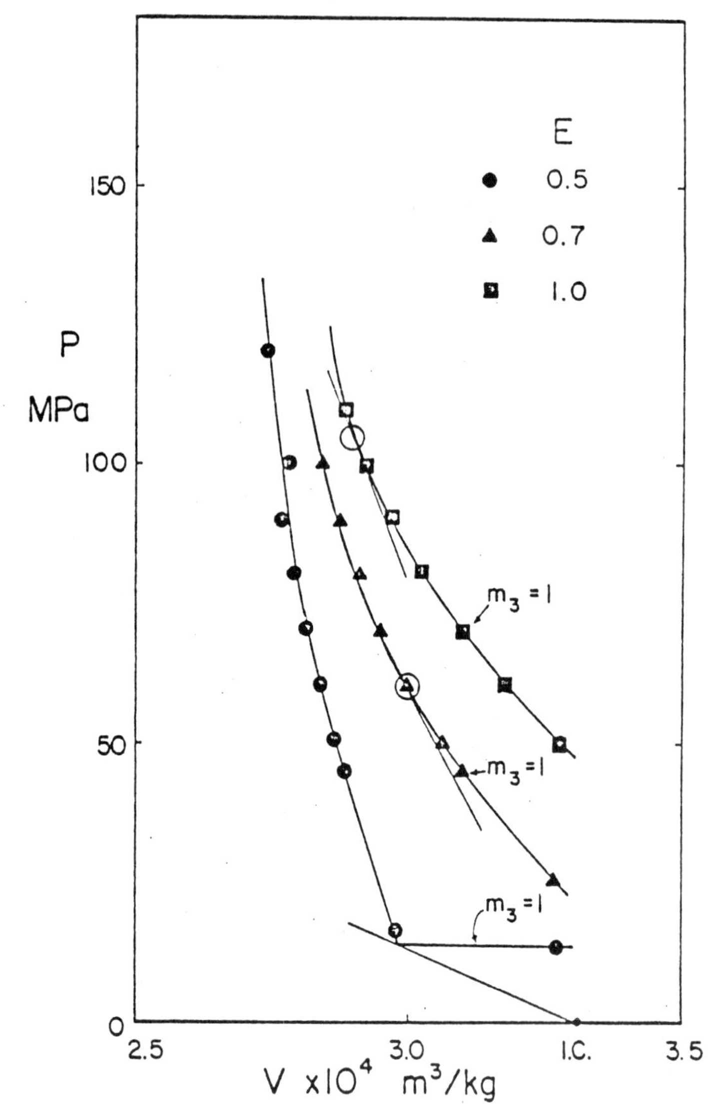

Fig. 9 Partial-fragmentation Hugoniot curves for a UO_2/Na mixture,[25] initially with equal volumes fuel, coolant vapor and coolant liquid. Mass fuel/mass coolant = 10. Fuel temp. = 3550K; liquid coolant temp. 1200K. This example shows C-J pressures of order 10^2 MPa for E = 1, in agreement with Ref. 1, but subcritical pressures if only half of the drop mass is fragmented at the sonic-velocity plane.

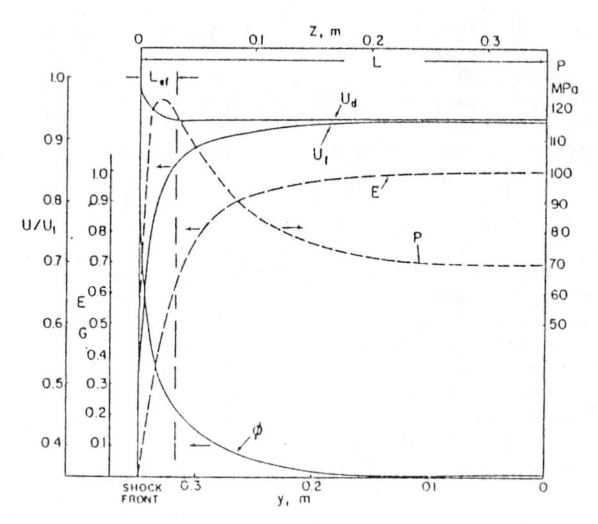

Fig. 10 Theoretical calculation of fragmentation zone behind plane shock in UO_2/Na mixture[32]. $\alpha_1 = 1/3$, $\epsilon_v = 0.5$; dimensionless breakup time $T_b = 0.9$; $C_D = 2$. Note that velocity equilibrium is never achieved, but that the relative velocity between fuel drops and coolant/fuel debris mixture is reduced to 5% of the relative velocity immediately behind the shock in an effective length, L_{ef}, of ~ 3 cm. Here α_d is the volume fraction of fuel drops ahead of the shock; ϵ_{v1} is the vapor fraction in the coolant ahead of the shock; L represents the length required for the relative velocity of fuel and coolant to be reduced to 0.5% of its value immediately behind the shock. φ is a dimensionless heat transfer.

the total momentum jump balance, and integrating the four mass and momentum differential equations backwards from a point where u_r/u_1 = 0.005 to the shock front, taken to be given by E = 0. The Reinecke-Waldman[26] expression for the rate of fragmentation was used. This expression was obtained empirically from air-water data in the boundary-layer stripping range, but a slope function appropriate to Taylor instability does not change the results materially, providing the same time scale, T_b, for fragmentation is used. It is seen that the zone length is unrealistically long for these examples if $T_b > 1.5$, which is indicated by gas-liquid experiments[26-28], as well as some mercury-water experiments[29,30]. However, other liquid-liquid experiments over a wide range of Bond numbers[31,53] give $T_b \sim 0.4$, which implies reaction zone lengths of 5-10 cm, which seems physically acceptable. This discrepancy cannot at present be considered to be fully resolved.

Scott and Berthoud[59] have independently solved the complete set of transport equations behind the steady shock, with slightly different assumptions. They assume no momentum transfer between the phases in the shock, which allows them to integrate forward from the shock through the reaction zone. The validity of this assumption is not clear, in view of the large change in relative velocity, as well as volumetric fraction, of the fuel and coolant across the shock. On the other hand, Sharon and Bankoff[25] do not make this assumption, which forces them to integrate backwards from the Chapman-Jouguet plane where the relative velocity between the unfragmentation portions of the fuel drops and the coolant/fuel debris mixture is nearly zero. The shock plane is then determined by the plane of zero fragmentation, where the pressure and velocities must be physically consistent with the shock jump balances. Nevertheless, somewhat similar results are obtained. An important assumption in both sets of calculations is that the fragmentation rate at any distance behind the shock is determined by the relative velocity between the fuel and coolant at that position, rather than solely by the relative velocity immediately behind the shock. In air/water single-drop fragmentation experiments the relative velocity is practically constant, but in a dense liquid-liquid dispersion the relative velocity can change rapidly. In addition, Scott and Berthoud point out that, in theory at least, the Chapman-Jouguet conditions cannot be satisfied until the velocities equilibrate (zero entropy generation rate).

Fishlock[60] and Jacobs[61] have performed quite different calculations, since their interest was in the escalation of the pressure shock wave from a small value (~ 10 bar) to a large value (~ 10^3 bar). Both used a three-equation (homogeneous flow) set in Lagrangian coordinates in order to make the problem tractable. Relative velocity and fragmentation effects must therefore be supplied as initial conditions, and the distortion of the original fuel-coolant packets due to their initially large, but time-varying, relative velocity is not considered. In the former case[60] the fragmentation rate is taken to be constant, determined by the initial conditions behind the shock. Sonic velocity conditions are not brought to bear. As might be expected, rapid escalation is a few centimeters is predicted. Similarly high pressures were predicted by Jacobs with different assumptions, leading to an inherently transient expulsion calculation. In this case an initial degree of fragmentation and a mixing time before heat transfer is cut off by vapor films are supplied as parameters.

The subject is clearly in need of further work, both theoretical and experimental. Foremost is the need for fragmentation rate data in freezing and non-freezing liquid-liquid systems, with and without vapor present, under well-controlled conditions. On the other hand, it can be argued that a detonation at supercritical pressures in UO_2/Na seems quite improbable for other reasons. Film boiling of subcooled sodium from molten UO_2 is unstable[33], but it has been suggested that the delay before collapse may be as long as 200 ms.[1] If the delay is much shorter (favored by small fuel drops, high relative velocity, low sodium temperature), subcooled nucleate boiling will result in rapid fragmentation[34]. The collapsing bubbles at the freezing interface produce a inwardly-pointing jet[35], which can exceed the yield strength of the growing UO_2 crust[36]. On the other hand, if the vapor film is stable for ~ 200 ms, an application of the Miles-Dienes nonlinear Taylor instability theory[37] for elastic-plastic solids indicates that the crust thickness will prevent fast breakup by Taylor instability[38]. This is summarized in the next section.

3.2 Effects of Solidification on Fuel/Coolant Taylor Instability

Cooper and Dienes[37] have given a model for fragmentation due to Taylor instability, based upon a method of generalized coordinates due to Dienes[93], which in turn stems from earlier work by Fermi[94] and Miles and Dienes[95]. The general theory makes no assumptions concerning the existence of a potential, and hence is applicable to dissipative media. When applied to plane periodic inviscid flow of an incompressible fluid (the inertia of the lighter fluid is ignored, to this approximation) potential flow can be assumed in a coordinate system in which acceleration is replaced by an effective gravity g(t). Because of the exponential growth, exp(n(k)t), of the Fourier component of the potential, $\tilde{\Phi}(k,t)$, during the early phase of instability growth, the Fourier integral for the potential is rapidly dominated by the fastest-growing wavelength, determined by $dn(k)/dk = 0$ at $k = k_o$. The potential is thus approximated by

$$\Phi(x,y,t) = \frac{\dot{q}(t)}{k^2} \cos kx \; e^{-ky} \tag{10}$$

where it is understood hereafter that k is evaluated at k_o. Here q(t) may be looked upon as a generalized coordinate, and the overdot signifies a time derivative. From this the explicit kinematic relation for the surface perturbation is obtained:

$$y(x,t) = k^{-1} \ln (1 + q(t) \cos kx); \quad 0 \le q \le 1 \tag{11}$$

which has the characteristic spike-and-bubble configuration of late-time Taylor instability as $q(t) \to 1$. The differential equation for q(t) is obtained from an energy balance over one wavelength, equating the rate of energy dissipation to the rate of change of kinetic and potential energy. The potential energy consists of "gravitational" and surface energy, and, if the denser material is frozen or freezing, elastic energy. When the deformation energy density exceeds the yield strength, there is energy dissipation in the plastic regime, where the integration over y is over the frozen region, as well as viscous dissipation. The thickness of the frozen region, d(t), is approximately given by the plane geometry result[23]:

$$d(t) = 2\lambda^* (\alpha_d^* t)^{\frac{1}{2}} \tag{12}$$

where α_d^* is the thermal diffusivity of the fuel drops, and λ^* is the solution of a transcendental equation[26].

The mixing time scale, t_{mix}, is of the order of 10-100 milliseconds while the time scale for instability growth is two or three orders of magnitude smaller. Consequently, the time for crust growth $t_{fr} \sim t_{mix}$ and the crust thickness, d, can be considered constant during the instability growth[92].

Cooper and Dienes found that the equation describing the early instability growth is given by the following linearized second order ODE:

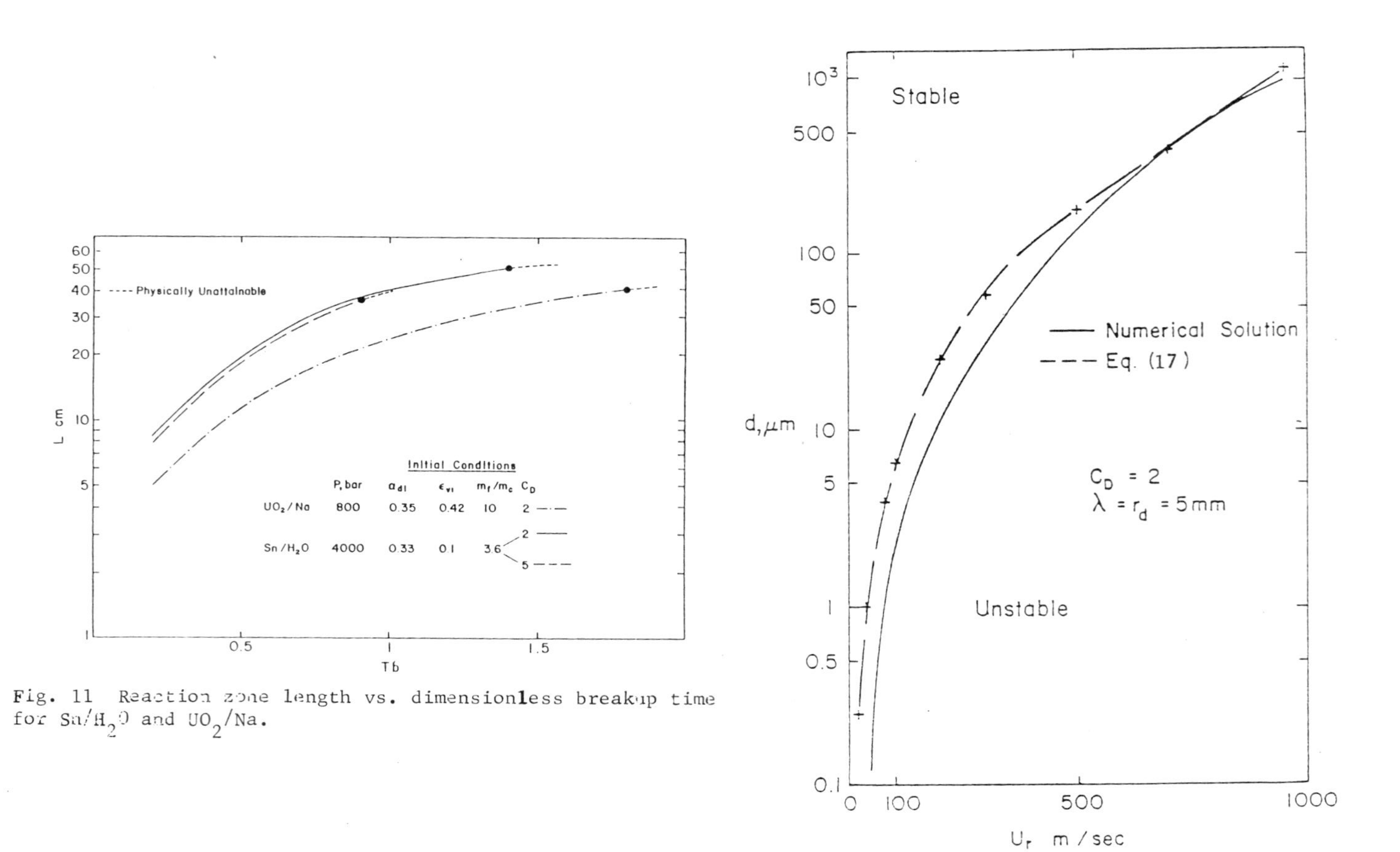

Fig. 11 Reaction zone length vs. dimensionless breakup time for Sn/H_2O and UO_2/Na.

Fig. 12 The effect of the frozen crust on the relative velocity required for the instability to grow[32].

$$\ddot{q} + 4k^2 \nu_d \dot{q} + [-\tilde{g}k + \frac{4k^2}{\rho_d}(1-e^{-2kd}) G H(\frac{Y_o}{2G} - q)] q$$

$$+ \operatorname{sgn}(\dot{q}) \frac{4k^2}{\rho_d}(1-e^{-kd}) H(q - \frac{Y_o}{2G}) = 0 \qquad (13)$$

where H(x) is the Heaviside step function, sgn(x) is the sign function, and

$$\tilde{g} = g - \frac{\sigma k^2}{\rho_d} \qquad (14)$$

Here g is the acceleration given by:

$$g = \frac{3}{8} C_D \frac{\rho_f u_r^2}{\rho_d}, \qquad (15)$$

k is the wave number ($2\pi/\lambda$), Y_o is the yield stress and G is the modulus of rigidity of the fuel (for UO_2 $Y_o = 10^9$ dynes/cm^2 and $G = 3.7 \cdot 10^{13}$ dynes/cm^2).

The initial conditions are taken to be:

$$q(0) = 0; \quad \dot{q}(0) = \pi^{-1} k u_{ro} = 2\lambda^{-1} u_{ro} \qquad (16)$$

Eq. (13), with the initial conditions (16), was solved numerically[32,92], assuming constant relative velocity ($u_r = u_{ro}$), $C_D = 2$ and $\lambda = r_d = 0.5$ cm. By varying the crust thickness, d, as a parameter for a given initial relative velocity, the critical crust thickness above which the instability will not grow can be determined. A plot of the critical crust thickness vs. the (initial) relative velocity is shown in Fig. 11.

If we neglect the effect of viscosity, the stability of Eq. (13) can be analytically determined by considering the trajectories in the $\dot{q}$ - q phase space[93]. The stability criterion becomes:

$$\dot{q}(0)^2 = \frac{k^2 u_{ro}^2}{\pi^2} \geq \frac{16 k^3 Y_o^2 (1-e^{-kd})^2}{\rho_d^2 \tilde{g}} \qquad (17)$$

and the critical crust thickness, d_{cr}, is:

$$d_{cr} = \frac{1}{k} \ln (1 - \frac{\rho_d u_{ro} \tilde{g}^{\frac{1}{2}}}{4\pi Y_o k^{\frac{1}{2}}}) \qquad (18)$$

As shown in Fig. 12 the numerical solution and Eq. (18) are in good agreement.

In all cases investigated for UO_2/Na mixture, the initial relative velocity was in the range of 100 - 200 m/s, so that the critical crust thickness is of the order of 10 μm. For an initial sodium temperature as low as 600K, $\lambda^* = 0.93$, and hence for $d_{cr} = 10$ μm, the corresponding fuel-coolant contact time is only 0.07 ms. Even if one assumes a constant early growth rate (up to 1 ms) of 2 cm/s (which may be not relevant to the present case, since it is associated with the crystallization-controlled advance of a solidification front into a highly supercooled liquid[96]), the required contact time is about 0.5 ms. Since the principal resistance to heat transfer is in the fuel, the same results apply even if contact is not made, but heat conduction across a thin vapor film takes place. Hence, we conclude that if the mixing time scale for UO_2/Na is larger than 0.1 ms, the crust will be thick enough to prevent the growth of interfacial waves and hence no hydrodynamic fragmentation can occur.

For carbide fuel (UC) the initial relative velocity can be of the order of 300 - 400 m/s ($\rho_d \sim 60$ μm (for UC $Y_o = 1.5 \cdot 10^9$ dyne/cm^2). The crust growth constant, λ^*, is 0.67, hence from Eq. (12) $t_{fr} \sim 0.4$ ms. Such rapid premixing on a coarse scale, for both UO_2 and UC fuels, seems to be very highly improbable in any large-scale event.

4. FRAGMENTATION MECHANISM

A variety of fragmentation mechanisms have been proposed for fuel-coolant interactions. One group can be classified as purely hydrodynamic, depending only on relative velocity between fuel and coolant, such as Taylor and Helmholtz instabilities, hydrodynamic deformation and boundary-layer stripping[10,24,31]. These would be the principal mechanisms which would be operative in a propagating detonation at supercritical pressures. A single supercritical detonation (7 kg of 790°C aluminum mixed with 6°C water in a strong metal vessel) has recently been reported[39], but details of the pulse width at supercritical pressures are not yet available. A second group consists of miscellaneous mechanisms such as thermal stress cracking[40,41] and gas bubble precipitation[42,43]. The former mechanism is probably too slow in terms of fuel-coolant mixing, while the latter mechanism is probably not important for UO_2/Na, in view of the low solubility of inert gases in molten ceramics. A third group may be broadly identified as boiling mechanisms, under such titles as violent boiling[44,52] compression waves[45]; bubble collapse[46,47]; jet penetration[36]; coolant entrapment[48]; and vapor blanket collapse[49]. It is seen that these titles are non-exclusive, so that more than one (or all) may be operative at once. A cyclic mechanism which has been observed in metal-water systems consists of the vaporization of entrapped coolant within the fuel mass into the surrounding subcooled coolant, followed by asymmetric collapse of the two-phase bubble which has been produced. This results in a coolant jet which penetrates the fuel, becomes superheated and vaporizes, repeating the cycle[50]. Liquid-liquid contact is necessary to trigger the cycle. An example of cyclic interaction of this type is shown in Fig. 13, where the fuel-coolant interaction (42 g. of 60/40 lead-tin alloy at $T_H = 372°C$ initially contained in a crucible below the surface of a column of water at 35°C within a vertical shock tube) was triggered by a weak shock (4 bars overpressure).

The ratio of the pressures behind and in front of the shock, P_2/P_1, determined whether a vigorous interaction was obtained (Fig. 14). This suggests that the degree of initial liquid-liquid contact, which is related to the rate at which the vapor blankets collapse, determines the strength of the subsequent interaction. The effect of the initial metal temperature on the maximum pressure in each cycle is shown in Fig. 15. There is no interaction for $T_i < T_m$, (Regions A and B); weak interaction when $T_m < T_i < T_{sn}$ (Region C), due to nucleate boiling from oxidized surfaces; strong interaction for $T_{sn} < T_i < T_{hn}$, which confirms the SN theory; and still stronger interaction for $T_{hn} < T_i < 1.0\ T_{cr}$ (Region D) with no cutoff within this region[7]. Earlier experiments, for which this shock tube was originally constructed, on destabilization of film boiling on a heated nickel tube[11] gave similar plots of the effect of metal temperature on the peak heat transfer, q_p, obtained from the slope of the heater temperature-time trace immediately after passage of the shock (Fig. 16). For the tube there is no rapid increase in surface area for $T_i > T_{sn}$ ($\sim T_{hn}$ for Freon), as will occur in liquid-liquid interaction. It is inferred that the principal mechanism for the initiation of fuel-coolant interaction is the local pressurization following vapor film collapse. Local liquid-liquid contacts for $T_i > T_{sn}$ produce local pressures of order 10^2 bar, which result in impulsive mixing. An approximation of this effect is the splash theory of Ochiai and

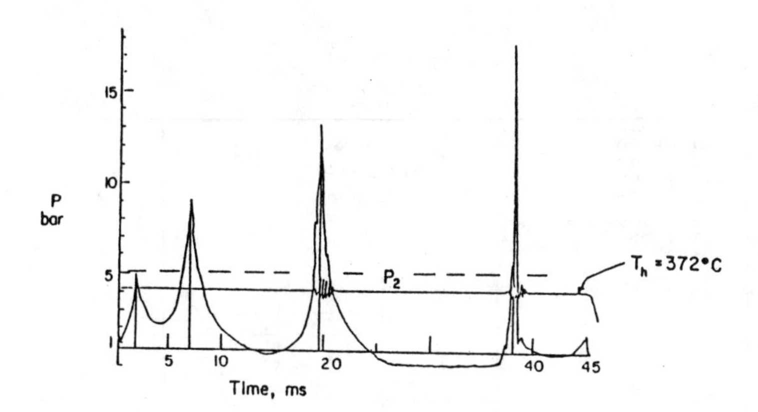

Fig. 13 Pressure and bulk metal temperature traces in shock-initiated cyclic interaction[32]. P_2 = 5.1 bar; P_1 = 1 bar; T_c = 35°C; τ = 38 μs.

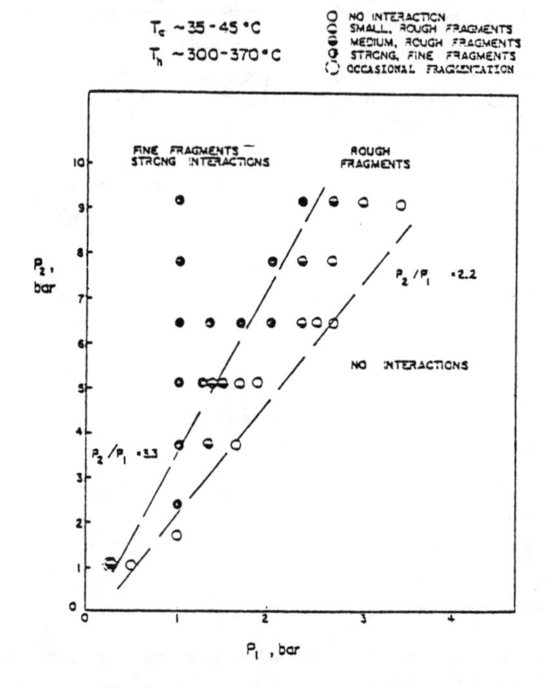

Fig. 14 Fragmentation regions as a function of the initial driver and test section pressures with water. τ = 38 μs. Scatter is principally due to random variation in vapor blanket mass at the instant of shock arrival.

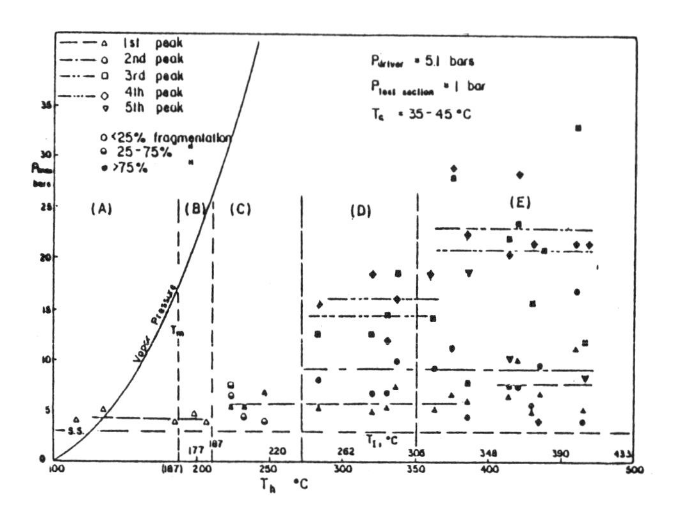

Fig. 15 Effect of the hot metal temperature on the peak pressure and fragmentation with water[7]. P_2 = 5.1; P_1 = 1 bar; T_c = 35-45°C; τ = 38 μs.

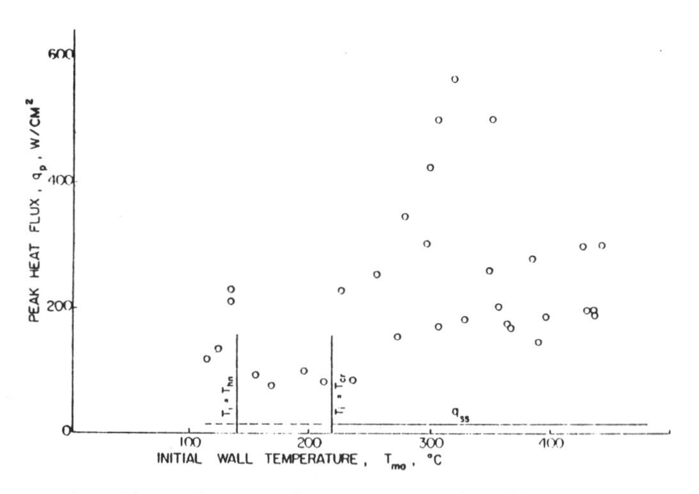

Fig. 16 Peak Heat Flux vs. Initial Wall Temperature for Freon-113 P_1 = 0.56 bar; P_2 = 4.2 bar; τ = 80 μs; T_ℓ = 22°C. q_{ss} refers to the steady-state heat flux before arrival of shock.

and Bankoff[12], which computes the annular splash jet by potential flow theory resulting from an impulsive pressure applied to a circular region on the liquid surface.

5. SOME LARGE-SCALE FCI STUDIES

To illustrate the complexity of the actual fuel-coolant interaction phenomenon, results of several large-scale studies of molten metal-water mixing are reviewed, one of which (Long[62]) is perhaps the earliest systematic study, while the others are more recent studies (Briggs[63]; Fry and Robinson[39]).

As a result of improved ingot casting methods in the aluminum industry in which large quantities of molten metal and water are in close proximity, Long[62] undertook a systematic study of explosion hazards in pouring molten aluminum in water. A total of 880 tests were carried out, in which the reference test, consisting of the sudden discharge of (22.7 kg) of molten aluminum through an 89 mm diameter hole into a clean degreased steel container partially filled with water at 12.8-25.6°C, always produced an explosion. A number of parameters were varied, with the following results:

1. No explosions occurred if the discharge hole diameter was less than 70 mm, or if the distance the metal dropped before reaching the bottom of the water container was increased from 1.2 m to 3 m. It was supposed that the metal stream breaks up during a long vertical fall, and this was checked by interposing a steel grid over the water. No explosions then occurred in 13 tests with a metal drop distance of 0.46 m.

2. As the water depth was increased, progressively higher metal temperatures were required. With a 0.15 m pool, explosions were obtained at a metal temperature of 670°C, while with a 0.25 m depth, a 750°C temperature was required. With a 0.76 m depth, a 900°C temperature was successful in producing an explosion. With 750°C aluminum temperature and a water depth of either 0.15 or 0.25 m explosions always occurred with a water temperature of 0-50° C, but never at 60-100°C.

3. Wetting agents (.01%) and soluble oils (0.5%) prevented explosions, while the use of a 15% NaCl solution enhanced the possibility of explosions. Grease, oil or paint on the bottom surface of the container prevented explosions, while lime, gypsum, rust or aluminum hydroxide coatings favored explosions. Similar results were obtained with molten magnesium.

4. Molten NaCl or KCl poured into the water always produced explosions, which were not prevented by a grease coating on the bottom, by soluble oil in the water or by a 0.5 m water depth.

Long concluded that three requirements must be fulfilled in order to have an aluminum-water explosion of the type studied in his test program:

1. Molten metal in considerable quantity must penetrate to the bottom surface of the container.

2. A triggering action must then occur.

3. Water depth, temperature and composition must have proper values.

The triggering action was ascribed to the sudden conversion to steam of a thin layer of water trapped below the incoming metal. The prevention of an explosion when using deep water, small amounts of metal, or metal dispersed into many small streams was related to the first requirement above. The inhibiting effect of paint, grease or oil was related to the second requirement, in that a thin water layer was not trapped on the poorly-wetted bottom surface. Long further hypothesized that hot water or the use of wetting agents stabilized the formation of vapor blankets around the metal particles, preventing rapid heat transfer. On the other hand, blanket formation was hindered by a salt solution. It was further concluded that chemical reactions between the aluminum and water played a minor role, since hot water or 0.5% soluble oil stopped the explosion, and since the coating on the bottom container was so important.

From these experiments one can deduce that a fairly energetic trigger is required to set off an aluminum-water explosion in the pouring mode of contact. Small differences in surface treatment were very important, and unless a layer of water was trapped by the aluminum at the bottom of the container, solidification took place quietly while the film boiling regime was still stable. However, the superheated entrapped water layer supplied a sufficiently large pressure pulse to mix the aluminum and water, producing rapid heat transfer and further pressurization. Similar experiments were performed with much better instrumentation by Briggs[33], who poured about 2 kg. of aluminum into water into a container with transparent walls. He was unable reliably to reproduce the "standard" explosion experiment of Long, possibly due to small differences in the methods of degreasing the container prior to the test.

These small differences, based on surface properties, argue strongly for the entrapment of coolant by the fuel as a triggering mechanism. In the latter experiments[33] the metal stream entered the water surrounded by a thick vapor film, reached the base of the vessel, and spread radially across the base and up the walls of the vessel. An interaction might be triggered a few hundred milliseconds after the initial entry of the metal into the vessel. Typically, the metal-water zone had spread about 100 mm up the wall of the vessel at the time of interaction, at which time detail disappeared in a small zone near the base of the vessel, spreading through the whole metal-water region. No clear front was observed but the disturbance involved a 100 mm region in about 0.5 ms, implying propagation at an average velocity of over 200 m/s. Usually small pressure pulses were observed before any visual change was seen on the film. There was a large range of maximum pressures (1 - 150 atm), with rise times and spike widths of 10 - 50 μ s, occurring 1 - 1.5 ms after the first indication of positive pressure in the vessel. On the other hand, Hess and Brondyke[84], who photographed experiments similar to those of Long, observed rapid volatilization of the organic coatings upon contact with the molten metal, which presumably prevented water entrapment. It thus seems likely that the interaction with the wall led to entrapment of some water, and hence triggered the disturbances.

Fry and Robinson[39] continued these experiments with about 20 kg. of aluminum or tin heated to temperatures up to 800°C and dropped into a tank of water. A wide range of temperature, geometry and triggering conditions was covered. Three types of interaction were observed:

i) Coherent interaction over a large region

ii) Localized interaction which did not propagate

iii) Incoherent interaction of a large region with a sequence of small interactions at different locations which might damp out, or continue until all the mixture had interacted

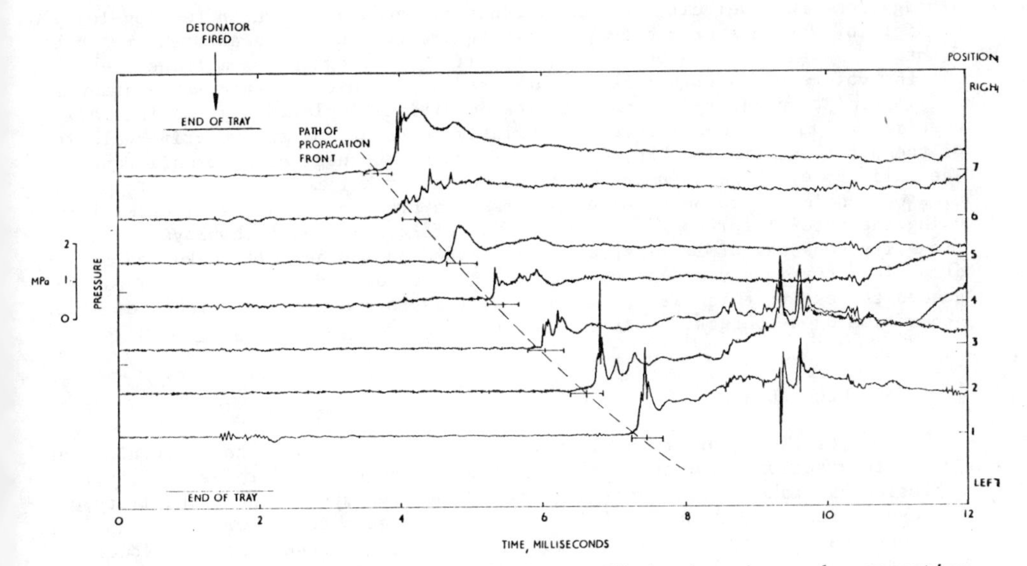

Fig. 17 Pressure records from test No. 107 showing observed propagation front.

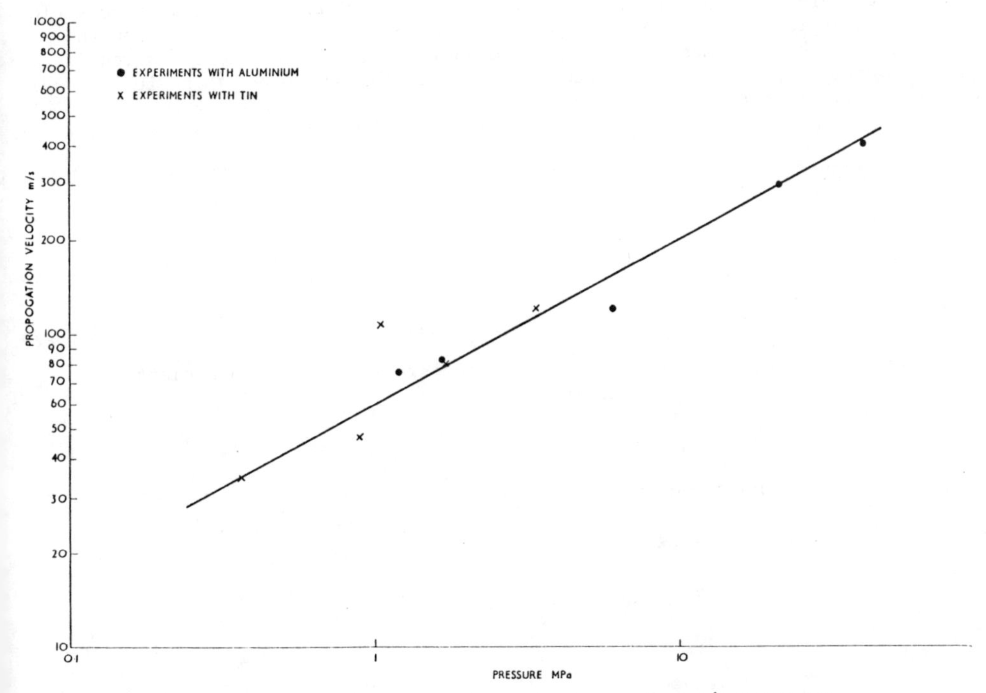

Fig. 18 Variation of propagation velocity with peak pressure.

Fig. 17 shows a propagating pressure front for 800°C tin poured into 85°C water in a narrow tank fitted with an array of pressure transducers. The interaction was triggered at one end by firing a detonator. It is seen that the propagation front velocity is nearly constant, and, as shown in the log-log plot of Fig. 18, varied approximately as the square root of the peak pressure of the front. This seems to agree with Eq. (4) if the following assumptions are made: 1) the volume of vapor per kg. of mixture in the initial state is constant for all initial temperature and mass ratios conditions employed; 2) the initial specific volume of the mixture is similarly constant; 3) the relative velocity of the water and tin leaving the shock front can be neglected for all runs. The validity of these assumptions needs to be further investigated. The breakup time was estimated to be of order 0.1 ms with breakup distances of order 1 cm, using the Patel-Theofanous[31,53] correction of Eq. (1). This employs a proportionality constant of 1.5 instead of 44, based on isothermal breakup of mercury drops in water in a vertical shock tube. As noted above, additional data are needed to resolve this discrepancy, as well as to examine the effects of vapor generation upon contact and of surface freezing.

6. CONCLUDING REMARKS

Most of the effort on nuclear reactor vapor explosions has been aimed at UO_2/Na interactions, for which it is not known whether a large-scale vapor explosion can take place. For the reasons outlined above, the probability of a steady detonation at supercritical pressure appears to be very low. However, the consequences of vigorous interactions with longer rise times and lower pressures must be explored for each accident scenario.

On the other hand, steel/water and lava/water explosions are well-known, so that there is no question that a vapor explosion can occur in a conceivable light-water reactor accident. However, a recent paper[51] contending, on the basis of risk assessments, that the probability of a steam explosion leading to containment failure must reach a large fraction of unity before it seriously impacts on the total risk, may be noted.

NOMENCLATURE

Subscripts

cr	Critical
d	(Fuel) drops
f	Fluid surrounding drops, consisting of liquid coolant and fragmented fuel particles
H	Hot liquid
hn	Homogeneous nucleation
i	Interface or phase index
m	Melting
o	Initial
r	Relative

sn	Spontaneous nucleation
1	Plane ahead of shock front
2	Plane behind shock front
C-J	Chapman-Jouguet plane
C_D	Drag coefficient
E	Mass fraction of fuel which has been fragmented
g	Acceleration
h	Enthalpy
k	Wave number
P	Pressure
q	Generalized coordinate, dimensionless amplitude
T	Temperature
T_b	Dimensionless breakup time = $(\rho_f/\rho_d)^{\frac{1}{2}} U_{ro} t_b / 2r_d$
t	Time
t_b	Breakup time
U	Internal Energy
u	Velocity
V	Specific volume
x	Distance parallel to the unperturbed interface
y	Distance normal to the unperturbed interface
Y	Yield stress
z	Distance coordinate behind shock front
α	Phase volumetric fraction
α_d^*	Thermal diffusivity of fuel drops
λ^*	Constant, Eq. (12)
ρ	Density
σ	Surface tension
Γ_f	Mass source function for fuel debris, Eq. (9)
Φ	Velocity potential function

ACKNOWLEDGMENT

This work was supported by the U.S. Department of Energy under Contract DE-AC02-76ET37210.M005.

REFERENCES

1. Board, S.J. and Caldarola, L. 1977. Fuel Coolant Interactions in Fast Reactors, in Thermal and Hydraulic Aspects of Nuclear Reactor Safety, O.C. Jones and S.G. Bankoff, eds., ASME, N.Y. 2:195.

2. Bankoff, S.G. 1978. Vapor Explosions - A Critical Review, Proc. Sixth Int. Heat Transfer Conference. 6:355.

3. Goldhammer, H. and Kottowski, H.M. 1976. CONF-761001. 2:1889.

4. Wright, R.W. et al. 1966. US AEC Rpt. STL-372-50.

5. Darby, K., Pottinger, R.C., Rees, N.J.M. and Turner, R.G. 1972. Proc. Int. Conf. on Eng. of Fast Reactors for Safe and Reliable Operation, Karlsruhe.

6. Segev, A. 1978. Thesis, Northwestern Univ., Evanston, Ill.

7. Sharon, A. and Bankoff, S.G. 1979. Fuel-Coolant Interaction in a Shock Tube with Initially-Established Film Boiling, ANS/ENS Int. Conf. on Fast Reactor Safety, Seattle, Wash.

8. Fauske, H.K. 1972. Trans. ANS. 15:813.

9. Fauske, H.K. 1973. Nucl. Sci. Eng. 51:95.

10. Board, S.J., Hall, R.W. and Hall, R.S. 1975. Nature. 254:319.

11. Inoue, A. and Bankoff, S.G. 1978. Destabilization of Film Boiling due to Arrival of a Pressure Shock, in Topics in Two-Phase Heat Transfer and Flow, S.G. Bankoff, ed. ASME, N.Y., p. 77.

12. Ochiai, M. and Bankoff, S.G. 1976. A Local Propagation Theory for Vapor Explosions, PNC-N251 76-12. 1:129.

13. Wakeshima, H. and Takata, K. 1958. J. Phys. Soc. Japan. 13:1398; 1959 p. 568.

14. Apfel, R.E. 1971. J. Chem. Phys. 54:62.

15. Blander, M., Hengsteberg, D. and Katz, J.L. 1971. J. Phys. Chem. 75:3613.

16. Skripov, V.P. and Pavlov, P.A. 1970. High Temp. 8:782.

17. Skripov, V.P. 1974. Metastable Liquids, John Wiley and Sons, N.Y.

18. Briggs. L.J. 1950; 1951. J. Appl. Phys. 21:721; J. Chem. Phys. 19:970.

19. Gunnerson, F.S. and Cronenberg, A.W. 1978. J. Heat Transfer. 100:734.

20. Henry, R.E., Fauske, H.K. and McUmber, L.M. 1976. Vapor Explosions with Subcooled Freon, PNC-N251 76-12. 1:231.

21. Dullforce, T.A., Reynolds, J.A. and Peckover, R.S. 1978. Interface Temperature Criteria and the Spontaneous Triggering of Small-Scale Fuel-Coolant Interaction, Culham Lab. Rpt. CLM-P517.

22. Armstrong, D.R., Testa, F.J. and Raridon, D., Jr. 1971. Argonne National Lab. Rpt. ANL-7890.

23. Amblard, M. and Berthoud, G. 1979. UO_2-Na Interactions - The CORECT-II Experiment, OECD-CSNI, Bournemouth, U.K.

24. Harper, E.Y., Grube, G.W. and Chang, I.D. 1972. J. Fluid Mech. 52:565.

25. Sharon, A. and Bankoff, S.G. 1978. Propagation of Shock Waves through a Fuel/Coolant Mixture, Topics in Two-Phase Heat Transfer and Flow, ed. by Bankoff, S.G., ASME, N.Y. p. 51.

26. Reinecke, W.G. and Waldman, G.D. 1970. Third Int. Conf. on Rain Erosion and Related Phenomena, Hampshire, England.

27. Simpkins, P.G. and Bales, E.L. 1972. J. Fluid Mech. 55:629.

28. Dabora, E.K. and Fox, G.E. 1972. Astronaut. Acta. 17:669.

29. Baines, M. and Buttery, N.E. 1975. CEGB Rpt. RD/B/N3497, Berkeley Nucl. Lab., U.K.

30. Baines, M. 1979. Hydrodynamic Fragmentation in a Dense Dispersion, OECD CSNI Spec. Mtg. on FCI, Bournemouth, U.K.

31. Patel, P.D. and Theofanous, T.G. 1978. Nature. 247:142.

32. Sharon, A. 1979. Thesis, Chem. Eng. Dept., Northwestern Univ., Evanston, Ill.

33. Henry, R.E. 1972. Trans. ANS. 15:420.

34. Ganguli, A. and Bankoff, S.G. 1978. Sixth Int. Heat Transfer Conf., Toronto, Canada. 5:149.

35. Plesset, M.S. and Chapman, R.B. 1971. J. Fluid Mech. 47:283.

36. Buchanan, D.J. 1973. Phys. D: Appl. Phys. 6:172.

37. Cooper, F. and Dienes, J. 1978. Role of Rayleigh-Taylor Instabilities in Fuel-Coolant Interaction, Los Alamos Scientific Lab., Rpt. LA-UR-77-1945.

38. Sharon, A. and Bankoff, S.G. 1979. Modeling of Steady Plane Detonation Waves, ANS/ENS Conf. on Fast Reactor Safety, Seattle, Wash.

39. Fry, C.J. and Robinson, C.H. 1979. Experimental Observations of Propagating Thermal Interactions in Metal/Water Systems, OECD-CSNI Spec. Mtg. on FCI, Bournemouth, U.K.

40. Cronenberg, A.W., Chawla, T.C. and Fauske, H.K. 1974. Nucl. Eng. Design, 30:443.

41. Zyszkowski, W. 1975. Int. J. Heat Mass Transfer. 18:271.

42. Epstein, M. 1974. Trans. ANS. 19:249.

43. Epstein, M. 1974. Nucl. Sci. Eng. 15:462.

44. Swift, D. and Baker, L. 1965. ANL-7152, p. 87.

45. Caldarola, L. and Kastenberg, W.E. 1974. On the Mechanism of Fragmentation During Molten Fuel-Coolant Interactions, CONF-740401-P3, p. 937.

46. Board, S.J. et al. 1972. Fragmentation in Thermal Explosions, CEGB RD/B N2423.

47. Vaughan, G.J., Caldarola, L. and Todreas, N. 1976. A Model for Fuel Fragmentation During Molten Fuel-Coolant Interactions, CONF-761001, 2:1879.

48. Brauer, F.E., Green, N.W. and Mesler, R.B. 1968. Nucl. Sci. Eng. 31:551.

49. Anderson, R.P. and Armstrong, D.R. 1977. R-22 Vapor Explosions in Nuclear Reactor Safety Heat Transfer, ed. by A.A. Bishop, F. Kulacki, O.C. Jones, Jr. and S.G. Bankoff, ASME, NY.

50. Bjornard, T.A. 1974. M.I.T. Nucl. Eng. Rpt. MITNE-163.

51. Duffey, R.B. and Lellouche, G.S. 1979. Fuel-Coolant Interactions in LWRs and LMFBRs. Relationships and Distinctions, OECD-CSNI Spec. Mtg. on FCI, Bournemouth, U.K.

52. Corradini, M.L. 1978. Heat Transfer and Fluid Flow Aspects of Fuel-Coolant Interactions, M.I.T. Nucl. Eng. COO-1781-12TR.

53. Theofanous, T.G., Saito, M. and Efthimiadis, T. 1979. The Role of Hydrodynamic Fragmentation in Fuel Coolant Interactions, OECD-CSNI Spec. Mtg. FCI, Bournemouth, U.K.

54. Briggs, A.J., Fishlock, T.P. and Vaughan, G.J. 1979. A Review of Progress with Assessment of MFCI Phenomenon in Fast Reactors Following the CSNI Spec. Mtg. in Bournemouth, April 1979, Proc. ANS/ENS Int. Mtg. on Fast Reactor Safety Tech., Seattle, Wash. III:1502.

55. Enger, T., and Hartman, D., Proc. Third Int. Conf. on Liquefied Natural Gas, Washington, D.C., September 1972.

56. Nakanishi, E., and Reid, R.C., Chem. Eng. Progress, Vol. 67, p. 36-41, December 1971.

57. Katz, D., and Sliepcevich, C., Hydrocarbon Processing, Vol. p.240-241 1971.

58. Courant, R. and Friedrichs, K.O. 1948. Supersonic Flow and Shock Waves, Interscience Publ., N.Y.

59. Scott, E. and Berthoud, G. 1978. Multiphase Thermal Detonation, Topics in Two-Phase Flow and Heat Transfer, ed. by Bankoff, S.G., ASME, N.Y.

60. Fishlock, T.P. 1979. Calculations on Propagating Vapor Explosions for the Aluminum/Water and UO_2/Sodium Systems, Fourth CSNI Specialist Mtg. on FCI in Nuclear Reactor Safety, Bournemouth, U.K.

61. Jacobs, H. 1979. Simulation of the CORECT II Experiment No. 18 with MURTI Computer Program, Ibid.

62. Long, G. 1957. Metal Progr. 71:107.

63. Briggs. A.J. 1976. Third Spec. Mtg. on Sodium Fuel Interactions, Tokyo, PNC N251, 76-12, 1:75.

64. Enger, T. and Hartman, D. 1972. Proc. Third Int. Conf. on LNG, Washington, D.C.

65. Nakanishi, E. and Reid, R.C. 1971. Chem. Eng. Progr. 67:36.

66. Katz, D. and Sliepcevich, C. 1971. Hydrocarbon Processing, 50:240.

67. Porteous, W. and Reid, R.C. 1976. Chem. Eng. Progr., 72:83.

68. Anderson, R.P. and Armstrong, D.R. 1973. AIChE-ASME National Heat Transfer Conference, Atlanta, Ga.

69. Farahat, M.M. 1971. Transient Boiling Heat Transfer from Spheres to Sodium, Ph.D. Thesis, Northwestern Univ., Evanston, Ill.; ANL-7909

70. Hall, W.B. 1977. ANL FCI Experts Mtg., OECD-Nucl. Energy Agency, Paris SINDOC(77)173.

71. Briscoe, F. and Vaughan, G.J. 1978. LNG/Water Vapor Explosions - Estimates of Pressures and Yields, UKRSR-176, U.K.A.E.A. Safety and Reliability Directorate, Culceth, Warrington, England.

72. Hicks, R.P. and Menzies, D.C. 1965. ANL-7120.

73. Waldram, K.L. 1974. M.S. Thesis, Northwestern Univ., Evanston, Ill.

74. Fauske, H.K., Conf-740401. 2:992.

75. Waldram, K.L., Fauske, H.K. and Bankoff, S.G. 1976. Can. J. Chem. Eng. 54:456.

76. Ochiai, M. and Bankoff, S.G. 1976. ANS-ENS Conf. on Fast Reactor Safety, Chicago, CONF-761001, IV, 1843.

77. MacInnes, D.A., Martin, D. and Vaughan, G.J. 1979. Fourth CSNI Specialist Mtg. on FCI in Nuclear Reactor Safety, Bournemouth, U.K.

78. Board, S.J. and Hall, R.W. 1976. Third Spec. Mtg. on SFI, Tokyo. PNC N251 76-12, 1:249.

79. Board, S.J., Hall, R.W. and Brown, G.E. 1974. CONF-740401-P2, 935.

80. Dullforce, T.A., Buchanan, D.J. and Peckover, R.S. 1976. J. Phys. D: Appl. Phys. 9:1295.

81. McCracken, G.M. 1973. UKAEA Safety Res. Bull. 11:20.

82. Wey, B.O., Hall, R.W., Board, S.J. and Baines, M. 1980. 1980 ICHMT Seminar on Nuclear Reactor Safety Heat Transfer, Dubrovnik. Yugoslavia.

83. Tso, C.P. and Tien, C.L. 1980. A Physical Model for Vapor Explosion, to be published.

84. Baines, M. and Bitter, N.E. 1979. CEGB RD/B/N4643.

85. Landau, L.D. and Lifshitz, B.M. 1959. Fluid Mechanics, Pergamon Press, Oxford.

86. Zeldovich, I.B. and Kompaneets, A.S. 1960. Theory of Detonation, Academic Press, N.Y.

87. Board, S.J. and Hall, R.W. 1973. Paper SNI 2/4, 2nd FCI Meeting, Ispra.

88. Hall, R.W., Board, S.J. and Baines, M. 1979. Observations of Tin/Water Thermal Explosions in a Long-Tube Geometry; Their Interpretation and Consequences for the Detonation Model, Paper FCI 4/20, 4th CSNI Specialist Meeting on FCI, Bournemouth.

89. Reynolds, J.A. Dullforce, T.A., Peckover, R.S. and Vaughan, G.J. 1976. SNI 6/2 at 3rd FCI Meeting, Tokyo.

90. Hall, R.W. and Board, S.J. 1977. Propagation of Thermal Detonations. Part 3: An Extended Model of Thermal Detonation, CEGB Report RD/B/N4085.

91. Wallis, G.B. 1969. One Dimensional Two Phase Flow, McGraw Hill Book Co., New York.

92. Sharon, A. and Bankoff, S.G. 1978. Propagation of Shock Waves through a Fuel/Coolant Mixture, Rpts. COO-2512-14 and -16, Chem. Eng. Dept., Northwestern Univ., Evanston, Ill.

93. Dienes, J.K. 1978. Phys. Fluids, 21(5):736.

94. Fermi, E. 1956. Los Alamos Report LA-1927.

95. Miles, J.W. and Dienes, J.K. 1966. Phys. Fluids, 9:2518.

96. Cronenberg, A.W. and Grolmes, M.A. 1975. Nuclear Safety, 16:683.

97. Henry, R.E., Gabor, J.D., et al. 1974. Large Scale Vapor Explosions, CONF-740401-P2 : 922.

98. Henry, R.E., Fauske, H.K. and McUmber, L.M. 1976. Vapor Explosion Experiments with Simulant Fluids, CONF-761001, II :1862.

21 NATURAL CONVECTION COOLING

by

D. Grand

CENG, Grenoble, France

One of the major concerns identified previously differentiating nuclear energy systems from those utilizing conventional fuel is the need to remove decay heat. This chapter is concerned with the removal of decay heat from the core of a Liquid Metal Fast Breeder Reactor (LMFBR) during a shut down consecutive to a loss of power supply to the coolant pumps. Natural convection provides an inherent, thus safe cooling process. While this chapter is mainly directed toward problems associated with the LMFBR, many of the considerations given apply equally well to LWR systems. The following specific points are covered:

- Description of the heat and flow paths
- Thermohydraulic phenomena in components (subassembly core, plenums, pipes)
- Efficiency of decay heat removal systems

1. INTRODUCTION

Under normal operating conditions, the heat generated in the core is extracted through the normal cooling circuits (primary circuit, intermediate circuit, water-steam circuit). An accidental condition considered is the complete loss of A.C. power supply to the plant. The coolant pumps stop, and the reactor trips. The residual energy production, the decay heat, must be removed without the help of the normal cooling circuit so that the temperature in the reactor vessel remains at a safe level.

Studies of these cooling conditions are taken into account in the conceptual studies of numerous plant projects. This lecture is devoted to the projects of Liquid Metal Fast Breeder Reactors.

Our intention is to illustrate some important thermohydraulic phenomena and give results whose utility is wider than the specific plant considered.

In this lecture we concentrate on results which may be useful in different plant concepts. It is divided into three parts :

- a schematic description of the systems, the normal heat removal system and the decay heat removal system, leading to a list of thermal and thermohydraulic problems.
- a review of the results available for some of the separate problems previously identified.
- an insight on the methods used to simulate, physically or analytically, the entire system. The influence of some parameters on the system's dynamics is then shown.

Reliability and safety considerations concerning the decay heat removal systems will not be discussed. This subject is treated in a session entitled "Decay Heat Removal System Reliability", of the international meeting on Fast Reactor Safety Technology (see references).

Previous reviews on the subject have been made by Singer et al - 1977 for pool type reactors, Additon and Parziale - 1977 for loop type reactors. Almost all references considered in this review are posterior to 1977 and most of them are from three meetings listed at the beginning of the references.

2. LIST OF THERMOHYDRAULIC PROBLEMS

2.1 Description of the system

A sketch of a LMFBR is shown in figure 1. The primary circuit includes the core, Intermediate Heat Exchangers (IHX) and primary pumps.

Under normal conditions, the heat is removed through the secondary loop from the IHX to the steam generators.

Under the accidental condition of a protected loss of flow, the loss of power to the pump causes the reactor to scram. However, there is a residual heat generation in the reactor which decreases with time. For example, decay heat values for Super Phenix, whose nominal power is 3000 MW, are given in table 1.

Time (s)	Decay heat (MW)
10^3	74.5
$5\ 10^3$	48.1
$5\ 10^4$	25.9
1 day	22
10 days	9

Table 1 - Decay heat versus time in Super Phenix (Debru et al - 1980)

This residual power may be removed from the primary circuit in different ways :

- through the normal heat removal system, i.e the secondary loops, steam generators, steam water loops, and to the condensers. The flow has to be maintained by natural convection in the secondary loops (figure 1). This requires heat rejection to the steam generators. Complementary heat sinks are sometimes provided on the secondary loops (see figure 9). However, as a safety requirement, the complete unavailability of the secondary loops has to be considered.
- the decay heat removal systems (DHRS), whose principle is shown in figure 2. A heat exchanger takes the heat from the primary circuit and an air cooled heat exchanger releases this heat directly to the atmosphere. The fluids used in these systems are either sodium or sodium-potassium eutectic. The fluid flow may be maintained by pumps, but as a safety requirement the DHRS must be able to function by natural circulation only.

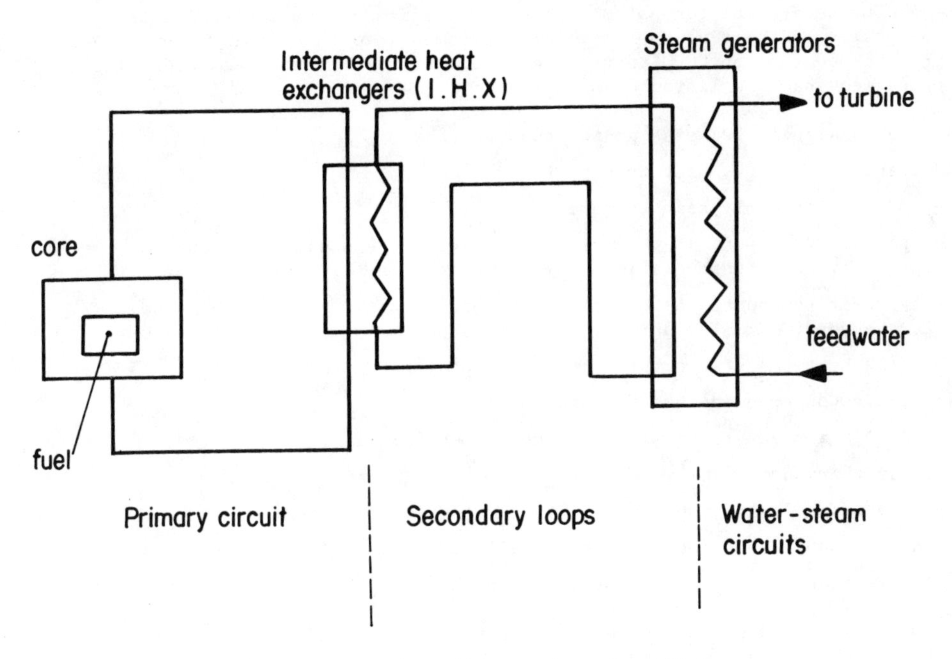

Fig. 1 - Normal cooling system of a LMFBR

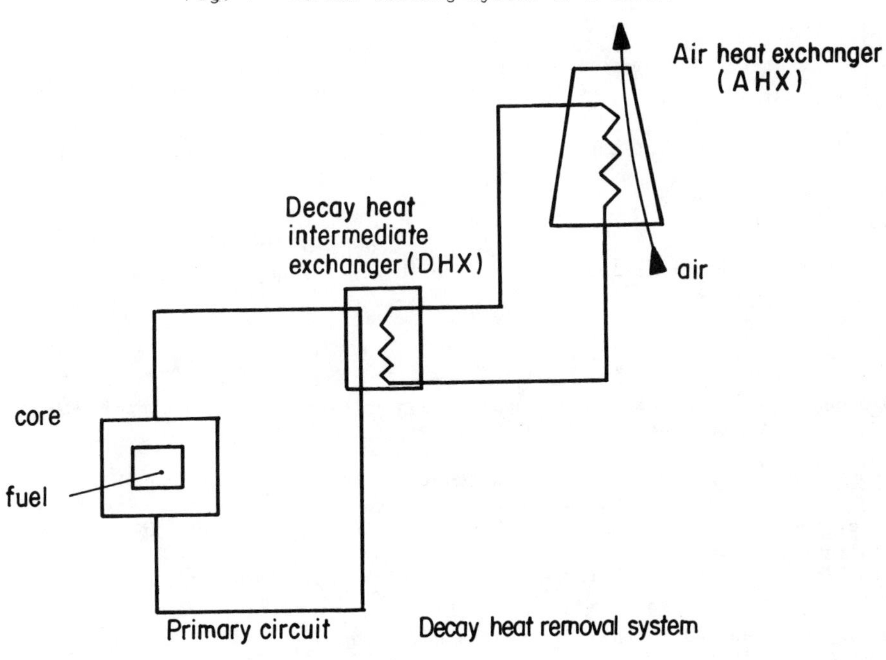

Fig. 2 - Decay Heat Removal System

2.2 Details of the heat path

The decay heat generated in the fuel region of the core transits by conduction to the clad surface. Then it comes into the primary circuit and is transported through DHRS to the atmosphere via different thermohydraulic processes that are listed below :

- the local heat transfer at the clad surface
- intra-subassembly heat transport
- inter-subassembly heat transport
- the heat transport in the exit plenum where the upward flow through the core arrives. Flow recirculations may be present in this volume.
- the heat transport in the heat exchanger of the DHRS, the decay heat exchanger (DHX). The primary coolant is on the shell side of the DHX.
- local heat transfer outside the tubes of the DHX (primary side)
- heat conduction through the pipe walls
- local heat transfer inside the tubes (DHRS circuit)
- heat transport in pipes
- convection in the tubes of the Air Heat Exchanger (AHX)
- heat conduction through the pipe wall
- convection of air on the shell side of the AHX

This completes the description of the heat flow from the core to the atmosphere. It involves three fluids :

- the sodium of the primary circuit
- the sodium (or sodium-potassium eutectic) of the DHRS
- atmospheric air

The circulation of these fluids is the object of the next paragraph.

2.3 Fluid flow

The flow in either loop is sustained by two means :

- residual pump head
- buoyant head

Pump head : After a loss of AC supply, the pump of the primary circuit can still deliver a pressure head :

- auxiliary motors, driven from sources independent of the AC supply, automatically engage the pump drive shafts as the primary pumps shut down. They are provided for in most of the projects. Although they are highly reliable, their failure is considered in the studies reviewed in this paper. The exact instant of their failure is, however, an important parameter for the onset of natural circulation.
- flywheels coupled to the pump shafts have stored mechanical energy and slow the flow coast down.

Buoyant head : The buoyant or thermal head of a thermosyphon is defined by the contour integral :

$$B \triangleq -\oint (\rho - \rho_c)\, g\, dz \qquad (1)$$

with $(\rho - \rho_c)$ the difference between local and reference densities

This can be applied to a very crude schematisation of the primary circuit given in figure 3.

In the fuel region (AB) and in the DHX (CD), we assume that the temperature varies linearly along the vertical axis Oz. It is further assumed that the temperature is constant along the legs BC and AD ($T_h > T_c$).

The density varies linearly with the temperature :

$$\rho - \rho_c = -\rho_c\, \beta\, (T - T_c)$$

By inspection of the temperature diagram, the result of the integration of equation (1) is easily found :

$$B = g\, \beta\, \rho_c\, (T_H - T_c)\, (Z_X - Z_f)$$

The buoyant head is proportional to the temperature difference and distance between centers of the heat exchanger and fuel region.

This result can easily be extended to the DHRS.

However, one must remember the simple temperature distributions that have been assumed in order to obtain this result. In reality and especially during transient evolutions, the temperature will not be uniform in the two legs BC and DA. Factors which influence the temperature distribution along BC for example, are :

- the heat exchange with the upper axial blanket of the core
- the degree of stratification or mixing of the upper plenum
- the thermal front propagation to the inlet of the DHX (in a plenum or in pipes depending on the design)

Flow resistances : The irreversible pressure losses which hinder natural circulation are :

- friction pressure loss in the bundle of a subassembly
- form losses inside and at the exit of the subassembly
- form losses at the inlet, outlet and inside the DHX
- friction losses in the tube bundle of the DHX
- pressure loss in the primary pump
- pressure loss in the core diagrid
- pressure loss in the subassembly inlet and in the diaphragms
- friction losses in pipes connecting some of these components

Moreover in the pool-type concept, there is a difference between the levels of the free surfaces of the inner and outer pools during nominal

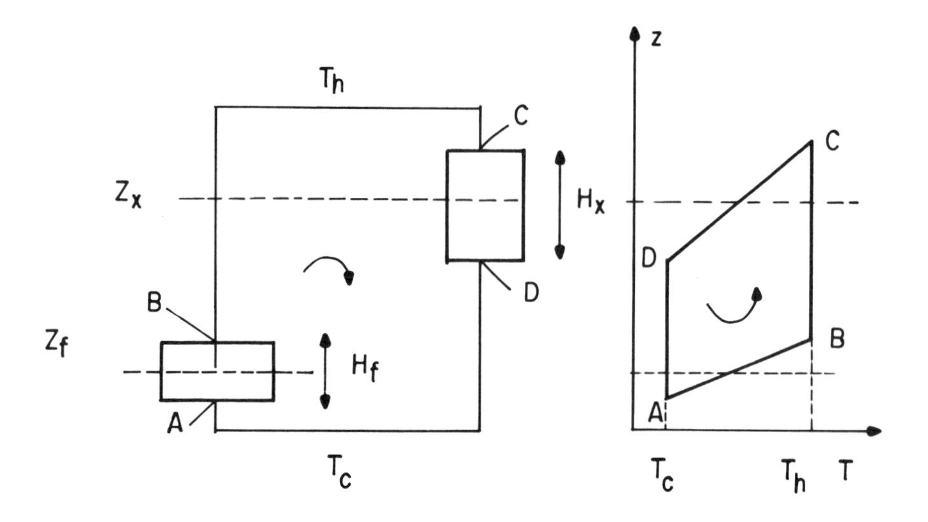

Fig. 3 - The primary circuit sketched as a thermosyphon

conditions. It corresponds to the pressure losses in the intermediate heat exchanger. In the initial period of a loss of flow, the decaying free surface level difference brings opposition to the buoyant head. The strength of this opposition increases with the volume of sodium which must flow out of the hot plenum in order to equalize the levels of the two plenums. It thus depends on the design of the pools. While it is taken into account in the initial period of the loss of flow for the design shown in figure 4.a (representative of PFR : Durham - 1976), it is negligible in designs like Super Phenix (figure 4.b).

3. THERMOHYDRAULIC PHENOMENA IN COMPONENTS

The preceeding chapter has raised a number of thermohydraulic problems which will now be discussed component by component.

3.1 The core

Low velocities characterize the flow conditions in the core during decay heat removal. Typical flow rate within the subassemblies is a small percent of the nominal flow rate. Under these conditions, buoyancy effects are expected to play an important role.

The flow and temperature distributions across the core will not be similar to those which occur under nominal conditions.

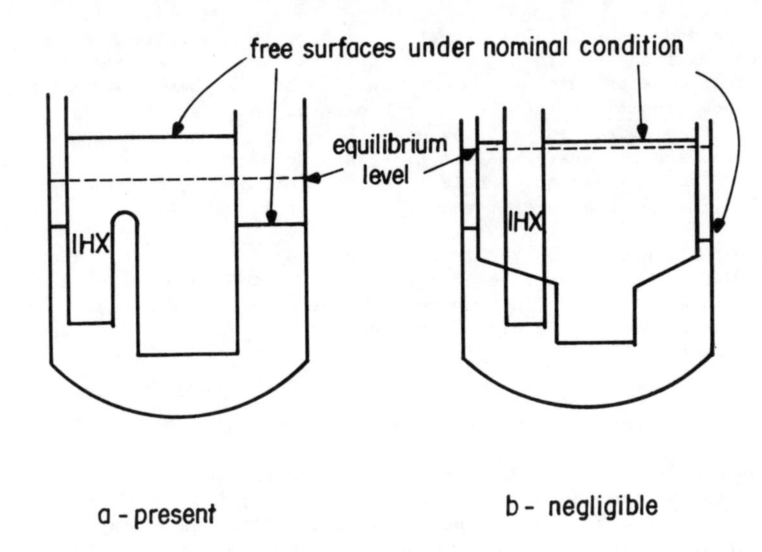

Fig. 4 - Opposition brought by the decaying free surface level difference. Influence of the design

The knowledge of flow and temperature distributions is important for the peak temperature in the core and as one element in the evaluation of the general circulation in the primary loop. An approach generally used in order to evaluate them is to consider different levels :

a) the core is divided into parallel channels, each one representative of a subassembly or of a set of subassemblies. These channels are coupled hydraulically by known pressures at both ends of the core, and thermally by conduction through the wrapper walls and by interassembly flow.

b) the bundle of the subassembly is divided into subchannels or sets of subchannels. In contrast to the preceeding case, there is no impermeable boundary between the subchannels and the coupling is more complex.

c) the subchannel is the smallest flow domain which can be considered in a complete calculation of the core. Thus laws must be provided for the shear stress and heat flux at the surface of the clad.

We now review results obtained for these different levels starting with the smallest.

3.1.1 Law for shear stress on the clad

In contrast to the nominal regime where the flow inside a subchannel is turbulent and driven by forced convection, it may be laminar or transitional during decay heat removal and may, moreover, be influenced by buoyancy.

In most studies the influence of buoyancy on the local shear has been considered as negligible. This assumption is based upon the fact that both temperature and length scales characteristic of the subchannel are in most cases too small to lead to significant buoyant forces. Pressure loss measurements made in water along a heated pin bundle (Namekawa et al - 1980) did not show significant influence of buoyancy. Local shear stress in a bundle is deduced from hydraulic tests at a low Reynolds number. For the case of a bundle with wire spacers, Engel et al - 1979 produced results showing a smooth transition from laminar to turbulent regime. This procedure is generally used for qualifying the pressure losses of the whole subassembly.

Analytical studies on the influence of buoyancy were undertaken by Yang - 1979 and Wang et al - 1980, for fully developed flow in bare rod arrays. Wang et al concluded that for CRBR core design the effect of buoyancy on the friction factor was negligible in the fuel assembly but was important in the blanket assembly.

3.1.2 Flow and temperature distributions inside a subassembly

Under nominal flow conditions, temperature differences exist at the exit of the heated section of a subassembly. They result from differences in the areas of peripheral and inner subchannels and additionnally may result from a power skew across the subassembly. These thermal differences tend to be flattened under DHR conditions - at least when steady state thermohydraulics can be assumed. This flattening results from :

- flow redistribution : the intervention of buoyancy tends to increase the velocity in the hottest parts of the subassembly to the detriment of colder ones. The higher vertical velocities in high power density regions and lateral cross flows which bring colder fluid to these regions tend to reduce the temperature differences. This flow redistribution can even lead to flow recirculation for low flow rates and large power skew (Khan et al - 1978).
- exchanges by thermal conduction : for low velocities in sodium, the thermal exchanges are dominated by the molecular transport. The longer transit times through the subassembly increase the efficiency of this effect.

These two effects can lead to strong reductions of the temperature differences across a subassembly.

The peaking factor during a loss of flow is thus substantially less than during nominal conditions. Figure 5 gives an estimate of this reduction. The peaking factor ΔT_h / $< \Delta T >$ (where ΔT_h is the temperature increase in the hot subchannel, and $< \Delta T >$ the average increase for the bundle) is given for subassemblies with uniform power distribution. Taking this reduction into account should lead to larger margins of safety limits. Thus important programs are devoted to this aspect : Leteinturier et al - 1980, Markley and Engel - 1978, Coffield et al - 1980, Singer et al - 1980.

3.1.3 Flow and temperature distribution for the core

The flow redistribution between parallel assemblies of the core is also expected to occur under decay heat condition. This effect and the longer transit time should also lead to a reduction of the peaking factor.

It has indeed been shown in EBR II tests (Singer and Gillette - 1977) : the use of temperature distributions evaluated at nominal conditions could

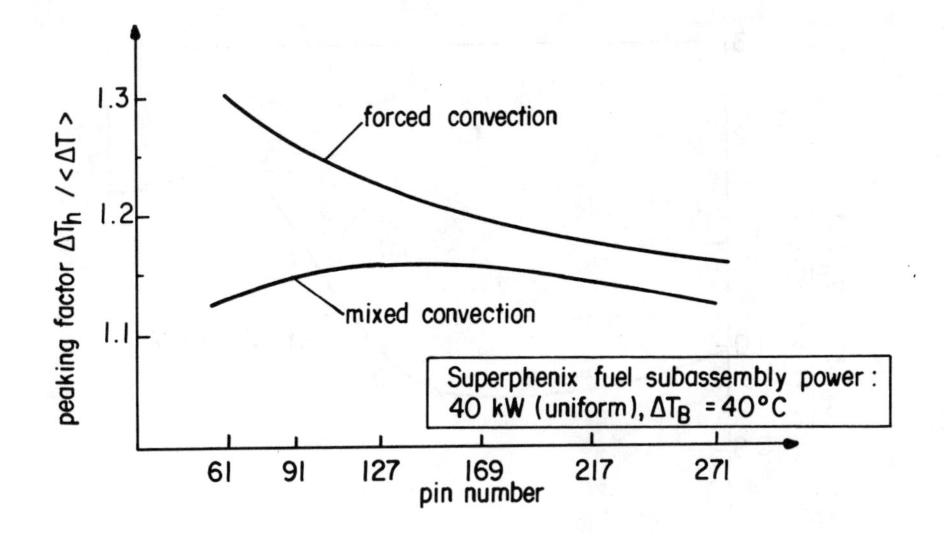

Fig. 5 - Evaluation of the peaking factor with and without the influence of buoyancy (from Leteinturier et al - 1980)

result in an overprediction of the maximum temperature rise by 45 to 65 % when applied to decay heat removal conditions.

Calculation tools are being developed for mixed convection in the core. Their elementary volumes are a set of subchannels : Coffield et al - 1980, Khan et al - 1978. Figure 6 shows a calculated flow redistribution in a core during a loss of flow. The flow factor (ratio of the mass flow rates in the hottest subassembly to the average mass flow rate) is plotted on the ordinate axis. It is normalized by its initial value (at t = 0). After 60 s, the flow factor rises to 1.2.

For extreme conditions, a flow recirculation may occur in the core, the coolant flowing upward in the fuel zone and downward through the blanket (Khatib et al - 1980 a.).

3.2 Upper plenum

Depending on the reactor's concept, the upper plenum's size and arrangement may be very different. However, some results exist which are of interest independently of the concept.

Thermal stratification of the plenum is one of them. Whereas a plenum is well mixed and quasi isothermal in nominal conditions, velocities found during DHR are generally too low to ensure mixing.

Two cases will be considered which depend on the sign of the buoyancy forces in the core's jet.

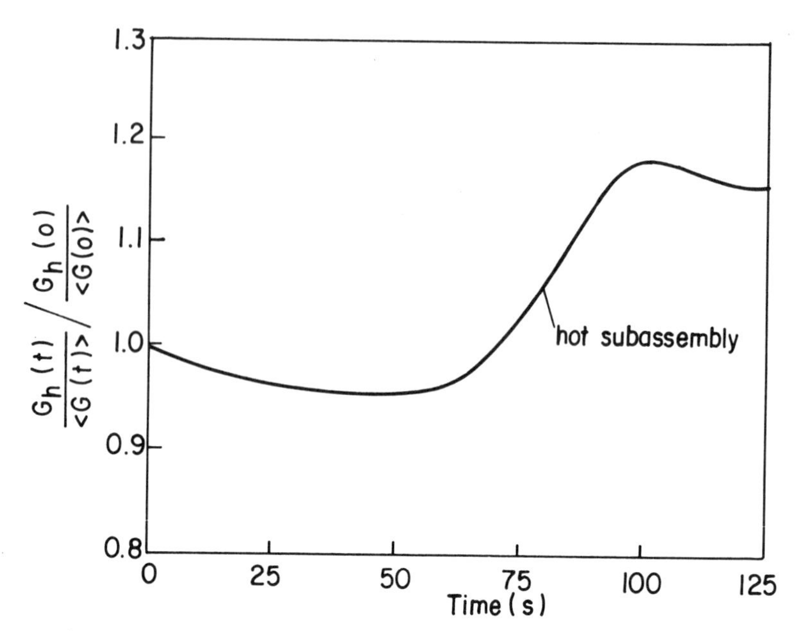

Fig. 6. Hot Subassembly Flow Relative to Reactor Flow Coastdown (Coffield et al - 1980)

a - Negative buoyancy : This situation may occur in the first instants of a protected loss of flow. The power to flow ratio decreases, causing the core outlet temperature to decrease. The jet flowing out of the core is denser than the fluid in the plenum. Depending on the pump run down, stratification of the upper plenum may occur. This is the case in a loop type reactor like FFTF for which small scale models and calculations have shown that the plenum is stratified : the denser flow coming from the core is forced downward and outward while a recirculation region develops in the upper zone. Figure 7.a gives a sketch of the flow pattern, obtained from Lorenz and Carlson - 1976. Yang - 1977 derived a correlation for the penetration height. Kinjo et al - 1980 looked at the influence of an inner barrel on the stratification.

b - Positive buoyancy : This situation may appear in long term cooling when the outer plenum is subjected to a heat sink (heat losses through the vessel, or presence of a decay heat exchanger). In this case, the jet coming from the core is warmer and lighter than the fluid contained in the plenum. Positive buoyancy tends to develop a plume above the core, centered along the core's centerline and narrower than the core's cross section. A recirculation zone develops between this plume and the heat sink (figure 7.b). This flow pattern is a schematic view of the flow visualizations reported by Astegiano et al - 1980, which were obtained in a small scale water model of the inner and outer pools of Super Phenix.

A narrow plume rising vertically along the core's centerline and a colder cross flow sweeping the peripheral subassemblies are also the two phenomena used by Anderson et al - 1980 to explain temperature distributions observed in PFR.

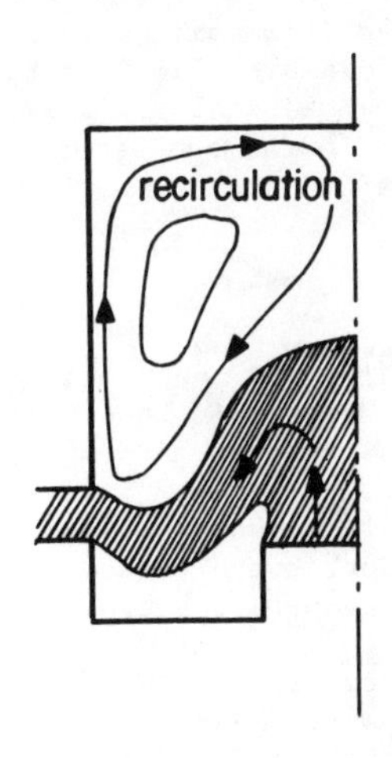

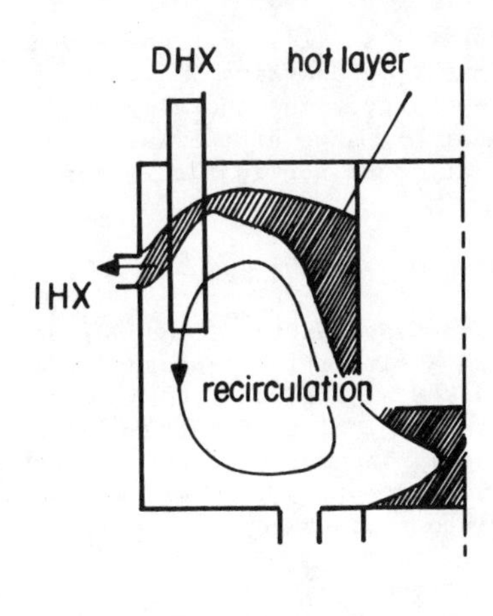

a - negative buoyancy b - positive buoyancy

Fig.7. Stratifications in the upper plenum. Defence upon the core's flow buoyancy. (a) from Lorenz and Carlson - 1976, (b) from Astegiano et al - 1980

3.3 Pipes

The piping system is an important component. In the loop type concept, the coolant of the primary circuit spends the longest time in the piping. This is not the case for the pool type concept however. In DHRS, piping is again the main component for the transit time.

The intervention of buoyancy at low flow rates is suspected in horizontal pipes. A colder fluid entering a warm pipe tends to slide under the warm sodium. As a result of this stratification effect, a temperature difference exists between the top and bottom of any cross section. Such a result should invalidate one dimensional models where a single value of the temperature is used in a cross section. In order to check this, Khatib et al - 1980 b made parallel one dimensional and three dimensional calculations for a particular transient at the inlet of a horizontal pipe. In this case, they concluded that the one dimensional representation gave results satisfactory enough for system simulation.

A systematic study has been undertaken in Grenoble in order to define the different flow regimes in a horizontal pipe (well-mixed, stratified, stratified with flow reversal).

3.4 Heat exchangers

To our knowledge, no publication relates the thermohydraulic behavior of a DHX or IHX under combined convection. If the exchanger is composed of a vertical tube bundle, some of the results obtained in subassemblies may be transposed for the shell side. Buoyancy and forced flow again act in the same direction, this time downward. However, the tube diameters are notably larger in the heat exchanger, and the buoyancy may influence the local laws for shear stress and heat exchange at the wall, in contrast to the subassembly where this effect was found negligible in most cases (§ 3.1.1).

3.5 Lower plenum

For a loop type reactor (SNR 300) Brukx et al - 1977 report experiments with sodium in a simplified representation of the reactor's lower plenum shown in figure 8.a. The downward flow represents the sodium from the breeder blanket. At the beginning of emergency cooling, it introduces a hot shock into the inlet plenum. Thermal coupling through the wall produces a heat flux from the annular to the central space. Buoyancy reinforces the forced flow. It has to be taken into account to explain the measured temperature evolution.

For a pool type reactor and when natural circulation is established, Astegiano et al - 1980 show from the small-scale water model of Super Phenix that the bottom of the plenum is filled with stratified fluid almost up to the primary pump inlet level. A positive result of this stratification is that the core's support structure is kept at a temperature lower than it would be with a mixed plenum (fig. 8.b).

3.6 Results concerning buoyant flows of liquid metals

The list of thermohydraulic aspects in components involved during decay heat removal is finished but not exhaustive. Complementary information can be found in reviews related to basic phenomena of natural or mixed convection in liquid metals : Welty - 1973, Sheriff and Davies - 1977, Grand and Vernier - 1979.

4. DECAY HEAT REMOVAL SYSTEM'S EFFICIENCY

After the presentation of the thermohydraulics of the components, the object of this paragraph is the dynamic behavior of the entire system. First we will give some characteristics of the DHRS implemented in reactors and an overview of the programs underway to improve the knowledge of these systems.

4.1 Characteristics of DHRS

The review will be restricted to the systems which produce a direct heat sink to the reactor hot pool. They are of two types (figure 9).

A) In-vessel heat exchanges which transfer heat from the hot pool to air heat exchangers.

B) Outer vessel cooling coils. They are installed outside the safety vessel in the vessel cavity. They remove heat radiated from the safety vessel. Secondary loops (on the right side of the dotted line in figure 9) may also provide decay heat removal. However, they are out of the scope of the present study.

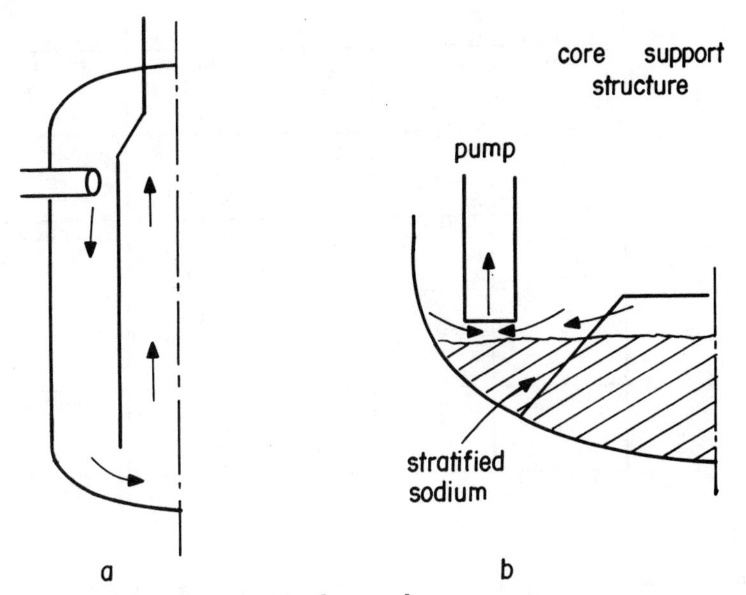

Fig.8. Stratifications in the lower plenum
(a) from Brukx et al - 1977, (b) from Astegiano et al - 1980

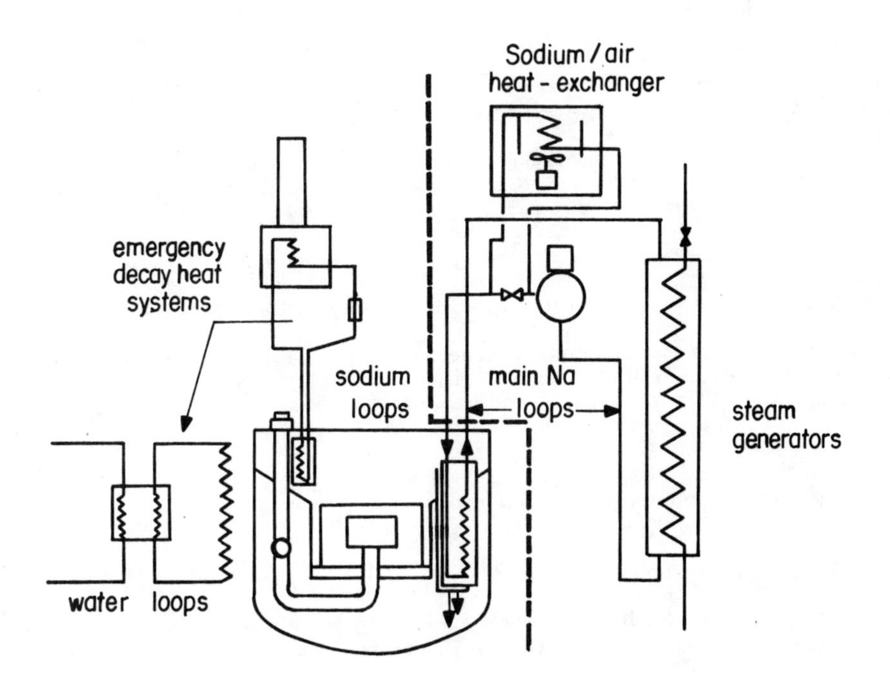

Fig. 9. Decay Heat Removal systems in Super Phenix
(Debru et al - 1980)

	type of DHR	number of loops	fluid	DHX's location	reference
CDFR (pool)	A	4	NaK	top of IHX	Broadley and Bland (1979)
Phenix (pool)	B		H_2O	Coils on safety vessel	Jubault and Carnino (1979)
Super Phenix (pool)	B	2	H_2O	Coils in vessel's cavity	Debru et al (1980)
	A	4	Na	DHX in uppler plenum	
SNR 300 (loop)	A	3	Na	DHX in uppler plenum	Roesgen et al (1980)

Table 1 - DHRS implemented in LMFBR's

Table 1 shows the redundency retained in those designs where multiple independent loops, sometimes of different nature, are installed.

4.2 Programs for system studies

Evaluation of the capacity of natural convection cooling of reactors needs experimental and analytical programs in order to refine the precision of predictions which may be made during conceptual studies.

A first approach is to take advantage of existing demonstration plants and study their behavior after simulated protected loss of flows. Such studies are reported for Phenix : Clauzon et al - 1976, PFR : Anderson et al - 1980.

Another approach relies on smaller scale experimental reactors EBR II (60 MW) : Singer et al - 1980, FFTF : Additon and Paziale - 1977, or on electrically simulated facilities, Ribando - 1980, Gabler et al - 1980.

The last approach relies on code development supported by detailed studies of particular components or overall comparison with the preceeding approaches : Agrawal et al - 1977, Mohr and Feldman - 1980, Dawson - 1980, Astegiano et al - 1980.

4.3 Some results on the system's dynamics

4.3.1 General considerations

The thermal evolution of the primary circuit is determined by numerous parameters related either to the design of the reactor or to the scenario considered. The main design parameters are listed in paragraph 2.3. To this list must be added the factors which influence the transit time through the components : The length of the pipes, the volume of the plenums,...

Parameters related to the scenario are for example :

- the time it takes for the DHRS to start up after the reactor trips
- the length of the period of pony motor operation
- the availability of secondary loops,...

A reference scenario is retained afterwards for the sake of simplicity :

- At time t = 0, the loss of AC supply causes :
  - shut down of primary pumps
  - reactor scram
- The auxiliary motors refuse to start and the pumps stop at $t = \tau_1$
- The secondary loops are unavailable, thus the IHX's do not produce any heat sink.
- The DHRSs become available sometime after the pumps stop (e.g. $t \sim 3\ \tau_1$). They work by natural circulation only.

In this scenario, flow is caused in both the primary circuit and DHRSs by natural convection only. This choice has been dictated in order to simplify the presentation. This scenario is highly improbable and severe since it assumes a total loss of heat sink through the IHX. However, inside this option, there are other scenarios which could lead to more severe conditions; e.g. in some circumstances the start up followed by the failure of some of the auxiliary motors.

A typical temperature history is shown in figure 10 taken from Gerner and Sciacca - 1980. The primary system's mean temperature rises slowly, reaches a maximum value for which the energy extracted by the DHRS balances the decay energy. For longer periods, this temperature would decrease since the reactor power drops continuously.

The core inlet slowly increases (it corresponds to a pool type design with a large inertia of the inlet plenum, a loop type reactor would show a steeper increase in the inlet temperature).

The core outlet shows a first peak at $t \sim \tau_1$ close to the origin and then rises slowly with the mean temperature, reaches a maximum and decreases.

The first peak in core outlet temperature corresponds to the time it takes for natural convection to be established. During this period, the mean temperature does not rise significantly. It is the short term cooling. This period will be discussed at greater length in the next paragraph.

4.3.2 Short term cooling

Figure 11 shows one example of the flow and temperature evolutions during this period. In all cases, the power drops much faster than the primary pumps' speed. Thus the outlet temperature first drops lower than the initial temperature. The initially isothermal outlet plenum is progressively filled with colder sodium which causes the stratification depicted in figure 7.a.

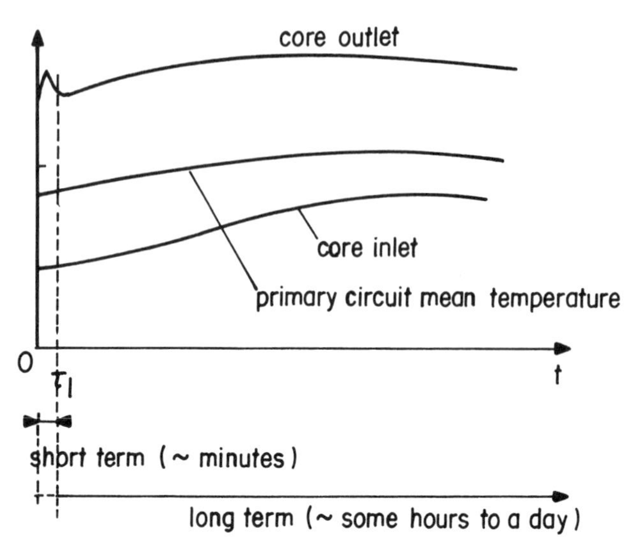

Fig.10. Temperatures' history during a protected loss of flow (from Germer and Sciacca - 1980)

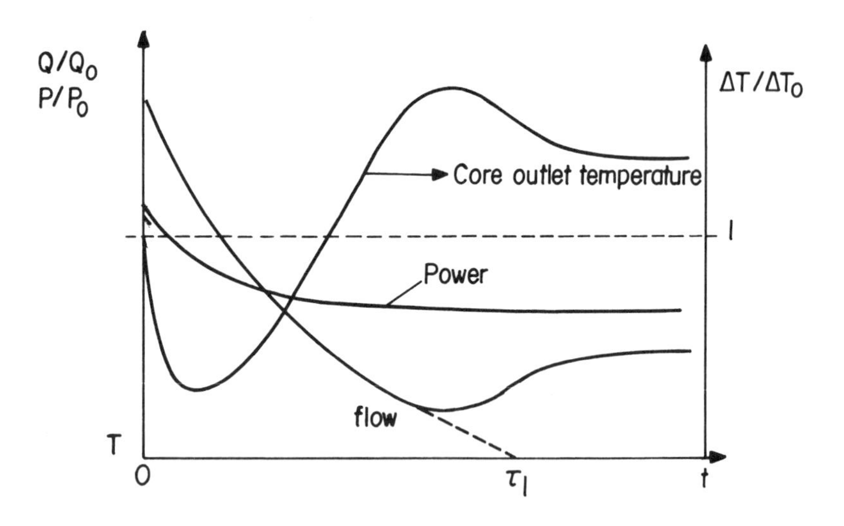

Fig. 11. Short term temperature and flow evolutions (Q_o, P_o, ΔT_o are nominal values of flow-rate, power and temperature increase in the core)

This cold sodium above the core outlet brings opposition to the flow.

With further decrease of the flow, the power to flow ratio first recovers and then surpasses its initial value. The core outlet temperature becomes higher than its nominal value. This helps to recover the buoyant head, and buoyancy starts to sustain the flow : the flow rate becomes higher than what would be imposed by the pump alone (dotted line).

After a minimum in the flow rate and a maximum in the core outlet temperature, steady state natural circulation tends to become established.

The core outlet's evolution is strongly related to the pump coastdown and the time when the pumps stop. Figure 12 shows the results of a parametric study made by Durham - 1976 for a pool type reactor like PFR. Results for three pump rundowns are given in figure 12. For both the short and long rundowns (1 and 3) there is a smooth transition from forced to natural convection. For the medium rundown (2), flow reversal takes place in the core. This is caused by two contrarying effects : the cold sodium formed in the initial period and the difference in level of the free surfaces (§ 2.3).

4.3.3 Start up of DHRS

The DHRS become available when the dampers closed during normal conditions are opened (some time after the primary pumps stop in the scenario retained here). The initial temperature distribution in the loop depends on many factors, such as the heat losses along the pipes and the temperature evolution imposed on the DHX by the primary coolant. In all cases, the buoyant head develops from the tubes of the AHX where the sodium (or NaK) is cooled down. The opening of the dampers must be programmed in such a way that the flow motion is initiated before the coolant temperature drops too low. Otherwise the sodium could freeze in the tubes of the AHX. The initiation of the flow motion is also dependent on the DHRS loop design.

To avoid any freezing problem, some DHRS loops are planned with NaK which is liquid at room temperature. Once the motion is initiated, flow oscillations may occur caused by the transport time along the DHRS loop and its coupling with the two other natural circulation loops (primary coolant and air) : Kesavan et al - 1980.

4.3.4 Long term cooling

This is the period which covers the number of hours during which the mean temperature of the primary circuit rises. This rise is controlled by the imbalance between the decay energy and the energy extraction through the DHRS. The length of this period also increases with the thermal inertia of the primary circuit. In the case of Super Phenix, it takes approximately one day to reach the maximum temperature (figure 13). In a loop-type, 1000 MW of electric power, Germer and Sciacca find the maximum after 3 hours.

The efficiency of the DHR is measured by the capacity to keep particular regions within temperature limits under which the strength of the materials can be assured. These materials are generally :

- the clad at the hottest point of the core

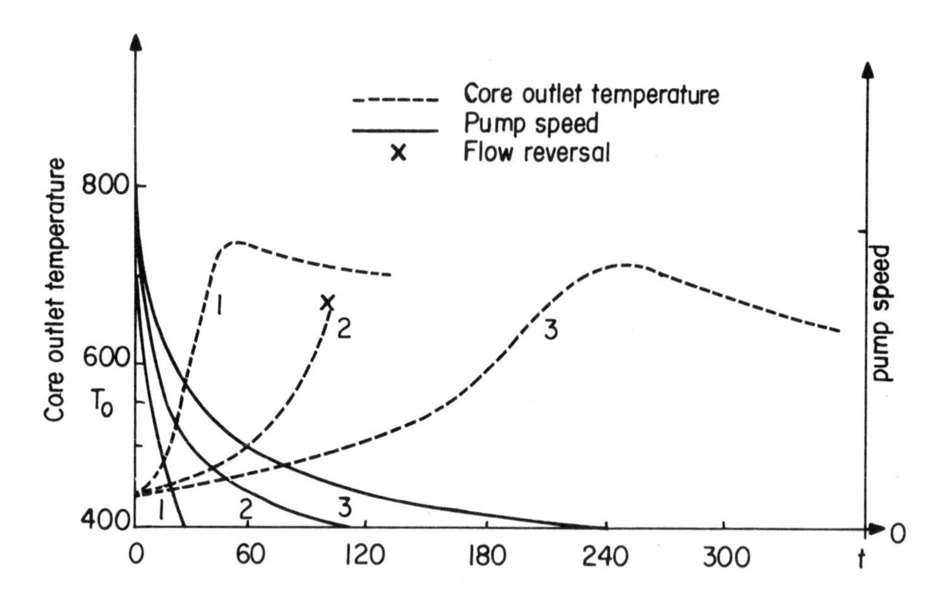

Fig.12. Influence of the pump run down on the onset of natural circulation (from Durham - 1976)

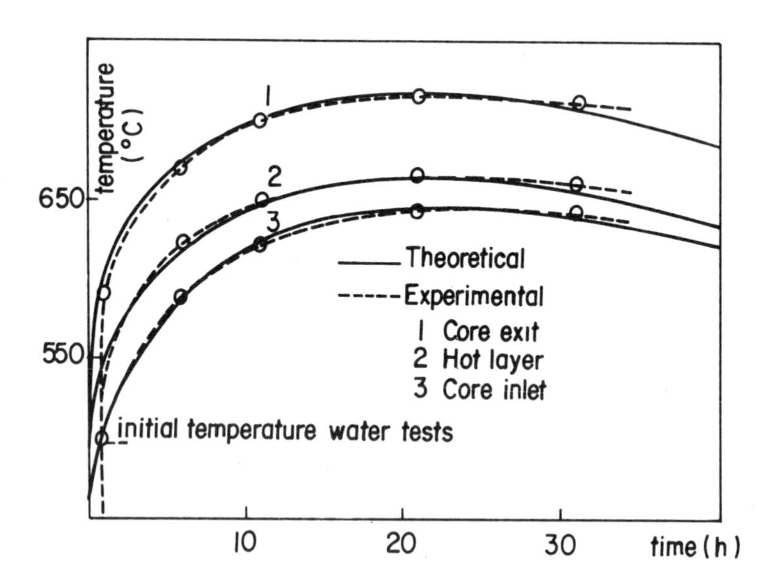

Fig.13. Long term temperature evolution in Super Phenix (Astegiano et al - 1980)

- the core support structure
- the suspension of the vessels in the roof slab

It thus implies a good knowledge not only of the mean temperature but also of the temperature distribution for such detailed points. This requires the thermohydraulic studies of the core and plena previously reviewed.

ACKNOWLEDGMENTS

The author is grateful for the helpful suggestions made by Drs Astegiano and Menant who read a preliminary version of this manuscript.

MEETING REFERENCES

Proceeding of the international meeting on Fast Reactor Safety Technology, Seattle, Washington, August 19-23, 1979, ANS publisher. (hereafter referred to as "International meeting on F.R.S.T., Seattle")

Summary report of the Specialist's meeting on Thermodynamics of FBR fuel subassemblies under nominal and non-nominal operating conditions, International Working Group on Fast Reactor, Karlsruhe, Germany, February 5-7, 1978, IAEA publisher. (hereafter referred to as "specialist's meeting of I.W.G.F.R., Karlsruhe)

Proceeding of the specialist's meeting on Decay Heat Removal and Natural Convection in FBRs, Brookhaven National Laboratory, Upton, N.Y, February 28-29, 1980. (hereafter referred to as "specialist's meeting on D.H.R., Upton")

REFERENCES

Additon, S.L. and Parziale, E.A. 1977. Natural circulation in FFTF, a loop type LMFBR. Symposium on the Thermal and Hydraulic aspects of nuclear reactor safety, Vol. 2 : Liquid Metal Fast Breeder Reactors. Winter Annual Meeting of the ASME, Atlanta, Georgia, November 27 - December 2, 1977.

Anderson, R., Dawson, C.W., Gregory, C.V., Lord, D.J. and Webster, R. 1980. Observations on coolant flow patterns in the PFR primary circuit during natural circulation experiments. Specialist's meeting on D.H.R., Upton.

Agrawal, A.K., Guppi, J.G., Madni, I.I., Quan, V., Weaver, W.L. and Yang, J.W. 1977. Simulation of transients in Liquid-metal fast breeder reactor systems. Nuclear Science and Eng., Vol. 64, p. 480-491.

Astegiano, J.C., Gesi, E. and Martin, R. 1980. Theoretical and experimental analysis of Super Phenix 1. Thermohydraulic problems in natural convection. Specialist's meeting on D.H.R., Upton.

Broadley and Bland 1979. CDFR Emergency Decay Heat Removal System. International meeting on F.R.S.T., Seattle, p. 1607-1616.

Brukx, J.F.L.M., Hansen, G.P.R. and Voj, P. 1977. Combined free and forced convection on transient temperature distributions in a scaled-down model of the inlet plenum of the SNR-300 reactor. Nuclear Technology, Vol. 33, 1, p. 5-16.

Clauzon, P.P., Coulon, P., Meyer-Heine, A. and Penet, F. 1976. Safety conclusions from start up tests and from the analysis of the core behaviour of Phenix. Proc. of Int. meeting on Fast reactor safety and related physics, Vol. 1, p. 61-75.

Coffield, R.D., Tang, Y.S., Killimayer, J.S. and Markley, R.A. 1980. Buoyancy induced flow and heat redistribution during LMFBR core decay heat removal. Specialist's meeting on D.H.R., Upton.

Dawson, C.W. 1980. A theoretical analysis of the establishment of natural circulation in the Dounreay prototype fast reactor. Specialist's meeting on D.H.R., Upton.

Debru, M., Lauret, P., Deckert, J. and Schneider, J.C. 1980. Decay Heat Removal in Super Phenix and related design basis plant conditions. Specialist's meeting on D.H.R., Upton.

Durham, M.E. 1976. Influence of reactor design on the establishment of natural circulation in a pool type LMFBR. J. Br. Nucl. Energy Soc., Vol. 15, p. 305-310.

Engel, F.C., Markley, R.A. and Bishop, A.A. 1979. Laminar, Transition and Turbulent parallel flow pressure drop across wire-wrapped spaced rod bundles. Nuclear Science and Engineering, Vol. 69, p. 290-296.

Gabler, M.J., Mills, J.C., Zweig, H.R. and Lancet, R.T. 1980. A test loop concept for verification of natural convection decay heat removal in FBRs. Specialist's meeting on D.H.R., Upton.

Germer, J.H. and Sciacca, F.W. 1980. The effect of system configuration and initiating conditions on natural convection in an LMFBR. Specialist's meeting on D.H.R., Upton.

Grand, D. and Vernier, Ph. 1978. Combined convection in Liquid metals. Turbulent Forced Convection in Channels and Bundles, Vol. 2, ed. by Kakaç, S. and Spalding, D.B., Hemisphere Publishing Corp.

Jubault, G. and Carnino, A. 1979. Fast Neutron Reactor Safety Reliability Analysis of PHENIX Decay Heat Removal Function. International meeting on F.R.S.T., Seattle, p. 2172-2182.

Khan, E.U., George, T.L. and Wheeler, C.L. 1978. Cobra and Cortran Code thermal-hydraulic models for LMFBR Core wide temperature distribution during a natural convection transient. Specialist's meeting of I.W.G.F.R., Karlsruhe.

Khatib-Rahbar, M., Guppy, J.G. and Agrawal, A.K. 1980 a. Hypothetical Loss-of-Heat-Sink and In-Vessel Natural Convection : Homogeneous and Heterogeneous Core Design. Specialist's meeting on D.H.R., Upton.

Khatib-Rahbar, M., Madni, I. and Agrawal, A.K. 1980 b. Impact of Thermal buoyancy in LMFBR piping systems, Specialist's meeting on D.H.R., Upton.

Kinjo, H., Wada, H., Okubo, Y. and Sawada, T. 1980. Development of computer codes for decay heat removal analysis in reactor vessel and primary loop system. Specialist's meeting on D.H.R., Upton.

Kesavan, K., Bradley, J.D. and Noe, D.E. 1980. A computer simulation of a pool-type LMFBR passive decay heat removal system. Specialist's meeting on D.H.R., Upton.

Leteinturier, D., Blanc, D., Menant, B. and Basque, G. 1980. Theoretical study and experimental investigation of mixed and natural circulation in LMFBR core subassemblies. Specialist's meeting on D.H.R., Upton.

Lorenz, J.J. and Carlson, R.D. 1976. An investigation of buoyancy-induced flow stratification in a cylindrical plenum. Proc. of the 1976 Heat Transfer and Fluid Mechanics Institute, Standford University press.

Markley, R.A. and Engel, F.C. 1978. LMFBR blanket assembly heat transfer and hydraulic test data evaluation. Specialist's meeting of I.W.G.F.R., Karlsruhe.

Mohr, D. and Feldman, E.E. 1980. A dynamic simulation of the EBR-II Plant during natural convection with the NATDEMO code. Specialist's meeting on D.H.R., Upton.

Namekawa, F., Mawatari, K., Tamaoki, T. and Makiura, R. 1980. Out-of-pile experiments for natural circulation decay heat removal in LMFBR assembly, power skew effects upon intra assembly hydraulic characteristics. Specialist's meeting on D.H.R., Upton.

Ribando, R.J. 1980. Comparison of numerical results with experimental data for single-phase natural convection in an experimental sodium loop. Specialist's meeting on D.H.R., Upton.

Roesgen, F., Duweke, M. and Timmermann 1980. Decay heat removal of the SNR-300 with special consideration of the natural convection effects. Specialist's meeting on D.H.R., Upton.

Sheriff, N. and Davies, N.W. 1977. Review of liquid metal natural convection relevant to fast reactor conditions. TRG report 2959, UKAEA, Northern Division.

Singer, R.M. and Gilette, J.L. 1977. Experimental study of whole core thermal-hydraulic phenomena in EBR-II. Nuclear Eng. and Design, Vol. 44, p. 177-186.

Singer, R.M., Grand, D. and Martin, R. 1977. Natural circulation heat transfer in pool-type LMFBR's. Symposium on Thermal and Hydraulic aspects of nuclear reactor safety, Vol. 2, Liquid Metal Fast Breeder Reactors, Winter Annual Meeting of the ASME, Atlanta, Georgia, November 27 - December 2, 1977.

Singer, R.M., Gilette, J.L., Mohr, D., Sullivan, J.E., Tokar, J.V. and Dean, E.M. 1980. Response of EBR-II to a complete loss of primary forced flow during power operation. Specialist's meeting on D.H.R., Upton.

Wang, S.F., Rohsenow, W.M. and Todreas, N.E. 1980. Subchannel friction factors for bare rod arrays under mixed convection conditions. Specialist's meeting on D.H.R., Upton.

Welty, J.R. 1973. Natural convection heat transfer in liquid metals. Progress report n° RLO-2227-T4-4, Oregon State University, Corvallis, 46 p..

Yang, J.W. 1977. Penetration of a turbulent jet with negative buoyancy into the upper plenum of an LMFBR. Nuclear Engineering and Design, Vol. 40, p. 297-301.

Yang, J.W. 1979. Buoyancy effects on pressure loss in rod arrays under laminar flow. ANS Trans., Vol. 32, p. 829-830.

22 BLOCKAGES IN LMFBR SUBASSEMBLIES

by
H.M. Kottowski
CEC – JRC, Ispra, Italy

It is commonly agreed that obstructions of the coolant flow in reactor subchannels might trigger a chain of fault propagation leading to destruction of the core. Experimental and analytical investigations on the formation of blockages and of the thermo-hydraulic effects of blockages in LMFBR fuel assemblies are in progress in the United States, Germany, France, Great Britain and Japan. Models are developed from experimental data obtained for blockages of various shapes and sizes, with and without pins, using water and sodium as coolant. Experiments where sodium boiling occurred behind blockages indicate that boiling is stable for a wide range of boundary conditions. This results could also be predicted by analytical models. Most of the experimental research was addressed to plane disk blockages. Flow obstractions due to pin bowing, swelling and particle agglomeration have not yet been adequately investigated.

This chapter will give a general review on the actual state of knowledge on the effects of partial flow blockages in LMFBR fuel assemblies. The majority of the blockage studies were simulation experiments focussed on the application to the different LMFBR development programs. The applicability of these studies to real reactor situation however seems limited and has to be analyzed from case to case.

1. INTRODUCTION

Channel blockages in fuel subassemblies of LMFBR's can cause due to cooling malfunction propagating clad melting, fuel melting, and, if not stopped lead to subassembly and core disassembling. In principle, blockages or flow obstructions can be initiated by swelling and bending of pins or by agglomeration of particles from failed pins or from other parts of the system. In modern reactors specifc design features prevent sudden and large blockages by fuel debris or other particles. Furthermore, destinction has to be made between grid and wire spaced subassemblies. Wire wrap spacers, for example, which are used for most of the LMFBR designs, minimize the possibility of debris agglomeration in the fuel assembly region. Nevertheless, the question of how blockages can affect the

safety and design of LMFBR's has been of sufficient importance to perform extensive experimental and theoretical research.
The effects of blockages depend on several factors; size and thermo-physical properties of the blockage, location of the blockage in the subassembly, design of the subassembly, fuel pin power and coolant flow velocity.

The first in pile blockage experiment in larger bundle geometry (37 pins with blockage extended over 11 pins) was conducted by KFK Karlsruhe and CEN/SCK Mol in the BR2 reactor and the first full scale grid bundle (SNR) out of pile blockage experiment also (169 pin) by the KFK Karlsruhe. All other blockage studies were run out of pile and with smaller bundles. Even if the experiments were run with sodium, essential features of the reactor could not be simulated, e.g. the axial and radial subassembly power repartion and the thermophysical properties of the blockages. Since the blockages were simulated by narrow steel plates fixed at various locations in the bundle, the question rises on how representative such experiments are and do the analytical models developed on the basis of these experimental results describe the effects of blockages sufficiently conservative. Even with the 37-pin in pile bundle experiment and the 169 full scale out of pile bundle experiment available, the extrapolation to full scale reactor bundle size and operation conditions represents a substantial jump. This is aggravated by the generally acknowledged lack of sufficient analytical tools to aid in the interpretation and extrapolation of small bundle results to full size. Nevertheless, the understanding of the thermohydraulics downstream a plane blockage has reached a reasonable level of maturity. Not adequate is however the understanding of blockage formation and flow and cooling mechanisms in axially extended blockages.

The following sections will present a review of theoretical blockage models and experimental results.

2. REVIEW OF THE THEORETICAL BACKGROUND

2.1. Phenomenological Background

The mechanisms and history of blockage formation in a subassembly are only phenomenologically known and there is no chance to obtain more than a so called most probable approach from simulation experiments. Bowing and swelling of pins, accumulations of debris represent different mechanisms and history and consequently different size, shape and physical property of blockage. The complex nature of possible blockages in subassemblies requires for the analytical approach at least a separation between the blockage itself within its geometrical limits and the space downstream the blockage. Experiments with grid spaced bundles (SNR) have shown that the extension of blockages generated by trapped debris are small in axial direction, but can be spread over a large radial area. Such type of blockage, however, is unlikely to occur in wire spaced bundles (FFTF, CRBR, Phenix).

For experimental reasons the majority of blockage studies done used flat steel plates mounted tightly or with a certain permeabil ty in a bundle geometry. The results of these experiments are qualitatively comparable to those in simple geometries without heater elements, e.g. circular disks set normal to flow in a duct. The circulation flow and temperature distribution downstream such symmetrical flow obstacles could be calculated in a two dimensional manner. The calculation of the complicate three dimensional flow and temperature fields, which one mets in a bundle geometry could not yet be resolved satisfactorily without simplifications and the help of experimental modelling.

2.2. Flow Pattern Behind Blockages

In case of recirculation flow destinction has to be made between recirculation zone and main flow. The flow around bodies normal to the flow depends on the characteristic Re-number. Essentially two typical flow pattern depending on the Re-number can be identified:

(a) At small Re-number, defined by the undisturbed velocity (U_∞) and the characteristic dimension of the obstacle (e.g. disk or tube diameter normal to the flow) the space downstream is dominated by Karman eddies ($60 < Re < 5000$).
(b) At Re-numbers ($Re > 5000$) the Karman eddies disappear and a stable recirculating dead water zone is building up (Fig. 1).

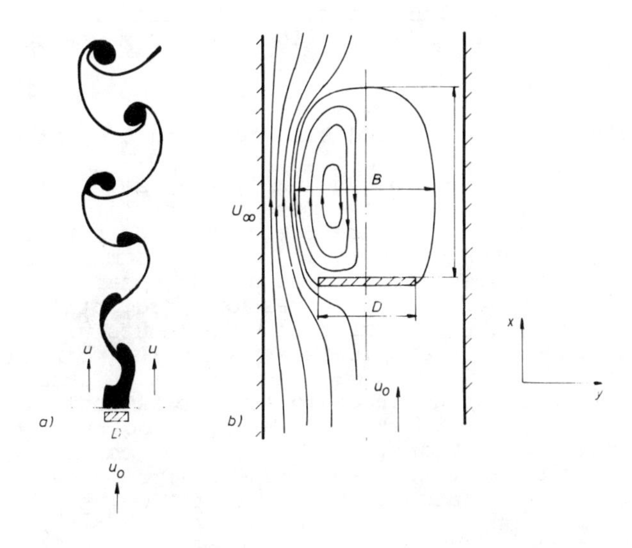

Fig. 1. Flow pattern behind plane impermeable blockage

The flow behind blockages in a bundle geometry, however, is not sufficiently described only by the "blockage Re-number", even if there is a general similarity of the flow pattern behind blockages in free flow and in the presence of rods. The presence of rods influences the cross flow, especially at small pitch over diameter ratios (p/d). With smaller p/d the cross flow diminishes and as a consequence the recirculation zone becomes longer. Beside the subassembly's geometry, other parameters were found to be of importance; the blocked fraction of the total flow area (β), the location of the blockage in the subassembly (central, wall or angle position), the Re -number of the undisturbed bundle flow as a criteria for the build up of a steady state recirculation zone. Basic investigations of Schleisiek [1], Basmer, Kirsch and Schultheiß [2] in a 169 bundle with p/d = 1,317 found that the size and shape of the recirculation zones are nearly independent of the blockage-Re - number and similar to that behind flow blockages of the same size in ducts without rods. They summarized the appearing flow patterns as follows:

- tight blockages in a (SNR) bundle geometry and turbulent flow involve steady state recirculation flow;
- no influence of the Re-number (Re_∞ and Re_B) on the shape and size of the recirculazion zone;
- at the same blocked fraction of the total flow area (β) and similarity of blockage shape, the length and volume of the recirculation zone increase when located in corner position;
- small residual flow lifts the recirculation zone from the blockage without destroying its nature. With increasing residual flow through the blockage the distance between recirculating flow area and blockage increases. For residual flow greater 15% of nominal undisturbed flow, the recirculation disappears and pulsating flow in the main flow direction occurs.
- An application of the results to bundles with larger p/d values is possible, whereas different flow pattern may appear at small p/d values.

2.3. Temperature Pattern Behind Blockages and Simulation Experiments

In the reactor case blockages are associated with temperature excursions and the aim of the research work is to provide tools to predict the temperature field (in and) behind blockages. Since experiments, even out of pile, with sodium are difficult to perform and the measurement of certain hydrodynamic parameters (e.g. local velocity vectors and pressure gradients) are uncertain or impossible to measure, simulation experiments with apparatus and liquides easy to use are indispensable. The main question to answer is the possibility to extrapolate simulation data to real reactor conditions. Using for example water as coolant simulant, the extrapolation to sodium will be limited too few flow and heat transfer conditions. Especially near to the wall, even if all other conditions are similar in sodium flow, the molecular heat conductivity dominates over turbulent flow, so that the temperature distributions become different. Kirsch [3], for example, could show that in a highly turbulent recirculation flow, when the wall effect of the heating rods can be omitted, the heat transfer term of molecular conduction becomes negligible and water experiments can simulate particular blockage conditions in LMFBR's. From modelling laws is known that the real temperature distribution can be calculated from measurements in simulation experiment when the dimensionless temperature distribution is equal for both cases. The basic requirements are [4]:

1. geometrical similarity
2. similarity of the differential equations describing the problem
3. similarity of boundary conditions (This condition cannot be fullfilled in case of water/sodium modelling. But since only steady state turbulent flow and time mean values are considered, the real boundary conditions can be neglected).

For general analysis, a two dimensional approach of the flow and heat transfer mechanisms is thought being appropriate. The basic tools are the continuity, momentum and energy equations for quasi steady state conditions (time mean values).
Velocity and temperature vectors:

$$u(x,y,t) = \bar{u}(x,y) + u'(x,y,t); \quad u' \ll \bar{u} \qquad (1)$$

$$T(x,y,t) = \bar{T}(x,y) + T'(x,y,t); \quad T' \ll \bar{T} \qquad (2)$$

Using these definitions in the continuity, momentum and energy equations and equivalent reference parameters in undistrubed flow for normalisation, one obtains with the normalized parameters the dimensionless equations for continuity, momentum and energy.
Dimensionless variables:

$$X = x/D, \; Y = y/D, \; \bar{U} = u/U_o, \; \bar{V} = v/U_o \qquad (3)$$

$$\bar{P} = (p - p_o)/\rho \cdot U_o^2 \, , \; \Theta = (T - T_E) \cdot \frac{\rho \cdot c_p \cdot U_o}{q'''_{max} \cdot D}$$

D : characteristic blockage length

U_o : velocity in undisturbed flow

T_E : temperature in undisturbed flow at blockage location

P_o : pressure in undisturbed flow upstream the blockage

$q'''_{max} \cdot F(x,y)$: equivalent volume heat source

$u'(x,y,t); \; T'(x,y,t)$: fluctuations

continuity eq.:

$$\frac{\partial \bar{U}}{\partial X} + \frac{\partial \bar{V}}{\partial X} = 0 \qquad (4)$$

momentum eq.:

$$\bar{U}\frac{\partial \bar{U}}{\partial X} + \bar{V}\frac{\partial \bar{U}}{\partial Y} = -\frac{\partial \bar{P}}{\partial X} + \frac{\partial}{\partial X}\left(\frac{1}{Re}\frac{\partial \bar{U}}{\partial X} - \overline{U'^2}\right) + \frac{\partial}{\partial Y}\left(\frac{1}{Re}\frac{\partial \bar{U}}{\partial X} - \overline{U'V'}\right)$$

$$\bar{U}\frac{\partial \bar{V}}{\partial X} + \bar{V}\frac{\partial \bar{U}}{\partial Y} = -\frac{\partial \bar{P}}{\partial Y} + \frac{\partial}{\partial X}\left(\frac{1}{Re}\frac{\partial \bar{V}}{\partial X} - \overline{U'V'}\right) + \frac{\partial}{\partial Y}\left(\frac{1}{Re}\frac{\partial \bar{V}}{\partial Y} - \overline{V'^2}\right) \qquad (5)$$

energy eq.:

$$\bar{U}\frac{\partial\bar{\theta}}{\partial X} + \bar{V}\frac{\partial\bar{\theta}}{\partial Y} = \frac{\partial}{\partial X}\left(\frac{1}{Re.\,Pr}\frac{\partial\bar{\theta}}{\partial X} - \overline{U'\theta}\right) + \frac{\partial}{\partial Y}\left(\frac{1}{Re.\,Pr}\frac{\partial\bar{\theta}}{\partial Y} - \overline{V'\theta}\right) \quad (6)$$

Applying these eq's to the following conditions:

- velocity on the walls: $u(R) = v(R) = 0$

$$U(R) = V(R) = 0 \qquad \text{(dimensionless eq.)} \qquad (7)$$

- assuming adiabatic wall: Normal to the wall the terms for the boundary conditions yield:

$$\left(\frac{\partial T}{\partial y^*}\right)_R = 0 \qquad y = \text{normal to the wall} \qquad (8)$$

$$y^* = y\,/D$$

$$\left(\frac{\partial\bar{\theta}}{\partial y^*}\right)_R = 0 \qquad \text{(dimensionless eq.)} \qquad (9)$$

$$\bar{U}_o = 1,\ \bar{V}_o = 0,\ \bar{\theta}_o = 0,\ \bar{P}_o = 0,\ Q = F(X,Y) = \frac{\dot{q}(x,y)}{\dot{q}_{max}} \qquad (10)$$

Since from these equations in the restrains of the boundary conditions only the Reynolds (Re) and Prandtle (Pr) numbers appear as independent dimensionless groups, one can conclude that the temperature distributions for the compared cases are similar if Re and Pr numbers or other equivalent dimensionless numbers are equal. These dimensionless groups in the equations, describing the flow and temperature in the wake (eq 4 to 10), are coefficients of terms describing molecular transport. The conclusion can be drawn that:

If the molecular transport terms can be neglected no factors depending on dimensionless groups appear in eq (5) and (6). The dimensionless temperature distributions are equal, if the boundary conditions are similar. This conclusion is only true when the flow and heat transfer in the recirculation wake is not influenced by the rod walls or subassembly housing. Kirsch [3], Carmody [5], Huetz [6] showed for sodium flow with blockage Reynolds numbers $Re_B > 10^6$ that the molecular heat transfer in the recirculation wake becomes less than 5% of the turbulent heat transport.

The results can be summarized as follows: With similar geometry and similar boundary conditions recirculating flow behind blockages in LMFBR-subassemblies can be simulated using water if the Reynolds number is sufficiently high and wall effects can be excluded.

2.4. Size and Shape of Recirculation Wakes

The almost equal kinematic physical properties of sodium and water permit simulation experiments to determine the size and shape of recirculation wakes downstream in LMFBR subassemblies in case of similarity of the blockage. Two methods to measure the length and shape are generally applied: (a) the kinematic method with tracers and (b) the pressure gradient measurement method normal to the recirculation flow velocity vectors. The pressure gradient method is more accurate. The theoretical background of this method will therefore be discussed. The flow normal to the main flow is described by the Navier Stokes equation:

$$\bar{u}\,\frac{\partial \bar{v}}{\partial x} + \bar{v}\,\frac{\partial \bar{v}}{\partial y} = -\,\frac{1}{\rho}\,\frac{\partial \bar{p}}{\partial y} + \left(\frac{\partial^2 \bar{v}}{\partial x^2} + \frac{\partial^2 \bar{v}}{\partial y^2}\right) - \left(\frac{\partial \overline{u'v'}}{\partial x} + \frac{\partial \overline{v'^2}}{\partial y}\right) \qquad (11)$$

In the central part of the recirculation wake the flow in the wake boundary can be assumed (Fig. 2) parallel to the main flow. If this is true, the terms $\bar{v}$, $\partial \bar{v}/\partial x$ and $\partial \overline{u'v'}/\partial x$ in the vicini ty of the main stream/wake boundary and the molecular momentum terms:

$$\nu\,\frac{\partial^2 \bar{v}}{\partial x^2} \quad \text{and} \quad \nu\,\frac{\partial^2 \bar{v}}{\partial y^2} \qquad (12)$$

(in case of no wall interference) can be neglected. Equation (11) yields then:

$$\frac{1}{\rho}\cdot\frac{\partial \bar{p}}{\partial y} = -\,\frac{\partial \overline{v'^2}}{\partial y} \qquad (13)$$

Introducing Prandtl's shearstress assumption:

$$-\,u'v' = \varepsilon\,\frac{\partial \bar{u}}{\partial y} \qquad (14)$$

With the fluctuation normal to the flow $v' \sim u'$, eq. 14 yields:

$$-\,v'^2 = \varepsilon^*\,\frac{\partial \bar{u}}{\partial y} \qquad (15)$$

The pressure gradient $\frac{\partial \bar{p}}{\partial y}$ can be determined from radial velocity measurements if ε^* is nearly independent of "y", which has been shown by Carmody [5] and Kirsch [3]. Equation (15) in (13) yields than:

$$\frac{1}{\rho}\,\frac{\partial \bar{p}}{\partial y} = \varepsilon^*\,\frac{\partial^2 \bar{u}}{\partial y^2} \qquad (16)$$

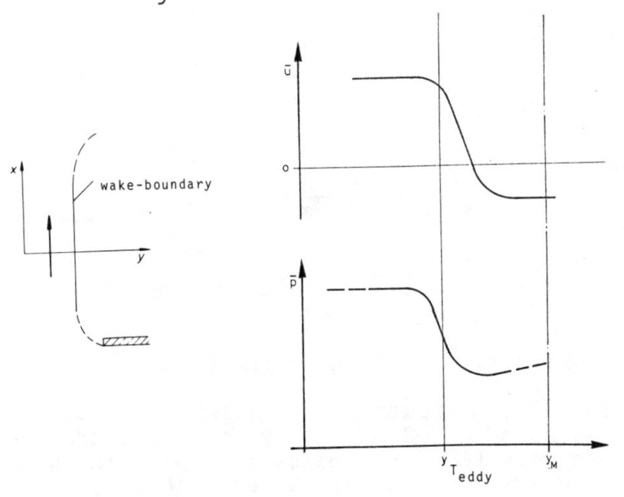

Fig. 2. Recirculation mechanism behind tight blockages

From the interpretation of equation (16) follows (Fig. 2): the local pressure normal to the main flow when passing through the boundary region will decrease and reach a minimum in the center of the eddy where the velocity $\bar{u}$ is zero. The turning point of the pressure curve corresponds with the point of maximum change of velocity gradient ($y_{t\ eddy}$). Carmody [5] and Winterfeld [7] for example could show that the boundary of the recirculation zone is best defined by the area which corresponds to the maximum pressure gradient (turning point). They made their measurements without rods, but Kirsch [3] and Schultheiß [2] could demonstrate that the influence of the rods, especially with a p/d ratio >1,25 is not aggravating. The extreme length of the recirculation zone is then, according to the definition, given by the axial distance where no radial pressure gradient exists. For the measurement of the wake length in sodium flow Schleisiek [1] profited from the fact that in the stagnation point in the extreme of the recirculation wake the axial temperature gradient becomes zero.

$$\left(\frac{\partial \bar{T}}{\partial x}\right)_{x=L} = 0 \qquad (17)$$

Schleisiek [1] could show that the main flow velocity had insignificant influence on the recirculation wake length (Fig. 3). He measured a length to width ratio, being independent of the main flow velocity, of 2,2 to 2,7 in sodium. From simulation experiments with water a ratio of 1,65 to 1,85 is reported [3].

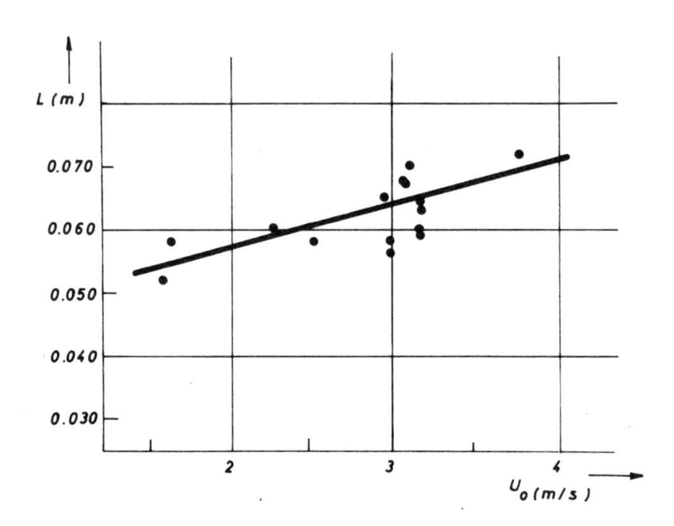

Fig. 3. Length of the recirculation zone

2.5. Mass Transport between Wake Area and Main Flow

The temperature distribution in a turbulent recirculation wake is dominated also in sodium flow by the mass transport from the wake into the main flow. Since the flow pattern in the recirculation is complex and can only be described by three dimensional computation methods, which are not yet available, significant simplifications are necessary for the development of analytical correlations. Blockage experiments with and without rods and in water and sodium suggest that at least for the steady state situation the one dimensional treatment of the flow and temperature pattern is accurate enough and provides results within the tolerances of the experimental measurements.

The following mode of mass transfer is commonly acknowledged:

(1) the driving force is the pressure difference between the downstream stagnation point and the blockage, which is governed by the pressure in the mean undisturbed flow around the blockage;
(2) a ficticious mean mass flow $\dot{m}_R$ proportional to the real one is assumed: $\dot{m} = K_1 \cdot \dot{m}_R$;
(3) for the pressure drop through the recirculation zone one takes into account the complex geometry (axial, radial flow and diversions) and that the real flow is bigger than the ficticious one and assumes the real pressure drop being proportional to the axial pressure drop with respect to L: $\Delta \overline{p}_R = K_2 \Delta \overline{p}_L$;
(4) the driving pressure between stagnation point and blockage is smaller than calculated from Bernoulli's equation and assumed proportional to it: $\Delta \overline{p} = K_3 \Delta P_B$;
(5) the recirculation length is proportional to the width of the blockage and assumed independent of the Re-number $L = K_4 . D$;
(6) radial pressure and temperature in the wake are assumed uniform.
Supposing the volume and the heat generation of the wake is known, the coolant residence time in the wake can be calculated:

From the energy balance for a blockage with N subchannels ($= \frac{N}{2}$ heating pins) one obtains:

$$M_R \cdot c_p \cdot \frac{\partial T}{\partial t} = \frac{N}{2} \cdot d \cdot \pi \cdot L \cdot \emptyset - \dot{m}_R \cdot c_p (T - T_o) \qquad (18)$$

$$\frac{\partial T}{\partial t} = \frac{N \cdot d \cdot \pi \cdot L}{2 M_R \cdot c_p} \cdot \emptyset - \frac{\dot{m}_R}{M_R} (T - T_o) \qquad (19)$$

$$M_R = N \cdot A_p \cdot L \cdot \rho \qquad (20)$$

($\emptyset$ = heat flux, $\overline{T}$ = mean coolant temperature over the cross section of the wake, T_o = mean coolant temperature at the stagnation point, d = rod diameter, $\dot{m}_R$ = mass flow through the wake, L = wake length, A_p = flow cross section of one subchannel).

Since blockages are supposed to develop over a long time the thermohydraulic pattern in the recirculation zone can be assumed stationary and $\partial \overline{T} / \partial t$ becomes zero. Eq (19) and eq (20) with $d_h = 8A_p/d$ yields:

$$\frac{M_R}{\dot{m}_R} = \frac{d_h}{4} \rho \, c_p \frac{(\overline{T}_{max} - T_o)}{\emptyset} = \tau \qquad (21)$$

τ is defined as the mean residence time of "M_R" in the wake.

The assumption made for equation (21) that the coolant enters the recirculation zone at the downstream stagnation point with the temperature T_o, flowing upstream in the center of the wake and leaving immediately above the blockage with the temperature T_{max} could be shown also in experiments with sodium [Fig. 14].

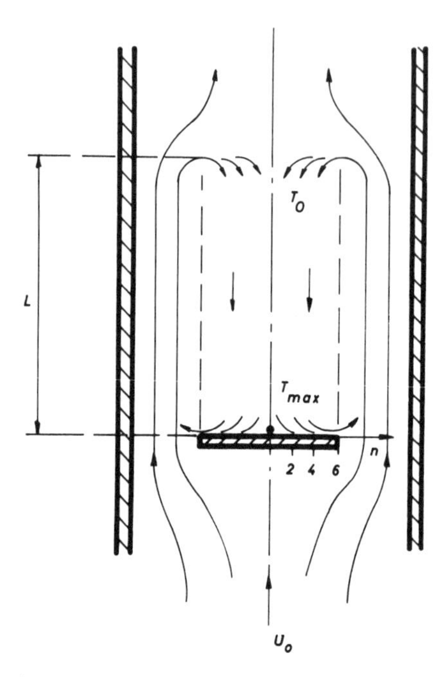

Fig. 4. Temperature mode in the wake

Equation (21) correlates the mean residence time with the peak coolant temperature and heat flux, but there is no information on the influence of the blockage size. To find a correlation between blockage size and residence time the simplified assumptions 1 through 6 are used:

Assumption 2 yields for the "upstream" velocity in the wake center:

$$V_R = \frac{K_1 \cdot \dot{m}_R}{\rho \cdot A_p \cdot N} \tag{22}$$

For the friction pressure drop follows for the upstreaming flow with assumption 3.

$$\Delta \bar{p}_R = \xi \cdot \frac{K_2 L}{d_h} \cdot \frac{\rho}{2} \cdot V_R^2 \tag{23}$$

ξ = friction factor for axial flow

With assumption 4 the driving pressure yields:

$$\Delta p = K_3 \cdot \frac{\rho}{2} \cdot (U_\infty^2 - U_o^2) = K_3 \cdot \frac{\rho}{2} \left(\frac{1}{p^2} - 1\right) U_o^2 \tag{24}$$

p : permeability = $\frac{A - A_B}{A}$

A : total bundle flow cross section

A_B: blocked flow cross section

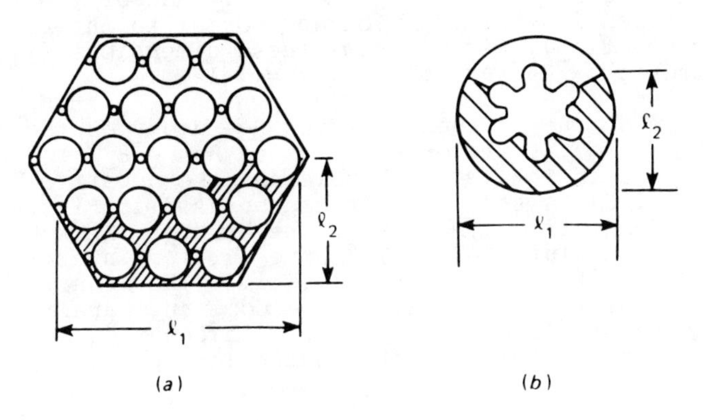

Fig. 5. Definition of the two orthogonal dimensions of (edge) blockages

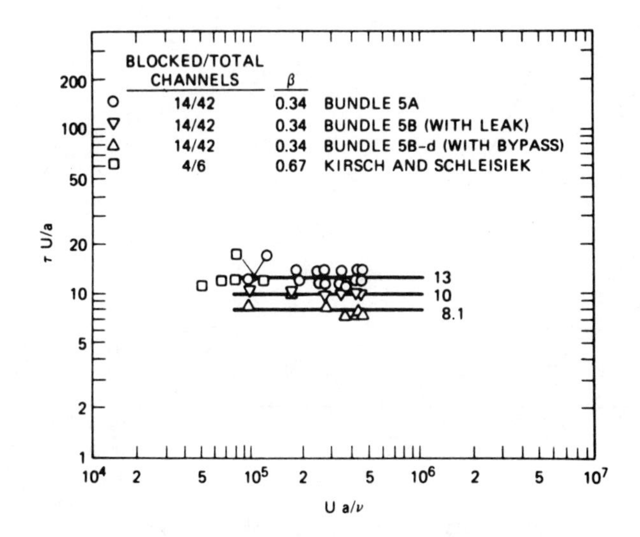

Fig. 6. Dimensionless residence time behind blockages

With $\Delta \bar{P}_R = \Delta \bar{P}$ and assumption 4, the residence time becomes a function of the size and nature of the blockage and undisturbed flow velocity

$$\tau = \frac{C}{U_o} \cdot \sqrt{\frac{\xi \cdot D^3}{d_h \cdot (\frac{1}{p^2} - 1)}} \; ; \; C = K_1 \cdot \sqrt{\frac{K_2 \cdot K_4^3}{K_3}} \tag{25}$$

C is the so called blockage constant and has to be determined for each geometry from simulation experiments. For general application and simplified presentation Han et al. proposed for the characteristic blockage dimension "a" the geometrical mean of the two longest orthogonal dimensions of the blockage normal to the flow (independent of its location). Fig. 5 shows these dimensions (l_1 and l_2) for an edge blockage [8] as follows: $a = \sqrt{l_1 \; l_2}$.

The question on the influence of the Re-number on the residence time is answered by the experiments of Han [8] and Schleisiek with blockages in the 19 rod bundle and the negative bundle respectively (Fig. 6). A stright line can be used to fit the dimensionless residence time for each of the bundles. Kirsch [3], Schleisiek [1] and Winterfeld [7] were further investigating the effect of viscosity on the residence time. They concluded that the velocity and not the viscosity is the dominant parameter, which justifies to omit the viscosity term in equation (21). The importance of the velocity is clearly shown in Fig. 7.

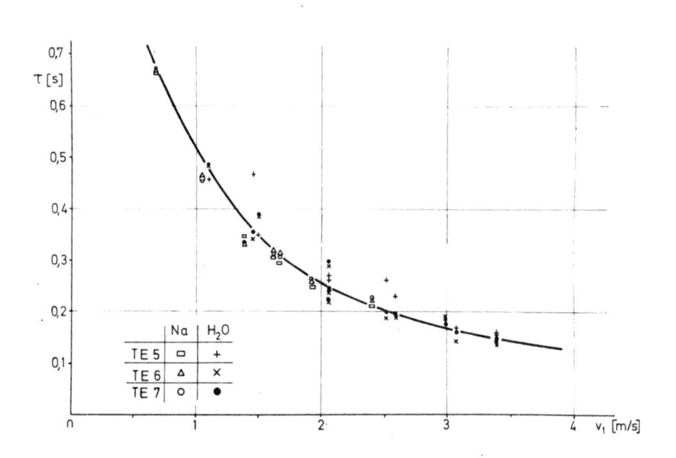

Fig. 7. Residence time vs velocity

The authors conclude that water simulation experiments provide reliable data for sodium flow in similar geometry and that the blockage constant "C" (eq 25) measured in water can be applied to sodium flow.

2.6. Heat Transfer

The heat transfer correlations for the recirculation flow are not unequivocal. Since the local flow Re-numbers are unknown, it is obvious that the choice of the blockage or the free flow cross section as characteristic length for the heat transfer in the wake zone is arbitrary and explains the differences in the correlations proposed by various authors. Han et al [8] calculated the Nusselt number from experimental data using the equation:

$$Nu = \frac{q'' \cdot d_h}{K \cdot (T_W - T_{wk})} \qquad (26)$$

with d_h = hydraulic diameter of the channel, T_W = wall temperature, T_{WK} = average coolant temperature in the wake. The so defined Nu-number for the heat transfer in the wake behind a 6 channel central blockage is shown in Fig. 8 as a function of the free flow Reynolds number. The Nu-number obtained for a 14 channel edge blockage (ORNL THORS bundle 5B) [8] is shown also in Fig. 8. The interesting point is that the Nu-number is approximately in the range of 0,65 to 3, which is much smaller than a value of 5 or 7 for laminar sodium flow in a bundle with no blockage. Although there is some scatter in the data points there is a clear indication that the Nu-number in the wake increases as the Re-number increases. A line is used to fit these results for $10^4 \leq Re_\infty = Re/(1-\beta) \leq 1,5 \cdot 10^5$ which corresponds to the equation:

$$Nu = 0,0043\ Re_\infty^{0,55} \qquad (27)$$

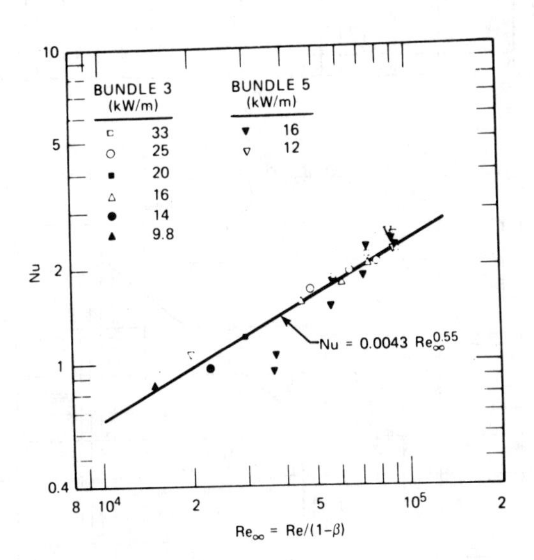

Fig. 8. Nusselt number for average heat transfer in a wake downstream plane blockages

Schleisiek calculated the Nu-number, using equation (26), for the wake behind the blockage in his negative bundle. The results are presented in Fig. 9 as Nu vs Peclet number. The wake-Nu-number, approximately 5, is much greater than that obtained in the wire-wrap-spaced 19-pin bundle of Han. The reason is not clear. The Nu-numbers in the wake obtained from water experiments are shown as a function of the Re-number (a) calculating the Re-number for the total free flow cross section and (b) calculating the Re-number for the narrowest free flow cross section at blockage level (Fig. 10). For comparison, the Colburn correlation (eq. 28) calculated with the characteristic length according to (a) is also shown in Fig. 10.

$$Nu = 0{,}025 \cdot Re^{0{,}28} \cdot Pr^{0{,}33} \qquad (28)$$

As long as no further results are available, equations 27 and 28 can be used within the limits established by the experiments performed.

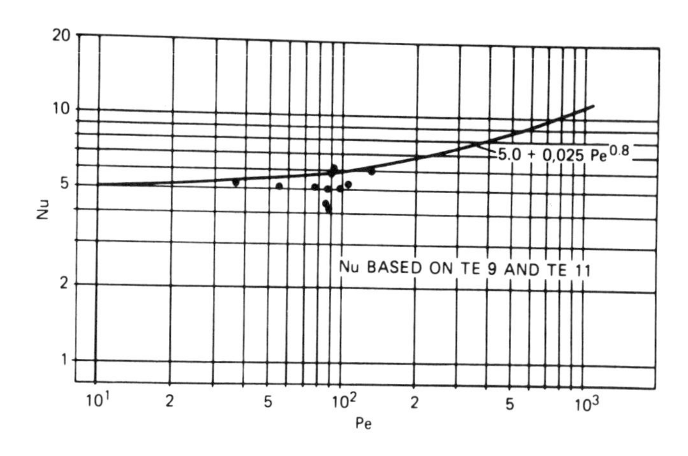

Fig. 9. Nusselt number behind blockages [1]

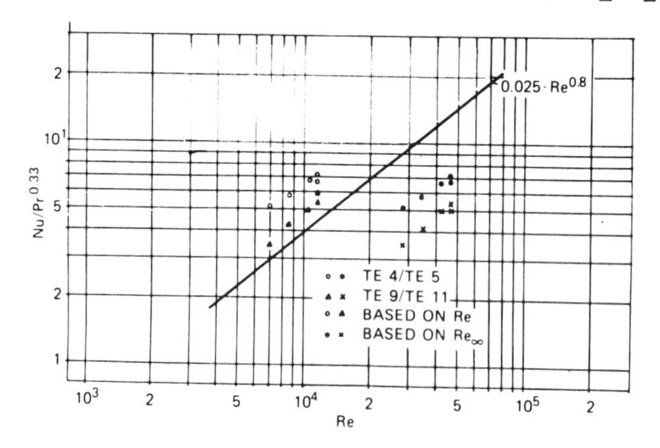

Fig. 10. Nusselt number vs Re-number behind blockages in water

2.7. Temperature Distribution in Recirculation Flow of a Rod Bundle

Theoretical and experimental work **showed:**

(a) the flow patterns behind blockages normal to the flow in rod bundle and simple geometry without rods are similar and comparable if the characteristic parameters are equal and the p/d ratio in the rod bundle is greater than 1,2.

(b) The similarity of the temperature distribution is not essentially affected substituting the heat flux from the rod walls by an equivalent volume heat source distribution to role out the wall effect.

This is established by measurements in different rod bundle geometries and comparison with simple geometry experiments in water and sodium for different Re-numbers and heat fluxes [1,3,8]. To demonstrate the good agreement of dimensionless temperature distributions for sodium and water the basic experiments of Schleisiek and Kirsch are quoted. Table 1 shows also dimensionless water temperature distributions measured in the test section shown in Fig. 11. (The last column shows an avereged dimensionless water temperature in order to compensate the measuring errors due to the less precise water temperature measurements). These data confirm the conclusion from the theoretical analysis that there is no (visible) influence neither of the Reynolds (Re_B) or Prandtle number nor of the heat flux.

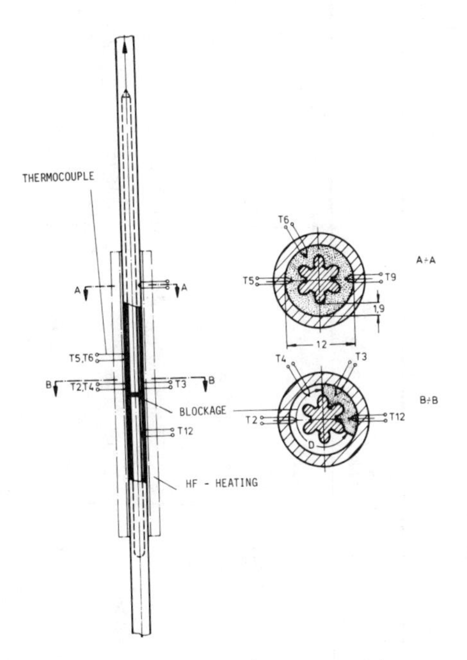

Fig. 11. Test section negative bundle [1]

Table I. Dimensionless Temperatures in the Test Section (Fig. 11)

Coolant		Na	H_2O	Na	H_2O	Na	H_2O	H_2O
Re_B	1	$0,81.10^5$	$0,81.10^5$	$1,64.10^5$	$1,62.10^5$	$2,05.10^5$	$1,98.10^5$	Average
$\dot{q}_f$	$W\ cm^{-2}$	57	129	133	129	133	159	
$\bar{\Theta}_2$	1	1,14	1,18	1,18	1,12	1,22	1,30	1,20
$\bar{\Theta}_4$	1	1,07	1,24	1,20	1,16	1,23	1,30	1,23
$\bar{\Theta}_5$	1	0,77	0,97	0,93	0,76	0,93	1,02	0,92
$\bar{\Theta}_6$	1	0,70	0,82	0,76	0,66	0,77	0,74	0,74
$\bar{\Theta}_7$	1	0,34	0,38	0,32	0,32	0,31	0,29	0,37

As shown by Kirsch [3] the turbulent eddy heat transport between subchannels in the wake can be calculated from measured temperature differences. Schleisiek made these calculations for the thermocouple positions T_3 and T_4 (Fig. 12) and correlated the dimensionless value $\varepsilon H/(U_o \cdot D)$ vs Re_B. As shown in Fig.12 the dimensionless eddy heat transport for water and sodium are equal and independent of the Reynolds number. This leads to the conclusion that the temperature patterns in recirculation wakes with sodium can be simulated with water experiments.

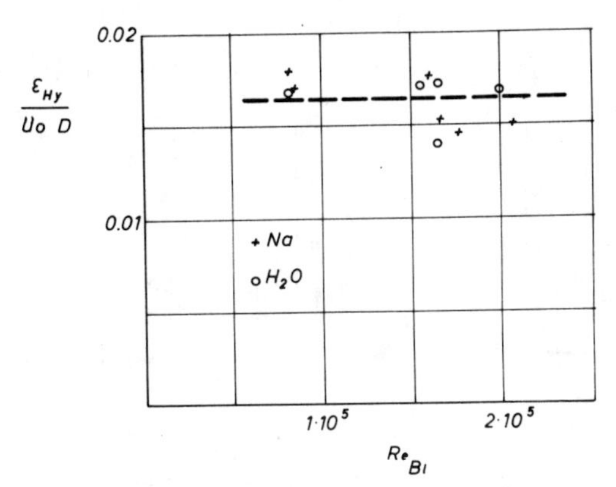

Fig. 12. Dimensionless eddy heat transfer vs Re -number

The discussion up to now was dealing for simplicity with central and plane blockages. There is in fact a strong influence of the location of the blockage in the duct and of its contour. The phenomenological investigations of Basmer [9], Kirsch [3] and Schultheiß [2] (Fig. 13) show especially the effect of blockage location. The length of the wake, measured between the blockage plate and the stagnation point, was found to be about 2 times the blockage diameter with the tendency to decrease with increasing blockage area. If the same blockage is moved to the wall or to the corner, the wake becomes considerably longer and asymmetric with the stagnation point moving to the wall. The wake length becomes up to 4 times the blockage diameter.

An important role plays the blockage permeability. Gregory and Lord investigated this parameter and found that at a relatively small residual flow through the blockage the recirculation disapperars (Fig. 14).

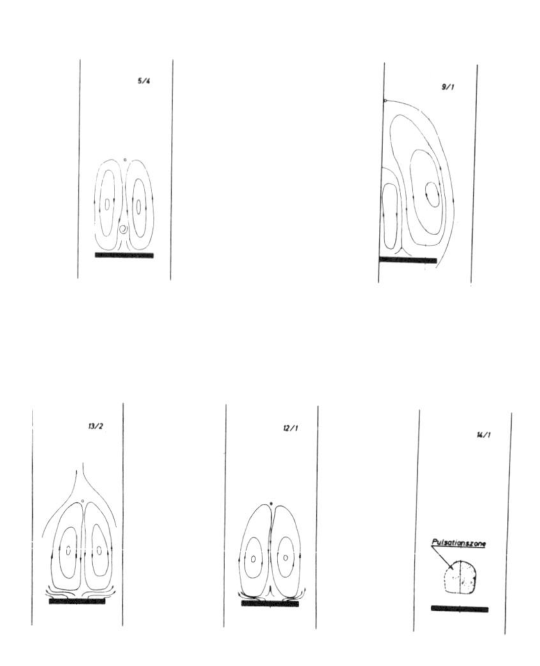

Fig. 13. Blockage flow pattern

Table 2

Test series	-	5/4	9/1	12/1	13/2	14/1
Blockage position	-	Fig. 13	Fig. 13	Fig. 13	Fig. 13	Fig. 13
Reynolds number Re	1	$3,0.10^4$	$2,32.10^4$	$3,34.10^4$	$3,32.10^4$	$3,13.10^4$
L/D ratio	1	1,69±0,07	2,5±0,05	1,60±0,04	1,60±0,04	-
Porosity (diameter of hols in the sub channel position)	mm	0	0	0,5	1,0	1,5
Residual flow "η"	1	0	0	0,008	0,054	0,162
Lifting of the wake from the blockage	mm	0	0	5	15	-

At a residual flow ratio $\eta = U_B/U$ of 0,2 for a porous plate normal to the flow, independent from other hydraulic or geometrical conditions, the recirculation wake disappears. (U_B = residual flow through the blockage, $U_\infty = U_0/(1-\beta)$. Since the flow pattern behind blockages in rod bundles are found not being disturbed significantly by the rods the application of Gregory's results to

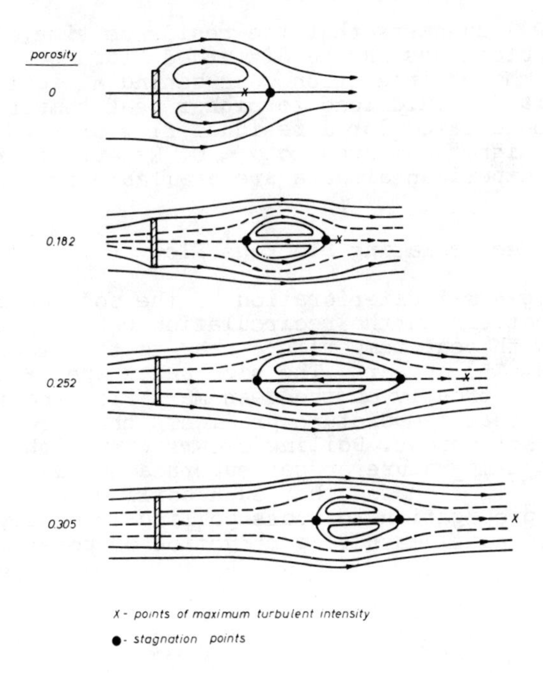

Fig. 14. Influence of porosity on the wake formation

rod bundle seem possible. The permeability of the blockage leads to a rediscussion of residence time and possible temperature peak in the zone behind the blockage.

For $\eta > 0,2$ the residual time of the coolant behind the blockage becomes:

$$\tau = \frac{L}{U_B} \tag{27}$$

Gregory et al. found for "L" as length of disturbed zone with $U_o < U_\infty$:

$$L = 3\ R\ (1 + \Delta) \qquad R = \frac{D}{2}\ \text{(blockage radius)} \tag{28}$$

Eq. (27) and (28) yield:

$$\frac{\tau\ U_\infty}{R} = \frac{3(1 + \Delta)\ (1 - \beta)}{\eta} \tag{29}$$

Δ ="length increment"as a function of residual flow

$\frac{\tau\ U_\infty}{R}$ is the dimensionless residence time which was found being constant for tight blockages.

$$\frac{\tau\ U_\infty}{R} = C\ (C \sim 15 \text{ from experimental results}) \tag{30}$$

Equation (29) suggests that the residence time of the coolant behind a porous blockages in the disturbed flowing zone could be greater than in the recirculation zone behind a tight blockage of the same size, which would lead to higher peak temperatures. Gregory et al. calculated for a residual flow of η = 0,2 a peak temperature 50% higher compared to η = 0. Except the report of Clare [10] no experimental data are available on this subject.

2.8. Boiling in Recirculating Blockage Flow

Due to the general deterioration of the cooling and the nature of the flow pattern in the recirculation wake a pronounced temperature field with peak temperature near to the stagnation zones of the wake (Fig. 15) appears. The size and shape of the isothermal planes depend strongly on the mean mass flow and size of blockage. Due to the great temperature gradients the hot zones become isolated stable hot spaces. Boiling occurs when such "hot spaces" reach saturation temperature or get superheated. Gast [11], De Vries et al. [12] showed that such boiling zones are extremely stable and a propagation to cross boiling is mostly due to further blockage increase, mass flow reduction or power excursion.

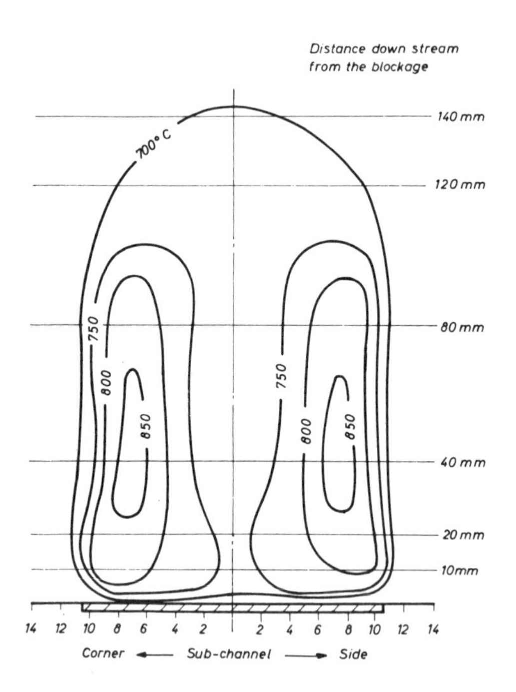

Fig. 15. Temperature pattern behind blockages

3. EXPERIMENTAL RESULTS AND EXTRAPOLATION TO REAL REACTOR CONDITIONS

Han and Fontana [8] summarized the state of the art of blockage experiments in a reviewing paper in 1977. This report is complete and gives a survey of the work done and results obtained in the various countries. Research details of a thermohydraulic fast breeder safety programme are known from US, DBNLUX (Germany, Belgium, Netherlands, Luxembourg), France, Japan and U.K. Due to the different subassembly designs (using wire wrapped spacing for the FFTF, Phenix, PFR, CRBR and grid spacing for the SNR and MONJU), complementary studies are available. The majority of the studies are out of pile and focussed on plane tight blockages. Inpile experiments with axially extended porous blockages were the Mol 7 C experiments in the frame of the SNR-programme.

3.1. SNR Thermohydraulic Blockage Studies

Extensive blockage investigations in simulated SNR fuel subassemblies have been performed by the Karlsruhe Nuclear Research Center in Germany, the Netherlands Energy Research Foundation (ECN) at Petten and the Centre d'Etude de l'Energie Nucléaire CEN/SCK at Mol.

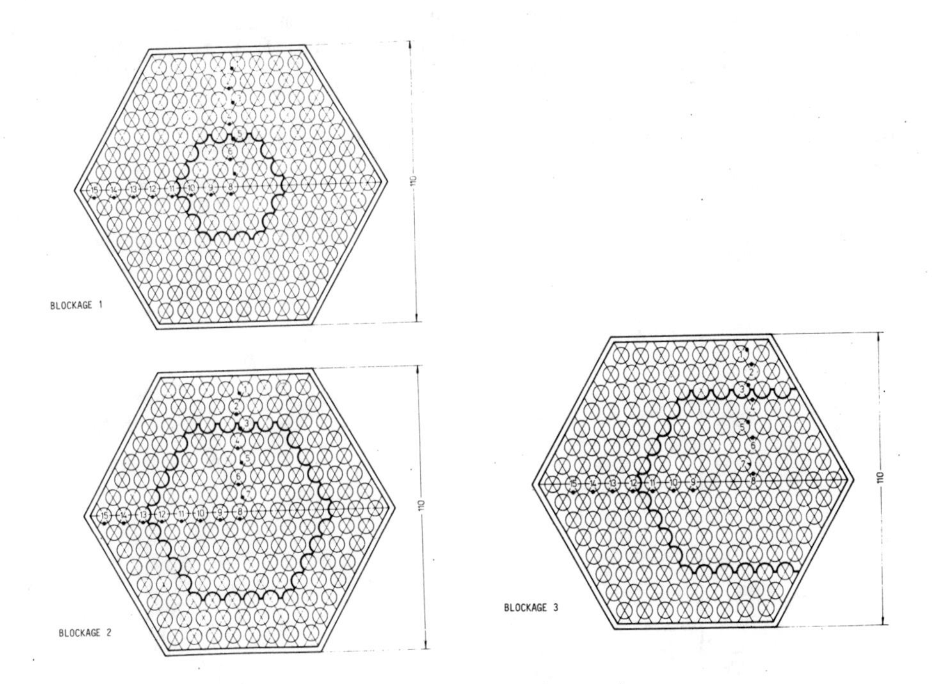

Fig. 16. 169 blockage experiment

Fig. 17. 169 blockage experiment

Following the experiments of Kirsch [3] who studied three kinds of blockages in a 169 -pin water bundle that has the same pin diameter and pitch as a SNR subassembly. Fig. 16/1 shows an internal blockage in the center with β = 0,147 (blocking 3 rings of flow channels). Fig. 16/2 shows a larger internal blockage in the center with β = 0,411 (blocking 5 rings of flow channels), and Fig. 17/3 shows the same blockage attached to one corner of the bundle with β = 0,47. All pins were uniformly heated with a length of 0,7 m. The blockage, a 5 mm thick plate, was located 0,1 m downstream the beginning of the heated section. On the basis of these experiments and the analysis discussed before, Kirsch et al. [3] and Schleisiek et al. [1] developed experimental tools to determine the wake temperature behind internal blockages. If the measurements in the water bundle can be extrapolated to sodium bundles the following conditions have to be satisfied:

$$(\bar{\Theta}_i)_{H_2O} = \left(\frac{\bar{T}_i - \bar{T}_B}{\bar{T}_{out} - \bar{T}_{in}}\right)_{H_2O} = \left(\frac{\bar{T}_i - \bar{T}_B}{T_o}\right)_{Na} = (\bar{\Theta}_i)_{H_2O} \qquad (30)$$

and with

$$\Delta T_o = \frac{N \cdot Xmax}{\dot{M} \cdot cp} \cdot \Delta x_{fuel}$$

$$(\bar{T}_i - \bar{T}_B)_{Na} = \frac{N \cdot Xmax}{\dot{M} \cdot cp} \cdot \Delta X_{fuel} \cdot (\bar{\Theta}_i)_{H_2O} \qquad (31)$$

(T_i : temperature in position "i", T_B : temperature in the free flow at blockage level, N : number of pins, $\dot{M}$: total subassembly mass flow, Δx_{fuel}: heating length, X_{max}: W/m = fuel power).

Fig. 18 shows an example of temperature distributions 20 and 40 mm behind a central blockage with β = 0,147 and a Re-number of 0,9 . 10^6.

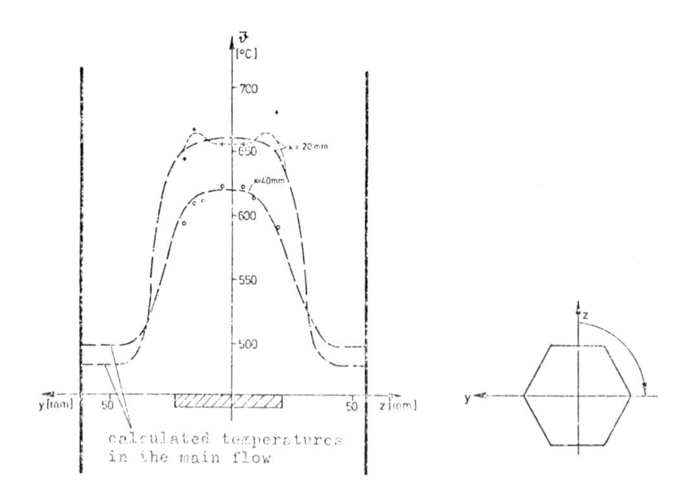

Fig. 18. Temperature distribution behind blockages

Schleisiek et al. calculated on the basis of these results the average wake temperatures behind the internal blockages in a peak-power SNR subassembly at two assumed locations, the core midplane and the outlet. The results are presented in Table 3 which shows that the average wake temperature should be of about 200 to 300°C below the sodium saturation temperature, and therefore sodium boiling behind such blockage is unlikely to occur under the flow and power conditions indicated.

Table 3. Average wake temperatures in an SNR fuel assembly at peak power with $\bar{T}_{in}$ = 380°C, $\bar{T}_{out}$ = 576°C, and $\dot{M}$ = 23,0 kg/s

	Core midplane		Core outlet	
q' (W/cm)	446		190	
β	0.147	0.411	0.147	0.411
T_B (°C)	478	489	567	598
T_{wk} (°C)	688 ± 14	781 ± 20	665 ± 6	723 ± 9
$T_{sat,B}$ (°C)	1000		950	

Basmer et al. [13] used Schleisiek's and Kirsch's method to extrapolate the temperatures measured in a water cooled bundle to a sodium cooled SNR subassembly and calculated the temperature peak behind blockages of various sizes and position in the bundle.

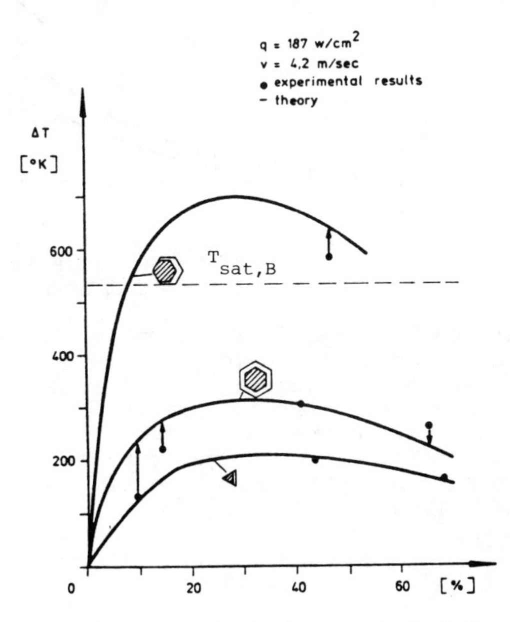

Fig. 19. Maximum temperature rise behind various blockages in peak power SNR subassembly

Fig. 19 shows their results $\bar{T}_{max} - \bar{T}_B$ behind the blockages at the midcore of a SNR subassembly at maximum power. The simbols on the curves indicate the blockage position. There are several important points shown. The temperature behind the edge blockage with $\beta > 0,10$ will exceed the local saturation temperature where sodium boiling might occur. For the central blockage the maximum calculated temperature rise $\bar{T}_{max} - \bar{T}_B$ is 310°C at $\beta \sim 0,3$, which is far below the saturation temperature. The lowest temperature rise was obtained in a 60° section of a bundle with 28 heater elements. Therefore great care must be taken to extrapolate these results to full scale bundle. For all these considerations a constant pressure head and sufficient bypass flow in the unblocked subassembly section is assumed. In order to assess the flow reduction due to the presence of a blockage, Basmer et al. measured pressure drops in an (unheated) unblocked subassembly and compared them with different blockages in the same subassembly. These results are shown in Fig. 20 for the percentage of velocity reduction vs blockage size and relative radial position of the blockage. They made their calculations for the highest and lowest powered subassembly of the SNR. The calculations show a higher flow reduction for the higher powered subassemblies.

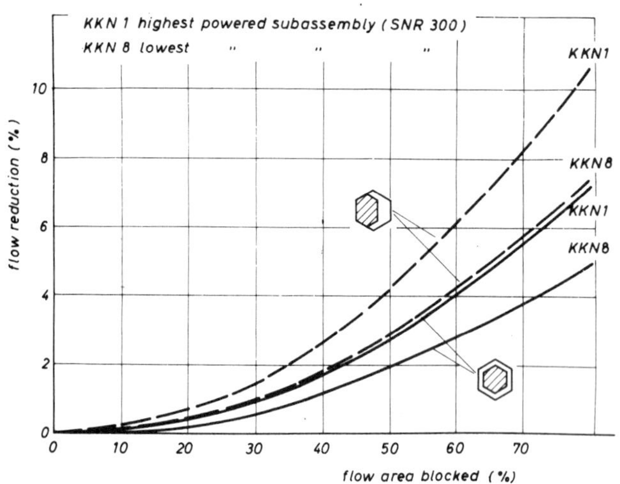

Fig. 20. Flow rate reduction as a function of blockage size in SNR subassembly

Note that the blockage of about 50% at corner position for the highest powered subassembly leads to a flow reduction of only 4%. The fact that a 4% flow reduction in the highest powered subassembly increases the mean outlet temperature of about 8°C, and as shown in Fig. 19 the peak temperature in the wake behind a corner blockage exceeds sodium saturation temperature at $\beta > 0,1$, the detection of blockages with already local boiling by monitoring only outlet temperatures might not be sufficient.

3.2. Sodium Boiling Tests in Blocked Bundle Geometry

KFK internal blockage in a 169 SNR simulated subassembly

Brook, Huber and Peppler [14] performed single phase and boiling tests behind a central 49% blockage in 169 sodium cooled bundle. The plane blockage plate was located axially 40 mm behind the beginning of the heated section. 91 rods were within the blockage boundary, 88 of which were uniformly heated. 3 were used for instrumentation (Fig. 16/2). One of the essential aims of these experiments was the verification of the modelling of sodium flow behind blockages by water experiments. Their single phase flow results are shown in Fig. 21 for temperature distributions at various levels behind the blockage as normalized temperature profiles:

$$\theta(m) = \frac{H \cdot (T - \bar{T}_B)}{T_{out} - T_{in}}$$

H : heated length

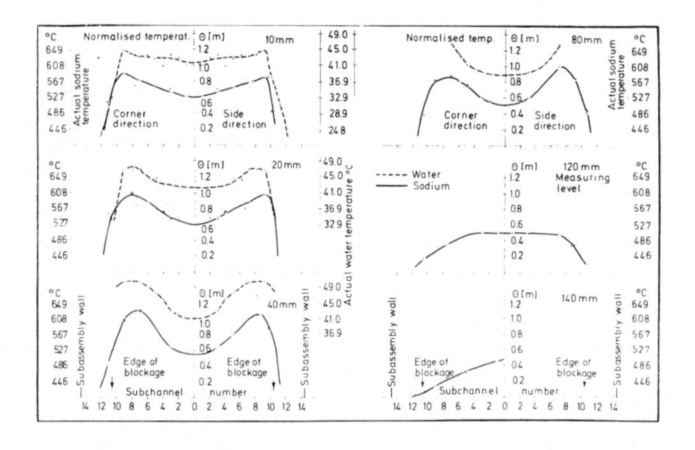

Fig. 21. Normalized temperature distribution behind 49% blockage in 169 SNR bundle

The maximum temperature is located at the slowly moving center of the wake as predicted from the water experiments. Fig. 21 shows further the temperature distributions obtained in corresponding water experiments. The agreement between water and sodium experiments is fairly good and confirms the applicability of water results obtained at similar conditions to predict the temperature distributions in fuel subassemblies. Brook et al. performed in the same bundle test section sodium boiling experiments keeping the power and inlet temperature constant while reducing the flow rate by throttling the inlet flow to a selected value until boiling was reached. Fig. 22 shows an example of measured temperature distributions for various mass flow rates with $\overline{T}_{in}$ = 590°C and $\dot{q}''$ = 132 W/cm² per heated rod. No sodium boiling occured at flow velocity V_1 = 3,7 m/sec, however as the velocity was reduced to 2,8 m/sec local boiling was detected. The local boiling zone increased when the velocity was decreased to 2,0 m/sec. They found that the boiling region in this velocity range was confined to the wake center and no propagation occurred. They concluded from their results that the recirculation flow pattern corresponded to the single phase behaviour and no flow instability was provoked by the boiling.

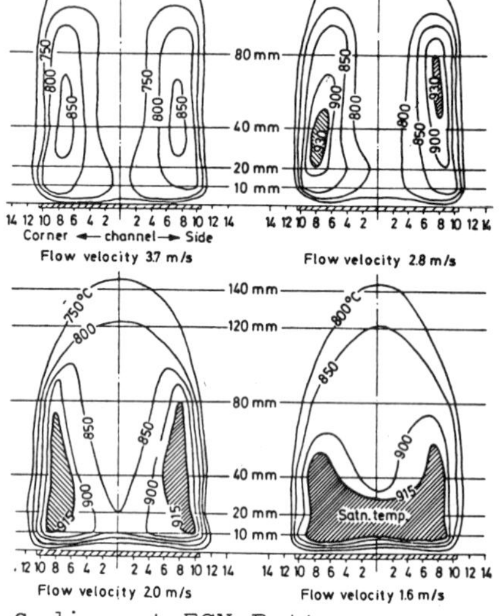

Fig. 22. Temperature distribution behind 49% blockage in a 169 SNR bundle

3.3. Local Boiling Investigations in Sodium at ECN Petten

Single phase and boiling experiments were performed by Brinkmann, de Vries, Dorr /15/ in a 28 rod test section simulating a 60° section of a SNR subassembly (rod diameter = 6 mm Ø; pitch = 7,9 mm). Fig. 23 shows the geometry of the test bundle, the location and size of the blockages in the test bundle and the simulated size of the real subassembly. Blockages of 68,5% and 34,5% of the cross section normal to the flow were investigated. The blockage, a metallic plate was located 0,1 m downstream the beginning of the heated section. The total length of the bundle was 1,0 m with a 0,40 m heated section. Devices for gas injection at blockage level were installed. Grid spacers were located 60 mm, 140 mm and 290 mm downstream the blockage.

Experiments were run with coolant flow rates between 0,25 and 4 m/sec. inlet temperature variations of 150 ÷ 600°C and heat fluxes $\dot{q}''$ = 5 to 160 W/cm². The evaluation of the experiments showed some dependence of the normalized temperature on the Re-number, (more than in the experiments of Schleisiek) and on the axial temperature gradient $(\overline{T}_{out} - \overline{T}_{in})/H$. Remarkable differences in the maximum normalized temperatures were observed between the 68,5% blocked bundle and the 34,5% blocked bundle. The corresponding

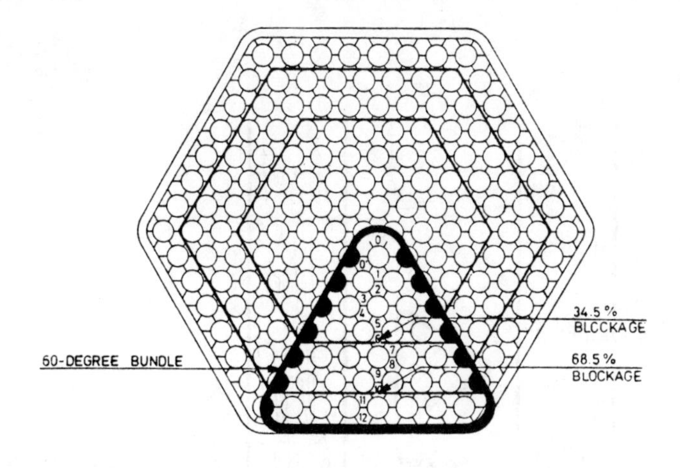

Fig. 23. 60° 28 rod bundle

measured values are 0,85 m, respectively 1,35 m for the smaller blockage. The authors indicate that the cross section geometry of the 60° bundle might have an influence and that they hesitate to extrapolate these values to SNR subassembly geometry. Nevertheless the non explained difference in the normalized temperature of the two blockage sizes Brinkmann et al. showed the strong influence of the spacers on the wake flow and temperature pattern. Fig. 24 shows the experimentally obtained flow distribution and normalized isotherms behind the 68,5% blockage. Separate areas with high temperature could be identified which prove the development of secundary flow circulations between the grids within the main wake (observed in the water experiments also).

The boiling experiments indicate three distinguishable boiling modes within the wake. The authors introduced for the indentification of the boiling characteristics the so called "boiling intensity" $\Delta T_{MS} = \bar{T}_{max} - T_{sat}$ (Fig. 25) as a measure of the heat virtually storred in the liquid phase wake above saturation and which can be transported to the main flow by evaporation and condensation. With increasing "boiling intensity" the boiling pattern in the wake changes from subcooled "nucleate boiling", "single bubble boiling" to "developed boiling".

As can be seen in Fig. 26 the boiling intensity $\Delta T_{MS} = \bar{T}_{max} - T_{sat}$ has to increase with increasing temperature rise in the wake in order to obtain single "bubble boiling" or "developed boiling". The authors estimate that under nominal LMFBR conditions a local temperature rise of more than 400°C is needed to obtain local boiling, and as seen from Fig. 26, single bubble boiling might occur at boiling intensities greater 100°C.

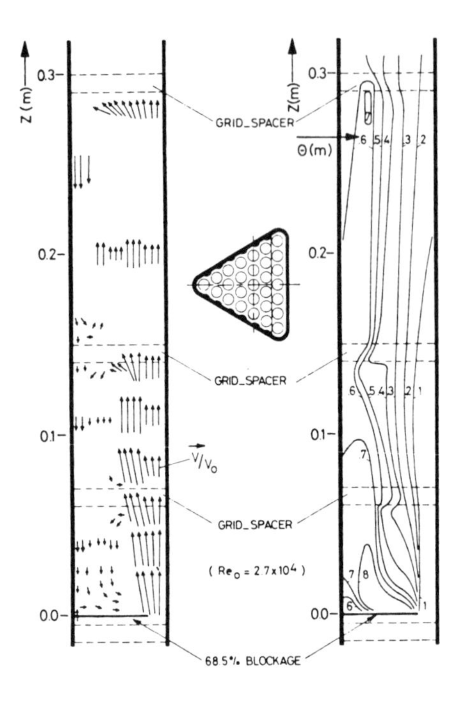

Fig. 24. Normalized isotherms and flow distribution in the 60° 28 rod bundle

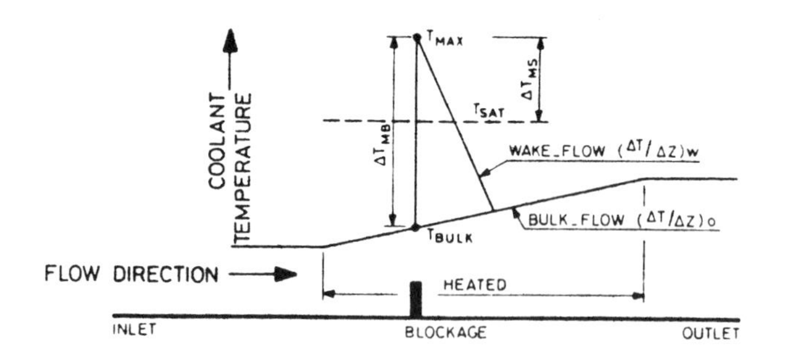

Fig. 25. Boiling mode characterisation

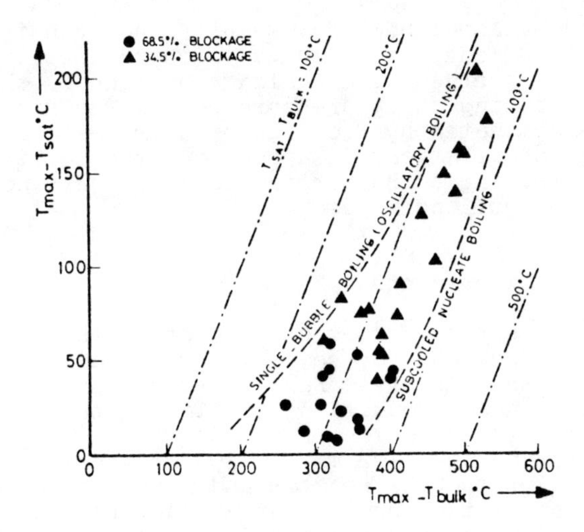

Fig. 26. Experimental results establishing boiling modes

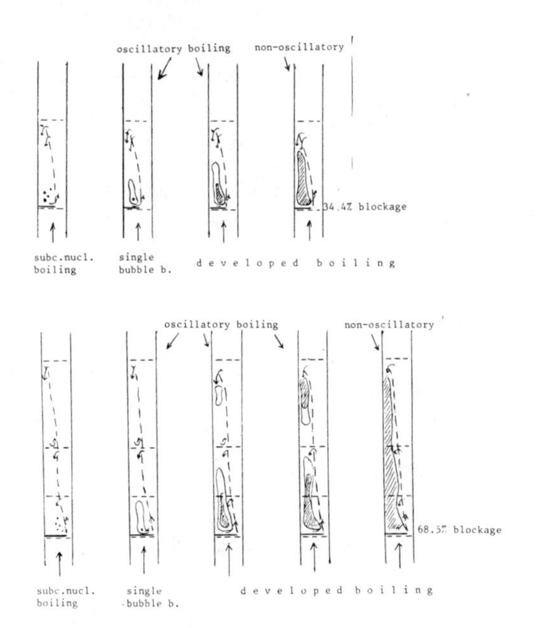

Fig. 27. Boiling regimes in the 60°-bundle with 34,4% and 68,5% blockages

The characteristic boiling pattern of the different boiling regimes are shown in Fig. 27. Note that boiling occurs also in the secondary wake at developed boiling. Even if the transitions between the main boiling modes are fluid, characteristic hydrodynamic pattern are observed. The nucleate boiling is characterized by cavitations causing sharp pressure peaks. The single bubble boiling regime is dominated by flow and pressure oscillations. At developed boiling, oscillations disappear and the boiling regime becomes steady state. The authors reported the occurrence of dryout at the developed boiling regime.

3.4. In Pile Mol 7C Experiments

Kramer et al /16/ performed experiments with a 37 rod bundle of fresh UO_2-fuel and for several days irradiated fuel in the BR.2 reactor at Mol. The test facility was designed for testing fuel pin bundles under conditions similar to those in the SNR-reactor. The in pile test section (Fig. 28) with a length of 8,2 m and a diameter between 122 mm and 300 mm was inserted into the central channel of the BR2 reactor and fixed at the reactor top nozzle. The test section housed the 37 fuel rod bundle which consisted of 30 fuel rods, one central dummy and 6 corner dummies (Fig. 29/1/2). The length of the rods was 800 mm with 400 mm fissile length. The U-235 enrichment was 90% for the inner row, 80% for the second row and 65% for the outer one,which produced a maximum uniform radial rod power of 400 W/cm. The 6 mm outer diameter rods were grid spaced with a pitch of 7,9 mm. The bundle contained a 40 mm axially extended porous blockage of 0,5 mm steel spheres located in the maximum power region. The blockage extended over six inner pins and part of the second row. The blockage boundary is made of a cage. In order to avoid impairment of the rods inside the blockage during the pretests additional cooling could be done through the central dummy. The sodium flowed downwards in the outer annulus and entered the test bundle and an annular bypass from below (Fig. 28).

Fig. 28 shows further the instrumentation of the test bundle

- 17 thermocouples (Ni/CrNi) at various locations of the test section
- 3 thermocouples (W/Re) in the center of the fuel rods
- 6 pressure gauges at bundle in- and outlet and expansion tank
- 5 pressure transducers for clad failure detection

- 4 acoustic noise detectors
- 2 flow meters
- molten fuel detection by 3 bi-axial cablesbelow the bundle
- delayed neutron detectors at the top of the test section.

Each fuel bundle was preirradiated for about 12 days corresponding to a burn up of 2,5 MWd/kg. During the preirradiation phase and out of pile simulation tests it was found that the interruption of the forced cooling in the blockage through the central dummy did not lead to boiling under nominal flow conditions.

The initial conditions for the so called transient test phase are illustrated in Table 4.

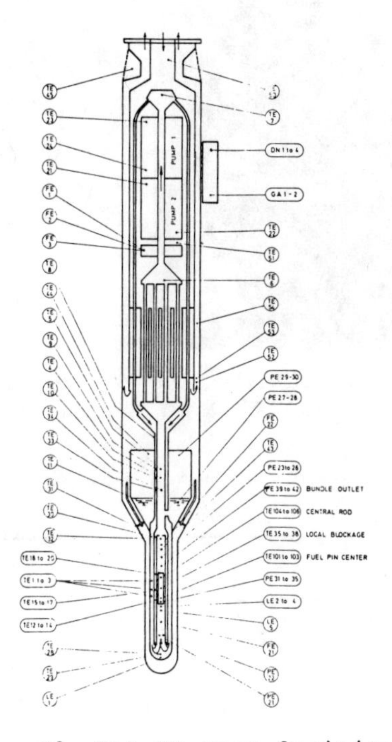

Fig. 28. Mol 7C test facility

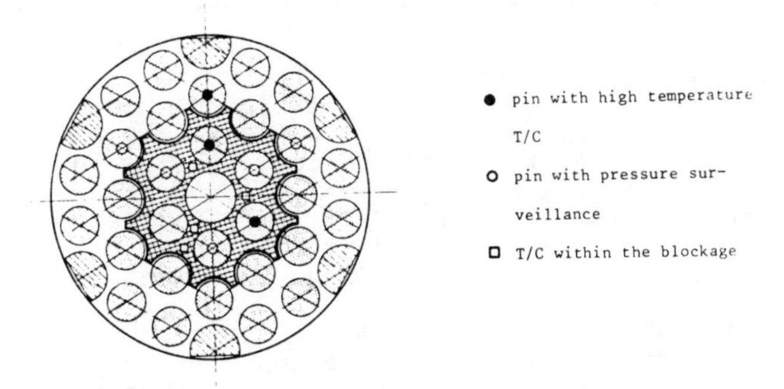

Fig. 29/1. Mol 7C test bundle

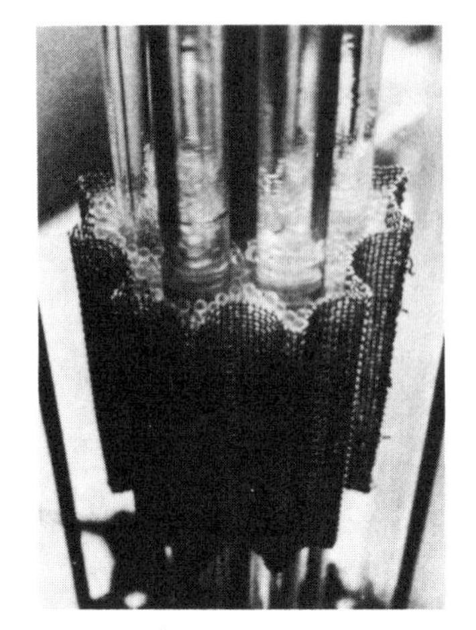
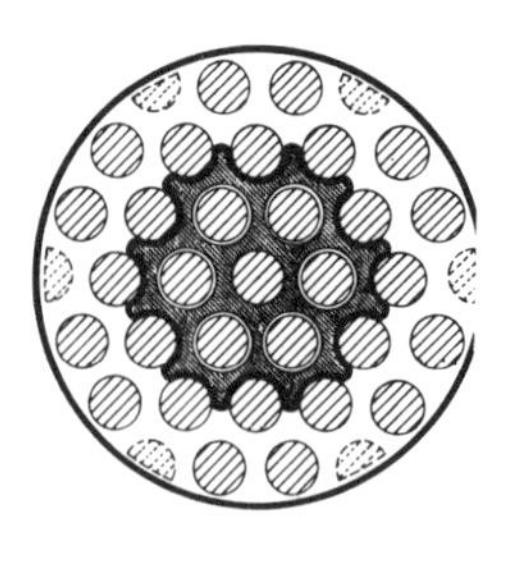

Fig. 29/2. Model of the Mol 7C test bundle with blockage

Table 4. Starting conditions for the transient test phase

	Dim.	Mol 7C/1	Mol 7C/2
Temperature bundle inlet	^{o}C	380	380
bundle outlet	^{o}C	730	740
Na-Flow	kg/s	0,9	0,9
Covergas pressure	bar	2	2
Max. linear rod power	W/cm	400	410
Max. burn up	MWd/kg	2,5	2,5
Gas pressure in the pins (operating conditions)	bar	60 and 53	12 and 10

To establish boiling conditions the nominal flow rate was reduced to half of its value so that the bundle outlet temperature increased to 730°C and 740°C respectively. The boiling phase was than initiated interrupting the cooling flow in the blockage (via the central dummy). Fig. 30 shows the registrations of the main parameters during the test 7C/1. After initiation of the transient the reactor was operated at full power for 49 min during the 7C/1 and 6 min during the 7C/2 experiment respectively.

The post test examinations showed considerable material relocations within the fuel bundle. The authors analyzed, for example, the increase of the bundle outlet temperature as a consequence of the formation of secondary blockages due to molten clad and

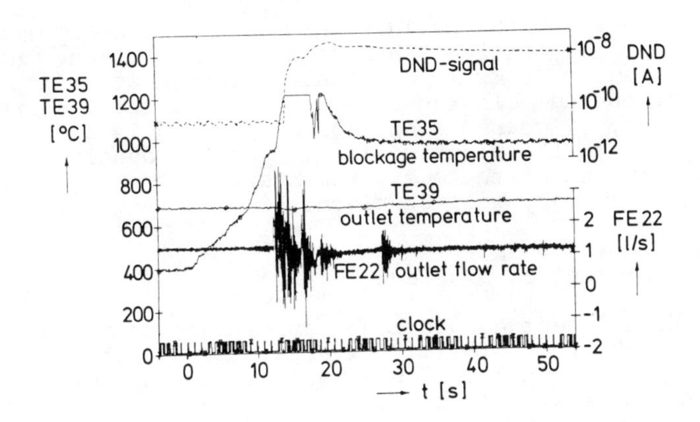

Fig. 30. Mol 7C/1 registration of the main parameters

fuel relocation. These experiments suggest the following at present only preliminary interpretation.

- At normal hydraulic and power opeation condition, a porous local blockage of up to 40 mm length and 35% of the flow cross section of a subassembly is coolable. No sodium boiling occurs within the blockage.
- Local boiling, dry out and fuel melting will occur at normal operation power within the porous blockage when the total mass flow is reduced to 50%.
- Fast failure propagation is unlikely to occur.
- The coolability of a distroyed blocked zone could be assured after melting and relocation at full power and 90% nominal mass flow.
- Pin failure could be detected immediately by DND instrumentation.

3.5. CRBR and FFTF Subassemblies

Han and Fontana /‾8_7 reported on investigations of the effects of blockages at the Thermal - Hydraulic Out of Reactor Safety (THORS) facility. The complete review of their work on impermeable blockage is given in /‾8_7. Their investigations were focussed, on the simulation of blockages in CRBR and FFTF wire-spcaed subassemblies. The test bundles consisted of 19 electrically heated rods. The rods had an outside diameter of 5,84 mm and were spaced by 1,42 mm diameter helical wire wraps of a 305 mm pitch, giving a distance between adjacent pin centers of 7,26 mm. The circular housing enclosed the 19 rods simulating a partial hydrodynamic section of an infinite rod array by having unheated rod segments attached to the housing wall. In vertical direction, beginning with the inlet, the sodium encountered a 406 mm long unheated section, then an about 533 mm long uniformly heated section and a 76 mm long unheated section before the exit. Three types of channel blockages have been studied: 6 channel internal blockages, 13-channels edge blockages and 14 and 24 channels blockages with the blockage located at the bundle inlet. A detailed description of these experiments is given in [8]. In the following a summary will be given.

3.5.1. 6 channel central blockage

The 6 channel central blockage corresponds to a blockage factor β = 0,117 of the total flow area. The stainless steel blockage plate (6,35 mm thick) was located 381 mm above the beginning of the heated section. Tests have been performed at uniform pin power in the range of 16 to 33 KW/m with a mean coolant velocity of 4 to 8 m/sec. Fig. 31 shows the test section.

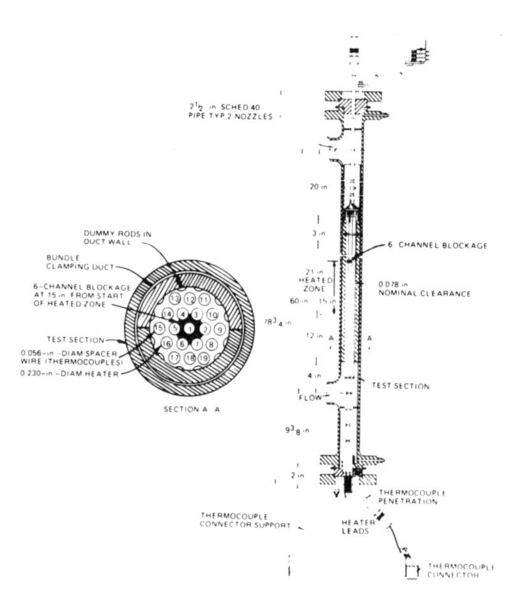

Fig. 31. Test Section of THORS-bundle 6 channel blockage

Han et al. assumed for the description of the recirculation wake the same model as shown by Schleisiek and Kirsch. Their measurements confirmed the results of Schleisiek and Kirsch that the coolant residence time and the wake length is independent of the Re-number and that there is only a moderate effect of the velocity. Fig. 32 shows the measured data of the 6 channel blockage single phase experiments, which generally can be extrapolated also to other tight plain blockages. From these experiments Han and Fontana developed a dimensionless maximum wake temperature correlation as a function of the free flow Reynolds number.

$$Y = K(\bar{T}_{max} - \bar{T}_B) / \dot{q}' = 1970\ Re^{-0,79} \qquad (32)$$

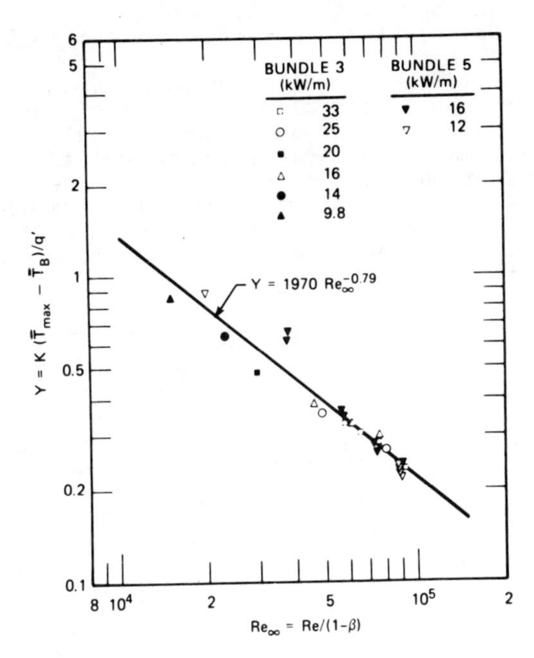

Fig. 32. Dimensionless maximum wake temperature vs Reynolds number

The reference temperature $\bar{T}_B$ is the mean bulk temperature at the blockage side. (The free flow Reynolds number is calculated from $U_\infty = U_O/(1 - \beta)$, U_O : velocity in the undisturbed bundle), $\dot{q}' \hat{=}$ W/m. This correlation is also confirmed by the 14 edge blockage results described later. The important result is that the maximum wake temperature rise ($T_{max} - \bar{T}_B$) decreases as the Reynolds number increases or as the power decreases which is physically reasonable. This behaviour of wakes has been described by Han and Schleisiek independently.

Although the quantitative extrapolation from the 19 pin bundle and 6 channel blockage to a full size CRBR or FFTF assembly is not known at present time, it is highly unlikely that at nominal flow and power conditions a 6 channel blockage will provoke boiling in the subassembly of the CRBR or the FFTF. This is because the maximum wake temperature increase of about 300°C behind the blockage is still to less to reach sodium boiling temperature (typical outlet temperature approximately 535°C, boiling temperature approximately 950°C). Hannus et al. [29] extended these experiments to investigate the boiling pattern behind blockages especially with respect to the propagation from local

boiling to cross boiling. They performed tests either reducing the mass flow in the bundle at constant rod power or varying the power while the flow was maintained steady. They obtained boiling behind the blockages at $\bar{T}_{in}$ = 560°C, $\check{q}'$ = 16,4 KW/m, bundle mass flow = = 1,23 m/sec (Test Nr. 3), and $\bar{T}_{in}$ = 488°C, q' = 14,8 KW/m (Test Nr. 4), bundle mass flow = 0,85 m/sec.
From the examination of the radial temperature distribution they concluded that there was no radial propagation of the boiling zone from the blocked channels to the unblocked channels. But for the experiments with low mass flow (e.g. 0,85 m/sec) they could not exclude that radial propagation of the boiling zone could have occurred downstream the blockage wake. From the measurements Hannus et al. judged that during their tests sodium dry out did not take place behind the blockage and that the boiling zone remained steady during the test periods.

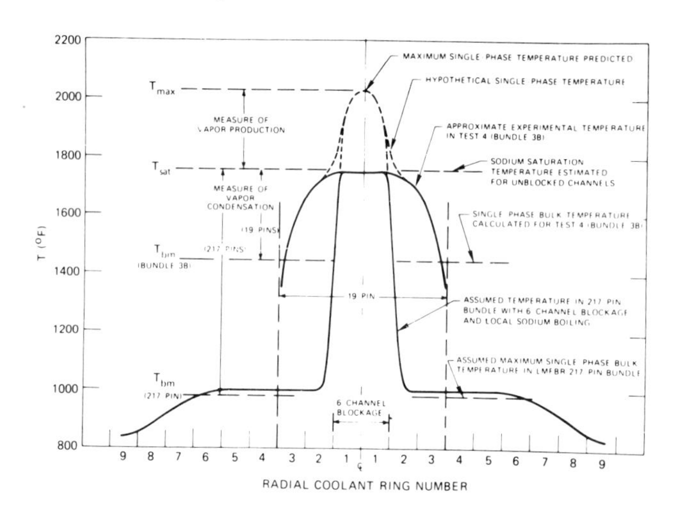

Fig. 33. Radial temperature profiles 25 mm downstream the blockage

In order to extrapolate the 19-rod bundle boiling results to that in a full size 217-rod fuel assembly, Hannus et al introduced a criterion to predict the boiling propagation capability in a bundle. They compare the boiling restraining force , which is the vapour-condensing capability of the system, and the driving force, which is the vapour-producing capability of the system. The boiling restraining force is produced by the local bundle mean subcooling $-(\bar{T}_B - T_{sat})$ and the boiling drivind force the maximum hypothetical superheat of the wake $(T_{max} - T_{sat})$. Fig. 33 illustrates the model and shows the temperature profile extrapolation to a 217 rod assembly, assuming a 6 channel blockage. Based on this idea Hannus et al. propose similar to Brinkmann and De Vries's suggestion a boiling propagation number to indicate the tendency of radial propagation of boiling from blocked channels into unblocked surrounding channels.

$$BP = (T_{max} - T_{sat}) / (T_{sat} - \bar{T}_B) \qquad (33)$$

The larger the number, the greater the tendency for propagation of local boiling. This method is a helpfull and valid tool to extrapolate from small scale simulation experiments to full size bundle geometry. The authors concluded from their experiments: Since there was no radial boiling propagation in the 19-rod bundle, it is very unlikely to have radial boiling propagation at similar conditions they investigated, in a CRBR fuel subassembly unbiased by some non prototypic features of the test section.
Assuming a 6-channel blockage at the outlet of a CRBR assembly and its maximum wake temperature rise being the same as that of test 4, they calculated the BP numbers shown in Table 5.

Table 5. Radial boiling propagation number for a 6-channel blockage

	Bundle size pins	$\bar{T}_{in}/\bar{T}_{out}$ (°C)	$\bar{T}_B$ (°C)	T_{sat} (°C)	T_{max} (°C)	B.P.
CRBR	217	388/535	535	950	1060	0,27
Test 3	19	560/874	800	965	1030	0,39
Test 4	19	488/883	790	960	1060	0,59

T_{max} was calculated from eq 32.

3.5.2. 13 channels edge blockage, 14 and 24 channels inlet blockage in a 19-rod bundle

In a 19 pin test section with the same pin diameter and pin array as for CRBR and FFTF fuel assembly a 3,175 mm thick stainless steel plate was mounted 102 mm above the beginning of the heated section which was 457 mm long blocking 13 flow channels (β = 0,34). Although the bundle was carefully built, a leak occurred between the blockage plate and the duct wall. In another bundle a leak passage on the two small parts of the blockage (Fig. 34) of 0,356 mm were intentionally built in. Tests have been performed in these bundles with a mean sodium velocity of 1 to 8 m/sec and an uniform power per pin of 10 to 16 KW/m and 27 KW/m per pin.

The dimensionlees maximum wake temperature calculated for the edge blockage case without leakage fits with the correlation developed for the 6-channel central blockage (Fig. 32). In case of even a very small leakage between the housing and the edge blockage the maximum wake temperature is substancially reduced. The authors conclude from these experiments that for a 13 channel edge blockage in a CRBR subassembly at maximum power ($\dot{q}'$=49 KW/m, K = 70 W/m°C Re = $7 \cdot 10^4$ and β = 0,03 ≙ 13 channels) the maximum wake temperature rise ($T_{max} - \bar{T}_B$) will be less than 200°C and not cause boiling at these flow and power conditions.

Fontana et al. investigated also in a 19 pin bundle the effect of inlet blockage on the temperature rise in the wake. The pin diameter and the pin pitch was the same as for the CRBR and FFTF. The pins had a heated section of 533 mm preceeded by an un-

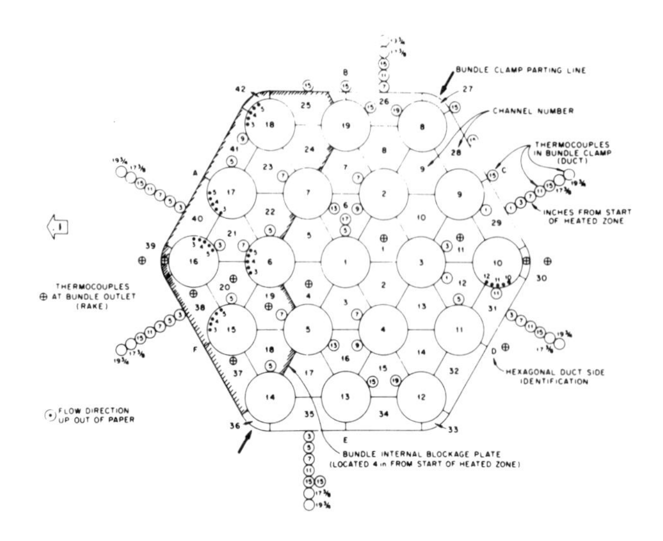

Fig. 34. 13 channels THORS edge blockage test section

heated length of 76 mm. A 1,59 mm stainless steel plate was located 76 mm upstream of the heated section blocking a) 14 central flow channels ($\beta = 0{,}25$) and b) 24 flow channels ($\beta = 0{,}46$). A series of tests were performed for sodium velocity in the range of 1 to 8 m/sec and uniform pin power in the range of 6 to 26 KW/m. For comparison Fontana et al. used the temperature measurements of the unblocked bundle. Fig. 35 shows the dimensionless temperature rise $(T - \bar{T}_{in})/(\bar{T}_{out} - \bar{T}_{in})$ along with the heated length. Fig. 36 shows their results for the corresponding dimensionless temperature rise with reference to the unblocked bundle of the 14 flow channel blockage. There is generally no significant difference between the temperature rise in the unblocked bundle and the 14 channel inlet blockage.

Fig. 37 shows the ratio of the dimensionless temperature rises for the 24 flow channel inlet blockage. This ratio is greater than 1 for the channels in the vicinity of the blockage and less than 1 for the unblocked flow channels due to the increased velocity in the unblocked channels. The maximum temperature rise ratio of 1,8 was measured 51 mm downstream of the beginning of the heated section. The temperature increase over the unblocked bundle at this location was only 4°C. Therefore, as pointed out by Fontana et al. the temperature increase in the blocked channels is very small even for the 24 flow channel blockage, which blocked approximately half of the total inlet flow cross section of the 19 pin bundle.

The effect of the blockage ceased about 150 mm behind the blockage which corresponds to a L/D ratio of 3.

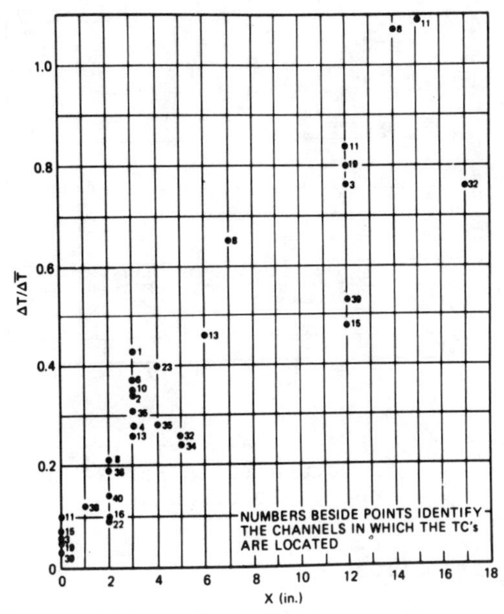

Fig. 35. Dimensionless temperature rise vs axial distance for unblocked tests

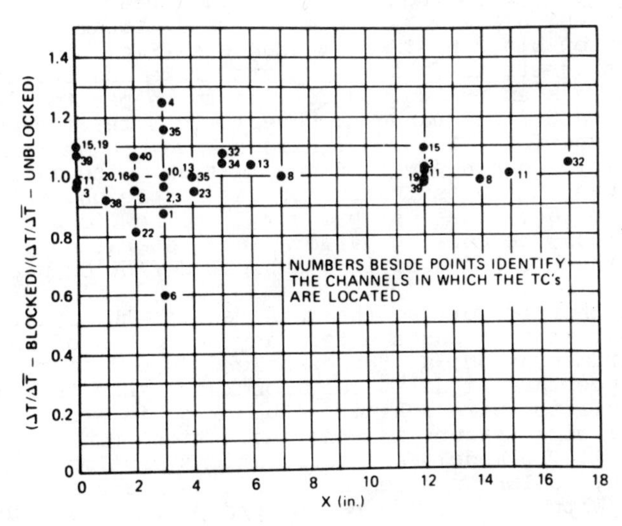

Fig. 36. Ratio of dimensionless temperature rise for 14 channel inlet blockage

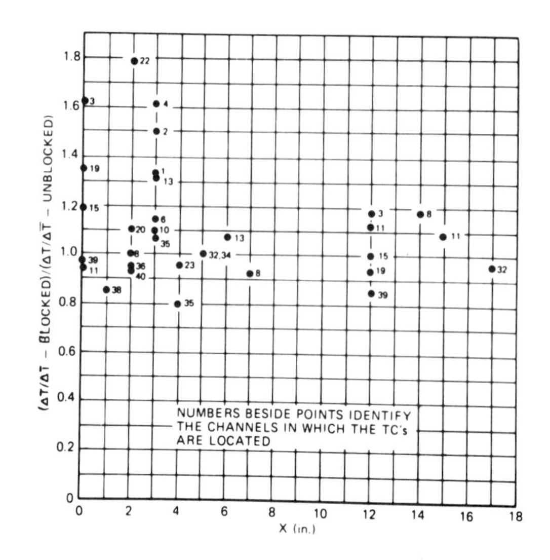

Fig. 37. Ratio of dimensionless temperature rises for 24 channel inlet blockage

3.6. Phenix Subassemblies

Menant et al. [17] performed blockage experiments in a 19 pin wire spaced bundles with central blockages. The pins had a diameter of 8.65 mm and were spaced by 1,29 mm wires wrapped on a helical pitch of 180 mm, giving a distance between adjacent pin centers of 9.95 mm as those in the Phenix reactor. A 2 mm thick stainless steel plate was located 50 mm upstream the 600 mm long heated section blocking 12 flow channels (Fig. 38). During the single phase tests the blockage plate broke and the two parts were removed as seen in Fig. 38 forming two separate displaced blockages of 4 and 8 flow channels. Nevertheless the mechanical defect extensive experimental investigations have been performed varying the mass flow and pin power. Since initially the main task of the test programme was the investigation of the single phase flow wake behind the blockage, the test programme was now focussed on the investigation of the boiling pattern and propagation of the boiling zone from the blocked area into unblocked flow channels.

The authors report as the main results the occurrence of early dryout in most cases before any detection of boiling by classical instrumentation; as for example flwo oscillations by flowmeters or pressure signals. They concluded that the formation of the recirculation wake behind the blockage is strongly biased by the helical wire spacers and the small tight pin to pin distance ($p/d = 1{,}15$). The models and correlations developed to calculate wake length, coolant residence time in the wake and heat transfer in bundles with a great p/d ratio ($p/d > 1{,}25$) seem not to be

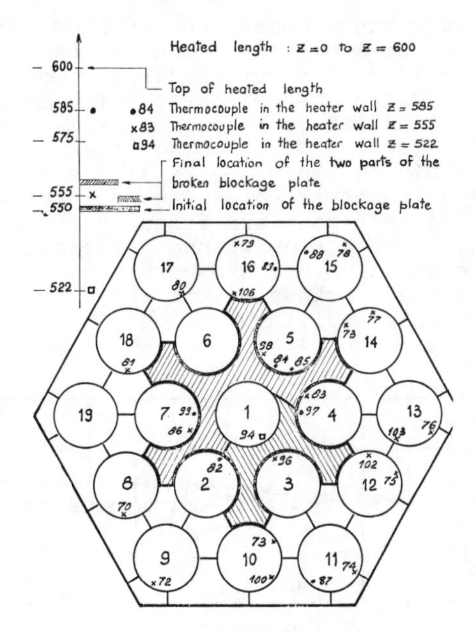

Fig. 38. 19 rod bundle 12 sub-channel blockage

applicable to Phenix type subassemblies. These results do not agree with those obtained in other laboratories with bundles of great pitch to diameter ratios. One of the reasons for the different results could be the extremely high heat flux of 500 W/cm per pin.

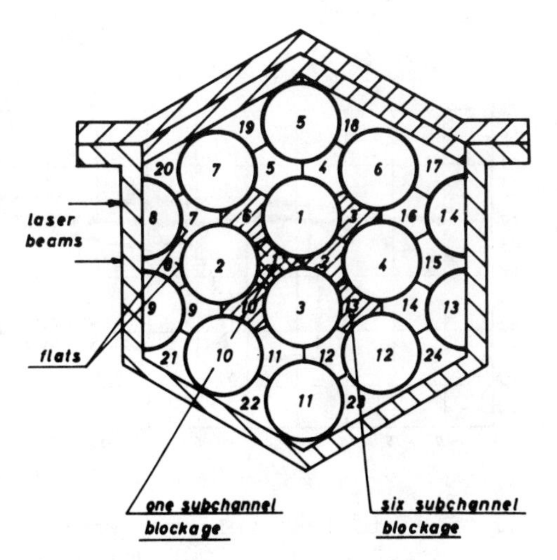

Fig. 39. Cross section simulation experiments p/d = 1,168

Simulation experiments measuring the flow pattern behind plain blockages in bundles of a pitch to diameter ratio of 1,168 were performed by Jolas /18/. For the measurement of the velocity distribution he applied the Doppler laser velocimeter. The test section was 3520 m long and 322 mm thick in the measuring section. The rods had 80 mm diameter and were made of transparent material and grid spaced with a pin to pin center distance of 93,4 mm. The array of the test section is shown in Fig. 39. 10 rods compose the central part and two half rods are fixed on the housing in the measuring section. 1 and 6-flow channel blockages were investigated. The 1-flow channel blockage showed a recirculation zone of 2 d_h length behind the blockage within the subchannel. Adjacent subchannels were nearly undisturbed.

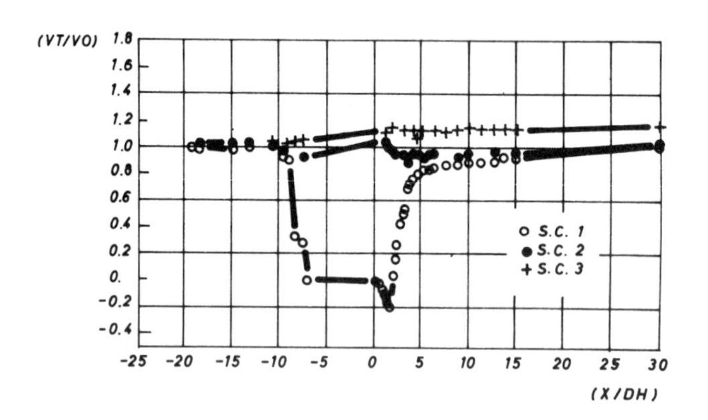

Fig. 40. Dimensionless velocity profile of one subchannel blockage Re= 30000

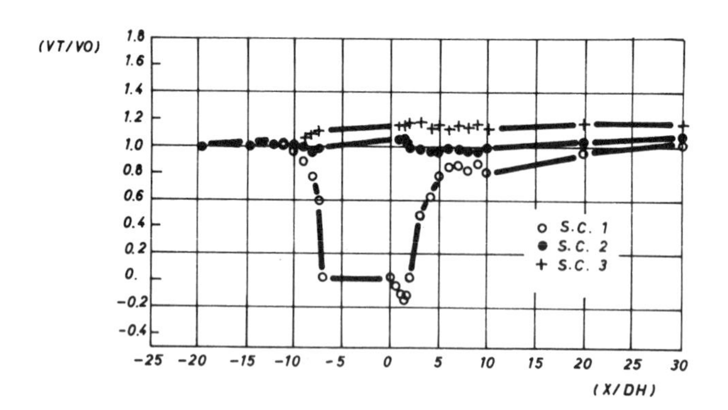

Fig. 41. Dimensionless velocity profile of one subchannel blockage Re = 90000

Fig. 40 and Fig. 41 show the dimensionless velocity profiles (local velocity to inlet mean bundle velocity) vs the dimensionless distance from the blockage X/d_h for two extreme Reynolds numbers. The Reynolds number has no influence on the length of the recirculation wake. In the 6-flow channel blockage a pronounced recirculation zone could be observed. The wake length to blockage diameter ratio was $L/D \sim 1,5$ which is in the range of values measured by Kirsch, Schleisiek and Fontana [3] [1] [8]. The velocity profiles behind the 6-flow channel blockage for free flow Reynolds numbers of 35000 and 90000 respectively are shown in Fig. 42 and Fig. 43. The measurements in subchannel 1 and 3 show clearly the flow inversion in the axis of the wake and the mass flow increase in the unblocked subchannels in the vicinity of the blockage.

The results of Jolas do not confirm Menant's observation that the recirculation wake behind a blockage in a tight bundle ($p/d = 1,168$) is not established. Jola's results show the occurrence of a recirculation zone similar to that in a free flow. The rods had no significant influence. Even if the p/d ratio was approximately the same as in Menant's experiments essential geometrical differences of the test sections have to be noted. The rod diameter was 80 mm instead of 8,65 mm, the smallest rod to rod distance 13,4 mm instead of 1,29 mm and the simulation experiments were done in a grid spaced bundle. These results suggest that the formation of the recirculation wake depends more on the real dimensions than on the p/d ratio.

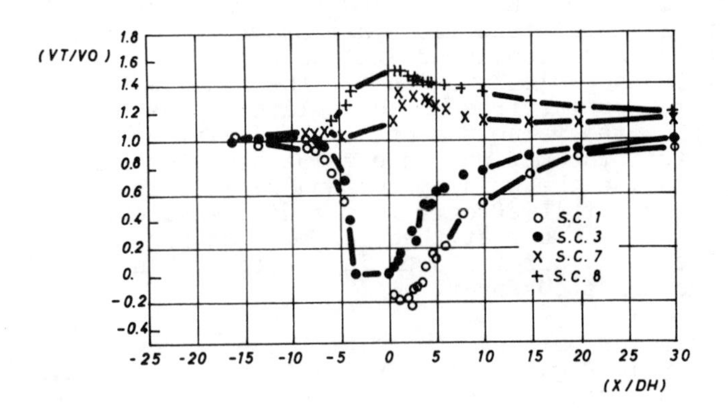

Fig. 42. Dimensionless velocity profile of 6 subchannel blockage, Re = 35000

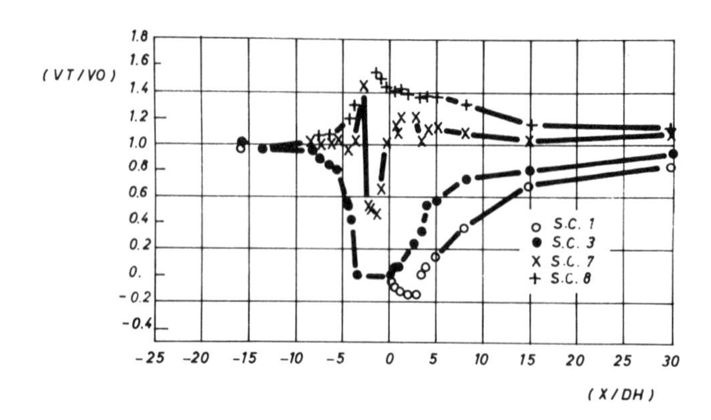

Fig. 43. Dimensionless velocity profile of 6 subchannel blockage, Re = 90000

3.7. Japanese LMFBR Studies

Daigo et al./ 19 / measured temperature distributions in a 7 pin sodium cooled bundle with a 6 channel blockage. Fig. 44 shows their test section, where a blockage plate of 0,5 mm was attached to the upstream side of a grid spacer. The pins were of 6,5 mm diameter with a pitch of 7,9 mm simulating the MONJU fuel assembly. The plate and grid spacers blocked 42% of the total flow area. Fig. 45 shows the circumferential pin wall temperature distributions at the blockage, and at 15 and 50 mm downstream of the blockage. The experimental conditions are shown in Fig. 45. The highest temperature was measured immediately behind the blockage and was about 52°C. Using Han's equation for the maximum wake temperature, the maximum temperature rise is calculated to be 36°C, which is about 30% below the value measured. The deviation is probably due to the fact that the maximum wake temperature in Daigo's experiment occurs inside the grid spacer and Han's correlation does not take into account this effect. From their measurements they calculated also the Nu-number at three axial locations downstream the blockage. The Nu-number is lower in the wake than outside the wake similar to Han's observation.

Their results obtained in an unblocked wire wrapped bundle are also shown in Fig. 46. The interesting point is that the type of spacers has little effect on the Nu-number in the unblocked bundle of this configuration. Note that the Nu-number measured by Daigo is even smaller than that measured by Fontana.

Uotani et al. / 20 / report on center and half bundle blockages in a sodium cooled 37 pin bundle at PNC's O-arai Engineering Center. 22 of the 37 pins where heater rods. Diameter and pitch of the rods simulated the japanese MONJU reactor bundle design. 24 subchannels corresponding to half of the 37 pin bundle were blocked. Design and overall dimensions are shown in Fig. 47. A 5 mm

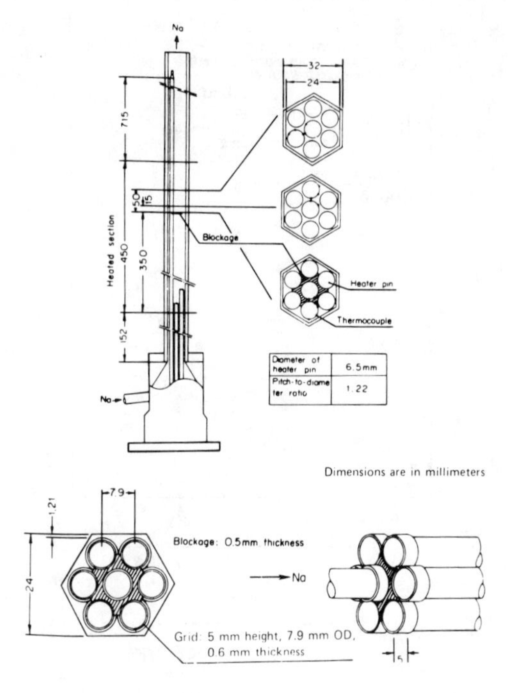

Fig. 44. 7-pin sodium cooled test section with 6 channel internal blockage

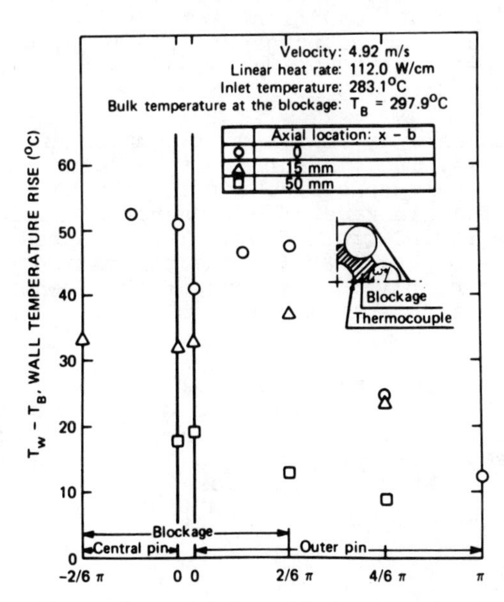

Fig. 45. Circumferential wall temperature distribution downstream the 6 channel blockage

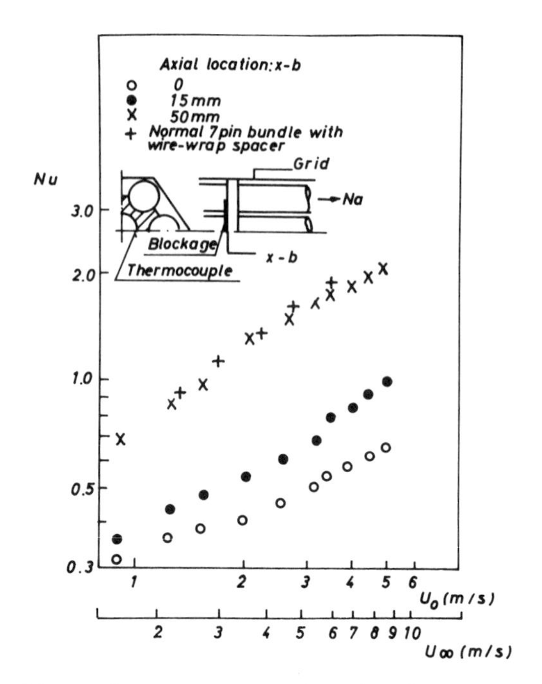

Fig. 46. Nusselt number behind the 6 channel blockage

thick stainless steel plate was located 308 mm downstream the beginning of the 460 mm long heated section below the honey comb grid spacer. The test conditions were as follows: velocity in the undisturbed bundle in the range of 0,48 to 3,49 m/sec, heat flux density per pin 1,0 to 53,96 W/cm^2 and inlet temperature fluctuations between 257°C to 299°C.

Fig. 48 shows a typical radial temperature field of pin surface temperatures. The pin surface temperature attained a local peak inside the grid spacer behind the blockage and decreased with distance from the blockage. The temperature field got disturbed notably due to radial cross flow. The measured residence time was about 1,4 times bigger than for the central 24 subchannel blockage (Fig. 49), but nearly independent of the Re-number. The 24 central subchannel blockage experiments confirmed Han's results.

Uotani et al. estimated the local maximum temperature rise evaluating the maximum to average temperature rise ratio in the wake region of the 37 pin bundle experiments and extrapolated to the full size bundle (Note the reservations mentioned by Han and Fontana).

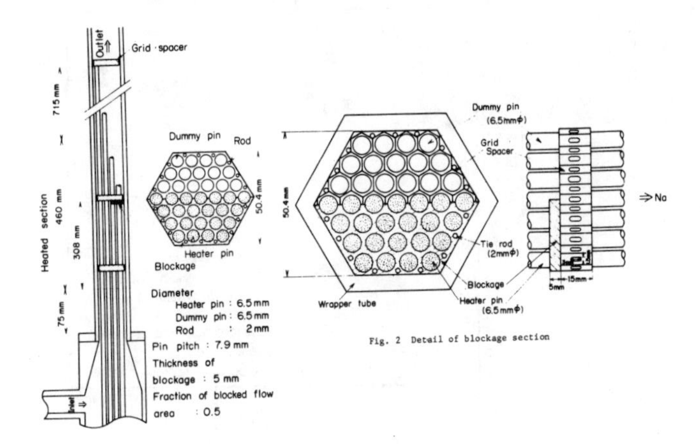

Fig. 47. 37 pin half bundle blockage experiment

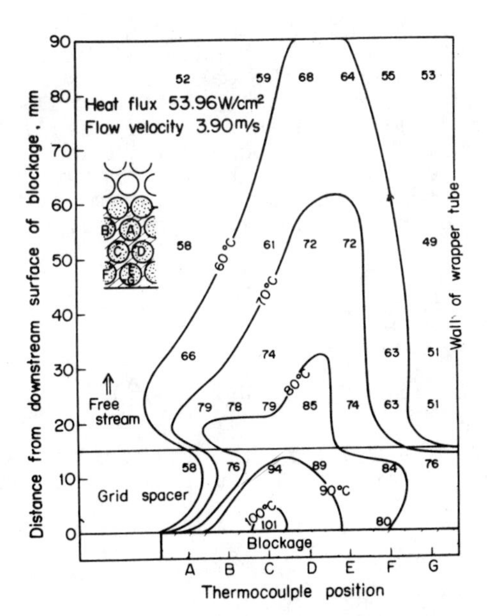

Fig. 48. Temperature behind the half bundle blockage in the 37 pin bundle

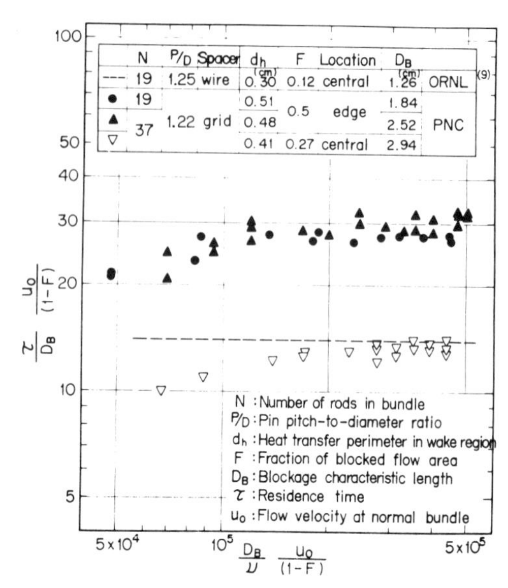

Fig. 49. Dimensionless residence time vs Re-number for central and half bundle blockages

Fig. 50 shows the maximum to average temperature rise ratio for both, the central and half bundle blockage, T_{max} was the maximum temperature measured in the wake. The ratio $(T_{max} - T_o)/(T^* - T_o)$ is independent of both the flow velocity and the heat flux. There is a small difference of the mean values between the central and the half bundle blockage (1,5 and 1,4 respectively). The difference is explained by the temperature fields due to the different flow patterns. Uotani et al. assume the temperature rise ratio also independent of the blockage size. Using eq (21), and eq (30):

$$\frac{d_h}{4} \cdot \rho \cdot c_p \cdot \frac{T_{max} - T_o}{\phi} = \tau \qquad (21)$$

$$\frac{\tau \cdot U_\infty}{R} = C \quad \text{with } U_\infty : U_o/(1-\beta) \qquad (30)$$

and $(T_{max} - T_o)/(T^* - T_o) = K$, (the maximum to average temperature rise ratio), they obtain for the full size 169 bundle and nominal operation conditions:

mass flow : U_o = 6 m/sec
heat flux : ϕ = 200 W/cm^2
hydraulic diam.: d_h = 0,41 cm

the maximum local temperature rise as a function of the blockage size. Fig. 51 shows the temperature rise for the two blockage types investigated. In both blockage cases the temperature increa-

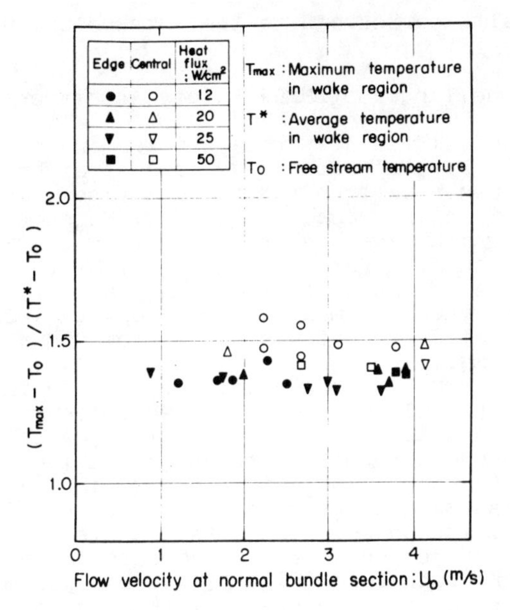

Fig. 50. Maximum temperature rise to average wake temperature rise vs undisturbed bundle velocity

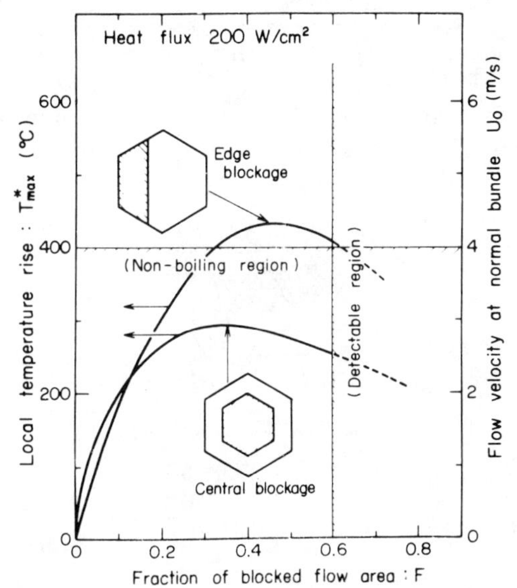

Fig. 51. Local temperature rise as a function of the blockage size in a full size MONJU subassembly

ses reaching a maximum at 30% blockage in the central area and at about 40% half bundle blockage. The calculation shows that for half bundle blockage local boiling might occur, whereas the peak temperature of the central blockage is far below boiling temperature. These results correspond to the extrapolations of Basmer [13].

3.8. Blockage Modelling for LMFBR Safety Assessment

The effect of different permeabilities on the temperature rise behind the blockage appears to be less straight forward. It was concluded that in general the temperature rises were comparable to or below those of an impermeable blockage. However specific cases were identified which gave higher temperatures. A 10% in temperature rise was calculated for an annular jet at the leading edge of the blockage. Also the central and annular jets combined gave a 5% increase above the impermeable blockage value. To support or reject these calculated predictions water flow visualisation experiments were performed. The complex flow distributions behind permeable blockages normally were visualized by the injection of air in the wake. In the following only the experiments of Robinson [21] as representative experiments will be discussed. The linear test section chosen to represent the subassembly is illustrated in Fig. 52. 11 stainless steel pins, 6,4 mm in diameter and 600 mm in length were assembled in a rectangular cross section with a pitch of 7,87 mm. The permeability was simulated by holes in the blockage plate. The residual flow was controlled and measured independently of the bulk flow (Fig. 53). The experimental investigations were performed in three stages with a range of residual flow in each case.

- Residual flow at the intersection of the blockage and the test section wall (referred as corner blockage)
- Residual flow near to the leading edge of the blockage
- Residual flow at the center of the blockage.

Photographic plates and cine film taken were used for the flow pattern evaluation.

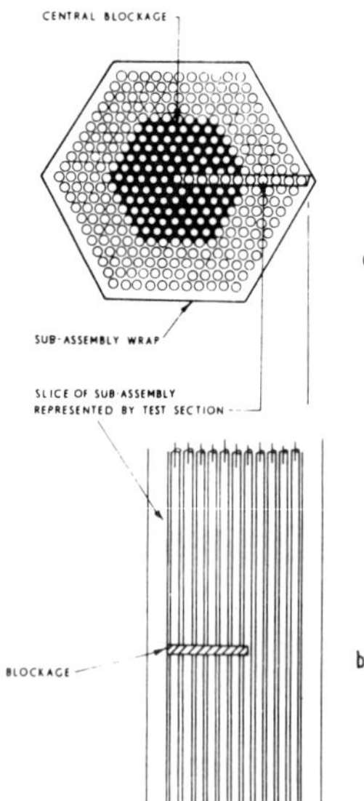

Fig. 52. 11 pin test section representing one row of a fast reactor subassembly of 325 rods

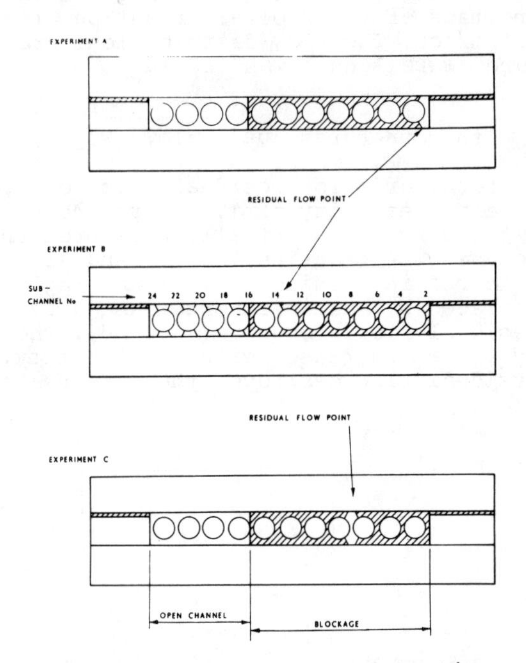

Fig. 53. 11 pin linear test sections with residual flow injections

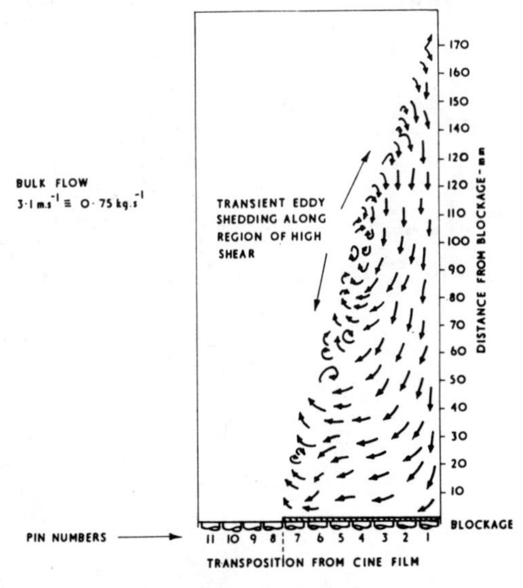

Fig. 54. Recirculation flow pattern of impermeable blockage

3.8.1. Impermeable blockage

With the blockage geometry shown in Fig. 53 and zero residual flow, records were made of the flow distribution at a bulk flow equivalent to 3,1 m/sec. The flow distributions are shown in Figure 54. The measured wake length was 160 mm.

3.8.2. Blockage with corner residual flow (Fig. 53 exp A)

The flow patterns for a low residual flow of 0,029 kg/sec corresponding to a permeability of 1,94% are shown in Fig. 55. The envelope of the wake is essentially unaltered from that of the impermeable blockage with transient eddies and slight increase in wake length. At increasing residual flow the single vortex wake structure was replaced by multiple vortex and a secondary vortex driven by the residual flow produced (Fig. 56). The axial length of the secondary vortex increased with permeability. With 6% permeability the residual flow destroyed the wake reattachment to the blockage.

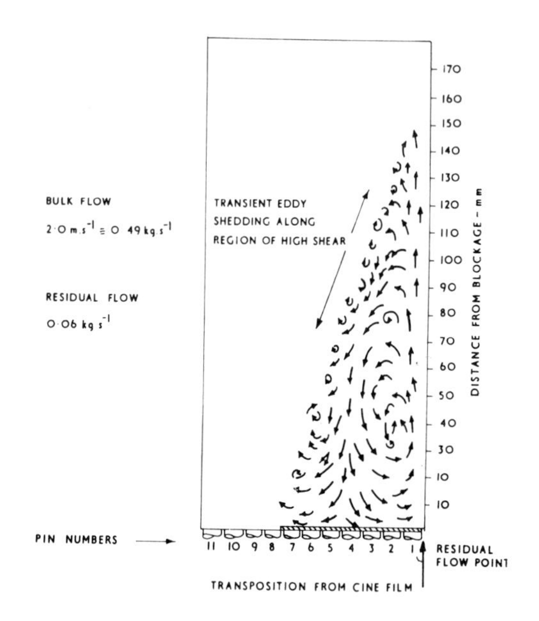

Fig. 55. Recirculation flow pattern at low residual flow (corner blockage)

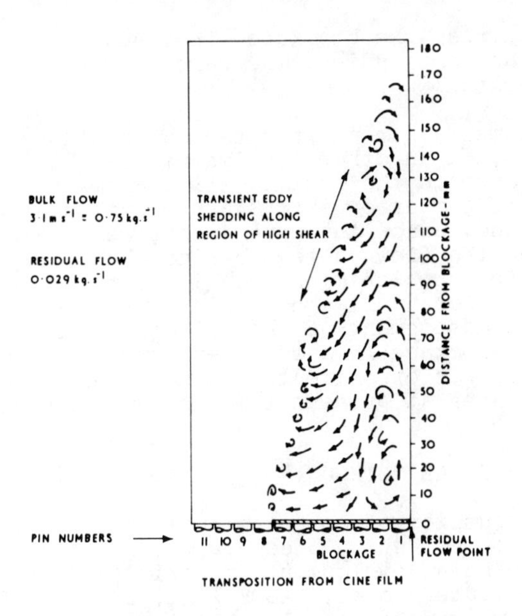

Fig. 56. Recirculation flow pattern at high residual flow (corner blockage)

3.8.3. Blockages with edge residual flow (Fig. 53, exp. B)

The residual flow in the blockage was repositioned in subchannel 14 (Fig. 53). For two different bulk flows (2 m/sec and 3 m/sec) a range of residual flows were investigated up to 11,5% of the main flow. In all cases the wake consisted of a single vortex. For a constant main flow velocity of 3 m/sec there was a monotonic decrease in wake length as the permeability increased. A 6,3% permeability produced a 25% reduction.

3.8.4. Blockages with central residual flow

The residual flow point was relocated to subchannel 8 (Fig.53). As the permeability increased a reduction in wake length occurred and a transition in the wake geometry from a single recirculation zone to one with more complex flow pattern was observed. With a low permeability the interaction between residual flow and the main wake recirculation flow is dominated by superimposed complex transient eddies. The residual flow is diverted diagonally across the pin bundle towards the region of high shear with a relatively low velocity remaining between the residual flow injection point and the tip of the blockage. An increase of residual flow to 4.6% of the main flow was sufficient to break up the recirculation flow of the main vortex. An elongated vortex was observed between the residual flow point and the channel wall with, on the other side of the residual flow injection point, a counter rotating secondary vortex downstream the blockage. The boundary of the wake was dominated by transient eddies. The wake length reduced as the permeability increased.

It becomes apparent from the observations made that small residual flows, in an otherwise impermeable blockage, can lead to a dramatic change in the recirculation geometry. These experimental findings were compared wiht "SABRE 1" calculations and good agreement was found. Since the experiments were carried out in a linear test section the flow pattern behind the blockage is biased by the wall effects which might stabilize the flow regime and lead to a wrong model for bundle geometries. Therefore, in order to make quantitative assessment of the effect of permeabilities in reactor subassembly blockages, heated experiments in rod bundles outgoing from these simple experiments are necessary. The knowledge of permeable blockages at present time is far from being adequate.

4. CALCULATIONAL METHODS

The discussion of computational tools predicting flow and temperature patterns behind blockages is not the aim of this paper. However, the main developments in this field will be indicated. There are computer codes, SABRE [22, 23, 24] COBRA [24], UZU [25], which numerically calculate the velocity and temperature field in the wake. The SABRE and COBRA codes have recently been compared with each other and with experimental steady state temperature measurements taken in the THORS facility at Oak Ridge (ORNL) [24]. COBRA III-C is in wide use in the United States, while SABRE is used by the UKAEA. The validation experiment was performed in a 19 pin guard heated bundle with a 6 subchannel blockage, called THORS bundle 3C. Both codes show good agreement with data in terms of radial temperature gradient. An example of temperature rise above inlet across the diametral transverse is shown in Fig. 57 plotted for COBRA, and for SABRE in Fig. 58. The results show temperature profiles, which are compared with experimental wire-wrap thermocouple data at the same axial position (circles represent measured data). The profiles are given at three axial locations, 432 mm, 483 mm and 533 mm from the beginning of heated section. Differences between data and code results are probably caused by pin bowing, which is at present time the limiting factor in validating coded empirical models.

Less elaborated at present time seems the UZU-code of the PNC-O-arai Center in Japan, which is a 3D code suited to calculate velocity and temperature fields in a bundle geometry in the wake behind blockages. Miyaguchi [25] published numerical results on velocity distribution and pressure drop calculations.and he compared the numerical results with the experimental data obtained in a 61 rod bundle with a 38% central blockage (Fig. 59).

Fig. 60 shows normalized calculated and measured velocity profiles at various distances from the blockage. In general, a fairly good agreement is obtained. Some discrepancies, however, appear between computed and measured results. The UZU-code calculates smaller values in the unblocked and higher values in the blocked region. The differences are probably caused by neglecting the grid spacer's effect in the code.

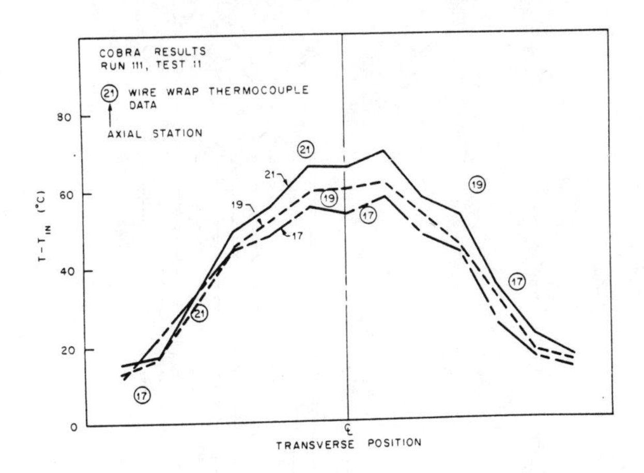

Fig. 57. Temperature profiles across diametral cross section (COBRA results)

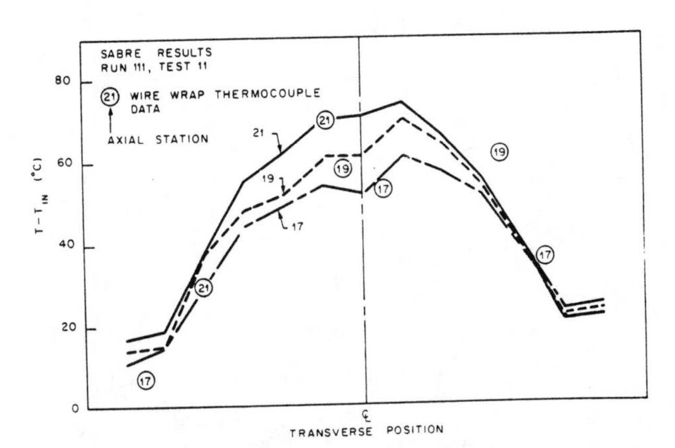

Fig. 58. Temperature profiles across diametral cross section (SABRE results)

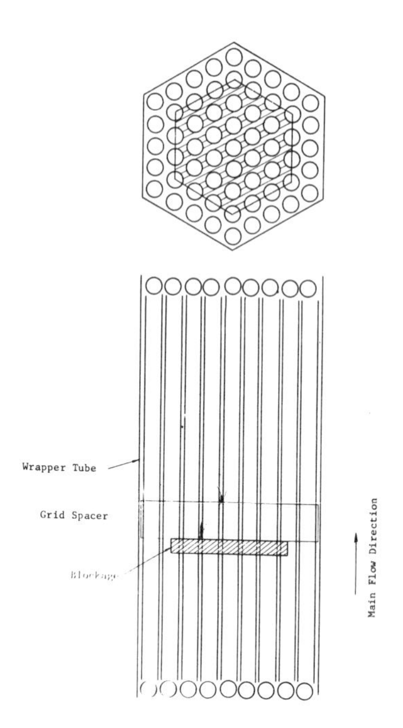

Fig. 59. 61 pin_blockage test section

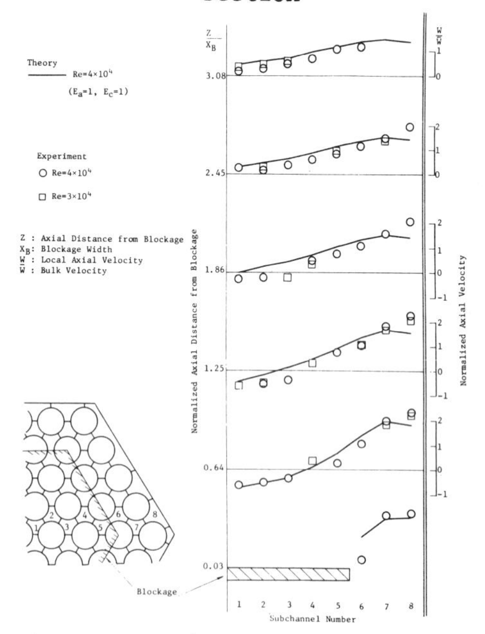

Fig. 60. Comparison of the UZU-code with measured data

5. CONCLUSIONS

In reviewing the studies presented in this paper, the following conclusions are obtained:

1. Blockages in a fuel subassembly were almost exclusively studied in small bundles with stainless steel plates located normal to the flow within the bundle structure.
2. Recirculating flow exists in the wake behind a blockage whether there is local boiling or not. The presence of the rod bundle structure within the wake has no or insignificant influence on the recirculation flow pattern at a pin pitch bigger 1,2.
3. Blockage porosity with residual flow disintegrates the recirculation wake at moderate residual flow rate. The flow pattern at low residual flow is very sensitive to the location of the porosity within the blockage.
4. For turbulent flow, with or without rods, the hydrodynamics of the wake behind a planar blockage is characterized by a "blockage constant" "C" which is nearly independent of the Reynolds number. It depends only on the bundle geometry. The blockage constant can be used to extrapolate the experimental results for one planar blockage type to similar blockages of different sizes in the bundle.
5. Although the presence of rods has no significant influence on the wake length or wake dimensions, it has however significant influence on the coolant residence time in the wake.
6. Water experiments can be used to obtain the residence time for other bundle geometries. Dimensionless wake temperature distribution, flow pattern and wake length can (for high turbulent flow) be extrapolated to similar blockages in sodium cooled bundles.
7. The maximum wake temperature rise behind the blockage depends on flow power conditions, size and location of the blockage.
8. The Nusselt number for sodium heat transfer in the wake is smaller than in normal channel flow. The most probable Nusselt number correlation has been empirically determined to be $Nu = 0{,}0043 \cdot Re_{\infty}^{0{,}55}$.
9. Experiments show that 24 subchannel blockages in a full size fast reactor subassembly is very unlikely being detected by means of classical instrumentation (assembly outlet temperature or flow measurements).
10. Boiling experiments have shown that boiling propagation from the wake into the unblocked area is very unlikely.
11. There is a range of blockage size which could cause local boiling without being detected by thermocouples located at subassembly outlet in a reactor.
12. There is a lack in analytical tools to interpret experiments and predict the temperature and flow pattern behind blockages of full fuel subassembly. Few codes are available discribing temperature and flow pattern behind planar blockages in test bundles. The most progressed codes are at present the SABRE, COBRA and UZU. These codes were compared with sodium experiments giving satisfactory results.
13. The SABRE code is also capable to treat local permeabilities with residual flow in small bundles with satisfactory agreement. Extrapolation to complex permeability and full size bundle is however questionable.
14. There is a great lack in analytical and experimental studies in axially extended porous blockages as possibly caused by particle agglomeration.

ACKNOWLEDGEMENTS

The author writes to express his gratitude to Mrs. C. Kind for preparing the typescript for this lecture in record time and Mr. A. Birke for the preparation of the graphs. He also acknowledges the continuing support of his colleagues and the programme management of the J.R.C. Ispra which made possible to preparate this lecture.

NOMENCLATURE

A_p subchannel flow cross section
A total bundle flow cross section
A_B blocked flow cross section
C blockage constant
D, a equivalent blockage diameter
R equivalent blockage radius
P_o pressure in undisturbed flow
$\bar{P}$ average pressure
p local average pressure
ΔP_B Bernouilli's pressure drop
ΔP_L pressure drop over the wake length
$T_{(x,y,t)}$ local time dependent temperature
$\bar{T}_{(x,y)}$ average local temperature
$T'_{(x,y,t)}$ local temperature fluctuation
T_o average temperature in undisturbed flow
$\bar{T}_E$ average inlet temperature
T_o temperature in the stagnation point of the wake
T_w rod wall temperature
T_{wk} average wake temperature at location T_w
M_R wake mass
$\dot{m}_R$ mass exchange rate between wake and main flow
$F_{(x,y)}$ heat transfer surface
$U = U_{(x,y,t)}$ local velocity in "X" direction
$u' = u'_{(x,y,t)}$ local velocity fluctuation
$V = V_{(x,y,t)}$ local velocity in "y" direction
$v' = v'_{(x,y,t)}$ local velocity fluctuation
$\bar{u} = u_{(x,y)}$ local mean velocity
V_R recirculation velocity
K_1, K_2, K_3, K_4 constants
t time

X	coordinate
Y	coordinate
d	rod diameter
d_h	hydraulic diameter
p	pitch
l_1, l_2	characteristic blockage lengths
Nu	Nusselt number
Re	Reynolds number
Pr	Prandtle number
$\dot{q}'$	heat flux per unit length
$\dot{q}''$	heat flux per unit surface
$\dot{q}'''$	heat flux per unit volume
τ	residence time
ξ	friction factor
$\varepsilon, \varepsilon^*$	turbulent heat transfer ratio
ν	kinematic viscosity
θ	dimensionless temperature
ϕ	rod power
ρ	specific weight
c_p	specific heat
β	blocked fraction A_B/A
K	heat conductivity
U_o	undisturbed flow upstream blockage
U_∞	undisturbed flow parallel to the wake
P	permeability $\frac{A - A_B}{A}$

REFERENCES

1. Schleisiek, K. Natriumexperimente zur Untersuchung lokaler Kühlungsströmungen in brennelementenähnlicher Testanordnungen. Bericht KFK 1914, 1974.

2. Schultheiß, G. Untersuchungen zur Bestimmung der Strömungsform im Totwasser hinter lokalen Kühlmittelblockaden in Stabbündeln. Atomwirtschaft-Atomtechnik (atw) 17, 1972, S 416.

3. Kirsch, D. Untersuchungen zur Strömungs- und Temperaturverteilung im Bereich lokaler Kühlkanalblockaden in Stabbündelbrennelementen. KFK 1974, 1973.

4. Kutateladze, S.S. Fundamentals of heat transfer. Adward Arnold Ltd, London 1963

5. Carmody, T. Establishment of the wake behind a disk. J. Basic Eng. 84, 4, 869, 1972.

6. Huetz, J. Lecture Nr. 2. Intern. Seminar on Heat Transfer in Liquid Metals, Trogir, September 1971.

7. Winterfeld, G. Rezirkulationsströmungen in Flamen Zeitschrift Flugwissenschaft. JS 10 Heft 415, 1962.

8. Han, J. T. et al.Blockages in LMFBR Fuel Assemblies Thermal and Hydrodynamic Aspects of Nuclear Reactor Safety, Volume 2, Liquid Metal Fast Breeder Reactors ASME 345 East 47th Street, New York, N.Y. 10017, 1977

9. Basmer P. et al. Phänomenologische Untersuchungen der Strömungsverteilung hinter lokalen Kühlkanalblockaden in Stabbündeln, KFK 1548, 1964.

10. Clare, A.J. A semi-empirical model for predicting maximum coolant temperature produced by partial blockages in fast reactor subassemblies, CEGB report RD/B/N 3809, 1976

11. Gast, K. Die Ausbreitung örtlicher Störungen im Kern schneller Natriumgekühlter Reaktoren und ihre Bedeutung für die Reaktorsicherheit, KFK-Bericht 1380, Mai 1971

12. De Vries, J.E., Dorr, B. Some aspects of local boiling behind a 70% central blockage within simulated LMFBR-subassembly. LMBWG-proceeding 7th LMBWG-Meeting, June 1-3 1977, Petten.

13. Basmer et al. Experiments on local blockages, LMBWG-Meeting Risley, October 1-3, 1975

14. Brook, A., Huber, F. Peppler, W. Temperature distribution and local boiling behind a central blockage in a simulated FBR subassembly. International Meeting on Fast Reactor Safety and Related Physics, October 5-9, 1976, Chicago III.

15. Brinkmann, K.J., De Vries, J.E., Door, B. Survey of local boiling investigations in sodium at Petten. ANS, Aug. 19-23, Seatle 1979

16. Kramer et al. Investigation of local cooling disturbances in an in-pile sodium loop BR2, ENS/ANS Meeting, October 16-19, 1978, Bruxelles.

17. Menant, B. Nucleation, dry-out and boiling behind a sodium tight local blockage in a 19 pin bundle with helical wire spacers. 7th LMBWG-Meeting, June 1-3, 1977, Petten.

18. Jolas, P. Experimental investigation on the flow around blockages in fast reactor assembly. October 11-13, 1978, SCK/CEN Mol, Belgium.

19. Daigo, Y. et al. Local temperature rise due to a 6-channel blockage in a 7 pin bundle, JAPF Nr-202, September 1975.

20. Uotani et al. Local flow blockage experiments in 37-pin sodium cooled bundles with grid spacers. LMBWG-Meeting Proceedings, October 11-13, 1978, SCK/CEN Mol, Belgium.

21. Robinson, D.P. Water modelling studies of blockages with discrete permeabilities in a 11 pin geometry. LMBWG-Meeting Proceedings, June 1-3, 1977, ECN Petten, Netherlands.

22. Potter et al. SABRE 1: A computer programme for the calculation of three dimensional flows in rod clusters, AEEW-R 1057, July 1976.

23. MacDougall, J.D., Lillington, J.N. The development of calculational methods for coolant boiling in rod clusters using the SABRE-code. LMBWG-Meeting-Proceedings, October 11-13, 1978, SCK/CEN, Mol, Belgium.

24. Dearing, J.F., Nelson, W.R., Rose, S.D. A comparison of COBRA III-C and SABRE-1 (wire-wrap-version) computational results with steady-state data from a 19-pin internally guard heated sodium-cooled bundle with a six-channel central blockage (THORS bundle 3C). International Meeting on Fast Reactor Safety Technology, August 19-23, 1979, Seattle, Washington.

25. Miyaguchi, K. UZU-Three dimensional analytical code for local flow blockage in a nuclear fuel subassembly. LMBWG-meeting proceedings, October 11-13, 1978, Mol, Belgium.

26. Rowe, D.S., COBRA III-C: A digital computer program for steady-state and transient thermal-hydraulic analyiss of rod bundle nuclear fuel elements, BNWL-1695, Battelle Northwest Laboratory, 1973.

27. Potter, R. et al. SABRE-1 (amendment 2) - A computer program for the calculation of three-dimensional flows in rod clusters. AEEW-R 1057, Rev. 2, United Kingdom Atomic Energy Authority, 1978.

28. Davies, A.L. and Wilkinson, S.A. The wire-wrap model in SABRE. AEEW-M 1588, United Kingdom Atomic Energy Authority, 1979.

29. Hanus et al.Steady-state sodium tests in a 19-pin internally guard heated simulated LMFBR fuel assembly with a six-channel internal blockage - Record of experimental data for THORS bundle 3C. ORNL/TM-6498, Oak Ridge National Laboratory, 1979.

23 SODIUM BOILING

by
H.M. Kottowski
CEC – JRC, Ispra, Italy

There has been considerable concern relative to possible LMFBR accidents as to whether or not the onset of sodium boiling could lead to the initiation of a CDA. This chapter describes relevant work in this area. Experimental measurements and analytical approaches are reported for basic sodium boiling features. Studies on incipient boiling, boiling flow pattern and boiling characteristics are discussed. Experiments with respect to reactor application are shown. The main topics of the chapter are:

- incipient boiling
- boiling flow pattern
- boiling characteristics
- two phase flow heat transfer and critical heat flux.

1. INTRODUCTION

The development of LMFBR's motivated extensive studies in sodium thermohydraulics of single phase flow and boiling two phase flow. Malfunction analysis with respect to power excursion or possible loss of coolant showed that if power excursion or loss of flow exceeds a certain threshold boiling will occur in the core. Under this aspect two questions have to be answered:

(a) which are the boiling features of sodium, and
(b) which are the predominant boiling characteristics of interest for reactor operation.

Since the single phase flow behaviour is dominated by viscosity, conductivity and turbulence effects, for single phase aspects one could refer to studies performed with water and other liquids and stress the studies on the few particular liquid metal features. Boiling of liquid metals, however, was found being different from the known water boiling behaviour. This fact required first of all to discover the main basic boiling features, as for example, boiling inception, boiling two phase flow pattern, boiling characteristics and parameters influencing these factors.

Basic experiments were carried out in many laboratories in countries with a fast reactor development programmé. The most involved or sponsoring laboratories are: KFK-Karlsruhe and Interatom in Germany, Grenoble and Cadarache in France, ECN Petten in Nether lands, CNEN-Casaccia in Italy, SCK/CEN Mol in Belgium and UKAEA Risley in England, J.R.C. Ispra in Italy, ANL Argonne and Brookhaven in USA, PNC-O-arai in Japan. Additional work on boiling was done at Universities in U.S. and UdSSR.

Design oriented work was mostly done at KFK Karlsruhe, Inter atom in Germany, Grenoble and Cadarache in France, PNC O-arai in Japan and ANL Argonne in USA.

The prime purpose of this paper is the discussion of the basic boiling features and characteristics and indicate some application cases.

2. BOILING NUCLEATION IN LIQUID METALS

2.1. Nucleation Model

Independent of the liquid, two distinct nucleation phases precede the formation of vapour nuclei:
(a) The formation of vacancies in the liquid structure and
(b) the formation of molecules from its surface into the vacancy space.

Though normally in the liquid structure a temperature dependent density of vacancies is present, the nucleation vacancy has to be large enough and the molecule diffusion into the void fast enough, otherwise the nuclei collapse.

The degree of nucleation difficulty is measurable and is indicated by the superheat. Non-metallic liquids have inherently sufficient molecular cluster defects or impurities so that high superheat is hard to occur. Liquid metals, however, have a regular cluster structure and are normally extremely pure that superheat of more than 100°C can be obtained under ordinary circumstan ces. The basic requirements to activate nucleus are:
- the thermodynamic equilibrium conditions for the existance of vacancies of critical size have to be satisfied and
- that during the local energy fluctuations in the liquid, energy peaks occur sufficiently high to produce vacancies.

The first condition is met by the saturation condition:

$$T = f(P_{oo}, \sigma, r) \quad (1)$$

(P_{oo} = system pressure)

The second criterium is dominated by statistical thermodynamic processes, which indicates the probability of void formation in the lattice structure of the liquid caused by statistical entropy fluctuation. According to the intensity of entropy fluctuations vacancies of various sizes are created. They collapse when a so-called critical threshold size is not exceeded. Vapour bubble for mation is only possible on vacancies of critical size[1,2,3].

Stranski and Knacke [3] developed a theoretical model of the probability of nucleation in homogeneous liquids which has been confirmed by Kaischew and Döring [2] [3].

$$I = Ng \left(\frac{6\,\sigma}{(3-b)\cdot\pi\cdot M}\right)^{1/2} \cdot \exp\left(-\frac{e}{K}\right) \quad [\text{events/cm}^3\text{sec}] \qquad (2)$$

$$b = \frac{2\,\sigma}{r_c Pr}$$

The model is based on the random thermodynamic interaction of momentum and energy exchange of molecules or clusters of molecules causing locally time-dependent energy fluctuations.

Since eq. 2 does not take into account the latent heat of evaporation, the correlation calculates only the number of vacancies created per time and volume unit. This so called homogeneous model is not applicable to nucleation on solid surfaces.

Knacke and Kottowski [2], [4] proposed a heterogeneous model which takes into account the wall effect. Knacke and others [1, 2, 4] proposed the following approach:

- Only the molecules or clusters of molecules in the liquid layer of the thickness of the critical radius size are involved in the nucleation process.
- The number of defects on which nucleation preferably will occur is determined by the plastic deformation during manufacturing processes ($N \sim 10^8$ defects/cm^2).
- The size distribution of the defects is assumed to be random distributed.
- The molecular oscillation frequency, which contributes to the vacancy formation is assumed to be $\nu = 10^{13}\ sec^{-1}$.
- The interfacial forces on the nucleation sites are dominated by dynamic physico chemical mechanisms, which change the nucleation conditions as a function of time and location

$$I = K \cdot \nu\, f(N) \cdot \exp(-A_K/kT) \quad [\text{events/cm}^2\text{sec}] \qquad (3)$$

$$A_K = A_{ko} \cdot f(A) \qquad (4)$$

$K = [N \cdot N_g \cdot \frac{\rho_1}{M}\,\delta]^{1/2}$; $\delta = f(r_c)$ = boundary layer thickness. $f(N)$ = nucleation cavity distribution. $f(A)$ = nucleation activation factor due to cavity shape and physico-chemical interface mechanisms. A_{ko} = maximum energy of vacancy formation of a nucleus with the curvature radius r_c. N_g = number of molecules per unit volume. M = molecular weigth. ρ_1 = liquid density. K = Stephan Boltzmann constant.

The prediction of nucleation is uncertain. It is essentially dominated by the complexity of physical and chemical processes. Kottowski [5] showed that there are mainly three terms contributing to the nucleation energy which on their part, even taken into account only the physical properties, are affected by many factors.

$$A_K = A_{ko} f(A) = A_\sigma + A_{s,1} + A_v$$

A_σ = work surface formation: $\int \Sigma\, dF_{liquid}$

$A_{s,1}$ = recovery work due to wetting retraction:
$- \int \xi_{s,1} \cdot dF_{spot}$

A_V = recovery work due to void expansion: $-\int Pdv$

Taking into account only the physical properties, the energy term "A_σ" of surface formation is determined by the type of the thermodynamic process of surface formation, adiabatic or isothermal, and by the temperature dependence of the physical properties. In case of an adiabatic process, the void liquid boundary temperature is decreasing which changes the thermodynamic nucleation conditions.

The formation of a void nucleus on the wall on its part is strongly influenced by the physico chemical wall-liquid interactions. The dominating factors are the cohesive forces on the liquid-solid interface and the solid surface conditions. This relationship is described qualitatively by the Dupré equation:

$$\frac{\xi_{s,l}}{\sigma_l} = \frac{1}{2}(\cos\vartheta + 1) + \frac{\pi\sigma}{l} \qquad (5)$$

$\pi\sigma$ stands for the superposing physico chemical surface factors, which might come e.g. from surface covering impurities. In case of a positive $\pi\sigma$ effect the wetting angle is apparently increasing.

The competitive mechanism between cohesive and adhesive forces is shown in Fig. 1 for an example of ideal surface as a function of the wetting behaviour. Note the strong influence of the wetting behaviour on the interfacial forces.

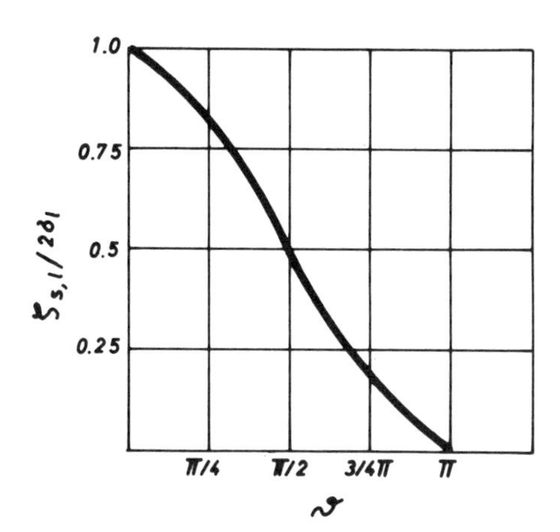

Fig. 1. Ratio of adhesive to cohesive forces vs ϑ

The quantitative value and its behaviour as a function of time can only be estimated.

More sensitive to surface roughness and interfacial boundary conditions is the term "A_V" of expansion work. Fig. 2 shows the possible instantaneous cross shape of a vacancy in a wall surface cavity. As can be seen the vacancy volume depends on the cavity geometry, wetting angle, surface tension forces and gravity forces. Eqs 6 to 9 describe the surface and volume work of the vacancy.

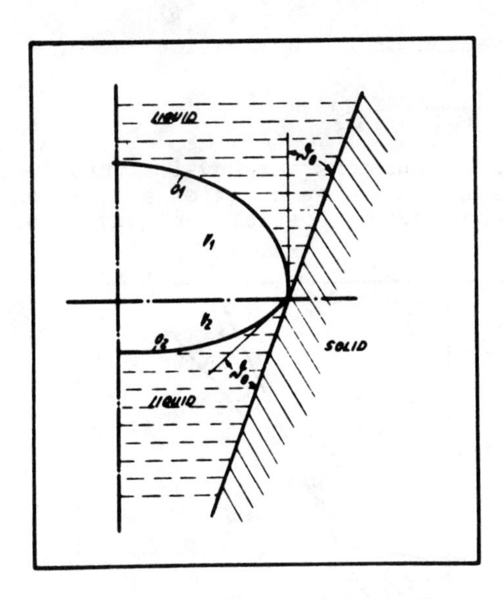

Fig. 2. Probable contour of nucleation vacancy

$$A_K = A_{1\sigma} + A_{1V} + A_{2\sigma} + A_{2V} \tag{6}$$

$A_{1\sigma}$ and A_{1V} are described by the equation

$$A_{1\sigma} + A_{1V} = 2\sigma \pi r^2 \left[1 - \sin(\vartheta - \frac{\beta}{2})\right] - \frac{2}{3}\sigma \pi r^2 \left[2 - 3 \sin (\vartheta - \frac{\beta}{2}) + \sin^3(\vartheta - \frac{\beta}{2})\right] \tag{7}$$

Whereas the terms $A_{2\sigma}$ and A_{2V} can only be solved by iterative integration of the contour of the liquid surface, which is described by equations 8 and 9.

$$\sigma_1 (\frac{1}{r_1} + \frac{1}{r_2}) + g(z - z_o) = 0 \tag{8}$$

With $\frac{1}{r_1} = - \frac{d\sin\varphi}{dx}$; $\frac{1}{r_2} = - \frac{\sin\varphi}{x}$; $\varphi = 90^o - \vartheta - \frac{\beta}{2}$

the definitive equation becomes:

$$\frac{d(x\sin\varphi)}{x \cdot dx} = 2 \left(\frac{2\sigma_1}{\rho g}\right)^{-\frac{1}{2}} (z - z_o) \tag{9}$$

($x = y$ because of the symmetry to the Z-coordinate).

Fig. 3 describes the surface contour of the work terms $A_{2\sigma}$ and A_{2V}.

The numerical evaluation of the activation factor

$$f(A) = \frac{A_K}{A_{Ko}} = \frac{A \;\; + A_V}{A_{Ko}} = H$$

for constant superheat and constant physical properties is shown in Fig. 4. It seems that in deep cavities activation of nuclei at constant wetting behaviour is possible only in a small range of cavity shape.

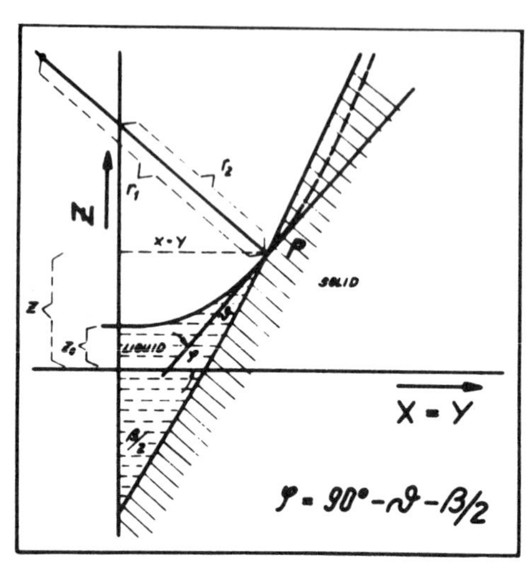

Fig. 3. Geometry parameters of the lower vacancy contour.

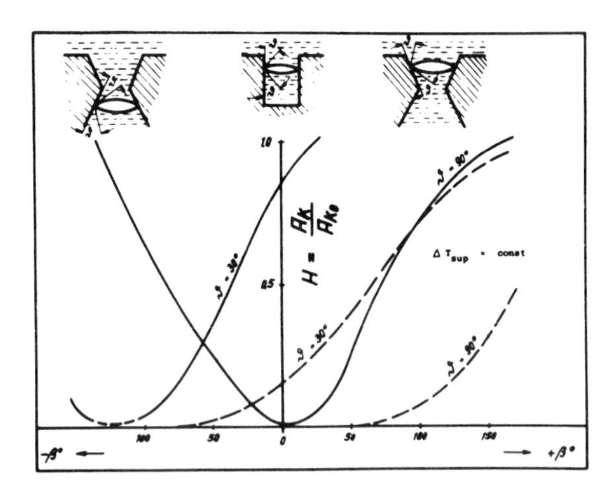

Fig. 4. Nucleation activation factor for deep cavities

Cavity shape and cavity distribution depend normally on the material properties and its surface treatment. These involve unpredictable factors for the numerical evaluation, which is more aggravating when taking into account the physico chemical factors which change with time. From these arguments it becomes obvious that the nucleation probability(equation (3))calculated with constant parameters gives even wrong qualitative information.

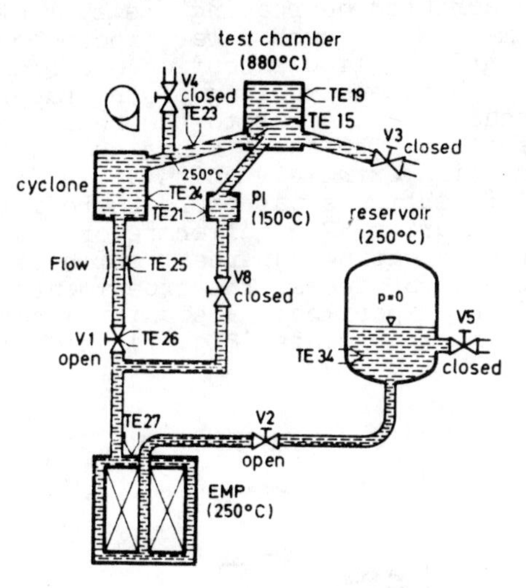

Fig. 5. Waiting time measurement test facility

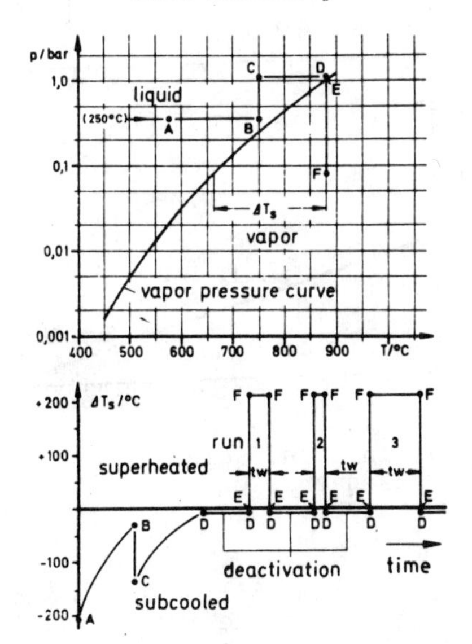

Fig. 6. Waiting time test history

Rietmüller /‾4_7performed nucleation waiting time experiments with well controlled and determined test conditions. The experiments were run in a test apparatus shown schematically in Fig. 5. After cleaning and evacuation, the apparatus was filled with sodium. Heating and pressurization were carried out according to curve A-B-C-D in Fig. 6. Reaching point D, the temperature and pressure were kept constant for 20 minutes. A predetermined superheat was than achieved decreasing the system pressure to point F in the diagramme. The in such a way imposed superheat ΔT_s remained constant until boiling inception occurred by itself. The measured time "tw" between imposing of the superheat and boiling inception was assumed equivalent to the reciprocal value of the nucleation probability. The experiments showed a large scatter in waiting time for which extremely high values were measured. The reproducibility of test runs was poor. Unpredictable was the influence of oxide impurity. Rietmüller reports a diminution of the waiting time with increasing superheat which is physically reasonable and the sequence of tests. The experiments run first show at the beginning of the experiments a significant shorter waiting time as for the later experiments (Fig. 7) which is not explainable.

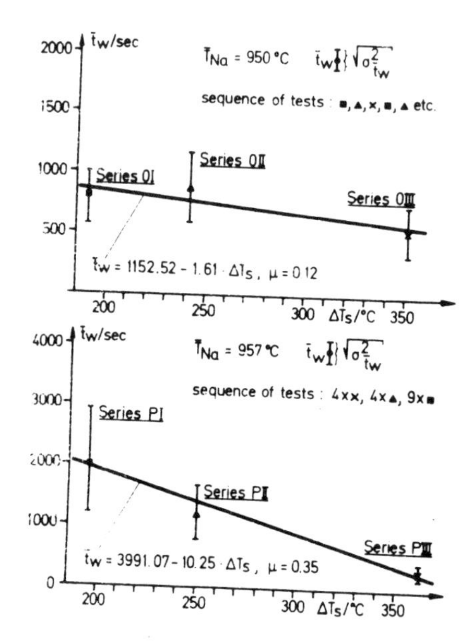

Fig. 7. Influence of the test sequence on the waiting time.

An example of waiting time histogrammes shows Fig. 8. (The CTP-temperature indicates the oxide impurity). As shown in the graph a change of 3°C of sodium temperature provoked a drastic change in the histogramme pattern and the mean waiting time changed of about 4000 hours.

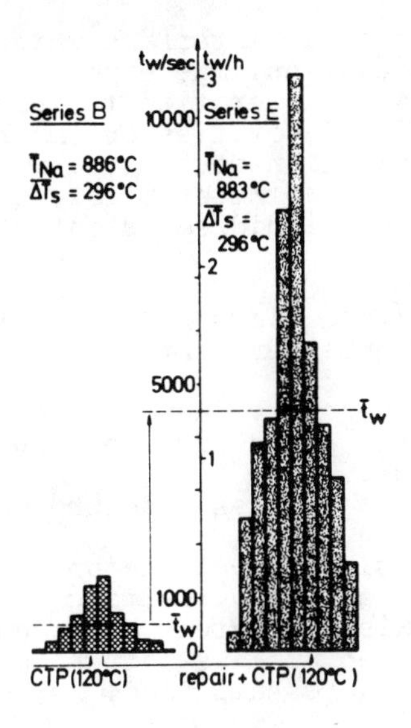

Fig. 8. Waiting time history

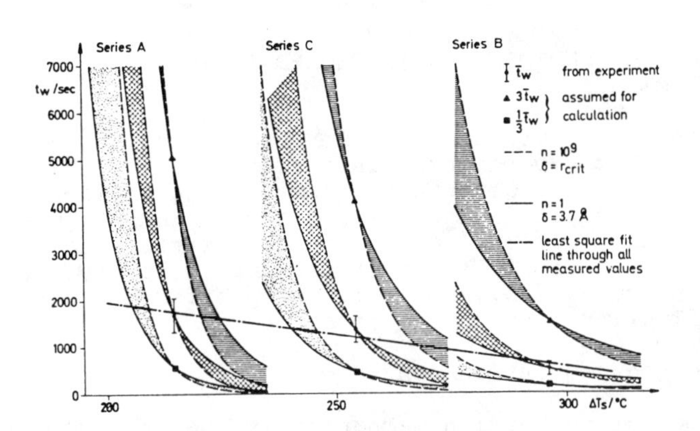

Fig. 9. Comparison between measured and predicted waiting time

A comparison between the measured and calculated values of "tw" is given in Fig. 9. The curves calculated with the heterogeneous nucleation probability equation using constant activation and nucleation site factors do not fit with the measured data. The measured flat curve of tw = f(ΔTs) demonstrates that the formation of boiling nuclei is determined essentially by physico chemical reactions on the liquid solid interface which over-role the mechanism of entropy fluctuation. The experiments show that even after several 1000 hours of operation the housing surface was not cleaned and that gas filled cavities or small unwetted spots existed on which boiling nucleation could occur. Note that in the energy term "A_K" of equation (3) the chemical energy of the physico chemical processes on the liquid/solid interface have to be taken into account.

Holz, Singer [6] and Class [5] developed a heterogeneous nucleation model based on experimental observations. Their basic assumptions are:

- Boiling nuclei are inherently (non wetted) existing cavities, cracks or clefts in the solid surface in which the wetting angle ϑ locally exceeds 90°.
- Corrosion and diffusion processes determine the time dependent wetting pattern.
- Contact angle hysteresis and wetting propagation in cavities determine the nucleation delay (waiting time effect).
- If a collective of existing latent boiling nuclei becomes superheated the most favoured nucleus, at which the thermodynamic equilibrium conditions are satisfied, is activated.

The onset of boiling at a certain superheat in this model is coupled with the existence of non activated nucleation sites. This model can only be compared with the entropy (A_K/T) fluctuation model taking into account the energy due to physico chemical reactions as equivalent nucleation energy.

3. SUPERHEAT MEASUREMENTS WITH REGARD TO EXPERIMENTAL AND PHYSICAL PARAMETERS

As shown in the previous chapter predictions of superheat are difficult to make. Experimental investigations in various laboratories discovered the following factors having an important influence on the incipient superheat: gas entrainment, O_2-impurity, surface finishing, material of the heating surface, operation history, heat flux ramp, temperature ramp, velocity, and conditioning of the surface by chemical treatment. Concerning several factors conflicting results are obtained.

3.1. Undissolved gas entrainment

The simplest nucleation mechanism is the passage of gas bubbles through a superheated region. A bubble circulating with the liquid metal with a radius greater than 10 μm will nucleate boiling at any superheat greater than a couple of degrees. The superheat depends on how much superheat can be built up before a gas bubble comes along.

3.2. Oxide impurity

The influence of disolved oxide on the superheat is shown in Fig. 10. A definite trend of decreasing superheat with increasing oxide ppm is observed. Based on experimental data an empirical correlation of wall superheat in °C vs oxide concentration in ppm for loop operation conditions was suggested:

$$\Delta T_s = \frac{163}{\log C} - 77 \ [^\circ C] \qquad (10)$$

$C = O_2$ concentration in ppm $= \frac{c}{c_o}$; $c_o = 1$ ppm

This purely empirical correlation represents the mean value of available experiments for non prepared stainless steel surfaces in the range of 5 to 60 ppm O_2 impurity and coolant velocities up to 3 m/sec.

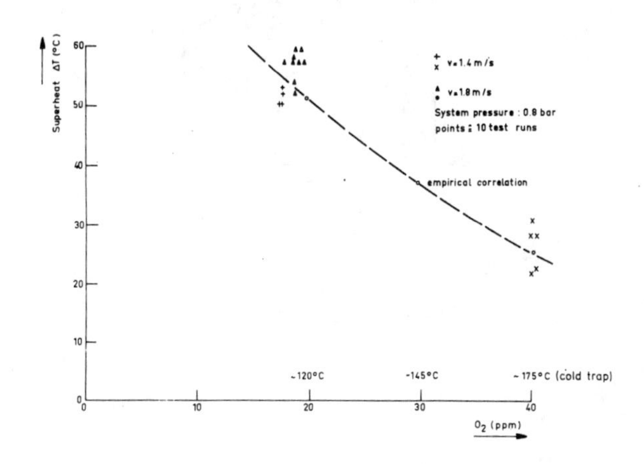

Fig. 10. Influence of oxide on wall superheat

3.3. Surface conditioning

The surfaces employed in Na boiling experiments are of the following types:
(a) as received, (b) refinished with grit abrasive, (c) mechanically drilled tube, (d) drilled by electro erosion, (e) drilled holes in the surface. The different surface preparation results in different surface roughness: (a) of different rms reading, (b) nature of roughness, and (c) definite cavities.

Definite results were obtained with drilled hole cavities. The trend of measured points follows the Laplace correlation. The scattering is in the limits of experimental accuracy of measurements. Fig. 11 shows measured data over a wide range of saturation temperature. Non unanimous are the results obtained with technical as received surfaces. Measurements are reported where heater pins "as received" were subjected to sodium flow and which changed the surface structure after 500 h of operation. The surface relief became similar to that of the "reference fuel cladding" of the LMFBR. A relation between the state of the technical surface structure and superheat could not be found.

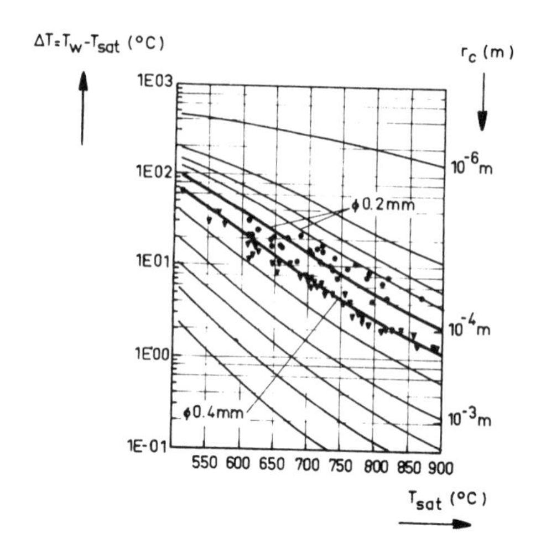

Fig. 11. Incipient boiling at artificial cavities

3.4. Material of the heating surface

Various materials have been tested in laboratory scale experiments by Dean, Rohsenow [7] and Schultheiß [8]. They performed experiments on well defined artificial cavities with nickel, molybdenum, iron and stainless steel. Fig. 12 shows an example on superheat measurements performed with various materials and an artificial cylindrical cavity of r_c = 0,168 mm for all test cases.

Dean and Rohsenow proposed an approximative correlation derived from the Laplace equation and based on experimental data:

$$\Delta T_s = \frac{5}{9} \frac{T_{sat}^2}{B} \log_{10} \left[1 + \frac{\sigma}{P_{sat} \cdot r_c} \right] \quad [^\circ C] \tag{11}$$

B = 9396,75 °R = empirical constant, σ = surface tension, T_{sat} = °R = saturation temperature, P_{sat} = saturation pressure, r_c = cavity mouth radius.

Since in equation (11) no term appears taking into account the wetting behaviour, which would represent the physical property of the heating surface material, the influence of the various materials on the amount of superheat seems negligeable. These results were confirmed by Schultheiß.

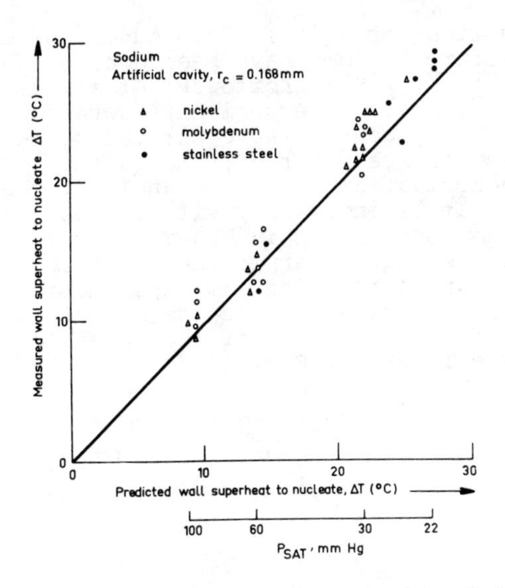

Fig. 12. Effect of wall material on superheat

3.5. Operation history

The operation history is believed by various authors to be a critical variable in determining the incipient wall superheat in connection with LMBR's operation.

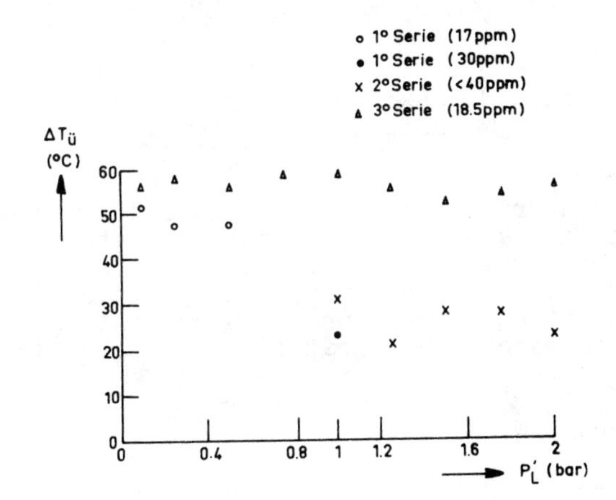

Fig. 13. Operation history effect (P_L' stands for the change of operation conditions)

Systematic investigations especially devoted to the effect of pressure and temperature history have been done by Holtz and Singer [6], Chen et al. [9], Kottowski et al. [10], Leonov and Prisnyakov [11]. Suitable experiments were done to give measurements of "ΔT_s" vs each of these parameters in term. No corresponding results are reported, though the same experimental procedures of deactivation have been applied. Chen et al. e.g. conclude that superheat is increasing with the increase of the operation pressure and subcooling in liquid flow conditions prior to boiling. Kottowski et al. observed the pressure temperature history effect only during the first 10 experimental runs. After that the statistical mean value and scattering of the measurements remained unchanged. The value of the superheat was only influenced by the oxide impurity, Fig. 13.

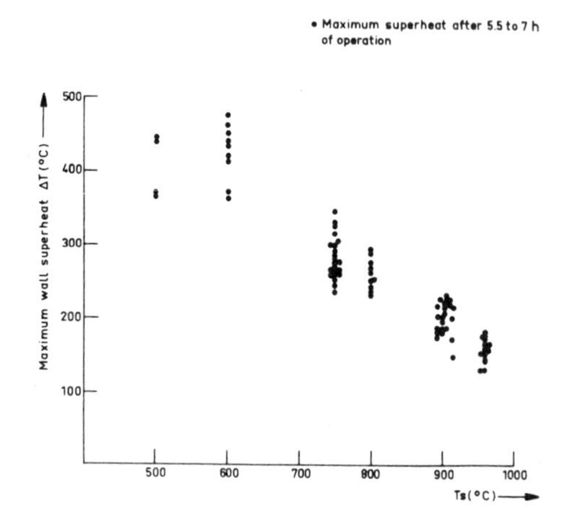

Fig. 14. Maximum wall superheat vs saturation temperature.

Leonev and Prisnyakov [11] conducted pressure temperature history experiments over a wide range of saturation temperature. The measurements show, Fig. 14, the known trend of decreasing superheat with increasing saturation temperature. From their own and other experiments they concluded an empirical correlation for the maximum threshold superheat of "as received" stainless steel surfaces:

$$\Delta T_s = 485 - 0{,}71\ (T_{sat} - 500)\ [°C] \quad (12)$$

The data measured by Leonev et al. are extremely high and do not correlate with most of the known measurements.

3.6. Velocity effect

The effect of velocity has received inconsistent experimental evidence. Data have been presented indicating, for example, incipient wall superheat may be decreased or unaffected by an increase in velocity. Most of the authors reported decreasing bulk superheat with increasing velocity, despite of the fact that the heat flux required to iniciate boiling had to be increased with velocity. As velocity and heat flux increase, the superheat characteristic is reported to change even to subcooled boiling nucleation (Fig. 15). We think that this is an apparent effect and a consequence of mass flow rate which is in accord with the heat balance.

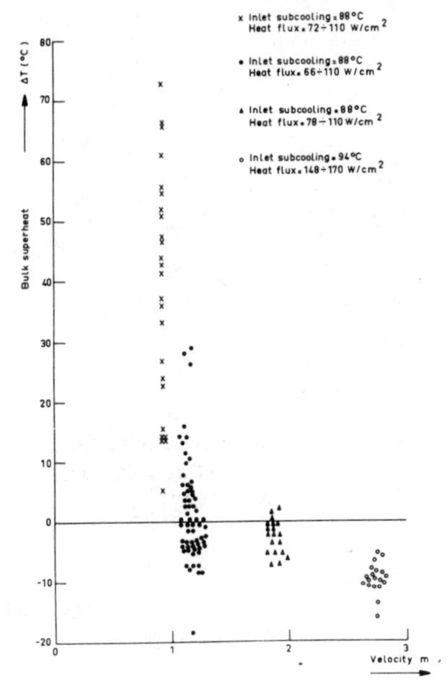

Fig. 15. Bulk superheat vs velocity.

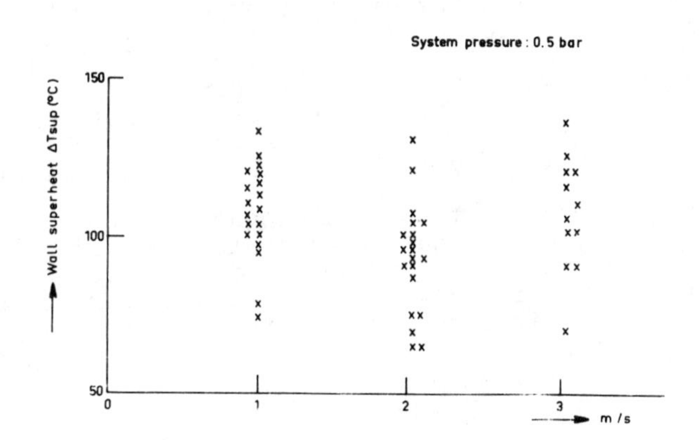

Fig. 16. Local wall superheat vs velocity.

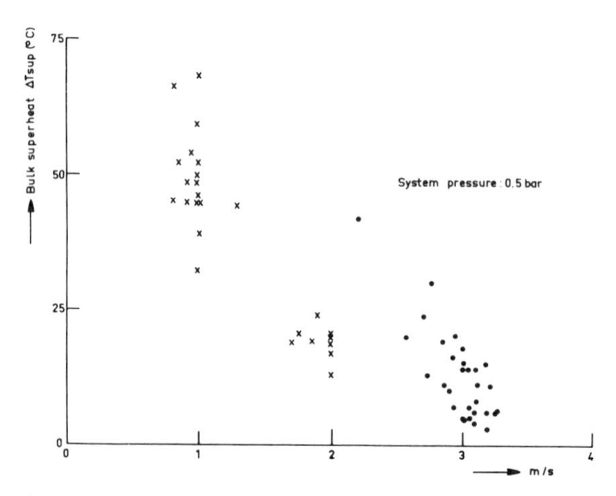

Fig. 17. Bulk superheat vs velocity.

Kottowski et al. performed well instrumented experiments which allowed the measurement of the location of boiling inception and the local superheat. The results are shown in Fig. 16. A change of the velocity up to 3 m/sec did not show any velcoity effect on the superheat, whereas the bulk superheat was observed to decrease as reported by other authors (Fig. 17). Kottowski et al. concluded that in the range of flow velocities of technical applications the velocity effect is negligeable.

3.7. Heat flux and temperature ramp effect

The effect of heat flux has received the most inconsistent and conflicting experimental evidence. For example, data have been presented indicating that incipient superheat may be decreased [11] increased or decreased [12] increased [6], increased or unaffected by an increase in the heat flux. Singer [13] made a critical review on the heat flux and temperature ramp effect and he demonstrated that the mobility of inert gas between the latent nucleation sites and the liquid can result in the observed behaviour. Since the solubility of inert gas in liquid alkali metals

increases with increasing temperature, gas will be lost from nucleation sites during the heating prior to and during boiling and might be gained by the sites during the cooling between the boiling runs. Since there is no way of direct measurement of inert partial gas pressure in the micro-cavity sites the heat flux effect must be treated with considerable caution.

4. BOILING CHARACTERISTICS

4.1. Boiling flow pattern

Due to the physical properties and operation conditions the two phase flow regime of boiling liquid metals is very inhomogeneous, especially as far as its space and time dependent pattern is concerned. A small vapour quality e.g. of 5% results already in a void fraction of about 90%. This feature influences strongly both the characteristics of steady state and transient boiling. Additionally, the transient boiling is further affected by the initial superheat and single phase flow thermo-hydraulic conditions. There is experimental evidence that in restricted geometry the boiling flow regime is characterized by single vapour bubbles which are separated by liquid plugs. The measurements in simple geometry show clearly this multi bubble flow pattern.

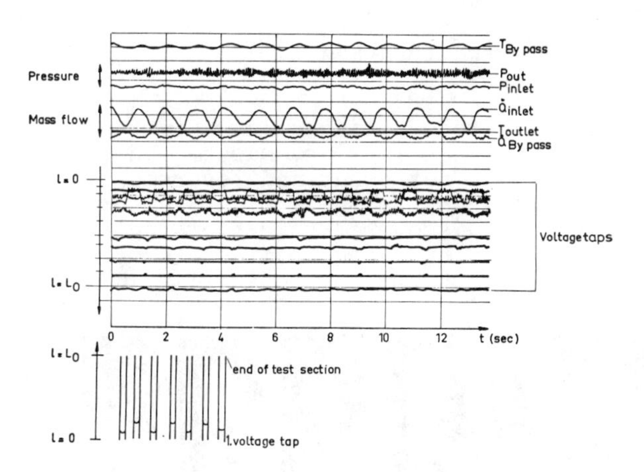

Fig. 18. Flow pattern registrations.

Fig. 18 illustrates boiling flow pattern measurements averaged over a tubular cross section. The signals of the voltage taps and EM flow meter enable the analysis of the flow pattern as a function of space and time. The signals show clearly bubble flow. With increasing heat flux the flow pattern is changing rapidly to slug, annular and droplet flow. A qualitative view of the flow pattern evolution with increasing heat flux at forced convection once through flow is illustrated in Fig. 19. As far as sodium boiling is concerned the local boiling and bubbly boiling regimes are negligeable due to the specific physical properties of liquid metals. These favour the slug flow, annular flow and mist flow regimes.

The best view of the boiling pattern in liquid metals is given by the X-ray pictures of Schmücker and Grigull [16]. The experiments were carried out with mercury for which wetting difficulties exist. It is highly probable that the boiling pattern

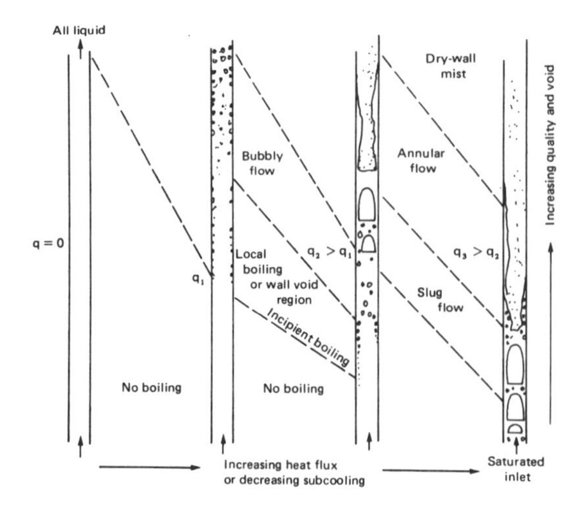

Fig. 19. Flow pattern model.

at low vapour quality and the boiling regime in the vicinity of the wall are biased by the non wetting. Due to the non wetting e.g., nucleate boiling will not occur and reversed annular flow, which means vapour annulus on the wall, might appear. The experiments of Schmücker et al. discovered this flow regime being extremely unstable. On the other hand, the wall unaffected central two phase flow can be assumed representative for liquid metal boiling.

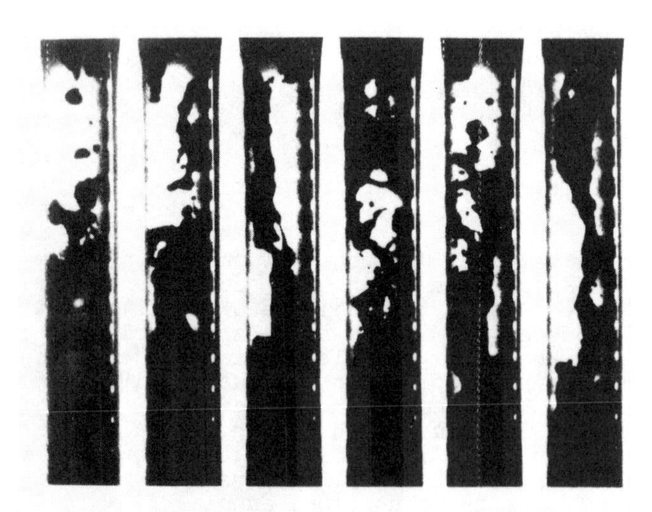

Fig. 20. X-ray photo of mercury boiling (low vapour quality)

A typical example of the boiling pattern at extremely low vapour quality is shown in Fig. 20. This flow is also representative for the boiling transition from liquid phase to two phase flow downstream boiling inception.

The picture shows the almost instantaneous occurrence of large bubbles and slugs. At higher vapour quality (about 5%) the liquid is lacerated mostly due to the vapour acceleration forces (Fig. 21). With increasing vapour quality the liquid is shredded to droplets and droplet size lumps (Fig. 22). The void fraction gets close to one and the vapour velocity reaches values close to sonic velocity. In the present experiment the vapour flow velocity was of about 150 m/sec and the slip ratio of about 100.

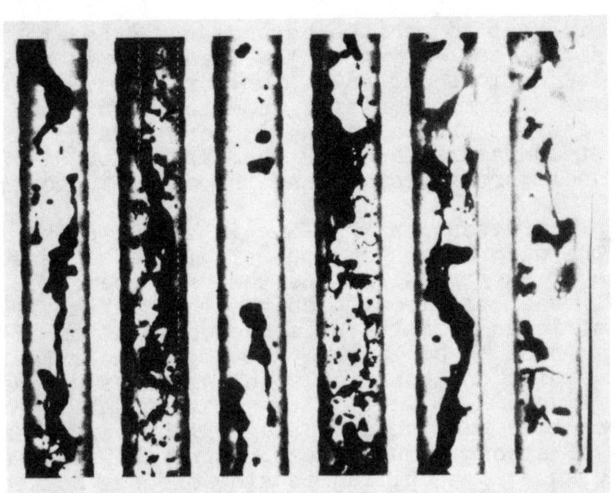

Fig. 21. X-ray photo of mercury boiling (about 5% vapour quality)

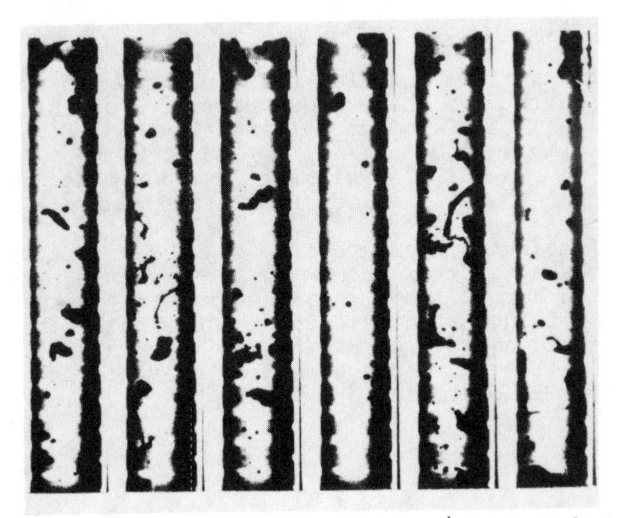

Fig. 22. X-ray photo of mercury boiling (droplet flow)

For comparison a flow pattern map obtained in water/gas two phase flow experiments is shown in Fig. 22a. The figure shows the various stages of characteristic flow regimes agreed upon for reference with increasing void fraction. When comparing this map with the X-ray pictures, the mercury boiling pattern can be identified as bubble or slug flow, churn flow and wispy-annular flow. Whether these flow pattern can be assumed being met in bundle geometry is at present time limited because of insufficient experimental results. The few steady state sodium boiling experiments in bundle geometry instrumented with void needles and voltage taps showed irregular passing of liquid slugs and vapour voids which can be interpreted as being caused by flow regimes shown in Fig.21 and Fig. 22.

Destinction has to be made between steady state and transient boiling. In simple geometry the flow transient is governed by single bubble ejections and multibubble regimes. The violence of the ejection is determined by the superheat. In a bundle geometry additional, possibly predominating factors have to be taken into account like temperature distribution, mass flow distribution across the bundle cross section and the overall geometry.

An example of transient boiling in simple geometry is shown in Fig. 23. The diagramme shows the single bubble growth at the initial stage of the transient and the disturbance of the steady state hydraulics of the loop. A characteristic feature of transients in restricted geometries is the oscillation of the system following the onset of boiling. A not negligeable factor of the flow pattern evolution, especially in case of slug and annular flow, is the liquid film left on the heating wall.
The at present time published bundle boiling experiments indicate that in case of strong temperature bukling the transient between local boiling and cross boiling is slow or does not occur. Quantitative data on boiling propagation as a function of the size of the bundle and of the temperature buckling are insufficient.

4.2. Two phase flow pressure drop

Boiling propagation and flow redistribution in a bundle depend strongly on the two phase flow pressure drop. At present time reliable results of sodium boiling two phase flow are only available for simple geometry [17], [18], [19], [20], [21]. For verification reasons also simulation experiments were performed in low pressure modelling loop facilities [22].

A typical measured pressure drop characteristic is shown in Fig. 24. The characteristic from A to B represents the single phase flow. Section B to C shows the steep rise in pressure drop associated with boiling. Point C shows the dry out point which experimentally could not be reached due to temperature limits and onset of instability. Such pressure drop characteristics were measured in various laboratories.

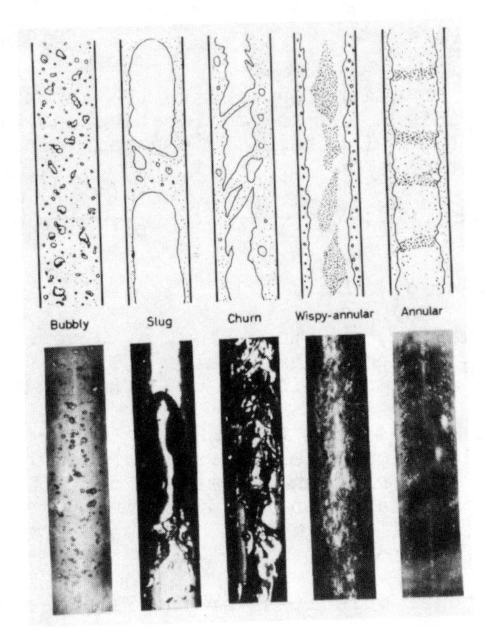

Fig. 22a. Boiling flow pattern map.

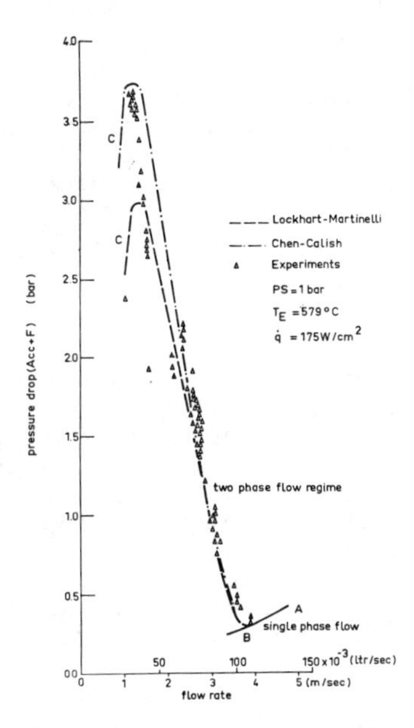

Fig. 24. Pressure drop characteristics.

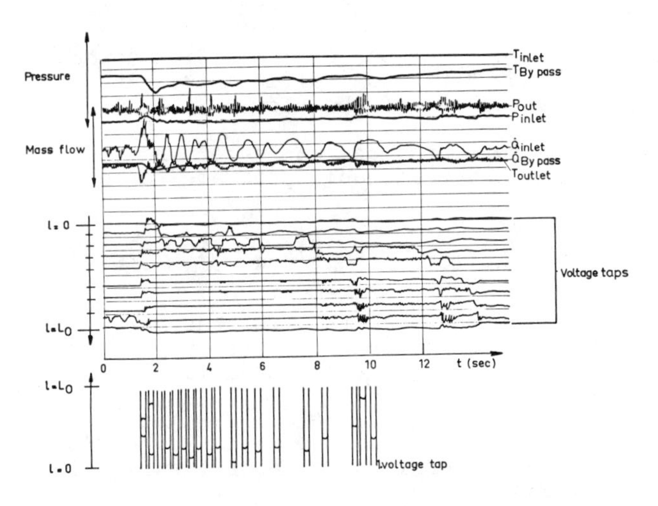

Fig. 23. Boiling transient.

It is further reported that bubble and annular flow were the predominant flow patterns at vapour quality values greater 1%. The overall two phase flow pressure drop is represented by the superposition of the pressure drop terms of friction, acceleration due to vapour quality increase and gravity. It can be written in the following form to illustrate the problem:

$$\Delta P_{2\varphi} = \Delta P_{2\varphi f} + \Delta P_{ac} + \Delta P_{h} \tag{13}$$

From the momentum and continuity equations applied to the vapour and liquid phase the pressure gradient due to acceleration and gravity can be calculated as a function of determinable parameters averaged over the flow cross section:

$$\Delta P_{ac} = \Delta(\alpha \cdot \rho_g \cdot v_g^2 + (1-\alpha)\, v_1^2);\quad V_g = \frac{\dot{m}_x}{\rho_g F};\quad V_1 = \frac{\dot{m}(1-x)}{\rho_1 (1-)F} \tag{14}$$

and $$\Delta P_h = g \cdot (\alpha \cdot \rho_g + (1-\alpha) \cdot \rho_1)\, \Delta z \tag{15}$$

α = void fraction, x = vapour quality, $\dot{m}$ = mass flow, ρ_g = vapour density, ρ_1 = liquid density, v_g = vapour phase velocity, v_1 = liquid phase velocity, z = axial coordinate, g = 9,81 m/sec^2.

The flow patterns observed suggest for the analytical approach the method of the separated flow model. Several procedures have been proposed to calculate the frictional pressure drop. The best known and probably most appropriate for low pressure systems is the approach of Lockhart Martinelli and Nelson. The basic premises upon which the separated flow model is based are:

- constant but not necessarily equal velocites for the liquid and the vapour phases,
- thermodynamic equilibrium between the phases, which means that the static pressure drop for the liquid phase must be equal to the static pressure drop of the vapour phase regardless of the flow pattern as long as no appreciable radial pressure gradient exists,
- the use of empirical correlations or simplified concepts to relate the two phase flow "friction multiplier" and "void fraction" to the independent variables of the flow.

From measured pressure drop one can determine the two phase flow $(\Delta P/\Delta z)_{2\varphi f}$ using any reasonable estimate for α in equations 14 and 15.

$$(\Delta P/\Delta z)_{2\varphi f} = \underset{\text{measured}}{(\Delta P/\Delta z)_{tot}} - (\Delta P/\Delta z)_{acc} - (\Delta P/\Delta z)_{h} \tag{16}$$

It is further commonly agreed that the two phase flow friction pressure drop is represented by the pressure drop of the liquid component as if flowing alone $(\Delta P/\Delta z)_1$ and a multiplication factor $\varnothing^2(x)$.

$$(\Delta P/\Delta z)_{2\varphi f} = (\Delta P/\Delta z)_1 \cdot \varnothing^2(X) \tag{17}$$

X is an empirical nondimensional variable of liquid and gas (vapour) component properties.
A number of empirical void fraction correlations are used to estimate (Table I). Of these correlations only those of Smith [37]

Baroczy [39] were obtained with liquid metals in mind. All others were originally derived for two phase flow of ordinary fluids (e.g. water-air, water-steam, etc.). The reader is refered to the original papers for the details of each correlation.
The most successful correlations applied are those one of Lockhart-Martinelli and Martinelli-Nelson ((a) and (b) in Table I).

Table I Correlation

(a) Martinelli and Nelson [38]
(b) Lockhart and Martinelli [19]
(c) Martinelli and Nelson [38]
(d) Homogeneous model
(e) Smith, Tek and Balzhiser [37]
(f) Baroczy [39]
(g) Levy [40]
(h) Lockhart and Martinelli [41]
(i) Baroczy [39]

Lockhart-Martinelli suggested a void fraction correlation based on their water-air experiments for liquid-gas flow which can also be applied to 1 bar liquid vapour flow.

$$\alpha = 1 - \frac{1}{\sqrt{1 + \frac{21}{X_{LM}} + \frac{1}{X_{LM}^2}}} \tag{18}$$

$$\text{with } X_{LM} = \left(\frac{1-x}{x}\right)^{0,9} \cdot \left(\frac{\rho_g}{\rho_l}\right)^{0,5} \cdot \left(\frac{\mu_l}{\mu_g}\right)^{0,1} \tag{18a}$$

x = gas or vapour quality, μ_g = vapour or gas viscosity, μ_l=liquid viscosity, X_{LM} = physical property variable.
For liquid-gas flow the frictional multiplier $\phi^2(X)$ in eq 17 becomes $\phi^2(X) = f(X_{LM})$ which is an empirical correlation established from experiments.
For 1 bar water steam flow Martinelli-Nelson modified eq 17 to the following expression:

$$\left(\Delta P/\Delta z\right)_{2\varphi f} = \left(\Delta P/\Delta z\right)_{lo} \cdot (1-x)^{1,75} \cdot \phi^2(X_{LM}) \tag{19}$$

with $(\Delta P/\Delta z)_{lo}$ = total flow rate (liquid and vapour) as liquid
x = vapour quality

These results were later modified for local value of:

$$(\Delta P/\Delta z)_{2\varphi f} \Big/ (\Delta P/\Delta z)_{lo} = \phi^2_{(X_{MN})} \tag{20}$$

$$\text{with } X_{MN} = \frac{1-x}{x} \left(\frac{\rho_v}{\rho_l}\right)^{0,571}_{sat} \cdot \left(\frac{\mu_l}{\mu_v}\right)^{0,142} \tag{20a}$$

as a function of local vapour quality and pressure, which takes into account the momentum term due to local evaporation and condensation. (ρ_v = vapour density, μ_v = vapour viscosity, sat = saturation). The reader is again refered for details to the original papers.

As far as low pressure liquid metal two phase flow is concerned, only the "local vapour quality" influence has to be considered (eq 17 and eq 19). For the determination of " α " either eq 18 or the empirical volume fraction correlation (eq 21) of Chen, Kalish, which is obtained from liquid metal measurements, can be applied

$$\ln(1 - \alpha) = 0,518 \cdot \ln(X_{LM}) - 0,0867 \left[\ln(X_{LM})\right]^2 - 1,59 \qquad (20b)$$

4.2.1. Discussion of experimental results

Lurie [23] performed steady state sodium boiling experiments varying the saturation pressure between 0,1 and 0,69 bar. Fig. 25 shows the evaluation of the measurements in terms of the frictional multiplier vs the turbulent Lockhart-Martinelli factor. The comparison between the measured data and the Lockhart Martinelli correlation:

$$\log \emptyset = \frac{- z + \quad z + 9,2}{4} \qquad (21)$$

$$\text{with } z = 2 \log X_{LM} + 0,176 \; X_{LM} + 0,382 \qquad (21a)$$

shows good agreement in the range of $0,1 < X_{LM} < 2$.

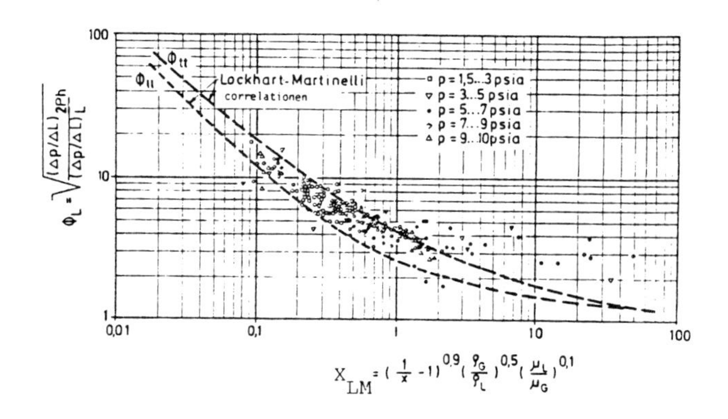

Fig. 25. Frictional multiplier vs X_{LM} (sodium boiling experiments: Lurie)

Chen and Kalish [24] investigated the pressure drop for potassium with and without vaporization. They obtained from their experiments as best fit correlation (Fig. 26):

tassium with and without vaporization. They obtained from their experiments as best fit correlation (Fig. 26):

$$\phi = e^{-A} \tag{22}$$

with $A = -1{,}59 + 0{,}518 \ln X_{LM} - 0{,}0867 (\ln X_{LM})^2$ (23)

Chen-Kalish's correlation shows in general the same trend as Lockhart-Martinelli's correlation, yields, however, 110% to 40% higher values with increasing X_{LM}.

As can be seen in Fig. 26 Kaiser et al. evaluated their experiments with respect to the two phase friction pressure drop and the total pressure drop. They suggest as frictional multiplier the correlation:

$$\phi = 8{,}2 \cdot X_{LM}^{-0{,}55} \tag{24}$$

in the range of $0{,}01 < X_{LM} < 2$.

Kaiser Peppler's correlation is of different nature compared to those of Chen-Kalish and Lockhart-Martinelli. It intersects Lockhart-Martinelli's correlation at $X_{LM} \simeq 1{,}5 \cdot 10^{-2}$ yielding higher values with increasing X_{LM} which diverge up to 100% at $X_{LM} = 1$. Chen Kalish's correlation yields up to 110% higher values for low X_{LM} and intersects Kaiser-Peppler's correlation at $X_{LM} \approx 6 \cdot 10^{-1}$, and yielding up to 40% lower values at $X_{LM} = 1$.

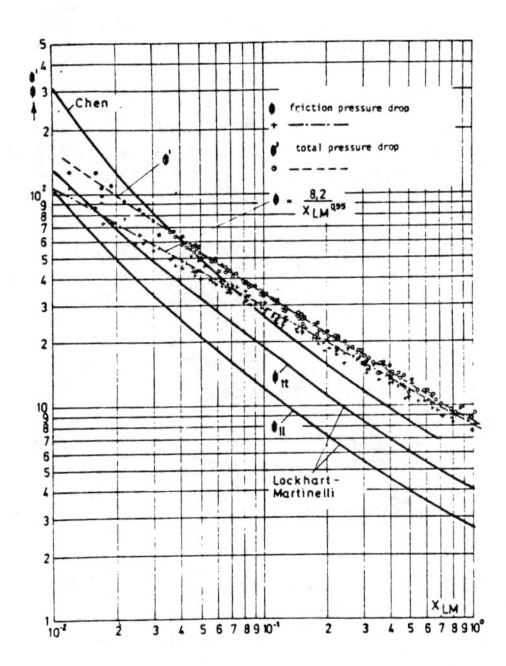

Fig. 26. Frictional multiplier vs X_{LM} (potassium boiling experiments: Chen-Kalish, sodium boiling experiments: Kaiser-Peppler).

Kottowski et al [25] performed steady state sodium boiling experiments in tubular and annular test sections. From their experimental results they evaluated as best squares fit fit of the data a ϕ correlation covering the range $7 \cdot 10^{-2} < X_{LM} < 30$ of the X_{LM} factor (Fig. 27):

$$\log \phi = 0.1046 (\log X_{LM})^2 - 0,5098 \log X_{LM} + 0,625 \qquad (25)$$

This correlation yield values of up to 20% below the Lockhart Martinelli correlation over the whole range of X_{tt}.

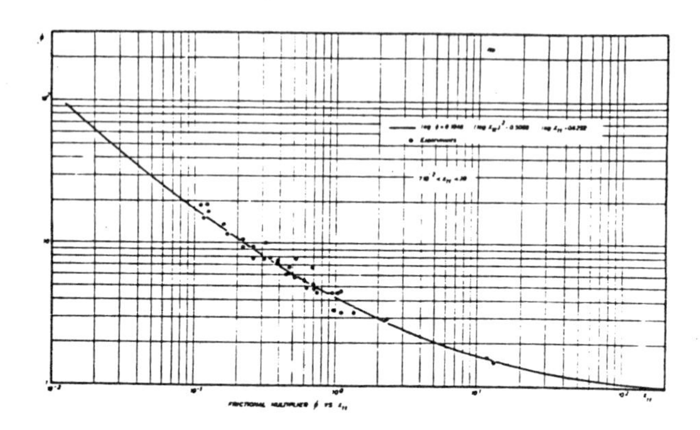

Fig. 27. Frictional multiplier vs X_{LM} (sodium boiling experiments: Kottowski-Savatteri)

The ϕ correlations developed on the basis of experimental results differe appreciably. At present time only a qualitative discussion on the reasons for these deviations is possible. The experiments were run in various laboratories with different test sections, different liquid metals and at different test conditions. The experiments available do not allow the valuation of the various factors at present time. On the other hand, the experiments of Lurie and Kottowski show that the friction multiplier for liquid metal boiling two phase flow can be predicted fairly good by correlation eq. 25.

4.3. Liquid metal boiling heat transfer

Before treating heat transfer to the two phase flow, an understanding of the correlation between various flow regimes and heat transfer is essential. Fig. 28 is a model pattern of the flow regimes and the heat transfer mechanisms as the vapour quality increases.

Fig. 28 shows in a simplified manner flow pattern changes that might occur in liquid metal boiling flow at an intermediate heat flux level. Bubbles are shown forming initially at active

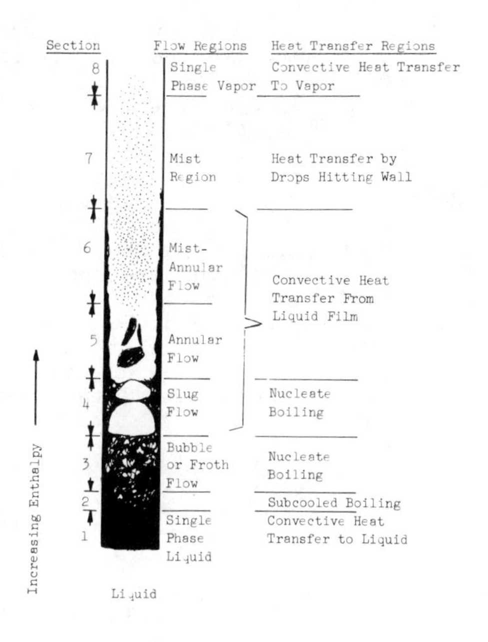

Fig. 28. Flow pattern and heat transfer modes.

sites along the wall and in the liquid itself. The presence of vapour bubbles or nucleus in the liquid provides other locations at which vaporization can occur if superheating of the liquid tends to exist away from the wall. Metallic liquids possessing high liquid thermal conductivities and requiring high wall superheats to nucleate bubbles can produce these conditions. Pressure drop along a flow channel which results in decrease of saturation temperature also tends to create bulk superheat if not accompanied by fluid vaporization or compensating heat loss. The heat transfer in this flow region is best represented by the nucleate boiling heat transfer mechanism. The axial extension of this flow regime is small and therefore negligeable. The slug flow regime represents a transition between all liquid flow and the annular - (churn) - mist flow configuration. This flow regime seems inherently unstable in that it leads to flow pulsation. The heat transfer in this flow regime is described by the nucleat boiling and convective heat transfer (Section 3 and 4 of Fig. 28).

The thin annular film in the annular and mist flow regime seems to preclude nucleate boiling due to the high conductivity of metallic fluids that evaporation occurs at the liquid vapour interface. One can argue that if the interface remains at saturation temperature, then it seems unlikely that sufficient superheat could be developed at the solid-liquid interface to produce nucleation.

At high heat fluxes the large vaporization rates could lead to film instabilities and complicate the orderly progression of the flow regimes. The heat transfer mechanism for these flow patterns, shown in section 5 and 6 of Fig. 28, seems best represented by convective heat transfer from the liquid film.

At sufficiently high heat flux levels the liquid film next to the wall could be destroyed long before the vapour quality reaches the level of section 7 (shown in Fig. 28). Such phenomena would be accompanied by a sharp reduction in the heat transfer coefficient and consequently in a jump in wall temperature. Such condition, and in case of regular progression of the flow regime where the liquid in the flow decreases (section 8, Fig. 28) to the point that the wall sees only vapour, is called the "critical flux" condition. This is ordinarily the case when an insufficient number of liquid droplets impinge on the wall and the droplets are repelled by virtue of a Leidenfrost type of phenomenon.

As illustrated in Fig. 28 various heat transfer mechanisms, linked to the flow pattern, exist. Since low vapour quality in metallic fluids produces high void fraction the following general trends for two phase heat transfer can be predicted.

- At constant heat flux, saturation pressure, and given wall roughness nucleate boiling tends to be suppressed with increasing vapour quality.
- If all other variables are held constant, lowering the pressure suppresses nucleate boiling.
- If all other variables are held constant, lowering the heat flux suppresses nucleate boiling.

As a consequence, the nucleate heat transfer in metallic fluids is of secondary importance, as far as technical application is concerned.

Although voluminous literature exists concerning two phase flow characteristics its application to metallic fluid remains however limited. On the other hand experimental results on steady state liquid metal boiling heat transfer are not adequate to explain the influence of some factors and tendency of the boiling characteristics observed.

4.3.1. Pool boiling

Experimental studies of boiling heat transfer on horizontal plates and tubes in pool boiling condition using mercury, potassium, sodium and sodium-potassium were performed by various authors covering the pressure range up to atmospheric pressure and heat flux up to about 150 W/cm^2.

Bonilla and workers [26] investigated the heat transfer on boiling sodium-potassium alloy on the surface of a horizontal plate at pressures of 2 to 760 mm Hg and on boiling potassium at pressures from 2 to 1500 mm Hg. The following empirical equations were proposed for boiling NaK-alloy:

$$\alpha = 236{,}9 \cdot P^{0{,}2} \cdot \dot{q}^{0{,}2} \text{ W/m}^2\text{°K} \quad (27)$$

and for potassium

$$\alpha = 0{,}05 \; P^{0{,}243} \cdot \dot{q}^{0{,}885} \text{ W/m}^2\text{°K} \quad (28)$$

(P in mm Hg, $\dot{q}$ in W/m^2)

Noyes [27] published sodium boiling heat transfer results obtained on horizontal tubes in the pressure range of 50 to 500 mm Hg. He suggested for sodium boiling the correlation:

$$\alpha = 26{,}7 \; \dot{q}^{0{,}58} \text{ W/m}^2\text{°K} \; (\dot{q} \text{ in W/m}^2) \quad (29)$$

No influence of pressure was observed.

Kovalev and Zhukov reported on sodium boiling studies on horizontal tubes. They deduced from their experiments the following heat transfer correlation as a function of pressure and heat flux:

$$\alpha = 0{,}84 \cdot P^{0{,}25} \cdot \dot{q}^{0{,}7} \text{ W/m}^2\text{°K} \quad (30)$$

(P in mm Hg in W/m^2)

Fig. 29 shows the measured heat transfer as a function of the heat flux: $\alpha = f(\dot{q})$. The different dots correspond to measurements at pressures of: (1) 7 mm Hg, (2) 10 mm Hg, (3) 20 mm Hg and (4) 35 mm Hg. The measuring points can be represented by a stright line with a slope of 0,7. In order to evaluate the dependence of the heat transfer on the pressure the experimental data were represented in coordinates of $\alpha/\dot{q}^{0,7}$ vs the pressure. Fig. 30 shows the measured data within the range of 7 to 500 mm Hg. The least squares fitted data are best approximated by a straight line with a slope of 0,25. The scattering of the data does not exceed ± 25%.

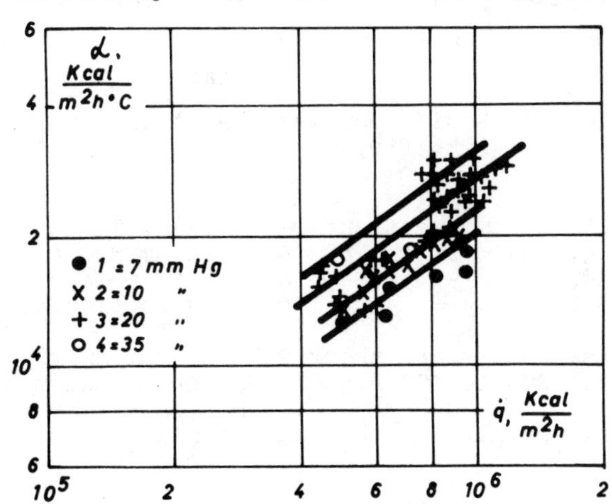

Fig. 29. Sodium boiling heat transfer $\alpha = f(\dot{q})$

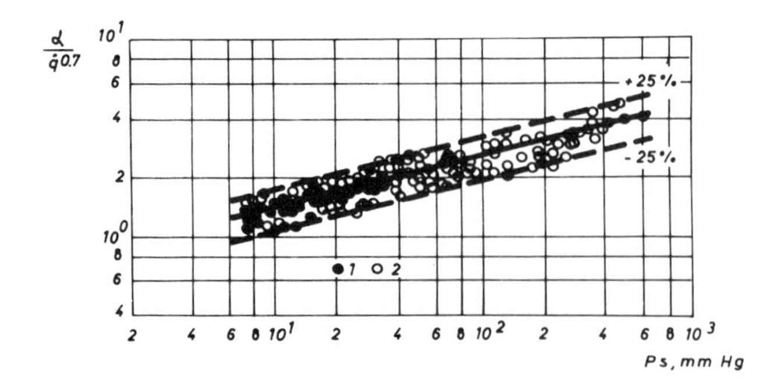

Fig. 30. Sodium boiling heat transfer
$= f(p)$

Kovalev and Zhukov found as influence of the tube diameter on the heat transfer $\alpha \sim 1/d^{0,25}$ which they neglected in their correlation. The transition from natural convection to boiling and the boiling results up to 350 mm Hg pressure are shown in Fig. 31.

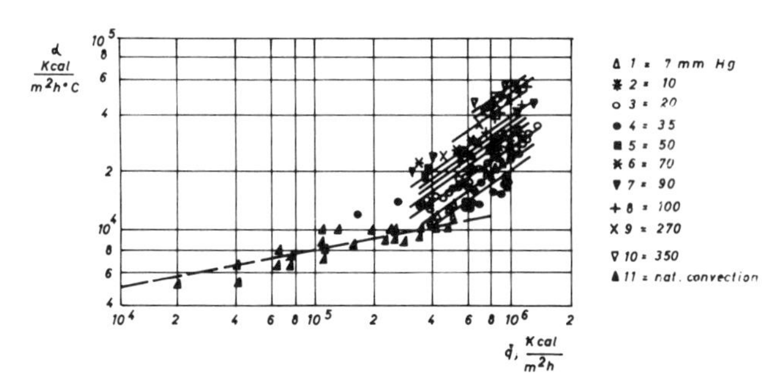

Fig. 31. Sodium boiling heat transfer
$= f(p, \check{q})$

The single test conditions were: (1) = 7mm Hg pressure, (2) = 10 mm Hg, (3) = 20 mm Hg, (4) = 35 mm Hg, (5) = 50 mm Hg, (6) = 70 mm Hg, (7) = 90 mm Hg, (8) = 100 mm Hg, (9) = 270 mm Hg, (10) = 350 mm Hg, (11) = natural convection without boiling.

Kovalev's results are about 20% lower than those of Noyes and Lurie and about 20% to 40% higher than the results of Hyman-Bonilla [30] and Kutateladze [29]. Hyman, Bonilla and Kutateladze evaluated their results in the dimensionless form $Nu = f(Gr.Pr^2)$. Bonilla and coworkers proposed the correlation:

$$Nu = 0,53\ (Gr\ .\ Pr^2)^{0,25} \qquad (31)$$

taking the tube diameter as characteristic length. Kutateladze and coworkers proposed a similar correlation:

$$Nu = 0,67\ (Gr \cdot Pr^2)^{0,25} \qquad (32)$$

Gr = Graßhoff number, Pr = Prandtle number

Fig. 32 shows the comparison of Bonilla's correlation (I), Kutateladze's (II) correlation and the data of Kovalev and Zhukov. Taking into account the experimental difficulties, a remarkably good agreement is found, between the correlations 30, 31 and 32.

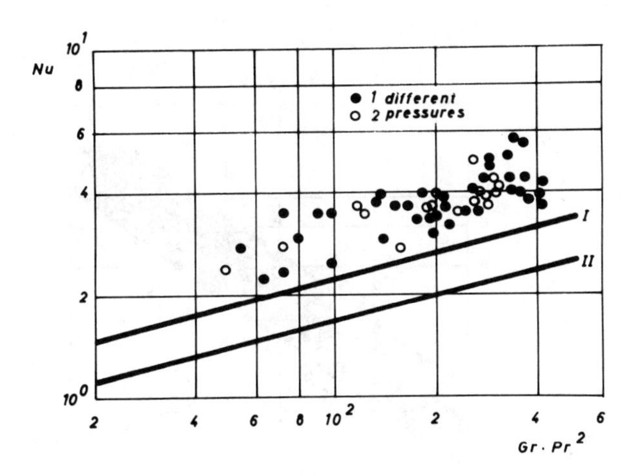

Fig. 32. Sodium boiling heat transfer $Nu = f(GrPr^2)$; (1=21,5mm tube, 2 = 29,6 mm tube)

4.3.2. Convective boiling heat transfer

Steady state heat transfer data are available from potassium and few sodium boiling experiments in tubular geometry. The suggested empirical correlations are mostly deduced from potassium boiling experiments. The complexity of the heat transfer mechanisms and the few results tolerate at present time the development of correlations as a function of the pressure and heat flux as independent variables. The influence of other variables like geometry and non uniformity of the heat flux is not yet established. This restricts the validity of the correlations to the range of experimental conditions the correlated data were obtained. Some representative work will be discussed in the following.

4.3.2.1. Boiling in tubes

Gorlov et al. performed experiments in Niobium alloy and Molybdenum tubes of 4 mm diameter and a length to diameter ratio l/d = 30 to 100. The pressure was varied between 0,15 to 17 bar, the heat flux $\dot{q}$ in the range of 10 to 180 W/cm^2 and the mass flow rate between 20 to 600 kg/m^{2}sec.

Gorlov et al. proposed on the basis of the measured data the best fitting correlation:

$$\alpha = 0{,}57\ \dot{q}^{0,7} \cdot p_s^{0,15} \quad W/m^2{}^oC \tag{33}$$

(P_s in N/m^2)

As seen from Fig. 33 the experimental data scatter within the range of ± 30% from the average value given by $\alpha/p_s^{0.15} = \dot{q}^{0,7}$. No influence of the tube material was observed.

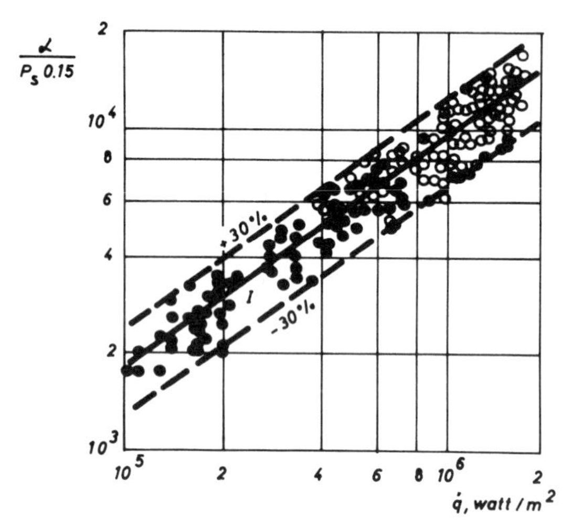

Fig. 33. Steady state boiling heat transfer of potassium in tubes.

The influence of the pressure is shown in Fig. 34 as a function of $\alpha/\dot{q}^{0,7} = f(Ps)$. The pressure dependence is best correlated by $\alpha \sim p^{0,15}$ within the range of $0{,}15 < Ps < 17$ bar. Furthermore, the authors report negligeable influence of the vapour quality on the heat transfer (Fig. 35).

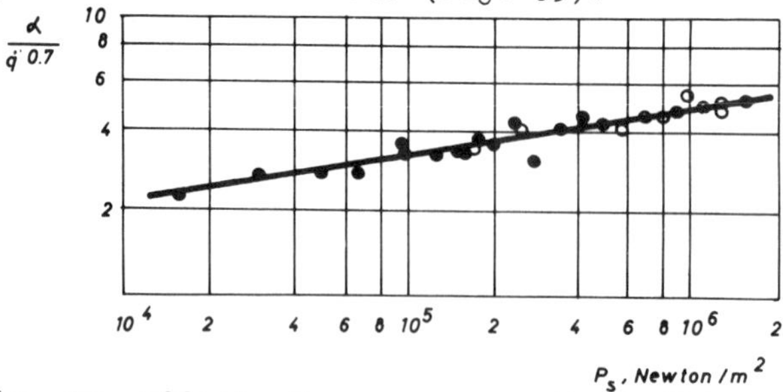

Fig. 34. Effect of pressure on the boiling heat transfer of potassium in tubes

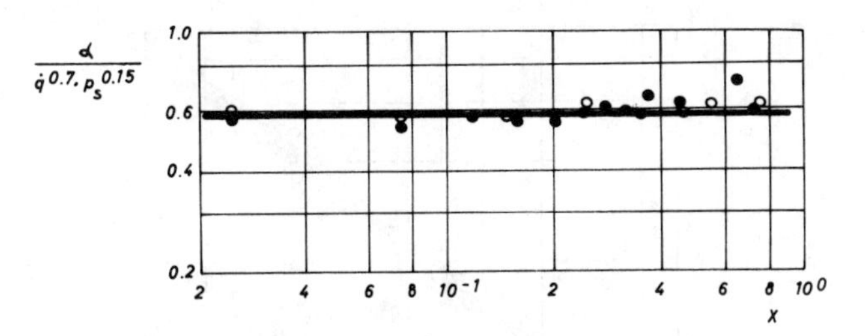

Fig. 35. Effect of vapour quality on the boiling heat transfer.

4.3.2.2. Boiling in coils

Alad'yev et al. [32] performed steady state boiling studies in helically coiled tubes. The experiments were run (a) in tubes of 10 mm diameter, 285 mm length, coil diameter D = 162 mm and a pitch of 73 mm, (b) in tubes of 4 mm diameter, D = 64 mm, a pitch of 60 mm and l/d = 242,5. The tubes were made of stainless steel and niobium alloy. The tests were run at pressures of 0,9 to 8 bar, heat fluxes of 3 to 51 W/cm^2 and mass flow rates of 14 to 205 kg/m^2sec. Alad'yev et al. propose as best fitting correlation:

$$\alpha_{coil} = 0{,}13\ \dot{q}^{0,85} \cdot p_s^{0,15} \quad W/cm^{2\,o}C \tag{34}$$

(P_s in N/m^2; $\dot{q}$ in W/m^2)

This correlation represents all measured data within $\pm$ 25%. The test results are shown in Fig. 36 where $\alpha/p_s^{0,15}$ is plotted vs the heat flux. The data are best correlated by $\alpha/p_s^{0,15} = \dot{q}^{0,85}$. The pressure dependence of the heat transfer in the pressure range of 0,8 to 8 bar is illustrated in Fig. 37.

Fig. 37 shows that the pressure effect is best represented by the exponent of n = 0,15, which is similar to that in tubes.

The interesting point to note is that the heat transfer in coiled tubes increases with increasing heat flux to power 0,85 instead of power 0,7 (in straight tubes). This might be caused by the superimposed radial secondary flow due to centrifugal forces. No change in the effect of pressure was observed. The similarity of boiling pattern and heat transfer mechanisms in pool boiling of potassium and sodium suggests the applicability of the correlations (33) and (34) also for sodium forced convection boiling.

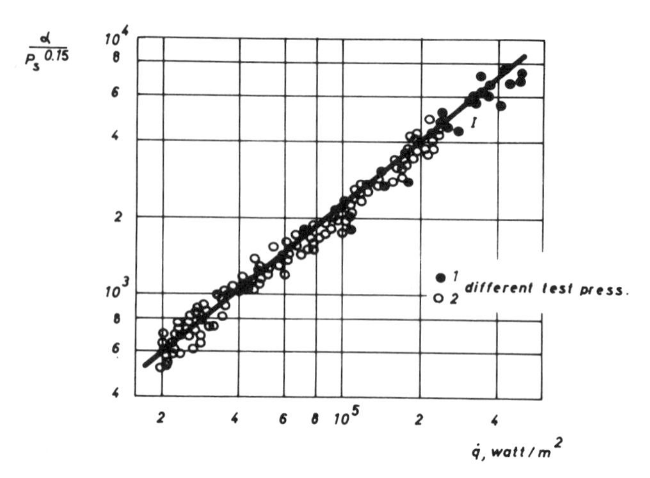

Fig. 36. Steady state boiling heat
transfer of potassium in coils

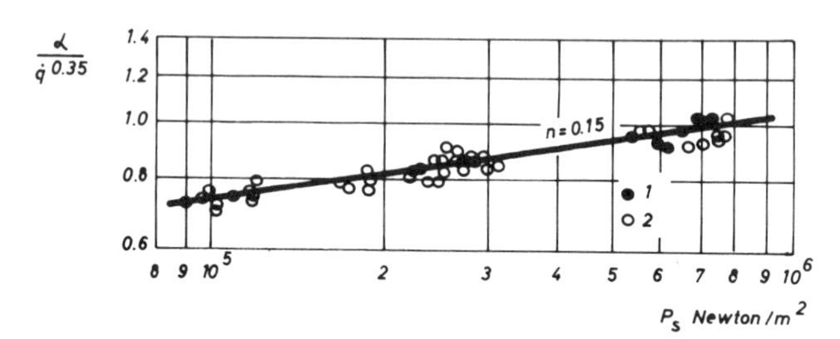

Fig. 37. Effect of pressure on the boiling heat transfer of potassium in coils (1=Niob-alloy coil with l/d=77÷232; 2 = SS coil with l/d = 23 ÷ 272)

4.3.3. Critical heat flux

About fifty equations have been proposed for predicting critical heat fluxes in various types of boiling regions and liquids. It has been pointed out by several investigators that for liquid metals at moderate pressure nearly all of these hydrodynamic baised correlations predict values of the critical heat flux that are much lower than the measured liquid metal results. In view of the foregoing, it appears that both hydrodynamic and liquid thermal transport properties must be considered in predicting the critical heat flux for metallic fluids. All attempts to incorporate thermal transport properties into burnout correlation have consisted of strictly empirical modifications of existing expressions. For liquid metals, these modifications raised the predicted value of the critical flux and corrected the pressure dependence to give somewhat better agreement with the experimental results. No experimental or theoretical work has yet been carried out, to my knowledge, which could enable to gain a more fundamental understanding of the extend to which thermal transport in the liquid influences burnout. Noyes, who studied sodium boiling, was the first to report liquid metal burnout results. He proposed a slight modification of Addoms [33] critical heat flux correlation to increase its applicability to metallic fluids

$$\frac{(\dot{q}/A)_c}{\lambda \cdot \rho_g (a \cdot g)^{1/3} (Pr/n)^{1/12}} = 1.19 \left[\frac{\rho_1 - \rho_g}{\rho_g}\right]^{0,56} \tag{35}$$

$(q/A)_c$ = critical heat flux in Btu/(hr) (sqft), λ = latent heat of vaporization, ρ_g = vapour density, ρ_1 = liquid density, a = void fraction, g = acceleration of gravity, Pr = Prandtle number, $n = g_g/g$, g_c = gravitational conversion ratio.

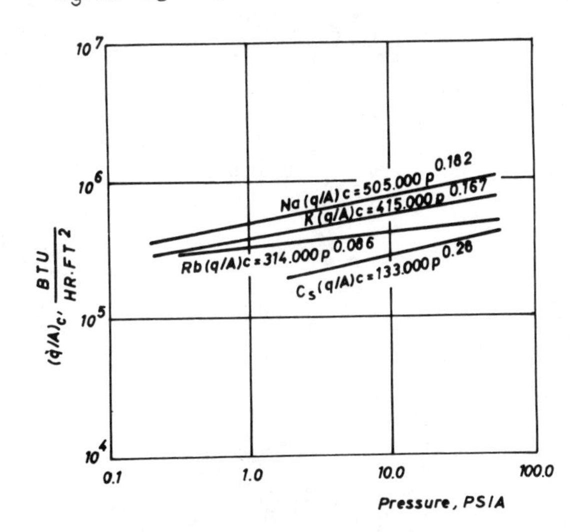

Fig.38. Summary of alkali burnout data

Correlating all liquid metal's critical data points for sodium, potassium rubidium and cesium, the following nondimensional equation resulted:

$$\frac{(\dot{q}/A)_c \cdot c_p \cdot \sigma}{\lambda^2 \rho_g K} = 1,18 \cdot 10^{-8} \left[\frac{\rho_1 - \rho_g}{\rho_g}\right]^{0,71} \tag{36}$$

where all properties are at saturation (Fig. 39). The above correlation was modified by Caswell [35] introducing the Prandtle number to the 0,71 power in order to include both metallic and nonmetallic data on a single line:

$$\frac{(\dot{q}/A)_c \cdot c_p \cdot \sigma}{\lambda^2 \rho_g \cdot K} \cdot Pr^{-0,71} = 1,02 \cdot 10^{-6} \left[\frac{\rho_1 - \rho_g}{\rho_g}\right]^{0,65} \tag{36a}$$

This correlation covers a range of Prandtle numbers from 0,004 to 11 and is recommended to predict the critical heat flux also for nonmetallic fluids (Fig. 40). Note that these liquid metal data are pool boiling measurements.

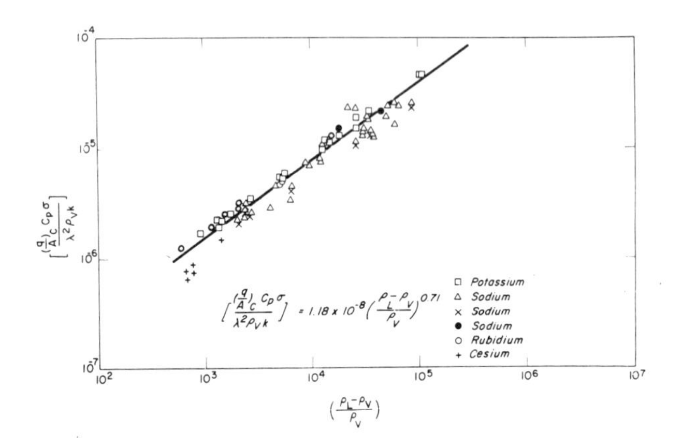

Fig. 39. Correlation of alkali metal burnout data.

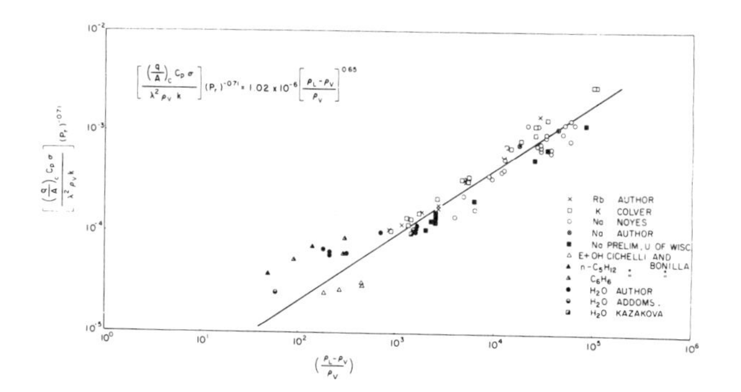

Fig. 40. Correlation of burnout data.

Forced convection critical heat flux results are reported by Gorlov et al. [31]. Experimental results were obtained in tubes of 4 mm diameter and up to 400 mm length. The boiling crises was approached by increasing the heat flux at constant mass flow or by decreasing the mass flow at constant heat flux. The mode of approaching the boiling crises had no influence on the critical heat flux. Gorlov correlated the measured results with the independent flow and geometry variables and proposed the following critical heat flux correlation within ± 15% of the measured data.

$$\dot{q}_c = 0{,}38 \cdot 10^6 \cdot \dot{Q}^{0{,}8}(1 - 2x)/(l/d)^{0{,}8} W/m^2 \quad (37)$$

$\dot{Q}$ = mass flow kg/m^{2}sec, x = inlet vapour quality of the heated length, l = tube length, d = diameter. Fig. 41 shows the experimental data and the line representing equation (37). No measurable effect of the pressure on the critical heat flux was observed.

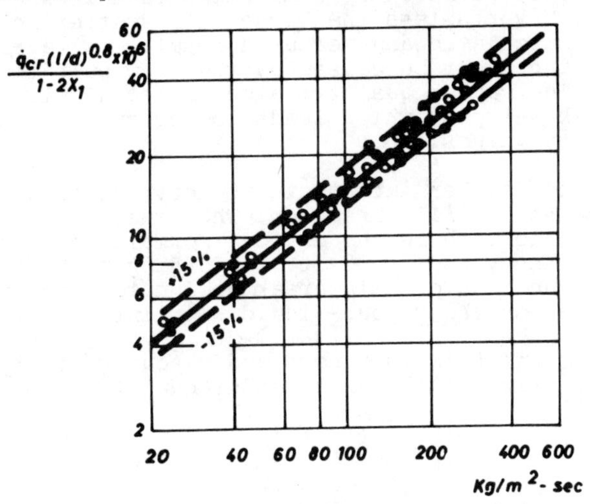

Fig. 41. Critical heat flux of boiling potassium at forced convection.

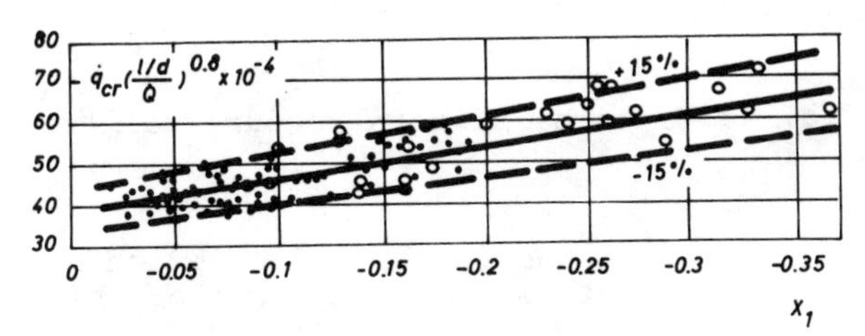

Fig. 42. Effect of subcooling on critical heat flux.

Gorlov investigated further the influence of subcooling on the critical heat flux. The results are illustrated in Fig. 42. The graph shows the tendency of increasing critical heat flux with increasing subcooling as observed for nonmetallic fluids. The subcooling refers to the inlet of the heated section in terms of negative vapour quality.

Gorlov's critical heat flux correlation was confirmed by Kaiser and Peppler [36]. Kaiser's data are shown in Fig 43 where Gorlov's correlation is compared with the experimental sodium results. Nevertheless the large data scattering the agreement between calculated and measured results is fairly good. The large scattering might be caused by the flow pulsation during the experiments. The vapour quality values were calculated with the pulsation peak velocity which leads to uncertainity in determining the quality values.

Kaiser and Peppler interprete the dry out as a consequence of film thinning and film break up. They measured at the exit of the heated section, where the dry out normally occurred, the film thickness and the vapour velocity. According to this model the dry out is a function of film break up due to evaporation and friction forces on the vapour-liquid film interface.

Fig. 44 shows film thickness and vapour velocity measurements as a function of the exit vapour quality.

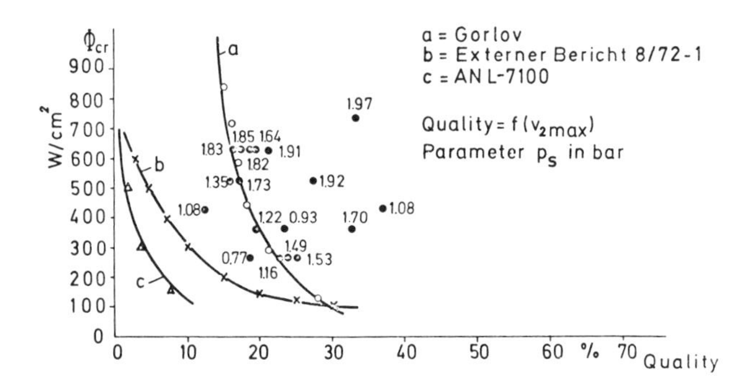

Fig. 43. Critical flux vs vapour quality.

No theoretical work has yet been done which could enable to predict the critical heat flux on the basis of this model.

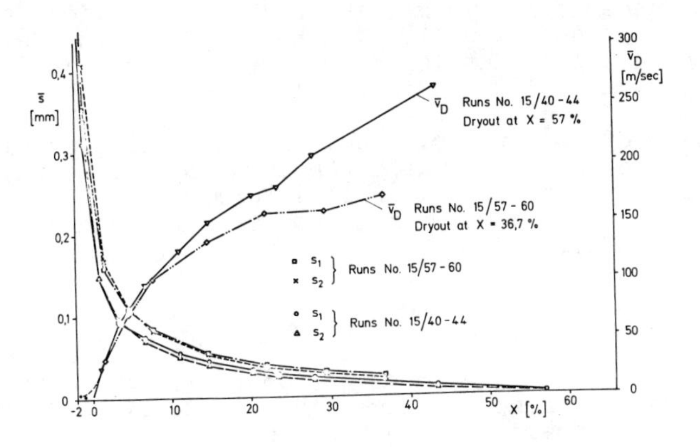

Fig. 44. Film thickness and vapour velocity vs vapour quality.

5. CONCLUSION

Reviewing the overall state of the art of sodium boiling the following conclusions are obtained:

1. There is experimental evidence that liquid metals inherently tend to superheat. Physico chemical processes and local hydrodynamic mechanisms lower the superheat. As far as reactor conditions are concerned the bulk superheat should not exceed 80°C. This does not exclude higher local superheat to occur.
2. The actual knowledge on boiling nucleation seems adequate. The experimental verification of the theoretical models fails in the last analysis due to experimental limitations. The empirical dynamic model of physico chemical processing being responsible for the nucleation seems realistic.
3. Good knowledge on flow pattern in simple and restricted geometry exists. There is still uncertainty on the flow pattern in complex geometry. Furthermore, destinction has to be made between steady state boiling flow regimes and transient flow regimes at boiling iniciation. The flow pattern at transient boiling is affected by the bulk superheat.
4. Two phase flow pressure drop characteristics of boiling liquid metals are known at present time only for steady state developed flow regimes. Results are available for simple geometry and small bundles. Comparison of measurements with metallic and non-metallic fluids show the great potential of non metallic liquids to simulate pressure drop characteristics of liquid metal two phase flow regimes provided that the flow pattern is reproduced.
5. The knowledge on steady state boiling heat transfer is not satisfactory. There is still a lack of systematically performed well instrumented experiments. On the other hand the known forced convection correlations can be used for engineering applications in the range of the experimental conditions the correlated data were measured.

6. The most uncertainty is met in predicting critical heat fluxes. It is questionable, whether pool boiling "critical flux" correlations developed from measurements in simple geometry, are applicable to bundle geometry.

The review of the present liquid metal boiling knowledge shows, even if not all work was taken into consideration, the lack of theoretical and experimental information in the field of pressure drop in complex geometry, heat transfer and critical heat flux in general.

ACKNOWLEDGEMENT

The author wishes to express his appreciation to his colleagues and the programme management of the C.C.R. Ispra which made possible to prepare this lecture. Special thanks are extended to Mrs. C. Kind for preparing the typescript in record time and Messrs. A. Birke and D. Droste for the preparation of the graphs.

NOMENCLATURE

A_{ko}	maximum nucleation work
A_K	nucleation work
A	surface formation work
A_v	volume work
$A_{s,1}$	wetting retraction work
N	constant (wall defects)
M	molecular weight
N	molecules per volume unit
K	constant
f(N)	wall cavity distribution
f(A)	nucleation activation factor
P	pressure
P_{sat}	saturation pressure
P_{oo}	local pressure
T	temperature
T_{sat}	saturation temperature
ΔT_s	superheat
Σ	surface formation energy
σ	surface tension
dF_{spot}	dry spot surface
dF_{liq}	liquid surface
X_{LM}	Lockhart Martinelli factor
X_{MN}	Martinelli Nelson factor

ϕ	fricitional multiplier
C	O_2-concentration in ppm
x	vapour quality
z	axial coordinate
π_σ	surface activation
K	Plank Boltzmann constant
v	velocity
ν	10^{13} sec^{-1} = molecular resonance frequency
ρ	density
r_c	critical nucleation radius
ϑ	wetting angle
β	representative cavity opening angle
ξ_1	cohesive forces
$\xi_{s,1}$	adhesive forces
g	gravity constant
δ	boundary layer

indices

2ϕ	two phase
l	liquid
g	gas or vapour
f	friction
h	hydrostatic height
ac	acceleration
σ	surface formation work
z	axial coordinate
s,sup	superheat
v	volume formation work
g_c	gravitational conversion ratio
λ	latent heat of vaporization
σ	surface tension
c_p	specific heat
K	heat conductivity
$(q/A)_c$	critical heat flux
α	void fraction
α	heat transfer coefficient
P_s	P_{sat}

REFERENCES

1. Kottowski, H.M. Mechanism of nucleation... Internal Heat Trans. Seminar, Heat Trans. in Liquid Met., Trogir, Sept. 6/11, 1971.
2. Knacke, O. Zur Keimbildung. Technische Mitteilungen, 58. Jg., Heft 3, 1965.
3. Knacke, O., Stranski, I.N. Mechanism of Evaporation. Progress in Metal Physics, 6, 1956.
4. Rietmüller, R. Results of delay time experiments and their application for checking of known nucleation models. Proceedings LMBWG-Meeting, June 1-3, 1977, Petten,Netherlands.
5. Class, G. Dynamisches Modell der heterogenen kollektiven Siedekeimbildung KFK 2007, Juli 1974.
6. Holz, R.E., Singer, R.M. A study of the incipient boiling of sodium, ANL-7608, October 1969.
7. Dean, S.N., Rohsenow, W.M. Mechanism and behaviour of nucleate boiling heat transfer to alkali metals, DSR-Report 76303-65, October 1969.
8. Schultheiß, G. Influence of cavities and oxide concentration on sodium superheat. KFK 874, EUR 4157 e., October 1968.
9. Chen, J.G. Incipient boiling superheats in liquid metals. J. Heat Transfer 90C, 303-312, 1972.
10. Kottowski, H.M., Grass, G. Influence on superheating by suppression of nucleation cavities and effect of surface microstructure on nucleation sites. ASME Meeting, New York, November 30, 1970.
11. Spiller, K.H. et al. Überhitzung und Einzelblasenejektion von stagnierendem Natrium. Akte 13, 245-251, 1968.
12. Petukov, B.S. et al. Study of sodium boiling. Proceedings of the International Heat Transfer Conference, Chicago, Vol.5, pp 80-91, A.I.Ch.E., New York, 1966.
13. Singer, R. An apparent effect of heat flux upon the incipient pool boiling of sodium. Trans. Am. Nucl. Soc. 13 (1), 367, July 1970.
14. Kottowski, H.M., Savatteri, C. Evaluation of sodium incipient superheat measurements with regard to the importance of various experimental and physical parameters. J.J. Heat Mass Transfer, Vol. 20, pp 1281-1300, 1977.
15. Collier, J.G. Convective boiling and condensation. Mac Graw Hill 07084402 X, 1972.
16. Schmücker, H., Grigull, U. Boiling of mercury in a vertical tube under forced flow conditions. Intern. Heat Trans., Seminar Heat Transfer in Liquid Metals, Trogir, Sept. 6/11, 1971.
17. Wirtz, P. Ein Beitrag zur theoretischen Beschreibung des Siedens unter Störfallbedingungen in natriumgekühlten schnellen Reaktoren. KFK 1958, Oktober 1979.
18. Grand, D., Letrebe, A. FLINT: A code for slow transients in simple channels. 7th LMBWG-Meeting, Petten, Holland, 1-3 June, 1977.
19. Lockhart, R.W., Martinelli, R.C. Proposed correlation of data for isothermal two-phase two-component flow in pipes. Chem. Eng. Progress, 1949.
20. Savatteri, C., Mol, M., Kottowski, H.M. Experimental results of Na-steady state boiling and comparison with boiling models. 7th LMBWG-Meeting, Petten, Holland, 1-3 June, 1977.
21. Kaiser, A., Peppler, W. Voross, L. Type of flow, pressure drop and critical heat flow of two-phase sodium flow. LMBWG, Grenoble, 1974.

22. Vlacher, P.T., Foulser, W., Robinson, P. Experimental and theoretical studies of forced convection boiling in the Winfrith low pressure water modelling. LMBWG, Risley, October 1-8, 1975.
23. Lurie, H. Steady state sodium boiling and hydrodynamics. NAA-SR-11586, 1966.
24. Chen, J.C., Kalish, S. An experimental investigation of two-phase pressure drop for potassium with and without net vaporization. Heat Trans. Conf. Paris-Versailles, Chapt.B 8.3, 1970.
25. Kottowski, H.M. et al. Steady state liquid metal boiling pressure drop characteristics. ANS Fast Reactor Safety, Chicago, 1970.
26. Madson, N., Bonilla, G.F. Heat transfer to sodium potassium alloy in pool boiling. Chemical Engineering progress, Symposium series, vol. 56, No. 30, 1960.
27. Noyes, R.C. An experimental study of sodium pool boiling heat transfer. Journal of Heat Transfer, Trans. of ASME Series C, No. 2, 1963.
28. Kovalev, S.A., Zhukov, V.M. Experimental study of heat transfer during sodium boiling under conditions of low pressure and natural convection. Progress in heat and mass transfer, Vol. 7, Pergamon Press, 1973.
29. Kutateladze, S.S., Borishansky, V.M., Novikov, I.I., Fedinsky, O.S., Heat transfer in liquid metals. Zhidkometallitcheskye Teplonosteli, Atomizdat, Moscow, 1958.
30. Hyman, S.C., Bonilla, C.J., Ehrlich, S.W. Chemical Engineering progress, Symposium Series, Vol. 49, No. 5, 1953.
31. Gorlov, I.G., Rzayev, A.I., Khudyakov, V.F. Boiling of potassium in pipes at high pressure. Heat Transfer - Soviet research, Vol. 7, no. 4, July August 1975.
32. Alad'yev, I.T., Petrov, V.I., Rzayev, A.I., Khudyyakov,V.F. Heat transfer in a sodium-potassium heat exchanger made of helically coiled tubes. Heat transfer - Soviet research, Vol. 8, Nr. 3, May June 1976.
33. Addoms, J.N. Sc.D. Thesis in chemical engineering. Massachussetts Institute of Technology, 1948.
34. Colver, C.P. A study of saturated pool boiling potassium up to burnout heat fluxes. 7th Heat transfer conference, Cleveland, Ohio, August 9-12, 1964.
35. Caswell. B.F. Critical heat flux determinations in saturated pool boiling. AFAPL-TR-66-85.
36. Kaiser, A., Peppler, W., Schleisiek, K. Flow pattern, pressure drop and critical heat flux with forced convection sodium. LMBWG-meeting Karlsruhe, Germany, November 6-7, 1972.
37. Smith, L.R., Tek, M.R., and Balzhiser, R.E., AIChE J.12, 50-8, 1966.
38. Martinelli, R.C. and Nelson, D.B., Trans. ASME 70, 695-702, 1948.
39. Baroczy, C.J., NAA-SR-8171, 1963.
40. Levy, S., J. Heat Transfer B2, 113-24, 1960
41. Lockhart, R.W. and Martinelli R.C., Chem. Eng. Progr. 45, 39-49.

24 EXPERIMENTAL METHODS IN TWO-PHASE FLOWS

by

O.C. Jones, Jr.

BNL – AUI, Upton, NY, USA

In virtually all the previous chapters, the technology described has been based on experimental observation. Experiments are used to provide the researcher with his first glimpse of certain behavior, thus providing him with a clue for further analytical modeling. Experiments then provide more detailed information upon which he may fix the values of unknown coefficients. They may also provide an independent check for completed analysis through the verification process. On the other hand they may simply be used to provide a "yes" or "no" answer to a question. In all cases, measurements must be obtained, the degree of sophistication depending on the need. It is the purpose of this chapter to describe techniques applicable to two-phase, gas-liquid systems.

In this chapter a detailed summary of measurement techniques which have achieved some degree of maturity is provided, not for the purpose of giving the most up-to-date research summary, but rather for indicating what has been relatively successful, and therefore achieved widespread usage. In this case, maturity is not necessarily an indicator of degree of understanding as many of the device-fluid interaction's are not well understood. This chapter attempts to indicate where improved understanding is desired and would be beneficial to improved use of a particular instrument.

The reader is not expected to have any detailed knowledge of the mechanics of multiphase flows, but a working feeling for electromechanical methods is taken for granted. Starting with a brief description which places the field in perspective from a scientific standpoint as well as from an application viewpoint, both electronic and mechanical devices for taking internal diagnostic measurements and devices for obtaining measurements from outside the flow field are discussed. Recent advances which are expected to have significant impact on the use or interpretation of classical methods are also covered.

1. INTRODUCTION

In the past 30 years, the advent of new chemical processing systems, modern power generation methods, and space propulsion devices have dealt increasingly with multiphase flows. This is especially true of nuclear reactor systems where off-normal and accident situations have been intensively studied to provide assurance of public safety under extreme conditions. In the future development of advanced energy systems, multiphase flows will play yet an increasing role due to the attractiveness of utilizing the latent heat of phase change to enhance the energy intensiveness of these systems. Currently, for instance, two-phase, gas-liquid flows are found in such diverse systems as ocean thermal energy conversion equipment, liquid metal magnetohydrodynamic generators, geothermal wells and turbine generating equipment, oil-gas pipelines, boilers, nuclear reactor systems, liquid metal blankets of fusion power systems, droplet combustors, distillation towers, turbomachinery, refrigeration systems, and coal liquifaction systems, to name just a few.

The overriding conclusion one draws from an examination of two-phase flow literature is that multiphase flow is such an exceedingly complex physical situation that practically no general analytical effort has had an impact on this field. Empiricism abounds. Experimental methods have been both borrowed from other fields and newly developed to allow examination of just one or two areas of the phenomena. Workers in general are in the position of the blind men attempting to describe an elephant by feel. One has hold of its trunk, another its leg, and yet a third the tail. Each arrives at a different conclusion due to his own vantage point, and none perceives the elephant as a whole.

In attempting to piece together the differing viewpoints, several facts come to mind. Like the experience of early workers in single-phase fluid mechanics, the overall picture remains elusive due to lack of insight into the intracies of the phenomena involved. When the experiments of J. Osborne Reynolds demonstrated the two major categories of single-phase flow fields, significant progress was then made in piecing together various conflicting results. Similarly, in two-phase flow, while people are cognizant of the vagaries of laminar versus turbulent conditions, few have paid more than lip service to other areas of demarcation. Flow regimes are of overriding importance. One would not expect the same equations to accurately describe pressure drop in both laminar and turbulent flow. No one would then expect a single equation to do the same for completely different two-phase regimes. And in fact virtually no distinction between two-phase flows of laminar or turbulent character has been seriously considered.

Many workers are still trying to treat this phenomenon as a single-phase fluid. It is not. Two-phase immiscible fluids are decidedly different. They behave differently. They are not generally microscopic intramixtures but are macroscopic conglomorations. Formulations and methodology based on single-phase technology can only be of limited utility. In the final analysis, new methods, new viewpoints, and new technology must be developed. We must view the entire field from as many points of view as possible in hopes of determining the true characteristics of two-phase flow.

Realizing that two fluids flowing together are different we must ask how they differ from a single fluid flowing in a pipe. First, one phase is usually lighter than the other, the result being diffusion of one fluid with respect to the other. This is due in part to Archimedes principal, a principal almost as old in concept as the study of fluid mechanics itself. This is basic. The relative movement of one fluid with respect to the other depends on the individual phase flow rates as related through the void fraction. No one would think of running a basic single-phase experiment without measuring the flow rates. But for two-phases, the void fraction is just as basic a parameter as the flow rates. No basic experiment in two-phase flow should be run without measuring the void fraction and phase flow rates where possible as an absolute minimum.

The need for determining the void fraction, that is the area or volume fraction of the flow occupied by the vapor phase, has beren widely recognized and has led to the development of numerous methods to obtain these measurements. Briefly, these methods may be divided into four general categories, depending on the spatial scale in relation to the duct size over which measurements are taken. The categories include:

a) Point average probe insertion techniques such as conductivity probes used by Neal [153], hot film anemometers used by Hsu, Simon, and Grahm [68], and impedance probes used by Bencze and Orbeck [63]. These will be discussed in Section 3.

b) Chordal average void measurements commonly obtained through the use of particle or photon beam attenuation techniques. These methods were used by Cravarolo and Hassid [154], Petrick [155], and Pike, Wilkinson, and Ward [156], and will be discussed in Section 5.

c) Area or small-volume average void fraction measurements by such means as impedance sensing devices similar to those employed by Orbeck [64], and Cimorelli and Premoli [66] (Section 3) or flow sampling devices (Section 4).

d) Large-volume average such as accomplished by Lockhart and Martinelli [151], and by Hewitt, King and Lovegrove [152], through the use of quick-closing valves at various points in a tube but not to be discussed hereing.

The methods listed above and the techniques employed are by no means exhaustive. Many will be discussed in what follows. The interested reader is directed to an excellent survey of the field of void fraction measurement compiled by Gouse [157] and published in April, 1964.

A second way in which two-phase flow differs from single-phase flow is due to the immiscibility of the phases, each separated from the other by interfaces. It can be shown that the existance of these interfaces, their number and location, are of fundamental importance in the analytical description of the movement of two-phase flows. Hydrodynamic and thermal transfer and relaxation phenomena in two-phase, gas-liquid flows, in addition to being dependent on the departure from equilibrium for phase interactions, are strongly dependent on the interfacial area density available for transfer to occur [1-3]. Such interactions include mass, momentum, and energy exchange for which both the transfer coefficients and the areas available for transfer must be known accurately in order to specify transfers based on driving potential [4]. For instance, closure of a problem involving mass transfer between the phases requires specification of the volumetric vapor generation rate, Γ_v, in order to predict the vapor volume growth rates. Vapor generation rates are usually calculated in general terms by means of a constitutive relation given by [1]

$$\Gamma_v = \frac{1}{A\Delta i_{fg}} \int_{\xi_i} \sum_{k=\ell,v} \vec{q}''_k \cdot \vec{n}_k \frac{dA_i}{dz} \tag{1}$$

where surface tension, shear, and relative kinetic energy effects are considered negligible. In this case the net heat flux to the interface is given by the summation of the normal phase fluxes $\vec{q}''_k \cdot \vec{n}_k$ at the point on the interface A_i having a density of dA_i/Adz. The interaction then sums the effects over all interfaces in A along the interfacial perimeter ξ_i. Similarly, specification of the momentum transfer between phases is important in prediction of the relative velocity between phases which is needed to couple void fraction to the quality in two-phase flows. The latter is especially important under low velocity situations such as with natural convective flows, or under conditions of countercurrent flows. The degree of momentum transfer and hence the relative velocity is very sensitive to the flow regime. In highly coupled flows such as bubbly and slug flows the transfer rates are high and relative velocities are low. Conversely, in weakly coupled flows such as in the annular or droplet regimes, momentum transfer is small and the relative velocities can be quite large [2]. In both cases, a knowledge of the interfacial area density is required. Thus basic experiments should also include some method of monitoring the passage and density of interfaces.

Is this sufficient? Certainly measurements of the gross flows in basic single-phase work were insufficient. Detailed velocity profiles were required as confirmation of theoretical models. It can be likewise assumed that both void and velocity profiles as well as interfacial area density profiles would be eventually required for analysis of two-phase flows. Likewise, different pressures are considered as essential measurements coupled with measurements relating to the frequency of void appearance. Lastly, as a minimum, certain characteristics of the flow must be identified as related to the structural behavior of the mixture, and levels of turbulance should be determined.

As in most other areas of scientific endeavour, purely theoretical analysis and prediction of the behavior of gas-liquid systems such as suggested above is clearly inadequate. Rather, experimental and analytical efforts must go hand-in-hand. While development of single-phase thermo-fluid measurement techniques has been intensively pursued during the last century, special problems are encountered in attempting to observe mixtures of fluids and it has been only over the last two to three decades that these problems have been considered. Measurements in two-phase, gas-liquid systems is very much still in the empirical stage. Researchers generally try to find something which "wiggles when tickled" and then try to calibrate and interpret the response. In short, it can be said that measurements in two-phase, gas-liquid systems carry all the complexities of similar measurements in single-phase systems plus additional difficulties, some of which are itemized below.

a) Interface deformation due to sensor presence causes temporal distortion of the true pointwise behavior. Unequal distortions are generally observed in going from gas-to-liquid rather than from liquid-to-gas.

b) Inadequate transference from liquid-phase response to gas-phase response due to film retention, causes bias in resulting interpretation unless precautions are taken.

c) Destruction of metastability due to sensor presence, which, for instance, causes cavitation on a thermal probe which may then record saturation rather than superheat conditions.

d) Inaccurate response of pressure tap lines may result due to liquid vapor interfaces caused by gas entrapment or flashing. The former can occur in two component systems even in steady flow conditions due to the normal fluctuations while the latter can occur due to rapid decompression in heated lines during transients.

e) Improper averaging of nonlinear signals occurring due to fluctuations, with or without consideration of damping factors, causes intepretation errors.

f) Errors due to distortion of signals due to the transmitting medium environment such as the shunting effects in sheathed, high temperature thermocouples.

g) Simple lack of understanding of the physical phenomena affecting a sensor response mass lead to misinterpretation of the results.

There have been several excellent summaries of two-phase measurement techniques written previously for steady state measurements [5-8] and for transient or statistical techniques [9-10]. The purpose of this summary is to combine the techniques developed to date into a coherent form suitable for a short, first graduate level, tutorial on modern methods.

This presentation will be divided into the following three sections:

1. Instream Electrical Sampling Techiques
2. Instream Mechanical Sampling Techniques
3. Global Sampling Techiques

2. TWO-PHASE, GAS-LIQUID FLOW PATTERNS

Since the entire thrust of this document is directed towards making measurements in two-phase flows, it is appropriate to briefly describe the basic configurations or patterns which these flows can attain.

Just as in single phase flows, there can be two different regimes within which different descriptions of momentum and heat transfer apply, so too in two-phase flows do we find laminar and turbulent situations. Unfortunately, at this stage, any distinctions made on the basis of these two classifications are purely qualitative. Rather, the major distinctions, which have been used, and which are probably the first order contributors to system behavior, are space-time phase distributions. These can probably be best described as a sequence of developing situations within a vertical tube evaporator.

As described by Collier [11] and shown in Figure 1, the changes with vertical upflow of an initially all liquid system are mainly due to "departure from thermodynamic equilibrium coupled with the presence of radial temperature profiles and departure from hydrodynamic equilibrium throughout the channel."

In the initial single-phase region the liquid is being heated to the saturation temperature. A thermal boundary layer forms at the wall and a radial temperature profile is set up. At some position up the tube the wall temperature will exceed the saturation temperature and the conditions for the formation of vapor (nucleation) at the wall are satisfied. Vapor is

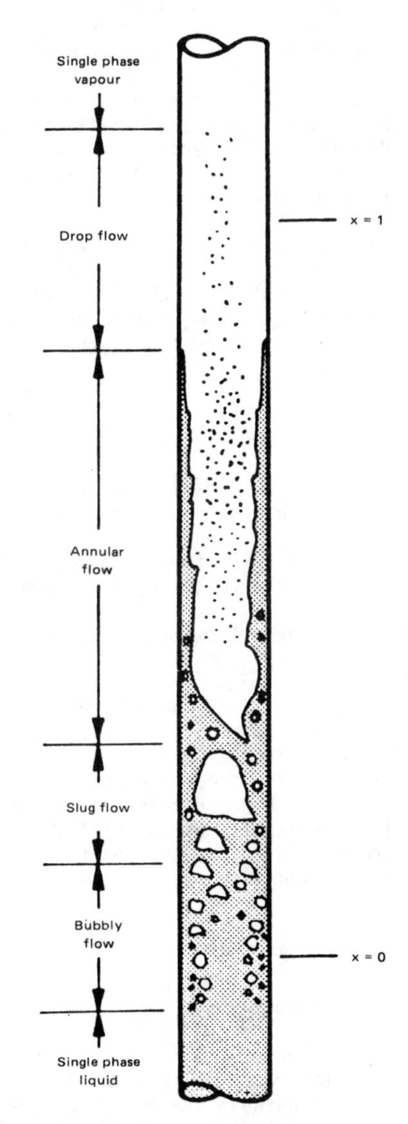

Figure 1 - Flow Patterns in a Vertical Evaporator Tube [11] (BNL Neg. No. 9-136-79)

formed at preferred positions or sites on the surface of the tube. Vapor bubbles grow from these sites finally detaching to form a "bubbly flow." With the production of more vapour due to continued heat addition, the bubble population increases with length and coalescence takes place to form slug flow which in turn gives way to annular flow as the elongated bubbles increase in length and merge further along the channel. Close to this point the formation of vapour at sites on the wall may cease and further vapour formation will be as a result of evaporation at the liquid film-vapour core interface. Increasing velocities in the vapour core will cause entrainment of liquid from the surface waves on the film which result in of droplets in the central vapor stream. The depletion of the liquid from the film by this entrainment and by evaporation finally causes the film to dry out completely. Droplets continue to exist, some of which may even be remnants of the upstream destructing of liquid slugs, and are slowly evaporated until only single-phase vapour is present. While this description presents a commonly perceived and encountered evolution of flow patterns in a heated duct, it should be kept in mind that these patterns may, in fact, be encountered in other situations and evolved differently.

From a researcher's viewpoint, then, for both analytical and experimental purposes, the phenomena of interest are generally associated with those detailed characteristics which distinguish one pattern from another. Of course, as in single phase flow, temperature, pressure, velocity and mass flow rates are of interest. In addition, specific items which deal with particular aspects include:

a) Bubbly flows: boiling inception, bubble size, bubble trajectory, bubble boundary layer thickness, liquid superheat, bubble agglomeration.

b) Slug flows: Slug lengths, bubble sizes, bubble spacings, film thicknesses, vapor entrainment, bubble velocity, slug velocity.

c) Annular flows: wave height, wave length, wave celerity, film thickness, film velocity, entrainment rate, de-etrainment rate.

d) Drop flows: dryout inception, drop size, drop trajectories, drop impingement.

In all cases, departures from mechanical or thermal equilibrium as evidenced by relative velocities of one phase with respect to the other and by differences between phase temperatures are important variables both for design and analysis as they can affect operating conditions of engineering equipment. Phase residence time fractions at a point in space and volume fractions at an instant in time as well as the distributions and alternate averages of each, are of overriding importance and interest. The balance of this report is devoted to describing methods of obtaining the measurements identified above.

3. INSTREAM SENSORS WITH ELECTRICAL OUTPUT

3.1. Conductivity Devices

The general principle of operation for conductivity sensing devices is for two electrodes to be immersed in a two-phase mixture. A potential difference is created between the two electrodes so that the current flow is a direct measurement of the conductivity of the fluid between the two electrodes. This current flow may be measured by means of a voltage drop across a calibrated resistor connected to some steady state or transient measuring device, such as an oscilloscope, oscillograph, or recording voltmeter. Within certain limits, the relative amounts of conducting and non-conducting fluid, in other words the amounts of liquid and gas, will determine the amount of current flow between the two electrodes. In this manner, current may be calibrated as a function of the vapor fraction and/or phase distribution between the two electrodes. (These devices may also be used in the impedance mode where the capacitive reactance is the primary variable but they will still be called conductivity devices). Such devices may be used for indications or measurements of void fraction, liquid level, film thickness, flow patterns and, with some modifications, can be arranged to give quantitative information on wave frequencies, heights and velocities.

Level probe. Figure 2 shows an early liquid level transducer, designed and built by Kordyban and Ranov [12], for use in their air-water experiments. This probe consisted of a pair of insulated copper wires inserted through a hypodermic tubing coated with insulating paint. The wire ends were stripped of insulation and attached to the outside of the hypodermic needle tubing, thus being electrically insulated from themselves and from the tubing. When a potential was placed between the two wires, the amount of current flow was then a mea-sure of the percentage of the sensing element covered by the conducting fluid. These transducers were used in series along the test section to measure the time interval of slug travel between sensing stations in order to obtain the velocitiy of slugs and variations along the length of the tube. The authors were also able to use the transducer to make qualitative measurements of flow regime within their horizontal tube. Figure 3 shows a typical output trace from a pair of transducers located 20 inches apart in the horizontal tube during conditions of slug flow. Not only was the passage of slugs immediately evident from the trace, but also when the two traces are compared they show that the flow regimes retained their identity along the 20 inch length so that slug velocities could immediately be measured. Kordyban and Ranov mention that "the main difficulty with the transducers . . . was the fact that the exposed copper electrodes deteriorated gradually, affecting the range and linearity of output and require frequent cleaning and recalibration." In addition, 60 cycle noise on the transducer output made measurements difficult.

Needle probes. The needle type conductance probe as designed by Solomon [13] was further developed and used by Griffith [14], Neal and Bankoff [15] and by Nassos [16]. This needle probe, shown schematically in Figure 4, consisted of a single 0.8 mm diameter steel wire completely insulated and bent at an angle of 90°, giving a point of $\sim$30 mm length. This needle formed the ungrounded conductor in a conductivity measurement system. The probe was inserted into a tube such that the point was aligned parallel to and opposing the flow streams. The non-insulated tip of the probe, when im-mersed in a conducting medium, would allow current to travel between the tip and a grounded electrode imbedded in the tube wall. Thus, the distribution of fluid between the probe tip and the grounded conductor on the wall determined the amount of electric current flow. In this manner, information on the distribution, number and velocity of bridges of liquid traveling in the tube, as well as qualitative information on flow re-gime, could be determined. This type of instrument seems also readily adaptable for the qualitative determination of the transition between slug flow and an-nular flow. Further, Neal and Bankoff describe a method of using this type of probe for the determination of local void fractions in mercury-nitrogen flow [15]. The non-wetting characteristics of mercury made nearly instantaneous make and break contacts possible, such that there was no surface tension effect on

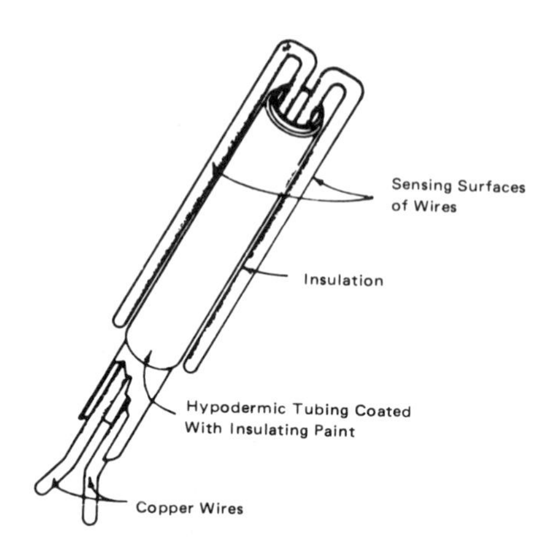

Figure 2 - Liquid Level Transducer of Kordyban and Ranov [12] (BNL Neg. No. 9-89-79)

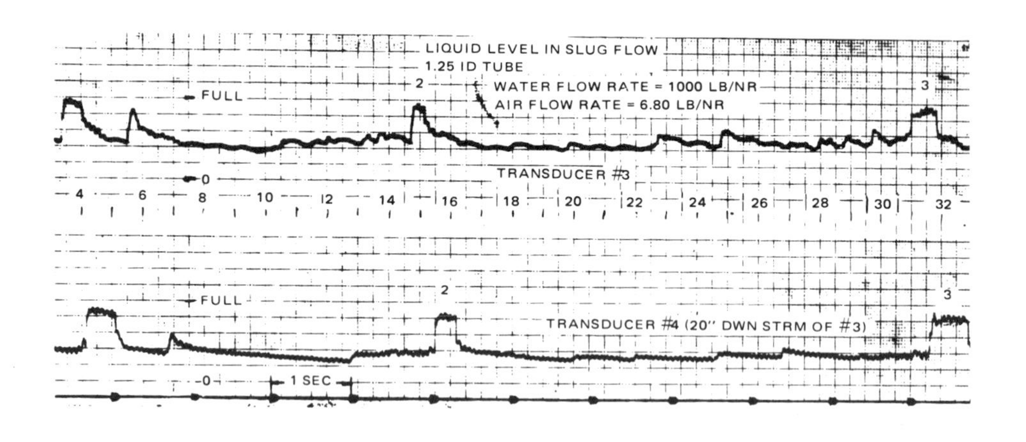

Figure 3 - Liquid Level Readings from Two Transducers at Different Axial Locations [12]. (BNL Neg. No. 9-93-79)

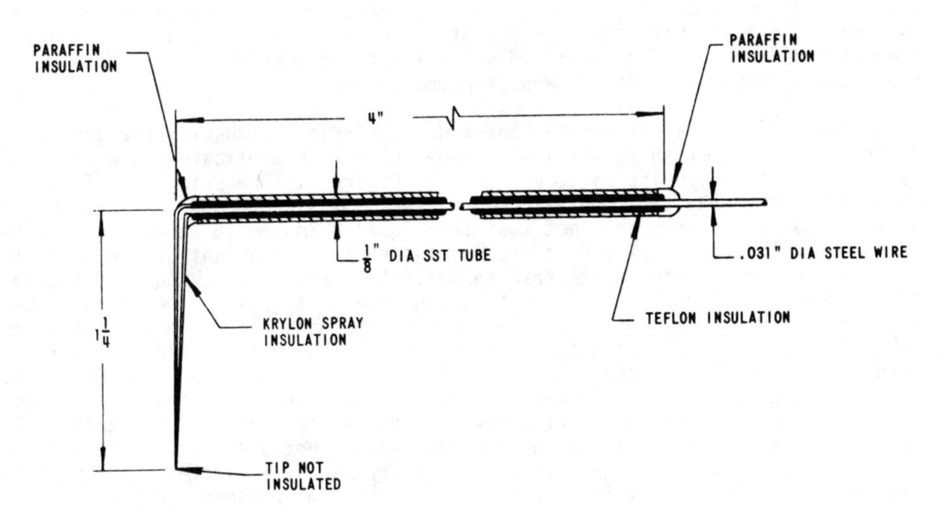

Figure 4 - Electrical Conductivity Probe Developed by Soloman [13]
(BNL Neg. No. 9-99-79)

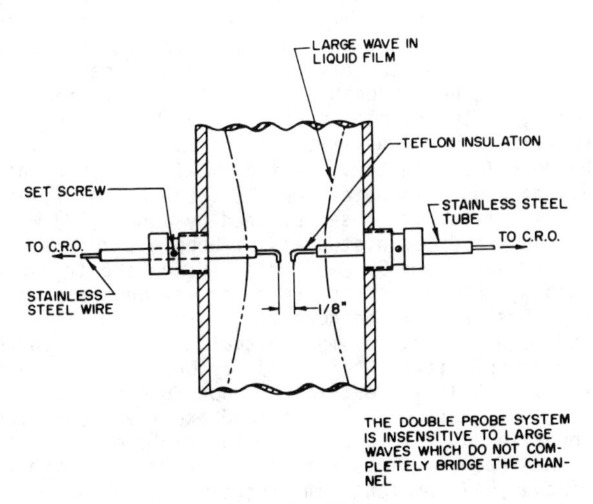

Figure 5 - Double Conductivity Probe Arrangement tried by Wallis [17]
(BNL Neg. No. 9-120-79)

void fraction measurements. Nassos, on the other hand, when attempting to adapt this to an air-water system, experienced difficulty due to the surface tension effects at low temperature [16], which today are well accepted. In other words, when entering a small void, the gas phase would deform, keeping the probe tip within the conducting medium. Thus, the measurements for vapor fractions were low. Nassos stated that some improvement was obtained by means of a separate triggering device but the calculated vapor fractions were still somewhat lower when compared with gamma-ray attenuation measurements.

Wallis [17] used a similar mechanism where, instead of having the ground electrode attached flush to the inside tube surface, a duplicate probe was inserted from the opposite wall, as shown in Figure 5. The principle of operation was exactly the same as that used in the Nassos probe and in Kordyban and Ranov's probe except for the fact that the liquid bridge must be between the two electrodes which can be placed at any point covering any amount of the channel diameter. An interesting fact arose in Wallis' experiments. When he compared data on the transition between slug flow and annular flow with the Nassos-type probe on the one hand, and his double probe on the other hand, in all cases having similar conditions in the test assembly and the same transition criteria, he found a definite discrepancy between the two sets of data as shown in Figure 6. The discrepancy was due to the way in which the slug-annular transition was defined, both conceptually and implicitly through the type of instrumentation used. Certainly if a liquid conduction path exists between the tube centerline and the wall, and if the signal is a pure binary signal regardless of the path, then there should be no difference between transition from one probe and another. However, if large globules of liquid float down the core of the pipe in annular flow, these would appear to be bridges to the dual sensor but not be seen by the single probe. Conversely, if the conductance path for the single probe is tortuous, and the signal not truly binary in nature, then subjectiveness on the part of the observer and arbitrariness must result in variations in one's estimation of transition over another. In spite of these drawbacks, a number of other workers have used both the single wire probe [18-23] and the double probe [24-26], mainly due to its simplicity.

Considerable other information may be obtained from conductivity devices. If local velocity is obtained, bubble diameter distributions can also be measured [48-50] as well as gas and liquid slug lengths [48]. Uga [51] obtained histograms of bubble sizes in the riser and downcomer sections of a BWR. However, he used average values of bubble velocities rather than instantaneous values corresponding to the dewetting signals so that his size distributions were really normalized inverse dry-time distributions where a constant value of velocity was the normalized factor. Similarly, Ibragimov et al. [52] used miniature traversing probes to obtain bubble frequency profiles in water-nitrogen flow similar to those obtained by Jones [53] with the anemometer. Sekoguchi et al. [49] reported histograms of bubble sizes using a double-tipped probe with tips 4 mm apart and 30 μm in dia. Using the transport time averaged over 10 observations, the bubble rise velocity was obtained which was then used to obtain the bubble size from the dewetting time. Similar to the method of Uga [51], this method requires the assumption that there is no correlation between bubble size and rise velocity. While this may be true for the > 1 mm bubble observed by Sekoguchi [49], it certainly is not in general. It is interesting to note, however, that these workers obtained excellent agreement between void residence time profile and the volume fraction profile calculated from the bubble diameter and frequency measurements, in agreement with the theoretical analysis by Serizawa [54] for long sampling times.

In their studies of thin film characteristics of annular two-phase flow, Telles & Dukler [55], Dukler [56], and Chu & Dukler [57] have used the straight

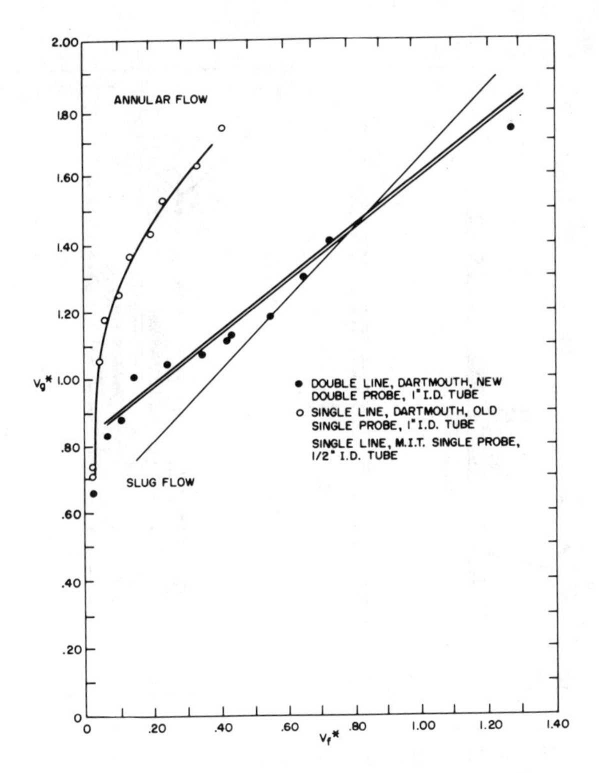

Figure 6 - Comparison of the Slug-Annular Flow Transition as Determined by the Single Conductivity Probe and by the Double Probe Method. (BNL Neg. No. 9-108-79)

electrical probe to determine information on wave structure. Such information has included probability density functions for wave height where the contributions due to the film substrate and due to large waves have been identified along with wave frequency. Also the spectral and cross spectral densities of film thickness fluctuations were determined.

Other statistical data can be obtained by using a double probe. Several authors tentatively described the granulometry of bubble flows with sophisticated models [51, 58, 54]. A bubble displacement velocity has also been looked for by the same authors and also by Lecroart & Porte [59], Kobayasi [40], and Galaup [50].

Transport methods have now begun to be utilized to obtain velocity information, although no author to date has specifically identified recognition that the correlated quantity is interfacial passage so that the velocities thus obtained are interfacial velocities. Serizawa [54], using a double-tip probe with tips 5 mm apart utilized both correlation and pulse-height methods to determine bubble velocity spectrums. Correlation of the outputs of the two probes after passing through Schmitt triggers provides a function which exhibits a well defined maximum at a time delay corresponding to the transport time

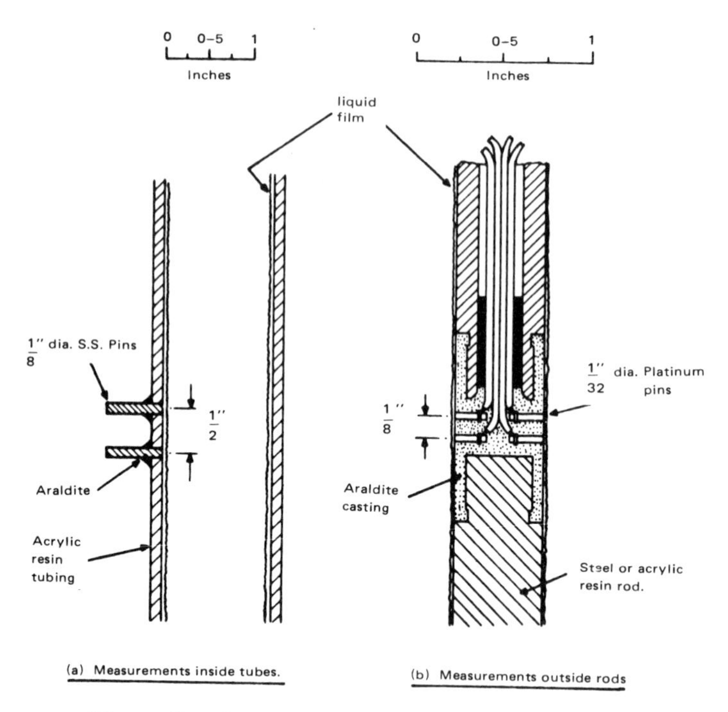

Figure 7 - Installation Characteristics of the Harwell Flush-Mounted Conductance Probes. (BNL Neg. No. 9-90-79)

between probes. Dispersion of the amplitude of the correlation function is representative of the probability distribution of this velocity, and hence of the fluctuations. Also, by using one probe signal as a starter and the other as a stopper, ramp functions are generated during the transport time which, when stopped and differentiated yield pulses whose heights are proportional to the transport time delay. Height analysis of this pulse train is assumed to yield the bubble velocity spectra.

Wall probes. A second type of conductivity instrument is a flush-mounted type developed at the Atomic Energy Research Establishment (AERE) at Harwell, England, and used in a number of studies in that Laboratory [27-32]. This device, which is shown in Figure 7, can be mounted for use in both tubes and rods. The operational principle is similar to that previously discussed, except for the fact that a high frequency alternating current (AC) carrier is used. Thus, when the probes are covered by a thin liquid film, the signal is amplitude-modulated by the film itself, such that the output is related uniquely to the film thickness. This type of probe was used to determine the transition between slug and annular flow and to measure film thicknesses in annular flow when the probes are calibrated. A similar type of sensing device was also used at AECL in Canada [33, 34] where a kicksorter circuit, shown in Figure 8, was used to ignore all waves of thicknesses less than a predetermined value. In this manner, a wave thickness spectrum may be determined. The signal can also be halfwave rectified as shown in Figure 9 and combined as shown in Figure 10,

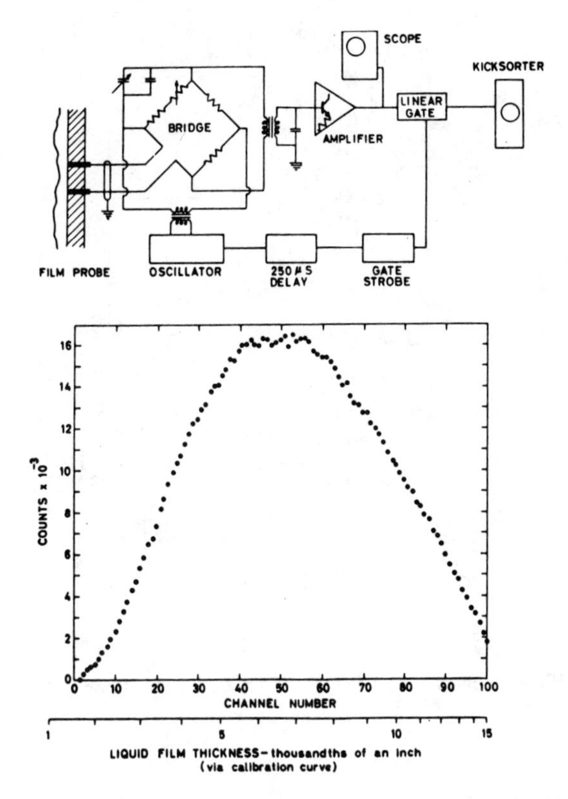

Figure 8 - Kicksorter Circuit and Representative Liquid Film Amplitude Distribution [33]. (BNL Neg. No. 9-97-79)

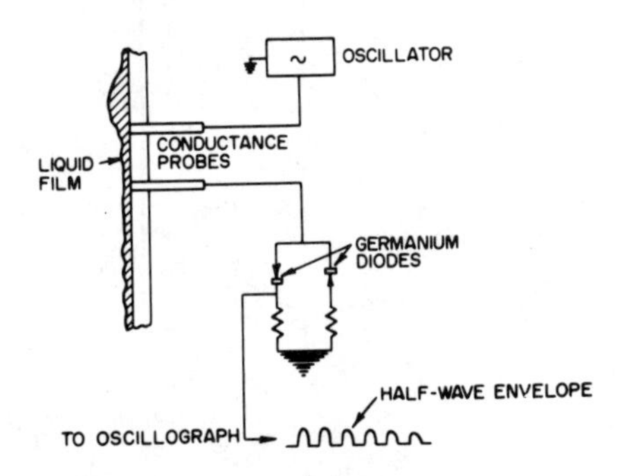

Figure 9 - Method of Half-Wave Rectification for Recording of Multiple Signals [35]. (BNL Neg. No. 9-113-79)

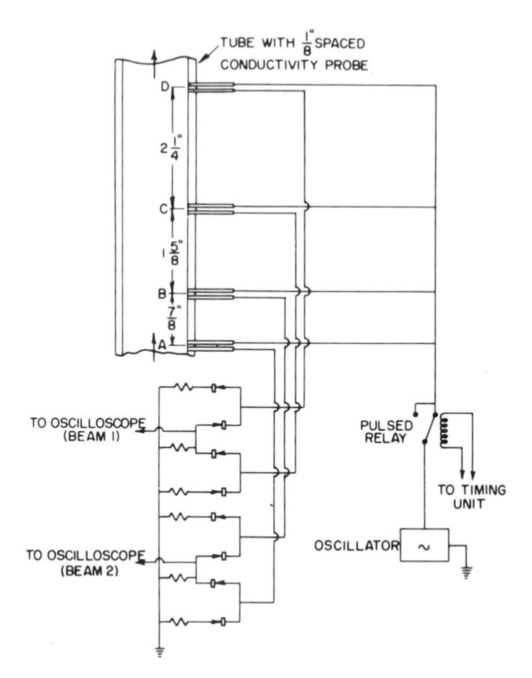

Figure 10 - Probes and Circuit for Multiple Probe Recording [35].
(BNL Neg. No. 9-105-79)

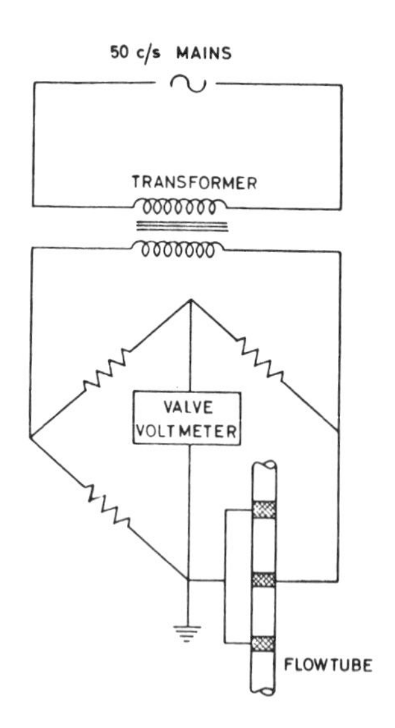

Figure 11 - Schematic of Film Conductance Method Developed at C.I.S.E.
(Ref. 36) (BNL Neg. No. 9-86-79)

such that the recordings from four separate measuring stations could be obtained simultaneously on, ray, a dual-beam oscilloscope [35]. Then, by comparing the outputs from probes at different stations, and keeping in mind that the flow patterns tend to retain their identity along the length of the tube, the velocities and frequencies of the waves may be determined.

Another type of surface conductance probe was developed at Centro Informazione Studi Esperienze (CISE) [36]. As shown in Figure 11, the tube is entirely surrounded at various axial locations by metal rings flush with the inner tube surface. Whenever a potential is generated between any two rings, the conductance will be a function of the amount of fluid between the rings. However, this type of instrument is insensitive in annular flow to rapid changes in film thickness and high, sharply peaked waves. Reference 37 describes a test assembly shown in Figure 12 which was used to compare both the AERE and CISE type conductance probes. In addition to insertions of both the CISE and AERE conductance devices, the test assembly contained quick-acting, simultaneously operated valves on both ends which enabled the air water mixture to be completely isolated so that a measurement of the total amounts of each phase could be made. The results from these tests are shown in Figure 13. In this figure, several things are immediately evident. First, the CISE probe gave measurements definitely lower than those produced by the AERE instrument. This is probably due to the fact that the CISE probe is not sensitive to sharp or high wave peaks thereby underestimating the average film thickness. This is evidenced by the approach of the CISE data to the AERE data with increasing film thickness, i.e., where the wave peaks start to make less and less difference in the overall film thickness. Secondly, the holdup method appears to show film thicknesses which are larger on the average than the AERE measurements. This is because the holdup method not only measures the liquid in the film itself, but also that the liquid which has been entrained in the gaseous core. In some instances, a large amount of the total liquid in the system is in the form of entrained droplets. Thus, any film thickness measurement which does not separate the entrained liquid from the film liquid must necessarily be overestimating the average film thickness. It is also probable that the AERE probes miss large waves thereby underestimating the average and instantaneous film thickness.

Problems encountered in the use of these instruments show that, as mentioned previously, wave peaks are not seen by the CISE instrument and, in some cases, not by the Harwell instrument. This may make the CISE instrument underestimate the average film thickness by as much as 40% in some cases. In addition, these instruments both require conducting mediums and as such may be difficult to use in systems having high resistivity water chemistry requirements. The AERE method requires an extremely, well-designed geometry mockup for calibration of the instrument. Any misalignments or nonconcentricities in the calibration unit itself can have a significant effect on the calibration.

An excellent review of conductance measuring devices is given in Reference 37. In general, the characteristics of such devices have restricted their use to studies where the continuous phase is a conducting medium, their major application being in film thickness studies. Bergles and his coworkers at Dynatech had some success in the use of needle-type probes as flow regime-detectors [60] but the previous discussion seems to lend some uncertainty to this application.

Conductivity probe summary. Conductivity devices of both the probe-type and the wall type have been used for many years for various purposes including local void fraction (residence time fraction), flow regime indication, film thickness wave celerity, etc., and when coupled with a velocity indication, bubble size

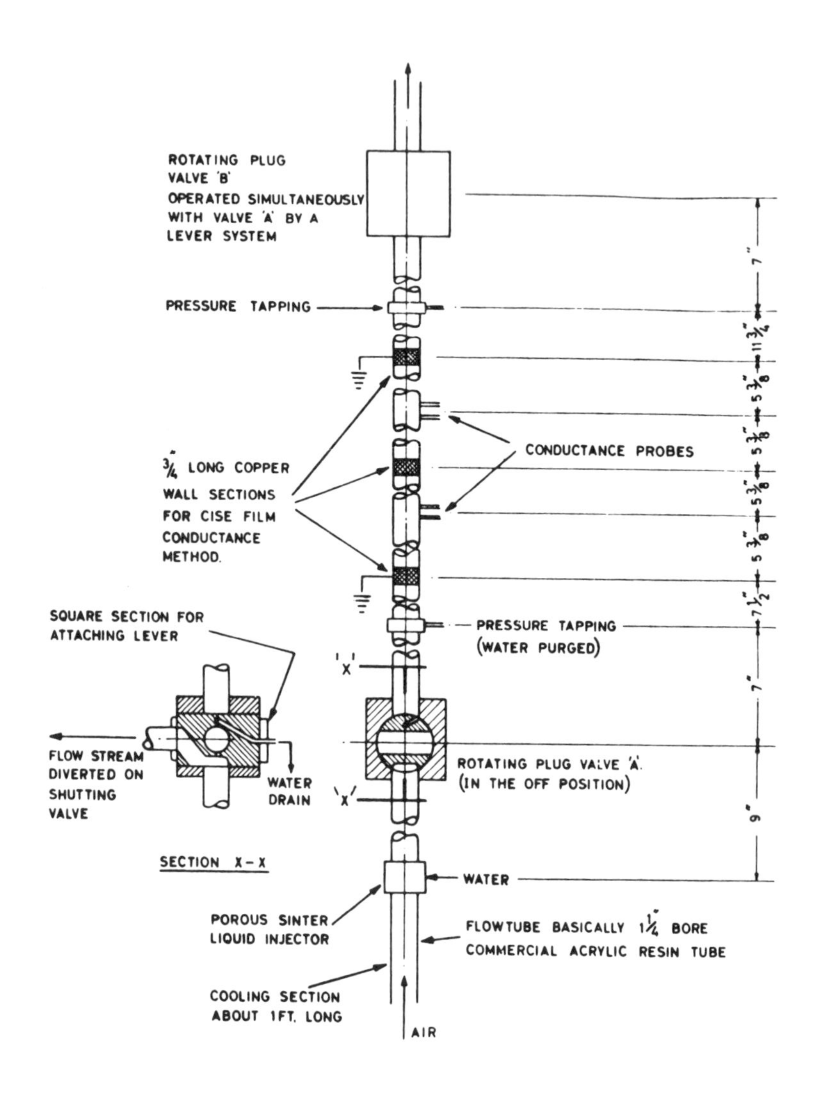

Figure 12 - Layout of Test Assembly used at Harwell to compare C.I.S.E. and Harwell Probes [37]. (BNL Neg. No. 9-88-79)

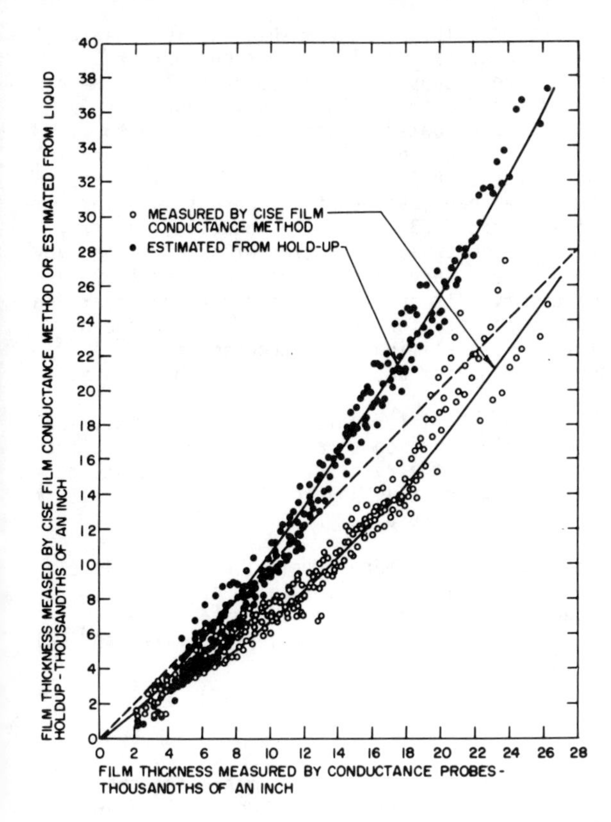

Figure 13 - Comparison of Data Obtained by C.I.S.E. and Harwell Conductance Probes [37]. (BNL Neg. No. 9-106-79)

and slug length distributions. Two used in tandem can also be interpreted to give a velocity indication. It should be understood that since the probe type device is basically a phase indicator, any transport information interpreted as a velocity between two points is really propagation of a phase delimiter or interface, ie.: interfacial velocity. As such, it is subject to all the effects which tend to produce errors at a single probe and may as well include selective distortion at one probe vs. another. In addition, interface velocities may or may not be a good indication of a phasic velocity, being subject to such things as biasing due to slip and evaporation.

General difficulties with electrical sensors include phase distortions due to probe insertion, electrochemical effects, nonrepresentative conductive liquid paths for electrical current, electrical field distortions, variable conductivity due to thermal or chemical effects, nonlinear response, and, if these aren't enough, inadequate physical interpretation of the results and resulting erroneous conclusions. Specifically, problems associated with this method include those due to probe wettability and surface tension, as well as bubble trajectory. Boundary contact times were noted by Sekoguchi et al. [45] to be 100-200 μs resulting in errors of up to 10% in void measurements and Jones [53] and Jones & Zuber [69] measured similar boundary times and errors with a 50 μm hot film anemometer. Generally speaking, more work is needed on this difficult problem, especially on the physical significance of the delay times measured with a double probe.

It has been seen that one of the principal features which differentiates the electrical circuits is the type of electrical supply, direct current or alternating current. With direct current supplies, electrochemical phenomena are encountered which obscure the desired signal unless low voltages are maintained. These, however, lead to complicated electronics and tend to yield poor signal-to-noise (s/n) ratio's. In addition, electrochemical deposits in low speed flows give alterations in the signals although at high velocities the sensors may remain clean. On the other hand, alternating current supplies generally eliminate electrochemical effects [38-40] while substituting stray capacitative effects. One must insure suitable separation of supply frequency from phenomena frequency. In some cases, very high frequencies over 1 MHz are required resulting in complex circuitry. It should be noted that Reocreux and Flamand [41] reported on a method using very low frequency, for high speed flows, to resolve this difficulty and yet eliminate electrochemical difficulties. Pseudo direct current behavior was obtained every half wave.

Finally, Tawfik [42] has shown that the interaction between the electrical field and the liquid field during bubble approach to a needle probe can affect the signal and hence the interpretation of the results. This points out the fact that a thorough knowledge of the interactive physics is very important in obtaining a good evaluation of experimental results.

In spite of the difficulties involved, the general simplicity of this class of measurement devices has resulted in their continued use in many fields, especially where qualitative diagnostic information is needed [43-46,54]. Continued development for improved understanding is, however, required and will no doubt be undertaken as the needs arise.

3.2. Impedance Void Meters

The second type of instream electrical measuring instruments are the impedance type void sensors [60]. While these may be considered a subset of the previous types, significant differences in geometry exist, and these devices are usually always used in the impedance mode relying on phase variations of the capacitative reactance. Several types have been designed and used. The first type is a concentric cylinder meter which, shown in Figure 14, consists of concentric, thin-walled short cylinders which are alternately connected to one of two electrodes. When connected to an alternating current supply, the relative impedance may be measured as a function of the vapor fraction. The void meter may be calibrated by means of a system such as is shown in Figure 15. Measured amounts of water and air are injected into the base of a vertical section. The water velocity is measured initially by means of a turbine meter. In the upper portion of the channel, another turbine meter is placed immediately downstream of the impedance void gage. Thus, the mixture velocity, coupled with an accurate knowledge of the air and water flow rates, may be used to calculate the void fraction. This calculation is then compared with the impedance of the void meter and used as a calibration point. In this manner a curve of relative impedance versus vapor fraction can be easily obtained. There does, however, seem to be a sensitivity on the void distribution between the plates as shown in Figure 16. This is in accordance with established theory [61,62]. Here there may be an effect amounting to as much as 15% or more depending on whether the voids are distributed horizontally or vertically with respect to the parallel plates of the void meter, and on the relative dielectric constants for the two phases.

Similar devices have been developed and used by a number of other investigators [63-67]. Marked effects due to void distributions at constant

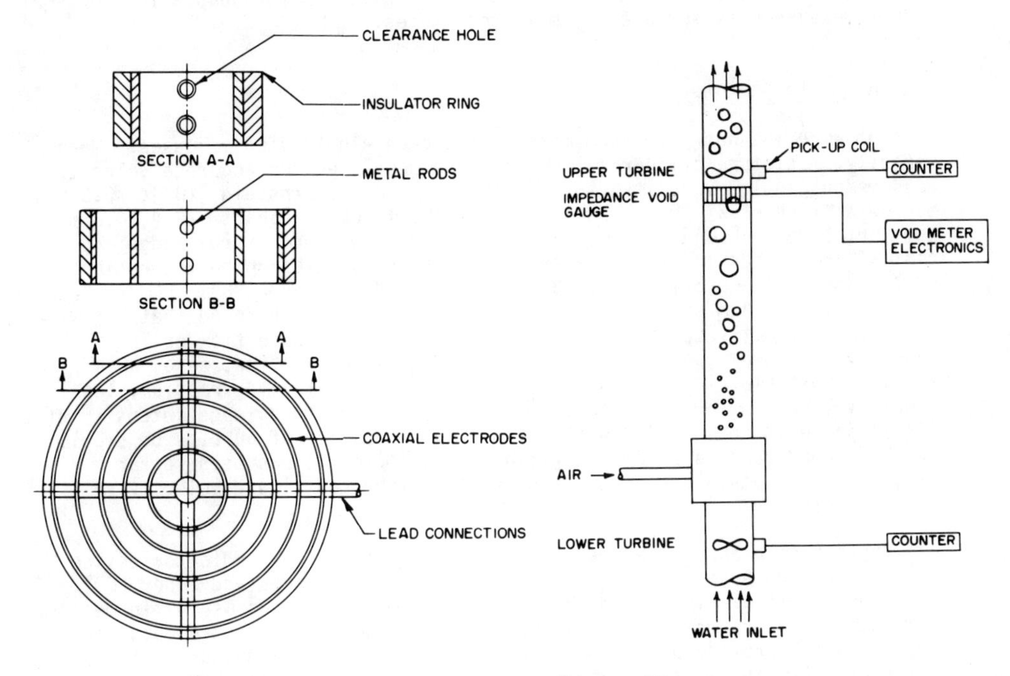

Figure 14 - Coaxial Impedance Void Meter [60]
(BNL Neg. No. 9-119-79)

Figure 15 - System for Calibrating Impedance Void Meter [60]
(BNL Neg. No. 9-121-79)

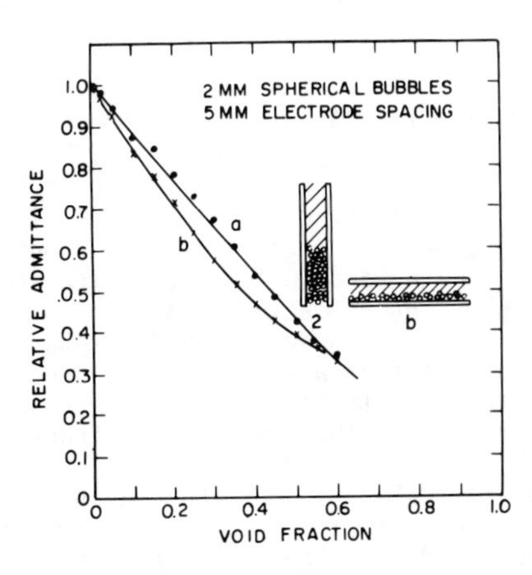

Figure 16 - Effect of Void Distribution on Impedance Void Meter Output [60]
(BNL Neg. No. 9-122-79)

void fraction are also confirmed by Cimorelli and Premoli [66]. In general these sensors are bulky and the methodology unsuitable for adaptation for small volume measurements for use in small geometries.

3.3. Hot Film Anemometer

Hot wire and hot film anemometers have been widely used in gases. More recently, miniature cylindrical probes have been used for accurate velocity measurement in low velocity (for instance, in water, Ornstein [181], <0.5 m/s; Morrow & Kline [182] <0.25 m/s in mercury; Hollasch & Gebhart [183] <0.6 m/s, and Hurt & Welty [184] , <0.05 m/s. The larger and more sturdy wedge and conical probes has enjoyed greater success at velocities up to the region of 5 m/s (Rosler & Bankoff [185], <3.7 m/s; Bouvard & Dumas [186], <5 m/s; Resch & Coantic [187], <4 m/s; Resch [188], <4 m/s). It has been found that hot-wire or hot-film anemometry can be used in two-component two-phase flow or in one-component two-phase flow with phase change. In the first case, an air-water flow, for example, it is possible to measure the local void fraction, the local liquid volume flux or instantaneous velocity and the turbulence intensity of the liquid phase in conjunction with the arrival frequency of bubbles or droplets. In the second case, a steam-water flow for example, it has been so far impossible to obtain consistent results on calibrated liquid velocity measurements.

The hot film anemometer was tried as a two-phase flow indicator device by Hsu, Simon, and Grahm [68], Jones [70], and later by Jones and Zuber [69]. The study by Hsu et al. indicated that this instrument could be useful in measuring all the parameters measurable by the previously described instruments and, additionally, provide both temperature and phase velocity measurements as well. The reason why this instrument has not received wide acceptance to date is because of its extreme fragility and short life time. To explain the principles of operation of this instrument, Figure 17 is used. The close-up view of the probe shows that it consists of a small diameter glass cylinder covered with a platinum coating and connected on either end to a copper lead wire. In operation, the resistance of the probe was set by means of an electrical current at a value corresponding to a desired probe temperature. If the probe resistance began to change from the preset value due to a change in temperature, the control unit would respond with a change in current to maintain a constant probe resistance. This setup is shown schematically in Figure 17 where the probe is one arm of a four-arm bridge. The output of the transducer was measured as a voltage drop across a calibrated resistor by an oscillograph.

A typical output trace for this instrument in different flow regimes taken from Reference 68 is shown in Figure 18. Here it is possible to compare the traces from bubbly flow, slug flow, and mist flow with low droplet concentration. As seen in this figure, the different flow regimes are readily discernible one from the other. The only possible conditions that might cause confusion are bubbly flow with large bubble concentration as opposed to mist flow with very rich droplet concentration. As seen in Figure 18, these two traces could look quite similar; however, one distinguishing feature would be the relative point on the scale at which the trace occurs. In bubbly flow, there would be expected to be a larger power dissipation on the probe than there would be in the mist flow due to the smaller void fraction.

Figure 19 shows the typical trace of a bubble as it passes the probe. From bottom to top in this figure, as the bubble approaches the probe, the vanguard consists of locally increased velocity as shown by increasing power dissipation.

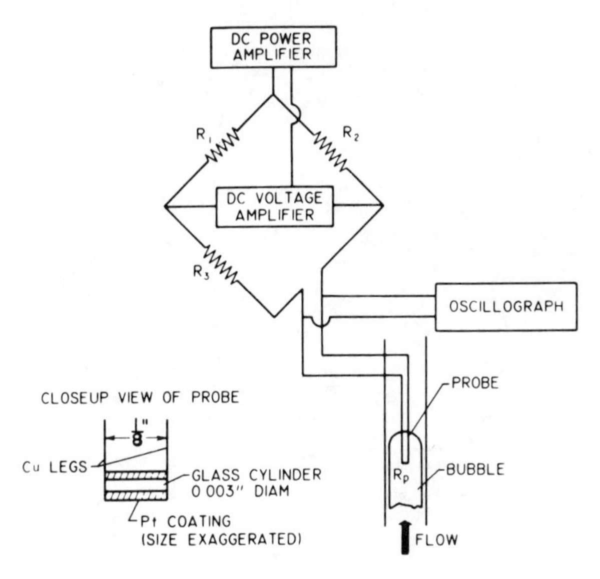

Figure 17 - Hot Film Anemometer System Used by Hsu, Simon, and Grahm [68]
(BNL Neg. No. 9-104-79)

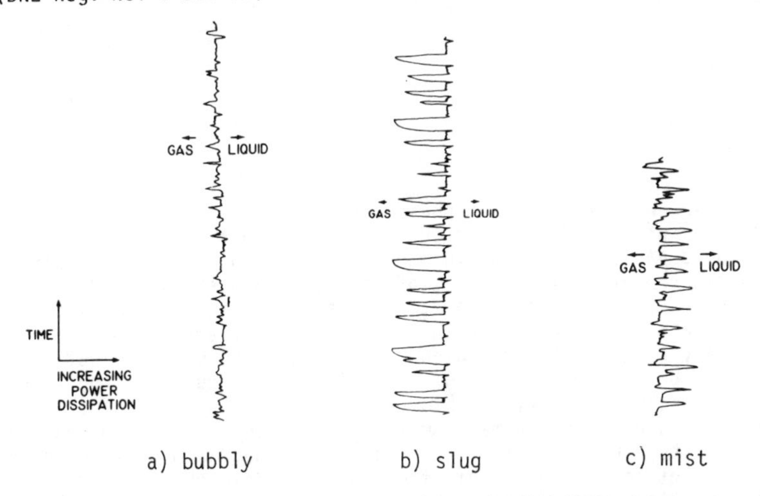

a) bubbly b) slug c) mist

Figure 18 - Typical output from the Hot Film Anemometer [68]
(BNL Neg. No's (a) 9-103-79; (b) 9-85-79; (c) 9-84-79

The entrance of the probe into the void is shown by a sharp decrease in power dissipation followed almost immediately by an increase as it enters the wake. The same effect happens again with another bubble and then, as the probe exits from the effect of the wake, the power dissipation decreases corresponding to the decreasing local velocities. In addition, since there is apparently a relatively rapid response of the transducer when entering a void, this instrument was expected to give good quantitative information on local vapor fractions.

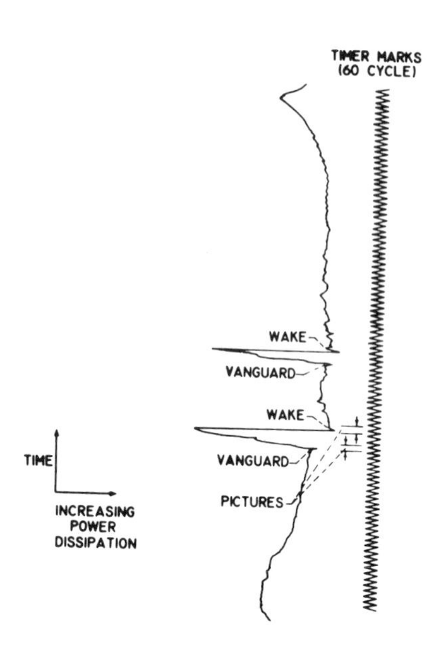

Figure 19 - Wake and Vanguard Effects Due to Bubble Passage by a Hot Film Anemometer [68] (BNL Neg. No. 9-96-79)

Measurements in two-component two-phase flow without phase change. Hot-wire anemometry has been used for measuring the concentration flux and the diameter histogram of liquid particles moving in a gas stream. Goldschmidt & Eskinazi [189,190] measured the arrival frequency of liquid droplets, 1.6 to 3.3 μm in dia, with a constant temperature anemometer and a cylindrical probe, 4.5 μm in dia. When the impaction frequency of the droplets is different from the energetic frequency range of the turbulent gas stream, the signal fluctuations due to impacts can be distinguished from the fluctuations due to turbulence. In experiments by Goldschmidt & Eskinazi [189,190], results showed that the ratio of the impaction frequency to the maximum impaction frequency was insensitive to the threshold amplitude of a discriminator used to produce a binary chain of pulses due to droplet impactions. This fact has also been observed by Lackme [191] for void fraction measurements with a resistive probe. Ginsberg [192] used the same technique to study liquid droplet transport in turbulent pipe flow. Goldschmidt [193] determined that the measured impaction rate is lower than the true value but proportional, and should thus be calibrated against another technique.

In determining droplet diameter histograms, Goldschmidt & Householder [72,194] theoretically found a linear relationship between particle diameter and cooling signal peak value which was verified experimentally for droplet diameters lower than 200 μm. Bragg & Tevaarverk [73], however, contradicted these results and concluded that the hot wire was unsuitable for this purpose. This conflict has yet to be resolved.

Time-averaged gas velocities as well as gas turbulent intensities were measured by Hetsroni et al. [204] in low concentration mist flow. The spikes due to the impingement of the liquid droplets on the hot-wire were eliminated with the help of an amplitude discriminator and a somewhat simpler electronic circuit than that of Goldschmidt & Eskinazi [189,190]. The resultant signal was used to obtain time averaged gas velocity and turbulent intensities. Chuang & Goldschmidt [195] employed the hot-wire as a bubble size sampler by theoretically investigating the nature of the signal due to the traverse of an air bubble past the sensor.

Despite several difficulties arising in droplet granulometry determination, the hot-wire has successfully been employed for studying the turbulent diffusion of small particles suspended in turbulent jets by Goldschmidt et al. [196]

Following the studies done by Goldschmidt in aerosols and by Hsu et al. [68] in steam-water flow, and the preliminary work of Jones [70], a thorough investigation of the hot-film anemometry technique in two-phase flow was carried out by Delhaye [77,78] using a conical constant temperature hot-film probe which has three major advantages over the cylindrical hot-film sensor: small particulate matter carried with the fluid does not attach to the tip, bubble trajectories are less disturbed, and the relatively massive geometry is less susceptible to flow damage at higher velocities. The maximum overheat resistance ratio of the probe of 1.05, (ratio of operating resistance to the resistance at ambient fluid temperature), was suggested by Delhaye [77,78] to avoid degassing on the sensor. This corresponded to a difference of 17°C between the probe temperature and the ambient temperature, significantly below saturation temperature. Jones [53], and Jones & Zuber [69] found little difference between resistance ratios of 1.05 and 1.10 insofar as degassing on their 50 μm dia cylindrical sensor was concerned, and chose the latter for increased sensitivity. Degassing in their system was found to occur in operation following failure of the 8000 Å-thick quartz coating over the platinum film. This failure occurred during forced resonant vibration of the sensor caused by vortex shedding at velocities over 1.5 m/s. Degassing caused the calibration to be unstable only at velocities less than $\sim$30 cm/s.

Delhaye [74-77] developed some rather sophistical anemometry techniques for measuring void fraction in large tubes with the rugged, conical, hot film probe. Using a multichannel analyzer, he obtained amplitude histograms of the voltage signal produced for a constant temperature sensor control system in a two-phase mixture. Referring to Figure 20, one may see that in the ideal case, (a), the anemometer voltage is either at the voltage corresponding to the presence of liquid at a given velocity, E_{ℓ}, or at the voltage corresponding to all gas, E_g, at the given gas velocity. The multichannel analyzer periodically determines the amplitude of the signal voltage. A block of memory locations is allocated where each location represents a different voltage level range of width ΔE. The first memory location represents the voltage range E_{min} to $(E_{min} + \Delta E)$ while the last in the block represents $(E_{max} - \Delta E)$ to E_{max}. At the time the voltage level is determined by the analyzer, one count is added to the memory location representing the voltage range within which the current amplitude was found to be. In this manner, each location or channel, will contain a count proportional to the fraction of the total sampling time during which the voltage level was within the range assigned to that channel. For the ideal case, then Figure 20 shows that the histogram on the right would contain N_{ℓ} counts for liquid voltage amplitude and N_g counts for gas voltage amplitude. The percentage of time the probe "sees" the gas would then be

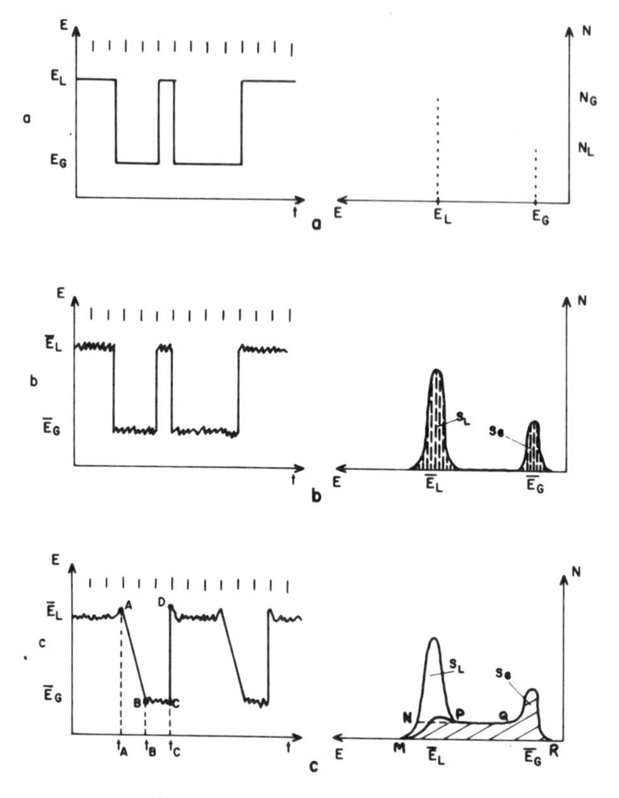

Figure 20 - Delhaye's Method for Local Void Fraction Using Multi-Channel Analysis [76] (BNL Neg. No. 9-94-79)

$$t_g = \frac{N_g}{N_g + N_\ell} \tag{2}$$

and could be taken as a measure of the local void fraction, α .

In the somewhat less ideal case where each voltage level has some fluctuations associated with it but still the switching from all one-phase to the other occurs instantaneously, (b). The amplitude histogram would show a cluster of vertical lines centered about E_ℓ and another cluster about E_g, the height of each line being the number of times the particular voltage level was encountered during periodic sampling. Thus, if all the lines clustered about E_g are associated with the gas, and all the lines clustered about E_ℓ associated with the liquid, the void fraction would be

$$\alpha = \frac{\sum N_{gi}}{\sum N_{gi} + \sum N_{\ell i}} \tag{3}$$

For this case it becomes convenient to begin thinking in terms of a continuous amplitude spectrum, N(E), such that in the limiting case where the voltage intervals vanish (3) becomes

$$\alpha = \frac{\int_{A_g} N(E)dE}{\int_{A_T} N(E)dE} = \frac{S_g}{S_g + S_\ell} \qquad (4)$$

where the void fraction becomes the ratio of the gas histogram area to the total histogram area.

In the real case, the anemometer signal is more as shown in Figure 20 (c) where a finite time is required for the sensor to dry out (A-B) before becoming characteristic of all gas (B-C). In some instances the gas residence time may be shorter than the dry out time and the gas voltage is never reached before the liquid appearance forces the voltage back to E_ℓ. Thus, the region between S_g and S_ℓ will contain some counts and the voltage amplitude histogram will appear as shown at the right of 20 (c). Thus, the gas-caused area of the histogram would be the cross-hatched area shown in the figure having some counts for all voltage levels up to approximately the maximum for all liquid.

Delhaye calculated the void fraction locally by application of Equation (4) where in practice

$$\alpha = \frac{A_{MNPQR}}{A_T} \qquad (5)$$

To a first approximation, then, the local void fractions were calculated as the ratio of the hatched area to the total area which then compared favorably with radiation absorption methods (γ-rays). Notice that trigger levels for S_g and S_ℓ affect the result. Delhaye adjusted both to get accurate comparison between the resulting line-averaged void fraction and an independent γ-ray measurement but developed no specific formula governing these settings. The liquid time-averaged velocity and the liquid turbulent intensity are calculated with the nonhatched area of the amplitude histogram (Figure 20c) and the calibration curve of the probe immersed in the liquid. The same method has extensively been used by Serizawa [54] for measuring the turbulent characteristics and local parameters of air-water two-phase flow in pipes.

A different processing method was proposed by Resch & Leutheusser [197] and Resch et al. [198] in a study of bubble two-phase flow in hydraulic jumps. The nonlinearized analog signal from the anemometer is digitally analyzed. A change of phase is recognized when the amplitude between two successive extremes of the signal is higher than a fluctuation threshold level ΔE. In this way the liquid mean velocities and turbulence levels were obtained along with bubble size histograms. ΔE was chosen to be in a plateau region of ΔE versus measured void fraction.

Jones [53] and Jones & Zuber [69] used a discriminator applied to the raw anemometer signal to obtain a binary signal representative of local void fraction but found the cutoff level needed to be adjusted depending on the local velocity to a point just below the minimum value for a liquid. Even though the threshold value was set at every point in the traverse, errors in averaged void

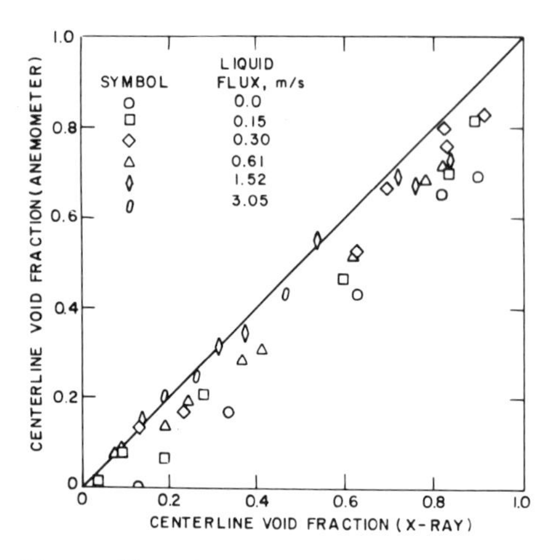

Figure 21 - Uncorrected Anemometer for Averaged Void Fraction (Jones [69]) (BNL Neg. No. 9-128-79)

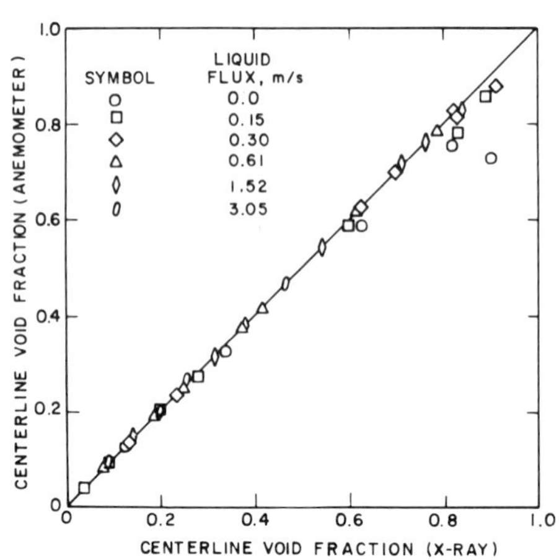

Figure 22 - Corrected Anemometer for Averaged Void Fraction per Equation (6) (Jones [69]) (BNL Neg. No. 9-129-79)

fraction were encountered when calibrated against an X-ray measurement. These errors were found to be dependent on the liquid volume flux and the mean void fraction.

Comparisons as shown in Figure 21 show significant errors exist, especially at low liquid throughput rates. By choosing a relationship between the corrected and measured void fraction as

$$\alpha_c(y) = \{1 + f_c(y)\}\, \alpha_m(y) + \alpha_{zc} \tag{6}$$

where

$$f_c(y) = C_1\ (1 - \alpha_m)\ (1 - \frac{4y^2}{s^2}) \tag{7}$$

and

$$\alpha_{zc} = \frac{C}{K_\alpha + j_\ell}\ (1 - \alpha_m)\ (1 - \frac{4y^2}{s^2}) \tag{8}$$

and averaging in y, a relationship between C_1 and the averages could be obtained as

$$C_1 = \frac{\bar{\alpha}_c - \bar{\alpha}_m - \bar{\alpha}_{zc}}{\overline{f_c \alpha_m}} \qquad (9)$$

Note that C and K_α for all data were found to be 0.0055 m/s and 0.028 m/s respectively, while the averages in (8) were found from the data. Good results were obtained as shown in Figure 22.

By counting the number of times the output of the discriminator changed from one level to another, Jones [2] also obtained local values for interface frequency. He also measured the liquid-volume flux directly by time averaging the linearized signal equal to the liquid velocity when the sensor was in liquid, and zero when the sensor was in gas. Liquid velocities were obtained by pointwise division of the measured liquid flux by the measured void fraction. The results were somewhat questionable, however, due to the cracking of the 8000 A-thick, quartz coating mentioned previously. No attempt was made to measure the turbulent fluctuations.

Serizawa [54] used a conical probe of much more sturdy construction and larger size, similar to that of Delhaye [77,78]. In bubbly and slug flow in air-water mixtures he used multichannel analysis techniques to obtain the frequency spectrum of the velocity signal including fluctuations up to $\sim$2 m/s. Ishigai et al. [199] used an anemometer to measure liquid film thicknesses.

Measurements in one-component, two-phase flow with phase change. The earliest paper on hot-wire anemometry in two-phase flow seems to have been published by Katarzhis et al. [200]. This preliminary and crude approach was followed by the work of Hsu et al. [68]. These authors, by comparing the signal with high-speed movies concluded that hot-wire anemometry was a potential tool for studying the local structure of two-phase flow, in particular for determining the flow pattern and for measuring the local void fraction. Hsu et al. [68] specified that in steam-water flow the only reference temperature is the saturation temperature. If water velocity measurements are carried out, the probe temperature must not exceed saturation temperature by more than 5°C to avoid nucleate boiling on the sensor. Conversely, if only a high sensitivity to phase change is desired, then the superheat should range between 5°C and 55°C causing nucleate boiling to occur on the probe when the liquid phase is present, and a resultant shift to forced-convective vapor heat-transfer when the vapor phase is present.

The low electrical conductivity of Freons enables bare wires to be used instead of hot-film probes. Shiralkar [201] used a 5 μm, boiling tungsten wire with a very short active length (0.125 mm) so that the whole active zone would generally be inside a bubble or droplet. Local void fraction was determined by an amplitude discriminator with an adjustable threshold level. For void fraction lower than 0.3 the threshold was set just under the liquid level whereas for high void fraction (0.8), it was set just above the vapor level. For void fractions ranging from 0.3 to 0.8 the threshold was set half-way between the liquid and vapor levels. The method was subsequently applied by Dix [202] and Shiralkar & Lahey [203].

Anemometer summary. The major advantages of the hot film sensor over the needle-type conductivity probe is that the sensor is self-contained not requiring a secondary electrode and is, therefore, not limited to use in specific flow regimes. This type of sensor appeared to be responsive to all flow conditions and apparently provides information on velocities and temperatures as well as void fractions. The major drawback with this type of sensor seemed to be the high initial cost for electronics and probes, and the general fragility of the sensing elements. In addition, an independent method must be utilized for calibration purposes which to date has been a chordal X-ray measurement of the void fraction. The line-averaged value obtained from the anemometer may then be compared with the X-ray measurement and parameters adjusted to yield suitable agreement.

3.4. Radio Frequency Probe

A relatively new development is the use of separate, small, transmitting and receiving antennas with radio frequency signals amplitude nodulated by the dielectric coefficient of the surrounding media [50,80-82]. The r-f probe developed at Brookhaven National Laboratory is shown in Figures 23 and 24 and consists of two 0.25 mm diameter insulated wires, with each wire encased in a 1 mm outside diameter stainless steel tube which were electrically connected to a common ground and acted as an electrical shield. The two shielding tubes themselves were encased in a larger stainless steel tube, which acts as a holder and provides rigidity. The sensitive part of the probe, the probe tip, was formed by extending the two insulated wires by $\sim$3 mm from the end of the shielding tubes. To prevent water from entering into the stainless steel tubes, each of the end connections were covered with a thin layer of epoxy including the tip of the two insulated wires which were also covered to insulate them from the surrounding media, water or air. When a d-c voltage was applied across one of the wires and the common ground, zero voltage was measured across the second wire and the ground. In operation of the probe, one of the wires was used as an emitter to which a sine wave was applied from a function generator. The second wire was used as a receiving antenna, and its output was fed directly to an oscilloscope or to a magnetic tape recorder after amplification. Similar r-f probes were previously described in the literature [50,81] but a systematic study of the response characteristics was not undertaken.

When a sine wave was applied to the transmitting antenna (input), the amplitude of the received signal (output) varied with the signal frequency of the input (from 100 Hz to 10^7 Hz) both in air and in water, Figure 25. Depending on the input frequency, the signal amplitude in water can be higher than the signal in air or vice versa . The r-f probe seems to act as a band pass filter. For a 500 kHz, 22.7 peak-to-peak sine wave input, the output voltage of the r-f probe was also observed to be dependent on the static immersion depth of the insulated nonshielded portion of the probe tip into the water. The output increased linearly with the immersion depth, reached a maximum and then decreased and leveled off at the all water signal level. An additional fact observed was that the output vs. input curve as presented in Figure 25 depends on the tube or pipe diameter in which the probe is immersed. Thus before undertaking any application of this probe for a specific geometry, a careful signal optimization with input frequency should be performed.

The probe with a 500 kHz, 22.7 v peak-to-peak sine wave input was also checked during the passage of bubbles with known velocities and lengths. Figure 26 presents the detailed output of the r-f probe during the passage of a bubble (obtained digitally with 20 μsec resolution). The output decreases from its water level to the air level with the penetration into the bubble and stays almost constant during the passage of the bubble. When the water impinges on the probe tip again, the signal increases, passes through a maximum, decreases, and then levels off at the steady air level. A possible explanation for this maximum was proposed by Fortescue [83] as being due to the additional capacitance of the water surrounding the insulated unshielded wires. Grounding the

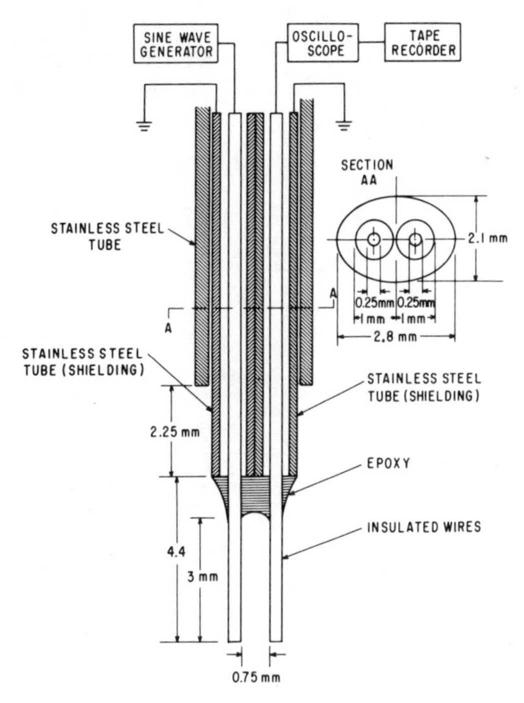

Figure 23 - Schematic representation of the r-f probe. (BNL Neg. No. 9-1493-78)

Figure 24 - r-f probe. (BNL Neg. No. 3-524-79)

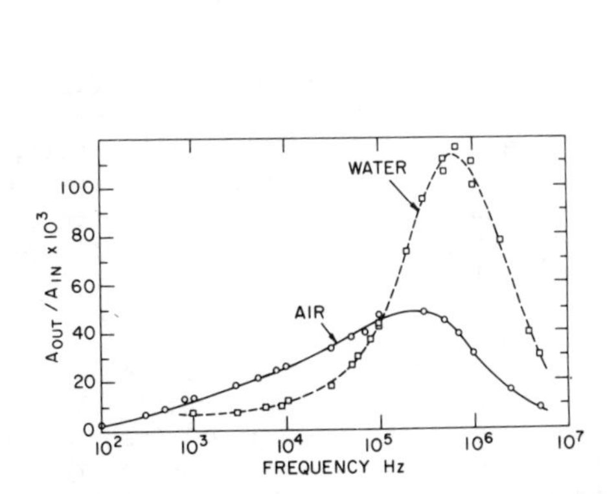

Figure 25 - Ratio of r-f probe output to input voltage level as a function of input sine wave frequency, for the probe tip in air and in water. (BNL Neg. No. 3-529-79)

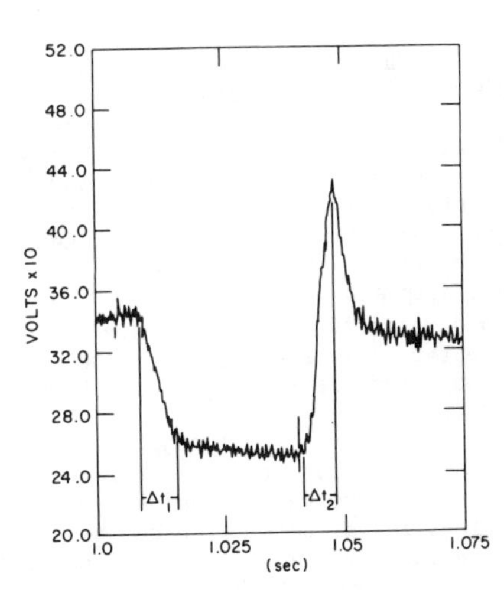

Figure 26 - Expanded output of the r-f probe during passage of the bubble. (BNL Neg. No. 3-526-79)

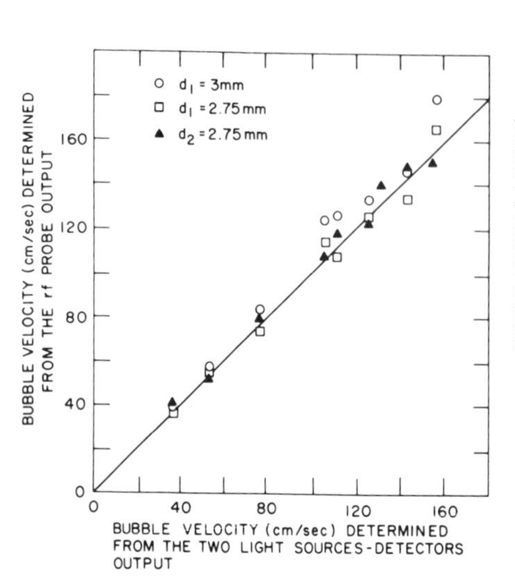

Figure 27 - Comparison of bubble velocity as determined by two independent methods, i.e, r-f probe and two light sources and detectors. (BNL Neg. No. 3-527-79)

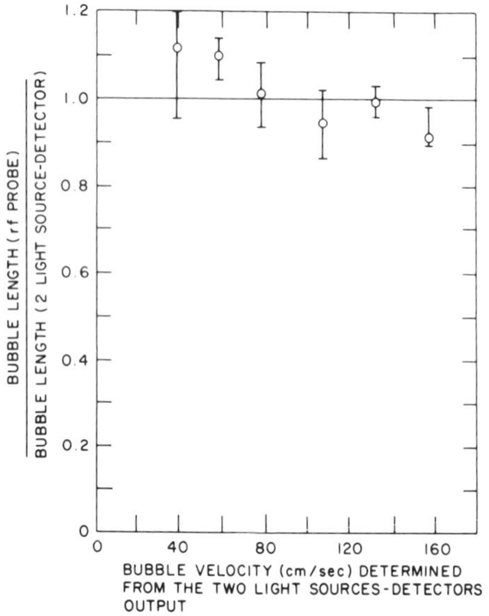

BUBBLE LENGTH (rf PROBE) / BUBBLE LENGTH (2 LIGHT SOURCE-DETECTOR)

BUBBLE VELOCITY (cm/sec) DETERMINED FROM THE TWO LIGHT SOURCES-DETECTORS OUTPUT

Figure 28 - Ratio of bubble length as determined by the r-f probe and the two light source detectors as a function of bubble velocity. (BNL Neg. No. 3-525-79)

water close to the tip by a separate copper wire was shown to eliminate this maximum.

In Figure 27, bubble velocities determined by the r-f probe are compared with velocities determined from the output of the two light-source detectors. Two penetration time intervals were measured from the r-f probe output (Figure 26), at t_1, and t_2. By considering a typical characteristic length of the sensitive part of the tip (3 mm and 2.75 mm), a bubble velocity was calculated. The actual dimension of the sensitive part of the tip was around 3 mm, (see Figure 24), but was difficult to determine exactly due to the geometry and construction. The bubble velocities determined by the two independent methods agree with each other within $\sim$10 percent. Thus with an r-f probe, the average bubble velocity can be determined from the passage time of either interface, air-water or water-air, along the insulated wires. This is true irrespective of the amplitude of the signal which may change with fluid state, purity, or test geometry.

The probe output levels for water and air did not change with the bubble velocity in the range considered (up to 160 cm/sec). Figure 28 depicts that the bubble sizes as determined by the two independent methods agree with each other with a maximum deviation of $\sim$10 percent. The bubble lengths recorded by the r-f probe are 10 percent higher at the low bubble velocities around 30 cm/sec. This fact may be due to surface tension effects during the penetration which become important at these low bubble velocities.

In summary, the r-f probe investigated has a relatively simple construction, and once tuned properly to the test geometry seems to provide information on both bubble sizes and velocities. More work, however, is needed to check the response of the probe in complex two-phase pipe flow conditions.

3.5 Microthermocouple Probes

The classical microthermocouple enables one to determine both steady and some statistical characteristics of the temperature. If combined with an electrical phase indicator, data regarding the local void fraction may also be obtained. Both are discussed below although the latter seems more appropriate in boiling two-phase flows.

Experiments on boiling heat-transfer include studies of temperature fluctuations near a heated surface with either pool boiling or forced convection.

A microthermocouple probe using wires 50 μm in diameters was used by Marcus and Dropkin [84] in measuring mean and fluctuating temperatures to evaluate the thickness of the superheated liquid layer in contact with a heated wall. The results, although timely, were somewhat inaccurate.

Bonnet and Macke [85] reported results obtained with a microthermocouple imbedded in a resin block in such a way that only 20 μm of the hot junction was in the flow. Unfortunately the size of the probe, 80 μm, produced a disturbance in the flow and its thermal inertia led to extra vaporization of the liquid on the sensor so that the significance of the signal was not clear.

Temperature profiles using a 125 μm diameter, chromel-alumel junction were measured by Lippert and Dougall [86] in the thermal pool boiling sublayer. According to the authors, this large diameter thermocouple data was shown to be reasonable by the results of tests in water, Freon-113 and methyl alcohol.

The interaction between bubbles and a microthermocouple, 25 μm in diameter was examined by Jacobs and Shade [87]. These authors, and also Van Stralen and Sluyter [88] were primarily concerned with the thermocouple response time. These investigations of the response time were augmented by Subbotin et al. [205] who examined the behavior of bubbles hitting different types of thermocouples.

Stefanovic et al. [89] verified the adequacy of a signal from a 40 μm diameter thermocouple by recording the impact of a bubble on the hot junction using high-speed movies. Amplitude histograms were obtained in pool boiling and in forced convection boiling. The authors separated steam and water temperature histograms by assuming that the predominant phase had a symmetrical distribution of temperature (Figure 29). The identical assumption was used by Afgan et al. [90,91].

Superheat layer thickness measurements were conducted in saturated and subcooled nucleate boiling by Wiebe and Judd [92] employing a 75 μm, chromel-constantan, microthermocouple. A time-average temperature was determined by integrating the temperature signal.

One of the first investigations into temperature profiles in forced convection boiling was carried out by Treschov [93]. The results appear less interesting than those obtained by Jiji and Clark [94] with a chromel-

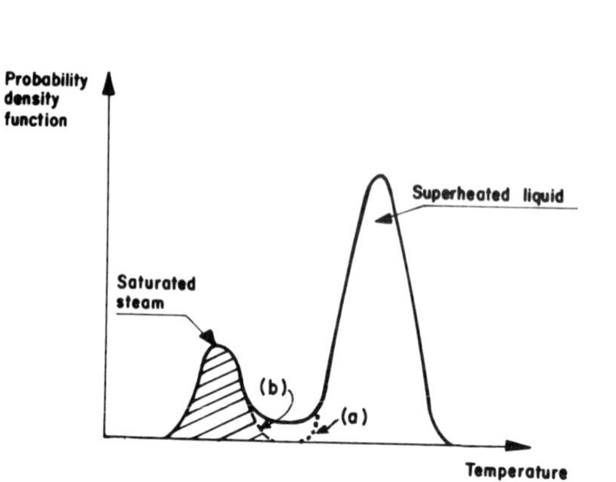

Figure 29 - Separation of steam and water distributions: (a) according to Stefanovic et al. [90,91]; (b) according to Barois [96]. (No BNL Neg.)

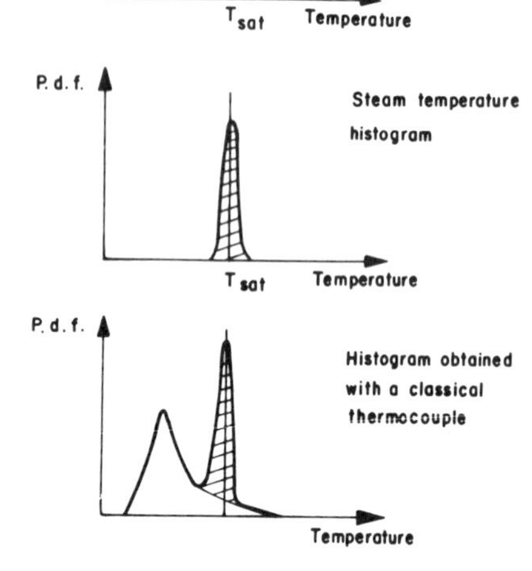

Figure 30 - Subcooled boiling Temperature histograms according to Delhaye et al. [98,99] (a) liquid temperature; (b) steam temperature (c) coupled classical temperature histogram. (No BNL Neg.)

constantan thermocouple, 0.25 mm in diameter. Despite the large size of their sensor, these authors succeeded in measuring an average temperature and average values indicative of the temperature extremes.

Local subcooled boiling, characterized by important nonequilibrium effects, was studied by Walmet and Staub [95] with the help of several local measurements: pressure, void fraction, and temperature. For temperature measurement the authors used a large, 0.15x0.2 mm copper-constantan thermocouple, and analytically related the measured value to the liquid temperature through the void fraction obtained using X-rays.

In his study of flashing flow of water, Barois [96] proposed to separate the distributions of steam and water by assuming that the steam temperature histogram was symmetrical (Figure 29), rather than the predominant phase as done by Stefanovic et al. [89] and Afgan et al. [90,91].

Although all the preceding works have contributed to a large extent to the understanding of the local structure of two-phase flow with change of phase, they have not provided any reliable statistical information on the distribution of the temperature between the liquid and the vapor phases.

The work done by Delhaye et al. [98,99] is based on the possiblity of separating the temperature of the liquid phase from the temperature of the vapor phase, and of giving the statistical properties of the temperature of each phase as well as the local void fraction. These workers used an insulated 20 μm thermocouple both as a temperature measuring instrument and as an electrical phase indicator (see section on electrical probes) by using a Kohlrausch bridge to sense the presence of a liquid conductor between the noninsulated junction and ground. The phase signal is used to route the thermocouple signal to two separate 1000 channel subgroups of a multichannel analyzer thus providing separate histograms of liquid and vapor tempratures as shown in Figure 30 for a subcooled boiling case. Comparison of these histograms with those in Figure 29 clearly shows the inconsistency in the assumptions of Stefanovic et al. [89], Afgan et al. [90,91], and of Barois [96].

Van Paassen [97] did a detailed study of the microthermocouple as a droplet size sampler, showing good agreement between theory and experiment in determining droplet sizes between 3 and 1188 μm. Detection frequencies of up to 1 kHz were obtained for small droplets.

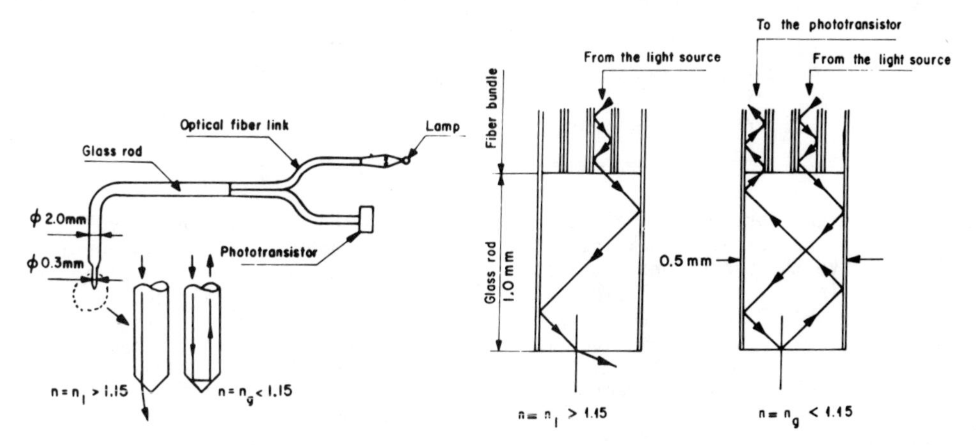

Figure 31 - Typical optical probe system. Miller & Mitchie [100,101] (BNL Neg. No. 9-127-79)

Figure 32 - Fiber bundle optical sensor. Hinata [104]. (ANL Neg. 900-5431)

3.6. Optical Probes

An optical probe is sensitive to the change in the refractive index of the surrounding medium and is thus responsive to interfacial passages enabling measurements of local void fraction or interface passage frequencies to be obtained even in a nonconducting fluid. By using two sensors and a cross-correlation method, information may be obtained on a transit velocity (Galaup [50]). A major advantage of such systems over others is the extremely high frequency response, limited electronically only by photoelectric electronics and photon statistics. The next best seems to be the conductivity probe having frequency response reported as high as 100-200 kHz [50].

Glass rod system. (Miller and Mitchie [100,101]; Bell et al. [102]; Kennedy and Collier [103].) This probe (Figure 31) consists of a glass rod 2 mm in diameter reduced to 0.3 mm at one end. The small tip of the rod is ground and polished to the form of a right-angled guide. The light from a quartz-iodine lamp is focused on one of the branched ends of the light guide. A phototransistor is located at the other branched end of the light guide.

Light is transmitted parallel with the rod axis towards the tip of the probe. When light beam strikes the surface at an angle of 45°, it emerges from the probe or is reflected back, depending upon the refractive indices of the surrounding material n and of the probe material n_0 according to Snell's law. Thus for a glass rod when $n_0 = 1.62$, if the incident internal angle between the light and the polished tip is 45°, light is reflected back along the rod if $n < 1.15$ and exists from the rod if $n > 1.15$. Table 1 gives possible combinations in which this probe can be successfully used.

Table 1. Liquid vapor systems where $n_g < 1.15$ and $n_\ell > 1.15$. Adequate for use of a 45°-tipped optic probe rod.

System	n_g	n_ℓ
Steam-water	1.00	1.33
Air-water	1.00	1.33
Freon-Freon Vapor	1.02	1.25

For signal processing, all the cited authors used a discriminator to transform the actual signal into a binary signal. A trigger level is set at a value above the background level corresponding to the case where the probe is immersed in the liquid. Miller and Mitchie [100,101] arbitrarily set the trigger level 10% of pulse amplitude above the all-liquid level obtained by comparing the value of the local void fraction at a given point to the volume void fraction measured with quick-closing valves. Bell et al. [102] set the trigger level half-way between the all-water and all-gas signal levels, while in a study of droplet jet flow, Kennedy and Collier [103] related the trigger level and the droplet time fraction with the sizes of the probe and of the droplets. It should be noted that significant variations in the results can sometimes be obtained with variations in the trigger level.

Fiber bundle system. (Hinata [104].) The basic element of this probe is a 30 μm diameter glass fiber which consists of a central core and an outer cladding Several hundred such elements are tied together in a Y-shaped bundle similar in appearance to that shown in Figure 31, with a light source and a phototransistor. The active end of this bundle is glued to a glass rod, 0.5 mm in diameter, 1 mm long, itself coated with a glass of lower refractive index (Figure 32). The extremity of the glass rod is ground and polished. The operation of this device is similar to that of the glass rod system. Hinata [104] obtained S-shaped curves of local void fraction measurement versus trigger level, with a plateau corresponding to a given value of the void fraction. He used the trigger level value corresponding to this plateau.

U-shaped fiber system. (Danel and Delhaye [105]; Delhaye and Galaup [106].) One of the major drawbacks of the glass rod and fiber bundle systems is the large dimension of the sensitive part of the probe (respectively, 0.3 and 0.5 mm). An alternative sensor configuration developed by Danel and Delhaye [105] has a distinct size advantage. This probe consists of a single coated optical fiber, 40 μm in diameter. The overall configuration is similar to that shown in Figure 33 with a miniaturized lamp and a phototransistor chosen for its high sensitivity. The active element of the probe is obtained by bending the fiber

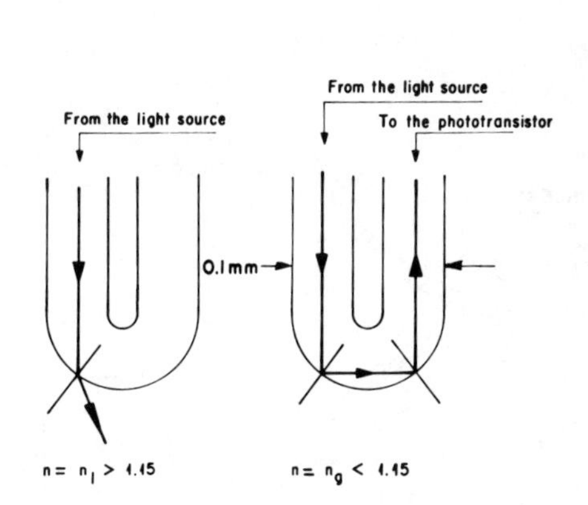

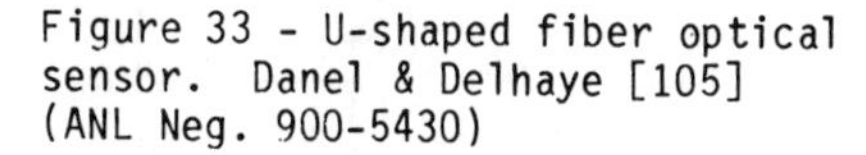

Figure 33 - U-shaped fiber optical sensor. Danel & Delhaye [105] (ANL Neg. 900-5430)

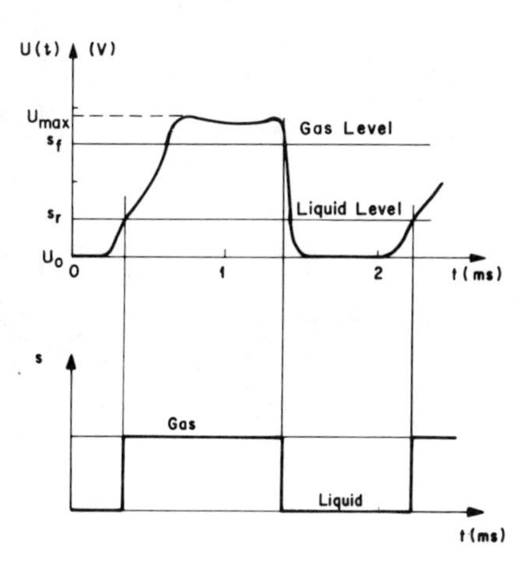

Figure 34 - Typical optical probe signal and discrimination method. Delhaye [107]. (ANL Neg. 900-5418)

into a U-shape and protecting the entire fiber, except the U-shaped bend, inside a stainless steel tube, 2 mm in diameter. The active part of the probe has a characteristic size of 0.1 mm as shown in Figure 33.

Signal processing for this system was taken one step further. A typical signal delivered by the probe is shown in Figure 34. The voltage U can be divided into a static component U_0 which was reported to vary with local void fraction, and a fluctuating component u.

Since U_0 corresponds to a sensitive part of the probe completely immersed in the liquid, the change in U_0 can be due to (a) the response time inherent in hydrodynamic and optoelectronic phenomena when the interface is pierced by the probe, and (b) the scattering of light by the bubbles surrounding the probe.

The fluctuating component constitutes the interesting part of the signal while the maximum value U_{max} corresponds to the sensor completely immersed in the gas and does not depend on the local void fraction. Signal analysis is accomplished through two adjustable thresholds, S_r and S_f which enable the signal to be transformed into a square-wave signal (Figure 34). Consequently, the local void fraction α is a function of S_r and S_f which are adjusted and then held fixed during a traverse in order to obtain agreement between the profile average and a γ-ray measurement of void fraction. Since the signal shape varies with void fraction, and since changes in S_r and S_f alter the result, local measurement of void fraction can be quite in error. This is true due to the compensating errors even though the average is in agreement with an independent measurement.

Wedge-shaped fiber system. All previously described methods suffer from difficulties associated with signal processing, hydrodynamic response time uncertainties, and fragility to which the U-shaped fiber system is especially susceptible. Signal processing at best seemed dependent on a global reference. The variable signal levels described by Delhaye and his coworkers [50,105,106]

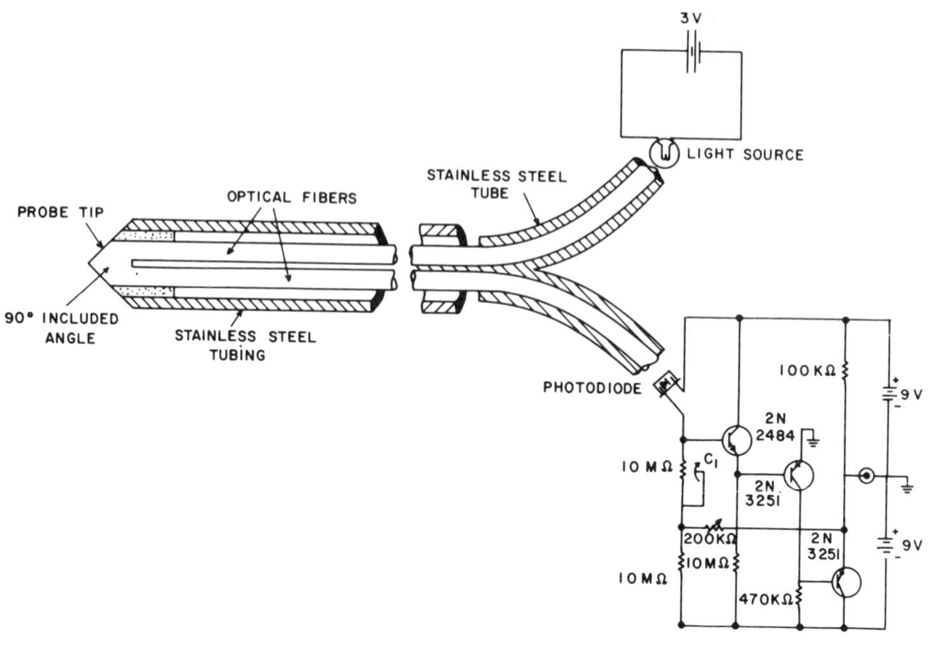

Figure 35 - Schematic representation of the optical probe.
(BNL Neg. No. 1-641-78)

can only yield an accurate local result at a specific void fraction. At other values of void fraction corresponding to different physical locations, values will be distorted due to differences between dryout and rewet response characteristics. The overall effect is a distortion in the void fraction profiles measured in a flowing mixture.

To circumvent these difficulties, a new design optical probe was developed by Abuaf, Jones, and Zimmer [107,108].

A schematic of the probe as developed is depicted in Figure 35, with the light source and amplifier circuit diagram used in the apparatus. Two .125 mm fibers were inserted into a 0.5 mm-O.D. stainless-steel tube. The two fibers were fused together at one end by means of a minitorch, forming a slightly enlarged hemispherical bead similar to that shown in Figure 33. This fused end of the fibers was then pulled into the tube and epoxied in place. The fibers were separated at the opposite end and encased in two pieces of stainless-steel tubing (0.25 mm O.D.). The ends of the fibers and the bifurcation were then epoxied for strength. The tip of the probe containing the fused end of the fibers was ground and polished at a 45° angle to the axes of the fibers, thus forming an included angle of 90° at the finished probe tip. The resultant probe geometry is extremely rugged, capable of withstanding considerable abuse without altering its characteristics, even to the extent of being dropped on its tip.

After grinding and polishing the free ends of the two fibers flat, one of them was placed in front of an incandescent light source (3 V), and the other in front of a Hewlett-Packard PIN photodiode (5082-4024). An amplifier with a design rise time of 20 μs [109] was used to enhance the output before going into the readout device (Figure 35). The electronic response of the system was

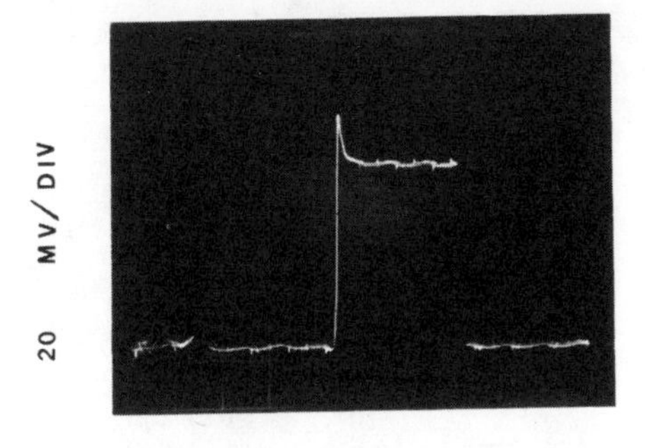

20 MS/DIV

U_B = 19 CM/SEC

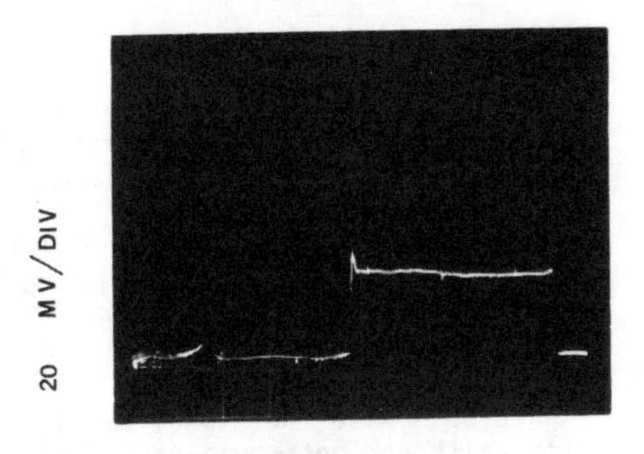

5 MS/DIV

U_B = 74 CM/SEC

Figure 36 - Typical oscillograms of the output. (BNL Neg. No. 9-111-77)

checked by means of a light-emitting diode (LED) placed in front of the probe tip. The LED output was modulated by using a signal generator so that rectangular light pulses of different spacings and widths were emitted, simulating the passage of bubbles. The rise time of the output was thus verified to be 20 μs as specified. The amplitude of the probe output did not change with the frequency of the input signal of the LED.

The hydrodynamic response of the probe to the passage of an interface or bubble was investigated as described in References 107 and 108, where single bubbles could be generated and forced past the probe while independent methods were used to determine the velocity. Typical oscillograms of the probe output during the passage of the bubble are presented in Figure 36. Here the probe output in mV is shown as a function of time for two cases where the bubble velocities were 19 and 74 cm/s, respectively. It was observed that when the tip was immersed in water the probe output was always zero without any artificial bias representing a significant improvement over previous optical probes. As the bubble hit the probe, the output was seen to increase, and after an overshoot, to level off to a certain steady value. At the end of the passage of the bubble the signal dropped to its original water level of zero. The bubble penetration time was clearly observed to be larger than the time it takes the probe tip to

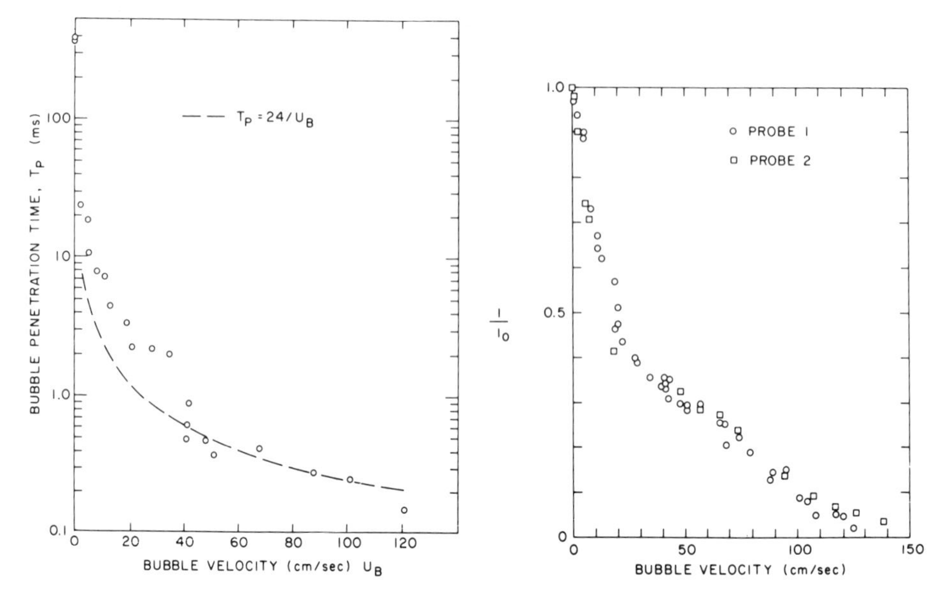

Figure 37 - Bubble penetration time vs. bubble velocity.(BNL Neg. No. 1-643-78)

Figure 38 - Probe signal at a given bubble velocity, I, divided by the steady air signal, I_0, vs. bubble velocity. (BNL Neg. No. 1-650-78)

be immersed in water, Figure 37. Also the signal amplitude decreased with increasing bubble velocity, although both bubbles had almost the same length (void fraction), Figure 38. These two effects were investigated in some detail. Results obtained with two different probes are presented in Figure 38 as a function of bubble velocity. Although the two probes had a steady air signal amplitude of 125 and 600 mV, respectively, the ratio I/I_0 follows the same consistent pattern.

A similar observation was noted by Miller and Mitchie, "With smaller bubbles and higher velocities. . . . The probe signal generated under these conditions. . . never reached maximum amplitude." [101]

An important conclusion that can be drawn from Figure 38 is that the optical probe is able to measure the local interface velocity as well as the local void fraction, after proper calibration within the velocity range observed.

A computer program was written to study the theoretical response of the output to hydrodynamic conditions at the tip of the probe by tracing individual rays of light from their source to the detector. The effects of varying liquid film thicknesses on the the probe tip was included. Comparison of the geometric effects for the probe of Danel and Delhaye[105] and the new design are shown in Figure 39 where the new geometry is seen to be dependent on probe tip angle.

For a possible explanation for the optical probe output behavior for various bubble velocities, it was proposed that a water film thickness left on the probe tip and increasing with the bubble velocity could explain the decrease in the signal intensity that was observed experimentally.

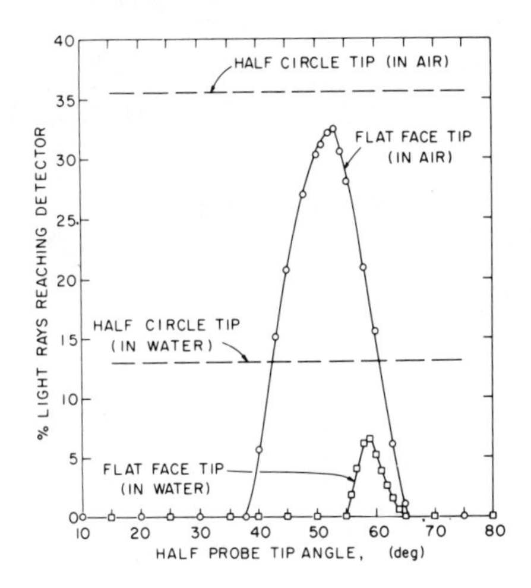

Figure 39 - Percent of light rays reaching the detector as a function of the half probe tip angle for the cases of the probe tip in air and in water. Comparison between the half circle and the flat face probe tips [108] (BNL Neg. No. 1-646-78)

Computer calculations were thus extended to study the signal attenuation that would be experienced when a variable film thickness is present at the probe tip. The water layer was assumed to increase linearly along the flat face. β_w is defined as the angle between the outside face of the water layer in contact with the air and the glass face of the probe tip in contact with the water layer. The attenuation of the rays during their passage from one media to another was not taken into account. Although the 52° half tip angle gave a higher signal in air when compared to the 45° half angle probe tip, the large angle tip (52°) was found to be strongly dependent on the water layer thickness left on the probe tip. In order to show this strong dependence of the probe output to the water layer left on the probe tip, we plot in Figure 40 the maximum angle β_{max} that can be sustained on a probe tip angle before the coherent light rays are refracted out and the signal is zero. Within the tip angle range of interest, $37° < \beta_t < 65°$, (Figure 39), the maximum angle β_{max} increases, and the sensitivity of the probe to the water layer thickness decreases for the lower values of tip angle.

It is known that when a cylinder is withdrawn from a liquid that the film thickness remaining on the cylinder increases with withdrawal velocity (White and Tallmadge [110]). This theoretical prediction was also checked experimentally for various liquids. The dimensionless film thickness (Tallmadge and White [111] was related to a Capillary Number,

$$N_{ca} = \frac{\mu U_w}{\sigma} \qquad (10)$$

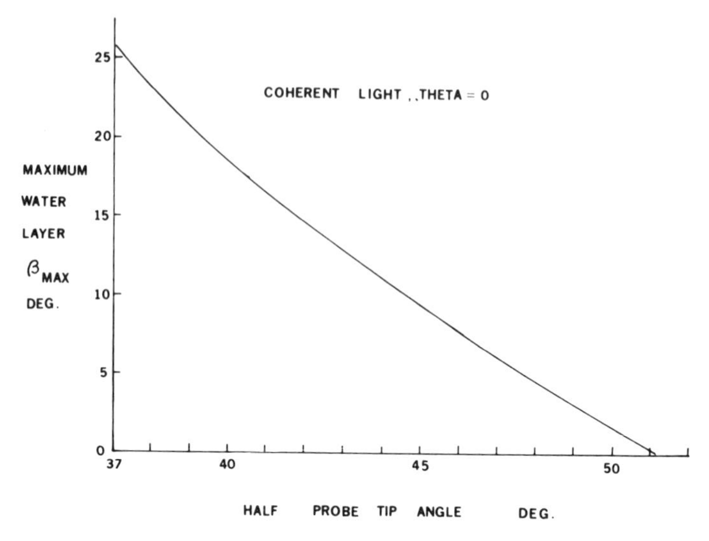

Figure 40 - Maximum thickness of water layer on the flat face probe tip for zero output [108] (BNL Neg. No. 3-1722-78)

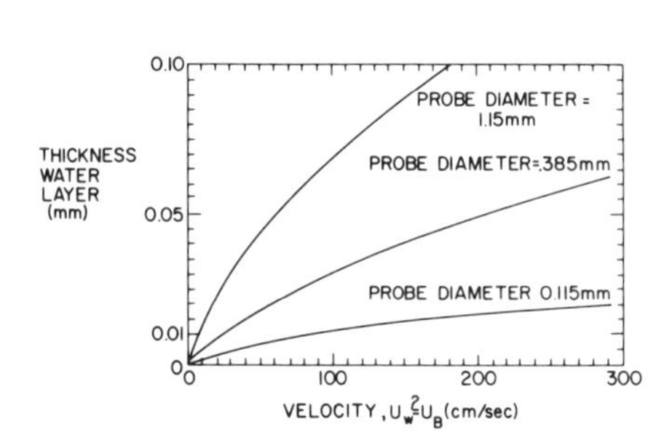

Figure 41 - Plot of water layer thickness vs. wire withdrawal velocity (Tallmadge and White [111]). (BNL Neg. No. 3-1723-78)

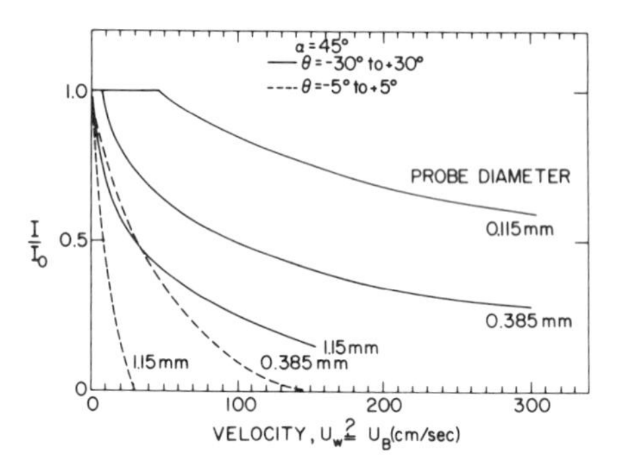

Figure 42 - Cross plot of normalized probe signal, I/I_0, vs. wire withdrawal velocity for three probe diameters and two values of the acceptance angle [108] (BNL Neg. No. 31-1719-78)

and a Goucher Number,

$$N_{Go} = \frac{R}{a} \tag{11}$$

where μ and σ are the viscosity and surface tension of the liquid, u_w is the withdrawal velocity, R is the wire radius, and a is the capillary length defined as $a = (2\sigma/\rho g)^{1/2}$. In addition to this, White and Tallmadge [112] observed that experiments conducted with distilled water provided film thicknesses almost twice those predicted by the theory. This fact is still unexplained. In any event, the results may be used to obtain Figure 41 showing the increase in film thickness with increasing velocity.

The two parts of the theory explaining the probe behavior thus include:

a) Amplitude of signal vs. water film thickness;

b) Water film thickness vs. interface passage velocity.

A combination of the two will thus yield the predicted variation of signal amplitude with interfacial velocity as shown in Figure 42, in good qualitative agreement with the observed behavior. Differences may perhaps be explained by the presence of nonlinear films, slanted optical surfaces instead of surfaces colinear with the probe direction, etc.

4. INSTREAM SENSORS WITH MECHANICAL OUTPUT

A large class of devices developed for taking measurements in two-phase flow have been designed for use primarily in annular flow. The single needle conductance probe when traversed toward a film has been used for measuring film thickness. The dual, flush mounted Harwell conductivity [27] probe and the ring-type CISE probes [36] were also designed to measure film thicknesses. Likewise, a number of devices with mechanical output have also been designed for measuring specific properties of annular flow. The major impetus is the general idea that the critical heat flux phenomena is predominantly an annular flow phenomena caused by dryout or disrupture of the liquid film.

In all of these sensors, the general principle is to remove from the flowing stream a certain portion of the fluid to determine the flow rate of either one phase or the other, usually the liquid phase. The sensor may be either stationary or may be designed with traversing mechanisms to determine the transverse distribution of the particular parameter under study. In some more intricate designs, the amount of suction is determined by matching the probe inlet static pressure with the channel static pressure in order to have minimum effect on the flow stream lines. The main development of these instruments may be attributed to the Atomic Energy Research Establishment (AERE) at Harwell, England, and the Centro Informazioni Studi Esperanzi (CISE) in Milan, Italy.

4.1. Wall Scoop

The wall scoop device was developed at CISE and was reported by Cravarolo and Hassid [113], and Adorni, et al. [114]. This device, shown in Figure 43 consisted of a movable scoop built into the wall of the tube being investigated. In their 2.5-cm diameter tube, the scoop could measure the integrated flow rates in the film region from 0.13mm to 2mm from the wall. To obtain a sample, the valve was opened until the static pressure just inside the scoop was identical to the static pressure at the same axial location on the opposite side of the tube. The theory is that by matching the static pressures, only minimal disturbance to the flow stream occurs and the sample is thus taken at the undisturbed, "or "isokinetic" conditions. In reducing the data to obtain local values of void fraction an assumption regarding the slip ratio must be made. Cravarolo and Hassid [113] assumed that gas in the film region would occur as

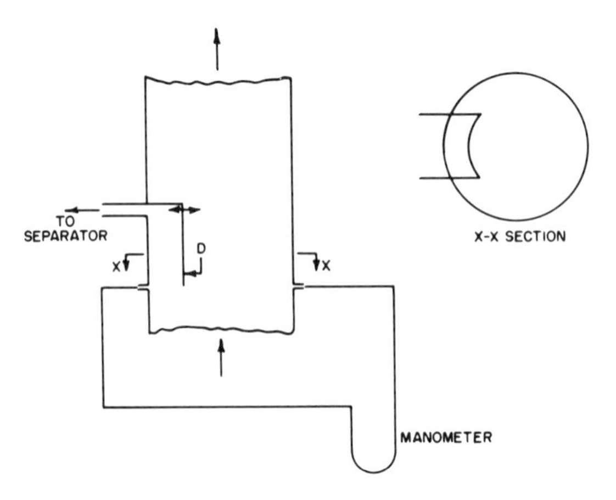

Figure 43 - Schematic of Wall Scoop used at C.I.S.E. [113]
(BNL Neg. No. 9-114-79)

small bubbles and that the slip ratios would thus be close to unity. The flow rates, coupled with the known area, also provided information regarding the velocities once the assumption of unity slip was been made. In any event, if the flow rates are accurately known, the phase volume fluxes may also be readily calculated without further assumptions.

One basic disadvantage of this type of instrument is that it disturbs the flow regime irrevocably. In other words, once the film is removed from the wall there is no method of putting it back in the same distribution. Other disadvantages are its limited size which tends to give erroneous readings for very thin or very thick films. In addition, while the method provides reasonable results in two-component flows, single component flows are more difficult to determine accurately due to phase change effects caused by heat losses and pressure drops, and under nonequilibrium conditions become virtually impossible to determine accurately.

Truoung Quang Minh and Huyghe [115] used a somewhat similar device wherein a circumferential slit was provided at a point in a tube. The axial width of the slit could be varied as shown in Figure 44(a). They found that the film flow rate measured would become almost independent of both slit width and differential pressure for widths over a certain minimum, and for differential pressures above some minimum. This behavior is shown in Figure 44(b). As discussed by Moeck [116] both this device and that designed at CISE can underestimate the film flow rate when large waves are present.

4.2. Porous Sampling Sections

Another device which has been used for annular flow studies is the porous sampling section, sometimes fabricated from a sintered porous plate and sometimes from a fine-mesh screen [117-127]. The purpose of this porous sample section is to remove the film of water selectively in annular or dispersed annular flow in order to measure the film flow rates. Typical of the devices used is that used at General Electric's Atomic Power Equipment Department [117] in an annular test section shown in Figure 45(a). Pressure is decreased in the plenum chamber behind the porous plate by opening of a sampling valve. In two-

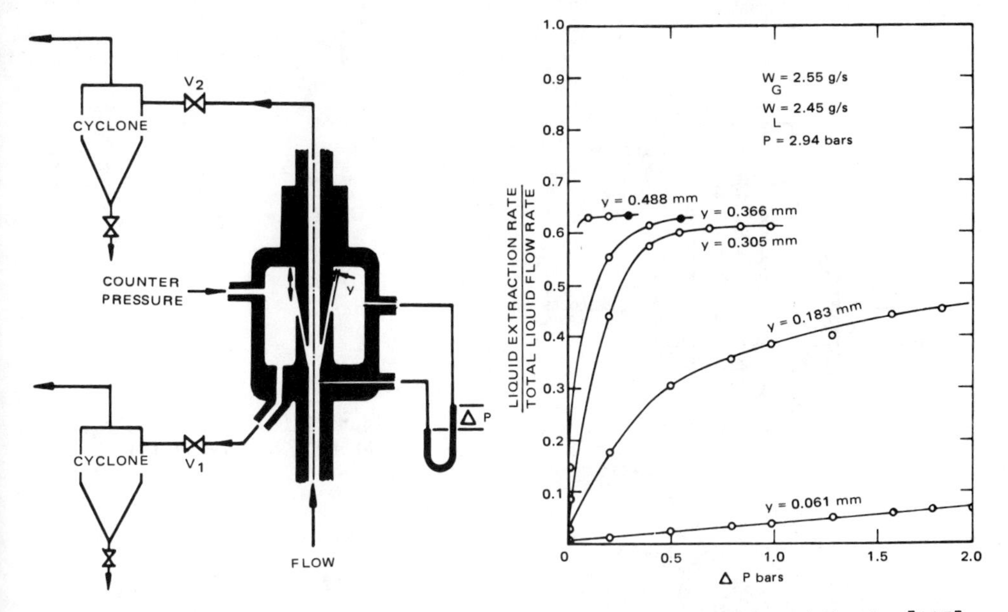

Figure 44 - Film Suction Device used by Truoug Quang Minh and Huyghe [115] a) left; b) right. (BNL Neg. No. 9-101-79)

component flow the phases are usually separated and measured separately while the single-component flow the mixture is condensed in a calorimetric device and a heat balance is applied to calculate the mass flow rates of the separate phase. In the case of two components, the liquid flow rate is usually plotted as a function of the pressure drop across the plate and curves similar to those for the wide gap slit sampling device shown in Figure 44(b) are usually obtained; that is, beyond some nominally small differential pressure the sample flow rate does not change for a wide range of differential pressures. The concept of this behavior is that if the sintered plate is sufficiently long, the entire film is removed with little pressure drop and thereafter only additional amounts of core gas are pulled through the porous plate. If sufficient pressure drop is applied, however, core-borne liquid would also begin to be extracted causing the liquid sample flow rate to increase again. The film flow rate in these cases is defined as the liquid flow rate independent of the pressure drop, the plateau value.

A second method of determining the film flow rate is to plot the liquid flow rate verses the gas phase flow rate. This method has also been the predominant method of determining the film flow in single component fluids. As shown in Figure 45(b), the liquid removal rate becomes quite insensitive to the vapor rate with increasing total sample flow [118]. The plateau in this case can also be quite flat. In some instances such as low quality or where the annular core is highly laden with liquid, the determination of the film flow rate is not so clean cut and various extrapolation schemes must be applied as shown in Figure 46. In addition to uncertainties, Singh et al. [125] showed that in single component flow, similar to two-component cases, the experimentally determined film flow rate was dependent on the length of the porous section due to deposition of core-borne droplets and due to increasingly more efficient capture of the liquid in large roll waves with longer plates.

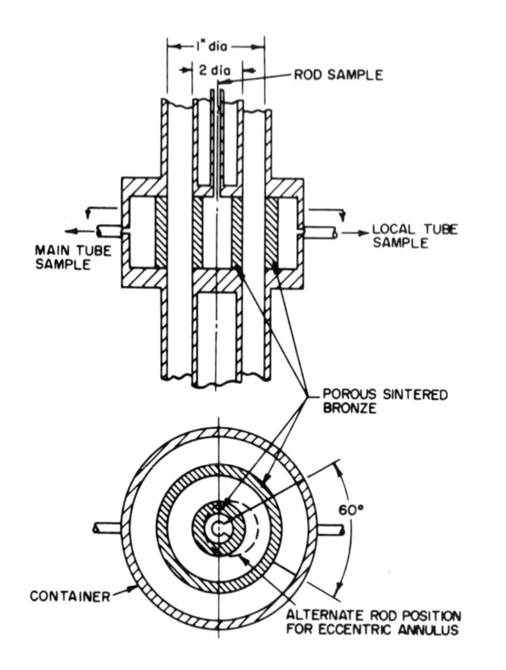

Figure 45(a) Porous plate
(BNL Neg. No. 9-109-79)

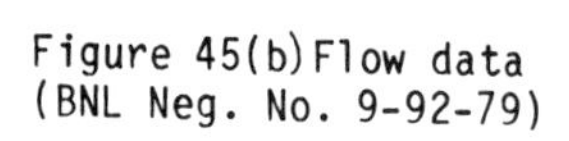

Figure 45(b) Flow data
(BNL Neg. No. 9-92-79)

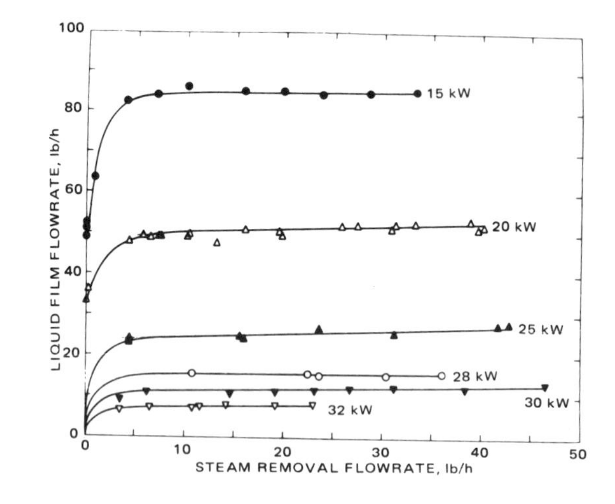

Typical Design and Results for a Porous Plate Film Flow Sampling Device
(References 117,118)

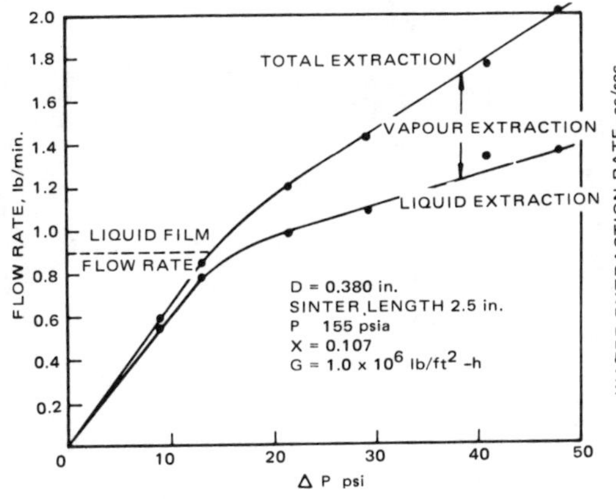

Figure 46(a) - Data of Staniforth et al. [127] (BNL Neg. No. 9-83-79)

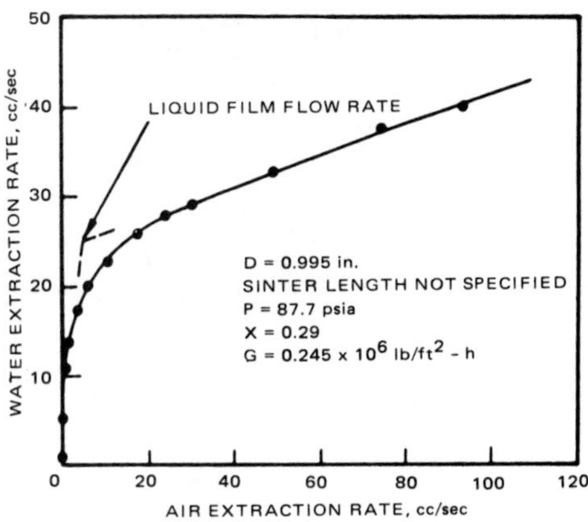

Figure 46(b) - Data of Schraub et al. [126] (BNL Neg. No. 9-82-79)

4.3. Isokinetic Sampling Probe

Where the two previously described mechanical sampling devices have been stationary or slightly moveable and restricted to measuring film flow quantities, a device was developed by CISE [128] for measuring component flow rates and, with the aid of a suitable assumption regarding slip, the void fraction at various locations of a tube cross section. This device has subsequently been used by Lahey, Shiralkar, and Radcliff [129], by Schraub [126,130], Adorni et al. [128], Todd and Fallon [131], and Jannsen [132]. All of these designs are similar in that they resemble a small Pitot-static probe as shown in Figure 47. In general, this probe is used in either of two distinct manners. The first manner is quite similar to the wall scoop in that flow is drawn off through the main body of the probe until the static pressure near the interior tip of the probe just equals the static pressure in the tube adjacent to the probe opening. Various degrees of sophistication can be employed in the design of such a probe from the very simple device used by Wallis and Steen [133], to that used by Burick, Scheuerman and Falk [134] based on a design by Dussord and Shapiro [135]. Ryley and Kirkman [136], in fact, added a momentum deflector with a floating impulse cage to combine momentum flux measurements with measurements of the mass flow rates of saturated steam and liquid in the exhaust section of a steam turbine. The second method of isokinetic probe operation simply utilizes the probe as a stagnation pressure metering device where reverse purging of one phase or the other, usually the gas, is used to provide a metering reference.

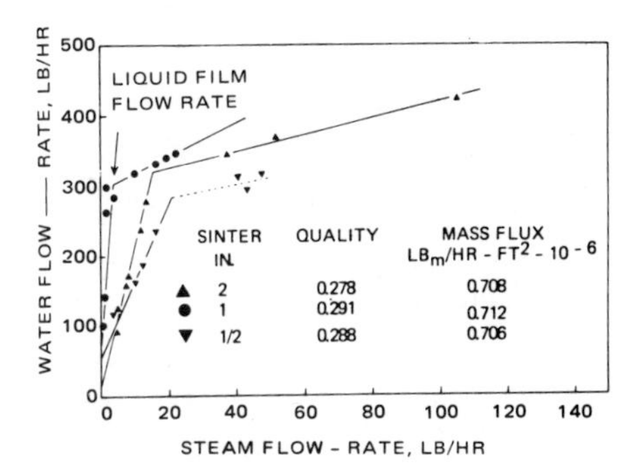

Figure 46(c) - Various Film Sampling Results of Different Investigations (BNL Neg. No. 9-87-79)

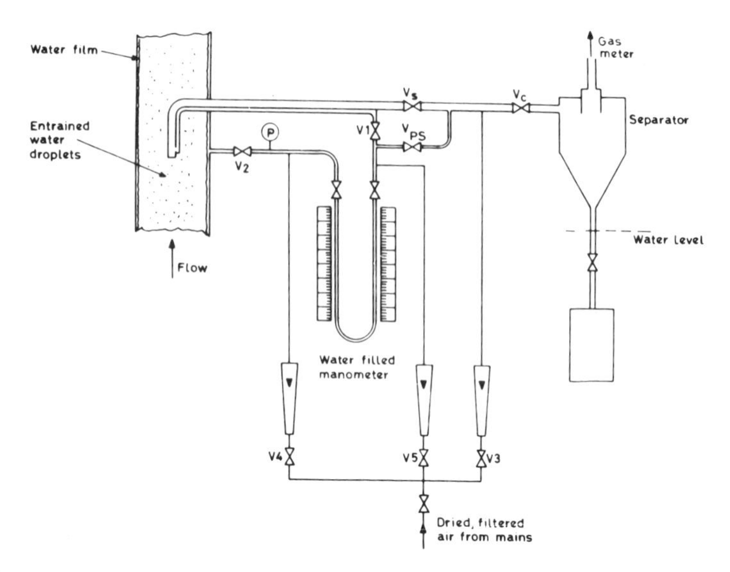

Figure 47 - Isokinetic Probe System as Designed by C.I.S.E. [128]
(BNL Neg. No. 9-102-79)

A number of problems exist with the use of the isokinetic probe, not the least of which is the necessity to compensate for the pressure losses between the probe tip entrance and the static pressure port. For small probes Schraub [130] outlines a method to sample at various conditions around the isokinetic condition and then to iterate on the proper conditions by checking the integrated flow profiles against the known values. This is good for two component systems with axisymmetric geometries and perhaps would also work in equilibrium single component systems where this check is possible. In planar or grossly two dimensional geometries, such a system would be extremely difficult. In addition, since the corrections are flow dependent, virtually every reading requires a different correction and profile distortion of measured results would occur.

Another problem lies with the fact that different assumptions lead to different results for the void fraction. The Italians [128] assumed a constant slip in the fluid approaching the probe whereas Schraub [130] assumed variable slip. Schraub [137] mentions, however, that only slightly differing results are obtained with different assumptions. In addition, Shires and Riley [138] show that if the probe is much larger than the dispersed phase, the vapor volume flow fraction is measured whereas if the reverse is true, the vapor volume fraction itself, α, is measured. In spite of these various problems many workers have used this instrument with varying degress of success, mainly because it appeared to be the only general class of instrument capable of providing the desired combination of measurements of phase velocities and void fraction.

A variation on the isokinetic probe has been used by Gill et al. [139] who demonstrated that for two component, slightly laden flows, the measured liquid flow rate is practically independent of sampled gas flow rate as shown in Figure 48. Thus, if only gas core liquid content in annular flow is desired, it is simply sufficient to have a slight amount of gas to obtain a representative liquid flow rate measurement.

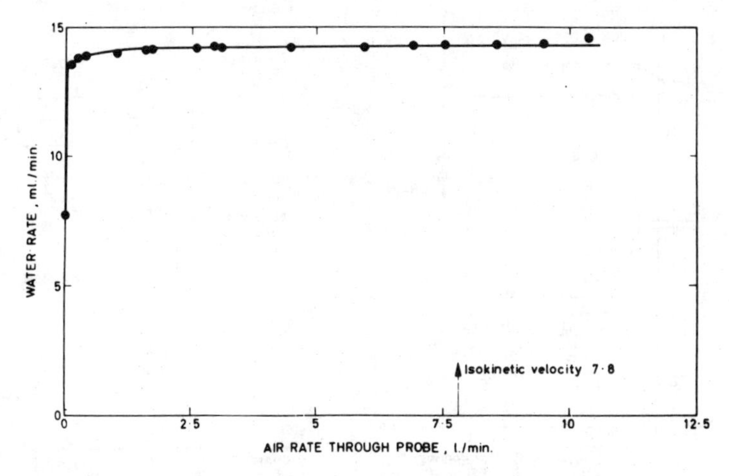

Figure 48 - Typical Results from an Isokinetic Sampling Probe When Used in an Air-Water System. Data of Gill, et al. [139] (BNL Neg. No. 9-98-79)

4.4. Wall Shear and Momentum Flux Measurement Devices

Various mechanical or electromechanical devices have been designed or adapted for the measurement of skin friction in two-phase flow. The latter type are included herein because these instruments are basically mechanical in nature, using ancillary electrical methods for readout purposes only.

Perhaps the most widely used device for directly measuring wall shear in single-phase flow is the Preston [140] tube which is simply a small right angled total pressure probe which is used when the mouth rests against the wall with its opening directed upstream. Preston's calibration as corrected by Patel [141] relates the dimensionless wall shear stress to the dimensionless dynamic head. This device was used successfully by King [142] with condensing annular-dispersed flow in a 1-inch tube, and by Jannsen [143] in a nine-rod bundle.

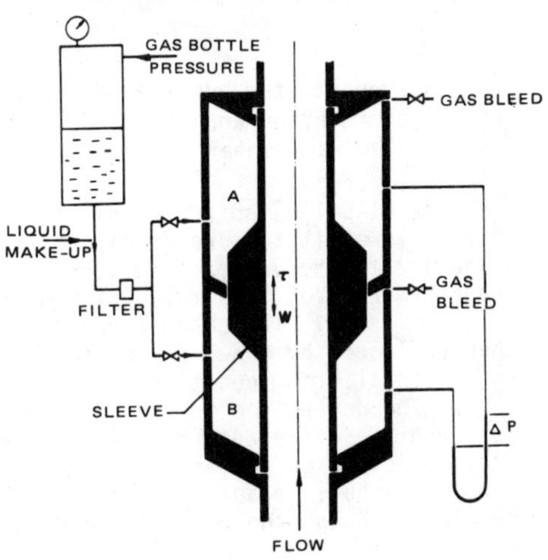

Figure 49 - Schematic Drawing of the Device Developed by Cravarolo et al. for Measurement of Shear Stress on the Wall of a Conduit. [144] (BNL Neg. No. 9-91-79)

Cravarolo et al. [144] devised a null-balancing wall sensor (shown in Figure 49) for measurement of the skin friction within an accuracy in non-fluctuating flows of $\sim$2%. In slug flow, however, they were unable to balance their device. By noting the pressure drop required to just begin to move the sleeve off the lower support, then doing the same to drive the sleeve down off

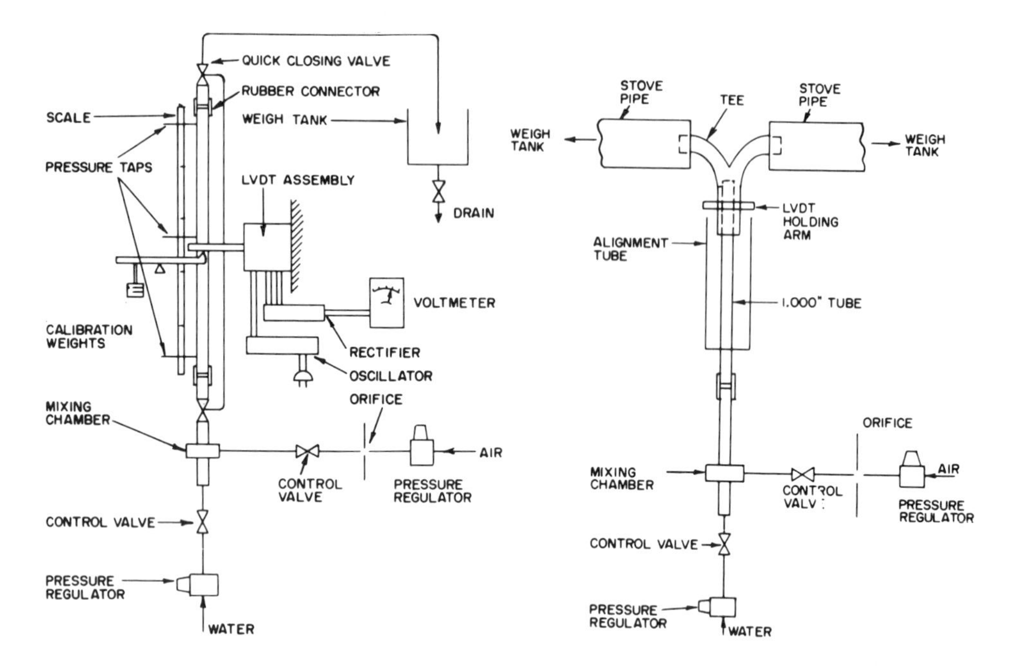

Figure 50 - Diagram of the Vertical Tube Apparatus Used by Rose. (Reference 146) (BNL Neg. No. 9-110-79)

Figure 51 - Schematic Diagram of the Exit Momentum Efflux Measurement System Used by Rose and by Andeen (References 146 and 147)(BNL Neg. No. 9-117-79)

the top support, the average value provided the shear stress without the effects of friction and inertia. Similar to the Preston tube, this device can give the wall shear for a relatively small area of a tube.

Rose [145,146] suspended his test section between two rubber connectors which allowed it to more freely in the vertical direction (Figure 50). Attached somewhere to the vertical section was a linear, variable, differential transformer (LVDT) assembly which detected small changes in position of the test section itself. The channel was first balanced in an equilibrium position and then deflected by known forces in order to obtain a calibration. Then, starting at the equilibrium position, flow of two-phase mixture is passed through the tube. The deflection of the tube is then due only to the shear stress on the wall and the difference in pressure between the inlet and the outlet of the test section acting on the end areas of the tube. A similar device was designed by King [142] but, because of hardware difficulties was never actually put into service although the method appears promising.

A device for measuring exit momentum flux was built and used by Rose, and subsequently by Andeen and Griffith [147]. This instrument, shown in Figure 51, is simply a device to change by 90° the direction of flow at the annular channel and measure the force required to do this. With an LVDT apparatus, similar to that described for his wall shear-stress instrument, Rose measures the forces needed to deflect the flow stream. Thus, the measurement of exit momentum flux, combined with a knowledge of the inlet momentum flux, and the integrated wall shear stress can result in accurate values for the component or pressure drop due to elevation and average channel vapor fraction.

5. OUT-OF-STREAM MEASURING DEVICES

This class of instruments consists of those whose measurement ability depends on the attentuation or reflection of electromagnetic radiation or atomic particles. Excellent reviews of light photography methods are given by Arnold and Hewitt [148], Cooper et al., [149], and by Hsu et al. [150], and will not be discussed herein. The majority of those methods are aimed at obtaining qualitative information. Indeed, visual examination of photographs has been one of the major methods of flow pattern determination to date. On the other hand, the attenuation of non-visable radiation has come to be the most widely used method of obtaining quantitative measurements of void fraction to date.

Of all the methods used, the most popular involves the attenuation of the strength of a concentrated beam of photons (γ- or X-rays). This method is popular mainly because measurements may be made of space averaged void fraction along a chord length without disturbing the fluid to a noticeable degree. This attentuation technique has generally been used to obtain average measurements of void fraction in steady systems [154-156] or systems with slowly varying transients ($<$ 10 HZ). Schroch [158], however, has had some success with more rapidly varying signals, and Jones [53,157,159] has reported a system capable of measuring transients in the millisecond range. More recently, high intensity, high energy systems have been developed and used for rapid transients in large pipes, driven by the emphasis on large scale nuclear safety tests [171,172].

There are basically four types of void measurement systems using attenuation techniques:

a) X-ray systems where a source of electromagnetic radiation in the general range of 25 - 60 kev is provided by an X-ray tube.

b) γ-ray system where the source is a radioisotope which emits photons with energies usually between 40 and 100 kev.

c) γ-ray systems which obtain a stream of electrons from a radioisotope with energies up to 10 mev.

d) Neutron systems where neutrons are supplied by a source in the range up to about 1 mev.

The general problems of attenuation methods are similar for all four systems and shall be discussed shortly. The basic differences between these systems lie in the differences in the attentuation laws and the hardware required to accomplish the measurement. Otherwise, the overall concepts are similar as discussed by Schrock [160]. The choice of a method then usually is dependent on the experimental constraints, cost, hardware availability, etc., rather than on the desirability of one system over another for reasons of accuracy or ease of application.

5.1. X-Ray and Gamma Ray Methods

From a source standpoint, only the X-ray system does not require a nuclear or radioisotopic source. Instead, X-rays are generated in a vacuum tube where a high potential is applied between a target of a specific material, and a heated filament. Electrons which are "boiled" off the filament are accelerated in the potential field toward the target. Target material may be of any element but is usually of high melting temperature and high atomic number such as Tungsten or Molybdenum. Since a large amount of heat is usually generated, these targets are sometimes made hollow to permit circulation of cooling water. About 99% of the electrons striking the target simply give off their energy in the form of heat upon being decelerated. The remaining 1% will give off a spectrum of electromagnetic radiation, X-rays, having energies from zero up to the maximum energy of an incident electron. In addition, since some electrons will knock bound electrons from atoms of the target material, other electrons will fall back into these vacancies giving off particular quanta of X-radiation characteristic of the material and the energy level vacated. This latter effect tends to produce a localized maxima in the X-ray energy spectrum at these characteristic energies. Filtering of the emitted X-ray beam may be used to remove lower energy radiation, producing a beam of X-ray of nearly monoenergetic characteristics near the characteristic energy.

The advantage of using X-rays lies in the lack of bulky source holders and the repeatability and long term stability of the source strength. The disadvantages lie in the short term unsteadiness in the source due to the alternating portion of the applied voltage. This problem has severely limited the capabilities of transient systems such as those used by Schrock [158] and Zuber et al. [161]. Jones, [157,159], however, used special filtering techniques in the high voltage side of his X-ray transformer supply to reduce the induced thermal ripple in the beam strength [160-162] to a negligible amount. In an attempt to minimize the effects of short term unsteadiness, many users have employed dual beam X-ray tubes where one beam was used for the test while the other served as a reference. In theory, then, the ratio of the intensities of the two beams removes the original source from consideration.

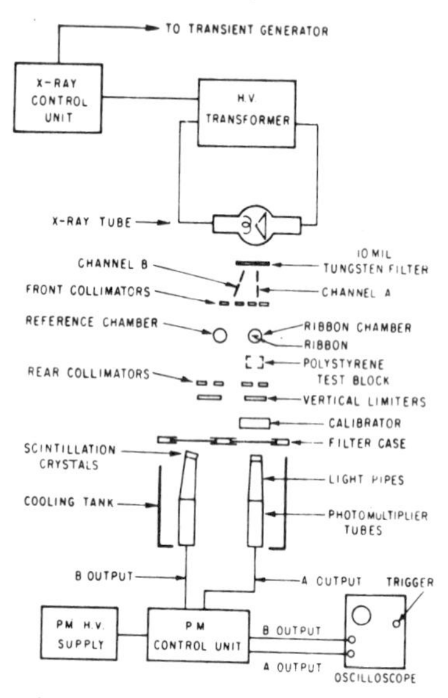

Figure 52 - Schematic of X-ray Densitometer System [160] (BNL Neg No. 9-100-79)

A schematic of such a system is shown in Figure 52 as taken from Reference [160]. This is a typical dual-beam system where, in this case, Schroch used a tungsten target operated at 100 kv with a tungsten filter. The result was a sharp resonant peak in the 50-70 kev range, corresponding to the lowest electron quantum level, $K\alpha$, of 58.5 kev, and the second, $K\beta$, level of 68 kev. Detection and measurement of each beam intensity was by means of a Thallium-activated, sodium-iodide crystal which absorbs X-rays wave length radiation and gives off radiation in the visible range in its place. While the crystal has some self absorption, the emergent intensity of the visible light is proportional to the incident X-radiation intensity. Light-piping was employed to direct the scintillated light to photomultiplier tubes which produced an electrical current proportional to the incident light intensity. The output from the two photomultipliers was then differentially amplified, and recorded. Additional problems in such a system include the inability to obtain well-matched photomultiplier tubes, differences in the fatigue or short term aging characteristics of the tubes, and problems in maintaining a stable excitation voltage for the pair of detectors.

For both X- and γ-rays, the intensity of a beam, I, of original intensity, I_0, is given by

$$I = I_0 e^{-\mu x} \tag{12}$$

where μ is the linear absorption coefficient and is dependent both on the material and the energy of the photons, and x is the thickness of the absorber. The major difference between the X-ray and γ-ray systems is the source. γ-rays are identical in nature to X-rays. Historically, however, γ-rays were usually extremely energetic, being obtained from radioisotopes and having wave lengths much shorter than those generally associated with X-rays. In general usage, however, the term now applies to any photons originating from a nucleus while the term X-ray is reserved for similar but extra-nuclear radiation. Since γ-rays are obtained from nuclear disintegration, the source strength is time-dependent. Thus, the half life of the source material becomes as important a consideration as the strength of the beam. Since the uncertainty (1 standard deviation) in a measurement of beam intensity is equal to the square root of the number of events measured, an intense beam from a highly radioactive source must be used to minimize observation time. The level of radioactivity measured in curies, (1 ci = 2.22×10^{12} disintegrations/min), is dependent on the half life, (time to disintegrate 50% of the material), and the mass of the element, (total number of atoms available for radioactive decay). A short half life material requires frequent recalibration whereas a long half life material of acceptable disintegration rate must be massive and necessarily hazardous to handle. In addition, most systems have been concerned with measuring voids in water-metal or freon-metal systems. Since the linear absorption coefficient for photon attenuation decreases more rapidly with increasing energy for water than for most metals, energies less than 100 kev are desirable from a sensitivity standpoint. The three radioisotopes which have found the most use are compromises based on the above factors and include Thulium-170, Samarium-145, and Gadolinium-153. The table below taken from Reference [160] gives the half life and γ-energies of these materials.

Table 2 - Popular isotopes for void measurements

Isotope	Half Life (days)	γ-Energies kev
Samarium-145	240	39-61
Gadolinium-153	240	42-72
Thulium-170	170	52-84

The advantages of γ-sources over X-rays are principally in cost, short term stability, and simplicity. Since short term stability is not a problem, dual beam systems become unnecessary. In addition, high voltage supplies and regulating systems are not required.

In general, X- and γ-ray systems have been the most popular void measuring technique used to date. Extensive discussion of the errors associated with these techniques was presented by Hooker and Popper [162], with exception of errors due to the fluctuations in nonlinear signal [157,159,163-166]. In the system described by this reference, a one-shot method was designed to measure the voids for the entire cross section of a 0.5 x 2.175-inch rectangular channel by collimating a thulium-170 source parallel to the wide plates of the channel. They identify errors due to the following sources:

a) Errors in the electronics system due to such characteristics as amplifier drift, photomultiplier gain sensitivity to small changes in the supply voltage, and temperature sensitivity of the sodium-iodide crystal.

b) Errors in measuring technique due to measurement of γ's which reach the detector by some path other than through the test section, strip chart reading errors, and errors due to calibration at non-test conditions.

c) Errors due to decay of the source.

d) Errors due to Preferential Phase Distributions

For a uniform distribution of voids, Hooker and Popper concluded that the maximum absolute errors in such a system is about 2.5% voids over the entire range of 0-100% voids. Similar conclusions have been reached for X-ray systems as well, and, in References 157, 159, 160, and 163 an estimate was given of absolute error of 1.7% voids is given for stationary measurement of local void fraction and 2.3% voids for local measurement during a transverse void profile scan.

The errors indicated above do not take into consideration additional errors due to void streaming effects in preferential void distributions, nor errors due to linear averaging of a nonlinear, fluctuating quantity. Hooker and Popper discuss the former errors associated with nonhomogeneous phase distributions. Basically these errors arise when liquid and vapor phases are separate, the limiting geometry possible being layers of liquid and gas whose planes are parallel to the beam of photons. In this case, the attentuated beam can be considered as two separate beams, one having passed through the gas and the other having passed solely through the liquid. The total intensity is, then, the sum of the individual intensities, each of which has been separately attentuated. The result is not the ideal exponential attenuation law normally employed in data reduction. Petric and Swanson [206] verified this source of errors and showed that typical one-shot measurement techniques could be off by up to 40%. Errors of the same order or larger were also predicted by the writer [159,163] to arise due to the fact that two-phase flow is by nature a nonstationary phenomena. Thus time averaging of a quantity nonlinearly related to the desired quantity, in this case void fraction, leads to significant errors when the void fraction is calculated from the average. These errors were verified by Harms and his co-workers [164-166] in consideration of their neutron experiments.

In the case of void streaming errors, these errors may be significantly reduced or eliminated by reduction of the beam size with respect to the voids, and also by reducing the length of the measurement path, thereby reducing the magnitude of the fluctuations in intensity. Problems associated with the fluctuation source of errors may be eliminated by linearizing the signal with respect to the desired quantity, or by sampling on time scales smaller than the fluctuation periods of interest. Many workers have used fine collimation to reduce photon streaming errors [161, 156, 167-169]. One has used linearlization to eliminate fluctuation errors [157,159] and, to the writer's knowledge, only one group has attempted to use alternative methods [164-166] (i.e., short term sampling), to circumvent this problem. In the latter case, however, sampling periods were not judged by this writer to be sufficiently short to eliminate all errors.

Quite recently, the γ-ray or X-ray methods have been extended to multibeam techniques where multiple beams, spreading radially from a single source or in parallel from separate sources, to individual detectors are used. These methods have been utilized most recently in the large scale testing programs being undertaken for nuclear safety studies. Single-source systems have been described by many workers such as Smith [170] who used five beams in a vertical plane plus two references to characterize phase distributions in a horizontal pipe during blowdown experiments. Cut-metal windows, (sometimes Berillium filled), were used for high pressure access while otherwise standard scintillation crystal and photomultiplier instrumentation methods were used. Similarly, Yborrando [171] utilized a γ-ray system having a single source. Lassahan *et al*. [172] recently summarized the technology of X-ray and γ-ray techniques. Such devices are useful compromises for discerning cross-sectional flow variations where flow stratification exists under transient conditions, but their usefulness where a high degree of accuracy is required is limited.

All but one multi-beam system to date have been static, non-traversing, systems usually Cs-137 or Co-60. As such, except in high water-to-metal mass thickness ratio systems, sensitivity tends to be limited and parasitic attenuation high. In a newly described system, however, Abuaf *et al*. [173], describes instead a parallel 5-beam, traversing system utilizing 5 separate Thulium-170 sources. The solid state detection and readout system has been previously described [174] utilizing Cd-Te crystal detectors requiring no elaborate thermal stabilization techniques and using standard nuclear instrumentation. The 84-kev resonance is used since it has the highest activation efficiencies of $\sim$3%, requiring only 30-35 Ci total source activity for each effective Curie at 84-kev. 84-kev source strengths of 1-5 Ci have been reported with activation up to 30 Ci or more expected. The major difficulty encountered was impurity activation, here-to-fore unencountered by other isotope researchers. It was found that any impurity having a highly efficient, high decay energy resonance would lead to difficulties necessitating unwieldy shielding. If impurity levels were maintained at less than 100 ppb, than 2.5 cm of lead would be sufficient at 30 Ci (84-kev) activation level to keep radiation levels at one foot at 3 mr/hr or less except in the beam. At high source strengths, however, the low energy edge yields an extremely sensitive beam with good transient response capabilities.

5.2. Beta-Ray Methods

In 1961, Perkins, Yusuf, and Leppert [175] made the following assessment of γ-attentuation methods:

> "...the γ-method is reasonably accurate for the determination of void volumes under the following conditions: (1) the voids are distributed in a homogeneous manner; (2) the test section offers

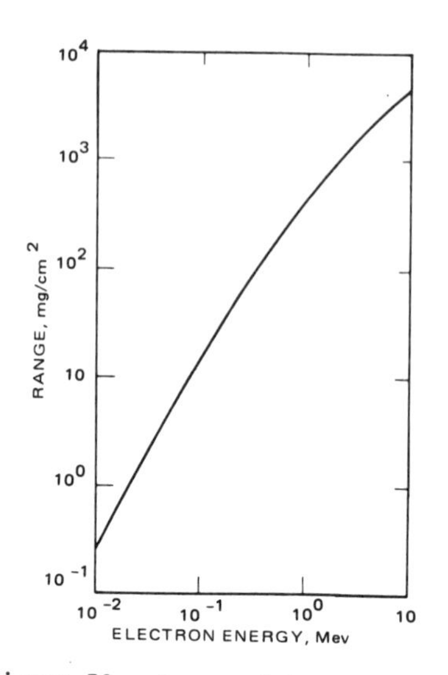

Figure 53 - Range of Beta Particles in Water [160]
(BNL Neg. No. 9-95-79)

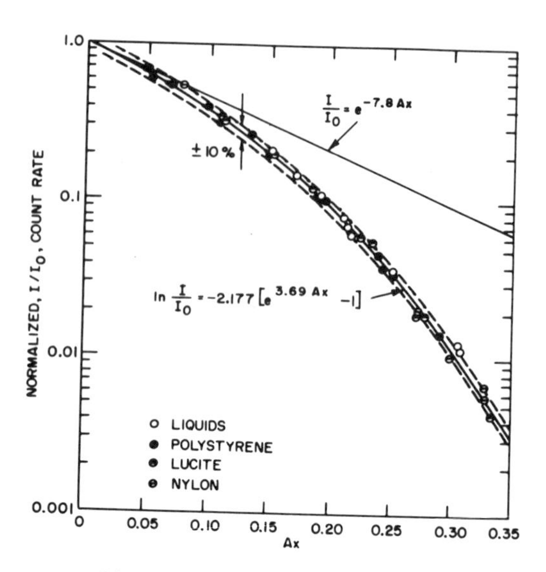

Figure 54 - Experimental Results for Absorption of Betas [175] (BNL Neg. No. 9-118-79)

a radiation path of greater than about 2500 mg/cm^2 (equivalent to one inch of water); and (3) the void fraction is greater than 25%. For smaller void fractions, smaller channels, or preferential distributions, the gamma technique may not offer sufficient accuracy."

While methods and accuracy of β-ray techniques have improved considerably since then, the methods they outlined are worthwhile as representative of the β-ray method of void measurement. Specifically, β-rays are really made up of a stream of electrons emitted at high energy from a radioisotope. The basic method of attenuation for electrons is different than for photons. Attenuation of X-rays and γ-rays is caused by photoelectric effects where X-rays are completely absorbed by ejecting electrons from the absorbing material, and by Compton or recoil scattering where only partial energy loss of a photon occurs with the electron ejection and a new photon of longer wavelengths is emitted. In both cases the electron density of the abosorbing medium is important because interaction is primarily with material electrons. Likewise, the electron density is important for the absorption of β's or free electrons. Just as when electrons decelerating in a target of an X-ray tube generate X-rays, so do they decelerate in any absorber. This loss of energy is caused by both electron interaction and nuclear interaction. The result is some X-ray production in a continuous spectrum of wavelengths called bremstrahlung. Up to a certain thickness of a material this absorption is exponential, similar to photon attenuation. However, over a certain thickness the absorption coefficient is nearly the same for all absorbing materials. The mass per unit area required to produce a given beta absorption is nearly independent of material type and is called the stopping distance or range of the material. The thickness required to stop a β-particle is then the range divided by the material density. The range of an electron, as seen in Figure 53, is necessarily dependent on the initial energy since absorption is mainly by collision. Thus, for water density of 1 gm/cc, a 2 mev electron will be stopped by about 1 cm of water.

While β-absorption is exponential for thin absorbers, Perkins, Yusuf, and Leppert [175] found that a more general law for β's was

$$\ln \frac{I}{I_0} = -2.177 \left\{ e^{3.69Ax} - 1 \right\} \tag{13}$$

where A is proportional to the equivalent linear absorption coefficient.

For small thickness (13) degenerates to

$$I = I_0 \, e^{-7.8Ax} \qquad x \sim 0 \tag{14}$$

These two equations are compared in Figure 54 and it is seen that exponential-like absorption occurs for values of Ax up to about 0.05.

There are two major advantages in using betas for void measurements where complete absorption doesn't occur. First, the sensitivity is greater. Perkins et al. [175] using a Strontium-90 source producing 2.26 mev betas found a sensitivity ratio between the full and empty case of 82:1 whereas for photons of about 60 kev the ratio would have been about 1.15:1. In addition, shielding is not a major obstacle since β's are absorbed readily in most dense materials. A major disadvantage however exists due to the advantage of high sensitivity. That is, due to the high attenuation rates, massive sources are required to obtain event rates at a test detector which would allow accurate transients to be measured. Since electrons are absorbed in the source itself, (self shielding), use of betas becomes impractical for this application.

5.3. Neutron Methods

Lastly, neutron attenuation has been used occasionally to measure void fraction in high pressure channels. Sha and Bonilla [176] employed a Sb-Be neutron source to measure voids within 3% in simulated rod bundle geometry. Sensitivities of over 20 times that of Tm^{170} gammas was obtained. More recently Moss and Kelly [177] measured film thicknesses in a heat pipe to an accuracy of 0.006 inches out of 0.125 inches while Harms and company [178-180] measured void fractions by the one-shot method to an accuracy of 6.5% between 3% and 70% voids. Neutrons being uncharged are not subject to coulomb scattering as are the charged particles including beta rays. If elastic scattering is the predominant mode of neutron slowing down, the lightest elements will tend to take on the largest percentage of the neutron current energy. Thus, in any given collision, neutrons are attenuated much more rapidly by the hydrogen atoms in water than by metallic atoms in, say, a test section wall, and so the sensitivity of void measurement approaches that of the β-ray equipment unhindered by the metal walls of a test section. For a water system it was found [180] that the attenuation law obtained was

$$I = I_0 B(x) e^{-\mu x} \tag{15}$$

where

$$B(x) = 1 + 1.011\ x^2 - 0.475\ x^4 + 0.488\ x^6 \tag{16}$$

Currently, however, the use of neutrons as a test vehicle is extremely limited due to the inaccessibility and general unavailability of nuclear reactors and the high cost of intense alternative sources.

6. SUMMARY

This summary has attempted to outline the basic methodologies used in making two-phase flow measurements today. Rather than provide an up-to-date research synopsis and critique carefully including every latest reference in characteristic sophisticated scientific manner, the writer has instead tried to review the fundamental methods which have found wide acceptance and utilization, and which have provided workers with insight into the mechanics of two-phase, gas-liquid flows. The difficulties of using each approach have been identified and important advances which are expected to lead to significant improvements listed. For the beginning researcher or for the engineer wishing to gain an overview, this is probably already too much.

Specifically, this summary has discussed the design, use, interpretation, and difficulties associated with the following techniques.

A) Instream Sensors with Electrical Output Including:

1) Conductivity Devices

a) Level Probe

b) Needle Probes

c) Wall Probes

2) Impedance Void Meters

3) Hot Film Anemometer

4) Radio Frequency Probe

5) Microthermocouple Probes

6) Optical Probes

a) Glass Rod System

b) Fiber Bundle System

c) U-Shaped Fiber System

d) Wedge-Shaped Fiber System

B) Instream Sensors with Mechanical Output Including:

1) Wall Scoop

2) Porous Sampling Sections

3) Isokinetic Sampling Probe

4) Wall Shear and Momentum Flux Measurement Devices

C) Out-of-Stream Techniques including:

1) X-ray and Gamma-ray Methods

2) Beta-ray Methods

3) Neutron Methods

ACKNOWLEDGEMENTS

This manuscript was originally prepared for the Minnesota Short Course on Fluid Mechanics Measurements, given at the University of Minnesota's Department of Mechanical Engineering during the week of September 10-13, 1979. Certain portions of this manuscript were extracted from the author's prior publications, especially Reference 9. This work was performed under the auspices of the U.S. Nuclear Regulatory Commission.

NOMENCLATURE

English

a	Capillary length
A, A_i	Area, Interfacial area
C, C_1	Coefficients (Eqn. 7-9)
d_1, d_2	r-f probe tip dimensions (Fig. 27)
E, E_{max}	Anemometer probe voltage, (maximum)
f_c	Correlation function (Eqn. 7)
g	Gravitational acceleration
i	Enthalpy
I	Photon intensity
j	volume flux
K	Correlation coefficient (Eqn. 8)
n	Index of refraction
N	Count
N_{Ca}	Capillary number
N_{Go}	Gocher Number
q''	heat flux
R, R_p	Radius, probe resistance
S	Area of PDF
t	Time

$\Delta t_1, \Delta t_2$	Time delays associated with d_1 and d_2.
U	Optical Probe Signal, velocity
x	Thickness
y	Distance from wall
z	Streamline coordinate

Greek

α	Vapor volume or area fractions
β	Vapor volume flow rate fraction
β_W	Water layer angle (optical probe)
μ	Viscosity or linear atenuation coefficient
Γ_v	Rate of vapor generation per unit volume
ξ_i	Interfacial perimeter in A
σ	Surface tension
ρ	Density

Subscripts

B	Bubble
c	Computed
g	Gas
i	Initial or interfacial
ℓ	Liquid
m	Measured
max	Maximum
v	Vapro
w	Withdrawal

REFERENCES

1. Jones, O. C. Jr. and Zuber, N., "Slug-Annular Transition With Particular Reference to Narrow Rectangular Duct," presented at the 1978 International Seminar for Momentum Heat and Mass Transfer in 2-Phase Energy and Chemical Systems, Dubrovnik, Yugoslavia, September 4-9, 1978.

2. Jones, O. C. Jr. and Zuber, N., "Interfacial Passage Frequency for Two-Phase, Gas-Liquid Flows in Narrow Rectangular Ducts," presented at The Institution of Mechanical Engineers Conference, September 13-15, 1977, and published in Heat and Fluid Flow in Water Reactor Safety, 5-10, I Mech E, London, 1977.

3. Jones, O. C. Jr. and Saha, P., "Non-Equilibrium Aspects of Water Reactor Safety," Thermal and Hydraulic Aspects of Nuclear Reactor Safety: Vol. 1. Light Water Reactors, O. C. Jones and S. G. Bankoff, Ed., ASME, 1977.

4. Jones, O. C. Jr. and Zuber, N., "Post-CHF Heat Transfer: A Non-Equilibrium, Relaxation Model," presented at the AIChE-ASME Heat Transfer Conference, Salt Lake City, Utah, August 15-17, 1977.

5. LeLourneau, B. W., and Bergles, A. E., Ed., Two-Phase Flow Instrumentation, ASME, (1969).

6. Hewitt, G. F., and Lovegrove, P. C., "Experimental Methods in Two-Phase Flow Studies," EPRI Report NP-118, (1976).

7. Delhaye, J. M. and Jones, O. C., "A Summary of Experimental Methods for Statistical and Transient Analysis of Two-Phase Gas Liquid Flow," Argonne Report ANL-76-75, (1976).

8. Hewitt, G. F., Measurements of Two-Phase Flow Parameters, to be published by Academic Press, (1978).

9. Jones, O. C., and Delhaye, J. M., "Transient and Statistical Measurement Techniques for Two-Phase Flows: A Critical Review," Int. J. Multiphase Flow, 3, 89-116, (1976).

10. Hsu, Y. Y., Ed., "Two Phase Flow Instrumentation Review Group Meeting," NUREG-0375, (1977).

11. Collier, J. G., Convective Boiling and Condensation, McGraw-Hill, New York (1972).

12. Kordyban, E. S., and Ranov, T., "Experimental Study of the Mechanism of Two-Phase Slug Flow in a Horizontal Tube," Winter Annual ASME Symposium on Multiphase Flow, Philadelphia, November, 1963.

13. Solomon, V. J., "Construction of a Two-Phase Flow Regime Transition Detector," MS Thesis, Mechanical Engineering Department, MIT, June, 1962.

14. Griffith, P., "The Slug-Annular Flow Regime Transition at Elevated Pressure," ANL-6796, November, 1963.

15. Neal, L. G., and Bankoff, S. G., "A High Resolution Resistivity Probe For Determination of Local Void Properties in Gas-Liquid Flow," AIChE J., 9. 4, 490, (1963).

16. Nassos, G. P., "Development of an Electrical Resistivity Probe for Void Fraction Measurements in Air-Water Flow," ANL-6738, June, 1963.

17. Wallis, G. B., "Joint US - Euratom Research and Development Program," Quarterly Progress Report, July, 1963.

18. Jannsen, E., "Two-Phase Flow and Heat Transfer in Multirod Geometries," GEAP-10214, April, 1970.

19. Haberstroh, R. E., and Griffith, P., "The Slug-Annular Two-Phase Flow Regime Transition," ASME Paper No. 65-HT-52, 1965.

20. Chevalier, H., Lakme, C. and Max, J., "Device for the Study of Bubble Flow within a Pipe," English Patent Application No. 36315/65, August, 1965.

21. Gardner, G. C., and Neller, P. H., "Phase Distributions in Flow of an Air-Water Mixture Round Bends and Past Obstructions at the Wall," Paper No. 12, IMechE Conference, Bristol, 1969.

22. Yu, H. S., and Sparrow, E. M., "Experiments on Two-Component Stratified Flow in a Horizontal Duct," ASME Paper No. 68-HT-14, August, 1968.

23. Jannsen, E., "Two-Phase Flow and Heat Transfer in Multirod Geometries, Eighteenth Quarterly Progress Report January 1 - March 31, 1970," GEAP-10214, April, 1970.

24. Akagawa, K., "Fluctuations of Void Ratio in Two-Phase Flow," Bul. JSME, 7, 25, 122, (1964).

25. Lafferty, J. F., and Hammitt, F. G., "A Conductivity Probe for Measuring Local Void Fraction in Two-Phase Flow," Nuc. Appl., 3, 317, (1967).

26. Jannsen, E., "Two-Phase Flow and Heat Transfer in Multirod Geometries," Fourth and Fifth Quarterly Progress Reports," GEAP-5056, January, 1966.

27. Collier, J. G., and Hewitt, G. F., "Film Thickness Measurements," ASME Paper 64-WA/HT-41, (1964).

28. Hewitt, G. F., and Lovegrove, P. C., "Comparative Film Thickness and Holdup Measurements," AERE-M-1203, April, 1963.

29. Hall-Taylor, N., and Hewitt, G. F., "The Motion and Frequency of Large Disturbance Waves in Annular Two-Phase Flow of Air-Water Mixture," AERE-R-3952, June, 1962.

30. Hewitt, G. F., and Wallis, G. B., "Flooding and Associated Phenomena in Falling Film Flow in a Tube," AERE-R-4022, May, 1963.

31. Hewitt, G. F., Kearsey, H. A., Lacy, P.M.C., and Pulling, D. J., "Burnout and Film Flow in the Evaporation of Water in Tubes," AERE-R-4864, March, 1965.

32. Butterworth, D., "Air-Water Climbing Film Flow in an Eccentric Annulus," AERE-R-5787, May, 1968.

33. Moeck, E. O., "The Design, Instrumentation, and Commissioning of the Water-Air-Fog Experimental Rig (WAFER)," APPE-1, Atomic Energy of Canada Ltd., January, 1964.

34. Wickhammer, G. A., Moeck, E. O., MacDonald, I.P.L., "Measurement Techniques in Two-Phase Flow," AECL-2215, October, 1964.

35. Hewitt, G. F., King, R. D., and Lovegrove, P. C., "Techniques for Liquid Film and Pressure Drop Studies in Annular Two-Phase Flow," AERE-R-3921, March, 1962.

36. Adorni, et al., "Experimental Data on Two-Phase Adiabatic Flow; Liquid Film Thickness, Phase and Velocity Distribution, Pressure Drop in Vertical Gas-Liquid Flow," EURAEC-150, (CISE Report R35) (1961).

37. Hewitt, G. F., and Lovegrove, P. C., "Comparative Film Thickness and Holdup Measurements in Vertical Annular Flow," AERE-M-1203, April, 1963.

38. Iida, Y., and Kobayasi, K., 1969, "Distributions of Void Fraction Above a Horizontal Heating Surface in Pool Boiling," Bull. J.S.M.E., 12, pp. 283-290.

39. Iida, Y., and Kobayasi, K., 1970, "An Experimental Investigation of the Mechanism of Pool Boiling Phenomena by a Probe Method," Heat Transfer, Vol. 5, pp. 1-11, Elsevier, Amsterdam.

40. Kobayasi, K., 1974, "Measuring Method of Local Phase Velocites and Void Fraction in Bubbly and Slug Flows," presented at the Fifth Int. Heat Transfer Conf., Round Table RT-1.

41. Reocreux, M. and Flamand, J. C., 1972, "Etude de l'utilisation des sondes resistives dans des ecoulements diphasiques a grande vitesse" CENG, STT, Rapport interne No. 111.

42. Tawfik, H., Alpay, S.A., and Rhodes, E., "Resistivity Probe Error Study Using a Two-Dimensioonal Simulation of the Electrical Field in a Two-Phase Media," Proc. 2nd Multiphase Flow and Heat Transfer Symposium-Workshop, Miami Beach, Florida, April 16-18, 1979.

44. Sheppard, J., private comnmunication.

45. Sekoguchi, K., Fukui, H., and Sato, Y., "Flow Characteristics and Heat Transfer in Vertical Bubble Flow," Proc. Japan-U.S. Seminar on Two-Phase Flow Dynamics, 107--127, Kobe, July 31-August 3, (1979)

46. Block, J., private communication.

47. Gouse, S.W. Jr., "Void Fraction Measurement," AD-600524, April, 1964.

48. Neal, L. S., and Bankoff, S. G., "A High Resolution Resistivity Probe for Determination of Local Void Properties in Gas-Liquid-Flow" A.I.Sh.E. J. 9, 49-54, (1963).

49. Sekoguchi, K., Fukui, H., Matsuoka T., and Nishikawa, K., "Investigation into the Statistical Characteristics of Bubbles in Two-Phase Flow," Trans. J.S.M.E., 40, 336, 2295-2310,(1974).

50. Galaup, J. P., "Contribution a l'etude des methodes de mesure en ecoulement diphasique." These de docteur-ingenieur, Universite Scientifique et Medicale de Grenoble, Institute National Polytechnique de Grenoble, (1975).

51. Uga, T., "Determination of Bubble Size Distribution in a BWR," Nucl. Engng Design, 22, 252-261, (1972).

52. Ibragimov, N., KH., Bobkov, V. P., and Tychinskii, N. A., "Investigation of the Behavior of the Gas Phase in a Turbulent Flow of a Water-Gas Mixture in Channels." Teplofiz. Vysok. Temp., 11, 1051-1061. Also High Temperature, 11, 935-944, Consultants Bureau, New York, (1973).

53. Jones, O. C., Statistical Considerations in Heterogeneous, Two-Phase Flowing Systems, Ph.D. Thesis, Rensselaer Polytechnic Institute, Troy, N.Y., (1973).

54. Serizawa, A., "Fluid-Dynamic Characteristics of Two-Phase Flow," Institute of Atomic Energy, Kyoto University, Ph.D. thesis, (1974).

55. Telles, A. S., and Dukler, A. E., "Statistical Characteristics of Thin, Vertical Wavy Liquid Films," I/EC Fundamentals, 9, 412-421, (1970).

56. Dukler, A. E., "Characterization, Effects, and Modeling of the Wavy Gas-Liquid Interface," Progress in Heat and Mass Transfer (Edited by Hetsroni, G., Sideman, S. & Hartnett, J. P.) Vol. 6, pp. 207-234. Pergamon, Oxford, (1972).

57. Chu, K. J., and Dukler, A. E., "Statistical Characteristics of Thin, Wavy Films," A.I.Ch.E. J., 20, 695-706, (1974).

58. Lecroart, H. and Lewi, J., "Mesures locales et leur interpretation statistique pour un ecoulement diphasique a grande vitesse et taux de vide," Societe Hydrotechnique de France, Douziemes journees de l'Hydraulique, Paris, 1972, Question IV, Rapport 7.

59. Lecroart, H. and Porte R., "Electrical Probes for Study of Two-Phase Flow at High Velocity," Presented at the Interna. Symposium on Two-Phase Systems, Haifa, Israel, (1972).

60. Wamsteker, A.J.J. et al., "The Application of the Impedance Method for Transient Void Fraction Measurement and Comparison with the X-ray Attenuation Technique," EURAEC-1109, June 1964.

61. Leung, J., The Occurrence of Critical Heat Flux During Blowdown with Flow Reversal, M.Sc. Thesis, Northwestern University, Evanston, Illinois, (1976).

62. Cimorelli, L., and Evangelisti, R., "The Application of the Capacitance Method for Void Fraction Measurement in Bulk Boiling Conditions," Int. J. Heat Mass Trans., 10, pp277, (1967).

63. Bencze, I. and Oerbeck, I., "Development and Application of an Instrument for Digital Measurement and Analysis of Void Using an AC Impedance Probe," KR-73, September 1964.

64. Oerbeck, I., "Impedance Void Meter," KR-32, November 1962.

65. Cimorelli, L., DiBartolomeo, and Premoli, A., "Void Fraction Measurement in a Boiling Channel Using the Impedance Method," RT-ING-(65) 7, Oct. 1965.

66. Cimorelli, L,. and Premoli, A., "Measurement of Void Fraction with Impedance Gage Technique," Energia Nucleare, 13, 1, 12, (1966).

67. Nielson, D. S., "Void Fraction Measurements in an Out-of-Pile High-Pressure Rig MK II-A by the Impedance Bridge Method," RISO-M-894, May 1969.

68. Hsu, Y. Y., Simon, F. F. and Grahm, R. W., "Application of Hot Wire Anemometry for Two-Phase Flow Measurement Such as Void Fraction and Slip Velocity," presented at the Two-Phase Flow Symposium at the Winter Annual ASME Meeting, Philadelphia, November 1963.

69. Jones, O. C. and Zuber, N., "Use of a Hot-Film Anemometer for Measurement of Two-Phase Void and Volume Flux Profiles in a Narrow Rectangular Channel," AIChE Sym. Ser., 74, 174, pp. 191-204, (1978).

70. Jones, O. C., "Preliminary Investigation of Hot Film Anemometer in Two-Phase Flow," TID-24104, November 1966.

71. Goldschmidt, V. and Eskmazi, S., "Two-Phase Turbulent Flow in a Plane Jet," ASME Paper No. 66-WA/APM-6, 1966.

72. Goldschmidt, V. and Householder, M. K., "The Hot Wire Anemometer as an Aerosol Droplet Size Sampler," Atmos. Environ., 3, 643, (1969).

73. Bragg, G. M. and Tevaarverk, J., "The Effect of a Liquid Droplet on a Hot Wire Anemometer Probe," Paper No. 2-2-19, presented at the First Symposium on Flow--Its Measurement and Control in Science and Industry, Pittsburgh, May 1971.

74. Delhaye, J. M., "Measurement of the Local Void Fraction Anemometer," CEA-R-3465(E), October 1968.

75. Delhaye, J. M., "Anemometer a Temperature Constante Etalonnage des Sondes a Film Chaude dans les Liquides," CENG Rapport TT No. 290, March 1968.

76. Delhaye, J. M., "Measure du Taux de Vide Local en Ecoulement Diphasique Eau-Air par un Anemometre a Film Chaud," CENG Rapport TT No. 79, October 1967.

77. Delhaye, J. M., "Theoretical and Experimental Results About Air and Water Bubble Boundary Layers," presented at the Novosibirsk Symposium, May 1968.

78. Delhaye, J. M., "Hot Film Anemometry in Two-Phase Flow," presented at the Symposium on Two-Phase Flow Instrumentation at the National Heat Transfer Conference, Minneapolis, August 1969.

79. Bergles, A.E., Roos, J.P., Bourne, J.G., "Investigation of Boiling Flow Regimes and Critical Heat Fluxes, Final Summary Report," Dynatech Corp. Report NYO-3304-13, 1968.

80. Abuaf, N., Swoboda, A. and Zimmer, G. A., "Reactor Safety Research Programs, Quarterly Progress Report," BNL-NUREG-50747, p. 175, 1977.

81. Abuaf, N., Feierabend, T.P., Zimmer, G.A., and Jones, O.C., "Radio Frequency (R-F) Prober for Bubble Size and Velocity Measurements," BNL-NUREG-50997, March 1979.

82. Abuaf, N., Feierabend, T.P., Zimmer, G.A., and Jones, O.C., "Radio Frequency (R-F) Prober for Bubble Size and Velocity Measurements," Rev. Sci. Inst., 50, 10, 1260-1263, Oct. 1979.

83. Fortescue, T., personal communication, 1978.

84. Marcus, B. D. and Dropkin, D., "Measured Temperature Profiles Within the Superheated Boundary Layer Above a Horizontal Surface in Saturated Nucleate Pool Boiling of Water," J. Heat Transfer, 87C, 333-341, (1965).

85. Bonnet, C. and Macke, E., "Fluctuations de temperature dans la paroi chauffante et dans le liquide au cours de l'ebullition nucleee," EUR 3162f. (1966).

86. Lippert, T. E. and Dougall, R. S., "A Study of the Temperature Profiles Measured in the Thermal Sublayer of Water, Freon-113, and Methyl Alcohol During Pool Boiling," J. Heat Transfer, 87C, 333-341, (1965).

87. Jacobs, J. and Shade, A. H., "Measurement of Temperatures Associated with Bubbles in Subcooled Pool Boiling," J. Heat Transfer, 91C, 123-128, (1969).

88. Van Stralen, S.J.D. and Sluyter, W. M., "Local Temperature Fluctuations in Saturated Pool Boiling of Pure Liquids and Binary Mixtures," Int. J. Heat Mass Transfer, 12, 187-198, (1969).

89. Stefanovic, N., Afgan, N., Pislar, V. and Jovanovic, L. J., "Experimental Investigation of the Superheated Boundary in Forced Convection Boiling," Heat Transfer, 5, Elsevier, Amsterdam, (1970).

90. Afgan, N., Jovanovic, L. J., Stefanovic, M. and Pislar, V., "An Approach to the Analysis of Temperature Fluctuation in Two-Phase Flow," Int. J. Heat Mass Transfer, 16, 187-194, (1973).

91. Afgan, N., Stefanovic, M., Jovanovic, L. J. and Pislar, V., "Determination of the Statistical Characteristics of Temperature Fluctuation in Pool Boiling," Int. J. Heat Mass Transfer, 16, 249-256, (1973).

92. Wiebe, J. R. and Judd, R. L., "Superheat Layer Thickness Measurements in Saturated and Subcooled Nucleate Boiling," J. Heat Transfer, 93C, 455-461, (1971).

93. Treschov, G. G., "Experimental Investigation of the Mechanism of Heat Transfer with Surface Boiling of Water," Teploenergetika, 3, 44-48, (1957).

94. Jiji, L. M. and Clark, J. A., "Bubble Boundary Layer and Temperature Profiles for Forced Convection Boiling in Channel Flow," J. Heat Transfer, 86C, 50-58, (1964).

95. Walmet, G. E. and Staub, F. W., "Electrical Probes for Study of Two-Phase Flows," Two-Phase Flow Instrumentation, edited by LeTourneau, B. W. and Bergles, A. E., pp. 84-101, ASME, (1969).

96. Barois, G., "Etude Experimentale de l'autovaporisation d'un Ecoulement Ascendant Adiabatique d'eau dans un Canal de Section Uniforme," These de Docteur-Ingenieur, Faculte des Sciences de l'Universite de Grenoble, (1969).

97. Van Paassen, C.A.A., "Thermal Droplet Size Measurements Using a Thermocouple," Int. J. Heat Mass Transfer, 17, 1527-1548, (1974).

98. Delhaye, J.M., Semeria, R., and Flamand, J.C., "Void Fraction, Vapor and Liquid Temperatures: Local Measurements in Two-Phase Flow Using a Microthermocouple," J. Heat Trans., 95C, pp365-370, (1973).

99. Delhaye, J.M., Semeria, R., and Flamand, J.C., "Mesure du Taux de Vide et des Temperatures du Liquid et de la Vapeur Ecoulement Diphasique Avec Changement de Phase a l;Aide d;un Microthermocouple," CEA-R4302, (1972).

100. Miller, N. and Mitchie, R. E., "Electrical Probes for Study of Two-Phase Flows," Two-Phase Flow Instrumentation (edited by LeTourneau, B. W. and Bergles, A. E.), pp. 82-88, ASME, (1969).

101. Miller, N. and Mitchie, R. E., "Measurement of Local Voidage in Liquid/Gas Two-Phase Flow Systems," J. Br. Nucl. Energy Soc., 9, 94-100, (1970).

102. Bell, R., Boyce, B. E. and Collier, J. G., "The Structure of a Submerged Impinging Gas Jet," J. Br. Nucl. Energy Soc., 11, 183-193, (1972).

103. Kennedy, T.D.A. and Collier, J. G., "The Structure of an Impinging Gas Jet Submerged in a Liquid," Multi-Phase Flow Systems, Inst. Chem. Engng. Symp. Ser. No. 38, II, Paper J4, (1974).

104. Hinata, S., "A Study on the Measurement of the Local Void Fraction by the Optical Fibre Glass Probe," Bull. J.S.M.E., 15, 1228-1235, (1972).

105. Danel, F. and Delhaye, J. M., "Sonde Optique pour Mesure du Taux de Presence Local en Ecoulement Diphasique," Mesures-Regulation-Automatisme, pp. 99-101, (1971).

106. Delhaye, J. M. and Galaup, J. P., "Measurement of Local Void Fraction in Freon-12 with a 0.1 mm Optical Fiber Probe," private communication (1975).

107. Abuaf, N., Jones, O.C., and Zimmer, G.A., "Response Characteristics of Optical Probes," ASME Preprint 78-WA/HT-3, August, 1978.

108. Abuaf, N., Jones, O. C., Jr. and Zimmer, G. A., "Optical Probe for Local Void Fraction and Interface Velocity Measurements," Rev. Sci. Instruments, 49, 8, 1090-1094, August, 1978.

109. Hsu, Y. Y., Ed., "Two Phase Flow Instrumentation Review Group Meeting," NUREG-0375 (1977).

110. White, D. A. and Tallmadge, J.A., "A Gravity Corrected Theory for Cylinder Withdrawal," AIChE J, 13, 4, pp. 745-750, 1967.

111. Tallmadge, J. A. and White, D. A., "Film Properties and Design Procedures in Cylinder Withdrawal," I&EC Process Design and Development, 7, 4, pp. 503-508, 1968.

112. White, D. A. and Tallmadge, J. A., "A Theory of Withdrawal of Cylinders from Liquid Baths," AIChE J, 12, 2, pp. 233-339, 1966.

113. Cravarolo, L. and Hassid, A., "Phase and Velocity Distribution in Two-Phase Adiabatic Dispersed Flow," CISE-R-98, August 1963.

114. Adorni, N., Alia, P., Cravarolo, L., Hassid, A. and Pedrocchi, E., "An Isokinetic Sampling Probe for Phase and Velocity Distribution Measurements in Two-Phase Flow Near the Wall of the Conduit," CISE-R-89, December 1963.

115. Truong Quang Minh and J. Huyghe, "Measurement and Correlation of Entrainment Fraction in Two-Phase, Two-Component, Annular Dispersed Flow," CENG Report No. TT-52, June 1965.

116. Moeck, E. O., "Measurement of Liquid Film Flow and Wall Shear Stress in Two-Phase Flow," Symposium on Two-Phase Flow Instrumentation, National Heat Transfer Conference, Minneapolis, August 1969.

117. Jannsen, E., "Two-Phase Flow and Heat Transfer in Multirod Geometries, Third Quarterly Progress Report April to July 1965," GEAP-4933, August 1965.

118. Hewitt, G. F., Kearsey, H. A., Lacy, P.M.C. and Pulling, D. J., "Burnout and Film Flow in the Evaporation of Water in Tubes," AERE-4864, March 1965.

119. Moeck, E. O., "Annular Dispersed Two-Phase Flow and Critical Heat Flux," AECL-3656, May 1970.

120. Hewitt, G. F. and Wallis, G. B., "Flooding and Associated Phenomena in Falling Film Flow in a Tube," AERE-R-4022, May 1963.

121. Hewitt, G. F., Kearsey, H. A., Lacy, P.M.C. and Pulling, D. J., "Burnout and Nucleation in Climbing Film Flow," AERE-R-4374, August 1963.

122. Hewitt, G. F., Lacy, P.M.C. and Nichols, B., "Transitions in Film Flow in a Vertical Tube," AERE-R-4614, April 1965.

123. Cousins, L. B., Denton, W. H. and Hewitt, G. F., "Liquid Mass Transfer in Annular Two-Phase Flow," AERE-R-4926, May 1965.

124. Butterworth, D., "Air-Water Climbing Film Flow in an Eccentric Annulus," AERE-R-5787, May 1968.

125. Singh, K., St. Pierre, C. C., Crago, W. A. and Moeck, E. O., "Liquid Film Flow Rates in Two-Phase Flow of Steam and Water at 1000 psia," AIChE J, 15, 1, 51, 1969.

126. Schraub, F. A., Simpson, R. L. and Jannsen, E., "Two-Phase Flow and Heat Transfer in Multirod Geometries: Air-Water Flow Structure Data for Round Tube, Concentric and Eccentric Annulus, and Nine-Rod Bundle," GEAP-5739, January 1969.

127. Staniforth, R., Stevens, G. F. and Wood, R. W., "An Experimental Investigation into the Relationship Between Burnout and Film Flow Rate in a Uniformly Heated Round Tube," AEEW-R-430, March 1965.

128. Adorni, N., Casagrande, I., Cravarolo, L., Hassid, A. and Silvestri, M., Experimental Data on Two-Phase Adiabatic Flow; Liquid Film Thickness, Phase and Velocity Distributions, Pressure Drops in Vertical Gas-Liquid Flow," CISE-R-35 (EUREAC-150), 1961.

129. Lahey, R. T., Jr., Shiralkar, B. S. and Radcliff, D. W., "Subchannel and Pressure Drop Measurements in a Nine-Rod Bundle for Diabatic and Adiabatic Conditions," GEP-13049, March 1970.

130. Schraub, F. A., "Isokinetic Sampling Probe Techniques Applied to Two-Component, Two-Phase Flow," ASME Paper No. 67-WA/FE-28, 1967. Also CEAP-5287, November 1966.

131. Todd, K. W. and Fallon, D. J., "Erosion Control in the Wet-Steam Turbine," Proc. I.M.E., 35, 180, 1965.

132. Jannsen, E., "Two-Phase Flow and Heat Transfer in Multirod Geometries: Second Quarterly Progress Report, January to April 1965," GEAP-4863, May 1965.

133. Wallis, G. B. and Steen, D. A., "Two-Phase Flow and Boiling Heat Transfer: Quarterly Progress Report for July to September 1963," NYO-10,488 October 1963.

134. Burick, R. J., Scheuerman, C. H. and Falk, A. Y., "Determination of Local Values of Gas and Liquid Mass Flux in Highly Loaded Two-Phase Flow," Paper No. 1-5-21 presented at the First Symposium on Flow--Its Measurements and Control in Science and Industry, Pittsburgh, May 1971.

135. Dussord, J. L. and Shapiro, A. H., "A Deceleration Probe for Measuring Stagnation Pressure and Velocity of a Particle-Laden Gas Stream," Jet Propulsion, p. 24, January 1958.

136. Ryley, D. J. and Kirkman, G. A., "The Concurrent Measurement of Momentum and Stagnation Enthalpy in a High Quality Wet Steam Flow," Paper No., 26, IME Thermodynamics and Fluid Mechanics Convention, Bristol, March 1968.

137. Schraub, F. A., "Isokinetic Probe and Other Two-Phase Sampling Devices: A Survey," presented at the Symposium on Two-Phase Flow Instrumentation, National Heat Transfer Conference, Minneapolis, 1969.

138. Shires, G. L. and Riley, P. J., "The Measurement of Radial Voidage Distribution in Two-Phase Flow by Isokinetic Sampling," AEEW-M-650, 1966.

139. Gill, L. E., Hewitt, G. F., Hitchon, J. W. and Lacy, P.M.C., "Sampling Probe Studies of the Gas Core in Annular Two-Phase Flow, Part 1; The Effect of Length on Phase and Velocity Distributions," Chem. Eng. Sci., 18, 525, 1963.

140. Preston, J. H., "Determination of Turbulent Skin Friction by Means of Pitot Tubes," J. Roy. Aero. Soc., 58, 109, 1954.

141. Patel, V. C., "Calibration of the Preston Tube and Limitations of its Use in Pressure Gradients," J. Fluid Mech., 23, 185, 1965.

142. King, C. W., "Measurement of Wall Shear Stress of a High Velocity Vapor Condensing in a Vertical Tube," PhD Thesis, University of Connecticut, 1970.

143. Jannsen, E., "Two-Phase Flow and Heat Transfer in Multirod Geometries; Eight Quarterly Progress Report, July to October 1966," GEAP-5300, November 1966.

144. Cravarolo, L., Giorgini, A., Hassid, A. and Pedrocchi, E., "A Device for the Measurement of Shear Stress on the Wall of a Conduit--Its Application in Mean Density Determination in Two-Phase Flow Shear Stress Data in Two-Phase Adiabatic Vertical Flow," CISE-R-82 (EURAEC-930), February 1964.

145. Rose, S. C., "Some Hydrodynamic Characteristics of Bubbly Mixtures Flowing Vertically Upwards in Tubes," ScD Thesis, MIT, September 1964.

146. Rose, S. C. and Griffith, P., "Flow Properties of Bubbly Mixtures," ASME Paper No. 65-HT-58, 1965.

147. Andeen, G. B. and Griffith, P., "The Momentum Flux in Two-Phase Flow," MIT-3496-1, October 1965.

148. Arnold, C. R. and Hewitt, G. F., "Further Developments in the Photography of Two-Phase Flow," AERE-R-5318, January 1967.

149. Cooper, K. D., Hewitt, G. F. and Pinchin, B., "Photography of Two-Phase Flow," AERE-R-4301, May 1963.

150. Hsu, Y. Y., Simoneau, F. J., Simon, F. F. and Grahm, R. W., "Photographic and Other Optical Techniques for Studying Two-Phase Flow," presented at the Symposium on Two-Phase Flow Instrumentation, National Heat Transfer Conference, Minneapolis, August 1969.

151. Lockart, R. W. and Martinelli, R. C., "Proposed Correlation of Data for Iso-Thermal Two-Phase, Two-Component Flow in Pipes," Chem. Eng. Progr. 44, 1944.

152. Hewitt, C. F., King, I. and Lovegrove, P. C., "Holdup and Pressure Drop Measurements in Two-Phase Annular Flow of Air-Water Mixtures," AERE-R-3764, June 1964.

153. Neal, L. G., "Local Parameters in Cocurrent Mercury-Nitrogen Flow," ANL-6625, January 1963.

154. Cravarolo, L. and Hassid, A., "Liquid Volume Fraction in Two-Phase, Adiabatic Systems," Energia Nucleare, 12, 11, 1965.

155. Petrick, M., "Two-Phase Air-Water Flow Phenomena," ANL-5787, March 1958.

156. Pike, R. W., Wilkinson, B., Jr., and Ward, H. C., "Measurement of the Void Fraction in Two-Phase Flow by X-ray Attenuation," AIChE J., 11, 5, 1965.

157. Jones, O.C., and Zuber, N., "The Interrelation Between Void Fraction Fluctuation and Flow Patterns in Two-Phase Flow," Int. J. Multiphase Flow, 2, 273-306, 1975.

158. Schrock, V. E. and Selph, F. B., "An X-ray Densitometer for Transient Steam Void Measurement," SAN-1005, March 1963.

159. Jones, O.C., "Determination of Transient Characteristics of an X-ray Void Measurement System for Use in Studies of Two-Phase Flow," KAPL-3859, February 1970.

160. Schroch, V. E., "Radiation Techniques in Two-Phase Flow Measurement," presented at the Symposium on Two-Phase Instrumentation, National Heat Transfer Conference, Minneapolis, August 1969.

161. Zuber, N., Staub, F. W., Bijwaard, G., and Kroeger, P. G., "Steady State and Transient Void Fraction in Two-Phase Flow Systems--Final Report for the Program on Two-Phase Flow Investigation," GEAP-5417, January 1967.

162. Hooker, H. H. and Popper, G. F., "A Gamma-Ray Attenuation Method for Void Fraction Determination in Experimental Boiling Heat Transfer Test Facilities," ANL-5766, November 1958.

163. Jones, O. C., "Procedural and Calculational Errors in Void Fraction Measurements by Particle or Photon Attenuation Techniques," KAPL-3361, October 1967.

164. Harms, A. A. and Forrest, C. F., "Dynamic Effects in Radiation Diagnosis of Fluctuating Voids," Nuc. Sci. Eng., 46, 408-413, 1971.

165. Harms, A. A. and Laratta, F.A.R., "The Dynamic-Bias in Radiation Interrogation of Two-Phase Flow," Int. J. Heat Mass Transfer, 16, 1459-1465, 1973.

166. Hancox, W. T., Forrest, C. F. and Harms, A. A., "Void Determination in Two-Phase Systems Employing Neutron Transmission," ASME Paper 72-HT-2, 1972.

167. Bestenbreur, T. P. and Spigt, C. L., "Study of Mixing Between Adjacent Channels in an Atmospheric Air-Water System," presented at the Two-Phase Flow Meeting at Winfrith, June 12-16, 1967. (See CONF-67065-6)

168. Martin, R., "Measurements of the Local Void Fraction at High Pressure in a Heating Channel," Nuc. Sci. Eng., 48, 125, 1972.

169. Spigt, C. L., Wamsteker, A.J.J. and von Vlaardingen, H. F., "Review of the Measuring, Recording and Analyzing Methods in Use in the Two-Phase Flow Programme of the Laboratory of Heat Transfer and Reactor Engineering at the Technological University of Eindhoven," Report WW016-R64 (EURATOM #III-17, Special TR#18), June 1964.

170. Smith, A. V., "A Fast Response Multi-Beam X-ray Absorpiton Technique for Identifying Phase Distributions During Steam-Water Blowdowns," J. Br. Nucl. Ener. Soc., 14, pp. 227-235, July 1975.

171. Yborrondo, Y. "Dynamic Analysis of Pressure Transducers and Two-Phase Flow Instrumentation," presented at the Third Water Reactor Safety Research Information Meeting, Washington, D.C., September 29-October 2, 1975.

172. Lassahn, G. D., Stephens, A. G., Taylor, J. D. and Wood, D. B., "X-Ray and Gamma-Ray Transmission Densitometry," presented at the International Colloquium on Two-Phase Flow Instrumentation, Idaho Falls, Idaho, June 11-14, 1979.

173. Abuaf, N., Zimmer, G. A., and Jones, O. C., private communication.

174. Zimmer, G. A., Wu, B.J.C., Leonhardt, W. J., Abuaf, N., and Jones, O. C., "Pressure and Void Distributions in a Converging-Diverging Nozzle with Non-Equilibrum Water Vapor Generation," BNL-NUREG-26003, April 1979.

175. Perkins, H. C., Jr., Yusuf, M., and Leppert, G., "A Void Measurement Technique for Local Boiling," Nuc. Sci. Eng., 11, 304, 1961.

176. Sha, W. T. and Bonilla, C. F., "Out-of-Pile Steam-Fraction Determination by Neutron-Beam Attenuation," Nuc. Appl., 1, 69, 1965.

177. Moss, R. A. and Kelly, A. J., "Neutron Radiographic Study of Limiting Planar Heat Pipe Performance," Int. J. of Heat and Mass Trans., 13, 3, 491, 1970.

178. Harms, A. A. and Forrest, C. F., "Dynamic Effects in Radiation Diagnosis of Fluctuating Voids," Nuc. Sci. Eng., 46, 408, 1971.

179. Harms, A. A., Lo, S., and Hancox, W. T., "Measurement of Time-Averaged Voids by Neutron Diagnosis," J. Appl. Phys., 42, 10, 4080 (1971).

180. Hancox, W. T., Forrest, C. F., and Harms, A. A., "Void Determination in Two-Phase Systems Employing Neutron Transmission," ASME Paper 72-HT-2, National Heat Transfer Conference, Denver, 1972.

181. Ornstein, H. L., "An Investigation of Turbulent Open Channel Flow Simulating Water Desalination Flash Evaporators," PhD Thesis, University of Connecticut, 1970.

182. Morrow, T. B. and Kline, S. J., "The Evaluation and Use of Hot-Wire and Hot-Film Anemometers in Liquids," Stanford University Report MD-25, 1971.

183. Hollasch, K. and Gebhart, B., "Calibration of Constant-Temperature Hot-Wire Anemometers at Low Velocities in Water With Variable Fluid Temperature," J. Heat Transfer, 94C, 17-22, 1972.

184. Hurt, J. C. and Welty, J. R., "The Use of a Hot-Film Anemometer to Measure Velocities Below 5 cm/sec in Mercury," J. Heat Transfer, 95C, 548-549, 1973.

185. Rosler, R. S. and Bankoff, S. G., "Large-Scale Turbulence Characteristics of a Submerged Water Jet," AIChE J, 9, 672-676, 1963.

186. Bouvard, M. and Dumas, H., "Application de la Methode du Fil Chaud a la Mesure de la Turbulence dans l'Eau," Houille Blanche, 3, 257-270, 1967.

187. Resch, F., "Etudes sur le Fil Chaud et le Film Chaud dans l'Eau," Houille Blanche, 2, 151-161, 1969.

188. Resch, F., "Etudes sur le Fil Chaud et le Film Chaud dans l'Eau," CEA-R3510, 1968.

189. Goldschmidt, V. W. and Eskinazi, S., "Diffusion de Particules Liquides dans le Champ Retardataire d'un Jet d'air Plan et Turbulent," Les Instabilites en Hydraulique et en Mecanique des Fluides, pp. 291-298. Societe Hydrotechnique de France, 8 emes journees de l'Hydraulique, Lille, France, 1964.

190. Goldschmidt, V. W. and Eskinazi, S., "Two-Phase Turbulent Flow in a Plane Jet," J. Appl. Mech., 33, 735-747, 1966.

191. Lackme, C., "Structure et Cinematique des Ecoulementes Diphasiques a Bulles," CEA-R3202, 1967.

192. Ginsberg, T., "Droplet Transport in Turbulent Pipe Flow," ANL-7694, 1971.

193. Goldschmidt, V. W., "Measurement of Aerosol Concentrations With A Hot-Wire Anemometer," J. Colloid Sci., 20, 617-634, 1965.

194. Goldschmidt, V. W. and Householder, M. K., "The Hot-Wire Anemometer as an Aerosol Droplet Size Sampler," Atmospheric Environment, 3, pp. 643-651, 1969.

195. Chuang, S. C. and Goldschmidt, V. W., "The Response of a Hot-Wire Anemometer to a Bubble of Air in Water," Turbulence Measurements in Liquids (edited by Patterson, G. K. and Zakin, J. L.), Univ. of Missouri, Rolla, Continuing Education Series, 1969.

196. Goldschmidt, V. W., Householder, M. K., Ahmadi, G., and Chuang, S. C., "Turbulent Diffusion of Small Particles Suspended in Turbulent Jets," Progress in Heat and Mass Transfer (edited by Hetsroni, G., Sideman, S., and Hartnett, J. P.), 6, pp. 487-508, Pergamon, Oxford, 1972.

197. Resch, F. J. and Leutheusser, J. H., "Le Ressaut Hydraulique; Mesures de Turbulence dans la Region Diphasique," Houille Blanche, 4, 279-293, 1972.

198. Resch, F. J., Leutheusser, H. J., and Alemu, S., "Bubbly Two-Phase Flow in Hydraulic Jump," J. Hydraul. Div. Am. Soc. Civ. Engrs., 100, No. HY1, Proc. Paper 10297, 137-149, 1974.

199. Ishigai, S., Nakanisi, S., Koizumi, T., and Oyabu, Z., "Hydrodynamics and Heat Transfer of Vertical Falling Liquid Films (Part I: Classification of Flow Regimes)," Bull. J.S.M.E., 15, 594-602, 1972.

200. Katarzhis, A. K., Kosterin, S. I., and Sheinin, B. I., "An Electric Method of Recording the Stratification of the Steam-Water Mixture," Izv. Akad. Nauk SSSR, 2, 132-136, A.E.R.E. Lib./Trans. 590, 1955.

201. Shiralkar, B. S., "Local Void Fraction Measurements in Freon-114 With a Hot-Wire Anemometer," NEDO-13158, General Electric Co., 1970.

202. Dix, G. E., "Vapor Void Fractions for Forced Convection With Subcooled Boiling at Low Flow Rates," PhD Thesis, University of California at Berkeley, 1971. Also General Electric Co., NEDO-10491, November 1971.

203. Shiralkar, B. S. and Lahey, R. T., Jr., "Diabatic Local Void Fraction Measurements in Freon-114 With a Hot-Wire Anemometer," ANS Trans., 15, No. 2, p. 880, 1972.

204. Hetsroni, G., Cuttler, J. M., and Sokolov, M., "Measurements of Velocity and Droplets Concentration in Two-Phase Flows," J. Appl. Mech., 36E, 334-335, 1969.

205. Subbotin, V. I., Sorokin, D. N., and Tsiganok, A. A., "Some Problems on Pool Boiling Heat Transfer," Heat Transfer, Vol 5, Elsevier, Amsterdam, 1970.

206. Petrick, M., and Swanson, B.S., "Radiation Attenuation Methods for Measuring Density in a Two-Phase Fluid," Rev. Sci. Inst., 29, 1079-1085, 1958.

25 THE ACCIDENT AT THREE MILE ISLAND

by

J.G. Collier & L.M. Davies
UKAEA, Harwell, UK

S. Levy
S. Levy Inc., Campbell, CA, USA

On 28th March 1979 what has been called the worst accident in the history of commercial nuclear power generation occurred at the Three Mile Island No. 2 reactor unit in Pennsylvania. This chapter describes the sequence of events which took place during the first 16 hours of the incident and considers the various fundamental heat transfer processes which occurred over this time period. There was extensive damage to the core of the reactor and some release of gaseous fission products from the station. However, the radiation doses received by the general public as a result of the exposure to this released radioactivity were so small that there will be no detectable adverse health effects. Nevertheless, the accident at Three Mile Island will have a profound effect upon commercial nuclear power in the United States and throughout the world. Attention is therefore drawn to particular items which might have been overlooked during the design and operation of the power plant.

1. INTRODUCTION

Heat transfer is an essential part of nuclear engineering in general and nuclear safety in particular. The events that occurred at Three Mile Island Unit 2 (TMI-2) nuclear plant on 28th March 1979 are therefore of considerable interest to the heat transfer community. Even after all the publicity given to the incident, it is difficult for those not actually involved with nuclear safety research to obtain a clear picture of the events. This chapter attempts to give a brief but succinct account of the incident, highlighting the features of particular interest to, and the implications for the heat transfer engineer.

2. THE PLANT

Three Mile Island in the Susquehanna River is about 10 miles south-east of Harrisburg, Pennsylvania. Two pressurised water reactors, each of approximately 900 MW(e) designed and built by the U.S. Babcock & Wilcox Co., are situated on the island. Unit No.1 had been in commercial service since 1974; Unit No.2 entered service in December 1978.

Figure 1 shows the main primary circuit components. The reactor vessel contains the nuclear fuel which is cooled by water circulating in two independent coolant loops. Two reactor coolant pumps in each loop force the water through the reactor vessel and the steam generators. The latter are shell and tube heat exchangers in which the hot high pressure water from the core flows through the tubes and generates steam in the shell on the secondary side. The steam produced in the steam generators is used to drive a conventional turbo-

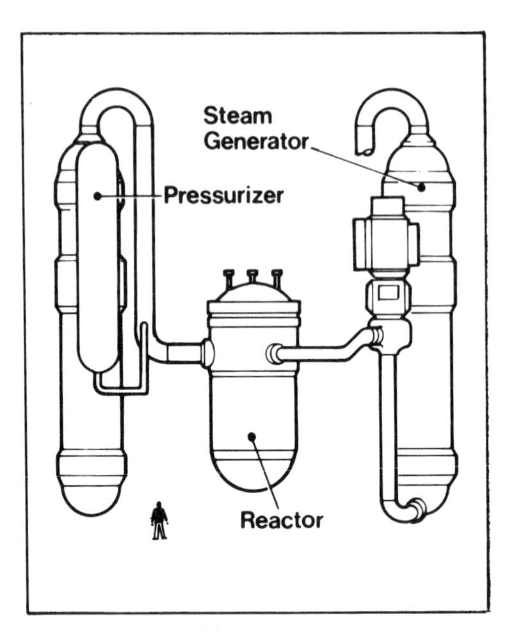

Figure 1: Main primary circuit components for B&W PWR

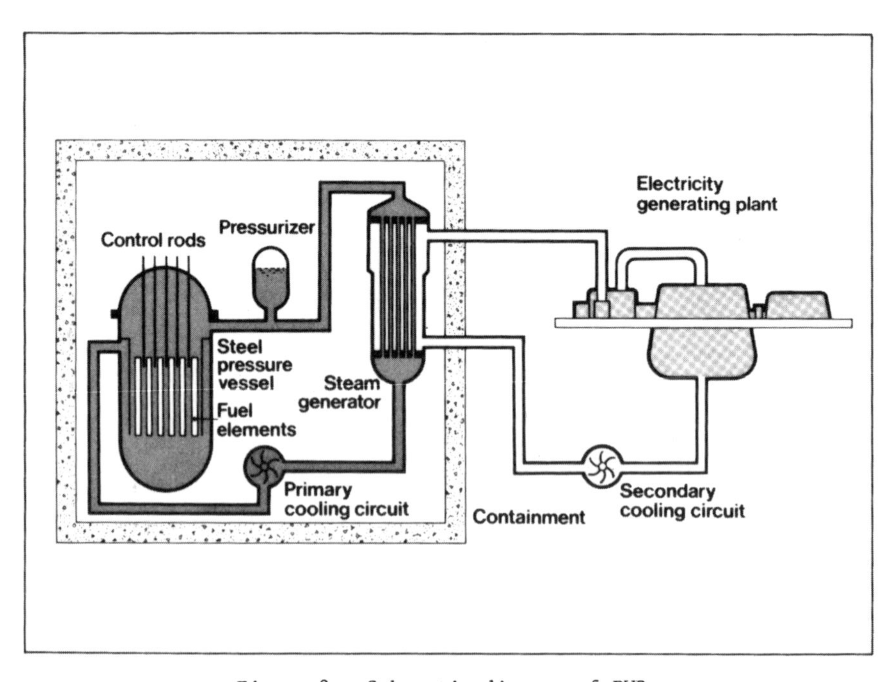

Figure 2: Schematic diagram of PWR

generator (Figure 2). A unique feature of the B&W design is that the steam generators are of the "once-through" type which means that they contain a comparatively low water inventory to sustain heat removal from the primary circuit in the event of a loss of feedwater. The fuel is in the form of enriched uranium dioxide pellets encased in zirconium alloy (Zircaloy) tubing. These fuel pins are arranged in clusters in the reactor core and are cooled by light (ordinary) water. There are nearly 37,000 such fuel pins in the reactor. The water also acts as a moderator to slow down the neutrons. To prevent boiling the coolant system is operated at high pressure (152 bar).

The pressuriser is an important component in the primary circuit. Since the volume of water changes with temperature, it is necessary to have a free space or "bubble" in the primary circuit to provide elasticity and to control the primary system pressure. The pressuriser is equipped with electrical heating which can generate steam and thus increase circuit pressure and with a cold water spraying system which can condense steam and thereby reduce pressure. Because steam and water are in equilibrium in the pressuriser it is the hottest part of the circuit. Under transient conditions steam can be blown off through a Power Operated Relief Valve (PORV) connected to the steam volume in the pressuriser. The tail pipe from the PORV is connected to the Reactor Coolant Drain Tank (RCDT) which has a volume of $28 m^3$. There is a cooling coil in this tank to cool down its contents. The pressure relief valves on the tank discharge to the reactor building sump through floor drains.

The primary coolant circuit, together with the steam generators, are housed in a reactor building which is made leak-tight - known as the reactor containment (Figure 2). The containment is designed to retain the entire contents of the primary circuit should a major leak occur.

The PWR is also equipped with a number of special additional safety systems. In the event of an accident involving leakage of the primary system coolant, the reactor can be cooled by a group of systems collectively known as the Emergency Core Cooling System (ECCS). The main components of this system are:

- High Pressure Injection System (HPIS) - which uses pumps which are capable of forcing borated water into the system at pressures up to and above the normal operating pressure. This system is automatically triggered when the primary coolant system pressure falls below 110 bar.

- Core Flooding System (CFS) - an arrangement of pressurised tanks which automatically inject cold borated water when the system pressure falls below 41 bar.

- Low Pressure Injection System (LPIS) - which operates at 28 bar. This system can draw water from a large tank of borated water or, alternatively, from the sump of the containment building, thus setting up a recirculating system which can be kept going indefinitely.

In the event of a large loss of coolant during an accident, there is a containment water spray system which contains dilute sodium hydroxide. The spray is triggered by high containment pressure and operates to cool down the contents of the containment and to remove any iodine leaking from damaged fuel.

The containment building can be isolated either manually or automatically during the course of an accident. In TMI-2 a high containment pressure is needed to trigger automatic isolation. (In some other plants ECCS operation triggers isolation.) The containment is also fitted with hydrogen recombiners to remove free hydrogen which might arise from radiolysis of the coolant or from a reaction between steam and the zirconium fuel cladding.

Adjacent to the containment building is the auxiliary and fuel handling building which contains, amongst other systems, the make-up and waste treatment plants.

The turbo-generator is housed in the turbine building together with the condensate and feedwater treatment plants. Feedwater for the secondary circuit is normally provided from the condenser. In the event of a fault which results in a loss of this secondary circuit feedwater flow, an auxiliary circuit is provided which can inject water directly into the steam generators. This auxiliary feedwater system has three separate pumps, two electrically driven and one driven by a steam turbine (Figure 3).

3. STATUS OF THE PLANT PRIOR TO INCIDENT

In the early hours of 28th March 1979 TMI-2 was operating under automatic control at 97% of its rated output of 2772 MW(t) (961 MW(e)). There was a persistent slight leakage from either the PORV or from the ASME code safety valves situated on the pressuriser (approximately 0.3 kg/s). Difficulties were being experienced at the time in transferring chemical treatment resins from an isolated condensate water system to a regeneration tank. This transfer is normally accomplished by injecting compressed air and demineralised water into the condensate polishing vessel.

What follows is an abbreviated and simplified account of the initiating incident and the sequence of events which followed. For ease of description the accident may be divided up into a series of arbitrary phases. Each phase is described with the aid of illustrations.

4. THE ACCIDENT SEQUENCE

4.1. Initiating incident

The efforts to transfer the clogged resins appear to have caused water to get into the service air system. Some water from the service air line appears to have entered the instrument air line causing the condensate polishing isolation valves to drift shut and the condensate booster pump to trip because of loss of suction pressure. In turn, the main feedwater pumps in the secondary circuit tripped resulting in an almost instantaneous trip of the main turbine. The time was 04.00.37.

4.2. Phase 1: Turbine Trip (0-6 minutes) (Figure 3)

The valves which allow steam to be dumped to the condenser opened and the auxiliary feedwater pumps started. The interruption in the flow of feedwater to the steam generators caused a reduction in the heat removal from the primary system. The reactor coolant system responded to the turbine trip in the expected manner. The reactor coolant pumps continued to operate to maintain coolant flow through the core. The reactor coolant system pressure started to rise because the heat generated by the core - which was still operating - was not being removed from the system at the required rate by the steam generators.

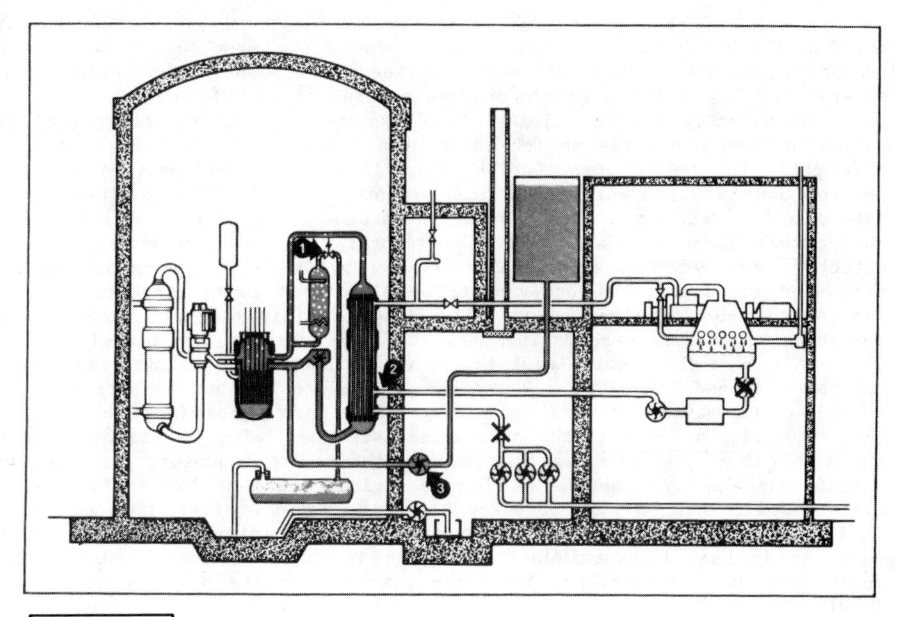

Figure 3: Phase 1 - Turbine Trip

Between 3 and 6 seconds after the start of the incident the Reactor Coolant System (RCS) pressure reached the PORV set point of 155 bar and the valve operated to relieve the pressure. This was, however, insufficient to reduce the pressure immediately and the RCS pressure continued to increase until at 8 seconds into the incident the pressure reached 162 bar - the set point for tripping the reactor. The control rods were driven into the core and the fission reaction stopped.

Thus, at this early stage all the automatic protection features had operated as designed and the reactor had been shut down. There was no further generation of heat as a result of fission but there was a continuing need to remove the heat generated by the decay of fission products in the fuel. The amount of decay heat generation decreases with time; the following table gives approximate values appropriate for TMI-2.

Time after shutdown	MW(t)
1 minute	97
1 hour	36
1 day	13
1 week	5.1
1 month	2.1

At 13 seconds the now reducing RCS pressure reached the set point for utomatic PORV closure (152 bar) (1)*. It failed to close. So started a equence of events where, to paraphrase an earlier instance at Concord some 200 ears earlier, the "shot was heard around the world".

These figures refer to the particular components identified in the illustrations

Formally, the plant had suffered a "small breach loss-of-coolant-accident". RCS water was being lost through the stuck open PORV. In the secondary circuit, all three auxiliary feedwater pumps were running, but the water level in the steam generators was continuing to fall and they were drying out. No water was being injected into the steam generators because of closed valves between the pumps and the steam generators - they had been closed some time prior to the incident (probably at least 42 hours earlier) for routine testing and had apparently been inadvertently left in that position. Status tags on other valves obscured some of the status lights for these valves. Thus, during this first crucial period the reactor coolant circuit was deprived of effective means of heat removal and could only dispose of the energy by blowing off water and steam. At one minute the difference in temperature between the hot and cold legs of the primary circuit was rapidly reaching zero, indicating the steam generators were drying out. The RCS pressure was also dropping. At about this time the liquid level in the pressuriser began to rise rapidly. At 2 minutes 4 seconds the RCS pressure had dropped to 110 bar and the ECCS system triggered automatically feeding borated water into the primary coolant system (3). The liquid level in the pressuriser was continuing to rise. Concern was expressed that the HPIS was increasing the water inventory in the primary circuit and the steam bubble in the pressuriser would be lost. In effect, the circuit would then be full of water or water "solid". (The HPIS shut-off head is 197 bar.) Subsequent analysis has shown that, initially, expansion of the water as it heated up and later, boiling in parts of the circuit displacing water into the pressuriser, were the reasons for the increasing pressuriser level.

One of the HPIS pumps was tripped at 4 minutes 38 seconds with the others continuing to be operated in a throttled condition.

4.3. Phase 2: Loss of Coolant (6 minutes - 20 minutes) (Figure 4)

At 6 minutes the pressuriser steam bubble was lost. The reactor coolant drain tank (RCDT) pressure started building up rapidly and at 7 minutes 43 seconds the reactor building sump pump switched on to transfer water from the sump to the miscellaneous waste hold-up tanks located in the auxiliary building. The isolation valves in this transfer line (which close automatically only when the reactor building exceeds 270 millibar and not on initiation of the HPIS as in some other PWR plants) were open and transfer of water from the reactor building to the auxiliary building was initiated.

At 8 minutes, the operators found that the steam generators were dry. Checks showed that the auxiliary feed pumps were running but that the valves were shut. The operator opened the valves (5) and the hot and cold leg temperatures of the RCS dropped as a result. "Hammering" and "crackling" was heard from the steam generators confirming that the auxiliary feed pumps were now delivering water to the steam generators.

The closed valves in the auxiliary feedwater circuit received a great deal of publicity immediately after the accident. It seems likely that the fact that the auxiliary feedwater was not available for the first 8 minutes of the accident did not, however, significantly affect the future course of the events which was largely determined by the stuck open PORV.

At 10 minutes 24 seconds a second HPIS pump tripped out, was restored but tripped out again to be eventually restarted at 11 minutes 24 seconds (6) but in a throttled condition. The balance between flow of water in from the HPIS, and

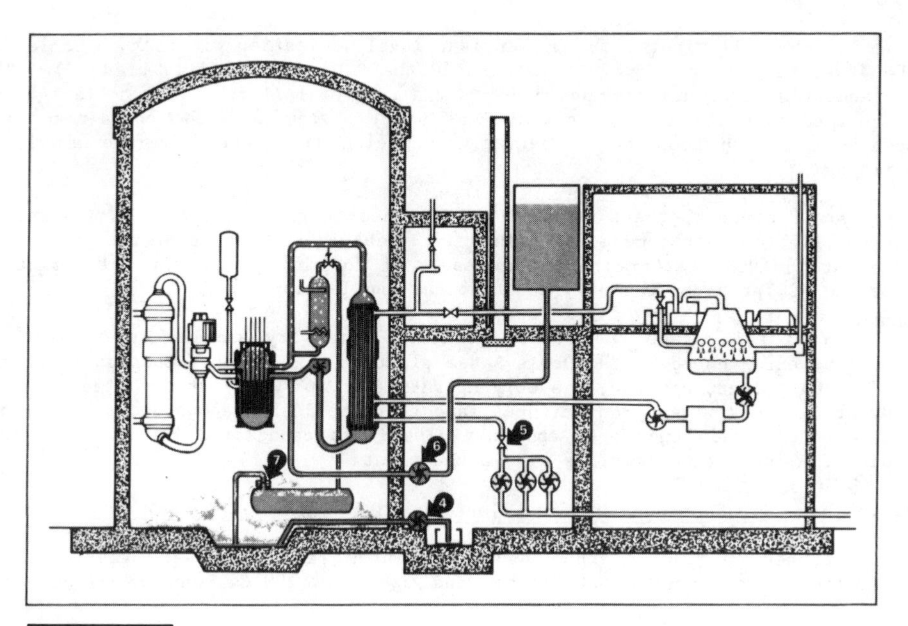

Figure 4: Phase 2 - Loss of Coolant

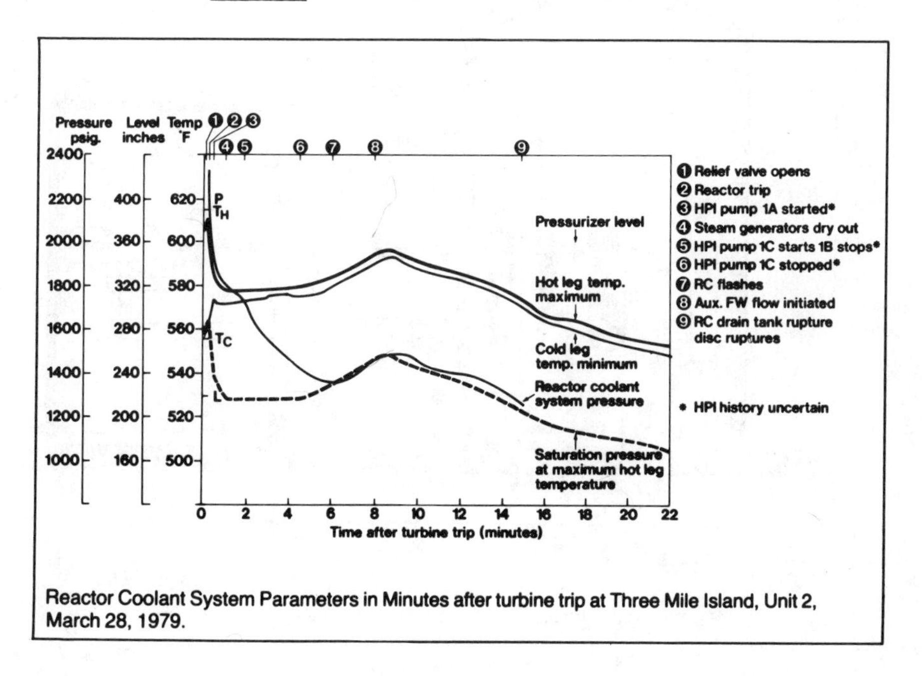

Reactor Coolant System Parameters in Minutes after turbine trip at Three Mile Island, Unit 2, March 28, 1979.

Figure 5

that out from the PORV, was such that there was a net outflow from the primary coolant system.

At about 11 minutes the pressuriser level indication was back on scale and the level was decreasing. At 15 minutes the RCDT rupture disc blew (7). The consequential pressure rise was seen in the reactor building. The coolant being discharged from the primary circuit was now emptying into the containment and because the sump pump was operating, was being transferred to the auxiliary building.

At 18 minutes, there was a sharp increase in gaseous activity measured by the ventilation system monitors. This was probably the result of the pressurisation of the RCDT rather than as a result of fuel failures. At this point in time the RCS pressure was only about 83 bar and falling.

Up to this stage the events at TMI-2 were very similar to a feedwater transient experienced at the Davis-Besse plant at Oak Harbour, Ohio in September 1977. That plant, however, was only operating at 263 MW(t) at the time when its PORV stuck open. Twenty-one minutes into that incident the operators determined that the PORV had stuck open and they closed its associated block valve, thus ending the incident. This was also a B&W plant.

4.4. Phase 3: Continued Depressurisation (20 minutes - 2 hours) (Figure 6)

Between 20 minutes and 1 hour the system parameters were stabilised at saturation conditions; about 70 bar and at about 290°C. At 38 minutes the

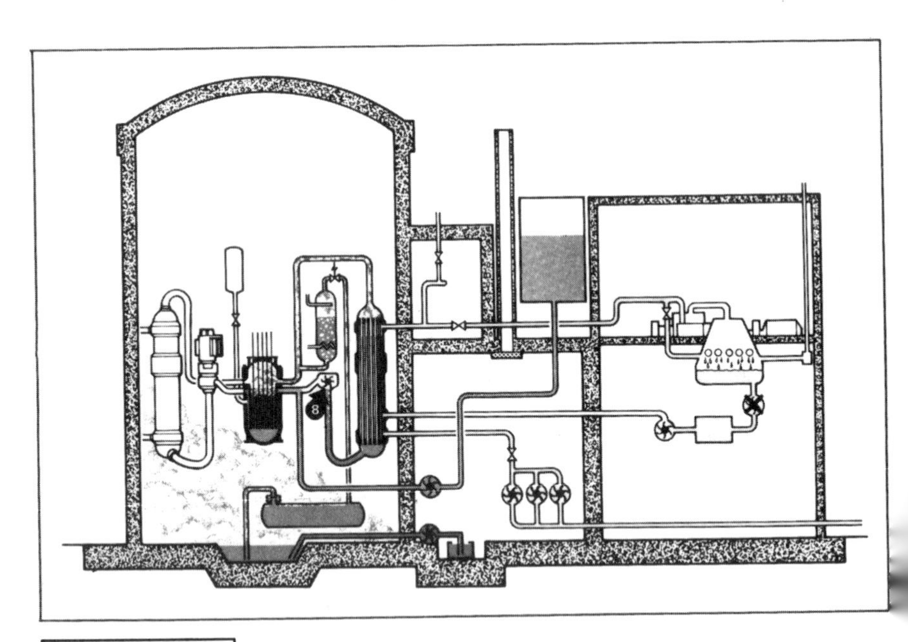

Figure 6: Phase 3 - Continued depressurisation

reactor building sump pumps were turned off after approximately 30 m^3 of water had been pumped to the auxiliary building. The amount of activity trarsferred was, however, relatively small since the transfer was secured before any significant failure of fuel.

At 1 hour 14 minutes the main reactor coolant pumps (RC pumps) in Loop B were tripped because of indications of high vibration, low system pressure and low coolant flow. The operators would normally be expected to take such action to prevent serious damage to the pumps and associated pipework. However, turning off the pumps in Loop B allowed the steam and water phases in that circuit to separate, effectively preventing further circulation in that loop.

At 1 hour 40 minutes the RC pumps in Loop A were tripped for the same reasons (8). One concern was that a pump seal failure could occur. The operating staff expected that natural circulation of the coolant would occur but because of the separated voids in both loops, this did not take place. Subsequent analysis has shown that about $^2/_3$ of the water inventory in the primary circuit had been discharged and that when the main coolant pumps were switched off the water level in the reactor vessel settled out at about 30 cm above the top of the core. Thus the core began a heat-up transient and it is these events which were the precursors of core damage.

4.5. Phase 4: The heat-up transient (2 hours - 6 hours) (Figure 7)

At 2 hours 18 minutes into the incident, the PORV blocking valve was closed by the operators (9). The indications of the position of the PORV were ambiguous to the operators. The control panel light indicates the actuation of

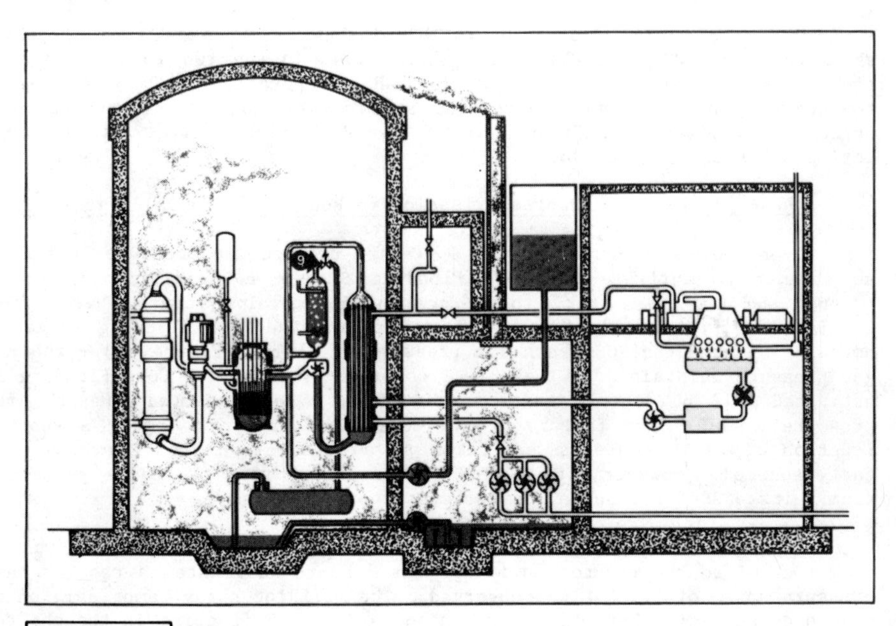

Figure 7: Phase 4 - The Heat-up Transient

the solenoid; there is no direct indication of valve stem position. Interpretation of high temperature in the pipework downstream of the valve was confused by the pre-existing leak through the PORV or one of the code valves mentioned earlier. However, it must be said that failure to recognise that a considerable loss of RCS inventory had occurred as a result of the stuck open PORV was a significant feature of the accident. Even at this point in time, however, a repressurisation of the RCS using the HPIS could probably have been successful in terminating the incident.

Following closure of the PORV the RCS pressure began rising. At 2 hours 55 minutes a site emergency was declared after high radiation fields had been measured in the reactor coolant let-down system. By this time a substantial fraction of the reactor core was uncovered and had sustained high temperatures. This condition would be expected to result in fuel damage, release of volatile fission products and the generation of hydrogen from a zirconium-steam reaction.

Attempts were made to re-start the RC pumps around this time. One pump in Loop B did operate for about 19 minutes but tripped out due to vapour binding and vibration alarms at 3 hours 13 minutes. Subsequent analysis showed that at about 3 hours 20 minutes a reactivation of the HPIS effectively terminated the initial heat-up transient. Peak fuel temperatures in excess of 2000°C were reached soon after 3 hours into the incident. A "general emergency" was declared at about 3 hours 30 minutes as a result of rapidly increasing radiation monitor readings in the reactor building, the auxiliary building and the fuel handling building. At 4 hours a detector shielded with 4 inches of lead and located in the containment dome was reading 200 R/h and, at 20 minute intervals, 600, 1000 and 6000 R/hr respectively.

Over the period from 4 hours 30 minutes to 7 hours into the incident attempts were made to collapse the steam voids in the two loops by increasing the system pressure and by sustained HPIS operation. These attempts to re-establish heat removal though the steam generators were unsuccessful and, moreover, involved significant use of the PORV block valve. This course of action was therefore abandoned.

4.6. Phase 5: Extended Depressurisation (6 hours - 11 hours) (Figure 8)

Over the next four hours the operators reduced the pressure in the RCS in an attempt to activate the Core Flooding System and establish heat removal through the low pressure (28 bar) decay heat removal system. This action was initiated at 7 hours 38 minutes (10) by opening the PORV blocking valve. At 8 hours 41 minutes the RCS reached a pressure of 41 bar (Figure 10) - the nominal gas pressure maintained in the core flood tank system. The Core Flooding System activated (11) but only a small amount of water was injected into the reactor pressure vessel. The failure of the core flood tanks to inject a substantial fraction of their contents was taken as an indication that the core was, in fact, covered. However, the design of the piping from the tanks prevents them from being used for ensuring satisfactory core coverage.

During the depressurisation a considerable volume of hydrogen was vented from the RCS to the reactor building. At 9 hours 50 minutes a reactor building pressure spike of 1.9 bar was observed. The building spray pumps came on within 6 seconds and were shut off after 6 minutes (12). This spike is thought to have been due to the ignition of a hydrogen-air mixture in a part of the reactor building.

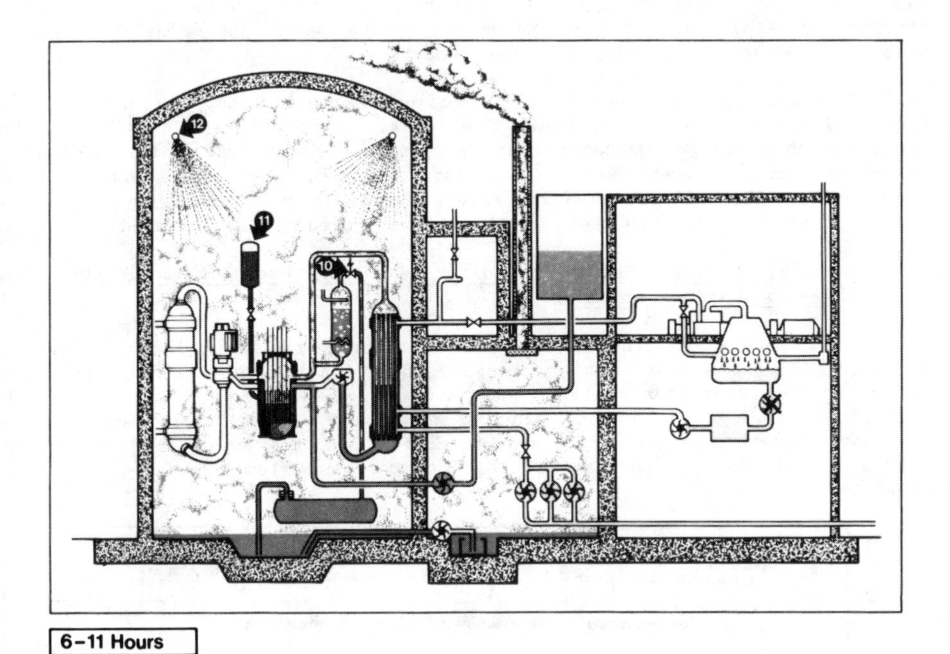

Figure 8: Phase 5 – Extended Depressurisation

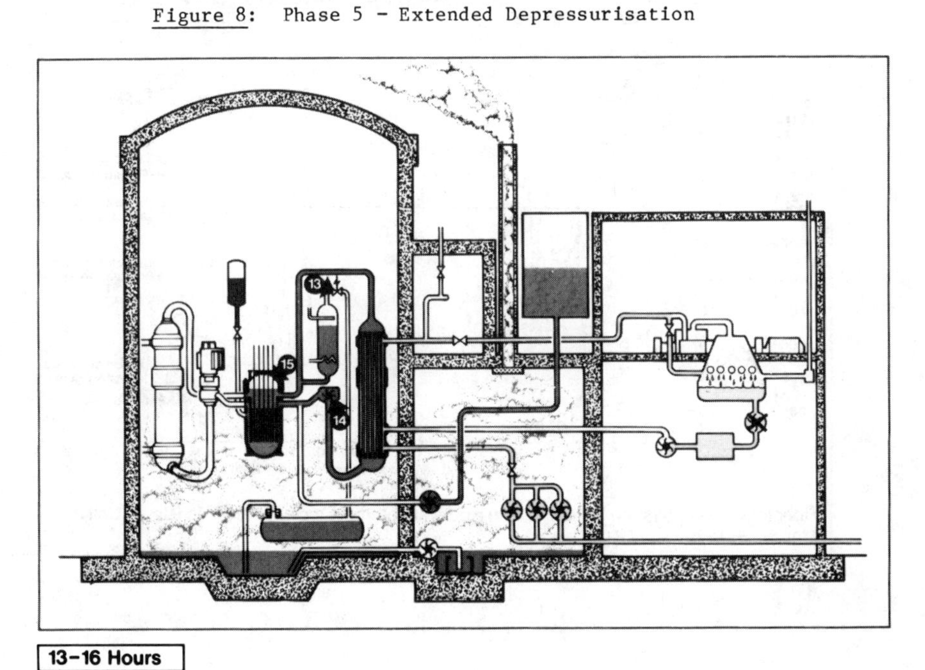

Figure 9: Repressurisation and Ultimate Establishment of Stable Cooling Mode

The depressurisation was unsuccessful in that the lowest pressure achieved was 30 bar and it remained there. Nothing tried could drive it lower and it obstinately remained above the maximum pressure at which the decay heat removal system can be brought into operation (28 bar).

With the operators unable to depressurise the RCS any further the PORV block valve was closed at 11 hours 8 minutes. Over the next two hour period there was no effective mechanism for removing the decay heat. The PORV block valve was kept closed during this period except for two short periods. Injection via the HPIS was at a low rate and was almost balanced by the outflow in the let-down line; both steam generators were effectively isolated.

4.7. Phase 6: Repressurisation and ultimate establishment of stable cooling mo
(13 - 16 hours) (Figure 9)

At 13 hours 30 minutes into the incident, the PORV block valve was reclosed (13) and sustained high pressure injection via the HPIS was initiated in order to repressurise the RCS and to allow the RC pumps to be restarted (14). At 15 hours 51 minutes an RC pump in Loop A was restarted and the hot leg temperature decreased to 293°C and the cold leg temperature increased to 205°C, indicating that there was flow through the steam generators.

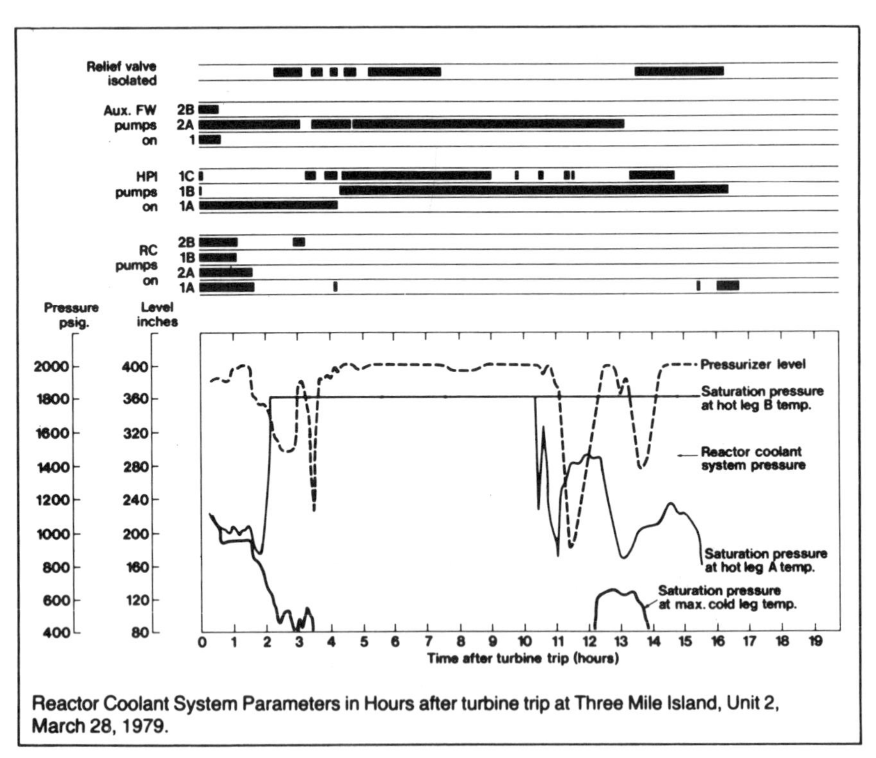

Reactor Coolant System Parameters in Hours after turbine trip at Three Mile Island, Unit 2, March 28, 1979.

Figure 10

4.8. Phase 7: Removal of the hydrogen bubble (Day 1 - Day 8) (Figure 11)

The voidage (about 28 m^3) (15) in the pressure vessel was composed mainly of incondensible hydrogen. The volume of this was decreased progressively (16) between day 1 and day 8 by degassing using the PORV (17) and let-down system. Recombiners were put into operation (18) to reduce the hydrogen level in the reactor building. Overspill water in the auxiliary building was pumped back into the reactor building (19).

As indicated earlier, the decay heat from the core reduces quite rapidly with time and about one month after the incident the RC pumps were turned off and the reactor placed in a natural circulation mode. This action was taken because the primary pumps inject some 4-5 MW(t) into the RCS which is considerably greater than the decay heat in the core (2 MW(t)) at that time.

5. CONSEQUENTIAL EVENTS AND CORE DAMAGE

The core was uncovered in part or wholly during three separate time periods. The first period started about 100 minutes into the incident. At least 1.5m of the core was uncovered for a period of about 1 hour and perhaps all the core for a period of about half an hour. This was a period of major core damage with significant zirconium/steam reaction and the production of considerable amounts of hydrogen; also a considerable amount of fission product gas was released from the fuel to the RCS. The second period of core uncovery occurred about 7½ hours into the event. About 1.5 m of the core was exposed for

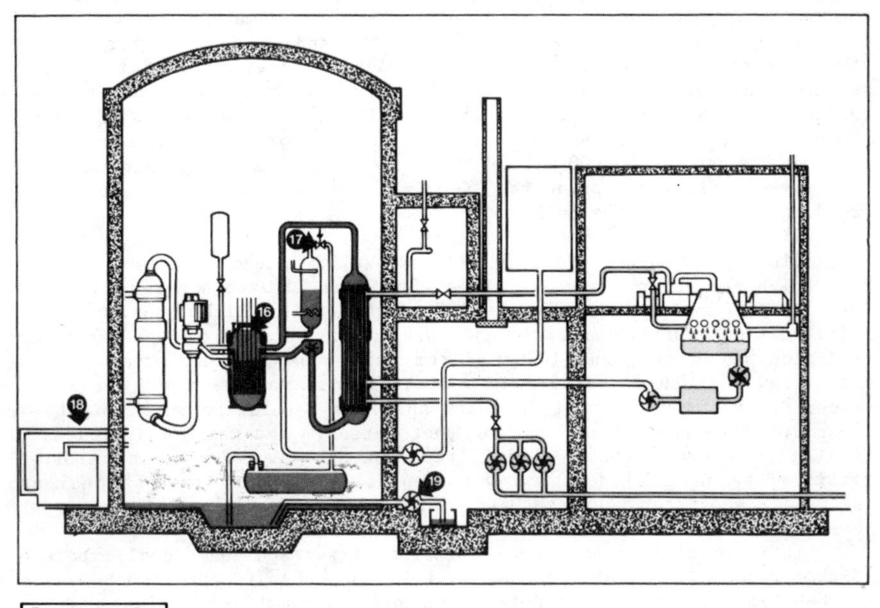

Day 1- Day 8

Figure 11: Removal of hydrogen bubble

a much shorter period and the fuel temperatures were probably considerably lower than for the first period. The third period occurred about 11 hours into the event when the water level decreased to between 2.1 and 2.3 m for a period of 1-3 hours. Again, significant fuel temperatures were reached during this period.

It has been estimated that about 30-40% of the total zirconium inventory was oxidised during the heat-up transient and that the upper third of the reactor core is seriously damaged. Even with pessimistic assumptions of low heat transfer coefficients of 2.8 W/m^2K, it is estimated that the fuel temperatures would have levelled off between 1750°C and 2600°C. It is more likely that the peak fuel temperatures reached in the region of 2000°C. 51 out of the 52 core thermocouples survived the incident and the values of temperature distribution have been used for a variety of calculations. For example, the flow resistance coefficient of the damaged core is estimated to be between 200-400 times normal.

Some 1500-2000 m^3 of active liquid will have to be disposed of from the containment building, and the active core material will have to be removed.

Some mention should be made of the radioactive species released within the containment during the incident. About 30 to 40% of the noble gases generated were released from the fuel. The core had only received a relatively low burn-up (approximately 4000 MWD/tonne). About 10-15% of the Iodine was also released from the fuel, together with smaller amounts of Strontium and Caesium. The iodine release would correspond to between 3 and 5 million curies of Iodine-131 and 0.2-0.3 million curies of Iodine-133. The actual discharges from the site included most of the Xenon-133 (approximately 13 million curies), together with comparable amounts of Krypton and just 16 curies of Iodine-131 (the filters on the exhaust appear to have retained most of the Iodine which reached the Auxiliary building). Liquid discharges were negligible - 0.24 curies of Iodine-131 and 12.4 curies of Tritium and 0.06 curies of mixed fission products. The remaining activity in the reactor building 1 year after the accident is about 50,000 Ci of ^{85}Kr in the atmosphere and 850,000 Ci of long-lived fission products (^{137}Cs, ^{90}Sr) contained in the water flooding the base of the building.

Finally, exposure off-site was very small indeed. Various estimates give collective radiation doses of between 2000 and 5000 person-rem up to April 7th 1979. The average value is 3300 person-rem, which provides an average individual dose of 1.5 millirem and this has to be assessed against a background radiation level of about 100 millirem per annum in that area. It has been calculated that this increase in radiation dose may produce 1 or 2 additional cancer deaths over the next 20 years and in a population which might expect at least 300,000 cancer deaths during that period. The maximum individual theoretical exposure was about 80 millirems and the maximum known exposure is 37 millirems by an individual who was identified as being present on nearby Hill Island for about ten hours on March 28th and 29th 1979.

It is possible that these estimates of average individual exposure may be further reduced when full account is taken of other factors, such as the effect of habitation, actual nuclide inventory, actual population and dosimetry instrument calibration.

6. POST MORTEM

Two weeks after the accident President Carter ordered an inquiry into the causes of the accident. This presidential commission, chaired by J.G. Kemeny, president of Dartmouth College, reported in October 1979. Whilst recognising that equipment failures had contributed to the accident, the commission concluded that the main cause was human failure.

Although inappropriate operator action was a major factor, many other factors contributed including deficiencies in training, lack of clarity in operating instructions, failure to apply the lessons learned from previous incidents and deficiencies in the design of the control room. The commission was critical of the nuclear industry in the United States in general and of the Nuclear Regulatory Commission in particular.

Since March 1979 the nuclear industry in the United States has made strenuous efforts to correct the deficiencies revealed by the TMI-2 accident. The Electric Power Research Institute has set up the Nuclear Safety Analysis Centre (NSAC) to carry out a detailed technical analysis of the accident and the Institute for Nuclear Power Operations (INPO) to up-grade the training and education of reactor operators and to improve and simplify reactor control room design. The NRC has completed a review of the lessons learned from the accident which can be applied to other light water reactor plants.

The reality is that the TMI-2 accident was a bad plant accident (it will cost at least $500M to put the plant right again), but the physical effect on the public was small compared with other man-made accidents. Indeed, worse accidents have occurred in the history of nuclear power. On 3rd January 1961 at the National Reactor Testing Station at Idaho, the SL1 reactor was almost totally destroyed as a result of a reactivity excursion incident killing three operators. The Windscale accident, which happened in the United Kingdom in 1957, released an estimated total of about 20,000 curies of Iodine-131 (compared with just 16 curies of Iodine-131 at TMI-2) to the atmosphere over a 24 hour period. It is in the nature of technology that accidents will happen, there will be design faults, equipment malfunctions and operator errors. It is the function of an effective safety philosophy to protect the public from the consequences of such events. The U.S. safety philosophy is based on "defence-in-depth" involving multiple barriers to prevent or limit the escape of radioactivity. The fact that the release of radioactivity was relatively small is confirmation that the safety philosophy is basically sound.

7. IMPLICATIONS FOR THE HEAT TRANSFER ENGINEER

It has been said that architects can hide their small mistakes behind bushes, that lawyers can put their big mistakes in jail, and that medical doctors can bury their worst mistakes. But engineers are expected to face up to all their mistakes and to use whatever lessons they might have learned in future applications. So it is with Three Mile Island, where an extraordinary heat transfer mistake was made. The fuel was allowed to overheat excessively when it was not supposed to. The result was not only the disabling of the $1000m power plant facility for several years but also a cost to the nuclear industry of between $10,000 and $20,000 million damage when one takes into account all the cleanup costs, plant changes, and delays that can eventually be traced back to this accident. This estimate does not include the very serious setback in public acceptance or perception of nuclear power.

After Three Mile Island, heat transfer engineers will best be served by calling for a return to fundamentals. As indicated earlier, the heat transfer process in a PWR under normal operation is quite straightforward. To be effective, this heat transfer process requires three simple things:

1. Enough water in the reactor core and primary circuit

2. A heat sink that can absorb the heat produced in the reactor

3. A way to transport the heat from the reactor core to the steam generator by moving the water from one to the other.

At Three Mile Island, all three of the above fundamentals were neglected.

Several important lessons can be learned from the recounting of events given earlier:

1. The pressurizer level may not tell what the water inventory is

2. When the availability of the heat sink is in doubt, always inject enough water to at least keep the reactor core covered.

3. Even with a heat sink available, natural circulation may not always occur, especially when non-condensable gases are present in the circuit.

The fundamental issue is not whether heat transfer coefficients or flow rates can be predicted or are known to 5, 10 or 20%, but rather whether there is any kind of heat sink, any type of recirculation flow, and enough water inventory. Once the fundamental problems are identified, many solutions can be found and implemented. It is not our intention to describe the many options available because they go beyond the fundamentals dealt with here. However, stopping at this point will fail to cover two other issues just as fundamental. The first question one might raise is how did this type of situation arise, especially when loss of water or coolant from a water-cooled nuclear reactor has been considered one of the most - if not the most - important accident to be dealt with in LWR plants. Water-cooled nuclear plants are supposed to be designed to handle any such event, and as a matter of fact, the fuel damage is to be quite limited even when the largest pipe breaks instantaneously at the same time as total loss of off-site power and another worst single failure are occurring. Also, to make sure that the plants can handle smaller breaks, one expects to get even less fuel damage for a small break as might be produced by a stuck-open relief valve together with a total loss of off-site power and another worst single failure.

Now, as shown in the following simple calculation,

• Large pipe failure statistics	10^{-4} to 10^{-6}/reactor year
• TMI Statistics	
• Feedwater transient	1/reactor year
• Failure of relief valve to close	5×10^{-2}
Total	5×10^{-2}/reactor year
• Human factors to match large pipe failure	2×10^{-3} to 2×10^{-5}

the probability of a large pipe failure is estimated to be of the order of 10^{-4} - 10^{-6} per reactor year or once every 10,000 - 1,000,000 reactor years. On the other hand, the probability of experiencing a loss of feedwater is about once per reactor year. The possibility of a relief valve's opening and failing to close during such an event is 5×10^{-2}, or the valve will fail to close once out of every 20 times. If we now assume that we will accept the same fuel damage for a relief valve's failing to close as for the very large pipe break, this leaves us with the need to handle other failures that might occur at the rate of 2×10^{-3} (or $10^{-4}/5 \times 10^{-2}$) to 2×10^{-5} (or $10^{-6}/5 \times 10^{-2}$) or at rates of once every 500 - 50,000 times. This means that the operators must be capable of performing at least at levels of 2×10^{-3} to 2×10^{-5}, which are well above the accepted levels of between 10^{-1} and 10^{-3}. To achieve this high level of operator proficiency, one must look at a very large number of operator actions and make sure to a high degree of certainty that no serious human mistake is made or left uncorrected for accidents involving loss of feedwater and a stuck-open relief valve. This was not done. Another option is to change the design so that the operator probability of taking the right action has been increased. This alternative was also not considered. Most probably a combination of the above two options would be best. The reason for not trying to do it all on the operator side will become clear when one realizes that such small loss of coolant accidents develop slowly and over long periods of time, thus giving the operator a substantial number of opportunities to intervene during the course of events.

Figure 10 shows the actions taken at TMI-2 that might have a serious impact on the heat transfer process and that involve operator interactions with relief valve, reactor pumps, high-pressure injection system, and the emergency flooding tanks. According to this chart, there were at least 70 important actions until core cooling was re-established as stable. It is important to realize that the operators could have tried more or fewer actions, different actions, or different scenarios. The design process, therefore, must recognize the need to examine enough such possibilities to obtain the high degree of confidence that the situation could be handled. Unfortunately, such a large set of events involving a variety of human-machine interactions was not looked at in enough detail in the design. Instead, the design emphasis was put on the very large pipe break because it is the most difficult to analyze, most difficult to design for, and also the most glamorous. The analytical tools developed for such large breaks take many hours to run on some of the largest computers available and even when such tools are simplified, they will not make it possible to look at the myriad of possibilities involving human interactions. The last fundamental lesson to be learned from Three Mile Island is therefore that we may need a set of design tools different from those developed to date. What may be needed is a simplified heat transfer model with good physics that can give an accurate and integral accounting of the water inventory and of the heat transferred to the sink, a model that is coupled to a good simulator of valves, pumps, instruments, controls, and human beings to test a variety of operator actions, scenarios and, equipment and instrument failures. This type of simulating interaction with human operators needs to be on a real-time basis because it involves human intervention and decision making. Thus, the last fundamental lesson to be learned from Three Mile Island is that when human-machine interactions play an important role in any piece of equipment, real-time analysis with real hardware simulation is most important.

In conclusion, it is hoped that this chapter may have helped convince heat transfer engineers that only by more emphasis on the fundamentals of the heat process, enough evaluation of the human role, and real-time analysis and simulation can we expect to avoid another Three Mile Island.

ACKNOWLEDGEMENT

This chapter draws upon the factual and derived information about the accident given in various U.S. Nuclear Regulatory Commission reports, in particular NUREG-0600 "Investigation into the March 28th 1979 Three Mile Island Accident by the Office of Inspection and Enforcement, August 1979". Reference was also made to the report prepared by the Nuclear Safety Analysis Center, NSAC-1 "Analysis of Three Mile Island - Unit 2 Accident' July 1979 and to the report of the Ad Hoc Population Dose Assessment Group dated May 10th 1979.

The chapter is based on articles published in Heat Transfer Engineering, Vol.1, No.3 and Vol.1, No.4 and on a paper presented by Sal Levy to the ASME Heat Transfer Division Luncheon, ASME Winter Annual Meeting, New York, December 6th, 1979.

INDEX